D0225223

Third Edition

CONSTRUCTION MATERIALS, METHODS, AND TECHNIQUES

Building for a Sustainable Future

WILLIAM P. SPENCE EVA KULTERMANN

DELMAR CENGAGE Learning™

Australia • Brazil • Japan • Korea • Mexico • Singapore • Spain • United Kingdom • United States

Construction Materials, Methods & Techniques, Third Edition
Spence & Kulterman

Vice President, Career and Professional Editorial: Dave Garza

Director of Learning Solutions: Sandy Clark

Senior Acquisitions Editor: Jim DeVoe

Managing Editor: Larry Main

Product Manager: Erin Cutis, Ohlinger Publishing Services

Editorial Assistant: Cris Savino

Vice President, Career and Professional Marketing: Jennifer Baker

Marketing Director: Deborah Yarnell

Marketing Coordinator: Mark Pierro

Production Director: Wendy Troeger

Production Manager: Mark Bernard

Content Project Manager: Michael Tubbert

Art Director: Casey Kirchmayer

Technology Project Manager: Joe Pliss

Compostitor: PrePress PMG

Library of Congress Control Number: 2009943503

ISBN-13: 978-1-4354-8108-4

ISBN-10: 1-4354-8108-9

Delmar
Executive Woods
5 Maxwell Drive
Clifton Park, NY 12065
USA

Cengage Learning is a leading provider of customized learning solutions with office locations around the globe, including Singapore, the United Kingdom, Australia, Mexico, Brazil, and Japan. Locate your local office at: **www.cengage.com/global**

Cengage Learning products are represented in Canada by Nelson Education, Ltd.

To learn more about Delmar, visit **www.cengage.com/delmar**

Purchase any of our products at your local bookstore or at our preferred online store **www.cengagebrain.com**

Printed in the United States of America
1 2 3 4 5 6 7 XX 14 13 12 10

Table of Contents

Preface

INTENT OF THIS BOOK

Construction Materials, Methods, and Techniques: Building for a Sustainable Future is designed for use in introductory courses on construction materials and methods, construction management, and architecture programs at primarily two-year and four-year schools. The text follows the logical progression of a construction project, with references to the 2004 edition of the MasterFormat™, developed by the Construction Specifications Institute and Construction Specifications Canada. Students are able to build a foundation of knowledge pertaining to construction materials, processes, and techniques as well as gain insight into the construction industry.

BUILDING FOR A SUSTAINABLE FUTURE

This third edition of *Construction Materials, Methods, and Techniques: Building for a Sustainable Future* has been supplemented to include additional topics that address the fundamental changes taking place in the construction industry today. Ever since the Earth Summit in Rio de Janeiro in 1992, much has been written about the subject of Sustainable Development. The concept was formulated by the World Commission on Environment and Development and is defined as "development that meets the needs of the present without compromising the ability of future generations to meet their own needs." Sustainable Construction seeks to mitigate the environmental impact of building practices by promoting strategies that will conserve natural resources, advance energy and resource efficiency, deal responsibly with waste, and create healthy environments.

This revision is an effort to integrate sustainable construction innovations holistically into all aspects of the larger construction context. Sustainable construction is no longer a unique means of building, but rather it is an essential part of how the industry will operate in the decades to come. New materials, building systems, and construction practices have been added to the existing knowledge in each chapter. In addition, case studies of individual buildings that exemplify sustainable construction materials and technologies have been added. This book has been thoroughly updated and revised to remain current with industry technologies and standards:

- The U.S. Green Building Council, LEED Rating system, and other green certification systems.
- Effective construction planning for efficient material use.
- Environmentally friendly building materials with emphasis on recycled content and materials that promote manufacturer, contractor, and occupant health.
- Energy-efficient environmental systems.
- Sustainable construction operations and practices.
- New organizations and resources actively providing metrics and research.

About the Cover

The cover image shows the headquarters building of Heifer International, a nonprofit organization dedicated to fighting global hunger and poverty. The project clearly demonstrates how thoughtful design and construction planning can result in a sustainable building that reduces environmental impacts and provides a healthy environment for its occupants. From site selection and landscape design, to passive solar energy and daylighting strategies, to the use of recycled content and natural materials and water conservation, the Heifer International Headquarters provides a showcase of the holistic integration of green building strategies.

HOW TO USE THIS BOOK

This text provides a detailed view of major construction methods and building systems and the vast array of materials and products provided by manufacturers supplying the construction industry.

The first part of the text gives a brief overview of different aspects of the construction industry. Pre-construction activities, the role of design professionals, various project delivery methods, and the MasterFormat™ are covered. A discussion of zoning and building codes is followed by information about some of the industry's major professional and technical organizations. A section on sustainable design and construction discusses the environmental impact of the construction industry and provides an introduction of the LEED and other sustainable building certification systems.

Early in the book is a discussion of the properties of materials. This discussion relates to all of the materials in the remainder of the book. New materials are developed each year, and the architect, contractor, and engineer must be informed of their properties in order to use these materials in the most effective and safe manner.

The remainder of this book is organized following selected divisions of the 2004 Edition of the MasterFormat™.

Division 1 addresses general requirements, including activities such as price and payment procedures, administrative requirements, and various other controls and requirements. See Appendix A for more details.

Division 3 presents a detailed study of the manufacture, types, characteristics, and properties of concrete. Consideration of the use of admixtures, proportions, water, mixing, and placing are included. Drawings and photos are used extensively to illustrate cast-in-place and precast concrete structural systems.

Division 4 includes detailed information on mortar, the key to satisfactory masonry construction. The materials and techniques involved with clay brick and tile, concrete masonry, and stone construction are explained in detail and generously illustrated.

Ferrous and nonferrous metals are presented in Division 5. Their characteristics, mechanical properties, and practical applications are discussed. Steel frame construction systems are illustrated.

One of the largest divisions in the book is Division 6, which covers the vast array of wood and plastic materials. Their properties, characteristics, and recommended applications are explained. Several chapters detail wood structural framing systems and the methods and materials of light wood frame construction. The remainder of this division is used to present information on paper and pulp product composites and plastic materials useful in building construction.

Insulating, waterproofing, and sealing buildings against the weather are covered in Division 7. Walls, ceilings, and floors need to be properly insulated and sealed against moisture penetration. Various bonding agents, sealers, and sealants are discussed. Bituminous materials are another form of waterproofing that also serve as surfaces for parking areas and roads. The roofing systems for residential and commercial building conclude this division.

The types, styles, methods of operation, and materials used for doors and windows are extensive. Many of the products available are illustrated in Division 8, as well as factory stock and custom-made storefronts. Glass is used extensively in door and window assemblies, so the types, properties, and uses of various glass products available are discussed. Finally, the exterior of commercial buildings is covered with some form of cladding system. An entire chapter is devoted to discussing and illustrating cladding systems.

Finishing the interior of a building involves the most diverse range of products in any area of construction. Division 9 includes interior finishes; decorative and protective coatings; gypsum, lime, and plaster materials and products; acoustical finishes and materials; and all types of finish flooring. The construction and finish of interior walls, partitions, and ceilings in residential and commercial buildings is included.

Division 10 covers some of the specialty products, such as visual displays, screens, grills, service walls, and identifying signs. Other products included as specialty items are shelving, partitions, fire protection devices, telephone enclosures, and toilet and bath accessories.

Division 11 includes items specified as equipment. Examples include ecclesiastical equipment, library equipment, and vending equipment. Furnishings form the final touch to a completed building, and Division 12 classifies these into seven major groups. Artwork, window treatments, and rugs and mats offer a variety of types and styles. Casework and furniture for use in commercial buildings form the largest group.

A most interesting assembly of special construction features is found in Division 13. This division discusses and illustrates a diverse offering, including air-supported and fabric structures, geodesic domes, and other pre-engineered assemblies.

Division 14 discusses and illustrates the variety of conveying systems available, such as conveyors, elevators, escalators, moving walks, and material-handling systems.

Division 21 organizes the many complex factors related to the operation and maintenance of fire-suppression systems, including installation, instrumentation, and control. Water, carbon dioxide, clean-agent, wet chemical, and dry chemical fire-extinguishing systems are included.

Division 23 is very large, including heating, ventilating, and air-conditioning systems. Piping, pumps, and fixtures for residential and commercial systems are included.

Division 26, Electrical Systems, is covered next. It carries the discussion from the generation and transmission of electrical power to the service entrances in residential and commercial buildings. Internal power distribution systems are illustrated, and extensive information on lighting is available. Equipment for controlling and operating electrical systems as well as equipment used for communication, such as alarm, television, public address, and other communication systems, is presented.

Division 28 covers the various electronic safety and security systems in residential and commercial buildings and related exterior areas where security is important. Video surveillance and personal protection systems are included, as are electronic detection and alarm systems for fire detection.

Division 31 encompasses earthwork, including site preparation, site clearing, soil treatments and stabilization, and factors related to excavation, foundations, and tunneling.

A master glossary at the end of the book provides an additional means of locating definitions. Appendix A provides details concerning the information in Division 1, General Requirements. Appendix B provides an extensive listing of U.S. and Canadian professional and trade organizations. These organizations provide information vital to the continual improvement of materials and construction techniques. They develop materials standards and installation specifications and even develop and publish building codes, manuals, and technical reports. Readers can reference this information to provide additional insight as the chapters in this book are read. Detailed metric information is available in Appendix C. This information is essential because construction is moving toward using the metric system in the future. Appendix D reports the weights of many construction materials. Appendix E gives the names and atomic symbols of selected chemical elements, and Appendix F provides data on the coefficients of thermal expansion of commonly used construction materials.

NEW TO THIS EDITION

- New Information on sustainable and green building materials, assemblies, and systems has been added to each chapter.
- Extensive and up-to-date, the text is highlighted with over 300 new color photos and illustrations that focus on the most current equipment, tools, and procedures.
- Construction Materials, Methods, and Techniques Boxes highlight specific ground-breaking materials and systems, providing more detailed information on the selected areas of interest.
- New Case Studies provide examples of the application of new construction materials and strategies. Each Case Study highlights a building that showcases real-world applications of sustainable building principles and strategies. Many of these projects have received a Green or LEED certification.
- Significant text revisions and edits make the material more interesting and accessible to today's students.

FEATURES OF THIS BOOK

This book includes many features to assist students as they progress through the chapters:

- Correlations to the MasterFormat™ are included at the beginning of each chapter, providing students with a quick reference to this important manual.
- Learning Objectives open each chapter to give students a framework for study to ensure full comprehension of the material.
- Key Terms are rendered in color throughout the text and listed at the end of each chapter, highlighting essential terminology. Complete definitions are provided in the end-of-book Glossary.
- Review Questions enhance students' understanding of key concepts.
- Activities encourage students to apply what they have learned in the chapter and build experience for actual on-the-job tasks.
- Additional Resources are listed at the end of each chapter and point students in the direction of organizations, periodicals, web links, and other references to further learning on selected topics.

SUPPLEMENTS

Spend Less Time Planning and More Time Teaching with Delmar, Cengage Learning's Instructor Resources to accompany *Construction Materials, Methods, and Techniques: Building for a Sustainable Future, 3rd*

Edition, preparing for class and evaluating students has never been easier!

This invaluable instructor CD-ROM is intended to assist you, as the instructor, in classroom preparation and management of a construction materials and methods course. Included within this electronic resource are tools that help reinforce the important building techniques introduced in the book, as well as provide the necessary materials for evaluation of student comprehension of critical concepts:

- An Instructor's Manual including Lesson Plans with correlating PowerPoint slides, Discussion Questions, and Answers to Review Questions help you prepare for class.
- PowerPoint Presentations highlight critical concepts in each chapter and include art from the book to enhance classroom lectures. PowerPoint presentations also correlate to the Lesson Plan in the Instructor's Manual, allowing for a seamless presentation of the content of the book.
- A Testbank in ExamView format includes over 1,000 questions and enables you to edit, delete, or add your own questions, as well as create your own tests using the questions provided. This flexible format makes this feature a handy tool for evaluating your students on the concepts presented in each chapter.
- An Image Library containing illustrations from the book enables you to supplement and enhance your classroom presentations.
- The ACCE Correlation Grid outlines the course objectives for construction materials and methods and correlates the objectives to the chapters in the book containing the supporting content. These course objectives are outlined in *Standards and Criteria for Baccalaureate and Associate Programs*, published by the American Council on Construction Education, an accrediting body for various 2-year and 4-year construction criteria.
- Link to *delmarlearning.com* and click on building trades to review other Delmar Learning titles available in the construction fields.
- Link to the American Council on Construction Education to learn more about the mission and the purpose of this accrediting body as well as to review Standards and the process of accreditation.

The use of these tools, along with *Construction Materials, Methods, and Techniques: Building for a Sustainable, Future, 3rd Edition*, will assist you as you guide your students down the path to success!

Order #: 1435481097

ABOUT THE AUTHORS

William P. Spence is Dean of the College of Technology as well as Professor of Construction Engineering Technology emeritus at Pittsburg State University, KS. He currently is active as a full-time author of over 30 construction technical books.

Eva Kultermann is a licensed architect and assistant professor at the Illinois Institute of Technology, College of Architecture. She has a background in construction and is currently involved in research in the area of energy- and resource-efficient construction technologies.

ACKNOWLEDGMENTS

A major factor in the organization, writing, and illustrating of this book was the help given by hundreds of manufacturer representatives. Representatives of many of the professional and technical organizations supporting the construction industry also made important contributions. A special note of appreciation is due to the consultants located at universities across the country for their assistance in reviewing the manuscript and the illustrations. In particular, I would like to thank the following reviewers:

James Freygang—Ivy Tech, South Bend, Indiana

Timothy L. Andera—South Dakota State University

Scott Waldinger—Westwood College, Chicago, Illinois

Craig Passley—Westwood College, Woodridge, Illinois

Finally, I dedicate this book to my wife, Bettye Margaret Spence, for her steady encouragement, editorial assistance, and the many hours she spent at her computer preparing the manuscript.

—William P. Spence

Disclaimer

The information presented in this book was secured from a wide range of manufacturers, professional and trade associations, government agencies, and architectural and engineering consultants. In some cases, generalized or generic examples are used. Every effort was made to provide accurate presentations. However, the author and publisher assume no liability for the accuracy of applications shown. It is essential that appropriate architectural and engineering staff be consulted and specific information about products be obtained directly from manufacturers.

DIVISION

00

Procurement and Contracting Requirements CSI MasterFormat™

001000 Solicitation

002000 Instructions for Procurement

003000 Available Information

004000 Procurement Forms and Supplements

005000 Contracting Forms and Supplements

006000 Project Forms

007000 Conditions of the Contract

009000 Revisions, Clarifications, and Modifications

(Image copyright Mearicon, 2009. Used under license from Shutterstock.com)

General Requirements CSI MasterFormat™

011000 Summary
012000 Price and Payment Procedures
013000 Administrative Requirements
014000 Quality Requirements
015000 Temporary Facilities and Controls
016000 Product Requirements
017000 Execution and Closeout Requirements
018000 Performance Requirements
019000 Life Cycle Activities

(Image copyright Pchemyan Georgiy, 2009. Used under license from Shutterstock.com)

CHAPTER 1

The Construction Industry: An Overview

LEARNING OBJECTIVES

Upon completion of this chapter the student should be able to:

- Gain an understanding of the scope of the construction industry.
- Understand the basic processes by which buildings come into being.
- Be familiar with the responsibilities of Owners, Architects, and Contractors.
- Get an overview of the makeup and organization of construction documents.
- Understand the role of the Construction Specifications Institute and the Master Format.
- Understand the process of construction contracts and procedures.

The construction industry is one of the largest business sectors in the United States, encompassing establishments engaged in the construction of buildings and other engineering projects, such as roadways and utility infrastructure. Construction with all of its related and supporting industries is vital to the economic health of the country, accounting for an annual average of five to ten percent of our gross domestic product. The Associated General Contractors of America, Inc. has described the construction industry as follows:

> *The construction industry is a brawny, hearty giant stretching to embrace all kinds of construction activity, from the erection of towering skyscrapers, construction of an interstate highway, or the establishment of a massive dam on a wilderness river, to major maintenance and alterations.* ***(Fig. 1.1)***

CONSTRUCTION: A DYNAMIC INDUSTRY

The construction industry can be divided into three major areas: building construction, heavy or infrastructure construction, and industrial construction (Figs. 1.2 through 1.4). **Building construction** involves the erection of a building on a piece of property. This includes residential, commercial, civic, educational, religious, and agricultural buildings among others. While many building projects consist of new construction, a vast amount of construction activity is concerned with the renovation, repair, or the provision of additional space for existing buildings. **Heavy construction** is the term used for larger infrastructure projects such as highways, bridges, canals, dams, subways, tunnels, utility piping systems, water control construction, and communications networks. Heavy construction projects are usually financed by governmental agencies and other institutions or are incorporated into master plans (such as universities) to serve the public good. **Industrial construction** refers to the building of large-scale manufacturing, processing, and chemical plants or utility generation installations. Regardless of the type of construction, each requires planning, financing, and observance of regulatory constraints.

Construction involves the coming together of skilled workers, architects, engineers, and a vast array of materials and equipment to execute a carefully conceived plan. Supporting them is a broad range of industries that manufacture the materials and components designated for a project. These can include suppliers of aggregate and cement products; manufacturers of lumber, structural beams, doors and windows, siding, roofing, and finish materials; and dealers of appliances and mechanical equipment. Many manufacturing endeavors involve the cutting, shaping, and assembling of various materials into a useful product. Each product area has a tremendous variety of materials and installation methods, requiring

Figure 1.1 The construction industry is a dynamic field, offering a wide range of challenging job opportunities. *(Image copyright Sculpies, 2009. Used under license from Shutterstock.com)*

Figure 1.2 New construction materials and methods are constantly being developed and those in the industry must keep abreast of new technologies. *(Courtesy H.H. Robertson)*

Figure 1.3 Heavy construction includes projects like the Falkirk Wheel in Scotland, which lifts boats from an upper level canal to a lower one. *(Courtesy Sean Mack)*

Figure 1.4 Heavy construction defines larger infrastructure projects such as this elevated bridge. *(Image copyright Aleksandar Kamasi, 2009. Used under license from Shutterstock.com)*

extensive technical expertise (Fig. 1.5). Consider just the electrical and mechanical components involved in servicing a building and the procedural knowledge required to design, manufacture, specify, and install them.

Construction materials and methods are constantly changing fields with new products and equipment being developed continuously. The people involved in construction need to be well educated and kept up-to-date as new materials and methods of construction are introduced. They need to maintain membership in professional organizations, attend conferences, and read professional journals to become aware of new technical developments and their proper application.

The industry is currently in the midst of profound changes caused by unprecedented environmental forces that are reshaping all areas of production. The most important challenge of the twenty-first century is to develop strategies for using natural resources in ways that will sustain the natural environment for generations to come. The importance of the construction industry instituting changes that promote both energy and resource efficiency cannot be overstated and will be discussed in greater detail in Chapter 2.

PRE-CONSTRUCTION ACTIVITIES

Owners, both private and public, plan buildings and other construction projects to accommodate desired functional and spatial requirements. Planning, designing,

Figure 1.5 The construction of this high-rise building is supported by a broad range of industries that manufacture the materials and components designated for the project. *(Courtesy Eva Kultermann)*

and completing construction of a large building project is a complicated undertaking. A vast array of planning decisions must be made at the onset of any construction project. These include an evaluation of the vision for a project, its scope, functional requirements, the determination of a site, budgets and financing, project scheduling, and assurance that regulatory constraints such as zoning and code regulations will be satisfied. Owners most often procure the services of an architect or an architect/engineering/contracting firm (AEC) to help in the development of project definition and feasibility, referred to as the pre-design phase.

The design of a building involves the utilization of space and all the factors that go with it. This includes physical requirements (such as a series of spaces or room for mechanical equipment), psychological requirements that reflect the attitudes and behavior of those using the space, and the need for those in the building to carry out the activities expected to occur in the building. In addition, factors involving the use of materials, energy efficiency, and economic utilization of space are considered. Today's developers are involved with treating the public right of way or contributing to public open space as part of the development process. In some cases, the project team may be required to create a scope of work that involves new sidewalks, curbs and gutters, or street lighting.

Pre-Design and Design Development

Once an architect or AEC firm has been identified, the pre-design portion of the work can begin. In this phase, the architect, other design consultants, and the owner mutually determine the goals and objectives of the project. Design sketches and feasibility studies are developed as a series of alternatives for approval by the owner. Early design efforts focus on establishing what is known as the design intent. **Design intent** is a statement that defines the anticipated aesthetic, functional, and performance characteristics of the finished building or project (Fig. 1.6). Functional and spatial requirements, including sizes and adjacencies of spaces, are defined and tabulated. The architect will conduct initial checks to insure that zoning and building code requirements can be met within the design concept. Surveys and drawings of existing conditions are obtained or prepared. The result of pre-design work will be a written program. A program is a written document that explains design intentions, controls, and standards for a project, including detailed space requirements and the types of equipment and systems to be used. This phase is complete when the owner and architect agree that the scope of work, anticipated construction cost, and time schedule are well defined.

During design development, initial design ideas are further elaborated into detailed drawings of the building indicating exact sizes and relationships between building elements. Architects use a system of orthographic drawings of a building or structure to simplify the graphic understanding of complex forms. The **floor plan** is a representation of a building viewed from above after a horizontal plane has been cut through it and the top portion removed (Fig. 1.7). A **building section** gives

Figure 1.6 The Chicago Center for Green Technology was planned with a design intent of achieving a healthy indoor environment, reducing energy use, and providing water efficient landscape features. *(Courtesy Eva Kultermann with thanks to the Chicago Department of the Environment)*

Figure 1.7 The floor plan is the most often referred to drawing in a set of construction documents, showing the relationship of spaces and overall building dimensions. *(Courtesy Eva Kultermann)*

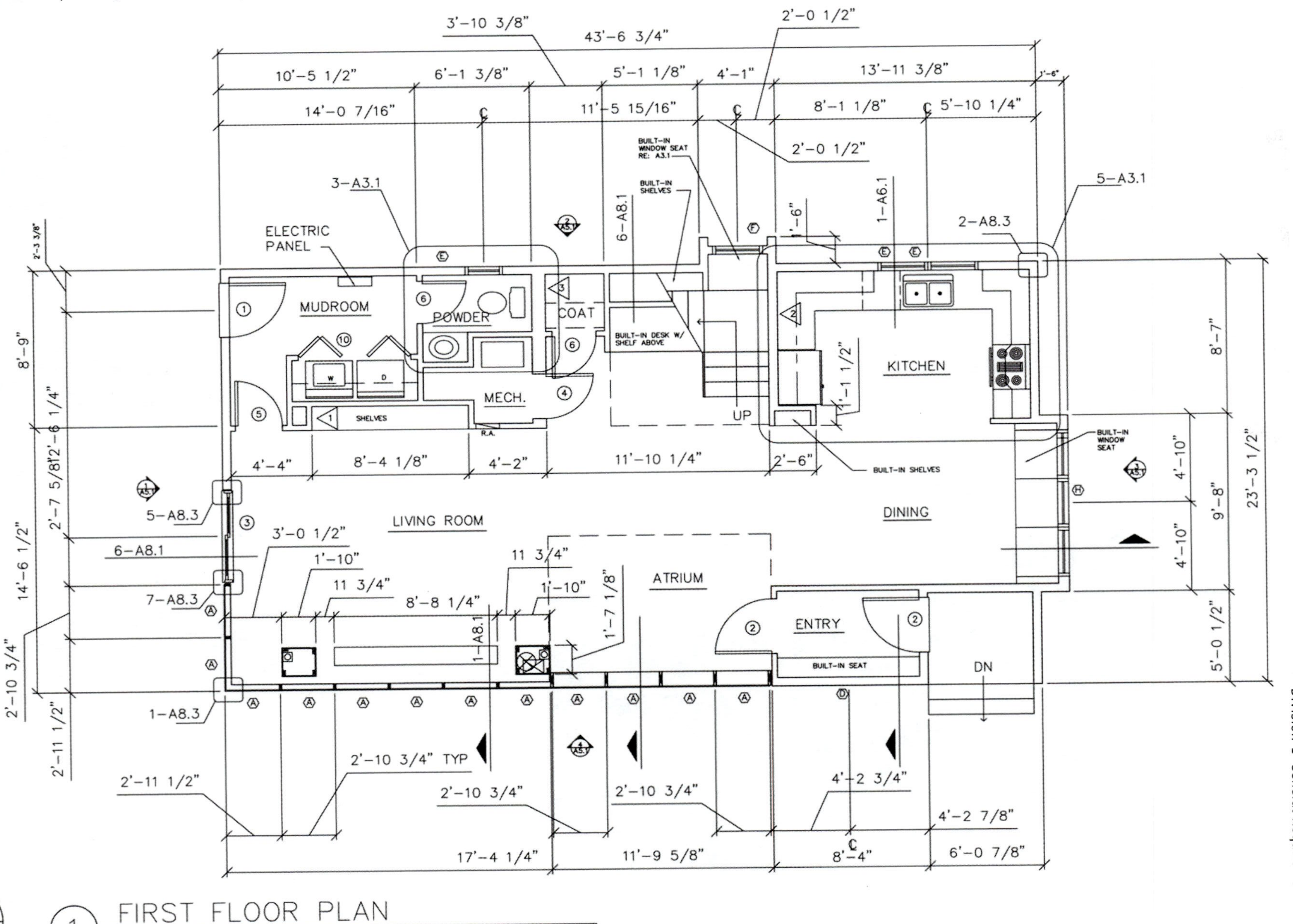

a view of a building after a vertical plane has been cut through it and the front portion removed (Fig. 1.8). An **elevation** drawing shows the exterior façade of a building, delineating geometries and the materials of construction (Fig. 1.9).

Because of the multitude of systems under consideration at this stage, the architect will normally hire the services of other design consultants in the development of systems. Structural, mechanical, electrical, plumbing, and civil engineers, fire protection consultants, interior and landscape designers, and cost estimating consultants are common on most projects of larger scale. Depending on the type of building being designed, other specialty consultants, such as traffic and parking professionals, lighting designers, or acoustical specialists, may be brought in to work with the team. During this stage of the work, most substantive decisions regarding the structural frame, mechanical systems, materials, and construction methods are determined. If a contractor for the project has been selected it is advisable to include them in the development of solutions for constructability and other technical aspects of the construction.

Figure 1.8 A building section graphically illustrates a cut through the building from foundation to roof, showing rooms and the materials of construction. *(Courtesy Eva Kultermann)*

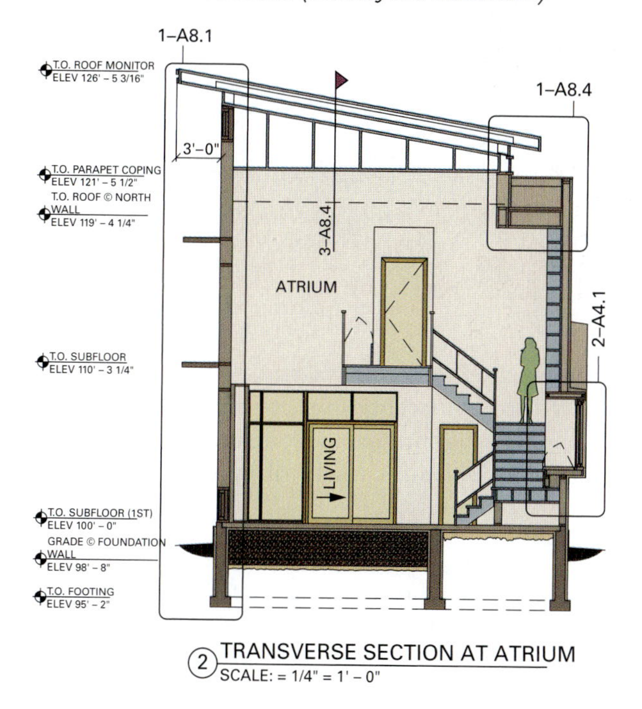

Figure 1.9 An elevation drawing shows the exterior façade of a building, delineating geometries, building heights and the materials of construction. *(Courtesy Eva Kultermann)*

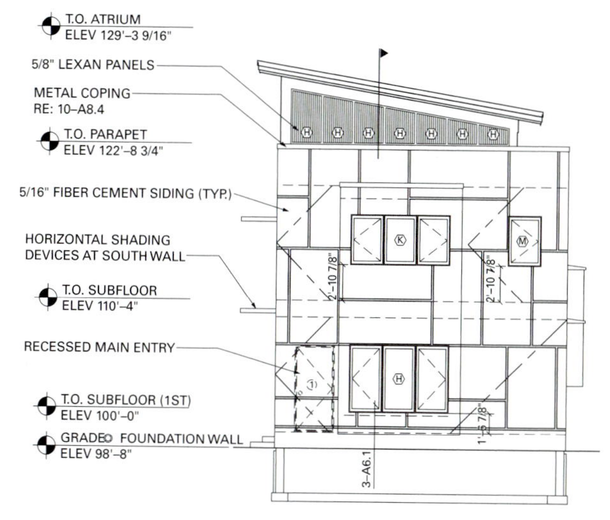

Construction Documents (CD)

Once all basic decisions have been approved by the owner, the architect and all consultants will commence with the final phase of the design process, the preparation of construction documents. **Construction documents** consist of two interdependent components—the drawings and the specifications. **Construction drawings**, sometimes referred to as working drawings, show the dimensional relationships between all aspects of the building: their form, sizes, and quantities. These drawings, usually computer generated, are dimensioned, indicating the physical relationship between components—their location, range, the materials used, and their colors and textures. Drawing sets are organized from the general, overall building plans, sections and elevations, down to the level of the detail, indicating fasteners and connections. Full drawing sets are collated according to discipline and include civil, architectural, structural, mechanical, electrical, plumbing (MEP), and fire protection sheets among others (Fig. 1.10). The drawings are the basis on which the contractor generates cost estimates and are used to guide the actual construction of the building on the site.

Since working drawings cannot give all of the details involved in describing the quality and performance characteristics of specific materials, construction methods, and project procedures, the construction documents include a written manual called the specifications. The **specifications** describe in writing more detailed information on the exact types of materials to be used and the ways in which construction processes are conducted. The drawings graphically indicate the relationship between elements, while the specifications spell out the specific types, qualities, and properties of those materials, and their installation.

Figure 1.10 Construction drawings are organized sequentially from the general to the detailed and in order of design discipline.

INDEX OF DRAWINGS

COVER	CODE MATRIX/SITE LOCATION
C1.1	SITE PLAN
ARCHITECTURAL	
A1.1	BASEMENT AND 1ST FLOOR DEMOLITION PLANS
A1.2	2ND FLOOR AND ROOF DEMOLITION PLANS
A2.1	BASEMENT AND 1ST FLOOR PLANS
A2.1	2ND FLOOR PLAN/DOOR AND FINISH SCHEDULE
A3.1	ENLARGED INTERIOR PLANS
A4.1	1ST AND 2ND FLOOR REFLECTED CEILING PLANS
A5.1	EXTERIOR ELEVATIONS
A6.1	BUILDING SECTIONS
A7.1	STAIR SECTIONS AND DETAILS
A8.1	ENLARGED WALL SECTIONS
A9.1	DETAILS
A9.2	SOLAR CHIMNEY DETAILS
A9.3	MILWORK AND TRIM DETAILS
STRUCTURAL	
S1.1	FOUNDATION PLANS
S1.2	1ST AND 2ND STRUCTURAL FLOOR PLANS
S1.3	STRUCTURAL DETAILS
MECHANICAL	
M1.1	1ST AND 2ND FLOOR MECHANICAL PLANS
M1.2	MECHANICAL DETAILS AND SCHEDULES
PLUMBING	
P1.1	PLUMBING RISER DIAGRAMS
P1.2	PLUMBING SCHEDULES AND DETAILS
ELECTRICAL	
E1.1	BASEMENT ELECTRICAL PLANS AND NOTES
E1.2	1ST AND 2ND FLOOR POWER PLANS
E1.3	1ST AND 2ND FLOOR LIGHTING PLANS

Specifications and the MasterFormat

Construction specifications for buildings and other projects are written using the MasterFormat. The MasterFormat system was developed by the Construction Specifications Institute (CSI) and Construction Specifications Canada. It is used by U.S. and Canadian construction companies and material suppliers, the McGraw-Hill Information Systems as a basis for their Sweet's Catalog Files, and the R.S. Means Company construction cost data publications. MasterFormat provides a standard for writing specifications using a system of descriptive titles and numbers to organize construction activities, products, and requirements into a standard order that facilitates the retrieval of information and serves as a means for all participants within the construction industry to communicate (Fig. 1.11). Since the design and completion of construction projects involves individuals in many technical fields, the ability to effectively communicate by having a standard sequence for identifying and referring to construction information is essential.

The numbers and titles in MasterFormat are divided into fifty basic groupings called divisions, thirty-four of which are active, with sixteen reserved for future expansion. Each division has a title and identifying number. Division one describes the general requirements of the contract, outlining administrative methods for a construction project, such as project management procedures and construction facilities and controls. Divisions two through nineteen, the *Facility Construction Subgroup*, deals mainly with the materials of construction for buildings. The *Facilities Services Subgroup*, divisions twenty through twenty-nine, cover mechanical, electrical, plumbing, fire protection, and communications equipment. Divisions thirty through thirty-nine, the

Figure 1.11 The 2004 MasterFormat is used by U.S. and Canadian construction companies and material suppliers to organize construction activities, products, and requirements into a standard order. *(Courtesy Construction Specifications Institute)*

Site Facilities Subgroup, are concerned with earthwork, transportation, and marine construction topics, while the final subgroup, *Process Equipment*, deals with larger industrial processes.

Divisions one through sixteen, which previously encompassed the entire MasterFormat, are primarily concerned with the construction of buildings. The division ordering is loosely arranged to follow the sequence of the construction process itself, helping to facilitate easy recall of where information is located. Division three, concrete, is the material from which most foundations are built. The divisions move on through major structural and cladding materials, to interior furnishings and components, and finally mechanical and electrical systems.

Within each division, specifications are identified in numbered sections, each of which covers a particular part of the work of the division. Each section is identified by three pairs of numbers, defining a further level of specificity. The first two digits are the same as the division number, and are referred to as level one numbers. Each of the following pairs of numbers defines a further level of detail, referred to as level two and three. Level two titles cover the broad categories and are the major section titles, while level three titles deal with more specific information. Each level three heading is further subdivided into three parts. Part one describes the general requirements for the category dealing with definitions, referencing standards, quality control, and warranties. Part two describes the actual materials and products, their physical properties, manufacturers, and performance requirements. Part three describes how and under what circumstances the product must be installed in the field. Within this system every building component can be easily found under its division and number.

In addition to the purely technical information, division 00 of the specifications outlines the guidelines for the pre-construction phase of the project. These include the General Conditions of the contract, a blueprint for the construction delivery process. Division 00 addresses such issues as how the bidding process will be conducted, what kinds of bonds and insurance the contractor is required to hold, and when the actual on-site construction may commence. It includes the bid forms that the contractor will use in submitting the proposal for construction. The divisions and sections of the 2004 MasterFormat are shown in Table 1.1.

Table 1.1 Level Two Numbers and Titles of the 2004 CSI MasterFormat™

Number	Title
Procurement and Contracting Requirements Group Divsion 00—Procurement and Contracting Requirements	
Introductory Information	
Procurement Requirements	
00 10 00	Solicitation
00 11 00	Advertisements and Invitations
00 20 00	Instructions for Procurement
00 21 00	Instructions
00 22 00	Supplementary Instructions
00 23 00	Procurement Definitions
00 24 00	Procurement Scopes
00 25 00	Procurement Meetings
00 26 00	Procurement Substitution Procedures
00 30 00	Available Information
00 31 00	Available Project Information
00 40 00	Procurement Forms and Supplements
00 41 00	Bid Forms
00 42 00	Proposal Forms
00 43 00	Procurement Form Supplements
00 45 00	Representations and Certifications
Contracting Requirements	
00 50 00	Contracting Forms and Supplements
00 51 00	Notice of Award
00 54 00	Agreement Form Supplements
00 55 00	Notice to Proceed
00 60 00	Project Forms
00 61 00	Bond Forms
00 62 00	Certificates and Other Forms
00 63 00	Clarification and Modification Forms
00 65 00	Closeout Forms
00 70 00	Conditions of the Contract
00 71 00	Contracting Definitions
00 72 00	General Conditions
00 73 00	Supplementary Conditions
00 80 00	*Unassigned*
00 90 00	Revisions, Clarifications, and Modifications
00 91 00	Precontract Revisions
00 93 00	Record Clarifications and Proposals
00 94 00	Record Modifications
Specifications Group	
General Requirements Subgroup	
Division 01—General Requirements	
01 00 00	Summary
01 11 00	Summary of Work
01 12 00	Multiple Contract Summary
01 14 00	Work Restrictions
01 18 00	Project Utility Sources
01 20 00	Price and Payment Procedures
01 21 00	Allowances
01 22 00	Unit Prices
01 23 00	Alternates
01 24 00	Value Analysis
01 25 00	Substitution Procedures
01 26 00	Contract Modification Procedures
01 29 00	Payment Procedures
01 30 00	Administrative Requirements

(Continued)

Table 1.1 Level Two Numbers and Titles of the 2004 CSI MasterFormat™ *(Continued)*

Procurement and Contracting Requirements Group Divsion 00—Procurement and Contracting Requirements	
01 31 00	Project Management and Coordination
01 32 00	Construction Progress Documentation
01 33 00	Submittal Procedures
01 35 00	Special Procedures
01 40 00	Quality Requirements
01 41 00	Regulatory Requirements
01 42 00	References
01 43 00	Quality Assurance
01 45 00	Quality Control
01 50 00	Temporary Facilities and Controls
01 51 00	Temporary Utilities
01 52 00	Construction Facilities
01 53 00	Temporary Construction
01 54 00	Construction Aids
01 55 00	Vehicular Access and Parking
01 56 00	Temporary Barriers and Enclosures
01 57 00	Temporary Controls
01 58 00	Project Identification
01 60 00	Product Requirements
01 61 00	Common Product Requirements
01 62 00	Product Options
01 64 00	Owner-Supplied Products
01 65 00	Product Delivery Requirements
01 66 00	Product Storage and Handling Requirements
01 70 00	Execution and Closeout Requirements
01 71 00	Examination and Preparation
01 73 00	Execution
01 74 00	Cleaning and Waste Management
01 75 00	Starting and Adjusting
01 76 00	Protecting Installed Construction
01 77 00	Closeout Procedures
01 78 00	Closeout Submittals
01 79 00	Demonstration and Training
01 80 00	Performance Requirements
01 81 00	Facility Performance Requirements
01 82 00	Facility Substructure Performance Requirements
01 83 00	Facility Shell Performance Requirements
91 84 00	Interiors Performance Requirements
01 85 00	Conveying Equipment Performance Requirements
01 86 00	Facility Services Performance Requirements
01 87 00	Equipment and Furnishings Performance Requirements
01 88 00	Other Facility Construction Performance Requirements
01 89 00	Site Construction Performance Requirements
01 90 00	Life Cycle Activities
01 91 00	Commissioning
01 92 00	Facility Operation
01 93 00	Facility Maintenance
01 94 00	Facility Decommissioning
Facility Construction Subgroup	
Division 02—Existing Conditions	
02 01 00	Maintenance of Existing Conditions
02 05 00	Common Work Results for Existing Conditions
02 06 00	Schedules for Existing Conditions
02 08 00	Commissioning of Existing Conditions
02 10 00	*Unassigned*
02 20 00	Assessment
02 21 00	Surveys
02 22 00	Existing Conditions Assessment
02 24 00	Environmental Assessment
02 25 00	Existing Material Assessment
02 26 00	Hazardous Material Assessment
02 30 00	Subsurface Investigation
02 31 00	Geophysical Investigations
02 32 00	Geotechnical Investigations
02 40 00	Demolition and Structure Moving
02 41 00	Demolition
02 42 00	Removal and Salvage of Construction Materials
02 43 00	Structure Moving
02 50 00	Site Remediation
02 51 00	Physical Decontamination
02 52 00	Chemical Decontamination
02 53 00	Thermal Decontamination
02 54 00	Biological Decontamination
02 55 00	Remediation Soil Stabilization
02 56 00	Site Containment
02 57 00	Sinkhole Remediation
02 58 00	Snow Control
02 60 00	Contaminated Site Material Removal
02 61 00	Removal and Disposal of Contaminated Soils
02 62 00	Hazardous Waste Recovery Processes
02 65 00	Underground Storage Tank Removal
02 66 00	Landfill Construction and Storage
02 70 00	Water Remediation
02 71 00	Groundwater Treatment
02 72 00	Water Decontamination
02 80 00	Facility Remediation
02 81 00	Transportation and Disposal of Hazardous Materials
02 82 00	Asbestos Remediation
02 83 00	Lead Remediation
02 84 00	Polychlorinate Biphenyl Remediation
02 85 00	Mold Remediation
02 86 00	Hazardous Waste Drum Handling
02 90 00	*Unassigned*
Division 03—Concrete	
03 01 00	Maintenance of Concrete
03 05 00	Common Work Results for Concrete
03 06 00	Schedules for Concrete
03 08 00	Commissioning of Concrete
03 10 00	Concrete Forming and Accessories
01 11 00	Concrete Forming
03 15 00	Concrete Accessories
03 20 00	Concrete Reinforcing
03 21 00	Reinforcing Steel
03 22 00	Welded Wire Fabric Reinforcing
03 23 00	Stressing Tendons
03 24 00	Fibrous Reinforcing
03 30 00	Cast-in-Place Concrete
03 31 00	Structural Concrete
03 33 00	Architectural Concrete
03 34 00	Low Density Concrete
03 35 00	Concrete Finishing
03 37 00	Specialty Placed Concrete
03 38 00	Post-Tensioned Concrete
03 39 00	Concrete Curing
03 40 00	Precast Concrete
03 41 00	Precast Structural Concrete

Table 1.1 Level Two Numbers and Titles of the 2004 CSI MasterFormat™ *(Continued)*

Procurement and Contracting Requirements Group Divsion 00—Procurement and Contracting Requirements	
03 45 00	Precast Architectural Concrete
03 47 00	Site-Cast Concrete
03 48 00	Precast Concrete Specialties
03 49 00	Glass-Fiber-Reinforced Concrete
03 50 00	Cast Decks and Underlayment
03 51 00	Cast Roof Decks
03 52 00	Lightweight Concrete Roof Insulation
03 53 00	Concrete Topping
03 54 00	Cast Underlayment
03 60 00	Grouting
03 61 00	Cementitious Grouting
03 62 00	Non-Shrink Grouting
03 63 00	Epoxy Grouting
03 64 00	Injection Grouting
03 70 00	Mass Concrete
03 71 00	Mass Concrete for Raft Foundations
03 72 00	Mass Concrete for Dams
03 80 00	Concrete Cutting and Boring
03 81 00	Concrete Cutting
03 82 00	Concrete Boring
03 90 00	*Unassigned*
Division 04—Masonry	
04 01 00	Maintenance of Masonry
04 05 00	Common Work Results for Masonry
04 06 00	Schedules for Masonry
04 08 00	Commissioning of Masonry
04 10 00	*Unassigned*
04 20 00	Unit Masonry
04 21 00	Clay Unit Masonry
04 22 00	Concrete Unit Masonry
04 23 00	Glass Unit Masonry
04 24 00	Adobe Unit Masonry
04 25 00	Unit Masonry Panels
04 27 00	Multiple-Wythe Unit Masonry
04 28 00	Concrete Form Masonry Units
04 30 00	*Unassigned*
04 40 00	Stone Assemblies
04 41 00	Dry-Placed Stone
04 42 00	Exterior Stone Cladding
04 43 00	Stone Masonry
04 50 00	Refractory Masonry
04 51 00	Flue Liner Masonry
04 52 00	Combustion Chamber Masonry
04 53 00	Castable Refractory Masonry
04 54 00	Refractory Brick Masonry
04 57 00	Masonry Fireplaces
04 60 00	Corrosion-Resistant Masonry
04 61 00	Chemical-Resistant Brick Masonry
04 62 00	Vitrified Clay Liner Plate
04 70 00	Manufactured Masonry
04 71 00	Manufactured Brick Masonry
04 72 00	Cast Stone Masonry
04 73 00	Manufactured Stone Masonry
04 80 00–04 90 00	*Unassigned*
Division 05—Metals	
05 01 00	Maintenance of Metals
05 05 00	Common Work Results for Metals
05 06 00	Schedules for Metals
05 08 00	Commissioning of Metals
05 10 00	Structural Metal Framing
05 12 00	Structural Steel Framing
05 13 00	Structural Stainless-Steel Framing
05 14 00	Structural Aluminum Framing
05 15 00	Wire Rope Assemblies
05 16 00	Structural Cabling
05 20 00	Metal Joists
05 21 00	Steel Joist Framing
05 25 00	Aluminum Joist Framing
05 30 00	Metal Decking
05 31 00	Steel Decking
05 33 00	Aluminum Decking
05 34 00	Acoustical Metal Decking
05 35 00	Raceway Decking Assemblies
05 36 00	Composite Metal Decking
05 40 00	Cold-Formed Metal Framing
04 41 00	Structural Metal Stud Framing
05 42 00	Cold-Formed Metal Joist Framing
05 43 00	Slotted Channel Framing
05 44 00	Cold-Formed Metal Trusses
05 45 00	Metal Support Assemblies
05 50 00	Metal Fabrications
05 51 00	Metal Stairs
05 52 00	Metal Railings
05 53 00	Metal Gratings
05 54 00	Metal Floor Plates
05 55 00	Metal Stair Treads and Nosings
05 56 00	Metal Castings
05 58 00	Formed Metal Fabrications
05 59 00	Metal Specialties
05 60 00	*Unassigned*
05 70 00	Decorative Metal
05 71 00	Decorative Metal Stairs
05 73 00	Decorative Metal Railings
05 74 00	Decorative Metal Castings
05 75 00	Decorative Formed Metal
05 76 00	Decorative Forged Metal
05 80 00–05 90 00	*Unassigned*
Division 06—Wood, Plastics, and Composites	
06 01 00	Maintenance of Wood, Plastics, and Composites
06 05 00	Common Work Results for Wood, Plastics, and Composites
06 06 00	Schedules for Wood, Plastics, and Composites
06 08 00	Commissioning of Wood, Plastics, and Composites
06 10 00	Rough Carpentry
06 11 00	Wood Framing
06 12 00	Structural Panels
06 13 00	Heavy Timber
06 14 00	Treated Wood Foundations
06 15 00	Wood Decking
06 16 00	Sheathing
06 17 00	Shop-Fabricated Structural Wood
06 18 00	Glued-Laminated Construction
06 20 00	Finish Carpentry
06 22 00	Millwork
06 25 00	Prefinished Paneling

(Continued)

Table 1.1 Level Two Numbers and Titles of the 2004 CSI MasterFormat™ *(Continued)*

Procurement and Contracting Requirements Group Divsion 00—Procurement and Contracting Requirements	
06 26 00	Board Paneling
06 30 00	*Unassigned*
06 40 00	Architectural Woodwork
06 41 00	Architectural Wood Casework
06 42 00	Wood Paneling
06 43 00	Wood Stairs and Railings
06 44 00	Ornamental Woodwork
06 46 00	Wood Trim
06 48 00	Wood Frames
06 49 00	Wood Screens and Exterior Wood Shutters
06 50 00	Structural Plastics
06 51 00	Structural Plastic Shapes and Plates
06 52 00	Plastic Structural Assemblies
06 53 00	Plastic Decking
06 60 00	Plastic Fabrications
06 61 00	Simulated Stone Fabrifications
06 63 00	Plastic Railings
06 64 00	Plastic Paneling
06 65 00	Plastic Simulated Wood Trim
06 66 00	Custom Ornamental Simulated Woodwork
06 70 00	Structural Composites
06 71 00	Structural Composite Shapes and Plates
06 72 00	Composite Structural Assemblies
06 73 00	Composite Decking
06 80 00	Composite Fabrifications
06 81 00	Composite Railings
06 82 00	Glass-Fiber-Reinforced Plastic
06 90 00	*Unassigned*
Division 07—Thermal and Moisture Protection	
07 01 00	Operation and Maintenance of Thermal and Moisture Protection
07 05 00	Common Work Results for Thermal and Moisture Protection
07 06 00	Schedules for Thermal and Moisture Protection
07 08 00	Commissioning of Thermal and Moisture Protection
07 10 00	Dampproofing and Waterproofing
07 11 00	Dampproofing
07 12 00	Built-Up Bituminous Waterproofing
07 13 00	Sheet Waterproofing
07 14 00	Fluid-Applied Waterproofing
07 15 00	Sheet Metal Waterproofing
07 16 00	Cementitious and Reactive Waterproofing
07 17 00	Bentonite Waterproofing
07 18 00	Traffic Coatings
07 19 00	Water Repellents
07 20 00	Thermal Protection
07 21 00	Thermal Insulation
07 22 00	Roof and Deck Insulation
07 24 00	Exterior Insulation and Finish Systems
07 25 00	Weather Barriers
07 26 00	Vapor Retarders
07 27 00	Air Barriers
07 30 00	Steep Slope Roofing
07 31 00	Shingles and Shakes
07 32 00	Roof Tiles
07 33 00	Natural Roof Coverings
07 40 00	Roofing and Siding Panels
07 41 00	Roof Panels
07 42 00	Wall Panels
07 44 00	Faced Panels
07 46 00	Siding
07 50 00	Membrane Roofing
07 51 00	Built-Up Bituminous Roofing
07 52 00	Modified Bituminous Membrane Roofing
07 53 00	Elastomeric Membrane Roofing
07 54 00	Thermoplastic Membrane Roofing
07 55 00	Protected Membrane Roofing
07 56 00	Fluid-Applied Roofing
07 57 00	Coated Foamed Roofing
07 58 00	Roll Roofing
07 60 00	Flashing and Sheet Metal
07 61 00	Sheet Metal Roofing
07 62 00	Sheet Metal Flashing and Trim
07 63 00	Sheet Metal Roofing Specialties
07 65 00	Flexible Flashing
07 70 00	Roof and Wall Specialties and Accessories
07 71 00	Roof Specialties
07 72 00	Roof Accessories
07 76 00	Roof Pavers
07 77 00	Wall Specialties
07 80 00	Fire and Smoke Protection
07 81 00	Applied Fireproofing
07 82 00	Board Fireproofing
07 84 00	Firestopping
07 86 00	Smoke Seals
07 87 00	Smoke Containment Barriers
07 90 00	Joint Protection
07 91 00	Preformed Joint Seals
07 92 00	Joint Sealants
07 95 00	Expansion Control
Division 08—Openings	
08 01 00	Operation and Maintenance of Openings
08 05 00	Common Work Results for Openings
08 06 00	Schedules for Openings
08 08 00	Commissioning of Openings
08 10 00	Doors and Frames
08 11 00	Metal Doors and Frames
08 12 00	Metal Frames
03 13 00	Metal Doors
08 14 00	Wood Doors
08 15 00	Plastic Doors
08 16 00	Composite Doors
08 17 00	Integrated Door Opening Assemblies
08 20 00	*Unassigned*
08 30 00	Specialty Doors and Frames
08 31 00	Access Doors and Panels
08 32 00	Sliding Glass Doors
08 33 00	Coiling Doors and Grilles
08 34 00	Special Function Doors
08 35 00	Folding Doors and Grilles
08 36 00	Panel Doors
08 38 00	Traffic Doors
08 39 00	Pressure-Resistant Doors
08 40 00	Entrances, Storefronts, and Curtain Walls
03 41 00	Entrances and Storefronts
08 42 00	Entrances

Table 1.1 Level Two Numbers and Titles of the 2004 CSI MasterFormat™ *(Continued)*

Procurement and Contracting Requirements Group Divsion 00—Procurement and Contracting Requirements	
08 43 00	Storefronts
08 44 00	Curtain Wall and Glazed Assemblies
08 45 00	Translucent Wall and Roof Assemblies
08 50 00	Windows
08 51 00	Metal Windows
08 52 00	Wood Windows
08 53 00	Plastic Windows
08 54 00	Composite Windows
08 55 00	Pressure-Resistant Windows
08 56 00	Special Function Windows
08 60 00	Roof Windows and Skylights
08 61 00	Roof Windows
08 62 00	Unit Skylights
08 63 00	Metal-Framed Skylights
08 64 00	Plastic-Framed Skylights
08 67 00	Skylight Protection and Screens
08 70 00	Hardware
08 71 00	Door Hardware
08 74 00	Access Control Hardware
08 75 00	Window Hardware
08 78 00	Special Function Hardware
08 79 00	Hardware Accessories
08 80 00	Glazing
08 81 00	Glass Glazing
08 83 00	Mirrors
08 84 00	Plastic Glazing
08 85 00	Glazing Accessories
08 87 00	Glazing Surface Films
08 88 00	Special Function Glazing
08 90 00	Louvers and Vents
08 91 00	Louvers
08 92 00	Louvered Equipment Enclosures
08 95 00	Vents
Division 09—Finishes	
09 01 00	Maintenance of Finishes
09 05 00	Common Work Results for Finishes
09 06 00	Schedules for Finishes
09 08 00	Commissioning of Finishes
09 10 00	*Unassigned*
09 20 00	Plaster and Gypsum Board
09 21 00	Plaster and Gypsum Board Assemblies
09 22 00	Supports for Plaster and Gypsum Board
09 23 00	Gypsum Plastering
09 24 00	Portland Cement Plastering
09 25 00	Other Plastering
09 26 00	Veneer Plastering
09 27 00	Plaster Fabrications
09 28 00	Backing Boards and Underlayments
09 29 00	Gypsum Board
09 30 00	Tiling
09 31 00	Thin-Set Tiling
09 32 00	Mortar-Bed Tiling
09 33 00	Conductive Tiling
09 34 00	Waterproofing-Membrane Tiling
09 35 00	Chemical-Resistant Tiling
09 40 00	*Unassigned*
09 50 00	Ceilings
09 51 00	Acoustical Ceilings
09 53 00	Acoustical Ceiling Suspension Assemblies
09 54 00	Specialty Ceilings
09 56 00	Textured Ceilings
09 57 00	Special Function Ceilings
09 58 00	Integrated Ceiling Assemblies
09 60 00	Flooring
09 61 00	Flooring Treatment
09 62 00	Specialty Flooring
09 63 00	Masonry Flooring
09 64 00	Wood Flooring
09 65 00	Resilient Flooring
09 66 00	Terrazzo Flooring
09 67 00	Fluid-Applied Flooring
09 68 00	Carpeting
09 69 00	Access Flooring
09 70 00	Wall Finishes
09 72 00	Wall Coverings
09 73 00	Wall Carpeting
09 74 00	Flexible Wood Sheets
09 75 00	Stone Facing
09 76 00	Plastic Books
09 77 00	Special Wall Surfacing
09 80 00	Acoustic Treatment
09 81 00	Acoustic Insulation
09 83 00	Acoustic Finishes
09 84 00	Acoustic Room Components
09 90 00	Painting and Coating
09 91 00	Painting
09 93 00	Staining and Transparent Finishing
09 94 00	Decorative Finishing
09 96 00	High-Performance Coatings
09 97 00	Special Coatings
Division 10—Specialties	
10 01 00	Operation and Maintenance of Specialties
10 05 00	Common Work Results for Specialties
10 06 00	Schedules for Specialties
10 08 00	Commissioning of Specialties
10 10 00	Information Specialties
10 11 00	Visual Display Surfaces
10 12 00	Display Cases
10 13 00	Directories
10 14 00	Signage
10 17 00	Telephone Specialties
10 18 00	Informational Kiosks
10 20 00	Interior Specialties
10 21 00	Compartments and Cubicles
10 22 00	Partitions
10 25 00	Service Walls
10 26 00	Wall and Door Protection
10 28 00	Toilet, Bath, and Laundry Accessories
10 30 00	Fireplaces and Stoves
10 31 00	Manufactured Fireplaces
10 32 00	Fireplace Specialties
10 35 00	Stoves
10 40 00	Safety Specialties
10 41 00	Emergency Access and Information Cabinets
10 43 00	Emergency Aid Specialties
10 44 00	Fire Protection Specialties

(Continued)

Table 1.1 Level Two Numbers and Titles of the 2004 CSI MasterFormat™ *(Continued)*

Procurement and Contracting Requirements Group Divsion 00—Procurement and Contracting Requirements	
10 50 00	Storage Specialties
10 51 00	Lockers
10 55 00	Postal Specialties
10 56 00	Storage Assemblies
10 57 00	Wardrobe and Closet Specialties
10 60 00	*Unassigned*
10 70 00	Exterior Specialties
10 71 00	Exterior Protection
10 73 00	Protective Covers
10 74 00	Manufactured Exterior Specialties
10 75 00	Flagpoles
10 80 00	Other Specialties
10 81 00	Pest Control Devices
10 82 00	Grilles and Screens
10 83 00	Flags and Banners
10 86 00	Security Mirrors and Domes
10 88 00	Scales
10 90 00	*Unassigned*
Division 11—Equipment	
11 01 00	Operation and Maintenance of Equipment
11 05 00	Common Work Results for Equipment
11 06 00	Schedules for Equipment
11 08 00	Commissioning of Equipment
11 10 00	Vehicle and Pedestrian Equipment
11 11 10	Vehicle Service Equipment
11 12 00	Parking Control Equipment
11 13 00	Loading Dock Equipment
11 14 00	Pedestrian Control Equipment
11 15 00	Security, Detention, and Banking Equipment
11 16 00	Vault Equipment
11 17 00	Teller and Service Equipment
11 18 00	Security Equipment
11 19 00	Detention Equipment
11 20 00	Commercial Equipment
11 21 00	Mercantile and Service Equipment
11 22 00	Refrigerated Display Equipment
11 23 00	Commercial Laundry and Dry Cleaning Equipment
11 24 00	Maintenance Equipment
11 25 00	Hospitality Equipment
11 26 00	Unit Kitchens
11 27 00	Photographic Processing Equipment
11 28 00	Office Equipment
11 29 00	Postal, Packaging, and Shipping Equipment
11 30 00	Residential Equipment
11 31 00	Residential Appliances
11 33 00	Retractable Stairs
11 40 00	Foodservice Equipment
11 41 00	Food Storage Equipment
11 42 00	Food Preparation Equipment
11 43 00	Food Delivery Carts and Conveyors
11 44 00	Food Cooking Equipment
11 46 00	Food Dispensing Equipment
11 47 00	Ice Machines
11 48 00	Cleaning and Disposal Equipment
11 50 00	Educational and Scientific Equipment
11 51 00	Library Equipment
11 52 00	Audio-Visual Equipment
11 53 00	Laboratory Equipment
11 55 00	Planetarium Equipment
11 56 00	Observatory Equipment
11 57 00	Vocational Shop Equipment
11 59 00	Exhibit Equipment
11 60 00	Entertainment Equipment
11 61 00	Theater and Stage Equipment
11 62 00	Musical Equipment
11 65 00	Athletic and Recreational Equipment
11 66 00	Athletic Equipment
11 67 00	Recreational Equipment
11 68 00	Play Field Equipment and Structures
11 70 00	Healthcare Equipment
11 71 00	Medical Sterilizing Equipment
11 72 00	Examination and Treatment Equipment
11 73 00	Patient Care Equipment
11 74 00	Dental Equipment
11 75 00	Optical Equipment
11 76 00	Operating Room Equipment
11 77 00	Radiology Equipment
11 78 00	Mortuary Equipment
11 79 00	Therapy Equipment
11 80 00	Collection and Disposal Equipment
11 82 00	Solid Waste Handling Equipment
11 90 00	Other Equipment
11 91 00	Religious Equipment
11 92 00	Agricultural Equipment
11 93 00	Horticultural Equipment
Division 12—Furnishings	
12 01 00	Operation and Maintenance of Furnishings
12 05 00	Common Work Results for Furnishings
12 06 00	Schedules for Furnishings
12 08 00	Commissioning of Furnishings
12 10 00	Art
12 11 00	Murals
12 12 00	Wall Decorations
12 14 00	Sculptures
12 17 00	Art Glass
12 19 00	Religious Art
12 20 00	Window Treatments
12 21 00	Window Blinds
12 22 00	Curtains and Drapes
12 23 00	Interior Shutters
12 24 00	Window Shades
12 25 00	Window Treatment Operating Hardware
12 30 00	Casework
12 31 00	Manufactured Metal Casework
12 32 00	Manufactured Wood Casework
12 34 00	Manufactured Plastic Casework
12 35 00	Specialty Casework
12 36 00	Countertops
12 40 00	Furnishings and Accessories
12 41 00	Office Accessories
12 42 00	Table Accessories
12 43 00	Portable Lamps
12 44 00	Bath Furnishings
12 45 00	Bedroom Furnishings
12 46 00	Furnishing Accessories
12 48 00	Rugs and Mats

Table 1.1 Level Two Numbers and Titles of the 2004 CSI MasterFormat™ *(Continued)*

Procurement and Contracting Requirements Group Divsion 00—Procurement and Contracting Requirements	
12 50 00	Furniture
12 51 00	Office Furniture
12 52 00	Seating
12 53 00	Retail Furniture
12 54 00	Hospitality Furniture
12 55 00	Detention Furniture
12 56 00	Institutional Furniture
12 57 00	Industrial Furniture
12 58 00	Residential Furniture
12 59 00	Systems Furniture
12 60 00	Multiple Seating
12 61 00	Fixed Audience Seating
12 62 00	Portable Audience Seating
12 63 00	Stadium and Arena Seating
12 64 00	Booths and Tables
12 65 00	Multiple-Use Fixed Seating
12 66 00	Telescoping Stands
12 67 00	Pews and Benches
12 68 00	Seat and Table Assemblies
12 70 00–12 80 00	*Unassigned*
12 90 00	Other Furnishings
12 92 00	Interior Planters and Artificial Plants
12 93 00	Site Furnishings
Division 13—Special Construction	
13 01 00	Operation and Maintenance of Special Construction
13 05 10	Common Work Results for Special Construction
13 06 00	Schedules for Special Construction
13 08 00	Commissioning of Special Construction
13 10 00	Special Facility Components
13 11 00	Swimming Pools
13 12 00	Fountains
13 13 00	Aquariums
13 14 00	Amusement Park Structures and Equipment
13 17 00	Tubs and Pools
13 18 00	Ice Rinks
13 19 00	Kennels and Animal Shelters
13 20 00	Special Purpose Rooms
13 21 00	Controlled Environment Rooms
13 22 00	Office Shelters and Booths
13 23 00	Planetariums
13 24 00	Special Activity Rooms
13 26 00	Fabricated Rooms
13 27 00	Vaults
13 28 00	Athletic and Recreational Special Construction
13 30 00	Special Structures
13 31 00	Fabric Structures
13 32 00	Space Frames
13 33 00	Geodesic Structures
13 34 00	Fabricated Engineered Structures
13 36 00	Towers
13 40 00	Integrated Construction
13 42 00	Building Modules
13 44 00	Modular Mezzanines
13 48 00	Sound, Vibration, and Seismic Control
13 49 00	Radiation Protection
13 50 00	Special Instrumentation
13 51 00	Stress Instrumentation
13 52 00	Seismic Instrumentation
13 53 00	Meteorological Instrumentation
13 60 00–13 90 00	*Unassigned*
Division 14—Conveying Equipment	
14 01 00	Operation and Maintenance of Conveying Equipment
14 05 00	Common Work Results for Conveying Equipment
14 06 00	Schedules for Conveying Equipment
14 08 00	Commissioning of Conveying Equipment
14 10 00	Dumbwaiters
14 11 00	Manual Dumbwaiters
14 12 00	Electric Dumbwaiters
14 14 00	Hydraulic Dumbwaiters
14 20 00	Elevators
14 21 00	Electric Traction Elevators
14 24 00	Hydraulic Elevators
14 26 00	Limited-Use/Limited-Application Elevators
14 27 00	Custom Elevator Cabs
14 28 00	Elevator Equipment and Controls
14 30 00	Escalators and Moving Walks
14 31 00	Escalators
14 32 00	Moving Walks
14 33 00	Moving Ramps
14 40 00	Lifts
14 41 00	People Lifts
14 42 00	Wheelchair Lifts
14 43 00	Platform Lifts
14 44 00	Sidewalk Lifts
14 45 00	Vehicle Lifts
14 50 00–14 60 00	*Unassigned*
14 70 00	Turntables
14 71 00	Industrial Turntables
14 72 00	Hospitality Turntables
14 73 00	Exhibit Turntables
14 74 00	Entertainment Turntables
14 80 00	Scaffolding
14 81 00	Suspended Scaffolding
14 82 00	Rope Climbers
14 83 00	Elevating Platforms
14 84 00	Powered Scaffolding
14 90 00	Other Conveying Equipment
14 91 00	Facility Chutes
14 92 00	Pneumatic Tube Systems
Divisions 15–19	**Reserved for Future Expansion**
Facility Services Subgroup	
Division 20	**Reserved for Future Expansion**
Division 21—Fire Suppression	
21 01 00	Operation and Maintenance of Fire Suppression
21 05 00	Common Work Results for Fire Suppression
21 06 00	Schedules for Fire Suppression
21 07 00	Fire Suppression Systems Insulation
21 08 00	Commissioning of Fire Suppression
21 09 00	Instrumentation and Control for Fire-Suppression Systems
21 10 00	Water-Based Fire-Suppression Systems
21 11 00	Facility Fire-Suppression Water-Service Piping
21 12 00	Fire-Suppression Standpipes
21 13 00	Fire-Suppression Sprinkler Systems
21 20 00	Fire-Extinguishing Systems

(Continued)

Table 1.1 Level Two Numbers and Titles of the 2004 CSI MasterFormat™ *(Continued)*

Procurement and Contracting Requirements Group Divsion 00—Procurement and Contracting Requirements	
21 21 00	Carbon-Dioxide Fire-Extinguishing Systems
21 22 00	Clean-Agent Fire-Extinguishing Systems
21 23 00	Wet-Chemical Fire-Extinguishing Systems
21 24 00	Dry-Chemical Fire-Extinguishing Systems
21 30 00	Fire Pumps
21 31 00	Centrifugal Fire Pumps
21 32 00	Vertical-Turbine Fire Pumps
21 33 00	Positive-Displacement Fire Pumps
21 40 00	Fire-Suppression Water Storage
21 41 00	Storage Tanks for Fire-Suppression Water
21 50 00–21 90 00	*Unassigned*
Division 22—Plumbing	
21 01 00	Operation and Maintenance of Plumbing
22 05 00	Common Work Results for Plumbing
22 06 00	Schedules for Plumbing
22 07 00	Plumbing Insulation
22 08 00	Commissioning of Plumbing
22 09 00	Instrumentation and Control for Plumbing
22 10 00	Plumbing Piping and Pumps
22 11 00	Facility Water Distribution
22 12 00	Facility Potable-Water Storage Tanks
22 13 00	Facility Sanitary Sewerage
22 14 00	Facility Storm Drainage
22 15 00	General Service Compressed-Air Systems
22 20 00	*Unassigned*
22 30 00	Plumbing Equipment
22 31 00	Domestic Water Softeners
22 32 00	Domestic Water Filtration Equipment
22 33 00	Electric Domestic Water Heaters
22 34 00	Fuel-Fired Domestic Water Heaters
22 35 00	Domestic Water Heat Exchangers
22 40 00	Plumbing Fixtures
22 41 00	Residential Plumbing Fixtures
22 42 00	Commercial Plumbing Fixtures
22 43 00	Healthcare Plumbing Fixtures
22 43 00	Healthcare Plumbing Fixtures
22 45 00	Emergency Plumbing Fixtures
22 46 00	Security Plumbing Fixtures
22 47 00	Drinking Fountains and Water Coolers
22 50 00	Pool and Fountain Plumbing Systems
22 51 00	Swimming Pool Plumbing Systems
22 52 00	Fountain Plumbing Systems
22 60 00	Gas and Vacuum Systems for Laboratory and Healthcare Facilities
22 61 00	Compressed-Air Systems for Laboratory and Healthcare Facilities
22 62 00	Vacuum Systems for Laboratory and Healthcare Facilities
22 63 00	Gas Systems for Laboratory and Healthcare Facilities
22 66 00	Chemical-Waste Systems for Laboratory and Healthcare Facilities
22 67 00	Processed Water Systems for Laboratory and Healthcare Facilities
22 70 00–22 90 00	*Unassigned*
Division 23—Heating, Ventilating, and Air-conditioning (HVAC)	
23 01 00	Operation and Maintenance of HVAC Systems
23 05 00	Common Work Results for HVAC
23 06 00	Schedules for HVAC
23 07 00	HVAC Insulation
23 08 00	Commissioning of HVAC
23 09 00	Instrumentation and Control for HVAC
23 10 00	Facility Fuel Systems
23 11 00	Facility Fuel Piping
23 12 00	Facility Fuel Pumps
23 13 00	Facility Fuel-Storage Tanks
23 20 00	HVAC Piping and Pumps
23 21 00	Hydronic Piping and Pumps
23 22 00	Steam and Condensate Piping and Pumps
23 23 00	Refrigerant Piping
23 24 00	Internal-Combustion Engine Piping
23 25 00	HVAC Water Treatment
23 30 00	HVAC Air Distribution
23 31 00	HVAC Ducts and Casings
23 32 00	Air Plenums and Chases
23 33 00	Air Duct Accessories
23 34 00	HVAC Fans
23 35 00	Special Exhaust Systems
23 36 00	Air Terminal Units
23 37 00	Air Outlets and Inlets
23 38 00	Ventilation Hoods
23 40 00	HVAC Air Cleaning Devices
23 41 00	Particulate Air Filtration
23 42 00	Gas-Phase Air Filtration
23 43 00	Electronic Air Cleaners
23 50 00	Central Heating Equipment
23 51 00	Breechings, Chimneys, and Stacks
23 52 00	Heating Boilers
23 53 00	Heating Boiler Feedwater Equipment
23 54 00	Furnaces
23 55 00	Fuel-Fired Heaters
23 56 00	Solar Energy Heating Equipment
23 57 00	Heat Exchangers for HVAC
23 60 00	Central Cooling Equipment
23 61 00	Refrigerant Compressors
23 62 00	Packaged Compressor and Condenser Units
23 63 00	Refrigerant Condensers
23 64 00	Packaged Water Chillers
23 65 00	Cooling Towers
23 70 00	Central HVAC Equipment
23 71 00	Thermal Storage
23 72 00	Air-to-Air Energy Recovery Equipment
23 73 00	Indoor Central-Station Air-Handling Units
23 74 00	Packaged Outdoor HVAC Equipment
23 75 00	Custom-Packaged Outdoor HVAC Equipment
23 76 00	Evaporative Air-Cooling Equipment
23 80 00	Decentralized HVAC Equipment
23 81 00	Decentralized Unitary HVAC Equipment
23 82 00	Convection Heating and Cooling Units
23 83 00	Radiant Heating Units
23 84 00	Humidity Control Equipment
23 90 00	*Unassigned*
Division 24	**Reserved for Future Expansion**
Division 25—Integrated Automation	
25 01 00	Operation and Maintenance of Integrated Automation
25 05 00	Common Work Results for Integrated Automation
25 06 00	Schedules for Integrated Automation

Table 1.1 Level Two Numbers and Titles of the 2004 CSI MasterFormat™ *(Continued)*

Procurement and Contracting Requirements Group Divsion 00—Procurement and Contracting Requirements	
25 08 00	Commissioning of Integrated Automation
25 10 00	Integrated Automation Network Equipment
25 11 00	Integrated Automation Network Devices
25 12 00	Integrated Automation Network Gateways
25 13 00	Integrated Automation Control and Monitoring Network
25 14 00	Integrated Automation Local Control Units
25 15 00	Integrated Automation Software
25 20 00	*Unassigned*
25 30 00	Integrated Automation Instrumentation and Terminal Devices
25 31 00	Integrated Automation Instrumentation and Terminal Devices for Facility Equipment
25 32 00	Integrated Automation Instrumentation and Terminal Devices for Conveying Equipment
25 33 00	Integrated Automation Instrumentation and Terminal Devices for Fire-Suppression Systems
25 34 00	Integrated Automation Instrumentation and Terminal Devices for Plumbing
25 35 00	Integrated Automation Instrumentation and Terminal Devices for HVAC
25 36 00	Integrated Automation Instrumentation and Terminal Devices for Electrical Systems
25 37 00	Integrated Automation Instrumentation and Terminal Devices for Communications Systems
25 38 00	Integrated Automation Instrumentation and Terminal Devices for Electronic Safety and Security Systems
25 40 00	*Unassigned*
25 50 00	Integrated Automation Facility Controls
25 51 00	Integrated Automation Control of Facility Equipment
25 52 00	Integrated Automation Control of Conveying Equipment
25 53 00	Integrated Automation Control of Fire-Suppression Systems
25 54 00	Integrated Automation Control of Plumbing
25 55 00	Integrated Automation Control of HVAC
25 56 00	Integrated Automation Control of Electrical Systems
25 57 00	Integrated Automation Control of Communications Systems
25 58 00	Integrated Automation Control of Electronic Safety and Security Systems
25 60 00–25 80 00	*Unassigned*
25 90 00	Integrated Automation Control Sequences
25 91 00	Integrated Automation Control Sequences for Facility Equipment
25 92 00	Integrated Automation Control Sequences for Conveying Equipment
25 93 00	Integrated Automation Control Sequences for Fire-Suppression Systems
25 94 00	Integrated Automation Control Sequences for Plumbing
25 95 00	Integrated Automation Control Sequences for HVAC
25 96 00	Integrated Automation Control Sequences for Electrical Systems
25 97 00	Integrated Automation Control Sequences for Communications Systems
25 98 00	Integrated Automation Control Sequences for Electronic Safety and Security Systems
Division 26—Electrical	
26 01 00	Operation and Maintenance of Electrical Systems
26 05 00	Common Work Results for Electrical
26 06 00	Schedules for Electrical
26 08 00	Commissioning of Electrical Systems
26 09 00	Instrumentation and Control for Electrical Systems
26 10 00	Medium-Voltage Electrical Distribution
26 11 00	Substations
26 12 00	Medium-Voltage Transformers
26 13 00	Medium-Voltage Switchgear
26 18 00	Medium-Voltage Circuit Protection Devices
26 20 00	Low-Voltage Electrical Transmission
26 21 00	Low-Voltage Overhead Electrical Power Systems
26 22 00	Low-Voltage Transformers
26 23 00	Low-Voltage Switchgear
26 24 00	Switchboards and Panelboards
26 25 00	Enclosed Bus Assemblies
26 26 00	Power Distribution Units
26 27 00	Low-Voltage Distribution Equipment
26 28 00	Low-Voltage Circuit Protective Devices
26 29 00	Low-Voltage Controllers
26 30 00	Facility Electrical Power Generating and Storing Equipment
26 31 00	Photovoltaic Collectors
26 32 00	Packaged Generator Assemblies
26 33 00	Battery Equipment
26 35 00	Power Filters and Conditioners
26 36 00	Transfer Switches
26 40 00	Electrical and Cathodic Protection
26 41 00	Facility Lightning Protection
26 42 00	Cathodic Protection
26 43 00	Transient Voltage Suppression
26 50 00	Lighting
26 51 00	Interior Lighting
26 52 00	Emergency Lighting
26 53 00	Exit Signs
26 54 00	Classified Location Lighting
26 55 00	Special Purpose Lighting
26 56 00	Exterior Lighting
26 60 00–26 90 00	*Unassigned*
Division 27—Communications	
27 01 00	Operation and Maintenance of Communications Systems
27 05 00	Common Work Results for Communications
27 06 00	Schedules for Communications
27 08 00	Commissioning of Communications
27 10 00	Structured Cabling
27 11 00	Communications Equipment Room Fittings
27 13 00	Communications Backbone Cabling
27 15 00	Communications Horizontal Cabling
27 16 00	Communications Connecting Cords, Devices, and Adapters
27 20 00	Data Communications
27 21 00	Data Communications Network Equipment
27 22 0	Data Communications Hardware
27 24 00	Data Communications Peripherial Data Equipment
27 25 00	Data Communications Software
27 26 00	Data Communications Programming and Integration Services

(Continued)

Table 1.1 Level Two Numbers and Titles of the 2004 CSI MasterFormat™ *(Continued)*

Procurement and Contracting Requirements Group Divsion 00—Procurement and Contracting Requirements	
27 30 00	Voice Communications
27 31 00	Voice Communications Switching and Routing Equipment
27 32 00	Voice Communications Telephone Sets, Facsimiles, and Modems
27 33 00	Voice Communications Messaging
27 34 00	Call Accounting
27 35 00	Call Management
27 40 00	Audio-Video Communications
27 41 00	Audio-Video Systems
27 42 00	Electronic Digital Systems
27 50 00	Distributed Communications and Monitoring Systems
27 51 00	Distributed Audio-Video Communications Systems
27 52 00	Healthcare Communications and Monitoring Systems
27 53 00	Distributed Systems
27 60 00–27 90 00	*Unassigned*
Division 28—Electronic Safety and Security	
28 01 00	Operation and Maintenance of Electronic Safety and Security
28 05 00	Common Work Results for Electronic Safety and Security
28 06 00	Schedules for Electronic Safety and Security
28 08 00	Commissioning of Electronic Safety and Security
28 10 00	Electronic Access Control and Intrusion Detection
28 13 00	Access Control
28 16 00	Intrusion Detection
28 20 00	Electronic Surveillance
28 23 00	Video Surveillance
28 26 00	Electronic Personal Protection Systems
28 30 00	Electronic Detection and Alarm
28 31 00	Fire Detection and Alarm
28 32 00	Radiation Detection and Alarm
28 33 00	Fuel-Gas Detection and Alarm
28 34 00	Fuel-Oil Detection and Alarm
28 35 00	Refrigerant Detection and Alarm
28 40 00	Electronic Monitoring and Control
28 46 00	Electronic Detention Monitoring and Control Systems
28 50 00–28 90 00	*Unassigned*
Division 29	**Reserved for Future Expansion**
Site and Infrastructure Subgroup	
Division 30	**Reserved for Future Expansion**
Division 31—Earthwork	
31 01 00	Maintenance of Earthwork
31 05 00	Common Work Results for Earthwork
31 06 00	Schedules for Earthwork
31 08 00	Commissioning of Earthwork
31 09 00	Geotechnical Instrumentation and Monitoring of Earthwork
31 10 00	Site Clearing
31 11 00	Clearing and Grubbing
31 12 00	Selective Clearing
31 13 00	Selective Tree and Shrub Removal and Trimming
31 14 00	Earth Stripping and Stockpiling
31 20 00	Earth Moving
31 21 00	Off-Gassing Mitigation
31 22 00	Grading
31 23 00	Excavation and Fill
31 24 00	Embankments
31 25 00	Erosion and Sedimentation Controls
31 30 00	Earthwork Methods
31 31 00	Soil Treatment
31 32 00	Soil Stabilization
31 33 00	Rock Stabilization
31 34 00	Soil Reinforcement
31 35 00	Slope Protection
31 36 00	Gabions
31 37 00	Riprap
31 40 00	Shoring and Underpinning
31 41 00	Shoring
31 43 00	Concrete Raising
31 45 00	Vibroflotation and Densification
31 46 00	Needle Beams
31 48 00	Underpinning
31 50 00	Excavation Support and Protection
31 51 00	Anchor Tiebacks
31 52 00	Cofferdams
31 53 00	Cribbing and Walers
31 54 00	Ground Freezing
31 56 00	Slurry Walls
31 60 00	Special Foundations and Load-Bearing Elements
31 62 00	Driven Piles
31 63 00	Bored Piles
31 64 00	Caissons
31 66 00	Special Foundations
31 68 00	Foundation Anchors
31 70 00	Tunneling and Mining
31 71 00	Tunnel Excavation
31 72 00	Tunnel Support Systems
31 73 00	Tunnel Grouting
31 74 00	Tunnel Construction
31 75 00	Shaft Construction
31 77 00	Submersible Tube Tunnels
31 80 00–31 90 00	*Unassigned*
Division 32—Exterior Improvements	
32 01 00	Operation and Maintenance of Exterior Improvements
32 05 00	Common Work Results for Exterior Improvements
32 06 00	Schedules for Exterior Improvements
32 08 00	Commissioning of Exterior Improvements
32 10 00	Bases, Ballasts, and Paving
32 11 00	Base Courses
32 12 00	Flexible Paving
32 13 00	Rigid Paving
32 14 00	Unit Paving
32 15 00	Aggregate Surfacing
32 16 00	Curbs and Gutters
32 17 00	Paving Specialties
32 18 00	Athletic and Recreational Surfacing
32 20 00	*Unassigned*
32 30 00	Site Improvements
32 31 00	Fences and Gates
32 32 00	Retaining Walls
32 34 00	Fabricated Bridges
32 35 00	Screening Devices
32 40 00–32 60 00	*Unassigned*
32 70 00	Wetlands

Table 1.1 Level Two Numbers and Titles of the 2004 CSI MasterFormat™ *(Continued)*

Procurement and Contracting Requirements Group Divsion 00—Procurement and Contracting Requirements	
32 71 00	Constructed Wetlands
32 72 00	Wetlands Restoration
32 80 00	Irrigation
32 82 00	Irrigation Pumps
32 84 00	Planting Irrigation
32 86 00	Agricultural Irrigation
32 90 00	Planting
32 91 00	Planting Preparation
32 92 00	Turf and Grasses
32 93 00	Plants
32 94 00	Planting Accessories
32 96 00	Transplanting
Division 33—Utilities	
33 01 00	Operation and Maintenance of Utilities
33 05 00	Common Work Results for Utilities
33 06 00	Schedules for Utilities
33 08 00	Commissioning of Utilities
33 09 00	Instrumentation and Control for Utilities
33 10 00	Water Utilities
33 11 00	Water Utility Distribution Piping
33 12 00	Water Utility Distribution Equipment
33 13 00	Disinfecting of Water Utility Distribution
33 16 00	Water Utility Storage Tanks
33 20 00	Wells
33 21 00	Water Supply Wells
33 22 00	Test Wells
33 23 00	Extraction Wells
33 24 00	Monitoring Wells
33 25 00	Recharge Wells
33 26 00	Relief Wells
33 29 00	Well Abandonment
33 30 00	Sanitary Sewerage Utilities
33 31 00	Sanitary Utility Sewerage Piping
33 32 00	Wastewater Utility Pumping Stations
33 33 00	Low Pressure Utility Sewerage
33 34 00	Sanitary Utility Sewerage Force Mains
33 36 00	Utility Septic Tanks
33 39 00	Sanitary Sewerage Structures
33 40 00	Storm Drainage Utilities
33 41 00	Storm Utility Drainage Piping
33 42 00	Culverts
33 44 00	Storm Utility Water Drains
33 45 00	Storm Utility Drainage Pumps
33 46 00	Subdrainage
33 47 00	Ponds and Reservoirs
33 49 00	Storm Drainage Structures
33 50 00	Fuel Distribution Utilities
33 51 00	Natural-Gas Distribution
33 52 00	Liquid Fuel Distribution
33 56 00	Fuel-Storage Tanks
33 60 00	Hydronic and Steam Energy Utilities
33 61 00	Hydronic Energy Distribution
33 63 00	Steam Energy Distribution
33 70 00	Electrical Utilities
33 71 00	Electrical Utility Transmission and Distribution
33 72 00	Utility Substations
33 73 00	Utility Transformers
33 75 00	High-Voltage Switchgear and Protection Devices
33 77 00	Medium-Voltage Utility Switchgear and Protection Devices
33 79 00	Site Grounding
33 80 00	Communications Utilities
33 81 00	Communications Structures
33 82 00	Communications Distribution
33 83 00	Wireless Communications Distribution
33 90 00	*Unassigned*
Division 34—Transportation	
34 01 00	Operation and Maintenance of Transportation
34 05 00	Common Work Results for Transportation
34 06 00	Schedules for Transportation
34 08 00	Commissioning of Transportation
34 10 00	Guideways/Railways
34 11 00	Rail Tracks
34 12 00	Monorails
34 13 00	Funiculars
34 14 00	Cable Transportation
34 20 00	Traction Power
34 21 00	Traction Power Distribution
34 23 00	Overhead Traction Power
34 24 00	Third Rail Traction Power
34 30 00	*Unassigned*
34 40 00	Transportation Signaling and Control Equipment
34 41 00	Roadway Signaling and Control Equipment
34 42 00	Railway Signaling and Control Equipment
34 43 00	Airfield Signaling and Control Equipment
34 48 00	Bridge Signaling and Control Equipment
34 50 00	Transportation Fare Collection Equipment
34 52 00	Vehicle Fare Collection
34 54 00	Passenger Fare Collection
34 60 00	*Unassigned*
34 70 00	Transportation Construction and Equipment
34 71 00	Roadway Construction
34 72 00	Railway Construction
34 73 00	Airfield Construction
34 75 00	Roadway Equipment
34 76 00	Railway Equipment
34 77 00	Transportation Equipment
34 80 00	Bridges
34 81 00	Bridge Machinery
34 82 00	Bridge Specialties
34 90 00	*Unassigned*
Division 35—Waterway and Marine Construction	
35 01 00	Operation and Maintenance of Waterway and Marine Construction
35 05 00	Common Work Results for Waterway and Marine Construction
35 06 00	Schedules for Waterway and Marine Construction
35 08 00	Commissioning of Waterway and Marine Construction
35 10 00	Waterway and Marine Signaling and Control Equipment
35 11 00	Signaling and Control Equipment for Waterways
35 12 00	Marine Signaling and Control Equipment
35 13 00	Signaling and Control Equipment for Dams
35 20 00	Waterway and Marine Construction and Equipment
35 30 00	Coastal Construction
35 31 00	Shoreline Protection

(Continued)

Table 1.1 Level Two Numbers and Titles of the 2004 CSI MasterFormat™ *(Continued)*

Procurement and Contracting Requirements Group Divsion 00—Procurement and Contracting Requirements	
35 32 00	Artificial Reefs
35 40 00	Waterway Construction and Equipment
35 41 00	Levees
35 42 00	Waterway Bank Protection
35 43 00	Waterway Scour Protection
35 49 00	Waterway Structures
35 50 00	Marine Construction and Equipment
35 51 00	Floating Construction
35 52 00	Offshore Platform Construction
35 53 00	Underwater Construction
35 59 00	Marine Specialties
35 60 00	*Unassigned*
35 70 00	Dam Construction and Equipment
35 71 00	Gravity Dams
35 72 00	Arch Dams
35 73 00	Embankment Dams
35 74 00	Buttress Dams
35 79 00	Auxiliary Dam Structures
35 80 00–35 90 00	*Unassigned*
Divisions 36–39	**Reserved for Future Expansion**
Process Equipment Subgroup	
Division 40—Process Integration	
40 01 00	Operation and Maintenance of Process Integration
40 05 00	Common Work Results for Process Integration
40 06 00	Schedules for Process Integration
40 10 00	Gas and Vapor Process Piping
40 11 00	Steam Process Piping
40 12 00	Compressed Air Process Piping
40 13 00	Inert Gases Process Piping
40 14 00	Fuel Gases Process Piping
40 15 00	Combustion System Gas Piping
40 16 00	Specialty and High-Purity Gases Piping
40 17 00	Welding and Cutting Gases Piping
40 18 00	Vacuum Systems Process Piping
40 20 00	Liquids Process Piping
40 21 00	Liquid Fuel Process Piping
40 22 00	Petroleum Products Piping
40 23 00	Water Process Piping
40 24 00	Specialty Liquid Chemicals Piping
40 25 00	Liquid Acids and Bases Piping
40 26 00	Liquid Polymer Piping
40 30 00	Solid and Mixed Materials Piping and Chutes
40 32 00	Bulk Materials Piping and Chutes
40 33 00	Bulk Materials Valves
40 34 00	Pneumatic Conveying Lines
40 40 00	Process Piping and Equipment Protection
40 41 00	Process Piping and Equipment Heat Tracing
40 42 00	Process Piping and Equipment Insulation
40 46 00	Process Corrosion Protection
40 47 00	Refractories
40 50 00–40 70 00	*Unassigned*
40 80 00	Commissioning of Process Systems
40 90 00	Instrumentation and Control for Process Systems
40 91 00	Primary Process Measurement Devices
40 92 00	Primary Control Devices
40 93 00	Analog Controllers/Recorders
40 94 00	Digital Process Controllers
40 95 00	Process Control Hardware
40 96 00	Process Control Software
40 97 00	Process Control Auxiliary Devices
Division 41—Material Processing and Handling Equipment	
41 01 00	Operation and Maintenance of Material Processing and Handling Equipment
41 06 00	Schedules for Material Processing and Handling Equipment
41 08 00	Commissioning of Material Processing and Handling Equipment
41 10 00	Bulk Material Processing Equipment
41 11 00	Bulk Material Sizing Equipment
41 12 00	Bulk Material Conveying Equipment
41 13 00	Bulk Material Feeders
41 14 00	Batching Equipment
41 20 00	Piece Material Handling Equipment
41 21 00	Conveyors
41 22 00	Cranes and Hoists
41 23 00	Lifting Devices
41 24 00	Specialty Material Handling Equipment
41 30 00	Manufacturing Equipment
41 31 00	Manufacturing Lines and Equipment
41 32 00	Forming Equipment
41 33 00	Machining Equipment
41 34 00	Finishing Equipment
41 35 00	Dies and Molds
41 36 00	Assembly and Testing Equipment
41 40 00	Container Processing and Packaging
41 41 00	Container Filling and Sealing
41 42 00	Container Packing Equipment
41 43 00	Shipping Packaging
41 50 00	Material Storage
41 51 00	Automatic Material Storage
41 52 00	Bulk Material Storage
41 53 00	Storage Equipment and Systems
41 60 00	Mobile Plant Equipment
41 61 00	Mobile Earth Moving Equipment
41 62 00	Trucks
41 63 00	General Vehicles
41 64 00	Rail Vehicles
41 65 00	Mobile Support Equipment
41 66 00	Miscellaneous Mobile Equipment
41 67 00	Plant Maintenance Equipment
41 70 00–41 90 00	*Unassigned*
Division 42—Process Heating, Cooling, and Drying Equipment	
42 01 00	Operation and Maintenance of Process Heating, Cooling, and Drying Equipment
42 06 00	Schedules for Process Heating, Cooling, and Drying Equipment
42 08 00	Commissioning of Process Heating, Cooling, and Drying Equipment
42 10 00	Process Heating Equipment
42 11 00	Process Boilers
42 12 00	Process Heaters
42 13 00	Industrial Heat Exchangers and Recuperators
42 14 00	Industrial Furnaces
42 15 00	Industrial Ovens
42 20 00	Process Cooling Equipment
42 21 00	Process Cooling Towers

Table 1.1 Level Two Numbers and Titles of the 2004 CSI MasterFormat™ *(Continued)*

Procurement and Contracting Requirements Group Divsion 00—Procurement and Contracting Requirements	
42 22 00	Process Chillers and Coolers
42 23 00	Process Condensers and Evaporators
42 30 00	Process Drying Equipment
42 31 00	Gas Dryers and Dehumidifiers
42 32 00	Material Dryers
42 40 00–42 90 00	*Unassigned*
Division 43—Process Gas and Liquid Handling, Purification, and Storage Equipment	
43 01 00	Operation and Maintenance of Process Gas and Liquid Handling, Purification, and Storage Equipment
43 06 00	Schedules for Process Gas and Liquid Handling, Purification, and Storage Equipment
43 08 00	Commissioning of Process Gas and Liquid Handling, Purification, and Storage Equipment
43 10 00	Gas Handling Equipment
43 11 00	Gas Fans, Blowers, and Pumps
43 12 00	Gas Compressors
43 13 00	Gas Process Equipment
43 20 00	Liquid Handling Equipment
43 21 00	Liquid Pumps
43 22 00	Liquid Process Equipment
43 30 00	Gas and Liquid Purification Equipment
43 31 00	Gas and Liquid Purification Filtration Equipment
43 32 00	Gas and Liquid Purification Process Equipment
43 40 00	Gas and Liquid Storage
43 41 00	Gas and Liquid Storage Equipment
43 50 00–43 90 00	*Unassigned*
Division 44—Pollution Control Equipment	
44 01 00	Operation and Maintenance of Pollution Control Equipment
44 06 00	Schedules for Pollution Control Equipment
44 08 00	Commissioning of Pollution Control Equipment
44 10 00	Air Pollution Control
44 11 00	Air Pollution Control Equipment
44 20 00	Noise Pollution Control
44 21 00	Noise Pollution Control Equipment
44 40 00	Water Treatment Equipment
44 41 00	Packaged Water Treatment
44 42 00	General Water Treatment Equipment
44 43 00	Water Filtration Equipment
44 44 00	Water Treatment Chemical Systems Equipment
44 45 00	Water Treatment Biological Systems Equipment
44 46 00	Sludge Treatment and Handling Equipment for Water Treatment Systems
44 50 00	Solid Waste Control
44 51 00	Solid Waste Control Equipment
44 60 00–44 90 00	*Unassigned*
Division 45—Industry-Specific Manufacturing Equipment	
45 08 00	Commissioning of Industry-Specific Manufacturing Equipment
45 11 00	Oil and Gas Extraction Equipment
45 13 00	Mining Machinery and Equipment
45 15 00	Food Manufacturing Equipment
45 17 00	Beverage and Tobacco Manufacturing Equipment
45 19 00	Textiles and Apparel Manufacturing Equipment
45 21 00	Leather and Allied Product Manufacturing Equipment
45 23 00	Wood Product Manufacturing Equipment
45 25 00	Paper Manufacturing Equipment
45 27 00	Printing and Related Manufacturing Equipment
45 29 00	Petroleum and Coal Products Manufacturing Equipment
45 31 00	Chemical Manufacturing Equipment
45 33 00	Plastics and Rubber Manufacturing Equipment
45 35 00	Nonmetallic Mineral Product Manufacturing Equipment
45 37 00	Primary Metal Manufacturing Equipment
45 39 00	Fabricated Metal Product Manufacturing Equipment
45 41 00	Machinery Manufacturing Equipment
45 43 00	Computer and Electronic Product Manufacturing Equipment
45 45 00	Electrical Equipment, Appliance, and Component Manufacturing Equipment
45 47 00	Transportation Manufacturing Equipment
45 49 00	Furniture and Related Product Manufacturing Equipment
45 51 00	Other Manufacturing Equipment
45 60 00–45 90 00	*Unassigned*
Divisions 46–47	**Reserved for Future Expansion**
Division 48—Electrical Power Generation	
48 01 00	Operation and Maintenance for Electrical Power Generation
48 05 00	Common Work Results for Electrical Power Generation
48 06 00	Schedules for Electrical Power Generation
48 08 00	Commissioning of Electrical Power Generation
48 09 00	Instrumentation and Control for Electrical Power Generation
48 10 00	Electrical Power Generation Equipment
48 11 00	Fossil Fuel Plant Electrical Power Generation Equipment
48 12 00	Nuclear Fuel Plant Electrical Power Generation Equipment
48 13 00	Hydroelectric Plant Electrical Power Generation Equipment
48 14 00	Solar Energy Electrical Power Generation Equipment
48 15 00	Wind Energy Electrical Power Generation Equipment
48 16 00	Geothermal Energy Electrical Power Generation Equipment
48 17 00	Electrochemical Energy Electrical Power Generation Equipment
48 18 00	Fuel Cell Electrical Power Generation Equipment
48 19 00	Electrical Power Control Equipment
48 20 00–48 60 00	*Unassigned*
48 70 00	Electrical Power Generation Testing
48 71 00	Electrical Power Generation Test Equipment
48 80 00–48 90 00	*Unassigned*
Division 49	**Reserved for Future Expansion**

(Source: The Numbers and Titles used in this product are from MasterFormat™, which is published by The Construction Specifications Institute (CSI) and Construction Specifications Canada (CSC), and used with permission from CSI, 2008. The Construction Specifications Institute (CSI) 99 Canal Center Plaza, Suite 300 Alexandria, VA 22314 800-689-2900; 703-684-0300 CSINet URL: http://www.csinet.org)

GreenFormat, a newly released version of the MasterFormat, was introduced to incorporate new environmentally friendly products and construction procedures that were previously not included. GreenFormat is a new CSI framework that provides designers, contractors, and building operators with basic information to help meet green building requirements. The new format is entirely web-based and gives building product manufacturers a process by which to describe the sustainable aspects of the products for incorporation in the MasterFormat sequence.

THE PROJECT DELIVERY PROCESS

Crucial for the successful delivery of a complex building project is the selection of a qualified contractor and an effective project delivery process. A project delivery method governs the conditions under which a construction project will be completed and defines the relationships between the owner, architect, or AEC firm and contractor.

Construction Contractors

The actual construction activity is performed by construction contractors. They are generally divided into general contractors and specialty contractors. **General contractors** (GC) assume the responsibility for the construction of an entire project at a specific cost and by a specified date. They are responsible for developing project schedules and sequencing and coordinating the work of all subcontractors. The general contractor also determines the actual methods and techniques of construction as well as implementing safety precautions on the building site. A general contractor signs contracts with **Subcontractors** who perform the required work within their technical areas, such as all the electrical, plumbing, roofing, bricklaying, carpentry, or concrete.

Specialty contractors do the work required in a limited area, like elevator or communication equipment installation for instance (Fig. 1.12). Both specialty and subcontractors are independent contractors employed by the general contractor to perform specific work at a specified cost. While subcontractors work independently on a site, bringing their own employees, supervisors, and tools to the job, their work is overseen by a project superintendent. The **superintendent** is the general contractor's on-site representative responsible for continuous field supervision, coordination, and completion of the work. The superintendent makes sure that the work proceeds according to the project schedule and that the activities of the various subcontractors working on the site will not interfere with one another.

Figure 1.12 General contractors assume the responsibility for the construction of an entire project at a specific cost and by a specified date. The general contractor signs contracts with subcontractors who perform the required work within their technical areas, such as this electrician. *(Courtesy International brotherhood of Electrical Workers)*

Selecting the Contractor

A number of different options are available in the selection of a contractor for a building project. For publicly financed projects construction contractors are generally sought through a process of **competitive bidding.** In this method qualified construction contractors are invited to bid on the project on a competitive basis with the contract often being awarded to the contractor that submits the lowest bid. A variation of this process used more on privately funded projects is called **invitational bidding,** wherein only preselected contractors are asked to provide bids on a project.

An owner may decide on a contractor with whom they have worked on previous projects without seeking other bids and agree upon a **negotiated contract.** A negotiated contract can have the advantage of bringing a contractor into the project in the pre-construction phase of the work, encouraging a team approach that incorporates design, construction, and budget planning with owner involvement in a cooperative and coordinated effort. This integration, which incorporates estimating and cost control throughout the design and development effort rather than at its conclusion, often facilitates the best overall solutions.

Most projects select a single contractor to oversee and complete the work on an entire project. The single prime contractor is responsible for all of the work including that which has been subcontracted. An alternative method is to divide portions of the work among more than one entity by employing the services of multiple prime contractors. Multiple prime contracts are often used on large and complex projects where a number of specialty contractors are engaged to complete different parts of the work. A high-rise building, for instance, may use one contractor for the foundations and site work, another for the building structure, and a third for the interior build out and finishes. In this case, each will enter into a separate contractual agreement with the owner. Multiple prime contracts are often used for projects that are fast tracked. **Fast tracking** a project shortens the overall construction time by beginning work on site prior to the completion of the construction documents. Construction documents for fast-track projects are often completed in sequenced packages. A first package may be for site preparation and foundations, a second for the structural frame, and a third for the installation of the exterior envelope. Careful coordination between the design team and multiple contractors is essential for a fast track project.

Project Delivery Methods and Types of Construction Contracts

A project delivery method determines how a construction project will be completed, delegates the responsibilities, rewards, and risks between participants, and regulates the relationship between the owner and the contractor. Division 00 of the written specifications, General Requirements, outlines the general conditions of the construction contract. This information provides the basis upon which the contractual relationships between all parties involved in a project will be administered. Standard forms for construction contracts which can be modified to suit conditions are available from the American Institute of Architects (AIA). A number of options exist to define the contractual relationship between the owner and contractor on a particular project. The three most common methods currently in use are:

- Design-Bid-Build
- Design-Build
- Construction Management (CM)

Design-Bid-Build

Design-Bid-Build is the traditional method of project delivery that moves sequentially from conception of a project through its construction. The owner contracts initially with an architect to define the project scope, perform pre-design services, and produce the construction documents. Once the documents are complete, the project is sent out for proposals to qualified contractors. Division 00 of the written specifications, Procurement and Contracting, gives instructions to bidders for how to prepare the proposal and bidding forms. The contract is usually awarded to the most qualified bidder who submits the lowest cost estimate. The general contractor solicits bids from subcontractors using the same bidding procedure. The project responsibilities are divided among the various team members with whom the owner establishes separate contractual relationships (Fig. 1.13). Because the design, bidding, and construction phases occur sequentially, the design-bid-build delivery process tends to produce longer overall project schedules. Design-Bid-Build contracts do, however, generally result in the lowest construction cost.

Design-Build

In the design-build contracting method, the owner contracts with a single party that completes all portions of the work from design through construction. Rather than having multiple contracts with the architect and contractor, the design-build method assigns a single entity with full responsibility (Fig. 1.14). The entity may be a construction company that has contracted with an architect or engineering firm for design services, or one that maintains design professionals as permanent

Figure 1.13 The traditional design-bid-build project delivery method. Solid lines in this image indicate the contractual relationships that are held between project participants while dashed ones indicate lines of communication.

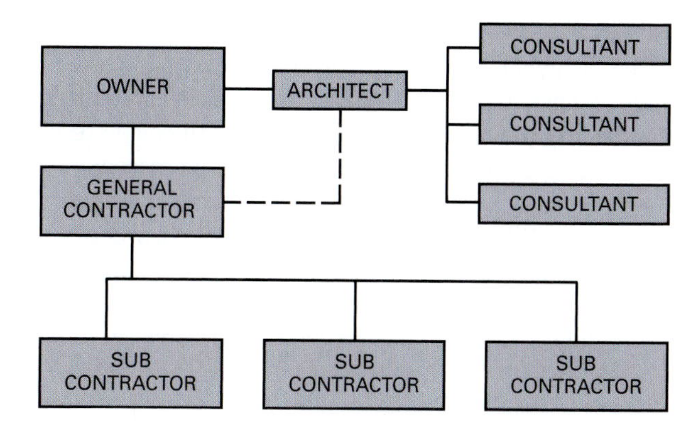

Figure 1.14 In the design-build delivery process the owner contracts with a single entity that provides both design and construction services for the project.

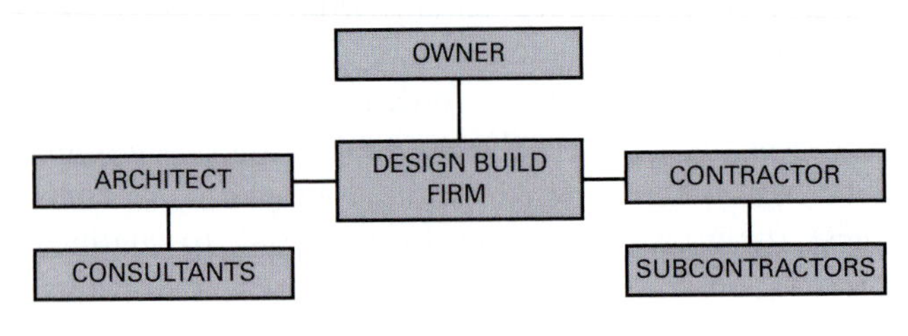

members of their staff. The design-build organization usually provides a bid price for both design and construction at the onset of a project, resulting in earlier and more tightly constrained cost controls. Because the bidding portion of the traditional design-bid-build delivery method is eliminated under this procedure, design-build contracts tend to produce shorter overall project schedules.

Figure 1.15 In the construction manager project delivery method, the owner contracts with a construction manager to oversee both the design and construction process. Solid lines in this image indicate the contractual relationships that are held between project participants while dashed ones indicate lines of communication.

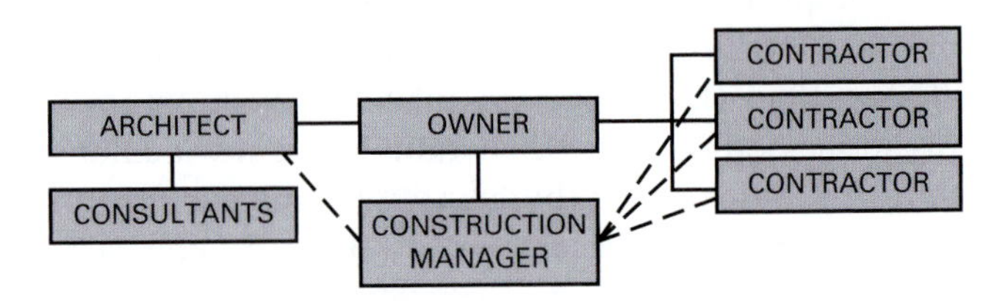

Construction Management (CM)

The owner may hire a **construction manager** to provide input during the design phase and oversight and administration of the bidding and construction phases (Fig. 1.15). The construction manager may act purely as an advisor, providing expertise throughout

Construction Methods

Subcontractor Agreements

Seldom does a general contractor have the personnel to complete all of the tasks required to construct a building. Many activities, such as mechanical, electrical, and plumbing (MEP) work, are handled by subcontractors employed by the general contractor. It is essential that the general contractor and subcontractor have a detailed written contract, stipulating what portion of the work is to done by which entity. Following are some of the basic topics covered in subcontractor agreements.

Contracts should include a schedule that states the starting, interim, and completion dates. Clauses outlining who is to pay for expenses caused by any delays are included. These may define what type of delay would constitute a breach of contract that would permit the general contractor to employ a new subcontractor. Who assumes responsibility for obtaining required permits and notifying inspectors when the job has progressed to benchmark points should be noted. Some statement regarding the failure of a subcontractor to complete the work to the required quality expected is necessary.

Of great importance is a clearly specified payment schedule. This includes the costs of materials and labor provided and the terms of payment. Materials suppliers for major portions of the work can be specified. Typically, a plumber supplies the pipe material, but the general contractor may prefer to purchase the plumbing fixtures. Drywall installers typically prefer to do the work but expect the general contractor to have the materials available when they appear on the job. If changes are to be made, the agreement should specify that the general contractor reviews these changes with the owner. The subcontractor should not proceed with changes unless they are approved by the general contractor.

All trades are subject to meeting safety standards, and even though the general contractor may have the overall responsibility, the contract should indicate that these safety standards must be met by all subcontractors. Another concern on all construction jobs is site cleanup. The contract must specify daily and final cleanup responsibilities in great detail.

All contracts should require that the subcontractor is responsible for having all the necessary licenses for his or her trade and for having a comprehensive, up-to-date insurance policy. Although legal differences exist across the country, it is advantageous to have the subcontractor agree to indemnify the general contractor if the subcontractor fails to pay workers compensation or injury claims and the general contractor is held responsible for paying them. The subcontractor should provide a written warranty for the work done so the owner can contact the subcontractor if something fails or needs repair.

the planning and construction of a project. Or the construction manager may act in an advisory capacity during design and sign a contract with the owner to act as general contractor once the construction documents are complete. In both cases, the construction manager becomes a representative of the owner in overseeing the work of both the architect and the contractor. The Construction manager's involvement from the onset of the project includes advising on issues of constructability during the design process. Construction management can be particularly advantageous on large, complicated projects where ongoing oversight can provide innovative and cost effective solutions to complex problems. A successful construction manager needs to have an education in both the technical and management portions of construction, along with a great deal of field experience.

THE CONSTRUCTION PROCESS

Prior to any work commencing on site, a **building permit** must be secured. The permit is a certificate issued by the local governing **authority having jurisdiction (AHJ)** authorizing the construction of a project after a thorough review of the construction documents to ensure compliance with local building, safety, and fire codes. The building permit must be posted in a clearly visible location until the project is completed (Fig. 1.16). Specialty subcontractors are responsible for securing permits to allow their own portion of the work to proceed. Most projects involve a general building permit, as well as electrical, plumbing, HVAC, and other permits. Certifications for heavy equipment or the use of cranes may also need to be secured. Inspectors employed with the jurisdiction will conduct periodic inspections of the work throughout the progress of construction to ensure that the installed work adheres to all applicable codes and standards. Permit fees are usually based on a percentage of the total construction costs. Any work involving obtaining utilities from the public right-of-way for the site and building typically require a Public Way construction permit obtained form the local government. A more thorough discussion of building codes follows in Chapter 2.

Pre-Construction Planning and Temporary Facilities

Once a building permit has been secured and the contract between the owner and contractor is signed, the actual on-site construction can begin. The owner will issue the contractor a formal notice to proceed, indicating that all pre-construction requirements have been met. Requirements for temporary facilities are outlined in division 01 of the contract documents. On larger projects the general contractor may be required to establish a field office on the construction site. Field offices are often housed in self-contained mobile trailers that are fully furnished and provided with modern communication equipment (Fig. 1.17). The field office provides a central command post for the day-to-day site activities, a repository for construction documents and project files, and an office for the project superintendant. Temporary utilities need to be brought to the site to provide the power and sanitation required by workers and equipment. Access for deliveries and equipment must be arranged along with designated space for the storage and staging of materials.

Many municipalities have instituted construction and demolition recycling ordinances requiring construction sites to recycle a percentage of the waste that is generated. This requires the contractor to identify a portion of the site to be used for the sorting and storage of construction waste materials to be picked up by a certified recycling agency. Materials such as wood, aluminum, steel, bricks and tile, and packaging materials can be easily recycled in today's expanding recycled-materials market.

Figure 1.16 A building permit is issued by the governing authority having jurisdiction that approves the construction of a project after a thorough review of the construction documents. The permit must be displayed throughout the duration of construction. *(Courtesy Eva Kultermann)*

A pre-construction conference, attended by the owner, architect, general contractor or construction manager, and invited subcontractors is held to communicate the ongoing management procedures of the project. These conferences establish efficient communication channels, discuss payment procedures, the sequencing of work, and set a schedule for ongoing progress meetings. The completed construction documents, both drawings and specifications, provide the guidelines under which the project will be completed, but a number of additional requirements and submittals must be generated to fully define all portions of the work.

Figure 1.17 A field office serves as a central command post for the day-to-day site activities, housing the construction documents and project files and providing an office for the project superintendant. *(Courtesy Eva Kultermann)*

Shop Drawings, Submittals, and Mock-ups

While the construction documents give detailed information as to how components and assemblies will be constructed, other more specific drawings are required on most projects. A **shop drawing** gives precise directives for the fabrication of certain components such as structural steel work, concrete reinforcing, or pre-cast concrete components. It is drawn to explain the fabrication and sometimes installation procedures of the items to the manufacturer's production crew or contractor's installation crews (Fig. 1.18). Shop drawings are substantially different from architectural drawings in style and content, including manufacturing conventions, and special fabrication instructions. They are generated by the manufacturer of the material, reviewed by the contractor, and submitted to the architect for final approval. The fast processing of shop drawings is crucial

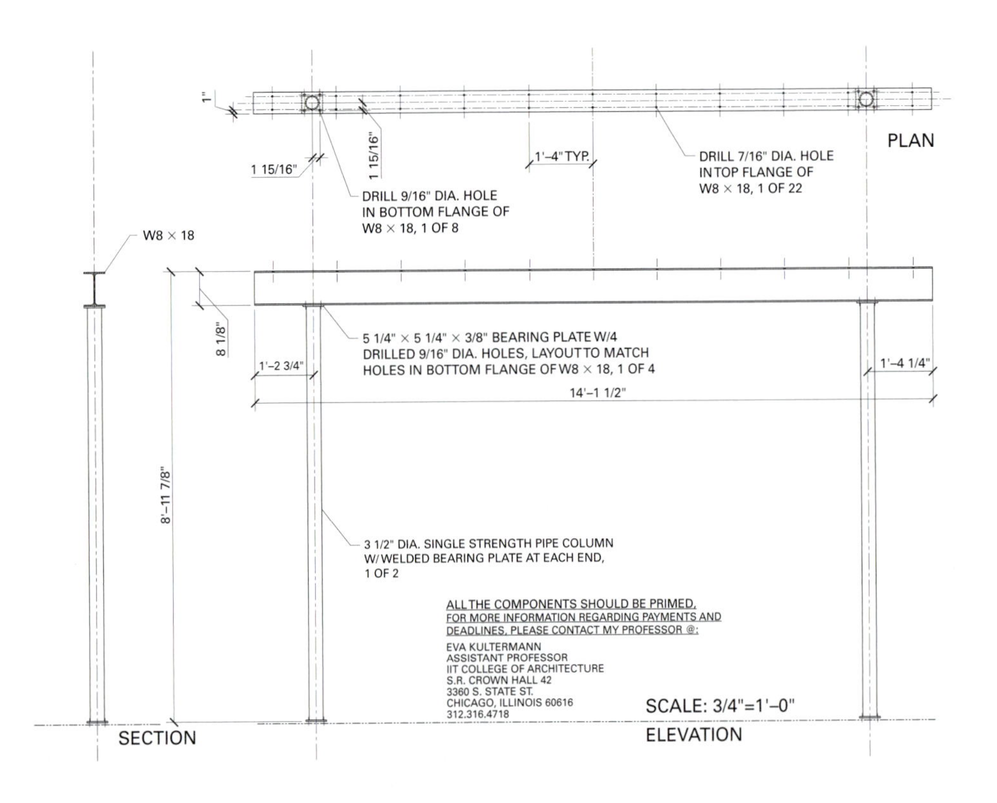

Figure 1.18 A shop drawing outlining the fabrication of a steel beam and column system. *(Courtesy Eva Kultermann)*

to the start of any project as the ordering of materials is often dependent on their completion.

The written specifications of the construction documents typically require the contractor to submit product data and material samples to the architect for final approval. The submittal process is an important and time-consuming step at the beginning of managing any construction project. Product data submittals are drawings, schedules, performance data, and brochures that give manufacturer's information on the characteristics of a material and allow the architect to verify that the product under consideration will satisfy the requirements listed in the specifications. Product samples may also need to be submitted to the architect for approval. A Product Sample is an actual physical example of a material that can be examined to assure that colors, textures, and other characteristics adhere to the original design intent.

Some contracts will ask for a sample of an entire building component, such as a wall panel, to be built for on-site evaluation. A construction **mock-up** is a full-size model of a proposed construction system that is built in order to judge the appearance of an assembly, examine its construction details, and test for performance under actual site conditions (Fig. 1.19). These models allow the contractor to investigate different means of achieving the level of workmanship required by the construction documents. Mock-ups often remain on the site throughout construction to provide a basis against which to judge the finished construction as it progresses.

Figure 1.19 An example of a full-scale mock-up used to study façade details on the Sacramento Civic Center. *(Courtesy Fentress Architects)*

Construction Observation

In addition to ongoing inspections by local building department officials, continuing inspections are conducted by representatives of the owner to ensure that the completed work complies with the guidelines set forth in the construction documents. In most contracts this responsibility is held by the architect, who will visit the site periodically and prepare written reports to document adherence to the project schedule and keep the owner updated on the progress of the work. Near the end of a project the contractor will request an inspection to document that the work has arrived at a point of substantial completion, indicating that the building is ready to be occupied by the owner. Some owners may also secure the services of an independent inspection firm to conduct testing and inspections in an effort to ensure quality control. If inspections find either materials or workmanship to be in non-compliance, the contractor is required to repair or replace the work at their own expense.

Contractor Requests for Information (RFI)

During the construction process the contractor may come upon situations for which the contract documents do not provide comprehensive directives. The request for information is used to obtain clarification that the contractor cannot access through document review or other research. A contractor could ask, for instance, if an alternate product model or manufacturer can be used instead of the one originally specified. In order to obtain the answer the contractor can submit a request for information to the architect. These requests often occur during the bidding phase, as the contractor examines the documents in detail for the first time. Requests for information must be processed quickly by the architect since the project schedule may be affected by unforeseen delays.

Construction Change Directive (CCD)

Almost every construction project encounters conditions that require a change to be made to the original contract documents. Changes may be required for a number of reasons, such as previously unknown conditions, design omissions or errors, or changes in the scope of the work. For instance, subsurface site conditions may be substantially different than those anticipated, requiring different or additional work and time from that set forth in the construction contract. A change is defined as any modification that will result in the project requiring more time or additional expense. A **construction change directive**, or *change order*, can originate with the architect or contractor. Standard forms available from the American Institute of Architects are used to describe the change, along with additional drawings and specifications when necessary. After completion the order must be approved by the owner and architect and becomes a modification of the construction contract. It authorizes the contractor to do the work and obligates the owner to cover the expense.

Project Close Out

Prior to all construction being concluded, a number of administrative benchmarks must be met to complete the contractual obligations between the owner and the contractor. When the contractor determines that the work is nearing completion, the architect or construction manager is asked to return to the site and conduct an inspection of substantial completion. The inspection occurs at a point when all major portions of the work have been completed but a number of final details have yet to be finished. A listing of the remaining items to be installed or repaired is called a **punch list**. The punch list is initially prepared by the contractor and then added to by the architect and engineering consultants during their inspections.

During project close out the contractor is obligated to collate and submit all product and equipment warranties for the installed work to the owner. Warranties are common for such items as windows, roofing, appliances and electrical, heating, and other mechanical equipment. The contractor also typically provides a warranty that the finished product is free of all defects and agrees that any work that requires it will be corrected within a one-year period. Operations and maintenance manuals are generated to educate building staff in the proper use and care of all building systems and equipment. Finally record drawings are completed, indicating where modifications were made to the original contract documents.

Before a final certificate of completion can be issued the inspecting authority must substantiate that all work is installed, complete, and that the quality of workmanship meets the specified standards. All equipment and appliances must have undergone functional testing and be working properly. The site must be thoroughly cleaned, all surplus materials and temporary equipment removed, and surrounding streets and sidewalks returned to their pre-construction condition. The contractor will apply to the local jurisdiction for a **certificate of occupancy**. The certificate is issued by the local building department indicating that the building is in compliance with locally adopted building codes and is in proper condition to be occupied.

Construction Safety

Throughout any construction project a thorough accident prevention and safety training program must be implemented and maintained. The construction industry has one of the most hazardous work environments in the country, accounting for hundreds of fatalities and countless injuries each year. In 1970, congress passed the Williams-Steiger Occupational Health and Safety Act (OSHA) that initiated the regulation of safety standards for the construction industry. OSHA develops safety guidelines and administers their enforcement through a system of recurring site inspections. OSHA Standards-29 CFR–1926 Safety and Health Regulations for Construction give extensive guidelines for construction means, methods, and materials handling. The standards cover every conceivable aspect of construction safety, from personal protective and life saving equipment (Fig. 1.20), tool and equipment safety, fire protection and prevention, materials storage, use, and disposal, to the signs, signals, and barricades that are required on a site.

The general contractor is responsible for administering safety procedures on the construction site. Detailed safety procedures are often specified in the general requirements division of the construction documents. An accident prevention program including training and education must be implemented at the start of any construction project. Workers must be trained in recognized safe working practices and the proper use of mandated personal protective gear. Copies of the current OHSA standards are required by law to be kept in the field office along with material data sheets outlining the hazards associated with certain construction materials.

Figure 1.20 Examples of common personal protective equipment used on the construction site.

All safety equipment and barriers must undergo regular inspections that are documented in a written log. With knowledge and expertise in construction safety, general contractors must provide the quality guidance that workers must have in order to prevent injury and the loss of life.

NEW DEVELOPMENTS IN THE DESIGN AND DELIVERY PROCESS

The development of multi-faceted software and data systems coupled with an increased desire to lessen the sometimes adversarial relationship between architects and contractors are resulting in new ways of approaching the construction delivery process. Two new tools are unifying the design and construction processes by redefining how responsibilities, rewards, and risks are allocated. These new methodologies are well suited to delivering high performance buildings with shorter project schedules, fewer disputes among team members, and a reduction in overall waste.

Integrated Design Process

Integrated Design Process (**IDP**) is a relatively new design approach that attempts to bring together all the diverse participants of a project at the beginning of the design phase. A project team including the owner, architect or AEC firm, engineers, energy and green design consultants, contractor, construction manager, material suppliers, facilities manager, and regulatory officials is established at the onset of building planning and design. The team meets for a brainstorming session at the beginning of a project and continues to work collaboratively in establishing design intent and functional requirements. By including all stakeholders early in the design process, many coordination issues can be solved holistically throughout the design and construction sequence. This approach considers the multi-faceted systems involved in a project as acting interdependently and seeks to identify synergies throughout the process. In general, integrated design is an approach that seeks to achieve high performance on a wide variety of well defined environmental and social issues while staying within budgetary and scheduling constraints. It follows the design through the entire project life, from predesign through occupancy and into operation.

Building Information Modeling

Efforts are currently under way to develop new ways of managing the complexity of the interdisciplinary information that goes into the design and construction phase activities of a building. The National Institute of Building Sciences (NIBS) has formed a committee whose aim is to optimize a computer software system that standardizes the design, construction, operation, and maintenance process for buildings by compiling all of the related information in a single electronic format. The software, known as **Building Information Modeling** (**BIM**), is set to revolutionize the way buildings are designed, constructed, and maintained. Architects and engineers use BIM to generate a three-dimensional model of a building with embedded links to other project information. In its current state BIM has a range of capabilities including: drawing, construction estimating and scheduling, interdisciplinary coordination, and generating fabrication protocols. BIM also runs analysis of structural design solutions, building code compliance, projected energy performance and consumption models, and daylight penetration (Fig. 1.21).

The use of the Building Information Modeling system results in enhanced efficiencies during the construction process itself. Sometimes referred to as "lean construction," the use of BIM has demonstrated increasing productivity and a reduction of waste throughout the project delivery process. Construction methods, sequences, and details can be studied using the modeling process, resulting in fewer mistakes, less changes, and ultimately shorter construction schedules. The software generates complete material takeoffs, eliminating the painstaking process of figuring materials by hand. By designing in a three-dimensional model, systems integration can be visualized and better coordination realized. Information from models can be fed directly to material fabricators, reducing the need for shop drawings and ensuring tight dimensional tolerances. Students of construction should become familiar with the Building Information Modeling software because its use in the construction industry will become widespread in the near future.

Figure 1.21

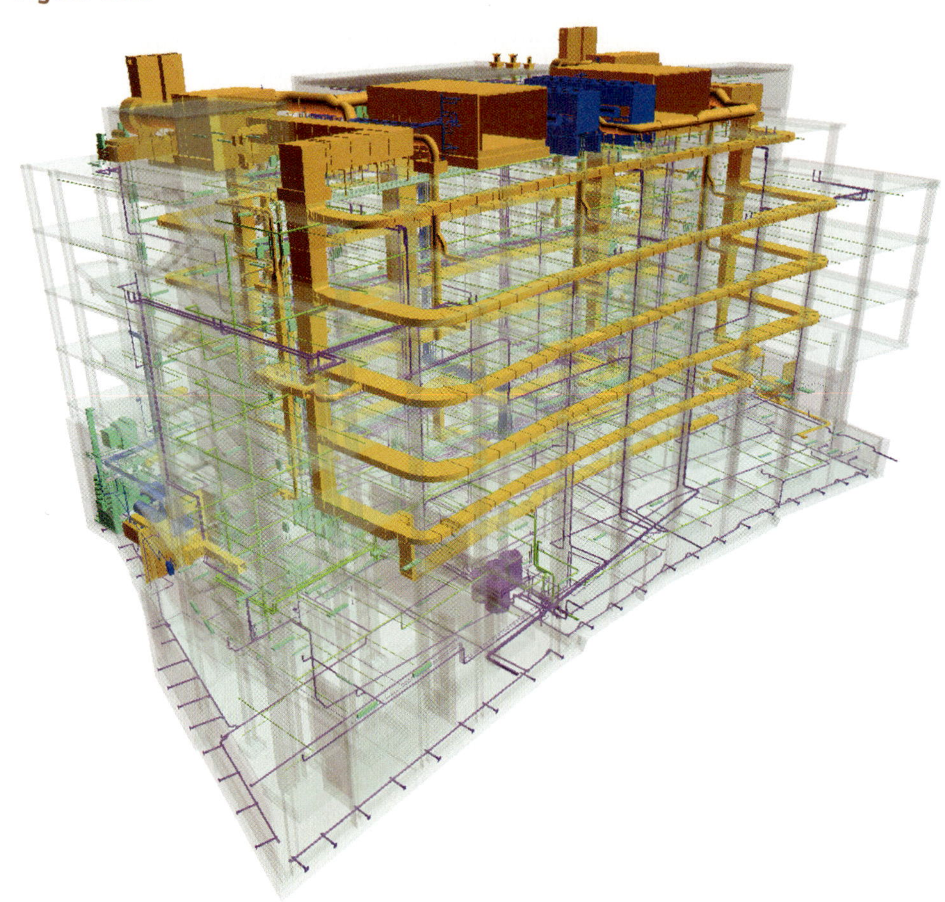

A BIM model illustrating the capability of visualization and analysis of various building systems. *(Courtesy of Mortenson Construction and McKinstry Company)*

Review Questions

1. What are the three major divisions of the construction industry?
2. What professionals are responsible for the preparation of construction documents?
3. What is the difference between construction drawings and specifications?
4. How does the MasterFormat help all participants in the industry to communicate?
5. What is the difference between a general contractors and subcontractors?
6. What are the three most commonly used types of construction contracts?
7. How are contractors selected to work on a construction project?
8. What is the purpose of a shop drawing?
9. When is a punch list generated and what does it contain?
10. What purposes are served by OSHA?
11. What are material data sheets?
12. What is IDP and how does it differ from the conventional design process?
13. How is BIM changing the way construction projects are designed and built?

Key Terms

Authority Having Jurisdiction
Building Construction
Building Information Modeling
Building Permit
Building Section
Certificate of Occupancy
Competitive Bidding
Construction Bid
Construction Change Directive
Construction Drawings
Construction Management
Contract Documents
Design Intent
Design-Bid-Build
Design-Build
Elevation
Fast-Track Project
Floor Plan
General Contractor
Heavy Construction
Industrial Construction
Integrated Design Process
Invitational Bidding
Master Format
Mock-up
Pre-design
Punch List
Shop Drawing
Specifications
Subcontractor
Superintendent

Activities

1. Contact the office of a local architect and ask them to conduct a tour of their facilities for the class. Request to see a completed set of construction documents while in the office.
2. Visit the Construction Specifications Institute Web site and investigate the on line forums for current discussions and developments.
3. Ask a local contractor and one of their subcontractors to address the class and explain the process of competitive bidding and the preparation of bid documents.

Additional Resources

The Architects Handbook of Professional Practice, AIA Press, Washington, D.C.

The Association of General Contractors, http://www.agc.org/

The Construction Specifications Institute, http://www.csinet.org/s_csi/index.asp

The Project Resource Manual, The CSI Manual of Practice, McGraw Hill Publishing.

Fisk, Edward, 1997, *Construction Project Administration*, Simon & Shuster, Upper Saddle River, New Jersey.

CHAPTER 2

Regulatory Constraints, Standards, and Sustainability

LEARNING OBJECTIVES

Upon completion of this chapter the student should be able to:

- Know the purposes served by zoning ordinances and building codes.
- Understand the function of the Americans with Disabilities Act.
- Become familiarized with professional and trade organizations and their contribution to the construction industry.
- Understand the impacts of the construction industry on global warming and resource depletion.
- Understand the concept of sustainable development and construction.
- Gain an understanding of Green Building rating systems.

A study of the materials of construction involves not only their technical aspects but also the influences on the use of a material by zoning ordinances, building codes, and various other standards established by trade and professional organizations. Standard writing organizations set forth additional best-practice guidelines that are frequently referenced in the building codes.

As the construction industry develops new practices in response to environmental issues, additional organizations have been formed to provide guidance and assessment systems for the design and construction of energy efficient, sustainable buildings. New standards that define the energy and resource implications of both construction materials and methods are entering the field every year. This chapter examines the variety of regulatory and additional supporting information that influences the construction industry.

THE REGULATORY ENVIRONMENT—ZONING AND CODES

Construction is regulated and supported by a variety of local, state, and federal authorities to ensure that buildings and infrastructure will be built to industry-determined safety standards. Architects manage code issues during the design phase, and contractors ensure that code mandated requirements are met during the construction process.

Zoning Ordinances

Zoning ordinances are local land-use regulations written into municipal law that regulate the degree of building development according to land use. These ordinances group buildings together into zoning districts that house similar activities in order to separate incompatible uses and promote cohesive city planning. Zoning districts typically define single and multi-family residential, business, park land, and industrial zones (Fig. 2.1). Within districts the ordinances set guidelines for the density of population, height and bulk of buildings, building coverage on a property in relation to open space, setbacks from property lines, and ancillary facilities, such as parking. In certain jurisdictions, zoning ordinances may also stipulate building style and the types of materials or colors that can be used on building exteriors. Zoning ordinances are usually based on a comprehensive master plan. A master plan is a long range strategy for the growth of an entire municipality over time. The goal of a master plan is to

Figure 2.1 This municipal zoning map defines residential, commercial, parkland, and industrial districts. *(Courtesy Village of Burr Ridge, Illinois)*

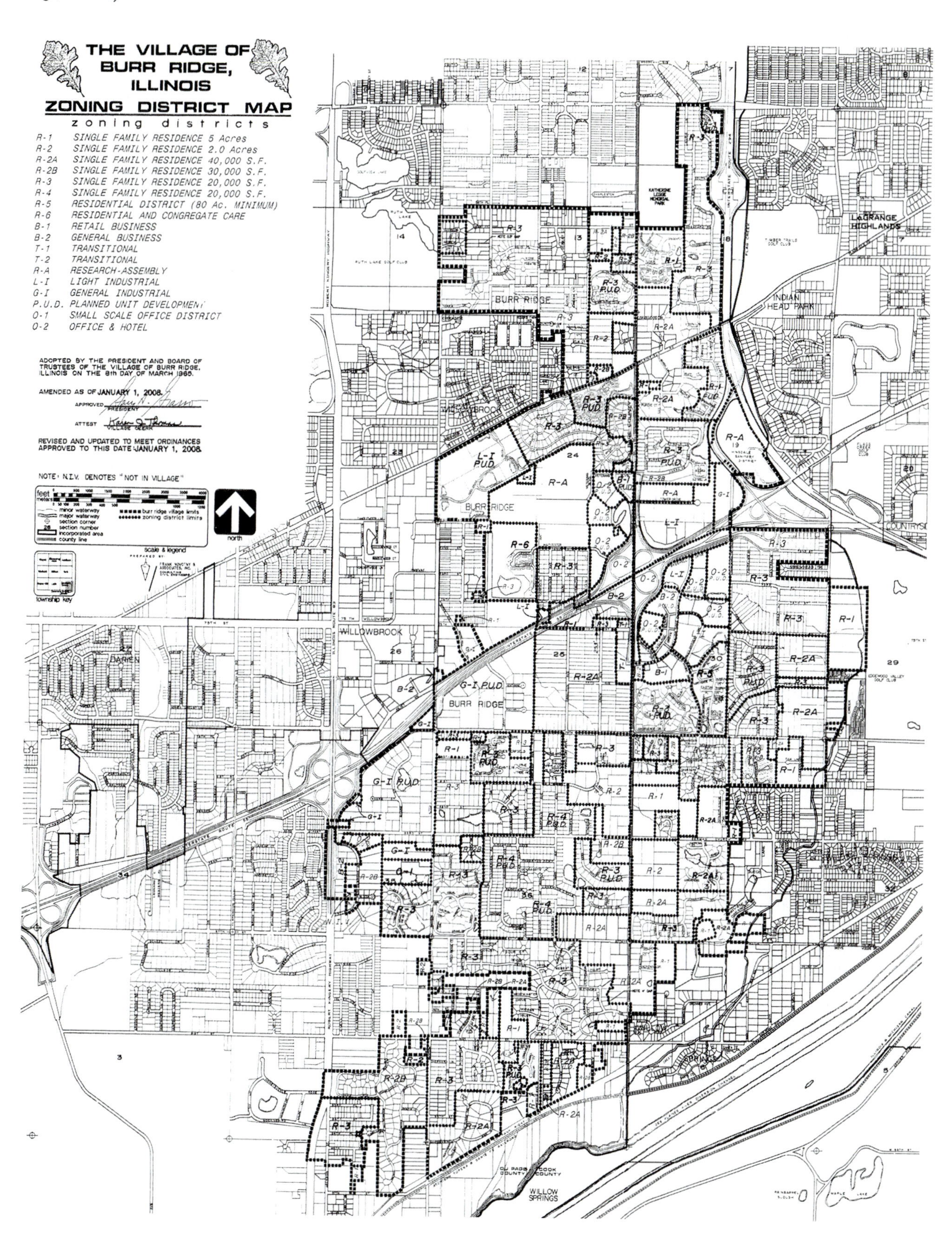

direct future growth of a municipality in terms of the provision of public services, transportation, and utility infrastructure. Zoning ordinances can be obtained from the local department of buildings or on their municipal website.

Building Codes

Building codes regulate the design and construction of buildings by providing minimum standards established to safeguard the life, health, property, and general welfare of the public. They dictate the design, means of construction, and maintenance of buildings within their jurisdiction. Their intent is to establish provisions for structural strength, sanitation, adequate light and ventilation, energy conservation, and safety from fire and other hazards. Building codes control not only the initial construction of a building but also any alteration, enlargement, repair, or demolition of buildings and other structures. The codes provide a basis upon which architects and engineers design buildings and material and equipment suppliers fabricate products. They promote the stability and safety of buildings by providing minimum construction standards and ensuring that they are adhered to through verification of compliance.

Building codes are legally binding documents and as such are voted upon and approved by the governing body of a municipality. They are adopted and administered by either local, county, or state authorities that review construction documents, issue building permits, and conduct periodic inspections of the work (see Chapter One). Most jurisdictions adopt what are called model codes, although some cities choose to write and maintain municipal codes that are tailored to address the specifics of local conditions.

The Development of Building Codes

Regulations regarding construction safety have been in use since the dawn of humanity. The ancient Babylonians decreed in writing that builders whose constructions led to occupant injury could be punished. The code of Hammurabi from the second century BC stated: "*If a builder build a house for some one, and does not construct it properly, and the house which he built fall in and kill its owner, then that builder shall be put to death.*" The development of American building codes originates with early efforts to prevent the outbreak of fires that were commonplace in the predominantly wood-built cities of the nineteenth century. The Chicago fire of 1871 in particular started a movement toward strict building regulations aimed at preventing the transfer of fire from one structure to another. Around the same time, the first laws intended to ensure adequate light and ventilation in residential buildings were enacted. While still concerned primarily with fire safety, the codes have expanded over the decades to include other security issues such as structural design and the safe exiting of a building.

Until 1995, most codes were written, maintained, and revised by four model code organizations, each adopted in different regions of the country. These organizations relied heavily on standards developed by groups such as the American Society for Testing and Materials (ASTM), the American Society of Heating, Refrigerating, and Air-conditioning Engineers (ASHRAE), and the National Fire Protection Association (NFPA). Local and state building ordinances typically adopted one of the four model codes: The Uniform Building Code (UBC) published by the International Conference of Building Officials (ICBO); the Standard Building Code (SBC) published by the Southern Building Code Congress International (SBCCI); the Boca National Building Code (BOCA/NBC) published by the Building Officials and Code Administrators International (BOCA); or the International Code (IC) published by the International Code Council (ICC).

Because of the difficulties associated with inconsistent codes regulating construction across different geographical regions of the country, the International Code Council moved to develop one, comprehensive model code by working cooperatively with the three model-code organizations. After years of intensive research, gathering input from design professionals, building contractors, material manufacturers, government organizations, and trade associations, the new *International Building Code (IBC)* was unveiled in 1997. The International Building Code provides a model code for adoption by all government agencies across the United States and other countries. The use of this code places all construction in the country under the same regulations related to design, materials, and construction techniques. The original four model codes are still available for adoption, however their use has diminished over the years. The International Code Council maintains a grouping of 13 codes called the I-codes, which govern not only the materials and construction of buildings but also their mechanical and environmental systems. In addition to the Building and Residential codes the ICC publishes and maintains the following model codes shown in Table 2.1.

Table 2.1 International Code Council Suite of Codes

International Fire Code—Deals specifically with fire prevention and suppression.
International Plumbing Code—Outlines requirements for water supply and waste systems.
International Mechanical Code—Relates to heating, ventilating, and air-conditioning systems.
International Fuel Gas Code—Concerned with fuel gas used for heating.
International Energy Conservation Code—Outlines requirements for energy performance.
International Performance Code—A rewriting of the ICC code as performance provisions.

Each code references national consensus standards, is comprehensive in itself, and all are compatible with each other. The codes are not static documents but are continuously updated, amended, and expanded for republication on a three-year cycle.

The International Residential Code (IRC)

The International Building Code regulates all structures within the jurisdictions where it has been adopted except one- and two-family dwellings. To deal with this residential building type, the ICC Board of Directors teamed with the National Association of Home Builders (NAHB) to form the ICC/NAHB Task Force. Input from this partnership led to the recommendation that the ICC develop and maintain a standalone residential code for one- and two-family dwellings and multiple single-family dwellings, such as townhouses. The task force developed a comprehensive residential code that is consistent with and inclusive of the scope and content of the existing codes promulgated by the BOCA, ICBO, and SBCCI. The resulting cooperative work led to the *International Residential Code for One- and Two-Family Dwellings*, first published in 2000 (Fig. 2.2).

The International Building Code (IBC)

The International Building Code establishes minimum regulations for commercial building systems using both prescriptive and performance related provisions (Fig. 2.3). A prescriptive code specifies exactly what type of design or material may be used and gives explicit guidelines for its implementation. A performance code lists the exact operating characteristics of a component and references industry standards to describe a minimum level of performance that an assembly must provide, without stipulating the exact materials or implementation strategies. Performance codes allow for a greater

Figure 2.2 The International Residential Code for One- and Two-Family dwellings is the result of an agreement between the International Code Council and several historic model codes. *(Courtesy International Code Council)*

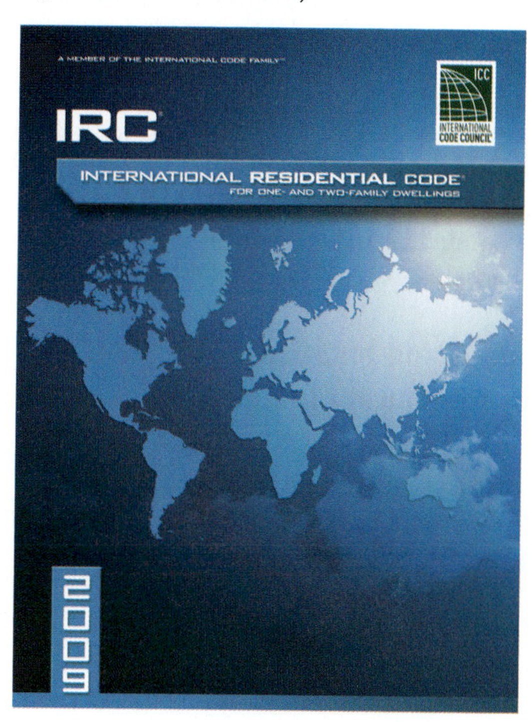

Figure 2.3 The International Building Code establishes minimum regulations for commercial building systems using prescriptive- and performance-based provisions. *(Courtesy International Code Council)*

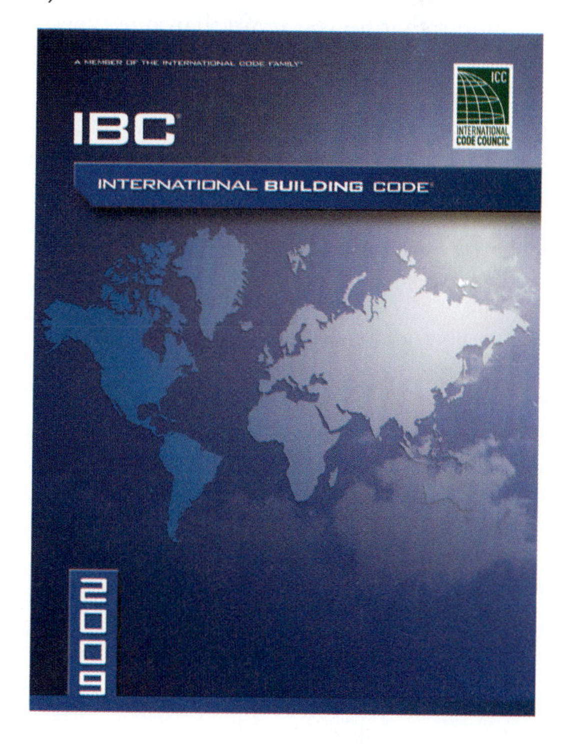

amount of flexibility in how requirements may be met. The written text of building codes are often technically complex and difficult to interpret. The designer, contractor, and material supplier must work closely with the local building official who has ultimate authority in how the wording of the code will be interpreted.

Many cities have municipal building codes that may be significantly different from the IBC or may exceed the IBC's requirements in certain instances. A check with the local government building department is required to understand what codes and standards must be followed.

Organization and Content of the International Building Code

The IBC begins with an administrative introduction that provides instructions as to how the code is to be adopted, administered, and enforced by a jurisdiction. Chapter One outlines the applicability of the code, clarifies the duties of building officials, dictates permitting and inspection procedures, and assigns fee schedules. A second chapter gives definitions for the terms used throughout the code book.

Use and Occupancy Classification Buildings and structures in the International Building Code are controlled by a classification system according to their intended function. Structures are classified with special requirements based on use and occupancy in one or more of the following groups (Table 2.2).

A building that houses more than one type of use is classified as mixed-occupancy and must conform to the code requirements of both types (or the one that is more restrictive). In addition to the main occupancy categories listed, the code specifies individual requirements for what are termed special uses, such as high-rise buildings or covered malls, that may require additional clarification. For instance, a building that houses hazardous materials will require supplementary fire protection equipment beyond that of normal occupancies. The occupancy group classification of a building governs much of the requirements covered in the remaining topics of the code, including allowable height and area and the type of construction that may be utilized.

Table 2.2 Occupancy Classifications

Group A	Assembly
Group B	Business
Group E	Educational
Group F	Factory and industrial
Group H	High hazard
Group I	Institutional
Group M	Mercantile
Group R	Residential
Group S	Storage
Group U	Utility and miscellaneous

Types of Construction The IBC classifies building construction according to five categories based on the types of materials used and their degree of combustibility. Type I and II construction are inherently non-combustible and utilize materials such as concrete and steel. Type III construction utilizes exterior building materials that are non-combustible with structural (load bearing) materials that may be of varying degrees of combustibility. Type IV construction defines exterior walls that are of noncombustible materials with interior building elements of solid or laminated wood without concealed spaces. This type is often used to designate heavy timber buildings. Type V is that type of construction in which the structural elements, exterior walls and interior walls, are of any material. Most wood frame buildings are designated as Type V construction. Each of these broad categories is further subdivided into A (protected) and B (unprotected). Protected describes a building in which all structural members have an additional fire-rated coating or cover that extends the fire resistance of the assembly. Unprotected means that all structural members of a building or structure have no additional fire-rated coating or cover. Protected buildings are generally allowed greater building heights and areas. Further subcategories account for the effects of a building being provided with an automatic sprinkler system.

Building Area and Height The height and area limitation for buildings of different construction types is governed by the intended use. Tables within the code specify the maximum floor areas allowed per occupant for various building types. Buildings of Type I and II construction are generally allowed unlimited areas due to their inherent incombustibility. The code allows for a certain amount of flexibility in determining building height and area by providing trade-offs between construction type, height, area, and the type of fire protection system utilized. Certain building types, such as low hazard industrial processes that require large areas and unusual heights, are exempt from area and height limitations. Code requirements must be balanced with local zoning ordinances that may be more or less restrictive.

Means of Egress A crucial part of the International Building Code is concerned with providing both safe ways for occupants to exit a building during a fire and safety to firefighters and first responders during emergency operations. The IBC defines a "**Means of Egress**" as: "A continuous and unobstructed path of vertical and horizontal egress travel from any occupied portion of a building or structure to a public way. A means of egress consists of three separate and distinct paths: the exit access, the exit, and the exit discharge" (Fig. 2.4). The required number and types of exits from a building, as well as the width of the egress path, vary according to building type, occupant load (the number of occupants), construction type, and whether a building has a sprinkler system or not. The code goes on to define the physical requirements for egress paths, including minimum head room, allowable projections, and the surface characteristics of the path. Travel distance, the distance a person must travel from any point in a building to a public way, is also specified. For most un-sprinkled occupancies, travel distances are set at two hundred feet. Specific guidelines are given for exit signage, illumination, and the configuration of stairs, doors, gates, and turnstiles.

Engineering Requirements Subsequent chapters of the IBC go on to define specific engineering guidelines for the structural adequacy of a building structure and enclosure. Chapters 14 and 15 outline requirements for the exterior envelope of a building, covering walls and roofs, including building openings, architectural trim, and projections, such as balconies and canopies. Chapters 16 through 18 set forth design criteria for different types of structural loading, structural tests and inspections, and the construction of foundation systems. The code defines two types of loading conditions that act on a structural system. The weight of the actual materials of construction, walls, floors and roofs, and finishes, is known as the **Dead load**. Conversely, **Live loads** are a result of occupancy and use, consisting of people, furnishings, and environmental forces such as wind, snow, and seismic (earthquake) activity. Chapters 19 through 26 give criteria for the use of specific materials, including concrete, aluminum, masonry, steel, wood, glass and glazing, gypsum board and plaster, and plastics. References to the ICC Electrical and Mechanical codes are given in lieu of detailed regulations for these systems. Final chapters of the code specify requirements for elevators, conveying systems, and other special constructions.

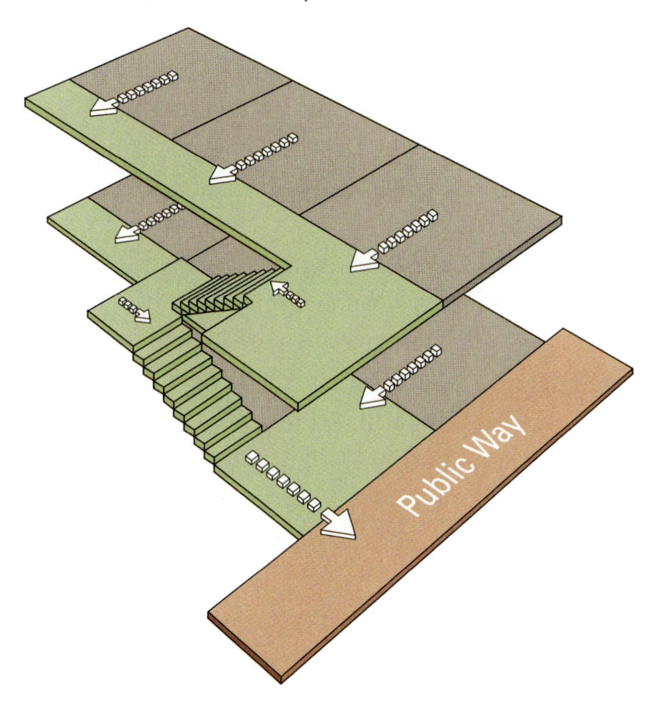

Figure 2.4 The IBC defines a means of egress as a continuous path of travel from any occupied portion of a building to a public way. *(Courtesy Eva Kultermann)*

Americans with Disabilities Act (ADA)/Universal Design

The Americans with Disabilities Act, enacted in 1992, is federal legislation that mandates buildings be made accessible to persons with both physical and cognitive disabilities. The ADA is not a code but a Federal law that is enforced by legal action through the courts. The Architectural and Transportation Barriers Compliance Board administers the ADA Accessibility Guidelines (ADAAG), a written and graphic set of compliance regulations outlining specific requirements for accessible parking, travel paths, doors, and fixtures (Fig. 2.5).

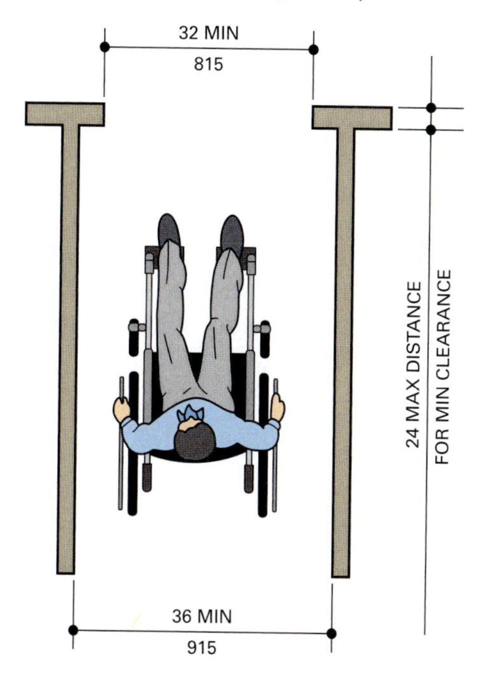

Figure 2.5 A graphic from the ADA Accessibility Guidelines showing the minimum clear width for a single wheelchair passage. *(Courtesy U.S. Department of Justice)*

The ADA includes provisions for persons with a variety of functional disabilities, including limitations in physical mobility, and visual, auditory, or mental sensory impairments. While building officials are not required to check for ADA compliance, many will make recommendations in cases where obvious violations are present. The ADA governs new construction but can be applied to existing buildings where access is not available.

In recent years, ADA guidelines have included issues in relation to the public right of way. Developers may be required to provide ADA ramps with detectable warning strips and sidewalks and drives with ADA compliant slopes. Local building departments should be consulted for the latest ADA regulations and standard details.

A new design movement that seeks to go beyond the barrier-free guidelines of the ADA is called universal design. **Universal Design** advocates the design of products and environments that are usable by all people, to the greatest extent possible, without the need for adaptation or specialized design. The movement encourages architects and product designers to make all spaces and fixtures accessible and comfortable to users with a variety of needs and abilities, including the disabled, elderly, women that are expecting, and those with temporary injuries. Universal design accommodates the widest range of individual abilities by creating spaces that are easy to understand regardless of the user's experience, knowledge, language skills, or current concentration level.

STANDARD DEVELOPMENT ORGANIZATIONS

In addition to the building codes, many professional, private, and governmental organizations make major contributions to the safe and efficient use of construction materials. These organizations support the construction industry by conducting research and testing of materials and components and widely disseminating the resulting information. Many of the provisions cited in the building codes reference standards produced by some of the following organizations.

Department of Housing and Urban Development

The purpose of the Department of Housing and Urban Development (HUD) is to promote the development of housing in the nation's communities. An important function is the stimulation of housing production through mortgage insurance and other federal subsidies. HUD publications are geared toward improving the quality of housing built and establishing more efficient land-planning standards (Fig. 2.6).

Figure 2.6 A guideline publication of the U.S. Department of Housing and Urban Development. *(Courtesy U.S. Department of Housing and Urban Development)*

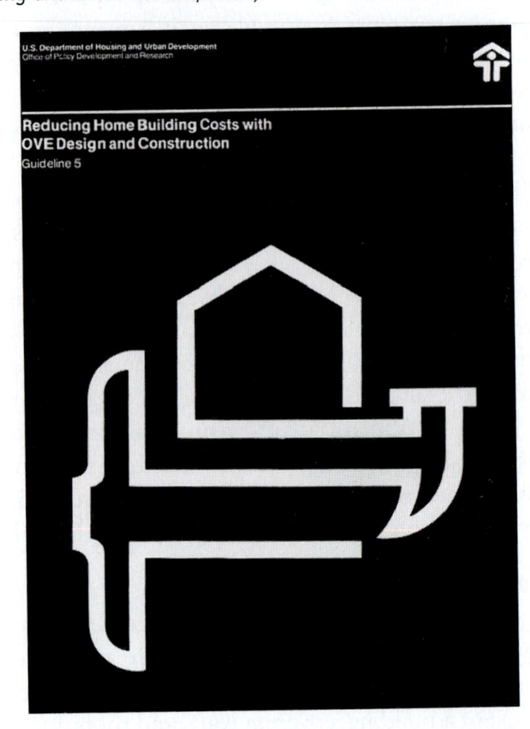

National Institute of Building Sciences

The National Institute of Building Sciences is chartered by the U.S. Congress as an authoritative national source of knowledge and advice on matters of building science and technology and as a leadership forum for the purpose of bringing together the building community to serve the public interest by:

- Promoting a more rational building regulatory environment
- Identifying and facilitating the development and use of new technology and processes
- Disseminating recognized technical information
- Facilitating building science and research
- Convening knowledgeable and affected interests to seek solutions to building community problems and needs

U.S. Department of Commerce Voluntary Product Standards

U.S. Department of Commerce (DOC) Voluntary Product Standards are developed under procedures published by the DOC in Part 10, Title 15, of the Code of Federal Regulations, Procedures for the Development of Voluntary Product Standards. The purpose of these standards is to establish nationally recognized requirements for products and to provide all concerned parties with a basis for common understanding of the characteristics of the products. The National Institute of Standards and Technology (NIST) administers the Voluntary Product Standards Program on a reimbursable basis. NIST also conducts research into sustainable construction methodologies and building automation systems.

Underwriters Laboratories Inc.

Underwriters Laboratories Inc. (UL) is a nonprofit organization that establishes, maintains, and operates laboratories for the examination and testing of devices and materials to determine their potential for hazards to life and property. UL conducts research to determine the probability of hazards to life and property resulting from the use of various materials, devices, products, equipment, assemblies, and systems. This is accomplished by scientific investigation, study, experiments, and tests. The organization works to publish standards, classifications, and specifications for materials, devices, products, equipment, constructions, and systems affecting such hazards. UL publishes yearly directories of manufacturers who have demonstrated an ability to produce products meeting UL requirements. They publish an extensive list of standards that have been accepted as "UL Standards for Safety." Products manufactured to these standards are permitted to carry the UL symbol indicating approval (Fig. 2.7). UL standards are designated by the initials "UL," followed by an identifying number. For example, UL943 covers ground-fault circuit interrupters.

American Society of Testing and Materials

The American Society for Testing and Materials (ASTM) is a nonprofit corporation formed for the development of standards on characteristics and performance of materials, products, systems, and services and for the promotion of related knowledge. The standards include test methods, definitions, recommended practices, classifications, and specifications. Standards developed by the society's committees are published annually in a 48-volume publication, *Book of ASTM Standards*. Section 4 contains five volumes that encompass the standards for the construction industry. ASTM specifications are designated by the initials ASTM followed by an identification number and the year of last revision. One example is *High-Strength Bolts for Structural Steel Joints*, ASTM A325-79.

Figure 2.7 Manufacturers whose products meet the safety standards of Underwriters Laboratories Inc. (UL) may be authorized by UL, subject to testing and evaluation, to use the UL mark on their products. *(Reproduced with permission from Underwriters Laboratories Inc.)*

ASTM also publishes *Standards Adjuncts*, which includes supporting material such as reference objects, charts, tables, and drawings used in conjunction with ASTM test methods. They are identified by a single letter of the alphabet followed by a code number and the word "adjunct." An example is C856 adjunct, *Petro graphic Examination of Hardened Concrete*. ASTM also publishes *Special Technical Publications*, which are books devoted to specific technical, scientific, and standardization topics. They are identified by the letters STP followed by a code number. An example is STP 691 *Durability of Building Materials and Components*.

American National Standards Institute

The American National Standards Institute (ANSI) is the coordinating organization for the national standards system in the United States. It is a federation of hundreds of companies, trade, technical, professional, labor, and consumer organizations. ANSI serves the nation in a variety of ways. Through cooperative efforts of its member organizations, ANSI's councils, boards, and committees coordinate the efforts of the hundreds of organizations in the United States that develop standards. ANSI helps identify what standards are needed and establishes timetables for their completion. It arranges for competent organizations to develop them or, if a standard-developing organization does not exist, it forms a group of persons who have the necessary competencies to develop the standard. ANSI provides and administers the only recognized system in the United States for establishing standards, regardless of the originating source, as American National (Consensus) Standards.

When it comes to standards development, ANSI is the official representative of the United States to the international community. It manages, coordinates, finances, and administratively supports effective participation in the International Organization for Standardization. ANSI helps govern ISO through membership on its Council Executive and Planning Committees. There are now more than 10,000 American National Standards in virtually every field and discipline. Standards deal with dimensions, ratings, terminology and symbols, test methods, and performance and safety specifications for materials, equipment, components, and products in the following fields: construction, electrical and electronics, heating, air-conditioning and refrigeration, information systems, medical devices, mechanical, nuclear, physical distribution, piping and processing, photography and motion pictures, textiles, and welding. An American National Standard is designated by code numbers beginning with the letters ANSI followed by letters indicating the organization that formulated the standard, such as ACI (American Concrete Institute). These letters are followed by an identification number and the date the standard was issued or released. A typical example is ANSI/ACI 517-70, *Practice for Curing Concrete.*

International Organization for Standardization (ISO)

The International Organization for Standardization (ISO) is the world body that coordinates the production of engineering and product standards for worldwide use. Each industrial nation in the world has a national standards group. In the United States, it is the American National Standards Institute (ANSI), and in Canada it is the Canadian Standards Association. Most nations are represented in ISO through their national standards organizations. The ISO publishes standards addressing a wide range of issues of interest to the construction industry (Fig. 2.8).

American Society of Heating, Refrigerating and Air-Conditioning Engineers (ASHRAE)

ASHRAE is an international organization with a mission of "advancing heating, ventilation, air conditioning and refrigeration to serve humanity and promote a sustainable world through research, standards writing, publishing and continuing education." It establishes standards for the uniform testing and rating of heating, ventilation, air conditioning, and refrigeration equipment. It also conducts related research, disseminates publications, and provides continuing education to its members.

Figure 2.8 The International Organization for Standardization (ISO) administers engineering and product standards for worldwide use and publishes numerous guides, such as this one addressing climate change. *(© ISO. With permission of the American National Standards Institute (ANSI) on behalf of the International Organization for Standardization (ISO))*

National Association of Home Builders of the United States

The National Association of Home Builders (NAHB) is a federation of approximately eight hundred state and local builders associations in the United States. NAHB established and staffs the National Housing Center in Washington, D.C., manages a National Housing Center Library, and publishes a number of newsletters and magazines. It issues reports pertaining to economic research and analysis and offers nationwide seminars, workshops, and conferences.

The NAHB Research Foundation is a subsidiary of the National Association of Home Builders. Its four divisions serve as a center of technical research for the residential and light-frame construction industries. The Laboratory Services Division performs tests on materials and components under contract from manufacturers. The Industrial Engineering Division designs and performs cost and economic studies of alternative construction techniques and materials. A Building Systems Division provides technical studies of whole functional systems in small buildings. The Special Studies Division initiates projects in new areas of interest to the residential construction industry.

Canadian Organizations

The following are a few of the Canadian organizations active in various aspects of the construction industry. The Canada Mortgage and Housing Corporation (CMHC) is an agency of the federal government responsible for administering the National Housing Act. This legislation is designed to improve the housing and living conditions in Canada. Among its responsibilities is conducting research into the social, economic, and technical aspects of housing and related fields. The Canadian Housing Information Centre (CHIC) is a part of the Canadian Mortgage and Housing Corporation. It provides thousands of publications related to all aspects of housing, building, and community development. The Canadian Wood Council (CWC) is a national federation of forest products associations. It is involved in developing and promoting technical product information to assist members of the specifying and regulatory community. It has an extensive listing of technical manuals and books.

Trade Associations

Other organizations are devoted to the advancement of knowledge about individual materials and methods used in construction. Some of these have developed guidelines that influence the content of the building codes and establish industry-wide standards. Trade associations are a major factor in materials development. A **trade association** is an organization whose membership comprises manufacturers or businesses involved in the production or supply of materials and services in a particular area. An example is the American Institute of Steel Construction. The goal of a trade association is to promote the interests of its membership. This includes dissemination of information on the proper use of the materials, products, and services of the members. Some trade associations support research into the proper use and improvement of construction materials and better construction methods. As they do this, they develop material specifications and performance procedures for a particular trade area. Often these become accepted national standards. Some trade associations are involved in programs of certification. Products made according to their specifications and standards are identified with a seal, label, or stamp. An example is the certification stamp on plywood issued by APA—The Engineered Wood Association (Fig. 2.9).

Following are brief descriptions of some of the associations involved with materials used in the construction industry.

The *American Concrete Institute (ACI)* is a technical and educational organization dedicated to improving the design, construction, manufacture, and maintenance of concrete products and reinforced concrete structures. ACI disseminates its information through conventions, seminars, publications, and volunteer committee work. The ACI series of technical publications is extensive, and one, *The ACI Manual of Concrete Practice*, is a major reference for the industry. ACI publishes several magazines.

Figure 2.9 This stamp on Plywood products indicates that APA—The Engineered Wood Association has certified that it meets its specifications. *(Courtesy APA—The Engineered Wood Association)*

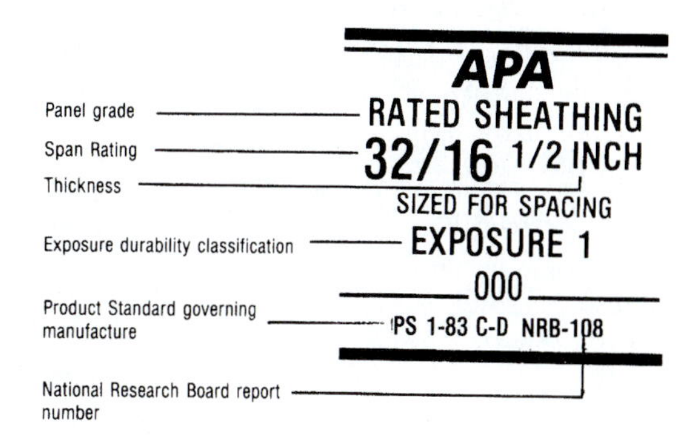

The *Portland Cement Association (PCA)* is a nonprofit organization dedicated to improving and extending the uses of Portland cement and concrete. It is supported by voluntary financial contributions by member companies. PCA serves as a clearinghouse for concrete design practices and construction methods. It has an extensive service program that offers publications, motion pictures, and slides depicting up-to-the-minute developments. It offers many educational programs and is involved in developing computer programs for construction-related activities. PCA provides research and development/engineering services that include troubleshooting field problems, engineering investigations, development services, and assistance with the manufacture of Portland cement.

The *American Institute of Steel Construction (AISC)* is a nonprofit association serving the fabricated structural steel industry. Its objectives are to improve and advance the use of fabricated structural steel through research and engineering studies to develop the most efficient and economical design of structures. The institute publishes manuals, textbooks, specifications, and technical books. The most widely used publication is the *Manual of Steel Construction* (Fig. 2.10).

The *Aluminum Association* is the primary source for statistics, standards, and information on aluminum and the aluminum industry. Members of the association produce almost all domestic primary aluminum ingots and most of the semi-fabricated mill products.

Figure 2.10 The *Manual of Steel Construction* gives extensive guidelines for the design of steel structures. *(Courtesy American Institute of Steel Construction)*

The association works to increase public awareness of aluminum, helps companies with operational problems, develops voluntary standards and technical data, collects marketing and statistical data, and works with product development and educational programs. It offers a variety of publications and audio-visual presentations. A most significant publication is the *Aluminum Association Construction Manual.*

The *Gypsum Association* has as its primary function the promotion of gypsum products. It serves as an information center by gathering and disseminating information pertaining to gypsum and gypsum products. Representatives of member companies contribute to the program of the association. Activities include mining and manufacturing, review of specifications incorporated into building codes, and research in such areas as fire testing, structural design, sound testing, and safety.

The *Asphalt Institute* works to advance the most efficient use of bituminous concrete through research, engineering, and educational activities. It assists in extending and developing the asphalt business. Basically it is an engineering organization with staff located in offices throughout the United States. The institute is active in educational activities and offers a wide range of technical publications and quarterly bulletins.

A number of organizations represent various aspects of the *wood industry.* Some of these are:

The *Southern Pine Inspection Bureau* establishes grades and inspects softwood lumber in the southeastern part of the United States.

The *West Coast Lumber Inspection Bureau* establishes grades and inspects Douglas fir, hemlock, cedar, and other softwood lumber in California and the western regions of Washington and Oregon.

The *National Hardwood Lumber Association* establishes grades and inspects hardwood lumber such as walnut, oak, birch, and maple.

APA—The Engineered Wood Association establishes grades and supervises the manufacture of softwood plywood. It offers an extensive variety of technical publications (Fig. 2.11).

The *American Institute of Timber Construction* performs research, testing, and structural analysis for solid and laminated timber construction.

Figure 2.11 This is one of the many technical publications of the Engineered Wood Association. *(Courtesy APA—The Engineered Wood Association)*

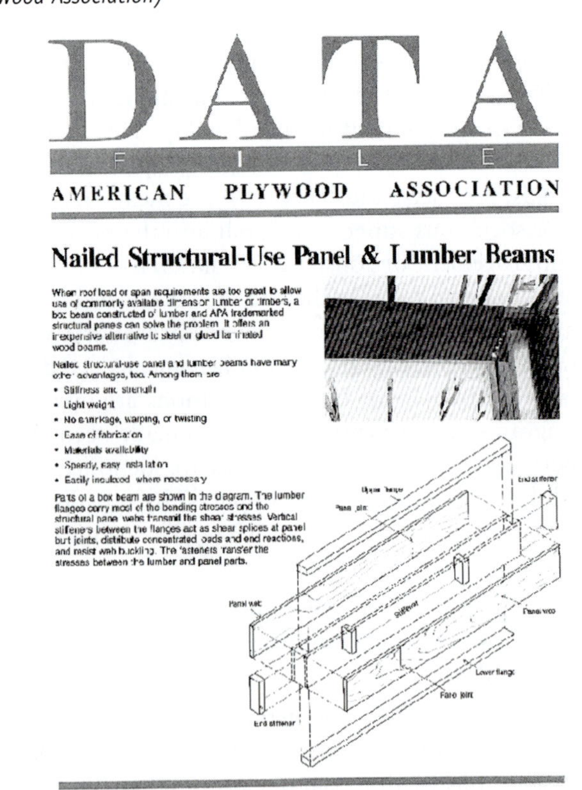

DATA FILE

AMERICAN PLYWOOD ASSOCIATION

Nailed Structural-Use Panel & Lumber Beams

When roof load or span requirements are too great to allow use of commonly available dimension lumber or timbers, a box beam constructed of lumber and APA trademarked structural panels can solve the problem. It offers an inexpensive alternative to steel or glued laminated wood beams.

Nailed structural-use panel and lumber beams have many other advantages, too. Among them are

- Stiffness and strength
- Light weight
- No shrinkage, warping, or twisting
- Ease of fabrication
- Materials availability
- Speedy, easy installation
- Easily insulated where necessary

Parts of a box beam are shown in the diagram. The lumber flanges carry most of the bending stresses and the structural panel webs transmit the shear stresses. Vertical stiffeners between the flanges act as shear splices at panel butt joints, distribute concentrated loads and end reactions, and resist web buckling. The fasteners transfer the stresses between the lumber and panel parts.

Addresses for a number of trade associations are listed in Appendix B.

SUSTAINABLE DESIGN AND CONSTRUCTION

Interest in the state of the environment and the future of our developing world has become a major topic of debate and discussion in all areas of our lives. For the first time in history we have the technology to monitor and confirm the problems facing the natural environment. This section will document some of the environmental issues that are relevant today, examine the impact of the construction industry on the environment, and look at new systems and technologies the design and construction industry can implement to limit future harm.

The Environmental Context

At the dawn of the twenty-first century, the global population reached the six billion mark and is expected to continue growing at an annual rate of 1.2 percent. In order to satisfy the needs of this ever-expanding population, natural resources are being depleted at unprecedented rates. Land that previously supported natural systems and agricultural production is coming under increasing development pressure. In the United States alone, about thirty-four million acres, an area roughly equal to the state of Illinois, was developed between 1982 and 2001. Sometimes referred to as **urban sprawl**, this unchecked land development places tremendous burdens on natural systems (Fig. 2.12).

Figure 2.12 This housing development in Colorado Springs is an example of the rapid, unchecked municipal growth known as urban sprawl. *(Courtesy David Shankbone)*

For the past half century, global energy consumption patterns have followed hand in hand with economic development. During this period, energy consumption doubled to keep up with the demands of the growing population and an ever increasing array of technological developments. The fossil fuels we use today are the result of millions of years of forces and pressures on biological materials, resulting in their being broken down into pure carbons (coal) and petroleum oils. These fossil fuels are a non-renewable resource; once they have been consumed they cannot be immediately replaced. Today, nearly all of the known petroleum reserves have been discovered and most are believed to be fully explored. Geologists predict that in the near future global oil producers will reach "peak production," a point after which securing more oil will become increasingly difficult and expensive.

The larger problem of the increased use of fossil fuels is the resulting impact on the environment. Scientists now believe that climate change, which for decades had been a hotly contested subject, is becoming a credible theory. Recent findings by the Intergovernmental Panel on Climate Change (IPCC), suggest that "*. . . there is new and stronger evidence that most of the warming observed over the last fifty years is attributable to human activities.*" As human societies have adopted increasingly automated lifestyles, the amount of heat-trapping gases present in the atmosphere has multiplied. The **greenhouse effect** is a process by which the earth's atmosphere traps solar radiation through the existence of gases, such as carbon dioxide (CO_2), water vapor, and methane, that allow incoming sunlight to pass through them but absorb the heat radiated back from the earth's surface (Fig. 2.13). The burning of fossil coal, oil, and natural gas, together with widespread deforestation, has raised the total amount of atmospheric CO_2 by thirty percent since the beginning of the industrial revolution (Fig. 2.14). Scientists now estimate that increased concentrations of CO_2 will continue to cause a global temperature increase, even if all additional emissions were eliminated today (Fig. 2.15). Global warming even within conservative estimates is predicted to have detrimental consequences for forest productivity, agricultural production, and air and water quality. Other long term consequences include a continued melting of polar ice, resulting in rising sea levels and more frequent extreme weather events. Dwindling reserves of fossil fuels and the release of greenhouse gases should be of particular concern to the people of the United States. With less than five percent of the global population, the U.S. consumes a staggering twenty-six percent of the world's energy supply and contributes more than thirty percent of the greenhouse gases thought to be responsible for climate change.

Figure 2.13

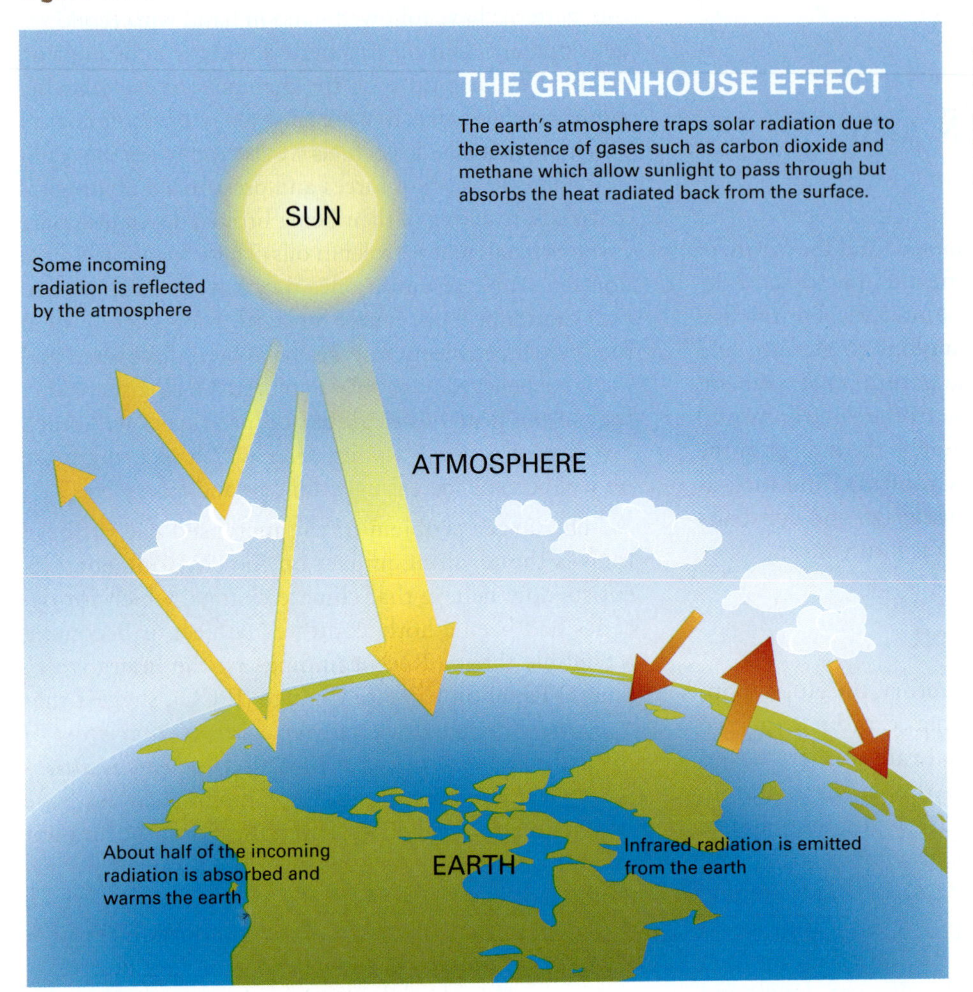

The Greenhouse Effect is a process by which the earth's atmosphere traps solar radiation causing a gradual increase in global temperatures. *(Courtesy Eva Kultermann)*

Figure 2.14 The dramatic rise in carbon emissions over the last two centuries is in part due to the activities of the construction industry. *(Courtesy Global Warming Art)*

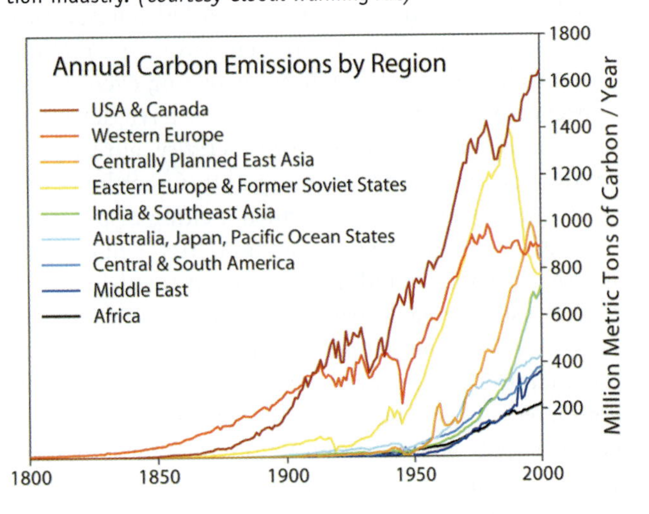

Figure 2.15 The increased burning of fossil fuels has caused global temperatures to rise, particularly in the last decade. *(Courtesy Global Warming Art)*

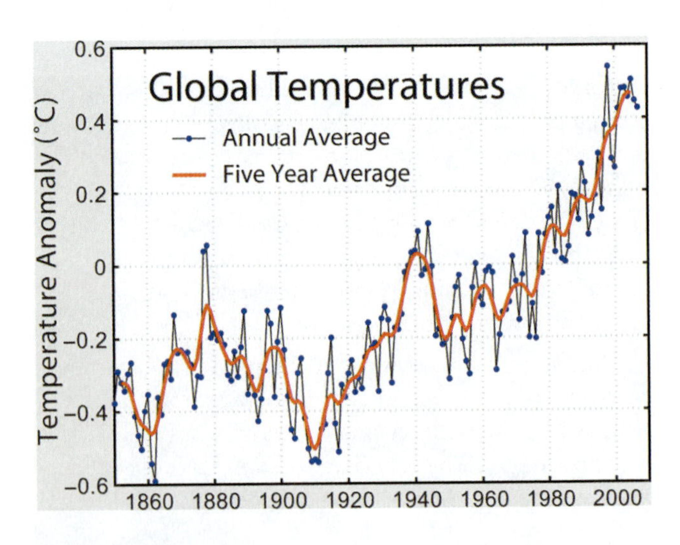

Natural materials obtained from the earth are essential to our modern life. Some of the raw materials required come from renewable resources, such as agricultural and forestry products, while others are nonrenewable, such as metals and minerals. A report by the United States Geological Survey (USGS) found that recent years have seen a decrease in the use of renewable resources and an increasing demand for nonrenewable resources, particularly in construction materials. Since 1900 the use of construction materials, such as stone, sand, and gravel, has soared. The world population doubled its consumption of non-fuel raw materials between 1970 and 1995, and estimates predict that this trend will continue. The large-scale exploitation of both renewable and nonrenewable materials, through mining and production activities, results in high greenhouse gas emissions that permeate Earth's water, air, and soil.

The Environmental Impact of the Construction Industry

Historically, buildings were built and managed in close harmony with nature. Materials were extracted locally from natural and renewable sources. Buildings were designed to be climate responsive, taking cues from long-established design methods that limited the amount of fuel required to operate them. Lighting came from the sun, and cooling was provided by building form and simple shading devices. Communities were organized in compact fashion, reducing the need for far-reaching infrastructure and transportation networks. At the onset of the twentieth century, however, a series of technological developments changed the way buildings were designed, built, and operated.

The advent of electricity allowed buildings to utilize artificial lighting as well as provide power to a dazzling array of new electrical appliances. By mid-century, electricity was being fed directly into homes all over the United States. The use of electricity to provide heat was expanded, and the fully electric home featured space heating and cooking and water heating. One of the biggest technical innovations to increase the amount of energy used in buildings was the development of the modern air conditioner. First utilized in movie theaters and industrial applications, the new invention quickly found its way into all types of buildings, fueling the ongoing pattern of high energy use.

Buildings, both commercial and residential, are a major contributor to energy and resource depletion. Each year, at least forty percent of the raw materials and energy produced in the world are used in the building sector, generating millions of tons of greenhouse gasses and solid wastes. Today, the environmental impact of buildings of all types accounts for nearly forty percent of our total primary annual energy consumption (Fig. 2.16) and around 65 percent of total electricity use. Combined, this translates into buildings contributing a solid thirty percent of total annual carbon emissions for the United States, with similar statistics for the rest of the developed world. In addition to the energy required to construct and operate our built environment, buildings consume other vital resources, including twelve percent of U.S. potable (drinkable) water. Building-related construction and demolition debris totals approximately 136 million tons per year, accounting for nearly 60 percent of total non-industrial waste generation in the U.S. (Fig. 2.17).

Figure 2.16 Buildings, both commercial and residential, account for thirty-nine percent of total annual U.S. energy consumption. *(Courtesy U.S. Energy Information Administration)*

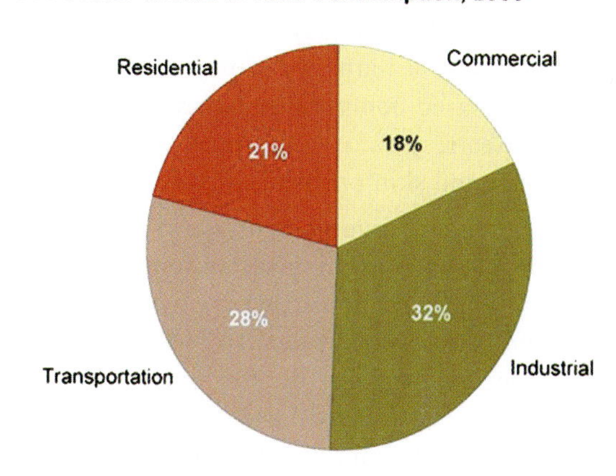

Figure 2.17 Building-related construction and demolition debris accounts for nearly 60 percent of total U.S. non-industrial waste generation each year. *(Courtesy Eva Kultermann)*

Other advances in material technologies have led to additional building-related problems, specifically, poor indoor air quality. A variety of factors have contributed to increasing complaints of discomfort associated with building occupancy. The past decades have seen a general trend of eliminating operable windows and relying solely on mechanical ventilation in larger commercial buildings. When mechanical systems are not properly maintained, the exchange of contaminated air with fresh air is reduced and bacterial agents can enter the air stream.

Indoor air quality was further compromised through the introduction of a vast number of new synthetic products to supply the increased demand for building materials. Pressed wood and fiber boards utilizing formaldehyde have since been found to be a major culprit of toxic off-gassing. Additional chemical concentrations found their way into buildings in the form of paints, carpeting, office equipment, cleaning agents, and pesticides. Gradually the air inside buildings has become a breeding ground for a number of maladies, including sick building syndrome. **Sick Building Syndrome** (SBS) is a blanket term often used to describe building-related complaints such as headaches; dizziness; nausea respiratory problems; eye, nose, or throat irritations; and skin problems, just to name a few.

With such a large percentage of global resources consumed in the building sector, understanding how to design and construct structures within environmental limits is a necessary step to reducing the negative environmental impacts of the construction industry. By designing and building with the environment in mind, and wisely combining products made of wood, steel, concrete, plastic and other materials, we can produce buildings that use energy wisely, use resources efficiently, and provide a healthier interior environment for occupants.

THE SUSTAINABLE DEVELOPMENT MOVEMENT

In response to the growing awareness of the depletion of natural resources, the impact of the burning of fossil fuels and other environmental concerns, governments and all areas of industry have been searching for ways to develop and implement long term solutions for our continued development. In 1987 the World Commission on Environment and Development (WCED) met to address growing concerns "*about the accelerating deterioration of the human environment and natural resources and the consequences of that deterioration for economic and social development.*" The commission originated the now common term **Sustainable Development**, defined as: "*... development that meets the needs of the present without compromising the ability of future generations to meet their own needs.*" Sustainable development is a far-reaching concept, encompassing social, economic and environmental concerns. A subset of this larger concern is the sustainable building movement.

Sustainable Design and Construction

A number of terms are used to describe **sustainable building**, including green building, high-performance building, and climate-responsive building. The American Society for Testing and Materials (ASTM) defines a Green Building as "*a building that provides the specified building-performance requirements while minimizing disturbance to and improving the functioning of local, regional, and global ecosystems both during and after its construction and specified service life.*" The organization goes on to advocate a design and construction process that emphasizes a reduction in the use of materials, resources, and energy (Fig. 2.18). Green Building seeks to achieve resource and energy efficiency, integrate healthy materials and buildings, and promote ecologically and socially sensitive land-use (Fig. 2.19).

The basic strategies that define sustainable building design and construction are outlined in Table 2.3.

While much of the work to be done in the creation of sustainable buildings rests in the domain of architects and engineers, the importance of considering

Figure 2.18 The Culver House, a multi unit residential development by Dirk Denison Architects, incorporates resource and energy efficiency and integrates healthy building materials. *(WAS, Antonio Petrov, Daniel Wolf)*

Figure 2.19 A diagrammatic view of the Culver House project pointing out the basic strategies that define sustainable building design and construction. *(Atilier Ten)*

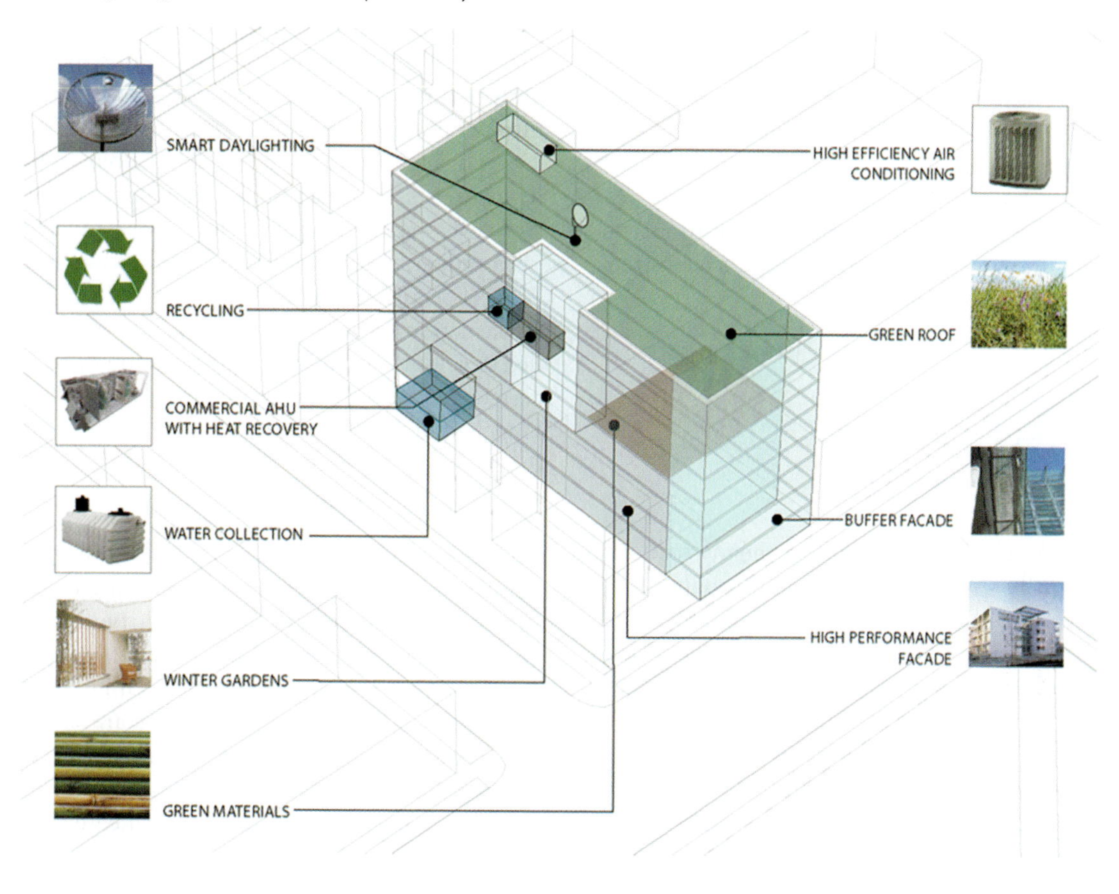

Table 2.3 Sustainable Building Design and Construction Guidelines

Site	Water
Develop building sites to restore and enhance natural eco-systems. Develop and restore previously disturbed sites instead of unspoiled rural or natural areas. Develop native and drought tolerant landscape systems. Minimize site paving and use pervious paving systems. Minimize construction-activity-related pollution.	Use water-efficient fixtures, appliances, and equipment. Minimize the use of potable water for landscape irrigation. Utilize rainwater harvesting for both building and landscape uses.
Energy	**Material Resources**
Promote energy conservation by avoiding energy-intensive operations. Use daylight and energy efficient electric lighting systems. Use renewable energy sources, such as solar, wind, and geothermal energy. Provide high levels of thermal insulation and an airtight building envelope. Use natural ventilation in place of mechanical ventilation. Assure that all installed systems are calibrated for optimal functioning. Utilize energy and heat recovery.	Reuse existing buildings and materials whenever possible. Design buildings for long service life and eventual recycling of components. Use renewable and recycled content materials. Practice waste management during construction to minimize waste. Use locally or regionally sourced materials to minimize transportation energy.
	Indoor Environment
	Use non-toxic, natural materials and finishes. Curtail sources of volatile organic compounds and other pollutants. Increase the use of natural daylight and ventilation. Provide high levels of natural ventilation.

sustainability during the construction process is crucial. Green building design is implemented on the construction site. Decisions made in the field regarding material procurement, site preservation during construction, equipment use, and the sequence and organization of the work are the responsibility of the contractor. By implementing a combination of traditional techniques and new technologies the construction industry can reduce and sometimes eliminate a great deal of the environmental damage associated with construction processes.

SUSTAINABLE BUILDING CERTIFICATION SYSTEMS

A number of organizations provide a means to both guide and quantify sustainable building through a recognized system of certification. Most are voluntary rating systems that are developed to be easily understood by all participants in the construction industry. In addition to certifying the environmental performance of a building, the systems allow for a way to compare and rate the success of different green building strategies. It is imperative for those involved directly in the construction of buildings to familiarize themselves with the basics of green building rating systems. Contractors are increasingly asked to participate in pre-construction planning teams in order to contribute their expertise to the design and development process. The implementation of sustainable technologies and green certification programs rests largely with those directly engaged in a building's construction.

The U.S. Green Building Council and LEED Rating System

The U.S. Green Building Council (U.S.G.B.C.) was founded in 1993 to identify and promote strategies with which the design and construction industry could advance the creation of energy- and resource-efficient buildings. The not-for-profit organization is made up of a diverse membership including architects, contractors, material manufacturers, financial and insurance firms, research institutions, and government organizations. Council members cultivate new standards, practices, and guidelines for the design, construction, and operating procedures of high performance, sustainable buildings. Working in a committee-based, consensus-driven process the U.S.G.B.C developed the **Leadership in Energy and Environmental Design (LEED)** green building rating system. The aim of LEED is to define principles for sustainable building by providing a common standard of measurement. LEED is a third-party certification program that provides a nationally accepted standard for the design, construction, and maintenance of high-performance buildings. The rating system promotes an integrated, whole building design process that stimulates green competition while also helping to raise public awareness of green building. After years of research and development, LEED version 1 was launched in 1998 with only a handful of projects achieving certification. After repeated review and adjustment, LEED NC (new construction) version 2.2 was released in 2006. Continued development and the incorporation of regional concerns has led to the current standard, LEED version 3, launched in 2009.

LEED NC Version 3

The LEED rating system provides a tool that simultaneously guides and evaluates a building over its entire life cycle, from project inception through building operation. Certification is based on achieving a certain number of points, or credits, for each building strategy that a project adopts. The standards are organized into five broad topics: Sustainable Sites, Water Efficiency, Energy and Atmosphere, Materials and Resources, and Indoor Environmental Quality. Each category contains a number of base criteria, or prerequisites, which must be achieved in order for other credits in the category to be counted. The collected LEED building strategies amount to 100 credits, with two additional categories, Innovation in Design and Regional Bonus Credits, bringing the total number of points to 110. The Innovation in Design category awards credits to projects that exhibit exceptional performance in achieving energy and resource efficiency. Regional and bonus credits can be awarded by local chapters to recognize specific regional issues. The U.S.G.B.C certifies individuals who have completed a comprehensive training program as LEED accredited professionals (LEED AP). The inclusion of a LEED AP on a project team to support the design and application process adds one credit toward certification. Depending on the number of points achieved, a project can be designated as follows: certified (40–49 credits); silver (50–59 credits); gold (60–79 credits); or platinum (80 credits and above), the highest rating possible.

The rating system is explained by a reference guide that lists in detail the intent, implementation, and submittal requirements and the technologies to be used for each individual credit. The basic intent identifies the main goal of the credit, usually with a one-sentence description. For instance, the prerequisite for the sustainable sites category, Construction Activity Pollution

Prevention, states the intent as: "Reduce pollution from construction activities by controlling soil erosion, waterway sedimentation, and airborne dust generation." The *Requirements* specify the criteria that must be used to satisfy the credit and lists accepted industry standards and tests that must be met to establish compliance. A section on *Submittals* lists the basic submittal requirements, templates, calculations, written summaries, drawings, and diagrams that must be sent to the U.S.G.B.C. for verification. Finally, a summary of references and standards cites the referenced standards on which the credit is based. The reference guide accompanies each credit explanation with a section entitled *Green Building Concerns* that gives a short narrative overview of the subject, listing environmental, economic, and community concerns.

Certifying a project is a three-step process consisting of project registration, submittal of technical documentation, and verification, which, if everything is in compliance, results in final certification by the U.S.G.B.C. LEED certification provides a number of tangible benefits including recognition of environmental stewardship by third-party certification, a plaque and official certificate for display, and a variety of marketing opportunities (Fig. 2.20). Certification under LEED adds to project costs, requiring additional consultants to manage the documentation of credits and perform energy modeling. Costs are currently estimated at anywhere between three to five percent of total project expenditures. The LEED system is growing in popularity with more federal and local municipalities that require certification for all public buildings within their jurisdiction.

Figure 2.20 LEED certification provides a number of benefits including recognition of environmental stewardship and a plaque for display. *(Courtesy Eva Kultermann)*

Table 2.4 shows the complete list of prerequisites and credits under the LEED for New Construction and Major Renovations Version 3 Rating System.

With the growing popularity of the LEED green building rating system, the U.S.G.B.C has been developing additional rating systems and standards that are explicitly tailored to specific building types. LEED now offers certification systems for major renovations, commercial interiors, core and shell, retail projects, schools, homes, and neighborhood developments among others.

Green Globes

The Green Building Initiative (GBI), a not-for-profit organization, defines its mission as: "to accelerate the adoption of building practices that result in energy-efficient, healthier, and environmentally sustainable buildings by promoting credible and practical green building approaches for residential and commercial construction." The Green Globes Rating System provides a third-party green building verification procedure that is administered entirely online. The system delivers an online assessment procedure that gives guidance for the design, construction, and management of a building. Point categories in the Green Globes system include project management, site, energy, water, resources, emissions and other impacts, and the indoor environment. The system is interactive and flexible, providing project teams with immediate feedback and a less costly certification option.

NAHB National Green Building Program™

The National Association of Home Builders (NAHB) and the International Code Council (ICC) have collaborated to establish a nationally administered green certification program for single and multifamily homes, residential remodeling, and site development projects. The ICC 700-2008 National Green Building Standard provides a flexible third-party rating system for residential green project certification, which will be used for the NAHB National Green Building Program.

The guidelines were developed through a public process that included an extensive review of local green homebuilder programs, existing energy-efficiency programs endorsed by NAHB, and leading life-cycle analysis (LCA) tools available for use in residential design and construction. Similar to the NAHB Model Green Homebuilding Guidelines, the main purpose is to provide a framework for builders to reduce the energy- and resource-related impacts of a home. Each topic addressed in the program has a point value attributed to it that is evaluated according to environmental impact, building science and best building practices, and ease of implementation.

Table 2.4 LEED for New Construction and Major Renovations, Version 3 Checklist

Sustainable Sites	Credits
SS Prerequisite 1: Construction Activity Pollution Prevention	Required
SS Credit 1: Site Selection	1
SS Credit 2: Development Density & Community Connectivity	5
SS Credit 3: Brownfield Redevelopment	1
SS Credit 4.1: Alternative Transportation: Public Transportation Access	6
SS Credit 4.2: Alternative Transportation: Bicycle Storage & Changing Rooms	1
SS Credit 4.3: Alternative Transportation: Low Emitting & Fuel Efficient Vehicles	3
SS Credit 4.4: Alternative Transportation: Parking Capacity	2
SS Credit 5.1: Site Development: Protect or Restore Habitat	1
SS Credit 5.2: Site Development: Maximize Open Space	1
SS Credit 6.1: Stormwater Design: Quantity Control	1
SS Credit 6.2: Stormwater Design: Quality Control	1
SS Credit 7.1: Heat Island Effect: Non-Roof	1
SS Credit 7.2: Heat Island Effect: Roof	1
SS Credit 8: Light Pollution Reduction	1
Water Efficiency	
WE Prerequisite 1: Water Use Reduction; 20% Reduction	
WE Credit 1.1: Water Efficient Landscaping: Reduce by 50%	2
WE Credit 1.2: Water Efficient Landscaping: No Potable Water Use or No Irrigation	2
WE Credit 2: Innovative Wastewater Technologies	2
WE Credit 3.1: Water Use Reduction: 30% Reduction	2
WE Credit 3.2: Water Use Reduction: 40% Reduction	2
Energy & Atmosphere	
EA Prerequisite 1: Fundamental Commissioning of the Building Energy Systems	Required
EA Prerequisite 2: Minimum Energy Performance: 10% New Bldgs, 5% existing	Required
EA Prerequisite 3: Fundamental Refrigerant Management	Required
EA Credit 1: Optimize Energy Performance	1–19
12% New Buildings or 8% Existing Building Renovations	1
16% New Buildings or 12% Existing Building Renovations	3
20% New Buildings or 16% Existing Building Renovations	5
24% New Buildings or 20% Existing Building Renovations	7
28% New Buildings or 24% Existing Building Renovations	9
32% New Buildings or 28% Existing Building Renovations	11
40% New Buildings or 36% Existing Building Renovations	15
44% New Buildings or 40% Existing Building Renovations	17
48% New Buildings or 44% Existing Building Renovations	19
EA Credit 2: On-Site Renewable Energy	1–7
1% Renewable Energy	1
5% Renewable Energy	3
9% Renewable Energy	5
13% Renewable Energy	7
EA Credit 3: Enhanced Commissioning	2
EA Credit 4: Enhanced Refrigerant Management	2

EA Credit 5: Measurement & Verification	3
EA Credit 6: Green Power	2

Materials & Resources

MR Prerequisite 1: Storage & Collection of Recyclables	Required
MR Credit 1.1: Building Reuse: Maintain 75% of Existing Walls, Floors, & Roof	1
MR Credit 1.2: Building Reuse: Maintain 95% of Existing Walls, Floors, & Roof	1
MR Credit 1.3: Building Reuse: Maintain 50% of Interior Non-Structural Elements	1
MR Credit 2.1: Construction Waste Management: Divert 50% from Disposal	1
MR Credit 2.2: Construction Waste Management: Divert 75% from Disposal	1
MR Credit 3.1: Materials Reuse: 5%	1
MR Credit 3.2: Materials Reuse: 10%	1
MR Credit 4.1: Recycled Content: 10% (post-consumer + 1/2 pre-consumer)	1
MR Credit 4.2: Recycled Content: 20% (post-consumer + 1/2 pre-consumer)	1
MR Credit 5.1: Regional Materials: 10% Extracted, Processed, & Manufactured Regionally	1
MR Credit 5.2: Regional Materials: 20% Extracted, Processed, & Manufactured Regionally	1
MR Credit 6: Rapidly Renewable Materials	1
MR Credit 7: Certified Wood	1

Indoor Environmental Quality

EQ Prerequisite 1: Minimum IAQ Performance	Required
EQ Prerequisite 2: Environmental Tobacco Smoke (ETS) Control	Required
EQ Credit 1: Outdoor Air Delivery Monitoring	1
EQ Credit 2: Increased Ventilation	1
EQ Credit 3.1: Construction IAQ Management Plan: During Construction	1
EQ Credit 3.2: Construction IAQ Management Plan: Before Occupancy	1
EQ Credit 4.1: Low-Emitting Materials: Adhesives & Sealants	1
EQ Credit 4.2: Low-Emitting Materials: Paints & Coatings	1
EQ Credit 4.3: Low-Emitting Materials: Flooring Systems	1
EQ Credit 4.4: Low-Emitting Materials: Composite Wood & Agrifiber Products	1
EQ Credit 5: Indoor Chemical & Pollutant Source Control	1
EQ Credit 6.1: Controllability of Systems: Lighting	1
EQ Credit 6.2: Controllability of Systems: Thermal Comfort	1
EQ Credit 7.1: Thermal Comfort: Design	1
EQ Credit 7.2: Thermal Comfort: Verification	1
EQ Credit 8.1: Daylight & Views: Daylight 75% of Spaces	1
EQ Credit 8.2: Daylight & Views: Views for 90% of Spaces	1

Innovation & Design Process

ID Credit 1–1.5: Innovation in Design	1–5
ID Credit 2: LEED Accredited Professional	1

Regional Bonus Credits

RB Credit 1–1.4: Region-Specific Environmental Priority: Region Defined	1–4

The standard contains six primary areas of concern:

Lot Preparation and Design

Resource Efficiency

Energy Efficiency

Water Efficiency

Occupancy Comfort and Indoor Environmental Quality

Operation, Maintenance, and Building Owner Education

The NAHB Research Center provides third-party certification for the program. There are four different levels of certification available to builders wishing to rate their projects: bronze, silver, gold, and emerald. A minimum number of points are required for each of the six guiding principles to ensure that green building strategies are addressed in a balanced and integrated way. After reaching the minimum requirements, an additional 100 points must be achieved by implementing any of the remaining line items.

The program is administered through a Green Scoring Tool available on the NAHB website. Contractors can compile information about the products and strategies used in a residential project. The tool specifies points that are awarded per the criteria of either the Guidelines or the Standard. The NAHB Research Center Green Certified mark identifies a project has been inspected at least twice by an independent verifier to confirm that every green point earned has been correctly implemented.

ENERGY STAR

ENERGY STAR is a green building certification program, administered by the United States Environmental Protection Agency (EPA), that is geared primarily toward the residential market, although it is available for commercial buildings as well. Established in 1992, the program was originally implemented to identify and promote the use of environmentally friendly products through a system of labeling. The ENERGY STAR label is widely used to certify appliances, office equipment, lighting fixtures, home electronics, and, more recently, residential, commercial, and industrial buildings. The program administers a third-party verification system to certify the energy performance of new homes and remodeling projects and publishes a series of builder's guides to help explain sustainable technologies. Unlike the LEED and Green Globes rating systems, certification under ENERGY STAR entails a series of site inspections during the construction process. These site visits are useful in identifying construction errors, allowing for correction to ensure full compliance with the ENERGY STAR guidelines. Buildings certified under ENERGY STAR receive an official certificate that identifies the contractor and third-party verification entity (Fig. 2.21).

Other Assessment and Certification Programs

In the United Kingdom, the British Research Establishment's Environmental Assessment Method (BREEAM) offers a verification standard that is similar to the LEED system in content and organization. The BREEAM rating system has been approved for use in Canada and several European countries. A slightly different assessment tool that deals specifically with a Life Cycle Assessment (LCA) of building materials is provided by the Canadian ATHENA system. LCA is a method of assessing the environmental impact of building materials and assemblies over the course of a building's full life span, from extraction and manufacturing of raw materials to their transportation, installation, and use, and finally through building maintenance and demolition. The effects of material choices are measured over a wide range of potential impacts, including resource depletion and pollution generation.

Figure 2.21 Buildings certified under ENERGY STAR receive an official certificate that is posted on a building's breaker box. *(Courtesy U.S. Environmental Protection Agency)*

AN ENERGY STAR® QUALIFIED HOME

Address:

Built by:

Verified by:

Date:

Optional information:

ENERGY STAR qualified homes are independently verified to meet strict energy efficiency guidelines set by the U.S. Environmental Protection Agency. Each home that earns the ENERGY STAR can keep 4,500 lbs of greenhouse gases out of our air each year

www.energystar.gov

Case Study

Heifer International Headquarters

Architect: Polk Stanley Rowland Curzon Porter Architects, Ltd. Contractor: CDI Contractors, LLC Location: Little Rock, Arkansas	Building type: Commercial office Size: 94,000 sq. feet Completed February 2006	Rating: U.S. Green Building Council LEED Platinum (52 points)

Overview

Heifer International is a nonprofit organization dedicated to fighting global hunger and poverty. It provides education in sustainable agriculture and gives donations of farm animals and plants to disadvantaged families around the world. The design intent of the new building was to provide a high performance building that would visibly communicate the sustainable building features utilized. The architects and construction team held frequent meetings early in the design process to communicate the implementation and documentation requirements of LEED criteria. The contractors played an active role in the process, suggesting alternative materials and construction methods to ensure compliance with sustainability goals.

The Heifer International Headquarters building was designed to clearly express its sustainable features. *(Courtesy Timothy Hursley)*

Site Design

The building occupies a former railroad switching yard that required environmental remediation to remove leftover petroleum and other chemicals. New parking, paved with permeable materials and marked by bio-swales between rows, leaves the majority of the site free for walking paths, native vegetation, and a constructed wetland. A **bio-swale** is a linear drainage ditch with gently sloped sides covered in vegetation designed to remove silt and pollution from surface runoff water. Storm water runoff is filtered through the bio-swales before being reused for landscape irrigation.

Energy

The building minimizes energy use in a variety of ways, each designed to be easily understood. Large expanses of clear glass ensure that sunlight will provide the majority of lighting required during working hours. The building's sixty-two foot width allows every employee access to light and views of the adjacent river and parkland. Electric lights are set on automatic dimmers in response to daylight levels, and occupancy sensors ensure that lighting activates only when needed. Extensive shading systems prevent unnecessary heat gain from reaching the interior. The building envelope, both opaque walls and glazing, utilizes high levels of thermal insulation. The completed building uses fifty-five percent less energy than a conventional office building of the same size.

Materials

Construction materials for the project were sourced locally to minimize the energy used in transporting them to the site. The materials for both the structural steel (ninety-seven percent recycled) and the aluminum and glass façade were fabricated by adjacent plants. The architects selected high recycled content, renewable, low-toxicity, and durable materials for the interior finishes, including flooring of bamboo and cork, insulation of recycled cotton and spray-on foam made from soybean byproducts, recycled brick, and

The steel structure and aluminum and glass facades were both fabricated by adjacent manufacturing plants, reducing energy use in the long-distance transportation of materials. *(Courtesy Timothy Hursley)*

countertops made of recycled glass. To ensure good indoor air quality, only materials with low levels of volatile organic compounds (VOCs) were utilized. During construction, the contractor recycled 75 percent of the building's construction waste, by weight.

Water

The roof shape allows rainwater to be captured in a 42,000-gallon water tower wrapped with a glass-enclosed fire stair. The water is reused in flushing toilets, in floor radiant heating installation, and in the building cooling system. Low-water-use plumbing fixtures, together with waterless urinals, greatly reduce the amount of potable water required to operate the building.

In addition to achieving the LEED platinum rating, the building has won numerous awards, including a U.S. Environmental Protection Agency award for Brownfield remediation (see Chapter Four). Heifer's headquarters provides a showcase for high-performance, energy-efficient, sustainable construction and an inspiration for other project teams.

Review Questions

1. Describe the purpose of local zoning ordinances.
2. What is the main intent of a building code?
3. What model building code is currently available for adoption in the United States and Canada?
4. What is the purpose of the Americans with Disabilities Act?
5. What is universal design?
6. What is the mission of the American Society of Testing and Materials?
7. What purposes are served by trade associations?
8. What are the environmental impacts of the construction industry?
9. What is urban sprawl?
10. What is meant by sustainable development?
11. What is the purpose of the LEED rating system?
12. What are the five broad topics addressed by the LEED rating system.
13. What are the purposes of the Energy Star program?

Key Terms

Bio-swale
Building Code
Dead Load
Greenhouse Effect
LEED
Live Load
Means of Egress
Sick Building Syndrome
Sustainable Development
Trade Association
Universal Design
Urban Sprawl
Zoning Ordinance

Activities

1. Visit a municipal zoning department and obtain a copy of local zoning districts for study.
2. Ask a local building inspector to address the class and review local building code requirements and the inspection process.
3. Contact one of the trade organizations and request information concerning its activities and publications.
4. Contact your local U.S. Green Building Council chapter to identify a LEED certified building in your area. Set up a tour for the class to visit the building.

Additional Resources

ENERGY STAR, http://www.energystar.gov/index.cfm?c=home.index

International Code Council, 2006 International Building Code

U.S. Green Building Council www.usgbc.org

United Nations World Commission on Environment and Development, (1987) *Our Common Future*, Oxford: Oxford University Press

Center for Universal Design, (2008) Universal Design principles, http://www.design.ncsu.edu/cud/

See appendix B for addresses of professional and trade associations and other technical information.

CHAPTER 3

Properties of Materials

LEARNING OBJECTIVES

Upon completion of this chapter, the student should be able to:

- Understand the parameters to be considered when selecting construction materials.
- Identify the properties of materials that must be known when specifying construction products.
- Discuss the technical aspects of mechanical, thermal, acoustical, and chemical properties.
- Understand the concepts of embodied energy and life cycle analysis.
- Understand the basic environmental issues related to the selection of construction materials.

As new materials are developed each year, the architect, contractor, and engineer must be informed of their properties in order to use these materials in the most effective and safe manner. For a material to be suitable for a particular application it must have predictable behavior. When the properties of the material are defined, the designer can predict its behavior and verify performance against that specified by codes and standards. This process requires that the design professional choose materials that adhere to certain requirements in a variety of considerations. Among these considerations are:

Availability—The material must be readily available, preferably within close proximity to the project site.

Manufacturing and processing—The environmental consequences of the extraction and manufacturing processes must be considered.

Strength—The material must carry design loads without excessive deflection and deformation.

Constructability—The material should be easily cut, shaped, and connected to other building materials.

Appearance—The material must conform to the aesthetics of the architectural design intent.

Environmental factors—Is the material renewable, recyclable, with a low embodied energy content?

Expense—The material must conform to specified budget constraints.

Durability—The material must be able to perform well over time and under a variety of environmental conditions.

MATERIAL GROUPS

For the purposes of discussing material properties, construction materials can be grouped into four broad categories: metals, non-metallic inorganic materials, polymers, and organic materials. **Metals** are refined from ores that have been extracted from the earth. Due to the addition of a variety of components, metals have a wide variation in properties. Most metals are ductile, strong, and are especially useful when tensile forces are expected to occur.

Non-metallic inorganic materials are also extracted from the soil. When they are refined they produce a variety of products. Typical inorganic materials include sand, limestone, glass, brick, cement, gypsum, mortar, and mineral wool insulation. Most are hard, rigid, brittle, and heavy, and are especially useful when compression forces are expected to occur.

Polymers consist of large molecules composed of repeating structural units connected by chemical bonds. While polymer in common usage suggests plastic, the term actually refers to a large class of natural and synthetic materials with a variety of properties and purposes. Polymers are used in construction applications such as pipes and conduit, wire and cable, textiles, adhesives, roofing, flooring and insulation among others.

Organic materials include wood, grasses, bitumens, and many synthetic materials based on a chemical compound containing carbon. Examples include wood and paper products, asphalts, rubber, and plastics.

These materials vary a great deal in their properties, and this influences how they can best be used in construction. The material properties that most influence decisions in construction can be divided into four categories: mechanical, thermal, acoustical, and chemical.

MATERIAL PROPERTIES

The fundamental properties of a material must be judged against the functional requirements of a given project. All materials respond to changes in applied load, temperature, and moisture content. A material must be able to perform under a variety of loadings, stresses, and environmental conditions. Compressive forces result from the weight of the building components, walls, floors, and roofs. Horizontal forces are the result of wind, subsurface water, soil, and earthquakes. Figure 3.1 illustrates the basic loads and forces that act upon a building structure.

Figure 3.1 A building structure is subject to various loads and forces.

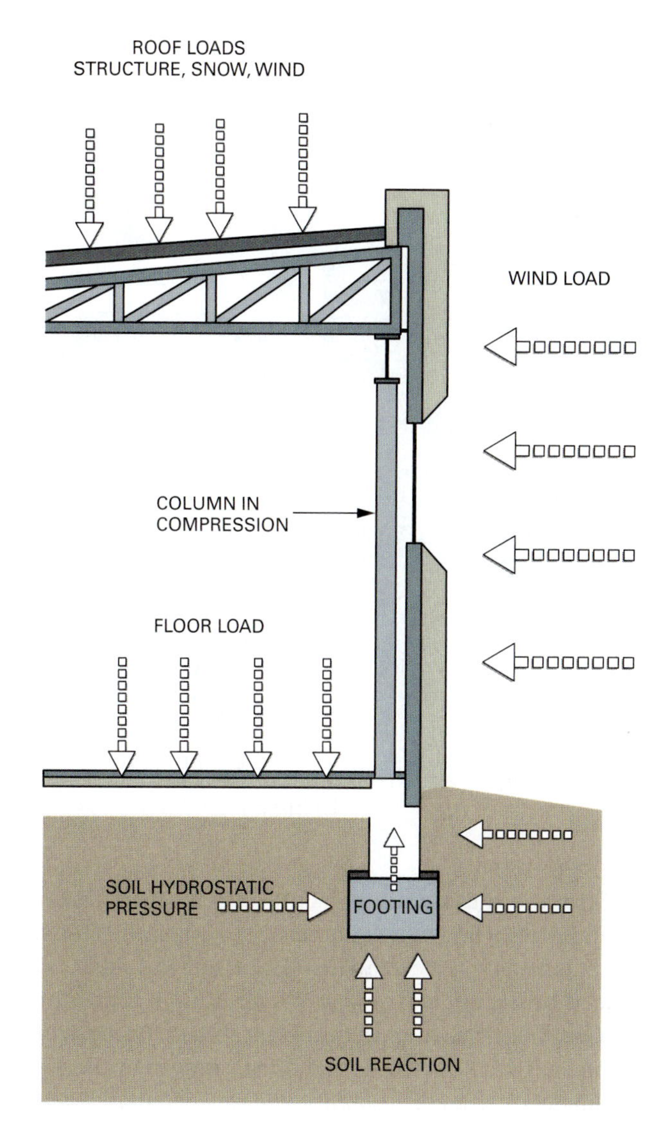

Mechanical Properties

Mechanical properties are a measure of a material's ability to resist a variety of mechanical forces. They are related to the response of a material to both static (continuous) and dynamic (intermittent) loads. Mechanical properties include tensile strength, compressive strength, shear strength, elasticity, ductility, hardness, impact resistance, fatigue strength, and permeability.

Stresses–Tension, Compression, Shear When a building load, or force, acts on a building component, such as a column, the component takes on an internal resistance to the force known as stress. As shown in Fig. 3.2, there are three possible types of stresses: tension, compression, and shear. Tensile stresses are created when forces pull on a member and tend to increase its length. Compressive stresses push on a member and tend to shorten it. Shear stresses produce forces that work in opposite directions parallel with the plane of the force, causing adjacent parts of a material to slide past one another.

How a material behaves under load and the amount of stress a member can withstand is determined through standardized tests, as specified by the American Society for Testing and Materials (ASTM). When a load is applied to a member it causes a stress-strain condition.

Stress and Strain Stress is the intensity of internally distributed forces that resist a change in the form of a body. It is measured in pounds per square inch (psi), mega-Newtons per square meter (MN/m^2), or mega-Pascals (MPa). Stress is equal to the load divided by the area upon which it is acting. Normal stress is indicated by the Greek letter sigma, σ.

$$\sigma \text{ (Stress)} = \frac{\text{Load}}{\text{Area}} = \text{psi, MN/m}^2\text{, MPa}$$

The formula applies to materials that have a force applied parallel with the axis of the member. It is important to know if a load is to be distributed over an area or concentrated on a small point. For example, if a 100 pound load is distributed over a two-inch square column, the stress is 25 psi. If the same load is concentrated on the end of a rod with a cross-sectional area of one inch^2 the stress would be 100 psi (Fig. 3.3).

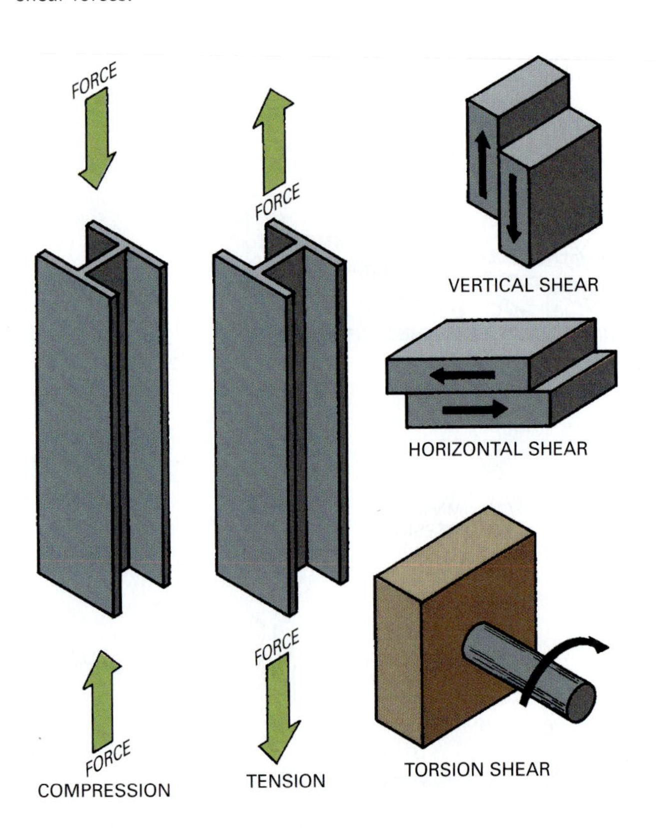

Figure 3.2 Stresses are created by tension, compression, and shear forces.

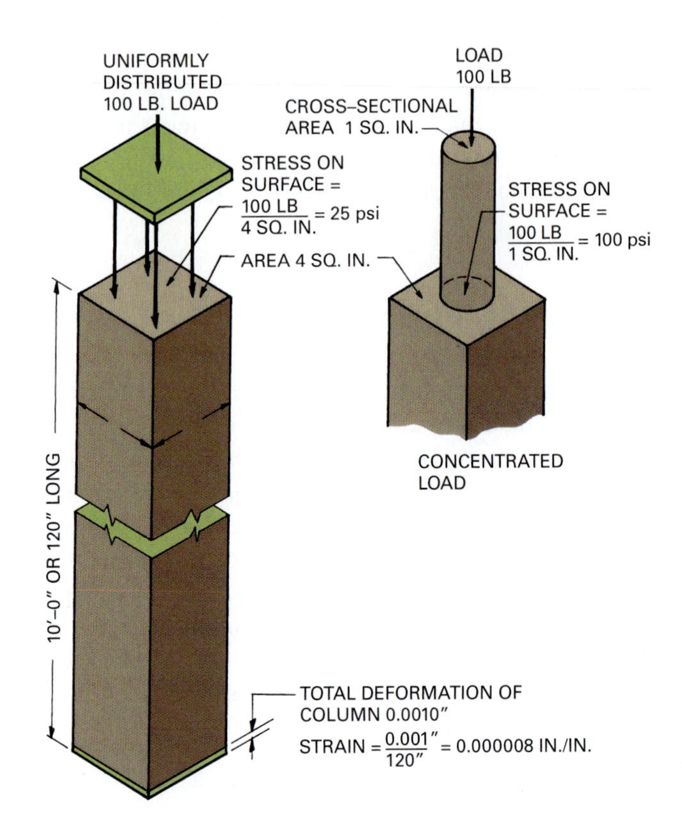

Figure 3.3 Columns are subjected to stress and strain forces.

Strain is the change per unit length in the linear dimension of a body that accompanies a change in stress. Strain is equal to the change in the original dimension divided by the original dimension. Strain is indicated by the Greek letter epsilon, ε, and is measured in inches of change per inch of the original dimension (in./in.) or in millimeters per meter (mm/m)—or it is given as a percentage. If strain is to be stated as a percentage of the dimensional change, the result is multiplied by 100.

$$\varepsilon(\text{Strain}) = \frac{\text{Change in the original dimension}}{\text{Original dimension}}$$

$$= \text{in./in. or mm/mm}$$

In Fig. 3.3, the deformation of the column is 0.0010 inches and the column is 120 inches long. The strain is 0.000008 in./in.

Stress-Strain Diagrams The behavior of a material under load can be represented on a stress-strain diagram (Fig. 3.4). A stress-strain diagram can be given for compression, tension, or shear. The data used to lay out the diagram are obtained from testing standard material specimens in a universal testing machine. The strain values are laid out on the horizontal axis (abscissa) and the stress values on the vertical axis (ordinate). An increase in stress will produce a proportionate increase in strain, up to the elastic limit. The elastic limit is the highest stress that can be imposed on a material without permanent deformation. This produces a straight line on the diagram from 0 to the elastic limit stress value.

Modulus of Elasticity The modulus of elasticity (E) is a proportional constant between stress and strain. For stresses within the elastic range, E, sometimes referred to as E-value, equals stress divided by strain. The modulus of elasticity is basically a measure of stiffness and rigidity. For beams, E-values are an indication of their resistance to deflection. The modulus of elasticity then defines the stiffness of a material, governs deflections, and influences buckling behavior. Table 3.1 lists E-values for selected construction materials. Remember that there is no single value of E for a particular material because of the many variations in the makeup of materials. For example, there are many types of plastics, each having its own stress-strain behavior. Within materials, such as steel, there are variations in the alloying ingredients used. Steel has high E-values and is used where loads must be carried with minimal deflection. The E-value of wood is higher than that of plastics but lower than that of concrete. Wood is useful in carrying light structural loads over short distances, as is evidenced in the use of residential joists and

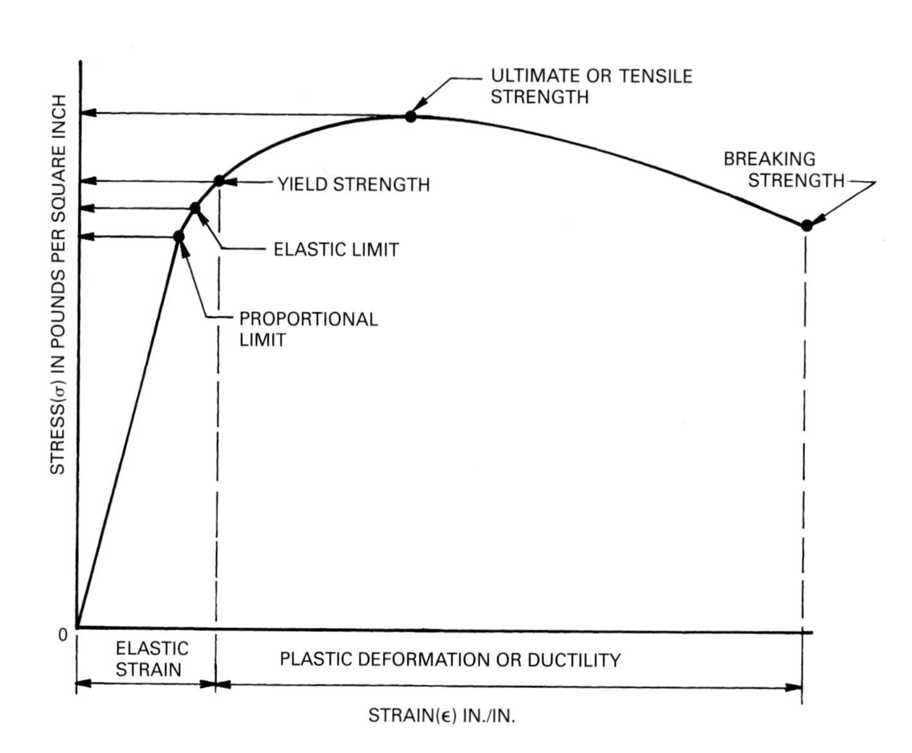

Figure 3.4 A typical stress-strain diagram. Note that this material sustained considerable strain between the ultimate strength and before it fractured.

Table 3.1 Modulus of Elasticity for Selected materials

Material Group	Material	10^6 psi
Metals	Aluminum alloys	25
	Cast iron, malleable	25
	Copper alloy	13–17
	Steel, carbon	29
Ceramic Materials	Brick	1.8–2.2
	Concrete	1.0–5.0
	Glass	9–11
	Stone	10–15
Organic Materials	Particleboard	0.9–2.0
	Plastics, molded	0.1–1.9
	Plastics, reinforced	0.1–2.2
	Wood, 2", visually graded	0.9–1.6

(These are approximate. Values will vary with specific samples.)

rafters. A comparison of the relative stiffness of selected construction materials is shown in Fig. 3.5.

Ductility **Ductility** refers to the ability of a material to be deformed plastically without actually breaking or fracturing. Copper, for example, is a very ductile material. When tested under tension, it stretches for some distance without breaking.

Hardness **Hardness** is a measure of the ability of a material to resist indentation or surface scratching. It is the result of several properties of a material, such as elasticity, ductility, brittleness, and toughness. Hardness is related to the tensile strength of the material. Generally, a material that has a high hardness rating also will have a high tensile strength. This is not true for all materials; concrete, for example, has high hardness and low tensile strength.

Impact Strength **Impact strength** is the ability of a material to resist a very rapidly applied load, such as the strike of a hammer. It is an indication of the toughness of the material. A material with high impact strength will absorb the energy of impact without fracturing. Impact strength is affected by strength and ductility. Metals that are strong and ductile have high impact strength. Ceramics are strong in compression but are brittle (lack ductility) and break under impact. In general, plastics are very ductile but low in strength and do not absorb impact well.

Fatigue Strength Resistance of a material to a cyclic load, one that varies in direction and/or magnitude, is called **fatigue strength**. This is illustrated by bending a wire back and forth until it breaks. Most materials are lower in fatigue strength than they are in tensile strength. Failures due to fatigue occur slowly, and most materials that fail due to fatigue offer some useful life before failure. This is an important factor to consider when the useful life of a product is established. Fatigue strength in buildings and other structures, such as bridges and elevators, is commonly caused by vibrations and rotational motion produced by machinery.

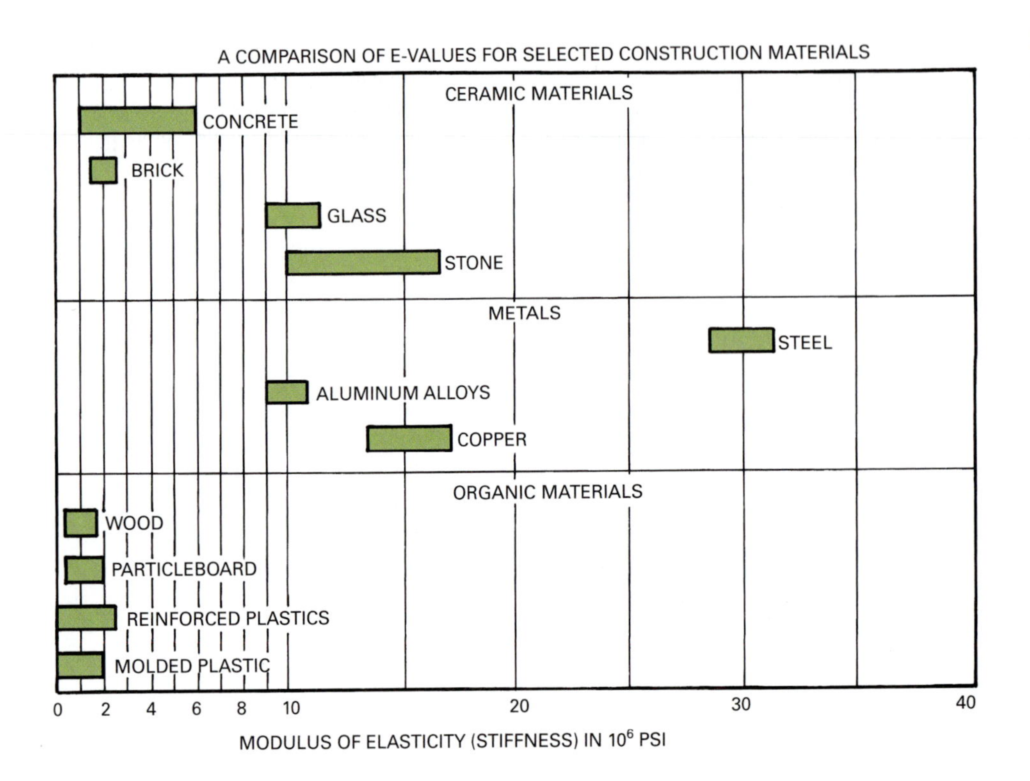

Figure 3.5 A comparison of modules of elasticity for selected construction materials.

Permeability The rate that water flows through a material is a function of the material's permeability. The permeability of a material is measured in units of permeance, called perms, and is typically referred to as the perm rating. A vapor retarder is defined as a material having a perm rating of 1.0 or less.

Thermal Properties

The thermal properties of a material are those that are related to the material's response to heat. When a material is subjected to a change in temperature it may expand, contract, conduct, or reflect heat. Construction materials can be classified either as insulators, materials that resist the transfer of heat, or conductors, materials that encourage the transfer of heat. Fibrous materials, such as rock wool, cotton, and cork, are good insulators, while most metals are good conductors.

Thermal Conductivity and Thermal Conductance

Thermal conductivity (k) denotes the ability of a material to transfer heat from an area of high temperature to an area having a lower temperature. Thermal conductivity is measured in the number of Btus (British thermal units) that can pass through one square foot of a material one inch thick when the temperature difference between the two surfaces is one degree Fahrenheit. It is expressed as (Btu.in)/(hr. ft.^{2} °F) or W/(m^2 °C). Metals tend to have the highest values of thermal conductivity. Non-metallic inorganics have much lower conductivities, and organic materials have the lowest (Table 3.2). Materials with low thermal conductivity tend to have excellent insulation qualities.

Thermal conductance (C), the coefficient of heat transfer, is used with products for which the unit thickness per inch used with the conductivity values is not a convenient measure. Conductance is the number of Btus per hour that will pass through one square foot of

Table 3.2 Thermal Conductivity (k) of Selected Materials

Material	(Btu/in.)/ (hr./ft.^{2}/°F) (English units)	W/(m^2°C) (metric units)
Polyurethane foam	0.11–0.14	0.016–0.020
Polystyrene foam	0.17–0.28	0.024–0.040
Softwoods	0.80	0.12
Most other plastics	1–2	0.14–0.28
Plate glass	5	0.72
Hardwoods	1.10–0.16	0.16–0.02
Brick	9	1.3
Concrete	12	1.7
Steel, stainless	100	14
Steel, carbon	300–500	43–70
Aluminum alloys	800–1600	115–230
Copper alloys	500–2500	70–360

(These values are approximate. Actual values will vary depending on the composition of the material.)

a material at a stated thickness in one hour per degree temperature difference.

Composite Thermal Performance There exists a single expression to describe the overall rate of heat transfer through a building assembly. The U-value is the overall coefficient of thermal expansion in Btu/(hr.ft^2 °F) or W/m^2 K. Thermal resistance or R-value is a measure of an individual material's resistance to heat flow, while U-value is used as a summary factor for the conductive energy potential of building assemblies. U-values are determined by calculating the thermal resistance of each part of an assembly, summing them to arrive at a total resistance, and taking their reciprocal.

$$U = \frac{1}{\Sigma R}$$

Both R- and U-values are frequently used in the building codes to specify the minimum thermal criteria for an exterior building component. The insulating potential of building materials such as insulation and windows is often identified by either their U- or R-value.

Changes of State Freezing and melting indicate the temperature at which a material is said to undergo a change of state. Since most construction materials are solid ("frozen") at normal service temperatures, a material's melting point is of greater importance. The **melting temperature** is that point at which a material turns from a solid to a liquid form. In general, materials with high melting temperatures, such as non-metallic inorganics, perform better at high temperatures. They also tend to retain their mechanical properties over a wider range of temperatures. The melting point of metals is below that of non-metallic inorganics, so metals do not perform as well under service conditions having high temperatures. This is why steel structural members must be protected with a fire-resistant material. Although steel will not burn, it does lose strength as its temperature rises. Organic materials perform the poorest in conditions of high temperature.

Heat Capacity The ability of a material to store and release heat is an important thermal property for building construction. Through natural heat transfer, heat always flows from hot to cold. Solid and liquid materials with a high **specific heat capacity** can store heat from the sun during the day and then discharge that heat by natural heat transfer during the night. Sometimes referred to as thermal mass, the process is an excellent way to provide low-cost heating and energy management. The subsequent transfer of heat occurs naturally as the material with an elevated temperature cools to the temperature of the surrounding environment, thus giving off heat. Heavy mass materials, such as masonry and concrete, have excellent thermal mass properties.

Heat can also be stored in a gaseous material, such as the air flowing through a central heating system, to provide thermal comfort in winter. The specific heat capacity of a solid, liquid, or gas is defined as the heat required to raise a unit mass of substance by one degree of temperature and is expressed in terms of Btu/(lb.°F) in the English system or Joules per kilogram per degree centigrade (J/(kg.°C)) in the metric system.

Acoustical Properties

Acoustics is the branch of physics that deals with the generation, transmission, and control of sound waves. It considers the ability of a material to either absorb or reflect sound waves within a room. The acoustical properties of interior finish materials directly affect occupants by influencing the quality of speech, music, and other audio sounds projected in a space. Acoustical materials that perform well as sound absorbers include soft materials such as fabrics, rigid but soft materials, and rigid but hard materials that have the exposed surface perforated with holes or slots of varying sizes and placement. Various construction materials can be tested to ascertain their sound control properties. A more detailed discussion is presented in Chapter 35.

Chemical Properties

The chemical properties of a material describe its tendency to undergo a chemical change or reaction due its composition and interaction with the environment. A chemical change can alter the original composition of a material and thereby affect its properties. Iron, for instance, has a tendency to oxidize or corrode under certain conditions. In addition to corrosion, other chemical properties are ultraviolet degradation and fire resistance.

Construction materials are chemically degraded by the environment in which they are placed. This degradation is called atmospheric corrosion. The most common form of atmospheric corrosion is oxidation. **Oxidation** is the reaction of a material with the oxygen in the atmosphere, such as iron rusting. Other corrosive chemicals are present in the air, water, and soil to which materials are exposed, including sulfate ions in groundwater, soils, and seawater; sulfur dioxide gas in the air; and alkali in soils.

Metals usually corrode due to electrochemical action. Corrosion occurs when minute amounts of electricity that occur naturally in the atmosphere or soil flow from one metal called an *anode* to a dissimilar metal called a *cathode* through a current-carrying medium (moisture) called the electrolyte. The anode deteriorates and the cathode remains unaffected by the electrochemical action. The electrolyte is usually a water or gas, such as carbon dioxide or sulfur dioxide. Metals are ranked by their tendency to be anodic or cathodic.

In Table 3.3, selected metals are ranked from those most anodic to those most cathodic. When a material near the top of the list, such as steel, is placed in contact with a material near the bottom of the list, such as copper, and an electrolyte is present, corrosion will occur. The metal that is destroyed (corrodes) is one that is high on the galvanic table. Metals that are cathodic are said to be nobler. Anodic materials are said to be less noble.

Metals that are widely separated on the galvanic table should not be in contact with each other. Those close together, such as cadmium-plated steel and aluminum, have practically no difference in electrolytic potential and, therefore, almost no galvanic action. Corrosion can be prevented by breaking up the electrical circuit through separation of dissimilar metals. This can be

Table 3.3 Galvanic Series of Selected Metals and Alloys

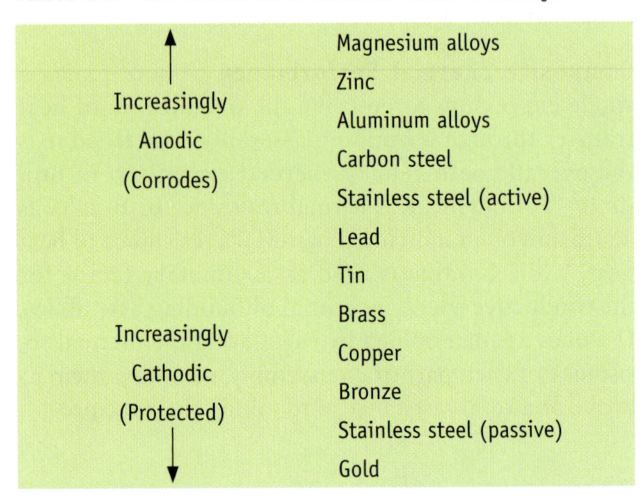

↑	Magnesium alloys
Increasingly Anodic (Corrodes)	Zinc
	Aluminum alloys
	Carbon steel
	Stainless steel (active)
	Lead
	Tin
	Brass
Increasingly Cathodic (Protected)	Copper
	Bronze
	Stainless steel (passive)
↓	Gold

Construction **Materials**

Composite Materials

Composite materials, or just composites, are engineered materials derived from two or more constituent materials with significantly different physical or chemical properties. They utilize the combination of two compounds, a matrix, and reinforcement. The matrix surrounds the reinforcement material, and the reinforcing imparts its special mechanical and physical properties to support the matrix properties. The combination results in material properties that can be engineered to produce a product or structure with optimum characteristics for a required application. Engineered composite materials are typically formed to their final shape by molding or tooling. Composites are used widely in the automobile, space, aeronautics, and construction industries.

An example of a historical composite is the common brick, made from mud and straw. The most visible composite today is the steel- and aggregate-reinforced Portland cement or asphalt concrete that paves our roadways. Another type is glass-fiber reinforced concrete. Glass fiber concretes are mainly used in exterior building façade panels and as architectural precast concrete. The material is well suited for creating complex shapes for building façades, canopies, or roofs and is less dense than steel.

A variety of composite materials are used for architectural cladding. *(Courtesy Image copyright Michel Stevelmans, 2009. Used under license from Shutterstock.com)*

done by painting or covering the metal surfaces with a material that will not conduct electricity and also protects the metal from other corrosive materials in the air or soil. Another means of preventing galvanic action is through the use of one metal in all parts of an assembly. For example, aluminum gutters can be hung with aluminum nails rather than copper nails.

Sometimes a metal, such as steel, is coated with another metal, such as zinc. The zinc, which is high on the galvanic series, is purposely sacrificed to save the steel. Aluminum and stainless steel are anodic and will corrode. However, they do develop some natural corrosion resistance when exposed to oxygen. The oxygen forms a transparent oxide film that protects the metal.

Ultraviolet Degradation The reduction in performance of some building materials is related to their exposure to ultraviolet light, or that portion of sunlight in the ultraviolet spectrum. UV radiation tends to break down solvents in many plastics, causing the material to become brittle, fade, and ultimately fail. Paint designed to be UV resistant is often used to prevent the adverse effects of continued exposure to direct sunlight. Data regarding a product's or material's resistance to ultraviolet light should be consulted during the selection process.

Fire Resistance Combustibility is of prime importance in the selection of building materials. A material of low combustibility is said to be fire resistant. For example, a firebrick in a fireplace must withstand temperatures of up to 2000 degrees Fahrenheit. An emphasis on the resistance to fire is a large component of building code constraints (see Chapter Two). Fires create indoor air-quality problems and can result in structural failure. The fire-related properties of specific materials are discussed in more detail in the various chapters related to each material.

ENVIRONMENTAL CONSIDERATIONS

The selection of environmentally positive materials is a central issue for sustainable design and construction. Materials selection is one of the most difficult problems to address in sustaining the built environment as we ask the question: What are green materials and how do we evaluate them? Environmentally preferable products (EPPs) are defined by Federal Government Executive Order 13101 as materials that have "a lesser or reduced effect on human health and the environment when compared to competing products that serve the same purpose." As industry provides an ever-increasing array of material options and new products with untested "green" performance claims, designers and contractors must be familiar with the factors that influence the environmental impact of a material. These factors include a material's embodied energy content, its impacts on natural resources and habitat degradation, and its potential for toxicity to humans and the environment.

Embodied Energy

Embodied energy is defined as the energy consumed by all processes associated with the production and use of a material or assembly, from the acquisition of natural resources to its final reuse or demolition (Fig. 3.6). Whereas the energy used in operating a building can be readily measured, the embodied energy contained within building materials is difficult to calculate. This energy use is often unknown and can only be fully quantified through a complete life cycle analysis (LCA). A **life cycle analysis** is a systematic procedure for compiling and analyzing the inputs and outputs of resources and energy and their associated environmental impacts directly attributable to the functioning of a product or service system throughout its life cycle. Life cycle assessment takes into account all of the resource and energy inputs that go into a material, while simultaneously calculating the resulting airborne emissions, solid and water borne wastes, and any other co-products and releases to arrive at a quantitative measure of embodied energy.

The life cycle energy cost of most building products starts with the extraction of raw materials. Here the energy consumed in the actual harvesting, mining, or quarrying of a resource, the building of access roads, and the transportation of raw resources to the mill or plant is tabulated. A second stage is initiated with the delivery of raw materials to the processing plant and ends with the finished product ready for shipment. The manufacturing stage accounts for the largest portion of embodied energy and emissions associated with the life cycle of a building product.

The on-site construction is like an additional manufacturing step where individual products and components come together in the assembly of an entire

Figure 3.6 The stages considered when determining the embodied energy content of a building material or component.

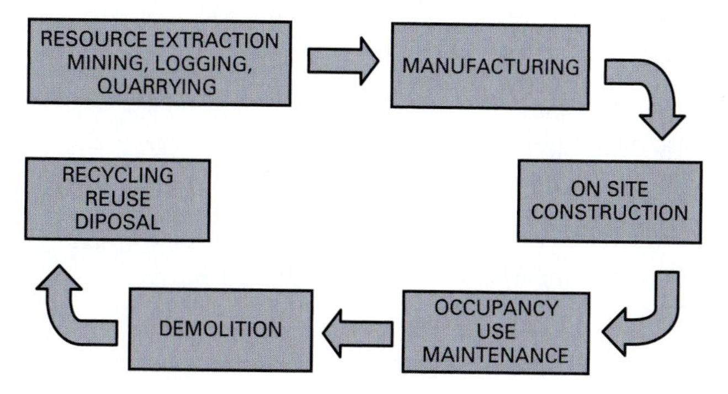

building. Energy is consumed in the transportation of individual products and assemblies from manufacturing plants to regional distribution centers, as well as in the use and transport of equipment, temporary heating, and electricity used during construction. Once the building is complete and occupied, operating energy is calculated, taking into account functions like heating, cooling, lighting, and water use.

Demolition marks the end of a building's life cycle, although it is not necessarily the end for all individual materials, some of which may face subsequent recycling or reuse.

Typically, embodied energy is measured as a quantity of non-renewable energy per unit of building material or component in either mega Joules (MJ) or gigajoules (GJ) per unit of weight (kg or ton) or area (square foot or meter) of a material. Table 3.4 gives embodied energy values for some common building materials. Published figures should be used with caution because values change depending on a variety of factors. For example, if a material is sourced locally, its embodied energy due to transportation will be lower than one sourced from a great distance.

Natural Resources/Habitat Degradation

The extraction, manufacture, and use of a construction material has profound effects on ecological systems and the supply of natural resources. To avoid habitat degradation, materials whose use results in reduced effects on erosion, salinity, vegetation loss, changes in the nutrient characteristics of soil, and aesthetic damage to landscapes are preferred. Impacts on water systems, such as the release of nutrients, salts, toxins, or suspended solids into aquatic systems, must also be considered.

Table 3.4 Embodied Energy of Selected Materials

Material	Embodied Energy in MJ/Kg
Fly ash (In concrete)	< 0.1
Gravel	0.2
Sand	0.6
Brick	2.7
Ceramics and quarry tiles	5.5
Glass	6.8
Wood	10.8
Steel	28.0
Cast iron	32.8
Asphalt	50.2
Polyethylene	79.5
EPDM Roofing	183
Aluminum (Not recycled)	207
Brass	239

Renewable Materials A **renewable material** is derived from a living plant, animal, or ecosystem that has the ability to regenerate itself. Renewable materials are preferred because, when managed properly, their supply will not be depleted. Most result in biodegradable waste and tend to have reduced net emissions of CO_2 across their life cycle compared to materials derived from fossil fuels.

Wood harvested from sustainably managed forests is a perfect example of the efficient use of a renewable material. It cleans the air and water as it matures (trees bind carbon during their growth), we are able to utilize nearly all of the raw material, its production uses low energy processes, and it is readily renewable and recyclable. Other examples of renewable materials include bamboo, cork, and various composite materials derived from agricultural products.

Recycled and Post consumer/Industrial/Agricultural Materials Materials that utilize the waste products from a variety of processes result in a reduced use of natural resources and energy consumption. An example of a post-industrial product is the use of fly ash, a glass-like powder recovered from the gases of coal-fired electricity production, as an inexpensive replacement for Portland cement in concrete. Other products use post-consumer waste in their production, such as synthetic polyester used for carpeting that is made entirely from recycled pop bottles. The LEED rating system awards credits for the use of environmentally preferred materials, as shown in Table 3.5.

Table 3.5 LEED Credits for Materials and Resources

MR Credit 1.1/2: Building Reuse: Maintain 75/95% of Existing Walls, Floors, & Roof
MR Credit 1.3: Building Reuse: Maintain 50% of Interior Non-Structural Elements
MR Credit 3.1: Materials Reuse: 5%
MR Credit 3.2: Materials Reuse: 10%
MR Credit 4.1: Recycled Content: 10% (post-consumer + 1/2 pre-consumer)
MR Credit 4.2: Recycled Content: 20% (post-consumer + 1/2 pre-consumer)
MR Credit 5.1/2: Regional Materials: 10/20% Extracted, Processed, & Manufactured Regionally
MR Credit 6: Rapidly Renewable Materials
MR Credit 7: Certified Wood

Toxicity to the Environment

Studies have found that humans now spend more than 80 percent of their time indoors. The EPA ranks indoor air quality (IAQ) as one of the most prominent environmental problems today. ASHRAE defines acceptable indoor air quality as air in which there are no known contaminants at harmful concentrations. Poor indoor air quality can result from inadequately maintained HVAC systems as well as the material makeup of interior finish materials. Many synthetic products containing volatile organic compounds (VOCs) tend to off-gas from interior materials, furnishings, and equipment. **Off-gassing** occurs when solid but chemically unstable materials evaporate at room temperature and slowly release contaminants into the air. Over 500 pollutants identified in very common building materials are known to cause asthma, allergies, cancer, and reproductive and nervous system disorders. Builders receive both short-term and long-term exposures to these chemical hazards from the off-gassing of solvent-based products and cleaners, as well as sawdust and dust particulates from construction materials.

Certain materials used frequently in the last few decades that carry risk of contamination include polyvinylchloride (PVC), used in flooring, wall coverings, plumbing, carpet backing, and electronics; formaldehyde products, such as plywood and particle board; synthetic carpets; and sheet vinyl. A number of organizations now provide official recognition and labeling that certifies the non-toxicity of building materials. More detailed information pertaining to the toxicity of specific materials is given in subsequent chapters.

Sustainable Construction Materials Assessment Criteria

There is a rapid increase in the types and numbers of materials that respond to environmental issues in the construction industry. Designers and builders should strive to specify materials that protect the natural environment, are renewable and recyclable, minimize resource and energy consumption, and are healthy and non-toxic. A complete life cycle assessment of proposed choices is recommended for a thorough understanding of the larger impacts of material selection.

Review Questions

1. List the parameters to be considered when selecting a construction material.
2. What are the major material groups?
3. What are the mechanical properties that define a material?
4. How does a tensile stress differ from a compressive stress?
5. What are the major thermal factors to consider when selecting a material?
6. Which of the three major material groups has the lowest thermal conductivity?
7. What factor describes the overall thermal performance of a building assembly?
8. What is the most common form of chemical degradation?
9. What stages in a material's life cycle are considered when calculating embodied energy?
10. What inputs and outputs does a Life Cycle Analysis of a material consider?
11. Give examples of renewable building materials.

Key Terms

Coefficient of Thermal Expansion
Compressive Stress
Embodied Energy
Environmentally-Preferable Products (EPPs)
Fatigue Strength
Good Indoor Air Quality
Hardness
Heat Capacity
Impact Strength
Life Cycle Analysis
Non-Metallic Inorganic
Melting Temperature
Metals
Modulus of Elasticity (E)

Off-Gassing
Oxidation
Polymer
Shear Stress
Strain
Stress
Tensile Stress
Thermal Conductance (C)
Thermal Conductivity (k)
Volatile Organic Compound

Activities

1. Prepare samples of various materials and test for compression and tensile strength. Keep an accurate record of what each sample contains and the conditions of its preparation. For example, prepare concrete samples, keeping a record of the ingredients, curing time, and how it was contained during curing. Vary certain ingredients in various samples. Test each for compression strength and compare the results to try to verify how the differences in mixture, curing time, and methods of curing influenced the strength. Similar tests can be made on other materials, such as various species of wood and different types of metals. Run some experiments on thermal conduction by applying heat to identical size samples of various materials and recording the time it takes to feel heat at the end opposite the end exposed to heat. Although this is rather crude, it will illustrate the property of thermal conduction.
2. Most schools have material testing equipment or even an entire testing laboratory. Have the instructor in this area help the class devise some more accurate, scientifically based experiments and help the class run these in the laboratory. For example, with the proper equipment, the class could possibly test a sample, such as a metal, and record the elastic limit, yield point, and breaking strength.
3. Try to set up a device that will subject a material, such as a metal, to constant rapid flexing and see which material has the best fatigue strength.

Additional Resources

American Society for Testing and Materials www.astm.org

The Athena Institute www.athenasmi.org

Shiers, H., Howard, S., Sinclair, M. (1996), *The Green Guide to Specification*, Oxon: GTI Specialist Publishers.

Wilson, F. (1984), *Building Materials Handbook,* New York, Van Nostrand Reinhold Company Inc.

DIVISION

02

Existing Conditions CSI MasterFormat™

(Image copyright Eimantas Buzas, 2009. Used under license from Shutterstock.com)

DIVISION 31

Earthwork
CSI MasterFormat™

311000 Site Clearing
312000 Earth Moving
313000 Earthwork Methods
314000 Shoring and Underpinning
315000 Excavation Support and Protection
316000 Special Foundations and Load-Bearing Elements
317000 Tunneling and Mining

(Image copyright Vitezslav Halamka, 2009. Used under license from Shutterstock.com)

(Courtesy Gomaco Corporation)

4 CHAPTER

The Building Site

LEARNING OBJECTIVES

Upon completion of this chapter, the student should be able to:

- Understand the preliminary steps that must be taken on the site before construction can begin.
- Be aware of the problems that may occur during excavation and techniques to control them.
- Be familiar with paving materials and design.
- Understand the types of forces developed by seismic activities.

Effective site design requires the application of ecologically based strategies to create projects that not only develop a site but also help repair and restore existing site systems. The coordinated efforts of architects, landscape architects, civil engineers, and construction managers are required to ensure that a structure will optimize site utilization, work well within existing natural systems, and minimize environmental damage.

Site work includes a wide range of preconstruction activities encompassing subsurface investigation, preparing the site for construction, dewatering excavations, shoring and underpinning, earthwork, installing various types of piping, water distribution, sewage, and drainage systems, and constructing site improvements, such as fences, walks and landscaping (Fig. 4.1). Building codes have sections devoted to site work and demolition. Site construction work is shown on site plans developed by an architect with the assistance of engineers specializing in the activities required.

Figure 4.1 Site preparation involves the work of a number of different construction trades. *(Image copyright Leah-Anne Thompson, 2009. Used under license from Shutterstock.com)*

SITE ASSESSMENT

Before any construction project can begin, the building site must be thoroughly analyzed and understood to ensure an efficient design and construction process. A series of preconstruction investigations must be conducted, including boundary and topographical surveys, subsurface soil testing, and environmental studies. An initial investigation may be conducted at the local municipal archives to collect historical data on a site.

Surveys

A boundary survey establishes the true property lines and corners of a parcel of land. Topographical surveys determine differences in elevation of all sections of a site and are used to inform building and landscape design and determine if a property is located in a flood zone. Lot surveys additionally locate natural and manmade features, such as buildings and drives, elevations, land contours, trees, and streams.

Subsurface Utility Engineering (SUE) surveys verify the location and condition of existing utilities on a site, including gas, water, and sewer lines. Civil surveys are conducted to determine the elevations of natural and manmade site features, drainage patterns, and other civil elevations. This information is used to ensure that the site will drain well and not cause ponding or flooding. All surveys must be conducted by state-licensed surveyors.

A **site plan** shows the placement of buildings, roads, and other improvements, including existing contours of the land and the desired finished contours. Locations of utilities (gas, electric, and water and sewer lines) are also indicated. Setbacks and easements must be located. The placement and size of parking areas, vehicle access routes, streets, curbs, and other details, such as fences, walks, and retaining walls, are also located. A typical site plan is shown in Fig. 4.2.

Environmental Assessment

An Environmental Site Assessment (ESA) is a report prepared for a building site that identifies potential or existing environmental contamination. A Phase I – ESA, defined by ASTM Standard E-1527-05, reviews the history of a site and addresses both the land itself as well as any physical improvements existing on a property, such as buildings, wells, or submerged tanks. The assessment looks for contamination to soil, subsurface materials, and the existence of hazardous substances, like asbestos within buildings.

A Phase II – ESA may include the Phase I portion of the assessment and also soil and ground water and surface sampling in suspect areas. Soil samples are collected using hand augers or a drilling rig and shipped to a laboratory according to standard industry methods. Ground water samples are collected from the borings or permanent monitoring wells located on a property. The Phase II – ESA report describes the soil borings completed, soil texture, soil and ground water analytical results, and presents the data in tabular format with a map illustrating the sampling locations and plan of site.

Sites found to contain environmental contamination are referred to as brownfield sites. The EPA defines a **brownfield site** as "a property, the expansion, redevelopment, or reuse of which may be complicated by the presence or potential presence of a hazardous substance, pollutant, or contaminant." Sites that contain hazardous contamination must be remediated prior to the commencement of new construction activities. The LEED rating system awards credits for the rehabilitation of brownfield sites as a positive alternative to the development of previously untouched land. Sites containing high levels of hazardous contamination are known as EPA superfund sites and require the services of specialty engineers.

Subsurface Investigation

A key factor in the design of any structure is the determination of the characteristics of the soil upon which it is built. This site investigation requires the testing of site-soil samples in a testing laboratory or by field-test procedures. The soil report is used by the architect and design engineers to determine how the building must be designed and constructed. Subsurface site conditions influence the design of foundations, retaining walls, paving, dewatering requirements, and other factors. The soil tests are performed on subsurface test borings, and the results are detailed in a soil test report. The report includes a detailed geologic description of the site subsurface conditions and gives recommendations based on the findings. Recommendations may pertain to site grading, foundation depth, paving design, retaining walls, and other aspects of the project.

Soil investigations for shallow depths (around 8 ft. or 2.4 m) can be made by digging an excavation and taking soil samples using a penetration method that involves driving a test rod into the soil and recording the number of blows it takes to penetrate a predetermined distance. For larger structures, soil samples are typically made by drilling test holes using a hollow-stem auger and wash-boring techniques. The auger is mounted on the end of a drill rod that rotates, causing the auger to penetrate the soil (Fig. 4.3). Water run into the pipes under pressure washes away waste material and lubricates the drilling process. The location, depth, and number of borings vary with site conditions and are based on the judgment of the soil engineer. The locations of boreholes are shown on a drawing of the site. The height of the water table can sometimes be noted in the borehole. The findings are reported using both written and graphic reports. A typical soils report drawing is shown in Fig. 4.4.

There are times when unexpected site conditions are identified after contracts have been signed and site work begins. For example, say a site is being excavated for a four-story masonry building and massive foundations from a previous building were found. Who should pay to excavate and remove the material. To prevent these kinds of potential problems, unexpected findings clauses can be written into contracts and thorough subsurface investigation should occur.

Figure 4.2 A typical site plan showing boundary lines, utility locations, site contours, and proposed improvements.

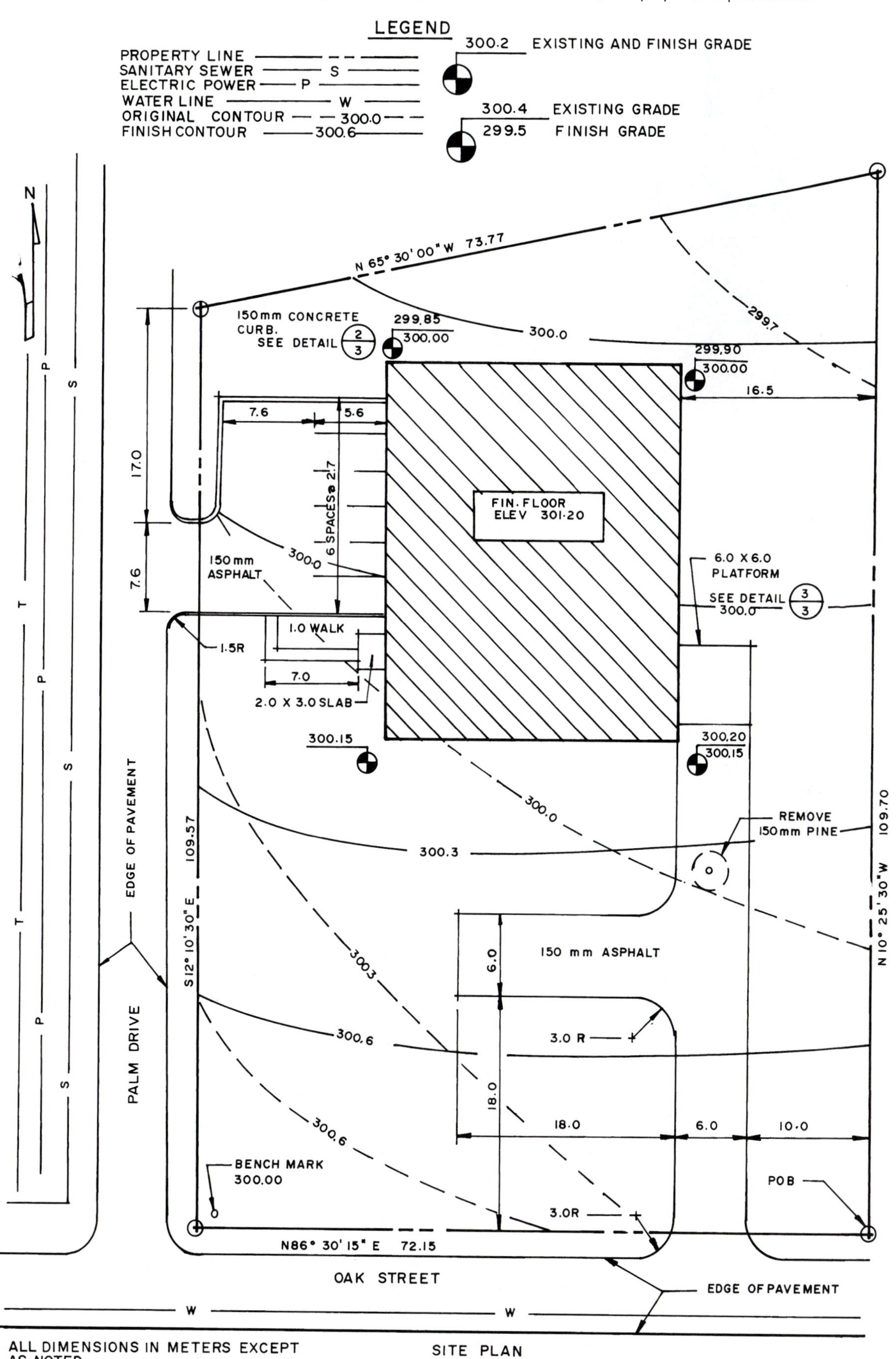

Figure 4.3 Deep soil test borings are made with a drilling rig. *(Courtesy Blastcube)*

SITE WORK ACTIVITIES

Site work includes many other activities, such as erosion control, dam construction, and tunneling. Tunneling involves excavating, ventilation, tunnel lining, grouting, and support systems. Foundations are a big part of site work and include spread footings, piles, and caissons (see Chapter 6). Building in marine areas may necessitate dredging, underwater work, and constructing seawalls, jetties, and docks. Utilities require site trenching for water, sewer, gas, oil, and steam distribution systems. The construction of ponds, reservoirs, and sewage lagoons may also be required. Finally, extensive site improvements are often needed in the form of walks, fences, irrigation systems, and landscaping.

Activities on a construction site should make every effort to conserve natural areas, prevent the escape of pollution from the site, and restore existing areas that may be damaged. Earthwork and other site disturbance should be kept as close as possible to the perimeter of built structures, roads and walkways, and utility trenches. Existing vegetation that is to remain must be protected during construction and redeveloped and restored at the end of a project. Topsoil must be carefully stripped, protected, and stockpiled for reuse. Discharge of toxic waste materials must be prohibited. The existence of flood zones must be acknowledged and decisions made concerning how they impact on the site.

Sedimentation and Erosion Control

Throughout the construction phase, activities that cause potential erosion and sedimentation problems must be controlled. Many municipalities are now requiring erosion and sedimentation control plans at the beginning of a project. Sediment is soil that has eroded and is transported by either water or wind. Erosion describes the washing or wearing away of soil through the actions of wind and water, as well as equipment and foot traffic. The construction manager must implement a Storm Water Pollution Prevention Plan (SWPPP) that conforms to best-management practices of the EPA's Storm Water Management for Construction Activities or local standards and codes.

Sedimentation and erosion control measures prevent the loss of soil by storm runoff and wind during the construction phase. Often, the civil engineer will identify areas prone to erosion and suggest strategies for soil stabilization. A cost-effective measure for preventing soil loss is the construction of a silt fence using a filter fabric media to capture sediment from storm water runoff. Permanent plantings or fast growing temporary grasses can be utilized to prevent the eroding of soil on the site. Structural controls include the placement of earthen swales to divert runoff water into temporary sediment basins that allow for a gradual settling of storm water.

Demolition

On some building sites, preparing for construction can require the demolition of existing structures. This may include the removal of buildings, parking areas, roads, and driveways; existing improvements above and below grade; as well as the clearing of trees and brush in areas designated for building. Demolition activities, such as removing and capping existing utilities and removing hazardous materials or contaminated soil, may also be required.

Requisite site work frequently involves demolishing or partially demolishing an existing building. Demolition work is subject to control by local building codes, and safety practices must be followed. Many municipalities maintain and enforce construction-waste regulations stipulating what portion of material must be recycled and approved means for disposing of remaining material. No demolition can occur until the construction documents are approved by the local building official.

Site Remediation

Remediation activities result in long-term benefits for building sites by preventing erosion, generating productive soils, restoring natural habitats, and removing toxic substances. On previously polluted

Figure 4.4 A typical soils report on the findings from test borings made at predetermined locations on the site.

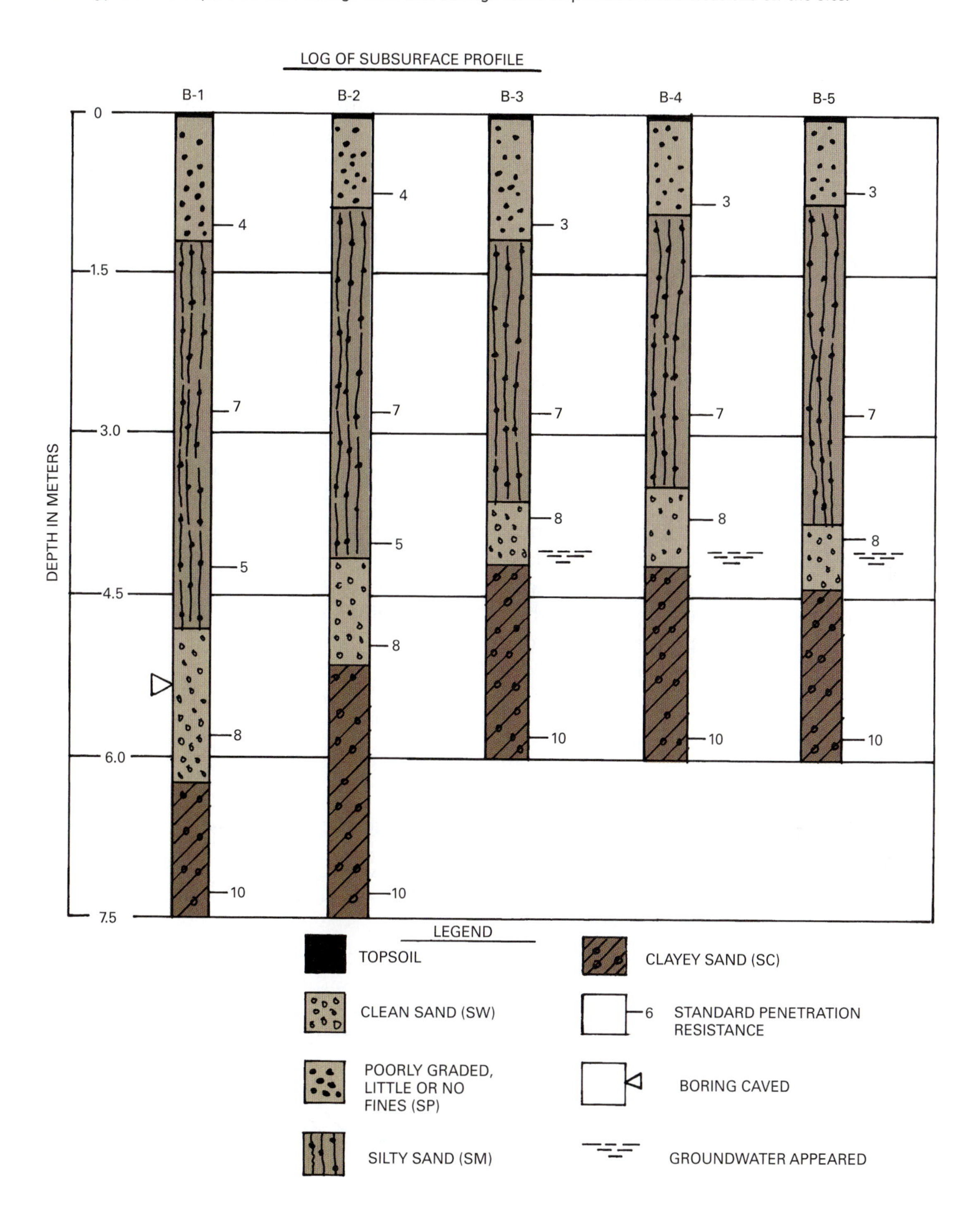

sites, restoration of structures, soils, and vegetation may be required. Both urban and rural sites can be problematic in a variety of ways, from minor damage and neglect to more complicated contamination. Depending on the type, extent, and toxicity of contamination found on a site, cleanup activities may include soil, surface water, and ground water remediation. Historically the main remediation strategy for damaged sites was the removal of contaminated material to special, contained landfills, a procedure known as dig and dump. More recently, means for remediating rather than dumping contaminated soil are being developed. New technologies include phytoremediation, which utilizes plants to metabolize organic contaminants. Another method, known as **bioremediation**, uses bacteria that are able to digest and remove toxic chemicals. The U.S. Environmental Protection Agency has a number of programs and resources to aid in the remediation of polluted sites and provides grants for assessment and cleanup, in addition to technical advice and training.

EARTHWORK

Earthwork involves a variety of activities, including grading, excavating, backfilling, compacting, laying base courses, stabilizing the soil, setting up slope protection, and establishing erosion control.

Grading

Grading is the act of remodeling the existing land form to provide a level area for a structure, create circulation paths, and create drainage and landscape features. Rough grading involves adjusting the level of the ground to facilitate the excavation and construction of a building. First, the topsoil is removed and stockpiled for reuse during finish grading. Finish grading occurs after the building and other improvements are complete, bringing the site elevations up to the levels stipulated on the site or grading plan. Removing or importing soil is costly. Cut-and-fill grading operations should be carefully balanced to avoid the need for transporting soils. Construction costs can be lessened by keeping earthwork activities to a minimum by utilizing site contours and avoiding large scale grading, which creates dust and the potential for erosion.

Excavation

Excavations are required for footings, foundations, underground utilities, and other aspects of a project. A wide variety of equipment can be used, including power shovels, backhoes, draglines, clamshells and cranes, trenching machines, wheel-mounted belt loaders, and bulldozers.

If foundations are shallow, they are usually dug with a backhoe. The depth of a foundation footing is determined by geographic location and its base must always be placed below the frost line. Excavations must reach to the depth at which tests have shown the soil to have the required load-bearing capacity. A building with a basement will have a large excavation and a considerable amount of soil must be removed and stockpiled.

Multistory buildings require extensive excavations that can reach several stories underground to access bedrock or other adequate supporting soil. If solid rock must be removed, the cost of excavation increases rapidly. Weak and thinly layered rock can be loosened with power shovels, tractor-mounted rippers, backhoes, or pneumatic hammers. Other rock formations can be fractured by explosives. The fractured rock can then be removed from the excavation by a front-end loader. Like soil, rock must be stockpiled or removed from the site.

Deep excavations present an ever-present danger of collapsing sides. The depth and type of soil and the location of the site influence what strategies should be used. If the excavation is not too deep and the site is large enough, excavation sides can be sloped. The angle of the sloped surface will vary with the type of soil. The **angle of repose** defines the steepest angle of a surface at which loose material will remain in place rather than sliding. This angle will range from 34° for cohesionless soil to 90° for stable rock. Deeper excavations can use a bench system, as shown in Fig. 4.5. The particles in cohesionless soils tend to dispersion, which can result

Figure 4.5 Ways of containing the sides of excavations to prevent collapse.

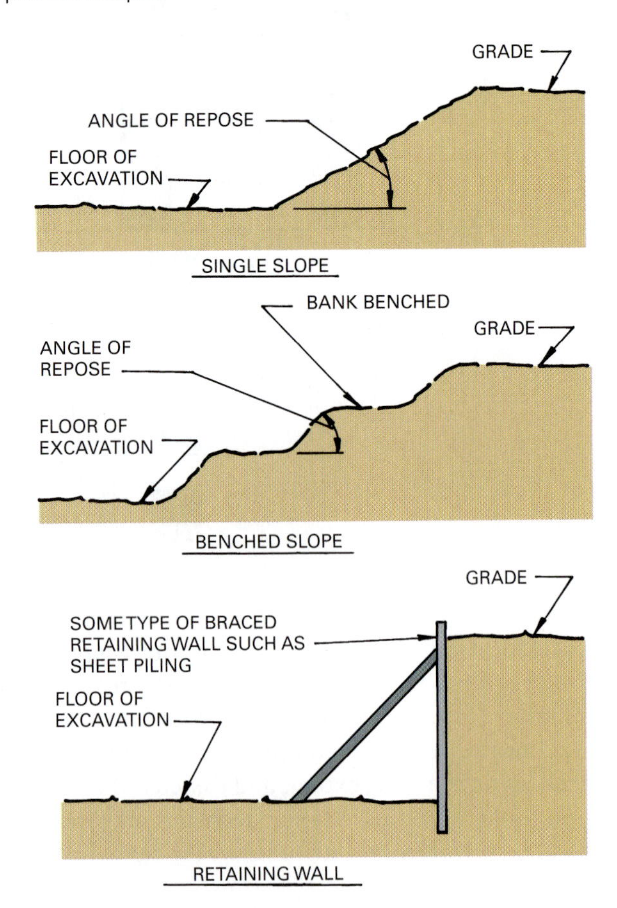

in slope failure unless the angle of repose is small. Cohesive soil grains tend to bond and provide good shear strength. Maximum allowable slopes for different materials are specified by OSHA regulations.

Sheeting

If the sides of the excavation cannot be sloped to provide protection from slides, some form of **sheeting** must be used. Sheeting, in the form of sheet piling, lagging, and slurry walls, is used to hold up the face of an excavation.

Sheet piling may be wood, aluminum, steel, or precast concrete, placed vertically into the ground (Fig. 4.6). While wood pilings are usually removed, steel and concrete sheet pilings are sometimes left in place after construction is finished. Another sheeting technique uses vertical steel piles, called soldier piles, that are driven into the soil. Horizontal wood lagging is then placed between them to contain the excavation (Fig. 4.7). Steel sheet pilings are typically driven into the earth before excavation begins (Fig. 4.8).

Since sheeting is subject to soil and subsurface water pressures, some form of bracing is needed as the excavation gets deeper. In shallow excavations, the sheeting may be driven deep enough below the bottom of the excavation so that bracing is not required. Vertical sheeting and narrow excavations can use horizontal beams, called wales, placed at regular intervals to resist soil pressures.

Wider excavations require the use of more substantial cross-lot bracing supported by vertical steel posts driven into the ground. Rakers are used on wide excavations where cross bracing would be impractical. The rakers are set on an angle and transfer the forces to a footing set in the bottom of the excavation (Fig. 4.9).

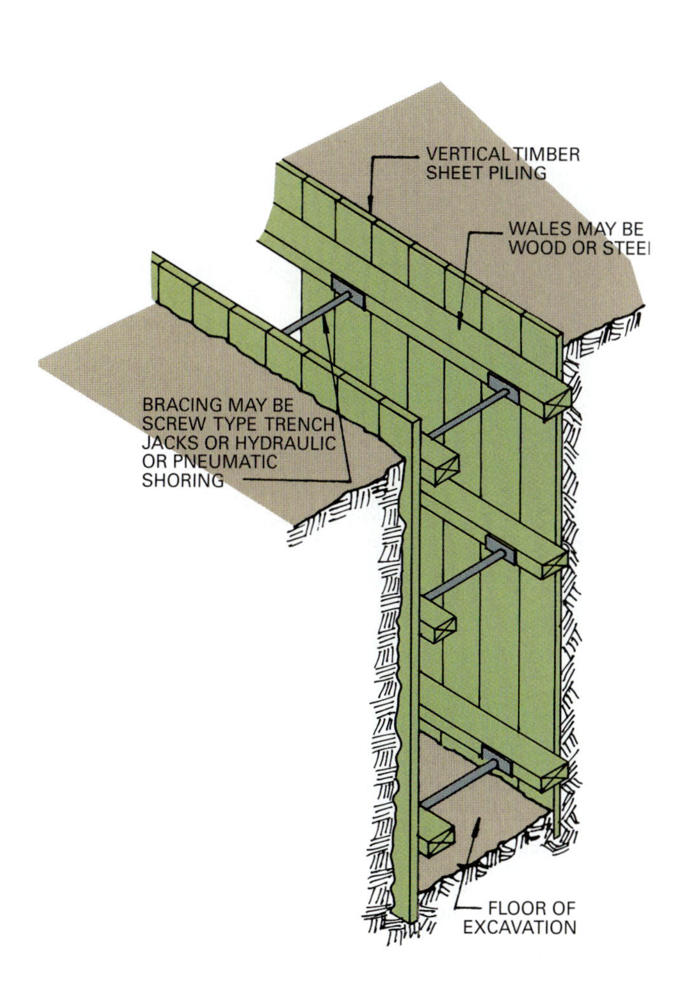

Figure 4.6 Vertical wood sheet piling can be used to support the sides of an excavation.

Figure 4.7 This form of sheeting uses steel soldier piles and horizontal timber lagging.

Figure 4.8 On shallow excavations, steel sheet pilings are driven before the excavation begins. *(Courtesy Eva Kultermann)*

Both types block the open space in the excavation and interfere with any work that occurs within it.

Some soils permit the use of tiebacks. Tiebacks are steel cables, or tenons, that are inserted into holes drilled through the sheeting and into the rock or subsoil. The drilled hole is filled with concrete grout. When the grout has set, the cables are tightened with hydraulic jacks and fastened to the wales to provide an excavation free of barriers (Fig. 4.10). A second type of tieback uses a screw anchor that is installed with the same rotary drilling equipment used in the grouted anchor construction. The screw anchor is rapidly installed and does not require the drilling of an anchor hole or the injection of grout. They are especially useful in loose sandy and clay soils where a drilled hole may collapse. Screw anchors can be withdrawn and reused.

A **slurry wall** serves as sheeting that protects the excavated area and becomes a part of the permanent foundation. It may be cast in place or built from precast concrete panels. The excavation is dug with a narrow clamshell bucket that establishes the width of the wall. To help prevent collapse, the excavation

Figure 4.9 Steel pilings on wide excavations can be braced with rakers and cross-lot bracing.

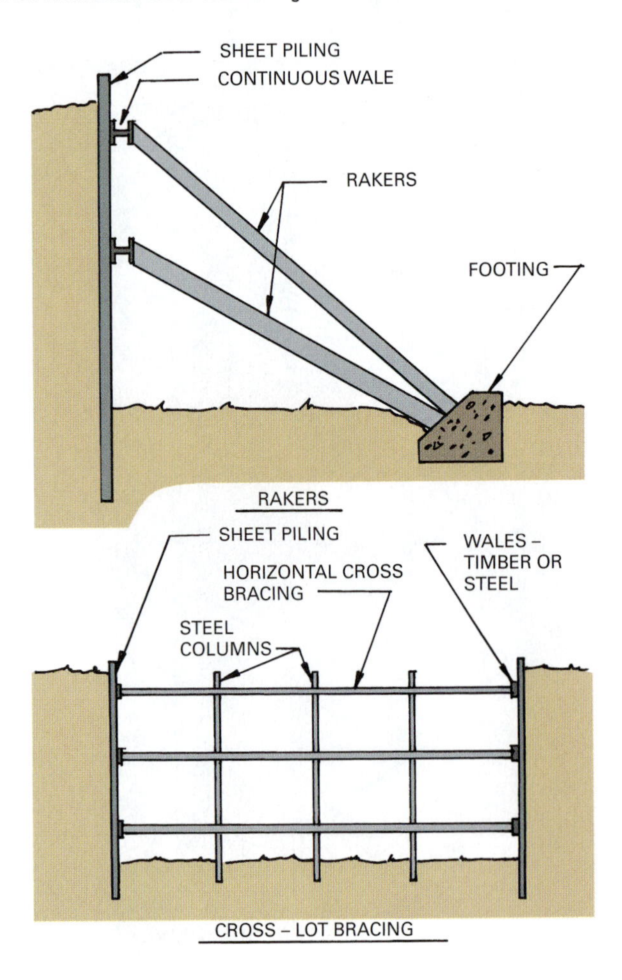

Figure 4.10 Sheeting can be supported by tiebacks anchored in stable earth or rock.

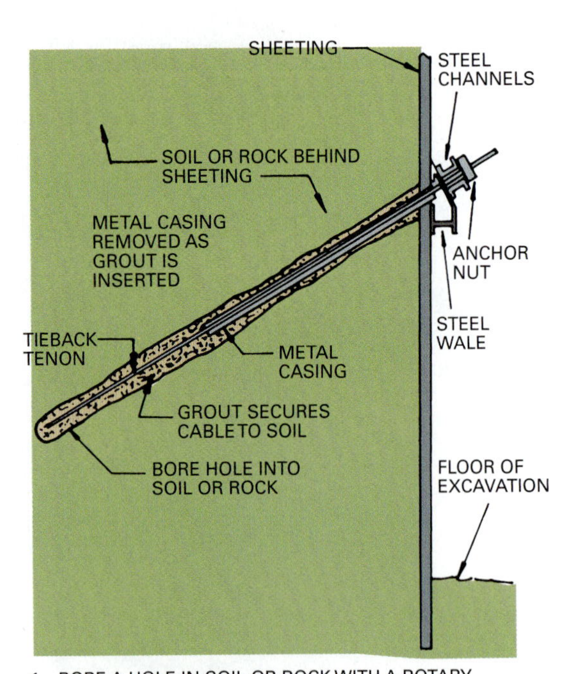

1. BORE A HOLE IN SOIL OR ROCK WITH A ROTARY DRILL. INSERT A METAL CASING TO KEEP THE SOIL FROM FILLING THE HOLE.
2. INSERT STEEL TIEBACK TENONS INTO THE HOLE. FILL THE HOLE WITH GROUT AS THE STEEL CASING IS REMOVED. LET THE GROUT SET.
3. WHEN THE GROUT HAS SET POST-TENSION THE TENON WITH A HYDRAULIC JACK AND ANCHOR TO THE WALE.

is filled with **slurry** composed of bentonite clay and water, which stabilizes the wall. The clam-shell bucket moves through the slurry to continue digging. After the required depth is reached, a welded cage of steel reinforcing is lowered into the cavity. Concrete is poured from the bottom of the excavation, filling it to the top. The slurry rises above the concrete before being pumped off. Once the entire wall is poured and has reached sufficient strength, the soil can be excavated from one side. As the excavation deepens, steel tiebacks are set in holes drilled through the wall and into the subsoil (Fig. 4.11).

A second technique is to insert precast concrete wall sections into the slurry instead of site casting the concrete wall. The surfaces of the precast sections that are on the excavation side are coated with a release agent to prevent the slurry from bonding to them. A tongue-and-groove joint or rubber gasket is used to seal adjacent sections. After the slurry remaining in the excavation has hardened, the soil can be dug away. The slurry on the back side of the wall remains and aids in waterproofing. The slurry on the excavated side falls free, leaving the surface of the precast units exposed.

Soil anchors are metal shafts, grouted into holes to stabilize sides of an excavation. This involves drilling holes into the soil embankment, removing the drill, and filling the hole with grout. Then a soil nail, typically a section of steel reinforcing bar, is pushed into the grouted hole. A centering device is attached to the nail so it remains in the center of the grouted hole. This is much like forming a cast pile that is made on an angle. (Additional information can be found in Chapter 6.)

Cofferdams and Caissons

Cofferdams are temporary watertight enclosures used either in water-bearing soil or directly in water. They prevent water from entering an interior area, allowing

Figure 4.11 The typical procedure for constructing a slurry wall.

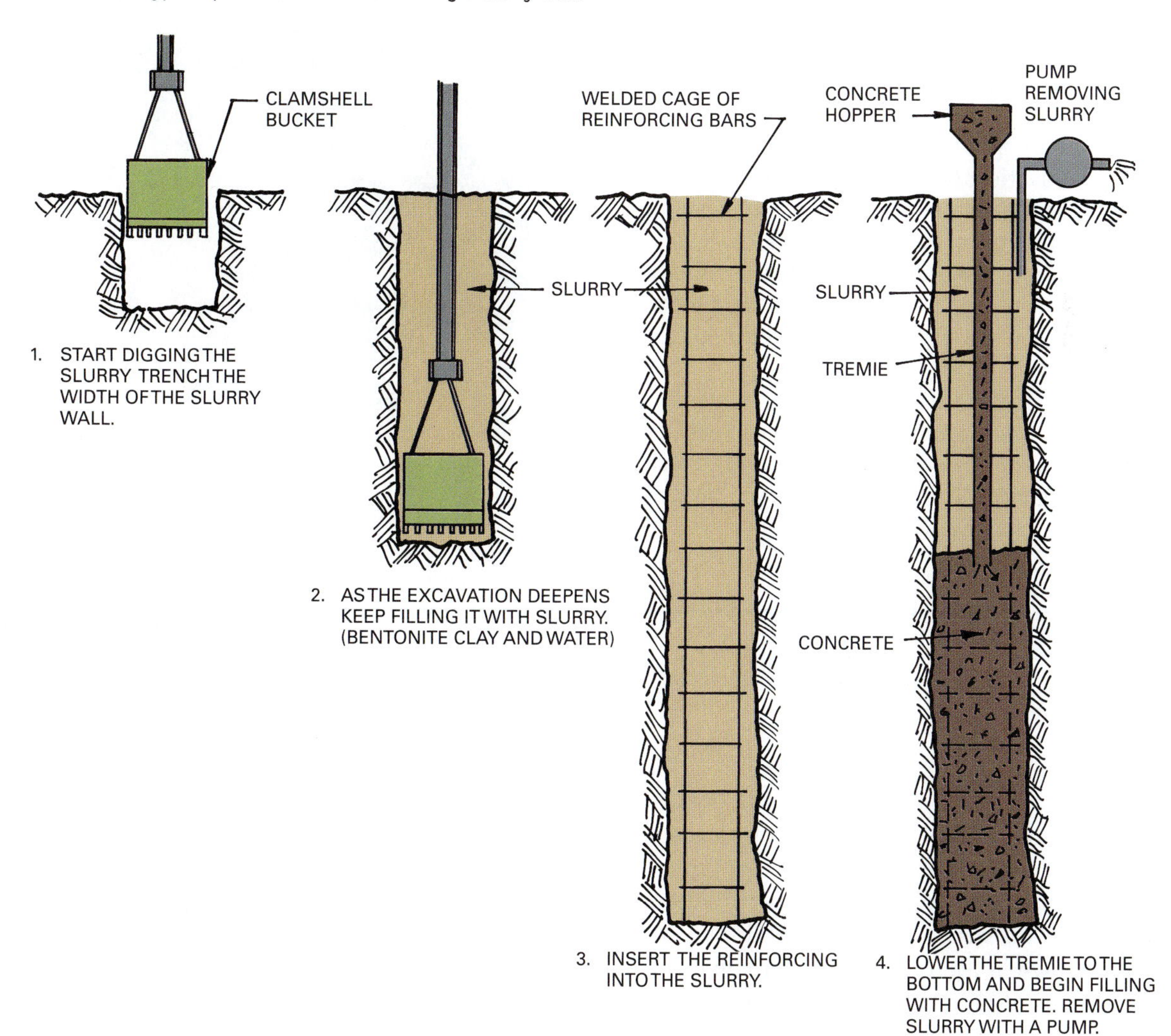

Construction Techniques

Blasting

Many parts of the country have considerable rock formations extending below the surface of the earth. Before a foundation, utility trenches, or other subsurface work can begin, it is sometimes necessary to blast out rock located in an excavation's path. Blasting requires the services of a qualified construction blaster who has a record of successful projects, is licensed as required by the state, and has adequate insurance coverage.

Drilling and blasting is expensive, and a careful analysis of the site must be made during the bidding phase of a project. When an area is known to contain rock, soil testing must be conducted to check for subsurface rock and determine whether blasting is needed prior to the start of construction. If rock is encountered once construction has begun, the material will have to be removed with air hammers, a time-consuming process. Subsurface investigations can reveal the presence of solid rock beneath a site. The project team may decide that structures and utility lines should be relocated to avoid the expense of blasting.

Explosives are employed in blasting for mining, roadwork, excavations demolition, and tunneling. They are made from materials that are extremely sensitive to stimuli from impact, friction, heat, or electrostatic sources of initiation. Explosives such as lead azide, lead styphnate, and TNT have a high rate of energy release and produce a forceful blast pressure.

Once subsurface conditions are known, a number of holes are drilled into the rock and then filled with explosive. Detonating the explosive will cause the rock to collapse and loosen, allowing for removal. Special care must be exercised if there are existing structures, such as buildings, wells, or submerged tanks, nearby. This may require more drilled holes and smaller charges or detonating fewer holes at one time. Heavy blasting mats can be utilized to cover the charged area and reduce the distribution of material.

Explosive material is inserted into drilled holes with funnels and covered with a protective mat made of recycled tires. *(Image copyright Zygimantas Cepaitis, 2009. Used under license from Shutterstock.com)*

construction to proceed in a dry and stable environment. Water is pumped from within the cofferdam, and pumps are maintained during construction to remove any leakage that may occur. Cofferdams are typically built using sheet piling, soldier beams with lagging, or as a double-wall structure. A typical double-wall cofferdam is shown in Fig. 4.12. The cofferdam extends through the areas of permeable water-bearing soil formations, through any impervious rock formations that have low bearing capacity, and on to solid bedrock that is used to support the foundation.

A **caisson** is a watertight shell in which construction work is carried out below water level. Caissons may be open or pneumatic. The top of the open caisson is exposed to the weather, and work is performed under normal atmospheric pressure. Pneumatic caissons are air- and watertight. They are open on the bottom so soil excavation can be accomplished. As an excavation

Figure 4.12 A cofferdam is used to provide a watertight area for the construction of footings and foundations.

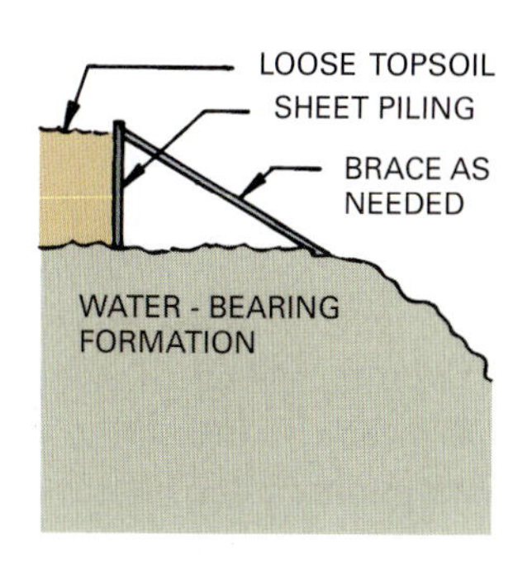

1. SHEET PILING IS USED TO HOLD THE LOOSE TOPSOIL FORMATION

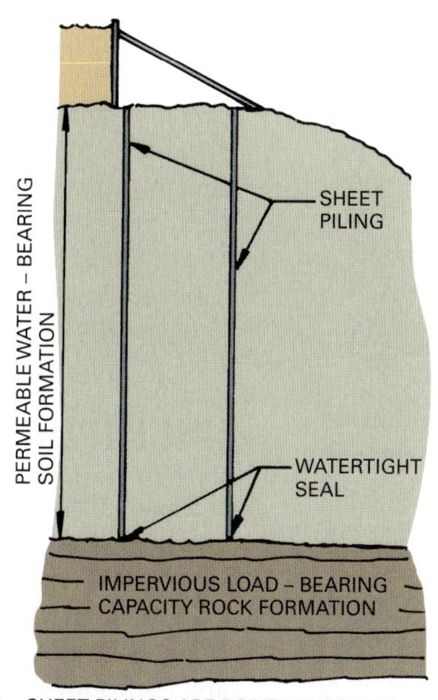

2. SHEET PILINGS ARE DRIVEN THROUGH THE PERMEABLE WATER - BEARING SOIL FORMATION TO IMPERVIOUS ROCK. THE PILING FORMS A WATER-TIGHT SEAL WITH THE ROCK.

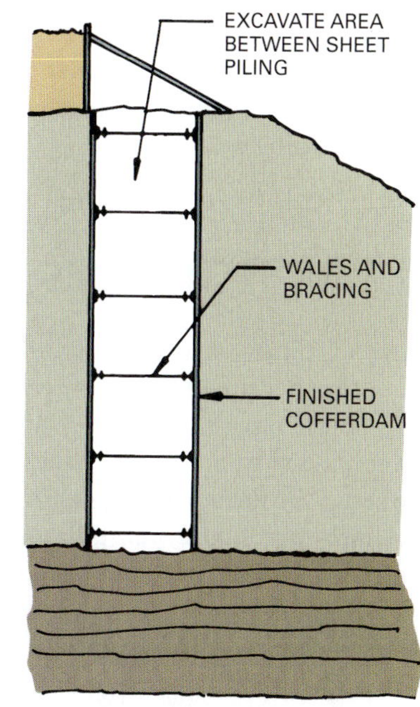

3. EXCAVATE THE AREA BETWEEN THE SHEET PILING AND BRACE WITH WALES AND CROSS BRACING AS THE EXCAVATION PROCEEDS

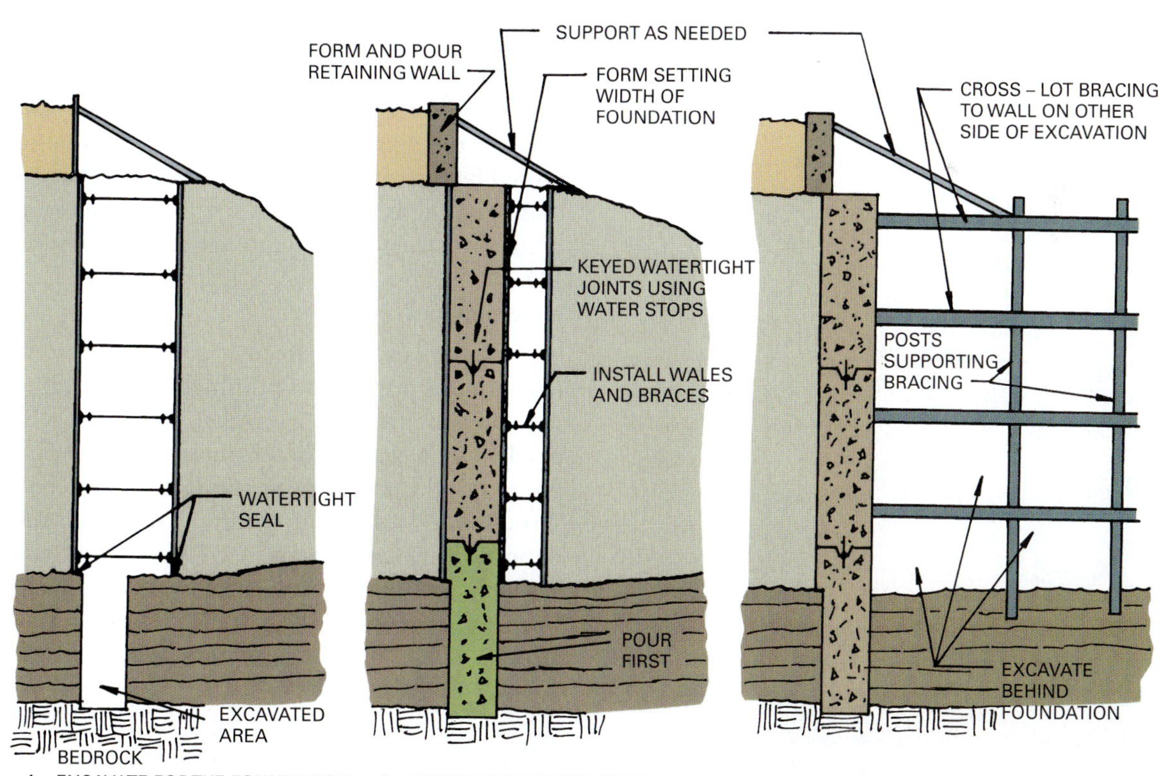

4. EXCAVATE FOR THE FOUNDATION THROUGH THE IMPERVIOUS ROCK DOWN TO BEDROCK. NO SHEET PILING NEEDED IN THIS AREA.

5. INSTALL THE FOUNDATION WALL FORM IN THE EXCAVATED AREA. INSTALL REINFORCING AND FILL WALL CAVITY WITH CONCRETE. POUR IN SECTIONS WITH WATER STOPS BETWEEN THE SECTIONS.

6. BEGIN EXCAVATION OF THE AREA BEHIND THE FOUNDATION. BRACE THE FOUNDATION AS THE EXCAVATION PROCEEDS. WHEN COMPLETELY EXCAVATED THE WALL CAN BE SUPPORTED WITH CROSS - LOT BRACING.

proceeds, the caisson is consistently filled with pressurized air to keep water from entering (Fig. 4.13). Workers must go through decompression after working in pressurized caissons to avoid suffering from the bends, a painful and sometimes fatal disorder. The structure forming the caisson is frequently left in place and filled with concrete, forming a caisson pile.

DEWATERING TECHNIQUES

Dewatering techniques involve lowering the level of subsurface water on a site to allow excavation to occur in a dry and stable environment. Dewatering usually begins before excavation and continues as the excavation proceeds. Subsurface water is removed using a system of pumps, pipes, and well points. A well point is a perforated unit placed at the bottom of a pipe that is driven into the soil. A series of well points connected to horizontal pipes above the ground, called headers, are driven in and around the area to be excavated. The headers are connected to centrifugal pumps that draw water up into the header pipes and eject it for drainage away from the excavation (Fig. 4.14). More than one series of well point, header, and pump assembly is usually required, especially as the excavation gets deeper.

Figure 4.13 A typical open-air caisson.

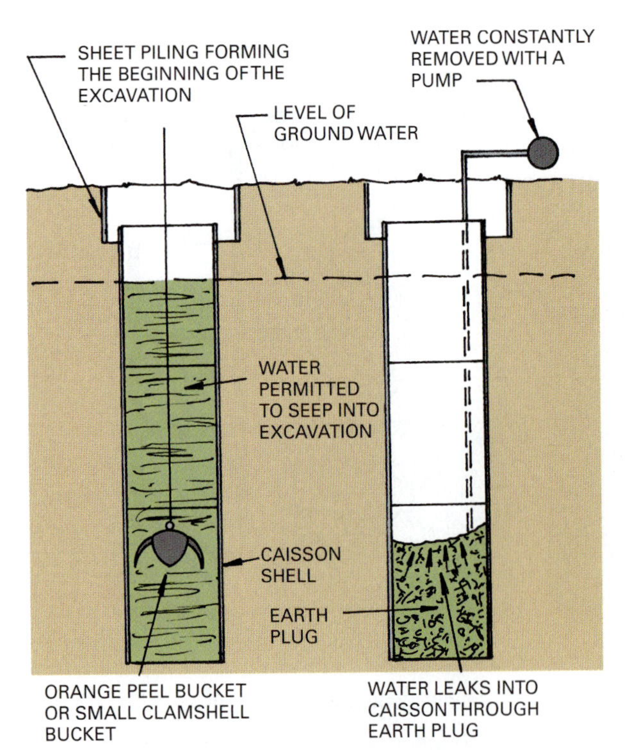

Water can also be kept out of an excavation by constructing a watertight wall. This could be a slurry wall or one constructed with sheet piling, as discussed under the Cofferdams heading. The wall must resist the hydrostatic pressure of the subsurface water and the pressure generated by the soil itself. It has to reach a depth where it penetrates a stratum of impermeable material so a watertight seal can be achieved. Pumps should be kept available to remove any water that may seep into the excavation.

UNDERPINNING TECHNIQUES

Occasionally, the excavation for a new building is so close to another building that it becomes necessary to support the existing building during construction by **underpinning**. This underpinning requires considerable engineering expertise, and special design considerations are needed for each situation.

The existing soil conditions must be evaluated to ensure that the new footings will be bearing upon a suitable stratum. There are various ways to underpin a foundation. Two frequently utilized techniques are the use of trenches dug below the foundation of the existing building or the use of needles.

The first technique uses trenches dug at intervals beneath the existing foundation. The building is supported by the remaining undisturbed soil. The required underpinning is installed in the trenched area and additional trenches are dug through the unexcavated area. This is repeated until the extant building has adequate underpinning.

Needles are heavy wooden timbers or steel beams that are run horizontally through the wall of a building and supported on each end. The spacing between needles in a wall is determined by an engineer. In Fig. 4.15 the building is supported by needles run through the foundation, then the area under the foundation is excavated and a new footing and foundation are built. Notice that this system uses sheeting under the building to prevent cave-in of the excavation wall.

PAVING

The most commonly used paving materials for roads and parking areas are asphaltic concrete or concrete. Finish paving occurs after a sub-base and base have

Figure 4.14 An excavation can be dewatered using a well point system.

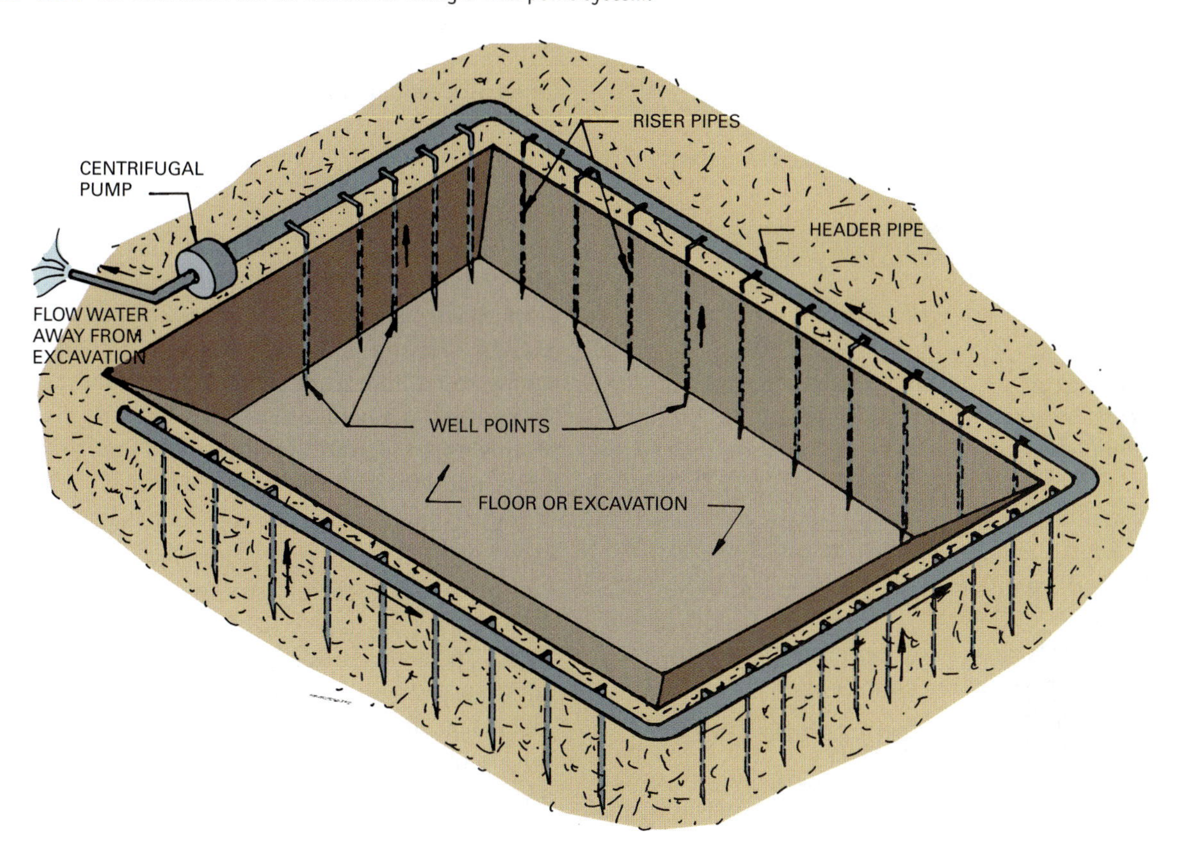

been laid. A typical situation with a flexible pavement, such as asphaltic concrete, is shown in Fig. 4.16. The lowest layer is the natural sub-grade, which is consolidated by compaction. The sub-base course, consisting of a natural soil of higher quality, is laid on top of the natural sub-grade. It provides additional support for distribution of loads imposed on the finished surface. The base course is laid over the sub-base and provides the surface upon which the finished wearing surface is laid. It is a granular base of high quality, such as crushed stone, gravel, slag, or some combination of these, sometimes mixed with sand. The material is often treated with bitumen to bind the materials. The sub-base and base courses are extended beyond the edge of the next layer to provide a cone disbursement for the loads.

Figure 4.17 shows a section through the base material for a typical rigid pavement, such as concrete. In this design the concrete slab has to be thick, reinforced, and strong enough to resist the imposed loads. The base material helps carry the loads, but the concrete slab actually carries most of the weight. The loads are transmitted over the entire width of the slab, which reduces the forces on the base materials.

Asphalt Paving

Asphalt concrete consists of asphalt cement and graded aggregates carefully proportioned and mixed in an asphalt plant at controlled temperatures. Four types of asphalt paving are in common use. They are made with either cutback or emulsified asphalt. The mix is transported to the site, spread by an asphalt paving machine, and rolled while it is still hot (Fig. 4.18). Cold-laid asphalt is similar to asphalt concrete but is laid when the asphalt liquid is cold.

Asphalt macadam is laid using a penetration method. A coarse aggregate is laid over the base, compacted to a smooth surface, and sprayed with an asphalt emulsion or hot asphalt cement. This is covered with fine aggregates and rolled, forcing the fine aggregate into the voids between the larger aggregate.

Old asphalt surfaces can be renewed by applying a surface treatment consisting of a 1 inch (25.4 mm) asphalt sealer. The treatment may consist of several layers of liquid asphalt covered with a mineral aggregate or a single layer of cold or hot asphalt mix that is plant-mixed and rolled after being machine laid. Other sealers used

Figure 4.15 A typical underpinning installation using needles to support the existing building.

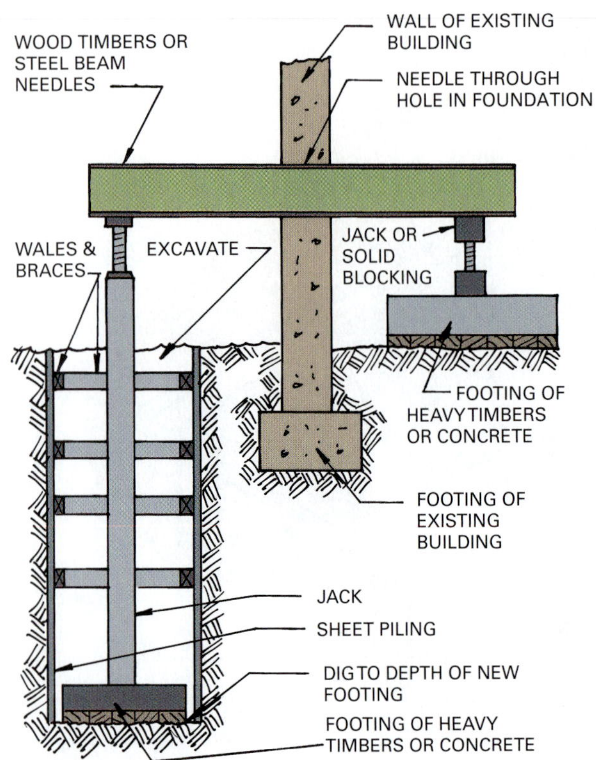

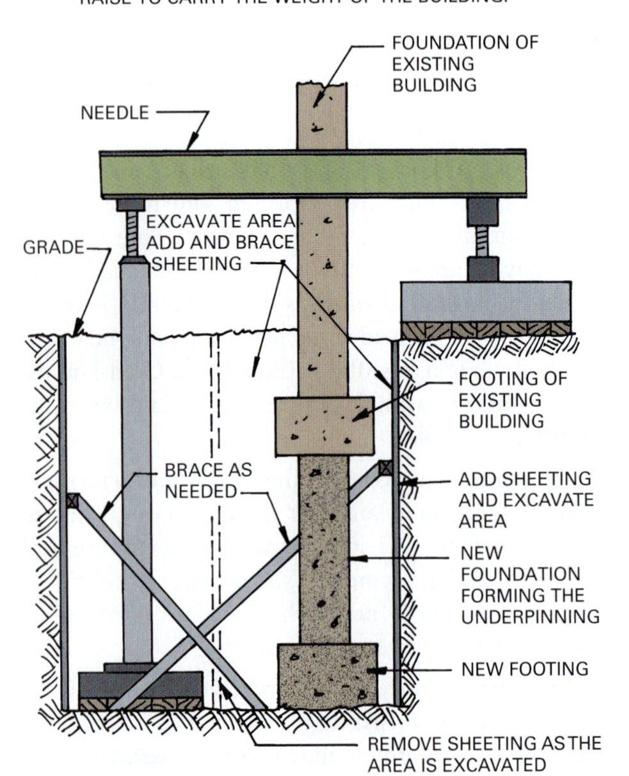

include a coal-tar-latex emulsion and vinyl and epoxy resin sealers. These have greater tensile strength than regular coal-tar sealers, and some are available in different colors. Additional information on bitumens can be found in Chapter 27.

Concrete Paving

Roads, driveways, and parking lots are often paved with concrete. The sub-grade, sub-base and base course preparation is much like that described for asphalt paving. Side forms are placed along the edges of the paving to contain the plastic concrete and act as rails upon which the concrete placing spreader rides. Some placing spreaders are designed to ride inside the side forms, and others ride outside the forms.

The paving machinery may be either form or slip-form types. The form type paver rides on the metal side forms, places the concrete, strikes it off, and consolidates it. Slip-form pavers place a slab without the use of side forms. The concrete is of a consistency that it develops sufficient strength to be self-supporting as it leaves the paver. The slip-form paver spreads, consolidates, and finishes the concrete slab (Fig. 4.19). Special machinery is used to create curb and gutter profiles (Fig. 4.20).

Information on concrete is presented in Chapter 7.

Pervious Paving

Materials that are pervious to water, such as gravel, crushed stone, open paving blocks, or pervious paving blocks, minimize runoff from a site and increase natural infiltration of storm water. Pervious materials can be used for drives, parking areas, walkways, and patios and can withstand both foot and heavy traffic. In order to ensure proper drainage, pervious systems will function best with a similarly porous sub-grade such as CA-17, a permeable backfill. A good solution for low-traffic walks is the use of interlocking pavers with open spaces that allow grass to grow in the openings while providing a surface of good load-bearing capacity (Fig. 4.21).

SEISMIC CONSIDERATIONS

Seismic forces are destructive forces caused by earthquakes. Although accurate prediction of earthquakes is difficult, certain **seismic areas** of the country are known to have subsurface conditions that make them possible earthquake zones, as shown in Fig. 4.22. These areas have

Figure 4.16 The base course and wearing surface for a typical flexible pavement.

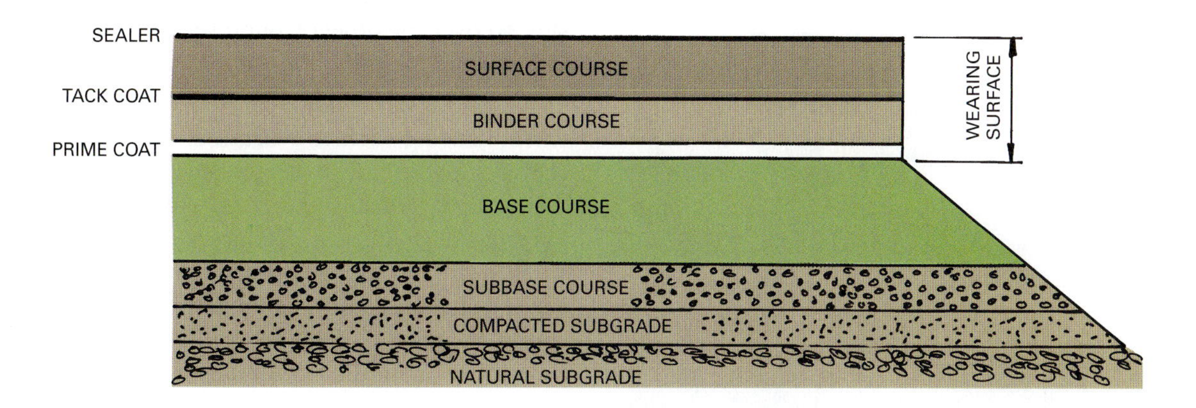

Figure 4.17 The base courses and concrete paving for a typical rigid pavement.

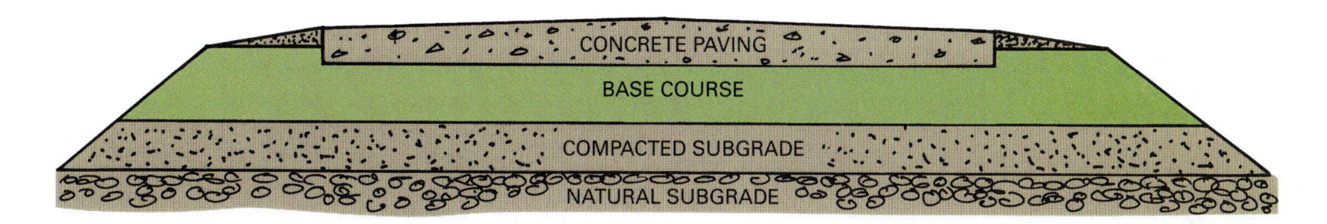

Figure 4.18 Laying asphalt paving. *(Image copyright V. J. Matthew, 2009. Used under license from Shutterstock.com)*

Figure 4.19 Laying concrete paving. *(Courtesy GOMACO Corporation)*

special building-code requirements regulating the design and construction of buildings and other structures.

An earthquake is caused by the slippage of the earth along a fault plane. A fault is a fracture in the earth's crust along which two sides of crust have slipped with respect to each other. Common movements include one crustal block moving horizontally in one direction while the block facing it moves horizontally in the other direction. This is referred to as a strike slip. In other cases, one block may move up vertically while the abutting block moves downward. This is referred to as a vertical slip. In some cases the disturbance along the fault may produce

Figure 4.20 Special machinery creates curb and gutter profiles. *(Courtesy GOMACO Corporation)*

Figure 4.21 A pervious paving system provides a hard surface and allows runoff water to percolate naturally into the ground. *(Courtesy Eva Kultermann)*

both vertical and horizontal movement. A thrust fault results when sections of rock press together, forcing one side up over the other. A blind thrust raises the surface into folded hills without breaking the surface.

Often these slippages of strata occur deep in the earth, causing surface vibration and shaking but no ruptures. These fault slippages generate ground motions that radiate vertically and horizontally in all directions. Earthquakes frequently produce ruptures of the ground surface as well as rolling and waving motions. The ground shakes as this occurs, and the shaking is what causes the most damage to buildings and other structures, rather than the more visible and dramatic ground rupturing (Fig. 4.23).

The magnitude of an earthquake is measured by the Richter scale, which is based on the maximum single movement recorded on a seismograph. The measurement

Figure 4.23 A collapsed apartment building after the 1994 Northridge earthquake. *(Image courtesy of Gary B. Edstrom)*

Figure 4.22 Seismic zone map of the United States.

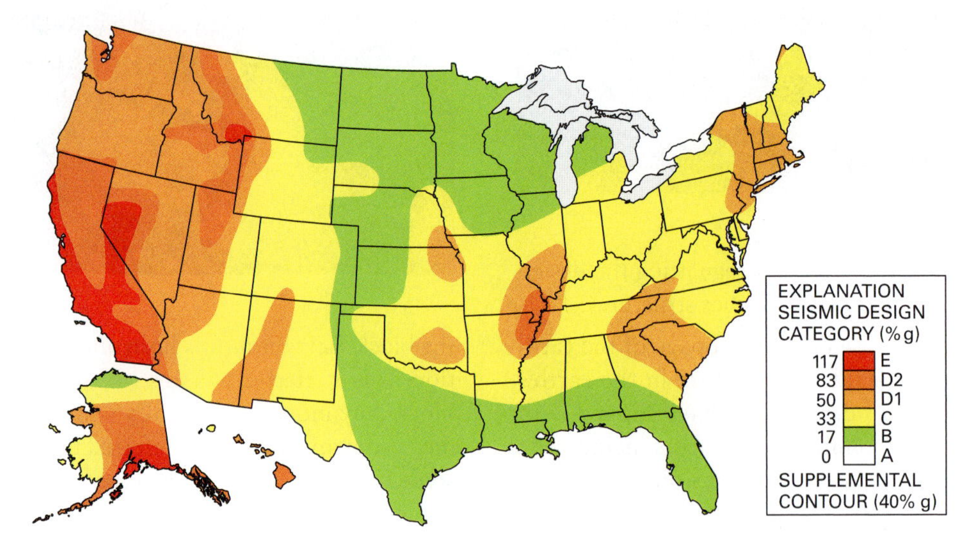

most widely used by seismologists is the moment magnitude. It is based on the size of the fault on which the earthquake occurs and the amount of earth slippage. The larger the fault and the larger the actual slip, the higher the moment magnitude of the earthquake. No buildings should be located over known active geologic faults. The types of destruction of buildings during an earthquake include ground rupture, ground shaking, liquefaction, differential settling or some combination of these (Fig. 4.24). Liquefaction occurs when the motions of the earthquake transform the soil into a semi-liquefied state much like quicksand. It occurs in areas with loose sands and silts that have high water tables.

The design of footings for buildings in seismic areas is specified by local codes and varies with the soil classification. Masonry construction must meet the requirements for seismic design set forth by the American Concrete Institute, American Society of Civil Engineers, and the Masonry Council. Seismic requirements for reinforced concrete are set forth by the American Concrete Institute and include any additional requirements specified by local codes. Structural steel seismic requirements are usually those set forth by the American Institute of Steel Construction, Inc., in their publication AISC Seismic Provisions for Structural Steel Buildings (plus other requirements specified by the local code). Seismic requirements for wood and timber construction follow the recommendations of the National Forest and Paper Association and additional local code specifications.

Figure 4.24 The effects of an earthquake can cause a variety of building failures. *(Courtesy U.S. Department of the Interior, Geological Survey)*

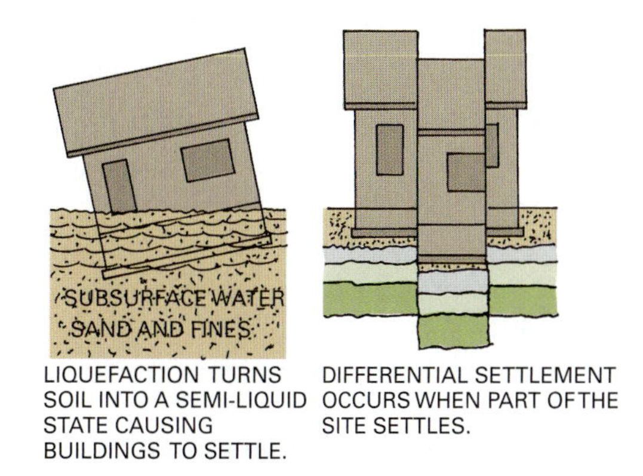

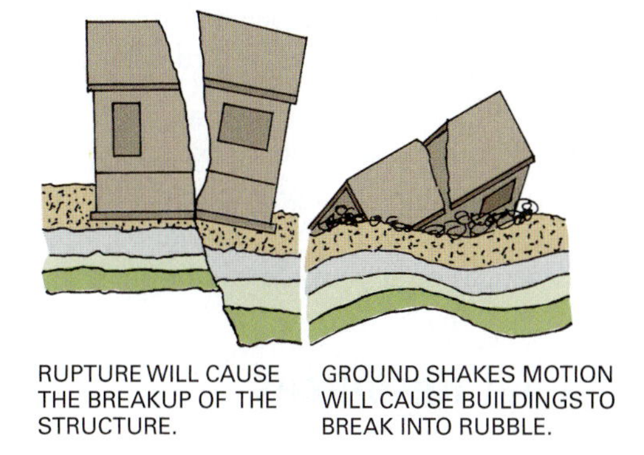

Review Questions

1. What information is shown on the site plan?
2. What types of activities might be required to prepare the site for construction?
3. What information is gained by making a subsurface investigation?
4. What types of technologies can be used for site remediation?
5. What type of work might be required under the earthwork section of a construction contract?
6. What are the various ways the sides of an excavation can be prepared to protect workers from caving earth?
7. What purpose does dewatering serve?
8. When would underpinning be necessary?
9. What are the layers of material used to pave a parking area with asphaltic concrete?
10. What is the difference in laying a concrete paving between form type pavers and slip-form pavers?
11. Describe what happens in the earth when an earthquake occurs.
12. What organizations provide information about designing structural systems for buildings located in an area of possible seismic activity?

Key Terms

angle of repose
bioremediation
brownfield site
caisson
cofferdam
dewatering
grading
seismic area
sheeting
site plan
slurry
slurry wall
soil anchors
underpinning

Activities

1. Secure site plans for recent projects from a local architect or contractor. Prepare a detailed list of the things shown on the plan. See if there were any environmental problems; if so, record what these were and the actions taken.
2. Visit local building sites where excavation is underway or just completed. Prepare a written report of the things you saw. Did you notice any especially good practices or any unsafe conditions?
3. Prepare sketches showing various ways the sides of excavations can be stabilized.
4. Examine the local building codes and list the requirements for site control of erosion, tree removal, and spoil banks.
5. If your locale is in an identified seismic area, cite the building code requirements specified because of possible seismic activity.

Additional Resources

Ambrose, J., *Building Construction: Site and Below-Grade Systems*, Van Nostrand Reinhold, New York.

Colley, B. C., *Practical Manual of Site Development*, McGraw-Hill, New York.

DeChiara, J., and Koppelman, L. E., *Time-Saver Standards for Site Planning*, McGraw-Hill, New York.

Publications from the Geo-Institute of the American Society of Civil Engineers, Reston, VA.

5 CHAPTER

Soils

LEARNING OBJECTIVES

Upon completion of this chapter, the student should be able to:

- Know the major types of soils and how they are classified.
- Be familiar with commonly used soil tests.
- Understand how factors related to soils can affect design decisions.
- Be aware of the various ways to stabilize soils.

Soils are created through the mechanical and chemical weathering of rock. Naturally occurring abrasive and mechanical forces involving heat and gravity wear down large masses of rock into smaller particles, producing gravels, sands, and fine silts. Mechanically weathered soil particles are three-dimensional in shape.

In clayey soils, less stable minerals in rocks produce very small, two-dimensional flake-like particles in crystalline form. An understanding of soils, their characteristics, and their properties is essential for those who design foundations. Building codes specify maximum design loads for various types of soils, and soil investigations on a site are required to produce information needed for design.

TYPES OF SOILS

Soils are classified by the sizes of their particles and their physical properties. Most soils are a mixture of the following five types of soils:

- **Gravel** is a hard rock material with particles larger than $\frac{1}{4}$ in. (6.4 mm) in diameter but smaller than 3 in. (76 mm).
- **Sand** consists of fine rock particles smaller than $\frac{1}{4}$ in. (6.4 mm) in diameter to 0.002 in. (0.05 mm).
- **Silt** is fine sand with particles smaller than 0.002 in. (0.05 mm) and larger than 0.00008 in. (0.002 mm).
- **Clay** is a very cohesive material with microscopic particles (less than 0.00008 in.).
- **Organic matter** is partly decomposed animal and vegetable matter.

Rock particles larger than 3 in. (76 mm) are called cobbles or boulders and are not classified as soil. Peat and other organic oils are insufficient in bearing strength to support structures.

SOIL CLASSIFICATIONS

The two commonly used soil classification systems are the Unified Soil Classification System (USCS) and the system of the American Association of State Highway and Transportation Officials (AASHTO).

Important to understanding soil classification are the liquid limit, plastic limit, plasticity index, and the shrinkage limit of soils. These states of soil consistency are indicated in terms of water content (Fig. 5.1). The **liquid limit** (**LL**) of a soil is the water content expressed as a percentage of the dry weight at which the soil will start to flow when subjected to a shaking test. The **plastic limit** (**PL**) of a soil is the moisture content percentage at which the soil begins to crumble when it is rolled into a thread ⅛ in. (3 mm) in diameter. The **plasticity index** (**PI**) is the difference between the liquid limit and the plastic limit and indicates the range in moisture content over which the soil will remain in a plastic condition. The **shrinkage limit** (**SL**) is the water content at which soil volume is at its minimum.

Test methods, practices, and guides for classifying soils and preparing test specimens are detailed in the American Society of Testing and Materials publication *ASTM Standards on Soil Stabilization with Admixtures*.

Figure 5.1 Soil becomes more fluid as its water content increases.

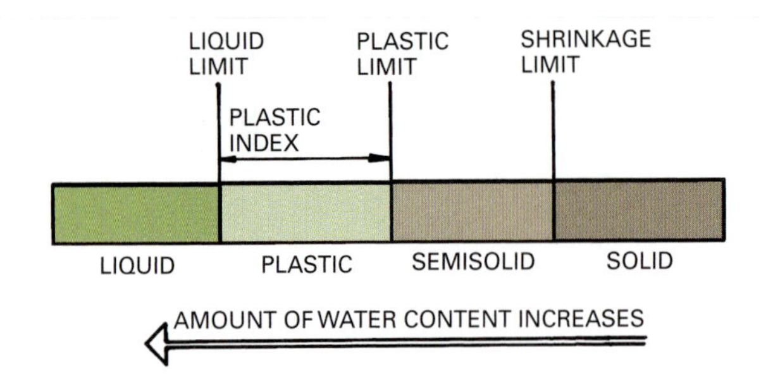

Unified Soil Classification System

The Unified Soil Classification System was developed by the U.S. Army Corps of Engineers to classify soils for use in roads, embankments, and foundations. Soils are classified according to the percentage of grain-size particles in each of the established soil grain sizes in the soil distribution index, the plasticity index, the liquid limit, and the organic-matter content. Soils are grouped in fifteen classes. Eight of these are coarse-grained soils that are identified by the letter symbols GW, GP, GM, GC, SW, SP, SM, and SC. Six classes are fine-grained and identified as ML, CL, OL, MH, CH, and OH. The one class of highly organic soils is identified as Pt. Soils on the borderline between two classes are given a dual classification, such as GP-GW. Descriptive details are given in Table 5.1.

American Association of State Highway and Transportation Officials System

The AASHTO soil classification system classifies soils according to those properties that affect their use in highway construction and maintenance. In this system the mineral soil is classified in one of seven basic groups, ranging from A-1 through A-7, on the basis of grain-size distribution, liquid limit, and plasticity index. Soils in the A-1 through A-3 groups are sands and gravels. They have a low content of fines (very small particles). Soils in the A-4 through A-7 groups are silts and clays and are therefore fine-grained soils. A special grade, A-8, is reserved for highly organic soils.

Table 5.1 Classification of Soils Using the Unified Classification System

Type	Letter Symbol	Description	Rating as Subgrade Material	Rating as surfacing Material
Gravel and gravelly soils	GW	Well-graded gravel; gravel-sand mixture; little or no fines	Excellent	Good
	GP	Poorly graded gravel; gravel-sand mixture; little or no fines	Good	Poor
	GM	Gravel with silt; gravel-sand-silt mixtures	Good	Fair
	GC	Clayey gravels; gravelly sands; little or no fines	Good	Excellent
Sand and sandy soils	SW	Well-graded sands; gravelly sands; little or no fines	Good	Good
	SP	Poorly graded sands; gravelly sands; little or no fines	Fair	Poor
	SM	Silty sands; sand-silt mixtures	Fair	Fair
	SC	Clayey sands; sand-clay mixtures	Fair	Excellent
	ML	Inorganic salts; fine sands; rock flour; silty and clayey fines	Fair	Poor sands with slight plasticity
Silts and clays with liquid limit greater than 50[a]	CL	Inorganic clays of low to medium plasticity; gravelly clays; silty clays; lean clays	Fair	Fair
	OL	Organic silts of low plasticity	Poor	Poor
Silts and clays with liquid limit less than 50[a]	MH	Inorganic silts; micaceous or diatomaceous fine sandy or silty soils; elastic silts	Poor	Poor
	CH	Inorganic clays of high plasticity	Very poor	Poor
	OH	Organic clays of medium to high plasticity; organic silts	Very poor	Poor
Highly organic soils	Pt	Peat and other highly organic soils	Unsuited for subgrade surfacing	Unsuited for material

[a]The liquid limit is the water content, expressed as a percentage of the weight of the oven-dried soil, at the boundary between the liquid and plastic states of the soil.
Courtesy U.S. Department of Interior, Bureau of Reclamation.

Table 5.2 The AASHTO System of Soil Classification

	A-1		A-2				A-3	A-4	A-5	A-6	A-7
	Sand and Gravel		Gravel or Silty or Clayey Sand								
Typical Material	**A-1-a**	**A-1-b**	**A-2-4**	**A-2-5**	**A-2-6**	**A-2-7**	**Fine Sand**	**Silt**	**Silt**	**Clay**	**Clay**
No. 10 sieve	50% max.										
No. 40 sieve	30% max.	50% max.					51% min.				
No. 200 sieve	15% max.	25% max.	35% max.	35% max.	35% max.	35% min.	10% max.	36% min.	36% min.	36% min.	36% min.
Fraction passing No. 40 sieve											
Liquid limit	6% max.	6% max.	40% max.	41% min.	40% max.	41% min.	—	40% max.	41% min.	40% max.	41% min.
Plasticity index	—	—	10% max.	10% max.	11% min.	11% min.	—	10% max.	10% max.	11% min.	11% min.

Source: Standard Specifications for Transportation Materials and Methods of Sampling and Testing, 2004, American Association of State Highway and Transportation Officials, Washington, D.C. Used by permission.

When laboratory tests are made on soil samples, the A-1, A-2, and A-7 groups can be further classified into groupings such as A-1-a, A-1-b, A-2-4, A-2-5. Details are shown in Table 5.2.

FIELD CLASSIFICATION OF SOIL

When it is not possible to have soil samples tested in a laboratory, a number of field tests can be made. Field testing involves separating soil particles according to size by screening them through sieves of various sizes. Frequently used sieve sizes are shown in Table 5.3. Following are generalized descriptions of field tests made by a soils engineer. More detailed information can be obtained from the U.S. Department of Interior, Bureau of Reclamation.

Table 5.4 shows the division of soil into various fractions based on the size of the soil particles. Although this is important when considering the properties of a soil, other properties, such as the shape of the particles (angular or sharp distinct edges versus subangular or rounded edges), also influence the ability of a soil to interlock particles.

Testing for Soil Coarseness

To test for coarseness, all soil particles larger than 3 in. in diameter are removed from the sample. The remaining particles are separated through a No. 200 sieve, which permits the smallest particles that can be seen by the naked eye to pass through. If more than 50 percent of the sample by weight does not pass through the sieve, the sample is a coarse-grained soil. The coarse sample is then divided into particles larger and smaller than $\frac{1}{4}$ in. (6 mm). If more than 50 percent of this sample by

Table 5.3 Some Frequently Used Standard Sieves

U.S. Standard	Opening Size	
Sieve Sizes	**Inch**	**Millimeter**
3 in.	3	76.2
1½ in.	1.50	38.1
¾ in.	0.75	19.0
No. 4	0.186	4.76
No. 10	0.078	2.00
No. 40	0.017	0.425
No. 100	0.006	0.150
No. 200	0.003	0.075

Table 5.4 Soil Size Fractions

	U.S. Standard
Constituent	**Sieve No.**
Cobbles	Above 3 in.
Gravel	
Coarse	3–¾ in.
Fine	¾ in.–No. 4
Sand	
Coarse	No. 4–No. 10
Medium	No. 10–No. 40
Fine	No. 40–No. 200
Silts and clays	Below No. 200

weight is larger than $\frac{1}{4}$ in. (6 mm), the sample is classified as gravel. If more than 50 percent by weight is smaller than $\frac{1}{4}$ in. (6 mm), the sample is classified as sand.

Dry Strength Test

A field test for dry strength, which is an indication of plasticity, begins with wetting a soil sample until its consistency approaches a stiff putty and molding it into a ball approximately 1 in. (25 mm) in diameter. After it dries, the sample is held it between the thumb and forefinger of both hands and squeezed. If it does not break, the soil is highly plastic and characteristic of clays. If the sample breaks but it is difficult to cause the sections to powder when rubbed between the fingers, it has a medium plasticity. If the sample easily breaks down into a powder, it has low plasticity.

Toughness Test

A toughness test ascertains the consistency of a soil sample near the plastic limit. All soil particles larger than the opening in a No. 40 (about $\frac{1}{64}$ in. or 0.4 mm) sieve are removed. A soil specimen in the shape of a $\frac{1}{2}$ in. (12.5 mm) cube is molded to the consistency of putty. Water may be added if necessary to get this consistency. If the sample is too sticky, the soil must be spread out in a thin layer to allow some moisture to evaporate. When the proper consistency is achieved, the soil is rolled into a thread about $\frac{1}{8}$ in. (3 mm) in diameter. The thread is folded and re-rolled repeatedly until it loses its plasticity and crumbles. When the soil crumbles, it has reached its plastic limit. The crumbled pieces are lumped together and kneaded until they crumble again. The tougher the thread is near the plastic limit and the stiffer the lump is when it crumbles, the more colloidal clay there is in the soil. Weakness of the thread at the plastic limit and easy crumbling of the lump indicates the presence of either inorganic clay with low plasticity or organic clay.

Shaking Test

The shaking test identifies the character of fines in a soil. After removing particles larger than No. 40 (about 1⁄64 in. or 0.4 mm) sieve size, a soil sample is formed into a ball about 3⁄4 in. (19 mm) in diameter. Enough water is added to make the sample soft but not sticky. The ball is placed in the open palm and shaken horizontally so it strikes against the palm of the other hand. The appearance of a glossy surface caused by water constitutes a positive reaction. As the sample is further shaken, the ball stiffens as the water gloss disappears. The ball will finally crack or crumble. Very fine sands give the quickest reaction. Inorganic silts give a moderately quick reaction, and clays give no reaction.

SOIL VOLUME CHARACTERISTICS

The volume of soil changes as it is excavated, hauled, placed, and compacted. When soil is excavated and transported, the grains loosen as air fills voids between them, and its volume is increased. Known as swell, this is measured as a percentage of the original volume and varies with different types of soil.

When soil is placed and compacted, the air between the grains is forced out and the soil's volume decreases. This is called shrinkage. The amount of shrinkage varies with the type of soil and is measured as a percentage of the original volume.

The volume of one cubic yard of soil in situ (undisturbed soil) is referred to as **bank measure**. When disturbed it will produce more than one cubic yard and when compacted, less than one cubic yard of soil. The weight and volume changes for typical types of soil are shown in Table 5.5.

Soil volume also changes due to fluctuations in moisture content. Clays are especially susceptible to changes in moisture. As moisture content increases, the soil expands, sometimes to the point of lifting a footing or slab. If the moisture decreases, the soil will shrink, sometimes causing a footing or slab to settle (Fig. 5.2).

SOIL WATER

The two major types of water present in soils are capillary water and gravitational water. Capillary water is retained in the microscopic pores of individual sand particles and enables them to bind together. Gravitational water flows within soil in the voids between particles. The water table, or **groundwater** level, is the level of the water below the ground surface. Since it is free flowing, water levels below the soil surface can be varied by pumping water from wells, by dry or rainy seasons, or by installing drain pipes to remove water from the construction area.

The moisture content of soil is an important factor during soil compaction. The density (p/cf or kg/m^3) varies with moisture content. A moisture content in optimal amounts enables soil to be compacted to its maximum density. This is called the optimum moisture content and is determined by laboratory testing of soil samples.

Another factor to consider when examining soil on a site is its permeability. Permeability measures the ability

Table 5.5 Typical Weights, Swell, and Shrinkage for Selected Types of Soils

	Weight[a]							
	Loose		Bank Measure		Compacted			
	lb./yd.^{3}	kg/m^3	lb./yd.^{3}	kg/m^3	lb./yd.^{3}	kg/m^3	Swell	Shrinkage
Clay, natural	2300	1300	2900	1720	3750	2220	38%	20%
Common earth, dry	2100	1245	2600	1540	3250	1925	24%	10%
Sand with gravel, dry	2900	1720	3200	1900	3650	2160	12%	12%
Sand, dry	2400	1400	2700	1600	2665	1580	15%	12%

[a]*Weights, swell, and shrinkage of actual samples may be larger or smaller.*

Figure 5.2 Expansive clay soils will heave when wet and settle when excessively dry.

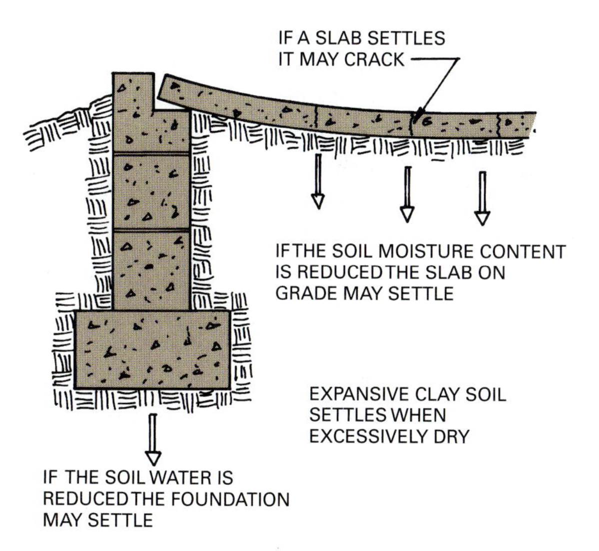

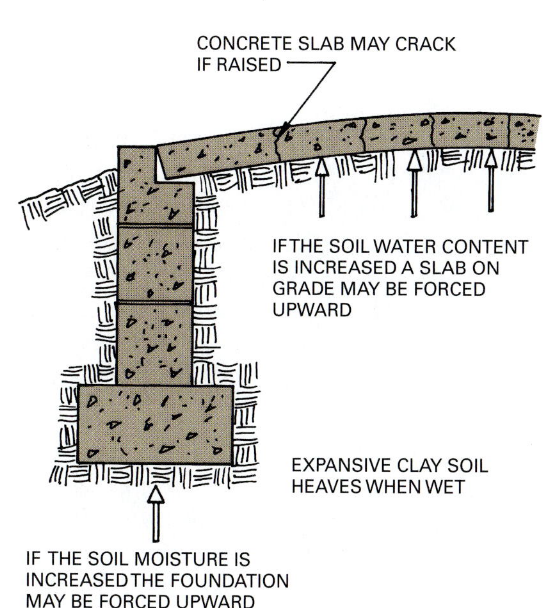

of water to penetrate through the soil. For example, clay has low permeability, whereas sand has a high permeability rating. Soil permeability is measured by testing samples with a permeameter.

SOIL-BEARING CAPACITIES

The required load-bearing capacities of the various types of soils are specified by local building codes. Table 5.6 shows the maximum allowable load-bearing values for supporting soils under spread footings placed at or near the surface of a site. In many cases it is necessary to test soil samples to determine the maximum allowable pressure that a soil can support. These surface values must be adjusted for deep footings and load bearing strata under piles as specified by the building code.

If the soil in the area of a structure is uniform in composition and the footing design is adequate, the building will remain stable. While some settlement may occur over time, it will be uniform. If the soil composition varies and is not compensated for in foundation design, the building could experience differential settlement. Other actions can also affect the stability of supporting soils. Extensive pumping of subsurface water, for example, can cause soil to settle because of a lowering of the water table.

Table 5.6 Presumptive Load-Bearing Values of Foundation Materials

	Load-Bearing Pressure	
Class of Material	lb./ft.^{2}	kg/m^2
Crystalline bedrock	12,000	58,560
Sedimentary rock	6,000	29,280
Sandy gravel or gravel	5,000	24,400
Sand, silty sand, clayey sand, silty gravel, and clayey gravel	3,000	14,640
Clay, sandy clay, silty clay, and clayey silt	2,000	9,760

Source: The BOCA National Building Code/1993.
Courtesy Building Officials and Code Administrators International, Inc.

Case Study

Chicago Center for Green Technology

Architect: Farr Associates Architecture and Urban Design	Building type: Educational Center, Commercial office	Completed 2003
Location: Chicago, Illinois	Size: 40,000 sq. feet (3,720 sq. meters)	Rating: U.S. Green Building Council LEED Platinum (37 points)

Overview

The Chicago Center for Green Technologies (CCGT) building is open for visitors to gain firsthand knowledge of how green buildings contribute to the good of the environment. The Center provides green technology educational resources to the public, including green building standards and construction guides, reference books, samples of green building materials, and staff and volunteers to guide the research process. The building also houses a recycling center, educating occupants and residents in everyday recycling strategies.

The structure is a renovation of an existing 1952 office building that was previously owned by a construction materials recycling company. The business was closed after it was discovered that the 17-acre site contained 600,000 cubic yards of illegally dumped debris stacked up to seven stories high. The Chicago Department of the Environment purchased and cleaned the site, recycling the discarded stone and concrete into public infrastructure projects throughout the city. The property was thoroughly investigated for toxic or hazardous wastes, and soil remediation was conducted.

Site Design

The landscaping components of the center were designed to showcase sustainable site-design technologies. A thorough sedimentation and storm water control system has been implemented. A constructed wetland and system of bioswales slows the volume of rainwater so that pollutants can settle out of the water before it reaches the ground or sewers. Native drought-resistant landscaping serves to filter and clean a majority of runoff directly on the site. All walks and drives were constructed using a system of permeable pavings. The paving matrix prevents contamination of nearby water systems by limiting urban storm water runoff. Rather than allowing storm water to flow over impervious surfaces, picking up contaminants along the way, all storm water is retained on site and slowly percolated back into the ground.

A large part of the roof at the Center for Green Technology consists of a green roof system utilizing sedum-based plantings that reduces the cooling load of the building while simultaneously protecting the roof's waterproof membrane. The planting species of the green roof absorb large portions of rainwater. The remaining storm water is collected in a series of cisterns and reused in landscape irrigation. This recycling of storm water for landscaping also helps reduce the center's water usage.

(Courtesy Eva Kultermann)

(Continued)

Energy and Materials

Environmental features of the building itself include: solar panels, recycled and healthy building materials, smart lighting, photovoltaic electricity, and a geothermal exchange system. The center has been found to use 40% less energy than a comparable code-compliant building of the same size. Asbestos existing in the building was removed and all interior materials were chosen for low toxicity content.

SOIL ALTERATIONS

After soil tests have been made and the soil's properties determined, the soil can be modified to improve its load-bearing characteristics. Commonly used processes to alter soil properties include compaction, stabilization, and dewatering.

Compaction

Compaction of soil refers to increasing its density by mechanically forcing the soil particles closer together. This expels the air present in the voids between particles. Soil density can also be increased by consolidation, which involves forcing the particles closer together through the removal of water from the voids between soil particles. Compaction produces immediate results. Consolidation requires longer time periods to increase the soil density.

The amount of compaction that can be obtained depends on the physical and chemical properties of the soil, its moisture content, the compaction method used, and the thickness of the soil layer being compacted. As compaction occurs, the air and water between particles is forced out of the soil and must be drained away. The soil engineer will use the results of soil tests to decide the best way to achieve the required compaction density.

There are many ways compaction can occur. These include a kneading action, vibration, static weight, explosives, and impact. Pneumatic-tired rollers are machines with multiple tires that provide compaction of the soil through a kneading action. The rows of tires are staggered to give wide coverage, and some units have wheels mounted to give a wobbly effect that increases the kneading action. The weight of pneumatic-tired rollers can be varied by adding ballast. They are effective on most soils but least effective on sands and gravels.

Tamping foot rollers use static weight to compact soil (Fig. 5.3). Tamping foot rollers are available in a range of foot sizes and shapes. As they pass over the soil the feet sink into it, compacting material below the surface. With repeated passes the feet do not penetrate as deep and eventually walk on the top surface. Tamping rollers are most effective on cohesive soils.

Another type of static-weight compactor has smooth steel wheels or drums (Fig. 5.4). It is best used to compact granular bases, asphalt bases, and asphalt pavements. It is not effective on cohesive soils because it tends to compact the surface, forming a crust over a loose, non-compacted subsurface.

A variety of compaction devices use vibration. Small hand-operated vibratory compactors are useful in

Figure 5.3 This compactor uses sheep's foot rollers to produce a tamping action.

Figure 5.4 Smooth wheel rollers are used to compact granular materials and asphalt pavements.

compacting areas where it may be difficult to fit larger equipment. Larger vibratory compactors include tamping foot rollers and smooth drum rollers. The vibratory action provides compaction in addition to static weight.

Impact rammers are used to provide compaction in very tight areas. The vertical movement of the rammer provides considerable compaction. Units are also available for mounting on the end of a backhoe boom.

Loose, saturated, granular soil can be compacted by subjecting it to a sudden shock and vibration. This causes the soil particles to fall into a denser pattern, displacing water from between them. The weight of the particles forces the water to flow from the soil. Explosives are typically used for this purpose. The spacing, depths, and sizes of the explosive charges are determined by experienced soil engineers.

Another way to increase the density of cohesionless soils is to use vibro-compaction. This employs a vibratory probe, a large-diameter tube with a vibrating device mounted on one end. The probe is lifted by a crane and driven into the soil by the weight of the tube and the vibrations. Figure 5.5 shows this process using a patented compacting probe. The probe is inserted in the soil by a powerful vibrator. The frequency and duration of vibration can be monitored to achieve maximum compaction for the prevailing soil conditions. The degree of compaction depends on the soil type, spacing of probes, frequency, and the duration of vibration.

Compaction of soil can also be achieved through a process called vibroflotation. As shown in Fig. 5.6, a probe with a powerful cylindrical vibrator penetrates the soil via vibration and is assisted by compressed air or water jets. When the probe reaches the desired depth, the space created by the vibroflot is filled with gravel by the simultaneous introduction of the material and withdrawal of the probe. Vibration during withdrawal ensures the efficient compaction of the added material and the neighboring soil.

Soil Stabilization with Admixtures

Test methods and specifications on soil stabilization with admixtures are detailed in the publication *ASTM Standards on Soil Stabilization with Admixtures*. Following are explanations of some of these. Soil remediation is discussed in Chapter 4.

Blending Soils Soils can be blended by mixing imported material with the original using a motor grader or disc. Power shovels or deep-cutting belt loaders can be used to cut deeply through the soils and provide consistent mixing.

Lime-Soil Stabilization Clays and silty clay soils can be stabilized through the addition of lime, which produces a chemical reaction. Clays expand and contract according to moisture content and can cause damage to the structures resting on them. A concrete slab, for instance, can heave and crack due to the expansion of clay. The addition of lime to clay reduces the amount of expansion and forms a moisture barrier that protects

Figure 5.5 Vibrocompaction consolidates soil by inserting a compaction probe with a powerful vibrator into the soil.

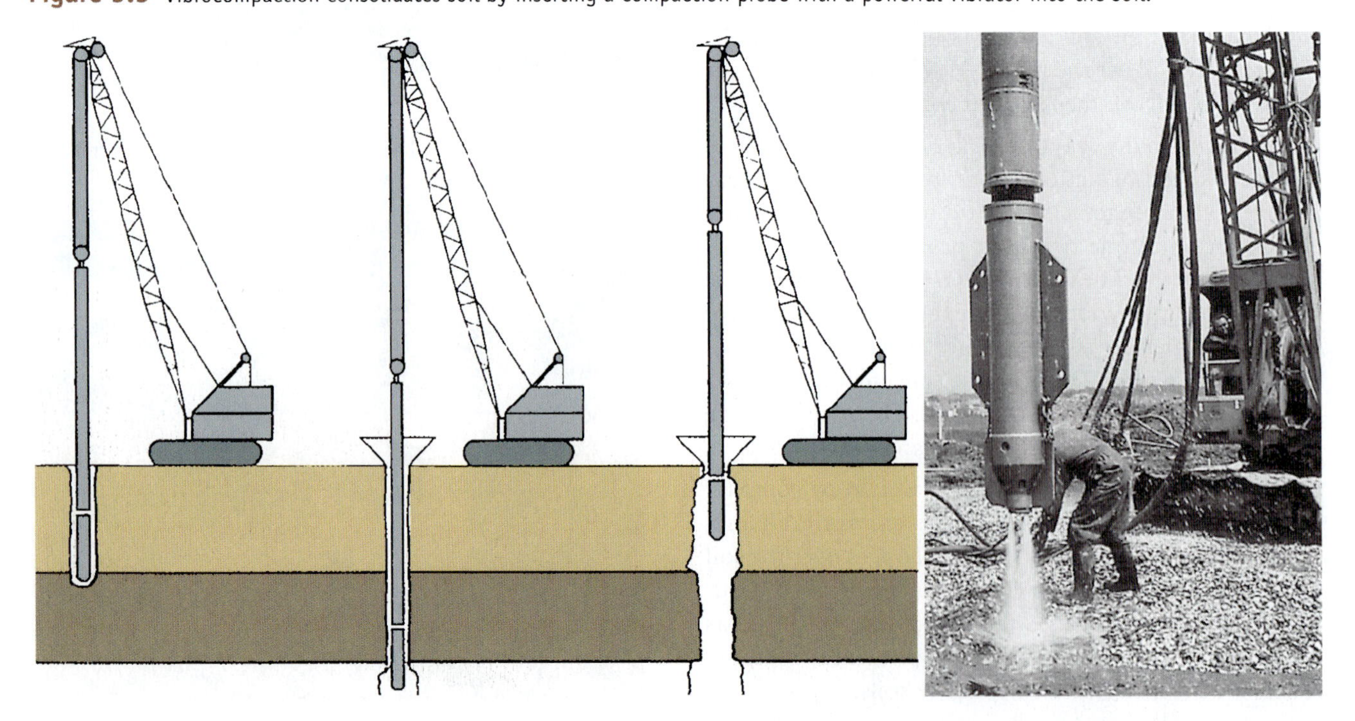

Figure 5.6 The vibroflotation compaction method inserts a vibrating probe into the soil. Penetration is assisted by jets of compressed air or water.

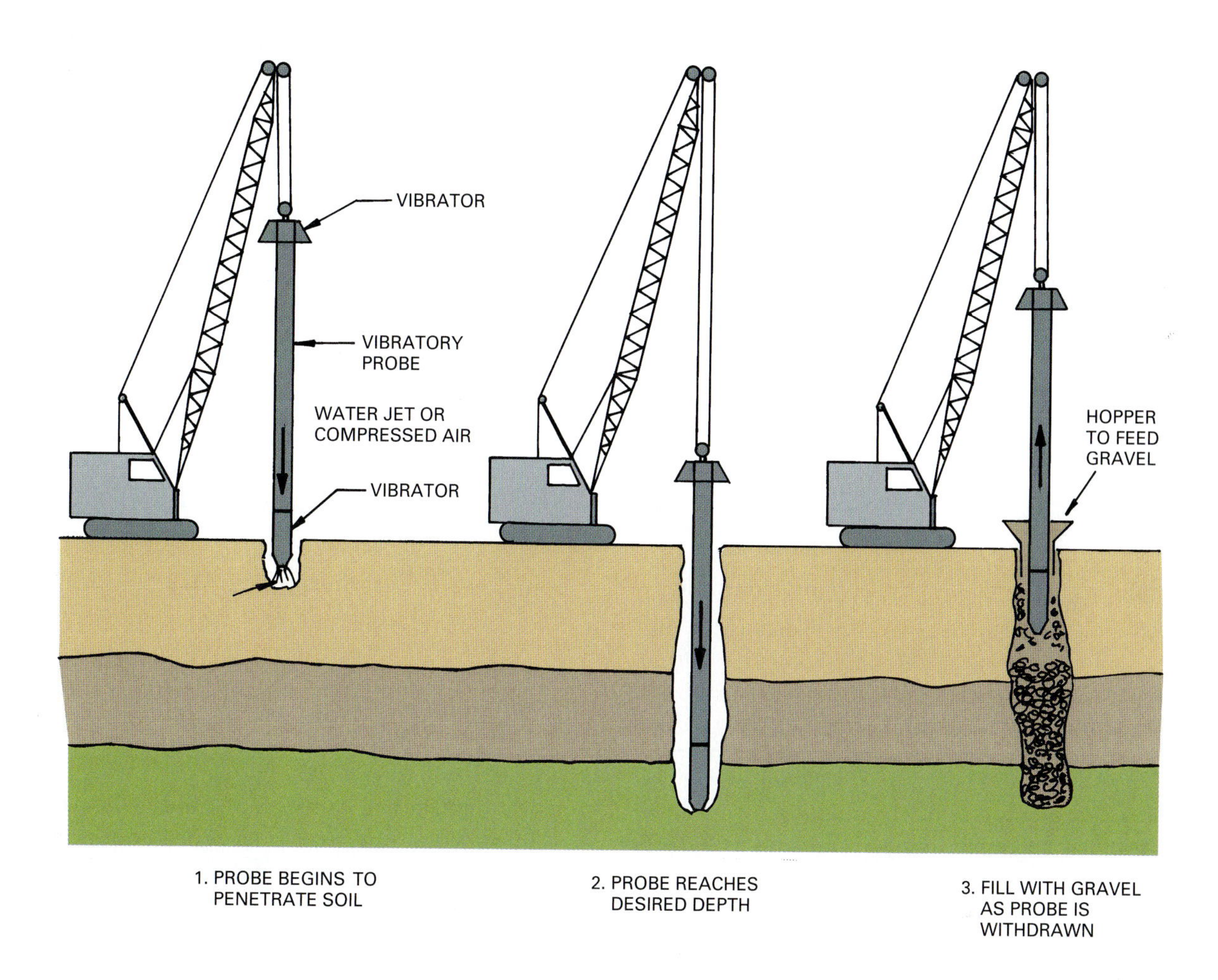

the expansive clay from subsurface water. In some cases, lime can eliminate the need to excavate and replace unsatisfactory soil. Typically, slaked, hydrated lime, or quicklime, is spread over the base material and blended into it with a pulverizing machine. Water may be added if conditions warrant. The blending layer is generally 12 to 18 in. (300 to 450 mm) deep and compacted to the required thickness. The compacted layer will continue to gain strength for many months.

Asphalt-Soil Stabilization Asphalts blended with granular soils produce a durable, stable soil that can be used as a finished surface for low traffic roads or a stabilized base course for higher-quality pavements. Some soils require the addition of fines with the asphalt to fill the spaces between the soil particles.

Cement-Soil Stabilization Predominantly granular soils that have minute amounts of clay particles can be stabilized by blending them with Portland cement. In some cases fly ash is used to replace some of the Portland cement. The cement is uniformly spread over the surface and blended with the soil using a pulverizing machine. After this has been accomplished to the specified depth, the surface is fine-graded and compacted. Water can be sprinkled on the surface during blending if the soil moisture content is low. Since Portland cement sets up quickly, compaction immediately after blending is required. Initial compaction can be accomplished with tamping or pneumatic rollers and followed by a steel smooth-wheel roller.

Salt-Soil Stabilization Coarse crushed rock salt can be used to stabilize well-graded clay or loamy soil containing some limestone fines. The salt can be blended in the soil in dry form or mixed with water to form brine. The base soil is pulverized to the specified depth, the salt added, and the mixture blended. This layer is then compacted, fine-graded, watered, and finished with a steel smooth roller. The salt and water form a bond with soil particles, creating a stable soil. It can take up to several weeks before the soil is cured.

Review Questions

1. How are soils classified?
2. What are the classes of soils?
3. How is the plasticity index of a soil determined?
4. What soil classification system is used by the U.S. Army Corps of Engineers?
5. What association has a soil classification system for use on highway construction?
6. Describe the field test for the dry strength of a soil.
7. What happens to the volume of a given amount of soil after it has been excavated?
8. What types of water are identified in soils?
9. Why are the load-bearing capacities of soil important?
10. In what ways can the properties of the soil be modified to improve its characteristics?
11. What admixtures are used to stabilize soil?

Key Terms

bank measure
clay
compaction
gravel
groundwater
liquid limit
organic matter
plastic limit (PL)
plasticity index (PI)
sand
shrinkage limit (SL)
silt

Activities

1. Have a representative of the state highway division visit the class and discuss local soils and their classifications.
2. Arrange for a visit to a local soils testing laboratory.
3. Secure soil testing sieves of various sizes and run some tests on local samples for coarseness.
4. Run dry strength and toughness field tests on soil samples from various sites.
5. Pack soil samples in a container of known volume, such as a quart or gallon. Ram each sample a pre-determined number of times. Then dump it in a loose condition into other containers and measure the increase in volume. Compute the percentage of swell for each sample. Although this is a crude test, it will possibly give some interesting results if the soil samples are chosen from a wide geographic area.
6. Visit a road construction site, observe the grading and compaction process, and report on the compaction and soil stabilization techniques being used.

Additional Resources

PCA Soil Primer, Portland Cement Association, Skokie, IL.

Standards on Soil Stabilization with Admixtures, American Society for Testing and Materials, Philadelphia, PA.

Standard Specifications for Transportation Materials and Methods of Sampling and Testing, American Association of State Highway and Transportation Officials, Washington, DC.

See Appendix B for addresses of professional and trade organizations and other sources of technical information.

CHAPTER 6

Foundations

LEARNING OBJECTIVES

Upon completion of this chapter, the student should be able to:

- Understand the factors to be considered when designing foundations.
- Identify various types of foundations.
- Discuss the features and construction of each type of foundation.

Buildings can be thought of as having three major parts: the superstructure, which is the portion of the building above grade; the substructure, the habitable portion below grade; and the foundation, which transmits the weight of the sub- and superstructure to adequate bearing below ground level. Foundations are classified as either shallow or deep **(Fig. 6.1)**. Shallow foundations, such as spread footings and slabs, transmit building loads directly beneath the substructure and close to existing grade. Deep foundations, such as caissons and piles, reach far below grade to access soils suitable for the required bearing. The choice of foundation systems is determined by the size and type of building to be constructed, the soil and subsurface water conditions, and building code requirements.

FOUNDATION DESIGN

The design of foundations begins as soon as the basic architectural concepts are developed. The architect is responsible for the design of the shape and size of the building, its orientation, space utilization, and the selection of materials. The structural engineer is involved with the structural frame of the building. The foundation engineer must work with both as foundation designs are developed. The final building is the result of the cooperative effort of these three design disciplines. Building codes require an investigation of subsoil conditions which may dictate the choice of foundation used. The size of the building, loads to be carried, conditions on the site, the climate, type of structural system to be used, presence of water, winds, potential earthquakes, and other such factors are all considered by the foundation engineer.

Foundation Loads

Foundations are designed to carry the loads imposed by the structure and any additional loads produced by occupancy. These loads are divided into dead loads and live loads. Dead loads are those, such as the weight of materials used to construct the building, that act continuously on the foundation. Live loads are those that are not constant, such as the weight of furniture, people, wind, or snow. Loads may be vertical or **lateral** (horizontal), **static** (fixed) or **dynamic** (moving), **concentrated** (on one spot) or **uniform** (spread evenly over a surface).

Foundation Types

Foundations are classified into three major types: spread, pile, and caisson foundations (Fig. 6.2). **Spread** or **mat foundations** transfer building loads to the soil through the footings at the bottom of a foundation wall or column. **Pile foundations** use long wood, concrete, or steel piles that are driven into the earth. They get their load-carrying capacity through frictional forces developed on the sides of the pile or from the end resting on load-bearing strata. **Caisson foundations** are formed by drilling holes in the earth and filling them with concrete and reinforcing steel. They get their load-carrying ability in the same manner as piles.

Figure 6.1 Deep foundations extend to rock or other satisfactory bearing surfaces. Light construction uses shallow foundations with footings just below the frost line of the earth.

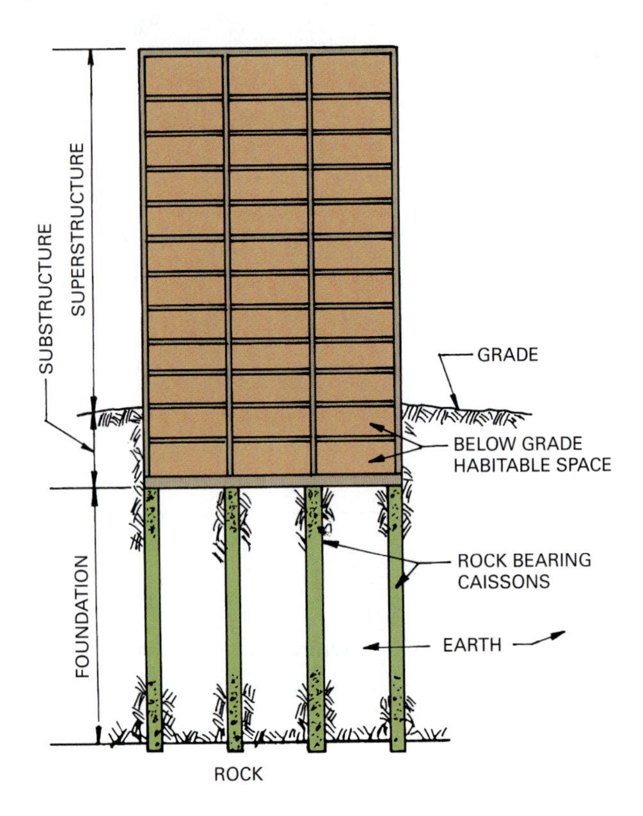

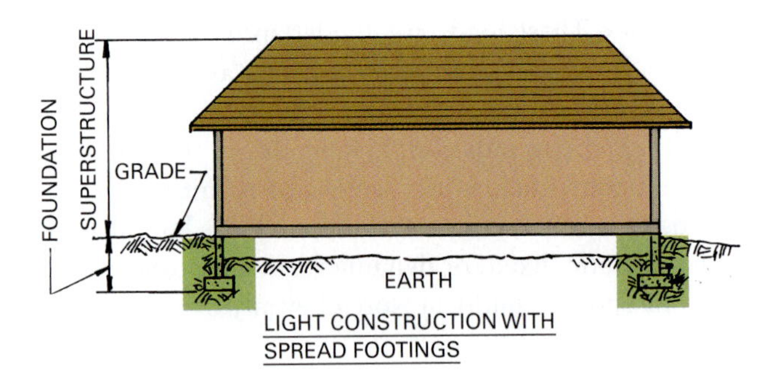

SHALLOW FOUNDATIONS

Shallow foundations use spread footings, slabs on grade, or mat foundations to distribute building loads over a large enough area of ground to not exceed the soil's bearing capacity. They are generally located not far below the surface of a site and are less costly than deep foundations. Following are the most commonly used shallow foundations.

Figure 6.2 Spread, pile, and caisson foundations.

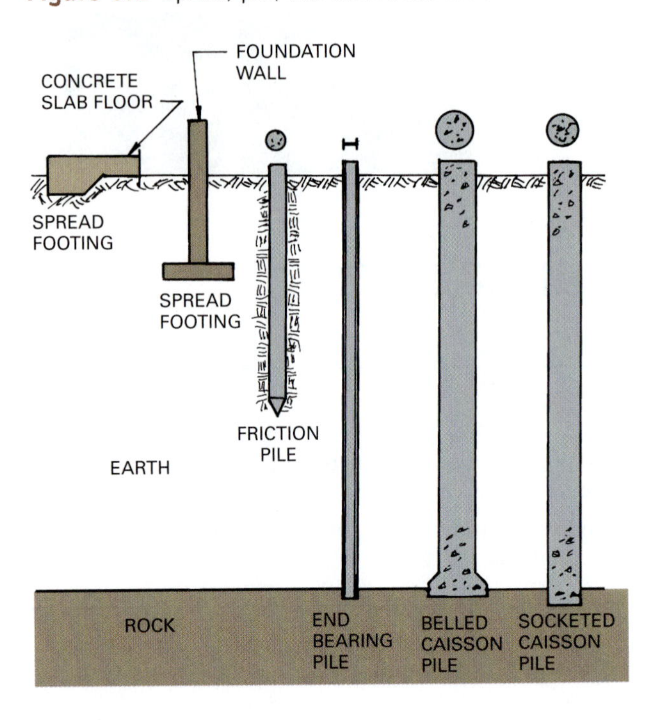

Spread Footings

The most common spread footing is a continuous wall or strip footing. This footing provides a stable base around the entire perimeter of a structure. Buildings with spread footings often incorporate interior **independent footings** to support point loads, such as columns or piers (Fig. 6.3). In both cases, the wider width of the footing base creates a large area to transfer the weight to the ground and prevent settlement. Both types require reinforcing with steel bars and must be placed on undisturbed soil. In some cases, a substrate of engineered soil is allowed if it is carefully designed to provide the required load-bearing capacity. Wall footings on sloped sites are usually staggered using horizontal steps (Fig. 6.4).

A **grade beam** is a reinforced concrete beam serving as the foundation wall (Fig. 6.5). Grade beam foundations support the building on a beam that is in turn supported on a series of piles or piers. Grade beam footings differ from continuous spread footings in how they distribute loads. The depth of a grade beam footing is designed to distribute loads to bearing points, while the width of a continuous spread footing is designed to transfer loads to the ground.

Slab and Raft Foundations

A widely used shallow foundation is a slab-on-grade, as shown in Fig. 6.6. There are many variations for this type of foundation. Slabs on grade are used widely in

Figure 6.3 Spread footings are used under foundation walls, piers, and columns.

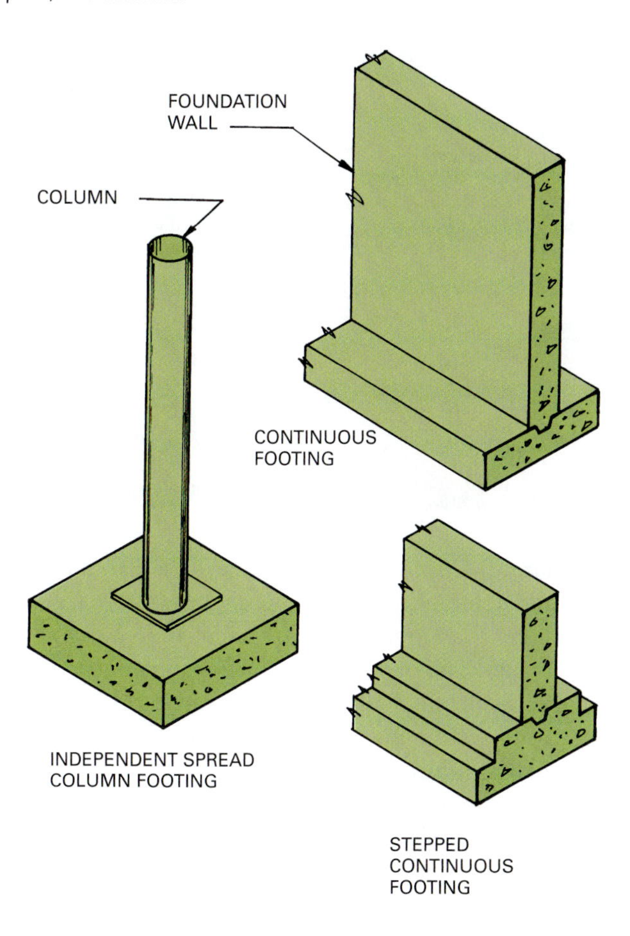

Figure 6.4 Typical stepped footings. *(Image copyright prism68, 2009. Used under license from Shutterstock.com)*

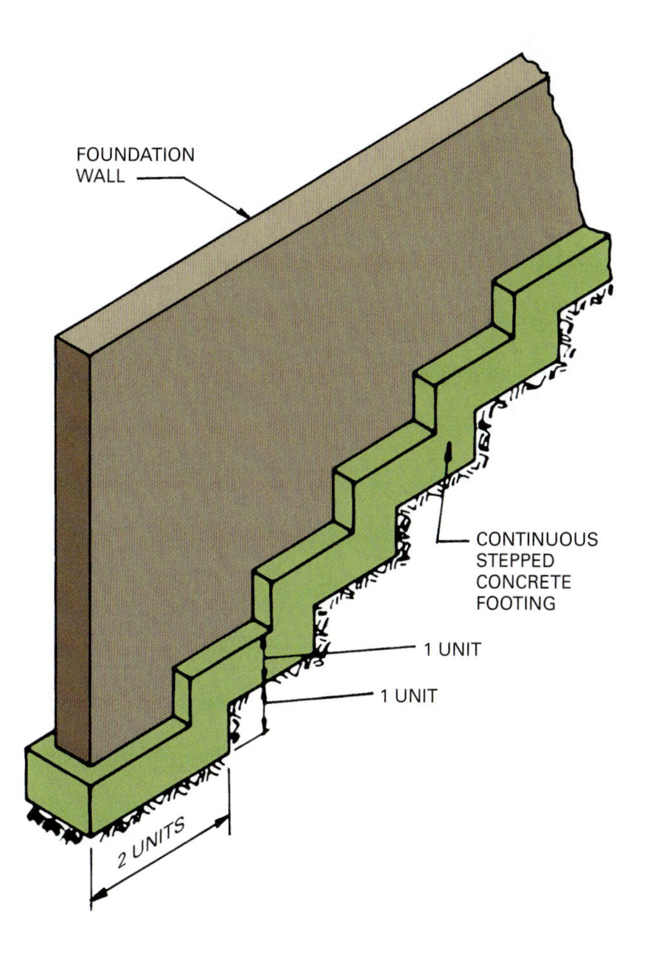

warm climates where the frost line is shallow and the surface soil has the necessary bearing capacity. The slab is provided with a thickened edge at its perimeter and at bearing points within the building to support interior load-bearing partitions and columns.

Raft or mat foundations are reinforced concrete slabs several feet in thickness that cover the entire footprint of a building (Fig. 6.7). The concrete mat spreads the weight of the building over the entire area below the building, thereby reducing the load per square foot on the soil. This is especially useful when the soil characteristics have low bearing capacity or vary across the excavated area. To avoid construction joints, the entire slab is poured monolithically and continuously. Another raft design uses a T-beam construction that has a solid poured slab and two-directional beams poured either above or below the slab (Fig. 6.8). When the beam is inverted, the slab becomes the basement floor. The beams' cavities are dug in the soil, which must be able to stand without caving in as the beams and slab are poured. When the beams are on top of the slab, the spaces between them must be filled before pouring a concrete floor.

Figure 6.5 Grade beams distribute loads to piles or piers.

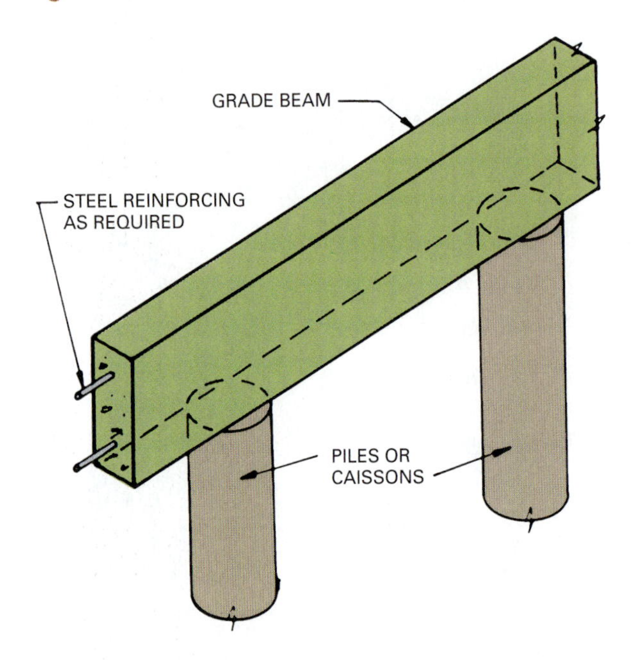

Figure 6.6 Slab on grade with thickened perimeter footing details.

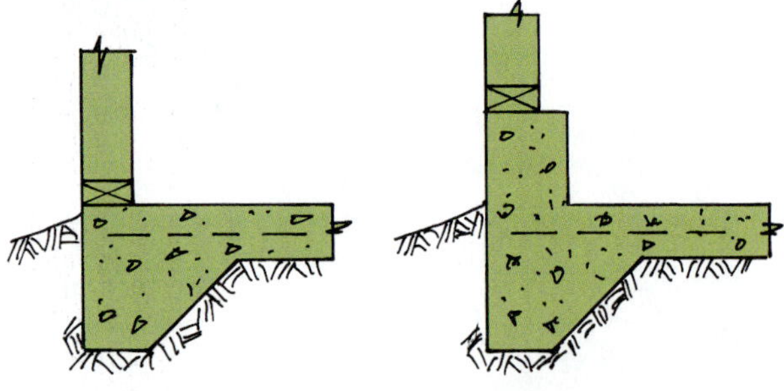

For heavy loading conditions, steel grillage footings incorporate steel beams to reinforce the footing (Fig. 6.9). A base layer of steel is sized to the area needed to support the load. The upper layer of beams is joined at right angles to the base level, and the entire construction is encased

Figure 6.7 A monolithically poured mat foundation. *(Courtesy Bill Bradley)*

in concrete. Combined footings are used to support walls and columns near the property line, where it is not possible for the footing to cross the line onto a neighboring property. The combined footing ties the outside row of columns to the next row within the building. If the outer row of columns were to be placed on the edge of individual footings, the loads would not be symmetrical, and the footings would tend to settle unevenly or rotate (Fig. 6.10). Cantilever footings are a form of combined footing that permits erection of wall columns near the edge of the footing. A beam rests on independent footings and cantilevers over the outer footing to support the column. The footings may be reinforced concrete or steel grillage.

Tieback Anchors

Tieback anchors are used to support sheeting for foundations. When used with sheeting, they reduce the need for bracing inside the excavation and are used to anchor foundations and slurry walls, as discussed in Chapter 4. **Tieback anchors** are either grouted or helical anchors. Grouted anchors are installed by drilling through the foundation or slurry wall into the soil and rock to the specified depth (Fig. 6.11). The steel tiebacks are inserted into the holes and the anchor zone area is filled with pressure-injected grout. After the grout hardens, the tiebacks are tensioned and secured to a steel chuck that maintains the tension. A typical construction detail is shown in Fig. 6.12.

PILE FOUNDATIONS

Pile foundations use long wood, steel, or concrete piles that are driven into the earth with a mechanical pile hammer. The pile hammer delivers a series of blows to the top of the pile, driving it to suitable bearing in the earth. Materials used for driven piles include timber,

Figure 6.8 Two types of T-beam construction.

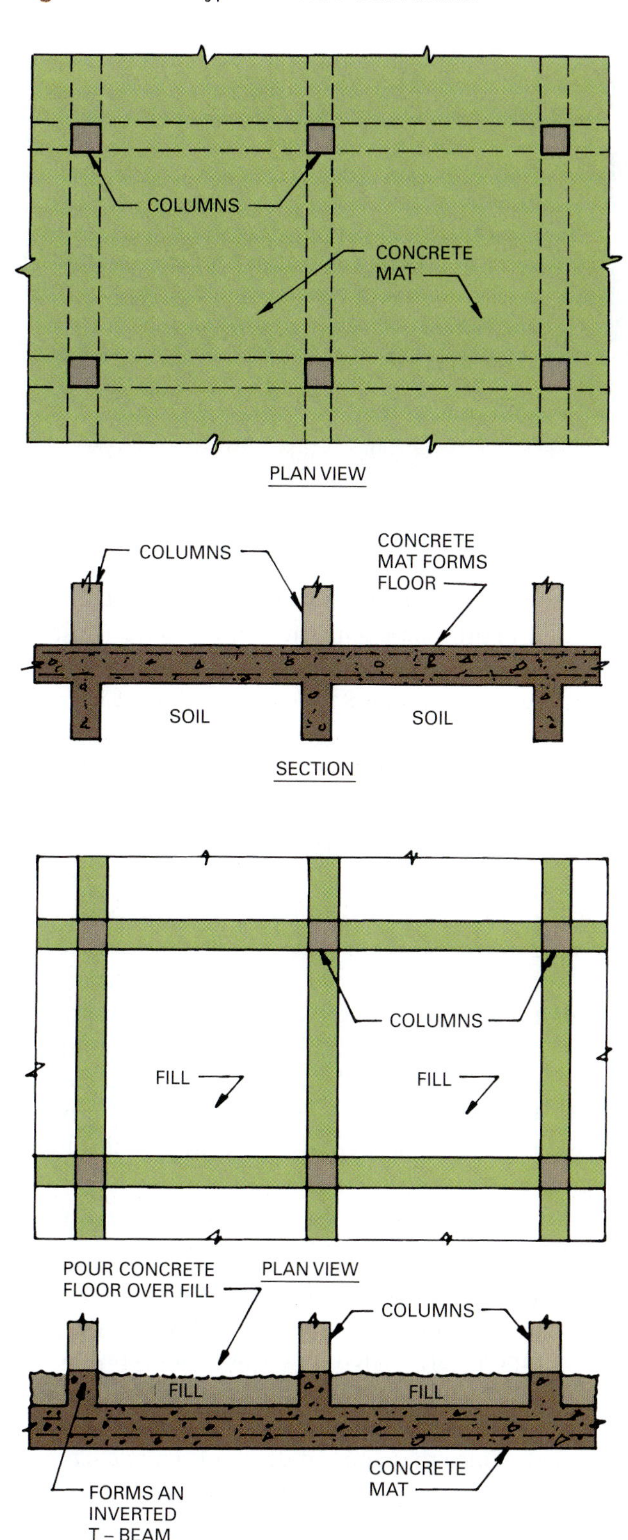

Figure 6.9 A steel grillage column footing.

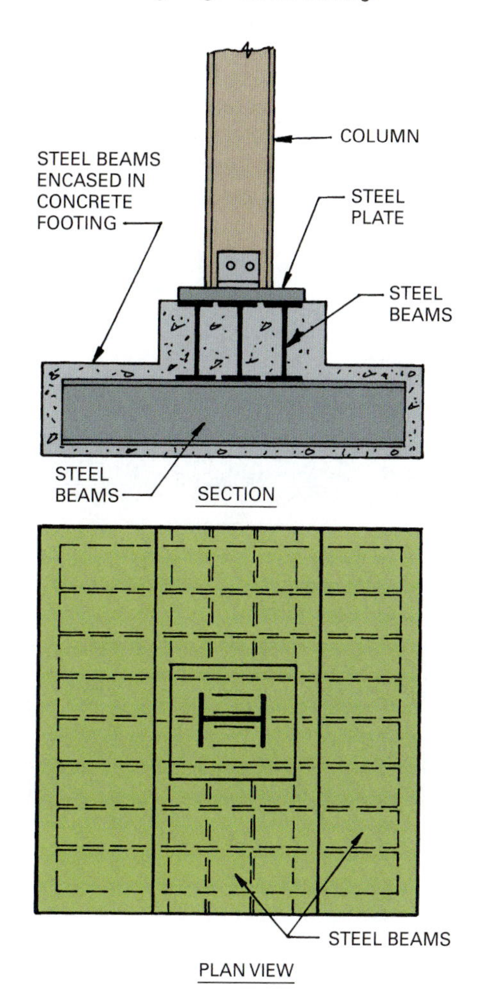

steel H-piles, pipe, precast concrete, and cast-in-place concrete with steel shells (Fig. 6.13).

Timber piles are tree trunks trimmed, stripped of bark, and treated with preservatives. They are driven with the small end down and may have a pointed metal shoe on the tip. The minimum tip is 6 in. (152 mm) in diameter. The top or butt end, typically with a minimum diameter of 12 in. (305 mm), is fitted with a metal pile ring to protect it during driving. Timber piles work best in soils relatively free of rock. Wood piles that reach rock or other load bearing strata transmit the load through the end and are called end-bearing piles. Those that get their load-bearing capacity from friction forces between the sides of the pile and the earth are called friction piles.

H-piles are HP structural steel shapes and are especially good for driving into soil containing rock or thin rock strata (Fig. 6.14). Since they have a small surface area, they cannot rely solely on friction but must be driven to end bearing on rock. They are available in widths from 8 to 14 in. (200 to 355 mm) and can be driven to depths of 150 to 200 ft. (45 to 61 m). These longer lengths are developed by driving a series of shorter sections and welding them to one another.

Pipe piles utilize heavy-gauge steel pipes that are driven with an open end. Small diameter piles generally have the end closed. As open end piles are driven, the earth inside is gradually removed to decrease resistance.

Figure 6.10 Footings close to a property line can be reinforced with a combined footing.

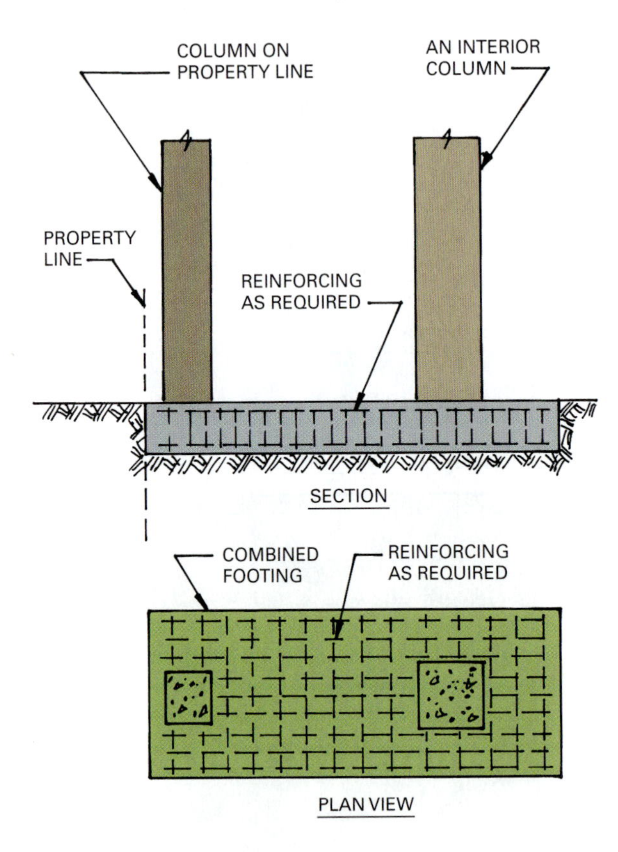

Figure 6.11 A typical detail showing the placement of tieback anchors.

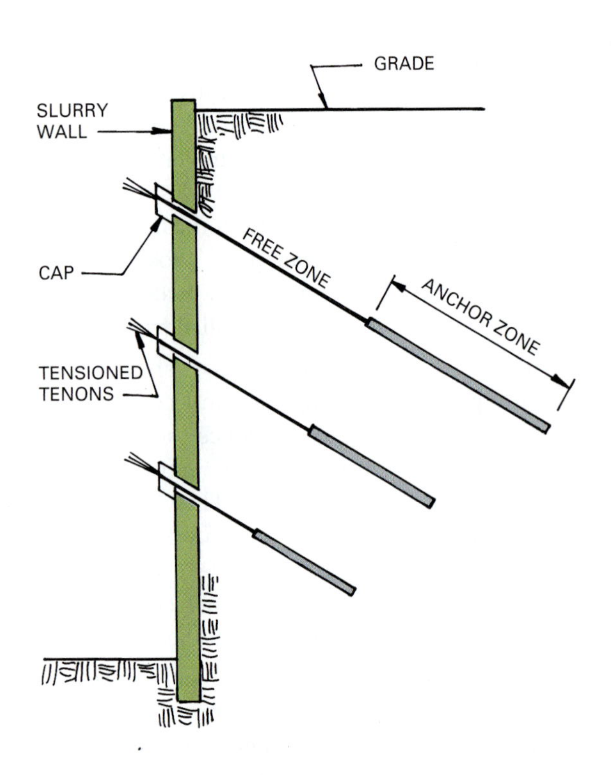

Figure 6.12 Helical tiebacks are screwed into the soil. *(Courtesy Midwest Construction)*

The soil can be removed by inserting a pipe and blowing it out or adding water to help loosen it. Once the required depth has been reached, concrete is placed inside the pipe, starting the fill from the bottom. Pipe diameters may be from 8 to 18 in. (203 to 457 mm) in width, with wall thicknesses from 1/4 to 1/2 in. (6 to 12 mm) (Fig. 6.15).

Precast and pre-tensioned concrete piles are made in a variety of ways. They may be cylindrical, square, or octagonal and are available either tapered or uniform in size for their entire length (Fig. 6.16). Although most are solid in cross section, some cylindrical types are hollow. The piles are reinforced with steel to withstand the installation stresses, as well as the building loads. Piles driven in clay soils have a blunt point, while a tapered point is used in sand and gravel. Precast concrete piles can be lengthened by welding together the reinforcing bars of two adjoining sections. Some are installed by jetting the excavation with water. A pipe is cast in the center of the pile and high pressure water is forced through it, blasting the soil as the pile penetrates it.

Cast-in-place piles can be cased with a metal shell, or they may be uncased without a shell. One type of pile uses a tapered steel shell driven with an internal steel mandrel. A mandrel is a steel core inserted into a hollow pile to reinforce the pile shell while it is being driven into the earth. Once the required depth of the pile has been reached, the mandrel is removed, and the shell is filled with concrete (Fig. 6.17). The shells come in 8 ft. lengths, and sections are added to increase the length as the pile is driven. The mandrel is used in areas with a high water table and where vibration may be problematic.

A shell-less uncased pile is formed by driving a steel shell with an interior driving core, removing the core, and filling the shell with concrete as the steel shell is removed from the earth. The Franki pressure-injected

Figure 6.13 Some commonly used piles.

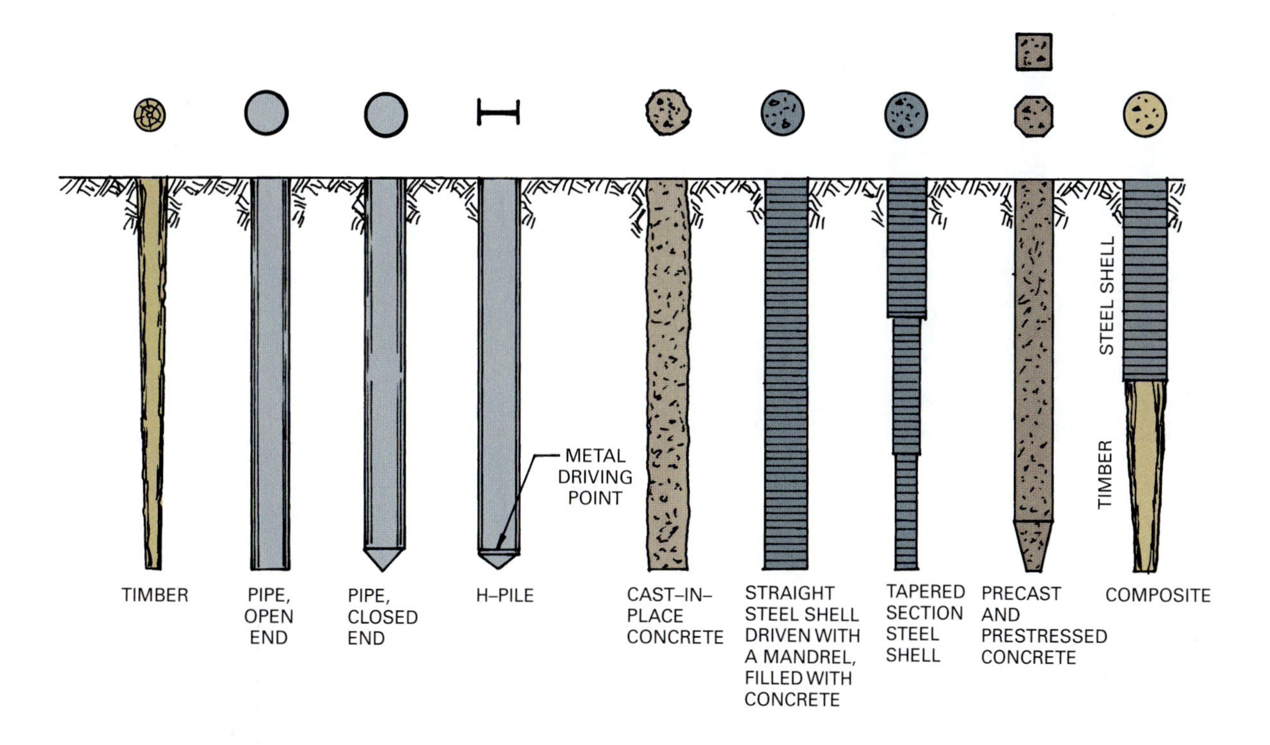

Figure 6.14 A pile driver driving an HP structural shape. *(Image copyright Albert H. Teich, 2009. Used under license from Shutterstock.com)*

Figure 6.15 Pipe piles of heavy gauge steel are driven with an open end. *(Courtesy Wikipedia, http://en.wikipedia.org/wiki/File:Driven_pipe_piles.jpg)*

footing construction procedure is shown in Fig. 6.18. It uses a steel casing into which zero-slump concrete is rammed to produce a pedestal footing.

A variety of composite piles are used to reduce costs or achieve other advantages. For example, a wood pile (low in cost) could be driven into the area below ground water and a concrete pile could be placed on top.

Structural foundation drawings will show where the piles are to be placed. Piles are usually grouped in clusters with a cast-in-place reinforced concrete pile cap.

Figure 6.16 Precast concrete piles. (*© David Sailors/CORBIS*)

Figure 6.17 Shells for cast-in-place concrete piles can be driven with a mandrel and filled with concrete after the required depth has been reached and the mandrel removed.

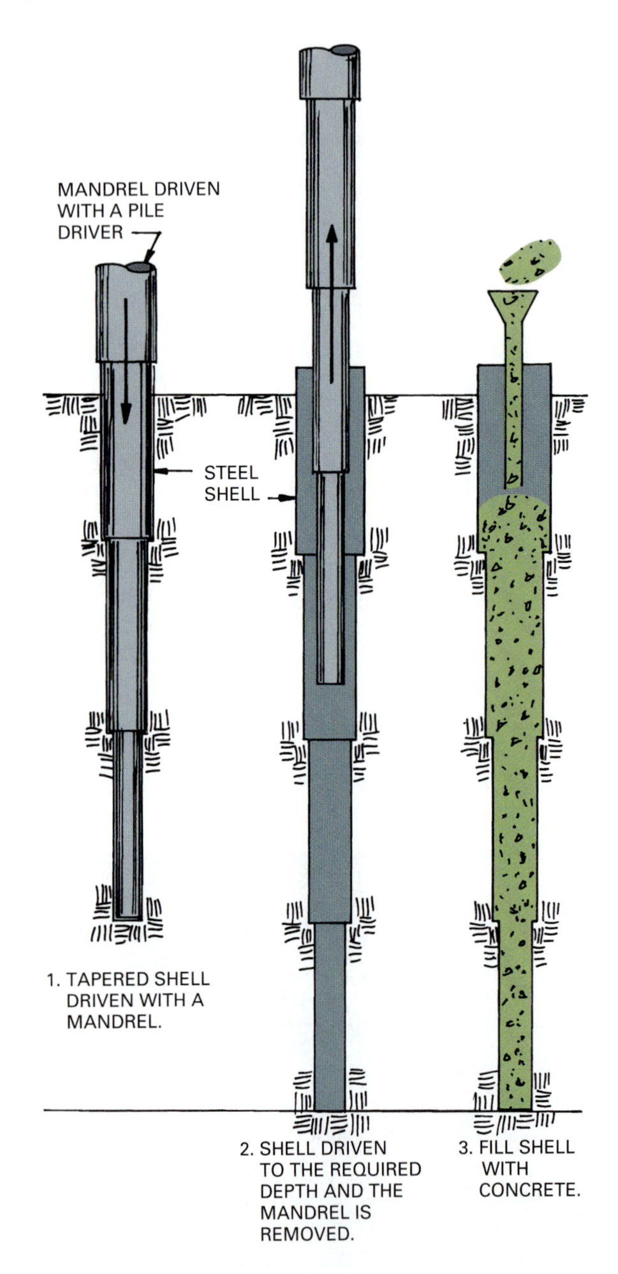

The cap distributes the load equally among the piles below the cap (Fig. 6.19). The footings can be supported by piles, or grade beams can be run over the pile caps to provide a foundation for construction of a wall. Mat foundations can also be supported on piles.

Individual pile caps and caissons are often interconnected by ties that carry tension and compression loads as specified by the building code.

Pile Hammers

Piles are driven into the earth with **pile hammers**. The most commonly used types include drop, single-acting steam, double-acting steam, diesel, vibratory, and hydraulic.

Drop hammers use a heavy weight lifted up vertical rails called leads (Fig. 6.20). When the weight reaches the top of the rail, it is released and falls onto the pile, driving it into the earth. Repeated cycles drive the pile to the required depth. A drop hammer can weigh from 500 to 3,000 lb. (226 to 1360 kg) and can be dropped from a height of 5 to 20 ft. (1.5 to 6.0 m), depending on the type of pile and soil conditions.

Single-acting steam hammers are mounted on top of the pile. The hammer (called a ram) is lifted by either steam or compressed air (Fig. 6.21). When the ram reaches the top of the stroke it is released and free falls onto the top of the pile. The steam raises the ram and the cycle repeats. Single-acting steam hammers are capable of forty or more blows per minute. Double-acting steam hammers operate in the same manner as single-acting hammers except that, on the down stroke, the steam pressure is applied to the backside of a piston, which increases the energy per blow. Hydraulic pile hammers operate much the same, but hydraulic fluid is used to move the ram.

Diesel pile hammers are self-contained driving units that include a vertical cylinder, a ram, an anvil, and the parts (fuel pump, injectors, fuel tank, etc.) that make up a diesel power unit. The ram is placed on top of the pile and released. As the ram falls, the fuel pump injects diesel fuel into the combustion chamber. As the ram continues to fall, the fuel is compressed and the heat generated causes an explosion. This drives the anvil hard against the pile and causes the ram to rise, thus starting a new cycle.

Vibratory pile hammers are held by a crane and mounted on the end of the pile (Fig. 6.22). A motor drives eccentric weights in opposing directions, producing vibrations that are transmitted to the pile. The vibrations agitate the soil, reducing the friction between the pile and the soil and allowing the pile to penetrate

Figure 6.18

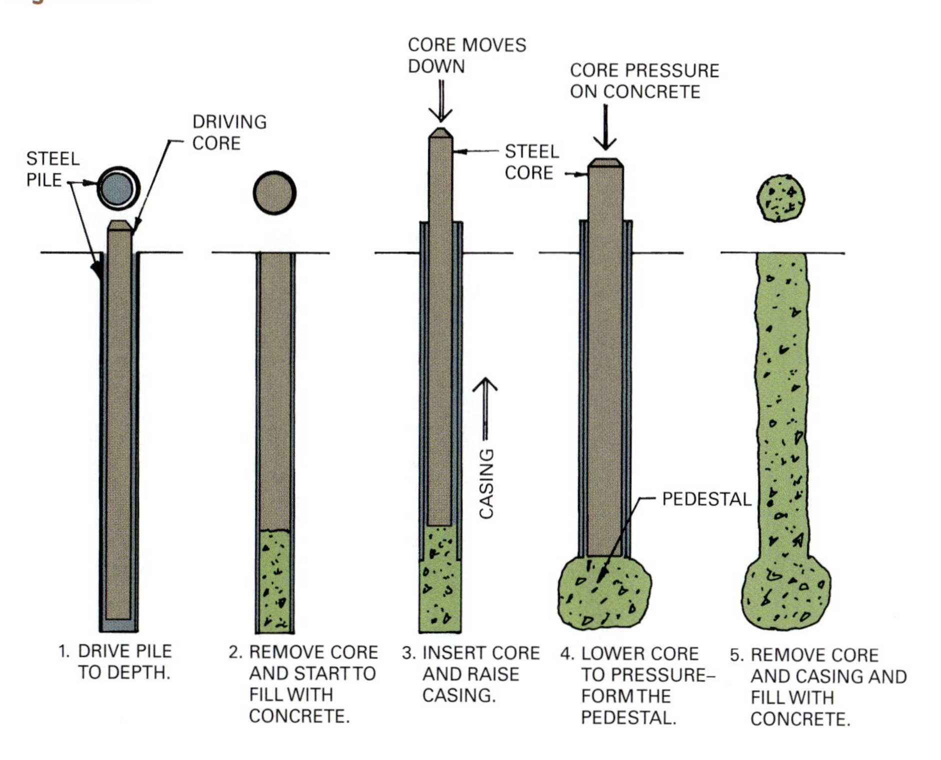

The Franki pressure-injected process produces a pile with a pedestal footing.

Figure 6.19 A pile cap supporting a column.

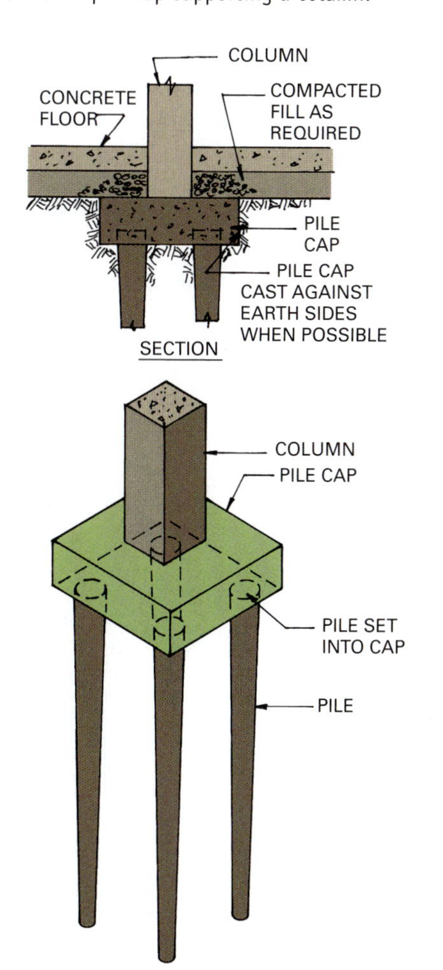

Figure 6.20 Piles are driven by pile hammers. *(Image copyright Richard Thornton, 2009. Used under license from Shutterstock.com)*

Figure 6.21 A diagram of how a steam pile hammer operates.

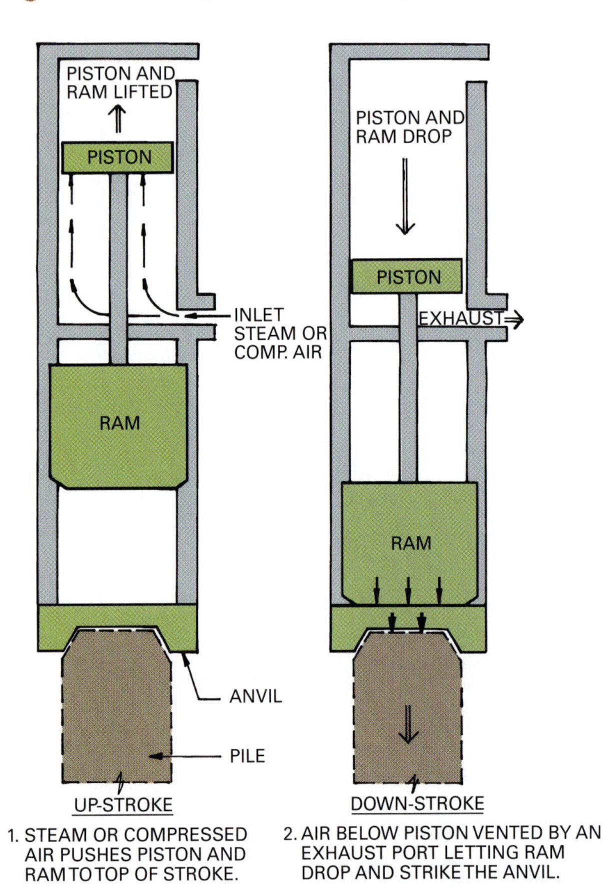

Figure 6.22 Vibratory pile hammers have a vibrating unit mounted at the end of the pile. *(Courtesy ZueJay)*

the soil. Vibratory pile hammers are especially effective for driving into water-saturated non-cohesive soils. They have difficulty in dry sand and will not penetrate tight, cohesive soils. They are also effective in driving sheet piling.

CAISSON FOUNDATIONS

As discussed in Chapter 4, a caisson is a watertight structure used to provide a dry working area for building foundations or structures below water. As previously mentioned, a *caisson foundation* is a pier that is drilled into the earth, filled with the required reinforcing steel, and poured with concrete (Fig. 6.23).

Caissons are end-bearing units drilled using large diameter augers. A steel casing is lowered into the hole as drilling continues. The soil must be able to

Figure 6.23 Caisson piles may rest on bedrock and utilize either belled or socketed ends.

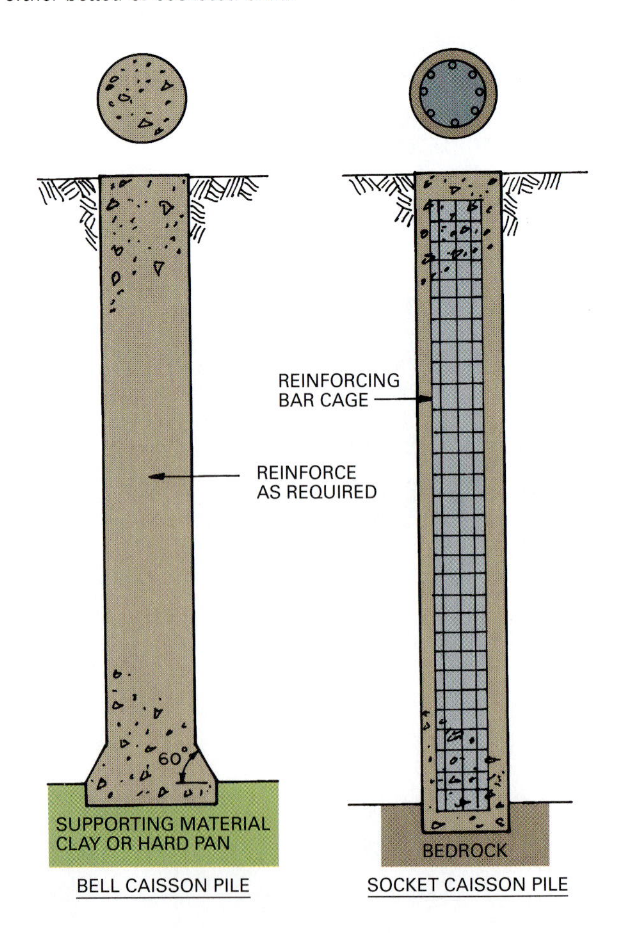

stand without collapsing when the casing is removed. Caissons are bored through the weaker surface soils until they strike rock or other acceptable bearing soil (Fig. 6.24). When rock is not encountered, the bottom is belled using a belling bucket on the end of the drill to provide a larger end-bearing surface. Shaft diameters can vary from 18 in. (460 mm) to 5 or 6 ft. (1.5 to 1.8 m) or more. Once the excavation is complete, the required reinforcing steel is lowered into the hole and the hole is filled with concrete. As it is filled, the steel casing is usually removed, although steel casings are sometimes left in place. The size and placement of reinforcing depends on building loads and the allowable soil-bearing capacity as specified by local codes.

An example of the use of caissons built under difficult conditions is illustrated by the section drawing in Fig. 6.25. This shows the construction of a bridge pier supported by six 2-meter-wide (6 ft. 6 in.) caissons. Notice the construction of a sheet-pile cofferdam and the watertight tremie concrete plug at the bottom. In this case, the caissons were 23 m (75 ft. 5 in.) deep with 7 m (72 ft. 1/2 in.) rock sockets. The tops of the caissons were 16 m below the mean water line.

Helical Pier® Foundation Systems

Helical Pier foundations use helical (screw) steel piers installed at intervals between the footing forms. They are installed by screwing the steel into the soil (Fig. 6.26). The torque required to drive each pier correlates to the bearing capacity of the soil below and is used to ascertain the anchor depth needed to support the required structural load. Helical Piers can also be used to underpin existing foundations that are experiencing settling by using a special support bracket.

Figure 6.24 A drilling rig with various augers and buckets used for caisson excavation. *(Image copyright Li Wa, 2009. Used under license from Shutterstock.com)*

Figure 6.25 A section through a bridge pier constructed on caisson piles.

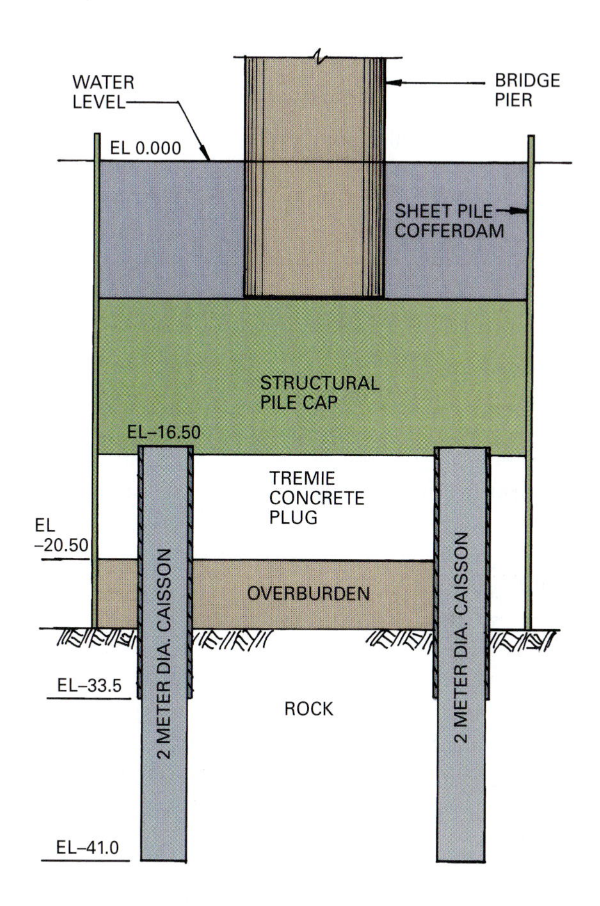

Figure 6.26 The Helical pier foundation uses steel helical piers drilled into the soil before the footing is poured.

Review Questions

1. What loads act upon a foundation?
2. What are the major types of foundations?
3. What types of piles are available?
4. What types of equipment are used to drive piles?
5. How are caisson foundations constructed?

Key Terms

caisson foundation
concentrated load
dynamic load
grade beam
independent footing
lateral load
mat foundation
pile foundation
pile hammer
spread foundation
static load
tieback anchors
uniform load

Activities

1. Secure foundation drawings from local architects and general contractors and identify the types of foundation specified, reinforcing, and other details. See if a log of test borings was made and identify the characteristics of the soils as the bore hole goes deeper.
2. Visit construction sites and observe the excavation, forms, piers, reinforcing, and other materials used to construct foundations.
3. If any buildings in your area are being built on piles, try to visit the site and take photos to illustrate the process, from locating the piles to establishing the finished elevation of the tops of the piles. What types of piles were being used?
4. When visiting foundation construction, report on any actions taken to keep the excavation dry.
5. As you visit construction sites, observe situations you think are not safe and prepare remarks so you can lead a class discussion on safety. (While on the site do not charge the supervisor or workers with allowing unsafe practices. Simply observe and remember.)

Additional Resources

Builders Foundation Handbook, U.S. Department of Energy, Oak Ridge, TN.

Design and Construction of Plain and Reinforced Concrete Masonry and Basement and Foundation Walls, National Concrete Masonry Association, Herndon, VA.

Other resources include: Many publications available from the U.S. Department of Housing and Urban Development, HUD USER, Rockville, MD.

See Appendix B for addresses of professional and trade organizations and other sources of technical information.

DIVISION

03

Concrete
CSI MasterFormat™

031000 Concrete Forming and Accessories
032000 Concrete Reinforcing
033000 Cast-in-Place Concrete
034000 Precast Concrete
035000 Cast Decks and Underlayment
036000 Grouting
037000 Mass Concrete
038000 Concrete Cutting and Boring

(Courtesy Portland Cement Association)

CHAPTER 7

Concrete

LEARNING OBJECTIVES

Upon completion of this chapter, the student should be able to:

- Identify the various types of Portland cement and cite the purposes for which they can be used.
- Be familiar with the properties of Portland cement and how they affect concrete.
- Discuss the role of water in concrete and its desired properties.
- Discuss the types of aggregate used in concrete and their desirable characteristics.
- Understand the purposes of various concrete admixtures and how they affect concrete.
- Understand what is required to produce concrete of differing properties for various applications.
- Be aware of the commonly used concrete tests and what they reveal about the mix.

Build Your Knowledge

For further information on these materials and methods, please refer to:

Chapter 10 Mortars for Masonry Walls
Chapter 12 Concrete Masonry
Chapter 34 Gypsum, Lime, and Plaster
Topic: Cement Board

Concrete is one of the most widely used construction materials and has a long history of use. Its constituent ingredients derive from a wide variety of naturally occurring materials that are readily available in most parts of the world. Concrete can be made by simple hand-mixing methods or in large quantities in a computer-controlled batching plant. It can be used in a wide range of applications and performs successfully for long periods of time.

Its beneficial properties include: high strength, durability, resistance to fire and insect damage, and good thermal mass properties. Concrete is an inexpensive construction material, but it has no form of its own or tensile strength. Forms must be built to shape it into useful structures, and reinforcing steel must be added to provide tensile strength.

Concrete is a solid, hard material produced by combining Portland cement, coarse and fine aggregates (sand and stone), and water in proper proportions (Fig. 7.1). The chemical reaction between the cement and water produces heat and a hardening of the mass.

PORTLAND CEMENT

Portland cement is a fine, pulverized material consisting of compounds of lime, iron, silica, and alumina. The exact composition of different types of Portland

Figure 7.1 Concrete components: cement, water, sand, and coarse aggregate next to a cut section of hardened concrete. *(Courtesy Portland Cement Association)*

cement varies, but the composition of Type 1, Normal, is representative (Table 7.1). The manufacture of Portland cement requires the combination of these elements in proper proportion under carefully controlled conditions. Current manufacturing techniques emit large amounts of Co_2 into the atmosphere, and research is underway to reduce the amount of Portland cement in concrete through the addition of other ingredients. (See "Construction Materials: Flyash Concrete" on page 116.)

The lime in Portland cement is commonly derived from limestone, marble, marl, or seashells. Iron, silica, and alumina are obtained from mining clays containing these elements. In some areas, iron furnace slag and flue dust are used, as are sand, chalk, and bauxite. These ingredients are crushed in a primary crusher and sent through a vibrating screen. Pieces falling through the screen are conveyed to storage silos while those that are too large are sent to a secondary crusher. After crushing, either a wet or dry process is used to produce the final cement.

The dry process grinds the raw materials to a powder, blends them in mixing silos, and moves them to a kiln. In the wet process, the ground materials are mixed with water to form a slurry, blended, and moved to a kiln. The kiln is a rotating cylinder operating at 2600–3000°F (1600–1780°C) that burns the materials into a clinker. The clinker is cooled, a small amount of gypsum added, and then it is ground into a powder so fine it can pass through a 40,000-openings-per-square-inch sieve. The gypsum serves to retard the curing process.

Most Portland cement is shipped in bulk, in railroad cars or large trailers designed especially for this material. Bulk cement is sold by the ton (2,000 lb. or 907.2 kg), and smaller quantities are bagged at a volume of one cubic foot (0.028 m3) weighing 94 pounds (42 kg). The entire process is illustrated in Fig. 7.2.

Table 7.1 Oxide Composition of Type 10 (ASTM Type I) or Normal Portland Cement

Oxide Ingredient	Range, %
Lime, CaO	60–66
Silica, SiO_2	19–25
Alumina, Al_2O_3	3–8
Iron, Fe_2O_3	1–5
Magnesia, MgO	0–5
Sulfur trioxide, SO_3	1–3

Courtesy Portland Cement Association

ASTM-Designated Types of Portland Cement

Cement is an easily molded temporary paste when mixed with water. After a short time it hardens, or sets, to form a rigid mass. The setting and hardening is called **hydration**. Hardening is not simply a process of evaporation or drying; hydration is the result of a chemical reaction between the Portland cement and water.

Portland cements are made following ASTM Designation C150, which defines eight types of cement. These are Type I, Normal; Type IA, Normal, Air-entrained; Type II, Moderate; Type IIA, Moderate, Air-entrained; Type III, High Early Strength; Type IIIA, High Early Strength, Air-entrained; Type IV, Low Heat of Hydration; and Type V, Sulfate Resisting (Table 7.2). In Canada these are specified as types 10, 20, 30, 40, and 50.

Type I: Normal Type I is a general-purpose Portland cement used whenever other special properties are not necessary. It is used for pavements, sidewalks, reinforced concrete structural members, bridges, tanks, water pipes, and masonry building units, among other things.

Type II: Moderate Type II cement is used when protection against moderate sulfate attack, found in some soils and groundwater, is required. It generates less heat by hydrating at a slower rate than Type I. This makes Type II useful in structures that have larger masses of concrete, such as abutments, piers, and large retaining walls.

Type III: High Early Strength Type III Portland cement provides higher strength in a shorter time period than Types I or II. It reduces the curing period, allowing for quick removal of forms, especially in cold-weather situations when freezing poses problems. High early strength normally reduces the required curing time to one week or less.

Type IV: Low Heat of Hydration Type IV reduces the rate and amount of heat generated by hydration. It develops strength slower than Type I and is mainly used for structures having large concrete masses, such as dams and nuclear plants.

Type V: Sulfate Resisting Type V Portland cement is used only in concrete that is exposed to severe sulfate conditions. This most frequently occurs in soils and groundwater in areas where sulfate concentrations are high.

White Portland Cement White Portland cement is a true Portland cement except that during the manufacturing process the components are controlled so the finished products will be white. It is used mainly for architectural purposes, such as in precast curtain-wall

Figure 7.2 The steps in the manufacturing of Portland cement. *(Courtesy Portland Cement Association)*

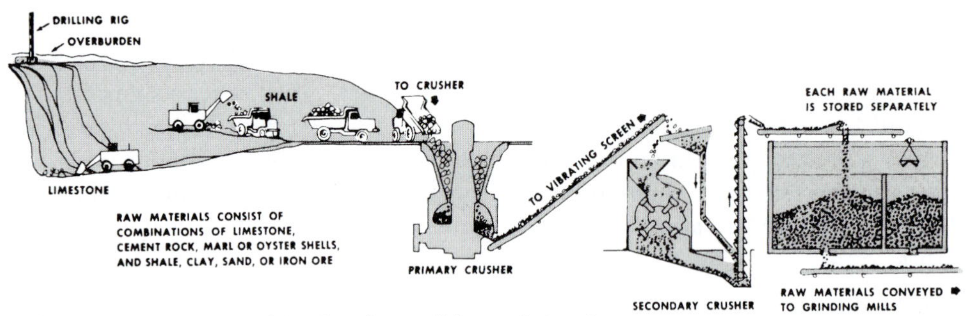

1. Stone is first reduced to 5-in. size, then to 3/4 in., and stored.

DRY PROCESS

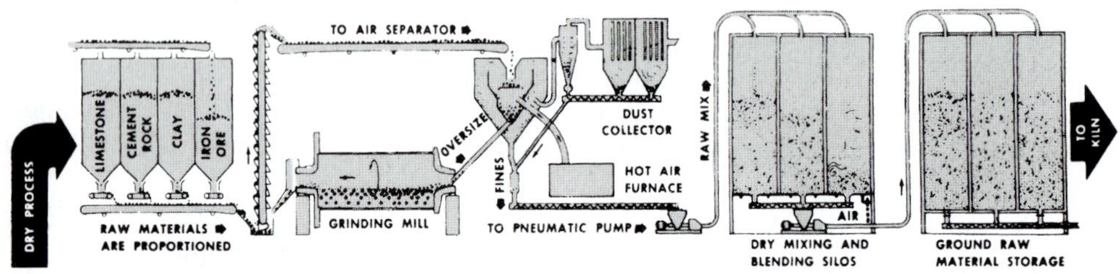

OR 2. Raw materials are ground to powder and blended.

WET PROCESS

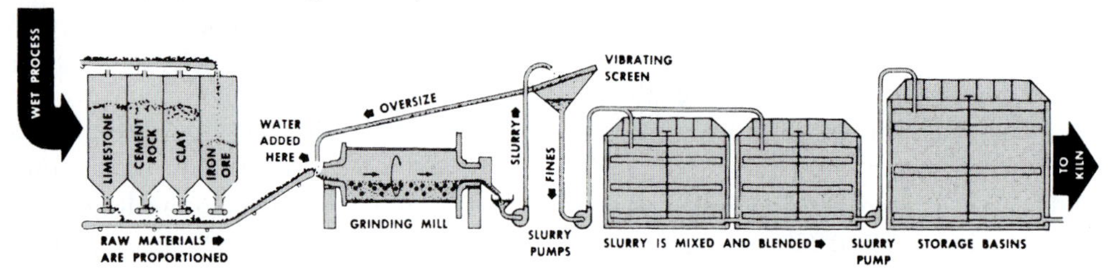

2. Raw materials are ground, mixed with water to form slurry, and blended.

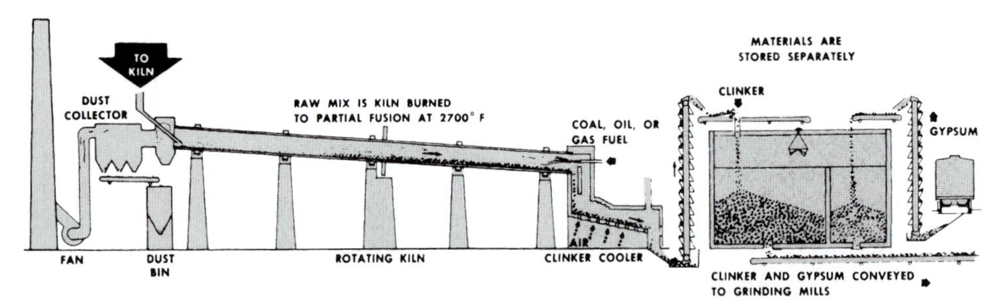

3. Burning changes raw mix chemically into cement clinker.

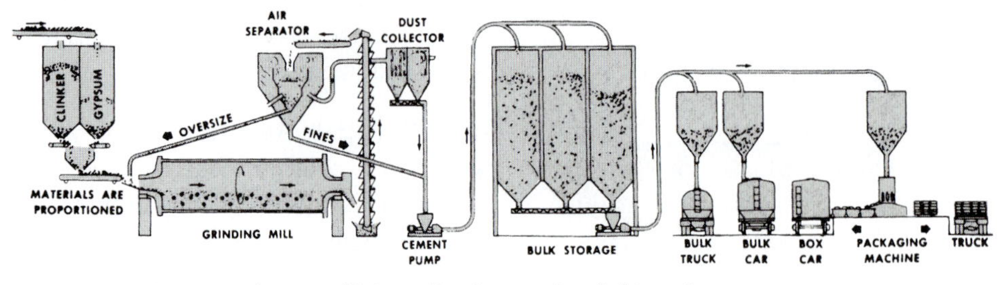

4. Clinker with gypsum is ground into portland cement and shipped.

Table 7.2 Types of Portland Cement

Canadian Designation	United States Designation	Type
10	I	Normal
20	II	Moderate
30	III	High early strength
40	IV	Low heat of hydration
50	V	Sulfate resisting
Air-Entrained Types		
10A	IA	Normal air-entrained
20A	IIA	Moderate air-entrained
30A	IIIA	High early strength air-entrained

Courtesy Portland Cement Association

and facing panels, terrazzo floors, stucco, finish-coat plaster, and tile grout.

Types IA, IIA, IIIA: Air-Entrained There are three types of **air-entrained Portland cement**: IA, IIA, and IIIA. These are similar to Types I, II, and III, except that small quantities of air-entraining materials are mixed with the clinker and gypsum.

Air-entraining encapsulates millions of microscopic air bubbles into the concrete mix (Fig. 7.3). This greatly improves the durability of concrete that is exposed to moisture and freeze and thaw cycles during winter. It also increases the concrete's resistance to surface scaling, which is caused by salts used to remove winter ice. Air-entraining improves the workability of concrete and reduces segregation of the aggregate, bleeding, and the amount of water needed.

Blended Hydraulic Cements Hydraulic cements are capable of hardening under water. There are two blended hydraulic cements: Portland blast-furnace-slag cements and Portland pozzolan cements.

ASTM C595 sets requirements for two types of Portland blast-furnace-slag cements: IS and IS-A (air-entrained). Blast-furnace slag is ground with the Portland cement clinker or separately ground and blended with Portland cement (Fig. 7.4). Blast-furnace slag content of these cements is 25 to 65 percent of the cement's weight.

The four types of Portland pozzolan cements are IP, IP-A, P, and P-A. The "A" indicates air-entraining. They are manufactured by intergrinding Portland cement clinker with a suitable pozzolan. **Pozzolan** is a siliceous and

Figure 7.3 A polished section of air-entrained concrete, as seen through a microscope. *(Courtesy Portland Cement Association)*

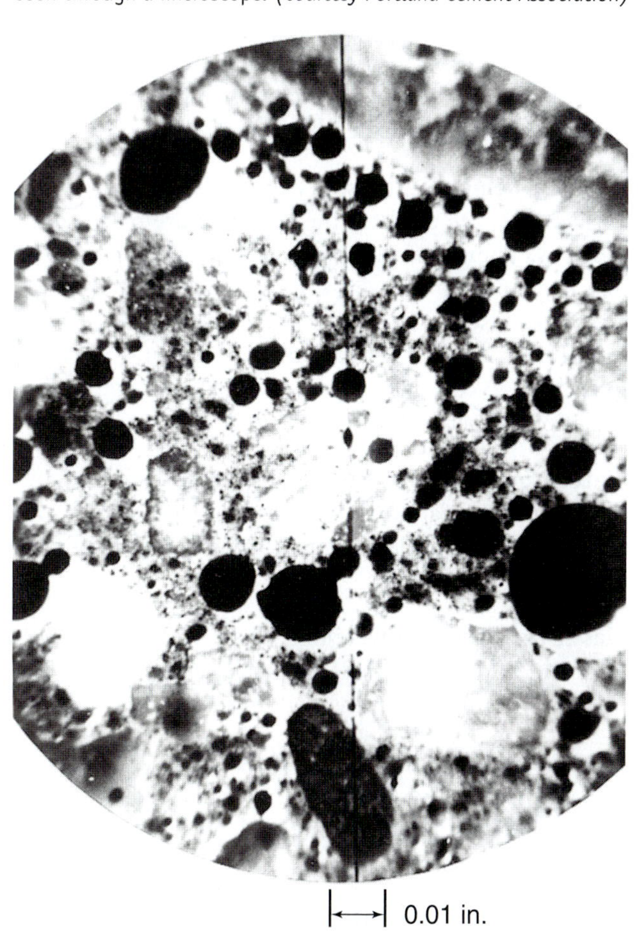

Figure 7.4 Blended cements use a combination of Portland cement or clinker and gypsum blended or interground with pozzolans, slag, or fly ash. *(Courtesy Portland Cement Association)*

aluminous material that chemically reacts with calcium hydroxide to form compounds possessing cementitious properties. Pozzolans may be manufactured or natural. Natural pozzolans include pumicite, volcanic ash, and volcanic tuff. Tuff is a porous rock formed by the consolidation of volcanic ashes and dust. Processed natural pozzolans include clay and shale that are burned or calcined in a kiln, then crushed and ground. All four types can be used for general concrete construction. Types P and P-A are used in massive concrete structures that do not require high early strengths.

Masonry Cements Masonry cements, covered under ASTM C91, are used for the manufacture of masonry mortars. They typically contain some of the following: Portland cement, slag cement, and hydraulic lime. They may also contain hydrated lime, limestone, chalk, calcareous shell, talc, slag, or clay. Mortar cements must have excellent workability, plasticity, and water-retention properties.

Waterproof Portland Cement Waterproof Portland cement has a small amount of calcium stearate or aluminum stearate added to the Portland cement clinker during grinding.

Plastic Cements Plastic cements have plasticizing agents added to Type I or Type II Portland cement during the milling operation. They are used in stucco and plaster.

Expansive Cement Expansive cement is hydraulic cement that expands during the early hardening period. The three types available are K, M, and S. They can be used to compensate for the effects of drying shrinkage.

Regulated Set Cement Regulated set cement is a hydraulic cement that can be compounded to have a set time from one to two minutes to as long as fifty or sixty minutes.

Properties of Portland Cement

The specifications for Portland cement place limits on its chemical composition and physical properties. Knowledge of these properties is necessary to evaluate the results of tests on the cement itself and on concrete made with the cement.

Fineness As cement increases in fineness, the rate of hydration increases, which accelerates the strength development of the concrete. The effects of greater fineness on strength are most apparent during the first five to seven days. The compound composition and fineness of Portland cements are listed in Table 7.3.

Soundness Soundness describes the ability of a hardened paste to retain its volume after setting. A major problem is delayed destructive expansion after the paste has hardened caused by excessive amounts of hard-burned free lime or magnesia. Cement can be tested following ASTM standards to determine soundness.

Table 7.3 Potential Compound Composition and Fineness of Cements

Type of Portland Cement	Compound Composition, Percentage				Wagner Fineness, m^2/g
	C_3S	C_2S	C_3A	C_4AF	
Type I	55	19	10	7	0.18
Type II	51	24	6	11	0.18
Type III	56	19	10	7	0.26
Type IV	28	49	4	12	0.19
Type V	38	43	4	9	0.19
White	33	46	15	2	

Courtesy Portland Cement Association

Consistency Consistency is the relative mobility of a fresh mixture, in other words, its ability to flow and its workability. Cement-paste tests for consistency are made using the tests specified by ASTM C191.

Setting Time Setting-time tests are made to determine how the cement paste undergoes setting and hardening during the first few hours. The initial set must not occur before the concrete surface finishing operations can be completed. After finishing, the final set should occur without unnecessary delay. Setting times of cement pastes and concrete do not correlate directly because of water loss, the surface on which the concrete has been poured, and temperature differences.

False Set False set results in the loss of plasticity without the development of much heat shortly after the concrete has been mixed. False set can be countered through the addition of water to the concrete mix. Chemical admixtures can also be added to delay its occurrence.

Compressive Strength Compressive strength, a major physical property of cement, is found by testing 2 in. (50 mm) cubes of mortar, as specified by ASTM C150. Compressive strength is influenced by the type of cement used, its compound composition and fineness. Due to variances in concrete mixtures, cement compressive strengths cannot be used to determine concrete compressive strengths. Actual concrete samples must be tested to determine the compressive strength of concrete. Compressive strengths of cements are shown in Table 7.4.

Heat of Hydration When cement and water chemically react, the heat generated is known as the heat of hydration. The amount of heat generated depends on the chemical composition of the cement. The rate of heat generated is affected by the fineness of the cement, its chemical composition, and the temperature during hydration. Structures with a large concrete mass may experience a significant rise in temperature unless the heat can be rapidly dissipated. If not controlled, it may create undesirable stresses. In cold weather the excessive heat can be beneficial because it helps to maintain favorable curing temperatures. The approximate amounts of heat of hydration that is generated are shown in Table 7.5.

Specific Gravity The specific gravity of Portland cement is about 3.15. Portland blast-furnace-slag and Portland pozzolan cements may have values as low as 2.90. The specific gravity is not an indication of the quality of cement, but is used when calculating mix designs.

Table 7.4 Relative Compressive Strength Requirements as Affected by Type of Cement[a]

Type of Portland Cement	Compressive Strength, Minimum, Percentage of Strength of Type 1 at 7 days			
	1 day	3 days	7 days	28 days
I	—	64	100	143[c]
IA	—	52	80	114[c]
II[b]	—	54	89	143[c]
IIA[b]	—	43	71	114[c]
III	64	125	—	—
IIIA	52	100	—	—
IV	—	—	36	89
V	—	43	79	107
IS	—	64	100	125
IS-A	—	52	80	100
IS(MS)	—	36	64	125
IS-A(MS)	—	27	50	100
IP	—	64	100	125
IP-A	—	52	80	100
P	—	—	54	107
P-A	—	—	45	89

[a]*ASTM Designations: C150-77 and C595-77.*
[b]*A strength reduction of one-fifth to one-third is allowed if the optional heat of hydration or the chemical limit on the sum of the C_3S and C_3A is specified. See ASTM specification.*
[c]*Optional specification requirement.*
Note: When suffixes MH or LH are added to any of last eight cement types listed, the strength requirement is 80% of the values shown.
Courtesy Portland Cement Association

Table 7.5 Approximate Relative Amounts of Heat Generated During Hydration[a]

Portland Cement Type	Percent of Hydration, Relative to Type 1, Normal
Type 1, Normal	100
Type 2, Moderate	80 to 85
Type 3, High early strength	up to 150
Type 4, Low heat of hydration	40 to 50
Type 5, Sulfate resisting	60 to 75

[a]*First seven days.*
Courtesy Portland Cement Association

Storing Portland Cement

Portland cement is moisture sensitive, so it must be protected from dampness. Sacked material must be stored on pallets, whether it is in a warehouse with a concrete floor or on a site in the open. Warehouse storage must be watertight, and bags must not touch the exterior walls. Bags should be packed closely to reduce airflow and covered with plastic or tarpaulins if stored for long periods. Bagged cement tends to pack if it is stored a long time. This can be corrected by rolling the bags on the floor or ground before opening them. Bulk cement is stored in watertight bins or silos. Dry low-pressure aeration or vibration should be used to make the cement flow better. When cement is loaded into bins or silos, it swells, so a unit will store only about 80 percent of its rated capacity.

WATER

Water used in making concrete should be clear and free of sulfates, acids, alkalis, and humus. Potable water from municipal water systems or wells provides water suitable for use. Water from lakes, ponds, or rivers should be carefully checked for suitability before use.

Water of questionable quality can be used for concrete if test mortar cubes have seven-day and twenty-eight-day strengths equal to 90 percent of samples made with drinkable water. These tests should be made following ASTM C109 specifications. ASTM C191 specifies how to test the samples to see if impurities in the water adversely shorten or lengthen the setting time.

Alkali carbonate and bicarbonate, chloride, sulfate, carbonates of calcium and magnesium, iron salts, inorganic salts, acid waters, and alkaline waters all have an effect on concrete. Additionally, sugar, silt, suspended particles of clay or fine rock, oils in suspension, and algae in water can make it unsuitable for concrete mixing.

Seawater containing up to 35,000 ppm (parts per million) of dissolved salts is generally suitable as mixing water for unreinforced concrete. Concrete made with seawater has a higher early strength than normal concrete and usually has lower strength after twenty-eight days. Seawater may be used for reinforced concrete, but the reinforcement must be protected from

Construction Materials

Fly Ash Concrete

Fly ash is a fine, glass-like powder recovered from the gases created by coal-fired electric power generation. U.S. power plants produce millions of tons of fly ash annually, much of it discarded in landfills. Fly ash can provide an economical and environmentally preferable substitute for the Portland cement used in concrete, brick, and block production. First discovered during the construction of the Hoover Dam, the material is now used frequently in concrete mixes.

Consisting mostly of silica, alumina, and iron, fly ash is a material that forms cement in the presence of water. When mixed with lime and water it forms a compound with properties similar to Portland cement. The circular shape of fly ash particles reduces friction and increases the concrete's consistency and workability. Improved workability means less water is needed, resulting in less segregation of the mixture and higher ultimate strength. The resulting concrete also tends to be denser, with a smoother surface and sharper detail. Fly ash can substitute for up to 20 to 35 percent of the Portland cement used to make concrete.

Class F fly ash reduces the risk of expansion due to sulfate attack, which can occur in certain soils or near coastal areas. Class C flyash is resistant to expansion from chemical attack, has a higher percentage of calcium oxide, and is more commonly used for structural concrete. Fly ash used in Portland-cement concrete is produced according to the requirements of ASTM C618, Standard Specification for Fly ash and Raw or Calcined Natural Pozzolan Class C Fly ash for use as a Mineral Admixture in Portland Cement.

As a recycled, post-industrial material, fly ash use in concrete offers environmental advantages by diverting material from the waste stream, conserving virgin materials, and reducing the energy investment in processing them. The use of fly ash is eligible for one recycled content credit in the LEED rating system, under the materials and resources category.

corrosion, and the concrete must contain entrained air. It is also necessary to check for sea salt on aggregates that were exposed to salt water. Not more than one percent of sea salt on aggregates is acceptable. The Environmental Protection Agency (and some state agencies) prohibit discharging untreated wash water from returned concrete or mixer washout operations into rivers, streams, and lakes.

AGGREGATES

Approximately 60 to 80 percent of concrete is made up of **aggregates**. The cost of concrete and its properties are directly related to the aggregates used. Natural aggregates are composed of rocks and minerals. Minerals are naturally occurring inorganic substances that have distinctive physical properties and a composition that can be expressed by a chemical formula. Rocks are usually composed of several minerals. For example, limestone is basically calcite (a mineral) with small amounts of quartz, feldspar, and clay (all minerals). The crushing and weathering of rock produces stone, gravel, sand, silt, and clay. Aggregates derived from the recycling and crushing of demolished concrete are also finding expanded use. Rocks and minerals commonly found in aggregates are shown in Table 7.6.

Aggregates must conform to ASTM specifications. They must be clean, strong, free of absorbed chemicals, and devoid of coatings of clay, humus, and other fine materials. Aggregates containing some shale, shaly rocks, soft or porous rocks, and some types of chert are not suitable because they do not weather well and can cause popouts on the exposed concrete surface.

Normal-weight concrete weighs about 135 to 160 lb. per ft.^{3} (2,150 to 2,550 kg/m^3). It uses sand, gravel, crushed stone, and air-cooled blast-furnace-slag aggregates, as specified by ASTM C33. Structural lightweight concrete weighs about 85 to 115 lb. per ft.^{3} (1350 to 1850 kg/m^3) and uses expanded shale, clay, slate, and slag aggregates. It is specified in ASTM C330. Insulating concrete, specified in ASTM C332, weighs about 15 to 90 lb. per ft.^{3} (250 to 1450 kg/m^3) and uses pumic, scoria, perlite, vermiculite, and diatomite aggregates. Heavyweight concrete weighs 175 to 400 lb. per ft.^{3} (2800 to 6400 kg/m^3) and uses barite, limonite, magnetite, ilmenite, hematite, iron, and steel slugs as aggregates. It is specified in ASTM C637.

Table 7.6 Rock and Mineral Constituents in Aggregates

Minerals	Pyrite	Sedimentary Rocks
Silica	Marcasite	Conglomerate
Quartz	Pyrrhotite	Sandstone
Opal	Iron Oxide	Quartzite
Chalcedony	Magnetite	Graywacke
Tridymite	Hematite	Subgraywacke
Cristobalite	Goethite	Arkose
Silicates	Ilmenite	Claystone, Siltstone,
Feldspars	Limonite	Argillite and Shale
Ferromagnesian	**Igneous Rocks**	Carbonates
Hornblende	Granite	Limestone
Augite	Syenite	Dolomite
Clay	Diorite	Marl
Illites	Gabbro	Chalk
Kaolins	Peridotite	Chert
Chlorites	Pegmatite	**Metamorphic Rocks**
Montmorillonites	Volcanic Glass	Marble
Mica	Obsidian	Metaquartzite
Zeolite	Pumice	Slate
Carbonate	Tuff	Phyllite
Calcite	Scoria	Schist
Dolomite	Perlite	Amphibolite
Sulfate	Pitchstone	Hornfels
Gypsum	Felsite	Gneiss
Anhydrite	Basalt	Serpentinite
Iron Sulfide		

For brief descriptions, see "Standard Descriptive Nomenclature of Constituents of Natural Mineral Aggregates" (ASTM C294).
Courtesy Portland Cement Association

Characteristics of Aggregates

The major characteristics for aggregates, listed in Table 7.7, are described below.

Resistance to Abrasion and Skidding Abrasion resistance is important when the aggregate is to be used in an area that is subject to heavy abrasive use, such as a factory floor. Aggregate for abrasion resistance is tested following ASTM C131 standards. To give skid resistance, the siliceous particle content of the fine aggregate should be 25 percent or more.

Resistance to Freezing and Thawing The freeze-thaw properties of an aggregate are important in concrete that will be exposed to a wide range of temperatures. Significant considerations are the porosity, absorption, permeability, and pore structure of the aggregate. Suitable aggregates may be chosen from those used in the past that have given good results. Unknown materials should be evaluated using ASTM tests.

Compressive Strength The strength of aggregates under compression is an important factor to consider when choosing materials. This is tested by standard compression tests on hardened concrete samples.

Shape and Texture of Particles The shape and texture of aggregate particles influence the properties of fresh concrete more than cured concrete. Rough-textured, angular, elongated particles require more water than do smooth, rounded aggregates. Therefore, angular particles require more cement to maintain the required water-cement ratio. Cement tends to bond better to angular particles than to smooth particles. This must be considered when flexural strength or high compressive strength is specified.

Specific Gravity Specific gravity is a measure of the relative density of an aggregate. It is a ratio of an aggregate's weight to the weight of an equal volume of water and is used to determine the absolute volume occupied by the aggregate. Most natural aggregates have specific gravities between 2.4 and 2.9.

Absorption and Surface Moisture The absorption and surface moisture conditions of an aggregate are tested in order to control the net water content of the concrete and determine suitable batch weights.

Moisture conditions are designated in four categories:

1. Oven dry. Completely dry and fully absorbent.
2. Air dry. Dry on the surface but with some interior moisture.
3. Saturated surface dry. Neither absorbing water nor contributing water to the mix. The surface is dry but the voids and interior of the aggregate are fully saturated.
4. Damp. Contains an excess of surface moisture.

Surface moisture can increase the bulk (volume) of average and fine sands more than coarse sands. Since moisture in sands delivered on a job can vary, it is necessary to weigh the aggregate rather than using volume measures when portioning concrete ingredients.

The test to determine the amount of moisture on fine aggregates is made according to ASTM C70, Surface Moisture in Fine Aggregate, which depends on the displacement of water by a known weight of moist aggregate. The density of the aggregate must be known to use this method. Another method, ASTM C566, Total Moisture Content of Aggregate by Drying, involves weighing a sample, heating it until dry, and reweighing it. The difference in weights is used to calculate the moisture content.

Table 7.7 Characteristics and Tests of Aggregates

Characteristic	Significance	Test Designation	Requirement or Item Reported
Resistance to abrasion	Index of aggregate quality; wear resistance of floors, pavements	ASTM C131 ASTM C295 ASTM C535	Maximum percentage of weight loss
Resistance to freezing and thawing	Surface scaling, roughness, loss of section, and unsightliness	ASTM C295 ASTM C666 ASTM C682	Maximum number of cycles or period of frost immunity; durability factor
Resistance to disintegration by sulfates	Soundness against weathering action	ASTM C88	Weight loss, particles exhibiting distress
Particle shape and surface texture	Workability of fresh concrete	ASTM C295 ASTM D3398	Maximum percentage of flat and elongated pieces
Grading	Workability of fresh concrete; economy	ASTM C117 ASTM C136	Minimum and maximum percentage passing standard sieves
Bulk unit weight or density	Mix design calculations; classification	ASTM C29	Compact weight and loose weight
Specific gravity	Mix design calculations	ASTM C127, fine aggregate ASTM C128, coarse aggregate ASTM C29, slag	—
Absorption and surface moisture	Control of concrete quality	ASTM C70 ASTM C127 ASTM C128 ASTM C566	—
Compressive and flexural strength	Acceptability of fine aggregate failing other tests	ASTM C39 ASTM C78	Strength to exceed 95% of strength achieved with purified sand
Definitions of constituents	Clear understanding and communication	ASTM C125 ASTM C294	—

Courtesy Portland Cement Association

Chemical Stability Chemically stable aggregates do not react chemically with cement, which could cause harmful reactions. Some aggregates contain minerals that do react with alkalies in cement, causing abnormal expansion and cracking in the concrete. Field records of aggregates provide evidence of their chemical stability. If this is unknown, a laboratory test based on ASTM specifications should be made.

Harmful Materials A number of harmful materials, such as silt, organic materials, coal, and soft rock particles, may be present in aggregate materials. A series of ASTM tests is used to identify these harmful materials in aggregate samples, and aggregate specifications limit the amount of these materials that may be present.

Grading Aggregates

Aggregates are graded into standardized sizes by passing them through a sieve, as specified by ASTM C136 and CSA A23.2.2. Sieves used for fine aggregates consist of seven sizes; there are ten sieve sizes for coarse aggregates (Table 7.8).

The limits used are specified for the percentage of material passing each sieve. Figure 7.5 shows the grading limits for fine aggregates. Notice that almost all of the particles of fine sand pass through the No. 4 sieve, but only 10 to 20 percent pass through the No. 100 sieve. The coarse sand graph shows that about 98 percent of particles pass through the No. 4 sieve and only about 3 percent pass through the No. 100 sieve. Each product is therefore an accumulation of the particles of various percentages passing through the various sieves.

When concrete mixes are designed, limits are specified for the percentage of material passing each sieve. This, plus specifying the maximum aggregate size, affects the relative proportions of aggregate, as well as the cement and water requirements, workability, economy, porosity, shrinkage, and durability of concrete. The best mixes are those that do not have a large deficiency or

Table 7.8 Fine and Coarse Sieve Sizes

Fine Sieve Sizes	
	3/8 in. (9.5 mm)
	No. 4 (4.75 mm)
	No. 8 (2.36 mm)
	No. 16 (1.18 mm)
	No. 30 (600 μm)
	No. 50 (300 μm)
	No. 100 (150 μm)
Coarse Sieve Sizes	
Size Number	**Nominal Size (sieves with square openings)**
1	3 1/2: to 1 1/2 in. (90 to 37.5 mm)
2	2 1/2 to 1 1/2 in. (63 to 37.5 mm)
357	2 in. to No. 4 (50 to 4.75 mm)
467	1 1/2 in. to No. 4 (37.5 to 4.75 mm)
57	1 in. to No. 4 (25.0 to 4.75 mm)
67	3/4 in. to No. 4 (19.0 to 4.75 mm)
7	1/2 in. to No. 4 (12.5 to 4.75 mm)
8	3/8 in. to No. 8 (9.5 to 2.36 mm)
3	2 to 1 in. (50.0 to 25.0 mm)
4	1 1/2 to 3/4 in. (37.5 to 19.0 mm)

Courtesy Portland Cement Association

Figure 7.5 Grading limits for fine aggregates. *(Courtesy Portland Cement Association)*

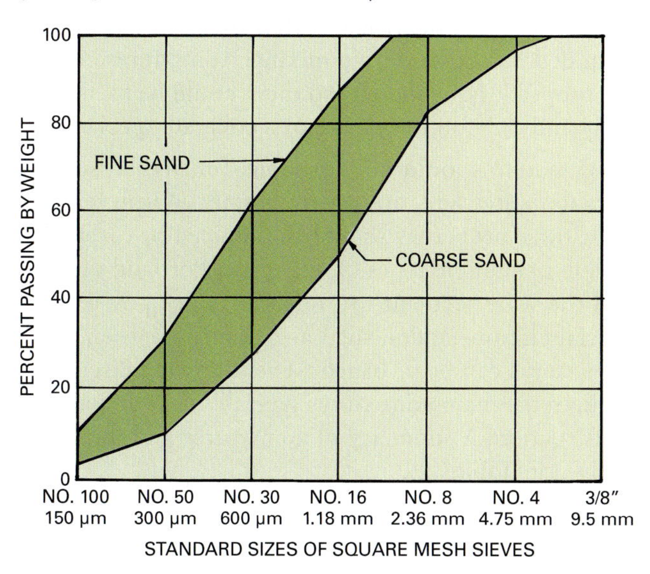

an excess of any size. Very fine sands are uneconomical, while very coarse sands produce a harsh, unworkable mix. A wide range of particle sizes is necessary to fill the voids between aggregates (Fig. 7.6). As the range of sizes fill the voids, voids constitute less volume, which decreases the cement needed.

Sand is graded using a layered set of sieves inside a shaker. Each sieve layer has a fine mesh. The sand remaining on each sieve can be weighed to get the percent of the sample for each sieve. Following is an example of a typical test using 1,000 g of sand.

Sieve	Weight of Sand Retained in Grams	Percent Retained	Percent Cumulative
4 (5 mm)	4	4	115.0
8 (2.5 mm)	115	11.5	115.5
16 (1.25 mm)	210	21.0	136.5
30 (630 mm)*	262	26.2	162.7
50 (315 mm)	201	20.1	182.8
100 (160 mm)	172	17.2	100.0
		Total	302.5

*micro millimeter

The fineness modulus of the sand sample can be found by adding the cumulative-percent-retained figures and dividing by 100. In the previous illustration, this total was 302.5, which gives a fineness modulus of 3.02. A range of 2.30 to 2.60 denotes fine sands; 2.61 to 2.90 medium; and 2.91 to 3.10 coarse. The sand in the example was coarse.

Structural Lightweight Aggregates

Lightweight aggregates are used to produce lightweight structural concrete. Lightweight concrete has a density ranging from 90 to 115 lb./ft.^{3} (1,440 to 1,850 kg/m^3), depending on the aggregate. Shale, slate, clay, and slag are generally used. Pelletized clay or crushed shale or slate is placed in a rotary kiln and

Figure 7.6 Separation of concrete aggregates into various sieve sizes gives a visual illustration of why control of aggregate gradation is important in maintaining uniformity in the concrete. *(Courtesy Portland Cement Association)*

heated to about 1,800°F (1,000°C). Gases in the material cause it to expand, forming small air cells in the pieces. Another process, known as sintering, involves crushing the aggregate, mixing it with a small amount of finely ground coal or coke, spreading it over a grate, and igniting the coke. Blowers increase the temperature of the burning, causing the material to expand and bond into large pieces. These are crushed, forming the smaller aggregate.

Expanded blast-furnace slag aggregate is produced by treating the slag with steam or water while it is in a molten state, causing it to expand and bond into large particles that are then crushed to the desired size.

Insulating Lightweight Aggregates

Perlite, diatomite, vermiculite, pumice, and scoria are commonly used as insulating lightweight aggregate. Perlite is siliceous volcanic rock that contains moisture. It is crushed and heated, causing it to expand and form a honey-comb structure. It is used for loose-fill insulation and plaster and concrete aggregate. Diatomite is a diatomaceous earth, composed of silicified skeletons of microscopic one-celled animals. Vermiculite is a hydrated magnesium-aluminum-iron silicate that occurs in thin layers with moisture between them. When vermiculite is crushed and heated, the layers expand, forming dead air cells. Pumice is a type of volcanic rock that is porous and lightweight. Scoria is a type of volcanic slag; scoria also is used to describe a form of blast-furnace slag. It is a cellular material that is crushed to form aggregate of various sizes.

Heavyweight Aggregates

Heavyweight aggregates are used to produce heavyweight concretes with densities of up to 400 lb./ft.^{3} (6400 kg/m^3) that are used for radiation shielding and other applications that necessitate great weight. Frequently used heavyweight aggregates include ferrophosphorus, barite, goethite, hematite, ilmenite, limonite, magnetite, and steel shot and punchings.

Testing Aggregates for Impurities

Aggregates must be free of clay, silt, and organic impurities. If present, these could influence the composition of the cement paste and, consequently, the strength and setting of the concrete. Organic impurities in fine aggregate are determined by ASTM C40, Organic Impurities in Sands for Concrete. Excess clay and silt in an aggregate can cause an increase in shrinkage, affect durability, and cause separation from the other aggregate particles. Aggregate specifications usually limit clay and silt passing through a No. 200 (75 mm) sieve to 2 or 3 percent for sand and less than one percent in coarse aggregate. Tests are made based on ASTM C117, Sieve in Mineral Aggregates by Washing. Testing for clay lumps is made following ASTM C142, Clay Lumps and Friable Particles in Aggregates.

Handling and Storing Aggregates

Aggregates should be stored in layers of uniform thickness. In stockpiles, each truckload should be discharged tightly against the previous load. Aggregate is removed from stockpiles with a front-end loader. If removing it from a tall, conical pile, the loader should move back and forth across the face of the pile so as to reblend sizes. Washed aggregates must have sufficient time in storage to drain to a uniform moisture content. Fine aggregates should not be dropped from a bucket or conveyor because the wind tends to blow away the very fine particles. Grades should be kept from intermingling by building dividers or storing in completely separate locations. Above all, aggregates must be stored and handled so they are not contaminated by unwanted substances, such as earth and leaves.

ADMIXTURES

Admixtures are ingredients added to concrete, other than Portland cement, aggregates, and water. They may be added before or during mixing. Admixtures change the properties of concrete, so they should be used sparingly and only on the advice of a concrete specialist.

Concrete should be workable, finishable, strong, durable, watertight, and wear resistant. Whenever possible, these properties should be obtained by careful selection of suitable types of aggregate, Portland cement, and the water-cement ratio. If this is not possible, or special circumstances, such as freezing weather, exist, admixtures can be of benefit. The following discussion outlines the main admixtures typically used in concrete construction. A summary of admixtures and their uses is given in Table 7.9.

Air-Entraining Agents

Air-entraining admixtures are used to entrain microscopic air bubbles in concrete. Entrainment can be produced by using air-entrained Portland cement or by adding an air-entraining admixture to concrete as it is being mixed. The entrained air bubbles are distributed uniformly throughout the cement paste

Table 7.9 Admixtures by Classification

Desired Effect	Type of Admixture	Material
Improve durability	Air entraining (ASTM C260)	Salts of wood resins Some synthetic detergents Salts of sulfonated lignin Salts of petroleum acids Salts of proteinaceous material Fatty and resinous acids and their salts Alkylbenzene sulfonates
Reduce water required for given consistency	Water reducer (ASTM C494, Type A)	Lignosulfonates Hydroxylated carboxylic acids (Also tend to retard set so accelerator is added)
Retard setting time	Retarder (ASTM C494, Type B)	Lignin Borax Sugars Tartaric acid and salts
Accelerate setting and early strength development	Accelerator (ASTM C494, Type C)	Calcium chloride (ASTM D98) Triethanolamine
Reduce water and retard set	Water reducer and retarder (ASTM C494, Type D)	(See water reducer, Type A, above)
Reduce water and accelerate set	Water reducer and accelerator (ASTM C494, Type E)	(See water reducer, Type A, above. More accelerator is added)
Improve workability and plasticity	Pozzolan (ASTM C618)	Natural pozzolans (Class N) Fly ash (Class F and G) Other materials (Class S)
Cause expansion on setting	Gas former	Aluminum powder Resin soap and vegetable or animal glue Saponin Hydrolyzed protein
Decrease permeability	Dampproofing and waterproofing agents	Stearate of calcium, aluminum, ammonium, or butyl Petroleum greases or oils Soluble chlorides
Improve pumpability	Pumping aids	Pozzolans Organic polymers
Decrease air content	Air detrainer	Tributyl phosphate
High flow	Superplasticizers	Sulfonated melamine formaldehyde condensates Sulfonated naphthalene formaldehyde condensates

Courtesy Portland Cement Association

and constitute 2 to 8 percent of the volume of the concrete. Air-entrained Portland cement has the air-entraining material ground into it during manufacture. If it is added to the concrete mix, it can be added before or during the mixing process. Numerous commercial air-entraining admixtures are manufactured from a variety of materials. Some ingredients used in air-entraining admixtures include polyethylene oxide polymers, fats and oils, sulfonated compounds, and detergents. Air-entraining admixtures are specified by ASTM C226.

Entrained air bubbles improve the durability of concrete, which increases resistance to damage due to freeze-thaw cycles and de-icers, which can cause scaling. Air-entraining gives improved workability during placement and superior water-tightness. It also improves resistance to sulfate attack from soil water and seawater. Another important feature is that properly proportioned air-entrained concrete requires less water per cubic yard than non-air-entrained concrete of the same slump, resulting in an improved water-cement ratio. Air-entrained concrete is used in cold climates where concrete, such as paving and architectural concrete, is exposed to the freeze-thaw cycle. It is also effective for concrete exposed to soil and water, where sulfate attack is possible.

Retarders

Retarding admixtures, or **retarders**, are used to slow the setting time of cement paste in concrete. They are often employed in hot weather, where hydration is accelerated by excessive heat. Without a retarder in hot weather, more water is required to achieve the desired slump, which produces lower-strength concrete. Retarding admixtures tend to reduce the water required, resulting in a better water-cement ratio and ultimately increased concrete strength. Retarders also help when it is necessary to pour large amounts of concrete or where placement is difficult. For example, if concrete must be pumped a considerable distance, retarders will enable it to be moved, placed, and finished before premature setting occurs. They are sometimes used in concrete mix trucks that have to travel an unusually long distance to a job. They also reduce increased temperatures caused by the heat of hydration in large concrete masses.

A variety of chemicals are used as retarders, and their use results in a reduction in strength during the first one to three days. Retarders can also cause shrinkage, which may cause cracking. Before retarders are used, tests should be made with job materials and conditions. Set times can often be slowed by other means. For example, higher air temperatures during hydration often cause increased hardening rates. A simple means for retarding set time in hot weather without admixtures is cooling the mixing water or aggregates or both. Set time can also be reduced by shading the concrete so it is not directly exposed to sunlight.

Water Reducers

Water-reducing admixtures reduce the amount of water needed to produce concrete of a given consistency. They can also be used to increase the amount of slump without requiring additional water. This makes for a lower water-cement ratio, resulting in greater concrete strength. Some water-reducing admixtures shorten set time and sometimes cause increased drying shrinkage. Lignin solfonic acids and metallic salts are common water-reducing agents.

Accelerators

An **accelerator** admixture speeds up the strength development of concrete. Strength development can also be accelerated by using Type III high-early-strength Portland cement, by increasing the amount of cement to lower the water-cement ratio, or by curing at higher temperatures. Accelerators are used in cold weather to develop strength faster in order to offset freeze damage.

Calcium chloride is a frequently used accelerator. It should be added to the concrete mix in solution rather than dry form. The amount added varies, depending on conditions and the desired set time, but never exceeds 2 percent of the cement's weight. Commercially available calcium chloride ($CaCl_2$) is produced in regular and concentrated flake types. Regular flake contains at least 77 percent $CaCl_2$, while concentrated flake has 94 percent $CaCl_2$.

Experience has shown that using 2 percent or less of calcium chloride has no significant corrosive effect on steel reinforcing materials if the concrete is of high quality. Calcium chloride is not recommended for use in (1) pre-stressed concrete, because of the possibility of corrosion; (2) in concrete containing embedded aluminum, such as electrical conduit; (3) in concrete subject to alkali-aggregate reaction or soils or water containing sulfates; (4) in nuclear shielding concrete; (5) in floor slabs to receive dry-shake metallic finishes; and (6) in hot weather.

Other commercial accelerators are made from chemicals other than calcium chloride. Some take a nonhygroscopic powder form and others are concentrated liquids. Since they are free of chlorides, they are useful in concrete in which steel is embedded. They produce a high early set, reducing normal set time from three hours to one hour, and the concrete develops higher-than-normal strength during the initial three-day curing period.

Pozzolans

A pozzolan is a siliceous and aluminous material that, when finely ground in the presence of moisture, chemically reacts with calcium hydroxide at ordinary temperatures to form compounds possessing cementitious properties. (See the earlier discussion under Blended Hydraulic Cements for more information.) Pozzolan materials are sometimes added to concrete to help reduce internal temperatures. This is especially helpful when pouring large masses of concrete. Some pozzolans are used to reduce or eliminate potential concrete expansion from alkali-reactive aggregates, while others improve resistance to sulfate attack.

Pozzolans can replace 10 to 35 percent of the cement. They substantially reduce the twenty-eight-day strength of concrete and require continuous wet curing and the maintaining of favorable temperatures for a longer period than that needed for normal concrete.

Workability Agents

If fresh concrete is harsh due to improper aggregate grading or incorrect mix proportions, agents can be added to improve workability (Fig. 7.7). **Workability** is a term used to describe the ease with which concrete can be placed and consolidated. Improved workability may be needed if the concrete requires pumping or placing in forms containing considerable reinforcing. If concrete needs a troweled finish, workability is important.

The addition of entrained air is the best workability agent. Some organic materials, such as alginates and cellulose derivatives, will increase slump. Finely divided materials can be used as admixtures to improve the workability of mixes deficient in aggregates passing through the No. 50 and No. 100 sieves. When added to mixes not deficient in fines, additional water is usually required. This may reduce strength, increase shrinkage, and adversely affect other properties of concrete. Fly ash and pozzolans used as workability agents must meet ASTM specifications. These also tend to reduce early strength and entrained air, so additional air-entraining admixture must be used.

Figure 7.7 Workable concrete should flow sluggishly into place without segregation. *(Courtesy Portland Cement Association)*

Superplasticizers

When cement and water mix, the wet cement particles can form small clumps that inhibit proper mixing of cement and water. This reduces workability and inhibits hydration. Superplasticizers are admixtures that coat the cement particles, causing them to break away from the lumps and disperse in water. Superplasticizers give each cement particle a negative charge, causing them to repel each other, thus providing more thorough dispersement. Superplasticizers can be used to:

- Reduce water and cement at a constant water-cement ratio, giving a concrete the same strength as a normal mix but reducing the amount of cement used.
- Produce normal concrete at normal water-cement ratios that is so workable it can be placed with little or no vibration or compaction and not have excessive bleed or segregation.
- Produce a concrete of higher strength by reducing the water content required while maintaining the normal cement content.

Some superplasticizers are formulated to give slightly accelerated or high early strength. Others are used with water-reducing admixtures. These tend to retard and lower early strength but increase ultimate strength. It is recommended that superplasticizers be added directly to the mix in the concrete mixer truck to enable the slump to be maintained during long hauls or delivery delays. Commonly used superplasticizers include a higher-molecular-weight condensed sulfonate naphthalene formaldehyde, sulfonated melamine formaldehyde, and modified ligninsulfonates, all of which are in liquid form. The effectiveness of superplasticizers has a short duration—thirty to sixty minutes—before the concrete has a rapid loss in workability.

Permeability-Reducing and Damp Proofing Agents

Sound, dense concrete that has a water-cement ratio of 0.50 by weight and is properly placed and cured will be watertight. Concretes that have low cement contents, a deficiency in fines, or a high water-cement ratio can have permeability reduced by adding permeability-reducing agents such as certain soaps, stearates, and petroleum products.

Permeability is a measure of the amount of water that passes through channels running between the outer faces of the concrete. The permeability-reducing agent reduces the flow of water through these channels. They should generally not be used in well-proportioned mixes because they increase the amount of mixing water required, thus increasing rather than decreasing permeability.

Other permeability-reducing strategies, which may avoid the need for a permeability reducing agent, can be employed. Concrete that is properly placed, compacted, and cured will have reduced permeability. Air-entraining admixtures also reduce permeability by increasing the plasticity and reducing the amount of water needed. In effect, air-entraining is an excellent permeability-reducing admixture. Superplasticizers also disperse cement particles throughout the mix and help reduce permeability.

Damp-proofing admixtures are used to reduce moisture that is transferred by capillary action. This occurs when one side of the concrete is exposed to moisture and the other to air, as in a slab on-grade. The surface exposed to air tends to dry. Capillary action occurs as moisture flows to a dry surface. The effectiveness of damp-proofing admixtures varies, and manufacturers' test data need to be studied.

There are other ways to damp proof concrete. Various types of film can be applied to the damp side. A typical example involves putting a plastic membrane on the ground below a concrete slab floor. Various coatings, such as asphalt, sodium silicate, and metallic-aggregate mixed with Portland cement, can also be applied. Above-grade concrete and masonry can be damp proofed by applying a coat of silicone sealer. This also protects against freeze-thaw cycles, efflorescence, weathering, and staining.

Bonding Agents

Fresh concrete must often be placed over a previously poured surface that has already set. When fresh concrete is poured over hardened concrete, the fresh concrete shrinks, breaking its bond to the hardened concrete. The hardened surface must be prepared so the fresh concrete will firmly bond to the aggregate in the hardened surface. The condition of the surface is of great importance. It must be dry, clean (free of dirt, dust, grease, paint, etc.), and the proper temperature should be maintained.

Bonding admixtures can be added to Portland cement mixtures or applied to the surface of old concrete to increase bond strength. These admixtures are usually water emulsions of certain organic materials, such as a liquid acrylic polymer, which may be added to the Portland cement with or without mixing water. The pores of the concrete absorb the water and the resin unites into a mass that bonds the two layers of concrete.

Bonding is also accomplished by exposing the surface of the aggregate in the hardened concrete, applying a cement paste slurry to the hardened surface, and immediately pouring the new layer of concrete over it. Another type of bonding agent that also forms a waterproof membrane is a two-compound moisture-insensitive epoxy adhesive applied to hardened concrete and immediately covered by fresh concrete.

Coloring Agents

Concrete can be colored by mixing pure, finely ground mineral oxides with dry Portland cement. Thorough mixing is necessary to produce a uniform color. Oxides added to normal Portland cement are usually limited to earthy colors and pastels because of the cost and graying effect of the cement. White Portland cement produces clearer, brighter colors and is preferred. Some color agents are compounded to provide water-reducing and set-controlling properties.

Concrete can also be colored by exposing the aggregate. Colored aggregates are spread on the surface of the freshly cast concrete and floated in place. At the proper degree of set, the un-hydrated paste is washed away. Color can also be placed on the surface of concrete before it sets. One way involves using natural mineral oxides of cobalt, chromium, iron, or ochres and umbers ground to fine powder. These are combined with Portland cement into a topping mix that is troweled over the surface of the concrete.

Another method uses synthetic oxides mixed with fine silica sand. This is spread over the unset but floated concrete surface. The mixture is then floated and troweled into the uncured surface and cured in the normal way. Colors can also be mixed with a metallic aggregate and dry Portland cement. This mixture is applied as a dry shake to the surface, then floated, troweled, and cured. Some dry shakes also give additional hardness to the surface.

Color can also be achieved by simply painting or staining the concrete surface after it has been completely cured and neutralized. Neutralization can occur naturally via aging and weathering or it can be accomplished with a neutralizing agent, such as zinc sulfate.

Hardeners

When a concrete surface is subject to heavy wear, such as that of a factory or warehouse floor, its life can be extended by using a liquid-chemical or dry-powder hardener. One form of chemical hardener is a colorless, nontoxic, nonflammable liquid containing magnesium and zinc fluosilicates with a wetting agent. The wetting agent reduces the surface tension of the liquid hardener, which makes it easier for it to enter the pores of the concrete. The hardening agent produces a chemical reaction with the free lime and calcium carbonates in the Portland cement.

This reaction densifies the surface, making it more wear resistant and impenetrable to many liquids and chemicals. The liquid hardener is applied to the surface of the cured concrete in several coats. Similar hardeners are made using Baume sodium silicate that reacts with the lime to form an insoluble crystal within the pores of the concrete.

Concrete surfaces can also be hardened using dry powder hardeners. One product uses quartz silica aggregates and alkali-fast inorganic oxides, which color the concrete. These two are mixed with Portland cement and plasticizing agents, giving a dry shake that is applied to freshly poured concrete. This provides a high-strength surface with color provided by the oxides. Another type uses a finely ground iron aggregate and, if desired, inorganic oxides for color. These are mixed with Portland cement and plasticizing agents and applied as a dry shake on the surface of freshly laid concrete.

Grouting Agents

Portland cement grouts are widely used for stabilizing foundations, filling cracks in concrete walls, filling joints, grouting tenons or anchor bolts, and in other applications. Grout properties can be altered using the various admixtures discussed in this chapter.

Gas-Forming Agents

Gas-forming agents are added to concrete or grout to cause a slight expansion in it before it hardens. Aluminum powder, one of several gas-forming agents in use, reacts with the hydroxides in hydrating cement and produces small hydrogen-gas bubbles. This also helps eliminate voids caused by settlement of the concrete or grout. A typical application consists of expanding the grout under a difficult-to-reach machine base or column to be certain the area is fully grouted. After concrete or grout has hardened, gas-forming agents will not overcome shrinkage. When they are used in large amounts, gas-forming agents produce lightweight cellular concrete.

BASICS OF CONCRETE

Concrete is made up of two parts: aggregates and a paste. The paste is a Portland cement and water mixture with some entrapped air that coats the particles of aggregate and hardens and binds them. The paste hardens due to hydration, a chemical reaction between the Portland cement and water. The aggregates used range from fine to coarse, with various degrees of size within each.

Figure 7.8 shows the range in proportions of materials used in typical designs for rich and lean mixes. Notice that the volume of both cement and water varies from 7 to 15 percent and 14 to 21 percent respectively. A small percentage consists of air, and the rest, the major volume, is aggregate. This means that the selection of aggregates is vital for the production of quality concrete. It is essential that a continuous graduation of particle size be maintained. Figure 7.9 shows a cross-section cut through a sample of hardened concrete. The paste must completely fill the spaces between the particles and coat each particle.

Water-Cement Ratio

If all conditions are held constant, the quality of the hardened concrete is determined by the water-cement ratio. The water-cement ratio is the ratio by weight

Figure 7.8 The range in proportions of materials used in concrete by volume. *(Courtesy Portland Cement Association)*

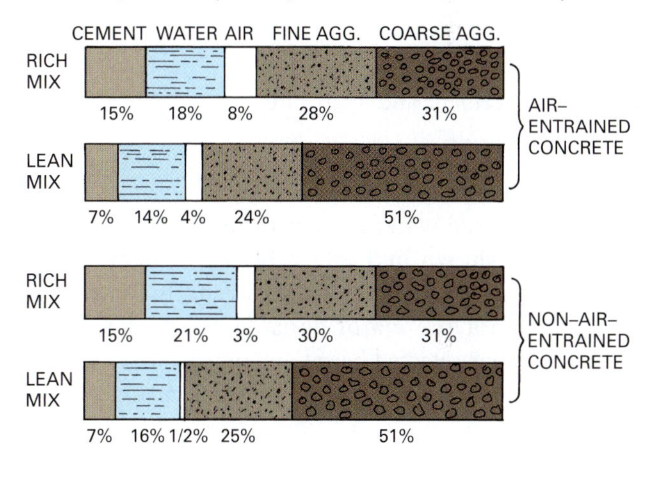

Figure 7.9 A cross section of hardened concrete showing how the cement and water ratio coats each particle of aggregate and fills the voids between particles. *(Courtesy Portland Cement Association)*

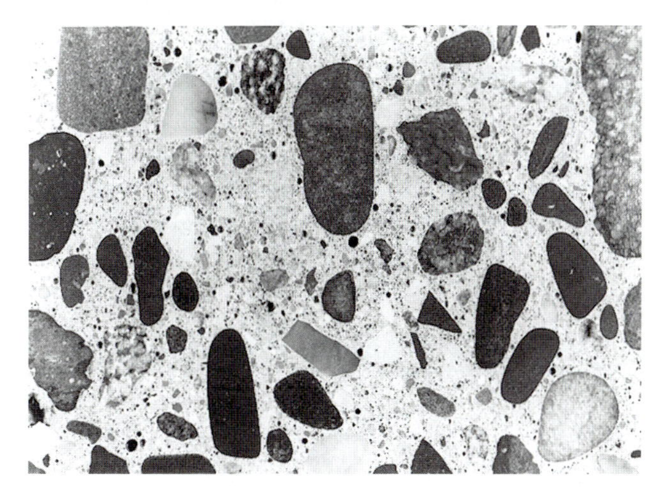

between the water and cement used to make paste. For example, a mix that uses 0.62 lb. of water per one pound of cement has a water cement ratio of 0.062/100, or 0.62. In metric measure, this would be 620 g of water per kilogram of cement, for a metric water cement ratio of 620. Stated another way, it is 62 percent of the weight of the cement.

The water-cement ratio should be the lowest value required to meet the design considerations. If too much water is used, giving a high water-cement ratio, the paste is thin and will be porous and weak after hardening. The effect of the water-cement ratio on the strength of concrete is illustrated in Fig. 7.10. Only a very small amount of water is needed for hydration to occur. A water-cement ratio of 0.31 will produce hydration, but usually more water is added so the concrete is workable and can be properly placed. The lower the water-cement ratio, the stronger the concrete (Table 7.10).

Concrete exposed to the elements must have the following for durability:

- Air entrainment
- Low water-cement ratio
- Quality cement and aggregate
- Proper curing
- Proper construction practices

The required water-cement ratios for various conditions are shown in Table 7.11. These data do not give consideration to relative strength. When durability is not the major consideration, the water-cement ratio is selected based on compressive strength. This often requires that tests be made using the actual job-site materials. If flexural strength is the basis

Figure 7.10 The graph demonstrates the effect of the water-cement ratio on the strength of concrete. *(Courtesy Portland Cement Association)*

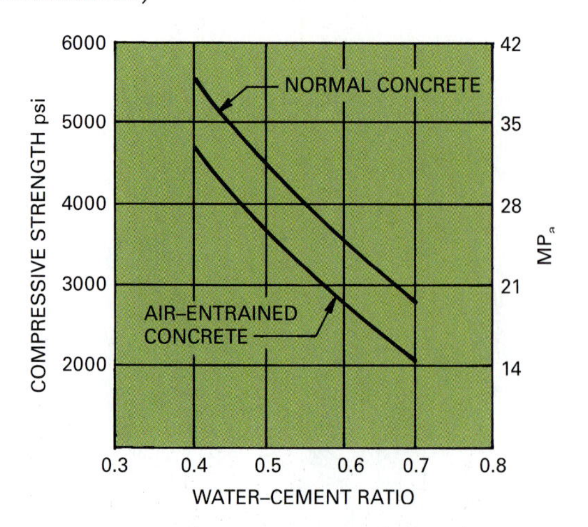

Table 7.10 Maximum Permissible Water-Cement Ratios for Concrete When Strength Data from Trial Batches or Field Experience Are Not Available

Specified Compressive Strength F'_c,psi[a]	Maximum Absolute Permissible Water-Cement Ratio, by Weight Non-air-entrained concrete	Air-entrained concrete
2,500	0.67	0.54
3,000	0.58	0.46
3,500	0.51	0.40
4,000	0.44	0.35
4,500	0.38	[b]
5,000	[b]	[b]

[a]*28-day strength. With most materials, the water-cement ratios shown will provide average strengths greater than required.*
[b]*For strengths above 4500 psi (non-air-entrained concrete) and 4000 psi (air-entrained concrete) proportions should be established by the trial batch method. 1000 psi ≃ 7 MPa.*
Courtesy Portland Cement Association

Table 7.11 Maximum Water-Cement Ratios for Various Exposure Conditions

Exposure Condition	Normal-Weight Concrete, Absolute Water-Cement Ratio by Weight
Concrete protected from exposure to freezing and thawing or application of deicer chemicals	Select water-cement ratio on basis of strength, work ability, and finishing needs
Watertight concrete	
In fresh water	0.50
In seawater	0.45
Frost-resistant concrete	
Thin sections; any section with less than 2-in. cover over reinforcement and any concrete exposed to deicing salts	0.50
All other structures	0.45
Exposure to sulfates	
Moderate	0.50
Severe	0.45
Placing concrete under water	Not less than 650 lb. of cement per cubic yard (386 kg/m^3)
Floors on-grade	Select water-cement ratio for strength, plus minimum cement requirements

Courtesy Portland Cement Association

for concrete design, tests are conducted to find the relationship between the water-cement ratio and flexural strength.

Minimum Cement Content

In addition to the specification of the water-cement ratio, the minimum cement content is also given. This ensures that the concrete will have good finishability, good wear-resistance, and good appearance (Table 7.12).

Aggregates

Aggregates must be properly graded and of proper quality. Grading pertains to the particle size and the distribution of particles. Properly graded aggregate produces the most economical concrete because it allows the use of the maximum-size coarse aggregate. This reduces the amount of water and cement required, and a reduction in cement reduces the cost.

The maximum size of aggregate used depends on the size and shape of the members being formed and the amount of reinforcing steel required. The maximum-size aggregate acceptable can be no more than one-fifth the narrowest dimension between the sides of the forms or three-fourths the clear space between steel, such as reinforcing bars, ducts, conduit, or bundles of bars. Aggregate in unreinforced slabs on the ground should not exceed one-third the slab thickness. For leaner concrete mixes, a finer grade of sand is used to improve workability. Richer mixes use a coarse grade of sand for greater economy.

Table 7.12 Minimum Cement Requirements for Concrete Used in Flatwork

Maximum Size of Aggregate, In.	Cement, lb. per Cubic Yard
1 1/2	470
1	520
3/4	540
1/2	590
3/8	610

1 in. ≃ 25 mm
100 lb./yd.^{3} ≃ 60 kg/m^3
Courtesy Portland Cement Association

Entrained Air

Entrained air should be used in all concrete paving regardless of the temperatures but especially for those exposed to freeze and thaw cycles. The required percentage of entrained air is shown in Table 7.13. The amount of entrained air required decreases as the maximum size of the aggregate increases. Entrained air reduces the amount of water required, producing a lower water-cement ratio.

Slump

Slump is a measure of the consistency of concrete, defining the ability of fresh concrete to flow. This is measured by the slump test explained later in this chapter. Slump is the decrease in height of a molded mass of fresh concrete that occurs immediately after it is removed from a standard metal slump cone. The higher the slump, or the more the sample lowers, the wetter the concrete mixture. A measure of slump can only be used to compare mixes of identical design. Slump is usually specified in the concrete specifications. Recommended slumps for concrete consolidated by mechanical vibration are shown in Table 7.14.

CONCRETE DESIGN

Concrete mixes are specially designed to provide the needed characteristics for a particular application. A mix designed for structural purposes will be higher in strength, while one intended for architectural finishes may have superior surface qualities.

Table 7.13 Recommended Average Total Air-Content Percentage for Level of Exposure

	Nominal Maximum Sizes of Aggregates					
Exposure	3/8 in. (10 mm)	1/2 in. (13 mm)	3/4 in. (19 mm)	1 in. (25 mm)	1 1/2 in. (40 mm)	2 in. (50 mm)
Mild	4.5	4.0	3.5	3.0	2.5	2.0
Moderate	6.0	5.5	5.0	4.5	4.5	4.0
Extreme	7.5	6.0	6.0	6.0	5.5	5.0

1 in. ≃ 25 mm
Courtesy Portland Cement Association

Table 7.14 Recommended Slumps for Various Types of Construction

Concrete Construction	Slump, In. Maximum[a]	Minimum
Reinforced foundation walls and footings	3	1
Plain footings, caissons, and substructure walls	3	1
Beams and reinforced walls	4	1
Building columns	4	1
Pavements and slabs	3	1
Mass concrete	2	1

[a]Slumps shown are for consolidation by mechanical vibration. May be increased 1 in. for consolidation by hand methods, such as rodding and spading.
Courtesy Portland Cement Association

Normal-Weight Concrete

The design of normal-weight concrete depends primarily on the required strength and durability. Consideration of workability, plasticity, and cost are factored secondarily.

As discussed earlier, the water-cement ratio is a basic premise used when designing normal-weight concrete. Based on research and experience, water-cement ratios for various applications can be recommended, as shown in Table 7.11. Another way mix design is accomplished involves using the absolute volume of material amounts. The absolute volume of a loose material, such as aggregate, is the total volume, including the particles and air spaces between them. This includes the absolute volume of the cement, aggregate, water, and trapped air.

$$\text{Absolute volume} = \frac{\text{weight of dry material}}{\text{specific gravity} \times \text{unit weight of water}}$$

For example, the absolute volume of 100 pounds of aggregate having a specific gravity of 2.5 would be:

$$\text{Absolute volume} = \frac{100 \text{ lb.}}{2.5 \times 62.5 \text{ (one cubic foot of water)}}$$

$$= 0.64 \text{ ft}^3$$

Publications of the Portland Cement Association give detailed instructions on concrete design by water-cement ratio and absolute volume.

Lightweight Insulating Concrete

The design of lightweight insulating concrete depends on the aggregate used and the desired compressive strength. The amount of water required varies greatly, depending on the circumstances. An air-entraining agent is recommended for some mixes. Detailed information is available from the Portland Cement Association.

Lightweight Structural Concrete

Lightweight structural concrete can be designed to produce structural members that are 25 to 35 percent lighter than members made with normal-weight concrete and have no loss in strength. Since the aggregate is cellular, and weights are different from normal aggregate, its design is usually derived from testing trial batches, experience, and other reliable test data.

CONCRETE TESTS

Hardened, cured concrete and freshly mixed concrete are tested to make certain they meet the written specifications for the concrete. This is especially important when working in various parts of the country because aggregates and water differ and the mix must be adjust accordingly.

Tests with Fresh Concrete

When testing fresh concrete, it is essential to take samples that are representative of the batch. Samples must be taken and handled following the specifications in ASTM C172. Except for slump and air-content tests, the sample must be at least one cubic foot (0.030 m^3) in volume. After it has been taken from the batch, the sample must be used within fifteen minutes and protected from sources of rapid evaporation during the test. Samples taken from the beginning and the end of a batch are not representative.

Slump Test Each load of transit-mixed concrete has a certificate listing its ingredients and their proportions. An on-site slump test is made to see if the required consistency has been achieved. A mix with a high slump may be too wet, and one with a low slump may be too stiff. The slump test is made following ASTM C143 specifications. A standard slump cone is 8 in. (200 mm) in diameter at the bottom and 12 in. (305 mm) high. The cone is placed on a flat surface and held still by standing on the foot support (Fig. 7.11). It is filled full and rodded twenty-five times with a 5/8 in. (16 mm) diameter, 24 in. (600 mm) long rod with a rounded tip. A second layer is poured and rodded as above, making certain the rod penetrates the surface of the layer below. After the top layer has been rodded, the excess concrete is struck off, leveling the top surface, and the mold is carefully lifted. The amount the concrete will slump is measured from the top of the cone (Fig. 7.12). For example, if the top of the concrete is 4 in. below the top of the cone, the slump is 4 in.

Another method for testing slump is the ball penetration test as specified by ASTM C360. The depth to which a 30 lb. (13.6 kg), 6 in. (150 mm) diameter hemisphere will sink into fresh concrete is measured.

Figure 7.11 A concrete sample is rodded during a slump test. *(Courtesy Portland Cement Association)*

Figure 7.12 The cone is lifted and the amount the concrete slumps below the top of the cone is measured. *(Courtesy Portland Cement Association)*

When calibrated for a particular set of materials, the results can be related to the slump. The concrete is placed in a container at least 18 in. (450 mm) square and at least 8 in. (200 mm) deep.

Unit Weight Test The unit weight test involves weighing a properly consolidated specimen in a calibrated container following ASTM C138 standards. It can determine the quantity of concrete produced per batch and give indications of air content.

Air-Content Test Methods for measuring air content include the pressure method (ASTM C231), the volumetric method (ASTM C173), and the gravimetric method (ASTM C138). The pressure method requires the sample be placed in a pressure air meter and subjected to an applied pressure (Fig. 7.13). The air content can be read directly. When lightweight aggregates are used, this method also compresses the air in them; therefore, tests by the pressure method are not recommended for concrete with lightweight aggregates.

The volumetric method measures air content by agitating a known volume of concrete in an excess of water. This method is suitable for concrete containing all types of aggregate. The gravimetric method uses the same test used for the unit weight test of concrete. The actual unit weight of the sample is subtracted from the theoretical unit weight, as determined from the absolute volumes of the ingredients, assuming no air is present. The mix proportions and specific gravities of the ingredients must be known. The difference in weight is given as a percentage as is the air content. This method requires laboratory control and thus is not suitable for on-site use. A quick check for air content can be made using a pocket-size air indicator. This suffices only for a quick on-site test and is not suitable for a standard ASTM test.

Cement Content Test The cement content test is used to determine the water and cement content of fresh concrete. The water-cement ratio has a major influence on strength. Therefore, this test gives an estimate of

Figure 7.13 A 1/4-cu ft. pressure-type air meter, ASTM C231. *(Courtesy Portland Cement Association)*

the strength potential without waiting for samples to harden and cure, which usually takes seven to twenty-eight days.

Strength Specimens Specimens of freshly poured concrete may be field-molded or laboratory-molded. The molding of test cylinders should be started within fifteen minutes after the specimens are obtained. Field-molded specimens (Fig. 7.14) should be made and cured as specified by ASTM C31 or AASHTO T23. Laboratory-molded specimens should be made and cured as specified by ASTM C192 or AASHTO T126. The size of the test cylinder depends upon the aggregate size. For example, a specimen with a maximum aggregate size of 2 in. (50 mm) or smaller can be made in a cylinder 6 in. (150 mm) in diameter.

The size of the mold for specimens used to test beams for flexural strength also varies depending upon the size of the aggregate. For example, a mold should be 6 in. × 6 in. (150 × 150 mm) in cross section for specimens with aggregates 2 in. (30 mm) or smaller. The cylinders are rodded and filled as specified.

Figure 7.14 Concrete cylinders being cast on the job site for compressive strength testing. *(Courtesy Portland Cement Association)*

After casting, the tops of the specimens should be covered with an oiled glass or metal plate, a special cylinder cap, or a plastic bag. The strength of the test specimen can be greatly affected by changes in temperature, jostling of the mold, and exposure to drying. Test specimens should be cast where they can be properly protected and movement is not necessary. Specimens taken and cured on the site in the same manner as the cast structure more closely represent the actual strength of the concrete in the structure at the time of testing. Detailed information is available from the Portland Cement Association.

Tests with Hardened Concrete

Specimens for strength tests of hardened concrete are made and cured according to ASTM C31 (in the field) and ASTM C192 (in the laboratory).

Compressive Strength Test After the test cylinder has cured as required it is ready for the compression strength test. The compression test is made according to ASTM specifications and is one of the most frequently required tests. The cylinders are placed in a compression testing machine (Fig. 7.15). As pressure is applied, the compressive strength is recorded up to the point when the cylinder fractures.

Figure 7.15 Compression strength testing of capped test cylinder. *(Courtesy Portland Cement Association)*

Flexural Strength Test The flexural strength test is used to determine the flexural, or bending strength, of concrete. The concrete sample is formed in a mold in the shape of a beam. Samples with aggregates up to 2 in. (50 mm) should have a minimum cross section of 6 × 6 in. (150 × 150 mm). Large aggregate samples should have a minimum cross-section dimension of three times the maximum size of the aggregate. The span of the test beam should be three times the depth of the beam plus two additional inches. A 6 × 6 in. (150 × 150 mm) beam would be 20 in. (508 mm) long. The mold is filled in two layers with one rodded stroke for every 2 in.^{2} (13 cm^2) of area. The top is struck flush with the mold and the sample is cured with controlled temperature and moisture. The cured specimen is tested as shown in Fig. 7.16. It is supported on each end, and pressure is applied to the midpoint until the specimen breaks. The ultimate flexural strength is read on a dial in pounds per square inch (kilopascals).

Abrasion Test The abrasion test is used to ascertain the resistance to wear of hardened concrete samples. A hardening admixture or surface coating is used with the sample concrete mix. The test is made on a machine that rolls steel balls under pressure in a circular motion on the surface of the specimen. The specimen is weighed before and after the test. The loss in weight determines the ability to resist abrasion (Fig. 7.17).

Freeze-Thaw Test Cured concrete specimens are placed in a freeze-thaw tester, which is a cabinet much like a freezer. It is run through a series of freeze-thaw cycles. The loss between the original weight and final weight of the specimen is used to determine which samples withstand the freeze-thaw cycle best.

Accelerated Curing Tests Accelerated curing tests are used when it is desirable to determine acceptance of structural concrete without the usual twenty-eight-day curing period. ASTM C684 has three methods for making accelerated strength tests.

Nondestructive Tests

Nondestructive tests are used to evaluate the strength and durability of hardened concrete. Commonly used tests are rebound, penetration, pull-out, and dynamic or vibration tests.

Figure 7.16 A concrete beam undergoing flexural strength testing. *(Courtesy Portland Cement Association)*

Figure 7.17 Test apparatus for measuring abrasion resistance of concrete. The machine can be adjusted to use either revolving disks or dressing wheels. *(Courtesy Portland Cement Association)*

Rebound Tests Rebound tests are made with a Schmidt rebound hammer. It measures the distance a spring-loaded plunger rebounds after striking the concrete surface. The reading is related to the compressive strength of concrete.

Penetration Tests The penetration method uses a Windsor probe, which is a power-activated gun that drives a hardened alloy probe into the concrete. The exposed length of the probe is measured and related by a calibration table to the compressive strength of the concrete. This leaves a small indentation in the concrete surface.

Pull-Out Tests A pull-out test requires that a steel rod with an enlarged end be cast in the concrete. A device used to pull the rod from the concrete measures the force required. This gives the shear strength of the concrete. It has the disadvantage of damaging the surface of the concrete.

Dynamic or Vibration Tests A dynamic or vibration test uses the principle that the velocity of sound in a solid can be measured by either recording the time it takes short impulses of vibrations to pass through a sample or determining the resonant frequency of a specimen. High velocities indicate a very good concrete while very low velocities indicate a poor concrete.

Emerging Trends in Concrete

Reactive Powder Concrete (RPC) is a high performance concrete able to achieve compressive strengths of up to 30,000 psi. Made by combining fine silica sand, cement, silica fume and crushed quartz with steel and synthetic fibers, RCP is highly ductile, extremely workable and relatively self-placing, requiring only minimal vibration.

Smog Eating Concrete uses cement that has been treated with titanium dioxide. The titanium dioxide sets off a photo-catalytic reaction with ultra violet light that accelerates natural oxidation and prevents bacteria and dirt from accumulating on a surface. The material can also break down nitrogen oxides emitted in the burning of fossil fuels.

Pervious Concrete is made with narrowly graded coarse aggregate, little or no fine aggregate, and a very low water/cement ratio. The result is a stiff, pebbly mixture with 15 to 25 percent of its volume composed of interconnecting pores through which water can flow. Pervious concrete is used for drives and walkways that allow water to percolate through a surface and thereby educe stormwater runoff.

Rammed earth, or pisé de terre, is a historic construction material that can be classified as a green building material because it utilizes locally available materials with little embodied energy (Fig. 7.18). Rammed earth walls are made by compressing a damp mixture of earth combined with sand, gravel, clay, and a small amount of Portland cement into conventional formwork. The damp material is poured into the forms 4 to 10 in. (100 to 250mm) at a time and is "rammed" with a pneumatically powered backfill tamper that compacts the mix to around 50 percent of its original height. Subsequent layers of the material are added and the process is repeated until the wall has reached the desired height. Like all solid masonry construction, a significant benefit of rammed earth construction is its excellent thermal mass qualities; it heats up slowly during the day and releases its heat during the evening. This can even out daily temperature variations and reduce the need for mechanical air conditioning and heating.

Figure 7.18 A rammed earth wall shows the layering of the construction process. *(Courtesy Eva Kultermann)*

Review Questions

1. What are the ingredients in concrete?
2. How does concrete harden?
3. What ingredients normally make up Portland cement?
4. What are the designations for the different types of Portland cement?
5. What is meant by air-entraining?
6. How does the addition of air-entraining materials to Portland cement improve the concrete?
7. What ingredients may be used in masonry cements?
8. How does the fineness of the cement affect the concrete?
9. Explain how Portland cement should be stored.
10. What natural aggregates are used in making concrete?
11. What aggregates should be avoided when making concrete?
12. What aggregates are used in insulating concrete?
13. What grains of sand are best for making concrete?
14. What are the commonly used admixtures?
15. What is the recommended water-cement ratio?
16. What five factors influence the quality of finished concrete exposed to the elements?
17. What can be done in hot weather to help retard rapid setting times?

Key Terms

accelerator
admixture
aggregate
air-entrained Portland cement
concrete
hydration
Portland cement
pozzolan
rammed earth
retarders
slump
water-cement ratio
workability

Activities

1. Visit a local concrete batch plant. Ask the supervisor to show the process for producing concrete and to explain instructions given to drivers of the delivery trucks.
2. If the concrete plant has a test lab, see if you can observe some of the tests made for the contractor.
3. Mix samples of concrete using various types of cement, varying the proportions of the ingredients and using different aggregates. Cure the samples and conduct compression tests. Compare the results and explain the differences that occur.
4. Prepare identical concrete samples and cure each the same number of days but under different temperatures (for example, at room temperature, above 100°F, and below freezing (use a freezer)). Test for compression strength. Examine the broken samples and report any differences in the appearance of the surfaces.
5. Prepare identical concrete samples, cure them under ideal conditions, and then expose them to various conditions to see how they perform. For example, immerse in salt water, fresh water, gasoline, oil, and other materials to which concrete is often exposed. Report your findings in a written paper.
6. On construction sites visited, observe how the aggregates are stored. Report the good and poor practices observed.
7. Prepare a list of admixtures and tell what purposes they serve.
8. Mix standard proportion concrete ingredients but vary the amount of water used. Measure the slump of each sample.

Additional Resources

Canadian Design and Control of Concrete, and other publications, Cement Association of Canada Headquarters, Ottawa, Ontario, Canada K1P 5Y7.

Concrete in Practice, National Ready Mixed Concrete Association, Silver Spring, MD.

The Contractors Guide to Quality Concrete Construction, Concrete Fundamentals, Cast-In-Place Walls, and numerous other technical concrete publications, American Concrete Institute, Farmington Hills, MI.

Design and Control of Concrete Mixtures, U.S. and Canadian metric editions, Portland Cement Association, Skokie, IL. Many other publications available.

Manual of Concrete Practice, and other publications, American Concrete Institute, Farmington Hills, MI.

Seismic and Wind Design of Concrete Buildings, International Code Council, Falls Church, VA.

8 CHAPTER

Cast-in-Place Concrete

LEARNING OBJECTIVES

Upon completion of this chapter, the student should be able to:

- Understand the processes for preparing, transporting, handling, and placing cast-in-place concrete.
- Discuss the finishes used on concrete surfaces.
- Know how concrete is cured.
- Cite various types of formwork used for cast-in-place concrete.
- Identify and describe types of concrete reinforcing materials.
- Describe cast-in-place on-grade concrete slabs and the types of joints commonly used.
- Explain how cast-in-place concrete walls, beams, and columns are formed, reinforced, and poured.
- Prepare sketches illustrating the various types of monolithically cast slab and beam floors and roofs.
- Describe the procedure for casting and erecting tilt-up concrete walls.
- Explain briefly what is meant by lift-slab construction.

Build Your Knowledge

For further study of these materials and methods, please refer to:

Chapter 7 Concrete

Topic: Basics of Concrete

Cast-in-place concrete is produced by setting wood, metal, molded plastic, or wood-fiber forms in place; placing reinforcing material in the forms; and pouring concrete over the reinforcing, filling the form. Cast-in-place concrete members must be designed by a professional engineer. The examples in this chapter are for illustration purposes only and are not intended to be used as design solutions.

An engineer can design cast-in-place concrete members in a wide range of sizes and shapes, with a variety of surface textures and colors. Although some concrete structural members can be precast and shipped to the site (see Chapter 9), cast-in-place parts of a structure generally are cast on-site. These include spread footings, foundation caissons, pilings, piers, slabs on-grade, and any members too large to precast and move to the site. Some designs have irregular shaped features that are difficult to precast and transport, so the pieces are cast-in-place. For example, construction of a dam requires huge intricate forms and massive amounts of concrete—all site-built and cast-in-place.

Cast-in-place concrete structural members usually are heavier than steel, wood, or precast concrete members, increasing the load on the foundation. Prefabricated steel, wood, and precast concrete members can be erected rapidly and in weather not suitable for cast-in-place concrete. There are continuing developments to make cast-in-place concrete faster and easier. A wide range of forms are available, in addition to equipment, such as concrete pumps and power finishing machines needed to speed up the process. Cast-in-place concrete is a widely used and effective construction material.

BUILDING CODES

Reinforced concrete structural members and pre-stressed concrete must be designed and constructed according to the provisions in the building code. This includes provisions to resist seismic forces if they are a factor in the area. Codes specify how to bend reinforcement, what the surface conditions must be, how to place the concrete in the forms, and what the coverage of the reinforcing within the concrete must be. For example, concrete that is cast against and will remain on the earth requires 3 in. (76.2 mm) of concrete cover over the reinforcement. Concrete walls, joists, and slabs not exposed to ground or earth typically require $\frac{3}{4}$ to 1 $\frac{1}{2}$in. (19 to 38 mm) of concrete over the reinforcing, depending on the diameter of the reinforcing bars. Codes also include specifications for placing concrete in corrosive environments, for thicknesses over reinforcing, fire protection, resistance to frost action, and for vertical and lateral loads.

PREPARING CONCRETE

Batching

Concrete is usually prepared in batches. A **batch** is the amount of concrete mixed at one time. The quantities of dry ingredient are usually weighed. Water and admixtures are specified by either weight or volume. The use of volume measurements is discouraged as they tend to be inaccurate because moisture in the aggregate changes the weight, which is not accounted for in volume measures. Aggregates, especially sand, tend to fluff when handled, so the actual volume of sand can vary from batch to batch. When concrete is produced by a continuous mixer, volumetric measure is used. The job specifications usually establish the percentage of accuracy allowed when measuring ingredients. These typically range from one to two percent, so accurate weighing facilities are essential.

Mixing

Concrete is mixed until it is uniform in appearance and all ingredients are evenly distributed. If an increased amount of concrete is needed, an additional mixer should be used, rather than overloading or speeding up those in operation. It is important to follow the manufacturer's recommendations and to keep the mixing blades clean. Bent or worn blades should be replaced.

Stationary Mixing

On a large job, the concrete is often mixed on-site using a stationary mixer. This can be a tilting or non-tilting type and may be manual, semiautomatic, or automatically controlled. Some mixers have data for various mix designs stored on computer programs. Generally, the batch is mixed one minute for the first cubic yard and an additional fifteen seconds for each additional cubic yard (0.35 m^3) or fraction thereof. Mixing time is measured from the moment all ingredients are placed in the mixer. All water must be added before one-fourth of the mixing time has elapsed. About 10 percent of the mixing water is placed in the drum before dry ingredients are added. The remaining water is combined uniformly with the dry ingredients, saving 10 percent for addition after all dry ingredients are in the drum.

Ready-Mix Concrete

Ready-mix concrete may be fully mixed in a central mixing plant and delivered to the site in a truck mixer that operates at agitating speed or in a special non-agitating truck (Fig. 8.1). The concrete may be partially blended in a central mixer and completed in a truck as it is moved to the site, or the dry ingredients may be placed in a transit truck mixer and the entire process done by the truck after the addition of water. When the entire batch is made in the truck mixer, seventy to one hundred rotations of the drum at the rotating speed specified by the manufacturer is enough to produce a uniform mix. All revolutions after one hundred should be at a slower agitating speed so as not to over-mix the batch. Concrete must be delivered and discharged within hours or before the drum has revolved three hundred times after the introduction of water.

Remixing

Fresh concrete in the drum tends to stiffen even before the concrete has hydrated to initial set. It can be used if remixing will restore sufficient plasticity for compaction in the forms. Under special conditions, a small amount of water can be added, but it must not exceed the allowable water-cement ratio, designated slump, and allowable drum revolutions, and it must be re-mixed at least half the minimum required mixing time or number of revolutions.

Figure 8.1 Central mixing in a stationary mixer of the tilting-drum type with delivery by a truck mixer operating at agitating speed. *(Courtesy Portland Cement Association)*

TRANSPORTING, HANDLING, AND PLACING CONCRETE

Before the fresh concrete arrives on the job, preparations for moving it to its point of placement must be complete. Delays in placing the concrete can cause a loss of plasticity. In addition, the method of moving the concrete must not result in the segregation of concrete materials. **Segregation** is the tendency of the coarse aggregate to separate from the sand-cement mortar. In some cases the heavy aggregate settles to the bottom and the sand-cement mortar rises to the top, producing unsatisfactory final results.

There are many ways to move and place concrete. As discussed earlier, truck agitators, truck mixers, and non-agitating trucks are used most frequently to bring concrete to sites. Concrete can also be mixed on-site with a stationary mixer. Concrete is moved about the site to points of placement with cranes using concrete buckets, barrows and buggies, chutes, belt conveyors, pneumatic guns, and concrete pumps (Fig. 8.2).

Moving Concrete

If transporting units are to be filled from a hopper, the concrete should pour straight into them. Concrete coming from a conveyor should be directed straight down by using deflector plates and a down pipe. If the concrete is moved with a chute, the chute should also be placed perpendicular to the receiving unit through a down pipe to prevent segregation of the mix. For many small jobs, such as pouring a garage floor or residential basement walls, wheelbarrows or manually pushed buggies can provide an adequate delivery of material. To speed up the flow of concrete, powered buggies can also be used.

Often a slab or foundation can be poured directly from the ready-mix concrete truck. If the truck can back up close enough, and the pour is not too high above grade, much of the pour can come directly from the truck (Fig. 8.3). In some cases, wheelbarrows or buggies may be used for part of a pour, with the rest pouring directly from the truck.

Multistory slabs, beams, and columns are poured from buckets lifted by cranes or from concrete pumps. The buckets are filled on the ground, lifted over the point of pour, and opened to deposit the concrete as shown in Fig. 8.4. Buckets are available in 2 yd.^{3} (1.5 m^3) capacities and may be round or square. They are opened by hand and poured from the bottom.

Concrete pumps are heavy-duty piston pumps that force concrete through a pipe ranging from 6 to 8 in. (152 to 203 mm) in diameter (see Fig. 8.2). They can place concrete over long distances, ranging up to about

Figure 8.2a A pump boom mounted on a mast can reach all points of placement. Concrete is supplied to the boom through a pipeline from a ground-level pump. *(Courtesy Portland Cement Association)*

Figure 8.2b A power buggy can move all types of concrete over short distances. *(Courtesy Portland Cement Association)*

Figure 8.3 Ready-mixed concrete discharged from a truck mixer to its final location. *(Courtesy Portland Cement Association)*

Figure 8.4 Ready-mixed concrete being elevated by bucket and crane to the top of a high-rise building. *(Courtesy Portland Cement Association)*

100 ft. (30.5 m) vertically and 800 ft. (244 m) horizontally. Pumps can also place concrete below grade, as required for foundations of multistory buildings, whose excavations must be several stories below grade. The maximum aggregate size for an 8 in. (203 mm) pipeline is 3 in. (75 mm); for a 7 in. (178 mm) pipeline it is $2\frac{1}{2}$ in. (64 mm); and for a 6 in. (152 mm) line it is 2 in. (50 mm). The pump requires an uninterrupted flow of plastic concrete to mitigate premature set and the production of unwanted joints in the slab, beams, or column. On the end of the discharge line is a choke that controls the flow of concrete. The pump and pipeline are thoroughly flushed after each use. Concrete is also moved with a conveyor, as shown in Fig. 8.5.

Placing Concrete

Concrete placement occurs after the base for on-grade pours is ready or the forms for walls, columns, and beams are erected and reinforcing is in place. Concrete should be placed continuously and as near as possible to its final location. In slab construction, work starts along one end, and each batch is discharged against the one previously placed. If the concrete is to be thick, like in a foundation wall, it should be placed in layers 6 to 20 in. (150 to 500 mm) deep for reinforced members and 15 to 20 in. (400 to 500 mm) deep for mass work. Each layer should be consolidated before a second layer is placed on it. It is necessary to work fast so the first layer is still plastic when the next layer is placed on it.

Consolidation is the process of compacting freshly placed concrete in forms and around reinforcing steel to remove air and aggregate pockets. Consolidation can be done by manually pushing a rod into the concrete or with a mechanical vibrator (Fig. 8.6). The vibrator should be lowered vertically into the concrete. It should not be used to move concrete along in a form. Rodding or vibrating should extend through the layer being consolidated and about 6 in. (150 mm) into the layer below.

Figure 8.5 A mobile conveyor belt carries a steady stream of concrete from ready-mix truck to the forms. *(Courtesy Portland Cement Association)*

Figure 8.6 A mechanical vibrator assures proper placement and compaction of concrete, even in heavily reinforced concrete members. *(Courtesy Portland Cement Association)*

It is important not to over consolidate. Too much consolidation tends to force heavy aggregate to the bottom and lighter cement paste to the top. The person doing the work judges the correct amount.

Placing Concrete in Cold Weather

Concrete placed in cold weather gains strength slowly. Fresh concrete must be protected from freezing. The critical period after which concrete is not seriously damaged by several freezing cycles depends on ingredients, conditions of mixing, placing, curing, and long-term drying. The concrete designer must consider the heat of hydration, the use of special cements and admixtures, and the temperature of the concrete, which is influenced by heating the aggregates and water. A formula is used to determine the temperature of fresh concrete. The final temperature of the combined ingredients should be well below 100°F (38°C)—most batches are kept within a 60°F to 80°F (15°C to 27°C) range. The temperatures of all batches should be about the same. If the overall temperature of the concrete exceeds 100°F (38°C), the concrete may flash set. When water is heated, most of the cement is not added until the bulk of the water and aggregates have been mixed in the drum. After placement, concrete must be protected from freezing. The length of time depends on the concrete mix and local conditions. Common methods of protection include insulated blankets and air heaters.

Placing Concrete in Hot Weather

In addition to maintaining low concrete temperature by cooling the aggregates and water, precautions must be taken to maintain a low temperature while placing concrete. This can involve shading and painting white the mixers, chutes, hoppers, pump lines, and other concrete handling equipment. Forms can be cooled with water, and the sub-grade can be moistened before the concrete is placed. Concrete must be transported from the mixing station to the point of placement as quickly as possible. In hot weather, the mix should be in place 45 to 60 minutes after mixing.

Pneumatic Placement

Concrete can be placed by pneumatically forcing a dry mixture of sand, aggregate, and cement through a hose and mixing it with water at a nozzle. This is referred to as pneumatically placed concrete, or shotcrete. It is used to form thin sections in difficult locations and to cover large areas. Shotcrete is ideal for placing concrete in free-form shapes, such as domes and shells; for applying protective coatings; and for repairing concrete surfaces. Typical applications include forming swimming pools, covering rock outcroppings along highways to prevent rock falls, and providing underground support, as in tunnel and coal mine shaft linings. A surface can be covered with metal lath and concrete sprayed over it. This can be done without the construction of expensive forms.

FINISHING CONCRETE

After concrete has been placed and consolidated, it is screeded (Fig. 8.7). **Screeding**, also called strike-off, involves removing excess concrete with a screed to bring it flush with the top of the form work. Immediately after strike-off, the surface is bull floated to lower high spots, fill low spots, and embed large aggregate that may be on the surface. A **bull float** has a long handle connected to a float (Fig. 8.8). A darby has a shorter handle and

Figure 8.7 Vibratory screed units reduce the work of strike-off while consolidating the concrete. *(Courtesy Portland Cement Association)*

Figure 8.8 A bull float is used to lower high spots and fill in low spots. *(Courtesy Portland Cement Association)*

is used for shorter distances. This work must be done before any bleed water appears on the surface. **Bleed** refers to water that rises to the surface very soon after concrete is placed in forms.

When the bleed water sheen has evaporated, the surface is ready for final finishing. Any finishing operation performed on the surface of the concrete while bleed water is present will cause it to scale and dust. Final finishing includes one or more of the following: edging, jointing, floating, troweling, and brooming.

Edging rounds off the edges of a slab to prevent chipping. **Jointing** forms control joints in a slab. A groove with the thickness of the slab is formed across it at intervals specified by the architect. Its purpose is to provide a weak spot where the slab can crack when stresses exceed the strength of the concrete. Control joints can be formed in wet concrete, sawed after concrete has hardened, or formed by inserting plastic or hardwood strips in the concrete (Fig. 8.9). Isolation and construction are other joints used. Isolation joints provide a space between a slab and a wall, allowing each to move without disturbing the other. Construction joints are formed where one pour ends and a joining one meets it (Fig. 8.10).

After concrete is edged and jointed it can be floated. This is done with a wood or metal handheld float or a finishing machine with float blades (Fig. 8.11). Floating embeds aggregate slightly below the surface, removes imperfections, compacts mortar at the surface for final finishing, and keeps the surface open, allowing excess moisture to escape. Marks left by edging and jointing are removed by floating. If they are wanted for decorative purposes, they need to be rerun after floating.

The final finish might be accomplished by troweling, brooming, or forming a pattern or texture in the surface. **Troweling** produces a hard, dense, smooth surface.

Figure 8.11 A power float, used after the bleed water sheen has evaporated and the concrete will sustain foot pressure with only slight indentation. *(Courtesy Portland Cement Association)*

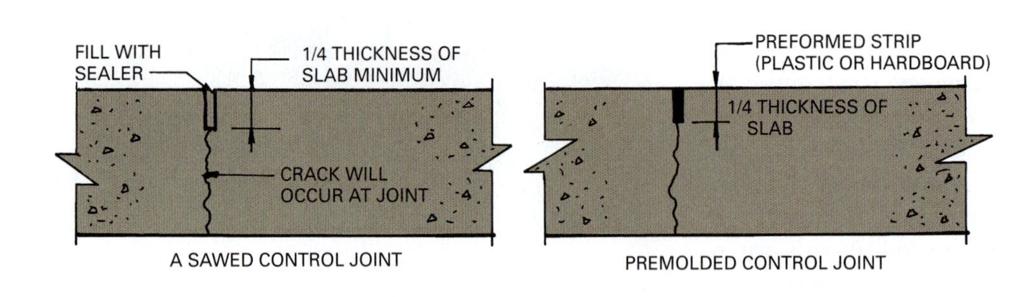

Figure 8.9 Control joints provide a controlled place for the slab to crack without being visible.

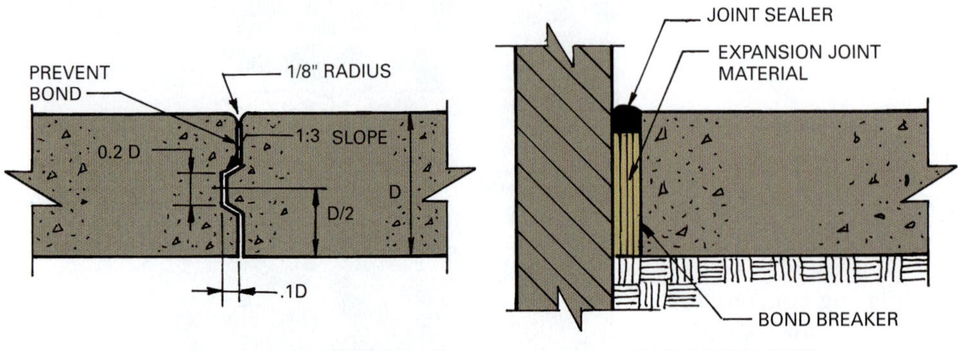

Figure 8.10 Construction joints occur where two pours meet. Isolation joints separate a pour from a wall, column, or other abutting form.

The trowel is a steel-bladed handheld tool. Brooming involves roughing the surface with a steel-wire or coarse-fiber broom. It provides a slip-resistant surface. Patterns can be formed in the surface by placing divider strips in the concrete. For example, it can be segregated to look like flagstones. For an exposed aggregate finish, aggregate can be embedded in the surface and excess paste washed away.

CURING CONCRETE

As explained in Chapter 7, adding water to Portland cement produces a chemical reaction called hydration. This reaction produces a hard cement paste that bonds the aggregate into a solid mass. Hydration continues for an indefinite period at a decreasing rate as long as water is in the mix and the temperature is favorable (73°F [23°C] is recommended). The material reaches design strength after 28 days of **curing**. Concrete should be protected, so moisture remains in the mix during the early hardening period and the temperature is maintained. Protected concrete that is kept moist for seven days has about twice the compressive strength of unprotected concrete (i.e., concrete exposed to air with no attempt to keep moisture in the mix). The curing process is essential to producing concrete members with the expected compressive strength.

The length of the curing period depends on the type of cement, the design of the mix, the strength required, the size of the member poured, and weather conditions.

Forms may be left in place and the exposed concrete surfaces kept moist. Curing compounds are sprayed on the surface to retard moisture evaporation. After the forms have been removed, the exposed concrete can be sprayed with curing compound. Exposed concrete can be covered with waterproof curing paper or plastic film to hold moisture in. Wet covering materials, such as burlap or moisture-retaining fabrics, can be placed over the concrete. They should be kept wet over the entire curing period. Continuously sprinkling the surface is an excellent method of curing. The sprinkling must be done so that the concrete is always wet.

FORMWORK

A variety of manufactured metal forms (steel and aluminum) are available in a range of sizes, along with assembly and bracing systems (Fig. 8.12). Plywood with wood or metal bracing is used for site-built forms. Although used for general wall construction, they are the main material for constructing more intricate and complex forms. An excellent material for carpenter-built forms is exterior high-density overlay plywood, although other plywood made with waterproof glue is also used. Solid lumber and Wafer board are also used in concrete form construction, (Fig. 8.13), and a number of molded plastic and waxed cardboard forms are also available.

Both wood and metal forms are generally assembled in inside/outside pairs. The face of the form is oiled or treated with a chemical release agent to facilitate easy stripping. When rebar is required, one form side is set and the reinforcing bars are placed inside before the second side is assembled (Fig. 8.14).

Figure 8.12 Foundation walls can be formed quickly using standard metal forms.

Figure 8.13 Site-built formwork made of plywood panels and lumber frames and bracing. *(Image copyright Mihai Simonia, 2009. Used under license from Shutterstock.com)*

Figure 8.14 A site-built plywood form assembly.

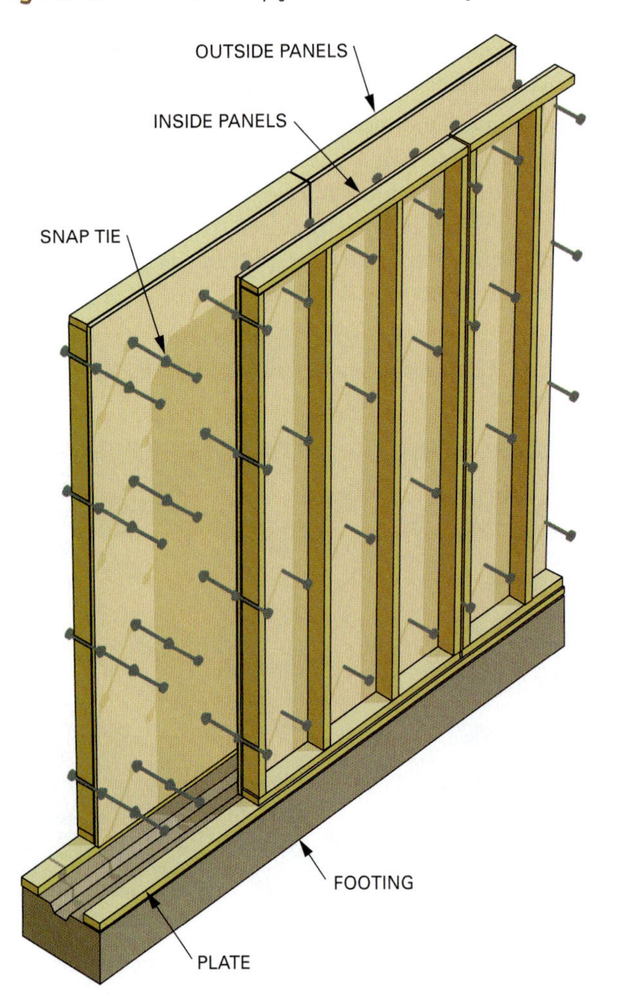

Snap ties hold forms together at the desired width. They support the forms' sides against lateral pressure from the wet concrete. Walers are used to provide additional reinforcing to form work. Once concrete has cured and the forms are removed, the snap tie ends are broken off just inside the surface of the concrete. The remaining holes are either grouted or capped.

The formwork sometimes must support the working deck in addition to the reinforcing steel, concrete placing and finishing equipment, and the weight of workers and equipment. It must be designed and erected to carry these vertical loads in additon to resisting lateral forces. The resident engineer or building inspector usually verifies that the form construction is safe.

The finish produced by forms ranges from the untreated surface left by the forms to one produced by form liners. Form liners are molded plastic sheets that have been modeled from actual concrete, masonry, or wood patterns. They are bonded to plywood sheets and secured inside a form, producing a textured surface (Fig. 8.15).

Figure 8.15 Form liners with textured surfaces can produce a variety of surface patterns. *(© Nathan Griffith/Corbis)*

Other forms are used to construct cast-in-place structural members. Columns are usually round, square, or rectangular.

Custom forms are offered by suppliers of concrete forms. One example is shown in Fig. 8.16. This rectangular column has heavily rounded corners, as does the spandrel beam that connects to the column. These forms were custom-made based on architectural designs. An example of one-sided formwork is shown in Fig. 8.17. This form uses vertical truss-like structural members against which the form sheets are placed. The bracing is set on footings and is positioned with a number of screw-adjusted feet.

Figure 8.16 A rectangular column and spandrel beam cast in custom-molded forms. *(Courtesy Molded Fiberglass Concrete Forms Co.)*

Circular formwork is shown in Fig. 8.18. The horizontal walers are curved metal units, and the form is supported with round steel braces bolted to the concrete floor.

An example of climbing formwork is shown in Fig. 8.19. After a section is poured and reaches sufficient strength, the form is raised to the next level and positioned, reinforcing is set in place, and the pour is repeated.

Figure 8.17 A one-sided form is supported by heavy tubular steel bracing. *(Image copyright Anyka, 2009. Used under license from Shutterstock.com)*

Figure 8.18 A plywood skin is used on this circular form work. Notice the curved walers and braces bolted to the concrete floor. *(Courtesy PERI Formwork Systems, Inc.)*

Large projects use a variety of forms. Figure 8.20 shows forms being installed for part of a dam construction, including climbing, curved concrete wall forms and forms for slanting concrete surfaces. Form construction and installation is a major consideration in large projects such as this.

Concrete floors and roofs are constructed in a number of ways. Figure 8.21 shows pans set for pouring a waffle floor. Reinforcing bars needed in the ribs run between the pans. The reinforcing for the floor is placed over the pans. Larger areas without pans form column heads cast over each column below the floor. These are discussed in greater detail later in this chapter.

Flying formwork is built in large sections that are lifted by crane to the next floor and reused (Fig. 8.22). The formwork is supported by metal posts, providing

Figure 8.19 Climbing formwork used to pour exterior wall panels is raised from one level to the next.

Figure 8.20 A concrete dam under construction. *(Courtesy Portland Cement Association)*

Figure 8.21 These pans with steel reinforcing are often used to form cast-in-place slabs with coffered ceilings.

Figure 8.22 Flying formwork tables with aluminum and timber joists. The tables are supported by shoes attached to previously poured columns and walls.

a strong, rigid unit. Reusability of the forms and bracing reduces costs by eliminating some of the labor required to strip and rebuild formwork on the next floor.

Another type of form used for residential and light commercial foundation construction is an insulated concrete form (ICF) with intergral stay-in-place insulation (Fig. 8.23). The form is made from expanded polystyrene in units that are notched to facilitate stacking to form the cavity for the cast-in-place concrete wall. The formwork is nonstructural, and although it remains in place after the wall is poured, the reinforced concrete is designed to carry the loads. The form provides insulation and acoustical values to the wall. It can be used to build walls above and below grade (Fig. 8.24).

Figure 8.23 Expanded polystyrene insulating concrete forms provide formwork and remain as insulation. *(Courtesy Portland Cement Association)*

CONCRETE REINFORCING MATERIALS

Concrete has no useful tensile strength of its own, so steel reinforcing with high tensile strength is added. Concrete and steel have about the same coefficient of thermal expansion, concrete bonds to steel, and steel is not corroded by concrete, so they work together

Figure 8.24 Insulated Concrete Forms (ICF) use plastic inserts as form ties that support the horizontal reinforcing bars. *(Courtesy Portland Cement Association)*

to provide an efficient structural system. The two commonly used steel reinforcing materials are reinforcing bars with associated hooks and stands, and welded wire reinforcement. A wide range of fibers also serve as concrete reinforcing.

Steel Reinforcing Bars

Reinforcing bars (also called rebars) are hot rolled steel rods that may be plain (smooth) or deformed. The deformed type has surface ridges, which provide better bonding to the concrete. The smooth type is used for special applications. The bars are available in 60 ft. (18.3 m) lengths and in eleven standard diameters. The metric bars are made in eight diameters (Table 8.1). The diameters are identified by the bar-size designation. Inch-size bar designations represent in. of bar diameter. For example, a No. 4 bar is $\frac{4}{8}$, or $\frac{1}{2}$, inch in diameter. Metric designations give the diameter in millimeters.

Reinforcing bars are manufactured to ASTM standards A615, A616, A617, and A706. They are made in grades 40, 50, 60, and 75. These refer to the minimum yield strength of the steel—40,000, 50,000, 60,000, and 75,000 psi (276, 345, 414, 517 MPa) (Table 8.2).

Reinforcing bars are made from three types of steel: rail steel, axle steel, and billet steel. The bars are available galvanized or coated with epoxy to prevent corrosion. Examples of the markings stamped into the rebar are shown in Fig. 8.25. The type of steel, bar size, grade mark, and identification of the production mill are given. Higher-strength bars are used where space is tight and smaller-diameter higher-strength bars provide the strength needed. Concrete columns and beams are members for which high-strength rebar is often used.

Table 8.1 Steel Reinforcing Bar Sizes

ASTM Inch-Size Steel Reinforcing Bars			
Bar Size Designation	Weight in Pounds Per Foot	Nominal Dimensions Diameter in Inches	Cross-Sectional Area in Square Inches
#3	0.376	0.375	0.11
#4	0.668	0.500	0.20
#5	1.043	0.625	0.31
#6	1.502	0.750	0.44
#7	2.044	0.875	0.60
#8	2.670	1.000	0.79
#9	3.400	1.128	1.00
#10	4.303	1.270	1.27
#11	5.313	1.410	1.56
#14	7.650	1.693	2.25
#18	13.60	2.257	4.00
ASTM Metric-Size Steel Reinforcing Bars			
Bar Size Designation	**Mass (kg/m)**	**Nominal Dimensions Diameter (mm)**	**Area (mm^2)**
#10M	0.785	11.3	100
#15M	1.570	16.0	200
#20M	2.355	19.5	300
#25M	3.925	25.2	500
#30M	5.495	29.9	700
#35M	7.850	35.7	1,000
#45M	11.775	43.7	1,500
#55M	19.625	56.4	2,500

The amount and placement of reinforcing steel for concrete structural members is specified on the structural drawings. The engineering drawings give the size, location, and bending information (Fig. 8.26). The steel fabricator cuts the bars to length and makes the required bends. The bends are specified by the engineer on the engineering drawings. Information pertaining to the bend designs can be obtained from the Concrete Reinforcing Steel Institute.

Table 8.2 Grades, Strengths, and Types of Reinforcing Steel

Type of Steel	Steel Type Symbol	Yield Strength (psi)	Tensile Strength (psi)	Yield Strength (MN/ m^2)	Tensile Strength (MN/ m^2)
ASTM A615					
Billet steel					
Grade 40	S	40,000	70,000	276	483
Grade 60		60,000	90,000	414	621
Grade 75		75,000	100,000	517	690
ASTM A616					
Rail steel					
Grade 50	R	50,000	80,000	345	552
Grade 60		60,000	90,000	414	621
ASTM A617					
Axle steel					
Grade 40	A	40,000	70,000	276	483
Grade 60		60,000	90,000	414	621

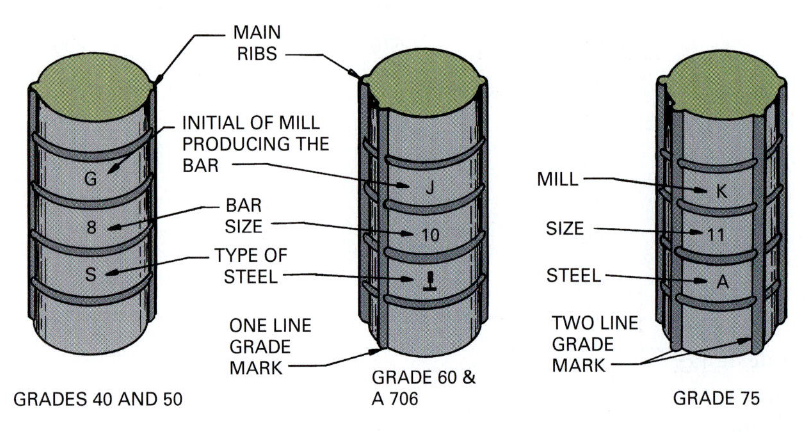

Figure 8.25 Markings stamped into reinforcing bars give the bar size, type of steel, and grade mark, as well as identifying the mill that produced the bar.

Figure 8.26 Examples of engineering drawings for cast-in-place structural members of a building.

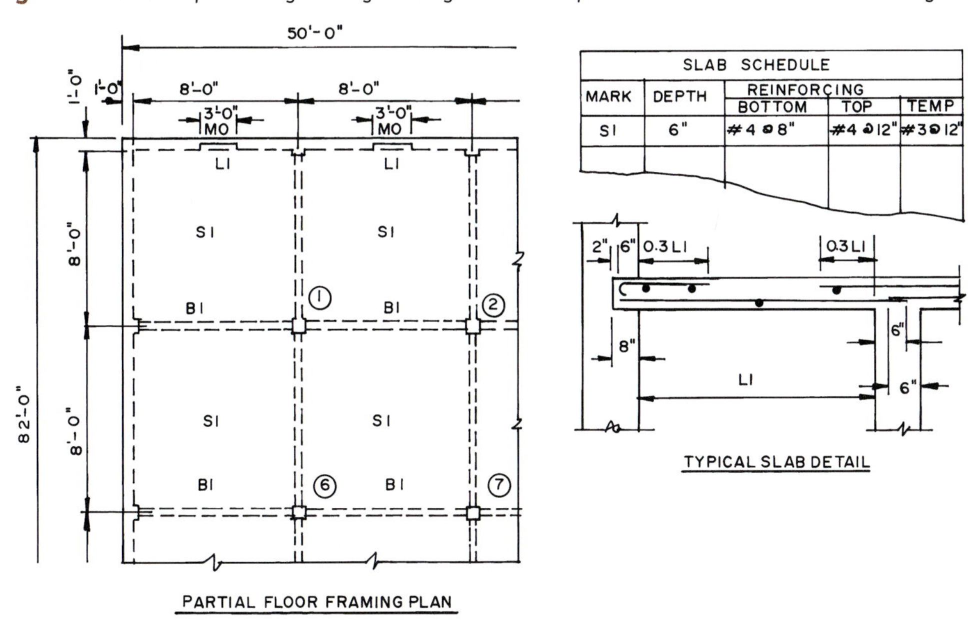

BEAM AND GIRDER SCHEDULE							
MARK	SIZE		BOTTOM		TOP	STIRRUPS	
	W	D	"A" BARS	"B" BARS		NO.- SIZE	SPACING FROM FACE OF SUPPORT
G1	12"	30"	3#8	3#8	2 #9 NON-CONTINUOUS ENDS 3# 8 AT COLUMNS #8 IN TOP LAYER	20#3	1@2", 3@6", 2@9", 2@10", 2@12"
B1	10"	20"	3#8	3#6	3# 10 AT COLUMNS	10#3	1@2", 2@12", 2@18"
L1	8"	12"	2#6	—	—	NONE	

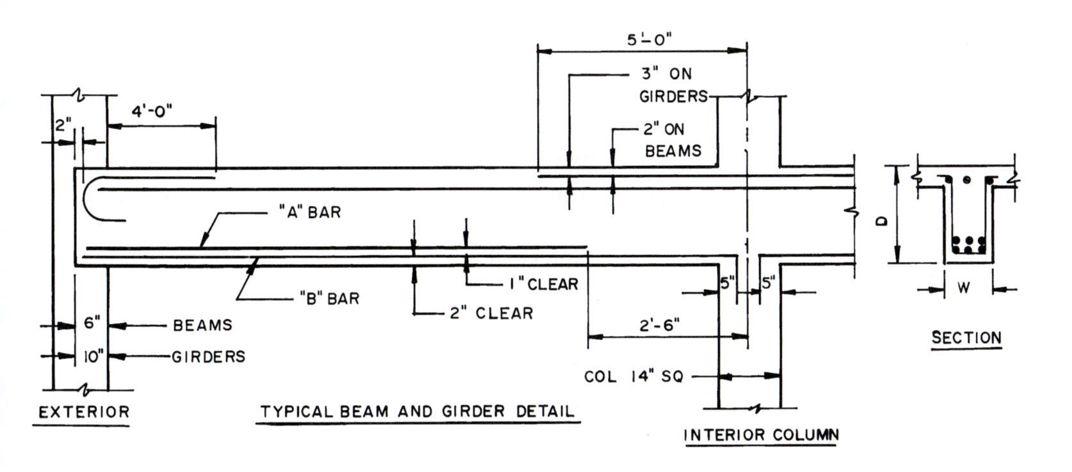

BENDING DETAILS							
MARK	SIZE	LENGTH	TYPE	A	B	C	D
B1	8	8'-6"	2	8	8'-0"		
S1	4	4'-6"	1	6	4'-0"		

B
A
TYPE 1

B
A
TYPE 2

B
K
C
H
D
TYPE 3

Bend types are standardized and identified by number. A few of these are shown in Fig. 8.27. Some of the standard hook bends are shown in Fig. 8.28. Reinforcing bars must be bent cold unless specific approval is given by the design engineer. The formed bars are wired together into bundles and tagged. The tag has the name of the fabricator, the address of the job, fabrication data, and the mark that locates their place on the structural drawing.

Some reinforcing is preassembled before being placed in the forms. Examples include column spirals, column ties, and footing bars (Fig. 8.29). The steel in beams usually involves a set of bottom bars and stirrups. The bottom bars resist tension forces that exist in the bottom of the beam. The stress is dissipated from the bars into the concrete through the bond between the concrete and the bars. The concrete in the top is under compression, and concrete is able to resist compression forces. The ends of the bottom bars are bent into hooks that help dissipate tension forces at the end of the beam. Tension forces remaining in the bearing ends of the beam are resisted by stirrups. Most applications use U-stirrups, but in some cases closed-stirrup ties are required (Fig. 8.30).

The bottom bars in the beam are raised above the bottom of the form with one of several types of bar supports. These can be bolsters or chairs. Bars in the top of the slab are supported with high chairs.

Figure 8.27 Some of the standard bends used with steel reinforcing bars. *(Courtesy American Concrete)*

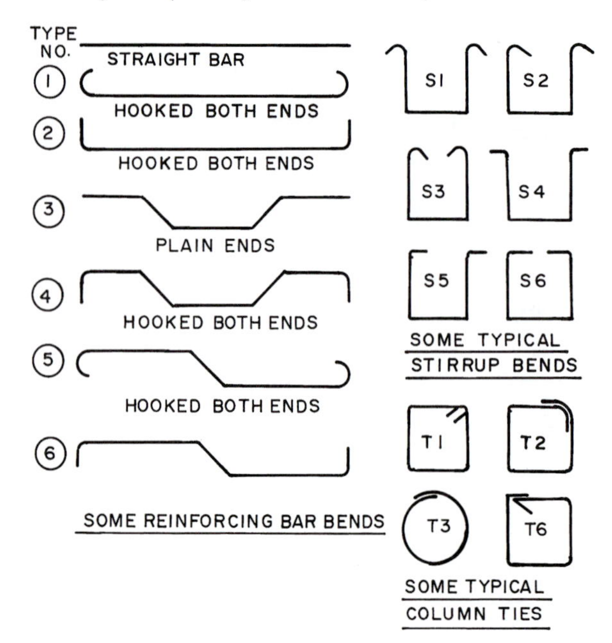

Bar supports are available in wire, precast concrete, reinforced cementitious fiber, and all-plastic types (Fig. 8.31). Reinforcement that rests directly on the ground can be supported by bar supports made from concrete, which become completely sealed from the earth by the concrete. Steel wire chairs tend to provide a passage for moisture and rust to reach the bottom reinforcing bars.

If a member is formed from several pours, the reinforcing bars are extended beyond the end of the pour and into the next one. The bars in the second pour overlap those from the first at a distance specified by codes. Figure 8.32 shows concrete columns with the steel extended. Connections between reinforcing bars in columns require that they be spliced end-to-end and secured with a mechanical splicing device or by welding. Connections should be made as specified by the structural engineer.

Welded Wire Reinforcement

Welded wire reinforcement (WWR), sometimes called welded wire fabric (WWF), is an assembly of steel reinforcing wires made from rod that is either cold-drawn or cold-rolled or both. The transverse wires are electrically resistance-welded to each of the longitudinal wires to form square or rectangular grids (Fig. 8.33). The material is available in rolls and sheets. The wires may be plain (W) or deformed (D). The plain welded wire bonds to concrete by the positive mechanical anchorage at each welded intersection. The deformed wire uses the deformations in the surface of the wire in addition to the welded intersections for bonding and anchoring. Welded wire reinforcement is widely used to reinforce concrete structures. The smaller diameter wires provide more uniform stress distribution and crack control. The selection of wire sizes by the engineer provides the needed cross-sectional area of the reinforcing steel.

In addition to uncoated wire, two coatings are available. One is a hot-dipped galvanized coating specified by ASTM A641 or A153. It is usually applied to the wire before welding. The other is an epoxy coating specified by ASTM A884. This is applied after the sheets have been welded.

The terms used when discussing WWR are building fabric, pipe fabric, and structural welded wire reinforcement. Building fabric includes products with wire sized up to W4.0. Wire spacing is generally 4 × 4 in. and 6 × 6 in. (101 × 101 mm and 152 × 152 mm). Sheets and rolls are available in building fabric styles. Pipe fabric is formed into cylindrical pipe forms (circular, elliptical, and arch types). Wire sizes go up to W12, generally

Figure 8.28 Standard hook forms used with steel reinforcing bars. *(Courtesy Concrete Reinforcing Steel Institute)*

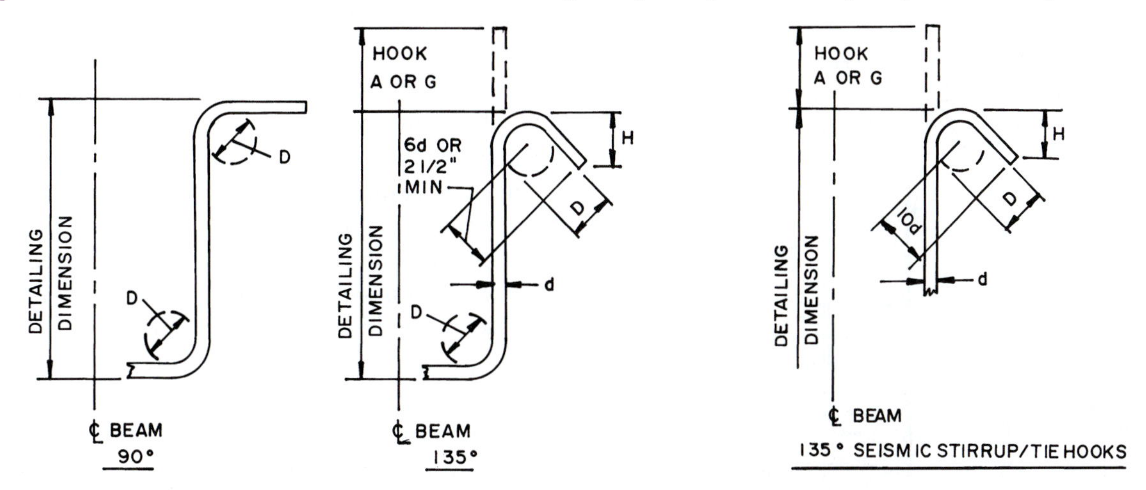

Bar Size	D, In.	Stirrup and tie hook dimensions, in.*		
		90-deg hook	135-deg hook	
		A or g	A or G	H, approx
#3	1 1/2	4	4	2 1/2
#4	2	4 1/2	4 1/2	3
#5	2 1/2	6	5 1/2	3 3/4
#6	4 1/2	1-0	7 3/4	4 1/2
#7	5 1/4	1-2	9	5 1/4
#8	6	1-4	10 1/4	6

Bar Size	D, In.	135 deg seismic stirrup/tie hook dimensions, in.*	
		135-deg hook	
		A or G	H, approx
#3	1 1/2	5	3
#4	2	6 1/2	4 1/2
#5	2 1/2	8	5 1/2
#6	4 1/2	10 3/4	6 1/2
#7	5 1/4	1-0 1/2	7 3/4
#8	6	1-2 1/4	9

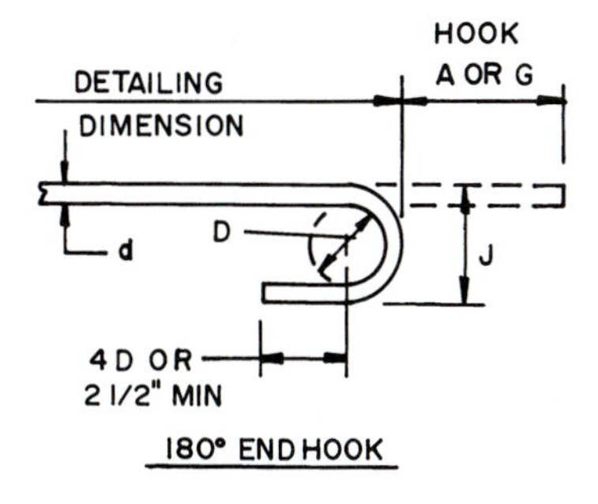

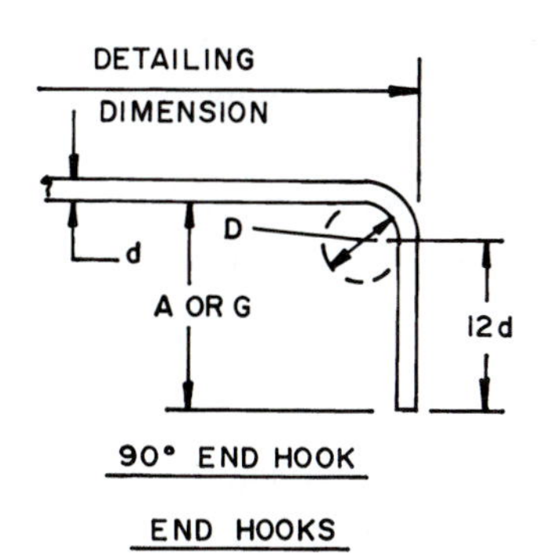

RECOMMENDED END HOOKS ALL GRADES				
Bar size	Finished bend diameter D, in.	180-deg hooks		90-deg hooks
		A or G, in.	J, in.	A or G, in.
#3	2 1/4	5	3	6
#4	3	6	4	8
#5	3 3/4	7	5	10
#6	4 1/2	8	6	1-0
#7	5 1/4	10	7	1-2
#8	6	11	8	1-4
#9	9 1/2	1-3	11 3/4	1-7
#10	10 3/4	1-5	1-1 1/4	1-10
#11	12	1-7	1-2 3/4	2-0
#14	18 1/4	2-3	1-9 3/4	2-7
#18	24	3-0	2-4 1/2	3-5

with spacing of 2 × 6 in. (50 × 152 mm), 2 × 8 in. (50 × 203 mm), and 3 × 6 in. (76 × 152 mm) in the standard styles. Pipe fabric usually comes in roll form. Structural welded-fabric-reinforcement wire sizes include anything over D4 or W4. They have a variety of wire spacings, from 3 in. to 18 in. (76 to 457 mm) in both directions. Generally, structural welded wire is furnished in sheet or mat form.

Figure 8.29 Reinforcing is often preassembled and placed into the form.

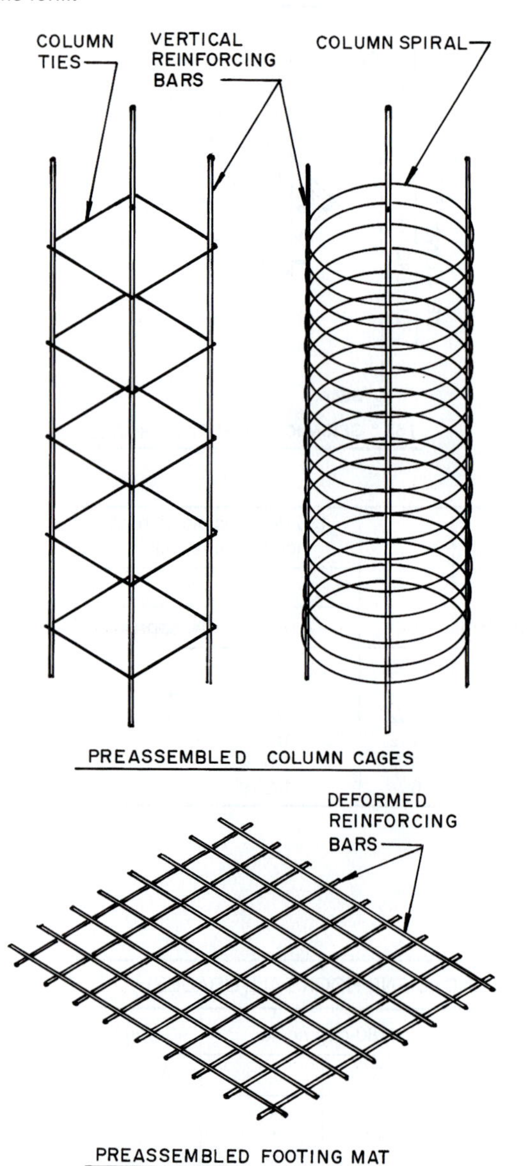

Welded wire reinforcement is specified by listing the longitudinal wire spacing, transverse wire spacing, longitudinal wire size, and the transverse wire size. An example for WWR style is 12 × 12 - W12 × W5. This means that the longitudinal and transverse wire spacing is 12 in. apart. The longitudinal wire type and area: W12 denotes a plain wire with an area of 12 in.^{2}/ft. The transverse wire type and area: W5 means plain wire, with an area of .05 in.^{2}/ft.

Longitudinal wire spacings are available in 2, 3, 4, 6, 8, 10, 12, and 16 in., and in some cases, spacing to 24 in. is available. Soft metric conversions are 51, 76, 102, 152, 203, 254, 305, and 406 mm. Transverse wire spacings are 3, 4, 6, 8, 12, and 16 in., with soft metric conversions of 76, 102, 152, 203, 305, and 406 mm. The designations for common styles of welded wire reinforcement are detailed in Table 8.3. Notice most sizes are available in both W- and D-type wires.

Detailed information relating to the manufacture, specifications, properties, design information, and building code requirements are available from the Wire Reinforcement Institute, Inc. The American Concrete Institute publication ACI 318, *Building Code Requirements for Reinforced Concrete*, contains design data information on welded wire reinforcement.

Welded wire sheets are shipped in bundles in quantities varying with the size and weight of the sheets. Typical bundles weighing 500 to 5000 lb. (227 to 2268 kg) are bound together with steel strapping. The bundles should never be lifted off a truck by the steel strapping. Lifting eyes can be specified when bundles are lifted by crane.

The engineer specifies the amount of reinforcement required and the correct placement of it within the slab or wall. The sheets must be placed on supports, or, in the case of walls, firm support spacers are used to maintain their position as the concrete is placed. The supports are usually concrete, steel, or plastic chairs, as discussed earlier. The amount of splice for sheets is also specified by the engineer. Slab-on-grade splices can generally be less than a structural splice, since the steel reinforcement is used primarily for crack control.

Fiber Reinforcement

In addition to steel reinforcing bars and welded wire fabric, a number of fibers are used to reinforce concrete. They are added to the concrete as it is prepared in the mixer. A number of manufacturers produce these products, and research will probably increase their effectiveness and use in the future. Under some conditions, they may reduce the amount of reinforcing bars required and replace welded wire fabric in some installations. Fibers also are used, along with rebar and welded wire, to produce more desirable properties in concrete.

The types of fibers available include glass fibers; polymeric (polypropylene, polyethylene, polyester, acrylic, and aramid); steel; asbestos; carbon; and natural fibers (wood, sisal, coconut, bamboo, jute, okwara, and elephant grass). Each of these fibers has different characteristics, and consultation with a concrete specialist should occur before using them. The design of the mix, strength, fatigue resistance, durability, shrinkage control, and other factors need to be considered.

In addition to employment in batch-mixed concrete, fiber reinforcement is also used in pneumatically placed concrete (shotcrete). The fiber is added with the

Figure 8.30 Typical reinforcing for a concrete beam.

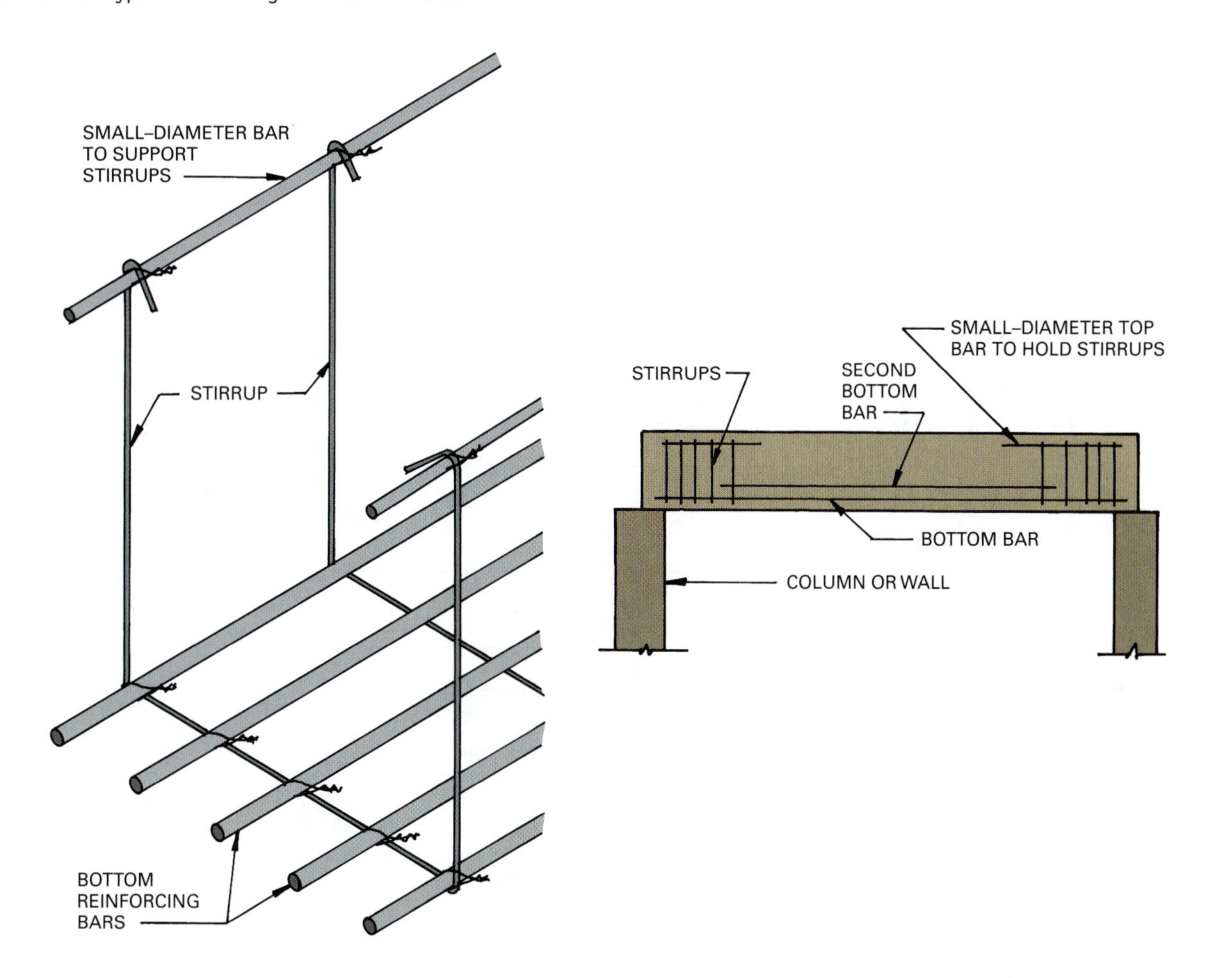

dry concrete mixture and fed through a hose to a nozzle where water is injected. The mix is then sprayed onto the desired surface. This is referred to as SFRC (sprayed fiber reinforced concrete).

CAST-IN-PLACE CONCRETE ELEMENTS

Casting On-Grade Slabs

On-grade concrete slabs require preparation of the slab bed. This varies with the type of soil, but often a base of compacted gravel is required over the soil. Some soils only require compacting before pouring the slab. To control moisture penetration, a plastic sheet is laid over the base. Sometimes 2 or 3 in. (50 to 75 mm) of compacted sand is placed over the plastic sheet. Rigid insulation should also be placed below the slab for conditioned spaces. Finally, the reinforcing is placed over this base and held the required distance above the surface with concrete bricks or bar supports (Fig. 8.34).

The common types of slabs include lightly reinforced and structurally reinforced. Lightly reinforced slabs are constructed with welded wire fabric, fiber reinforcement, or both, and depend on the earth for total support. This helps hold together surface cracks that occur during curing. Structurally reinforced slabs contain steel reinforcing bars and often welded wire fabric and/or fiber reinforcing. The design of the reinforcing and thickness of the slab varies with the loads to be carried. Some designs depend on the base for support, while the reinforcing helps control tensile stress. Other designs have sufficient reinforcing so that the slab can extend from one support, such as a foundation, to another without depending on the earth for support. These slabs are designed by a professional engineer.

Joints are necessary when building concrete slabs on-grade to help control cracking, reduce the size of the pour, and separate the slab from surfaces where it should not bond. Control joints are used to provide a weakened place in the slab where it can crack without causing problems, preventing cracks from running across slabs at all angles. Control joints are spaced 15 to 20 ft. (4.6 to 6.1 m) apart. They are formed

Figure 8.31 Types and sizes of reinforcing bar supports. *(Courtesy Concrete Reinforcing Steel Institute)*

TYPICAL TYPES AND SIZES OF WIRE BAR SUPPORTS

SYMBOL	BAR SUPPORT ILLUSTRATION	BAR SUPPORT ILLUSTRATION PLASTIC CAPPED OR DIPPED	TYPE OF SUPPORT	TYPICAL SIZES
SB	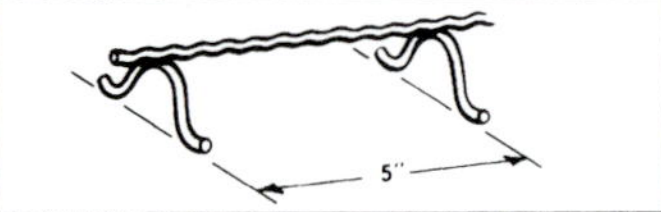 5"	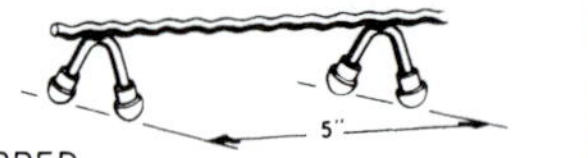 CAPPED 5"	Slab Bolster	¾, 1, 1½, and 2 inch heights in 5 ft. and 10 ft. lengths
SBU*	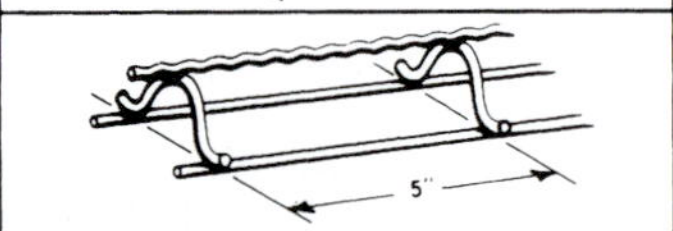 5"		Slab Bolster Upper	Same as SB
BB	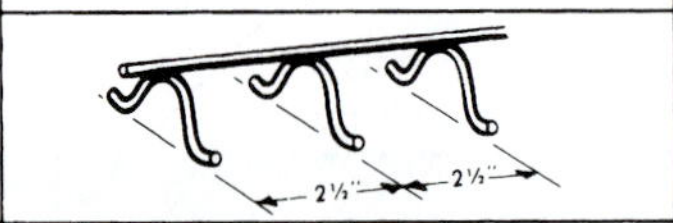 2½" 2½"	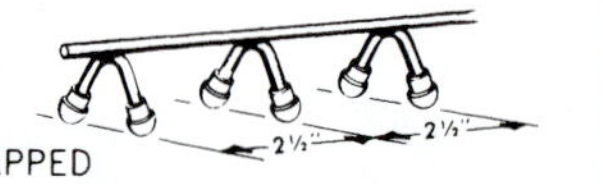 CAPPED 2½" 2½"	Beam Bolster	1, 1½, 2, over 2" to 5" heights in increments of ¼" in lengths of 5 ft.
BBU*	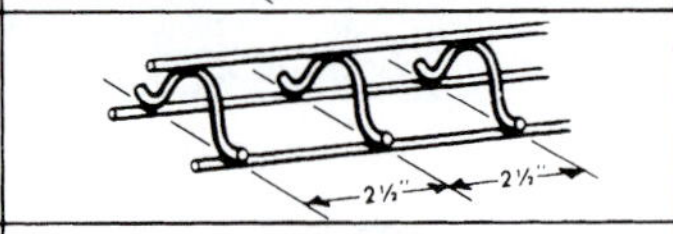 2½" 2½"		Beam Bolster Upper	Same as BB
BC	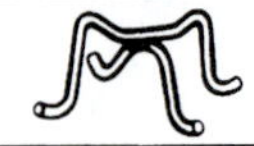	DIPPED 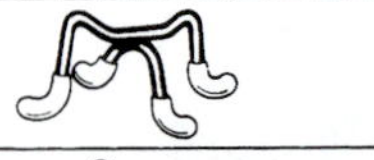	Individual Bar Chair	¾, 1, 1½, and 1¾" heights
JC	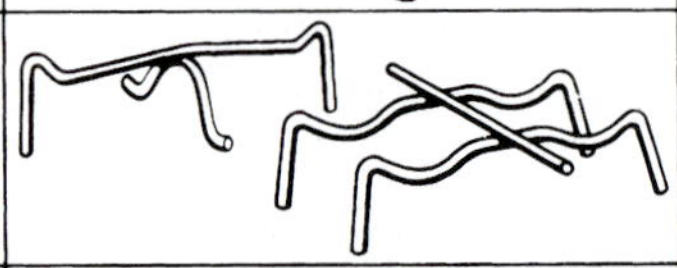	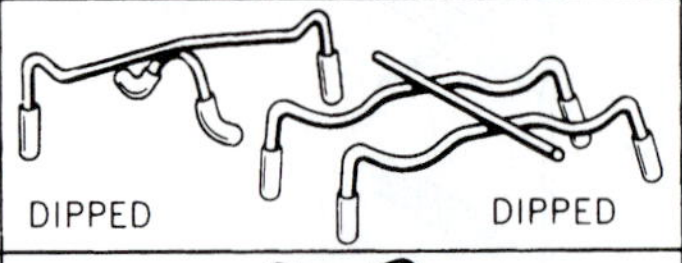 DIPPED DIPPED	Joist Chair	4, 5, and 6 inch widths and ¾, 1 and 1½ inch heights
HC	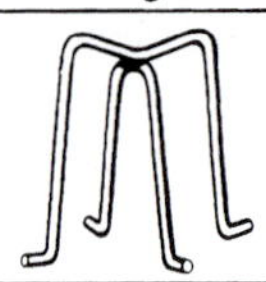	CAPPED 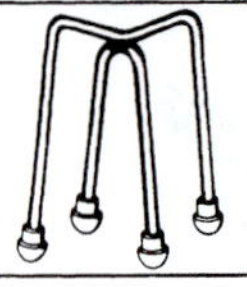	Individual High Chair	2 to 15 inch heights in increments of ¼ inch
HCM*	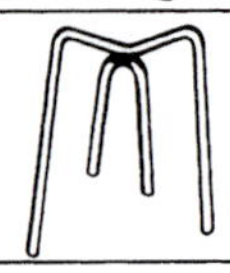		High Chair for Metal Deck	2 to 15 inch heights in increments of ¼ in.
CHC	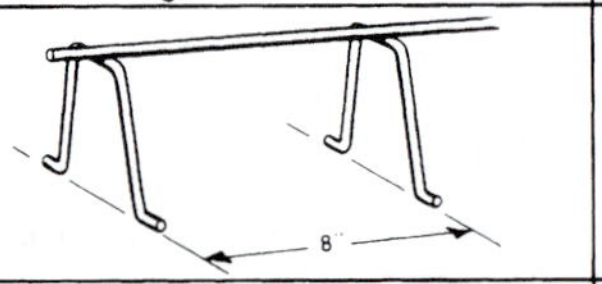 8"	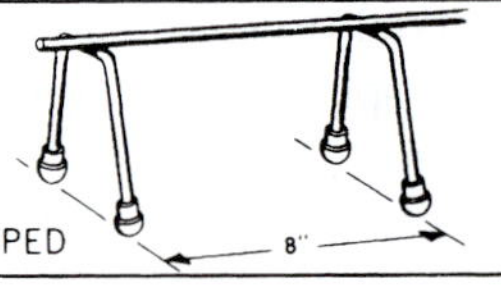 CAPPED 8"	Continuous High Chair	Same as HC in 5 foot and 10 foot lengths
CHCU*	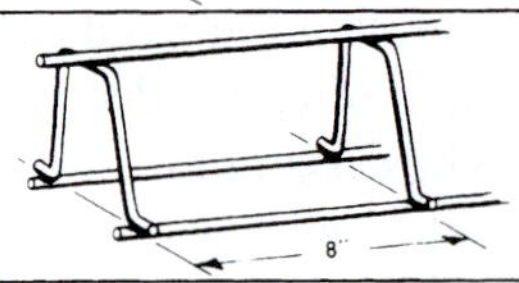 8"		Continuous High Chair Upper	Same as CHC
CHCM*			Continuous High Chair for Metal Deck	Up to 5 inch heights in increments of ¼ in.
JCU**	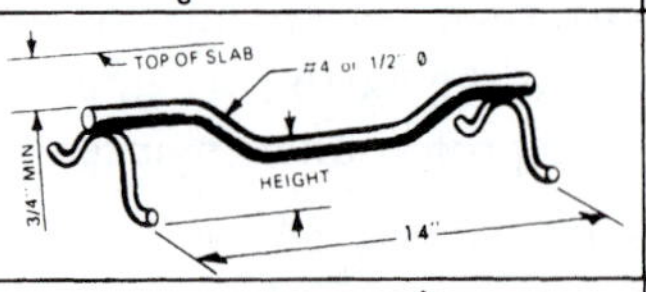 TOP OF SLAB #4 or 1/2" Ø 3/4" MIN. HEIGHT 14"	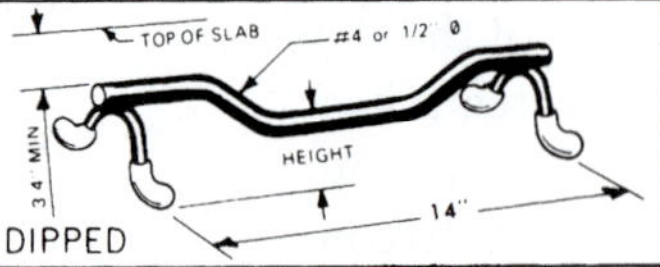 TOP OF SLAB #4 or 1/2" Ø 3/4" MIN. HEIGHT 14" DIPPED	Joist Chair Upper	14" Span Heights −1" thru +3½" vary in ¼" increments
CS	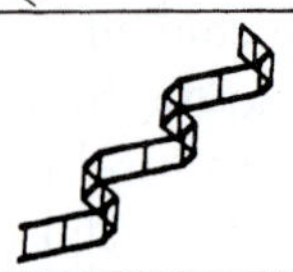		Continuous Support	1½" to 12" in increments of ¼" in lengths of 6′-8"

*Usually available in Class 3 only, except on special order.
**Usually available in Class 3 only, with upturned or end bearing legs.

Figure 8.32 Reinforcement in columns extends for subsequent pours. *(Courtesy Portland Cement Association)*

by sawing into the slab after it has begun to harden or by placing molded strips in the concrete before it hardens.

Construction joints are used to separate a large area to be poured into smaller, more manageable areas. They also help prevent cracking because each joint serves as a control joint, allowing for expansion and contraction within a large slab.

Isolation joints are used to keep a slab from bonding to some abutting part of the building. This permits each part to move independently. Typical isolation joints are $\frac{1}{8}$ to $\frac{1}{4}$ in. thick (3 to 6 mm) asphalt-impregnated fiber or molded plastic strips (Fig. 8.35).

Reinforcing Cast-in-Place Concrete Walls

Reinforced concrete cast-in-place walls may rest on a continuous concrete footing below grade, such as a basement wall, or extend above grade, forming the exterior or interior walls of a building. When the footing is poured, metal dowels are inserted that project above the top of the footing. Some footings have a key cast into them that forms a tie at the bottom of the wall (Fig. 8.36). One side of the wall form is set on the footing and braced. The vertical bars are wired to the dowels. The top of each vertical bar is wired to a horizontal bar,

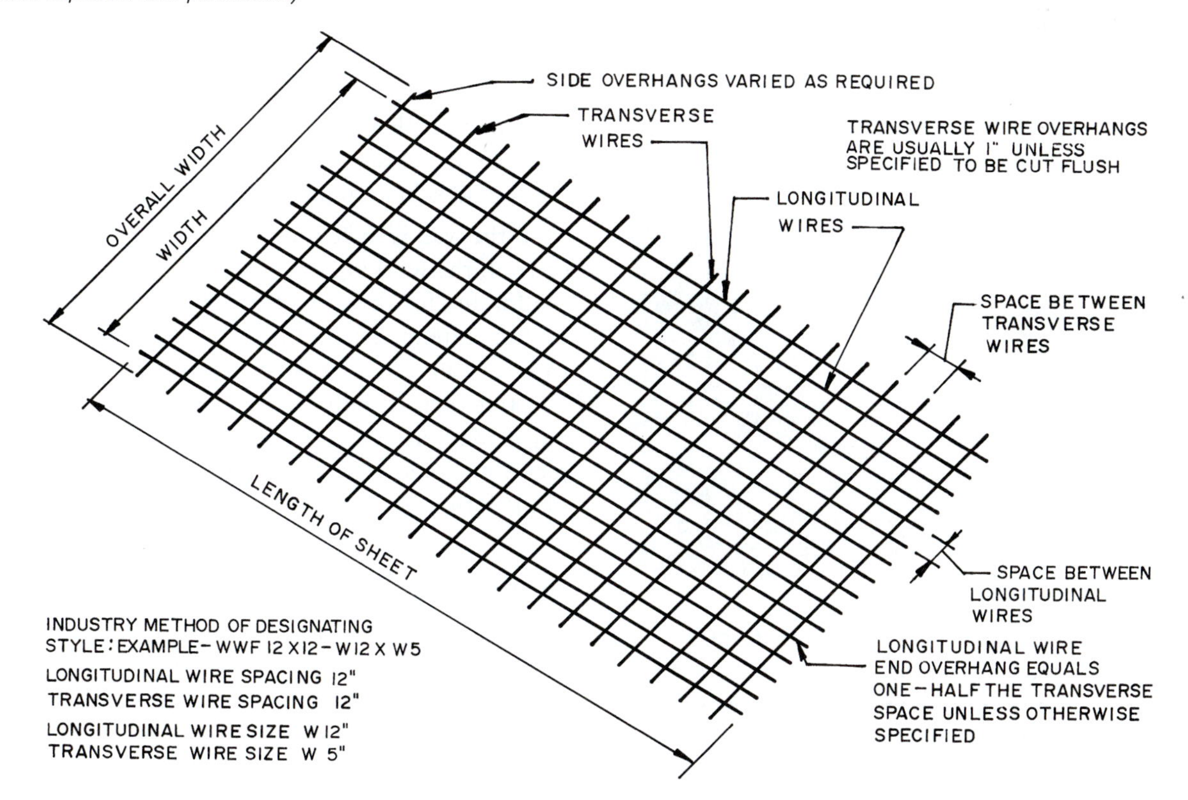

Figure 8.33 Identification and parts of a sheet of welded wire fabric. *(Courtesy American Society for Testing and Materials. Reprinted with permission.)*

Table 8.3 Common Styles of Welded Wire Fabric

Yield Strength (min.) (fy in psi)	Style Designation (W = Plain, D = Deformed)	Steel Area (in.²/1′-0″) Longit.	Trans.	Metric[a] Style Designation
65,000	4 × 4-W1.4 × W1.4	.042	.042	102 × 102 MW 9.1 × MW 9.1
(W only)	4 × 4-W2.0 × W2.0	.060	.060	102 × 102 MW 13.3 × MW 13.3
	6 × 6-W1.4 × W1.4	.028	.028	152 × 152 MW 9.1 × MW 9.1
	6 × 6-W2.0 × W2.0	.040	.040	152 × 152 MW 13.3 × MW 13.3
	4 × 4-W2.9 × W2.9	.087	.087	102 × 102 MW 18.7 × MW 18.7
	6 × 6-W2.9 × W2.9	.058	.058	152 × 152 MW 18.7 × MW 18.7
70,000	4 × 4-W/D 4 × W/D 4	.120	.120	102 × 102 MW 25.8 × MW 25.8
(W & D)	6 × 6-W/D 4 × W/D 4	.080	.080	152 × 152 MW 25.8 × MW 25.8
	6 × 6-W/D 4.7 × W/D 4.7	.094	.094	152 × 152 MW 30.3 × MW 30.3
	12 × 12-W/D 9.4 × W/D 9.4	.094	.094	304 × 304 MW 60.6 × MW 60.6
72,500	6 × 6-W/D 8.1 × W/D 8.1	.162	.162	152 × 152 MW 52.3 × MW 52.3
(W & D)	6 × 6-W/D 8.3 × W/D 8.3	.166	.166	152 × 152 MW 53.5 × MW 53.5
	12 × 12-W/D 9.1 × W/D 9.1	.091	.091	304 × 304 MW 58.7 × MW 58.7
	12 × 12-W/D 16.6 × W/D 16.6	.166	.166	304 × 304 MW 107.1 × MW 107.1
75,000	6 × 6-W/D 7.8 × W/D 7.8	.156	.156	152 × 152 MW 100.6 × MW 100.6
(W & D)	6 × 6-W/D 8 × W/D 8	.160	.160	152 × 152 MW 51.6 × MW 51.6
	12 × 12-W/D 8.8 × W/D 8.8	.088	.088	304 × 304 MW 56.8 × MW 56.8
	12 × 12-W/D 16 × W/D 16	.160	.160	304 × 304 MW 103.2 × MW 103.2
80,000	6 × 6-W/D 7.4 × W/D 7.4	.148	.148	152 × 152 MW 95.5 × MW 95.5
(W & D)	6 × 6-W/D 7.5 × W/D 7.5	.150	.150	152 × 152 MW 48.4 × MW 48.4
	12 × 12-W/D 8.3 × W/D 8.3	.083	.083	304 × 304 MW 53.5 × MW 53.5

[a]These are soft conversions from the inch sizes. Courtesy Wire Reinforcement Institute

Figure 8.34 Forming for a concrete slab with reinforcing in place.

which maintains the spacing. Now the other horizontal bars are located and wired to the vertical bars, forming a grid. The bars usually are wired together at every second or third bar. This forms a reinforcing wall mat. The mat is secured to the top of the form so that it remains in a vertical position during pouring. Sometimes the mat is formed flat on the ground and then lifted and placed on the footing with a crane. If the design calls for a second mat, the mat closest to the form is completed first, then the second is built in the same manner.

If the wall is to be topped with a concrete slab floor or roof, the working drawings will show the required reinforcing for the connection. Typically, vertical reinforcing extends beyond the top of a wall and is bent to be cast into the slab. Corners between walls are also tied together. Usually some form of a hook or elbow bar is used, as shown in Fig. 8.37. After all steel is in place, it should be rechecked before the enclosing side of the form is set and braced.

Reinforcing Cast-in-Place Beams

A simple single-span cast-in-place beam rests on end supports. The typical arrangement for placing the reinforcing was shown in Fig. 8.30. The stirrups are held in place with a small-diameter top bar or the top member of a truss bar.

Figure 8.35 Commonly used construction joints.

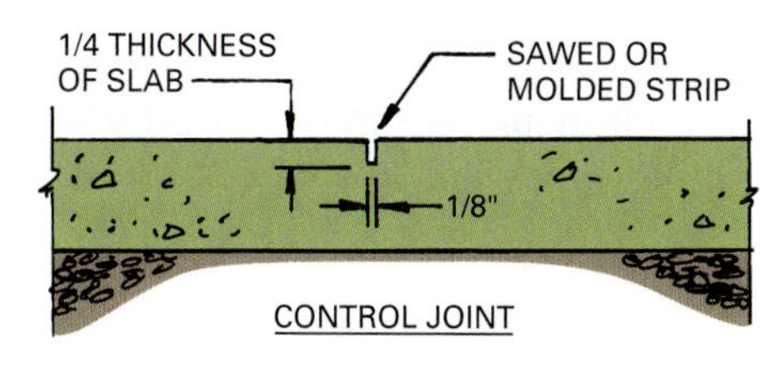

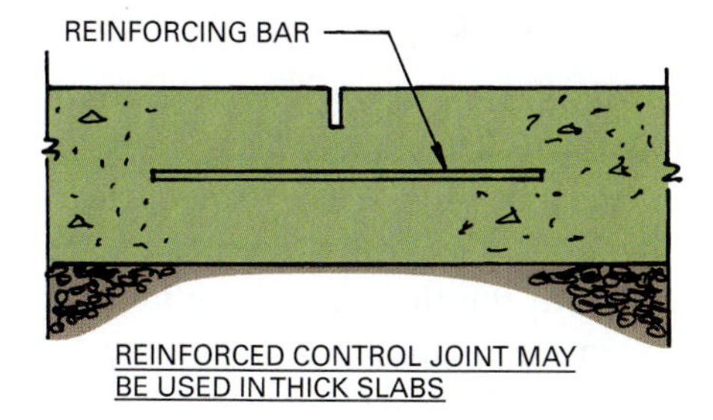

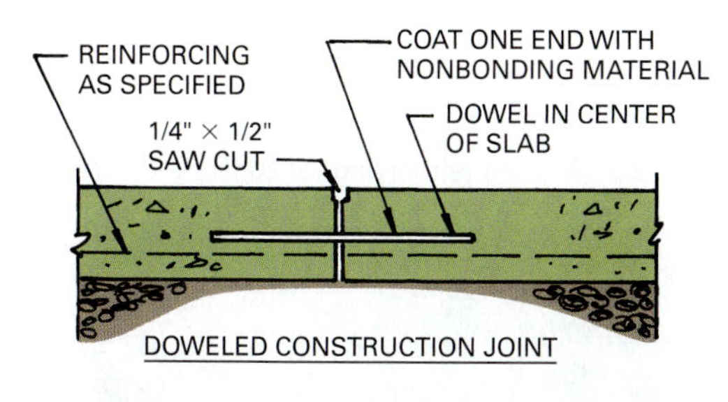

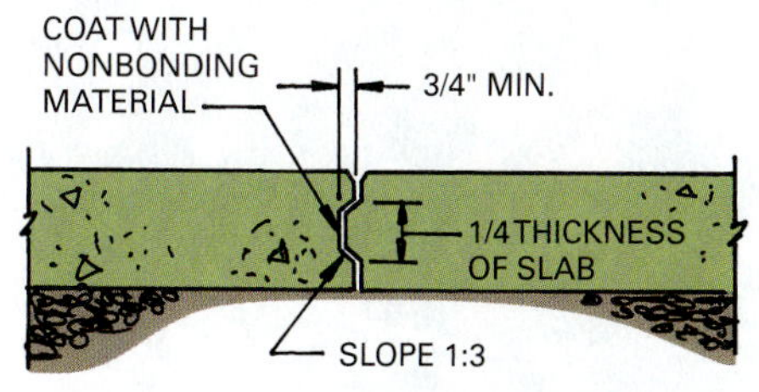

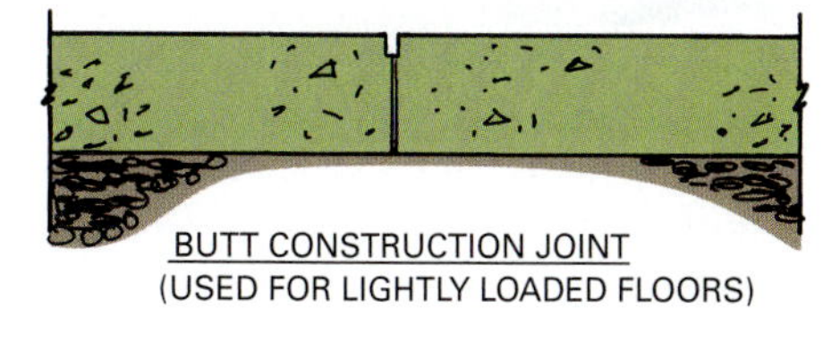

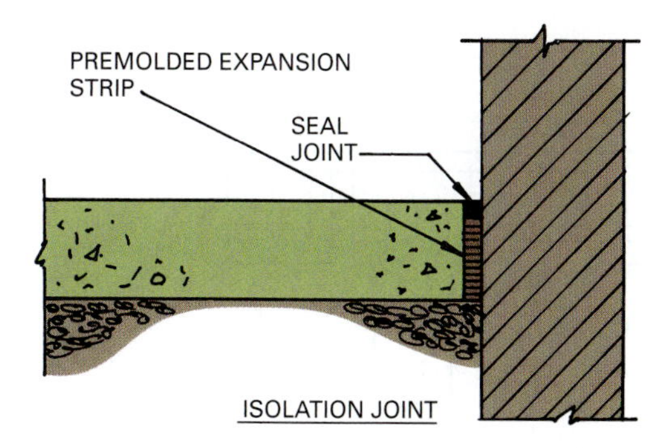

Continuous beam casting is common in structures with a cast-in-place concrete structural system. The beam ties together the structure from column to column. The bottom of the beam at mid span is in maximum tension, and this force dissipates toward the ends of the beam. The stirrups at the end transfer these tension forces to the concrete. Over the column, the top of the beam is subject to tension forces due to bending, so appropriate top bars are placed over each column (Fig. 8.38).

Reinforcing Cast-in-Place Columns

Cast-in-place concrete columns are typically round, square, or rectangular. They are reinforced with vertical bars, which help carry compressive loads and resist tension forces from lateral loads, such as those caused by a high wind. The vertical bars may be arranged in square, rectangular, or circular patterns. Either pattern can be used for square and circular columns.

Tied columns have vertical bars tied together with small-diameter smooth steel bars placed horizontally and wired to the vertical bars (Fig. 8.39). These ties hold the vertical bars in position during the pour and help them resist outward buckling when under load. Spiral columns act in the same manner as tied columns except they also restrain the concrete inside them. Rather than cracking, this enables the column to bend or bow under load.

CAST-IN-PLACE CONCRETE FRAMING SYSTEMS

A typical cast-in-place concrete framing system utilizes cast-in-place columns, one-way or two-way concrete slabs, and cast-in-place joists, beams, and girders. The exterior can be finished in a number of ways, including precast concrete spandrel panels. The flat slab concrete floor and roof slabs are among the simplest to form, reinforce, and pour. A generic example is shown in Fig. 8.40.

Several types of reinforced concrete framing systems are used with cast-in-place concrete construction. In some cases the beams and girders are cast with the floor or roof slab. The two commonly used reinforced concrete floor and roof systems are the flat slab and the flat plate. The flat slab is a concrete slab reinforced in two or more directions and supported by columns with dropped panels and capitals, which enlarge the columns

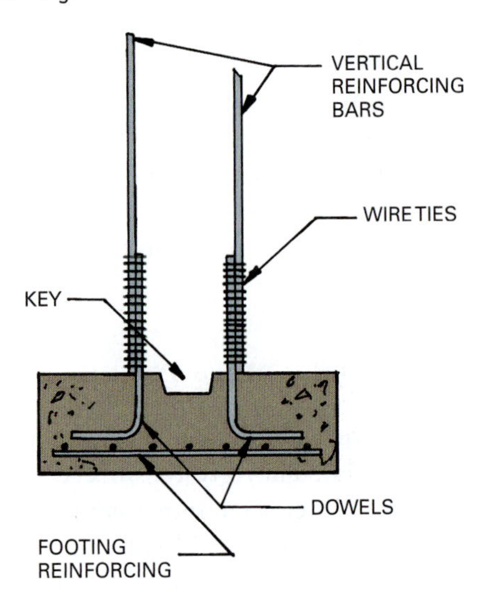

Figure 8.36 Vertical reinforcing bars are tied to dowels cast in the footing.

at the top, or by beams or joists. When two-way reinforcement is used in the slab, it is called a two-way flat slab. A flat plate floor or roof is much like the flat slab. Its reinforcing runs in two directions, but it is supported by columns that are not enlarged where they meet the slab.

General Construction Procedures

After the foundation is in place, the load-bearing walls and columns are poured. When they have cured enough to carry the floor or roof load, the forms for the slab are built. Generally, the beams are formed and poured with the floor, resulting in a monolithic unit. The forms are supported with temporary joists and beams of wood or metal, which are supported by

Figure 8.37 Typical corner reinforcing for cast-in-place walls.

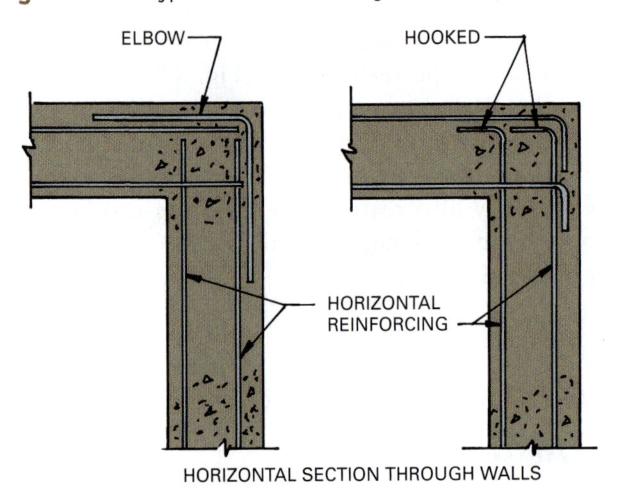

Figure 8.39 Pre-tied reinforcing for concrete columns.

Figure 8.38 Typical reinforcing for cast-in-place beams.

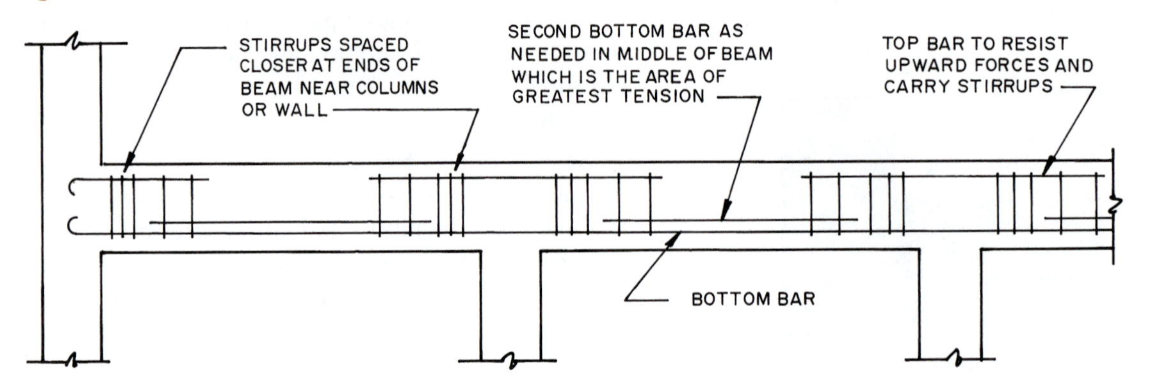

Figure 8.40 A typical flat plate cast-in-place structure. *(Reproduced with permission from the* Building Systems Integrated Handbook, *Richard Rush, ed., Butterworth-Heiman Publisher, Stoneham, Mass, 1986.)*

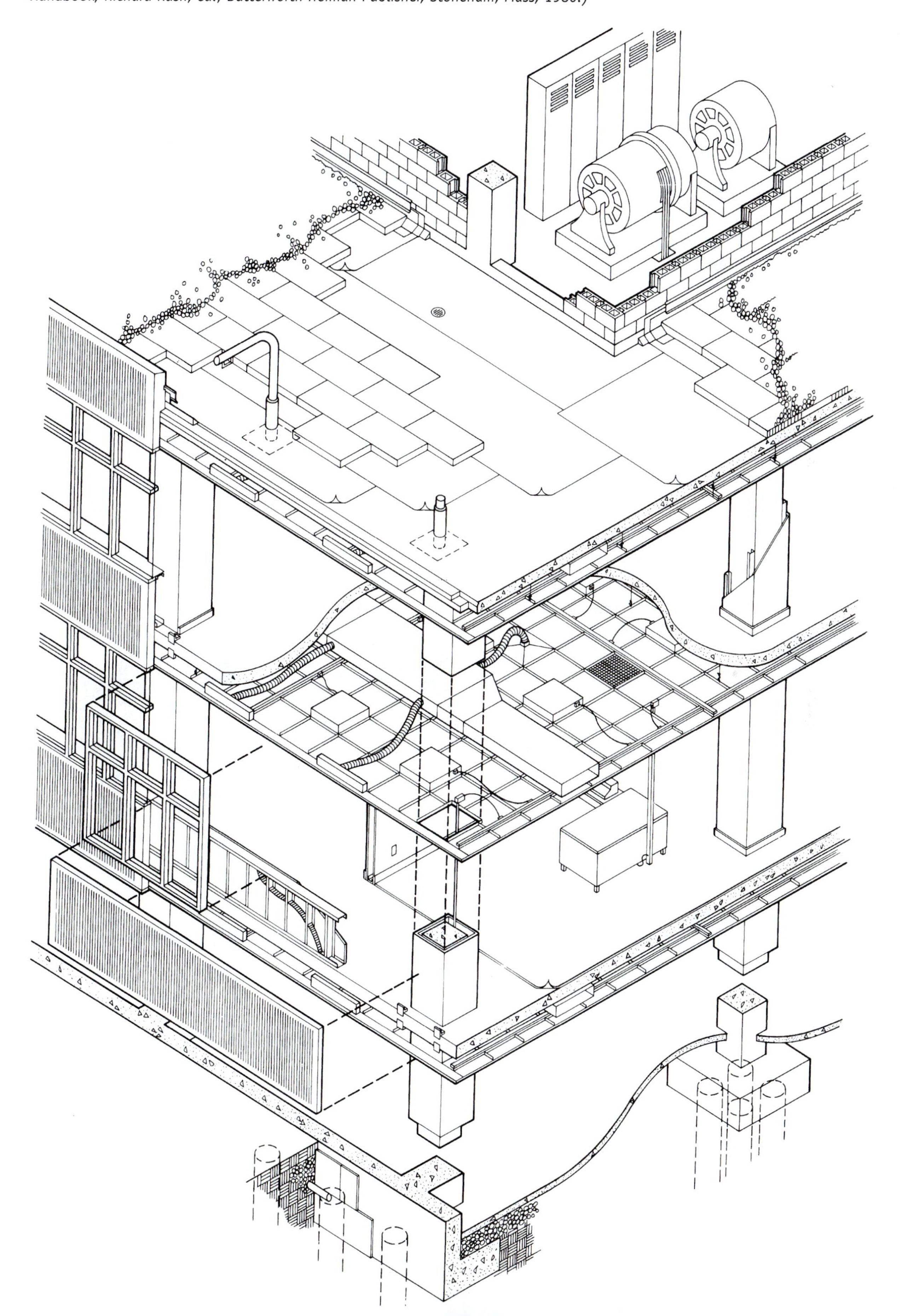

temporary shores. A shore is a column whose length can be adjusted. The entire temporary form support system must be designed by an engineer, so it can carry the required loads.

Sharp corners on concrete members are hard to cast and tend to break away if struck, leaving a ragged edge. Therefore, wood or plastic inserts are placed in the form to produce a beveled or rounded corner. The interior surfaces of the forms are coated with a form-release compound, which prevents the concrete from bonding to the form and aids in removing it with minimum damage to the concrete member and the form. Forms are expensive and must be preserved for reuse. Finally, the reinforcing steel is placed as specified by the engineer's placing drawings.

Before the pour begins, the entire assembly must be inspected and approved. As mentioned earlier, usually the beams, girders, joists, and floor or roof slab are poured monolithically. The exposed surface of the slab is finished as specified. If additional floors are to be built above the one poured, there will be reinforcing protruding through the slab to be joined to the reinforcing in the columns on the next floor or hooked to lay into the roof slab.

After the structure has reached sufficient strength, the formwork for the next floor is constructed. The formwork for the lower floor is stripped, and the beams and slabs are re-shored with temporary posts to provide added support.

Figure 8.41 One-way solid slab with beam construction has the main beams on one axis.

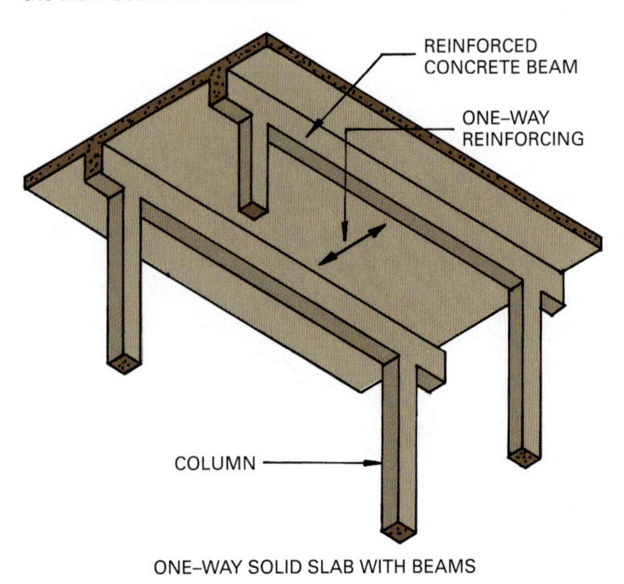

Figure 8.42 A one-way flat slab with beams and girders has a slab that spans the beams that are supported by girders.

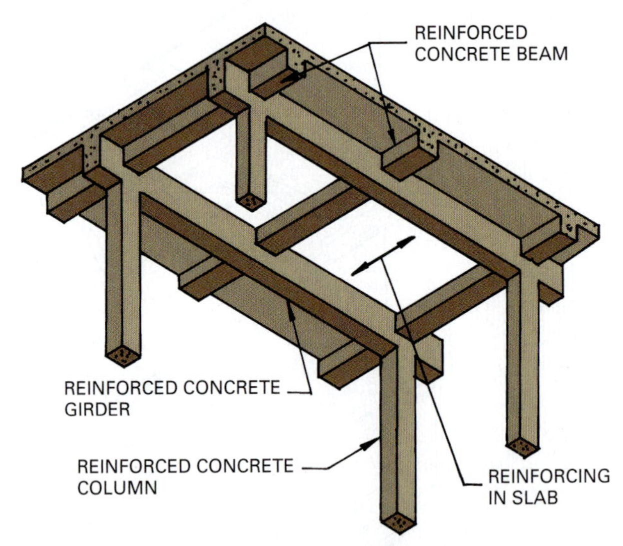

One-Way Flat Slab Floor and Roof Construction

The one-way flat slab systems in common use include a one-way solid slab, one-way flat slab with beams and girders, and one-way flat slab with joists and beams.

One-Way Solid Slab Construction One-way solid slab construction has the slab supported on two sides by beams or load-bearing walls, as shown in Fig. 8.41. The main beams are on one axis, and the slab spans the distance between them. The reinforcing bars or pre-stressed tendons are placed perpendicular to the supporting walls or beams.

One-Way Flat Slab with Beams and Girders Another one-way slab construction is shown in Fig. 8.42. It utilizes a one-way flat slab with beams and girders. The construction follows the same procedures described for one-way solid slab construction, but this slab is supported by cast-in-place concrete beams that are supported by cast-in-place girders. The girders rest on columns. The slab spans the distance between beams. Usually the girders, beams, and slab are cast monolithically.

One-Way Flat Slab with Joists and Beams Another widely used system is a one-way flat slab with joists and beams (Fig. 8.43). The joists span the beams and support

Construction Materials

Strip-Applied Waterstops

Waterstops in concrete construction are used for providing construction, contraction, and expansion joints as well as for creating the perimeter of any penetrations in a concrete wall. Strip-applied waterstops are typically $\frac{1}{2} \times \frac{3}{4}$ in. (25 × 19 mm), although some are larger. They are made from a variety of materials, which have varying characteristics and therefore different applications and installation procedures. Types commonly available include bituminous strips, bentonite-based strips, hydrophilic rubber strips and vinylester gaskets, and ethylene vinyl acetate strips. The concrete worker must know the proper way to install each type, including the method of bonding each to concrete. The engineer designing the concrete structure must choose the product best suited for the conditions that will exist.

Bituminous strips are typically some blend of refined bituminous hydrocarbon resins and plasticizing compounds reinforced with inert mineral fillers. They resist fresh and salt water and acids but not oil or oil byproducts. They are bonded with an asphalt primer.

Bentonite-based strips are a blend of sodium bentonite clay and butyl rubber or sodium bentonite clay with various binders and fabrics. Various grades are available that resist exposure to salt water, fresh water, and certain chemicals. They are bonded to concrete with a water-based latex adhesive. One type has an adhesive back that is simply pressed onto the concrete surface.

Hydrophilic rubber strips are made from various types of rubber that have been modified to make them hydrophilic. "Hydrophilic" refers to the product having a strong affinity for water. These strips swell when exposed to water and shrink to normal size when dry. They can resist fresh and salt water and various chemicals. Epoxy or polyurethane sealants are used to bond them to concrete.

Hydrophilic vinylester gaskets are made from a vinylester that is hydrophilic. Like hydrophilic rubber strips, they expand when wet and shrink when dry. They are available in several compositions, one for fresh water, another for salt water, and a third for selected chemicals. The concrete worker can secure these to the concrete with a xylene-based glue, nails, or screws.

Ethylene vinyl acetate strips are a closed-cell cross-linked foam produced from ethylene vinyl acetate. It resists exposure to oil-based products, fresh and salt water, and selected chemicals. The concrete worker bonds these strips to concrete with a trowel-applied structural grade epoxy adhesive.

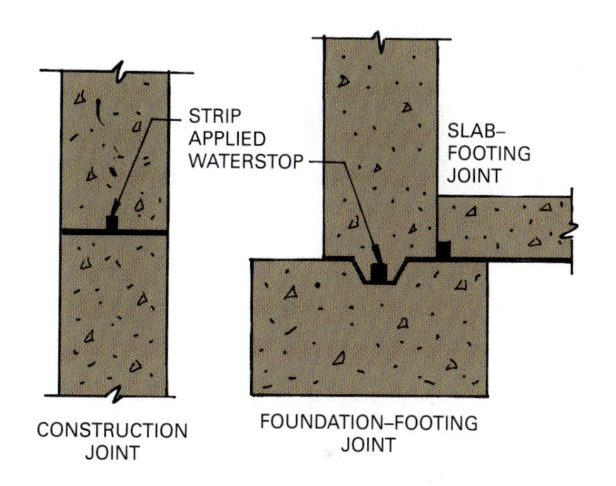

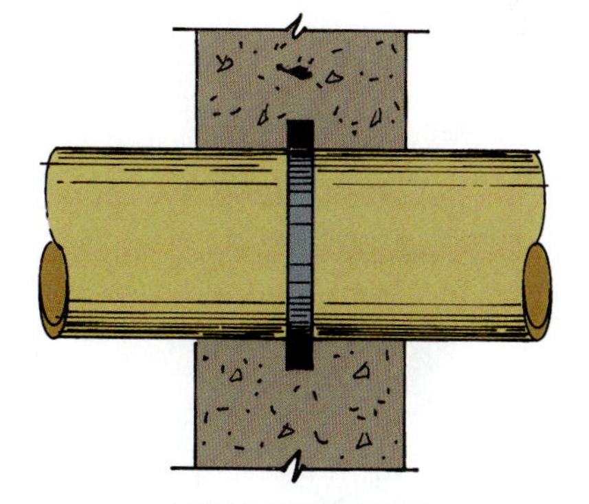

the slab. The beams are supported by the columns. This is often used when the span between beams is not large. As the span increases, the thickness of the slab increases, which adds to the total weight.

The beams, joists, and slab are cast monolithically. The joists are formed with molded fiber-glass or metal domes that rest on the temporary framing and shoring. When the forms are removed, the exposed ceiling appears, as shown in Fig. 8.44. This can be treated in many ways. For example, it can be painted, left natural, sprayed with acoustical plaster, or covered with a suspended ceiling.

Figure 8.43 A one-way slab with joists and beams has joists that span the distance between beams.

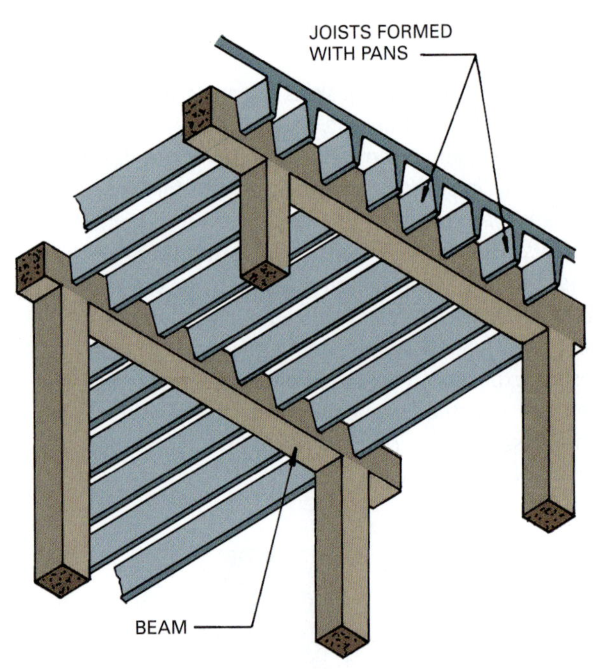

Two-Way Flat Slab Floor and Roof Construction

As mentioned earlier, the two-way flat slab has reinforcing running in two (perpendicular) or more directions. The systems in general use are the two-way flat slab, two-way solid slab, two-way flat plate, and the two-way joist (or waffle) slab. This construction enables a building structure to be designed in nearly square bays.

Two-Way Flat Slab Construction Two-way flat slab construction utilizes a flat slab supported by thickening the slab at each column with a drop panel or a drop panel and a capital (Fig. 8.45). The slab reinforcing runs in two directions. The increased thickness at each column helps resist the shear forces at the top of the column. This system is useful for buildings that will have heavy loads. In buildings designed for lighter loads, the thickness of the slab above the column can be reduced. A simplified partial-plan view and sections are shown in Fig. 8.46.

Figure 8.44 The beams, joists, and flat slab are cast monolithically using domes to form the joists and the base for the floor.

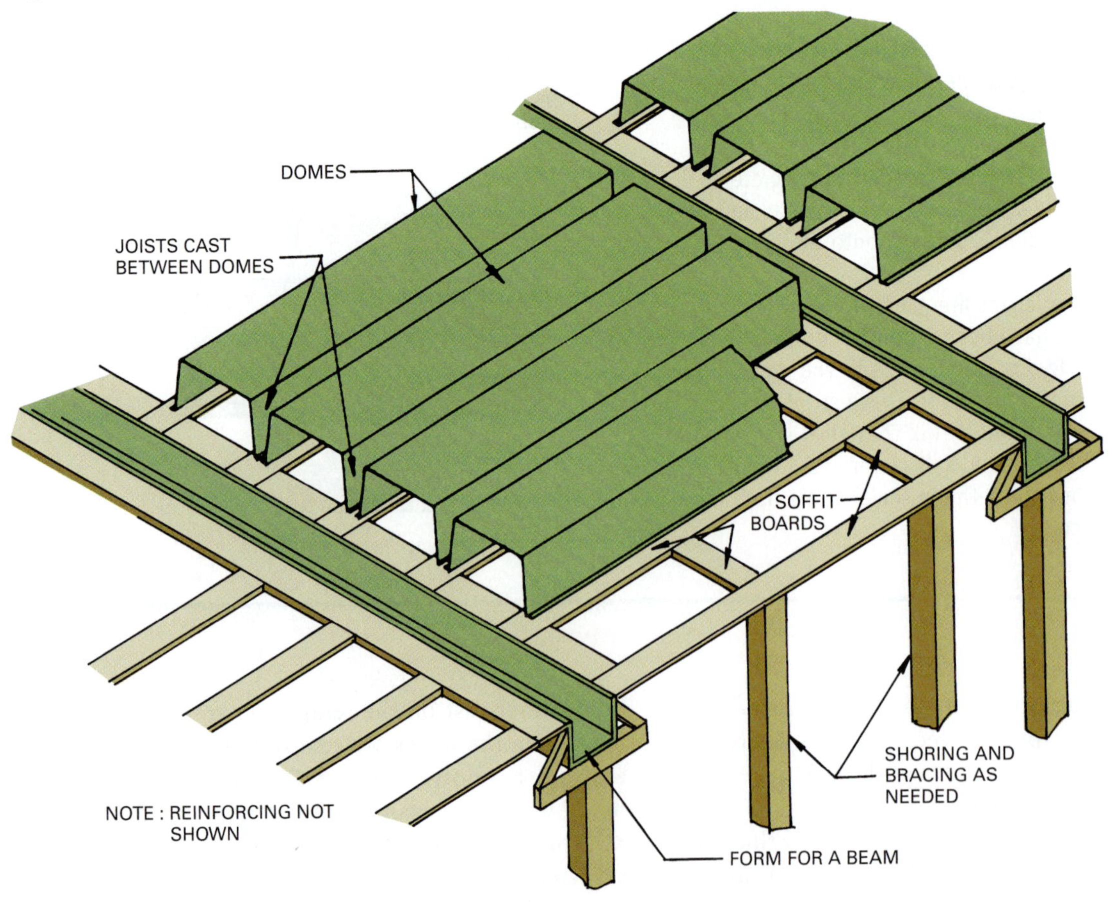

Figure 8.45 Two-way flat-slab construction with columns, using drop panels and a column capital.

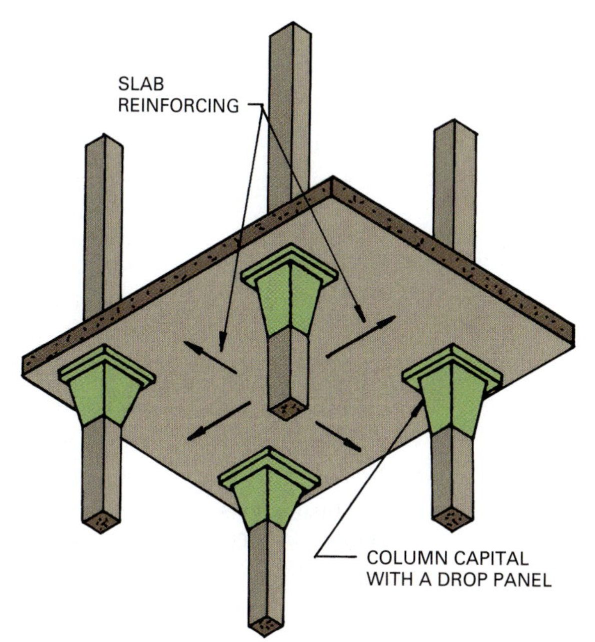

The slab reinforcing is divided into column strips and middle strips. Each is about one-half span. The column strips carry the bend forces developed in the area over the column, while the middle strips reinforce the slab in the area between columns. The slab reinforcement is shown on the engineering drawings. The drop panel usually has no additional reinforcing.

Two-Way Solid Slab The two-way solid slab has the flat slab supported by beams that run between columns (Fig. 8.47). The slab and beams are cast monolithically. This construction can be used for buildings designed to carry heavy loads. The slab is reinforced in two directions.

Two-Way Flat Plate The two-way flat plate uses the same general construction as the flat slab, but it has no drop panel or capital on the top of each column. The design is very similar to the flat slab.

Two-Way Joist (Waffle) Flat Slab The two-way joist framing system consists of a series of concrete joists cast at right angles to each other to form a grid. The slab spans the distances between the joists and both are poured monolithically (Fig. 8.48). The system is designed around a series of columns, usually equally spaced. The joists are formed by domes placed on a wooden deck that is supported by shoring. Areas over the columns are left open so a solid concrete slab area called a column head can be poured.

Tilt-up Wall Construction

Tilt-up concrete construction is a form of site-cast concrete. Wall panels are cast in a horizontal position near the building or, most frequently, on the concrete on-grade floor slab. A bond-breaking agent is placed on the slab so the wall panel will not adhere to it. The surfaces of the panel can be finished in many ways: textured (from a form liner); washed to expose embedded aggregate; troweled smooth; or painted.

After the wall panel has developed sufficient strength, it is lifted into position with a crane and secured to the footing (Fig. 8.49). The panels are temporarily braced until all wall and roof structural members are in place. The panels are secured to columns, which may be precast concrete, steel, or concrete that is cast-in-place after the walls are up.

Panels generally range in thickness from about 5 $\frac{1}{2}$ to 7 $\frac{1}{2}$ in. (140 to 190 mm) and may contain openings for windows and doors; they can be several stories high. A key to success is the selection and placement of steel reinforcing, which not only must serve when the wall is erect and carrying a roof or floor load but also must withstand the bending loads generated when the panel is tilted and lifted into place. Pick-up hangers must be carefully located in the panel so the crane can attach to and lift the wall. An experienced engineer must design the panels.

Generally, the wall panels are load bearing and will support floors and roofs. Typical decking includes hollow-core units, single or double tees, and metal open-web joists.

Panels have $\frac{3}{4}$ in. (18 mm) chamfers around the outside edges. Sharp corners are hard to cast and break off easily when hit. The wall panels usually are not connected to each other but have a $\frac{3}{4}$ in. (18 mm) space sealed with backer rods and a weatherproof sealant. A backer rod is a flexible, compressible rod made of foam plastic. It is placed in a joint to limit the depth to which a sealant can enter the joint. The wall panels are tied together by the roof framing and floor slab. Each panel acts as an independent member. A typical wall-floor connection is made by leaving out a section of the floor along the wall when the on-grade slab is cast. The wall panels may be supported by continuous footings, isolated footings, or caissons. They are aligned with dowels or metal angles. The void between the bottom of the panel and the top of the footing is filled with grout.

Figure 8.46 A partial drawing showing typical details for a two-way flat slab with drop panels.

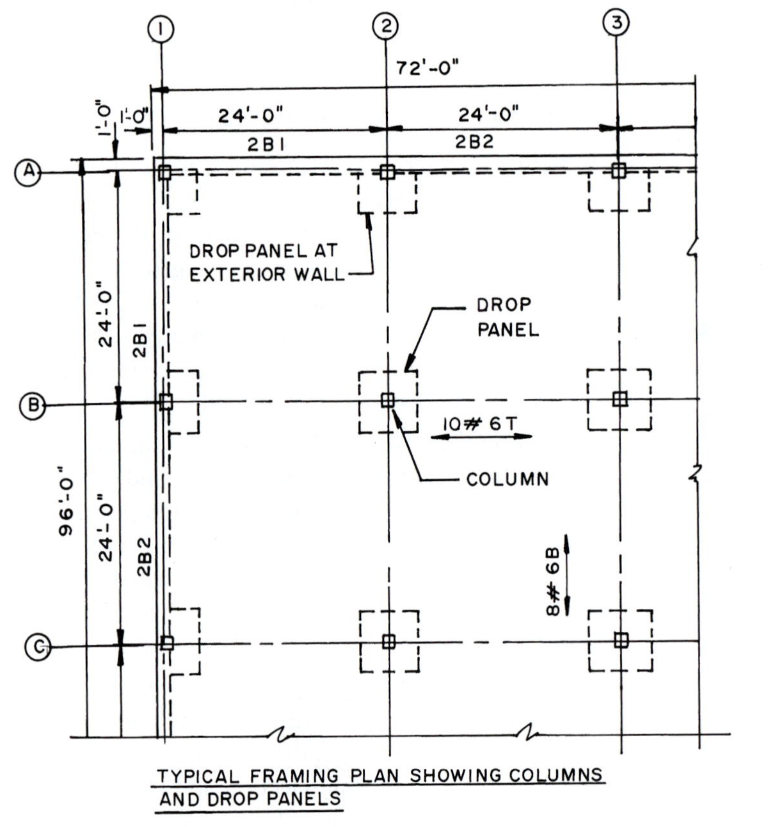

NOTE: DIMENSIONS ON THESE DRAWINGS ARE FOR ILLUSTRATION PURPOSES ONLY.

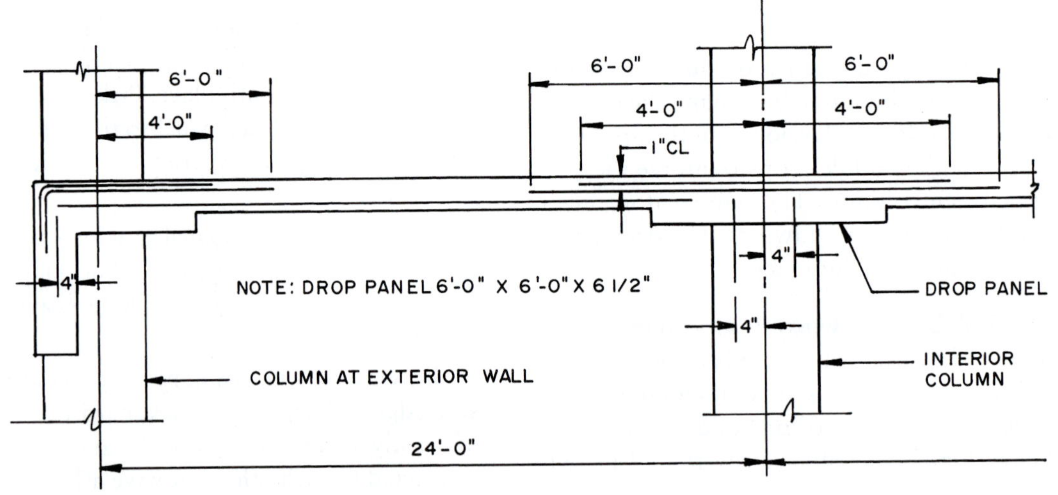

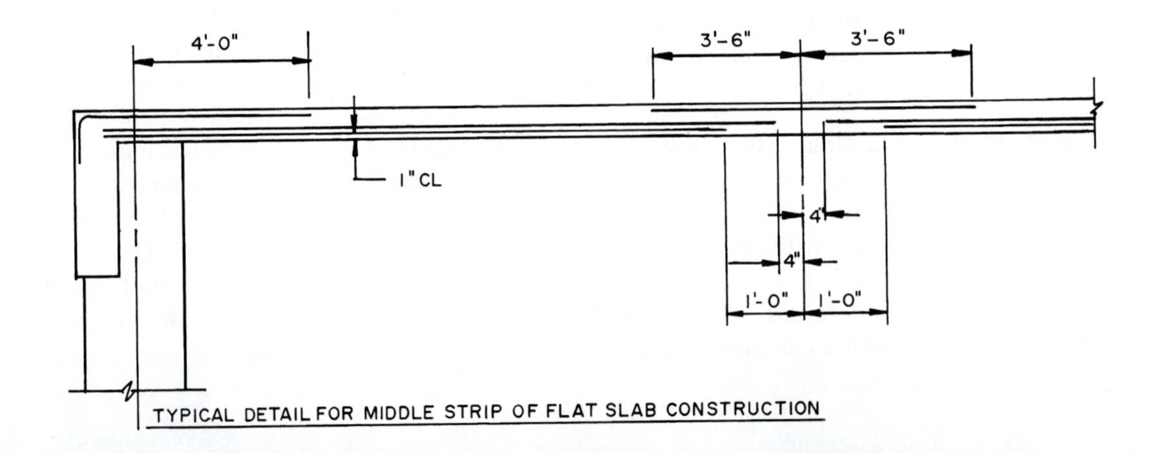

Figure 8.47 A two-way solid slab and beam is supported by beams running between the columns.

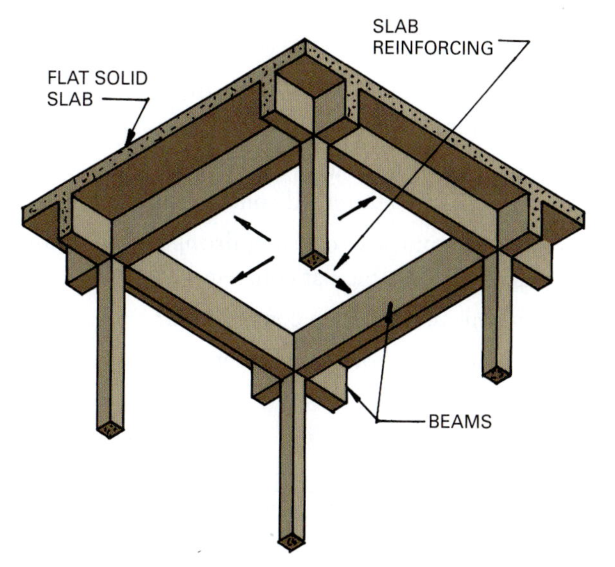

Figure 8.48 The two-way joist framing has cast-in-place joists run at right angles to each other creating a waffle-like ceiling.

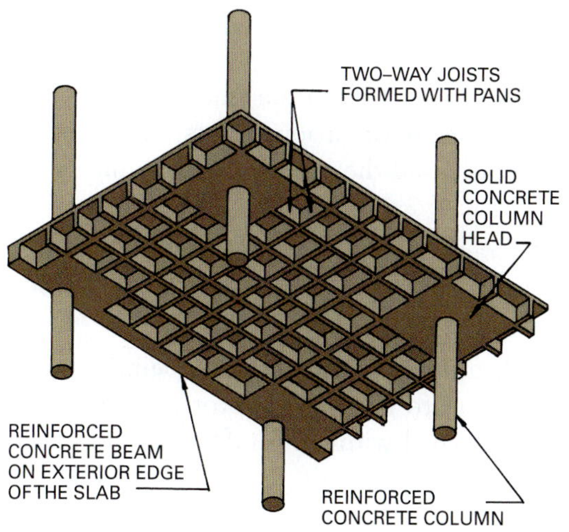

Figure 8.49 A tilt-up wall being lifted into place. *(Courtesy Portland Cement Association)*

Lift-Slab Construction

Lift-slab construction involves casting a reinforced concrete slab for each floor of a building and the roof, one on top of the other. The ground floor on-grade slab is cast first. It is coated with a bond release compound that prevents the next slab from bonding to it. Slabs are cast for each floor and for the roof, if it is to be a concrete slab. Each is coated with a bond-release compound. The slabs, usually two-way flat slabs, can be reinforced concrete flat slabs or pre-stressed. The slabs are lifted by hydraulic jacks mounted on the tops of the columns. They slide up to the desired elevation and are secured to each column by welding, bolting, or a series of pins. This is repeated for each floor.

Slip-forming is used to cast-in-place tall structures, such as grain elevators and stairwells. It involves constructing formwork, which is then lifted by jacks supported by vertical steel rods. The forms are filled with concrete and reinforcing bars as they are raised. During the duration of the lift, the concrete gains sufficient strength to carry the weight of the newly poured concrete. Keys to success are a proper concrete mix and the equipment to lift it to successive levels as the structure increases in height.

Review Questions

1. How long can ready-mix concrete remain in the mixer and still be used?
2. What devices are used on-site to move concrete to point of placement?
3. How is concrete consolidated?
4. What may happen if the temperature of a concrete mix exceeds 100°F (38°C)?
5. Why are freshly poured concrete slabs bullfloated?
6. What is the difference between an isolation joint and a construction joint?

7. What is meant by hydration?
8. What are the curing methods in common use?
9. What is the purpose of form liners?
10. How is the tensile strength of concrete beams increased?
11. What two styles of reinforcing bars are used?
12. What is the diameter in inches of a No. 8 bar?
13. What purpose do stirrups serve when installing reinforcing bars in a form?
14. How can you decide how many inches of concrete are needed to give fire protection to reinforcing bars?
15. What styles of wire are used to make welded wire reinforcement?
16. What types of coatings are used on welded wire reinforcement?
17. What types of fibers are used in the production of fiber-reinforced concrete?
18. What is the difference between lightly reinforced and structurally reinforced concrete slabs?
19. What is the purpose of using dropped panels and capitals when casting flat concrete slabs?
20. What is meant by tilt-up construction?
21. What is unique about the slip-forming technique?

Key Terms

batch
bleed
bull float
cast-in-place concrete
concrete pump
consolidation
curing
hydration
jointing
screeding
segregation
troweling

Activities

1. Visit a construction site where a cast-in-place concrete structural frame is being erected. Prepare a report giving full details about what you saw. For example, what types of forms and reinforcing were used? How were they pouring the concrete, and was it site-mixed or from a batch plant? How often are concrete samples taken from the delivery truck and on-site mixed batches? Who does the testing? What type of structural system was used?
2. Review the local building codes and list briefly the regulations relating to cast-in-place concrete structural systems.
3. Observe the pouring of floor and roof slabs and report on the use of various joints observed, the surface finish, and the protection used during the curing phase.
4. Collect small samples of concrete-reinforcing materials. Prepare a display and lead a discussion on reinforcing techniques.
5. If any tilt-up wall construction is being done in your area, try to get some photos and prepare a sketch of a panel with the reinforcing and panel-lifting devices.

Additional Resources

ACI Building Code Requirements for Structural Concrete and Commentary, American Concrete Institute, Farmington Hills, MI.

Builders Foundation Handbook, U.S. Department of Energy, Oak Ridge, TN.

Building Code Requirements for Structural Concrete (2005), Construction Practices (2004), *Essential Requirements for Reinforced Concrete Building (2002)*, ACI Manual of Concrete Practice (2005), Farmington Hills, MI.

Cast-In-Place Home Building, Portland Cement Association, Skokie, IL. Many other publications available.

Chemical Admixtures for Concrete, *Cold Weather Concreting*, *Guide for Consolidation of Concrete*, and many other concrete technical publications, American Concrete Institute, Farmington Hills, MI.

Structural/Seismic Design Manual—Volume 3, International Code Council, Falls Church, VA.

Other resources include: Publications of the Wire Reinforcement Institute, Findley, OH.

Technical publications from the Concrete Reinforcing Steel Institute, Schaumburg, IL.

See Appendix B for addresses of professional and trade organizations and other sources of technical information.

9 CHAPTER

Precast Concrete

LEARNING OBJECTIVES

Upon completion of this chapter, the student should be able to:

- Describe and identify the major groups of precast concrete units and the members within each.
- Explain how precast units are manufactured.
- Cite the advantages and limitations of using precast concrete structural units.
- Explain the differences between pre-stressed and non-pre-stressed precast concrete units.
- Explain the differences between pre-tensioned and post-tensioned structural concrete units.
- Describe the various types of precast concrete slab units and their applications.
- Be aware of the types and typical sizes of precast concrete columns, beams, girders, and wall panels.
- Discuss the procedure for erecting precast concrete units and the types of connections used.

Build Your Knowledge

For further study of these materials and methods, please refer to:

Chapter 12 Concrete Masonry

Structural precast concrete units can be used to form both structural and non-structural elements of a building, including columns, beams, girders, floor and roof slabs, and exterior and interior wall panels (Fig. 9.1). They are cast under factory-controlled conditions and moved to the job site for assembly.

Precast concrete units can be classified into two major groups: precast structural concrete units and precast architectural units. Typical precast structural units include floor and roof slabs, beams, girders, columns, and wall panels. Precast architectural concrete units include wall panels and other features that may or may not be structural. Precast units are cast in beds, which are permanent forms made of metal, wood, or fiberglass (Fig. 9.2). The quality of the surface of the form produces the finished surface of the precast unit. Both pre-tensioned and post-tensioned units can be produced.

ADVANTAGES OF PRECAST CONCRETE UNITS

Precast construction is widely used because it affords a number of advantages when compared with cast-in-place concrete. These advantages include the following:

1. Casting takes place in automated facilities where an experienced crew produces units under controlled conditions.
2. The control of materials, mixing, and placing the concrete is rigid, producing higher-quality concrete. Samples of the concrete are taken regularly and tested per ASTM standards. Typically, 5000 psi (35 MN/m^3) concrete is used.
3. Units are finished and cured under carefully controlled conditions. The concrete is mechanically vibrated in the form and cured. To speed up production, units are covered with insulation blankets and steam cured (autoclaved). When Type III, high early strength, Portland cement is used, the units can be removed from forms after twenty-four hours (Fig. 9.3).

Figure 9.1 A structure system of precast concrete members. (*Reproduced with permission from* The Building Systems Integration Handbook, *Richard Rush, ed., Butterworth-Heinmann Publishers, Newton, Mass., 1986.*)

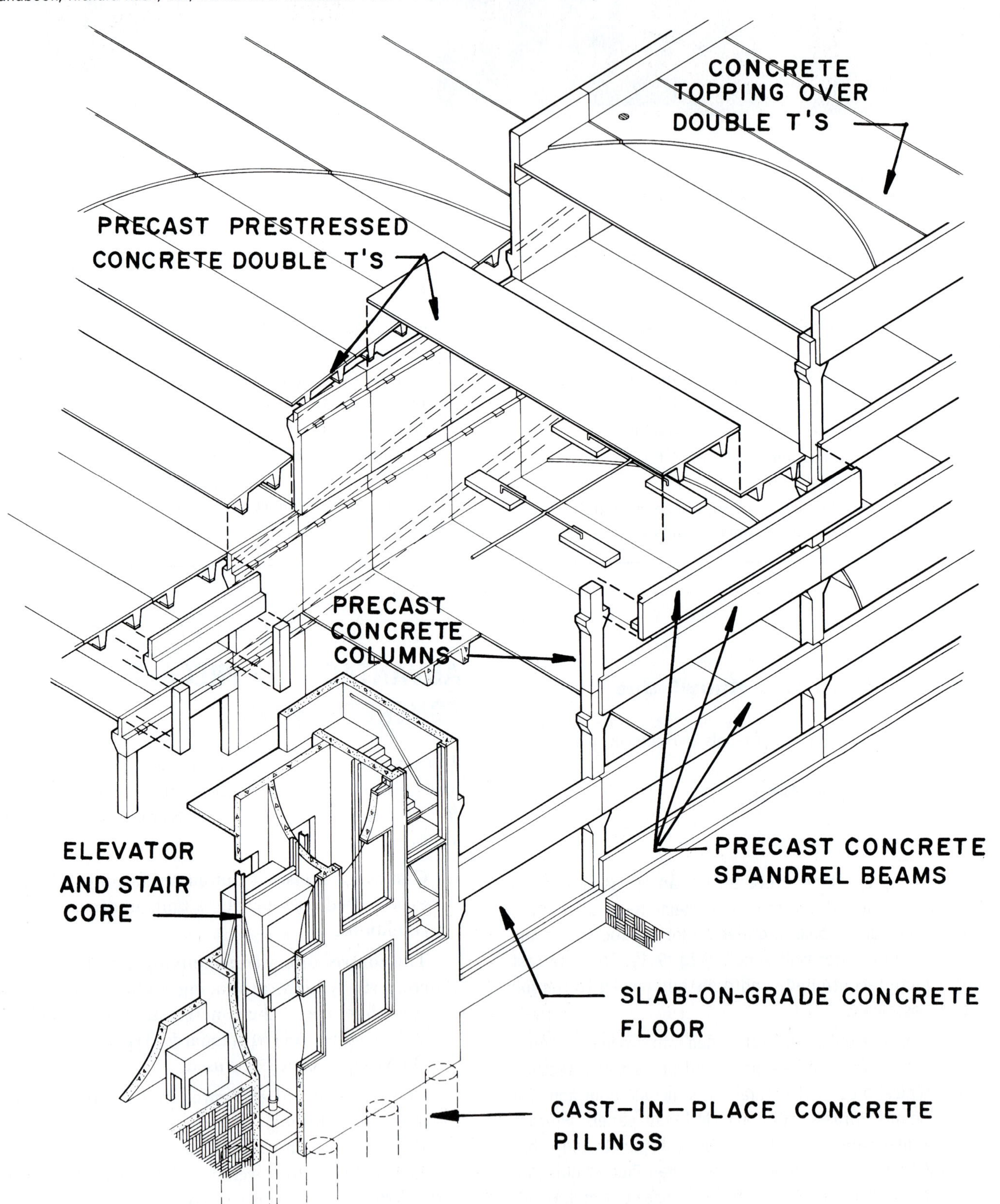

Figure 9.2 A cast concrete unit being removed from the casting bed. *(Courtesy Portland Cement Association)*

Figure 9.3 Precast units are removed from the casting bed and stored until they have sufficient strength to be moved to the site. *(© Construction Photography/Corbis)*

4. **Forms are situated on the ground, making the placing of steel reinforcing, casting, and curing faster and easier.**
5. **The forms are used repeatedly, reducing costs. However, many jobs require custom-built forms. Special surface finishes and textures can also be produced.**
6. **Generally, for the same spans and loads, precast units can be smaller and lighter than cast-in-place units.**
7. **Members can be cast and stored until needed, speeding up the erection process (Fig. 9.3). They are erected with a crane, in much the same manner as structural steel.**
8. **Inclement weather does not slow precast construction as easily as it does cast-in-place jobs.**

After the units have achieved adequate strength, they are delivered to the job site, usually by truck. Their size is limited only by what can be moved by truck or railroad. Highway regulations restrict load sizes and weights. Extra-large precast units can be cast by moving forms to the job and casting on-site. Another limiting factor is the condition of a building site. Space must be available to deliver and unload members and must allow for lifting them into place. Careful handling is necessary so that surface finishes are not damaged.

BUILDING CODES

Building codes have extensive specifications pertaining to the design and installation of precast concrete members. In addition to engineering design data, the codes refer to areas such as reinforcements, connections, lifting devices, fabrication, shop drawings, tendon anchorage, and grout. The codes also specify the requirements for meeting fire codes.

NON-PRE-STRESSED AND PRE-STRESSED PRECAST UNITS

Non-pre-stressed units are cast in molds in a plant, then cured and shipped to the job site. They are reinforced in the same manner as cast-in-place concrete and are not under tension forces (Fig. 9.4). Some large or unusual beams, girders, and many columns and lintels are cast this way. The architect will work with the casting company to establish the design, secure examples of the proposed finish for approval, and receive evidence that the units will meet quality standards.

PRE-STRESSED PRECAST CONCRETE

Pre-stressed concrete units have stresses introduced before they are placed under a load. There are two types: pre-tensioned and post-tensioned.

Figure 9.4 A typical reinforcing cage for a precast non-pre-stressed concrete beam.

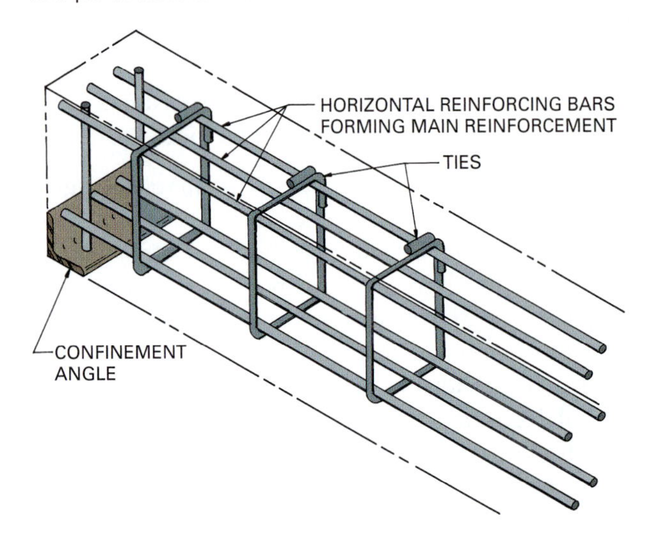

Figure 9.5 A cast-in-place concrete beam will deflect under load and possibly crack on the bottom because concrete cannot resist the tension forces.

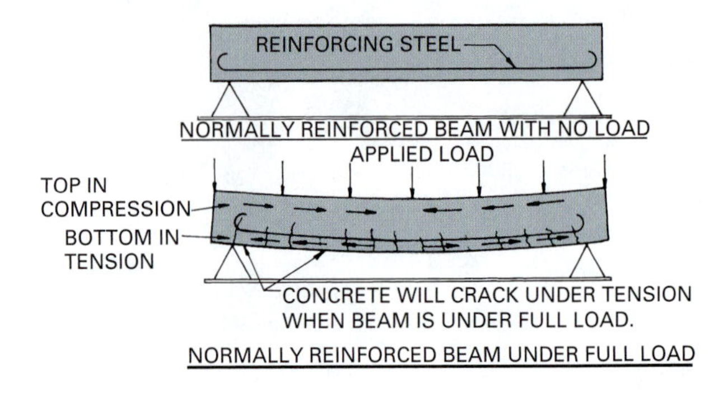

Figure 9.6 A pre-stressed concrete beam has a slight camber, which places the concrete in compression. When under load the concrete is still in compression.

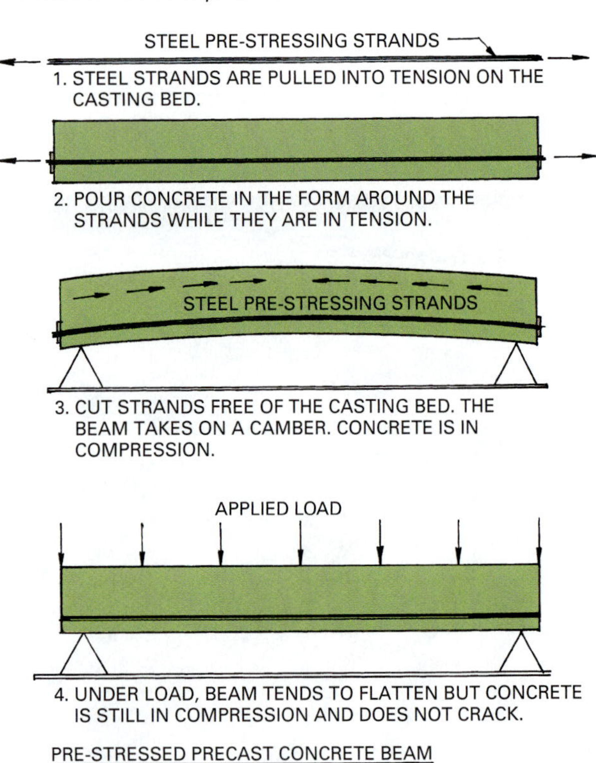

Pre-Tensioned Units

Pre-tensioning is used to produce units that are standardized in size or when enough identical units are needed to make it economically possible to cast them. They are cast in forms, some as long as 800 ft. (250 m) or more. High-tensile-strength steel reinforcing strands are stretched from one end of the form to the other. One end is anchored to a large fixed abutment. The other end passes through the abutment on the opposite side and has hydraulic stressing equipment fastened to it. The desired design stress is applied to the cable by pulling it. When the design tension is reached, the cable is clamped to the abutment.

After the strand is stressed, the concrete is placed in the form. As it hardens it bonds to the strands. After twenty-four hours, samples of the concrete are taken and tested. If they have attained the required strength, the strands are cut from the abutments. The resulting forces produced by the strands pre-stress the concrete. The units are lifted from the bed and moved to storage, from which they are shipped to job sites. The forms are now ready for laying up another pre-tensioned member.

A comparison of non-pre-stressed (Fig. 9.5) and pre-stressed (Fig. 9.6) concrete shows that the non pre-stressed unit has no **camber** (arch) and deflects slightly under load, but the pre-tensioned member has some camber due to the tension in the strands. When under load, a pre-tensioned member tends to settle in a level position. The stressed strands produce high compressive stresses in the lower part of the unit and a tensile stress in the upper part. As the unit is loaded these stresses are reduced.

Post-Tensioned Units

Post-tensioning applies stresses to the concrete unit after it has been cast and hardened. It is applied primarily with cast-in-place concrete but is sometimes used to tension precast concrete units.

Post-tensioning involves casting the unit using normal reinforcement and, in addition, running several post-tensioning strands through the unit in such a way that the concrete does not bond to them as it sets. The strands may be greased or placed in thin-walled metal tubes to prevent bonding (Fig. 9.7). After the concrete has hardened, the post-tensioning strands are anchored firmly on one end and a hydraulic jack is used on the other to apply tension to the strand. Once the desired tension is reached, this end is also anchored (Fig. 9.8). The steel strands may be left unbonded. If they are in steel tubes, they can be bonded by filling the space between the strands and the tube with high-pressure grout. Figure 9.9 shows the structural framing and tenons in tubes on a post-tensioned concrete structure.

Figure 9.7 Post-tensioned precast concrete members develop camber by applying tension after the member is cast and cured.

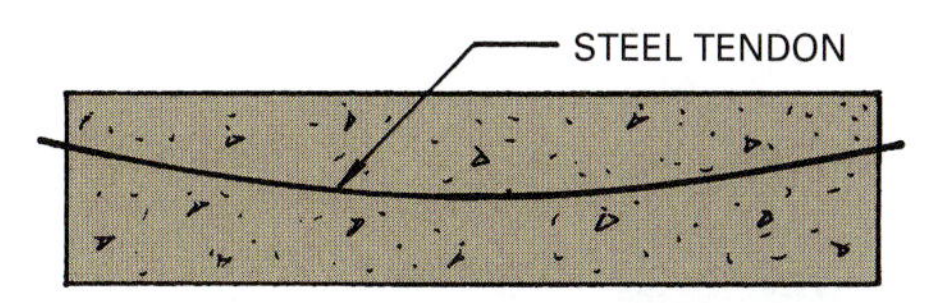

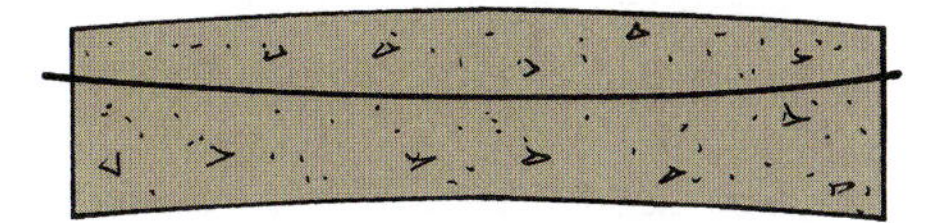

Figure 9.8 This concrete beam is being post-tensioned with a hydraulic jack tensioning the cables in each tendon. The steel protruding on the top of the beam will be embedded in the concrete deck to be poured on top of the beam. *(Courtesy Portland Cement Association)*

Figure 9.9 A post-tensioned concrete structure. *(Reproduced with permission from* The Building Systems Integration Handbook, *Richard Rush, ed., Butterworth-Heinmann Publishers, Newton, Mass., 1986.)*

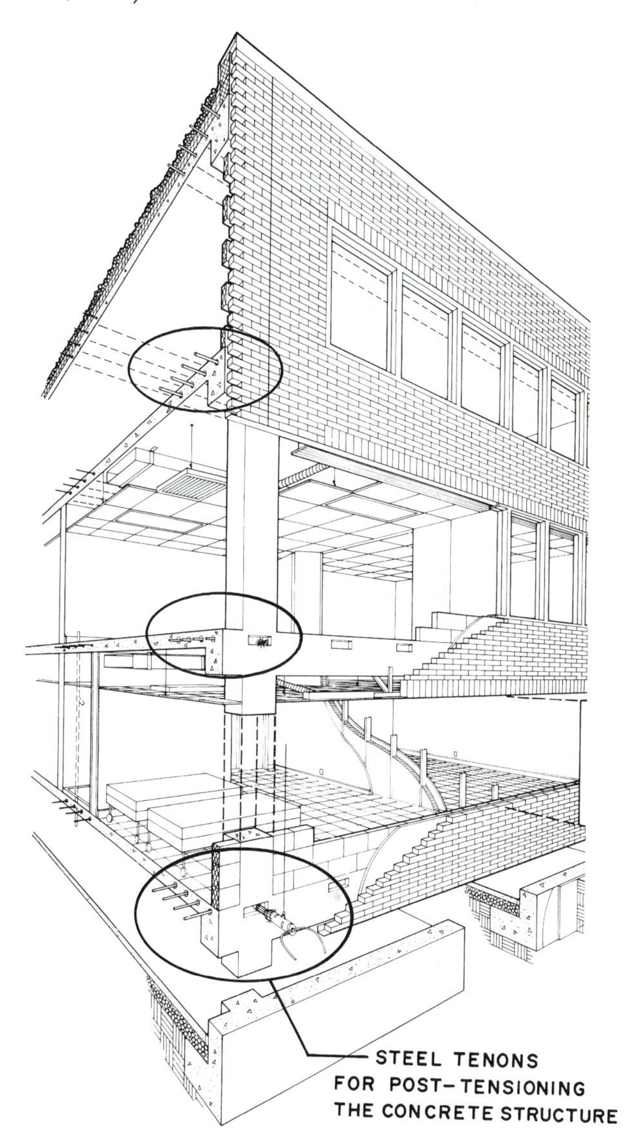

PRECAST CONCRETE ELEMENTS

Precast Concrete Slabs

The major types of slabs used for floor and roof construction are solid flat slabs, tongue-and-groove planks, channel slabs, hollow core, double tees, and single tees (Fig. 9.10). Some are pre-stressed and others precast with un-tensioned reinforcing.

Solid slabs and channel slabs are used for short spans and minimum slab depth. Spans of 25 ft. (7.6 m) with thicknesses of 2 to 8 in. (50 to 200 mm) and widths of 8 to 12 ft. (2.4 to 3.7 m) are common. Slabs made thicker than this to span longer distances become very heavy and are not economical.

Figure 9.10

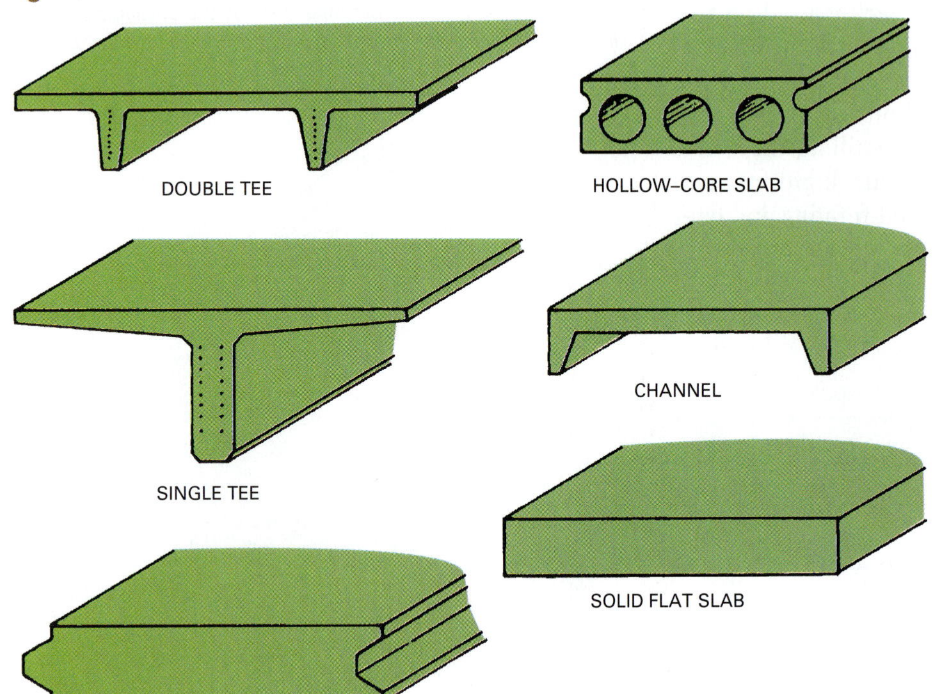

Precast concrete units used for floors and roof decks.

Hollow-core slabs are used for spans ranging up to 40 ft. (12 m) in length and are 6 to 12 in. (150 to 300 mm) thick. The hollow core, created by removing concrete from an area of the slab that has little influence on strength, lightens the member. A topping of 2 in. (50 mm) thick concrete is often applied after installation to increase structural performance and fire rating.

Double tees and single tees are used to span the longest distances. Single tees come in widths of 6, 8, 10, and 12 ft. (1.8, 2.4, 3.0, and 3.7 m) and depths of 16 to 48 in. (406 to 1219 mm). Double tees come in widths of 4, 8, 10, and 12 ft. (1.2, 2.4, 3.0, and 3.7 m) and depths from 10 to 40 in. (254 to 1016 mm). Standard lengths range to over 100 ft. (30.5 m), although lengths over 60 ft. (18.3 m) present special transportation problems.

Tees are cast with a rough or smooth surface on top. The rough surface is used when a concrete topping is to be applied over them. This topping is usually 2 in. (50 mm) thick. Structural continuity across the units can be provided by placing steel reinforcing bars in the topping.

Normal weight and lightweight concrete are used to cast these units. Lightweight concrete is more expensive than normal weight, but it reduces the weight on the structure. Since double tees do not require temporary support to prevent tipping, they are easier and more economical to erect than single tees.

Precast Concrete Columns

Precast columns are usually combined with precast concrete beams, forming a post-and-beam structural framework. Most precast concrete columns are reinforced conventionally. If they are pre-stressed, this is done mainly to reduce stresses on the column during transportation and handling. Columns may be rectangular or square. Square sizes range from 10 to 24 in. (254 to 619 mm). Rectangular columns vary, depending on the design. Columns up to 60 ft. (18 m) in length can be transported. Those above that size need special arrangements. Some columns have **haunches**, projections used to support beams, on two or three sides (Fig. 9.11). Concrete used may range from 5,000 to 11,000 psi (48 to 76 MPa).

Beams and Girders

Precast pre-stressed concrete beams and girders can be used in any building in which precast construction is desired. They are made in several standard shapes, and special designs can be produced (Fig. 9.12). The L-shaped and inverted tee beams provide a bearing surface for precast floor units, such as single and double tee units. This allows lower overall height than does resting floor and roof units on top of rectangular beams.

Rectangular beams range in depth from 18 to 48 in. (457 to 1,219 mm) and are 12 to 36 in. (305 to 914 mm) wide. Inverted tee and L-beams range in depth

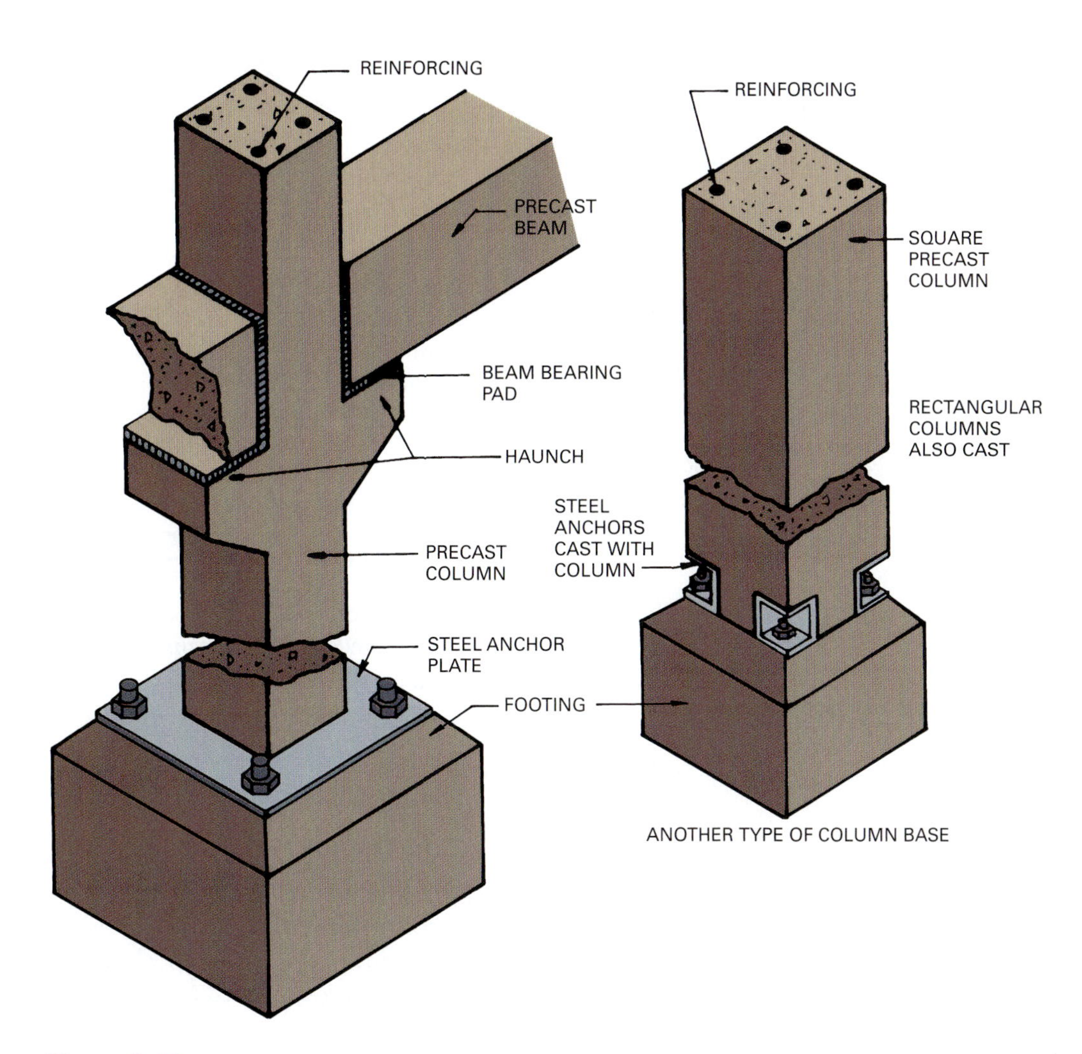

Figure 9.11 Typical precast concrete columns.

Figure 9.12 Typical precast pre-stressed concrete beams.

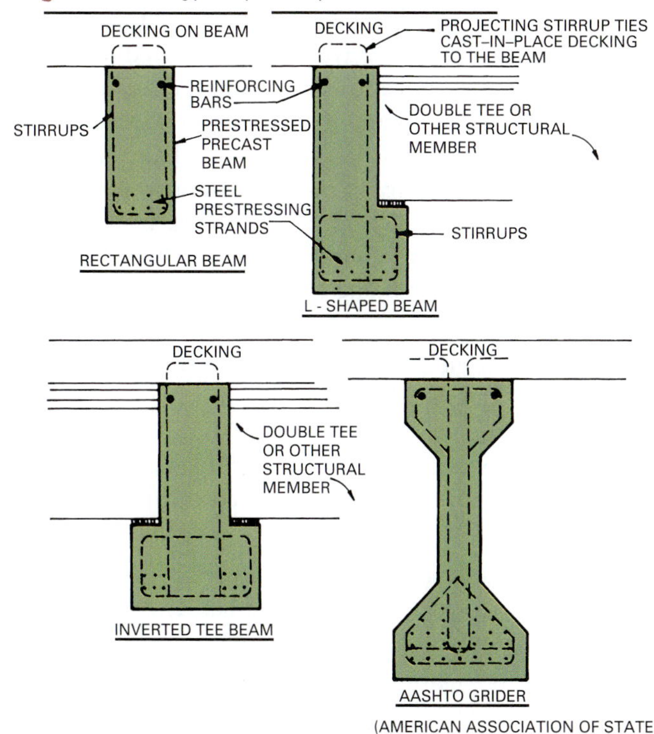

from 18 to 60 in. (457 to 1,524 mm) and in width from 12 to 30 in. (305 to 762 mm). Beam ledges are usually 6 in. (152 mm) wide and 12 in. (305 mm) deep. Beams are available in increments of 2 or 4 in. (50 or 100 mm).

Precast Concrete Wall Panels

Precast concrete wall panels may be pre-stressed or conventionally reinforced. Their design varies considerably. For example, they may be load-bearing or non-load-bearing, they may have a flat, ribbed, or other surface configuration, and they may be hollow, solid, or of a sandwich construction. They can be used in connection with a precast concrete framing system, cast-in-place concrete, or a steel framing system. Flat panels can be two stories high, and ribbed panels can be cast up to four stories high. Panels with openings are not pre-stressed.

Solid panels are typically 3 $\frac{1}{2}$ to 10 in. (89 to 254 mm) thick. Sandwich and hollow-core panels vary from 5 $\frac{1}{2}$ to 12 in. (140 to 305 mm) thick. Ribbed panels are made in thicknesses from 12 to 24 in. (305 to 610 mm). Most panels are designed with an 8 ft. (2.4 in.) width (Fig. 9.13).

Multistory load-bearing panels can be cast with a haunch or other type of support for carrying floor and roof slabs (Fig. 9.14).

Figure 9.13 A wide variety of precast panels can be manufactured, including ones with openings for doors and windows.

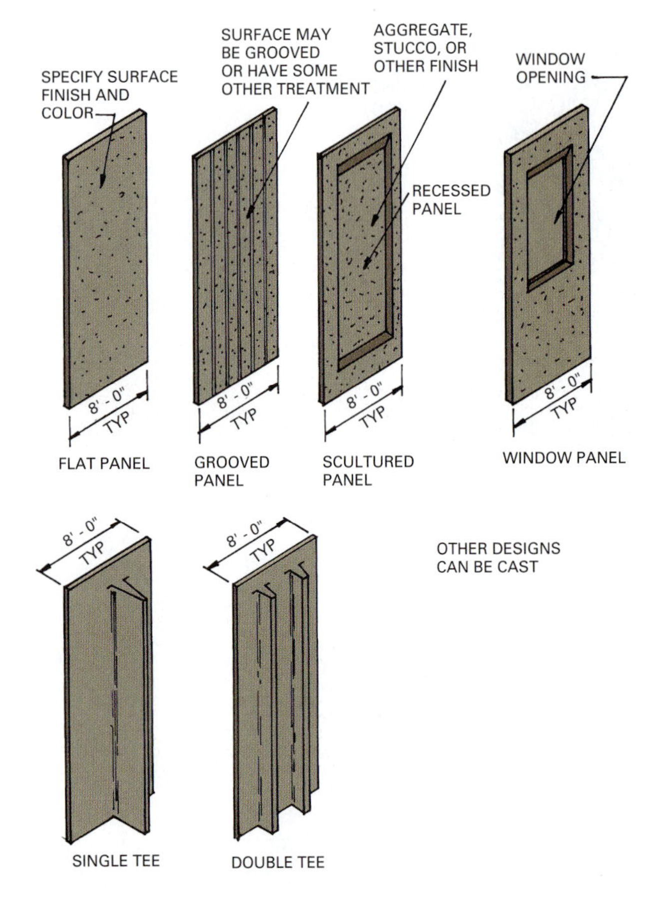

Figure 9.14 Multistory precast concrete panels can be cast with a haunch to carry floor and roof decking.

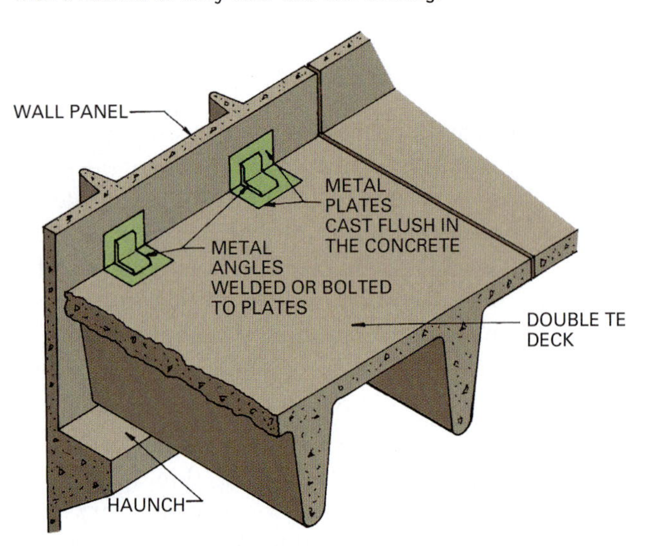

Double-Wall Precast Concrete Building System

A European technology for residential construction is now available in the U.S. The system uses a double-wall insulated precast concrete sandwich panel for both walls and floors. Common wall thickness is 8 in., consisting of two wythes of reinforced concrete 2-$\frac{3}{8}$ in. thick, sandwiched around 3-$\frac{1}{4}$ in. of high R-value insulating foam. The two wythes of the interior and exterior concrete layers are held together with steel trusses.

CONNECTING PRECAST UNITS

The types of connections for precast units include bolt connections, welded connections, post-tensioned connections, and doweled connections.

Types of Connections

Bolt connections speed up erection and allow final alignments to be made after the crane has placed the member. Generally $\frac{1}{2}$, $\frac{3}{4}$, and 1 in. (12, 18, and 25 mm) bolts with national coarse threads are used. Steel washers may be required under certain conditions. Bolts should be tightened to the recommended torque.

Welded connections are strong and easy to make on-site. They should be made following details on erection drawings, which include the size, type, length of the weld, type of electrode, and required temperatures. Welded connections are made before a unit is released by the hoisting device. This tends to tie up the hoisting device for longer periods than for bolted connections. The amount of heat generated may cause the concrete to crack and the metal to become distorted. Large or long, continuous welds should be used sparingly, and the welding temperature should be carefully controlled.

Post-tensioned connections are made using either bonded or unbonded tendons. Bonded tendons are installed in holes cast in the member and bonded to the member after tensioning by filling the hole with grout. Unbonded tendons are not grouted in place but are provided with an organic coating to prevent corrosion.

Doweled connections use reinforcing bars grouted into dowel holes in a precast member. The strength of the connection depends on the dowel diameter, its depth in the member, and the bond developed by the grout. A number of manufactured dowel systems use a sleeve with the dowel (Fig. 9.15).

Figure 9.15 An example of a doweled connection made using a metal sleeve.

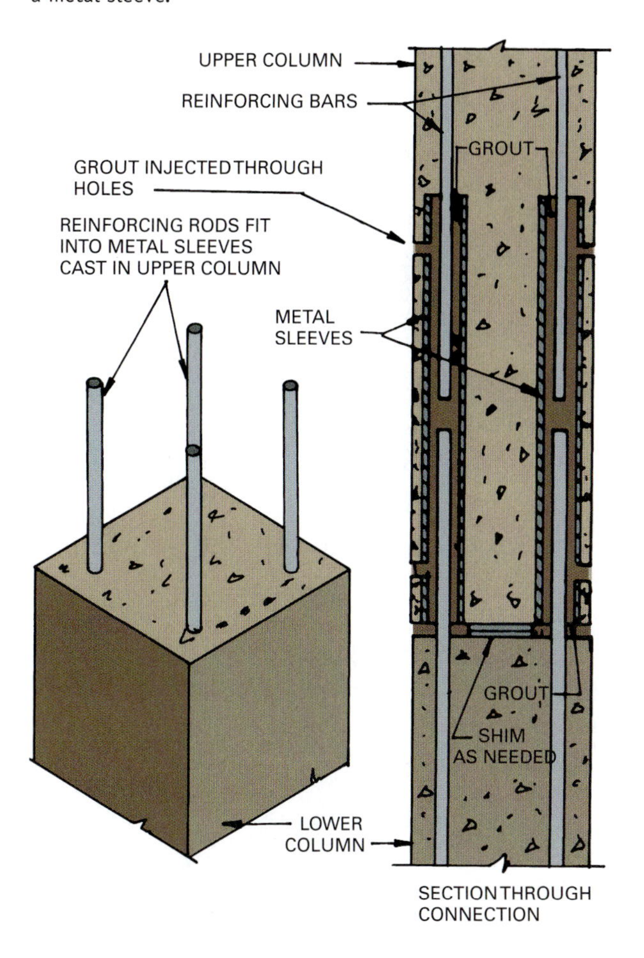

Grout, Mortar, and Drypack

Grouts and mortars are used to transfer loads between members. They are also used to fill noncritical voids. Load-bearing grouts should be non-shrinking, and grouts manufactured for this specific purpose should be used (Fig. 9.16). **Drypack** describes a method of placing grout. The dry grout material has only enough water added to produce a stiff granular mix, which is packed into an opening.

Connection Details

The following detail drawings are typical of connections used and are for general illustration purposes only. All member sizes, shapes, reinforcing, and connections must be designed by a professional engineer to meet the requirements of each specific situation. A typical column-to-column connection is shown in Figs. 9.17 and 9.18. The column can be shimmed as needed to get it plumb.

Figure 9.16 Non-shrinking load-bearing grout is placed below precast columns and precast walls. *(Courtesy Precast/Pre-stressed Concrete Institute)*

A typical column-to-foundation connection is shown in Fig. 9.19. Anchor bolts are cast in the foundation. The column base plate rests on leveling nuts, which are used to get the column plumb. The area below the base plate is filled with grout.

Typical beam-to-column connections include the use of dowels or haunches. Figure 9.20 shows metal dowels that are cast with the column and fit into holes cast in the beam. Figure 9.21 shows precast columns with haunches designed to support the beam. Typically, the beam is connected to the column by welding an angle to metal anchor plates cast into the beam and column.

Figure 9.17 A typical column-to-column connection.

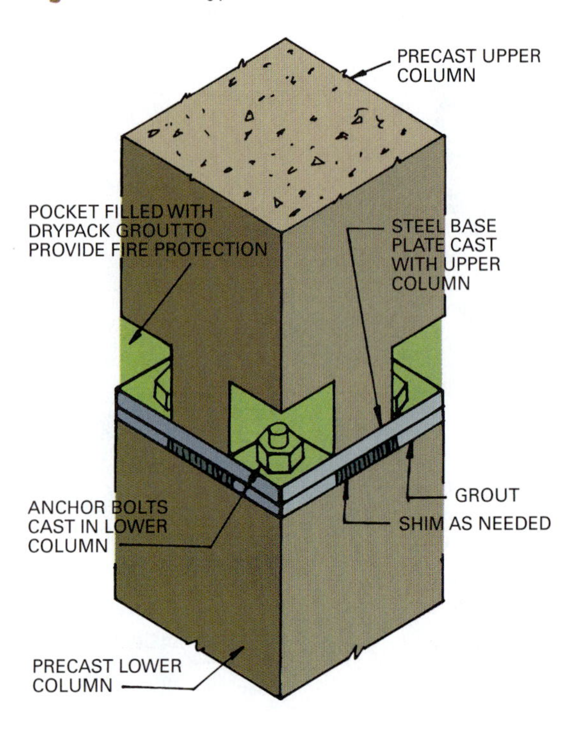

Connections of precast walls to a foundation are shown in Fig. 9.22. This typically involves metal plates that are welded or an anchor rod that is post-tensioned. One way to connect a precast concrete interior wall to a floor in multistory construction is shown in Fig. 9.23. The wall has a metal anchor plate cast in it with a pocket above. An anchor bolt in the wall below is secured to the wall above with a nut.

There are a number of ways to make floor and roof panel connections. The installation of hollow core deck units to a beam is shown in Fig. 9.24. Reinforcing rod is cast in concrete between butting units. Additional rods may be placed in the cores on either side of the joint. Figure 9.25 shows a detail for connecting hollow-core units to an interior precast concrete wall.

One method for installing single and double tees is to use a precast L-shaped beam that is supported on column haunches (Fig. 9.26). The L-beam is connected to the column by welding metal plates cast into each unit. The tee member is welded to the L-beam. Another technique involves supporting inverted tee beams on a precast column haunch and placing the precast tee decking units as shown in Fig. 9.27. A section through this construction shows the details in Fig. 9.28.

Figure 9.18 This column is anchored to the foundation with bolts located within the base of the column.

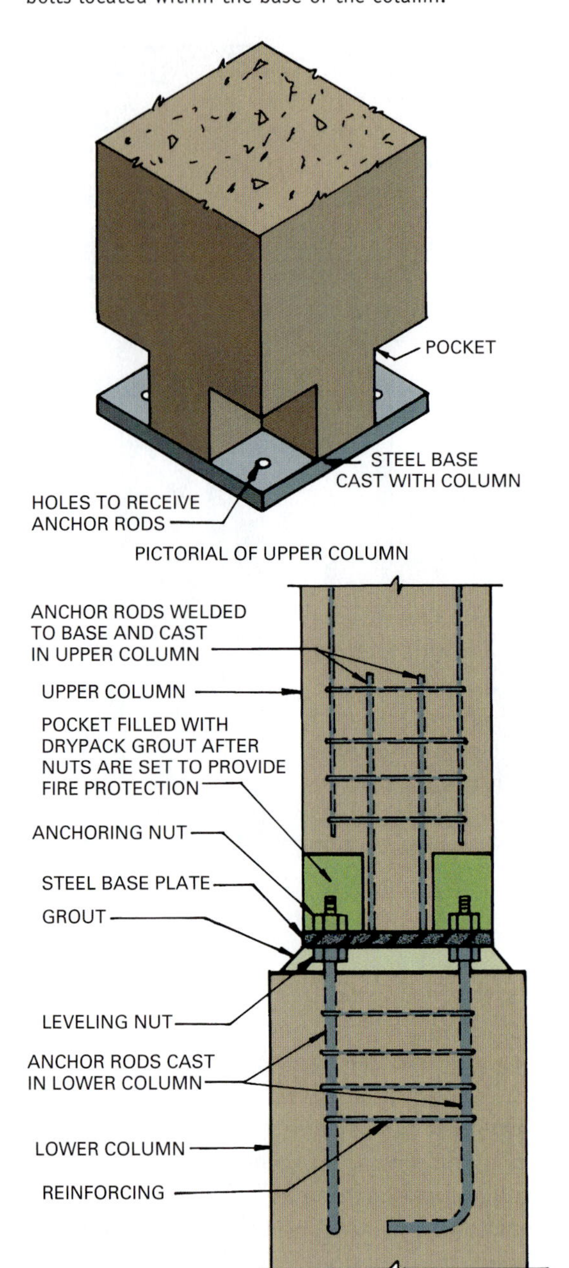

ERECTING PRECAST CONCRETE

Planning

After the structural design work has been completed and reviewed, the plans are sent to the manufacturer that will cast the units. The work is planned using standard molds when possible and custom-built molds as needed.

Figure 9.19 This column is anchored to the foundation with bolts located outside the base of the column.

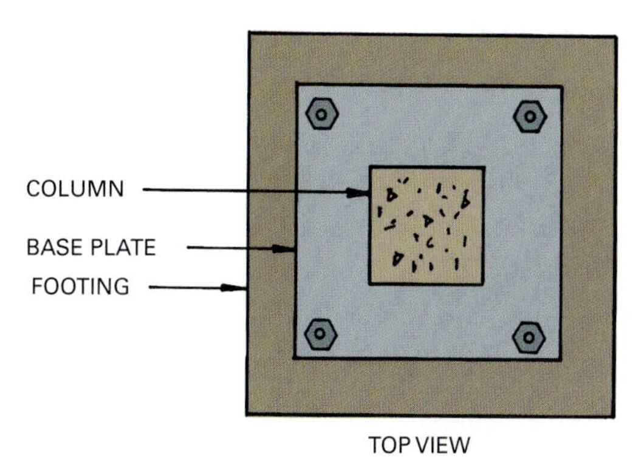

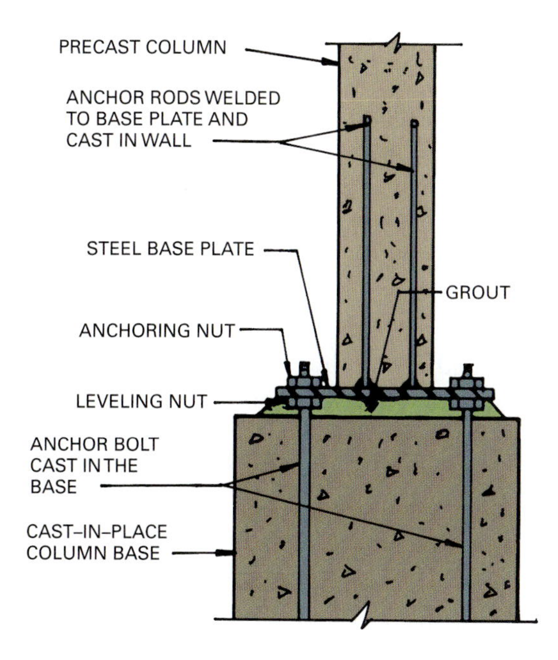

Figure 9.20 A typical connection for a precast rectangular beam to a precast column.

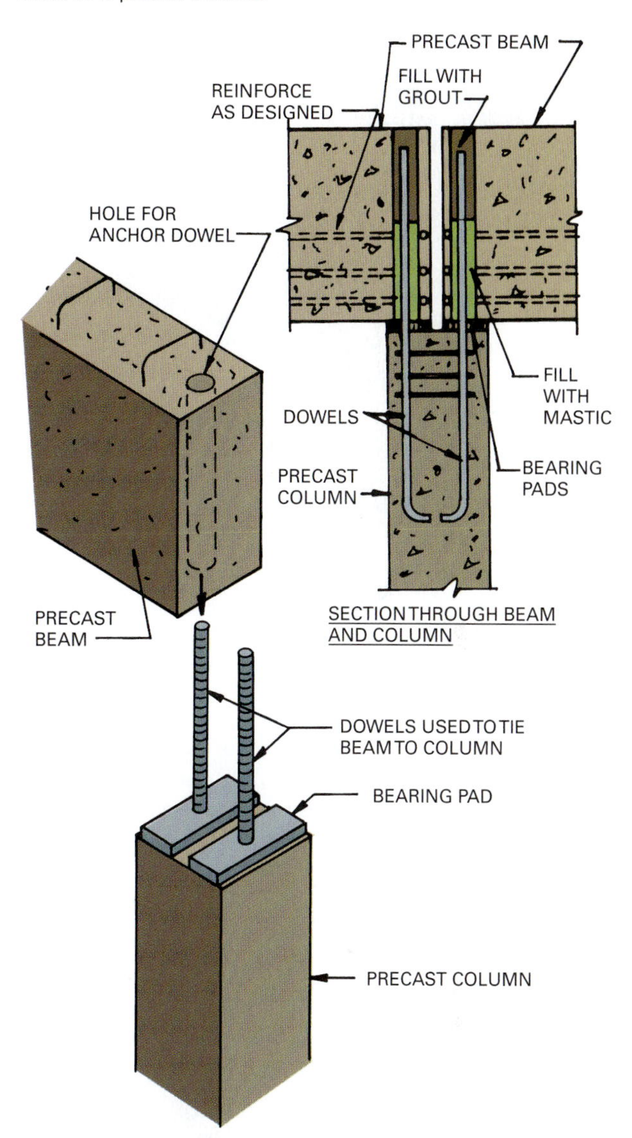

A major factor is the sequencing of the manufacture and delivery of members to the job site. A production and delivery schedule must be planned so the units can be delivered to the site as they are needed. Adequate time for the units to cure and gain sufficient strength to permit movement to the job site must be included.

Consideration of the erection procedures also begins before the units are cast. The design engineer must discuss the use of lifting hardware and connections with the erector. Precast units often have lifting devices cast in them. Their location must be worked out by the designer so the unit can be lifted without damage and leveraged into the position needed for connection.

Precast units are erected and connected as shown on the erection drawings. The erector must comply with the specified dimensional tolerances, hardware sizes, weld lengths, and torque on bolts. Connections should be planned so workers can make them from a stable work platform. Planning should include provision for activities such as welding, post-tensioning, and grouting. Temporary bracing may be required during construction. Final adjustment and alignment of precast structural units can then be made without the need for a crane to support them.

The erection procedure should include plans to maintain the stability and equilibrium of the structure during construction as well as after the structure is completed. For example, when an L-shaped or inverted tee ledger beam is subjected to eccentric loading, the beam experiences torsion and may tilt slightly on the column below. The connections should be designed to resist this motion.

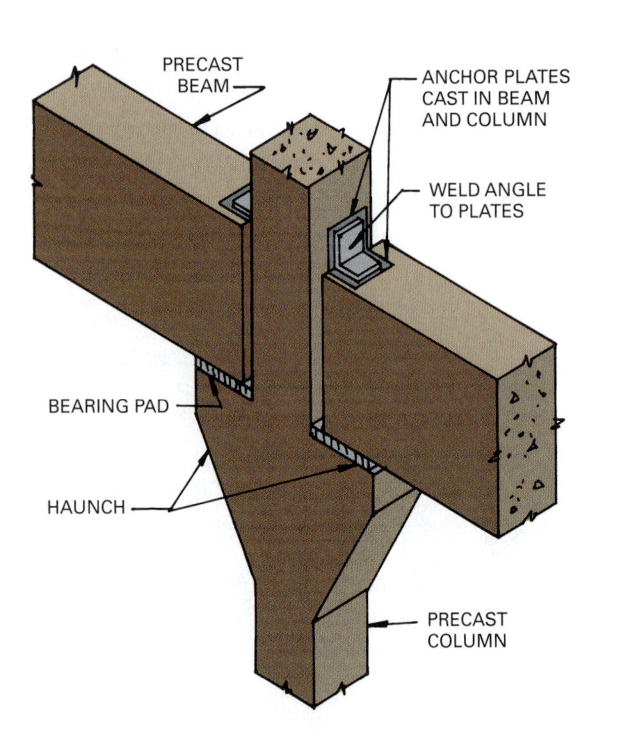

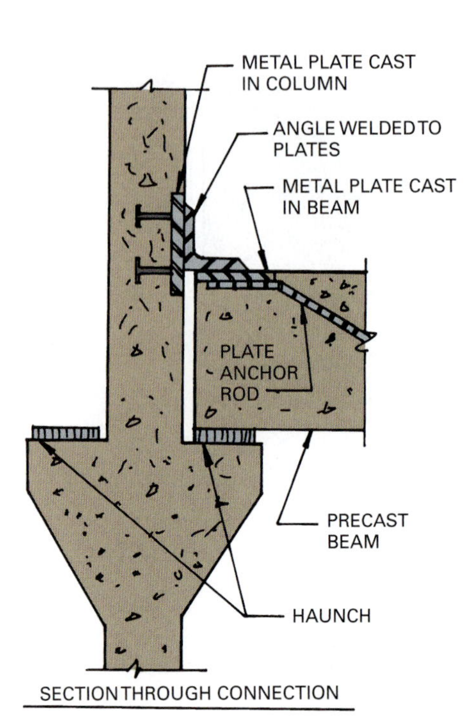

Figure 9.21 Precast rectangular beams are supported on column haunches.

Figure 9.22 Two ways to connect a precast wall to the foundation.

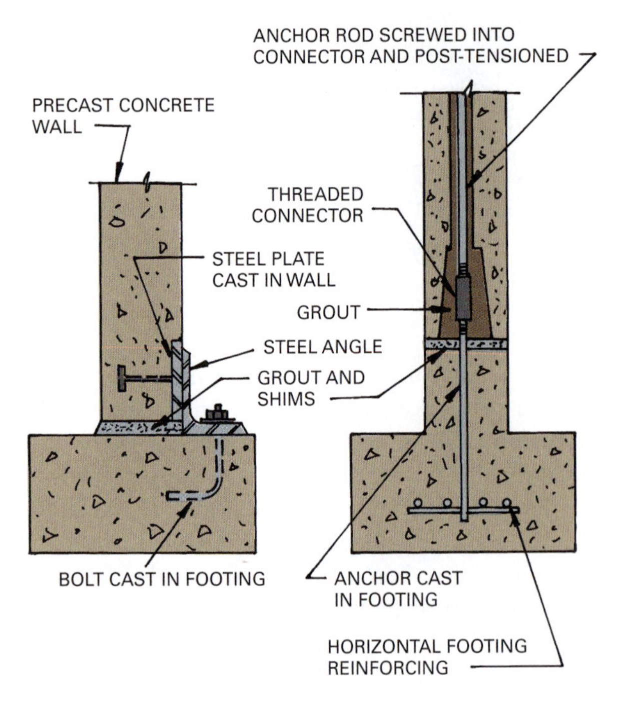

Transporting Units to the Site

Deliveries of precast members are usually made by truck (Fig. 9.29). The distance to the job; the type, size, and weight of members; the type of vehicle required; the weather; road conditions; and other such factors must be considered. At the site, proper equipment for unloading must be available. Finally, the members must be carefully stored until needed to avoid structural damage.

Units are placed on the site so they can be reached easily when needed and as near as possible to their destination points. The erection contractor will have made arrangements to have the necessary hoisting equipment available to unload arriving members and lift them to the point of erection.

Rigging

Rigging refers to the equipment, cables, slings, and other hoisting equipment used to lift precast units into place (Fig. 9.30). Pre-stressed members must never be lifted by their center only, and they should not be lifted or stored upside down. They must be kept in the upright position in which they will be installed. If lifting devices are not in the member, it can usually be safely lifted with slings placed near each end.

Placing Precast Units

The foundations and footings for precast structures are often cast-in-place and must be carefully planned. Tolerances are tight, and assembly will halt if tolerances

Figure 9.23 Anchor bolts can be used to connect precast interior walls when hollow-core decking is used.

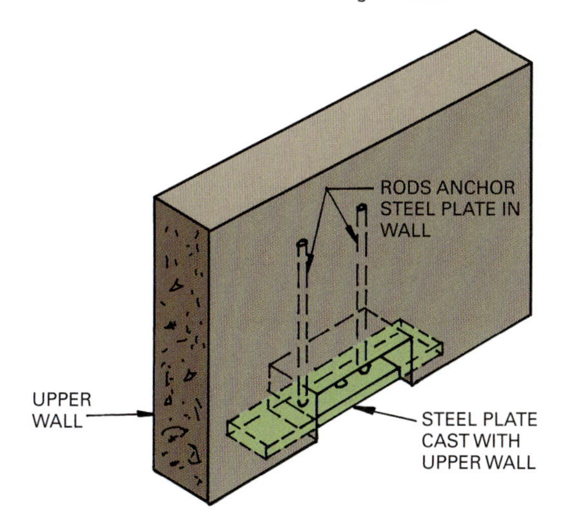

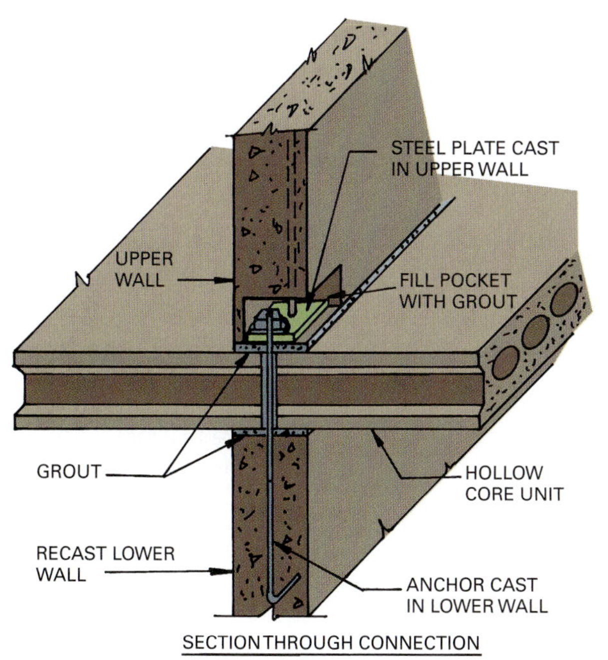

have not been carefully maintained. A structure may be built using precast columns and beams. The columns are lifted and placed with a crane, then plumbed and grouted with drypack grout.

Columns are joined by beams (Fig. 9.31) and the floor decking is set in place. Floor decking often consists of hollow-core units (Fig. 9.32).

Framing Plan

An example of a floor **framing plan** for a small building using hollow-core units is shown in Fig. 9.33. The specific way these are drawn varies, but the information shown is consistent. Single or double tees are also often used for floor decking. Double tees are preferred to single tees because they do not require bracing during erection (Fig. 9.34).

Figure 9.24 A typical detail for installing hollow-core units on a precast rectangular beam.

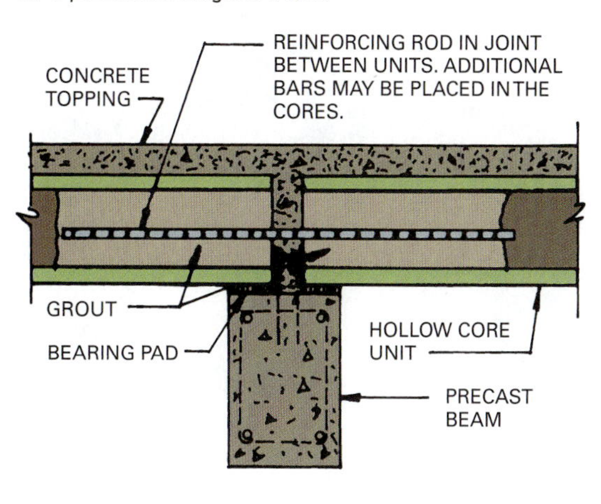

Figure 9.25 A typical detail for supporting hollow-core units on a precast interior wall.

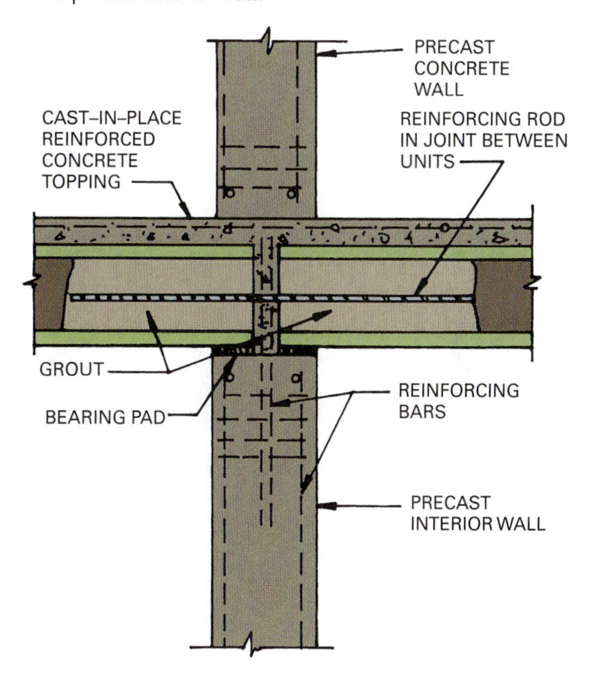

Some buildings use precast structural walls instead of columns. These are erected on foundations and braced until members abutting them are in place (Fig. 9.35).

If precast exterior wall panels are used to enclose the building, they are lifted in place and secured to the structural frame (Fig. 9.36).

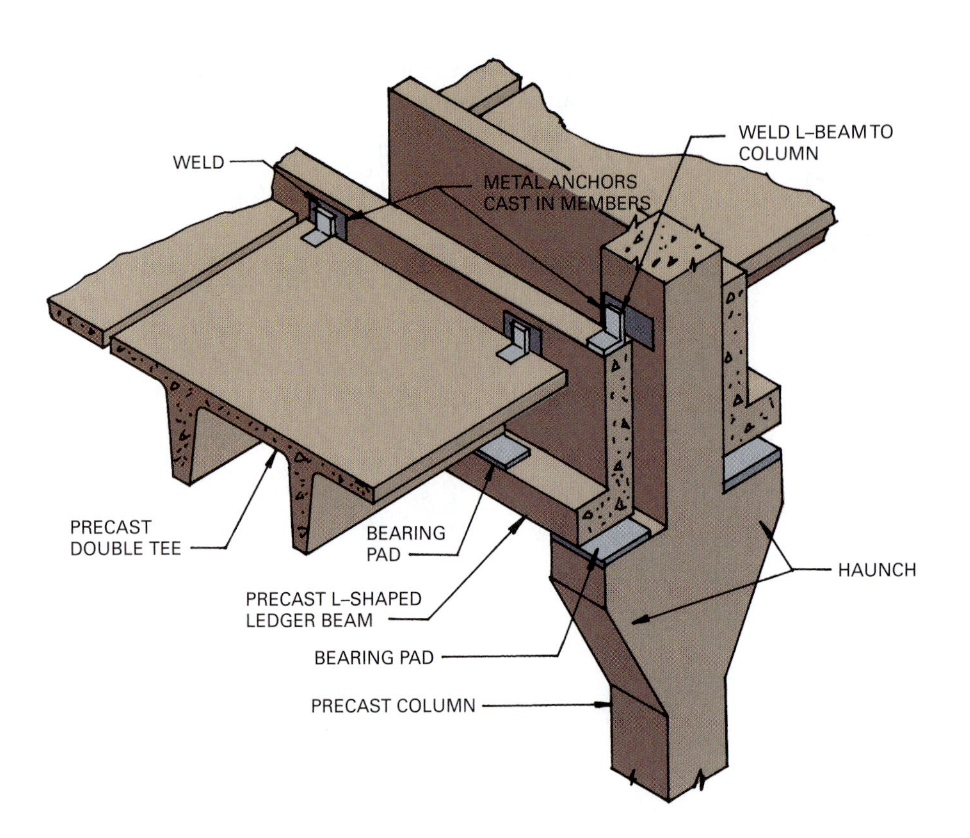

Figure 9.26 Double tees can be supported on a ledger beam that is seated on a haunch of a precast concrete column.

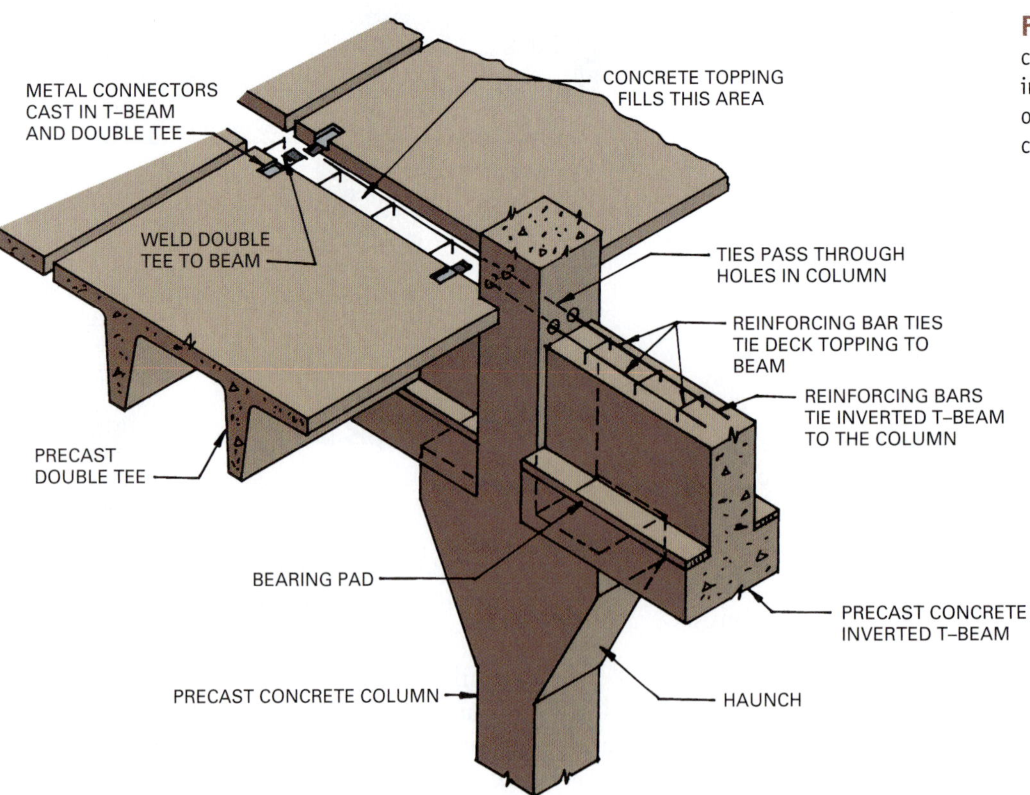

Figure 9.27 Double tees can be supported by precast inverted tee beams that rest on the haunches of precast concrete columns.

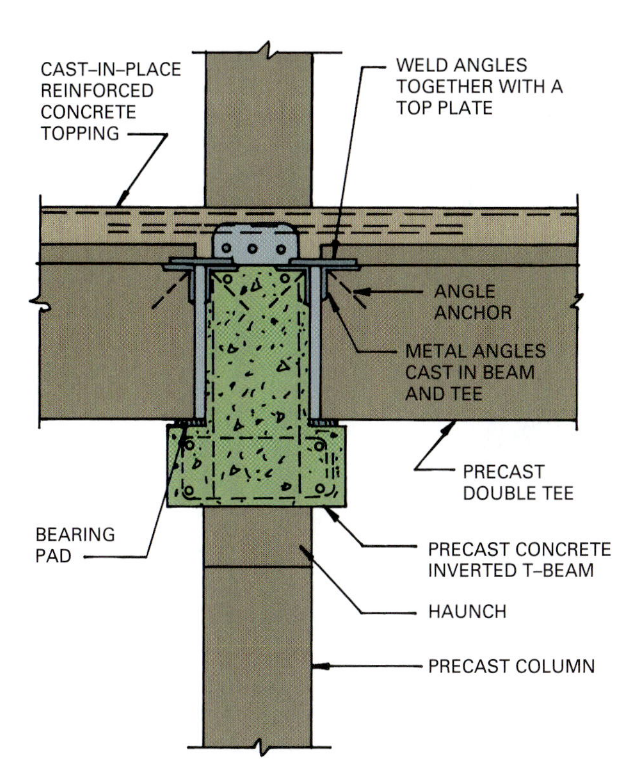

Figure 9.28 A section through a precast inverted tee beam supporting double tees.

Figure 9.29 Most precast units are transported to the site by truck and set in place with a crane. *(Courtesy Precast/Pre-stressed Concrete Institute)*

Figure 9.30 Precast units are lifted into place with a crane and proper cables, slings, and other hoisting devices. *(Courtesy Precast/Pre-stressed Concrete Institute)*

Figure 9.31 Precast pre-stressed beams are connected to the columns. *(Courtesy Precast/Pre-stressed Concrete Institute)*

Figure 9.32 This precast hollow-core floor decking is secured to the beams. *(Courtesy Precast/Pre-stressed Concrete Institute)*

Figure 9.33 A framing plan for a building with hollow-core floor units. Included are drawings showing construction details and a schedule giving details about the units required.

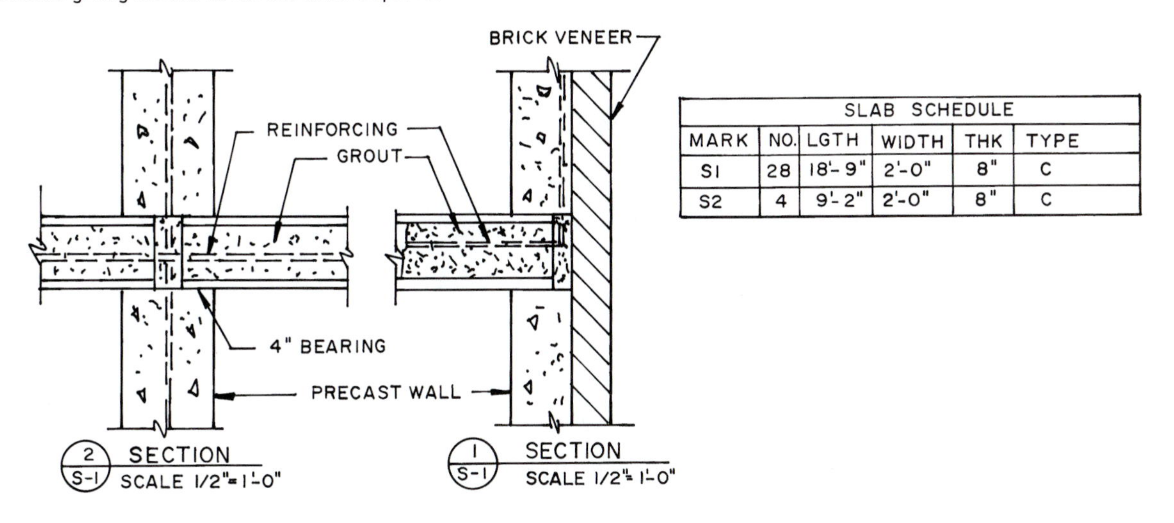

SLAB SCHEDULE					
MARK	NO.	LGTH	WIDTH	THK	TYPE
S1	28	18'-9"	2'-0"	8"	C
S2	4	9'-2"	2'-0"	8"	C

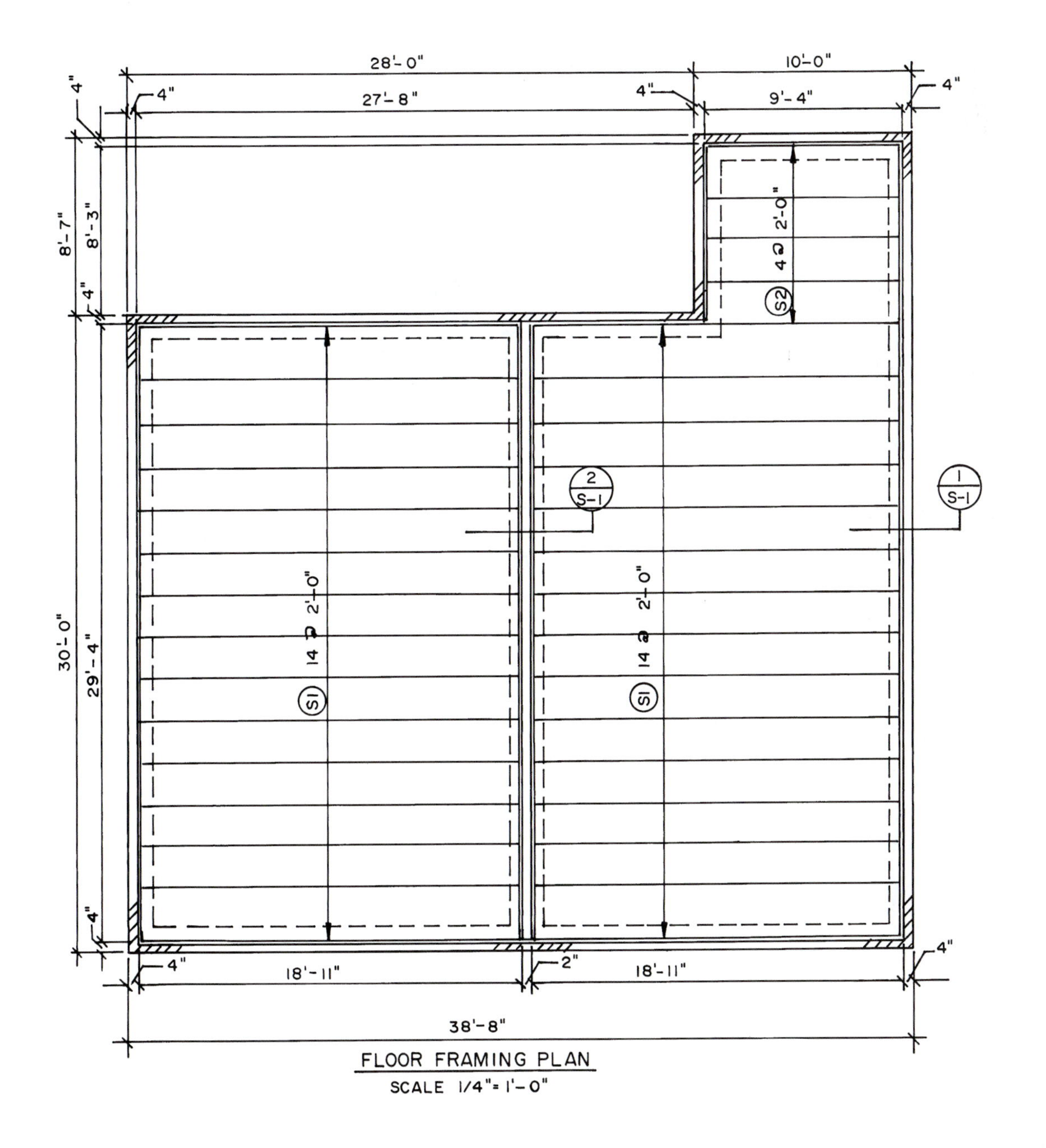

Figure 9.34 The underside of a deck built using precast double tees. *(Courtesy Precast/Pre-stressed Concrete Institute)*

Figure 9.35 Precast interior and exterior walls are connected to a foundation or beam. *(Courtesy Precast/Pre-stressed Concrete Institute)*

Figure 9.36 Precast exterior wall panels are connected to the structural frame. *(Courtesy Precast/Pre-stressed Concrete Institute)*

Review Questions

1. Give some examples of precast concrete structural units.
2. How do non-pre-stressed structural units differ from pre-tensioned units?
3. What does camber refer to?
4. What are the commonly used types of precast pre-stressed floor and roof slabs?
5. Which type of precast pre-tensioned slabs span the greatest distances?
6. What is the main value in pre-stressing precast concrete columns?
7. What is the difference between post-tensioning and pre-tensioning?

Key Terms

camber
double tees
drypack
framing plan
grout
haunch
non-pre-stressed units
post-tensioned
pre-stressed units
pre-tensioned
rigging

Activities

1. If there is a precasting plant within reasonable driving range, it is well worth the effort to make a visit. Ask the supervisor to explain and show the process from beginning to end. Take photos and prepare an exhibit detailing the process.
2. Explain how a precast plant prepares the concrete and how the quality control tests are taken.
3. See if you can get the structural plans for a precast concrete structure from a local architect or contractor. Prepare a report citing the size and spacing of columns, beams, and girders. What type of floor and roof decking were specified, and what are the sizes of each? Is there anything else on the plan that you would like to report?
4. Make a model with wood pieces representing precast concrete beams, girders, columns, wall panels, and decking.
5. Examine the local building code and briefly cite the specific areas related to precast concrete structures.
6. If you are fortunate enough to be able to visit the site of a precast concrete structure being erected, take photos and notes describing what you see. For example, what types of connectors are used? How are members lifted into place? What safety devices are the workers wearing?

Additional Resources

Architectural Precast Concrete, and numerous other publications, Precast/Pre-stressed Concrete Institute, Chicago, IL.

PCI Design Handbook, and numerous other publications, Precast/Pre-stressed Concrete Institute, Chicago, IL.

Other resources include: Many related publications available from the American Concrete Institute, Farmington Hills, MI.

Numerous publications from the Portland Cement Association, Skokie, IL.

See Appendix B for addresses of professional and trade organizations and other sources of technical information.

DIVISION

04

Masonry
CSI MasterFormat™

042000 Unit Masonry
044000 Stone Assemblies
045000 Refractory Masonry
046000 Corrosion Resistant Masonry

CHAPTER 10

Mortars for Masonry Walls

LEARNING OBJECTIVES

Upon completion of this chapter, the student should be able to:

- Describe the various materials used to produce mortar.
- Describe the ingredients used to manufacture masonry cements.
- Identify and cite the uses of the types of standard mortars.
- Discuss the desirable properties of plastic mortar.
- Discuss the properties of quality hardened mortar.
- Be aware of the frequently used mortar tests.
- Specify how to cure masonry mortar laid in cold weather.

Build Your Knowledge

For further information on these materials and methods, please refer to:

Chapter 7 Concrete

Mortar is the bonding agent used to join masonry units into an integral structure. In addition to bonding the units together, it also must (1) seal the spaces between the units so they are not penetrated by air or moisture; (2) tie steel reinforcement, ties, and anchor bolts into walls; (3) provide a design of lines of color and shadows; and (4) allow for adjustment of the slight variations that occur in masonry units.

MORTAR COMPOSITION

Mortar is composed of **cementitious materials**, clean carefully graded mortar sand, clean water, and sometimes a coloring agent or admixture. The actual composition of mortars varies according to intended use. For general masonry applications, mortars may contain Portland cement or masonry cement, sand, hydrated lime or lime putty, and water. Mortars for special applications, such as prefabricated masonry units that must be moved into place, require additives that increase compressive and tensile strength and have greater bonding capabilities.

Cementitious Materials

Cementitious materials provide the bonding ingredient in mortars. They are made according to ASTM (American Society for Testing and Materials) and CSA (Canadian Standards Association) specifications, as shown in the following list:

Masonry cement—ASTM C91
(Types M, S, or N), CSA A8 (Types S or N)
Portland cement—ASTM C150
(Types I, IA, II, IIA, III, or IIIA), CSA A5 (10, 20, 30, or 50)
Blended hydraulic cement—ASTM C595
(Types IS, IS-A, IP, IP-A, I(PM), or I(PM)-A) (Pozzolan modified Portland cement), CSA A362 (Type 10S)
Hydrated lime for masonry purposes—ASTM C207
(Types S, SA, N, or NA)
Quicklime for structural uses (for lime putty)—ASTM C5

Detailed information on various cements is given in Chapter 7. Most mortars are now made with mortar cement. Masonry cement mortars are made by combining masonry cement, clean carefully graded sand, and enough clean water to produce a plastic, workable mix.

Masonry Cements

Masonry cements are hydraulic cements used in mortars for masonry construction. They produce a mortar that has greater plasticity and water retention than Portland cement. Typically they may include Portland cement, slag cement, blast-furnace slag cement, Portland-pozzolan cement, natural cement, and hydraulic lime. Manufacturers may include chalk, clay, talc, calcareous shell, or limestone to add the properties desired.

Masonry cements meet the specifications found in ASTM C91. This report specifies three types of masonry cement: Type N, Type S, and Type M. They are used to produce mortars: Types N, O, S, and M, as specified in ASTM C270. Type N masonry cement is used in Type N and O mortars. It can be blended with Portland cement for use in Type S and M mortars.

Other Mortar Ingredients

Lime is used to help stabilize the volume of mortar by controlling shrinking and expansion. Masonry sand may be natural or a manufactured product that meets the requirements of ASTM C144, the Standard Specification for Aggregate for Masonry Mortar. In Canada the standard is CSA A82.56, Aggregate for Masonry Mortar. Since sand makes up the major portion of a mortar, the quality of the finished mortar depends heavily on clean, quality sand. The gradation requirements for sand specified in ASTM C144 are shown in Table 10.1.

Water used to produce mortar must be clean and free of acids, alkalies, and other organic materials. Water containing soluble salts will cause the mortar to effloresce. **Efflorescence** appears on the finished masonry wall as a white powdery substance that discolors the mortar and masonry units.

Table 10.1 Aggregate Gradation for Masonry Mortar

	Gradation Specified, Percent Passing ASTM C144[a]	
Sieve Size No.	**Natural Sand**	**Manufactured Sand**
4	100	100
8	95 to 100	95 to 100
16	70 to 100	70 to 100
30	40 to 75	40 to 75
50	10 to 35	20 to 40
100	2 to 15	10 to 25
200	—	0 to 10

[a]*Additional requirements: Not more than 50% shall be retained between any two sieve sizes, nor more than 25% between No. 50 and No. 100 sieve sizes. Where an aggregate fails to meet the gradation limit specified, it may be used if the masonry mortar will comply with the property specification of ASTM C270 (Table 2).*
Courtesy American Society for Testing and Materials

Mortar can be colored by adding various organic pigments. White mortar is made with white masonry cement or white Portland cement, lime, and white sand. Colored mortars are made with white masonry cement and pigments, which are typically some type of mineral oxide compound, such as iron, chromium, cobalt, or manganese oxides. Carbon black is used to produce a dark gray or black mortar.

Admixtures may sometimes be used to alter the properties of a mortar. However, the possibility of their creating unforeseen problems is great. Admixtures usually are not used unless laboratory tests have been made to verify the changes they will have on the mortar. For example, certain admixtures may affect the hydration process and hardening properties.

Table 10.2 lists some of the commonly available admixtures (modifiers), their benefits, and possible problems they may cause. More information on admixtures is detailed in Chapter 7.

TYPES OF MORTAR

Four types of mortar are specified for use in the United States. These are M, S, N, and O, as specified by ASTM C270. The Canadian Standards Association Standard A179 specifies two types, S and N.

Mortar types are identified by either proportion or property specifications. Table 10.3 lists the specifications for mortar proportion indicated by the combinations of Portland cement or blended cement, masonry cement, hydrated lime or lime putty, and the aggregate. Blended cement is a combination of Portland cement and masonry cement. Mortar types, classified under property specifications, are based on compressive strength, water retention, and air content (Table 10.4). This data is developed by testing 2 in. (50.8 mm) cubes of hardened mortar to find the compressive strength. The compression test is discussed later in this chapter.

Type M mortar is a high-strength mortar with a compressive strength of 2,500 psi (17 MPa). It has better durability than the other types. Type M is recommended for masonry below grade in contact with earth and for conditions of severe frost action. It is also used for reinforced masonry.

Type S mortar is a medium-high-strength mortar with a compressive strength of 1,800 psi (12.5 MPa). It is often permitted in wall construction instead of Type M because it has almost the same allowable strength and can be used above or below grade. Type S

Table 10.2 Admixtures—Benefits and Concerns

Admixture	Primary Benefits	Possible Concerns
Air-entraining	Freeze-thaw durability, workability	Effect on compressive and bond strengths
Bonding	Wall tensile (and flexural) bond strength	Reduced workability, bond strength regression upon wetting, corrosive properties
Plasticizer	Workability, economy	Effect on hardened physical properties under field conditions
Set accelerator	Early strength development	Effectiveness at cold temperatures, corrosive properties, effect on efflorescence potential of masonry
Set retarder	Workability retention	Effect on strength development, effect on efflorescence potential of masonry
Water reducer	Strength, workability	Effect on strength development under field conditions with absorptive units
Water repellent	Weather resistance	Effectiveness over time
Pozzolanic	Increase density and strength	Effect on plastic and hardened physical properties under field conditions
Color	Esthetic versatility	Effect on physical properties, color stability over time

Courtesy Portland Cement Association

Table 10.3 Proportion Specifications for Mortar

United States—ASTM C270						
	Parts by Volume					
		Masonry cement type				
Mortar type	Portland cement or blended cement	M	S	N	Hydrated lime or lime putty	Aggregate[a]
	1	—	—	1	—	$4\frac{1}{2}$ to 6
M	—	1	—	—	—	$2\frac{1}{4}$ to 3
	1	—	—	—	$\frac{1}{4}$	$2\frac{13}{16}$ to $3\frac{3}{4}$
	$\frac{1}{2}$	—	—	1	—	$3\frac{3}{8}$ to $4\frac{1}{2}$
S	—	—	1	—	—	$2\frac{1}{4}$ to 3
	1	—	—	—	Over $\frac{1}{4}$ to $\frac{1}{2}$	.
	—	—	—	1	—	$2\frac{1}{4}$ to 3
N	1	—	—	—	Over $\frac{1}{2}$ to $1\frac{1}{4}$	.
	—	—	—	1	—	$2\frac{1}{4}$ to 3
O	1	—	—	—	Over $1\frac{1}{4}$ to $2\frac{1}{2}$	.

Canada—CSA A179M					
	Parts by Volume				
		Masonry cement type			
Mortar type	Portland cement	S	N	Hydrated lime or lime putty	Aggregate[a]
S	—	—	1	—	$2\frac{1}{4}$ to 3
	$\frac{1}{2}$	—	1	—	$3\frac{1}{2}$ to $4\frac{1}{2}$
	1	—	—	$\frac{1}{2}$	$3\frac{1}{2}$ to $4\frac{1}{2}$
N	—	—	1	—	$2\frac{1}{4}$ to 3
	1	—	—	1	$4\frac{1}{2}$ to 6

[a]The total aggregate shall be equal to not less than $2\frac{1}{4}$ and not more than 3 times the sum of the volumes of the cement and lime used.

Notes: 1. Under both ASTM C270, Standard Specification for Mortar for Unit Masonry, and CSA A1 79, Mortar and Grout for Unit Masonry, aggregate is measured in a damp, loose condition and 1 cu ft. of masonry sand by damp, loose volume is considered equal to 80 lb. of dry sand (in SI units 1 cu m of damp, loose sand is considered equal to 1,280 kg of dry sand).

2. Mortar should not contain more than one air-entraining material.

Courtesy Portland Cement Association and the Canadian Standards Association. Canadian material presented with the permission of the Canadian Standards Association; material is reproduced from CSA Standard A179-94 (Mortar and Grout for Unit Masonry), which is copyrighted by CSA, 178 Rexdale Blvd., Etobicoke, Ontario M9W 1R3. Although use of this material has been authorized, CSA shall not be responsible for the manner in which the information is presented, nor for any interpretations thereof. This material may not be updated to reflect amendments made to the original content. For up-to-date information, contact CSA.

Table 10.4 Property Specifications for Laboratory-Prepared Mortar[a]

United States				
Mortar specification	**Mortar type**	**Minimum 28-day compressive strength, psi**	**Minimum water retention, %**	**Maximum air content, %[b]**
	M	2,500	75	12[c]
	S	1,800	75	12[c]
ASTM C270	N	750	75	14[c]
	O	350	75	14[c]
Canada				
Mortar specification	**Mortar type**	**Minimum compressive strength, MPa**		**Minimum water retention, %**
		7-day[d]	**28-day**	
	S	7.5	12.5	70
CSAA179	N	3	5	70

[a]*The total aggregate shall be equal to not less than 2¼ and not more than 3½ times the sum of the volumes of the cement and lime used.*
[b]*Cement-lime mortar only (except where noted).*
[c]*When structural reinforcement is incorporated in cement-lime or masonry cement mortar, the maximum air content shall be 12% or 18%, respectively.*
[d]*If the mortar fails to meet the 7-day requirement but meets the 28-day requirement, it shall be acceptable.*

has better workability and more **water retention** than Type M. It is used for reinforced and unreinforced masonry and has high tensile bond strength.

Type N mortar is a medium-strength mortar with a compressive strength of 750 psi (5 MPa). It is used for above-grade general construction involving exposed masonry load-bearing walls for which compressive strength and lateral strength requirements are not high.

Type O mortar is a medium-strength mortar with a compressive strength of 350 psi (2.5 MPa). It is used for general interior purposes, such as for non-load-bearing walls with a compressive strength that does not exceed 100 lb. per sq. in. It must not be in contact with soil or exposed to freezing conditions.

A guide for selecting masonry mortars for various uses above and below grade is given in Table 10.5.

Sources of Mortar

Mortars are available ready-mixed, dry-batch, and on-site mixed. Ready-mixed mortar is prepared at a central batch plant and transported to the job site, where a slump test is made. If required, additional water is added to get the desired slump. This mortar contains a retarding, set-controlling admixture that keeps the mortar workable and in a plastic condition for more than twenty-four hours. The mix is delivered by a ready-mix truck and stored on-site in large tubs. Ready-mixed mortar must meet the requirements of ASTM C1142, Specifications for Ready-Mixed Mortar for Unit Masonry.

Dry-batch mortar has the required ingredients blended and packaged in bags that are delivered to the site and stored in a sealed hopper. When mortar is needed, the dry ingredients are moved to an on-site mixer and the required water is added. Dry-batching provides control over the proportions of the various ingredients in the mortar. Premixed dry mortar ingredients are also available packaged in waterproof bags.

On-site mixed mortar can be made by storing dry ingredients in a silo-type mixer in which an auger mixing device receives the correct proportions of dry ingredients and blends them together. The dry ingredients are stored in separate chambers, from which they are fed to the mixer. The water is then injected from a pressurized water source, such as a city water main, and the mixer produces the finished mortar. The process is usually computer controlled. The mortar is discharged into tubs or wheelbarrows for distribution on-site. This produces a very accurately proportioned mortar (Fig. 10.1).

For small jobs, the ingredients typically are delivered to the site as separate items. The mortar cement is in bags, and the sand is in bulk. The sand must be protected from moisture and dirt. The mortar is often proportioned by placing shovelfuls of sand into a small power-operated mixer and adding the required number of bags of mortar cement with water (Fig. 10.2). The gasoline-powered mixer produces the mortar, which usually is moved to the masons by wheelbarrow. The proportions of the ingredients can vary considerably, as can be evidenced after mortar has cured and joints over

Table 10.5 Guide for the Selection of Masonry Mortars (United States)[a]

Location	Building segment	Mortar type	
		Recommended	Alternative
Exterior, above grade	Load-bearing walls	N	S or M
	Non-load-bearing walls	O[b]	N or S
	Parapet walls	N	S
Exterior, at or below grade	Foundation walls, retaining walls, manholes, sewers, pavements, walks, and patios	S[c]	M or N[c]
Interior	Load-bearing walls	N	S or M
	Non-load-bearing partitions	O	N

[a]Adapted from ASTM C270. This table does not provide for specialized mortar uses, such as chimney, reinforced masonry, and acid-resistant mortars.
[b]Type O mortar is recommended for use where the masonry is unlikely to be frozen when saturated or unlikely to be subjected to high winds or other significant lateral loads. Type N or S mortar should be used in other cases.
[c]Masonry exposed to weather in a nominally horizontal surface is extremely vulnerable to weathering. Mortar for such masonry should be selected with due caution.
Courtesy American Society for Testing and Materials

Figure 10.1 Sand and cement stored in a computer-controlled silo are mixed, discharging the desired mix and amount of mortar. *(Courtesy of Ennstone, Inc.)*

Figure 10.2 Small quantities of on-site mortar can be mixed in a portable mixer. *(Courtesy of Bon Tool Company)*

a wall do not have a uniform shade. To be most effective, proportions should be carefully measured and the power mixer run from three to five minutes. Undermixing produces a non-uniform mortar, and overmixing may reduce a mortar's strength.

PROPERTIES OF PLASTIC MORTAR

The properties that affect mortar in a plastic condition are workability and water retention.

Workability

Workability is a term used to describe the condition of a mortar that will spread easily, cling to vertical surfaces, extrude easily from joints but not drop off, and permit easy positioning of masonry units (Fig. 10.3). It is a condition of the mortar that is influenced by other properties, such as consistency, water retention, setting time, weight, adhesion, and penetrability. Masons judge workability by observing how the mortar slides or adheres to their trowel.

Figure 10.3 Mortar must be workable, adhere to vertical surfaces, and extrude from joints without falling off.

Water Retention

Mortar must have good **water retention** to resist rapid loss of mixing water to the air or an absorptive masonry unit. Precipitous mix water loss causes mortar to stiffen, making it difficult to achieve a good bond or watertight joint. Mortar that has good water retentivity remains workable, enabling masons to properly place masonry units. Low-absorption masonry units may float when placed on a mortar with too much water retentivity. This causes the mortar joint to bleed. Water retentivity is increased by entrained air, very fine aggregate, or cementitious materials.

Mortar Flow Test The water retention limit of mortar is measured by both initial flow and flow following laboratory suction tests, as described in ASTM C91. The initial flow test is made by placing a truncated cone of mortar with a 4 in. (100 mm) diameter on a metal flow table. The table is mechanically dropped $\frac{1}{2}$ in. (12 mm) twenty-five times in fifteen seconds. The mortar will flow into an enlarged circular shape. The diameter of this shape is measured and compared with the original diameter. Allowable initial flow should be in a range of 100 to 115 percent.

The flow-after-suction test is used to determine flow after loss of water to an absorbent masonry unit. The mortar sample is placed for one minute in a vacuum device that removes some water. The mortar is then tested as described above and the flow measured. The flow after suction should be in a range of 70 to 75 percent.

PROPERTIES OF HARDENED MORTAR

The properties essential to quality hardened mortar are bond strength, durability, compressive strength, low volume change, appearance, and rate of hardening.

Bond Strength

Bond strength refers to the degree of contact between mortar and masonry units and also to the **tensile bond strength** available for resisting forces that tend to pull masonry units apart.

The degree of contact between masonry units is essential to watertight joints and tensile bond strength. Good bond strength requires workable, water-retentive mortar, good workmanship, full joints, and masonry units with a medium rate of absorption (**suction rate**).

Tensile bond strength is necessary to withstand forces such as wind, structural movement, expansion of clay masonry units, shrinkage of mortar or concrete masonry units, and temperature changes. Tensile bond strength is tested by bonding together samples of concrete masonry units with mortar, curing the mortar, and pulling the units apart on a tensile testing machine (Fig. 10.4).

The major factors affecting the bond strength are (1) the characteristics (strength) of the masonry units; (2) quality of the mortar; (3) workmanship of the mason; and (4) curing conditions.

Figure 10.4 The tensile-bond-strength test measures the extent of a bond between mortar and masonry unit. *(Courtesy Portland Cement Association)*

Bond is high on textured surfaces and low on smooth surfaces. Suction rates of masonry units influence bond. Concrete masonry units tend to retain moisture after curing and have relatively low suction rates. Some clay bricks have very high suction rates and, unless wetted before use, will pull water from the mortar, resulting in a poor bond. After bricks are wetted, surfaces should be permitted to dry before use.

Mortar flow influences tensile bond strength. As the water content increases, the bond strength increases. This indicates that it is wise to use the highest water content possible while retaining a workable mortar. As water content increases, mortar compressive strength decreases. Bond strength takes precedence over compressive strength.

Good workmanship requires a minimum of elapsed time between spreading the mortar and placing the masonry unit. Some water in the mortar will evaporate and some will be sucked away by the masonry unit on which it is placed, leaving insufficient water to form a good bond on the next masonry unit. After placing a unit on mortar and getting its initial alignment, it should not be moved, tapped, or slid in any way. This can break the initial bond, which cannot be reestablished. The mortar must be replaced if this happens.

Good curing conditions require the maximum amount of water possible be in the mortar, because it is needed for hydration. The laid units should be covered with plastic to retain moisture while curing. Under severe dry conditions, it may be necessary to keep the wall wet with a fine mist spray for several days. Walls must also be protected from freezing with insulating blankets.

Durability

Mortar needs the durability to withstand weathering forces. Frost or freezing will not damage mortar joints unless they leak and are water soaked. High-compressive-strength mortar usually has good durability. Air-entrained mortar also provides protection against freeze-thaw cycles.

Compressive Strength

The compressive strength of mortar depends mainly on the type and quantity of cementitious material used in the mortar. It increases as cement content increases and decreases as air-entrainment, lime, or water content increases. The compressive strength of mortar is found by testing standard cured 2 in. (50.8 mm) square cubes in a laboratory compression testing machine following ASTM C270 standards. Mortar can be tested in the field using ASTM C780 standards.

The compressive strength of a wall depends not only on the mortar but also on the masonry unit, design of the structure, workmanship, and curing.

Low Volume Change

Mortars have low volume change. The actual shrinkage during curing of a mortar joint is minuscule.

Appearance

The mortar joints in a wall should have a uniform color or shade. Each batch of mortar should have exactly the same proportions of ingredients. Time of tooling also causes variances in the shade of a joint. If mortar is tooled when fairly hard, a darker shade will occur; if fairly soft, then tooling produces a lighter shade. White mortar cement should be tooled with a glass or plastic joint tool because a metal tool darkens the joint.

Any pigments added to color the mortar must be carefully measured. Typically, the color is premixed with enough mortar cement to do the entire job, rather than one batch at a time. A joint's color and shade is also affected by atmospheric conditions, admixtures, and moisture content of masonry units.

Rate of Hardening

The rate of hardening of a mortar due to hydration is the speed at which it develops the strength to resist an applied load. Mortar that hardens very slowly can delay construction because it will not support brick courses laid above it. Mortar with a very high rate of hardening can be difficult for the mason to use. The rate of hardening must be consistent from batch to batch, allowing sufficient time to place masonry units and tool joints.

Loss of water can also influence the stiffening of mortar. This is of particular concern in hot weather. In high temperatures, masons frequently lay fewer masonry units on shorter beds before stopping to tool joints.

COLORED MORTAR

Architects frequently specify the color of mortar. White mortar is made using white masonry cement or white Portland cement, lime, and sand. Colored mortars are made using white masonry cement or white Portland cement and adding color pigments, colored sand, or colored masonry cements (Fig. 10.5). The final color achieved is the result of blending these ingredients. Trial batches are made and cured until the desired color is developed.

Color pigments are a form of mineral oxide, such as iron, manganese, chromium, and cobalt oxides. Carbon black is used to produce a dark gray to black mortar.

Following are recommended pigments.

Color Used*	Pigment	Maximum Amount Used*
Gray to black	Carbon	3
Green	Chromium oxide	10
Blue	Cobalt oxide	10
Reds, yellows, browns, blacks	Iron oxide	10

**Maximum amount of pigment as a percent of the weight of the Portland cement.*

Figure 10.5 A colored mortar being made from white masonry and colored sand.

MORTAR IN COLD WEATHER

When mortar is placed and cured, its temperature should be kept in the range of 60°F to 80°F (15.7°C to 26.9°C). The water in the mortar is needed for hydration (the chemical reaction between masonry cement and water). This leads to the mortar's hardening, which slows or stops if the cement paste in the mortar drops below 40°F (4.5°C).

Mortar can be laid when air temperature is above 40°F (4.5°C) using normal procedures. When air temperature is below 40°F (4.5°C), the mortar water requires heating. However, the temperature of the mortar should never exceed 120°F (50°C) because higher temperatures will cause the mortar to set up too fast, resulting in a loss of compressive and bond strength. If air temperature falls as the mortar is laid, the minimum mortar temperature should be 70°F (21°C). The recommended mortar temperatures for various air temperatures are given in Table 10.6 and Table 10.7.

The finished laid masonry should be covered when work stops. When air temperatures top 40°F (4.5°C),

Table 10.6 Recommendations for All-Weather Masonry Construction[a]

	Construction Requirements	
Air temperature, °F	**Heating of materials**	**Protection**
Above 100 or above 90 with wind velocity greater than 8 mph	Limit open mortar beds to no longer than 4 ft. and set units within one minute of spreading mortar. Store materials in cool or shaded area.	Protect Wall from rapid evaporation by covering, fogging, damp curing, or other means.
Above 40	Normal masonry procedures. No heating required.	Cover walls with plastic or canvas at end of work day to prevent water entering masonry.
Below 40	Heat mixing water. Maintain, mortar temperatures between 40°F and 120°F until placed.	Cover walls and materials to prevent wetting and freezing. Covers should be plastic or canvas.
Below 32	In addition to the above, heat the sand. Frozen sand and frozen wet masonry units must be thawed.	With wind velocities over 15 mph, provide windbreaks during the work day and cover walls and materials at the end of the work day to prevent wetting and freezing. Maintain masonry above 32°F using auxiliary heat or insulated blankets for 24 hours after laying units.
Below 20	In addition to the above, dry masonry units must be heated to 20°F.	Provide enclosure and supply sufficient heat to maintain masonry enclosure above 32°F for 24 hours after laying units.

[a]Adapted from recommendations of the International Masonry Industry All Weather Council and requirements of ACT530 1/ASCE 6/TMS 602. (References 4 & 13).

Table 10.7 Canadian Protection Requirements[a]

Mean daily air temperature, °C	Protection
0 to 4	Masonry shall be protected from rain or snow for 48 h.
−4 to 0	Masonry shall be completely covered for 48 h.
−7 to −4	Masonry shall be completely covered with insulating blankets for 48 h.
−7 to below	The masonry temperature shall be maintained above 0°C for 48 h by enclosure and supplementary heat.

Note: The amount of insulation required to properly cure masonry in cold weather shall be determined on the basis of the expected air temperature and wind velocity and the size and shape of the structure.
Data provided by Canadian Standards Association

walls should be covered with plastic sheets to protect them from rain or light snow. Below 40°F (4.5°C), the walls should be covered with plastic or canvas to prevent their freezing or becoming wet. Some masons use insulated plastic or canvas-covered blankets. When air temperatures fall below 32°F (0°C), walls must be covered with insulated blankets and a source of heat provided to keep the mortar from freezing for at least twenty-four hours.

SURFACE-BONDING MORTARS

A variety of surface-bonding mortars are available from various manufacturers. These are applied by trowel, brush, or spray to any masonry surface and provide a base to which plaster, stucco, concrete, and cement-based paints will adhere. Surface bonding is the application of a cement mortar, reinforced with glass fibers, to both surfaces of concrete block walls laid up without mortar. This bonds them into a solid wall and provides a waterproof coating. Tests indicate that surface-bonded walls are as strong in bending flexure as walls laid with conventional mortar joints. If the surface between the blocks is not flat and smooth, vertical compressive strength is reduced.

Review Questions

1. What are the basic materials used to produce mortar?
2. What is meant by the workability of a mortar?
3. Why are mortar water-retention properties important?
4. What steps are performed to make a mortar flow test?
5. What does a flow-after-suction test reveal?
6. What is meant by the tensile bond strength of mortar?
7. What type of masonry surfaces produce the best mortar bond?
8. How does the water content in mortar influence bond and compressive strengths?
9. How does air-entrainment influence the compressive strength of mortar?
10. How does the volume change in mortar as it cures affect a mortar joint?
11. What things can cause a change in the shade of a mortar joint?
12. What are the types of mortar used in the United States and Canada?
13. Which type of mortar has the highest strength and where is it used?
14. Where would you use the lowest-strength mortar?
15. Which type of mortar is used for general construction of exposed masonry load-bearing walls?
16. What materials can be used to produce colored mortar?
17. How can you keep prepared mortar from having excessive moisture loss?
18. What should you do with mortar that becomes stiff from hydration while waiting for use?
19. What is the lowest temperature at which mortar can be laid before requiring protection from freezing?

Key Terms

cementitious materials
efflorescence
mortar
mortar flow
suction rate
tensile bond strength
water retention

Activities

1. Prepare a number of mortar samples using different variations of ingredients, including Portland cement mortar and masonry cement mortar. Cure each under identical conditions and test their compressive strength.
2. Get a supply of clay masonry and concrete masonry units. Mix mortars of different consistencies and use them to lay up a small wall. Observe and record the ease or difficulty in using the mortar and maintaining the desired-size mortar joint. After the walls have cured, try to separate the masonry units. Do some seem to have greater bonding strength?
3. Prepare several samples of mortar, varying the water content slightly. Then conduct an initial mortar flow test and report your results.
4. Prepare several samples of mortar using various ingredients. After they have cured, make compression tests and record your results.

Additional Resources

Cement, Lime, Gypsum, Volume 04.01, American Society for Testing and Materials, West Conshohocken, PA.

Other resources include: Numerous publications from the Portland Cement Association, Skokie, IL.

Publications from the Canadian Standards Association, Etobicoke, Ontario, Canada.

CHAPTER 11

Clay Brick and Tile

LEARNING OBJECTIVES

Upon completion of this chapter, the student should be able to:

- Describe how clay bricks are made.
- Identify various clay masonry products and explain how they are typically used.
- Be aware of the grades, types, and classes of clay masonry products.
- Use the properties of clay masonry products to select materials for various applications.

Build Your Knowledge

For further study on these materials and methods, please refer to:

Chapter 36 Interior Walls, Partitions, and Ceilings
Topics: Structural Clay Tile Partitions, Ceramic Wall Tile Finishes

Chapter 37 Flooring

Clay brick is made from surface or deep-mined clays that have the necessary plasticity when mixed with water to permit molding to desired shapes. The clay must have the tensile strength to hold a shape while in a plastic condition and contain clay particles that will fuse together when subjected to high temperatures. Properly manufactured brick is fire resistant. It is made in small units, providing the designer an opportunity to create a variety of designs and patterns.

CLAYS

Clay is found in three forms: **surface clay,** shale, and fireclay. Surface clays reside near the surface of the earth and are strip-mined. Most bricks are made from surface-mined clays. **Shales** are clays that have been subjected to high pressures, causing them to be relatively hard. **Fireclays** are found at deeper levels and have more uniform physical and chemical properties. Fireclays can withstand higher temperatures and are used where these will occur, such as in the lining of a fire-place.

Clays contain a variety of materials, but they are predominantly silica and alumina, with smaller amounts of metallic oxides and other ingredients. Clays are divided into two classes, calcareous and non-calcareous. **Calcareous clays** contain about 15 percent calcium carbonate and have a yellow color when burned. **Non-calcareous clays** contain silicate of alumina, feldspar, and iron oxide. The iron oxide causes them to take a buff, red, or salmon color when burned. The color varies depending on the amount of iron oxide.

MANUFACTURING CLAY BRICKS

The manufacture of clay bricks requires a seven-step manufacturing process: (1) winning and storage, (2) processing the raw materials, (3) forming the bricks, (4) drying, (5) glazing (if required), (6) burning and cooling, and (7) drawing and storage of the finished bricks (Fig. 11.1).

Winning and Storage

"**Winning**" is a term used to describe the mining of clay. Most bricks are made from surface-mined clays dug from open pits. Fireclays are obtained from underground mines. The clays are moved by truck or rail to a plant where they are crushed and shifted to storage piles.

Figure 11.1 The flow of materials in the modern brick manufacturing process.

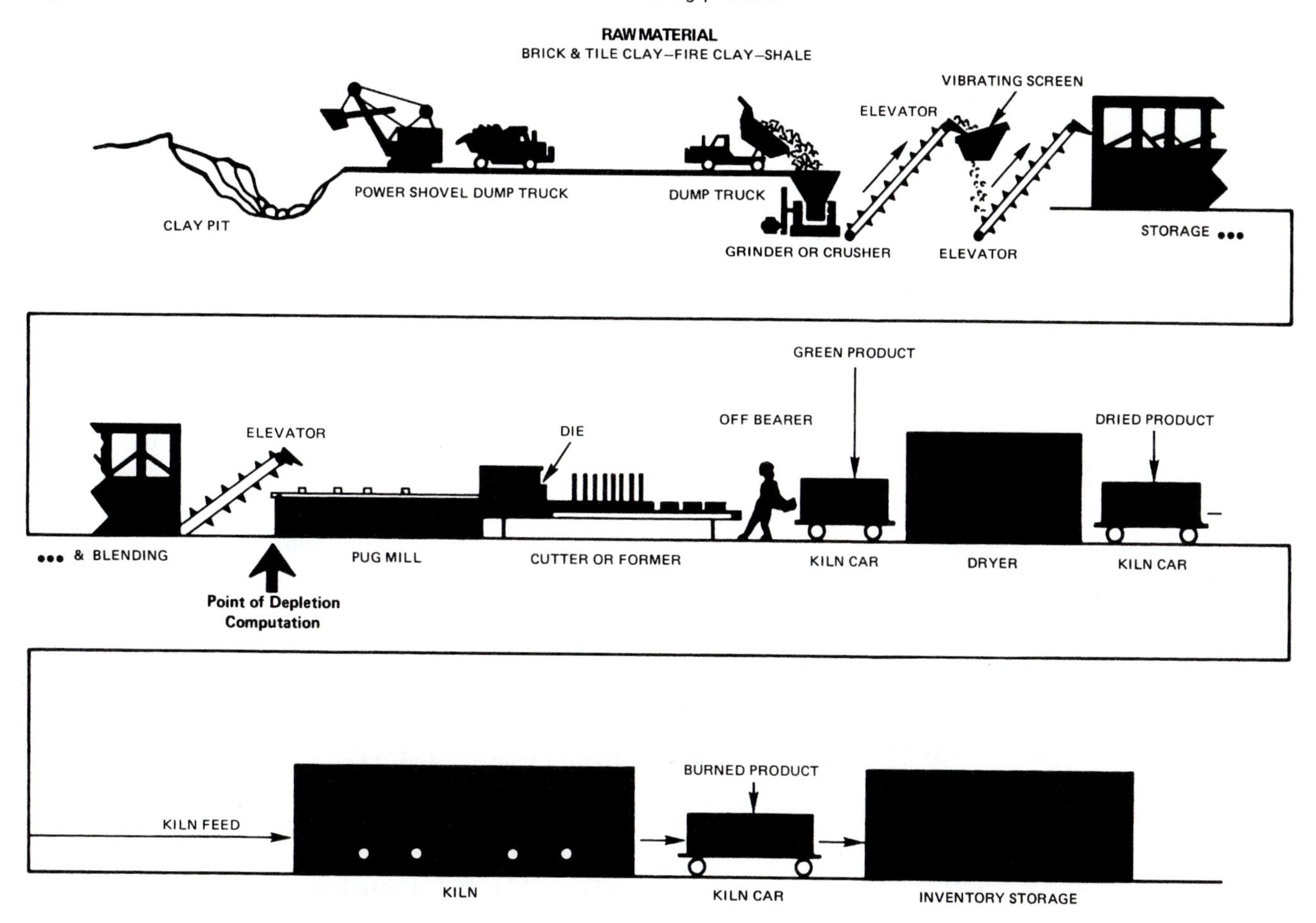

Processing the Raw Materials

Clays are removed from various storage piles and blended to produce the desired chemical composition and physical properties. The blended clays are then moved to crushers where stones are removed and the clay lumps are reduced to a maximum of about 2 in. (50 mm) in diameter. This material is then moved by conveyor to grinders where it is ground to a fine powder and passed over vibrating screens. Material too large to go through the screens is sent back to the grinder. The fine material is placed in storage.

Forming the Bricks

The three major methods for forming bricks are the soft-mud process, the stiff-mud process, and the dry-press process.

The **stiff-mud process** is most widely used. It is a high-production procedure that passes clay of 12 to 15 percent moisture through a vacuum that removes air pockets. The clay is then forced by an auger through a die, producing a continuous column of the desired size and shape (Fig. 11.2). As the clay leaves the die, a surface texture is applied by attachments that scratch, brush, roll, or in some way leave markings on the brick's face. Common textures include smooth, matt, rugs, barks, stippled, sand-molded, water-struck, and sand-struck.

The column then passes through a wire cutter that trims the bricks to size, they move via conveyor to an

Figure 11.2 The stiff mud process forces clay through a die and produces a column of brick that is cut to size with a wire cutter.

inspection area, and imperfect bricks are removed and returned for reprocessing. The good bricks are placed on drier cars for transfer to a drier kiln.

The **soft-mud process** is used for brick making from clays having too much natural water to permit the use of the stiff-mud process. The bricks are shaped in molds lubricated with water or sand (Fig. 11.3). Bricks formed in water-lubricated molds are called water-struck, while those made in sand-lubricated molds are called sand-struck. Water-struck bricks have a relatively smooth finish. Sand-struck bricks have a matte textured surface.

The **dry-press process** is used with clays having 10 percent or less moisture. The mix is formed into bricks in steel molds under high pressure.

Drying

The moisture content of green bricks (unfired, newly formed bricks) varies depending on the clay and the process used. Once formed, the bricks are placed in a low-temperature drier kiln for one to two days. The temperature and humidity are carefully controlled to prevent rapid shrinkage and possible cracking. They are then glazed, if required, or moved directly to high-temperature kilns.

Glazing

Some bricks have a ceramic glaze applied to one or more surfaces after the brick has been dried. **Glaze** is a sprayed coating of mineral ingredients that melts and fuses to the brick when subjected to the required temperature. The glaze forms a smooth, glasslike coating and is available in a wide range of colors.

Figure 11.3 In the soft-mud process, the mixture is placed in molds, formed, removed, and placed in the kiln.

Burning and Cooling

Burning involves raising the temperature of dried bricks to a predetermined level. The two common types of kilns in use are a periodic kiln and a tunnel kiln. The periodic kiln is filled with bricks stacked so air can circulate between them. The temperature in the kiln is raised, held, and lowered. The tunnel kiln is a long, narrow structure through which the bricks move on cars. The bricks enter the kiln on one end and are fired and cooled as they move through to the other. The burning process for both methods takes from 40 to 150 hours, depending on desired end results.

The burning process itself involves several stages: water-smoking, dehydration, oxidation, vitrification, flashing, and cooling.

Water-smoking removes free water by evaporation and requires temperatures up to 400°F (204°C). Dehydration removes additional moisture and requires temperatures ranging from 300 to 1800°F (150 to 980°C). Oxidation temperatures range from 1000 to 1800°F (540 to 980°C) and vitrification from 1600 to 2400°F (870 to 1315°C). These last processes transform clay into a solid, ceramic material. Flashing, if required, follows at this point. It is accomplished by adjusting the fire to reduce the atmosphere in the kiln (resulting in insufficient oxygen to support combustion). This produces a variation in the colors and color shading of the bricks.

As bricks are burned they undergo considerable shrinkage. This is taken into account when they are cut to size or molded. The higher the burning temperatures, the more the shrinkage and the darker the color of the brick. Therefore, dark bricks are usually slightly smaller than light-colored bricks. Some size variation is always possible, and slight distortion caused by the burning process is normal.

After bricks have been burned and flashed as required, a cooling period of forty-eight to seventy-two hours begins (Fig. 11.4). The rate of cooling affects a brick's color and controls cracking and checking.

The color of brick is related to the chemical composition of the clay or shale used and the temperature during the burn. The iron in clay turns red in an oxidizing fire and purple in a reducing fire. As mentioned earlier, the higher the temperature of the burn, the darker the color.

Figure 11.4 Burned brick leaves the kiln for drying.

Drawing and Storage

After the cooling stage is complete, bricks are removed from the kiln, sorted, graded, and moved to storage. Often they are stacked on wood pallets for loading by a forklift. Each pallet load is wrapped in plastic to keep the bricks dry.

STRUCTURAL CLAY MASONRY UNITS

Structural clay masonry units are classified as solid masonry and hollow masonry.

Solid Masonry

Bricks are classified as **solid amasonry** if they have cores whose area does not exceed 25 percent of the gross cross-sectional area of the brick. The cores (holes) help the drying and burning of the unit and reduce its weight.

Bricks are made in a wide range of types and sizes. Checking with a brick supplier is sometimes necessary to verify the size of the unit to be supplied. The commonly manufactured modular and non-modular bricks are shown in Fig. 11.5.

Modular bricks are those whose actual size plus a mortar joint can be assembled on a standard unit or module. The modular unit for inch-size bricks is 4 in. The actual size plus the thickness of a mortar joint ($^1/_2$ in.) is the nominal size. The actual and nominal sizes for modular inch bricks are shown in Table 11.1. Brick sizes are specified by three dimensions: width, thickness, and length, given in that order. Figure 11.6 shows the application of the modular size of a standard inch modular brick. Using the 4 in. module, three courses of brick produces an 8 in. module. The actual sizes of non-modular inch bricks are shown in Table 11.2.

Figure 11.5 Some of the commonly used types of modular and non-modular bricks.

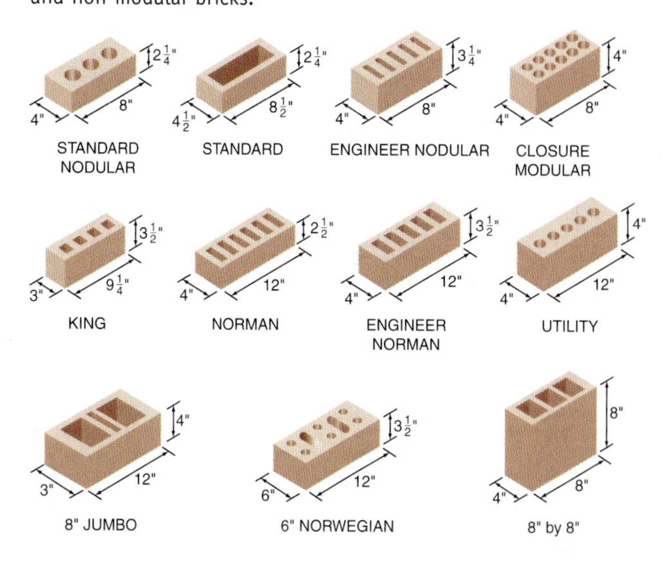

Table 11.1 Modular Inch-Size Common Brick Sizes

Unit Name		Actual Dimensions[a]	Nominal Dimensions[b]	Joint Thickness	Modular Courses
Modular	w	$3\frac{1}{2}$	4		
	h	$2\frac{1}{4}$	$2\frac{2}{3}$	$\frac{1}{2}$	3C = 8″
	l	$7\frac{1}{2}$	8		
Engineer modular	w	$3\frac{1}{2}$	4		
	h	$2\frac{3}{4}$	$3\frac{1}{5}$	$\frac{1}{2}$	5C = 16″
	l	$7\frac{1}{2}$	8		
Closure modular	w	$3\frac{1}{2}$	4		
	h	$3\frac{1}{2}$	4	$\frac{1}{2}$	1C = 4″
	l	$7\frac{1}{2}$	8		
Roman	w	$3\frac{1}{2}$	4		
	h	$1\frac{5}{8}$	2	$\frac{1}{2}$	2C = 4″
	l	$11\frac{1}{2}$	12		
Norman	w	$3\frac{1}{2}$	4		
	h	$2\frac{1}{4}$	$2\frac{2}{3}$	$\frac{1}{2}$	3C = 8″
	l	$11\frac{1}{2}$	12		
Engineer norman	w	$3\frac{1}{2}$	4		
	h	$2\frac{3}{4}$	$3\frac{1}{5}$	$\frac{1}{2}$	5C = 16″
	l	$11\frac{1}{2}$	12		
Utility	w	$3\frac{1}{2}$	4		
	h	$3\frac{1}{2}$	4	$\frac{1}{2}$	1C = 4″
	l	$11\frac{1}{2}$	12		

[a]Actual size unit as manufactured.
[b]Specified unit size plus intended joint size.
Courtesy Brick Institute of America

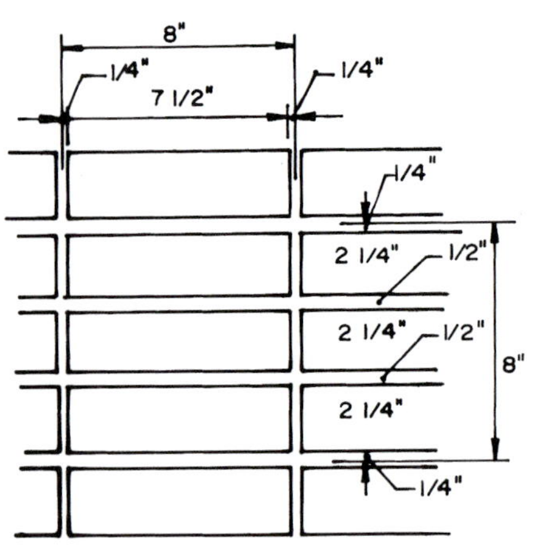

STANDARD MODULAR BRICKS SIZED FOR 1/2" MORTAR JOINT

NOMINAL 4" x 2 2/3" x 8"

ACTUAL 3 1/2" x 2 1/4" x 7 1/2"

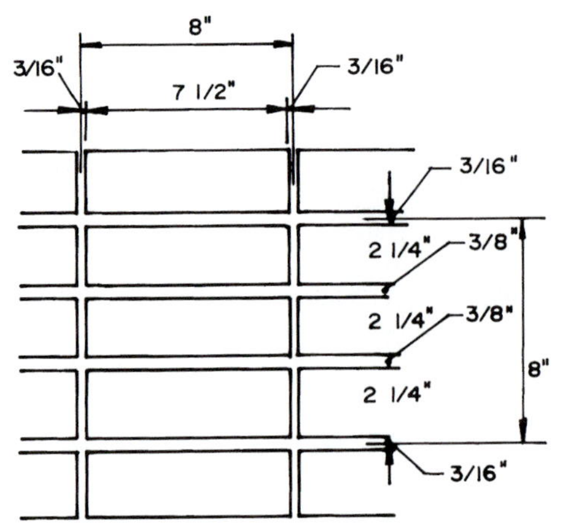

STANDARD MODULAR BRICKS SIZED FOR 3/8" MORTAR JOINT

NOMINAL 4" x 2 2/3" x 8"

ACTUAL 3 5/8" x 2 1/4" x 7 1/2"

Figure 11.6 Modular bricks are sized for an 8 in. module with $\frac{1}{2}$ or $\frac{3}{4}$ in. mortar joints.

Table 11.2 Specified Inch-Size Non-modular Common Brick Sizes

Unit Name		Actual Dimension[a]
Standard	w	$3\frac{5}{8}$
	h	$2\frac{1}{4}$
	l	8
Engineer standard	w	$3\frac{5}{8}$
	h	$2\frac{13}{16}$
	l	8
King	w	3
	h	$2\frac{3}{4}$
	l	$9\frac{5}{8}$
Queen	w	3
	h	$2\frac{3}{4}$
	l	8

[a]Anticipated manufactured dimension.
Courtesy Brick Institute of America

The actual size of modular metric brick is that of the manufactured product. The modular size is the size stated when specifying the brick. For example, when ordering an 89 × 57 × 190 mm actual-size modular metric brick, 90 × 57 × 190 mm is specified. The nominal size is the modular size plus the 10 mm mortar joint.

Hollow Masonry

Hollow clay masonry comprise units whose net cross-sectional area in the plane of the bearing surface is not less than 60 percent of the gross cross-sectional area of that face. A hollow brick may have a cored area from 25 to 40 percent of the gross cross-sectional area of the bearing surface.

Structural clay facing tiles are hollow clay units whose cores exceed 25 percent of the gross cross-sectional area (Fig. 11.7). They are used in load-bearing and non-load-bearing walls and are available with a glazed or an unglazed finished surface. A smooth, colored glaze is most frequently used, but they also come in matte, speckled, and mottled finishes. Surfaces with rough textures or those covered with small-diameter holes of varying sizes provide acoustical treatment. They must meet or exceed ASTM requirements for imperviousness and resistance to fading and scratching and have zero flame spread and toxic fumes. They can be made to provide for radiation protection and installed over wood and metal stud walls and any type of masonry or concrete wall (Fig. 11.8).

Figure 11.7 A wide variety of structural clay facing tile is available. *(Courtesy Stark Ceramics, Inc.)*

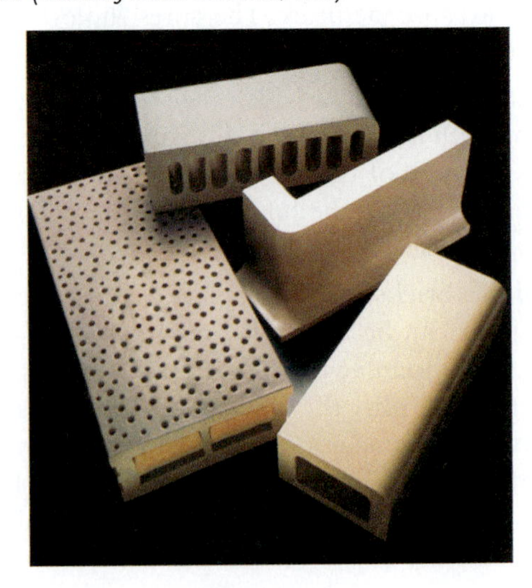

Figure 11.8 Typical installation details for structural clay facing tile.

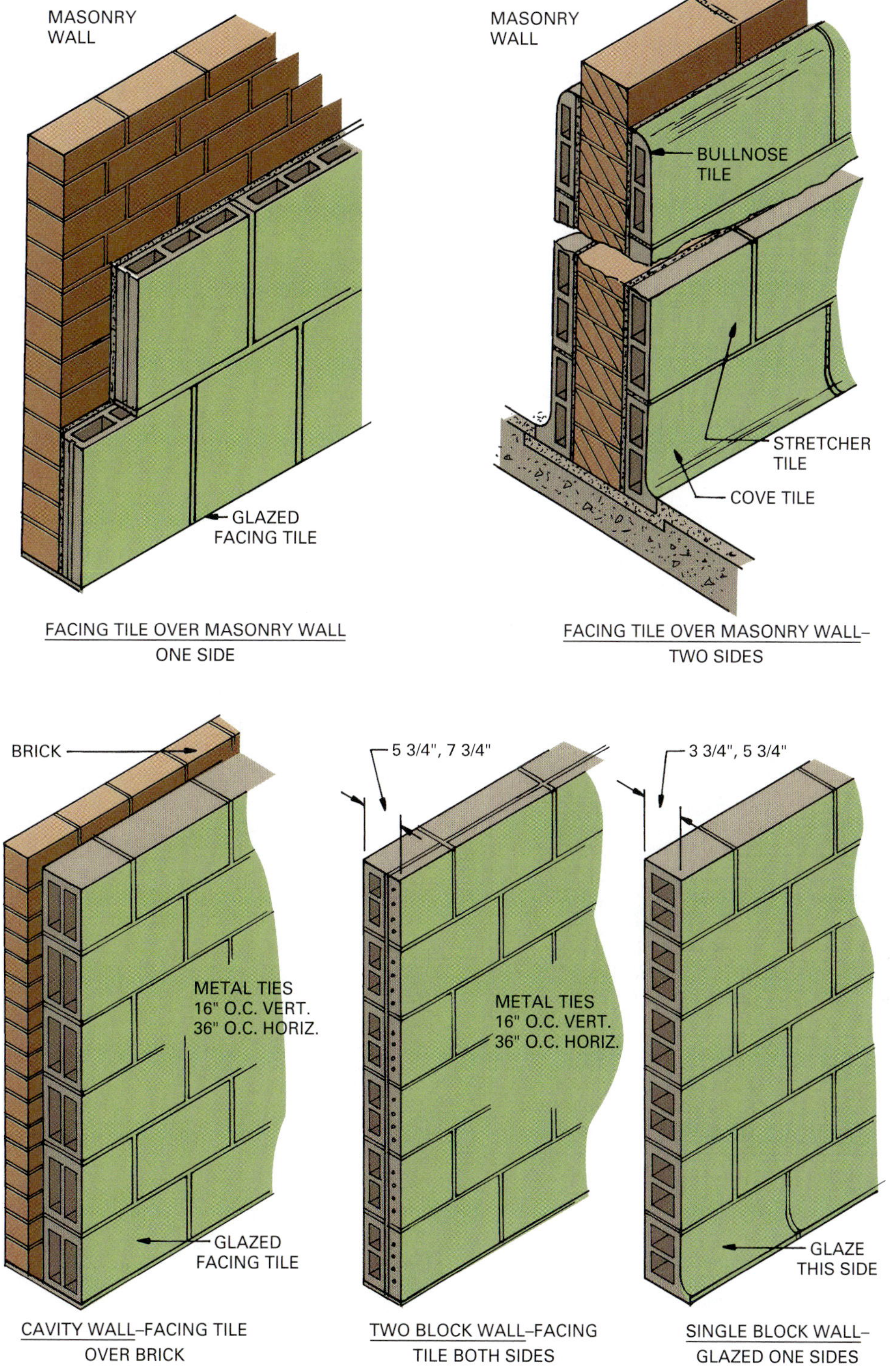

MATERIAL SPECIFICATIONS FOR CLAY MASONRY

The specifications for clay masonry units have been developed by the American Society for Testing and Materials (ASTM). When specifying clay masonry units, the appropriate ASTM specifications should be included. The ASTM specifications for brick, hollow brick, and structural clay tile are based on the weathering index.

Weathering Index

The **weathering index** reflects the ability of clay masonry units to resist the effects of weathering. It is the product of the number of freeze cycle days coupled with the annual winter rainfall. A **freeze cycle day** is a day when the temperature of the air rises above or falls below 32°F (0°C). Winter rainfall is measured in inches between the first and last killing frosts in the fall and spring. The weathering index ranges from 0 to more than 500. Regions rated higher than 500 are considered severe weathering regions, those between 50 and 500 are moderate regions, and areas below 50 are negligible regions (Fig. 11.9).

Grades, Types, and Classes

Building (common), facing, and hollow brick are available in a variety of grades based on the degree of weathering and appearance. (Tables 11.3 to 11.5).

Figure 11.9 The weathering index is an indication of the ability of clay masonry units to resist the effects of weather. *(Courtesy Brick Institute of America)*

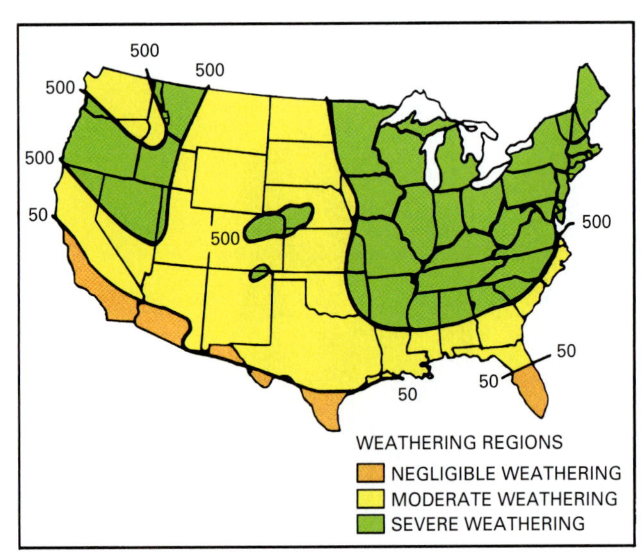

Weathering Indexes in the United States

Table 11.3 Grades and Uses of Solid Building Brick (ASTM C62)

Grade (based on weathering index)	Use
SW Severe weathering	For wet locations below grade where bricks may be frozen, such as in foundations
MW Moderate weathering	For vertical masonry surfaces exposed to the weather in relatively dry conditions where freezing can occur
NW Negligible weathering	For use as backup or interior masonry where no freezing occurs

Courtesy American Society for Testing and Materials

Table 11.4 Grades, Types, and Uses of Solid Facing Brick (ASTM C216)

Grade (based on weathering index)	Use
SW Severe weathering	Masonry in wet locations, in contact with the ground, and subject to freezing
MW Moderate weathering	Exterior walls and other exposed masonry above grade where freezing can occur
Type (based on appearance of the finished wall)	**Use**
FBX	High degree of physical perfection, minimum variation in color, minimum variation in size
FBS	Wider color range and size variations than permitted in type FBX
FBA	Nonuniform in size, color, and texture

Courtesy American Society for Testing and Materials

Table 11.5 Grades, Types, and Uses of Hollow Brick (ASTM C652)

Grade (based on weathering index)	Use
SW Severe weathering	High degree of resistance to disintegration by weathering when brick may be permeated with water and frozen
MW Moderate weathering	Moderate degree of resistance to frost action where brick is not likely to be permeated with water when it is exposed to freezing temperatures
Type (based on appearance of the finished wall)	**Use**
HBS	Visible interior and exterior walls where graded variations in color and size than specified for HBX are acceptable
HBX	Visible interior and exterior walls where a small variation in color and size are acceptable
HBA	Nonuniform in size, color, and texture
HBB	Color and texture are not a consideration Size variation greater than specified for HBX is acceptable

Courtesy American Society for Testing and Materials

Load-bearing structural clay tile is available in two grades, LBX and LB, as specified by ASTM C34 (Table 11.6). Load-bearing tile will carry building loads in addition to its own weight.

Non-load-bearing structural clay tile is available in one grade, NB, as specified by ASTM C56 (Table 11.7). Non-load-bearing tile carries only its own weight.

Table 11.6 Grades and Uses of Structural Clay Load-Bearing Tile (ASTM C34)

Grade (based on weathering index)	Use
LBX	Tile exposed to the weather and as a base for the applications of stucco
LB	Tile not exposed to frost or earth May be used in exposed masonry if covered with 3 in. or more of other masonry

Courtesy American Society for Testing and Materials

Table 11.7 Grade and Use of Structural Clay Non-Load-Bearing Tile (ASTM C56)

Grade	Use
NB	Non-load-bearing walls, partitions, fireproofing, and furring

Courtesy American Society for Testing and Materials

Table 11.8 Grades and Uses of Unglazed Structural Clay Facing Tile (ASTM C212)

Grade	Use
FTX	Exposed masonry with minimum variation in color and dimensions, smooth face, mechanically perfect
FTS	Smooth or rough textured, with moderate absorption and variation in dimensions, medium color range, and minor surface finish defects

Courtesy American Society for Testing and Materials

Unglazed structural facing tile is available in two grades, FTX and FTS, as specified by ASTM C212. These grades are based on face-shell thickness and factors affecting the appearance of the finished wall (Table 11.8).

Ceramic glazed facing brick and structural clay facing tile are available in two grades, S and SS, and two types, I and II, as specified by ASTM C126 (Table 11.9).

Pedestrian and light traffic paving brick is available in three classes, SX, MX, and NX, and three types, I, II, and III, as specified by ASTM C902 (Table 11.10). It is used on patios, walkways, floors, and driveways.

Table 11.9 Grades, Types, and Uses of Ceramic Glazed Structural Clay Facing Tile (ASTM C126)

Grade	Use
S Select	For masonry with narrow mortar joints (1/4″)
SS Select sized or ground edge	For masonry where the face dimension variation is small
Type	**Use**
I Single-face units	Where only one finished face is to be exposed
II Two-faced units	Where two opposite finished faces are to be exposed

Courtesy American Society for Testing and Materials

Table 11.10 Classes, Types, and Uses of Pedestrian and Light Traffic Paving Block (ASTM C902)

Grade (based on weathering index)	Use
SX	Where brick may be frozen when saturated with water
MX	Where resistance to freezing is not necessary
NX	Interior use when an effective sealer or water-resistant surface coating will be applied
Type (based on traffic)	**Uses**
I	Where brick will be exposed to extensive abrasion, as in driveways
II	Where brick will be exposed to intermediate traffic, as on floors in stores
III	Where bricks will be exposed to low traffic, as in residences

Courtesy American Society for Testing and Materials

PROPERTIES OF CLAY BRICK AND TILE

Finished clay units vary considerably in their physical properties. The types of raw materials used and the effects of the production process greatly influence these properties. The most important properties are compressive strength, durability, absorption, color, and texture.

Compressive Strength

Compressive strength depends on the clay, how the units are made, and the temperature and duration of firing. In general, plastic clays used in the stiff-mud process produce units with higher compressive strengths. Higher burn temperatures produce higher compressive strengths in almost any clay or process used. Bricks vary in compressive strength from 1,500 psi (10.35 MN/m^2) to more than 5,000 psi (34.50 MN/m^2).

The compressive strength in a wall of brick depends not only on the compressive strength of the brick but also

Case Study

University of Florida Rinker Hall

Architect: Croxton Collaborative Architects, P.C.

Contractor: Centex Rooney Construction Co., Inc.

Location: Gainesville, Florida

Building type: Higher Education

Size: 47,300 sq. feet (4,390 sq. meters)

Completed March 2003

Rating: LEED Gold (39 points)

Rinker Hall is home to the University of Florida's College of Design and School of Building Construction, providing classrooms, construction teaching labs, and administrative facilities for 1,600 students and faculty. The new building incorporates a range of green building features and in 2004 became the first building in Florida to achieve a LEED Gold rating from the U.S. Green Building Council. Located on the site of a former parking lot, Rinker Hall was designed to maximize daylighting, collect rain water (used for flushing toilets), and features a white roof to reflect heat instead of absorbing it.

For the school of building construction, great care was taken in the selection and procurement of materials for the new building. Material selections were based on stringent prerequisites, including: proximity of manufacturing, recycled content, renewable-resource content, durability, low-maintenance attributes, low toxicity, and potential for recycling and reuse at the end of the building's life. A material proximity and "optimum pathways" exercise involving construction students supported the design team in the procurement of building materials.

When an existing building on the campus was demolished, its wall brick was carefully recycled for reuse on the new building. The brick salvaged from the demolished Hume Hall provided historical significance and a rustic appearance in addition to low cost. Salvaged brick is about $30 to $70 less expensive per 500 bricks than new bricks. The other main advantage to salvaging brick is the reduced volume of waste that otherwise goes to a landfill.

For bricks to be effectively recycled, they must be properly cleaned. If old mortar is not completely removed from a brick's surface, the new bond can be negatively impacted. Removing bricks in mortar can be easy or difficult, depending on how hard the mortar is. Typically, bricks at least 50 years old are used, otherwise the mortar is too difficult to remove and reuse. For the Rinker Hall project, students volunteered their time to clean the bricks and store them on pallets for later use.

Walls made from salvaged brick are generally less durable than those composed of new brick masonry units. A lime mortar rather than one of Portland cement is recommended for laying reclaimed brick. Lime mortar is composed of lime and sand and is generally low in salt content that can cause efflorescence on the brickwork. Lime mortar is also highly plastic and more likely to achieve a good bond with porous brick.

Reused brick on the project was utilized in retaining walls and other non-critical load-bearing elements of the structure. The project allowed students to gain first-hand experience in sustainable construction principles as applied to planning, design, and reconstruction of the built environment.

(Courtesy Brisban H. Brown, Jr., Ph.D., PE, University of Florida)

on that of the mortar. For example, a brick may have a compressive strength of 4,000 psi, but when it is combined with a mortar, the allowable compressive strength will be limited by the mortar's compressive strength.

Durability

Durability refers to the ability of a clay masonry unit to resist damage from freeze and thaw cycles and moisture subjection. Durability is a result of fusion of the clays during burning. High burn temperatures tend to produce a harder, more durable unit.

Absorption

Clay units absorb a certain amount of water. This property affects the bond strength of mortar to brick. The properties of the clays, the process used to make the brick, and the burn temperature affect absorption. In general, plastic clays and high-temperature burns produce units that have lower absorption. Units made with the stiff-mud process usually have lower absorption than those made with other processes. The rate at which a clay unit absorbs moisture is called the **suction rate**. Suction refers to the tendency of a brick to take up moisture in pores and small openings in its surface by capillary action. It refers to surface water, not moisture that penetrates the brick itself. Suction affects the bond strength of mortar to brick. The strongest bond results when units have a suction rate that does not exceed 0.7 ounces (20 grams) of water per minute. If a brick has a greater suction rate than this, it should be wetted and the surface allowed to dry before the brick is laid.

The initial rate of absorption (suction) can be determined in a laboratory using ASTM 67 testing procedures. The masonry unit is immersed in $\frac{1}{8}$ in. (3 mm) of water for one minute. It is removed and weighed. The gain in weight from its dry condition is an indication of the initial rate of absorption (IRA) and is measured in grams per minute per 30 in.^{2} The IRA values range from 1 to 50 or more.

Color

The color of clay masonry units depends on the clays used, burning temperature, and the method of controlling color during the burn. Burned clays vary widely in color, from creams and buffs through reds and purples. Iron oxide, commonly found in clay, is the dominant oxide that influences color. It produces a red unit in an oxidizing fire and becomes purple if burned in a reduced atmosphere (called flashing). Under-burning produces a salmon-colored unit that is softer, has lower compressive strength, and a higher absorption rate than red bricks. Bricks that are over-burned are called clinkers and tend toward dark red to black when made from clays high in iron oxide. Buff clays are used to produce bricks ranging from yellow-brown to dark tan.

Since clay composition varies, the color of produced units has some variation within each unit. The application of surface coatings, such as glazes, enables units to be produced with almost any color desired. Chemicals that produce a range of colors when they vaporize can be introduced into the clay.

Texture

The surface texture of a finished clay masonry unit is produced by the surface of the die or mold used to form the unit or by attachments that cut, scratch, roll, or in some other way alter the surface as the clay unit leaves the die. Some of the standard textures are smooth, matte (with horizontal or vertical markings), barks, rugs, sand-molded, stippled, water-struck, and sand-struck (Fig. 11.10).

Heat Transmission

The ability of clay brick walls to facilitate or resist the transmission of heat directly influences the surface temperature of interior walls and rooms. In most applications, resistance to heat transmission is very important. In others, such as the storage and conduction of solar heat, the ability to transmit heat is important. Selected examples of heat transmission (U value) and heat resistance (R value) are given in Table 11.11. These figures vary for different types of brick because of the differences in both clay used and unit density.

Figure 11.10 Clay masonry units are available in a wide variety of colors and textures.

Table 11.11 Coefficients of Heat Transmission and Resistance of Solid and Cavity Clay Brick[a]

Brick	Transmission (U)	Resistance (R)
4″ (100 mm) solid brick	2.27	0.44
6″ (150 mm) solid brick	1.52	0.66
8″ (200 mm) solid brick	1.14	0.88
12″ (300 mm) solid brick	0.76	1.32
10″ (250 mm) cavity brick	0.34	2.96

[a]*Values reflect an average rating. Get actual values from the manufacturer of the masonry units to be used.*

Fire Resistance

The fire resistance of a material is an indication of its ability to prevent materials behind it from igniting. Fire resistance is stated in hours and is usually called the fire rating. These ratings are developed by testing units under actual fire conditions set up in a laboratory. Most floors, walls, and ceilings are made up of several materials, so fire ratings of these assemblies are also known. Table 11.12 lists the fire ratings for selected clay masonry units.

CERAMIC TILE

Ceramic tile includes wall tile, mosaic tile, quarry tile, and paver tile. It is made in a variety of thicknesses for both molded types and flat tiles. Tile is used widely to provide durable, waterproof surfaces.

Ceramic Tile Grade Marking and Certification

There are three grades of ceramic tile. Standard grade is the best and meets the highest performance specifications.

Table 11.12 Fire Resistance of Selected Brick and Tile Walls

Solid Brick	Fire Rating in Hours
4″ (100 mm) brick	1
6″ (150 mm) brick	3
8″ (200 mm) brick	4
Hollow Core Brick	
8″ (200 mm) brick wall	2
8″ (200 mm) brick wall plastered both sides	4
Structural Facing Tile	
4″ (100 mm) tile wall plastered one side	1
6″ (150 mm) tile wall	2
8″ (200 mm) tile wall plastered both sides	4

Second grade meets all specifications but may have facial defects noticeable from a distance of ten feet. Decorative thin wall tile grade meets all specifications except breaking strength. Ceramic wall tile, mosaic tile, quarry tile, paver tile, and special-purpose tile are shipped in sealed cartons with their grade indicated.

Types of Ceramic Tile

Ceramic wall tiles are fired clay tiles that are usually glazed and widely used on interior surfaces, such as walls, floors, showers, and countertops. They are available in a variety of shapes and sizes (Fig. 11.11). They are not expected to withstand excessive impact or be subject to freezing and thawing. High-gloss and matte finishes are commonly used and surfaces may be smooth or textured. A wide range of colors is available, including solid, multi-shaded, and speckled variations. Much tile is imported from foreign manufacturers. Individual tiles are often supplied mounted on a backing sheet so multiple tiles, properly spaced, can be applied to a wall at one time (Fig. 11.12).

Ceramic mosaic tile may be either natural clay or porcelain in composition. Mosaic tiles are smaller than wall tiles and come in square or hexagonal shapes. Standard sizes are listed in Fig. 11.11. These individual tiles are assembled on a backing material forming sheets from 9 × 9 in. (225 × 225 mm) to 12 × 24 in. (300 × 600 mm).

Ceramic tile can be installed over wood or metal studs, furring, or any masonry or concrete wall. Quarry tiles are hard, burnt unglazed or glazed clay tiles that are red, brown, or buff in color, depending on the clay. Quarry tiles generally are from $\frac{1}{2}$ to $\frac{3}{4}$ in. (12 to 20 mm) thick. A number of trim tiles are made for use with quarry tiles. These include straight base, cover base, double bull nose, and internal and external corner base units. Paver tile is the same as quarry tile but thinner, ranging from $\frac{3}{8}$ to $\frac{5}{8}$ in. (9.5 to 16 mm). Paver tiles are used in areas with lighter loads and less traffic.

OTHER CLAY TILE PRODUCTS

Clay tile is a vitrified clay product, such as clay pipe, used for drainage and sewer systems.

Roof tile is burnt clay used for finished roofing (Fig. 11.13). Tiles are made in flat, plain shingles, single-lap tiles, and interlocking tiles. They are fashioned in various styles, reflecting the countries where they have been used for centuries. Among these styles are Spanish, French, Roman, and English.

Figure 11.11 Commonly manufactured ceramic mosaic, glazed wall, and quarry and paver tile.

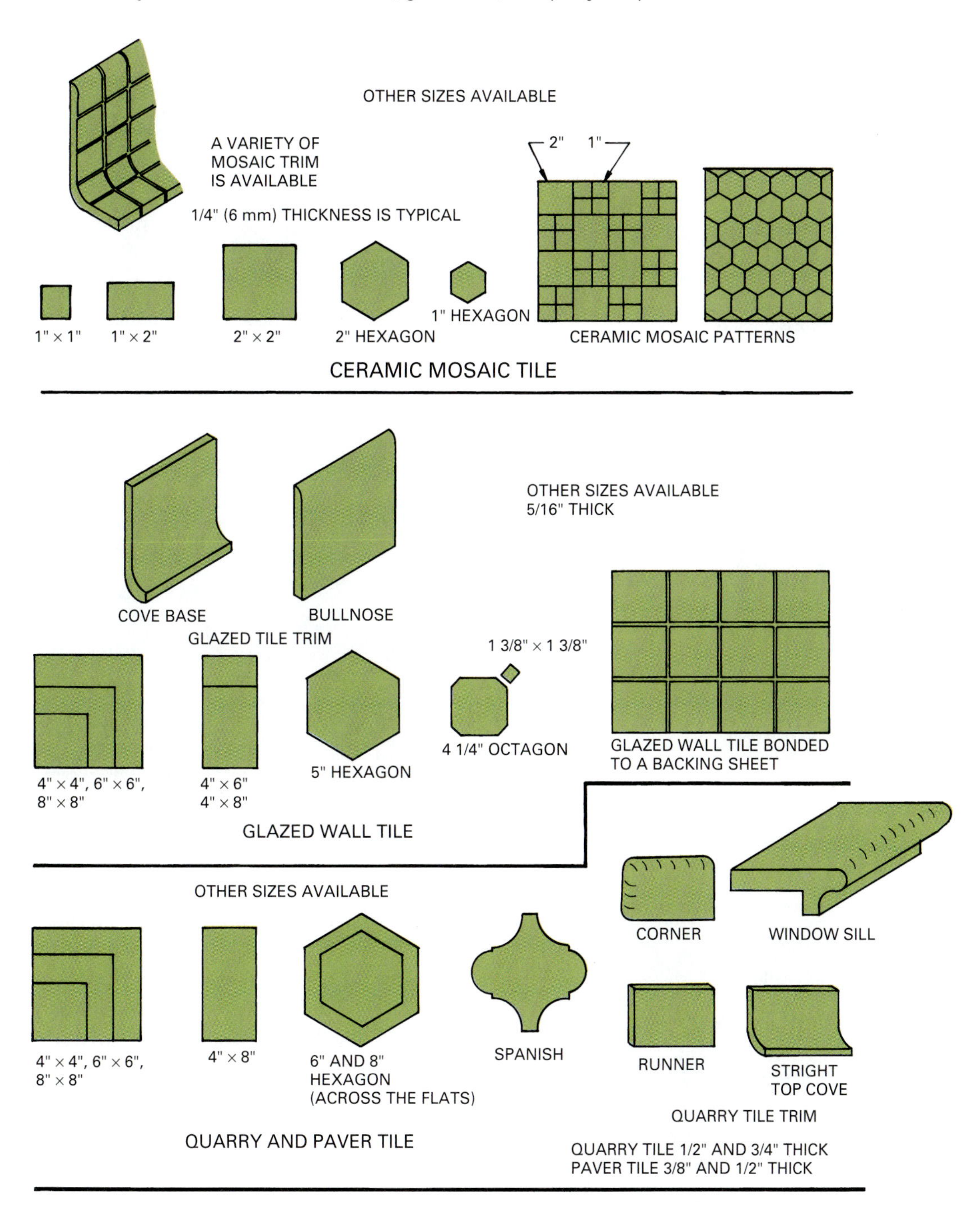

Architectural Terra-Cotta and Ceramic Veneer

Architectural terra-cotta is a hard-fired clay that has been used for hundreds of years for decorative purposes. Modern terra-cotta products are extruded and molded or pressed to shape. Most of these products are custom made to an architect's designs and specifications. Figure 11.14 shows a sun-shading façade panel made from architectural terra-cotta.

Architectural terra-cotta is a red earth color when left unglazed. It is available with a smooth, plain surface that may be unglazed, or it can have a transparent glaze, a non-lustrous glaze that gives a satin or matte finish, a ceramic color glaze, or a polychrome finish. Sculptural and decorative reproductions of classic architectural ornamentation can be made from terra-cotta and are generally handmade in special molds.

A machine-made unit having a flat or ribbed back and a flat face is called ceramic veneer.

Figure 11.12 Glazed tile are assembled with a backing sheet, permitting the installation of large areas at a time. (*Courtesy American Olean Time Company)*

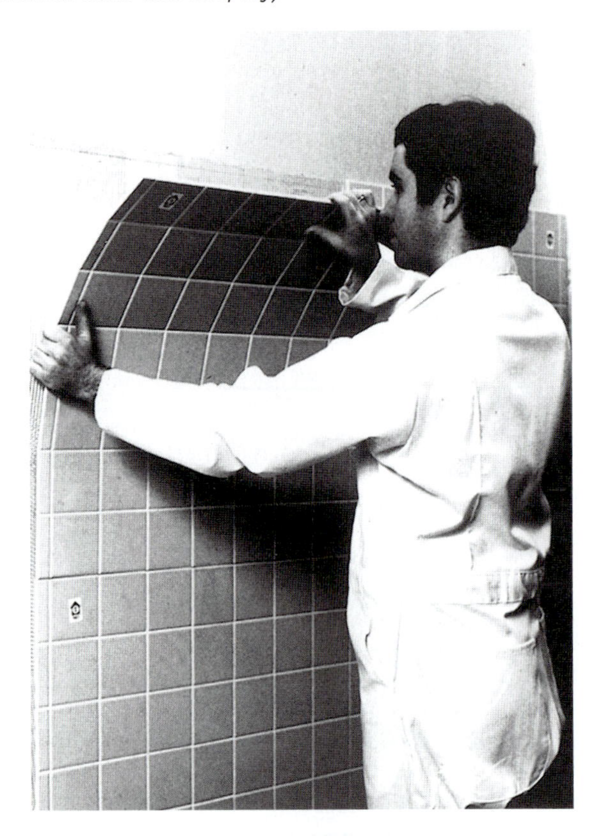

Figure 11.13 Clay roof tile has been used for centuries to provide a fireproof, durable roofing material.

Figure 11.14 Architectural Terra-Cotta used as a screen over a glazed façade.

Ceramic veneer may be an earth red if unglazed but may have a transparent ceramic glaze, a non-lustrous glaze with a satin or matte finish, or a ceramic color glaze, either solid or a mottled blend of colors. A polychrome finish is used, which means two or more colors are applied to separate areas and each color is burned separately. The surface may be smooth, scored, combed, or roughened. Ceramic veneer is installed by mortar adhesion or through a system of mechanical anchoring.

Review Questions

1. What are the forms in which clay is found?
2. How do fireclays differ from surface clays?
3. How do the colors of calcareous and non-calcareous clays differ?
4. What methods are used to form bricks?
5. What are some of the common surface textures used on face bricks?
6. Why are the temperature and humidity carefully controlled when drying bricks?
7. How long does the burning process take?
8. What stages do bricks go through in the burning process?
9. How does the burning temperature influence the size and color of bricks?
10. What are the classifications of structural clay masonry units?
11. How can you identify solid masonry clay units?
12. What is the module used when making standard bricks?
13. What is the actual size of a standard clay brick?
14. What is the allowable hollow cored area for a hollow brick?
15. What two factors are considered when establishing the weathering index?
16. What are the grades of common building brick?
17. What is the range of compressive strength for common bricks?
18. What unit of suction provides conditions for the best mortar bond?

Key Terms

architectural terra-cotta
burning
calcareous clays
dry-press process
fireclays
freezing cycle day
glaze
hollow clay masonry
non-calcareous clays
shales
soft-mud process
solid masonry
stiff-mud process
suction rate
surface clays
weathering index
winning

Activities

1. Collect samples of clay masonry products for use in the classroom. Label each, giving as much information about their properties as you can collect.
2. Design and run some tests on samples of clay masonry units, such as compression tests, hardness tests, and moisture absorption tests.

Additional Resources

Architectural Graphic Standards, the American Institute of Architects, John Wiley & Sons, New York, NY.

Jaffe, R., *Masonry Basics*, the Masonry Society, Boulder, CO.

Technical Notes on Brick Construction on CD-ROM, the Brick Industry Association, Reston, VA.

Numerous publications from the Portland Cement Association, Skokie, IL.

Other resources include:

Publications from the Brick Industry Association, Reston, VA.

See Appendix B for addresses of professional and trade organizations and other sources of technical information.

CHAPTER 12

Concrete Masonry

LEARNING OBJECTIVES

Upon completion of this chapter, the student should be able to:

- Understand how concrete masonry units are manufactured.
- Use information about the physical properties of concrete masonry units when making material selection decisions.
- Recognize the many types of concrete masonry units and be able to choose those suitable for various applications.

Build Your Knowledge

For further study on these materials and methods, please refer to:

Chapter 7 Concrete
Chapter 37 Flooring

Concrete masonry units are manufactured in a wide range of standard sizes and custom-designed architectural units. They are one of the most widely used modern construction materials, finding use in both structural and nonstructural applications.

MANUFACTURE OF CONCRETE MASONRY UNITS

Concrete masonry units are made of a relatively dry mix of Portland cement, aggregates, water, and, in some cases, admixtures. The dry materials are carefully weighed and moved to a mixer. The mixer adds the required water and agitates the batch for a predetermined time. The mixed batch is discharged into the hopper of a block machine. Here it is fed into molds and consolidated by pressure and vibration. The freshly molded "green blocks" are moved from the block machine on steel pallets to a curing rack.

The curing rack full of green units is moved to a low-pressure steam kiln or an autoclave for hardening. In a low-pressure steam **kiln**, the green units are allowed to attain an initial set before steam is introduced. This takes from one to three hours with the temperature kept at 70°F to 100°F (22°C to 38°C). Then steam is introduced, providing heat and moisture. The temperature is gradually raised to 150°F to 180°F (66°C to 82°C), depending on the composition of the concrete. This condition is maintained for ten to twenty hours until the units reach the required strength. After removal, the blocks are stored in a protected condition and achieve almost full strength in two to four more days.

An **autoclave** uses high-pressure steam. The molded units are placed in the autoclave and allowed to set for two to five hours. Then they are gradually heated with saturated steam under a pressure of 150 psi (1,035 kPa) for two to three hours. Once the maximum temperature (350°F (178°C)) and pressure are reached, the units soak for five to ten hours. Finally, pressure is gradually released over a thirty-minute period and the units go to storage. They can be used twenty-four hours after leaving the autoclave. These units have greater stabilization against volume changes caused by moisture than the low-pressure units.

PHYSICAL PROPERTIES OF CONCRETE MASONRY UNITS

The physical properties of concrete, discussed in Chapter 7, apply to concrete masonry units as well. The properties of the units are determined by cement paste and aggregates. Differences that do exist between concrete and concrete masonry units are caused by different mix compositions, methods of consolidation, and curing processes. Concrete masonry units are generally made with less cement per cubic yard and a lower water-cement ratio. The aggregate is finer, with $\frac{3}{8}$ in. (10 mm) being the largest size. Concrete masonry units have a large volume of void spaces between aggregate particles, while normal concrete should have no voids.

The important properties to consider when specifying concrete masonry units are weight, compressive strength, water absorption, and the coefficient of thermal expansion (Table 12.1). The required properties are established by national building codes and ASTM standards.

The weight of a concrete masonry unit varies with the design of the block and the mix used to make it. It is necessary to know weights so the dead loads of a structure can be calculated.

Compressive strength data give a means of determining a unit's ability to carry loads and withstand structural stresses. The strength varies depending on the wetness of the mix. Wetter mixes give the highest strength but cause difficulties in manufacturing the units. The manufacturer develops units and tests them so the user will know the compressive strength of the products available. The compressive strengths, given in Table 12.1, are based on the gross bearing area of the block, including the core spaces. Compressive strength of the net area (actual surface, excluding the cores) is 1.8 times the values shown.

Table 12.1 Properties of Concrete Blocks with Various Aggregate

Customary Units			
Aggregate	**Compressive Strength in psi (gross area)**	**Coefficient of Thermal Expansion per °F × 10^{-6}**	**Water Absorption (lb./ft.^{3} of concrete)**
Sand and gravel	1,200–1,800	5.0	7–12
Limestone	1,100–1,800	5.0	8–12
Air-cooled slag	1,100–1,500	4.6	9–13
Expanded shale	1,000–1,500	4.5	12–15
Cinders	700–1,000	4.5	12–18
Expanded slag	700–1,200	4.0	12–18
Pumice	700–900	4.0	13–20
Scoria	700–1,200	4.0	12–18
Metric Units			
Aggregate	**Compressive Strength in kg/cm^2 (gross area)**	**Coefficient of Thermal Expansion mm/mm/°C × 10^{-6}**	**Water Absorption (kg/m^3)**
Sand and gravel	84.4–127	9.0	128–190
Limestone	77.3–127	9.0	128–190
Slag	77.3–105.5	8.3	144–208
Expanded shale	70.3–105.5	8.1	192–240
Cinder	49.2–70.3	4.5	192–288
Expanded slag	49.2–84.4	7.2	192–288
Pumice	49.2–63.3	7.2	208–320
Scoria	49.2–84.4	7.2	192–288

Figure 12.1 The compressive strength of a wall made from concrete masonry units depends on the type of mortar and the strength of the concrete masonry units. *(From "Masonry Structural Design for Buildings," Department of the Army and Air Force)*

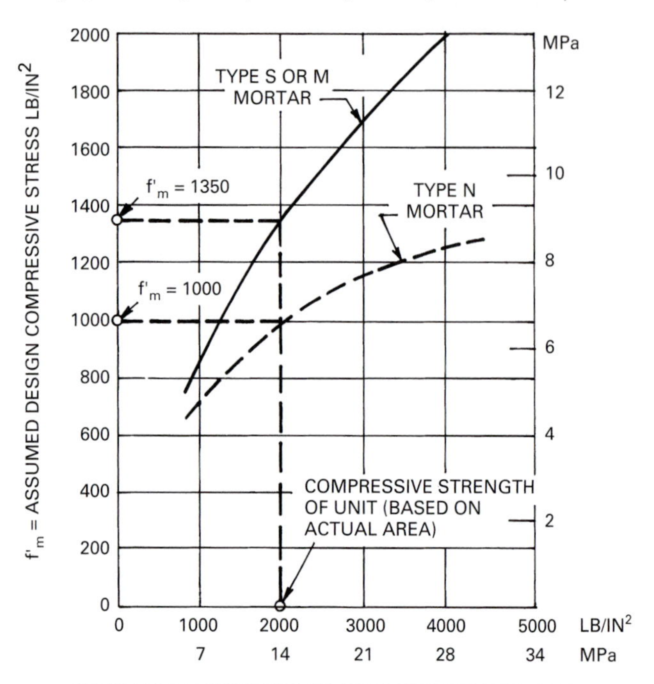

Using the graph in Fig. 12.1, design compressive data can be obtained if the compressive strength of the concrete masonry unit (based on actual area) and the type of mortar are known.

Tensile strength is about 5 to 10 percent of compressive strength. Flexural strength is about 15 to 20 percent of compressive strength, and the modulus of elasticity ranges from 300 to 1,200 times the compressive strength.

Water absorption varies with the density of the concrete masonry unit. Absorption is a measure of the pounds of water per cubic foot of concrete. Units made with a dense aggregate have much lower absorption rates than those made with lightweight aggregate. High water absorption is not acceptable for many applications. Units having a large number of interconnected pores and voids have a high absorption rate (also called suction rate). Units with a high suction rate and high absorption by aggregate have a high permeability to water, air, and sound, and are more likely to be damaged by freezing. Units with unconnected pores, such as those found in lightweight aggregate and in the air-entrained cement paste, provide for sound absorption, reduce thermal conductivity, and minimize permeability to water.

Units used for exterior walls that are not to be painted should be low absorption units. Painting and other waterproof coatings greatly reduce water absorption. It should be remembered that concrete masonry units ought to be in an air-dry condition (in equilibrium with the surrounding air). If they are used before this occurs, they will shrink slightly as moisture is lost, causing tensile and shearing stresses that may cause cracking.

The coefficient of thermal expansion is used to calculate the amount of expansion that can be expected as temperatures change. Concrete masonry units expand when heated and contract when cooled. These changes are reversible, and the unit returns to its original size after the temperature returns to the point at which the change started. Since the aggregate makes up about 80 percent of a unit, the coefficient of expansion of the aggregate is a major factor influencing expansion and contraction of the concrete masonry unit.

Other properties that are a result of those just discussed include insulation value, coefficient of heat transmission, sound-absorbing properties, and fire resistance.

The insulation value of units made with porous, lightweight aggregate is better than those using denser material. The units with lightweight aggregate also have a lower coefficient of heat transmission. Heat transmission, U, is stated in British thermal units (Btu) per hour per square foot per degree Fahrenheit for each degree difference in temperature between the air on the cool and warm sides of a wall. Concrete masonry walls are poor insulators and good heat transmitters. Therefore they require the addition of insulation materials to be energy efficient (Table 12.2).

Concrete masonry units resist the transmission of sound. Hollow block made with lightweight aggregate provides the best sound isolation. The addition of a plaster interior or exterior finish increases this property. Concrete masonry units that have an open, porous surface absorb sound better than denser units with

Table 12.2 Coefficients of Heat Transmission for Selected Concrete Masonry Units

Hollow Masonry			
Size	**Aggregate**	**Btu/hr/ft²/°F**	**Watts/m²/°C**
8″	Limestone	0.53	3.0
8″	Sand and gravel	0.53	3.0
8″	Cinders	0.37	2.1
8″	Expanded clay, slag, or shale	0.31	1.8

smooth surfaces. Painting the surface fills these pores and reduces sound-absorbing properties. Acoustical concrete block units are manufactured to combine a sound-deadening liner panel with a concrete unit with open slots on the face (Fig. 12.2).

Building codes are very strict on required fire-resistance ratings for various parts of a building. Products such as concrete masonry units must be carefully tested before they are given a fire-resistance rating. The rating specifies the number of hours the material can be exposed to a flame before it fails. The rating varies depending on the aggregate. Plaster on a concrete unit is an effective way of increasing its fire resistance. A method for estimating the fire-resistance rating of concrete masonry is shown in Table 12.3.

TYPES OF CONCRETE MASONRY UNITS

Concrete masonry units, often called concrete blocks, are made in a wide range of sizes and types. Many are made based on a special design prepared by an architect. They can be divided into three classifications: concrete brick (solid units), concrete block (hollow and solid units), and special units.

Modular Sizes

Concrete masonry units are made in modular sizes and are specified by their nominal or modular size. They are produced on a 4 in. module. The metric modular unit is 100 mm.

Figure 12.2 The sound-absorbing qualities of these acoustical blocks improve the acoustical properties of rooms, such as auditoriums. *(Courtesy of Trenwyth Industries, Inc.)*

The nominal or modular size is the theoretical size without allowance for a mortar joint. The actual size is $\frac{3}{8}$ in. less than the nominal size, so the actual unit plus a $\frac{3}{8}$ in. mortar joint gives the modular size (Fig. 12.3). Concrete units are also related to standard bricks. For example, as shown in Fig. 12.4, the unit is two bricks wide, three brick courses high, and two bricks long. This allows for the mortar joints between the standard modular bricks. A standard modular brick is $4 \times 2 - \frac{3}{8} \times 8$ in. nominal.

True metric modular concrete blocks' modular size is $200 \times 200 \times 400$ mm. Actual modular size is $190 \times 190 \times 390$ mm ($7\frac{1}{2} \times 7\frac{1}{2} \times 15\frac{3}{8}$ in.). American metric modular blocks soft-converted from inch sizes are (actual size) $194 \times 194 \times 397$mm ($7\frac{5}{8} \times 7\frac{5}{8} \times 15\frac{5}{8}$ in.), which is quite similar to a true metric block. The mortar bed is 10 mm. See Figs. 12.3 and 12.4.

Concrete Brick

Concrete bricks are made either solid or with a depressed area called a frog. The frog reduces the weight. They are laid with a $\frac{3}{8}$ in. mortar joint. A standard modular brick is $4 \times 2\frac{3}{8} \times 8$ in. nominal. Some block manufacturers make other sizes.

Slump Block

Slump block is made with a concrete mix that permits the unit to slump a little when it is removed from the mold. This produces units having irregular faces and some differences in height and surface texture. They give a rustic appearance when laid into a wall.

Concrete Block

A wide variety of types of concrete blocks are available, many designed for special uses. Some are shown in Fig. 12.5. They are produced in three major groups: solid load bearing, hollow load bearing, and non-load-bearing units. A solid load bearing unit is one whose cross sectional area in every plane parallel to the bearing surface is not less than 75 percent of the gross cross sectional area measured in the same plane. A hollow concrete block is one whose cross sectional area in every plane parallel to the bearing surface is less than 75 percent of the gross cross sectional area measured in the same plane.

Many units are made with two rather than three cores. The two-core unit has the advantage of having an increase of the face shell thickness and the center web. This increases strength, reduces cracking due to shrinkage during curing, produces a lighter unit, reduces

Table 12.3 Estimated Fire Rating in Hours for Concrete Masonry

	Minimum Equivalent Thickness in Inches for Fire Ratings in Hours			
Type of Aggregate	**1 hour**	**2 hours**	**3 hours**	**4 hours**
Pumice or expanded slag	2.2*	3.2	4.0	4.7
Expanded shale, clay, or slate	2.6	3.6	4.2	5.1
Limestone, cinders, or unexpanded slag	2.7	4.0	5.0	5.9
Calcareous gravel	2.8	4.2	5.3	6.2
Siliceous gravel	3.0	4.5	5.7	6.7

How to Figure Equivalent Thickness for Cored Blocks

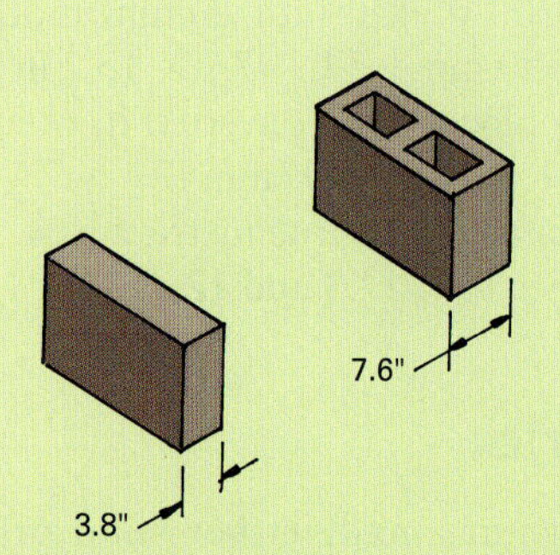

Equivalent thickness is the solid thickness that would be obtained if the same amount of concrete contained in a hollow unit were re-cast without core holes.

Calculating Estimated Fire Resistance. Example: A 7.6-in. (actual size) hollow masonry wall is constructed of expanded slag units reported to be 55%* solid. What is the estimated fire resistance of the wall? (modular units)

Eq Th $= 0.55 \times 7.6$ in. $= 4.2$ in.
From table: 3-hr fire resistance requires 4.00 in. Use 3-hr est. resistance.

	Minimum Equivalent Thickness in Millimeters for Fire Ratings in Hours			
Type of Aggregate	**1 hour**	**2 hours**	**3 hours**	**4 hours**
Pumice or expanded slag	55.9	76.2	101.6	119.4
Expanded shale, clay, or slate	66.0	91.4	106.7	129.5
Limestone, cinders, or unexpanded slag	68.6	101.6	127.0	149.9
Calcareous gravel	71.1	106.7	134.6	157.5
Siliceous gravel	76.2	144.8	114.8	170.2

How to Figure Equivalent Thickness for Cored Blocks

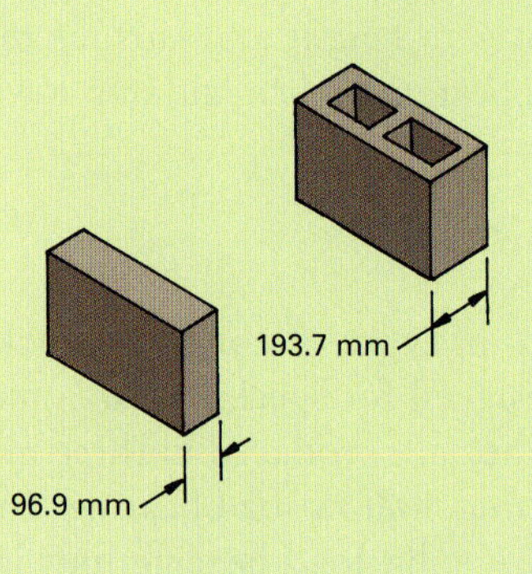

Equivalent thickness is the solid thickness that would be obtained if the same amount of concrete contained in a hollow unit were re-cast without core holes.

Calculating Estimated Fire Resistance. Example: A 193.7-mm (actual size) hollow masonry wall is constructed of expanded slag units reported to be 55%* solid. What is the estimated fire resistance of the wall? (modular units)

Eq Th $= 0.55 \times 193.68$ mm $= 106.43$ mm. From table: 3-hr fire resistance requires 101.6 mm. Use 3-hr est. resistance.

(Source: "Fire Safety with Concrete Masonry," 35C, with permission, National Concrete Masonry Association)

heat concentration in the wall, and permits the cores to line up vertically so reinforcing, plumbing, and electrical runs can be made inside the wall.

The stretcher unit is most commonly used in foundation and wall construction. The most common widths are 4, 6, 8, 10, and 12 in., with 8 in. height and 16 in. length. Half blocks are also available. Corner blocks are used to make an exposed corner and for piers and pilasters. They are made with one or both ends flush. A corner return block permits a full block face to appear on one wall with a recess to help turn the corner.

Header blocks have a recess that holds the header unit in a masonry bonded wall. Pilaster blocks are used when the design calls for pilasters to reinforce a wall. They also provide a bearing surface for beams. Some are made in halves and others as a single unit. They may have rebar and concrete in the core if greater strength is necessary.

Figure 12.3 Standard inch and metric concrete blocks.

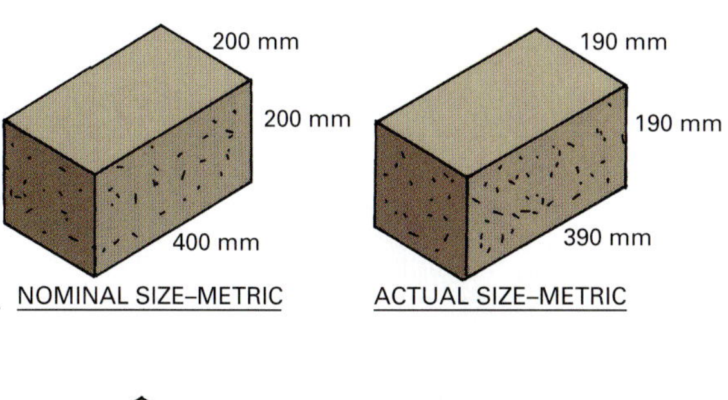

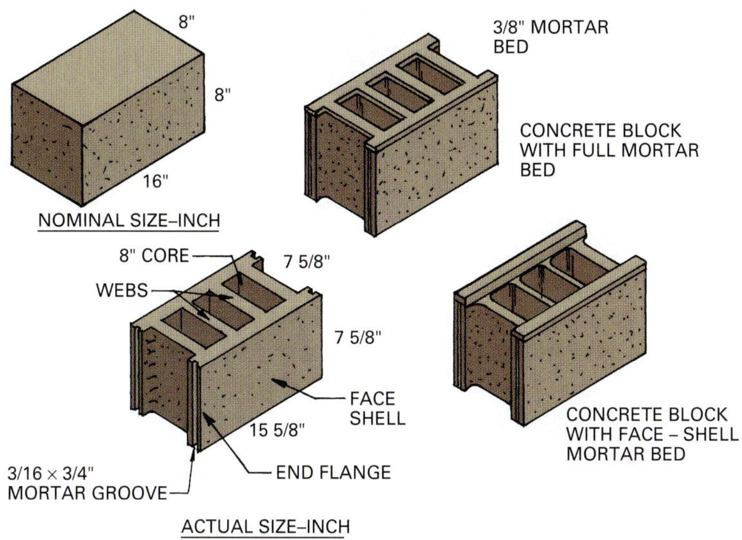

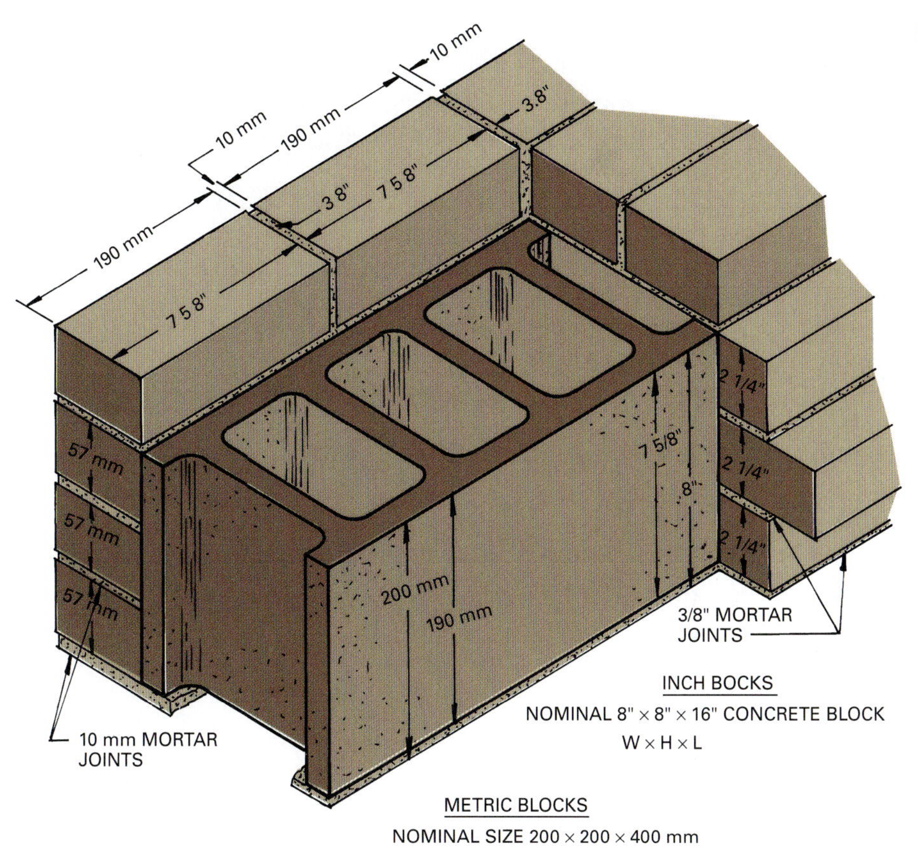

Figure 12.4 Standard brick and blocks are compatible in height and width.

Figure 12.5 Shapes of commonly available concrete masonry units.

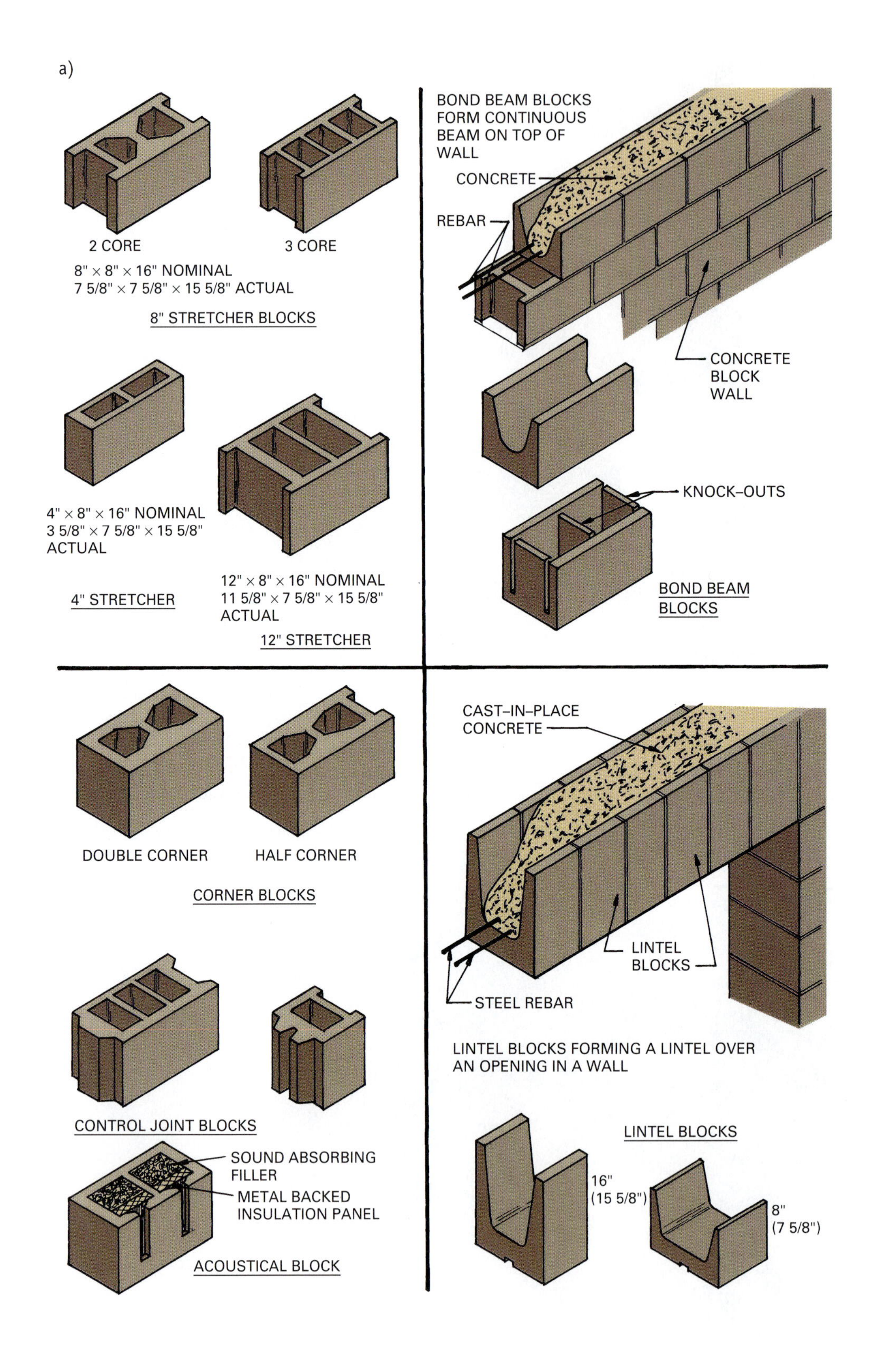

Figure 12.5 *Continued*

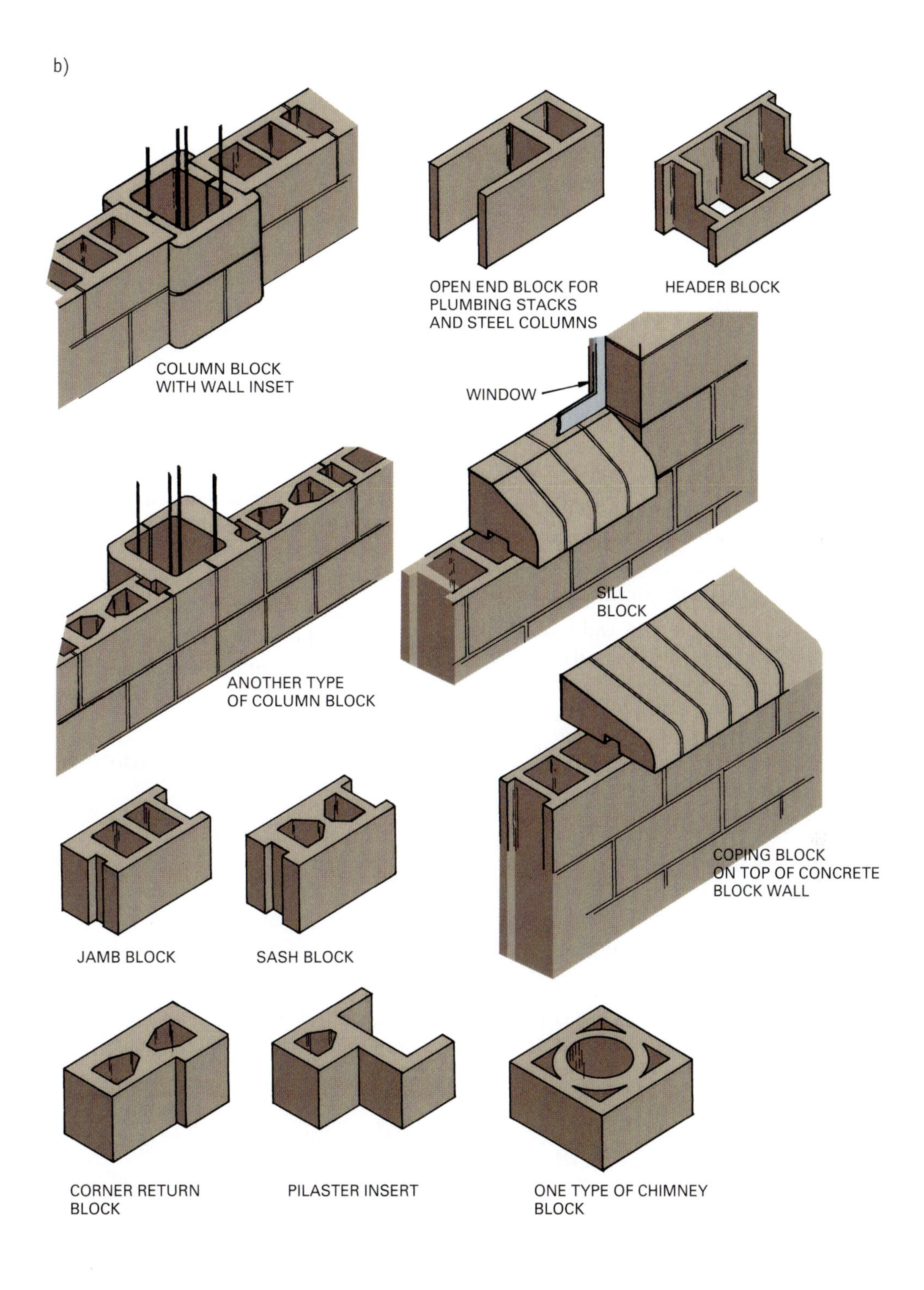

Control joint blocks are used when vertical shear control joints occur in a wall. This relieves stresses in long runs of masonry walls. Bond beam blocks are filled with concrete and reinforcing bar to form a continuous reinforced concrete beam along the top of a wall. They also are used to span above lintels and below sills. Sash blocks and jamb blocks are used where windows are to be installed. One style is used for wood windows and the other for metal windows. A joist block permits a floor joist to rest on the wall. The end is covered with the concrete wing, so from the exterior it appears as a normal block wall. Open-end units are used to provide a vertical cavity for running plumbing or electrical conduit or to enclose a steel beam.

Lintel blocks are used to form a steel reinforced concrete lintel over an opening in a wall, such as a window opening. They can also be used to form bond beams. More commonly, precast concrete lintels are used above and below openings in concrete block walls.

Sill and coping blocks are used to provide a finished cap. Sill blocks are placed below the window to direct water away from the wall. Coping blocks are placed on top of the exterior wall to seal out moisture.

Partition blocks are available in 4 and 6 in. thickness and are used for non-load-bearing walls. Chimney blocks are used to quickly form a concrete surround around a fireclay flue lining. Solid top blocks are used where a solid, coreless surface is needed, such as on the top of a foundation.

Split-face blocks are concrete units that are split in half lengthwise after they have hardened. This produces a rough, irregular surface that serves as the exposed face (Fig. 12.6). They may be hollow or solid units. Color variations are produced by using aggregates of various colors and by putting mineral colors in the mix.

Faced blocks have the exposed surface covered with a ceramic glaze or a plastic overlay, or they are polished by grinding the surface smooth. Decorative blocks are made with many different face designs. Typically they are pierced or recessed to produce an unusual wall surface. A decorative privacy wall is an example. Common sizes are 8×16 in. and 12×12 in. Special designs can be made by the block manufacturer to meet the requirements of an architect.

Special units are custom made blocks designed by an architect to face a project for which a special surface treatment is desired. The design possibilities are unlimited. Block manufacturers can produce almost any design, provided a mold can be made to fashion the unit. Decorative walls can be produced by using standard concrete masonry units and varying the bond pattern (Fig. 12.7).

Concrete Retaining Walls

Precast concrete retaining wall units are available in sizes ranging from rectangular units for small residential walls to large, multi-piece units suitable for high walls and large retaining walls, like those used for highway construction. They offer an architect the choice of units of various sizes, colors, and textures to blend architectural appearance with the design of a house. They provide great flexibility in creating a unique outdoor living environment.

Autoclaved Aerated Concrete Block (AAC)

Autoclaved aerated concrete block uses Portland cement mixed with lime, silica sand, or recycled fly ash, water, and aluminum powder. The mix is poured into a mold where the reaction between aluminum and concrete causes microscopic hydrogen bubbles to form, expanding the mix to about five times its original

Figure 12.6 Split face block adds dimension and texture to a wall.

Figure 12.7 A variety of brick bonds, positions, and orientations can create unique designs.

volume. After evaporation of the hydrogen, aerated concrete is cut to size and formed by steam-curing in a pressurized chamber (an autoclave). The result is a non-toxic, airtight material with excellent thermal, fire, and acoustical-resistance properties that can be used for wall, floor, and roof panels and blocks, and lintels to provide structural capacity.

MATERIAL SPECIFICATIONS

The American Society for Testing and Materials has specifications for four types of concrete masonry units. These are used by block manufacturers to produce quality blocks (Table 12.4).

Grades

Load-bearing concrete masonry units are available in two grades, S and N (Table 12.5). Grade S units are restricted to above-grade applications and, when used for exteriors, must have a protective coating. Grade N units can be used above and below grade and may be exposed to the weather or moisture penetration. When used below grade, protective coatings are recommended and often are required by building codes.

Types

Two types of concrete masonry units are specified by ASTM C90, Type I and Type II. Type I units are manufactured with specific limits on moisture content. Type II has no moisture control limits. Moisture limits on block manufacturing have the advantage of minimizing shrinkage after units have been laid. Shrinkage can cause cracking of the wall. Uses of the various grades and types are shown in Tables 12.6 and 12.7.

Weights

ASTM C90 establishes three weights of concrete masonry units: normal, medium, and lightweight. Normal weight units are made from concrete weighing more than 125 pcf (2,000 kg/m^3); medium weight uses concrete weighing 105 to 125 pcf (1,680 to 2,000 kg/m^3); and lightweight uses concrete 105 pcf (1,680 kg/m^3) or less. Due to the weight of concrete masonry units, erection is usually limited to ten courses or rows per day because mortar needs time to set and gain strength before additional weight is placed on it.

Table 12.4 ASTM[a] Standards for Concrete Masonry Units

Standard	Type of Unit
ASTM C55	Concrete building brick, solid veneer, and split block
ASTM C90	Hollow load-bearing concrete masonry units
ASTM CI29	Hollow non-load-bearing concrete masonry units
ASTM CI45	Solid[b] load-bearing concrete masonry units

[a]*American Society for Testing and Materials.*
[b]*Units with 75% or more net area.*

Table 12.5 Grades, Types, Weights, and Uses of Concrete Masonry Units (ASTM C90)

Grade	Use
N	General use above and below grade
S	Above grade only in walls not exposed to weather or with weather-protective coating if exposed to the weather
Type	**Use**
I	Moisture controlled during manufacture
II	Non-moisture-controlled units
Weight	**Use**
Normal weight	Manufactured using concrete weighing more than 125 pcf (2,000 kg/m^3)
Medium weight	Manufactured using concrete weighing between 105 and 125 pcf (1,680 to 2,000 kg/m^3)
Lightweight	Manufactured using concrete weighing 105 pcf (1,680 kg/m^3) or less

Table 12.6 Grades and Uses of Solid Load-Bearing Concrete Block, ASTM C145 and Hollow Load-Bearing Concrete Block, ASTM C90

Grade	Use
N-1, N-2	Exterior walls above or below grade that may be exposed to moisture and weather, and for interior walls and backup
S-1, S-2	Above-grade walls not exposed to weather or exterior walls if covered with weather-protective coatings

Table 12.7 Grades and Uses of Concrete Building Brick, ASTM C55

Grade	Use
N-1, N-2	On interior and exterior walls where high strength and resistance to moisture and frost action are required
S-1, S-2	Where moderate strength and resistance to moisture and frost action are required

Concrete Roof Tile

Concrete roof tile consists of a mix of cementitious materials, such as Portland cement, sand, hydraulic cements, fly ash, pozzolans, fine aggregates, and pigments, producing a durable, lightweight tile that is fire resistant. Some have an iron-oxide pigment added to the mix to produce the desired color. Others are colored by coating the face of the tile with a slurry of thin cement mixed with an iron-oxide pigment (Fig. 12.8).

Figure 12.8 Concrete tiles provide a roof covering with long life and excellent fire resistance. *(Courtesy Monier)*

Review Questions

1. What are the two ways concrete masonry units are cured?
2. What advantages are there to autoclaved concrete masonry units?
3. What two things influence the properties of concrete masonry units?
4. How does the tensile and flexural strength of concrete masonry units compare with compressive strength?
5. What affects the water absorption of concrete masonry units?
6. How is the insulation value of concrete masonry units improved?
7. What are the three major classifications of concrete masonry units?
8. What is the design module used when sizing concrete masonry units?
9. What are the grades established for load-bearing concrete masonry units?

Key Terms

autoclave
concrete masonry
kiln
special units

Activities

1. If you are near a plant that produces concrete masonry units, arrange for a visit. Before the visit prepare a list of specific things to look for and ask questions about, such as the actual proportions of the concrete used, curing procedures, and tests used to ensure quality control of products produced.
2. Secure a supply of concrete blocks and lay up a short wall and turn a corner. Vary the amount of water and mortar cement in several batches of mortar and prepare a written report citing what this does to the finished wall before and after curing.

Additional Resources

Beall, C., *Masonry*, Creative Homeowner Publishers, Upper Saddle River, NJ.

Building Code Requirements for Masonry Structures (ACI 530/ ASCE 5/ tms 402), American Concrete Institute, Farmington Hills, MI.

Concrete Masonry Handbook for Architects, Engineers, Builders, Portland Cement Association, Skokie, IL 50077. Many other publications available.

Ramsey, C. G., Sleeper, H. R., and Hoke, J. R., *Architectural Graphic Standards*, the American Institute of Architects, John Wiley & Sons, Hoboken, NJ.

Other resources include:

Numerous publications from the National Concrete Masonry Association, Herndon, VA.

See Appendix B for addresses of professional and trade organizations and other sources of technical information.

CHAPTER 13

Stone

LEARNING OBJECTIVES

Upon completion of this chapter, the student should be able to:

- Discuss the characteristics and uses of various stones used in building construction.
- Identify the various types of commercially available stone.
- Understand the processes used to quarry and work stone to make it useful for building construction.
- Select stone for a project based on the requirements of the job.

Build Your Knowledge

For further information on these materials and methods, please refer to:

Chapter 37 Flooring

For centuries, stone was used as a material to build structural load-bearing walls. Today it is used mainly as a veneer or facing, which greatly reduces the weight of a building while still enabling a designer to take advantage of stone's beauty as a finish material **(Fig. 13.1)**. **Rock** is solid mineral matter, occurring in individual pieces or large masses, such as an outcropping. **Stone** is rock that is quarried and shaped to size for construction purposes.

BASIC CLASSIFICATIONS OF ROCK

Rock is divided into three basic categories, depending on its origin: igneous, sedimentary, and metamorphic.

Igneous Rock

Igneous rock is formed, usually deep in the earth, when a molten material changes from a liquid to a solid state. Commonly used forms include granite, serpentine, and basalt.

Granite Granite is an igneous rock containing crystals or grains of visible size. It consists mainly of quartz, feldspar, mica, and other colored minerals. Colors include black, gray, red, pink, brown, buff, and green. It is hard, strong, non-porous, and durable. Granite is one of the most permanent building stones. It can be used under severe weather conditions and in contact with the ground. Granite can be finished with a range of surface textures, from rough to highly polished. Granite is used for windowsills, cornices, columns, floors, countertops, and wall veneers.

Figure 13.1 Stone is used as a finish material on the interior and exterior of buildings. *(Courtesy Eva Kultermann)*

Serpentine Serpentine is an igneous rock named after its major ingredient. It ranges from olive green to greenish black, has a fine grain, and is dense. Since some types deteriorate due to weathering, its major uses are on interiors. It can be cut into thin sections, $\frac{7}{8}$ to $\frac{11}{4}$ in. (22 to 32 mm), and is used for paneling, window-sills, stools, stair treads and risers, and landings.

Basalt Basalt is an igneous rock that ranges in color from gray to black. It is fine-grained and is used mainly for paving stones and retaining walls.

Sedimentary Rock

Sedimentary rock is formed of materials (sediments) deposited on the bottom of bodies of water or on the surface of the earth. Major types include sandstone, shale, and limestone.

Sandstone Sandstone is a sedimentary rock composed of sand-sized grains cemented together by naturally occurring mineral materials, such as silica, iron oxide, and clay. Quartz grains predominate in sandstone used for building construction. The two most familiar forms are brownstone, used mainly in wall construction, and bluestone, used for paving and wall copings. Colors include gray, brown, light brown, buff, russet, red, copper, and purple.

Since sandstone's hardness and durability depend on the cementing material, there is a wide range in weight and porosity. Sandstone is used for wall facing panels and can have a variety of surface finishes, including chipped, hammered, and rubbed.

Shale Shale is a sedimentary rock derived from clays and silts. It forms in thin laminations and tends to be weak along planes. It is not suitable as a concrete aggregate because it may expand in the presence of water, causing concrete failure. Shale that is high in limestone is ground into small particles and used in making cement, bricks, and tiles. It is predominantly gray in color but can be found in hues ranging from black to red, yellow, and blue.

Limestone Limestone is a sedimentary rock composed mainly of calcite and dolomite. There are three types of limestone. Oolitic is a calcite-cemented calcareous stone formed from shells that is uniform in composition and structure. Dolomitic limestone consists mainly of magnesium carbonate and has a greater compressive strength than oolitic. Crystalline limestone consists mainly of calcium carbonate crystals. It has high tensile and compressive strengths.

Limestone is used for building stones and is available as dimension (cut), ashlar, and rubble stones. It is used for paneling, veneer, window stools and windowsills, flagstone, mantels, copings, and facings (Fig. 13.2). It is also pulverized to form crushed stone aggregate and burnt to produce lime.

Metamorphic Rock

Metamorphic rock is either igneous or sedimentary rock that has been altered in appearance, density, and crystalline structure by high temperature and/or high pressure. Major types used in construction include marble, quartzite, shist, and slate.

Marble Marble is a metamorphic rock made up largely of calcite or dolomite that has been re-crystallized. There are a number of types of marble, with colors varying from white through gray and black. The presence of oxides of iron, silica, graphite, carbonaceous matter, and mica produce other color variations, including red, violet, pink, yellow, and green. Marble is used for wall panels and column facings, as well as window stools, windowsills, and floors. Some types are used on building exteriors, and others are limited to interior applications. Its surface can be ground to a fine, polished condition.

Quartzite Quartzite is a metamorphic rock that is often confused with granite. It is a variety of sandstone composed mainly of granular quartz that is cemented by silica, producing a coarse, crystalline appearance. It has high tensile and crushing strengths and is available in

Figure 13.2 A single rough-cut slab of stone accentuates one façade of a modern building. *(Courtesy Eva Kultermann)*

brown, buff, tan, ivory, red, and gray. Quartzite is used for building stone, gravel, and aggregate in concrete.

Schist Schist is a metamorphic rock generally made up of silica with smaller amounts of iron oxide and magnesium oxide. Its color depends on mineral makeup, but blue, green, brown, gold, white, gray, and red are common. It is commonly available in rubble veneer and flagstone and is used for interior and exterior wall facing, patios, and walks.

Slate Slate is a hard, brittle metamorphic rock consisting mainly of clays and shales. Its major ingredients are silicon dioxide, aluminum oxide, iron oxide, potassium oxide, magnesium oxide, and, sometimes, titanium, calcium, and sulfur. Slate is found in parallel layers, which enables it to be cut into thin sheets.

Slate is produced in three textures: sand-rubbed, honed, and natural cleft. It is cut into three types: roof tiles, random flagging, and dimension slate (cut to size) (Fig. 13.3). It is commonly used for interior and exterior wall facing, flooring, flagstones, countertops, coping, and windowsills and window stools.

TYPES AND USES OF STONE

Commercially, stone is used in several types: rubble stone, rough stone, monumental stone, dimension stone, flagstone, broken and crushed stone, and stone powder and dust.

Rubble stone consists of irregular fragments from a quarry that have one good face. The pieces are irregular in shape and sized usually in pieces 12 in. (300 mm) by 24 in. (600 mm) that are cut and fitted by a mason.

Figure 13.3 Slate is a long-lasting and durable roofing material. *(Courtesy Zureks, http://en.wikipedia.org/wiki/File:St_Fagans_Tannery_7.jpg)*

Rough building stone, sometimes called fieldstone, occurs in naturally found rock masses. The stone is generally used in its natural shape (Fig. 13.4).

Dimension stone, also referred to as cut stone, is cut to size at a stone mill and shipped to site. Its surface may be rough, as occurs when it is split, or polished. It is used as veneer on interior and exterior walls, floors, copings, stair treads, and for other similar applications. Ashlar, a form of dimension stone, is a cut rectangular stone with square corners and faces that is smaller than other dimension stone.

Flagstone is thin, flat stone from $\frac{1}{2}$ to 4 in. (12 to 100 mm) in thickness. Flagstones laid over a concrete base are usually $\frac{3}{4}$ to 1 in. (18 to 25 mm) thick. If laid over a sand or loam base, pieces $1\frac{1}{4}$ to $1\frac{1}{2}$ in. (31 to 37 mm) thick are required. Its surface may be left rough or polished. Random flagstones, with the exception of minor shaping, are left in their natural shape. Trimmed flagstones are random, shaped pieces with several edges sawed straight. Trimmed rectangular flagstones are pieces with four sides sawed, forming square and rectangular pieces. Typical patterns for laying flagstones are shown in Fig. 13.5.

Broken and crushed stone includes irregular shapes and crushed pieces of one type of stone that are graded for hardness and size and used as aggregate in concrete for surfacing roads and driveways and as aggregate for surfacing fiberglass asphalt shingles and built-up

Figure 13.4 Fieldstone is laid in a random pattern, forming a textured surface. *(Courtesy Eva Kultermann)*

Figure 13.5

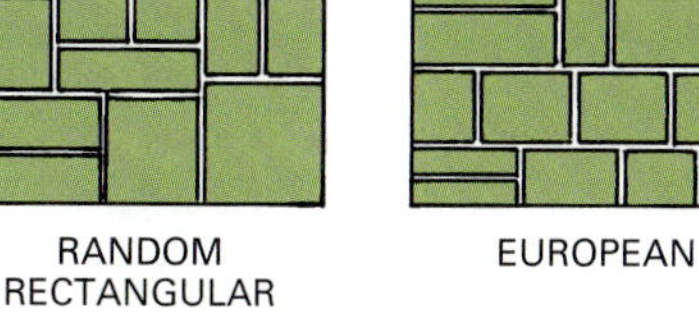

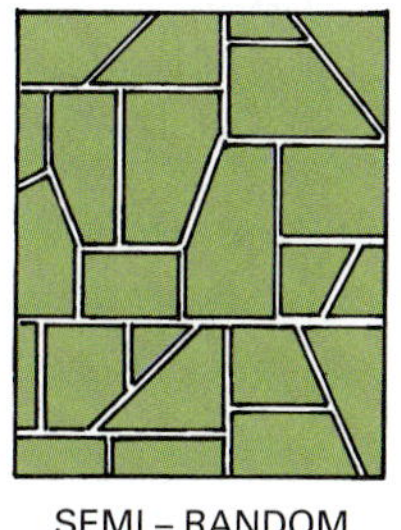

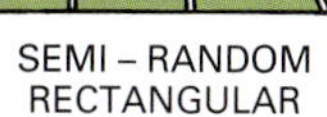

Some of the commonly used patterns for laying flagstone.

roofing. Stone powder and stone dust are used as fill-in paints and asphalt paving surfaces.

QUARRYING AND PRODUCING BUILDING STONES

A quarry is an excavation from which stone used for building is taken via blasting or cutting (Fig. 13.6). Broken stone is produced by blasting the rock. Larger pieces can be re-broken or cut into smaller units for use as an exterior finish material. The rest is crushed and sorted into various sizes for use as aggregate.

Most stone used in building construction is dimensional and produced by cutting large blocks from a quarry, often with a channeling machine, which makes an incision from one to three inches wide. Some machines use a rotating chisel cutter, and others use a wire that runs over pulleys and moves a quartz sand cutting agent over the stone, producing a saw-type cut.

Large blocks are removed from the quarry to a mill, where they are cut to the sizes and thicknesses needed.

Figure 13.6 Large blocks of rock are moved from quarry to mill to be cut into desired shapes and sizes. *(Image copyright Oleg - F, 2009. Used under license from Shutterstock.com)*

The architectural drawings specify the shape and size of each stone. Holes are drilled in each block, as indicated for lifting and anchoring it in place.

CHOOSING STONE

When selecting the stone to be used in a building, an architect must consider a variety of parameters, including cost, strength, durability, hardness, grain and color, and texture and porosity (Fig. 13.7).

The ease with which a stone can be quarried and shaped influences economic considerations. Soft stones have lower production costs. Accessibility is a second cost factor, because costs rise as stone transportation distances increase.

If stone is exposed to detrimental weather, it must have sufficient durability to withstand freeze-thaw cycles and erosive conditions. Hardness is very important in stones used for floors, steps, patios, and other areas exposed to traffic. Grain and color are considered when the appearance of stone in its finished position is decided. Colors vary widely and must be specified. The texture of the stone has a great influence on the finished appearance. Fine-grained stones can have a smoother, polished surface, while coarse-grained stones present a more open face. Porosity pertains to the ability of stone to resist moisture penetration. Porous rock tends to permit some minerals to dissolve and stain the exposed face. It also is not durable and will be damaged by freeze-thaw cycles.

Figure 13.7 Stone selected for carving is carefully chosen for proper grain, color, durability, and hardness. *(Courtesy Bybee Stone Company Inc.)*

Review Questions

1. What is the main use for stone in buildings built today?
2. What are the three basic categories of rock?
3. How is igneous rock formed?
4. What colors of granite rock are commonly available?
5. In what part of a building is serpentine rock used?
6. How is sedimentary rock formed?
7. What materials bind together the sand-sized grains forming sandstone?
8. Shale that has high limestone content is used to make what masonry materials?
9. What are the three types of limestone?
10. What are the main differences in the three types of limestone?
11. What are the major types of metamorphic rock used in building construction?
12. In what colors is marble found?
13. Where is marble commonly used in building construction?
14. Where is shist used in construction?
15. In what textures is slate produced?
16. What type of machine is used in a quarry to cut rock into manageable sizes?
17. What factors must a designer consider when choosing stone for a building?

Key Terms

igneous rock
metamorphic rock
rock
sedimentary rock
stone

Activities

1. Collect samples of various types of stone typically used in building construction. Write a report citing the characteristics of each type. Your building-materials supplier can help with samples.
2. Test selected stone samples for compressive strength. Note variances between samples of the same stone and between the different types of stone. Report why you think these variances may have occurred.

Additional Resources

Dimension Stone Cladding: Design, Construction, Evaluation, and Repair, American Society for Testing and Materials, West Conshohocken, PA.

Other resources include:

Publications of the Building Stone Institute, Purdys, NY.

Publications of the Cast Stone Institute, Winter Park, FL.

Publications of the Indiana Limestone Institute of America, Bedford, IN.

See Appendix B for addresses of professional and trade organizations and other sources of technical information.

CHAPTER 14

Masonry Construction

LEARNING OBJECTIVES

Upon completion of this chapter, the student should be able to:

- Select the type of masonry bearing wall best suited for a particular situation.
- Understand the purpose of expansion and control joints in masonry construction.
- Recognize the types of brick and concrete masonry wall constructions and the properties associated with them.
- Explain the advantages and disadvantages of the various mortar joints.
- Be familiar with the process used to lay brick and concrete masonry units.
- Evaluate the quality of stone wall construction.
- Recognize proper construction of structural clay-tile walls.

Build Your Knowledge

For further study on these materials and methods, please refer to:

Chapter 10 Mortars for Masonry Walls
Topic: Types of Mortar
Chapter 11 Clay Brick and Tile
Topic: Structural Clay Masonry Units
Chapter 12 Concrete Masonry
Chapter 13 Stone

The engineer designing load-bearing masonry buildings must consider magnitude and direction of all forces acting on a building. This includes live loads, dead loads, lateral loads, and forces caused by temperature change, impact, or unequal foundation settlement.

Masonry wall construction can consist of clay brick, concrete masonry units, stone, or structural tile. Load-bearing walls carry floor and roof loads on the exterior and interior of a building **(Fig. 14.1)**. Non-load-bearing masonry walls are used for interior partitions and on exterior building skins to protect from the elements. Masonry units provide acoustical control, fireproof construction, and are highly durable. Masonry walls are easy to design and construct and often produce a more economical building than do other construction methods. They can be used to construct low-rise buildings without lateral bracing. Walls above certain heights require lateral bracing. Masonry is a heavy material, limiting its use in high-rise buildings. The weight requires extensive footings and produces compressive stresses and can cause buckling. Lateral forces are controlled by various design features, such as building internally reinforced masonry walls.

MASONRY BEARING WALLS

A masonry load-bearing wall may be unreinforced or reinforced, solid masonry or cavity type. It may be a composite wall, using several types of units, such as clay brick over concrete blocks.

Reinforced masonry is used when the flexural, compressive, and shear stresses exceed those permitted for partially reinforced or unreinforced masonry. Building codes specify the amount of steel reinforcing required according to use and loading.

Cavity wall construction is generally used for exterior walls because they can control moisture penetration and be insulated. Solid masonry walls do not have these advantages and are therefore generally used today for interior partitions only.

Figure 14.1 Masonry can provide structural support or be used as a veneer for interior and exterior finishes. This building has masonry load-bearing walls supporting a bar joist roof. *(Reproduced with permission from The Building Systems Integration Handbook, Richard Rush, ed., Butterworth-Heinmann Publishers, Newton, Mass., 1986.)*

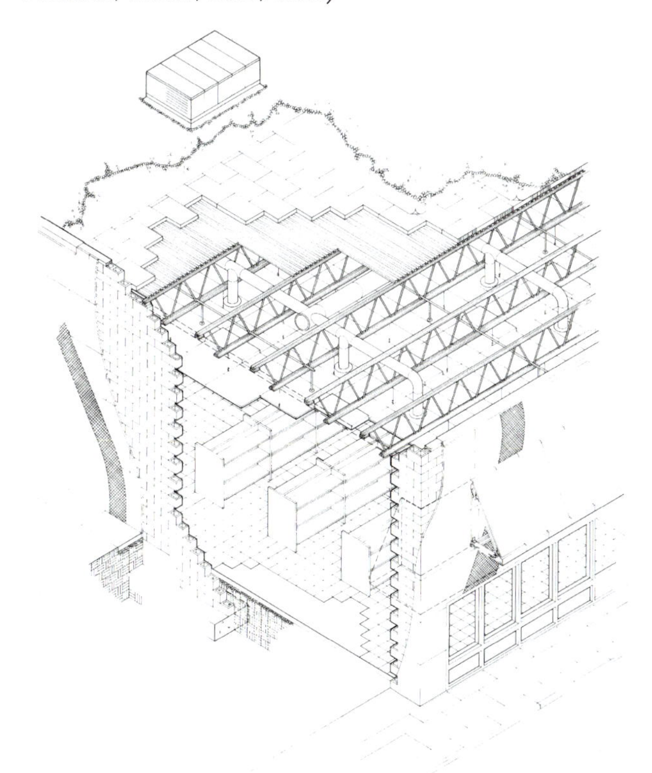

Composite masonry walls have an exterior veneer of a quality masonry unit, such as brick, tile, or stone, and the hidden interior portion of the wall is built from a more economical unit, such as concrete block. The engineer designing the wall must consider how differences in thermal expansion, moisture absorption, and load-bearing capabilities will affect the wall.

CONTROL AND EXPANSION JOINTS

Within masonry walls, movements can occur that produce stresses that may cause cracking. These stresses can be created by expansion and contraction due to changes in temperature or moisture content, structural movements due to settling of the foundation, and concentration of stresses at openings in the wall. Masonry walls often use units of different materials. These expand and contract at different rates and can create stresses. Examples include brick veneer over concrete block and the use of metal lintels over openings. The locations of control and expansion joints must be carefully determined as part of the wall design.

Expansion joints create a space between adjacent parts of masonry construction and permit a limited amount of movement. The actual spacing varies with the type of masonry unit, size of the area, and the reinforcing used. Vertical expansion joints are placed near corners or where walls change direction. Horizontal joints are placed above masonry walls that butt structural frames or on the bottom of floor or roof structures (Fig. 14.2).

Figure 14.2 Expansion joints used in masonry construction.

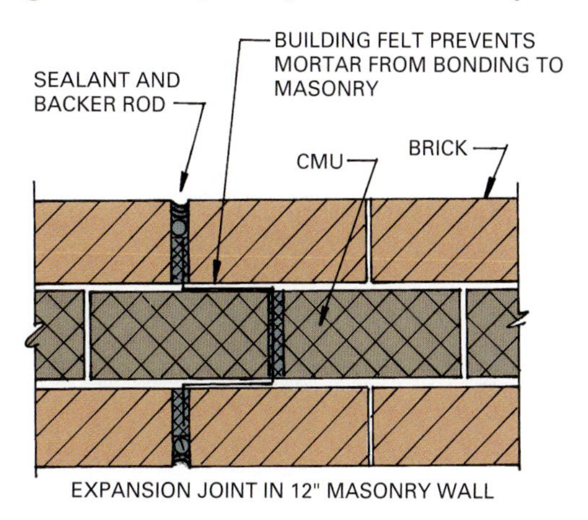

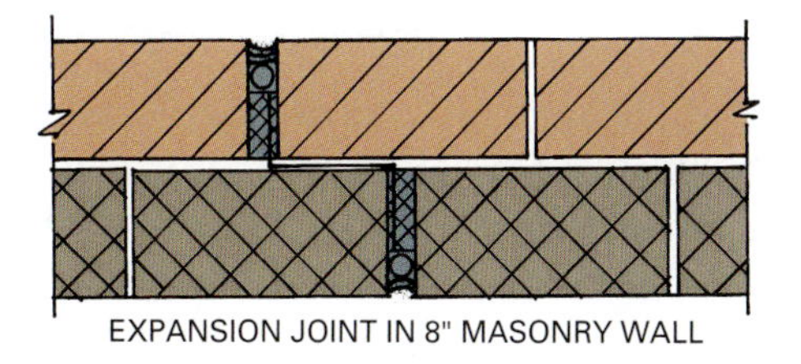

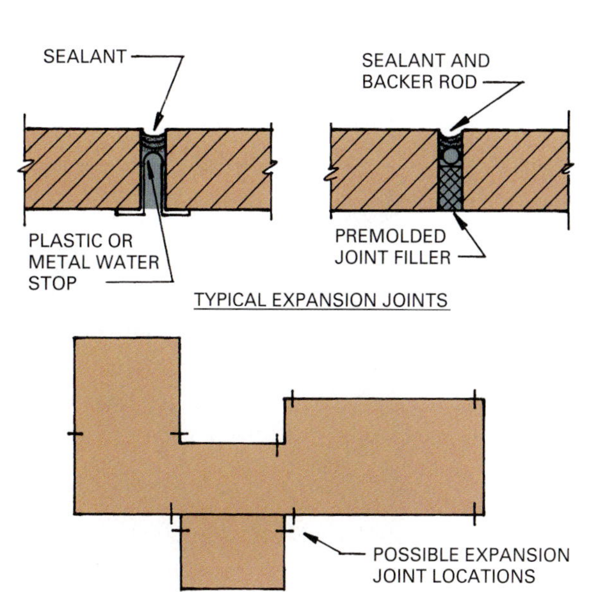

Control joints provide tension relief between parts of a masonry wall that may change from their original dimensions. One major contributor is the fact that masonry walls tend to contract after laying and drying. The tensile strength of the wall tries to resist these stresses, and the wall cracks if its strength is exceeded. This cracking is controlled by using carefully placed control joints and adequate steel reinforcing in the wall.

Control joints are continuous vertical joints built into a masonry wall as the units are laid. The locations of control joints must be carefully determined as part of the wall design. Generally, they are located where there are changes in wall height or wall thickness and at openings; wall intersections; returns in U, T, and L-shaped buildings; and junctions between walls and columns (Fig. 14.3).

Various types of control joint construction methods are shown in Fig. 14.4. Control joints must be carefully sealed with a specified elastic joint sealer. Regular inspection and maintenance of the seal is required to keep the wall watertight. Control joints are generally not used below grade because they can develop leaks. Temperatures below grade are more constant, so expansion and contraction are usually not an issue.

Figure 14.3 A typical control joint in a masonry wall and recommended layout locations.

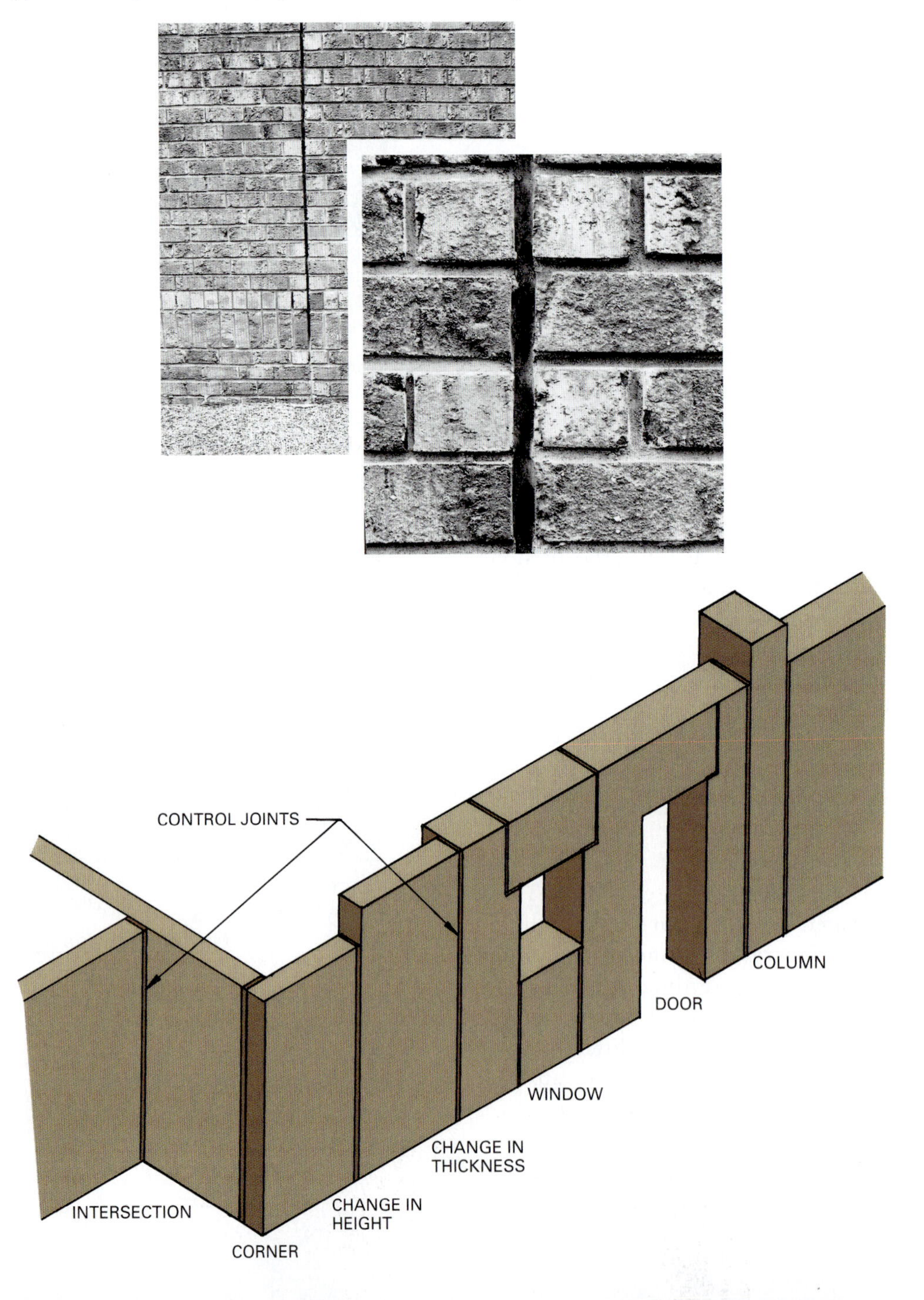

Figure 14.4 Typical control joint construction.

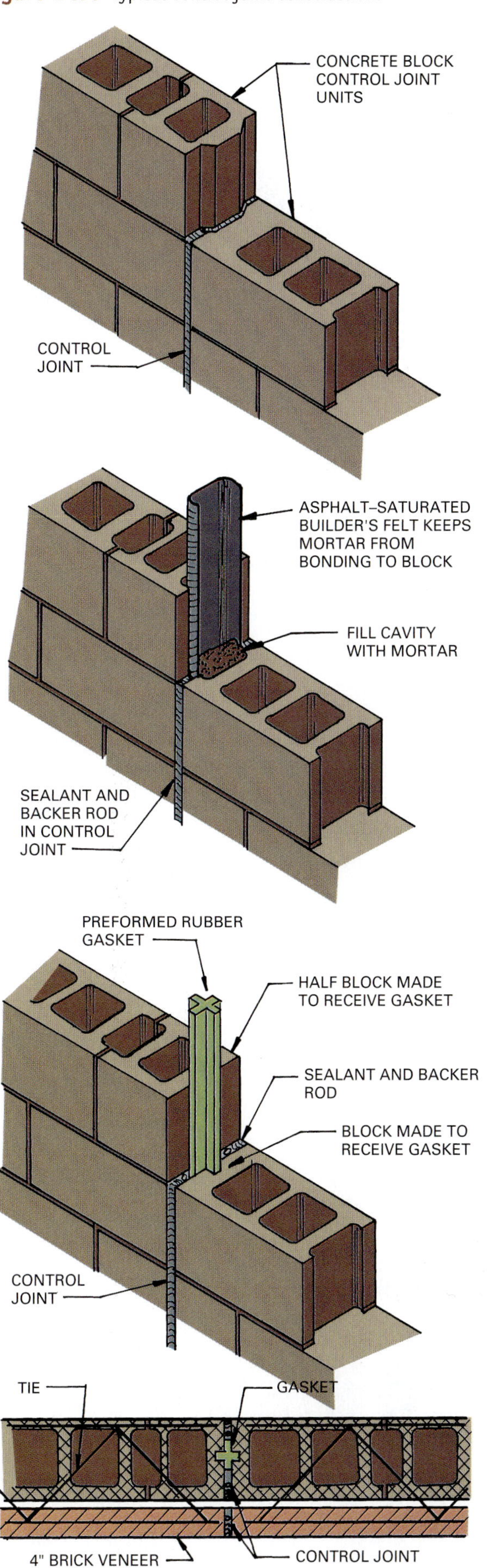

BRICK MASONRY CONSTRUCTION

Figure 14.5 shows commonly used solid and cavity masonry wall constructions. In cavity walls, the hollow cavity tends to collect moisture that is controlled by providing for its drainage. Cavity walls require flashing at the base with weep holes located just above the bottom of the flashing (Fig. 14.6). Moisture that enters the cavity moves to the bottom of the wall and is directed out of the building through the weep holes.

Figure 14.5 Typical brick masonry wall constructions.

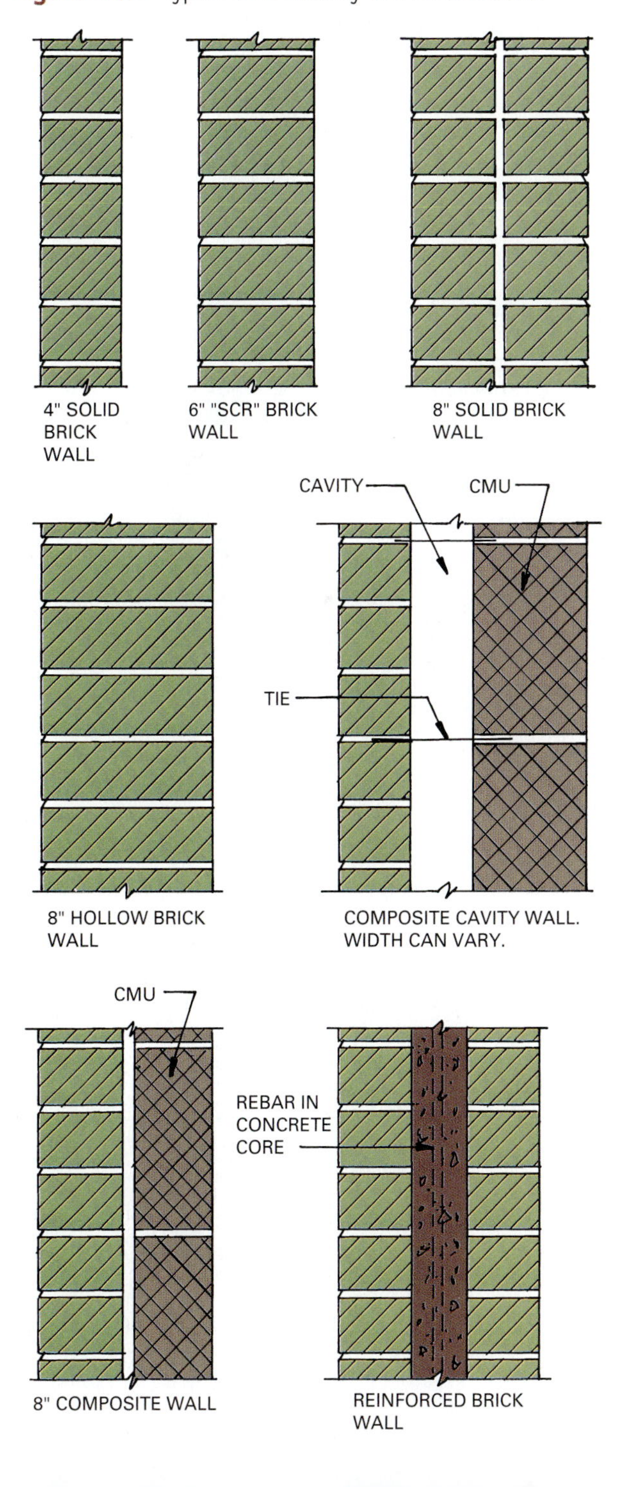

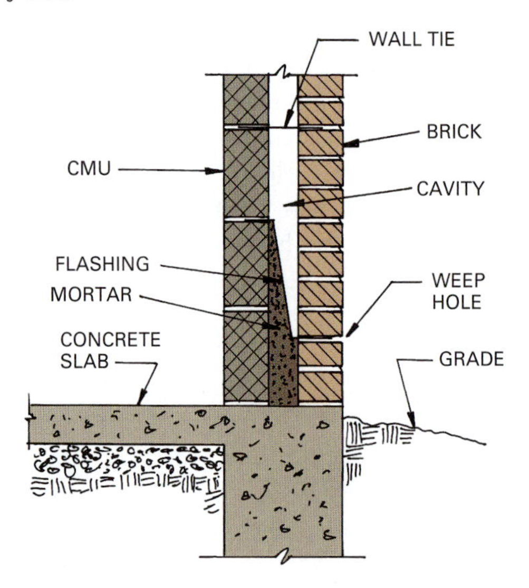

Figure 14.6 A detail used to control moisture inside a brick cavity wall.

Figure 14.7 Weep holes allow moisture to escape from the cavity.

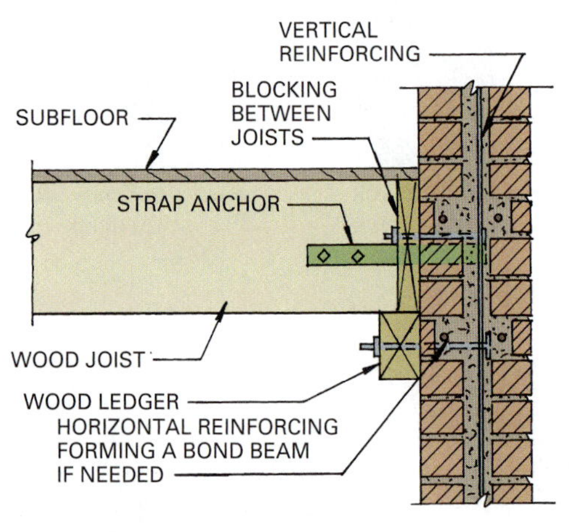

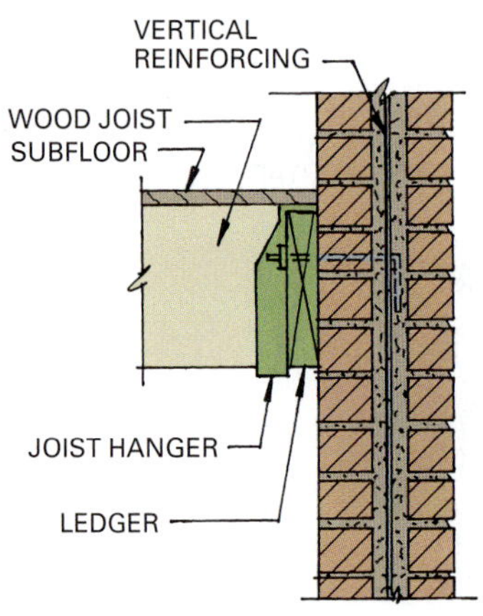
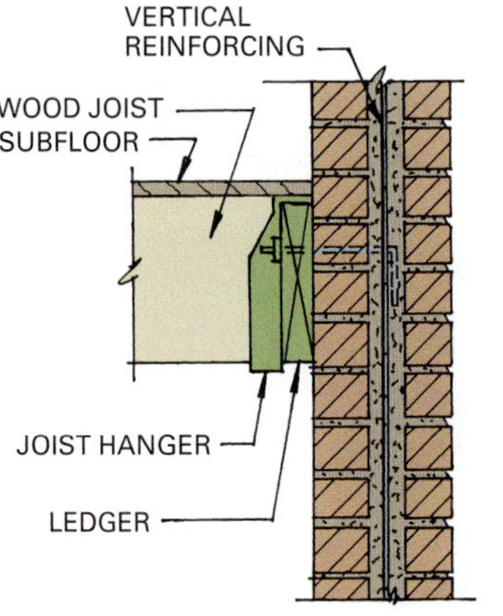

VERTICAL REINFORCING
HORIZONTAL REINFORCING FORMING A GRADE BEAM IF NEEDED
SUBFLOOR
WOOD JOISTS
BLOCKING BETWEEN JOISTS
TYPICAL DETAIL WHEN JOISTS RUN PARALLEL WITH THE WALL

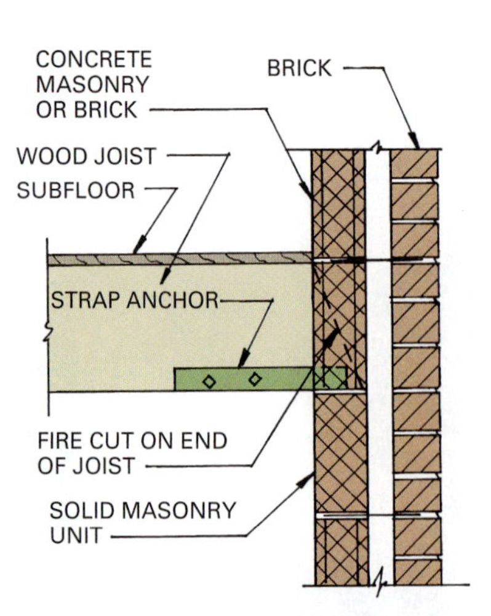

Figure 14.8 Commonly used ways to frame wood joists and beams to brick masonry walls.

Weep holes are openings left in the mortar joint. They can be formed with pieces of plastic or metal pipe laid in the joint, by omitting mortar in places along the joint, or by placing oiled rope in the joint and pulling it out after the mortar begins to set. Weep holes are often spaced 18 to 24 in. (457 to 508 mm) on center (Fig. 14.7).

The flashing is copper or plastic sheet material. Aluminum flashing is not used because it has an unfavorable chemical reaction with mortar. The masons lower a wood strip with wires into the cavity below the area where bricks are being laid. The strip catches mortar accidentally dropped in the cavity and is raised up and cleared. This keeps mortar from filling the area above the flashing and closing the weep holes. Moisture penetration of brick walls can also be controlled by parging the width between the two courses of brick with mortar or grout.

Some examples of brick bearing walls supporting floor and roof joists and decking are shown in Figs. 14.8 through 14.11. Examples of roof construction are shown in Fig. 14.12.

Figure 14.9 Several ways to frame steel joists or beams to brick masonry walls.

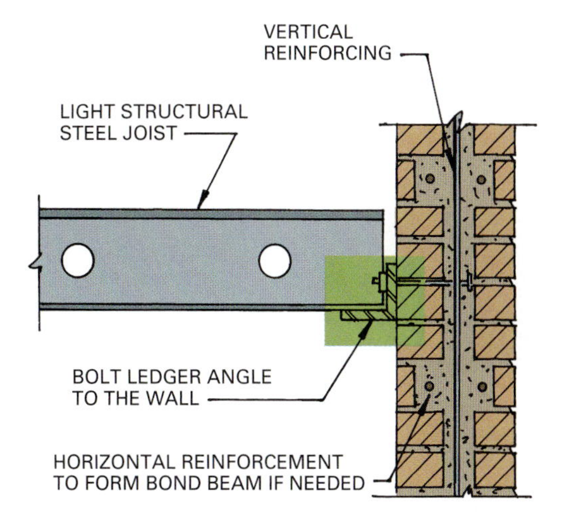

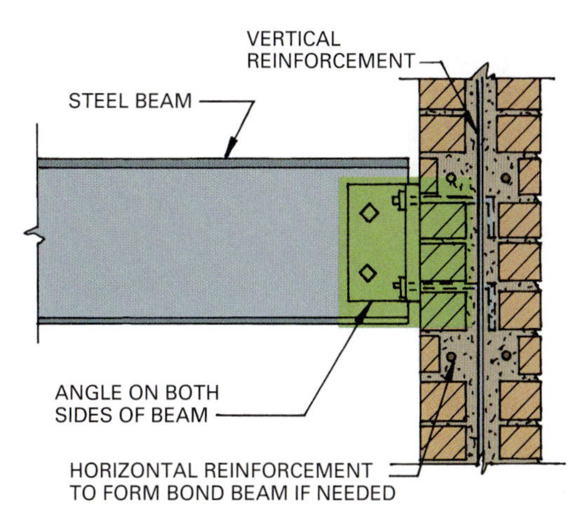

Openings in brick masonry walls are spanned with a lintel. Lintels are constructed with steel angles, precast concrete, or concrete masonry lintel blocks (Fig. 14.13). Reinforced brick lintels use steel reinforcing that is bonded to the masonry to form a beam. The choice and design depend on the span of the opening, the load to be carried, and the desired architectural expression. Precast concrete and concrete masonry lintels generally carry the loads imposed on the wall, and steel lintels

Figure 14.10 Two ways to support cast-in-place concrete decks with brick masonry walls.

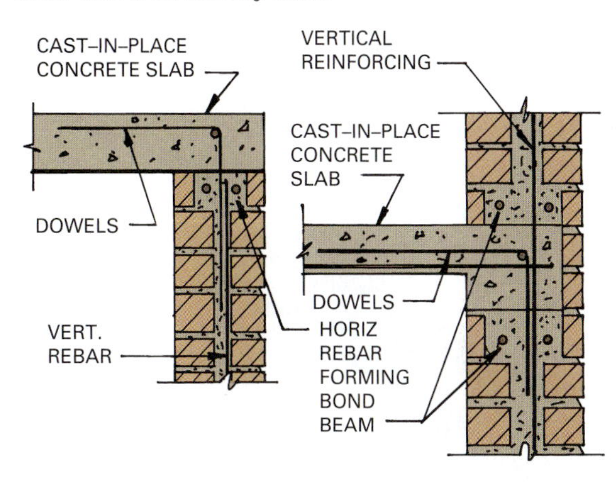

Figure 14.11 Open-web joists are anchored to masonry walls with metal straps set in the mortar joint.

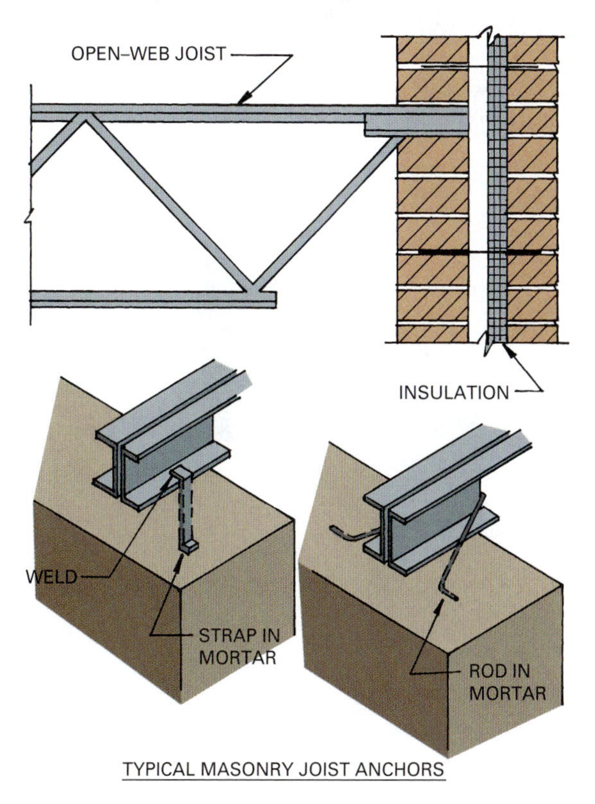

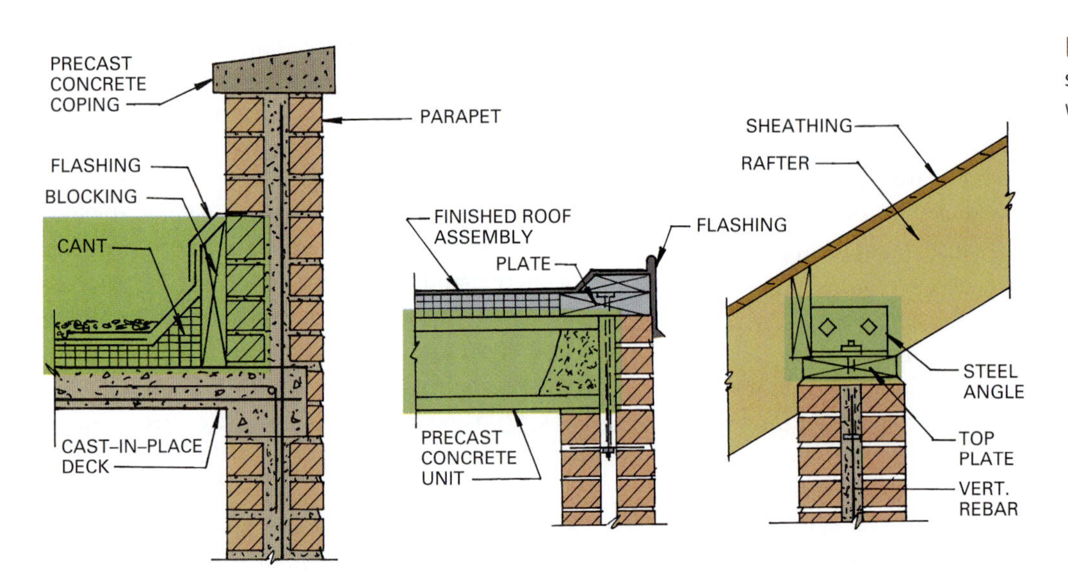

Figure 14.12 Examples showing roof construction with brick masonry walls.

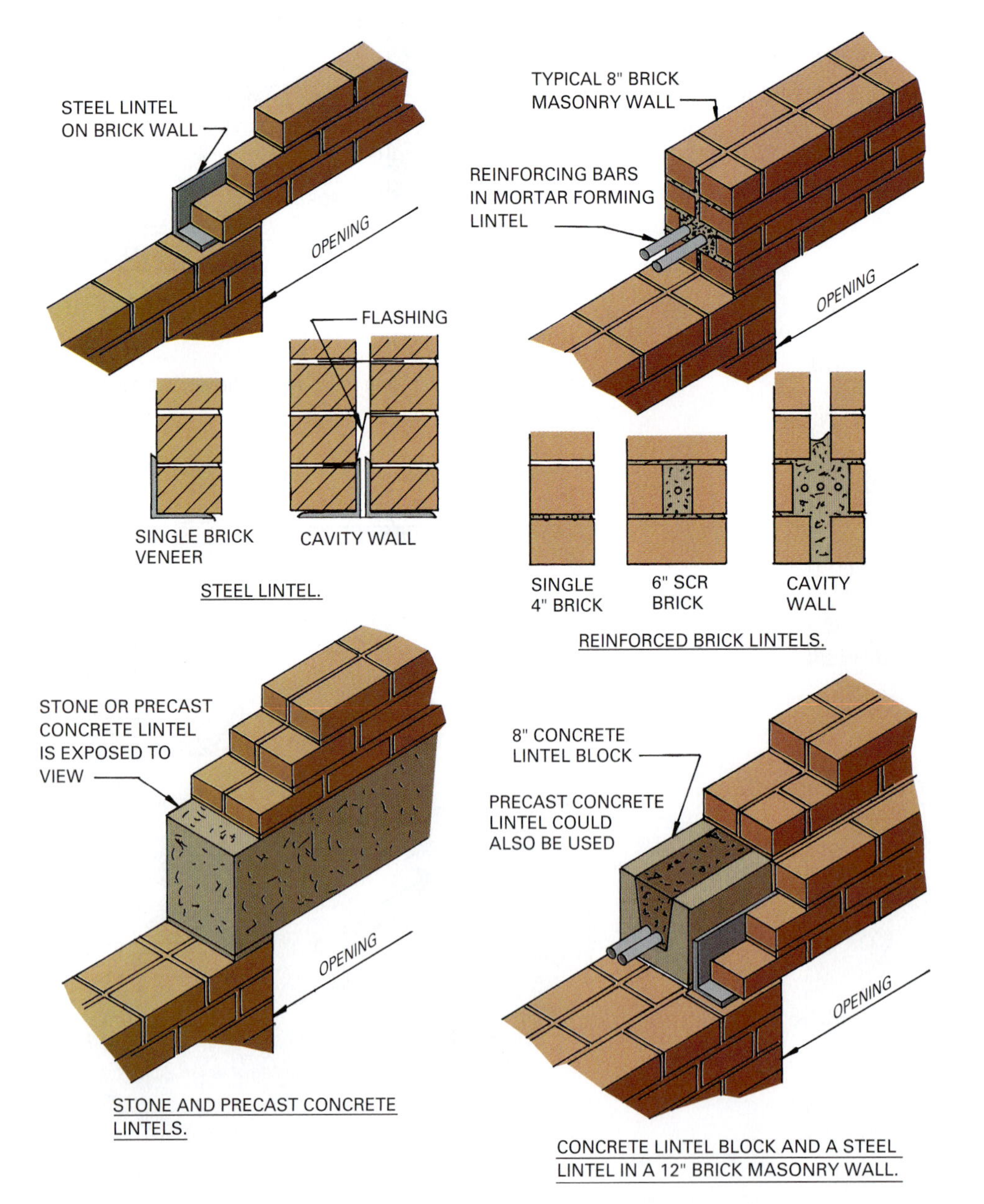

Figure 14.13 Types of lintels used with brick masonry construction.

carry the face brick. Some classic architectural styles use exposed concrete or stone lintels.

Brick masonry columns are typically designed for construction that doesn't require brick cutting. The interior core has reinforcing bars and is filled with concrete (Fig. 14.14). Pilasters are built in the same manner as columns but are an integral part of a load-bearing wall.

Brick masonry walls are reinforced using various types of metal ties (Fig. 14.15). The mortar and reinforcing work in tandem to provide the required structural bond. A solid wall two bricks wide can simply use header bricks to tie the two parts together. The header bricks also provide an interesting bond pattern. A cavity wall uses metal ties set in mortar. Brick veneer over a concrete block or sheathed wood-frame wall uses metal ties and straps. Brick walls veneered over tile units can use header bricks or metal ties.

Grouted reinforced brick masonry walls use vertical and horizontal reinforcing bars in the wall cavity, as shown in Fig. 14.16. The amount and placement of the reinforcing are engineering decisions. The reinforcing increases the strength of the wall to resist lateral loads and reduce buckling. A typical layout for a wall with openings is shown in Fig. 14.17.

Figure 14.14 Reinforced brick masonry pilaster and column construction.

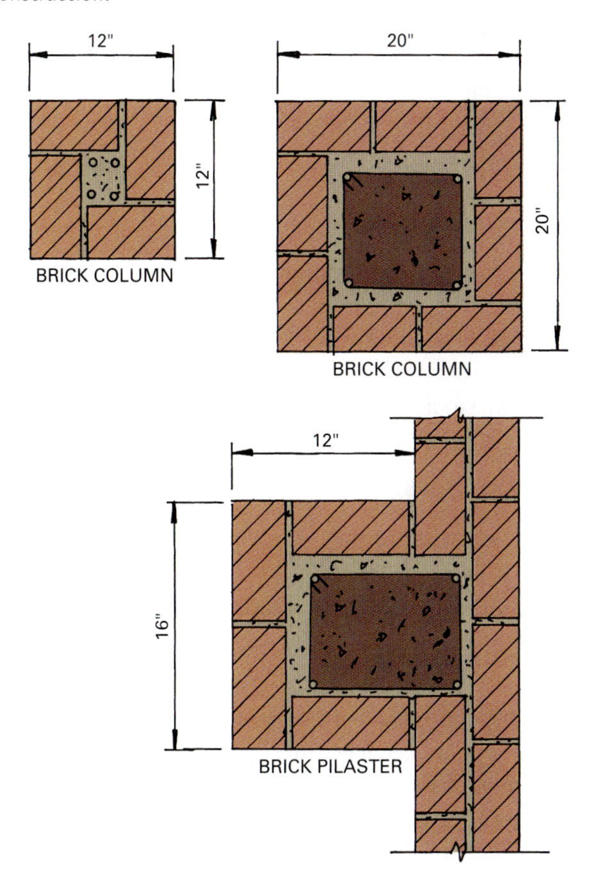

Insulation

Solid masonry walls are poor insulators, readily conducting heat and cold. To improve the energy efficiency of a masonry wall, insulation is placed either in its cavity or on its interior face. Rigid insulation consists of rigid foam sheets, adhered to the masonry with layers of acrylic polymer stucco, sometimes reinforced with an applied fiberglass mesh. This forms an attractive hard exterior finish. Since the masonry is completely hidden, less costly units may be used (Fig. 14.18).

Cavity walls permit the placing of insulation inside them. If rigid insulation is used, it is bonded to the inside face of the masonry units forming the inside wall. Another way to insulate interior surfaces is to construct back up stud walls with batt insulation.

The cores of concrete block walls can be filled with a granular insulation. The webs permit considerable energy loss, so rigid insulation should be applied to the exterior or interior surface.

BUILDING A BRICK MASONRY WALL

Part of the design process involves specifying brick bond patterns and types of mortar joints. Bricks are also laid in several directions, providing some structural and decorative properties. The terms used to describe brick in various positions are shown in Fig. 14.19.

Brick Bond Patterns

Brick bond patterns commonly used for standard and oversize bricks are shown in Fig. 14.20. Notice the running bond consists entirely of stretchers. The English bond alternates courses of headers and stretchers. The common bond employs a header course every sixth course.

Types of Mortar Joints

Mortar joints can take several forms. The appearance of a finished wall depends on the type of joint employed. Figure 14.21 shows joints in common use. The mortar joint is finished by troweling or tooling. Troweling involves striking off excess mortar with a mason's trowel. This does not produce the most watertight joint. Weathered, concave, and V-joints resist leakage the best. They are called *tooled joints*. Tooled joints result from compressing mortar into a joint and against the faces

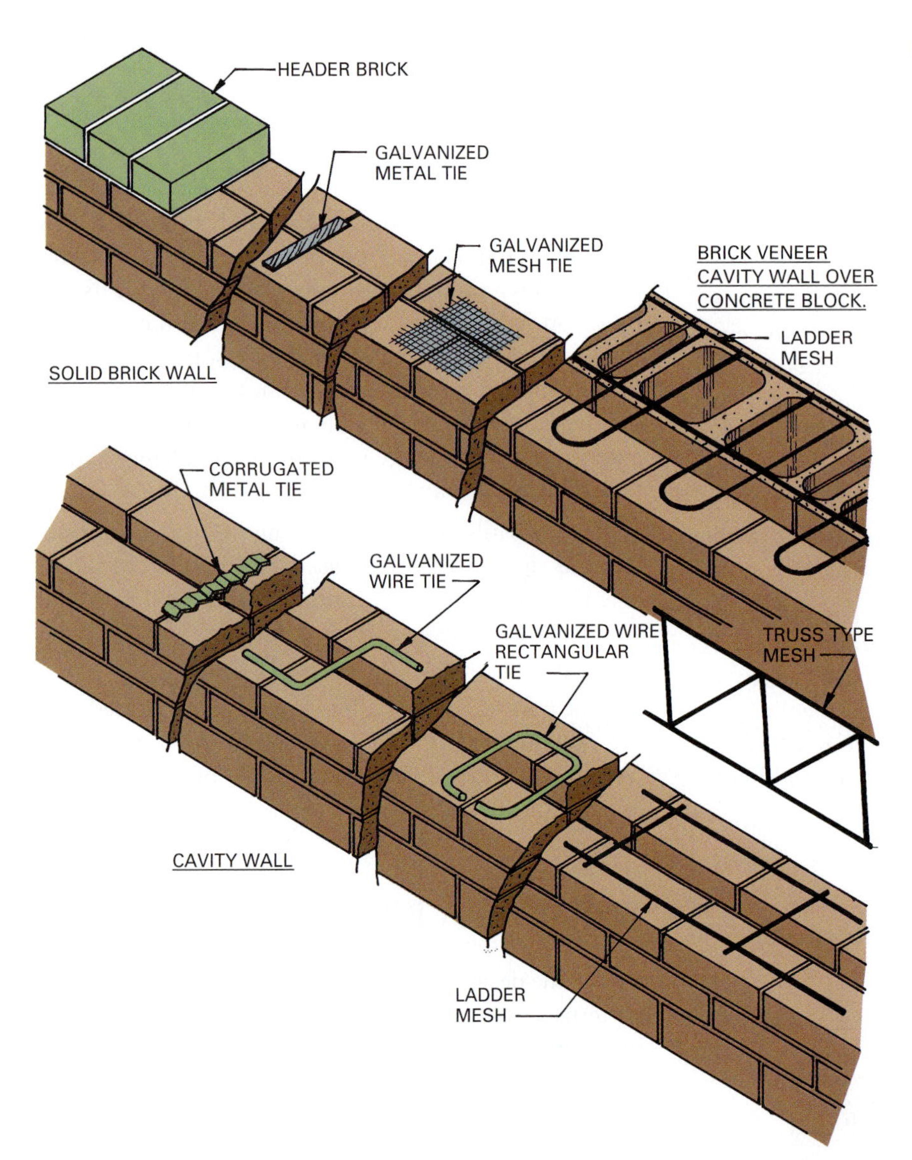

Figure 14.15 Brick masonry walls use various types of metal wall ties.

Figure 14.16 A grout-filled reinforced brick masonry wall with horizontal and vertical rebars.

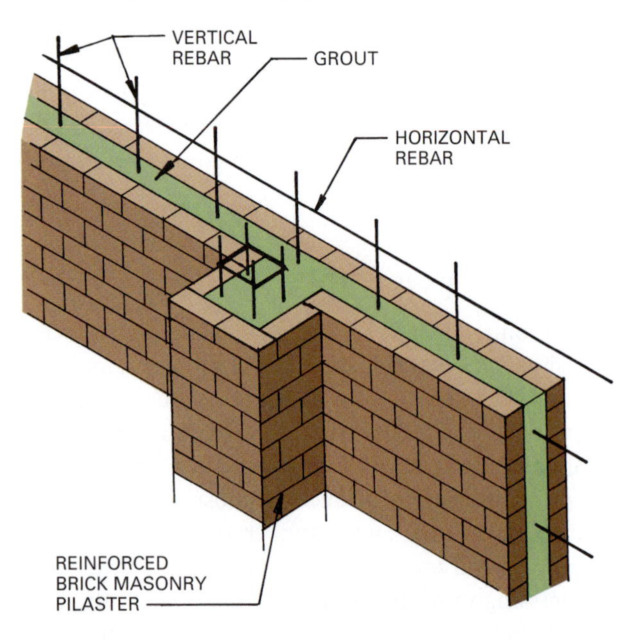

of adjacent units. Ruled, flush, and flush and rodded joints are not as watertight. Extruded, beaded, struck, and raked joints are most likely to eventually let moisture penetrate a wall. Information on types of mortar is given in Chapter 10.

Laying Bricks

When possible, masonry walls are sized to avoid the need for cutting bricks (Fig. 14.22). A $\frac{3}{8}$ to $\frac{1}{2}$ in. (9.5 to 13 mm) joint is commonly used. Each unit length then equals a brick plus a mortar joint. (Brick sizes are discussed in Chapter 11.)

The basic process for laying a brick masonry wall is shown in Fig. 14.23. The outer edge of a wall is located on its footing by snapping a chalk line. A first brick course without mortar is laid to determine if brick cutting is necessary to close the wall. A brick is laid and leveled at each end of the wall. Working from the ends

Figure 14.17 Typical reinforcing in brick masonry load-bearing walls that have openings.

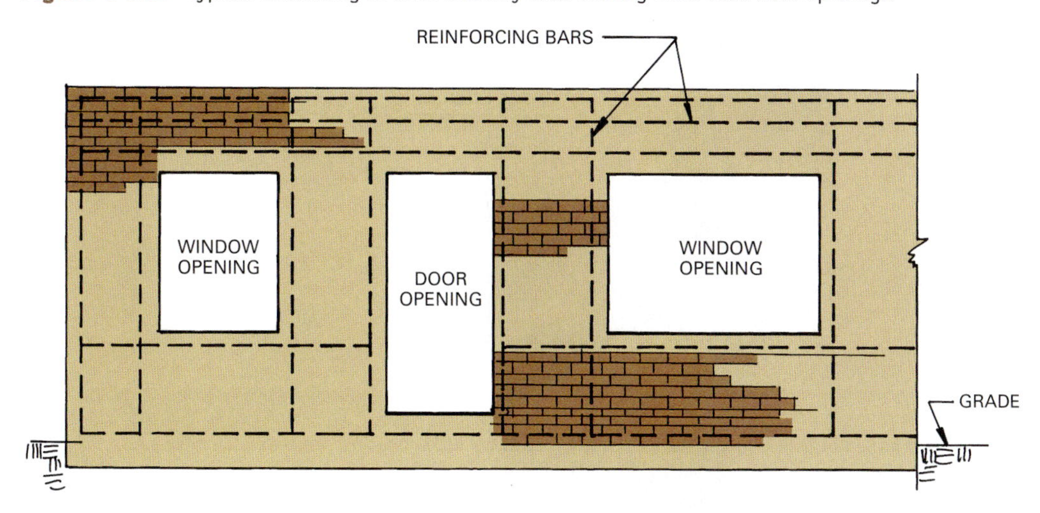

Figure 14.18 Rigid foam insulation enhances the energy efficiency of masonry construction.

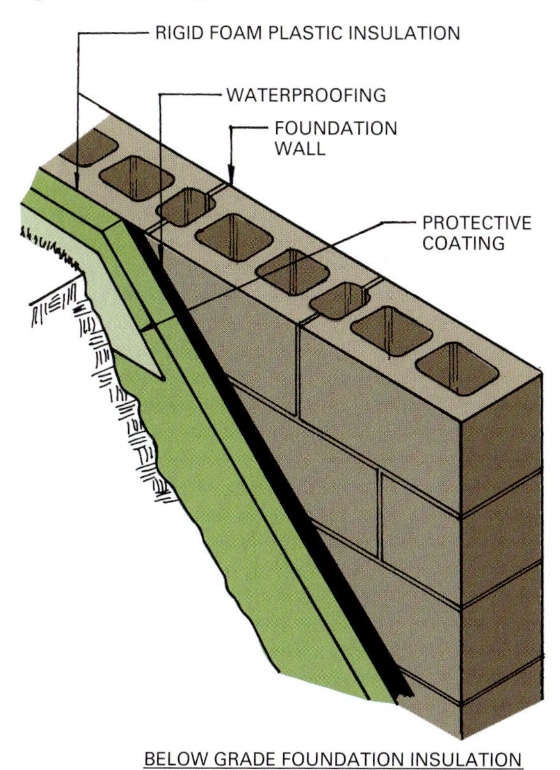

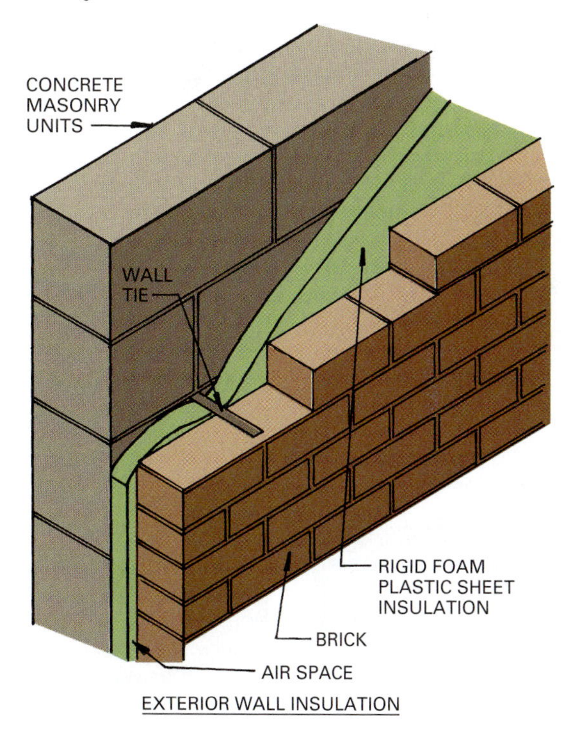

toward the center, the first course is laid in mortar, making certain it is straight and level. Then several courses are laid up in a pyramid fashion on each corner. A chalk line is run for each subsequent course, and the center bricks are laid in mortar. After several courses are laid up, the bricks are again raised at each corner and then filled in toward the center.

A story pole with the height of each course marked on it is used to check the height of each course as it is laid. Due to the weight of the masonry units, progress is usually limited to ten courses per day to allow the mortar to set and develop strength.

Figure 14.24 shows the basic steps generally followed to lay brick. The mason must work rapidly yet

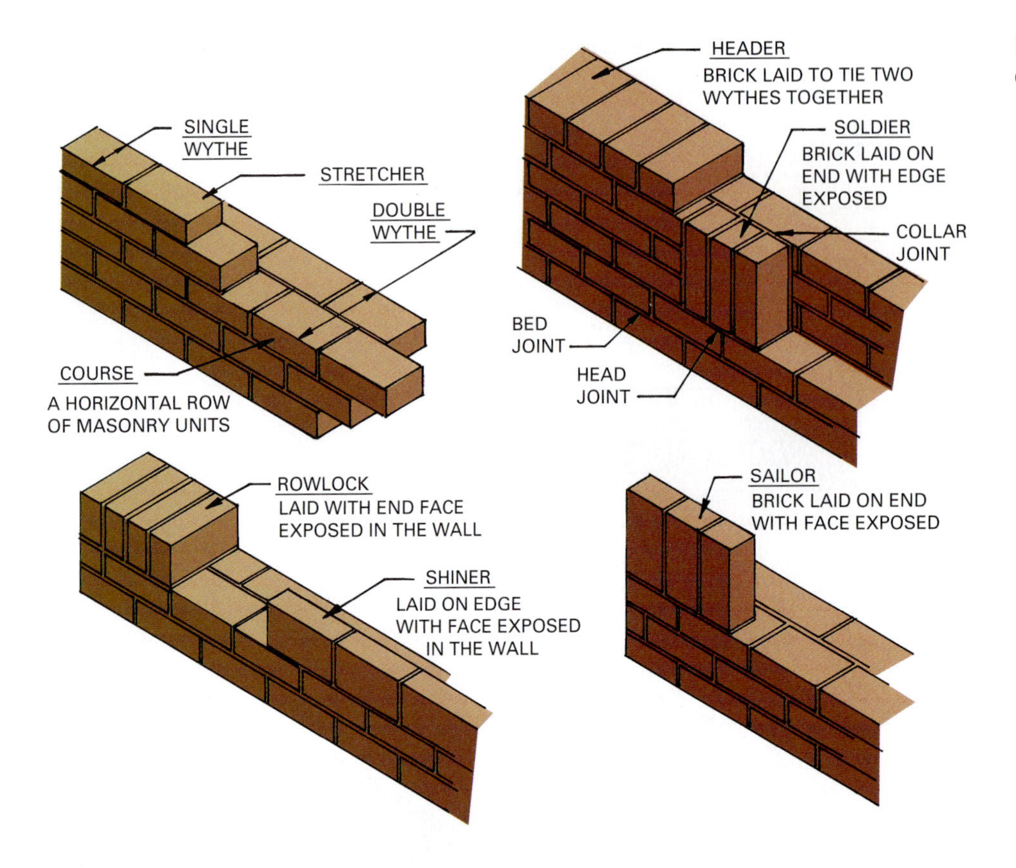

Figure 14.19 Terms used to describe brick positions in a wall.

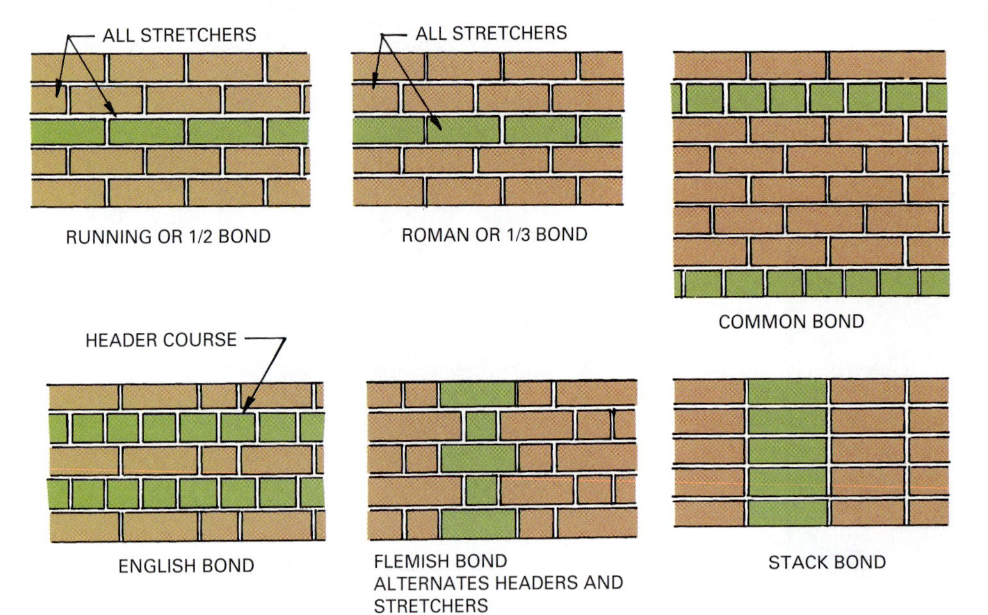

Figure 14.20 Brick bond patterns used with standard and oversize brick masonry units.

carefully in placing the bricks and maintaining a uniform mortar line. Because bricks are laid from each corner to the center of a wall, closure is made by placing the final brick in the last opening. Each course must be carefully executed to assure that the closure brick can be laid without requiring its being cut.

After the mortar has set sufficiently, the joints are tooled. Tooling forms the joints, as shown in Fig. 14.21. Various-shaped tooling tools are used to shape the joints. After tooling, the wall is lightly scrubbed with a fiber brush to remove loose mortar particles. Some weeks after the wall has cured, it is usually washed with muriatic acid (HCl) and thoroughly rinsed with clean water. This removes any remaining mortar stains on the brick face.

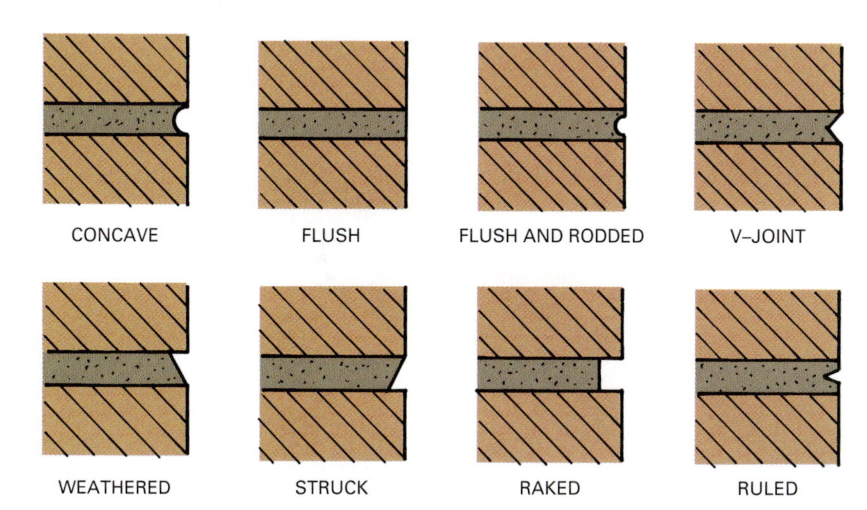

Figure 14.21 Commonly used mortar joints.

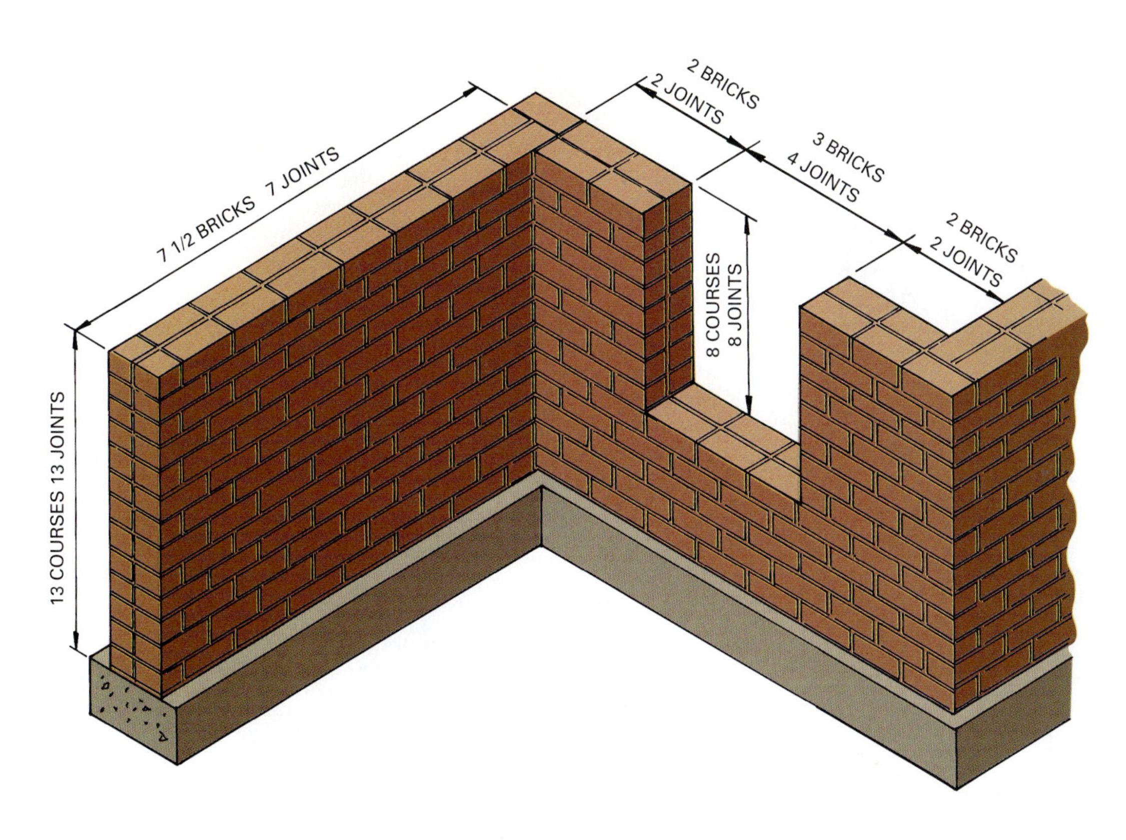

Figure 14.22 Masonry wall lengths are designed to fit the module of the masonry units, reducing the need to cut individual units.

MASONRY ARCHES

Masonry **arches** are a form of curved construction used to span an opening, usually consisting of wedge-shaped blocks called *voussoirs*. When rectangular blocks are used, the mortar between them assumes a wedge shape. Arches vary, from flat horizontal openings, through semicircular and semielliptical forms, to pointed configurations.

A masonry arch may be constructed using soldier courses, two or three rowlock courses, or alternating row-lock and soldier courses (Fig. 14.25). The block arch uses compressive strength of brick or stone to span an opening and carry imposed loads. The horizontal thrust generated is resisted by the masonry mass of the wall against which it butts. Two equal arches close together provide opposite but equal thrust, thus resisting the horizontal forces.

Figure 14.23 A typical procedure for laying a masonry wall.

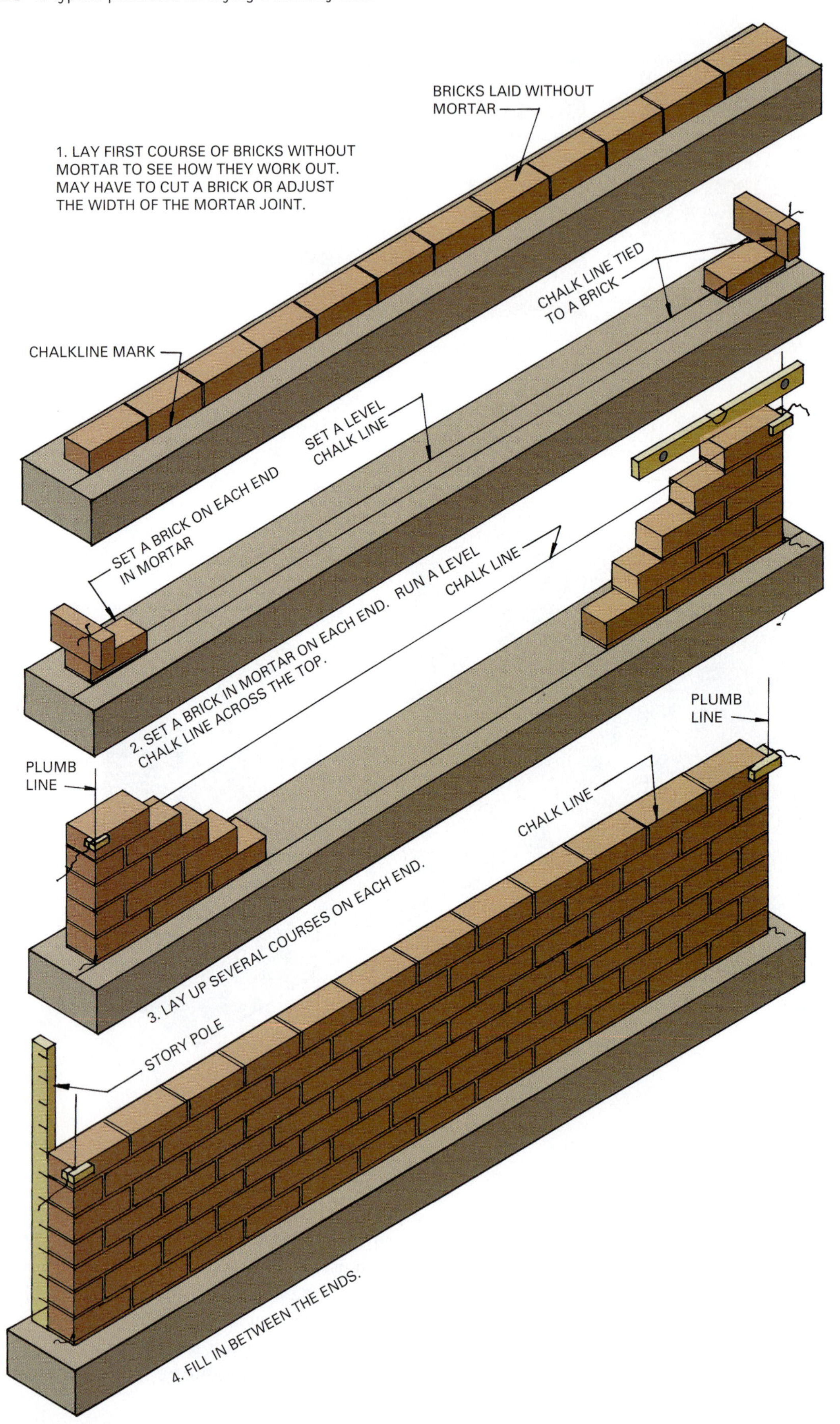

Figure 14.24 The basic steps for laying bricks.

1. Lay a trowel of mortar on the top of the bricks and spread evenly.

2. Cut off mortar projecting over the edges of the bricks.

3. Lay mortar on the end of a brick.

4. Place the brick on the mortar bed.

5. Press down and against the adjacent brick to close the head joint. Check the brick for levelness and correct height.

6. Use a level string to check alignment.

Figure 14.25 An arch formed by triple rowlock courses of brick. *(Courtesy Eva Kultermann)*

The technical terms used to describe the parts of arches are given in Fig. 14.26. Types of arches include Roman, Gothic, elliptical, segmental, Tudor, and jack (Fig. 14.27).

Figure 14.26 Terms used to identify parts of an arch.

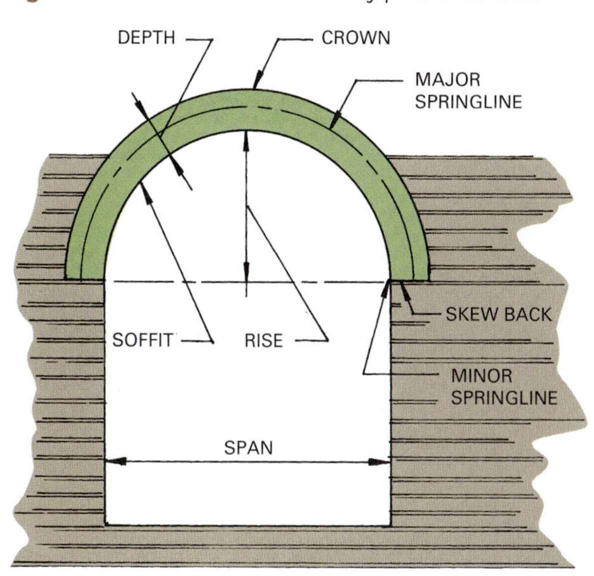

CONCRETE MASONRY CONSTRUCTION

Some of the frequently used concrete masonry wall constructions are shown in Fig. 14.28. Since the composition of materials used to make concrete masonry varies, their compression strength, fire resistance, coefficient of thermal expansion, and other properties must be known so the designer can specify the type of construction required.

Construction details for load-bearing concrete masonry foundations are shown in Fig. 14.29. Interior concrete masonry walls may be load-bearing or non-load-bearing, as shown in Fig. 14.30. Floor joists may rest on top of a concrete masonry wall, be set into it, or be supported by ledgers bolted through the wall (Fig. 14.31). Wood joists must bear at least 4 in. on the foundation. Some examples of roof construction are shown in Fig. 14.32. Openings are spanned using steel, precast concrete, or concrete lintel blocks, as shown in Fig. 14.33.

Concrete masonry columns are designed using standard-size blocks to reduce the need for cutting units to special lengths. The interior core is filled with concrete and specified vertical reinforcing. Customized concrete-masonry-column units with hollow centers are available in some areas. Pilasters are built in the same manner as columns but form part of the wall. Pilasters are used to stabilize long concrete masonry walls and provide a larger bearing surface for beams and girders resting on a wall. They are spaced along a wall at specified intervals and may have a reinforced concrete core.

Figure 14.27 Types of masonry arches.

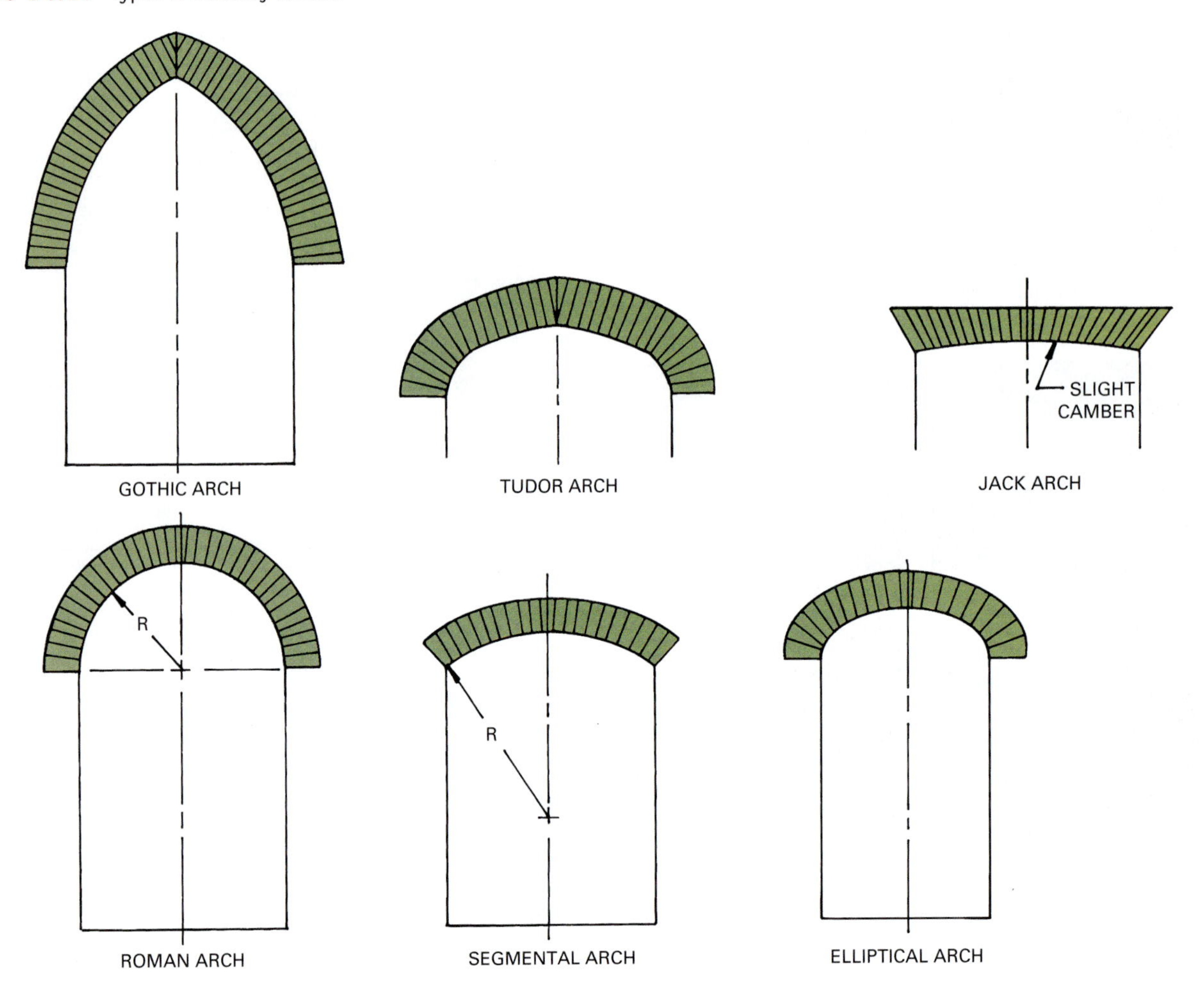

BUILDING A CONCRETE MASONRY WALL

When possible, a designer sets foundation wall lengths so standard-size concrete masonry units can be used. These includes full blocks (16 in.) and half blocks (8 in.). A $\frac{3}{8}$ in. or $\frac{1}{2}$ in. mortar joint is common. Figure 14.34 shows a typical example. The process for laying block is basically the same as for brick. First, blocks without mortar are spaced along a footing to establish spacing and to determine if some blocks need cutting.

The basic steps for laying up a concrete masonry wall are shown in Fig. 14.35. The mason begins by laying a full mortar bed on the footing and laying up each corner several courses high. The height of the courses is checked with a story pole. A chalk line is run the length of the wall to establish the level height of the first course, which is laid on a full mortar bed. Courses above it are usually laid without placing mortar on the webs. Mortar is placed on the face shells, and each course is laid from the corners to the center of the wall. As the blocks meet, closure must be made, as shown in Fig. 14.36. The walls are checked for level and plumb with long mason's levels. After the mortar has set properly, joints are tooled to the desired contour, and mortar crumbs are scrubbed off with a soft brush.

STONE MASONRY CONSTRUCTION

Stone laid in mortar may be rubble or ashlar masonry. **Rubble** refers to stone found naturally in irregular shapes and sizes. **Ashlar** masonry uses units cut into squared shapes. Stone may be laid in a random or coursed pattern. The coursed pattern maintains horizontal lines, while the random pattern uses irregular courses (Fig. 14.37). One type of rubble stone masonry,

Figure 14.28 Concrete masonry wall constructions.

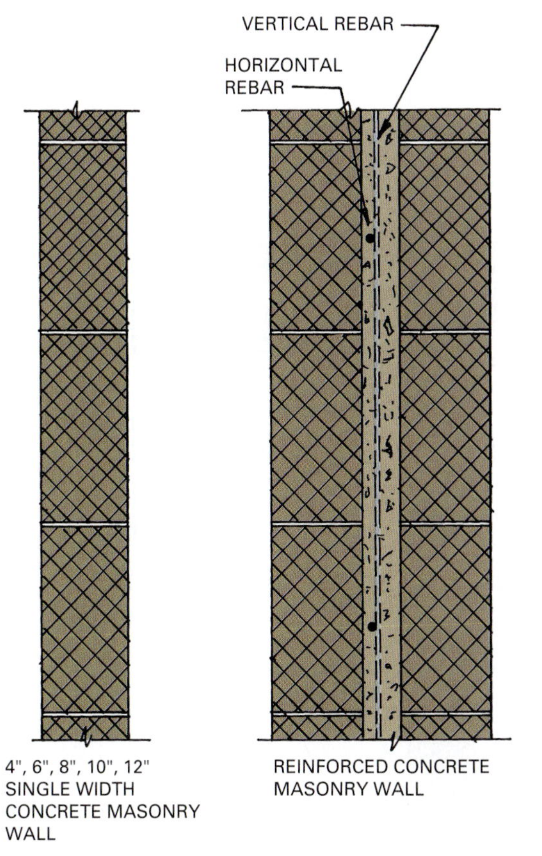

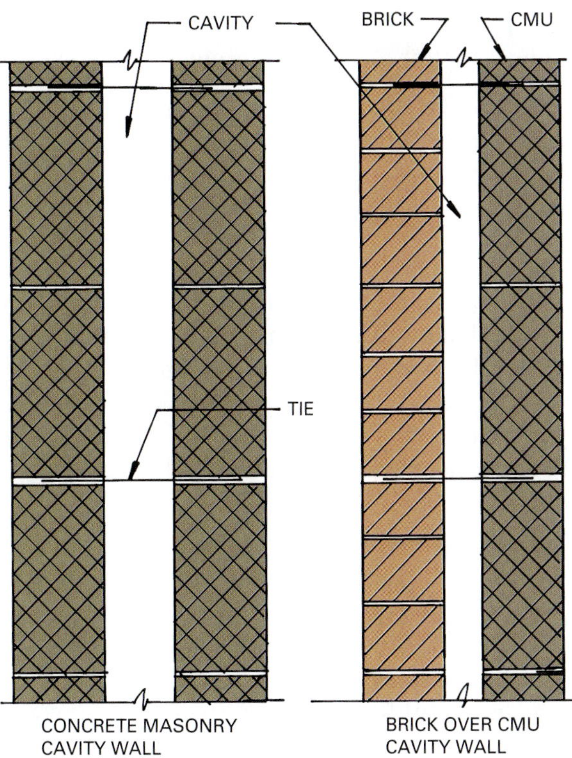

Figure 14.29 Concrete masonry units are widely used for foundation construction.

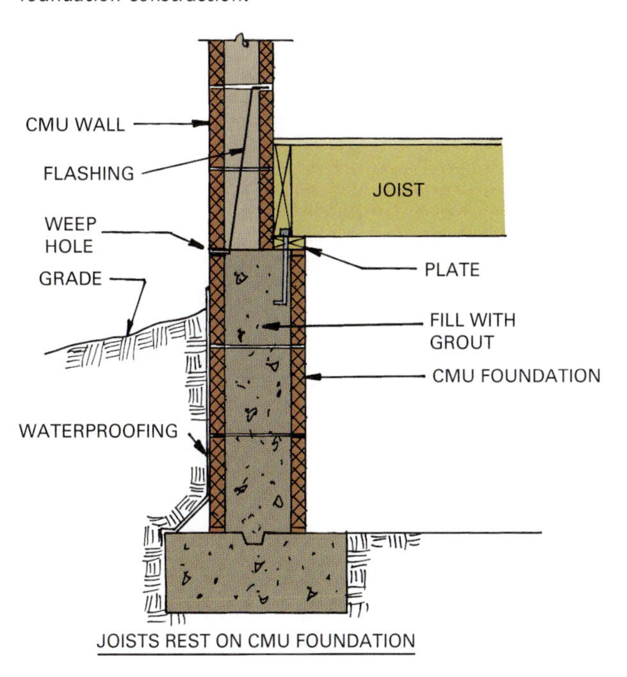

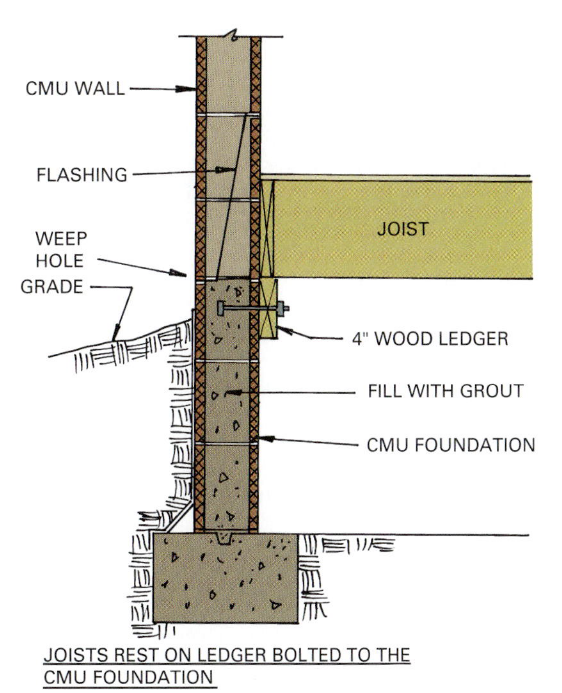

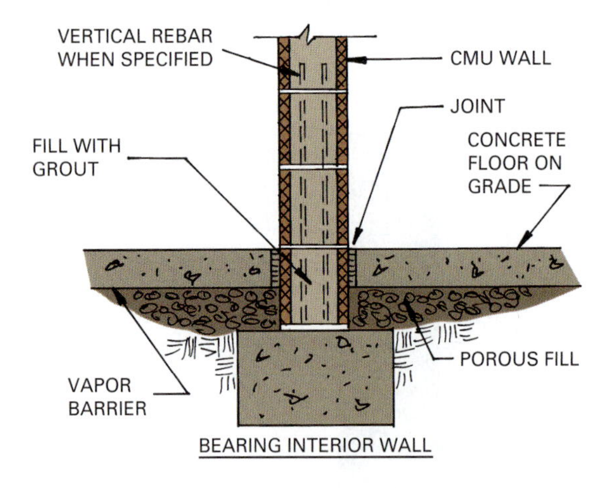

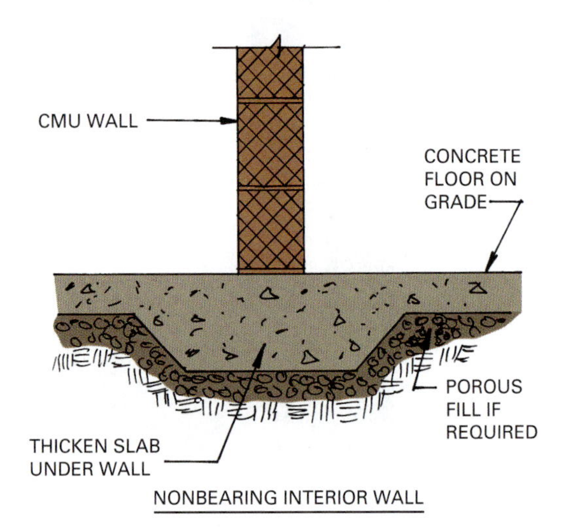

Figure 14.30 Load-bearing and non-load-bearing interior concrete block wall details.

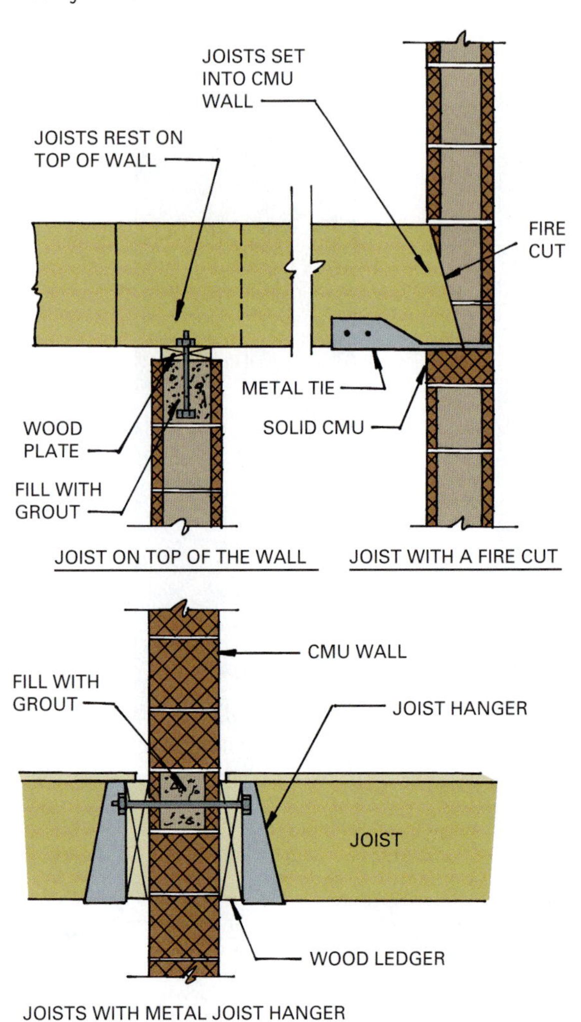

Figure 14.31 Ways to support floor joists using concrete masonry foundations and walls.

ledgerock, is a rock formed naturally into thin layers of varying thicknesses.

When stone is laid, the grain in each unit should run in a horizontal direction. The stone is stronger in this position and tends to better resist weathering. Rubble stone is more difficult to lay because a mason must select pieces that fit open spaces and work in with the shapes of adjoining stones. Sometimes it is necessary to trim stone with a hammer. The resulting mortar joints are irregular and often quite large. Squared ashlar stone is easier to lay than rubble but requires the mason to select stones that will fit together as needed. Since the stones are squared, the mortar joint can be of relatively uniform width over the entire wall. Joints in ashlar masonry are usually $\frac{3}{8}$ to $\frac{3}{4}$ in.

Large squared blocks are called *dimension stone*. They usually require a hoist to set them in place.

Mortar joints in ashlar masonry are usually raked after setting and allowed to set thoroughly. Then the raked space is filled with mortar (called *pointing*) and tooled to the specified shape. Great care is taken to avoid getting mortar on the face of the stone. The face is cleaned with a mild soap and soft brush and flushed clean with water.

Stone masonry is usually applied as a veneer over a backup wall, composed of concrete masonry units, for example. Metal ties are used to bond the stone veneer to the backup wall (Fig. 14.38), much the same as with brick masonry construction. Anchors should be chromium-nickel, stainless steel, or a zinc alloy. Building codes often ban the use of galvanized steel. Copper, brass, and bronze ties may cause some staining. Construction details for several types of stone masonry veneered wall construction are shown in Fig. 14.39.

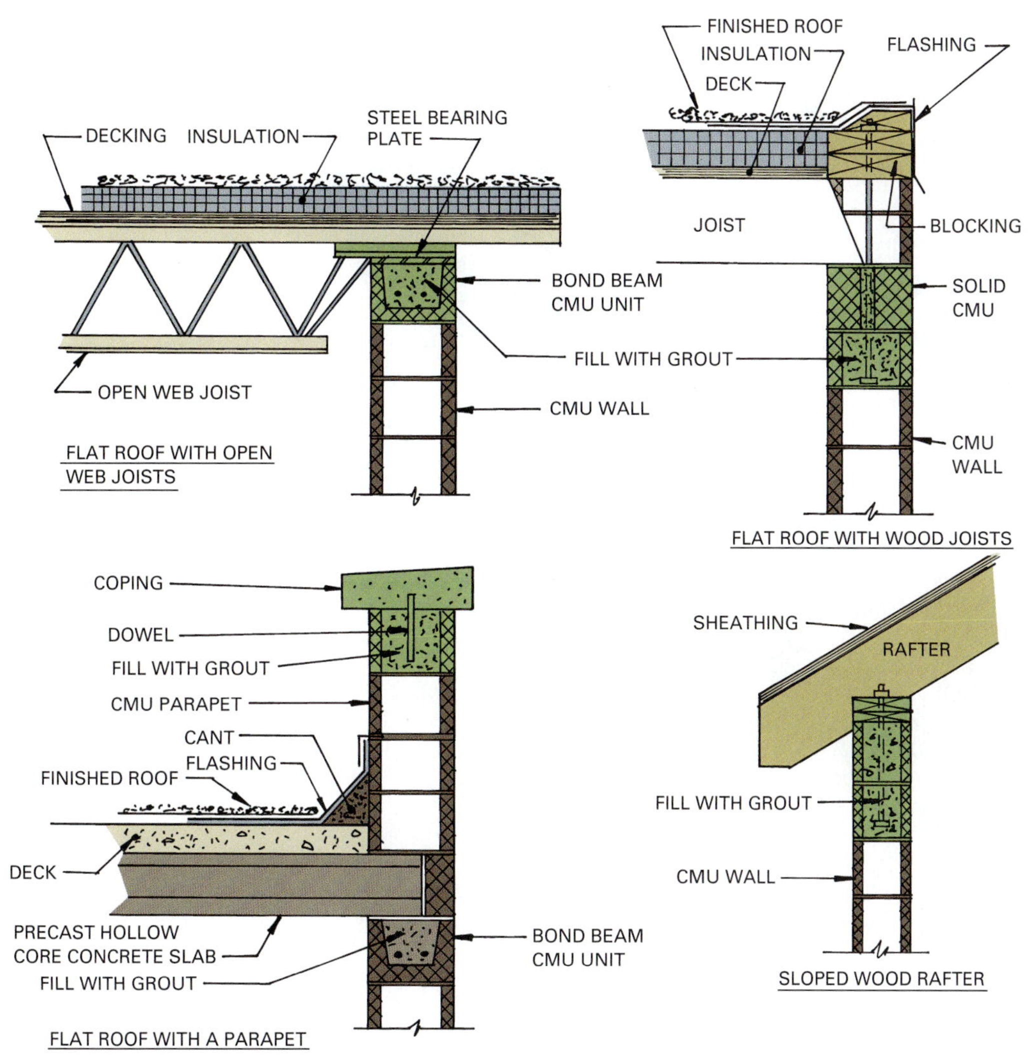

Figure 14.32 Typical roof details for concrete masonry wall construction.

Some stone is cut into large, thin panels that are scored to accept some form of metal tie that secures them in place to a backup wall. The manufacturers of various types of stone panels have engineered systems for securing panels in place. A number of connection methods are used. The first veneer panel on a base can be set on the foundation or other supporting masonry, or the panel can overhang the supporting structure and be secured with metal connectors, as shown in Fig. 14.40. As panels are placed up the wall, a metal connector is used to secure them to the backup wall. Typically, metal connectors tie the stone veneer to a column. Preassembled column covers are available bonded with high-strength epoxy adhesives.

Stone veneered walls usually use stone windowsills (Fig. 14.41) and various types of decorative moldings (Fig. 14.42), balusters (Fig. 14.43), and modillions and consoles. A modillion is a scroll supporting the corona (overhanging member of a cornice) under a cornice. A console is a decorative vertical scrolled bracket that projects from a wall to support a cornice, door, or window. Balusters are vertical members in a stair rail used to support the handrail.

STRUCTURAL CLAY TILE CONSTRUCTION

Structural clay tile may be load-bearing or non-load-bearing. Load-bearing units may form an unfaced wall, or they may be used as a backup wall and faced with another material. Non-load-bearing tile walls are used as interior partitions, fireproofing barriers, and masonry screens.

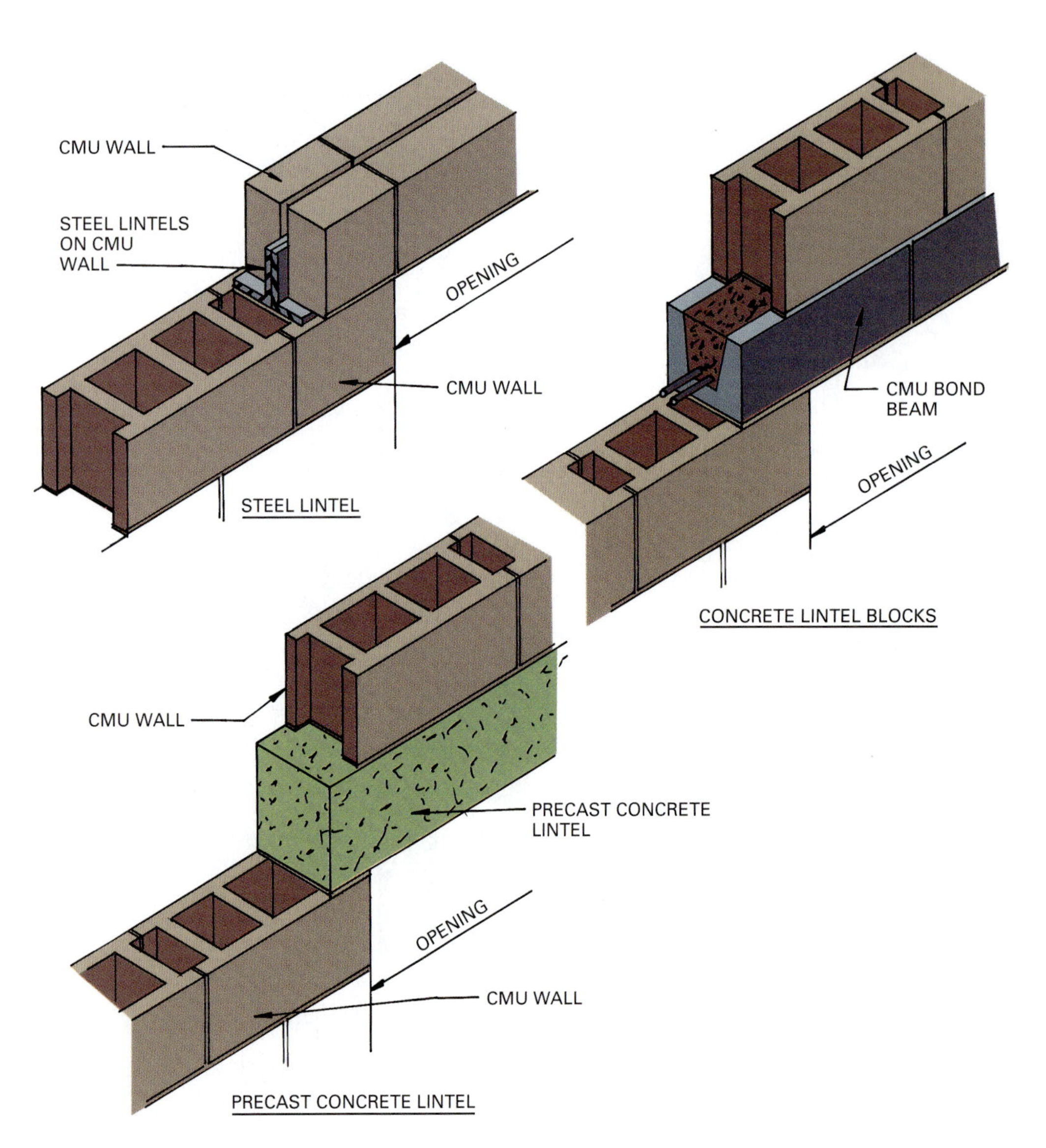

Figure 14.33 Lintels used in concrete masonry construction.

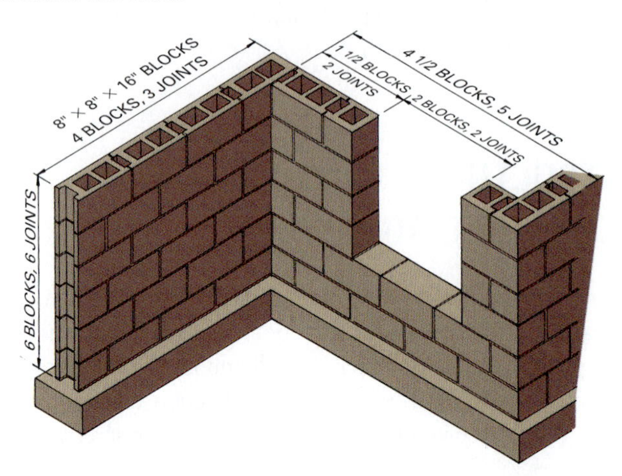

Figure 14.34 Concrete masonry walls are designed to use standard-size units.

Examples of commonly used structural tile construction are shown in Fig. 14.44. It should be noted that there are many possible variations, depending upon expected loads, and desired surface characteristics. Some tile units are utility block used to construct backup walls only, while others have glazed exposed faces available in a variety of colors and textures.

Cavity walls of structural tile use flashing and weep holes as described for brick construction. They may be insulated with rigid plastic foam insulation. Floor joists and rafters are also installed much like those on brick and concrete masonry. Structural clay tile can be used as a veneer over a backup wall of a different material, such as concrete masonry units or stud walls. Openings in structural clay tile are usually spanned using reinforced tile lintel units.

Figure 14.35 Steps for laying concrete masonry units. *(All photos courtesy Portland Cement Association)*

1. Lay the first course on a full mortar bed.

2. The blocks for the first corner are laid.

3. The corner blocks are laid up several courses.

4. The height of each course is checked with a story pole.

5. Blocks are laid up on each corner and worked toward the center of the wall.

6. The first course has been laid across the footing and the corners have been raised several courses. The mason is laying the second course.

Figure 14.36 Procedure for closure of a course of concrete blocks. *(Courtesy Portland Cement Association)*

1. A level is used to check the wall for plumb.

2. A level is used to check the wall for level.

3. The mortar is tooled to the desired contour when it has properly set.

4. Mortar crumbs are removed with a soft brush after tooling is finished.

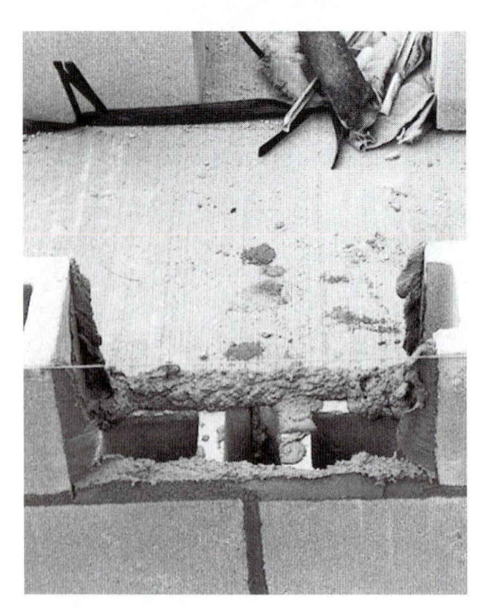

5. The closure opening has mortar buttered on all sides.

6. The closure block has mortar on the ends and is slid into the opening. Excess mortar is cut away.

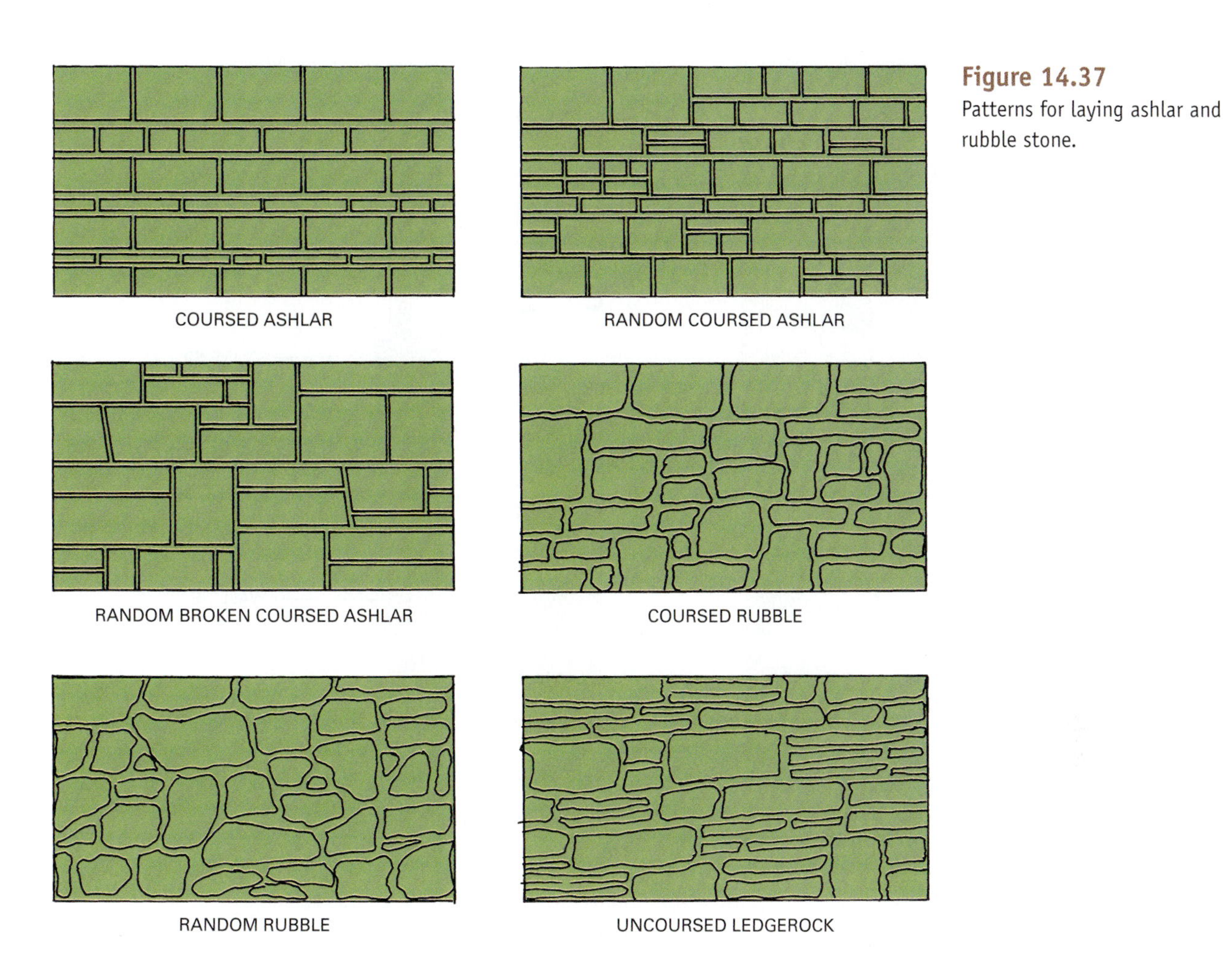

Figure 14.37 Patterns for laying ashlar and rubble stone.

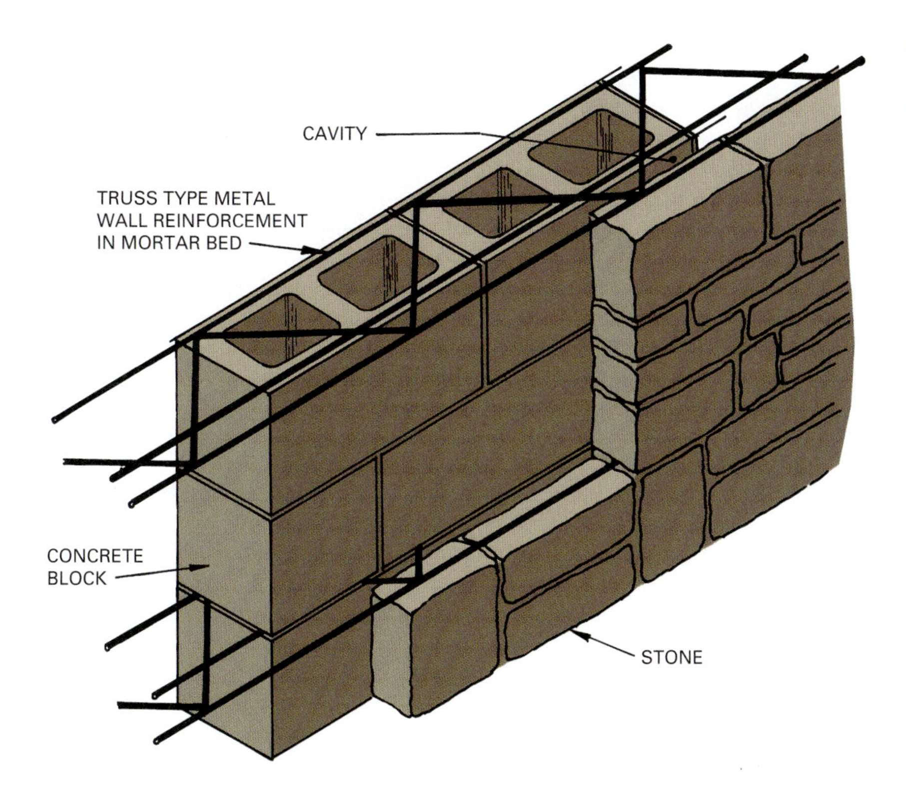

Figure 14.38 Stone masonry veneer is laid over a backup wall and connected to it with metal ties.

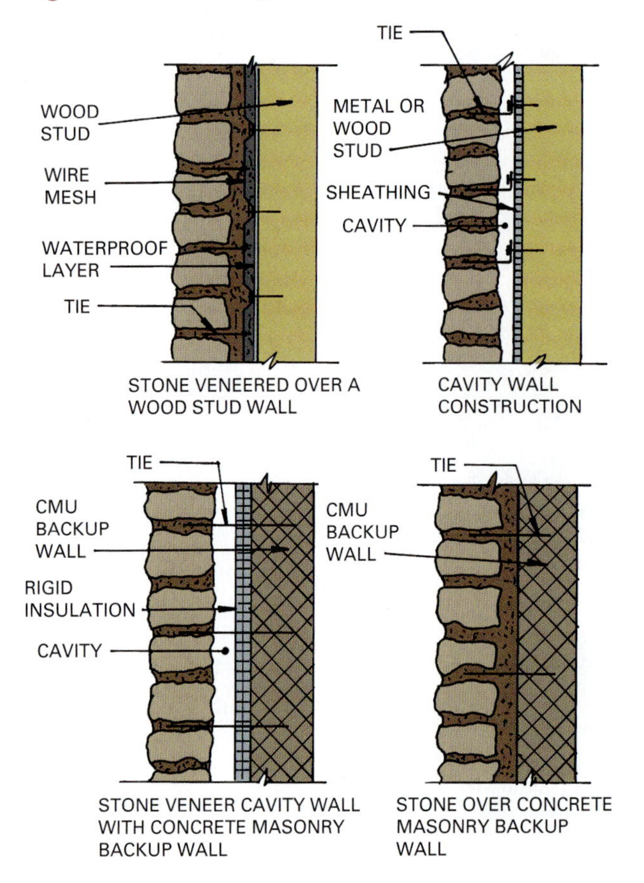

Figure 14.39 Some typical stone-veneer wall constructions.

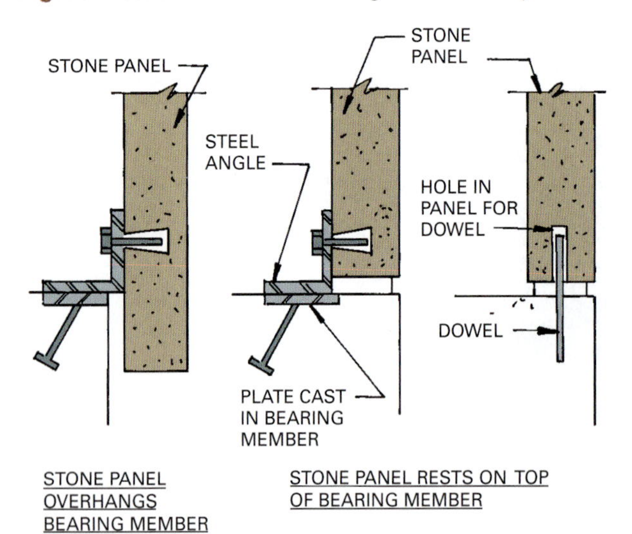

Figure 14.40 Details for anchoring vertical stone panels.

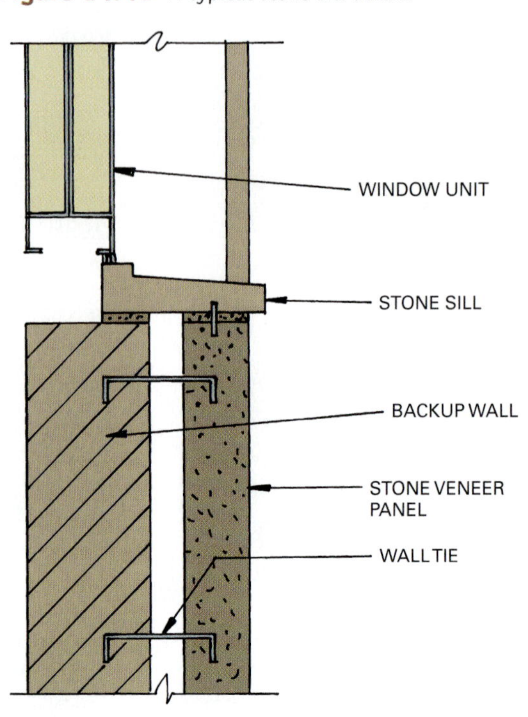

Figure 14.41 A typical stone sill detail.

Figure 14.42 Examples of moldings and trim that can be produced in stone.

Figure 14.43 Stone balusters can be made in a variety of profiles. *(Image copyright AGITA, 2009. Used under license from Shutterstock.com)*

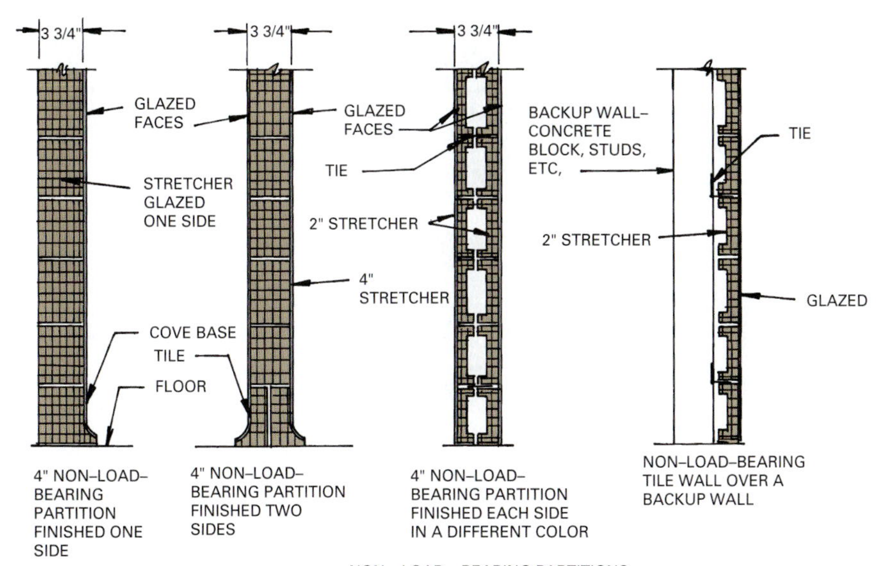

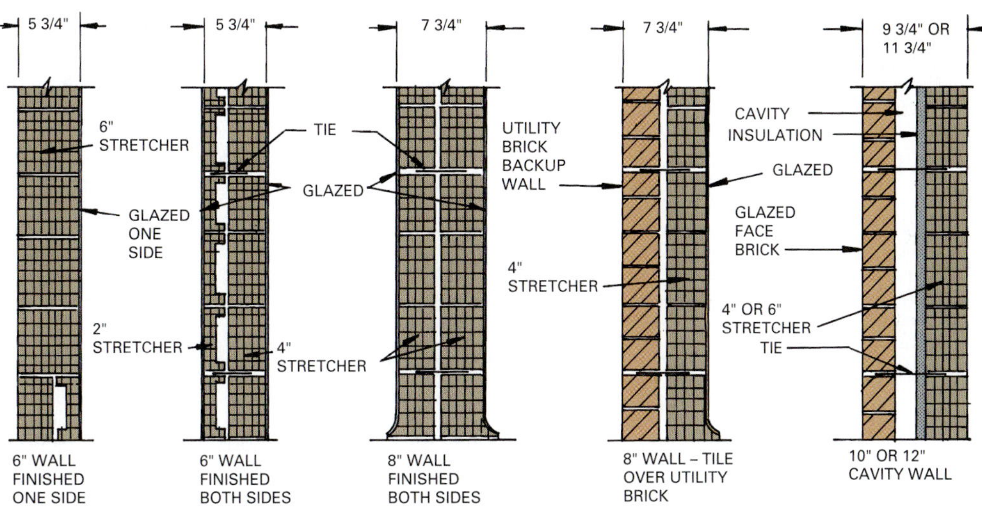

Figure 14.44 Typical wall and partition construction using structural tile.

BUILDING A STRUCTURAL CLAY TILE MASONRY WALL

As with brick and concrete masonry, structural clay tile walls are designed to use standard-size units that avoid custom cutting. The face sizes of a standard stretcher block are $5\frac{1}{16}$ in. high by $11\frac{3}{4}$ in. long ($5\frac{1}{2}$ in. nominal) and $7\frac{3}{4}$ in. high by $15\frac{3}{4}$ in. long (8×16 in. nominal). Buildings are often designed to meet these sizes, although other special units are also available.

High-strength mortars are used with structural tile. Local codes and information from tile manufacturers are used when specifying mortar for structural clay tile. Cavity walls utilize metal ties to integrate tile with its backup material, similar to how brick and concrete masonry are handled.

Corners can be built in several ways. Typical corner constructions and sills at openings can be fashioned using special bull-nose units, as shown in Fig. 14.45.

Figure 14.45 Several ways to construct corners and openings with structural tile.

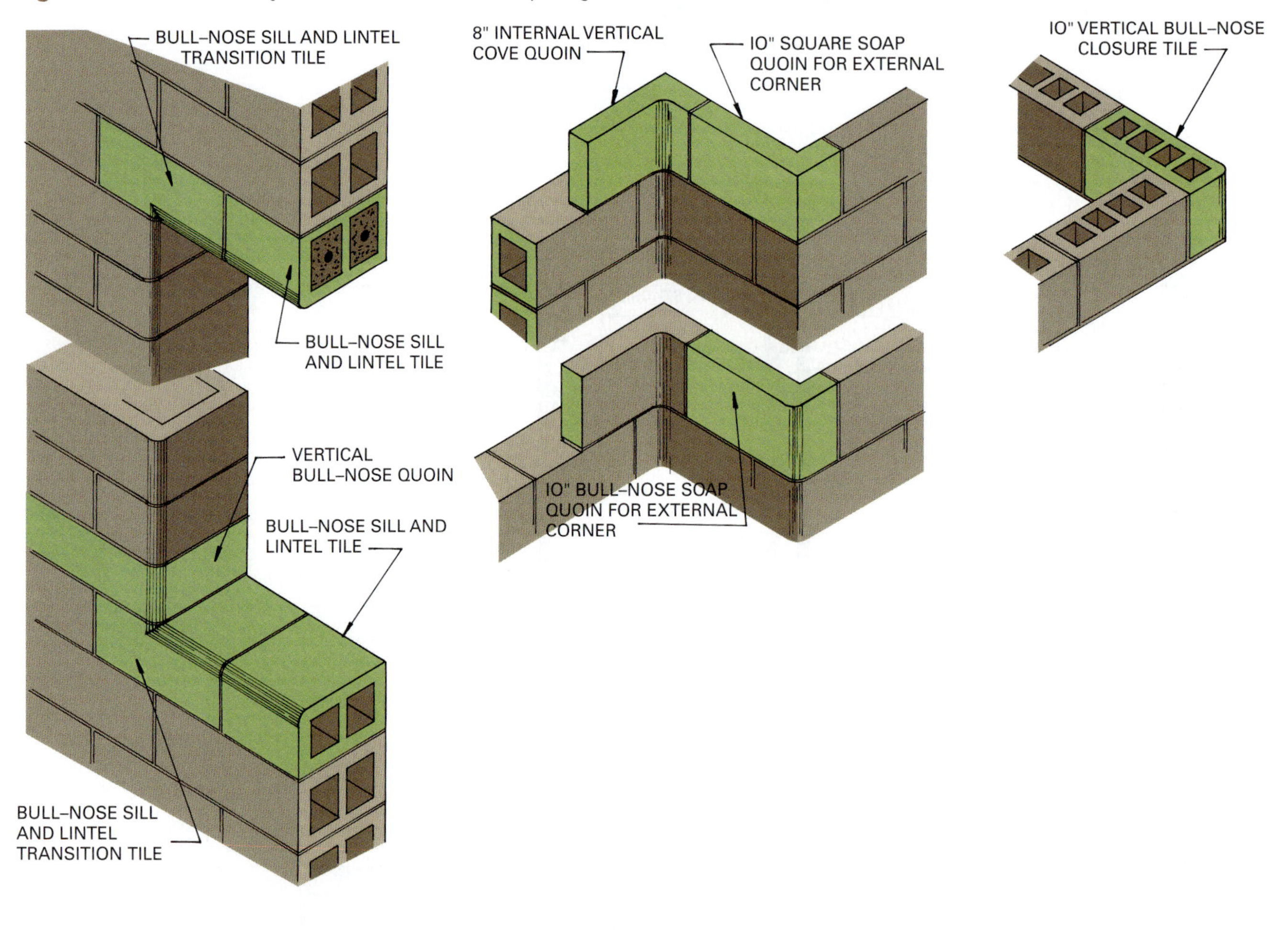

Review Questions

1. What are the commonly used types of masonry load-bearing walls?
2. Why are control and expansion joints required in masonry walls?
3. Why are weep holes used?
4. What can be done to reduce moisture penetration through masonry walls?
5. What types of metal flashing can be used with masonry wall construction?
6. How are masonry units supported over openings in a wall?
7. How can the heat and cold conduction of masonry walls be reduced?

8. What are the commonly used brick patterns used in wall construction?
9. What are the advantages and disadvantages of the various types of mortar joints?
10. What size mortar joint is commonly used?
11. What is the difference between rubble and ashlar stone used for wall construction?
12. What types of structural clay tile are used for wall construction?

Key Terms

arch
ashlar stone
brick
brick bond
brick veneer
cavity wall
concrete masonry unit
console
control joint
corona
expansion joint
lintel block
modillion
pilaster
reinforced masonry
rubble
story pole
structural clay tile
trowel
weep hole

Activities

1. Check your local building code and make a record of requirements pertaining to the construction of masonry load-bearing walls.
2. Using standard 8 × 8 × 16 in. concrete blocks, design a building so it will be approximately 10 ft. wide and 25 ft. long. Size the walls so it is not necessary to cut any concrete blocks. If the wall should be about 10 ft. high, how many blocks will it take to lay the back and one side?
3. Secure a quantity of clay bricks and mortar mix. Prepare a story pole using $\frac{1}{2}$ in. mortar joints. Lay up a length of wall about 6 ft. long and 4 ft. high. Try making several types of finished mortar joints.
4. Collect samples of all the masonry units you can find. Most clay brick manufacturers have sample boards with thin sections of the bricks they make glued onto a strong backing.
5. Pick an exterior wall of a brick building and measure its width and height. Calculate the number of standard and oversize bricks in the wall. Get the local cost of bricks and figure the cost of the masonry units in this wall. Refer to Chapter 11 for sizes of masonry units.

Additional Resources

ASTM Standards on Masonry, American Society for Testing and Materials, West Conshohocken, PA.

Beall, C., *Masonry Design and Detailing*, the Masonry Society, Boulder, CO.

Chrysler, J., and Amshein, J. E., *Reinforcing Steel in Masonry*, the Masonry Society, Boulder, CO.

Concrete Masonry Handbook, Portland Cement Association, Skokie, IL.

Jaffe, R., *Masonry Basics*, and many other technical publications, the Masonry Society, Boulder, CO.

Masonry Code and Specifications, the Masonry Society, Boulder, CO.

Masonry Designers Guide, the Masonry Society, Boulder, CO.

NCMA Masonry Design Software, National Concrete Masonry Association, Herndon, VA.

See Appendix B for addresses of professional and trade organizations and other sources of technical information.

DIVISION

05

Metals
CSI MasterFormat™

052000 Metal Joists
053000 Metal Decking
054000 Cold-Formed Metal Framing
055000 Metal Fabrications
057000 Decorative Metal

(Image copyright Dwight Smith, 2009. Used under license from Shutterstock.com)

CHAPTER 15

Ferrous Metals

LEARNING OBJECTIVES

Upon completion of this chapter, the student should be able to:

- Explain the processes for mining and processing iron ore and for producing pig iron and steel.
- Develop knowledge of the properties of ferrous metals to consider when making material selection decisions.
- Be familiar with the various steel identification systems and the Unified Numbering System for Metals and Alloys.

Build Your Knowledge

For further information on these materials and methods, please refer to:

Chapter 3 Properties of Materials
Topic: Mechanical and Chemical Properties
Chapter 8 Cast-in-Place Concrete
Topic: Concrete Reinforcing Materials
Chapter 17 Steel Frame Construction
Chapter 20 Wood and Metal Light-Frame Construction
Topic: Steel Framing
Chapter 28 Roofing Systems
Chapter 30 Doors, Windows, Entrances, and Storefronts
Chapter 36 Interior Walls, Partitions, and Ceilings

Ferrous metals are those in which the chief ingredient is the chemical element iron (ferrum). Iron (chemical symbol Fe), mixed with other minerals, is found in large quantities in the earth's crust. To be useful, iron must be extracted from mined ore, have impurities removed and ingredients added to alter its properties, and then be formed into usable products.

Ferrous metal products are widely used in the construction industry. They are a major construction material, and architects, engineers, and contractors should be familiar with the various types, their properties, and the proper applications for each. Although a ferrous metal product may fail, this usually does not occur because it is a poor material but because the type of ferrous metal chosen for a particular application was incorrect. When the properties of a ferrous metal are known, its performance can be accurately determined during the engineering design process.

IRON

Iron is found in large quantities in the earth's crust. Pure iron, free from impurities and other elements, is ductile and soft but generally not strong enough for structural purposes. It has good magnetic properties but oxidizes (rusts) easily and does not resist attack by acids and some chemicals. For commercial purposes, iron must have **alloying elements** added to improve its characteristics. Iron can be hardened by heating and rapid cooling, and can be made more workable by **annealing**, that is, heating it then allowing it to cool slowly.

The typical commercial iron product is called *pig iron*. It contains 3 to 5 percent carbon and traces of other elements, such as manganese, sulfur, silicon, and phosphorus. Pig iron is the base material used to produce various types of iron and steel. Iron is used for a wide range of purposes in construction. It is the main ingredient in cast iron, steel, stainless steel, and iron alloys. Iron particles may be used as an abrasive for sandblasting and sometimes as aggregate for specialized concretes. They also form the basis for some color pigments.

Construction Techniques

History of Ferrous Metals

Ferrous metals had little use in construction until the late eighteenth century, when cast iron found limited use as a structural material. By the early nineteenth century, wrought iron and cast iron were used for structural purposes—but not extensively, because cast iron was brittle, and the production of wrought iron was limited and expensive.

A breakthrough occurred in the 1850s when the Bessemer process for removing impurities from molten iron was developed. Among the steps in the purification process was the reduction of the carbon content that makes iron brittle. The Bessemer process involved blowing an air stream into a vessel holding molten iron. This enabled steel, a greatly improved material, to be produced quickly and in large quantities, thus reducing its cost. Steel was also easy to form into useful products because of its ductile and malleable characteristics. In the late 1860s, the open-hearth steelmaking process was developed in Europe. This provided another source of steel, one of the most valuable and widely used construction materials. Other methods used to make steel include electric processes and the basic oxygen process.

Mining and Processing Iron Ore

Iron is found in rock, gravel, sand, clay, and mud. It is mined in open pit and underground mines. Common iron-bearing minerals are pyrite (FeS_2), siderite ($FeCO_3$), and hematite (Fe_2O_3), which contain up to 70 percent iron. Jasper and taconite rock contain 20 to 30 percent iron.

Surface ores are mined, loaded onto trucks, trains, and ships, and moved to blast furnaces. If the iron content of the ore is less than 50 percent, it is too costly to ship any distance. These lower-content ores are processed on-site by a **beneficiation** process that removes some of the unwanted elements, leaving an ore with a high iron content. Beneficiation involves grinding ore to remove unwanted elements and then increasing the size of the ground particles by a process called *agglomeration*.

Agglomeration involves pelletizing the ore particles, which are then easier to ship than finely ground ore. Iron ore dust produced during this process is recycled through sintering. **Sintering** consists of fusing iron ore dust with coke and fluxes into a clinker that is high in iron content.

Ores that have very high iron content are made into pellets and briquettes containing more than 90 percent iron. These pellets are so pure they are not used in pig iron production but go directly into steelmaking. The method of producing these pure pellets is called *direct reduction*.

Producing Iron from Iron Ore

Iron ore is converted into pig iron in a blast furnace by smelting and reduction. **Smelting** is a process in which the ore is heated, permitting the iron to be separated from impurities with which it may be chemically or physically mixed. **Reduction** is a process that separates the iron from oxygen with which it is chemically mixed.

The Blast Furnace The blast furnace separates the iron from the waste materials and sinters the ore and flue dust. A very large blast furnace is shown in Fig. 15.1. It is computer-controlled, fed by a conveyor, and operates continuously.

Coal, oil, and natural gas are the commonly used fuels in a blast furnace. Most operate on coke, which is produced from coal. The coke not only provides heat to melt ore but also helps separate iron from its oxides.

A **flux** is a mineral added to molten ore in a blast furnace. The flux combines with impurities and forms a **slag** that floats on top of the molten material. Limestone and dolomite are typical basic fluxes, while sand, gravel, and quartz are acid fluxes.

Figure 15.1 This very large blast furnace is capable of producing 8,000 tons of molten iron per day. *(Courtesy Wikipedia)*

The heart of the blast furnace is a tall, cylindrical shaft, 150 to 200 ft. (45.8 to 61 m) tall, and about 30 ft. (9.2 m) in diameter at the base. It is usually smaller in diameter at the top. It is lined with a refractory brick made of a material, such as magnesia, that withstands temperatures approaching 3,000°F (1,662°C) (Fig. 15.2).

A charge of ore, hot coke (fuel), and limestone (flux) are loaded at the top. A blast of hot air is injected at the bottom. As the air works its way up through the charge, oxygen in the hot air combines with the hot coke. This causes combustion, which produces the heat necessary to melt the ore. As the gases from combustion pass through the heated ore, a chemical reaction occurs that frees iron from its oxide. The gases are removed from the top of the furnace and pass through cleaners, which retain the dust. The cleaned hot air passes through heat exchangers that in turn heat new, incoming air. A blast furnace must operate continuously.

Molten pig iron settles to the bottom of a furnace and is tapped off every few hours. At this point, it is impure and brittle, containing about 4 to 5 percent carbon. This high carbon content is what makes pig iron brittle and not as useful as steel. It has no ductility and cannot be rolled into useable products, like beams or studs. Pig iron must undergo additional processing in a steelmaking furnace to reduce carbon content. The impurities, flux, and oxygen combine to form a slag on top of the molten iron, which is periodically tapped off. Slag is used as an aggregate for concrete, loose fill, and road surfacing.

The molten pig iron is moved to furnaces called *mixers* where it is blended to equalize the chemical composition and kept in a liquid state until used. The molten iron sometimes goes into a casting machine where it is molded into ingots (pigs) and allowed to cool. In other operations, it is moved directly to basic oxygen steelmaking furnaces. In this case, energy is saved because the iron is

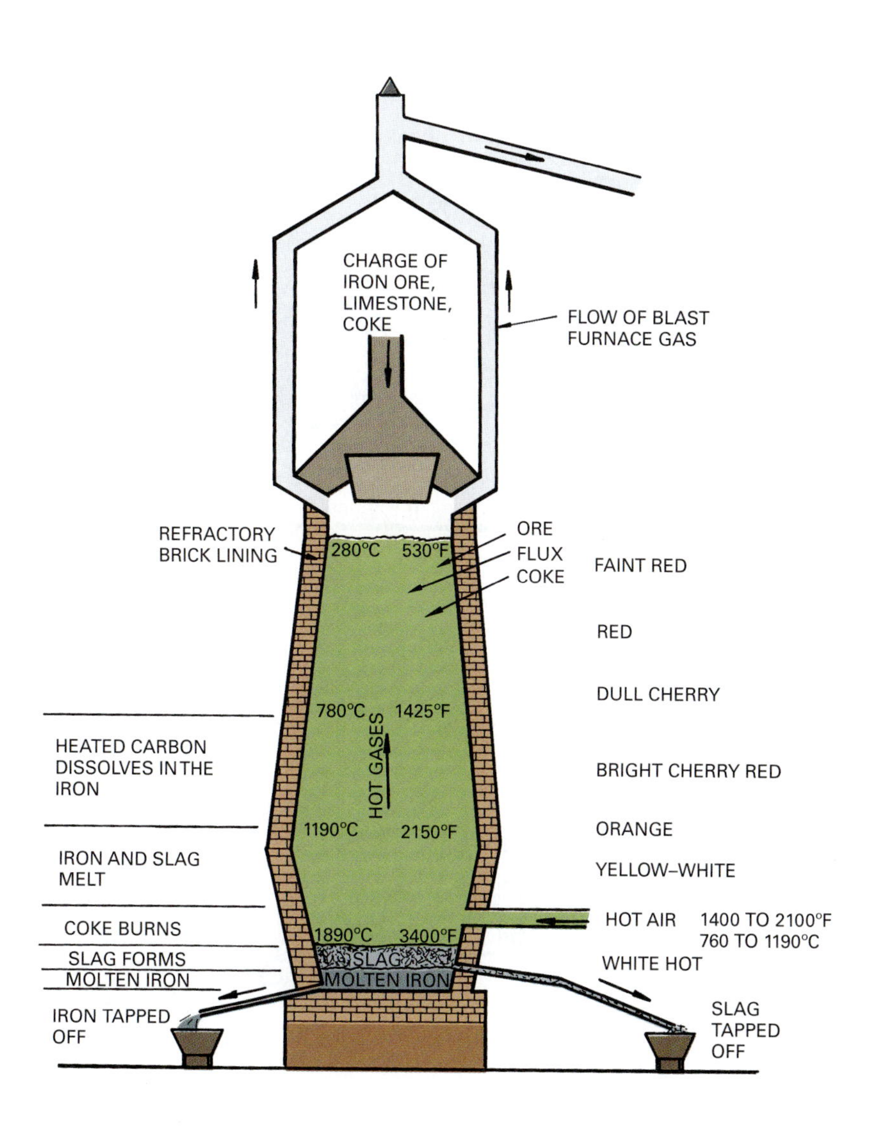

Figure 15.2 A blast furnace receives a charge at the top and hot air at the bottom, causing combustion that frees the iron from the oxide-forming pig iron.

already in molten form. A diagram of the total process is shown in Fig. 15.3.

Pig iron ingots are usually made to specifications based on their end use. The elements that are carefully controlled include carbon, sulfur, silicon, phosphorus, and manganese.

Cast Irons

Iron containing almost no carbon is identified as a wrought iron. It is a mixture of low-carbon iron and a large amount of slag. Wrought iron is soft, tough, and ductile (easily worked). Ingot iron is a very low-carbon iron that has no slag and is also tough, ductile, and soft.

Cast irons have carbon contents above 1.7 percent and include white, gray, and malleable types. White cast irons have low silicon content and are cooled rapidly. They are hard and brittle and have few applications for construction uses. Gray cast irons are produced by increasing silicon content and cooling the molten metal slowly. They are tougher and softer than white cast iron and gray in color. They may have additional elements, such as nickel, copper, and chromium. Gray cast irons are widely used for all types of castings, such as sewer pipe, ornamental railings, and decorative lamp posts.

Figure 15.3 The steelmaking process: preparing ore; processing it into steel; steel beam production. *(From "Steelmaking Flowline" by permission of the American Iron and Steel Institute)*

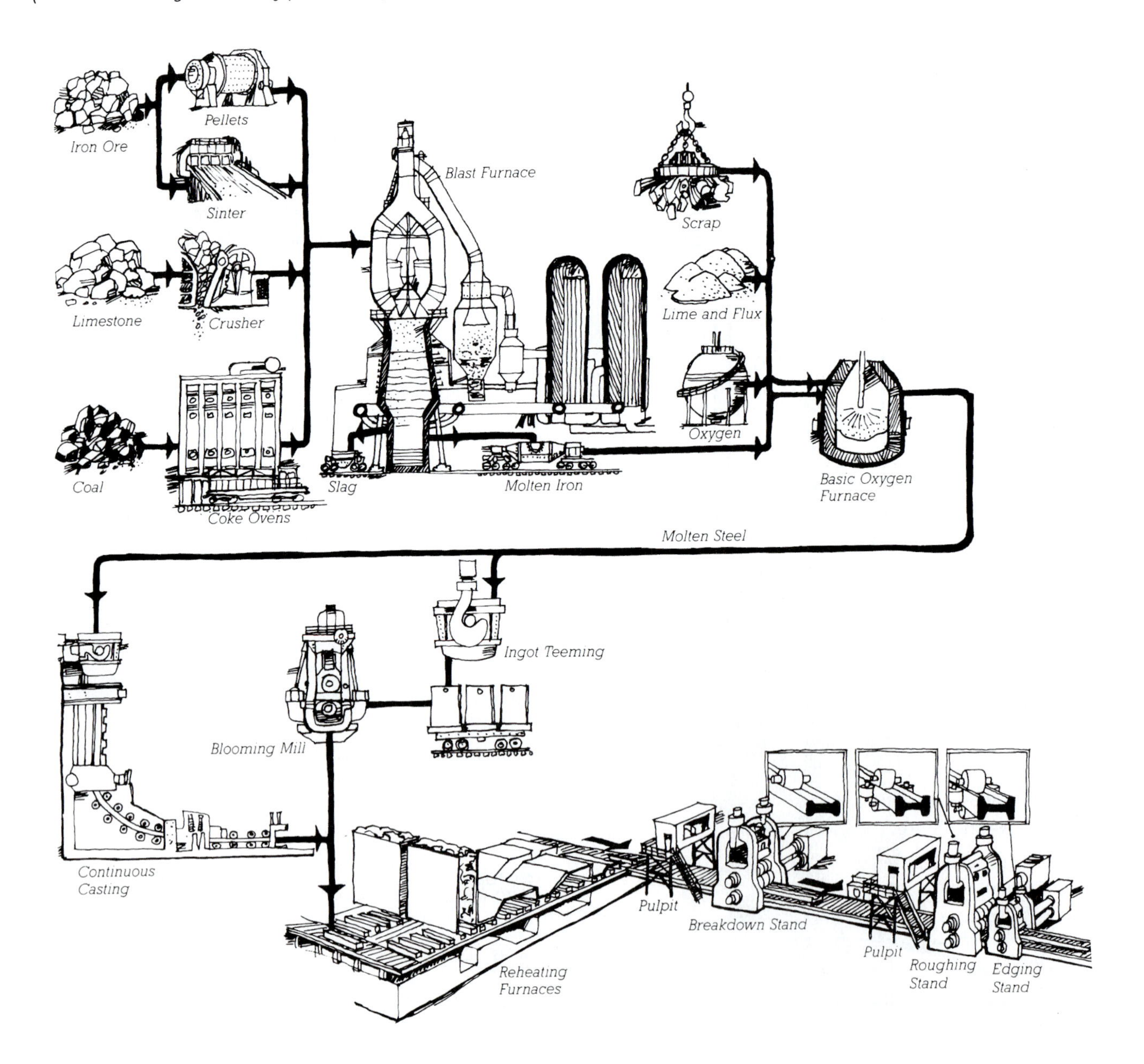

Malleability is a property of metal that allows it to be formed mechanically—by rolling or forging, for example—without fracturing. Malleable cast iron is produced by reheating white cast iron, holding the required temperature for a long time, and then cooling it slowly. It is not as brittle as white and gray cast iron because it has a lower carbon content and greater ductility. Malleable cast iron is used for hardware and other cast items requiring toughness and breakage resistance.

STEELMAKING

The production of steel entails controlling impurities in pig iron, adding alloying elements as needed to produce required properties, and lowering pig iron's carbon content. This involves combining molten pig iron, scrap metal, and fluxes in a steelmaking furnace. The molten steel is then cast into ingots or run through a strand-casting machine. Steel is produced using a basic oxygen process or with an electric furnace.

Basic Oxygen Process

The basic oxygen furnace has a large pear-shaped vessel lined with refractory material. The charge, consisting of molten pig iron, metal, scraps, and fluxes, is added at the top. A jet of high-purity oxygen is shot into the vessel through a water-cooled lance. The heat of the molten pig iron is great enough to start a chemical reaction between the oxygen and the carbon and other impurities. This oxidation produces the heat necessary to melt the charge. The slag and molten steel are tapped off. A modern basic oxygen furnace operates behind huge pollution control doors that remain closed during the steelmaking process. This enables dust and gases generated during the process to be collected by a pollution-control system. A pour of molten steel can be seen in Fig. 15.4.

Figure 15.4 Workers pour molten iron in the basic oxygen process that uses a jet of high-purity oxygen plus the heat of the molten pig iron to start the procedure. *(Image copyright Oleg - F, 2009. Used under license from Shutterstock.com)*

Electric Steelmaking Processes

High-grade steels, such as stainless, tool, heat-resisting, and alloy, are generally produced in electric furnaces. These furnaces can develop the high temperatures needed to produce required conditions. Electric furnaces use arc radiation and electric resistance to current flow to produce the needed high temperatures. High-purity oxygen, which oxidizes impurities, is injected when necessary. Since the process can be carefully controlled, there is less loss of alloying elements. The furnaces are charged with iron, scrap metal, fluxes, and alloying elements.

The two types of electric furnaces are the electric arc furnace and the induction furnace. The electric arc furnace can produce steels in large quantities. The induction furnace is used to produce smaller quantities of special grades of steel that require the use of expensive alloying elements.

The Electric Arc Furnace The electric arc furnace has a circular steel shell similar to a large cooking pot. Several cylindrical electrodes project into the furnace from the top. A high-voltage electric current is introduced through the electrodes, causing an arc to pass between them and producing the heat needed to melt the charge. The furnace tilts and the steel pours off under the slag (Fig. 15.5).

The Induction Furnace This furnace has a cylindrical vessel made of magnesia and insulated with refractory materials. Outside the vessel are windings of copper tubing through which high-voltage alternating current passes, creating an induction current resisted by the charge. This resistance generates the heat needed for melting the charge. The vessel tilts and the steel pours off below the slag.

Casting the Steel When steel is ready to be poured from a furnace, it is either cast into ingot molds or run through a **strand-casting machine** (Fig. 15.6). Ingot molds can be several feet in diameter and six to eight feet high. The molds are coated inside to prevent surface damage from steel. After an ingot begins to harden, it is removed from the mold and taken to a soaking pit where the molten steel inside solidifies. Strand-casting machines produce a continuous ribbon of steel that begins to harden as it passes through a series of rollers (Fig. 15.7). When it has hardened, it reaches a horizontal conveyor and is cut to required lengths. This process takes about thirty minutes.

Figure 15.5 White hot steel pours from a 35-ton electric arc furnace. *(Courtesy Alfred T. Palmer)*

MANUFACTURING STEEL PRODUCTS

Steel products are manufactured by rolling, extruding, cold-drawing, forging, and casting. Items produced by casting are referred to as cast steel products and all of the others are called *wrought products*. The most frequently used products are produced by hot-rolling and cold finishing. Cast steel products are made by pouring molten steel into sand molds. The molten steel goes through a process to alter its properties to suit the purpose served by the finished product.

Most wrought products are produced by hot-rolling and cold-rolling. The ingots fashioned in steel mills are used to produce slabs, blooms, and billets, which can be seen in Fig. 15.8. Slabs, blooms, and billets are semi-finished forms of steel. When semi-finished steel is fabricated directly into a finished product or into a product used to produce a finished product, it is considered finished steel.

Figure 15.6 Molten steel can be cast into a continuous slab by a strand-casting machine. *(Courtesy American Iron and Steel Institute)*

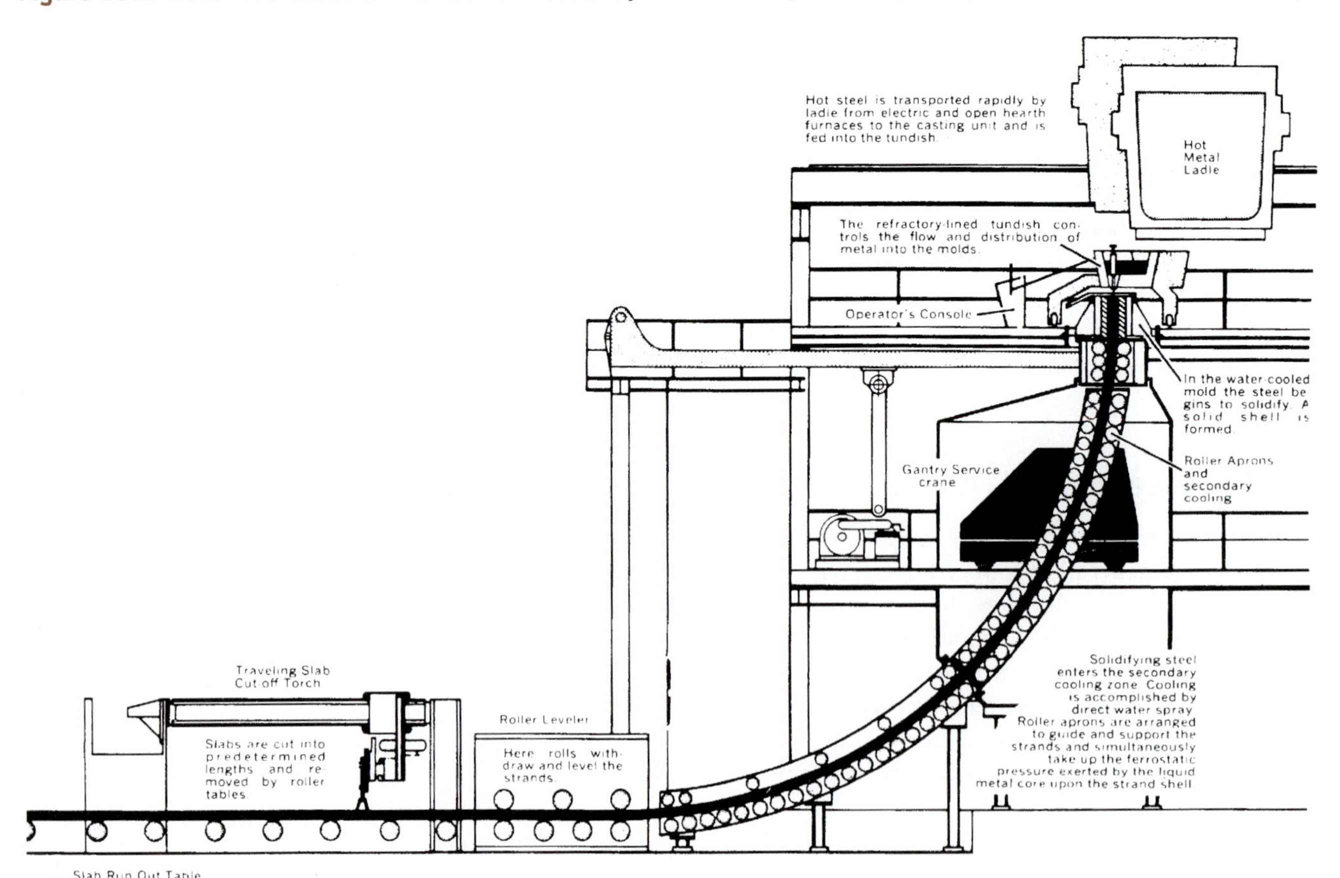

Figure 15.7 This is a steel strand produced by a continuous casting machine. The slab is cut into predetermined lengths by an automatic torch machine. *(Image copyright Oleg-F, 2009. Used under license from Shutterstock.com)*

Slabs are large rectangular semi-finished steel pieces made in a number of sizes. Their width is typically more than twice their thickness. The thickness is usually 10 in. (254 mm) or less. A bloom is an oblong of semi-finished steel with a square cross section that is usually larger than 6 in. (152 mm) square. A continuous bloom caster is shown in Fig. 15.9. Notice the size of the machine and the working environment. Billets are also oblongs of semi-finished steel, but they are smaller in cross section and longer than blooms.

Figure 15.8 lists typical finished products that use these semi-finished steel products. Notice that plates and skelp (skelp is used to produce pipe and tubing) are rolled from slabs. Finished steel plates resting on cooling beds can be seen in Fig. 15.10. The plate stock enters a four-tiered finishing stand where it is rolled to the final desired length and thickness by passing back and forth through finishing stand rollers.

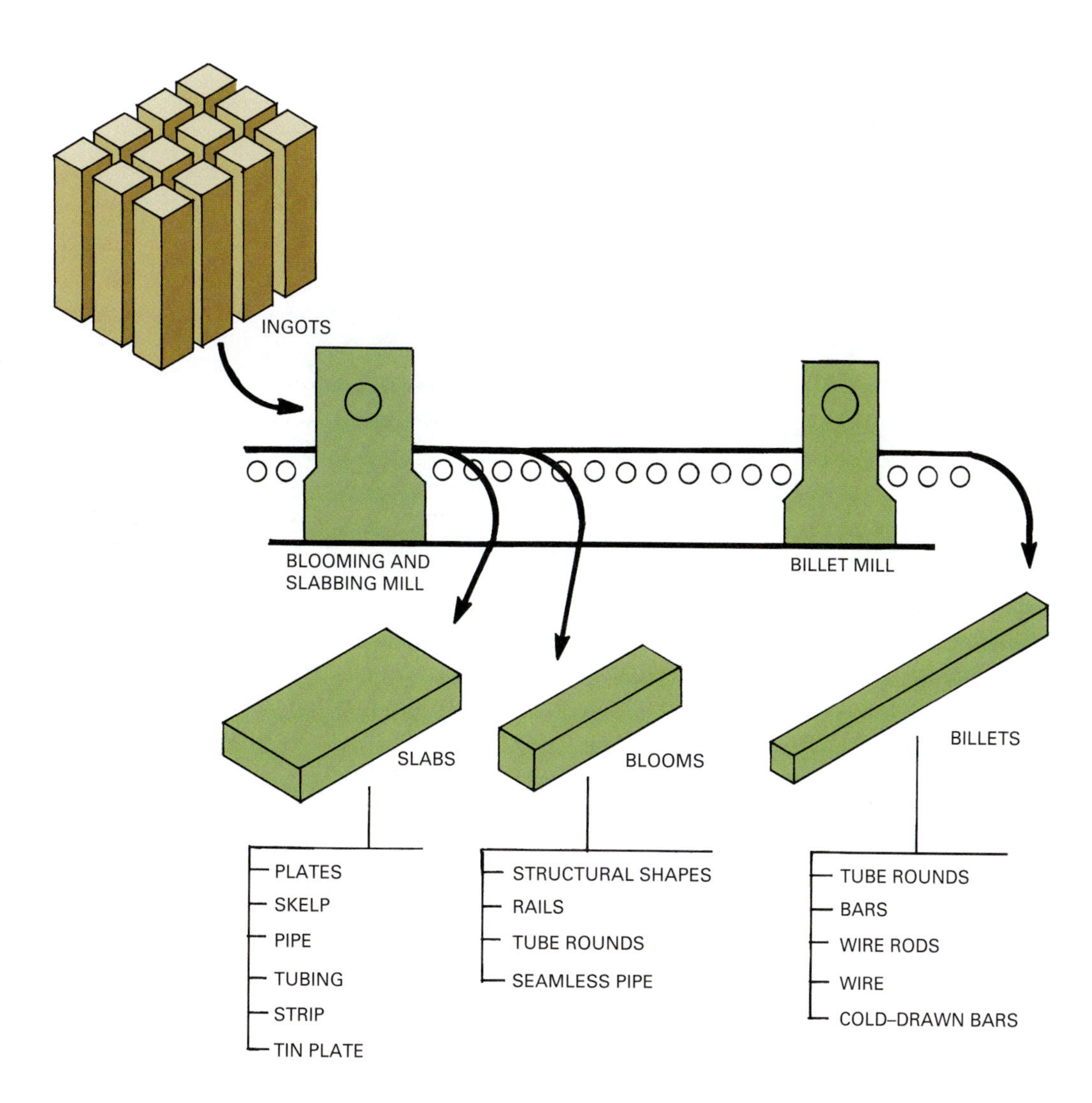

Figure 15.8 The steps in producing ingots into semi-finished steel products, which are then used to produce finished steel products.

Figure 15.9 A multiline continuous caster moving slabs after they have been cut to length. *(Image copyright Tischenko Irina, 2009. Used under license from Shutterstock.com)*

Figure 15.10 These perfectly formed steel plates are resting on a cooling bed. They remain there until their temperature drops below 300°F (150°C), after which they move to the marking area. *(Courtesy Bethlehem Steel Corporation)*

Figure 15.11 These structural steel beams have been roll-formed hot and are being transferred to a "walking" cooling bed before they are cut to the desired lengths. *(Courtesy Bethlehem Steel Corporation)*

Figure 15.12 Steel scrap stockpiled for reprocessing in a steel mill.

All types of structural shapes including rods, rails, strip, and seamless pipe are manufactured using blooms. Finished structural beams are shown in Fig. 15.11. One beam has just left the hot-rolling process and will move to cooling beds on the side. Billets are used to produce solid steel bars and rounds, wire rods and wire, seamless pipe, and cold-drawn bars.

Steel Recycling

Steel is one of the most widely recycled materials in the world (Fig. 15.12). The steel industry has been engaged in recycling for more than a century because it is economically advantageous. It is less costly to recycle steel than to mine iron ore and process it to form new steel. Steel loses none of its inherent physical properties during the recycling process, and the energy and material requirements for recycling are far less than those for iron ore refining. Today 60 to 90 percent of all steel products contain recycled content, and the energy saved by recycling reduces the industry's annual energy consumption by about 75 percent.

STEEL IDENTIFICATION SYSTEMS

Nationally used metal and alloy numbering systems are administered by societies, trade associations, and individual users and producers of metals and alloys.

Among these are numbering systems by the American Society for Testing and Materials (ASTM), SAE/Aerospace Materials Specifications (AMS), American Welding Society (AWS), American Iron and Steel Institute (AISI), Society of Automotive Engineers (SAE), Federal Specifications, Military Specifications (MIL), and the American Society of Mechanical Engineers (ASME). Following are brief descriptions of several of these systems.

The American Iron and Steel Institute (AISI) and the Society of Automotive Engineers (SAE) series of identifying numbers for carbon and alloy steels use four- or five-digit numbers. The first two digits indicate type of steel (carbon or alloy) and the last two digits indicate carbon content. For example, in AISI 1030, the 10 indicates a carbon content of 0.30 (actual range is 0.28 to 0.34). An AISI 4012 indicates a molybdenum steel alloy composed of 0.15 to 0.25 molybdenum. It has 0.09–0.14 carbon, 0.75–1.00 manganese, 0.035 phosphorus maximum, 0.040 sulfur maximum, and 0.15–0.35 silicon.

The American Society for Testing and Materials (ASTM) sets standards for steel designations using an arbitrary number to indicate chemical compositions and specify minimums for strength and ductility. These specifications regulate specific chemical elements that directly affect the fabrication and erection of the steel. This makes it possible for various proprietary steels of different chemical compositions to conform to ASTM performance standards. Structural steels used in construction are designated by ASTM standards. Typical steels used for construction purposes are identified in Table 15.1.

One system, the Unified Numbering System for Metals and Alloys (UNS), was developed jointly by the Society of Automotive Engineers, Inc., and the American Society for Testing and Materials. It provides a means for describing the composition of several thousand metal designations and cross-references the systems of the organizations just mentioned.

The Unified Numbering System for Metals and Alloys

The Unified Numbering System (UNS) provides the uniformity required for indexing, record keeping, data storage and retrieval, and cross-referencing of metals. The UNS is not a specification but is used as an identifier of a metal or alloy for which controlling limits have been established in specifications published elsewhere.

The UNS has eighteen series of designations for metals and alloys, seventeen of which are active. Each UNS designation has a single-letter prefix followed by five digits. The letter is used to identify the family of metals, such as S for stainless steels and A for aluminum. The prefixes are listed in Table 15.2.

Table 15.1 Identification Number, Type, and Yield Point of Selected Structural Steels

Unified Number	ASTM Number	Type	Yield Point, ksi[a]
K02600	A36	Carbon steel	36
K11510	A242	High-strength low-alloy, corrosion resistance 5 to 8 times carbon steel	42–50
K11630	A514	High-yield-strength, quenched and tempered alloy steel	90–100
K02703	A529	Carbon steel	42
K02303	A572	High-strength low-alloy, niobium-vanadium, structural quality, corrosion resistance four times carbon steel	42–65
K11430	A588	High-strength low-alloy, corrosion resistance four times carbon steel	42–50
K01803	A633	Normalized high-strength low-alloy steel	42–60
K01600	A678	Quenched and tempered steel	50–75
K12043	A852	High-strength quenched and tempered alloy steel	70

[a]*Kips per square inch (ksi). One ksi equals 1,000 lb. per square inch.*

Table 15.2 The Unified Numbering System (UNS) for Identifying Metals and Alloys

Series Designation	Family of Metals
Axxxxx	Aluminum and aluminum alloys
Cxxxxx	Copper and copper alloys
Exxxxx	Rare earth and similar metals and alloys
Fxxxxx	Cast irons
Gxxxxx	AISI and SAE carbon and alloy steels
Hxxxxx	AISI and SAE H-steels
Jxxxxx	Cast steels (except tool steels)
Kxxxxx	Miscellaneous steels and ferrous alloys
Lxxxxx	Low melting metals and alloys
Mxxxxx	Miscellaneous nonferrous metals and alloys
Nxxxxx	Nickel and nickel alloys
Pxxxxx	Precious metals and alloys
Rxxxxx	Reactive and refractory metals and alloys
Sxxxxx	Heat and corrosion-resistant steels (including stainless), valve steels, and iron-base "superalloys"
Txxxxx	Tool steels, wrought and cast
Wxxxxx	Welding filler metals
Zxxxxx	Zinc and zinc alloys

Courtesy Society of Automotive Engineers

The significance of the digits can vary with the UNS series. Therefore, their meaning for each series can serve a different purpose. Effort has been made to relate UNS digits to those established by related trade organizations. In the published UNS manual, Metals and Alloys in the Unified Numbering System, UNS numbers are cross-indexed with systems of other organizations whose documents describe materials that are the same as or similar to those covered by UNS numbers. These systems are identified by the letters for each organization, shown in Table 15.3. The base elements in the specification of composition of a metal are identified by standard chemical symbols. Some of these are shown in Appendix E.

STEEL AND STEEL ALLOYS

Steel is a term generally applied to plain carbon steels that are alloys of iron and carbon (with a carbon content of less than 2 percent). Other steel products include stainless and alloy steels.

Plain Carbon Steels

Plain carbon steels have iron as the major element (more than 95 percent), but they also contain impurities, such as sulfur, nitrogen, and oxygen. Other elements may be present as residual impurities or as alloys added to change the properties of the steel. These can include phosphorus, nickel, aluminum, copper, silicon, and manganese.

Table 15.3 Numbering Systems Cross-Referenced to UNS

Cross-Reference Prefix	Specifying Organization
AA	(Aluminum Association) numbers
ACI	(Steel Founders Society of America) numbers
AISI	(American Iron and Steel Institute) including SAE (Society of Automotive Engineers) numbers (carbon and low-alloy steels)
AMS	(SAE/Aerospace Materials Specification) numbers
ASME	(American Society of Mechanical Engineers) numbers
ASTM	(American Society for Testing and Materials) numbers
AWS	(American Welding Society) numbers Federal Specification Numbers
MIL	(Military Specification) numbers
SAE	(Society of Automotive Engineers) "J" numbers

Courtesy Society of Automotive Engineers

The properties of carbon steel are varied not only by the elements that are added to alter their chemical composition but also by the type of mechanical and heat treatment used in producing the steel. For example, during production, carbon steel could be hot- or cold-rolled, cast, cooled slowly, or cooled rapidly, all of which influence final properties.

Control of carbon content is the major factor in establishing the properties of carbon steel. As the carbon content increases, so does the strength and hardness, while ductility decreases. The differences in the amount of carbon among steel types is very small, usually hundredths of a percent. The percentage of carbon in several types of carbon steel is shown in Table 15.4.

Carbon steels contain other elements, such as manganese, phosphorus, and sulfur—amounts vary with the chemical design of the metal. For example, AISI/SAE 1008 carbon steel contains 0.10 percent maximum carbon, 0.30 to 0.50 percent manganese, 0.040 percent maximum phosphorus, and 0.050 percent maximum sulfur. Carbon steels are used for many construction products, including structural shapes, bars, sheet and strip products, plate, pipe, tubing, wire, nails, rivets, and screws. Carbon steel is also used to produce cast products, typically from medium-grade carbon steel.

Table 15.4 Carbon Content of Carbon Steels

Type of Steel	Percent Carbon Content	Characteristics
Extra soft grade	0.05–0.15	Ductile, soft, tough Used for wire, rivets, pipe, sheets
Mild structural grade	0.15–0.25	Ductile, strong Used for boilers, bridges, buildings
Medium grade	0.25–0.35	Harder than mild structural Used for machinery and general structural purposes
Medium hard grade	0.35–0.65	Harder than medium grade Resists abrasion and wear
Spring grade	0.85–1.05	Strong and hard Used to make springs
Tool steel	1.05–1.20	Hardest and strongest of carbon steel

The castings are often heat treated to relieve internal strain developed during the casting process.

Alloy Steels

An alloy steel contains one or more alloying elements other than carbon (such as chromium, nickel, and molybdenum) that have been added in amounts exceeding a specified minimum to produce properties not available in carbon steels. These elements give particular physical, mechanical, and chemical properties to the steel. Stainless steels, specialty steels, and tool steels are not considered alloy steels, although they do contain alloying elements.

Standard alloy steels generally have elements of carbon, manganese, and silicon occurring naturally. Alloy steel that has one additional alloying element is identified as a single alloy steel. Adding two alloying elements produces a double, or binary, alloy steel, and three elements produce a triple, or ternary, steel. The major alloying elements and the property changes they produce can be found in Table 15.5.

Table 15.5 demonstrates that alloying elements are used to improve properties, such as hardness, performance of the material at high and low temperatures, strength, ductility, workability, wear resistance, electromagnetic properties, and electrical resistance or conductivity. Architects and structural engineers are especially concerned with those alloys having increased strength, wear resistance, and resistance to **corrosion**, expansion, contraction, and ductility.

Table 15.5 Properties Imparted to Steel Alloys by Alloying Elements

Element	Properties
Aluminum	A deoxidizer used to control the grain size within the structure of the steel, promotes surface hardening
Boron	Increases the depth of hardness
Carbon	Increases hardness, strength High percentages contribute to brittleness
Chromium	Increases hardness, corrosion, and wear resistance
Cobalt	Hardens or strengthens the ferrite, resists softening at high temperatures Used in high-speed tool steels
Copper	Increases resistance to atmospheric corrosion and increases yield strength
Manganese	Increases strength and resistance to wear and abrasion
Molybdenum	Increases corrosion resistance, raises tensile strength and elastic limit, reduces creep, improves impact resistance
Nickel	Increases elastic limit and internal strength Increases strength and toughness in heat treated steel In some steels increases hardness, fatigue, and corrosion resistance
Niobium	Retards softening during tempering operations, increases resistance to creep at high temperatures, increases ductility and impact strength
Phosphorus	Increases corrosion resistance and strength
Silicon	Increases hardenability, strength, and magnetic permeability in low-alloy steels
Sulfur	Improves machining properties Especially useful in mild steels
Titanium	Prevents intergranular corrosion of stainless steels, is a deoxidizer, increases strength in low-carbon steels
Tungsten	Used in tool steels to promote hardness In stainless steels helps maintain strength at high temperatures
Vanadium	Improves resistance to thermal fatigue and shock
Zirconium	Inhibits grain growth and is a deoxidizer

There are many alloy steels specified by the various organizations mentioned earlier. Their uses range widely and include such types as alloy steel electrode, alloy steel welding wire, alloy steel, high-strength low-alloy steel, and others. Some of these have direct applications to products used in construction. Examples include high silicon content, which improves magnetic permeability, making it useful in transformers, motors, and generators. Nickel improves toughness, so nickel alloys are employed on cutting tools used in rock-drilling machines and air hammers. Since alloy steels are more expensive than carbon steels, they are typically not used for structural members unless special requirements exist, such as high temperature resistance or very high strengths. They are used in the manufacture of power tools and heavy construction equipment. Some common uses for carbon and alloy steels are shown in Table 15.6.

Types of Structural Steel

Steel specified for structural purposes is of major importance to architects and engineers. It has low-to-medium carbon content. The American Institute of Steel Construction (AISC) publication, "Code of Standard Practice for Steel Buildings and Bridges," includes those things that must be specified in construction documents, including columns, beams, trusses, bearing plates, and various fastening devices and connectors. The AISC publication, "Specification for Structural Steel Buildings," details information on structural steels for use in building construction.

Table 15.6 Construction Uses for Selected Carbon and Alloy Steels

UNS Designation	ASTM Designation	Construction Use
K02600	A36	Structural steel members
K20504	A53	Welded and seamless pipe
K11510	A242	Structural members
K02303	A572	Structural members
K11430	A588	Structural members
K02703	A529	Structural members
K01803	A633	Structural members
K03000	A500	Cold-formed, welded, and seamless tubing
K10600	A678	Quenched and tempered steel plate
K03000	A501	Hot-formed, welded, and seamless tubing
K02706	A325	High-strength bolts
K03900	A490	High-strength alloy steel bolts
	A307	Machine bolts and nuts
	A328	Sheet piling
KJ0300	A27	Cast steel items
J31575	A148	High-strength cast steel items
	615	Billet steel bars for reinforcing concrete
	616	Rail-steel bars for reinforcing concrete
	606, 607	Sheet steel

Courtesy Society of Automotive Engineers

Structural steels fall into four major classifications:

1. Carbon steel (ASTM A36, A529, UNS K02600)
2. Heat-treated construction alloy steel (ASTM A514, UNS K11630)
3. Heat-treated high-strength carbon steel (ASTM A633, A678, A852, UNS K01803, K01600, K12043)
4. High-strength low-alloy steel (ASTM A242, A572, A588 UNS K11510, K02303, K11430)

Carbon steels must meet maximum content requirements for manganese and silicon. Copper requirements have minimum and maximum specifications. There are no other minimums specified for other alloying elements. Heat-treated construction alloy steels have more stringent alloying element specifications than carbon steel. They produce the strongest general-use structural steel.

Heat-treated high-strength carbon steels are brought to desired strength and toughness levels by heat treating. Heat treating refers to the process of heating and cooling metals to produce changes in the physical and mechanical properties.

High-strength low-alloy steels are a group of steels to which alloying elements have been added to produce improved mechanical properties and greater resistance to atmospheric corrosion. Their carbon range is typically from 0.12 to 0.22 percent. The percent of alloying agents varies with the manufacturer of the steel and it is available from manufacturers under specific trade names. ASTM has specifications to cover all the trade name steels. One type, known by the general name **weathering steels**, is of special interest. They are designed for architectural exterior applications where exposure to the atmosphere causes the steel to form a natural, rust-colored, self-healing oxide coating. It is never painted but left to protect and color the exterior material.

High-strength low-alloy steels can be worked by many metal processing operations, including hot- and cold-forming, punching, shearing, gas cutting, and welding. In construction, alloy steels find use in high-strength bolts, cables used in pre-stressed concrete, sheet and strip material, plates, various bars and structural shapes, wire, tubing, and pipe.

Stainless and Heat-Resisting Steels

Stainless steels have outstanding corrosion and oxidation resistance at a wide range of temperatures. Heat-resisting steels maintain their basic mechanical and physical properties when subjected to high temperatures. Typically, standard stainless steels have 10 to 25 percent chromium.

The chromium alloying element gives stainless steel its corrosion-resistance qualities and produces a thin, hard, invisible film over the surface, which inhibits corrosion. Nickel and manganese increase strength and toughness and make the material easier to fabricate.

Other alloying elements that may be used to produce desired characteristics include zirconium, titanium, sulfur, silicon, selenium, phosphorus, molybdenum, and niobium. Refer to Table 15.5 for additional information on how each of these alters the characteristics of a product. The chemical compositions of several types of stainless steel frequently used in construction are shown in Table 15.7. Notice the small ranges allowed within each alloying element in each type of stainless steel.

Classifying Stainless and Heat-Resisting Steels

Stainless and heat-resisting steels are classified into three major groups based on their chemical composition and their reaction when heat treated: ferritic steels, martensitic steels, and austenitic steels (Table 15.8).

Ferritic stainless steels have a chromium content of about 16 to 18 percent and a low carbon content of 0.12 maximum. These are non-hardenable steels, meaning

Table 15.7 Chemical Compositions of Stainless Steels Used in Construction

UNS Designation	AISI/SAE Designation	Chemical Composition (percent)						
		Fe	Cr	Ni	Mn	C	Mo	Si
S20100	201[a]	67.51–73.51	16.0–18.0	3.5–5.5	5.5–7.5	0.15 max.		1.00 max.
S20200	202[a]	63.51–70.01	17.0–19.0	4.0–6.0	7.5–10.0	0.15 max.		1.00 max.
S30100	301[b]	70.84–74.84	16.0–18.0	6.0–8.0	2.0 max.	0.15 max.		1.00 max.
S30200	302[b]	67.84–70.84	17.0–19.0	8.0–10.0	2.0 max.	0.15 max.		1.00 max.
S30400	304[b]	66.28–70.77	18.0–20.0	8.0–10.5	2.0 max.	0.15 max.		1.00 max.
S31600	316[b]	61.83–68.83	16.0–18.0	10.0–14.0	2.0 max.	0.10 max.	2.0–3.0	1.00 max.
S43000	430[b]	79.81–81.05	14.0–18.0		1.0 max.	0.12 max.	0.75–1.25	1.00 max.

[a]*Contains maximum 0.06% phosphorus, 0.03% sulfur, 0.25% nitrogen.*
[b]*Contains maximum 0.04% phosphorus, 0.03% sulfur.*
Courtesy American Iron and Steel Institute

Table 15.8 AISI/SAE and UNS Designations and Characteristics of Heat-Resisting and Stainless Steels

Grain Structure	Chief Alloying Elements	Series	AISI/SAE Designation	UNS Designation	Characteristics	Steel Type
Martensitic (ferromagnetic and hardenable by heat treatment)	Chromium 4%–6%	500	502	S50200	Retain their mechanical properties at high temperatures	Heat-resisting steels
	Chromium 11.50–13.50%	400	410	S41000	Moderate corrosion resistance, high strength and hardness	
Ferritic (ferromagnetic and nonhardenable)	Chromium 16%–18%	400	430	S43000	Very good corrosion resistance, particularly at high temperatures	
Austenitic (nonmagnetic and hardenable by cold working)	Chromium 17%–19% Nickel 8%–10% Manganese 2.0% Max.	300	302	S30200	Excellent corrosion resistance, high strength and ductility Suitable for many fabrication techniques	Stainless steels
	Chromium 17%–19% Nickel 4%–6% Manganese 7.5%–10%	200	202	S20200	Excellent corrosion resistance at high temperatures, high strength and toughness	

Courtesy American Iron and Steel Institute

they cannot be hardened by heat treating. Ferritic steels have excellent corrosion resistance. Although they have especially good corrosion resistance at high temperatures, they do lose strength under these conditions. They are used when high temperatures exist, corrosion resistance is important, and when a low coefficient of thermal expansion is helpful.

Martensitic stainless steels have a chromium content of 11.50 to 13.50 percent and a carbon content of 0.15 percent maximum. They can be hardened by heat treating and worked hot, forged, or formed cold. They are typically used when hardness, strength, and abrasion resistance are critical, such as in a steam turbine.

Austenitic stainless steels have a chromium content of about 17 to 19 percent, a nickel alloying element of 8 to 10 percent, and a maximum carbon content of 0.15 percent. They are nonmagnetic and harden when worked cold. At high temperatures they have high strength and are very tough and corrosion resistant. They have a high coefficient of thermal expansion, which must be taken into consideration by an engineer when specifying this material's use. Of all of the stainless steels, they have the least resistance to corrosion by attack of sulfur gases. Typical uses include curtain wall panels, railings, and door and window finished surfaces (Fig. 15.13). Some types are used in the manufacture of food preparation equipment and various appliances in hospitals.

Designations for Stainless and Heat-Resisting Steels The American Iron and Steel Institute (AISI) designations for heat-resisting and stainless steels are based on a three-digit numbering system ranging from 200 to 500, as shown in Tables 15.7 and 15.8. The first digit indicates the group, and the last two digits indicate the type within the group. Some types have the prefix TP, which refers to tubular grades.

Figure 15.13 The 630 foot high St. Louis Gateway Arch is clad in type 304 stainless steel. *(Image copyright Mehmet Dilsiz, 2009. Used under license from Shutterstock.com)*

Stainless Steel Uses in Construction Typical applications for stainless steels frequently specified by architects and engineers are shown in Table 15.9. Of the ferritic stainless steels, Type 430 is most commonly used in construction for such applications as column covers, trim, grills, and gutters. Of the austenitic stainless steels, Types 301, 302, and 304 are used for applications such as storefronts, doors, windows, and railings. Types 201 and 202 are used for the same applications as 301 and 302. Other uses include drinking fountains, kitchen sinks, flatware, and cooking utensils (Fig. 15.14).

Table 15.9 Stainless Steels Used in Construction

UNS Designation	AISI Designation	Typical Applications
S30100	301	Gutters, trim, flashing, household and industrial appliances
S30200, S30400	302, 304	Storefronts, curtain walls, doors, windows, railings, household and industrial appliances
S31600	316	All exterior marine uses, on seacoast building exteriors, in chemical, petroleum, and paper manufacturing facilities
S43000	430	Exterior applications in areas exposed to salt water atmosphere, column covers, trim, grills, gutters
S20100	201	Gutters, trim, flashing, household and industrial appliances
S20200	202	Storefronts, curtain walls, doors, windows, railings, household and industrial appliances

Figure 15.14 This elevator uses a stainless steel finish for durability and ease of maintenance. *(Image copyright Adam Radosavljevic, 2009. Used under license from Shutterstock.com)*

STEEL PRODUCTS

Steel of various types is used extensively in construction, ranging from something as small as a nail to huge beams and columns. Following are a few of the products in common use.

Structural Steel Products

Rolled structural steel shapes are used extensively as structural members. The commonly used shapes are shown in Fig. 15.15. These include S (standard beam) shapes; W (wide flange) shapes; WWF (welded wide flange) shapes; M (miscellaneous) shapes; H-sections; HP (bearing pile) shapes; C (channel) shapes; ST (structural T) shapes cut from M, W, and WWF shapes; L (angle) shapes with equal and unequal legs; HSS (hollow structural section); square and rectangular shapes; pipe, bar (flat, round, and square), and plate (Fig. 15.16). Shapes are specified on architectural drawings, as shown in Table 15.10. The S beam has narrow flanges that taper at a 17 percent slope. The W beam has wide

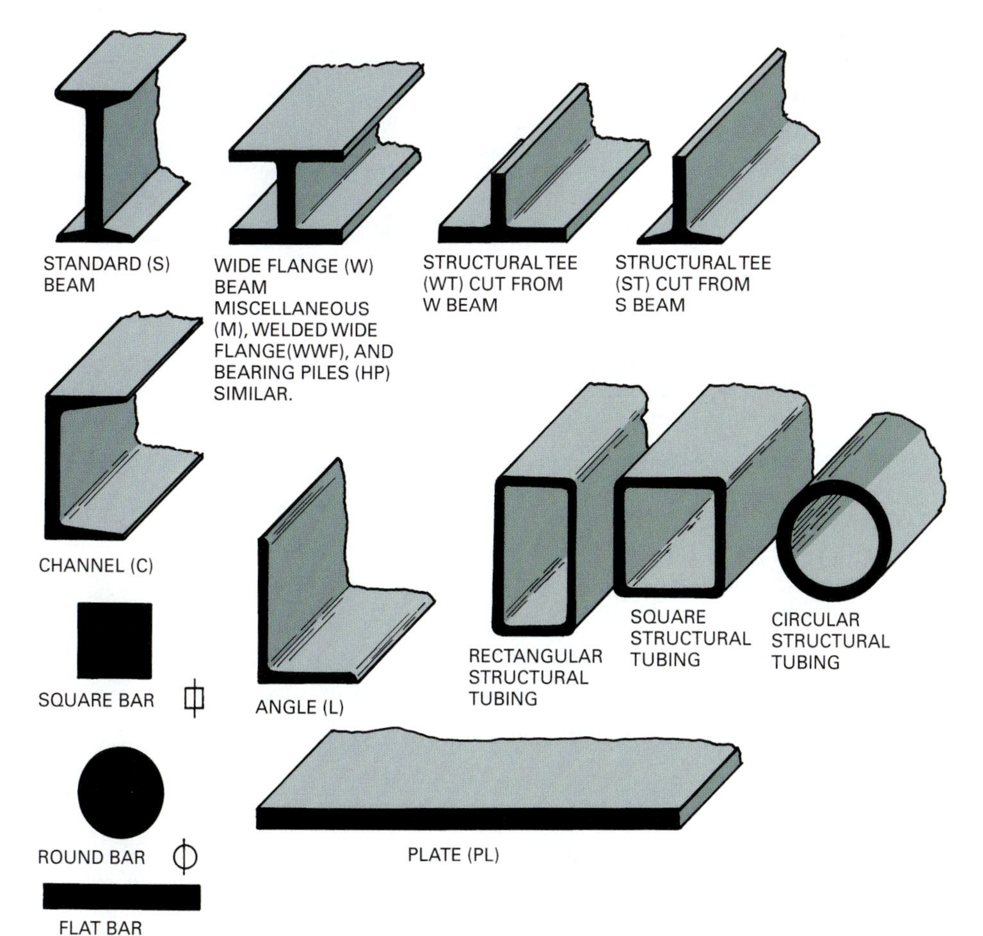

Figure 15.15 Commonly available structural steel shapes.

Figure 15.16 Structural steel beams are available in a wide range of sizes. *(Courtesy Bethlehem Steel Corporation)*

flanges that have very little slope. The WWF shapes are made by welding flanges to the web, while S and W beams are rolled to shape from a single piece of steel.

Open-web steel joists are a widely used steel product. They are lightweight and produced by welding structural steel shapes, such as angles and bars, into a Warren truss unit. They are manufactured in short-span, long-span, and deep long-span series. The short-span joists are manufactured to span clear openings from 8 ft. (2.4 m) to 60 ft. (18.3 m) and in depths from 8 in. (203 mm) to 30 in. (762 mm). The long-span joist series is heavier, made to span clear openings from 25 ft. (7.6 m) to 96 ft. (29.2 m) and depths from 18 in. (457 mm) to 48 in. (1219 mm). The deep long-span series spans a clear opening from 89 ft. (27 m) to 144 ft. (43.9 m) and in depths from 52 in. (13 mm) to 72 in. (1829 mm). They are used to support floor and roof loads. The openings in the web permits the passage of plumbing, heating ducts, and electrical runs (Fig. 15.17).

Figure 15.17 These open-web joists have corrugated steel decking on top. *(Courtesy Eva Kultermann)*

Table 15.10 Symbols and Abbreviations for Structural Steel Members

Structural Shape	Symbol	Order of Presenting Data	Sample of Abbreviated Note
Square bar	⌷	Bar, size, symbol, length	Bar 1 ⌷ 5′ – 9″
Round bar	⏀	Bar, size, symbol, length	Bar $^5/_8$ ⏀ 7′ – 8″
Plate	PL	Symbol, thickness, weight	PL $^1/_2$ × 24
Angle, equal legs	L	Symbol, leg 1, leg 2, thickness, length	L 2 × 2 × $^1/_4$ × 9′ – 6″
Angle, unequal legs	L	Symbol, long leg, short leg, thickness, length	L 4 × 3 × $^1/_4$ × 6′ – 2″
Channel	C	Symbol, depth, weight, length	C 6 × 10.5 × 12′ – 7″
American Standard beam	S	Symbol, depth, weight, length	S 10 × 35.0 × 13′ – 6″
Wide flange beam	W	Symbol, depth, weight, length	W 16 × 64 × 20′ – 2″
Structural tee	ST, WT	Symbol, depth, weight	ST 10 × 50
Pipe	Name of pipe	Name, diameter, strength	Pipe 3 X-strong
Miscellaneous	M	Symbol, depth, weight, length	M 14 × 17.2 × 19′ – 0″
Bearing pile	HP	Symbol, depth, weight, length	HP 14 × 117 × 36′ – 0″
Welded wide flange	WWF	Symbol, depth, weight, length	WWF 14 × 84 × 15′ – 6″

Sheet Steel Products

Many products are made by rolling them to shape from flat steel sheets. The thickness of steel sheets is given by gauge numbers, as shown in Table 15.11. Common among these are roofing, siding, decking, and light-gauge steel framing systems.

Many styles of steel roof and siding systems are available. A variety of coatings are used to protect the steel surfaces. Typical coatings include galvanizing, zinc-aluminum alloy with a covering paint, siliconized polyester in a variety of colors, fluoropolymer paint finish, and weathering copper coating. A few of the available wall panel configurations are shown in Fig. 15.18.

Corrugated steel floor decking is used with structural steel framing. This decking usually spans 6 to 15 ft. (1.8 to 4.6 m) and has a site-cast concrete topping (see Fig. 15.10). Cellular floor decking provides openings in floor slabs for running electrical systems (Fig. 15.19).

Steel roof decking may have a site-cast lightweight concrete or gypsum topping, or it may be covered with some form of insulation board and finished roofing system, such as tar and gravel. Examples of decking products are shown in Fig. 15.20.

Light-gauge steel framing systems use studs, joists, channels, and runners to frame wall and floor systems. Metal studs are used in the same manner as wood studs. They have metal runners for top and bottom plates. Metal joists are welded to the top runner and have a perimeter channel for a header. The floor and roof deck can be metal or plywood (Fig. 15.21). Additional information can be found in Chapter 20.

Expanded steel mesh is made by slitting metal sheets and stretching them to form diamond-shaped openings. It is used on gratings, decks, partitions, as a base for troweled stucco and plaster application, and in many other applications. Expanded metal mesh is usually designated by the width of the mesh opening and the gauge of the steel sheet. For example a $\frac{3}{4}$-16 expanded metal mesh has diamond openings $\frac{3}{4}$ in. (18 mm) wide on 16-gauge material. The length of the diamond is approximately twice the width, or 1 $\frac{1}{2}$ in. (38 mm) in this example.

Metal lath is a form of expanded steel that has a flat or ribbed mesh $\frac{3}{8}$ in. (9 mm) in height. It is used as a base for plaster and expanded metal corner bead. A wide variety of other expanded, slit, and woven products are available for both utilitarian and decorative purposes (Fig. 15.22).

Table 15.11 Standard Thicknesses for Some Basic Steel Sheets

Minimum Thickness	
Equivalent Inches	**Millimeters**
0.3937	10.0
0.3543	9.0
0.3150	8.0
0.2756	7.0
0.2362	6.0
0.2165	5.5
0.1969	5.0
0.1890	4.8
0.1772	4.5
0.1654	4.2
0.1575	4.0
0.1496	3.8
0.1378	3.5
0.1260	3.2
0.1181	3.0
0.1102	2.8
0.0984	2.5
0.0866	2.2
0.0787	2.0
0.0709	1.8
0.0630	1.6
0.0551	1.4
0.0472	1.2
0.0433	1.1
0.0394	1.0
0.0354	0.90
0.0315	0.80
0.0276	0.70
0.0256	0.65
0.0236	0.60
0.0217	0.55
0.0197	0.50
0.0177	0.45
0.0157	0.40
0.0138	0.35

Other Products

Welded wire fabric is used to reinforce concrete slabs. Several commonly used sizes are listed in Table 15.12. More information is given in Chapter 8. Inch sizes of welded wire fabric are designated by two numbers and

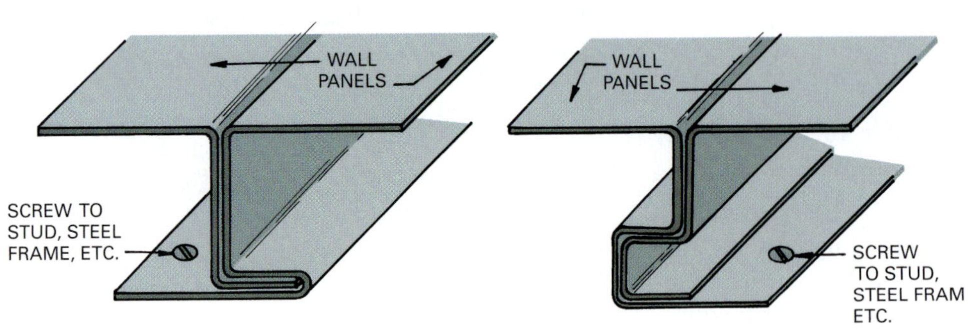

Figure 15.18 Metal siding is available in a variety of configurations and colors.

two letter-number combinations. The first two numbers, for example, 6 × 6, give the spacing of the wires in inches. The first number is the spacing of the longitudinal wires. The second number gives the spacing of the transverse wires in inches. The first letter-number combination gives the type and size of the longitudinal wire and the second for the transverse wire. "W" indicates a smooth wire, and the number following gives its cross-sectional area. A "D" instead of a "W" indicates the use of a deformed wire. The area is given in hundredths of an inch per foot. For example, a W8.0 wire is a smooth wire with a cross-sectional area of 0.08 in.^{2}

Longitudinal wires are spaced at 2, 3, 4, 6, 8, and 12 in. Transverse wires are spaced 4, 6, 8, and 12 in. apart. Metric welded wire fabric sizes are given in millimeters. Typical sizes are also shown in Table 15.12. The first two numbers represent the wire spacing in millimeters. The last two numbers indicate the type of wire and give the cross-sectional area in square millimeters (mm^2).

Reinforcing bars are placed in concrete members to improve the tensile strength of concrete (Fig. 15.23). They are made in accordance with ASTM requirements for yield and tensile strength, cold bend, and elongation. Three grades are available: structural, intermediate, and hard. Structural grade has the lowest yield point and tensile strength, and hard has the highest. The deformed bar has surface projections that provide a better bond to concrete. They are available in sizes 3 through 18. The number indicates the diameter in eighths of an inch. For example, a No. 5 bar has a diameter of $\frac{5}{8}$ in. Metric sizes are given in numbers shown in Table 15.13. A variety of accessories can hold reinforcing bars in place until concrete is poured. They are made from steel wire. Detailed information is available in Chapter 8.

Many fasteners used in construction are made from steel. Commonly used fasteners include bolts, nails, rivets, and screws. Standard steel bolts are available in a wide range of sizes and head types (Fig. 15.24). High-strength steel bolts are used where tensile strength is important, such as in structural steel framing. Additional details are given in Chapter 17. Commonly used nails are shown in Fig. 15.25. Some nails are specified in length by inches and others by the term "penny" (d). Penny lengths are shown in Table 15.14.

Figure 15.19 Cellular steel floor decking is used in multistory construction. *(Courtesy H.H. Robertson, A United Dominion Company)*

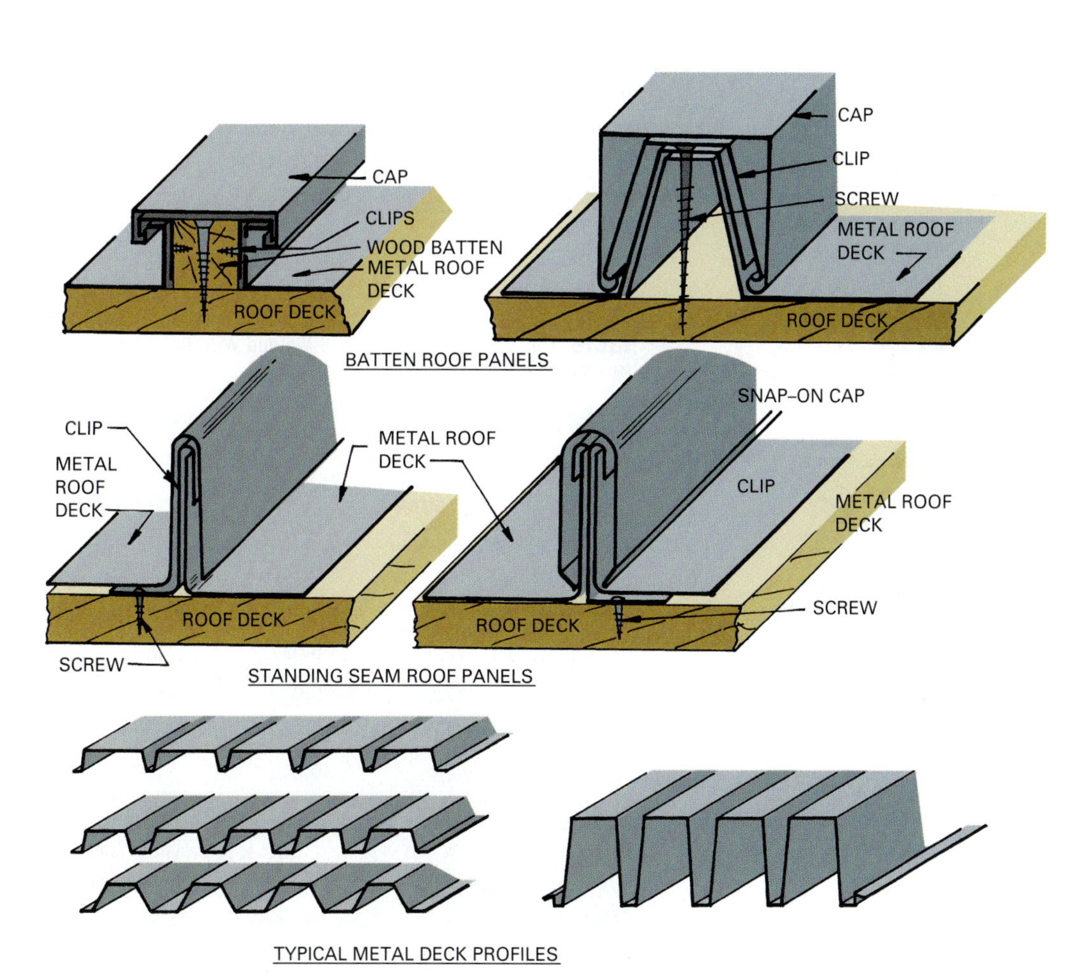

Figure 15.20 Several of the many types of steel roof decking available.

Most wood screws used have flat, round, or oval heads. Their lengths are specified in inches and their diameters with wire gauge numbers. A variety of slots are used in the heads (Fig. 15.26).

Rivets are permanent fasteners used to join two materials, such as two or more sheets of steel or a beam to a column. Standard rivets are available in a variety of diameters and head shapes. They are used when the back side of the material through which the rivet protrudes is accessible, allowing a head to be formed on it. Some rivets' heads are formed while the steel is cold. Others require heating before forming the back head. Blind rivets serve when a joint is accessible from only one side. After the rivet is inserted, a head is formed by a drive pin, a pull stem, or a chemical expansion in the protruding end.

Hundreds of other fastening devices are commercially available. These include concrete anchors, self-drilling fasteners, set screws, washers, eye bolts, U-bolts, and many types of hooks. Numerous other steel products are used in building construction, such as locks, hinges, gutters, and flashing.

TESTING METALS

Engineers and architects must know the capabilities of potentially usable metals. These are determined by standardized testing that follows the procedures of the American Society for Testing and Materials (ASTM). Some of the more frequently used tests include analyses of hardness, tensile strength, and fatigue.

Hardness testing determines the resistance of a metal to penetration. A Tinius Olsen Air-O-Brinell metal hardness tester is shown in Fig. 15.27. To conduct a test, the operator adjusts the air regulator until the desired Brinell load in kilograms is indicated. The specimen is placed on the anvil and the operator pulls the plunger valve. The plunger, which has a hardened ball on the bottom, impacts the metal and forms an impression in its surface. The plunger valve depresses and the plunger and ball retract from the metal. The impression diameter is measured, using optical instruments to provide an automatic and precise reading. A computer makes it possible for the operator to customize test parameters

Figure 15.21 Construction details of lightweight steel framing. *(Courtesy Marino Industries Corporation)*

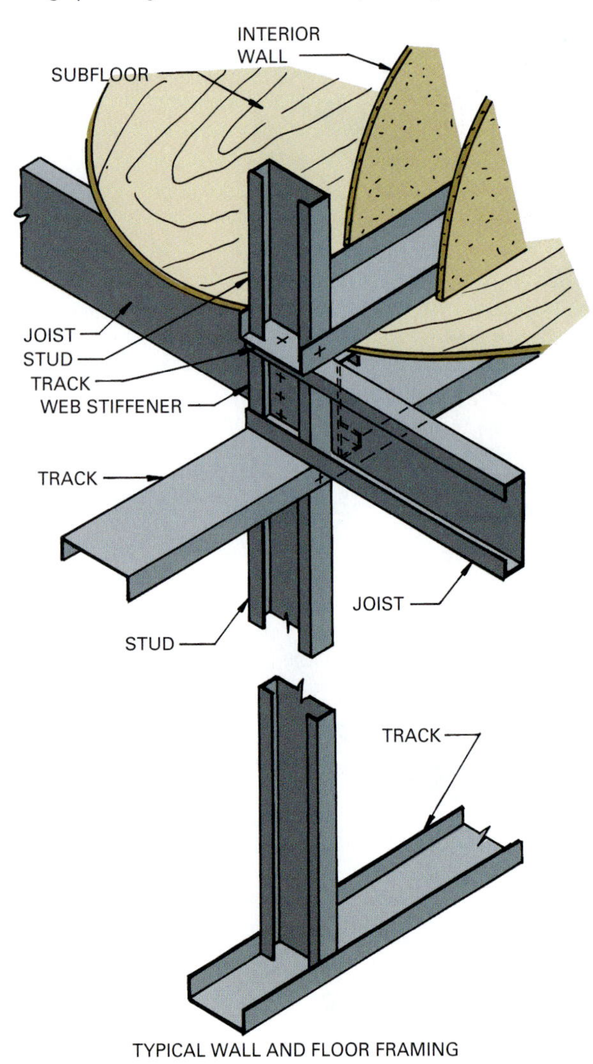

and test report formats. Test data can be recalled by individual test number, lot identifier, heat number, date, time, or other specific data.

Tensile testing determines the mechanical properties of a metal when it is subjected to a force that tends

Table 15.12 Selected Sizes of Welded Wire Reinforcement (WWR) (Inch and Metric)

Inch Sizes		
	Steel Area (in.²/ft.)	
Wire Style Designation (W or D)	Longitudinal Wire	Transverse Wire
4 × 4-W1.4 × W1.4	0.042	0.042
4 × 4-W2.0 × W2.0	0.060	0.060
4 × 4-W2.9 × W2.9	0.087	0.087
6 × 6-W1.4 × W1.4	0.028	0.028
6 × 6-W2.0 × W2.0	0.040	0.040
6 × 6-W2.9 × W2.9	0.058	0.058
Metric Sizes		
	Steel Area (mm²/m)	
Wire Style Designation (MW or MD)	Longitudinal Wire	Transverse Wire
102 × 102-MW9 × MW9	88.9	88.9
102 × 102-MW14 × MW14	127.0	127.0
102 × 102-MW19 × MW19	184.2	184.2
152 × 152-MW9 × MW9	59.3	59.3
152 × 152-MW14 × MW14	84.7	84.7
152 × 152-MW19 × MW19	122.8	122.8

Figure 15.22 A variety of utilitarian and decorative metal mesh products are available. *((a) Image copyright Marilyn Barbone, 2009. Used under license from Shutterstock.com (b) Image copyright Vartanov Anatoly, 2009. Used under license from Shutterstock.com (c) Image copyright J. Helgason, 2009. Used under license from Shutterstock.com (d) Image copyright Happy Alex, 2009. Used under license from Shutterstock.com)*

Table 15.13 Metric and Inch Reinforcing Bar Sizes and Diameters

Metric Bar Number	Diameter (mm)	Imperial Equivalent Bar Number	Diameter (inches)
10	9.5	3	.375
13	12.7	4	.500
16	15.9	5	.625
19	19.1	6	.750
22	22.2	7	.875
25	25.4	8	1.000
29	28.7	9	1.128
34	34.3	10	1.270
36	35.8	11	1.410
43	43.0	14	1.693
57	57.4	18	2.257

Table 15.14 Lengths and Diameters of Common, Box, Casing, and Finishing Nails

		American Steel Wire Gauge Number		
Size	Length (in.)	Common	Box and Casing	Finishing
2d	1	15	15 1/2	16 1/2
3d	1 1/4	14	14 1/2	15 1/2
4d	1 1/2	12 1/2	14	15
5d	1 3/4	12 1/2	14	15
6d	2	11 1/2	12 1/2	13
7d	2 1/4	11 1/2	12 1/2	12 1/2
8d	2 1/2	10 1/4	11 1/2	12 1/2
9d	2 3/4	10 1/4	11 1/2	12 1/2
10d	3	9	10 1/2	11 1/2
12d	3 1/4	9	10 1/2	11 1/2
16d	3 1/2	8	10	11
20d	4	6	9	10
30d	4 1/2	5	9	
40d	5	4	8	

to pull it apart. The tensile test conducted on a Tinius Olsen Universal Testing Machine determines the ultimate strength, yield point, yield strength, and factors related to ductility. It is available in capacities up to 600,000 lb. (2,000 kN). The test data is recorded by a four-channel digital indicating system. The testing sample is gripped with pinion-type wedge grips in the

Figure 15.23 Reinforcing bars and welded wire fabric provide tensile strength for concrete members. *(Image copyright Kenneth William Caleno, 2009. Used under license from Shutterstock.com)*

Figure 15.24 Standard steel bolts.

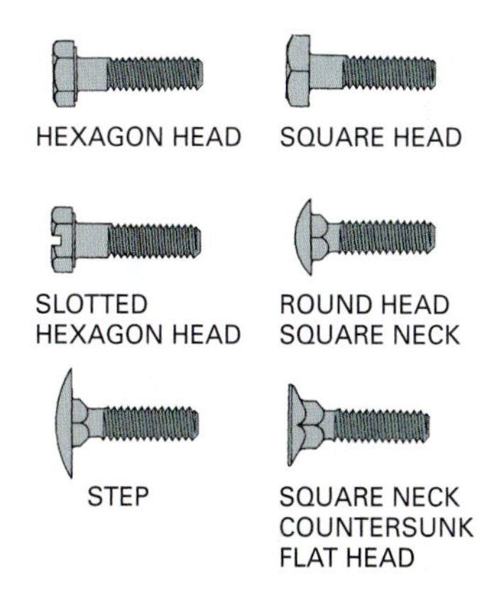

Figure 15.25 Commonly used nails.

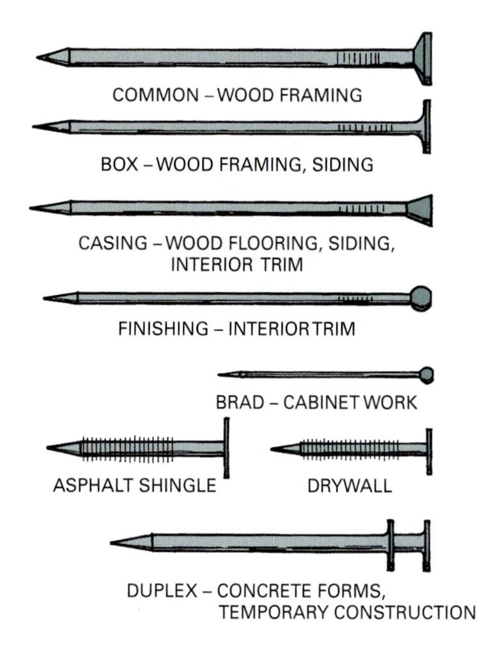

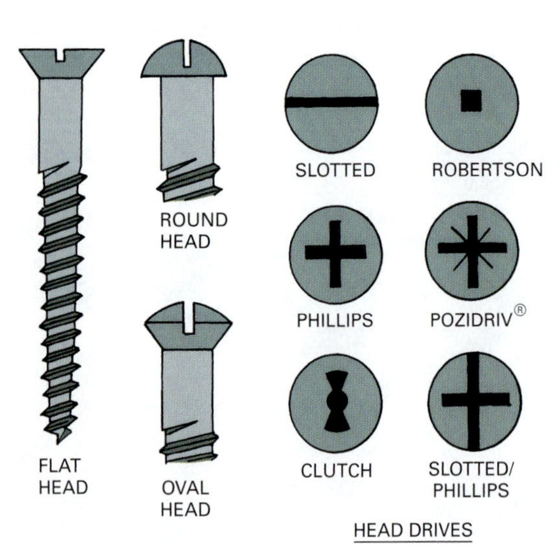

WOOD SCREWS		
DIA.	DECI. EQUIV.	LENGTH (INCHES)
0	.060	1/4–3/8
1	.073	1/4–1/2
2	.086	1/4–3/4
3	.099	1/4–1
4	.112	1/4–1 1/2
5	.125	3/8–1 1/2
6	.138	3/8–2 1/2
7	.151	3/8–2 1/2
8	.164	3/8–3
9	.177	1/2–3
10	.190	1/2–3 1/2
11	.203	5/8–3 1/2
12	.216	5/8–4
14	.242	3/4–5
16	.268	1–5
18	.294	1 1/4–5
20	.320	1 1/2–5
24	.372	3–5

Figure 15.26 Common head types used on wood screws.

Figure 15.27 The Tinius Olsen Model DS/AOB No. 2 Air-O-Brinell metal hardness tester with semiautomatic and automatic optical diameter measurement system. *(Courtesy Tinius Olsen, Inc.)*

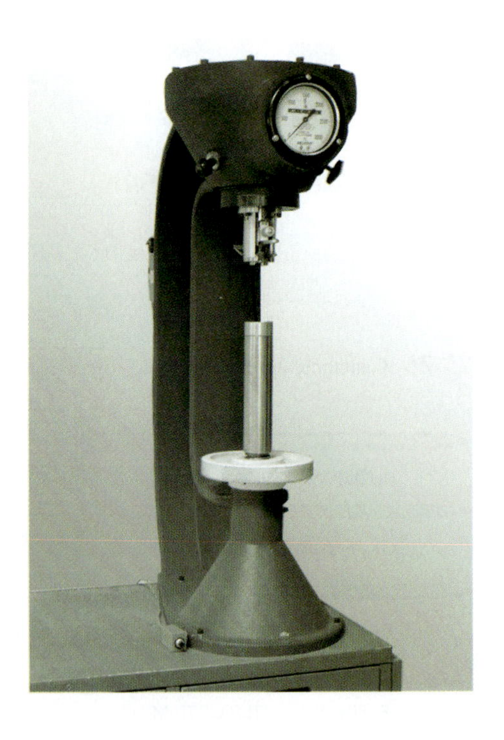

crossheads. Once the sample is secure in the grips, the lower crosshead moves down, exerting tension on the material. Data are recorded and the stress-strain curve values displayed.

Fatigue testing determines the stress level a metal can withstand without failure when subjected to an infinitely large number of repeated alternating stresses. Typical tests include subjecting a specimen, such as a beam, to rotating motion while under a bending movement. Another test involves bending the metal back and forth without rotating it.

Review Questions

1. What is the key element that influences the properties of iron?
2. What process is used to remove unwanted elements from freshly mined iron ore?
3. What fuels and fluxes are commonly used in a blast furnace?
4. What uses are made of slag removed from a blast furnace?

5. What are the commonly produced cast irons?
6. What are the main properties of white cast iron?
7. Why is malleable cast iron less brittle than white cast iron?
8. How does the alloying element nickel affect the properties of steel?
9. If steel has a carbon content less than 0.90 percent, how does this affect the properties of the steel?
10. How do the alloying elements copper and cobalt influence the properties of steel?
11. What are the three basic steelmaking processes?
12. How does carbon steel differ from alloy steel?
13. What are the two outstanding properties of stainless steel?
14. Explain the purpose of each part of the AISI steel designation code.
15. Why does stainless steel resist corrosion?
16. What are the three groups of stainless and heat-resisting steels?
17. Identify the following structural steel shapes: W, S, HP, ST, HSS, C, M.
18. How is welded wire fabric used in building construction?
19. What tests are used on metal to determine its hardness?
20. What does a tensile test show?
21. What does a fatigue test show?

Key Terms

agglomeration
alloying elements
annealing
beneficiation
cast iron
corrosion
ferrous metals
flux
malleability
oxidation
pig
pig iron
reduction
sintering
slag
smelting
strand-casting machine
toughness
weathering steel

Activities

1. Collect representative samples of ferrous metals and prepare a sheet for each one, identifying the type and its properties.
2. Make a tour of your campus and compose a list of all the applications of ferrous metals you can find. Take photos, if possible.
3. Test samples of ferrous metals for tensile strength and hardness and keep a record of your findings for each sample. Compare your findings with those of others who tested the same materials.
4. Find products made from ferrous metals produced by various manufacturing processes, such as casting and extruding.

Additional Resources

Metals and Alloys in the Unified Numbering System, and other related publications, American Society of Automotive Engineers, Inc., Warrendale, PA, and the American Society for Testing and Materials, West Conshohocken, PA.

Other resources include:

Numerous publications related to metals, American Society for Testing and Materials, West Conshohocken, PA.

Publications of the Wire Reinforcement Institute, Hartford, CT.

See Appendix B for addresses of professional and trade organizations and other sources of technical information

CHAPTER 16

Nonferrous Metals

LEARNING OBJECTIVES

Upon completion of this chapter, the student should be able to:

- Understand the effects of galvanic corrosion and how to design to eliminate it.
- Select nonferrous metals for a wide range of applications.
- Know the properties of nonferrous metals and how these will influence a material's performance.

Build Your Knowledge

For further study on these materials and methods, please refer to:

Chapter 3 Properties of Materials
Topic: Mechanical and Chemical Properties

Chapter 28 Roofing Systems
Topic: Roofing Materials

Chapter 45 Plumbing Systems
Topics: Piping, Tubing, Fittings

Nonferrous metals are those containing little or no iron. In other words, all metals other than iron and steel are nonferrous. Some commonly found in construction projects include aluminum, copper, lead, tin, and zinc. Nickel is used mainly as an alloying element. Titanium, once used in aircraft, aerospace, and military applications, is now employed as a building cladding material.

GALVANIC CORROSION

Both ferrous and nonferrous metals are subject to **galvanic corrosion**. In the presence of an electrolyte (moisture in the atmosphere), dissimilar metals in contact will corrode more rapidly than similar metals in contact. For example, an aluminum gutter secured with copper nails produces galvanic action at the point of contact. The presence of moisture in the atmosphere sets up an electrolytic action that causes the aluminum to corrode. The metal that is higher on the table of electrolytic corrosion potential will corrode more (Table 16.1). When aluminum is plated with zinc and

Table 16.1 The Galvanic Series

Electrolytic Potential	Metals and Alloys
High potential (anode +)	Magnesium Magnesium alloys Aluminum (pure and several cast and wrought alloys) Zinc Cadmium Aluminum (wrought 2024 and 356.0 cast) Iron or steel
Electric current flows from positive (+) to negative (−)	Cast iron Stainless steel Lead Tin Nickel (active) Brass Copper Bronze
Low potential (cathode −)	Chromium stainless steel Silver Titanium Platinum Gold

electrolytic action corrosion occurs, the zinc corrodes and protects the aluminum. When steel is coated in zinc, it is called **galvanized**. When it is not possible to avoid contact between dissimilar metals, they can be given a coat of unleaded paint or separated with a plastic or other non-conducting material. A joint can also be caulked or otherwise sealed to keep out moisture, thus eliminating the electrolyte.

ALUMINUM

Aluminum (chemical symbol Al) is a versatile material used widely in building, and the construction industry is one of its largest consumers. Aluminum is lightweight, having a specific gravity of only 2.7 times that of water and approximately one-third that of steel. Pure aluminum melts at 1,220°F (665°C)—considerably lower than the melting point of other structural metals. It also is relatively weak as far as mechanical properties are concerned. Aluminum elastically deforms about three times more than steel under comparable loading. It can be strengthened by alloying, cold-working, or **strain hardening**. Aluminum alloys do not lose ductility or become brittle at cryogenic (low) temperatures.

Aluminum is a good conductor of electricity. Compared to copper wire of the same diameter, aluminum's conductivity is roughly 65 percent that of copper.

Mining Aluminum

Aluminum is found in most rocks and clays, but concentrations of the aluminum oxide content must be about 45 percent to be viable for economic extraction. Aluminum ores are called **bauxites**. Most ores are secured via open-pit mining. After extraction, the ore is crushed, washed, screened, ground, and dried. The embodied energy impacts of aluminum production are high, though they are offset by its inherent recyclability. Aluminum is widely recycled, and reprocessing the material requires considerably less energy.

Refining Bauxite Ore

Aluminum is refined using a two-step process. The first, the Bayer process, produces a very pure alumina (Al_2O_3). Alumina is an oxide of aluminum in crystal form. The second step reduces the alumina to a metallic aluminum, which is about 99 percent pure. This process is referred to as the Hall-Héroult process. It is named after the two men who developed the electrolytic method of aluminum production.

The Bayer Process

The ground dried bauxite is mixed in a digester with soda ash, crushed lime, sodium hydroxide, and hot water. Live steam and mechanical agitators stir the mixture (Fig. 16.1). The mixture is then pumped to

Construction Materials

History of Aluminum

People probably were using alum, one of the aluminum compounds found in nature, as early as 500 BC. These naturally occurring compounds were used as astringents (a substance that contracts the tissues of the body), to fix dyes in cloth, and in the tanning of animal skins.

The Romans used a natural potassium sulfate they called alumen. This was purified into crystalline alum around AD 1200, and by the 1500s alum was produced from the clay in which it occurred.

The actual production of aluminum as we know it today happened much more recently—in fact, it is the most recently developed and used metal in construction. Various attempts were made by German, French, and Danish chemists and scientists to isolate aluminum. Around 1850, one was successful in isolating a small quantity of the metal by decomposing anhydrous aluminum chloride with potassium.

In 1855, an aluminum ingot was exhibited at the Paris Exposition. It was a rarity worth more per pound than gold. Emperor Napoleon III commissioned Henri Sainte-Claire Deville to find a way to manufacture large amounts of aluminum to be used for army equipment. He successfully produced a few tons but not the huge commercial quantities required.

In 1886, an American, Charles Martin Hall, and a Frenchman, Paul Louis-Toussaint Héroult, independently developed the production method used today, in which the alumina is dissolved in molten cryolite and decomposed electrolytically. Since that time, industry has begun to produce large amounts of aluminum. It first appeared in construction products around 1925. Applications have expanded since then to include hundreds of internal and external uses.

Figure 16.1 The first step in producing aluminum, the Bayer process, is to separate the alumina from the bauxite ore.

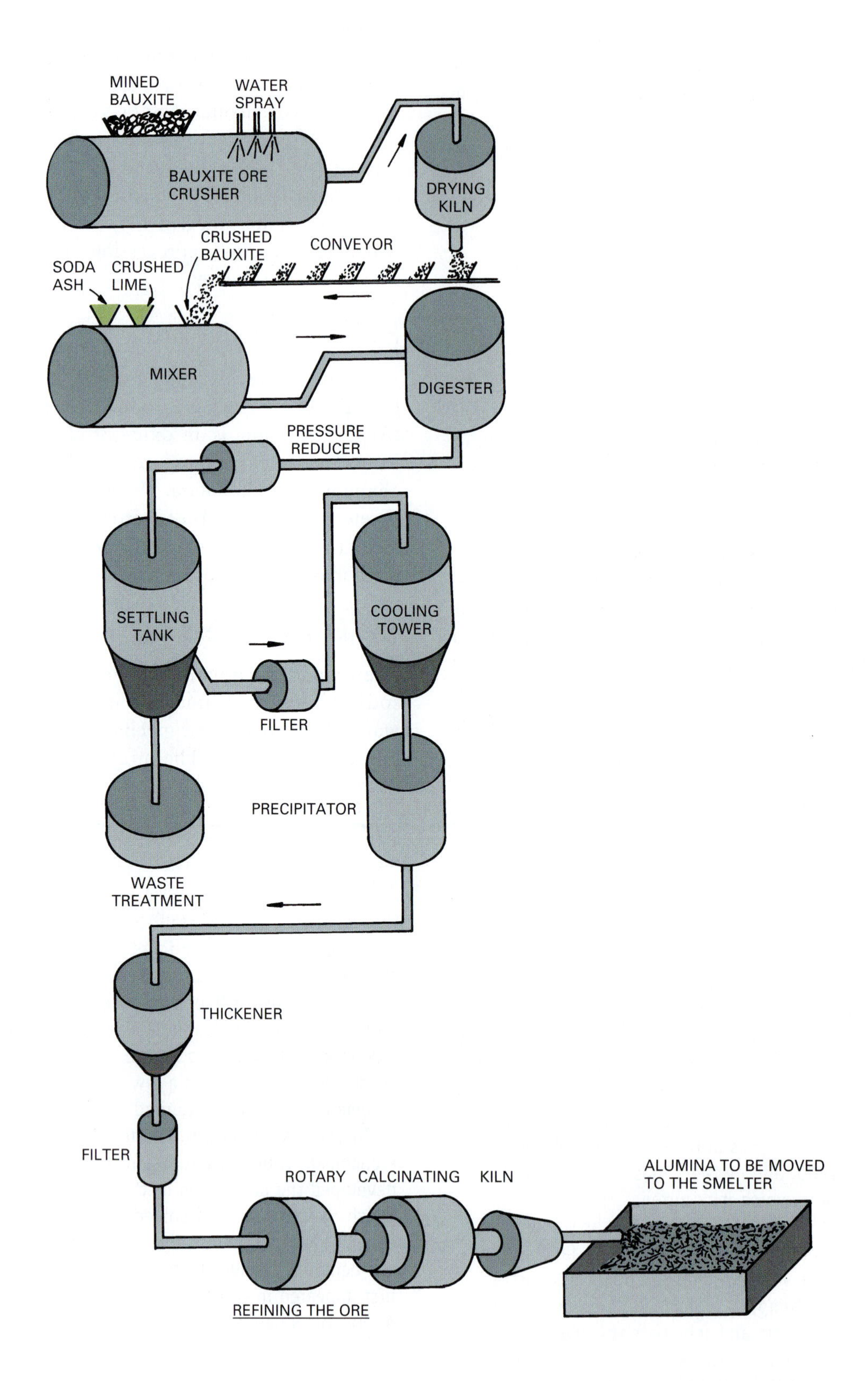

digester tanks where, under high pressure and the injection of steam, it is churned. The chemical reaction forms sodium aluminate, and the insoluble impurities form a waste material. The mixture flows through a pressure-reducing tank into a settling tank where waste is removed. The sodium aluminate solution passes through a filter to a cooling tower and into a precipitator where aluminum hydrate is added. Compressed air agitates the concoction, and cooling continues, allowing the sodium aluminate to precipitate as aluminum hydrate. This is pumped into filter tanks that separate the aluminum hydrate from the solution. It is then calcined in a rotary kiln operating at about 2,000°F (1,100°C), which produces the alumina used to produce the metallic aluminum in the second step, the Hall-Héroult electrolytic process.

THE HALL-HÉROULT ELECTROLYTIC PROCESS

Aluminum is produced from the oxide **alumina** by a reduction process (Fig. 16.2). **Reduction** is an electrolytic procedure that uses a carbon-lined vessel containing molten cryolite and alumina. An electric current is passed through this liquid via large carbon anodes suspended in it. When the current hits the liquid, molten aluminum separates and settles on the bottom of the vessel, where it is siphoned off. Pure aluminum may be cast in molds for use in products requiring its properties or sent to a holding furnace where alloying elements are added to alter its physical properties.

Figure 16.2 The second step in producing aluminum is the reduction of alumina using the Hall-Héroult process.

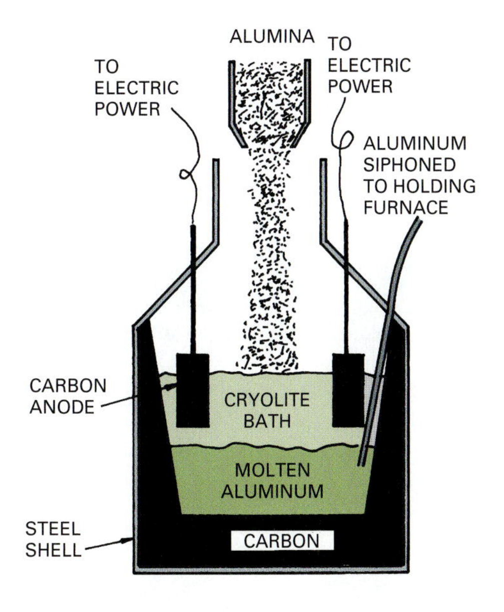

Aluminum Alloys

Produced aluminum is between 99.5 and 99.9 percent pure. In this form it is relatively soft and ductile and has a tensile strength of around 7000 psi (48,258 kPa). For many products, aluminum must have alloying elements added to alter its physical properties.

Aluminum alloys have two classifications: wrought and casting. **Wrought alloys** are those that are mechanically worked by processes, such as forging, drawing, extruding, or rolling, to form sheet material. **Cast alloys** are those used to produce a product for which the molten metal is cast in a finished shape, such as a grille, in a sand mold or permanent mold.

Wrought Aluminum Alloy Classifications The major alloying elements added to aluminum are manganese, copper, magnesium, silicon, and zinc. A wide variety of aluminum alloys are identified by a code system developed by the Aluminum Association, Inc. Each wrought aluminum alloy is specified by a four digit code number, as shown in Table 16.2.

The first digit denotes the alloy series and indicates the major alloying element. The second digit in the 1xxx series represents a modification of impurity limits, and in the 2xxx through 9xxx series it represents a modification of the alloy. The last two digits in the 1xxx series indicate aluminum purity above 99 percent. For example, a 1050 specifies an aluminum containing 0.50 percent more aluminum than the minimum 99 percent. In the 2xxx through 9xxx series, the last two digits are arbitrary numbers identifying the alloy in the series. Examine Tables 16.3 and 16.4 for a breakdown of these classification systems.

Table 16.2 Wrought Aluminum and Aluminum Alloy Designation System

Aluminum 99.00% Minimum 1xxx	
Aluminum Alloys Grouped by Major Alloying Elements	
Copper	2xxx
Manganese	3xxx
Silicon	4xxx
Magnesium	5xxx
Magnesium and silicon	6xxx
Zinc	7xxx
Other elements	8xxx
Unused series	9xxx

Courtesy the Aluminum Association

Cast Aluminum Alloy Classifications Cast alloys are specified by a three-digit number followed by a decimal (Table 16.5). The first digit identifies the alloy series and the second and third digits the specific alloy or purity. The decimal indicates whether the alloy composition is for final casting (.0) or for ingot (.1 or .2).

Unified Numbering System Designations

The Unified Numbering System (UNS) is described in detail in Chapter 15. It uses a five-digit number with a letter prefix to classify and identify various types of aluminum and aluminum alloys specified by other organizations, such as the Aluminum Association (AA), American Society for Testing and Materials (ASTM), Society of Automotive Engineers/Aerospace Materials Specifications (AMS), Military Specifications (MS), and Federal Specifications (FS). The UNS pulls together those materials with like specifications from these organizations and gives them a unified number. Table 16.6 gives a few selected examples to illustrate how this system combines them.

Table 16.3 Breakdown of 1xxx Aluminum Designations

1 X XX		
Indicates 99% pure commercial aluminum	0 indicates no special impurity control	Indicates pure aluminum content
	1–9 indicates special specific impurity controls	above 99% in hundredths of a percent

Table 16.4 Breakdown of 2xxx–9xxx Aluminum Designations

X X XX		
1 commercially pure	0 indicates original alloy developed	in 1xxx series indicates impurity limits
2–9 indicates major alloying element as shown in Table 16.2	2–9 indicate modifications of the original alloy	2xxx through 9xxx series—arbitrary numbers identifying the alloy in the series

Table 16.5 Cast Aluminum and Aluminum Alloy Designation System

Aluminum 99.00% minimum	1xx.x
Aluminum Alloys Grouped by Major Alloying Elements	
Copper	2xx.x
Silicon, with added copper and/or magnesium	3xx.x
Silicon	4xx.x
Magnesium	5xx.x
Zinc	7xx.x
Tin	8xx.x
Other elements	9xx.x
Unused series	6xx.x

Courtesy the Aluminum Association

Temper Designations

Temper is the degree of hardness and strength imparted to a metal by a process, such as **heat treating** or cold-working. Tempering refers to the process that renders the proper degree of hardness and elasticity to a metal so it can be used for an intended purpose.

Wrought aluminum alloys fall into two classes: heat-treatable and non-heat-treatable. **Heat-treatable alloys** are those whose strength characteristics are improved by heat treating. Their strength is also improved by adding alloying elements, such as copper, zinc, silicon, and magnesium.

Non-heat-treatable alloys have alloying elements added that do not cause an increase in strength when heat treated. The strength of non-heat-treatable alloys depends on elements such as iron, magnesium, manganese, and silicon. These alloys are strengthened by cold-rolling or **strain hardening.**

Some temper specifications apply only to cast aluminum, and others apply only to wrought aluminum. Alloys specified as F, H, or O can be hardened by cold-working and may or may not be non-heat-treatable. Heat-treatable aluminum alloys use the T and W designations.

The specification of an aluminum alloy requires a designation of temper or metallurgical condition. The following temper designation system for aluminum

Table 16.6 Selected Examples of Aluminum Alloy UNS Designations

UNS Designations	Material	Numbers Used by Various Organizations
A02400	Aluminum foundry alloy, casting	AA 240.0, AMS 4227
A91035	Wrought aluminum alloy, non-heat-treatable	AA 1035
A92011	Wrought aluminum alloy, heat-treatable	AA 2011, ASTM B210, FS QQ-A-22513, SAE J454

Courtesy Society of Automotive Engineers

alloys was developed by the Aluminum Association, Inc. The **temper designation** consists of letters and numbers placed after an alloy number and separated from it by a dash.

Temper designations include:

F (as fabricated): No special control over thermal or work-hardening conditions is used.

Q (annealed): Wrought products have been heated to effect recrystallation, which produces the lowest strength. Cast products are annealed to improve stability and ductility.

H (strain hardened): Wrought products are strain hardened through cold-working. The H is followed by one or two digits. See Table 16.7.

W (solution heat-treated): The alloy is heated to about 1,000°F (542°C) and then quenched.

T (thermally treated): The product has been heat treated and then strain hardened. The T is followed by one or two digits. See Table 16.8.

Some applications for aluminum alloys are shown in Table 16.9. Table 16.10 gives some detailed uses of aluminum in construction.

Aluminum Castings

The conditions involved with the production of aluminum castings influence their physical properties. The metallurgist has to consider not only the characteristics of an alloy and the use to which the product will be put but also how it will be cast. In general, aluminum cast in permanent (metal) molds is stronger than sand-cast products. The cost of producing metal molds is high but can be justified when many products need casting and the molds can be reused. A sand-casting mold produces only one casting.

Heat-treated alloys typically are stronger and have greater ductility than non-heat-treatable alloys. After casting, heat-treatable alloy castings are subjected to a very high temperature (but below the alloy's melting point), then quenched in cold or hot oil or water and left to age at room temperature. This can help control the characteristics of the alloy.

Aluminum Finishes

As mentioned earlier, when aluminum is exposed to oxygen it forms a natural protective **oxide layer**, so under ordinary circumstances no protective coating is necessary. However, various finishes are applied to aluminum products to improve their appearance and usefulness. The Aluminum Association specifies three categories of finishes: mechanical, chemical, and coatings. Aluminum can also have an **anodized** finish or be left natural. The flow chart in Fig. 16.3 illustrates these finishes and shows their relationships.

Table 16.7 Subdivision of H Temper: Strain Hardened

First Digit Indicates Specific Treatment:
H1—Strain hardened only
H2—Strain hardened and partially annealed
H3—Strain hardened and stabilized
H4—Strain hardened and lacquered or painted
Digit (0–8) Indicates the Degree of Strain Hardening as Identified by a Minimum Value of the Ultimate Tensile Strength[a]:
0—Annealed
2—Tempers whose ultimate tensile strength is midway between 0 and 4.
4—Tempers whose ultimate tensile strength is midway between 0 and 8.
6—Tempers whose ultimate tensile strength is midway between 4 and 8.
8—Tempers whose ultimate tensile strength exceeds that of 8 by 2 ksi or more.
1, 3, 5, 7—Tempers whose ultimate tensile strength falls between those defined above.

[a]*The tensile strength for each number is specified by Aluminum Association tables.*
Courtesy the Aluminum Association

Table 16.8 Subdivisions of T Temper: Thermally Treated

First Digit Indicates Specific Sequence of Treatments:
T1—Naturally aged after cooling from an elevated temperature shaping process
T2—Cold-worked after cooling from an elevated temperature shaping process and then naturally aged
T3—Solution heat treated, cold-worked, and naturally aged
T4—Solution heat treated and naturally aged
T5—Artificially aged after cooling from an elevated temperature shaping process
T6—Solution heat treated and artificially aged
T7—Solution heat treated and stabilized (overaged)
T8—Solution heat treated, cold-worked, and artificially aged
T9—Solution heat treated, artificially aged, and cold-worked
T10—Cold-worked after cooling from an elevated temperature shaping process and then artificially aged
Second Digit Indicates Variation in Basic Treatment:
Examples:
T42 or T62—Heat treated to temper by user
Additional Digits Indicate Stress Relief:
Examples:
TX51—Stress relieved by stretching after solution heat treating
TX52—Stress relieved by compressing after solution heat treating or cooling

Courtesy the Aluminum Association

Table 16.9 Typical Applications of Aluminum Alloys

AA Alloy Series	AA Typical Alloys	UNS Designation	Typical Applications
Typical Applications of Non-Heat-Treatable Aluminum Alloys			
AA Alloy Series	**AA Typical Alloys**	**UNS Designation**	**Typical Applications**
1xxx	1350	A91350	Electrical conductors
	1060	A91060	Chemical equipment, tank cars
	1100	A91100	Sheet metal work, cooking utensils, decorative
3xxx	3003	A93003	Sheet metal work, chemical equipment, storage tanks
4xxx	4043	A4043	Welding electrodes
	4343	A94343	Brazing alloy
5xxx	5005	A95005	Decorative and automotive trim, architectural and anodized, sheet metal work, appliances
	5050	A95050	
	5454	A95454	
	5456	A95456	
5xxx (3% Mg)	5083	A95083	Marine, welded structures, storage tanks, pressure vessels, armor plate, cryogenics
	5086	A95086	
	5454	A95454	
	5456	A95456	
Typical Applications of Heat-Treatable Aluminum Alloys			
AA Alloy Series	**AA Typical Alloys**	**UNS Designation**	**Typical Applications**
2xxx (Al-Cu)	2011	A92011	Screw machine products
	2219	A92219	Structural, high temperature
2xxx (Al-Cu-Mg)	2014	A92014	Aircraft structures and engines, truck frames and wheels
	2024	A92024	
	2618	A92618	
4xxx	4032	A94032	Pistons
6xxx	6061	A96061	Marine, truck frames and bodies, structures, architectural furniture
	6063	A96063	
7xxx (Al-Zn-Mg)	7004	A97004	Structural, cryogenics, missile
	7005	A97005	
7xxx (Al-Zn-Mg-Cu)	7001	A97001	High-strength structural and aircraft
	7075	A97075	
	7178	A97178	

Courtesy the Aluminum Association

Table 16.10 Specific Construction Product Applications

UNS Designation	AA Alloy Number	Use
A93003	3003	Flashing, ducts, garage doors, curtain wall panels, grilles, louvers, siding, shingles, termite shields
A91235	1235	Vapor barriers, insulation
A04430	B443.0	Cast products, such as hardware, curtain wall castings, architectural letters
A96063	6063	Extruded products, such as curtain wall, door, and window frames; grilles, mullions; railings; thresholds

The Aluminum Association classifies finishes with a letter and a two-digit number. Mechanical finishes are designated with “M” and chemical finishes with “C.” Coatings are designated by letters denoting a particular type, and anodic coatings use the letter A. The specifications and designations for aluminum finishes are given in Tables 16.11, 16.12, and 16.13.

Natural Finishes

Natural finishes may be controlled or uncontrolled. Uncontrolled finishes are imparted to wrought and cast products by the surface conditions of the rollers, molds, or extruding dies used to form them. Hot-rolled products have a brighter surface. Controlled finishes result

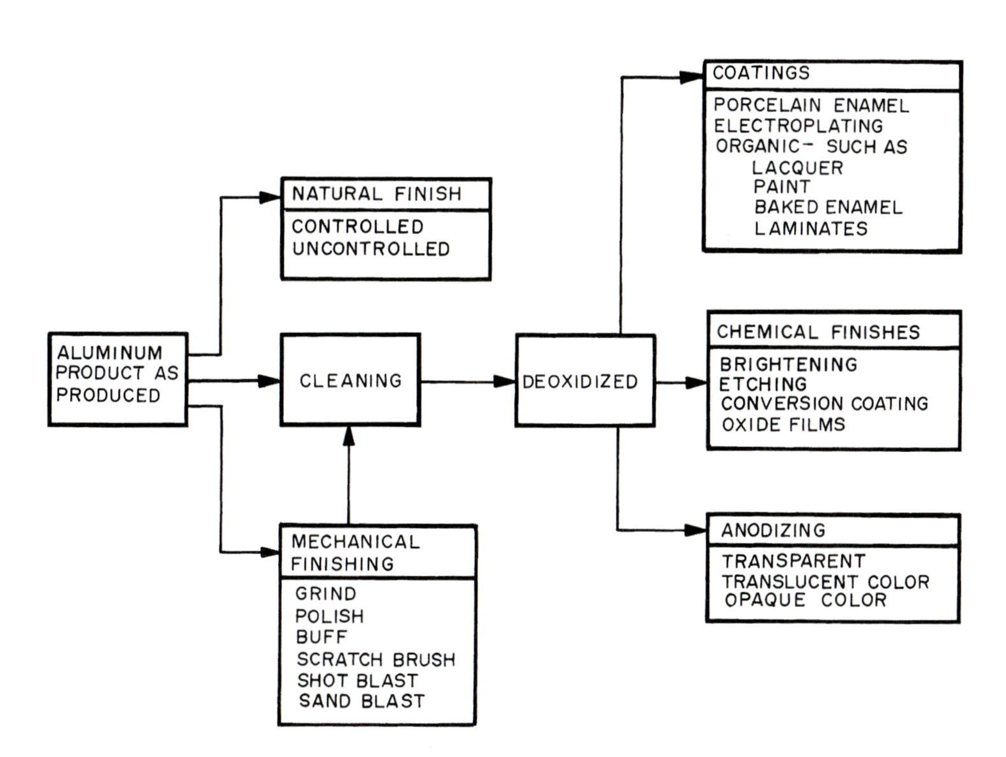

Figure 16.3 The finishing processes that can be used on various aluminum products.

from varying the smoothness of rollers or mold surfaces. A sheet can have a controlled finish on one side, effected with smooth rollers, and an uncontrolled finish on its other side. Sheets may be embossed using rollers with surface designs, as in the case of imitation-wood-grain aluminum siding.

Mechanical Finishes

Mechanical finishes are used to alter surface appearances of aluminum products. They are usually produced before surfaces are cleaned to receive other finishing processes. Surfaces may be buffed with a coarse abrasive or receive a high luster polish via a fine abrasive. They may be ground with a dry grinding wheel. This is often necessary to clean up ridges on castings. It produces a rough, scratched finish. Rotating wire brushes produce scratched surfaces. The scratches can vary from very fine to coarse, depending on the size of the wire in the brushes. A matte surface is produced by blasting with sand or steel shot. Round pieces of abrasive cloth or steel wool can be rotated against a surface to produce a series of concentric circles. Code designations for mechanical finishes are given in Table 16.11.

Chemical Finishes

Chemical finishes are produced by the reaction of an aluminum surface to various chemicals. Conversion coatings are a major chemical finish. They prepare surfaces for the bonding of paints, organic coatings, and laminates. A natural oxide film does not always provide an adequate bonding surface. A frosty surface can be etched with chemicals. Designs can be etched on surfaces by masking all areas except those to be etched. Chemical oxide films produce surfaces with greater corrosion resistance. Aluminum can be plated in a process called zincating, which produces a thin zinc coating. This coating protects aluminum from galvanic action and also prepares surfaces for electroplating. Aluminum use can produce surfaces highly reflective of heat and light. These mirror-like finishes result from chemical brightening. Code designations for chemical finishes are given in Table 16.12.

Anodic Finishes

A very important, widely used finish on aluminum is an anodized electrolytic oxide layer. Anodized films can be used as finishes on surfaces to be painted. Anodized coatings are most often employed as a finished protecting layer. A film is less than 0.1 mil thick, while a coating is 0.1 mil or thicker.

Following is a typical procedure for the anodizing process.

1. Alkaline and/or acid cleaners are used to remove grease and dirt from a surface.
2. Next, the surface receives an etching or brightening pretreatment. Etching involves producing a matte

Table 16.11 Mechanical Finishes on Aluminum

Type of Finish	Designation[a]	Description	Examples of Methods of Finishing[b]
As fabricated	M10	Unspecified	
	M11	Specular as fabricated	
	M12	Nonspecular as fabricated	
	M1X	Other	To be specified.
Buffed	M20	Unspecified	
	M21	Smooth specular	Polished with grits coarser than 320. Final polishing with a 320 grit using peripheral wheel speed of 30 m/s (6,000 feet per min). Polishing followed by buffing, using tripoli based buffing compound and peripheral wheel speed of 36 to 41 m/s (7,000 to 8,000 feet per min).
	M22	Specular	Buffed with tripoli compound using peripheral wheel speed 36 to 41 m/s (7,000 to 8,000 feet per min).
	M2X	Other	To be specified.
Directional textured	M30	Unspecified	
	M31	Fine satin	Wheel or belt polished with aluminum oxide grit of 320 to 400 size; peripheral wheel speed 30 m/s (6,000 feet per min).
	M32	Medium satin	Wheel or belt polished with aluminum oxide grit of 180 to 220 size; peripheral wheel speed 30 m/s (6,000 feet per min).
	M33	Coarse satin	Wheel or belt polished with aluminum oxide grit of 80 to 100 size; peripheral wheel speed 30 m/s (6,000 feet per min).
	M34	Hand rubbed	Hand rubbed with stainless steel wool lubricated with neutral soap solution. Final rubbing with No. 00 steel wool.
	M35	Brushed	Brushed with rotary stainless steel wire brush, wire diameter 0.24 mm (0.0095 in.); peripheral wheel speed 30 m/s (6,000 feet per min); or various proprietary satin finishing wheels or satin finishing compounds with buffs.
	M3X	Other	To be specified.
Nondirectional textured	M40	Unspecified	
	M41	Extra fine matte	Air blasted with finer than 200 mesh washed silica or aluminum oxide. Air pressure 310 kPa (45 psi); gun distance 203 to 305 mm (8 to 12 inches) from work at 90° angle.
	M42	Fine matte	Air blasted with 100 to 200 mesh silica sand if darkening is not a problem; otherwise aluminum oxide type abrasive. Air pressure 207 to 621 kPa (30 to 90 psi) (depending upon thickness of material); gun distance 305 mm (12 inches) from work at angle of 60 to 90°.
	M43	Medium matte	Air blasted with 40 to 50 mesh silica sand if darkening is not a problem; otherwise aluminum oxide type abrasive. Air pressure 207 to 621 kPa (30 to 90 psi) (depending upon thickness of material); gun distance 305 mm (12 inches) from work at angle of 60 to 90°.
	M44	Coarse matte	Air blasted with 16 to 20 mesh silica sand if darkening is not a problem; otherwise aluminum oxide type abrasive. Air pressure 207 to 621 kPa (30 to 90 psi) (depending upon thickness of material); gun distance 305 mm (12 inches) from work at angle of 60 to 90°.
	M45	Fine shot blast	Shot blasted with cast steel shot of ASTM size 70 to 170 applied by air blast or centrifugal force. To some degree, selection of shot size is dependent on thickness of material since warping can occur.
	M46	Medium shot blast	Shot blasted with cast steel shot of ASTM size 230 to 550 applied by air blast or centrifugal force. To some degree, selection of shot size is dependent on thickness of material since warping can occur.
	M47	Coarse shot blast	Shot blasted with cast steel shot of ASTM size 660 to 1,320 applied by air blast or centrifugal force. To some degree, selection of shot size is dependent on thickness of material since warping can occur.
	M4X	Other	To be specified.

[a]The complete designation must be preceded by AA—signifying Aluminum Association.
[b]Examples of methods of finishing are intended for illustrative purposes only.
Courtesy of the Aluminum Association

Table 16.12 Chemical Finishes on Aluminum

Type of Finish	Designation[a]	Description	Examples of Methods of Finishing[b]
Nonetched cleaned	C10	Unspecified	
	C11	Degreased	Organic solvent treated.
	C12	Inhibited chemical cleaned	Inhibited chemical type cleaner used.
	C1X	Other	To be specified.
Etched	C20	Unspecified	
	C21	Fine matte	Trisodium phosphate, 22–45 g/l (3–6 oz per gal) used at 60–71°C (140–160°F) for 3 to 5 min.
	C22	Medium matte	Sodium hydroxide, 30–45 g/l (4–6 oz per gal) used at 49–66°C (120–150°F) for 5 to 10 min.
	C23	Coarse matte	Sodium fluoride, 11 g/l (1.5 oz) plus sodium hydroxide 30–45 g/l (4–6 oz per gal) used at 54–66°C (130–150°F) for 5 to 10 min.
	C2X	Other	To be specified.
Brightened	C30	Unspecified	
	C31	Highly specular	Chemical bright dip solution of the proprietary phosphoric-nitric acid type used, or proprietary electrobrightening or electropolishing treatment.
	C32	Diffuse bright	Etched finish C22 followed by brightened finish C31.
	C3X	Other	To be specified.
Chemical coatings[c]	C40	Unspecified	
	C41	Acid chromate-fluoride	Proprietary chemical treatments used producing clear to typically yellow colored surfaces.
	C42	Acid chromate-fluoride-phosphate	Proprietary chemical treatments used producing clear to typically green colored surfaces.
	C43	Alkaline chromate	Proprietary chemical treatments used producing clear to typically gray colored surfaces.
	C44	Non-chromate	Proprietary chemical coating treatment employing no chromates.
	C45	Non-rinsed chromate	Proprietary chemical coating treatment in which coating liquid is dried on the work with no subsequent water rinsing.
	C4X	Other	To be specified.

[a]*The complete designation must be preceded by AA—signifying Aluminum Association.*
[b]*Examples of methods of finishing are intended for illustrative purposes only.*
[c]*Includes chemical conversion coatings.*
Courtesy the Aluminum Association

surface with hot solutions of sodium hydroxide. This process removes minor surface imperfections, and a thin layer of aluminum. Brightening produces a mirror-like surface with a concentrated solution of phosphoric and nitric acids. These smooth the surface through chemical reaction.

3. The third step is the actual anodizing process, in which the anodic film is built and combined with the aluminum by passing an electric current through an acid electrolyte bath immersing the aluminum. The coating thickness and finished surface characteristics can be carefully controlled.
4. Coloring the anodized surface can occur in several ways. One method is to combine the coloring with the actual anodizing process (step 3), which simultaneously forms and colors the oxide cell wall in bronze and black shades. This produces a more abrasion-resistant coating than other methods but is more expensive because it requires more electricity.

A second coloring procedure involves a two-step electrolytic coloring process. After aluminum is anodized (step 3), it is immersed in a bath containing an inorganic metal salt and subjected to an electric current that deposits the metal salt at the base of the pores. Color depends on the metal salt used (for example, tin, copper, nickel, cobalt). This method provides the greatest variation of colors. A third process involves organic dyeing. This produces vibrant colors that are highly weather resistant.

Table 16.13 Anodic Coatings on Aluminum

Type of Finish	Designation[a]	Description	Examples of Methods of Finishing[b]
General	A10	Unspecified	
	A11	Preparation for other applied coatings	3 μm (0.1 mil) anodic coating produced in 15% H_2SO_4 at 21° ± 1°C (70°F ± 2°F) at 129 A/m^2 (12 A/ft.^{2}) for 7 min, or equivalent.
	A12	Chromic acid anodic coatings	To be specified.
	A13	Hard, wear and abrasion resistant coatings	To be specified.
	A1X	Other	To be specified.
Protective and Decorative Coatings less than 10 μm (0.4 mil) thick	A21	Clear coating	Coating thickness to be specified. 15% H_2SO_4 used at 21° ± 1°C (70°F ± 2°F) at 129 A/m^2 (12 A/ft.^{2}).
	A211	Clear coating	Coating thickness—3 μm (0.1 mil) minimum. Coating weight—6.2 g/m^2 (4 mg/in.^{2}) minimum.
	A212	Clear coating	Coating thickness—5 μm (0.2 mil) minimum. Coating weight—12.4 g/m^2 (8 mg/in.^{2}) minimum.
	A213	Clear coating	Coating thickness—8 μm (0.3 mil) minimum. Coating weight—18.6 g/m^2 (12 mg/in.^{2}) minimum.
	A22	Coating with integral color	Coating thickness to be specified. Color dependent on alloy and process methods.
	A221	Coating with integral color	Coating thickness—3 μm (0.1 mil) minimum. Coating weight—6.2 g/m^2 (4 mg/in.^{2}) minimum.
	A222	Coating with integral color	Coating thickness—5 μm (0.2 mil) minimum. Coating weight—12.4 g/m^2 (8 mg/in.^{2}) minimum.
	A223	Coating with integral color	Coating thickness—8 μm (0.3 mil) minimum. Coating weight—18.6 g/m^2 (12 mg/in.^{2}) minimum.
	A23	Coating with impregnated color	Coating thickness to be specified. 15% H_2SO_4 used at 27°C ± 1°C (80°F ± 2°F) at 129 A/m^2 (12 A/ft.^{2}) followed by dyeing with organic or inorganic colors.
	A231	Coating with impregnated color	Coating thickness—3 μm (0.1 mil) minimum. Coating weight—6.2 g/m^2 (4 mg/in.^{2}) minimum.
	A232	Coating with impregnated color	Coating thickness—5 μm (0.2 mil) minimum. Coating weight—12.4 g/m^2 (8 mg/in.^{2}) minimum.
	A233	Coating with impregnated color	Coating thickness—8 μm (0.3 mil) minimum. Coating weight—18.6 g/m^2 (12 mg/in.^{2}) minimum.
	A24	Coating with electrolytically deposited color	Coating thickness to be specified. Application of the anodic coating, followed by electrolytic deposition of inorganic pigment in the coating.
	A2X	Other	To be specified.
Architectural Class II[c] 10 to 18 μm (0.4 to 0.7 mil) coating	A31	Clear coating	15% H_2SO_4 used at 21°C ± 1°C (70°F ± 2°F) at 129 A/m^2 (12 A/ft.^{2}) for 30 min, or equivalent.
	A32	Coating with integral color	Color dependent on alloy and anodic process.
	A33	Coating with impregnated color	15% H_2SO_4 used at 21°C ± 1°C (70°F ± 2°F) at 129 A/m^2 (12 A/ft.^{2}) for 30 min, followed by dyeing with organic or inorganic colors.
	A34	Coating with electrolytically deposited color	Application of the anodic coating followed by electrolytic deposition of inorganic pigment in the coating.
	A3X	Other	To be specified.

Table 16.13 Anodic Coatings on Aluminum *(Continued)*

Type of Finish	Designation[a]	Description	Examples of Methods of Finishing[b]
Architectural Class I[c] 18 μm (0.7 mil) and thicker coatings	A41	Clear coating	15% H_2SO_4 used at 21°C ± 1°C (70°F ± 2°F) at 129 A/m² (12 A/ft.²) for 60 min, or equivalent.
	A42	Coating with integral color	Color dependent on alloy and anodic process.
	A43	Coating with impregnated color	15% H_2SO_4 used at 21°C ± 1°C (70°F ± 2°F) at 129 A/m² (12 A/ft.²) for 60 min, followed by dyeing with organic or inorganic colors, or equivalent.
	A44	Coating with electrolytically deposited color	Application of the anodic coating followed by electrolytic deposition of inorganic pigment in the coating.
	A4X	Other	To be specified.

[a]*The complete designation must be preceded by AA—signifying Aluminum Association.*
[b]*Examples of methods of finishing are intended for illustrative purposes only.*
[c]*Aluminum Association Standards for Anodized Architectural Aluminum.*
Courtesy the Aluminum Association

One other coloring process is described as interference coloring. It involves modification of the pore structure produced in sulfuric acid. It results in lightfast colors ranging from blue to green and yellow to red.

Anodic coatings are classified into four groups: General, Protective and Decorative, Architectural I, and Architectural II. Coatings that are less than 0.1 mil thick are classified as General. Protective and Decorative coatings are less than 0.4 mil thick. These two classes are used for general industrial applications. The architectural classes are used on materials exposed to weather and wear. Details can be found in Table 16.13.

Architectural I coatings are recommended for exterior use where they will receive no regular maintenance. They are also used for interior purposes where extra protection is needed. They must be over 0.7 mil in thickness and weigh more than 27 mg per in.²

Architectural II coatings are recommended for interior applications not expected to receive heavy wear and for exterior uses where a product receives regular maintenance. They must have a thickness from 0.4 to 0.7 mils and weigh 17 to 27 mg per in.² Code designations for architectural classes can be found in Table 16.13.

Coatings

Many aluminum products are painted to provide additional protection or for surface decoration. These products, such as aluminum gutters, are factory painted and delivered to site finished and ready to install. The factory-applied finish is electrostatically sprayed or roller applied with an organic paint and then passed through an oven where it is baked to a hard, uniform finish. The paint is flexible and will not crack if the product is bent or formed on the job.

Porcelain enamel coatings are produced by using a vitreous inorganic material that is bonded to the metal by fusing it at high temperatures. The coating is very resistant to corrosion, durable, and available in a wide range of colors. It is widely used on curtain wall panels, as seen on high-rise buildings, where it is difficult to provide regular maintenance.

Aluminum can be electroplated with chromium, copper, and other materials. If chromium is electroplated, the aluminum must first be zincated or plated with copper, brass, or nickel. Chromium plating gives a mirror-like surface and provides resistance to abrasion. Copper plating requires that the aluminum be zincated. It is mainly used where electrical connections must be soldered to an aluminum product. Tin and brass plating also provide good soldering surfaces. Zinc and cadmium plating improve corrosion resistance.

Aluminum surfaces can also have laminated finishes. This involves bonding another material, such as vinyl or polyvinyl chloride films, to the surface with an adhesive. Usually the aluminum surface must be chemically cleaned to provide the strongest bond.

Aluminum surfaces that serve as high-quality mirrors are treated by electropolishing. This is an anodic smoothing of the surface and requires high-purity aluminum. Finished surfaces usually have a final anodic protective coating applied.

Protecting a Finished Aluminum Product

Aluminum products on a construction site need protection from damage during delivery, storage, and installation. Following are several coatings used for this purpose.

The best and most expensive protective coatings consist of paper or plastic sheet material bonded to the product with an adhesive that permits easy removal, leaving little or no sticky residue behind. If on-site protection is needed, masking tape, the kind used by painters, can suffice. Sometimes a coat of clear lacquer is sprayed on the aluminum's surface. Methacrylate lacquers chalk off after several years, but applying an automobile wax or a polish can afford added protection.

Routine Maintenance of Aluminum Surfaces

When cleaning aluminum surfaces, the type of protective film must be considered. A natural aluminum will develop its own oxide film that, when scrubbed or rubbed with abrasive cleaners or steel wool, will be damaged. Eventually, another oxide layer will form. However, unless the surface is stained, abrasive cleaners should be avoided. Likewise, anodized surfaces, plated surfaces, and other coatings can suffer damage that is not easily repaired.

The simplest cleaner is clear water. Surfaces can be washed with water and liquid soaps, then rinsed. If some discoloration needs removing, a mild polish, such as liquid auto polish, is acceptable. Liquid non-scratch cleaners, the kinds used on plastic bath fixtures, are suitable as well. Cleaners containing abrasive materials, such as kitchen cleansers, will scratch a surface. Liquid or paste wax is a good final protective coating.

Joining Aluminum Members

Aluminum members can be joined by any of the standard fastening techniques. Mechanical fasteners include screws, bolts, rivets, and a variety of specially designed products. Sheet stock can be joined by stitching, which involves sewing sheets together with aluminum wire. Aluminum members can be attached by welding, brazing, and soldering. This includes gas, arc, resistance, and inert gas-shielded arc welding. Adhesive bonding is a technique gaining wide usage. A variety of adhesives can be used, depending on materials and design considerations. Bonding produces a strong joint and increases the design possibilities for fusing aluminum and forming laminates.

Aluminum Products

Aluminum is a major construction material that finds use in many forms. Following are examples of some of the most common uses.

One major application for aluminum is in the production of metal doors and windows (Fig. 16.4). The members used for this purpose are extruded from an aluminum alloy, such as 6063. A wide range of interior and exterior aluminum doors and windows are made using framed glass units encased in an extruded aluminum frame. These can be panelized, flush, louvered, and have glass lights. A variety of other aluminum extrusions are available (Fig. 16.5).

Specifications developed by the Architectural Aluminum Manufacturers Association (AAMA) have been adopted by the American National Standards Institute (ANSI). These specify minimum frame

Figure 16.4 Aluminum windows are widely used in residential and commercial construction. *(Courtesy Acorn Building Components)*

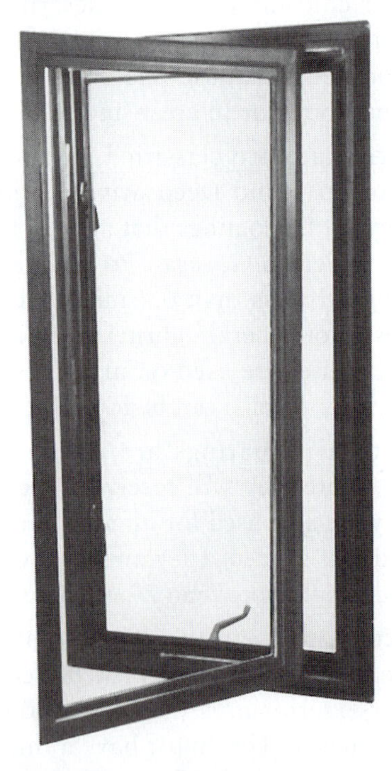

Figure 16.5 A detailed view of extruded aluminum shapes. *(Courtesy Wikipedia, http://en.wikipedia.org/wiki/Image:Extruded_aluminium_section_x3.jpg)*

strength, thickness, corrosion resistance, air infiltration, water resistance, wind load capacity, and condensation resistance. Units meeting these specifications have the AAMA seal attached.

Aluminum is used for residential and commercial siding and roofing. The specifications are developed by the AAMA. Residential siding is cold-rolled to shape and comes in several widths. Horizontal, vertical, and panelized siding sheets are available (Fig. 16.6). The panels may or may not have a backer board. Backer boards can be made of fiberboard or foamed plastic, both of which provide some insulation value.

The same material is used for soffits, fascia, flashing, gutters, and frieze boards (Fig. 16.7). Panels are secured to wood framing with aluminum nails.

Figure 16.6 Aluminum siding is weather resistant and adds insulation value with backer boards. *(Image copyright V. J. Matthew, 2009. Used under license from Shutterstock.com)*

Figure 16.7 Aluminum soffits, fascias, and gutters require no maintenance. *(Image copyright Amy Walters, 2009. Used under license from Shutterstock.com)*

Aluminum panels for siding and roofing commercial buildings are made in a variety of designs. Aluminum alloy 3004 is often used, and the panels are one-third the weight of steel. Common finishes include mill, unpainted, painted, and stucco embossed.

Aluminum curtain wall systems include preformed insulated wall panels that are integrated with aluminum windows to form a weather-tight, maintenance-free exterior surface. They are available with anodized or factory-applied baked enamel finishes. More information is given in Chapter 31. A wide array of roof accessories, such as skylights, roof hatches, and smoke and fire vents and louvers, are also made with extruded and sheet aluminum.

Aluminum structural shapes are rolled in much the same configurations as those discussed for structural steel members. These shapes include S and W beams, channels (in a variety of special forms), tees, zees, bulb angles, round and square tubes, pipe, plate, and rods of varying aspects.

Aluminum is used for large-diameter electrical conductors, various sheet metal applications, and for situations involving corrosion, like in food-processing and chemical plants.

COPPER

Copper (chemical symbol Cu) is a nonmagnetic reddish-brown metal with excellent electrical and thermal conductivity. It has the highest conductivity properties of all commonly used metals except silver. It is ductile and malleable and easily worked. When alloyed, it offers a wide range of properties, making it a very valuable and widely used material in construction.

Copper and copper alloys are used to produce a myriad of construction products. Although copper products are initially more expensive than aluminum, they have properties that make them less costly in the long run, such as resistance to corrosion and other damaging conditions.

History of Copper

Copper is one of the first metals humans used—to make tools and utensils and for various art and decorative purposes. In some parts of the world it is found naturally in almost pure form, and early humans could beat it into desired shapes in a cold condition.

Archaeological discoveries, such as hammered copper specimens, indicate that copper likely was used as early as 4500 BC (and possibly earlier). It is clear that by 3000 BC the Egyptians were making utensils, simple tools, and ornaments from copper. In 2500 BC or so, the Egyptians discovered how to make bronze, and the use of copper and bronze spread over most of the Mediterranean area.

Copper is named after the island of Cyprus, where large quantities were produced by 3000 BC. It provided a major supply for the Romans, and the island was frequently conquered by others because of the copper. Copper was first called cuprium, then cuprum, and, in English, became "copper."

Copper artifacts have been found in China and India dating from about 2500 BC, and bronze items were made around 1770 BC. Bronze, an alloy of copper and tin, became so widely used that the period is referred to as the Bronze Age.

Through the years, with improvements in mining, refining, and alloying, copper has become one of the major metals used in construction products and tools.

Properties of Copper

The most important properties of copper and copper alloys as used in the construction industry are electrical and thermal conductivity, resistance to corrosion, wear resistance, ductility, and high temperature performance. Copper has a relatively low tensile strength, about 32,000 psi, which can be improved by heat treating, cold-working, and alloying. High electrical conductivity makes it a good material for electrical wiring and parts in devices that conduct electricity. Its good thermal conductivity makes it useful in heat transfer situations, and its corrosion resistance properties make it ideal for plumbing pipe, gas lines, and components exposed to the atmosphere or corrosive chemical elements. Its ductility properties make it a material easily bent, stretched, stamped, machined, and otherwise formed into useful products. Copper has a melting point of 1,981°F (1,083°C) and a coefficient of thermal expansion of 0.0000168/°F (0.0000093/°C).

One unique characteristic of copper: when exposed unprotected to the atmosphere, it will organically develop a green coating over a period of years. This coating, called patina, provides a natural protection from additional corrosion. It takes many years of exposure to the atmosphere to build up a fully developed patina coating.

Production of Copper

Much of the copper mined in the United States is low-grade sulfate ore. There are various processes for producing metal copper from the ore. Figure 16.8 shows mined ore being crushed and pulverized in a ball mill. The crushed ore is ground to a powder by a grinder and flows to a concentrator where a flotation method is used. Here the pulverized ore is mixed with water, oil, and a foaming agent and agitated by air. The copper sulfate particles collect in the foam on the surface, and waste materials settle to the bottom and are removed. The froth flows to a reverberatory furnace where it is smelted to eliminate sulfur and certain metal impurities caused by oxidation. This produces a top layer of copper that is tapped off and moved to a converter.

The converting process is much like the Bessemer process used to produce steel. In the converter, the iron in the remaining material is oxidized, forming a layer of slag that is removed. The sulfur is oxidized, forming sulfur dioxide gas, and drawn off the converter. The remaining material, called blister copper, is about 99 percent pure. The blister copper is processed in a refining furnace, reducing its oxygen content and forming fire-refined copper. This fire-refined copper can be used to produce copper products by rolling, extruding, drawing, and casting.

The purest copper is produced by electrolytic processes. Fire-refined copper can be cast in molds that become anodes in an electrolytic process. The electrolytic process utilizes an acid-proof tank containing an electrolyte, warm diluted sulfuric acid, and copper sulfate. The fire-refined copper anodes are suspended in the tank. Thin sheets of pure copper are suspended between them and serve as cathodes. The system receives an electric current, causing the copper in the anodes to dissolve and bond to the cathodes. This produces copper approximately 99.9 percent pure. The cathodes are melted and used to produce the base metal for a wide variety of copper alloys.

After the copper has been fire-refined or purified by the electrolytic process, it is formed into sheets, bars, tubes, and wires used to produce various products. Examples of these products are shown in Fig. 16.9 and 16.10.

Classifying Copper and Copper Alloys

Copper and copper alloys are specified by the Unified Numbering System for Metals and Alloys (UNS). They are identified by a five-digit number code preceded by the letter C. Wrought materials are assigned UNS numbers from C10000 to C79999. Cast alloys are numbered from C80000 to C99999. A summary of these can be found in Table 16.14.

Coppers are numbered from C10100 to C15999 and are pure or nearly pure, having a minimum copper content of 99.3 percent or higher. The coppers in this series are very similar in chemical composition. The

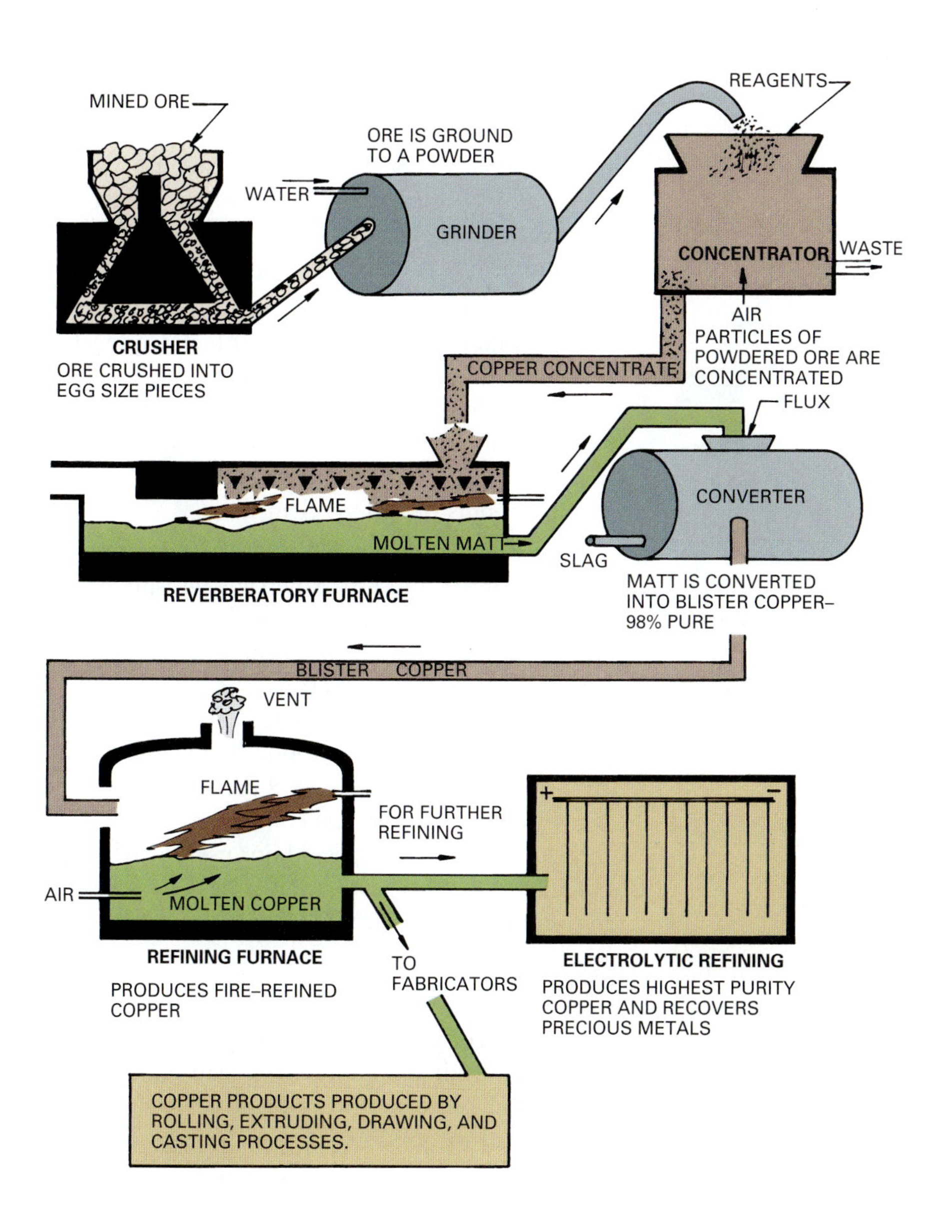

Figure 16.8 The production process to produce copper from copper ore.

Figure 16.9 A variety of copper fittings are produced from copper ore. *(Courtesy Torsten Bätge, http://en.wikipedia.org/wiki/Image:Kupferfittings_4062.jpg)*

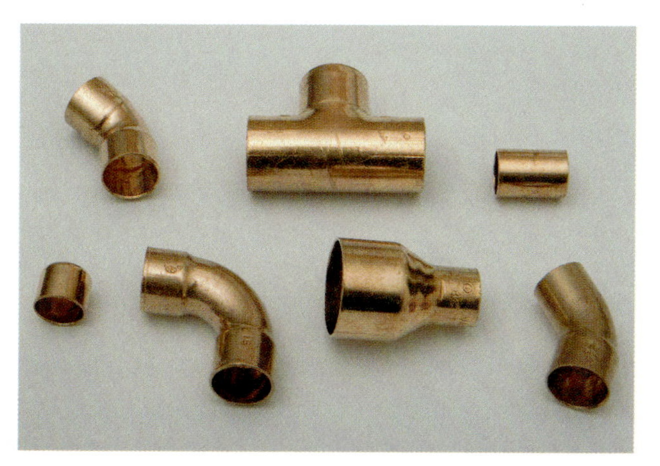

Figure 16.10 Copper plumbing tube wound on large-diameter reels for storage and shipment. *(Image copyright Scott Milless, 2009. Used under license from Shutterstock.com)*

Table 16.14 UNS Designations for Coppers, Brasses, and Bronzes

Wrought Coppers and Copper Alloys	
C10000–C15760	Copper
C16200–C19900	High copper alloys
Cast Coppers and Copper Alloys	
C80100–C81200	Copper
C81400–C82800	High copper alloys
Wrought Brasses and Brass Alloys	
C21000–C28000	Brasses
C31200–C38500	Copper-zinc-lead alloys
C40400–C48600	Copper-zinc-tin alloys
Cast Brass and Brass Alloys	
C83300–C83810	Brasses
C84200–C84800	Copper-tin-zinc and copper-tin-zinc-lead alloys
C85200–C85800	Copper-zinc and copper-zinc-lead alloys
C86100–C86800	Manganese bronze and leaded manganese bronze alloys
C87300–C87800	Copper-silicon alloys
Wrought Bronzes and Bronze Alloys	
C50100–C54400	Copper-tin-phosphous alloys
C55180–C55284	Copper-phosphorus and copper-silver-phosphorus alloys
C60800–C64210	Copper-aluminum alloys
C64700–C66100	Copper-silicon alloys
C66400–C69710	Other copper-zinc alloys
C70100–C72950	Copper-nickel alloys
C73500–C79800	Copper-nickel-zinc alloys (nickel silvers)
Cast Bronzes and Bronze Alloys	
C90200–C91700	Copper-tin alloys
C92200–C92900	Copper-tin-lead alloys (leaded tin bronzes)
C93100–C94500	Copper-tin-lead alloys (high leaded-tin alloys)
C94700–C94900	Copper-tin-nickel alloys
C95200–C95900	Copper-aluminum-iron and copper-aluminum-iron-nickel alloys
C96200–C96900	Copper-nickel-iron alloys
C97300–C97800	Copper-nickel-zinc alloys (nickel silvers)
C98200–C98840	Copper-lead alloys
C99300–C99750	Special alloys

Courtesy Society of Automotive Engineers

numbering system identifies the production or refining processes used to produce the copper. For example, oxygen-free copper is C10200, electrolytic tough pitch copper is C11000, and phosphorus-deoxidized high residual phosphorus copper is series C12200. Coppers containing less than 0.7 percent of specified alloying constituents include tellurium-bearing copper (C14500) and zirconium copper (C15000).

The numerical designations indicate the type of copper or copper alloy and the alloying elements and impurities. Comprehensive handbooks from the Copper Development Association give specific details.

Copper or copper alloys must be identified by their UNS number. Over the years, some frequently used types of copper were given trade names, as shown in the tables that follow. The trade name does not indicate the composition of the material and sometimes is used to describe similar materials that have slightly different amounts of various elements.

Deoxidized copper (UNS C122200) contains 99.9 percent copper and 0.025 percent phosphorus. It has better forming and bending qualities than electrolytic copper and resists embrittlement at high temperatures. It is used for water and refrigeration piping, oil burner service, and in sheets and plates where welding constitutes the major joining method. Electrolytic tough pitch copper (UNS C11000) is 99.9 percent copper and has high electrical and thermal conductivity. It can be easily formed into useful shapes. Major uses include electrical conductors of all types, as well as flashing, gutters, roofing, and forgings (Fig. 16.11).

Copper Alloys

The major copper alloying elements are tin, aluminum, zinc, nickel, silicon, manganese, lead phosphorus, and beryllium. Following is a brief discussion of each. Additional information about brasses and bronzes is given later in this section. The UNS designations shown are for wrought products. Designations for cast products are listed in Table 16.14.

Figure 16.11 Copper gutters and downspouts are used on high-quality construction. *(Image copyright Rolf Klebsattel, 2009. Used under license from Shutterstock.com)*

High copper alloys (UNS C16200–C19199) are wrought alloys having specified copper contents from 96 to 99.3 percent. Cast high-copper alloys have a minimum of 94 percent copper. Wrought high-copper alloys have very high electrical and thermal conductivity, almost matching that of pure copper. However, they are much stronger than pure copper, which increases the number of possible uses. They also have good corrosion resistance.

Brasses (UNS C20000–C49999) are copper alloys with zinc as the major alloying element. Other elements, such as lead, phosphorus, nickel, silicon, iron, and aluminum, may be added in small specified amounts. They are extremely useful and find application in many products.

Bronzes (UNS C50000–C66399) are copper alloys in which neither nickel nor zinc is used as a major alloying element. The various types are used for electrical contacts and in corrosion-resistant applications.

Miscellaneous copper-zinc alloys (UNS C66400–C69999) are often referred to as manganese or nickel bronzes. Typically, the major alloying element is zinc, so they are much like some of the brasses.

Copper-nickel alloys (UNS C70000–C72999) contain from 3 to 33 percent nickel. Other elements may be added to improve corrosion resistance and strength. They are used in marine applications because of their outstanding capacity to resist corrosion. Typical applications include use in heat exchangers, condensers, piping, valve and pump parts, and relay and switch springs. Alloys with more than 50 percent nickel are called Monel® metals and form a separate class. Monels retain high strength at elevated temperatures.

Copper-nickel-zinc alloys (UNS C73000–C79999) are referred to as silver nickels because of their color. Zinc is the principle alloying element and nickel is secondary, although other elements may be added to alter their properties. They have good electrical and mechanical characteristics, good corrosion resistance, and are used for fasteners and various electrical components.

Copper Alloy Finishes

Copper alloys are available with a variety of finishes. Some are supplied by the mill that manufactures the copper alloy, and others are supplied by the company fabricating the copper into a specific product or stock shapes. These finishes use the same three classifications used for aluminum products. They include mechanical, chemical, and coatings. A summary of coatings is given in Table 16.15. Notice the Copper Development Association finish designation. An "M" before the identifying digits denotes mechanical finishes, a "C" designates chemical finishes, and coatings use three-digit numbers.

Table 16.15 Copper Alloy Finishes

Finish	Copper Development Assn. Finish Designation
Mechanical	
As fabricated	M10 series
Buffed	M20 series
Directional textured	M30 series
Non-directional textured	M40 series
Patterned	M4X (specify)
Chemical	
Cleaned only	C10 series
Mane dipped	
Bright dipped	
Conversion coatings	C50 series
Coatings	
Organic:	
Air dry	060 series
Thermo-set	070 series
Chemical cure	080 series
Vitreous	
Laminated	L90 series
Metallic	

Courtesy Copper Development Association

Care of Copper and Copper Alloys

New copper products, especially sheet stock, have a bright, shiny, light-brownish color. A protective coating, such as a clear lacquer, is needed to maintain this color. When left exposed to the atmosphere, it will first turn a darker brown and eventually take on a permanent light-green patina. This is considered highly desirable from an appearance standpoint and can be produced by artificial means if the natural aging is too slow.

Copper can be washed with liquid soap and water. If scrubbed with an abrasive cleaner, the brown or green color will be damaged and require time to return.

Uses of Copper

Copper and copper alloys are excellent for outdoor uses, such as siding, roofing, flashing, guttering, and screen wire (Fig. 16.12). Alloys are used extensively for plumbing pipe in residential and commercial structures and in the manufacture of plumbing fittings, such as valves, drains, and faucets. Sewage treatment plants and industrial plants, such as chemical processing installations, utilize copper for many purposes, including lining vessels that contain corrosive materials. Various types of

Figure 16.12 Corrugated copper roofing being installed on a dome structure. *(Image copyright David P. Lewis, 2009. Used under license from Shutterstock.com)*

hardware and fasteners, such as nails, screws, and bolts, are made from copper alloys. A major use is in electrical wire, electrical conductors, and parts in electrical appliances that conduct electricity.

BRASS

Brasses (UNS C20000–C49999) are copper alloys having zinc as the principal alloying element, but variations are produced by adding small quantities of other elements. Zinc improves strength and ductility and produces changes in color. Adding lead improves machinability, and tin enhances strength, hardness, workability, and ductility.

Brasses are hardened by cold-working. However, hardness is also influenced by alloy composition. The compositions of several wrought brasses and brass alloys used in construction are shown in Table 16.16. Similar information is available for cast brasses and brass alloys. Notice that zinc is the major alloying element, with lead and iron present in much smaller amounts. The chemical symbols used to identify the various elements in all metals are shown in Appendix E.

Brasses fall into three general classes. White brasses contain less than 55 percent copper and are hard and brittle. They are used for cast products and cannot be hammered or worked without breaking. Alpha brasses contain 63 to 95 percent copper and are the easiest type to work. They are used to make radiator parts, springs, grilles, and decorative plating (Fig. 16.13).

Alpha-beta brasses contain 55 to 63 percent copper. They are stronger than alpha brasses and can be worked hot. They are used where strength is important, such as for rivets and screws. Table 16.14 gives the Unified Numbering System designation for wrought and cast brasses. Brasses are very important metals and are used in construction almost as much as copper.

Table 16.16 Composition of Selected Wrought Brasses and Brass Alloys[a]

UNS Designation	Descriptive Name	Major Alloying Elements in Percent[b]					Other Named Elements
		Cu	Zn	Pb	Fe	Sn	
Copper-Zinc Alloys (Brasses)							
C22000	Commercial bronze	89.0–91.0	REM[c]	0.05 max.	0.05 max.		
C23000	Red brass	84.0–86.0	REM	0.05 max.	0.05 max.		
C26000	Cartridge brass	68.5–71.5	REM	0.07 max.	0.05 max.		
C28000	Muntz metal	59.0–63.0	REM	0.03 max.	0.07 max.		
Copper-Zinc-Lead Alloys (Leaded Brasses)							
C31400	Leaded commercial bronze	87.5–90.5	REM	1.3–2.5	0.10 max.		0.7 Ni
C35000	Medium-leaded brass	60.0–63.0	REM	0.8–2.0	0.15 max.		
C37700	Forging brass	58.0–61.0	REM	1.5–2.5	0.30 max.		
C38500	Architectural bronze	55.0–59.0	REM	2.5–3.5	0.35 max.		
Copper-Zinc-Tin Alloys (Tin Brasses)							
C44300	Admiralty, arsenical	70.0–73.0	REM	0.07 max.	0.06 max.	0.8–1.2	0.02–0.06 As
C46400	Naval brass, uninhibited	59.0–62.0	REM	0.20 max.	0.10 max.	0.5–1.0	
C48500	Naval brass, high lead	59.0–62.0	REM	1.3–2.2	0.10 max.	0.5–1.0	

[a]*These are only a few of the many types of brass and brass alloys available.*
[b]*Cu copper, Zn zinc, Pb lead, Fe iron, Sn tin, Ni nickel, As arsenic.*
[c]*Remainder for the difference between elements specified and 100 percent.*
Reproduced with permission from Standards Handbook, *parts 5 and 6, Copper Development Association*

Figure 16.13 A decorative door utilizes embossed brass sheets. *(Image copyright CJPhoto, 2009. Used under license from Shutterstock.com)*

History of Brass

The Romans may have been the first to deliberately add zinc to copper to produce brass. They used it to manufacture coins and jewelry. In the Middle Ages, the production of brass in Europe was a major industry. It was widely used to cast religious objects, candlesticks, locks, and plates to remember the dead. These were engraved and contained words and various figures. Companies started producing brass utensils, lamps, and other household items. Nineteenth century cannons were massive brass castings and eventually replaced by steel.

Plain Brasses (Copper-Zinc Alloys)

Copper-zinc alloys are sometimes referred to as plain brasses. Several that find use in products related to construction are red brass, commercial bronze, cartridge brass, and muntz metal. The alloying elements for each can be found in Table 16.16.

Red brass contains 85 percent copper, 15 percent zinc, and small amounts of lead and iron. It has excellent resistance to corrosion and higher ductility and strength than copper. It is used for plumbing pipe, handrails, balusters, stair posts, tubing, and hardware.

Commercial bronze contains 90 percent copper and 10 percent zinc. It has good ductility and cold-working properties. It is used for screws, forgings, and some types of hardware.

Cartridge brass contains 70 percent copper and 30 percent zinc. It has the best strength and ductility of all the brasses and is easily worked cold. It is widely used where copper products require extensive fabrication, such as stamping or deep drawing. It finds use in the manufacture of electric sockets, reflectors, rivets, and heating units.

Muntz metal contains 60 percent copper and 40 percent zinc, which is the greatest percent of zinc alloyed with the copper. Muntz metal has low ductility but high strength and is used for things like sheet stock and exposed architectural features.

Leaded Brasses (Copper-Zinc-Lead Alloys)

Lead is added to brass to make it easier to machine. Leaded brasses are used in a variety of construction products, including architectural bronze, forging brass, and medium-leaded brass. The alloying elements for each can be found in Table 16.16.

Architectural bronze has the least copper and most lead of the three mentioned above. Widely used for forgings and products produced by machining, it contains 55 to 59 percent copper, 41 to 45 percent zinc, and 2.5 to 3.5 percent lead. Architectural bronze is used for decorative grilles, handrails, architectural trim, and hardware.

Forging brass contains about 60 percent copper, 38 percent zinc, and 2 percent lead. It has great plasticity when hot and is therefore used often for forgings. Because of its good corrosion resistance, it is used for plumbing and hardware items.

Medium-leaded brass contains about 62 percent copper, 34 percent zinc, and 2 percent lead. It is used when good machining properties are required, such as for keys, parts of locks, plaques, and various scientific instruments.

Tin Brasses (Copper-Zinc-Tin Alloys)

When tin is alloyed with copper, zinc, and other elements, the alloy has additional properties not present in plain brasses. The two types of tin brasses are admiralty and naval brasses.

Admiralty brass contains about 71 percent copper, 28 percent zinc, 1 percent tin, and traces of lead, iron, and arsenic. These alloying elements improve strength and ductility and, most important, increase resistance to corrosion. Admiralty brass is widely used in the manufacture of condenser and heat-exchanger plates, various tubes, and equipment in chemical and electrical power plants, as well as products needing resistance to seawater.

Naval brasses fall into several categories used for wrought products. These include uninhibited, arsenical, medium-leaded, and high-leaded. The copper content for all naval brass ranges from 59 to 62 percent. There is a wide range in the amounts of lead and tin alloying elements. Naval brasses are used in chemical, steam power plant, and marine equipment. They can withstand the corrosive effects of materials to which they are typically exposed.

BRONZE

Technically, the term "bronze" has been used to identify a product that is 90 percent copper and 10 percent tin. However, other alloying elements are now added to produce a wider range of materials in the bronze family of metals that have varied properties. Bronze now refers to copper alloys having elements of silicon, aluminum, manganese, and others. They may or may not have zinc. Some of the products in Table 16.16 are referred to as bronze but are actually brass alloys. These include commercial bronze, architectural bronze, and Muntz metal.

The composition of a few selected wrought bronzes and bronze alloys are shown in Table 16.17. The phosphor bronzes contain approximately 89 percent copper. C51800 has 4 to 6 percent tin and 0.1 to 0.35 percent phosphor. C53400 has more lead and is referred to as leaded phosphor bronze. Aluminum bronzes have 6 to 7.5 percent aluminum, while low silicon bronze has no tin or lead but 0.8 to 2.0 percent silicon. Cast bronzes have similar elements. Table 16.14 gives the Unified Numbering System designations for bronzes. Bronze is used for various cast products and hardware, including screws, washers, nuts, bolts, and weather stripping (Fig. 16.14).

LEAD

Lead (chemical symbol Pb) is a soft, heavy metal with good corrosion resistance that is easily worked. Its ability to resist penetration from radiation is a unique feature.

History of Lead

Archaeological findings indicate that lead may have been used as early as 5000 to 6000 BC. Through the centuries, it found use in pottery glazing and as a solder

Table 16.17 Composition of Selected Bronzes and Bronze Wrought-Type Alloys[a]

UNS Designation	Pevious Trade Name	Major Alloying Elements in Percent[b]										Other Named Elements
		Cu	Pb	Fe	Sn	Zn	P	Al	Mn	Si	Ni	
Copper-Tin-Phosphorus Alloys (Phosphor Bronzes)												
C51800	Phosphor bronze	REM[c]	0.02 max.	—	4.0–6.0	—	0.10–0.35	0.01				
Copper-Tin-Lead-Phosphorus Alloys (Leaded Phosphor Bronzes)												
C53400	Phosphor bronze B-1	REM	0.8–1.2	0.10 max.	5.0–5.8	0.30 max.	0.03–0.35					
Copper-Aluminum Alloys (Aluminum Bronzes)												
C61300	Aluminum bronze	REM	0.01 max.	2.0–3.0	0.20–0.50	0.10 max.	—	6.0–7.5	2.0 max.	0.10 max.	0.15 max.	0.15 P max.
C64200	—	REM	0.05 max.	0.30 max.	0.20 max.	0.50 max.	—	6.3–7.6	0.10 max.	1.0–2.2	0.25 max.	0.15 As max.
Copper-Silicon Alloys (Silicon Bronzes)												
C65100	Low silicon bronze B	REM	0.05 max.	0.8 max.	—	1.5 max.	—	—	0.7 max.	0.8–2.0		

[a]*These are only a few of the many alloys available.*
[b]*Cu copper, Pb lead, Fe iron, Sn tin, Zn zinc, P phosphorus, Al aluminum, Mn manganese, Si silicon, Ni nickel, As arsenic.*
[c]*Remainder for differences between elements specified and 100 percent.*
Reproduced with permission from Standards Handbook, *parts 5 and 6, Copper Development Association*

Figure 16.14 This bronze casting shows great detail, indicating that the metal has excellent casting qualities. *(Image copyright Casper Voogt, 2009. Used under license from Shutterstock.com)*

for joining copper. It was used in China for money and as an alloying element in early bronzes. The Romans made wide use of lead pipes for water distribution, not realizing lead poisoning results from its ingestion. By 800 BC, lead was used in the manufacture of glass and, by AD 1100, as roofing and cames, the latticed framework that holds glass pieces in stained glass windows.

Properties of Lead

Other advantageous properties of lead are its high density and weight, softness and malleability, low melting point, 620°F (327°C), and good electrical conductivity. Potentially disadvantageous is lead's low strength and lack of elasticity.

Production of Lead

The major source of lead is the mineral galena (or lead sulfide). Other sources include cerussite (lead carbonate) and anglesite (lead sulfate). Lead ores frequently contain zinc, and some have gold, silver, and other metals.

Lead-bearing ore is first crushed and ground into fine particles. Metal-bearing material is separated from the rock particles via flotation. Flotation involves mixing water, oils, and chemicals with the ground ore. As air is blown into the mix from the bottom, the lead-bearing particles are wetted by the oil and float to the top in a froth of air bubbles. The lead particles are drawn off and the waste material (gangue) settles to the bottom and is removed.

The concentrated ore is roasted in the air, which changes the lead sulfide to lead oxide. Sulfur escapes as sulfur dioxide gas and is recovered and made into sulfuric acid. The lead oxide is then smelted in a blast furnace, which causes the lead to settle to the bottom. Any gold or silver settles with it. Waste materials float to the top, forming slag, and are removed. The lead is then processed to separate the gold and silver.

Grades of Lead

Of the several grades of lead, chemical lead, desilverized lead, and corroding lead find use in some construction applications. Chemical lead and desilverized lead are used for pipes, sheets, and alloys. Corroding lead is used for white lead, red lead, and litharge (used in the manufacture of batteries, pottery, lead glass, and ink).

Lead Alloys

Lead is alloyed with antimony to improve hardness and strength. However, many other elements are also added. Among these are arsenic, nickel, zinc, copper, iron, and manganese.

Unified Numbering System Designations

Lead and lead alloys are designated by the UNS numbers L50001 through L59999. For example, L50121 is described as a solder alloy containing 98.0 percent lead, and L50770 is a battery grid alloy, lead-calcium, containing 99.6 percent lead.

Uses of Lead

Lead pipes and tank liners are used in installations that process highly corrosive materials, but they are never used for piping to carry drinking water. Since lead is a good self-lubricant, it is used when high-pressure lubricating is necessary. Lead solder works for electrical connections because it is a good conductor, but it is not used on water pipe connections. Hard solders have antimony added. High-temperature solders are alloyed with silver and are generally referred to as silver solder.

The use of lead in solder for joining copper water pipes has been banned because of the possibility of increasing lead content in water, which causes lead poisoning. Lead pipes are not used to move drinking water for the same reason. This danger is especially present when soft or distilled water is used. However, lead pipe and lead-lined tanks have high corrosion resistance and can find use in industrial production applications, such as in the chemical manufacturing industry. Sheet lead is used for roofing, flashing, and spandrels in areas where there is severe industrial air contamination or on seacoasts.

Another interesting use for lead occurs in the form of lead azide. Lead azide is easily exploded by an electrically heated wire, so it is used in the manufacture of blasting caps, which set off other explosives.

Lead works in products such as adhesives, caulking, pigments, glazing, pipes and tanks, compounds, and protective coatings over steel and copper. It is an element added to produce brass, bronze, certain asphalt products, glazes, various fusible alloys, glass, porcelain enamel, iron, steel, certain plastics, solder, rust-resistant pre-painting primer coatings, tin. It also serves as an additive in certain wood preservative preparations. As mentioned in Chapter 32, the use of lead in paint has been banned to avoid the danger of children ingesting peeling pieces of paint.

Architects specify lead for waterproofing, soundproofing, reduction of vibration, and radiation shielding. Following are more detailed examples of two lead products in order to provide a closer look at its elements and properties.

Lead strip and sheet products are made from lead that is almost 100 percent pure and from an alloy with about 7 percent antimony, which improves strength and stiffness. Seams can be folded, soldered, or folded and soldered, as shown in the roof details in Fig. 16.15. Soldering is done with a 50 percent lead, 50 percent tin solder.

Red lead, a lead oxide, is a widely used primer applied generally to steel before it receives final coats of finish paint. It is available dry, in a paste, or as a liquid paint. Dry red lead is available in grades A, B, and C. Grade C has the highest lead content. Red lead paste in an oil is available in grades B and C. Again grade C has a higher percentage of lead.

Figure 16.15 Typical details used when installing lead and terneplate roofing and flashing.

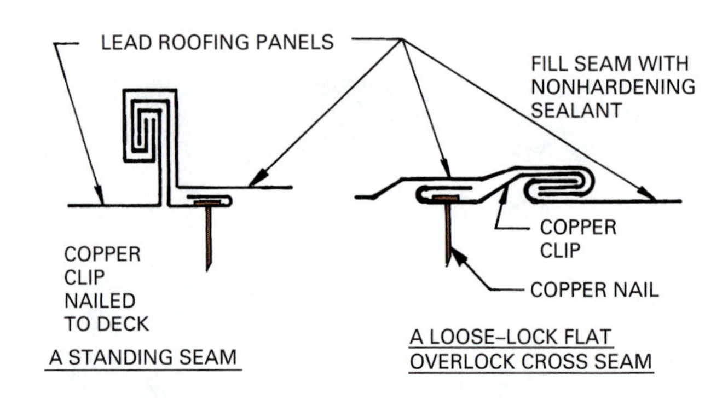

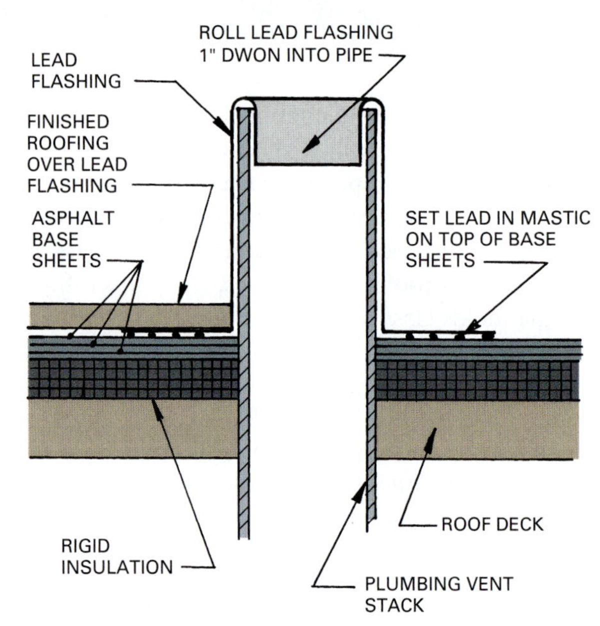

SOLDERS

Solders are nonferrous metals used to join metals in a waterproof joint and to make secure electrical connections.

Properties

Solders have low melting temperatures, typically around 375°F (192°C) to 595°F (315°C). However, some high-temperature solders range up to about 740°F (396°C). The low melting point enables a solder to join metals without melting them; therefore, the solder has no alloying action with the metals being joined.

Solders have little shear, tensile, or impact strength, so they are used where there will be no load on a joint or on joints that have other fasteners, rivets, bolts, or interlocking seams to carry loads or other stresses.

Types of Solder

Solders are mainly an alloy of lead and tin with small amounts of other elements included. The four major classifications of solder are tin-lead, tin-lead-antimony, silver-lead, and a variety of special alloys.

Tin-lead solders are general purpose alloys used for joining metals. The tin-lead composition varies from an alloy with 70 percent lead and 30 percent tin through a series of ten or so combinations in which the amount of lead increases and the amount of tin decreases. The maximum lead solder has 90 percent lead and 10 percent tin. Tin-lead solders have less than one percent antimony. Typical 50-50 tin-lead solder is most commonly used.

Tin-lead-antimony solders have from 1 to 2 percent antimony, and of the five commonly available types, all have more lead than tin. Tin content varies from 20 to 40 percent.

Silver-lead solders contain more than 97 percent lead, from 0 to 1.25 percent tin, traces of antimony, and a silver content of 1.5 to 2.5 percent. They have melting points of about 580°F (307°C) and produce fairly strong corrosion-resistant joints. They are used for soldering copper and brass using a torch heating method. Special purpose solders are designed for a specific application, such as soldering copper roofing, or when a high-temperature solder is required. They typically contain various amounts of lead, zinc, silver, and cadmium.

Fluxes

A metal surface must be clean and free of oxides for solder to bond to it. In addition to mechanically cleaning the surface (washing, buffing, brushing, etc.), any oxides on the surface must be removed. Fluxes are materials used to remove these oxides. Three general types of fluxes are available: neutral, corrosive, and non-corrosive.

Neutral fluxes are mild and used on easily soldered metals, such as copper, lead, brass, and tin plate. They typically take the form of a mild acid that is wiped on a surface and need not be removed before soldering begins.

Corrosive fluxes include salt-type and acid-type. They are more effective for cleaning surfaces than neutral fluxes, but it is important that they, along with any oxide residue, be removed. Otherwise their corrosive action continues to attack the base metal being soldered. Corrosive fluxes (often called acid fluxes) cannot be used on electrical connections.

Noncorrosive fluxes tend to be used only on easily soldered metals and electrical connections. Typically they contain rosin as the main flux. Noncorrosive fluxes are good for soldering electrical connections because they will not cause wires to corrode.

TIN

Tin (chemical symbol Sn) is produced from the ore containing the mineral cassiterite, which is a tin oxide. Since there is little cassiterite in North America, tin has to be imported from Malaysia, Brazil, Russia, Indonesia, Thailand, China, and Bolivia. Ore usually contains little tin, so an extensive refining process is required.

Properties and Working Characteristics of Tin

Tin is a soft metal that is malleable, ductile and blue-white in color. It is corrosion resistant when exposed to air and moisture. It has a low melting point of 450°F (232°C) and can be cast. It will take a high polish and has properties enabling it to coat other metals. Because tin is soft and malleable, it can be worked by rolling, spinning, extrusion, and casting.

Production of Tin

After ore is mined, impurities must be removed to get a concentrated ore. This involves a series of mechanical and chemical processes for crushing and cleaning the ore. Magnetic separators and screening operations are used to separate the ore and the tailings.

The concentrated cassiterite is refined, producing a high-purity metal of at least 99.8 percent tin. Several different processes are used, but all involve smelting ore in a furnace where crude tin is separated from slag. The crude tin is then heated to a predetermined temperature at which impurities with higher melting temperatures than tin remain. This process is called liquidation. This partly pure tin is drawn off and heated above its boiling point. As it is stirred, additional impurities rise to the surface and are drawn off. This continues until most impurities are removed, leaving an almost pure tin.

Unified Numbering System Designations

Tin alloys are designated in the Unified Numbering System by the numbers L13001 through L13999. For example, UNS L13630 is a lead-tin solder containing 37 percent lead and 63 percent tin.

Uses of Tin

Since tin and tin alloys have high corrosion resistance and excellent coating ability, they are used extensively to provide protective coatings on other metals, especially steel. One example is the coating on cans used to store food for retail. Tin is also used as an alloying element in other metals. Its low melting point makes it useful in some solders. It finds applications in mirrors, hardware, and fusible alloys. Tin compounds are used in the production of glazes, glass, and porcelain enamel.

TERNEPLATE

Terneplate is a mixture of lead and tin applied to copper-bearing steel sheet or stainless steel sheet to produce a corrosion-resistant coating. Tin is added because lead alone will not alloy with the iron. Terneplate sheets are available in two types: short terne and long terne.

Short terne, used for roofing, is available with stainless steel or copper-bearing steel bases. The stainless steel type uses UNS S30400 stainless steel in 26 and 28 gauge thicknesses. Copper-bearing steel terneplate is made in 26, 28, and 30 gauge thicknesses. Both types use a coating on both sides consisting of 75 to 80 percent lead and 20 to 25 percent tin.

Long terne is used for various industrial purposes and is available in 14 to 30 gauge low-carbon steel with a coating on both sides consisting of 75 to 87 percent lead and 12 to 25 percent tin.

Uses of Terneplate

Short terne, often called roofing terne, is used for finished roofing, gutters, downspouts, and flashing. Terneplate roofing is installed with batten and standing seams. Flat-locked and horizontal seams can be used when it is installed over wood sheathing.

Long terne is used for fireproof doors and frames, other fireproofing items, and roofing. The various types are available in sheets and rolls. Terne on steel should be painted on both sides after installation to seal any pinholes. If pinholes remain in terne-coated steel, corrosion results. Frequently, a mill applies a red iron-oxide primer, but a high-quality finish coat of paint is required over this. Terne-coated stainless steel does not require a primer or painting.

Production of Terne Sheets

Sheets are chemically cleaned by passing them through a dilute solution of sulfuric or hydrochloric acid. Then they are fluxed by a heated solution of zinc chloride. Finally, the sheets pass through a molten solution of lead and tin. The terne-coated sheet is then moved between smooth metal rollers to produce the finished surface. After cooling, it is cleaned to remove any traces of oil. It is then ready for priming.

TITANIUM

Titanium (chemical symbol Ti) is found in large quantities in the earth's surface. It is a very light, strong, ductile, silvery metal and is one of the most common elements.

History

The element titanium was discovered in England around 1790. It was in the form of black sand, which, upon analysis, yielded a white metallic oxide: titanium dioxide, a compound found in nature that produces a white powder when refined. This powder is used today in the paint industry. In 1910, a laboratory experiment proved that titanium in metallic form could be produced. It was not until 1946 that a process was patented that enabled titanium to be produced in commercial quantities. By 1950, an industrial-type titanium was available.

Properties

Titanium has low electrical conductivity and a low coefficient of thermal expansion. It is also paramagnetic, meaning that when placed in a magnetic field it possesses magnetization in direct proportion to the field's strength. It has a melting point of 3,300°F (1,820°C) and a coefficient of thermal expansion of 0.0000085/°C.

Two important characteristics of titanium are its high strength-to-weight ratio and its ability to resist corrosion by salt water and the atmosphere. These have an important impact on its use in various products.

Working Characteristics

Pure titanium is easier to use than titanium alloys; however, in general, all types can be fabricated using standard manufacturing processes. These include machining, welding, riveting, drilling, punching, hot- and cold-rolling, extruding, forging, and drawing.

Production of Titanium

Titanium is produced by the Kroll process, named after the person who developed it. The process is chemically driven, using magnesium in an inert atmosphere of helium or argon to produce the reduction of the titanium tetrachloride. These elements react, releasing magnesium chloride, which is distilled off. Left is a sponge metal that is crushed and melted into ingots. The ingots (similar to pig iron ingots) are used to produce titanium and titanium alloys that are processed into various products.

Titanium Alloys

The strength of titanium varies depending on the purity of the metal. The higher the purity the weaker the metal; therefore, various elements are added to produce titanium alloys with greatly increased strength. Typically, vanadium, molybdenum, aluminum, iron, chromium, and manganese are used as alloying elements.

Unified Numbering System Designations

Titanium alloys are designated in the Unified Numbering System by the numbers UNS R50001 through R59999. For example, R56210 is a titanium alloy containing 90.2 percent titanium.

Uses of Titanium

The major demand for titanium comes from aircraft and aerospace industries and the military. Its strength and light weight make it a desirable material for these applications. It can also be formed in sheet, strip, pipe, and tube products, and it can be forged and cast. Titanium is beginning to find uses in construction as cladding, flashing, and guttering (Fig. 16.16). Since titanium has a high melting point, it also works as a structural material.

NICKEL

Nickel (chemical symbol Ni) is a silver-colored metal mainly used as an alloying element. It provides increased resistance to atmospheric and chemical corrosion and increases an alloy's strength.

History

Nickel in various alloyed forms was used by prehistoric man because of its presence in the meteoric iron used to produce tools. Other metallic items from China and Asia Minor have been found to contain nickel, as well as copper and other alloying elements. In the mid-1700s a crude nickel was first isolated from an ore in England, and by the early 1800s, refined nickel was being mass produced. Canada and New Caledonia are the largest producers of nickel-bearing ore.

Figure 16.16 The Guggenheim Museum in Bilboa, Spain, uses a titanium cladding. *(Courtesy Wikipedia, http://en.wikipedia.org/wiki/Image:GuggenheimBilbao.jpg)*

Properties

Nickel is resistant to strong alkalis and many acids. It has good resistance to corrosion and oxidation and is strong and tough. It has a melting point of 2,651°F (1,455°C), a coefficient of thermal expansion of 0.000013/°C, and is magnetic up to 680°F (360°C).

Working Characteristics

Nickel can be fabricated using most of the commonly used processes, such as hot- and cold-rolling, extruding, bending, forging, and spinning. It can be soldered, brazed, joined by some welding processes, or joined with mechanical fasteners.

Production of Nickel

Several processes are used to produce nickel from ore. The most recent is the Hybinette process developed in Canada. Ore is processed using the Bessemer process and the molten metal goes to a cooling chamber. It is then crushed, ground, and magnetically separated. The result is a nickel-copper platinum alloy that is treated by an electrolysis process to separate the nickel, copper, and platinum.

Nickel Alloys

A major use for nickel is as an alloying element. The alloying of nickel to other metals provides increased ductility, corrosion resistance, strength, hardness, and toughness. Nickel alloyed to nonferrous metals improves electrical resistance and magnetism and helps control expansion. An examination of steel alloys in Chapter 15 demonstrates that nickel is a commonly used alloying element. It is also widely used in Monel® metals and aluminum alloys. Monel alloy is about 66 percent nickel and 34 percent copper. Inconel 600® is a special nickel alloy containing about 75 percent nickel, 15 percent chromium, and 7 percent iron. Since it has excellent oxidation resistance, it is used in food-processing and chemical industries. Other nickel alloys include those used for electrical resistance coils, magnetic and nonmagnetic alloys containing iron, alloys used in the production of glass that are designed to have a high coefficient of thermal expansion, and copper-nickel alloys used for products exposed to marine conditions.

Unified Numbering System Designations

Nickel and nickel alloys are designated in the Unified Numbering System by the numbers N02001 through N99999. For example, UNS N02250 is a commercially pure nickel alloy having 99.0 percent minimum nickel.

Uses of Nickel

In addition to its use as an alloying element in ferrous and nonferrous metals, nickel is also an excellent material for electroplating and electroless plating. Electroplating is the process of depositing a coating of metal on another metal by electrolysis. Electrolysis is a method of plating a material by chemical means in which the piece to be plated is immersed in a reducing agent that, when catalyzed by certain materials, changes metal ions to metal, forming a deposit on the surface of the piece. Nickel is used in electric heating elements, lamp filaments, plumbing fittings, and hardware (Fig. 16.17).

ZINC

Zinc (chemical symbol Zn) is a bluish-white metal that is brittle and has low strength. It is often referred to as a white metal and is widely used as a protective coating over steel to prevent corrosion.

Figure 16.17 Solid brass is used to produce a variety of hardware. *(Image copyright JR Trice, 2009. Used under license from Shutterstock.com)*

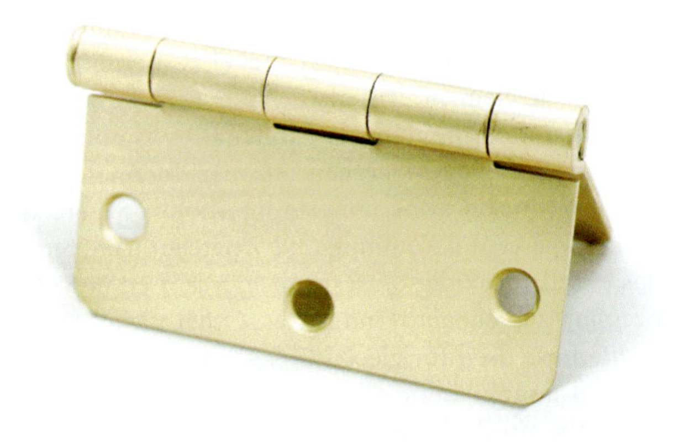

Properties

Zinc has low strength and is brittle. Although it can be damaged by alkalis and acids, it resists corrosion by water and forms a protective oxide when exposed to air. Zinc has a melting point of 787°F (419°C). It is also subject to creep. Its tensile strength can be greatly increased by cold-working and alloying.

Working Characteristics

Since zinc is a soft material, it can be hot- and cold-rolled, drawn, extruded, cast, and machined. It can be joined by welding, soldering, and various mechanical fasteners.

Production of Zinc

Zinc is extracted from zinc blende (sphalerite) ore. The mined ore is crushed and ground, and the ore particles are separated from the rock. Elements present in the ore, such as copper, lead, and iron sulfides, are separated by a flotation process. This involves dividing the elements in the finely grained ore by floating them on a liquid. The floating capacity of the elements varies. Lead and copper sulfates will float off the top of the liquid. Added chemicals cause the zinc sulfide to float, enabling it to be taken off. This concentrated zinc material is dried and ready to be refined into the metal zinc.

The electrolytic process involves roasting the zinc concentrate and removing soluable parts with a weak sulfuric acid solution. The solution is filtered to remove some of the other metals. Finally, this solution is moved to electrolytic tanks where cathodes of pure aluminum and anodes of lead or lead-silver are lowered into the tank and an electric current is passed between them through the solution. Pure zinc is attracted to and plates the cathodes. The layer is removed, melted, and poured in slabs for processing into various applications.

The vertical furnace method involves mixing coking coal briquettes and dried zinc concentrate in the top of a vertical furnace. They are then heated until the zinc concentrate vaporizes. The vapor is removed and condensed at the temperature at which the vapor turns into a solid. Solid zinc slabs are then ready for use in producing various products.

Zinc Alloys

Zinc alloys used for die casting consist of about 95 percent zinc and 4 percent aluminum and magnesium. Some copper may be present.

Unified Numbering System Designations

Zinc and zinc alloys are designated in the Unified Numbering System by the numbers Z00001 through Z99999. For example, Z13001 is identified by the name "zinc metal" and contains 99.90 percent zinc at minimum.

Uses of Zinc

Zinc's major use is in forming a protective coating over steel to prevent rust (Fig. 16.18). This is referred to as galvanizing. Galvanizing involves placing the steel to be coated into a bath of molten zinc, which bonds to its surface. It is important that the coating be free of imperfections, such as pinholes, that permit moisture to reach the steel and cause it to rust. Both galvanized sheet and strip material are available.

Since zinc and zinc alloys have low melting temperatures, they are easy to cast and are used for some types of hardware and plumbing items. They are usually die-cast and finished by polishing or plating with chromium, brass, or other materials.

Figure 16.18 Galvanized steel beams. *(Courtesy Phasmatisnox, http://en.wikipedia.org/wiki/Image:Galvanizing02.JPG)*

Zinc also finds use as an alloying element in brasses. Various zinc compounds serve in the production of paper, plastics, ceramics, rubber, abrasives, paint, and other products. Zinc is also used for specialized products in which corrosion resistance is important, such as anchors, flashing, screws, nails, expansion joints, and corner beads. Solid zinc strip material is used to produce a wide range of products, such as low-voltage buss bars, cavity wall ties, electric cable binders, electric motor covers, grading screens, and roofing and fascia material.

Zinc is high on the galvanic table of electrolytic potentials. This means it can be used to coat a material lower on the table to protect that material from galvanic action. The zinc is sacrificed, thus protecting the coated metal.

Zinc Galvanizing Processes

Zinc sheets, strips, coils, and wire are galvanized in a continuous process. The processes in use include electrogalvanizing, hot-dip galvanizing, metallic spraying, and sherardizing.

Electrogalvanizing involves placing cleaned steel or iron in an electrolyte solution of zinc sulfate. The electrolytic action deposits a layer of zinc on the material. The thickness of the coating can be controlled, but it is limited. Typical thicknesses range from 0.0001 to 0.0005 in. (0.0025 to 0.0127 mm). Hot-dip galvanizing involves immersing clean steel in a bath of molten zinc. It is a semiautomatic process. Metallic spraying involves coating sheet iron or steel by applying a fine spray of molten zinc. It can be applied after an installation is complete, thus also coating bolts, rivets, and welds. **Sherardizing** is a process in which cleaned iron or steel is placed in a container filled with zinc dust. The temperature in the container is raised and the objects to be coated are tumbled in the dust. The heated zinc bonds to the metal, forming a thin coating. This is usually used for small parts and provides minimum protection.

Review Questions

1. How does the specific gravity of aluminum compare with that of steel?
2. How does aluminum compare with copper as a conductor of electricity?
3. How is alumina produced from processed aluminum hydrate?
4. What is the purity of newly produced aluminum?
5. What do the aluminum temper designations F, O, and H mean?
6. What does solution heat-treating mean?
7. When galvanic corrosion occurs, what forms the electrolyte?
8. How can galvanic corrosion be reduced or eliminated?

9. What types of finishes are used on aluminum products?
10. What is a natural, controlled finish?
11. What is an anodized finish?
12. What are the types of anodic coatings used in industrial and architectural applications?
13. How are aluminum surfaces painted in a factory?
14. What is a porcelain enamel coating?
15. What is laminated finish?
16. How can aluminum members be joined?
17. What are the major properties of wrought copper alloys?
18. What are the two most widely used copper alloys?
19. What is copper patina?
20. What type of ore contains lead sulfide?
21. What are the major properties of lead?
22. What are the major properties of tin?
23. What are the major properties of zinc?

Key Terms

alumina
anodize
bauxite
cast alloys
galvanic corrosion
galvanize
heat-treatable alloys
heat treating
non-heat-treatable alloys
nonferrous metals
oxide layer
reduction
strain hardening
temper designation
wrought alloys

Activities

1. Collect samples of each of the commonly used nonferrous metals. Identify and list the properties of each.
2. How many uses of nonferrous metals can you find around your campus? Make a list of locations and types of material.
3. Test samples of a variety of nonferrous metals for tensile strength and hardness. Compare your findings with those of others who made the same tests.
4. Cite the location and name of a product made from nonferrous metals that were extruded, drawn, rolled sheet, and cast.

Additional Resources

Aluminum Design Manual, and many other technical publications related to aluminum, the Aluminum Association, Arlington, VA.

Metals and Alloys in the Unified Numbering System, and other related publications, American Society of Automotive Engineers, Inc., Warrendale, PA, and the American Society for Testing and Materials, West Conshohocken, PA.

Other resources include:

Numerous publications related to metals, American Society for Testing and Materials, West Conshohocken, PA.

Publications of the Copper Development Association, New York.

See Appendix B for addresses of professional and trade organizations and other sources of technical information.

CHAPTER 17

Steel Frame Construction

LEARNING OBJECTIVES

Upon completion of this chapter, the student should be able to:

- Become acquainted with drawings prepared for structural steel buildings.
- Describe the procedure for erecting the structural steel frame of a building.
- Be familiar with the fastening techniques used to join structural steel members.
- Describe fire-protection procedures for structural steel members required by building codes.
- Discuss steel framing systems using manufactured components.

Build Your Knowledge

For further study on these materials and methods, please refer to:

Chapter 15 Ferrous Metals
Topics: Steel and Steel Alloys, Steel Products

Chapter 20 Wood and Metal Light Frame Construction
Topic: Steel Framing

Chapter 36 Interior Walls, Partitions, and Ceilings

Steel-framed buildings utilize a skeleton frame construction in which the walls, floors, and roof are supported by a structural framework of steel beams, columns, girders, and related structural elements **(Fig. 17.1)**. Steel frames are strong, lightweight, and durable structures. Some designs rely partly on the skeleton frame for support while utilizing other systems, such as wall-bearing construction, where walls (masonry, panelized, etc.) form part of the overall structural system. The interior walls in a skeleton-framed building are non-load-bearing, permitting greater freedom in the layout of interior spaces. Frame members are manufactured to design sizes, transported to the construction site, and erected.

The design of a structural frame requires extensive engineering analysis. Design considerations include live and dead loads, as well as lateral loads caused by earthquakes, wind, rain, and snow. Soil and hydrostatic pressures act horizontally below grade and must also be considered. The stresses to which steel is subjected while in use must be considered as the design process continues. Provisions also must be made for temporary stresses that occur during construction, such as those resulting from supporting a crane. Allowable working stresses are regulated by building codes. Detailed information is available from the American Institute of Steel Construction (AISC). Design information for light-gauge steel structural members can be obtained from the Steel Joist Institute and the American Iron and Steel Institute.

The economic use of structural steel members is another design consideration. The spacing of columns influences the span of beams and girders. Spans that are too short or too long result in the uneconomical use of framing members. The design of connections and the securing methods for rivets, bolts, or welding are other crucial design considerations.

The procedure for the design and construction of structural steel-framed buildings includes the engineering design of the structure, the preparation of shop drawings, the manufacture of structural members, and the erection of these structural members on-site.

STRUCTURAL STEEL DRAWINGS

Several types of drawings are required for steel frames, including engineering design drawings, shop drawings, and erection plans. **Design drawings** are prepared by

Figure 17.1 This steel-framed building has a cellular steel floor through which electrical wiring can be run. Notice the sprinkler system and suspended ceilings. *(Reproduced with permission from* The Building Systems Integration Handbook. *Richard Rush, ed., Butterworth-Heinman Publishers, Newton, Mass., 1986)*

Figure 17.2 A partial engineering design drawing contains the data needed by the structural detailer.

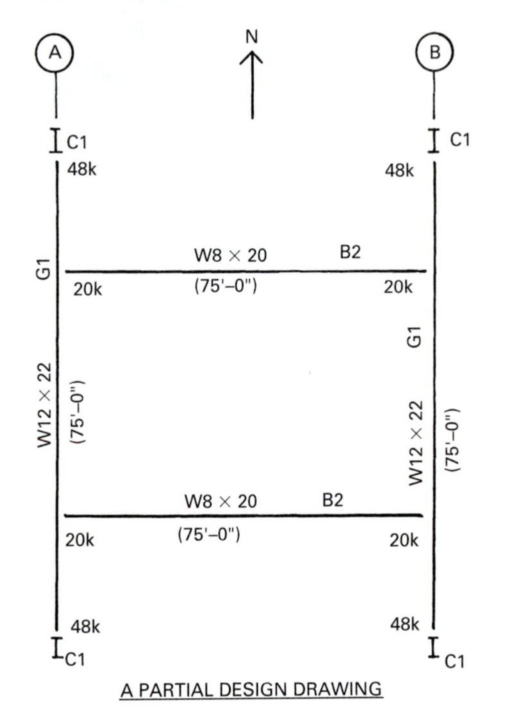

GENERAL NOTES

SPECIFICATIONS: LATEST AISC EDITION
MATERIAL: ASTM A36
FASTENERS: 3/40 A325 IN BEARING TYPE CONNECTIONS
THREADS IN SHEAR PLANES
CONNECTIONS: PER AISC MANUAL AND DEVELOP
INDICATED END REACTIONS

a structural engineer. They indicate the type of construction and give data on shears, loads, moments, and axial forces that must be resisted by each member and all connections. Figure 17.2 is a partial drawing that shows the elevation of a beam as (75' 0") above an established site datum, such as the finished floor. The sizes of the beams are given, and the forces they must resist are indicated in k (kips). Notes are used to provide additional information that allows a structural detailer to prepare shop drawings for the various members.

Shop drawings contain the necessary information for the fabrication of each member. They specify the size of the member and the exact location of holes for connections. Clearances must be allowed for so erection can proceed without interference between joining members. A typical example of a shop drawing is shown in Fig. 17.3, which gives specifications and the mark (B1) identifying the beam.

Erection plans are assembly drawings similar to the design drawings. They are used on-site to direct the placement of each member in its desired location. An example is shown in Fig. 17.4. Each piece, or any subassembly of pieces, and its assigned shipping and assembly mark is noted. The erection plan also includes details showing the anchor bolt locations in the foundation. Anchor bolts are used to secure columns to the foundation and are placed in the concrete as the foundation is poured, so their location is critical. Engineering design drawings are basically the same as erection drawings but have more design detail and, in some cases, can be used as erection drawings. Typically, they are small and crowded with details, so larger-scale erection drawings are usually provided with only the information needed to identify and locate each member.

Each steel member is identified by an erection or piece mark, usually a letter and number giving the part a unique identification. For example, "B2" means beam number 2, "C2" means column number 2, and "G2" means girder number two. All identical members use the same mark. Marks can also include numbers indicating

Figure 17.3 A typical shop drawing of a beam with shop-welded connectors that will be bolted on-site. *(Courtesy Bethlehem Steel Corporation)*

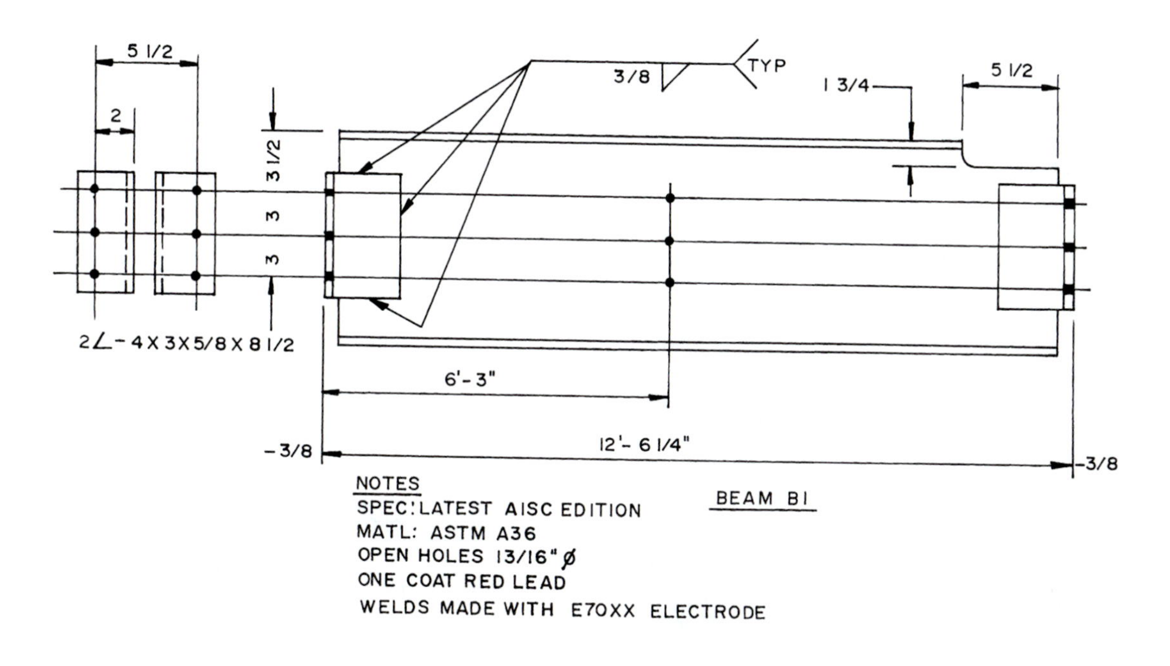

Figure 17.4 An erection drawing of the roof framing that identifies each structural member by a mark and shows their locations.

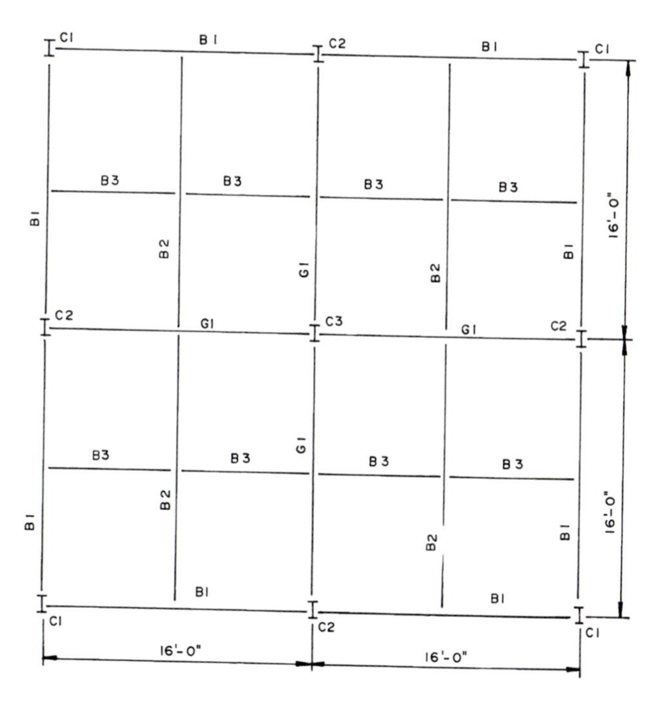

the floor on which a member will be used. For example, "B2 (3)" means beam B2 will be erected on the third floor. A column designation, such as "C2 (2–4)," means that column 2 is intended for the building's second tier, including the second to fourth floors.

Structural steel members are delivered on-site with a primer coat and numbered as shown on the erection plan. Connectors are also primed and numbered. If a steel member will be encased in concrete it usually is not primed. This allows better bonding between the concrete and the member. After the steel is erected, some field painting is required, for example, in areas where welding occurred. Welds are chipped free of any formed slag before they are painted.

THE ERECTION PROCESS

Finished members are delivered on-site in the order in which they are needed during erection. The erecting contractor is responsible for setting in place each member of the frame and securing it as required by the design drawings. As the steel arrives on-site, members must be placed as near to their point of erection as possible in the order they will be needed.

After verifying the correct placement and elevation of the anchor bolts, an erection crew lifts the first level of columns with a ground-level crane and places them over the bolts cast in the foundation. Columns in multistory buildings are generally two stories high (Fig. 17.5). The columns have steel base plates that distribute the load on the column over a larger area of the foundation. The base plate is usually welded to the column when it is manufactured. There are several

Figure 17.5 Columns on multistory buildings are usually two stories high. *(Image copyright Stephen Finn, 2009. Used under license from Shutterstock.com)*

Figure 17.6 Typical base connections for columns.

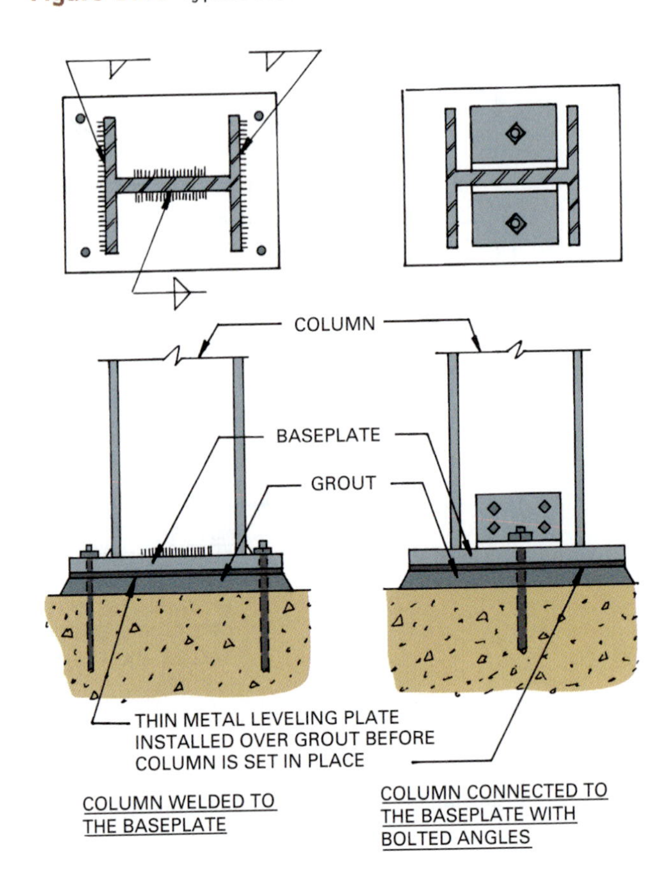

methods for setting columns. Smaller columns have a steel plate leveled over a bed of grout before erection (Fig. 17.6). Larger columns have leveling nuts on the base plate that are adjusted to plumb the column. Then grout is worked below the base plate. A column with stiffener plates is shown in Fig. 17.7. Very large plates required for large, heavy columns may have the base plate leveled and grouted before the column is welded to it. Large-diameter holes may be specified near the plate's center so grout can be forced below it. For additional examples, consult the publications of the American Institute of Steel Construction.

Once columns are in place, girders and beams are lifted by crane and secured to the columns with the required connections. Once the first level is complete, the next level of columns is erected (Fig. 17.8). The second level of two-story columns is lifted with a crane and connected to the columns below with a splice plate (Fig. 17.9). Column splices may be bolted or welded. A crane sets the second-level columns, beams, and girders. As the floors are framed, specified decking is installed (Fig. 17.10). Stringent safety regulations are observed that mandate the use of safety nets, railings, and the protection of openings.

As the building rises beyond the reach of the ground-based crane, a tower crane can be used to reach upper levels. It might be installed outside a building's frame, as shown in Fig. 17.11, or erected inside the structure, in an elevator shaft or a special, temporary opening planned for this purpose. The height of most cranes can be increased by adding additional sections.

Another lifting device for multistory construction is a guy derrick. It consists of a mast, a boom that pivots at the base of the mast, hoisting tackle, and supporting guy lines. The guy derrick can lift itself to floors constructed above it. The hoisting operation is controlled by a power winch located on grade. As the derrick rises to higher levels, the winch remains on the ground. A typical construction procedure is shown in Fig. 17.12. Since the derrick is mounted on the structural frame, the loading must be considered as the design is developed. When the frame is completed, the derrick is disassembled and lowered to the ground with smaller winches.

Figure 17.7 Stiffener angles can be used in welded construction.

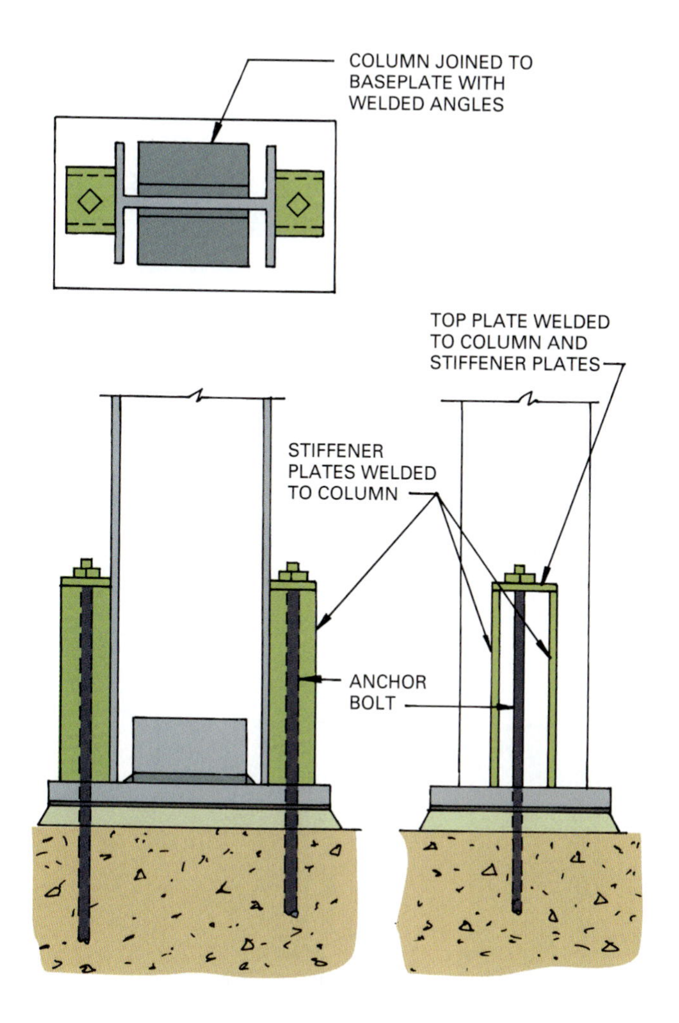

Figure 17.8 A second-story beam being placed and readied for connection. *(Image copyright Dwight Smith, 2009. Used under license from Shutterstock.com)*

FASTENING TECHNIQUES

Steel members in a building frame can be joined by riveting, bolting, or welding. In most cases, a combination of all of them is used.

Rivets used to connect structural steel members must have properties that enable them to handle the shear, bending, and other stresses developed by the frame. Rivets are made from high-strength carbon and alloy steels according to ASTM dimensional and chemical specifications.

Rivets are installed in holes drilled or punched in framing members. They are heated to a white heat, inserted into the holes of the members to be joined, and a head is formed by a pneumatic hammer that also produces an additional head on the opposite side (Fig. 17.13). When the rivet cools, it shrinks, shortens in length, and pulls the members together. The compression and friction developed between the joined members assist in resisting shear and tensile stresses in the joint.

Bolts are used more commonly than rivets (Fig. 17.14). Those used to join steel frame members are either carbon steel, as specified by ASTM A449, ASTM A325 or A490 high-strength bolts. ASTM A449 bolts are used in bearing connections in which their lower strength is adequate. ASTM A325 and A490 high-strength bolts are used for friction connections. They have higher shear strength than carbon A449 bolts and high tensile strength. Bolts are identified by head and nut markings (Fig. 17.15). Bolts used in friction connections must be tightened to a minimum of 70 percent of their ultimate tensile strength. Bolts can be tightened with calibrated torque wrenches, which develop the specified tension values in the connector. They have an automatic cutoff and a gauge that registers the tensile stress developed. Another way to determine tension is the turn-of-the-nut method. Here, the nut is turned until tight and then turned an additional, specified amount, such as one-third to one full turn.

Tightening can also be accomplished with a direct tension indicator. Load indicator washers—metal washers with rounded protrusions on one face—are placed under the bolt head. As the nut tightens they flatten. The degree of flattening is measured with a feeler gauge and correlated with the amount of tension developed (Fig. 17.16). Still another method uses tension-control bolts, as shown in Fig. 17.17. These bolts have a splined end that is inserted in a power wrench that grips both the nut and the splined end. When the specified torque is reached, the splined end twists off at the torque control groove.

Figure 17.9 Typical wide flange column splices.

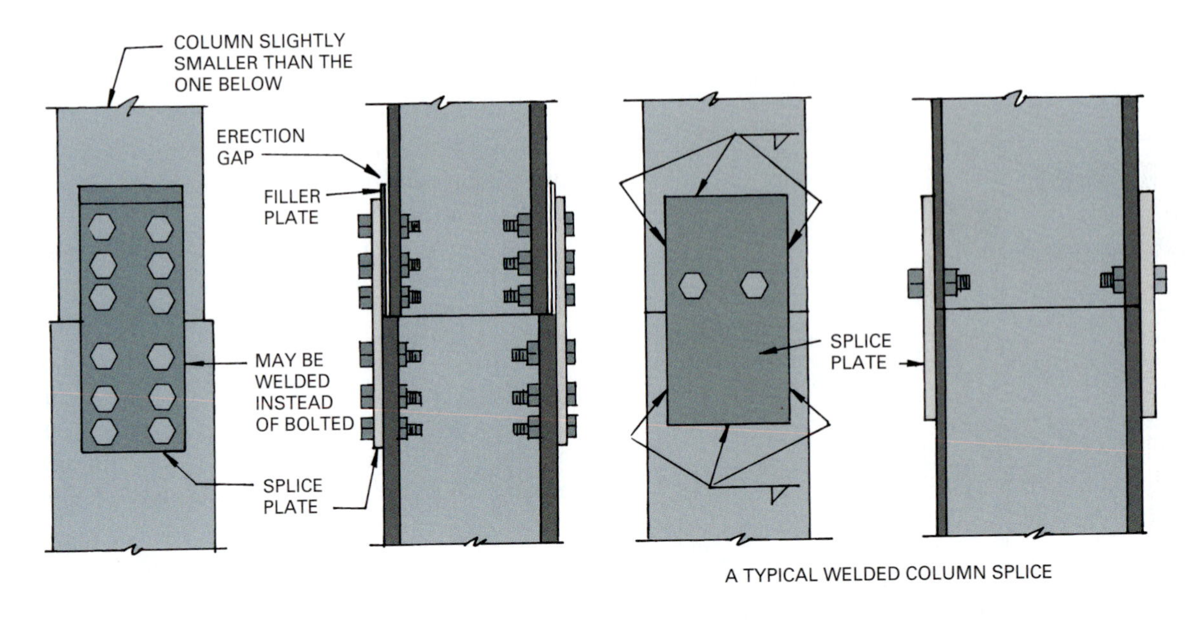

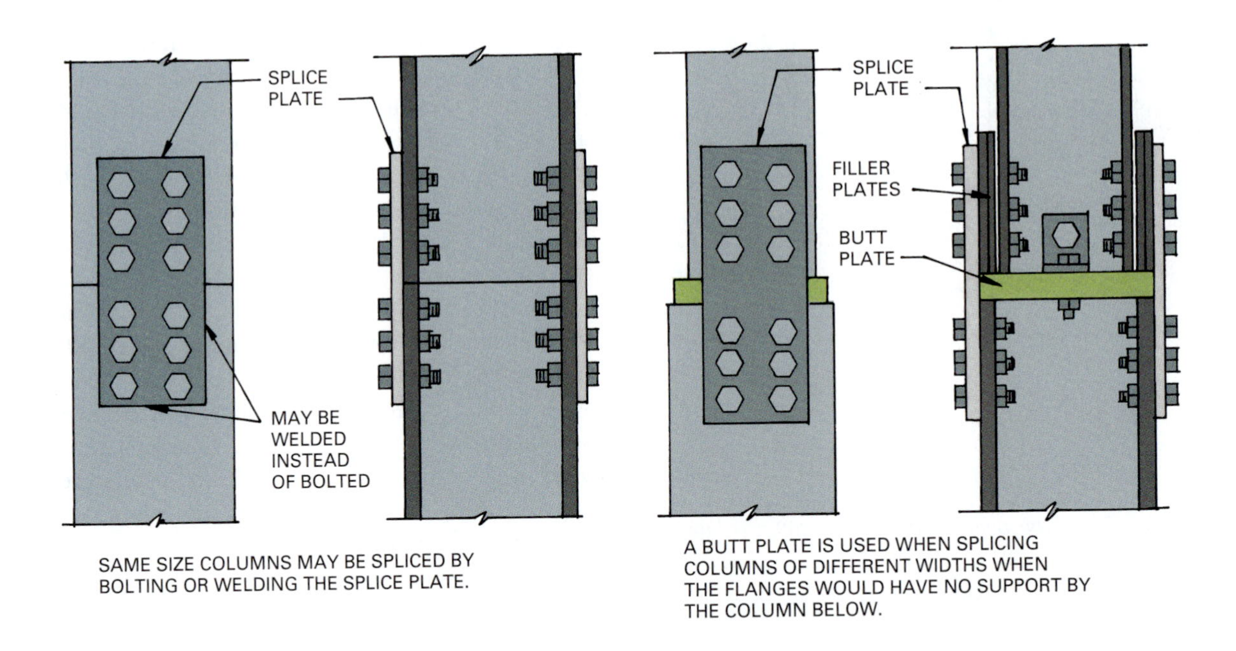

Plain smooth hardened washers may be used to distribute the load of a bolt or nut over a larger area. The hardened washer is used to prevent **galling,** which can render the tension calibration ineffective. **Galling** is the wearing or abrading of one material against another under extreme pressure. Bolt holes are usually $\frac{1}{16}$ in. (1.6 mm) larger in diameter than the bolt, but some field conditions may require a larger allowance. The size of oversize holes must be approved by a structural engineer.

The amount of threads that fall within the inside of the hole in the connection influences the strength of the connection. High-strength bolts have shorter threads than other types, so there is a minimum encroachment of threads into the grip. However, the bolt length must also provide sufficient protrusion beyond the connection to hold a washer and the nut. To achieve higher strengths in bearing connections, the number of threads that fall within the grip must be less than the thickness of the outside metal piece.

Figure 17.10 Steel decking is welded to beams and purlins, forming a base for the finished floor or roof. *(Courtesy Vulcraft)*

Figure 17.11 Tower cranes are used to reach the upper levels of multistory buildings. *(Image copyright Iurii Konoval, 2009. Used under license from Shutterstock.com)*

Welding is the third method used to fasten structural steel members. Occasionally, welding and bolting are used together on connections. Welded beam connections consist of the same general types used for bolted and riveted connections. Connecting plates may be riveted or bolted to a member in the fabrication shop and welded to a connecting member on-site or welded in the shop and bolted in the field (Fig. 17.18). When field connections are welded, several bolts or clamps hold the members together while the field welding occurs. Welds in a fabrication shop are made by clamping a connector to a member and welding it in place. A beam detail showing a shop-welded, field-bolted connection is shown in Fig. 17.19. Welding symbols indicate the size and type of weld. Common weld symbols are made up of several parts, as shown in Fig. 17.20. Some frequently used penetration welds are shown in Fig. 17.21. To keep the molten welding rod material from running out the bottom of a connection, a backing bar or backing bead is placed below it before a weld is started.

Although there are several types of welding processes, electric arc welding is generally used for structural steel connections. This involves passing an electric current through a metal electrode (the welding rod) to the piece to be welded. The electrode is kept slightly above the piece, which causes an arc to jump the gap. This produces considerable heat, which melts the electrode and a small area of the members being welded. As the electrode moves along the line of the weld it leaves a continuous bead of metal that fuses the members

Figure 17.12 The guy derrick is positioned on the building's highest level and can be elevated further as the height of the building increases.

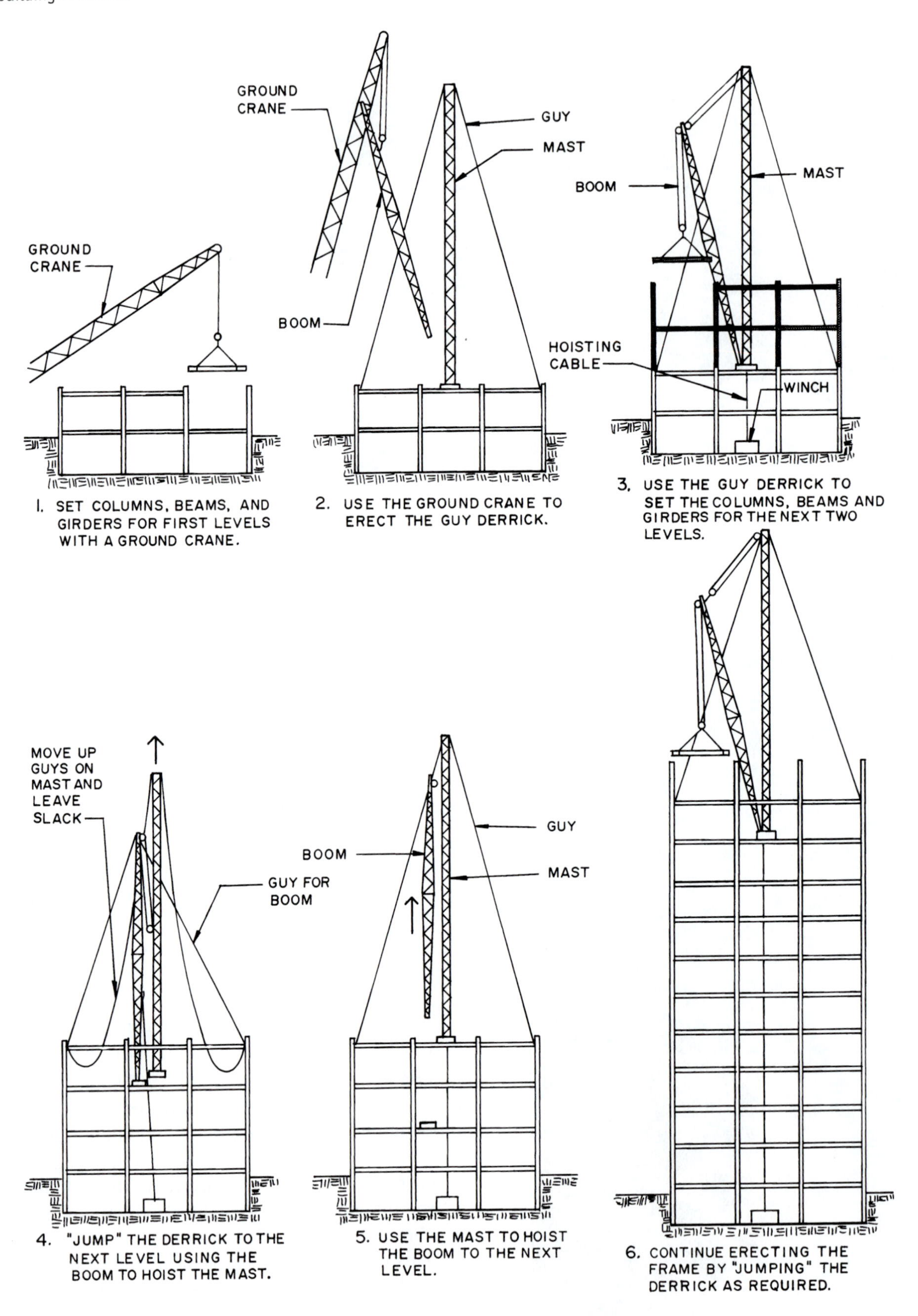

Figure 17.13 Structural steel connections can be joined using rivets.

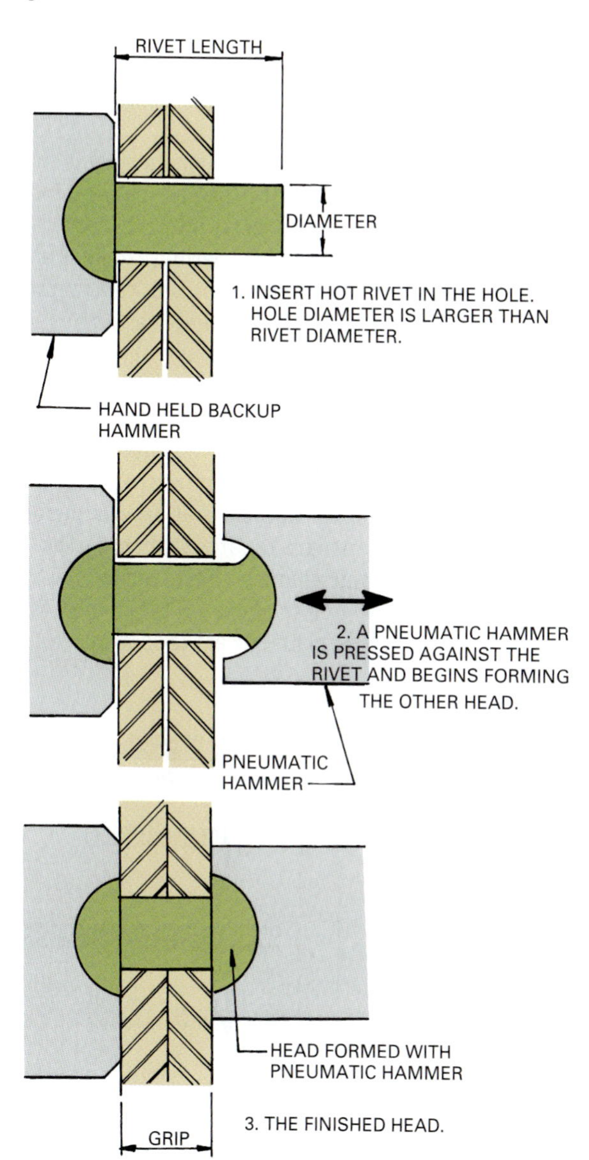

Figure 17.14 High-strength steel bolts are tightened with a pneumatic impact wrench. *(Courtesy Bethlehem Steel Corporation)*

Figure 17.15 High-strength steel bolts have ASTM identification markings on heads and nuts.

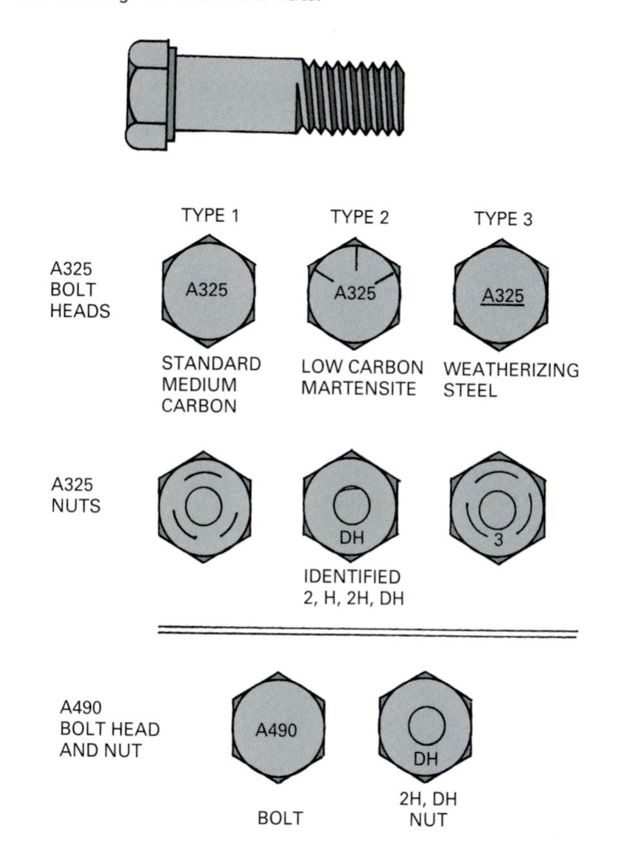

Construction **Methods**

Fasteners for Metal-Building Construction

In metal-building construction, two commonly used fasteners are designed to join metal-to-metal and metal-to-wood connections for sheet metal panels. Metal-to-metal fasteners include drilling fasteners and self-tapping fasteners. The drilling fasteners, shown in Fig. A, have a tip that drills the correct-sized hole through the sheet metal followed by threads that tie the pieces together. The fastener must have a high surface hardness and a head that withstands the use of a power impact driver. Drilling fasteners are available in carbon steel and several classifications of stainless steel and in many sizes and lengths. Manufacturers should be consulted for this information.

The self-tapping fasteners, shown in Fig. B, are installed in holes that are drilled or punched in the sheet metal. The size of the hole must correspond to those specified by the manufacturer for each of the screws they produce. Self-tapping fasteners are available in carbon steel and stainless steel. Consult a manufacturer for available sizes.

Typical sheet-metal-to-wood fasteners, available in carbon steel, are shown in Fig. C. They are power driven through predrilled holes in the metal into the wood. They are available with protective coatings, such as a polymer film and zinc-and-chromate finish. Manufacturers will paint heads a desired color. These fasteners come in lengths up to 2 in. (50.8 mm).

An examination of head types shows that a hexagon head is used for fasteners driven with socket-type power impact tools. Smaller types have slotted heads and also are power driven. A variety of washer faces are available. Some have metal washers that provide metal-to-metal contact. Others have a water-sealing gasket that is typically polyvinyl chloride (PVC), which is a plastic, and ethylene-propylenediene (EPDM), which is a synthetic rubber. Other materials are used also.

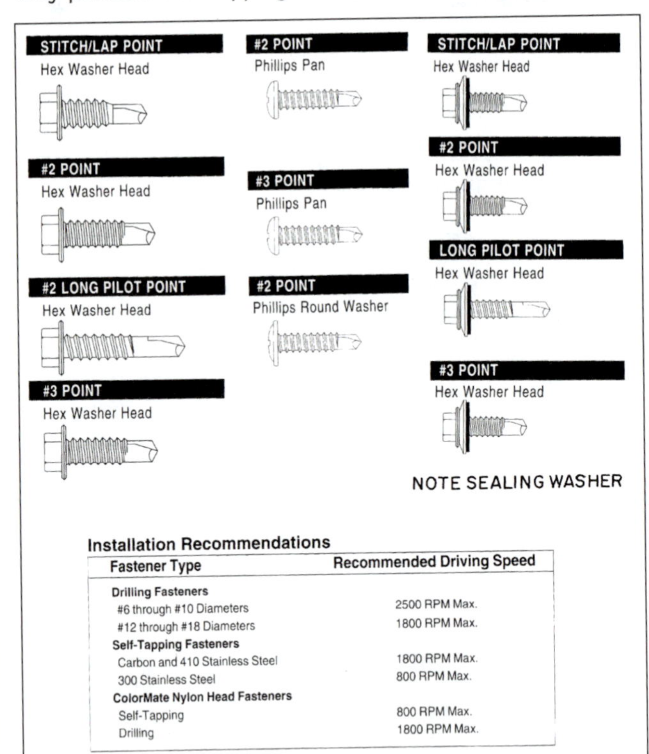

Installation Recommendations

Fastener Type	Recommended Driving Speed
Drilling Fasteners	
#6 through #10 Diameters	2500 RPM Max.
#12 through #18 Diameters	1800 RPM Max.
Self-Tapping Fasteners	
Carbon and 410 Stainless Steel	1800 RPM Max.
300 Stainless Steel	800 RPM Max.
ColorMate Nylon Head Fasteners	
Self-Tapping	800 RPM Max.
Drilling	1800 RPM Max.

Figure A

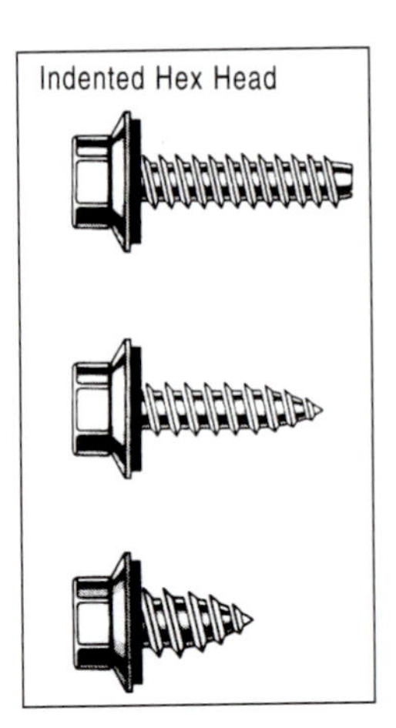

Figure B

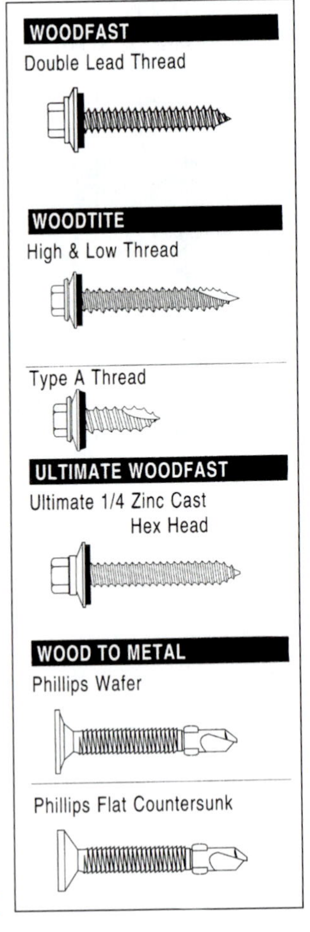

Figure C

Figure 17.16 Load indicator washers can be used to verify that a bolt is under the proper tension.

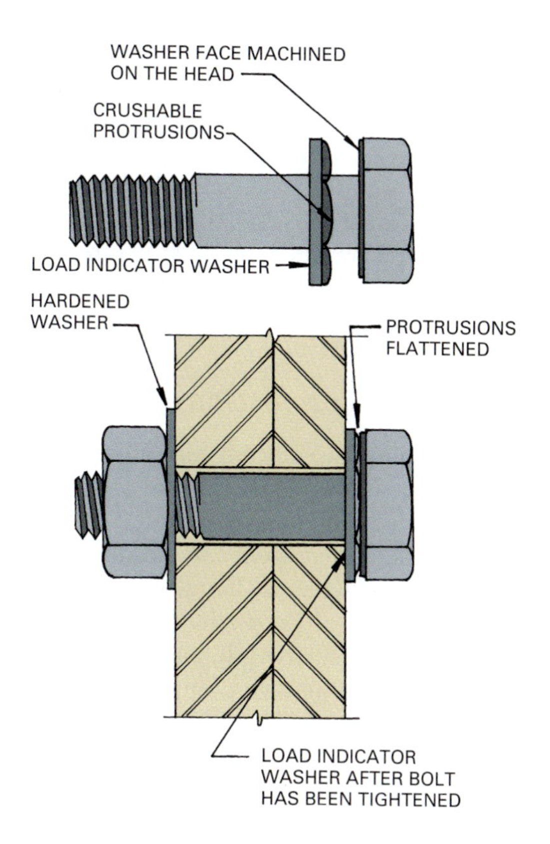

together as it cools (Fig. 17.22). Large welds require several beads laid on top of each other.

A structural engineer determines the size, length, and type of welds to be used. Welded connections in critical locations require that each weld be inspected. There are a number of techniques available for checking welds for flaws.

THE STEEL FRAME

The members making up a steel frame are joined with metal connections and riveted, bolted, or welded. The connections typically are angles, tees, or plates. The connections transmit vertical (shear) forces and bending moments (rotational forces) from one member to the other. **Shear stress** is produced when vertical loads are applied to a member, creating a deformation in which parallel planes slide relative to each other so as to remain parallel. A **moment** is the property by which a force tends to cause a body to which it is applied to rotate about a point or line, as shown in Fig. 17.23. **Bending moment** is the moment that produces bending at a section of a structural member. The various connections illustrated in the next section of this chapter are identified as shear connections, moment connections, or connections controlling both.

Figure 17.17 Connections can be tensioned as specified by using tension-control bolts, which have a spline that shears when the proper tension is reached.

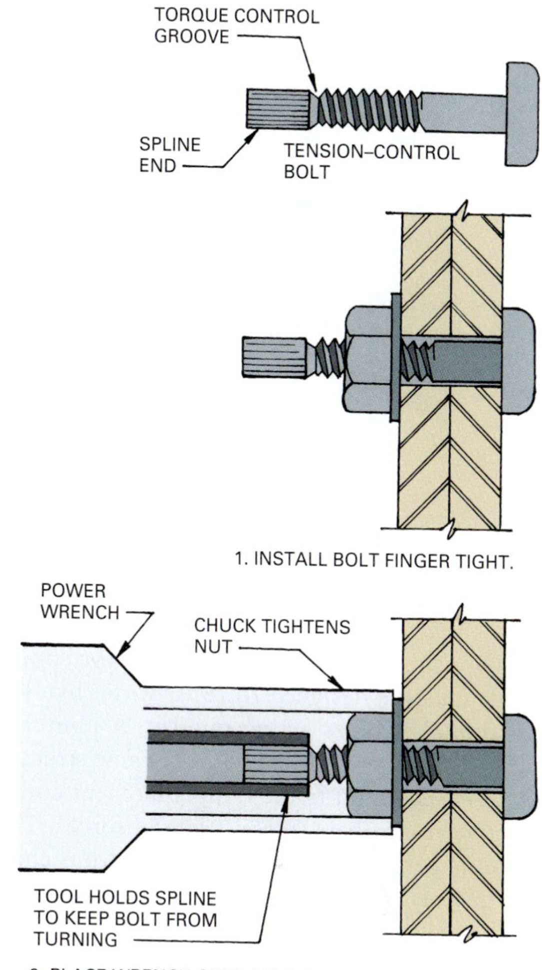

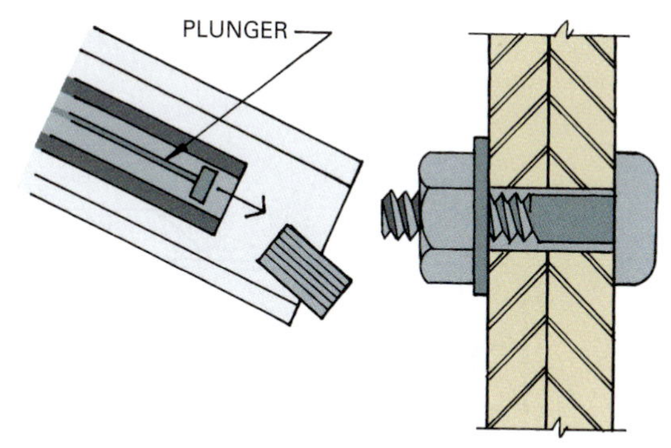

Figure 17.18 This angle connector has been welded to a beam during fabrication and welded to a column during erection.

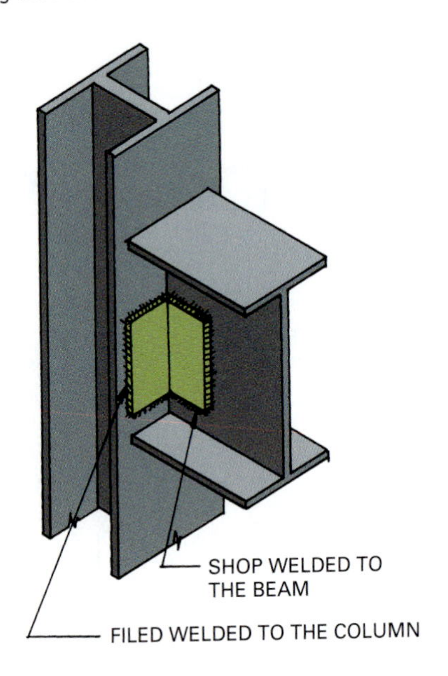

Frame Stability

A building using structural steel framing (or concrete or wood) must be designed to resist wind, live load, and earthquake forces acting upon it. A number of strategies exist to provide stability to frame structures (Fig. 17.24). The rigid frame utilizes moment connections between members to resist lateral forces. These connections prohibit adjacent members from changing the geometry between them. Braced frames use diagonal members to provide triangulation that braces the framing from deformations. Used often in tall buildings, a rigid core acts as a vertical beam that is stabilized with a deep foundation. The core is frequently built of reinforced concrete and houses stairs, elevators, and service chases. Flat shear walls of concrete or steel, stiffened in both directions, can be used to stabilize a structural frame. For very tall, slender buildings, hollow tubes provide maximum efficiency in resisting wind forces. The bundled tube concept combines the inherent strength of individual tubes in a bundled configuration (Fig. 17.25). Mega frames describe structures in which large members constitute the overriding lateral support of a building.

Types of Beam Connections

The most commonly used beam connections are either framed or seated (Fig. 17.26). **Framed connections** join a beam to an adjoining member with connectors, such as angles, that are secured to the beam web. **Seated connections** have the flange of a beam resting on a seat, such as a metal angle. When connections are designed, the type, strength, and size of fasteners; strength of the base material; and the end reaction of the beam are all considered. Usually, one side of a framed connector is joined to a beam in a fabrication shop and the other side is field-connected to a column or adjoining beam.

Simple, Rigid, and Semi-Rigid Connections Moment connections are used to transfer forces in beam flanges to a column. This force transfer does not preclude the need for shear connections used to support the beam. Framed, seated, and end-plate connections are

Figure 17.19 A detailed drawing of a beam with connections shop-welded to it and prepared for field-bolting to a column or girder.

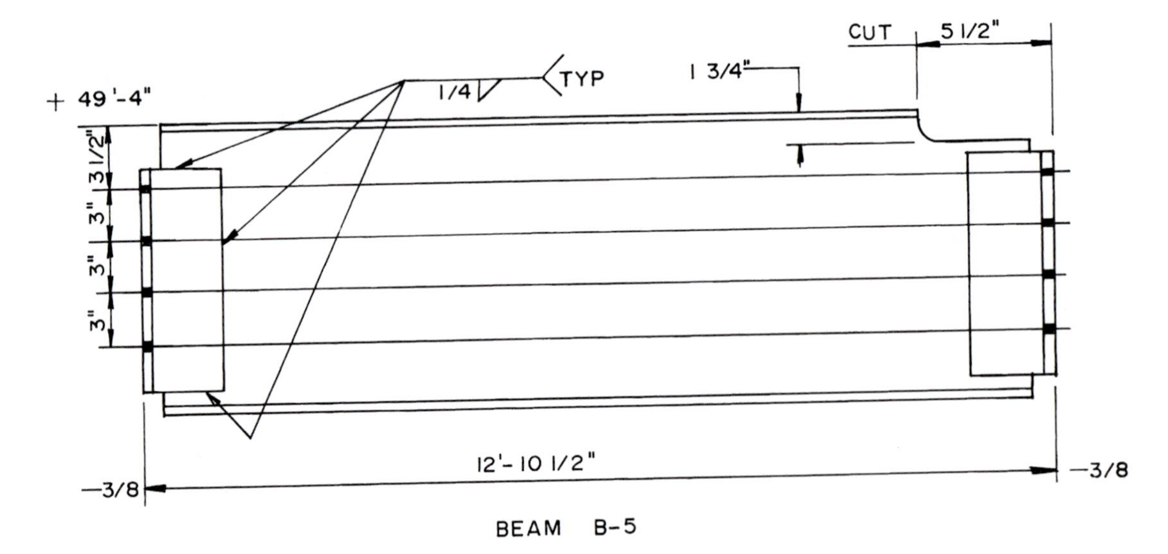

Figure 17.20 Standard symbols for welded connections.

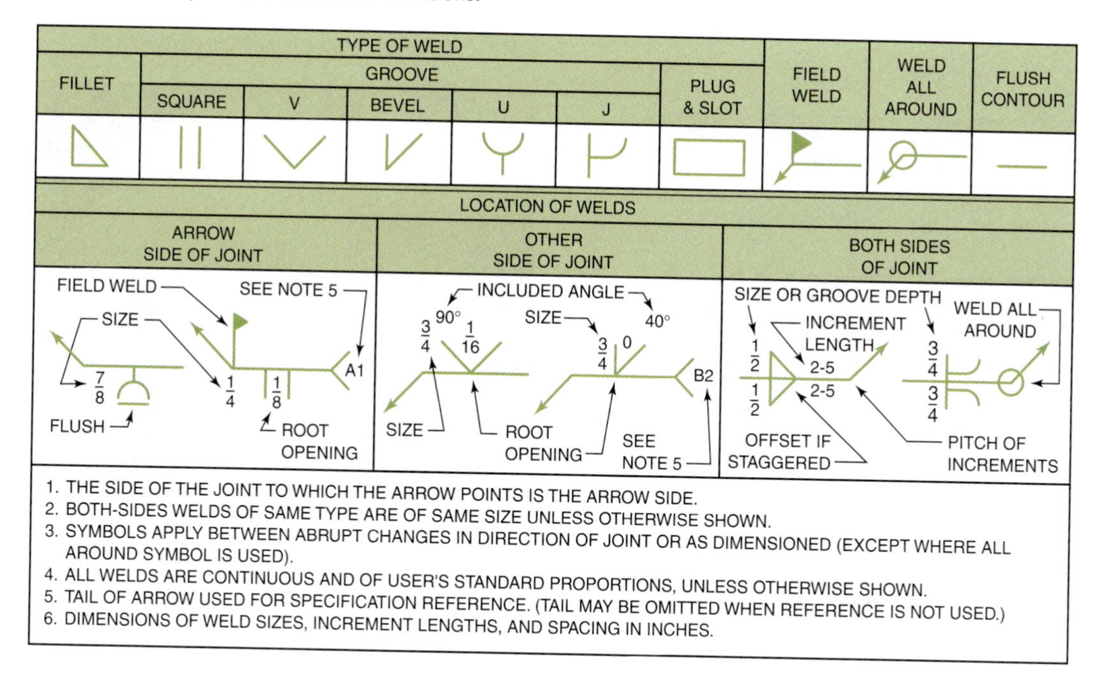

considered shear connections. Moment connections have flange stresses developed independently of the shear connections.

Simple connections provide for shear only and require additional support to stabilize the structural frame, such as diagonal bracing or shear panels. Rigid connections provide for shear and have sufficient rigidity to hold the original angles between connected members (moment). Rigid-frame construction does not require additional bracing.

Semi-rigid connections provide for shear, and, although not as unyielding as rigid connections, they do possess a known moment-resisting capacity.

Welded Connections Examples of welded beam-to-column and combinations of welded and bolted connections are shown in Fig. 17.27. Load capacity depends on the connector, type of weld, base material, and strength of electrode material. AISC tables indicating the sizes and capacities of angle connections are based on the use of E70 electrodes. Welded column splicing is illustrated in Fig. 17.11 on page 313.

Bolted and Riveted Connections Two types of connections are possible using bolts or rivets. A bearing connection is one in which the fasteners bear against the sides of the holes in the connections. The rivet or bolt bears the stress. All riveted and some bolted connections are bearing type.

Friction connections are those in which a fastener can be sufficiently tightened to clamp the connected parts together under high pressure. The shearing force on the connection is resisted by the friction between the connected parts rather than by the bolts. Only high-strength bolts are used for this type of connection.

Examples of bolted beam-to-column details are shown in Figs. 17.28 and 17.29. Riveted connections use similar details. A partial framing drawing indicating a beam-to-beam connection and design information is pictured in Fig. 17.30. A typical beam-to-beam connection is shown in Fig. 17.31. Notice the top of the beam butting the girder has a clearance cut to facilitate flushness. The notch (called a cope, block, or cut) allows clearance around the flange.

Channels are often used between beams to support decking, like on a roof. A typical connection between channels is shown in Fig. 17.32. Many special connections require careful analysis by a structural engineer. Not all designs fit the rather standard connections illustrated to this point. A variety of structural configurations and connections are possible, as illustrated in Fig. 17.33.

DECKING

Several types of floor and roof decking—for example, metal decking—and precast concrete units are used with structural steel framing.

Figure 17.21 Penetration welds commonly used on structural steel construction. Penetration welds extend all the way through a material.

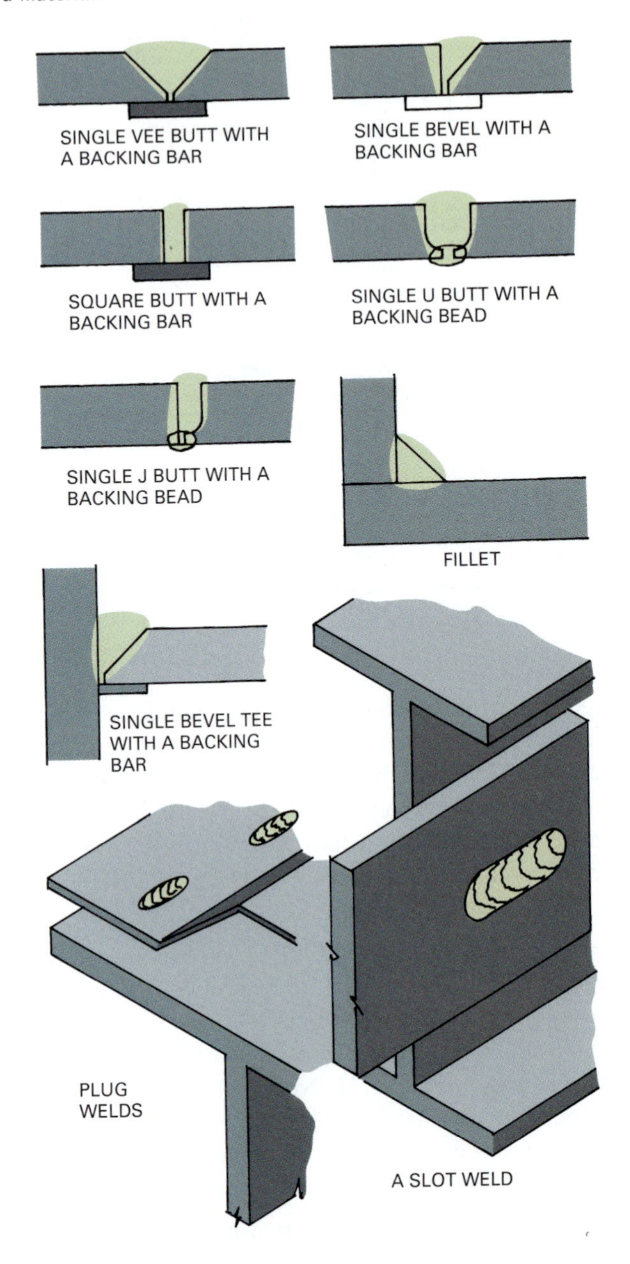

Figure 17.22 The welding arc merges the molten electrode core with the molten base material. *(Image copyright Lngdu, 2009. Used under license from Shutterstock.com)*

Figure 17.23 Structural members and connections are subject to shear forces and positive or negative bending moments.

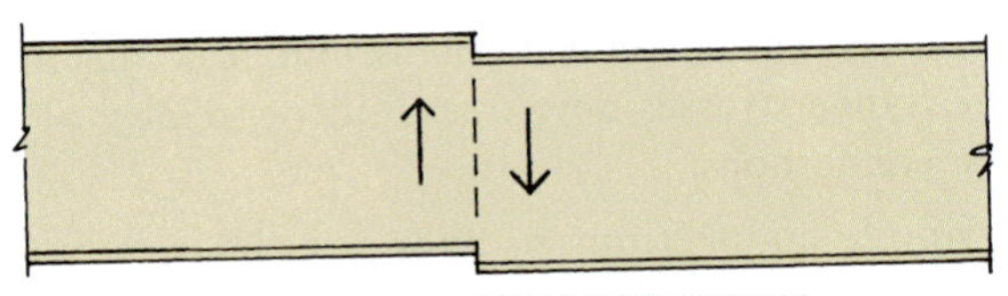

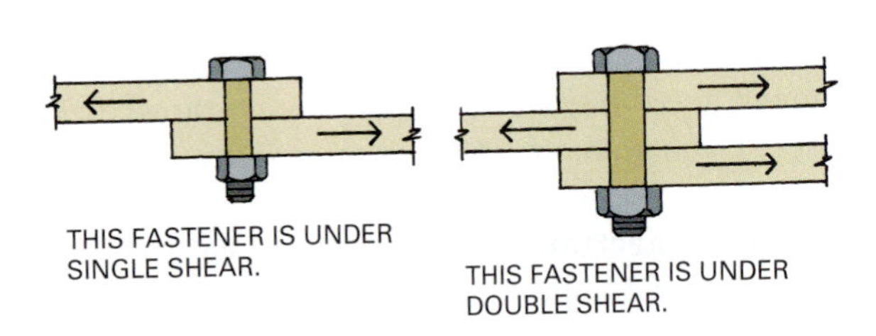

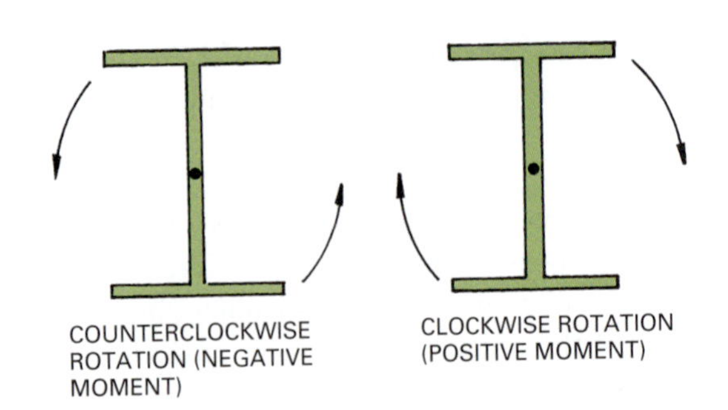

Figure 17.24 Typical methods for stabilizing a structural frame.

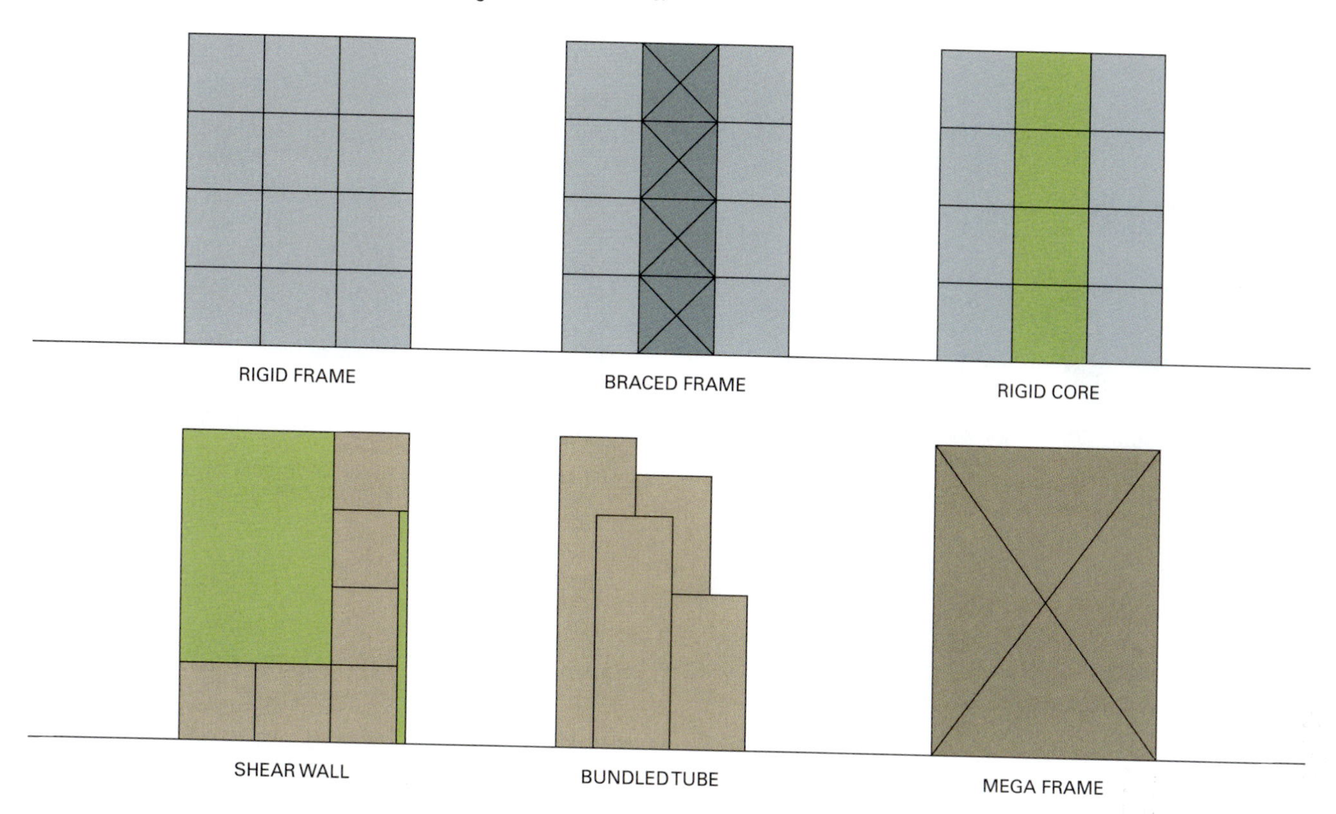

Figure 17.25 The Sears Tower is a classic example of the bundles tube concept. *(Image copyright EugeneF, 2009. Used under license from Shutterstock.com)*

Figure 17.26 Typical framed and seated connections.

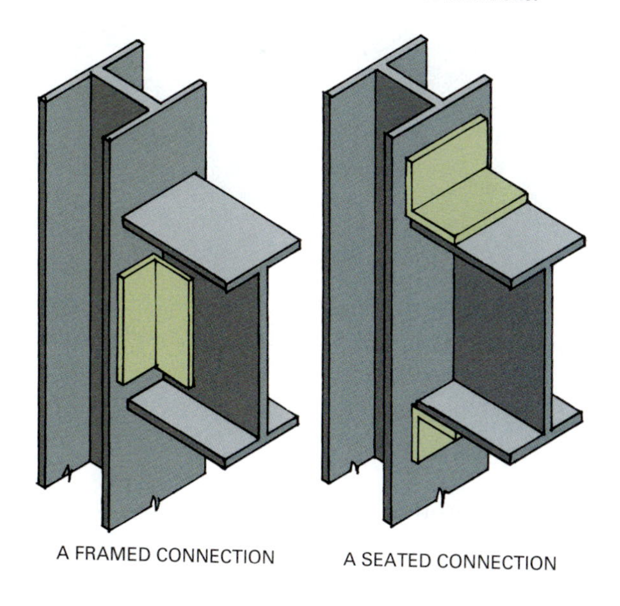

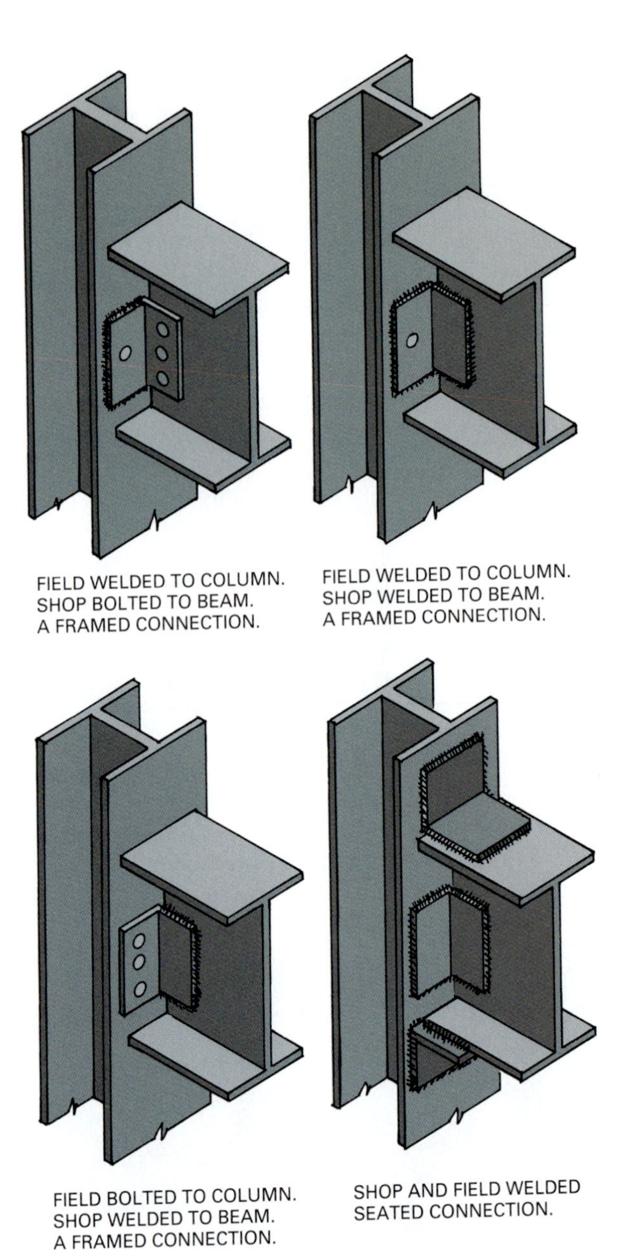

Figure 17.27 Typical welded beam-to-column framed, seated, and end-plate connections.

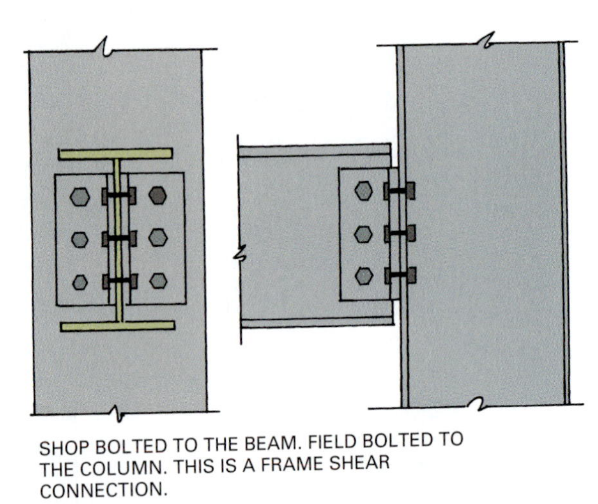

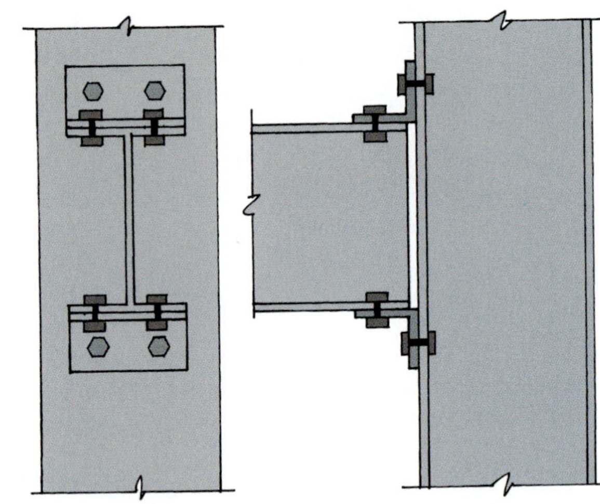

Figure 17.28 A typical bolted beam-to-column framed shear connection.

Metal Decking

Sheet and cellular steel decking are illustrated in Chapter 15. Sheet decking is available in various thicknesses of metal and depths of corrugation (Fig. 17.34). Designers consider spans between supporting members and load-carrying requirements when selecting a product. Metal roof decking is often covered with both rigid insulation and the finish roofing material. Under these conditions, it must carry all of the imposed loads.

Composite metal decking provides a base on which to pour a concrete slab (Fig. 17.35). The reinforced slab accounts for most of the structural strength, but the decking provides tensile reinforcing via bonding. Bonding occurs through perforations in the metal or by wire fabric tack-welded to the decking. Sometimes a designer specifies the use of metal **shear studs** that are welded to the beams and protrude up into the concrete slab (Fig. 17.36). These tie the steel beam and the concrete slab together, which increases the load-carrying capacity of the steel beam. Properly installed metal roof decking can serve as shear diaphragms against wind, seismic forces, and other lateral loads.

Cellular steel-floor raceway systems consist of metal decking with wiring raceways and a structural concrete slab on top. The cellular raceways provide space to run electrical wiring, telephone lines, and data cables. These connect to a large main header duct that has a removable cover for lay-in wiring. Inserts provide access

Figure 17.29 One type of bolted moment-resisting connection.

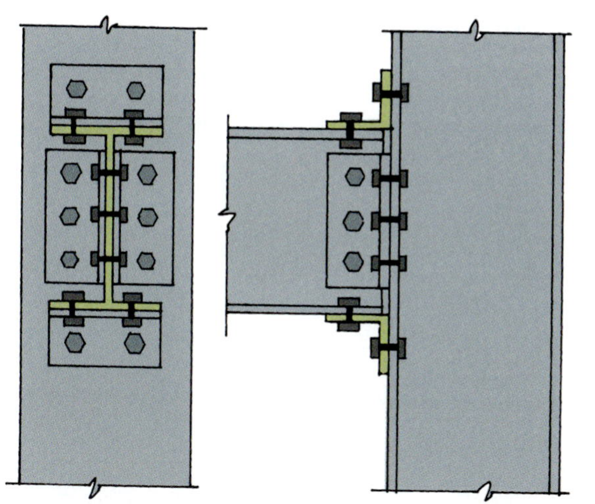

Figure 17.30 Design information for a beam-to-beam connection.

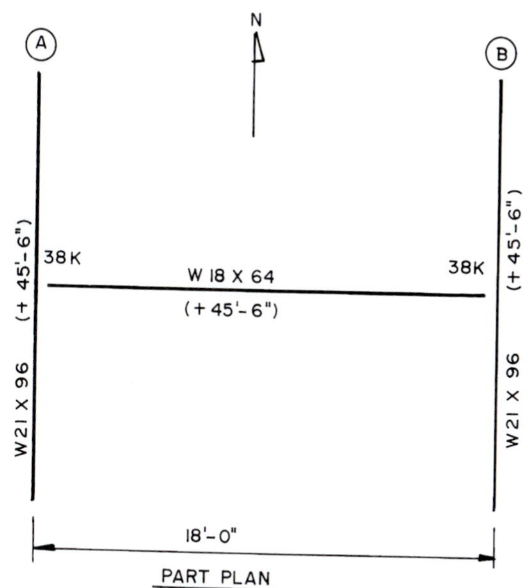

Figure 17.31 A typical bolted beam-to-beam connection. *(Image copyright Mark Winfrey, 2009. Used under license from Shutterstock.com)*

Figure 17.32 A typical bolted channel connection.

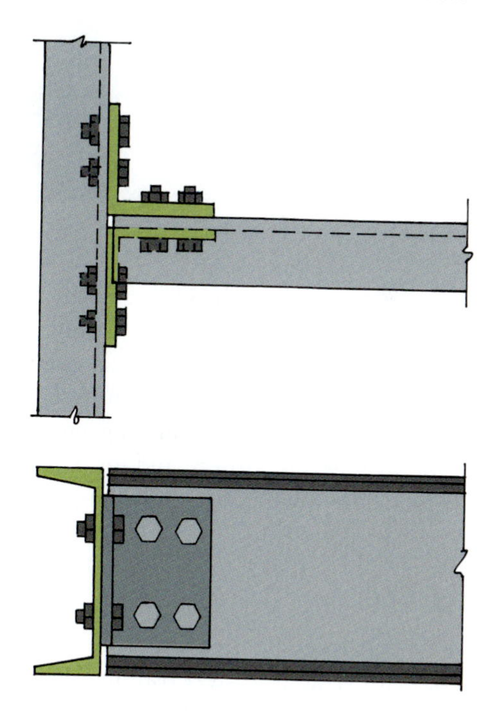

Figure 17.33 An example of a special connection design. *(Image copyright Elena Elisseeva, 2009. Used under license from Shutterstock.com)*

through the concrete slab to the wiring in the raceway (Fig. 17.37). The cellular system provides a fire-resistant barrier between floors, serves as a form for pouring the concrete deck, and provides tensile reinforcement for the concrete floor slab. Metal decking is generally plug-welded to the steel beams, girders, and joists. The edges are usually joined by welding or self-drilling screws.

Concrete Decking

A number of precast concrete decking units are available for use with structural steel framing. Detailed in Chapter 9, they include hollow-core units and precast slabs and channels of various designs. Long-span precast concrete roof and floor units are generally prestressed. Short-span and cast-in-place members have reinforcement.

Precast slabs and channels are situated on the structural steel framing. They form decks and, in some cases, ceilings (Fig. 17.38). They are secured to beams and joists by welding metal plates that are cast in them to the steel structure or by using clips, as shown in Fig. 17.39.

Figure 17.34 Typical types of metal decking.

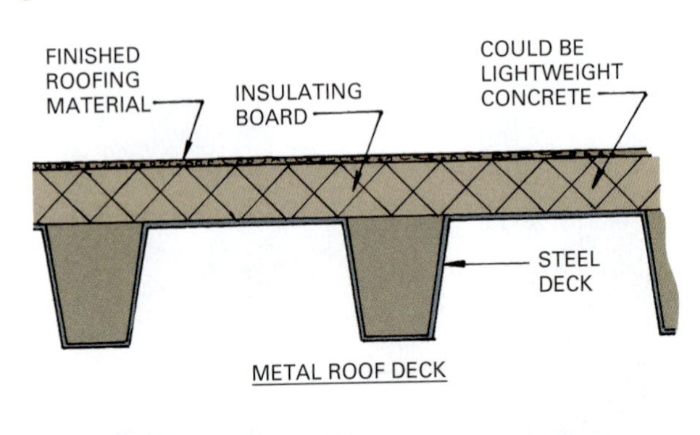

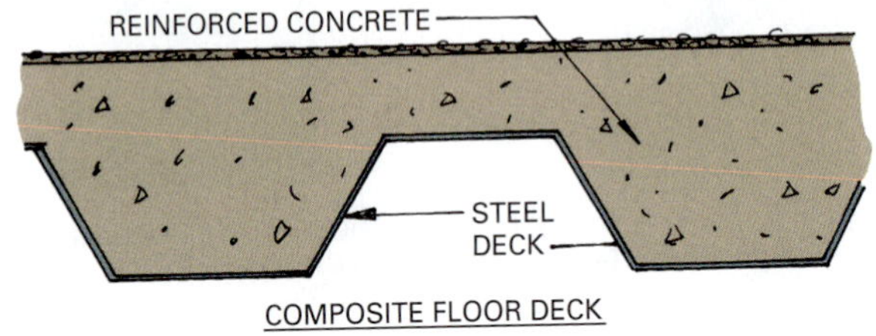

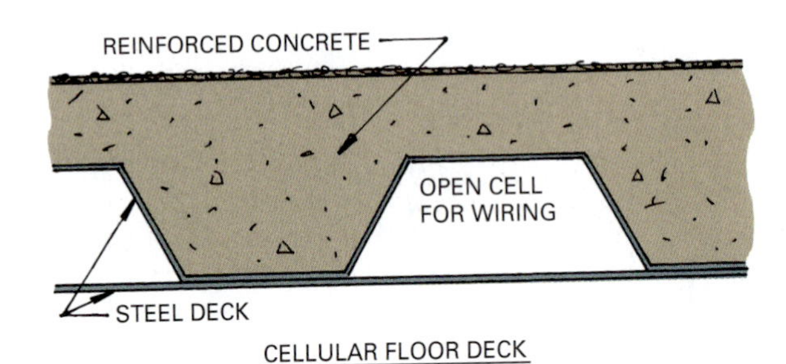

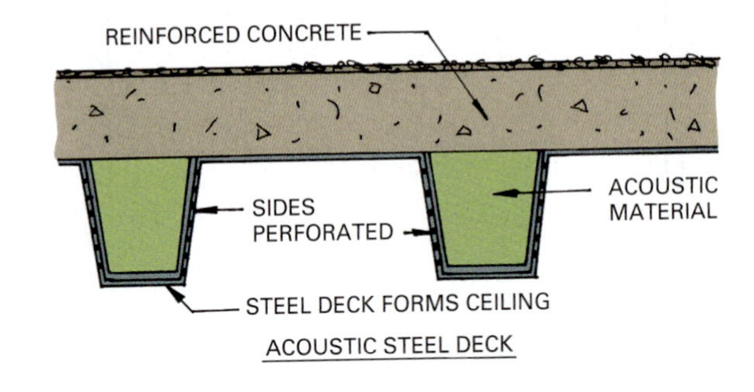

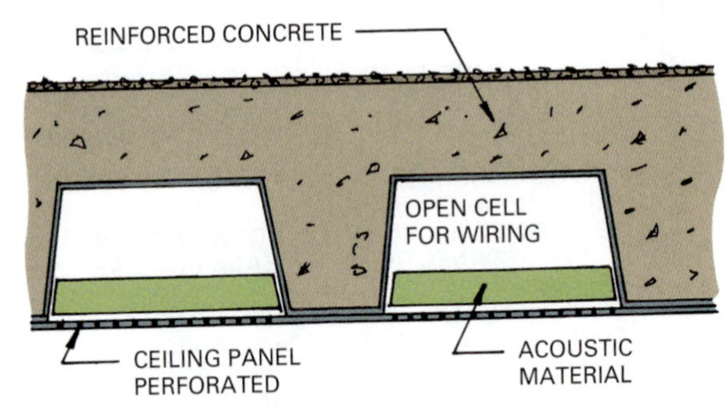

Figure 17.35 Composite metal decking serves as a base for a concrete slab floor or roof. *(Courtesy Vulcraft)*

Precast hollow-core decking units, used for long spans, are placed on the steel framing with a crane. For better bonding with topping materials, their top surface is often left rough. The topping helps tie precast units together and increases a roof's or floor's structural integrity. Reinforcing can be cast in the topping.

STEEL TRUSSES

A number of steel trusses are used for construction purposes. Included are roof trusses, open-web joists, and space frames. A truss is a coplanar (forces operating in the same plane) assembly of structural members joined at their ends. The diagonals form triangles, producing a rigid framework. Triangulation is the concept behind the design of a truss. A triangle is the only multisided geometric figure that is rigid and will not move or rotate (unless a structural member bends or a connection fails). This principle is illustrated in the design of the bridge shown in Fig. 17.40.

Roof trusses act like beams and support a roof and all imposed loads. Common types of floor and roof trusses are shown in Fig. 17.41. A typical steel

Figure 17.36 Shear studs tie the steel structural member to the concrete slab. *(Drawing Courtesy Vulcraft) (Photo Courtesy Bethlehem Steel Corporation)*

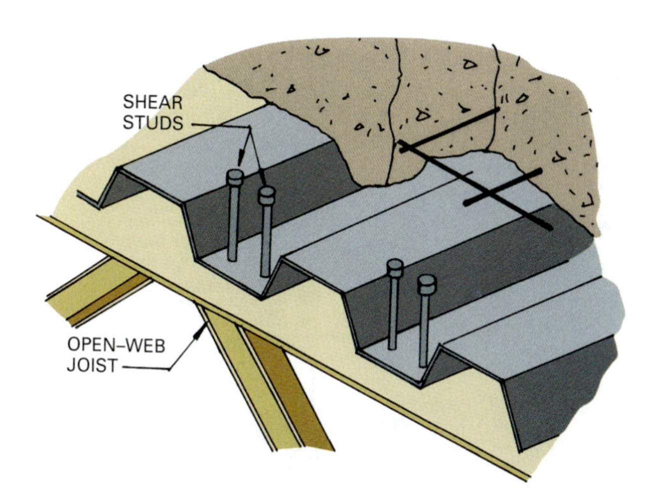

Figure 17.37 A steel raised floor system for installation of electrical, heating and plumbing distribution. *(Photo courtesy Jonathan Lamb, http://en.wikipedia.org/wiki/Image:Tile-lifter-in-use-raised-floor.jpg)*

Figure 17.38 These lightweight precast concrete deck panels provide a structural deck. *(Courtesy H.H. Robertson)*

Figure 17.39 Precast deck planks can be secured to steel framing with metal clips.

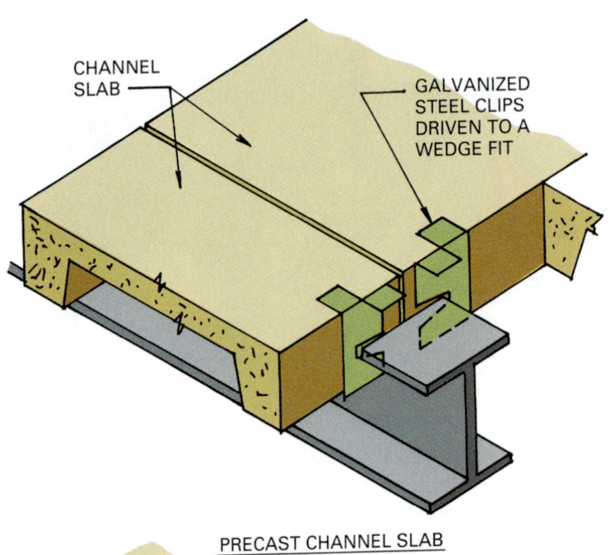

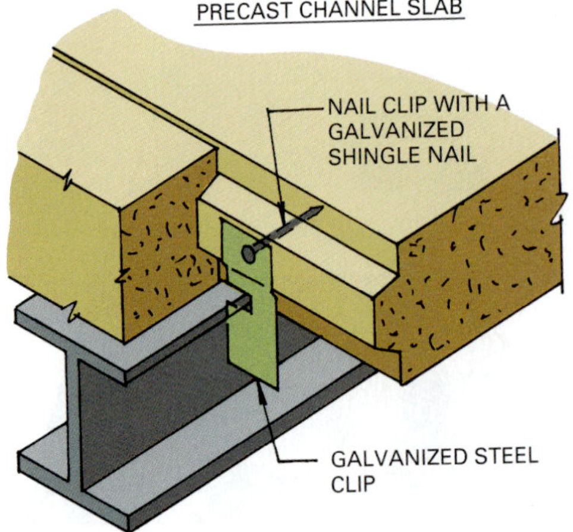

Figure 17.40 The structure of this bridge uses the principle of triangulation. *(Image copyright Khafizov Ivan Harisovich, 2009. Used under license from Shutterstock.com)*

roof truss consists of top and bottom chords and a system of webs. The slope of the top chord can vary as desired, and it can be parallel to the bottom chord for flat roofs. Parallel chord trusses are used for floors and flat roofs and as major structural beams and girders. A typical trussed-roof framing structure includes trusses, purlins (horizontal members that connect trusses), and a deck. As shown in Fig. 17.42, large multi-floor steel trusses can carry the loads of floors, roofs, and ceilings.

Open-web joists are a type of truss. They are lightweight structural members made from steel angles and bars (Fig 17.43). They, like other trusses, are end bearing and can span long distances in floor and roof constructions.

Space frames are a three-dimensional truss system. They provide a structural frame that spans in several directions and over large areas with a minimum number of vertical supports. Space frames are generally

Figure 17.41 Commonly used truss configurations.

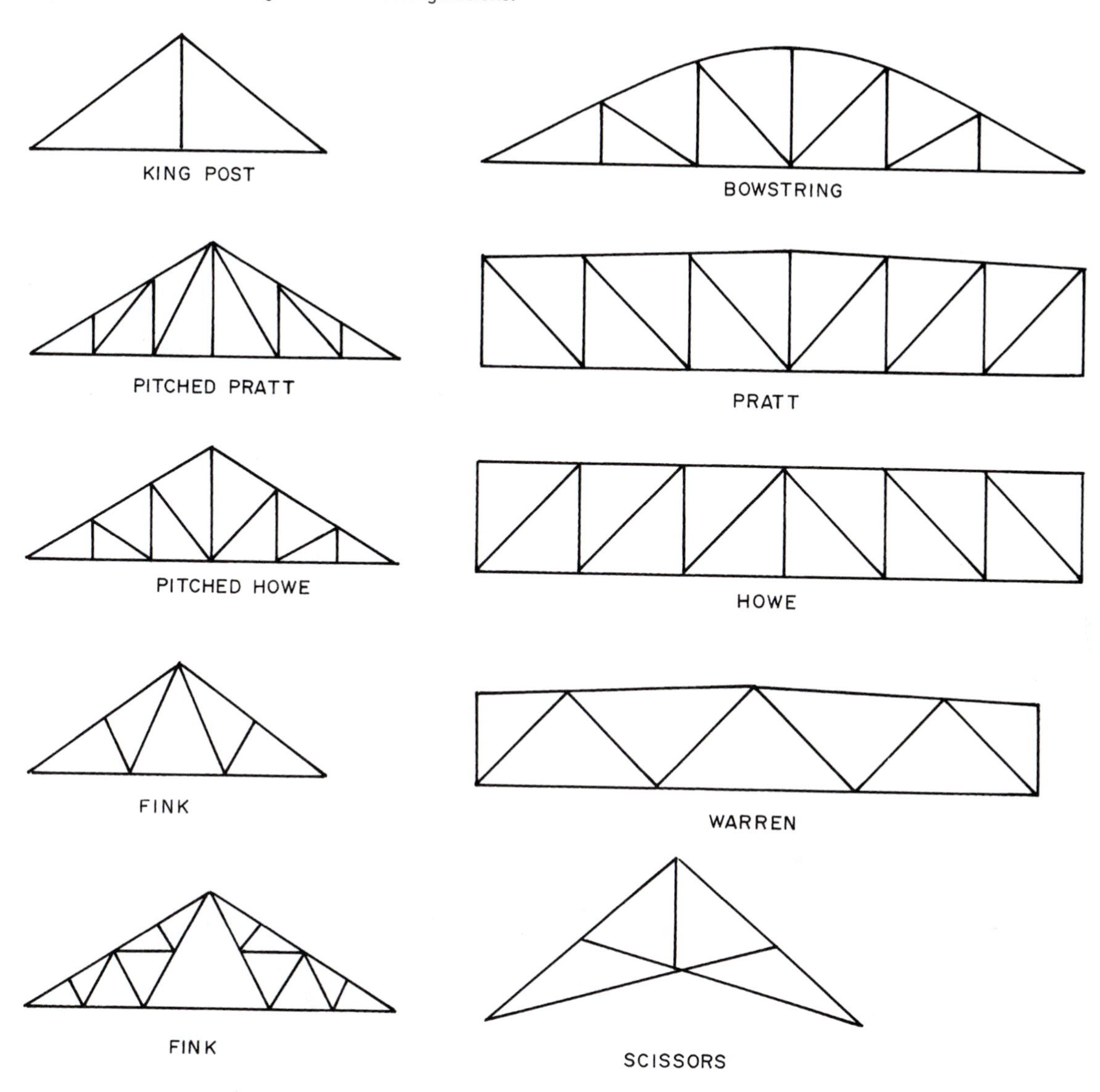

made from tubular members (Fig. 17.44). However, wide flange beams and tee members are also used. The attractiveness of the assembly is lost if it must be covered to meet fire codes. Some codes do not require fire protection if the space frame is more than 20 ft. (6.1 m) above the floor.

TENSILE STRUCTURES

Tensile structures and cable net structures are framed using high-strength cables of cold drawn steel suspended between supporting members and secured with cable stays (Fig. 17.45). These lightweight structures can span large distances in an almost unlimited number of shapes and curvatures. Cable net structures may be covered in a variety of different materials to provide a weather tight enclosure. Glass or polycarbonate panels can be fixed to cables with mechanical connections and sealed with rubber gaskets. Fabric coverings can be used to provide a lightweight and translucent enclosure for tensile steel structures (Fig. 17.46).

LIGHT-GAUGE STEEL FRAMING

Light-gauge structural steel shapes are formed from flat cold-rolled pieces of carbon steel. Gauge thicknesses range from No. 12 to No. 20. Some shapes are formed from a single steel sheet, but others have several forms shape-welded together. They are available in nailable and non-nailable types and are either

Figure 17.42 Multistory trusses provide lateral bracing and carry floor loads in the Hong Kong and Shanghai Bank Tower. *(Image copyright Norman Chan, 2009. Used under license from Shutterstock.com)*

Figure 17.43 Open-web joists are lightweight truss-type structural members that can span long distances. *(Courtesy Eva Kultermann)*

Figure 17.44 Space frames span long distances and permit the integration of electrical and mechanical systems within the structure. *(Photo by Andrew Dunn, http://commons.wikimedia.org/wiki/Image:LibeskindSpaceFrameTower.jpg)*

Figure 17.45 Tensile or tensioned fabric structures use a network of steel cables in sleeves in the fabric covering to form the structural support and exterior finish. *(Courtesy Eva Kultermann)*

galvanized or primed with zinc chromate. Load-carrying capacities and spans should be obtained from a manufacturer. Light-gauge metal is widely used in commercial, industrial, and residential construction (Fig. 17.47).

Figure 17.46 The tensioned fabric structure at the Denver Airport provides a lightweight, translucent enclosure system. *(Photo by David Benbennick, http://en.wikipedia.org/wiki/Image: Denver_International_Airport_terminal.jpg)*

Figure 17.47 Lightweight steel studs and joists are widely used in commercial and residential construction. *(Image copyright Mark Winfrey, 2009. Used under license from Shutterstock.com)*

Typical construction details for light-gauge steel structural members are shown in Fig. 17.48. The floors can consist of steel decking with a concrete topping or precast concrete planks. Roofs can be any decking material available. Stud walls can support lightweight steel or open-web joists and various roof trusses or metal rafters. The finished exterior can be any material typically used.

PRE-ENGINEERED METAL BUILDING SYSTEMS

The building frames of pre-engineered metal building systems are constructed from a variety of structural steel and light-gauge steel framing members. The vertical members are supported on cast-in-place concrete footings and designed to carry specified loads. The company that manufactures and sells the systems distributes them through a network of franchised dealers. The dealers work with customers and architects to produce the required structural systems. Figure 17.49 shows the assembly of a pre-engineered building using rigid frames as the main structure and enclosing materials supported by light-gauge steel members. The complete assembly, including the steel frame, girts, purlins, and bracing, provides the needed strength. The common types of pre-engineered buildings available include truss-type, rigid frame, post-and-beam, and sloped roof.

Truss-type buildings use steel columns and open-web joists and girders to form a roof (Fig. 17.50). Joist girders span the long distances between steel columns. Smaller open-web joists serve as truss purlins between joist girders.

Rigid frame buildings, as shown in Fig. 17.51, are available with low-sloped and high-sloped roofs. The roof decking is supported by light-gauge steel Z-shaped purlins spanning the rigid frames. The ceiling is insulated with long insulating blankets laid over the purlins. The insulation is typically a fiberglass blanket encased in a vinyl fabric covering.

Post-and-beam framed buildings are built using posts on the interior to support a roof and divide the building into standard-sized bays. A large building is made up of several bays with posts inside its enclosed area. This roof has a small amount of slope. If greater slope is desired, the roof is framed with trusses or a rigid frame is used. The roof decking is supported by light-gauge metal purlins that span roof beams, trusses, or rigid frames.

Tubular steel trusses are used to carry loads over long distances and provide an attractive, massive appearance. The truss shown in Fig. 17.52 rests on large steel columns and carries both roof and curtain wall loads.

Figure 17.48 Construction details used with light-gauge steel framing. *(Courtesy Dale/Incor, Inc.)*

Construction details

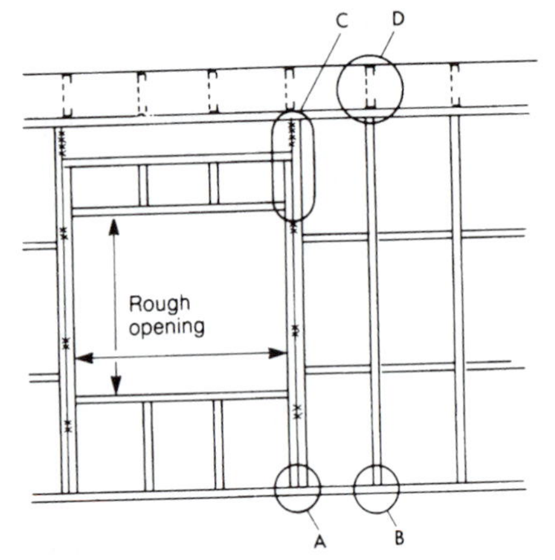

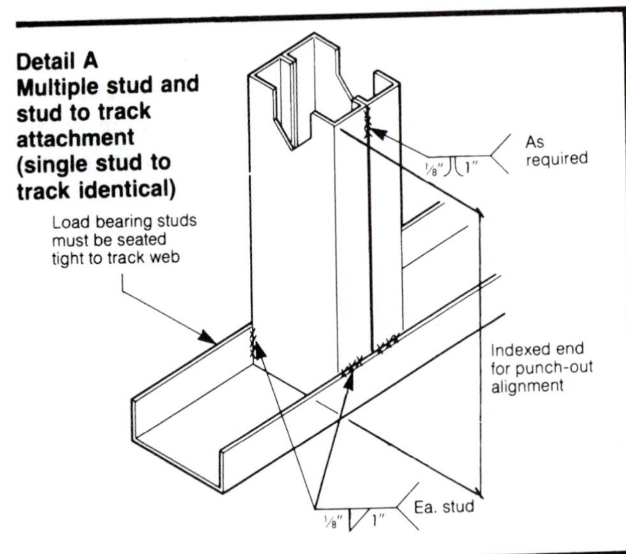

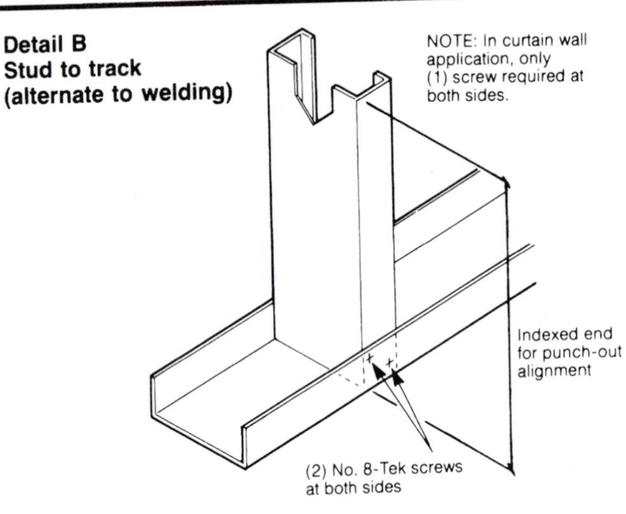

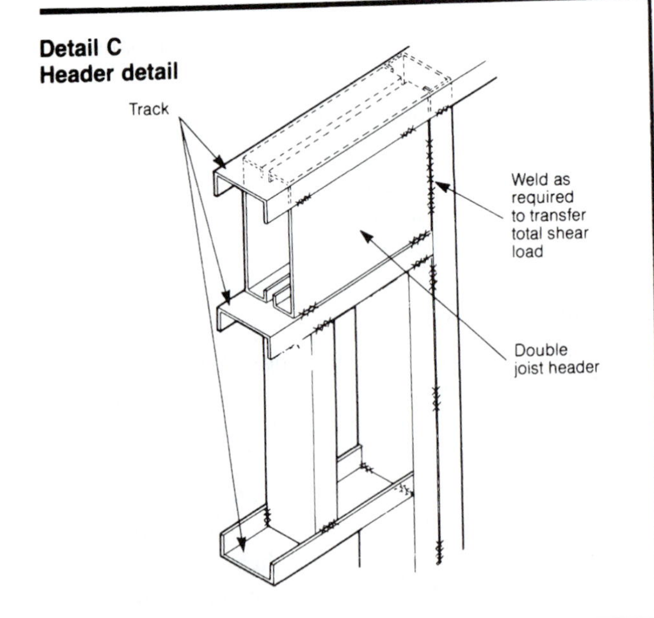

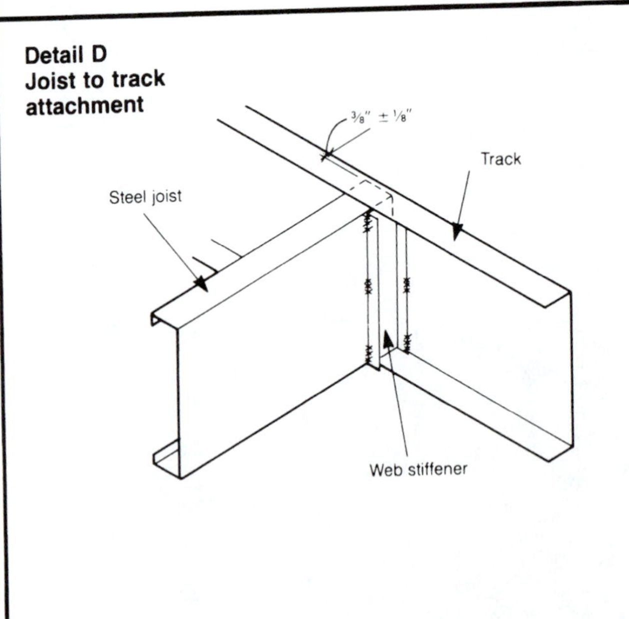

Figure 17.48 *Continued*

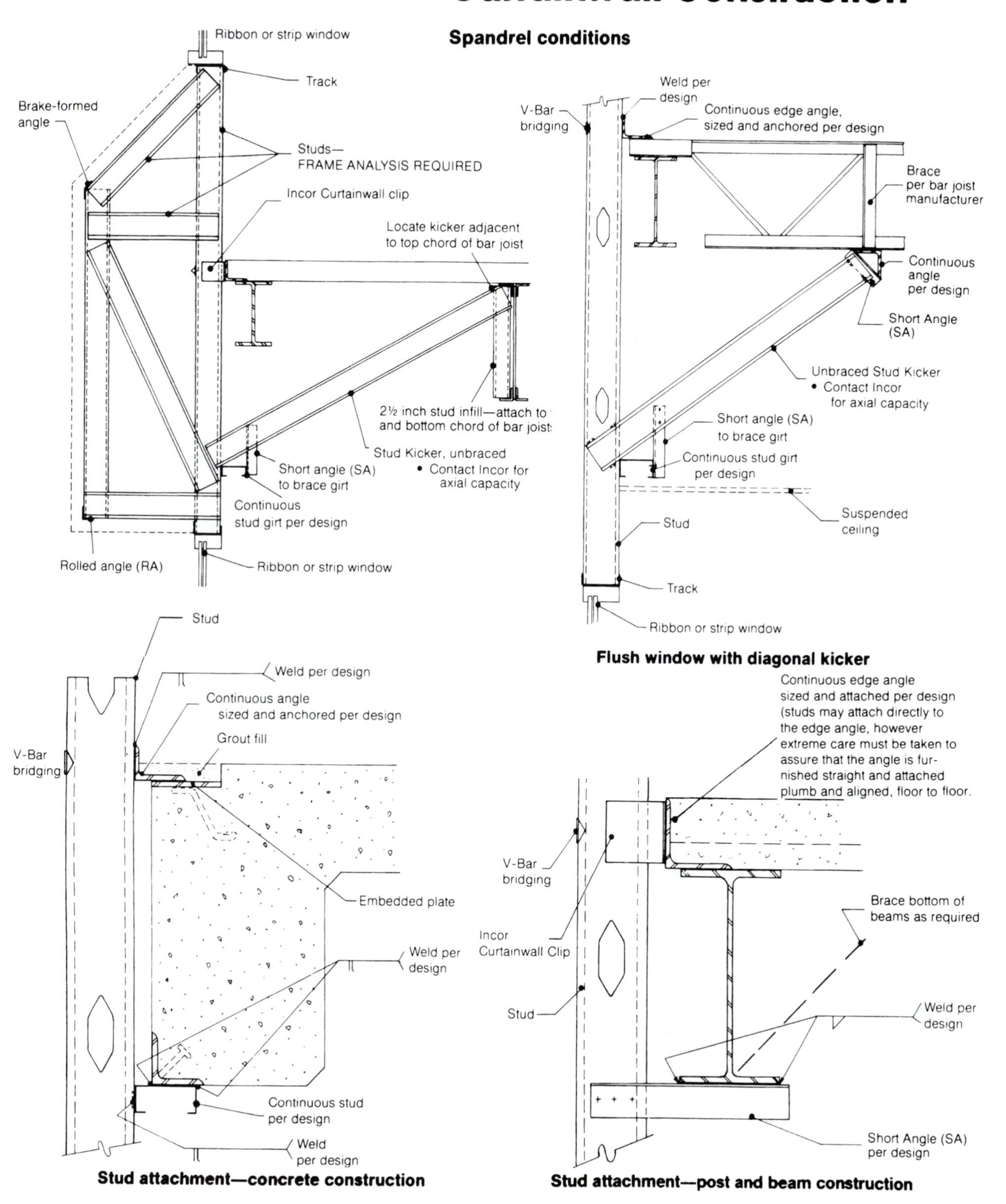

Figure 17.49 This pre-engineered metal building uses factory-assembled rigid frames for wall and roof structural members. Z-purlins, running perpendicular to the rigid frames, support the metal roof. *(Reproduced with permission from* The Building Systems Integration Handbook, *Richard Rush, ed., Butterworth-Heinmann Publishers, Newton, Mass., 1986)*

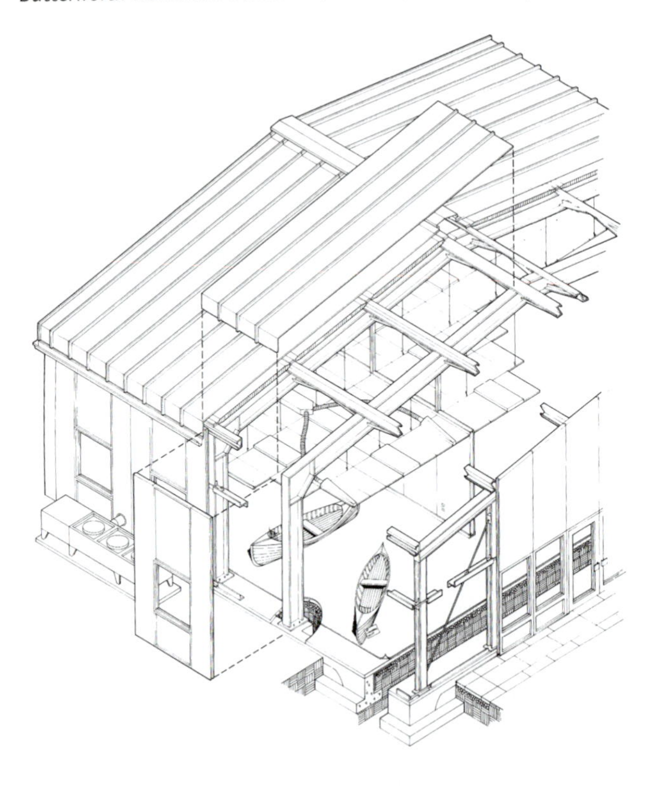

Figure 17.50 A pre-engineered truss-type building uses steel columns, joist girders, and open-web joists. *(Image copyright Kenneth V. Pilon, 2009. Used under license from Shutterstock.com)*

Figure 17.51 These rigid frames support metal purlins upon which metal roof decking will be installed. *(Image copyright Gerald Bernard, 2009. Used under license from Shutterstock.com)*

Figure 17.52 These tubular trusses carry both roof and cladding loads. *(Image copyright Mark Mendenhall, 2009. Used under license from Shutterstock.com)*

EXTERIOR WALL SYSTEMS

A number of metal exterior wall systems are available. Some are single thickness metal panels with ribs formed in them for strength and decorative purposes. Other types have insulation in a hollow-core panel. This insulation may consist of a foamed core, rigid insulation board, or some form of fibrous insulation blanket. Methods for joining panels vary depending on the manufacturer (Fig. 17.53).

Exposed exterior surfaces can have a colored coating, such as fluoropolymer over a smooth or embossed metal face (Fig. 17.54); a stucco-type finish, such as an applied fiber-reinforced polymer to the metal; panels with a stone aggregate bonded to the metal; or a kiln-fired clay brick surface.

Figure 17.53 Several types of metal wall panels used on pre-engineered buildings.

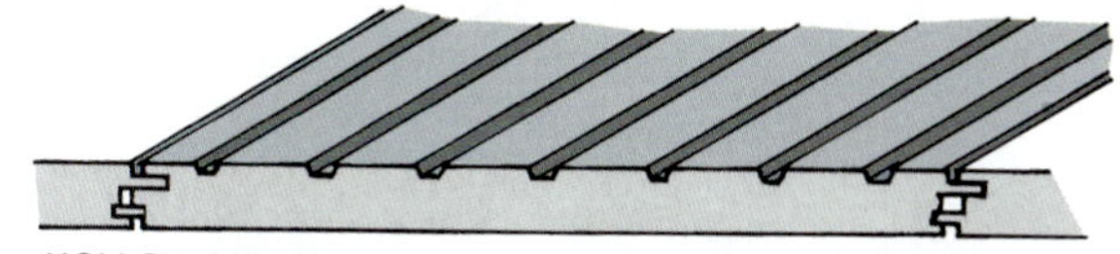

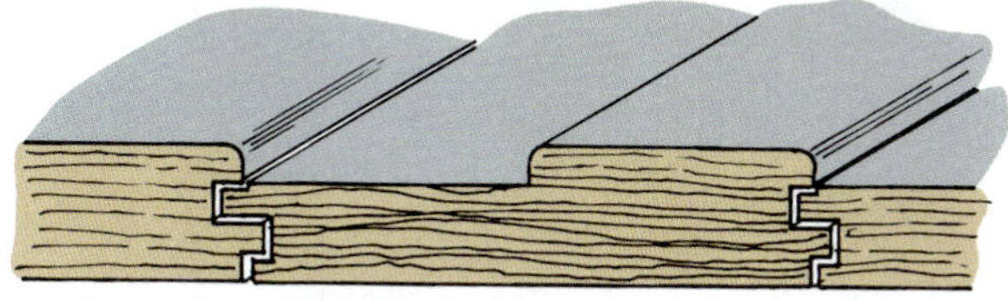

Figure 17.54 An example of a metal wall panel. *(Image copyright Thomas Fredriksen, 2009. Used under license from Shutterstock.com)*

FIRE PROTECTION OF THE STEEL FRAME

Structural steel is an incombustible material that will not melt even during a building fire. But when subjected to sustained extreme heat, its properties, including strength, are affected. This can lead to steel column and beam failure during a prolonged fire. To protect a building's occupants and structural integrity, codes require coating certain steel frames with a fire-resistant material. These requirements depend on types of construction, building heights, floor area, occupancy, fire protection systems, and building location. Many materials can be used to protect structural steel, including concrete, tile, brick, stone, gypsum board, gypsum blocks, fire-resistant plasters, sprayed-on mineral fibers, intumescent fire-retarding coatings, liquids, and flame shields.

Unprotected mild steel loses about half its room temperature strength at temperatures exceeding 1000°F (542°C). Fire resistance ratings are given as the number of hours a material can withstand fire exposure, as specified by standard test procedures. Most standard fire tests on structural steel members are conducted by the National Institute of Standards and Technology and the Underwriters Laboratories.

Insulating concrete can be used to protect steel members. The amount of protection depends on the thickness of the cover, the concrete mix, and the method of support. Lightweight concretes—those with aggregates such as perlite, vermiculite, expanded shale and slag, pumice, and sintered flyash—repel fire better than normal concretes due to greater resistance to heat transfer and a higher moisture content. Other insulating materials provide greater protection and are much lighter than concrete.

Masonry units, such as concrete block, brick, gypsum block, and hollow clay tile, are often used to enclose structural steel members. In addition to a material's protection, those with hollow cores can be filled with insulation (such as vermiculite) or mortar to increase their fire-protection qualities. Both concrete and masonry protection is heavy, limiting its use (Fig. 17.55).

Lightweight plaster troweled over a metal lath is a lightweight fire protection material. Lightweight plaster using perlite or vermiculite aggregates has good insulating qualities.

Fire-resistant sheet products are secured mechanically to structural steel members. Those with hard surfaces, such as gypsum board, can also be painted and serve as a finished interior surface (Fig. 17.56).

Sprayed cementitious materials, such as gypsum plaster with perlite, vermiculite, or other insulating materials, are sprayed directly on structural steel (Fig. 17.57). The steel must be clean and primed to receive a coating. Since abrasion can easily damage these coatings, they are used in unexposed areas and covered with a durable protective material. Thick applications (over 2 in. or 50 mm) require that a metal mesh be used.

Mineral fiber slabs are also used to enclose structural steel members. Mineral fibers have excellent heat flow

Figure 17.55 Structural steel can be fire protected by concrete or masonry encasement.

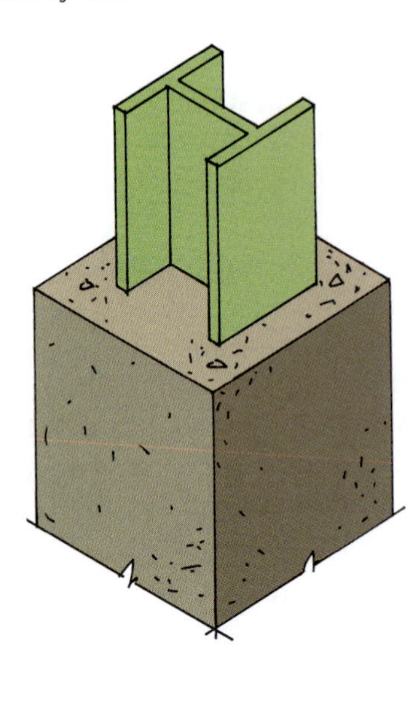

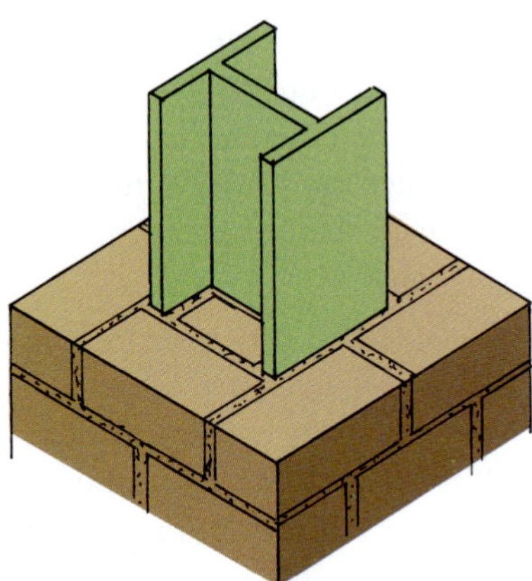

retardation properties and can withstand temperatures above 1000°F (542°C). This type of coating is easily damaged and requires a protective covering if vulnerable to possible impacts or weather.

Intumescent coatings are a sprayed-on mastic fire-retarding coating. They dry to a hard, durable finish, much like paint. When exposed to heat, the coating expands, increasing its thickness and forming an insulation blanket. These coatings are available in various colors and can serve as a finished surface.

Liquid-filled columns are used to reduce steel's heat. Tube- or box-type columns are filled from a water main that replenishes water lost due to fire heat. Vent valves release the steam produced. Pumps are sometimes

Figure 17.56 Examples of gypsum board enclosures used for fire protection. Precast channel roof decking is secured to the structural frame.

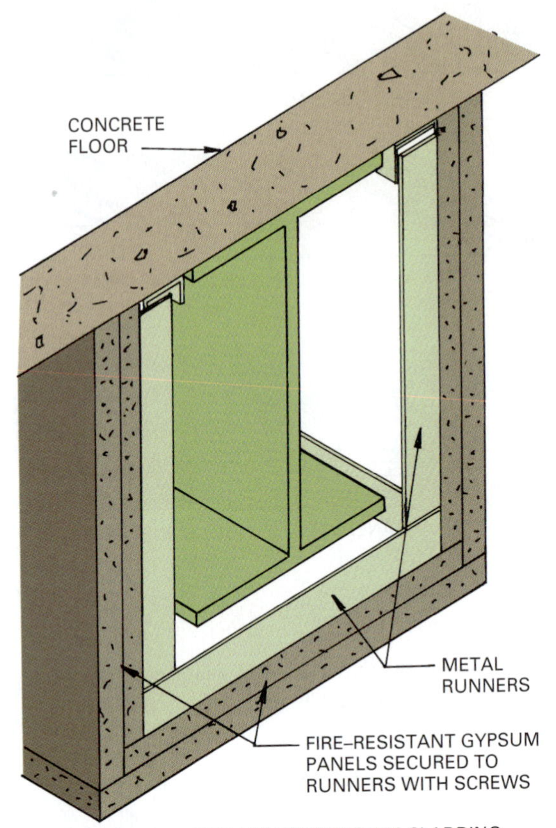

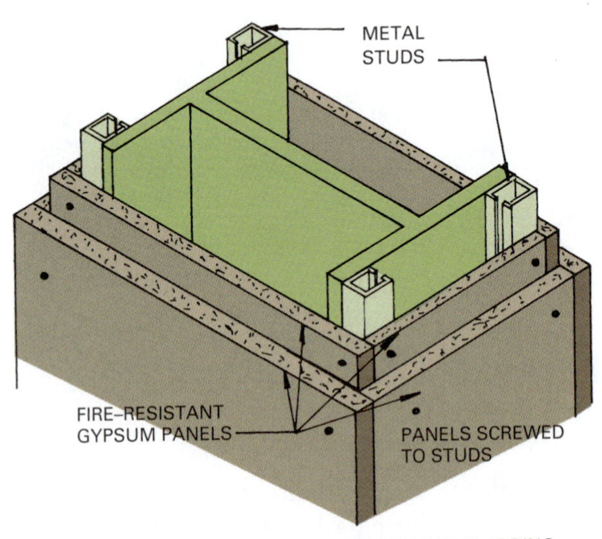

Figure 17.57 Cementitious material can be sprayed on structural steel assemblies for fire protection. *(Courtesy Eva Kultermann)*

used to maintain circulation within the system. Antifreeze is added if the columns are exposed to freezing temperatures.

Flame shields are metal barriers that deflect flames and reflect heat away from exterior structural steel members. The installation of ceilings that provide needed fire protection for floors and roofs constitutes another procedure. Designs depend on different situations, but plastered ceilings, fire-rated dropped ceilings, acoustic tiles of various types, and drop-in ceiling panels all provide various degrees of fire protection (Fig. 17.58).

Figure 17.58 Typical fire-rated ceilings used to protect structural steel framing.

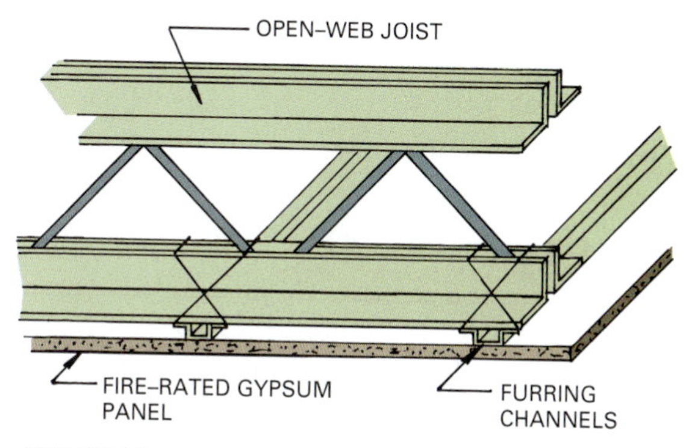

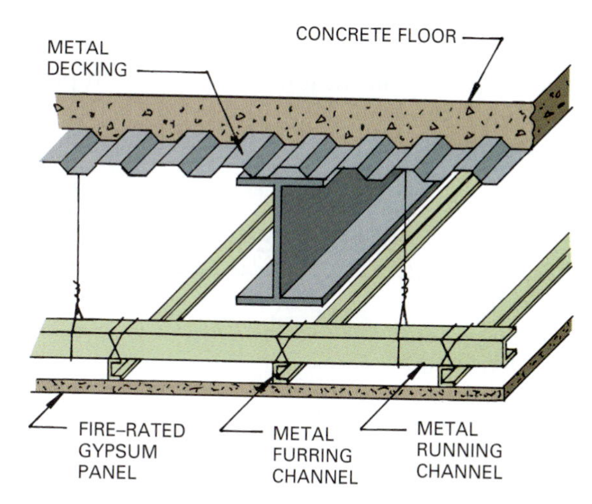

Review Questions

1. What are the purposes of the various types of drawings used for the design and erection of steel-framed buildings?
2. How are structural steel members lifted for installation as the erection process proceeds?
3. What are the commonly used fastening methods for joining structural steel members?
4. How can specified tension values be ensured when bolting structural steel members?
5. What techniques can be used to provide for the stability of a structural steel frame?
6. What are the two types of beam connections?
7. What is the difference between bearing-type and friction-type riveted connections?
8. What techniques are used to provide fire protection for structural steel framing?
9. What types of decking are used with structural steel framing?
10. What is a truss?
11. What type of structural system is used to support tensile structures?

Key Terms

bending moment
design drawings
erection plan
framed connections
galling
moment
seated connections
shear stress
shear studs

Activities

1. Contact an architect or general contractor and try to secure drawings of steel-framed buildings. Examine the framing and connection drawings. Note the system used to identify members and connections.
2. Select one method for fireproofing steel framing listed in your local building code and prepare a sketch showing details for beam and column fireproofing.
3. Visit a site where a steel-framed building is under construction. It can be a multistory or factory-manufactured metal building. Prepare a report with necessary sketches to show types of connections used.
4. Examine local building codes and describe each type of construction in which structural steel framing may be used.

Additional Resources

Detailing for Steel Construction, American Institute of Steel Construction, Chicago, IL.

Design Guide for Anchored Brick Veneer Over Steel Studs, Western States Clay Products Association, Los Angeles, CA.

Manual of Steel Construction, American Institute of Steel Construction, Chicago, IL.

Standard Specifications, Load Tables and Weight Tables for Steel Joists and Girders, and other related publications, Steel Joist Institute, Myrtle Beach, SC.

Structural/Seismic Design Manual—Volume 3, International Code Council, Falls Church, VA.

Waite, T., and NAHB Research Center, *Steel-Frame House Construction*, National Association of Home Builders, Washington, DC.

Other resources include:

Extensive number of publications and free downloads at *www.aisc.org*, American Institute of Steel Construction, Chicago, IL.

Numerous technical Fact Sheets related to metal building construction, North American Insulation Manufacturers Association, Alexandria, VA.

See Appendix B for addresses of professional and trade organizations and other sources of technical information.

DIVISION 06

Wood, Plastics, and Composites CSI MasterFormat™

061000 Rough Carpentry
062000 Finish Carpentry
064000 Architectural Woodwork
065000 Structural Plastics
066000 Plastic Fabrications
067000 Structural Composites
068000 Composite Fabrications

(Image copyright Ba Tu, 2009. Used under license from Shutterstock.com)

CHAPTER 18

Wood, Plastics, and Composites

LEARNING OBJECTIVES

Upon completion of this chapter, the student should be able to:

- Understand the structural composition of trees and identify species used in construction.
- Make decisions pertaining to the influence of defects in lumber on various applications.
- Know standard inch and metric lumber sizes.
- Use information about lumber grades to secure products that serve an intended purpose.
- Use data from the In-Grade Testing Program.
- Use information about properties of wood when selecting species.
- Apply the structural properties of wood when designing structural members.
- Take action to protect wooden structures from damage by insects and moisture.

Build Your Knowledge

For further study on these materials and methods, please refer to:

Chapter 19 Products Manufactured from Wood
Chapter 20 Wood and Metal Light-Frame Construction
Chapter 21 Heavy Timber Construction
Chapter 23 Paper and Paper Pulp Products
Chapter 37 Flooring

Wood is a natural organic material. It is unique among construction materials in that it is a renewable resource. Carefully managed timber farms and natural wild growth provide a continuing source of wood. Wood in the form of **lumber** and timbers is one of the most familiar construction materials. In addition, wood is used to produce a variety of reconstituted products, such as plywood, particleboard, and hardboard. Wood is also used in the manufacture of paper and cardboard. Wood fibers also provide a source of nitrocellulose for the manufacture of explosives.

Because there are many different species of trees, there is a wide variation in the properties of wood. As a result, wood is useful for many applications, both structural and decorative. It is important to understand the properties of a wood species before selecting it for a particular application.

TREE SPECIES

Several hundred species of wood grow throughout the world. Some are abundant and continuously farmed, while others are rare and endangered. Some species, because of their properties, availability, and abundance, are used commonly in construction. Woods are divided into two classes: hardwoods and softwoods (Fig. 18.1). This division is based on botanical differences and not on their actual hardness or softness. The major forest areas providing construction lumber in the United States and Canada are shown in Fig. 18.2.

Softwoods

Much of the production of wood for commercial use involves the softwoods class. Softwoods are used for structural framing, sheathing, roofing, subflooring, siding, trim, and millwork. **Softwoods** are referred to as coniferous, trees that bear cones, and, with a few exceptions, have needle-like leaves that stay green all year long.

Figure 18.1 Woods are divided into two classes, deciduous (hardwoods) and coniferous (softwoods).

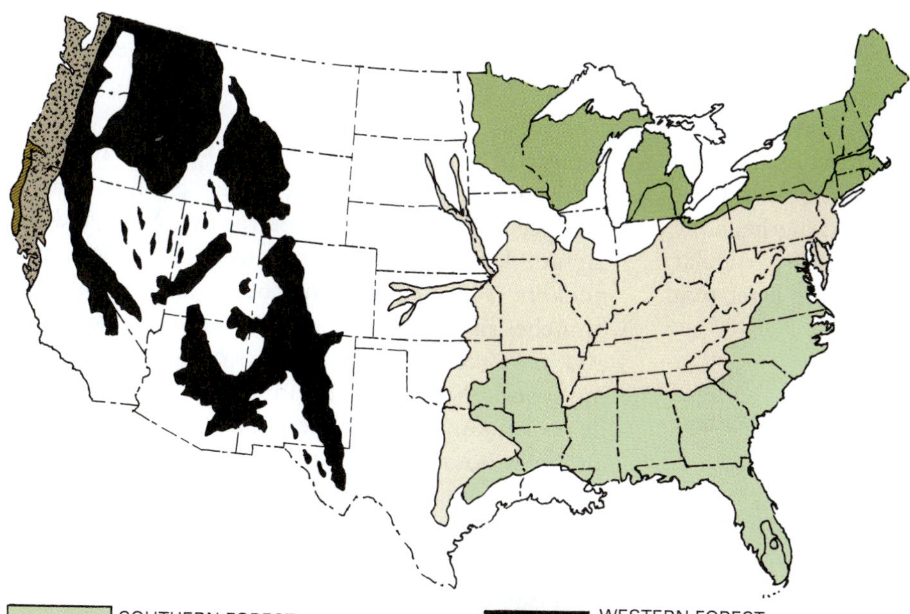

SOUTHERN FOREST
LOBLOLLY PINE, SHORTLEAF PINE, SLASH PINE, LONGLEAF PINE, GUM, HICKORY, CYPRESS, OAK, PECAN, ASH.

CENTRAL FOREST
GUM, HICKORY, OAK, YELLOW POPLAR, BLACK WALNUT, ELM, MAPLE, RED CEDAR.

NORTHERN FOREST
EASTERN SPRUCE, NORTHERN HEMLOCK, WHITE PINE, JACK PINE, FIR, MAPLE, BEECH, BIRCH.

WESTERN FOREST
LODGEPOLE PINE, PONDEROSA PINE, SUGAR PINE, IDAHO WHITE PINE, DOUGLAS FIR, WHITE FIR, RED CEDAR, LARCH, SPRUCE, ASPEN, ELM, ASH, COTTONWOOD.

WEST COAST FOREST
SITKA SPRUCE, WESTERN RED CEDAR, WEST COAST HEMLOCK, DOUGLAS FIR, ALDER, PONDEROSA PINE.

REDWOOD REGION
REDWOOD, DOUGLAS FIR.

Figure 18.2 Major forest regions in the United States and Canada.

Hardwoods

Hardwood, or deciduous, trees are broad-leaved and shed their leaves in the winter. Hardwoods are more expensive than softwoods and are used in cabinetry, furniture making, paneling, interior trim, and flooring.

THE STRUCTURE OF WOOD

Wood uses energy from the sun, water, and extracted carbon dioxide through the process of photosynthesis. Trees are anchored deep into the ground by their root systems. The tap root draws minerals and water from the ground. Side roots grow out around the tree, and feeder roots grow off the side roots. They absorb soil water and minerals in the ground to provide nutrients for the tree. Water and minerals flow up into the sapwood (xylem). The leaves use sunlight and carbon dioxide to change water and minerals into food. This food flows down into the tree through the inner bark (phloem).

Each part of a tree serves a specific purpose in the tree's growth and development (Fig. 18.3). At the center is a small core called the pith. During the early years of growth, the pith helps support the stem and feed the tree. As the tree matures, the pith ceases to function. Next to the pith is the **heartwood**, which is hard, mature wood that forms the largest part of the trunk. It strengthens the tree and displays the color associated with a particular species. The next layer, **sapwood**, is a living layer that carries water and food throughout the tree. It is soft, usually light in color, and contains more moisture than heartwood. The heartwood and sapwood form growth rings that can be seen clearly when a tree is cut down. The light, soft rings develop in the spring, when the tree grows rapidly, and are referred to as springwood. The dark, hard rings form in the summer, when growth is slow, and are called summerwood. When a tree is cut down, its approximate age can be determined by counting the hard summerwood annual rings.

Next to the sapwood is the cambium layer. It forms new cells that are either xylem or phloem. Sapwood and heartwood are formed with xylem cells. On the outside of the cambium layer are the phloem cells, which form the inner bark layer. Phloem cells in the inner bark carry food to the roots. As more phloem cells are formed, the outer layer of the inner bark changes into outer bark, a layer that protects the tree. Within the tree are medullary rays that run perpendicular to the growth rings. They carry food and water from the cambium layer to the interior of the tree.

Figure 18.3 A cross section of a tree illustrating its structure. *(Courtesy of Western Wood Products Association)*

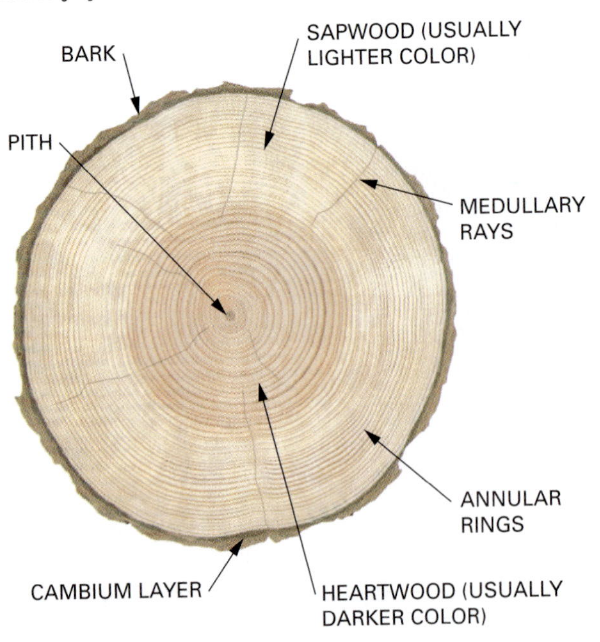

WOOD DEFECTS

A piece of wood may have natural defects that occur while a tree grows or seasoning defects that are produced as the wood is dried for use. Natural defects include knots, shake, wane, insect holes, and pitch pockets (Fig. 18.4). Knots occur where imbedded branches in tree trunks are cut. Knots weaken wood and are a factor lumber graders consider when judging wood. Pitch is the accumulation of sap or resin in pockets in a tree. It occurs in softwoods and usually does not cause major problems. Shake can result when a tree is racked or bent, as in a wind storm. Shake appears as small cracks running with or across annual rings. Wane refers to the absence of wood or the presence of bark along the edge of a board. Insect holes are caused by boring insects eating their way into wood.

Seasoning defects include warp, checks, stain, honey combing, and case-hardening. Warp refers to any variation in a board's shape beyond a flat, true surface. Warp develops as wood loses moisture and can be prevented by proper seasoning. Cup, bow, crook, and twist are common forms of warping (Fig. 18.5).

Checks are small cracks perpendicular to growth rings on the surface or ends of boards. Wood sometimes develops a stain on its surface after it is cut into lumber. Stains may be brown, green, or blue and generally do not influence the wood's strength. Honeycombing refers to cracks in a board's interior. They occur if lumber is

Figure 18.4 Knots in softwood lumber can loosen and drop out. *(Image copyright Michele Perbellini, 2009. Used under license from Shutterstock.com)*

not seasoned properly. When a board's outer surface is drier and subjected to more stress than its interior, case-hardening can result. Again, this is caused by improper seasoning.

SEASONING LUMBER

Seasoning refers to the process of converting wood from a **green lumber** to a finished lumber state through drying, which results in a recommended and desired moisture content. Lumber-seasoning methods include air drying, kiln drying, dehumidification, and solar drying, with air and kiln drying being the most common. Seasoning requirements for softwoods, as specified by the Southern Pines Inspection Bureau, are given in Table 18.1.

Air-Dried Lumber

Air drying involves stacking boards with stickers (wood strips) inserted between layers (Fig. 18.6). This permits air to circulate and aid drying. Air drying is a slow process and does not provide control over drying duration. On hot, dry, windy days, wood will season too quickly, and checking and warping may occur. On cool days with high humidity, wood dries very little. It is difficult to regulate desired moisture contents when air drying. Some dimension lumber and lower grades of softwood lumber are often air dried. Structural timbers are too large for quick air drying, so they are often shipped green.

Moisture content of lumber air dried for construction purposes should be in the 15 to 19 percent range in the United States. In some parts of the country, such as the Southwest, it can be slightly lower than 15 percent, while in the moist Pacific Northwest, it may be closer to 20 percent. Lumber grading rules for the various

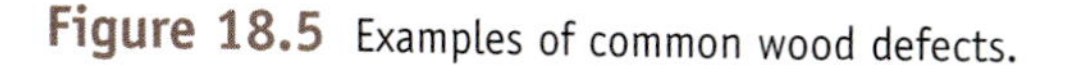

Figure 18.5 Examples of common wood defects.

Table 18.1 Seasoning Requirements for Softwoods[a]

Items (Nominal)	Moisture Content Limit	
	Maximum (Dry)	Kiln-Dried (KD or MC15)
D&btr grades		
1" & 1 1/4"	15%	12% on 90% of pieces 15% on remainder
1 1/2", 1 3/4" & 2"	18%	15%
Over 2" not over 4"	19%	15%
Over 4"	20%	18%
Paneling		
1"		12%
Boards		
2" and less and dimension 2" to 4"	19%	15%
Decking		
2" thick	19%	15%
3" and 4" thick		15% on 90% of pieces 18% on remainder
Heavy dimension		
Over 2" not over 4"	19%	15%
Timbers		
5" and thicker	23%	20%

[a]*As specified by the Southern Pine Inspection Bureau.*

Figure 18.6 Lumber is stickered and stacked for air drying. *(Courtesy of Weyerhauser Buisiness)*

softwoods specify maximum moisture contents for each grade regardless of how they are dried.

Hardwoods are usually air dried for a period of time, then kiln dried. It is difficult to get moisture contents of hardwoods low enough for furniture making and cabinetry by air drying only. Normally, a moisture content of 6 to 8 percent is required for these purposes, necessitating kiln drying.

Kiln-Dried Lumber

Kiln drying involves stacking lumber in the same manner for air drying inside a kiln. In the kiln, temperature, humidity, and air circulation are controlled to carefully reduce a wood's moisture content. Air temperatures in the kiln reach 180° F (82° C), with an equally high relative humidity (Fig. 18.7). Since both temperature and humidity are controlled, lumber can be quickly dried to any desired moisture content. Most woods take less than two weeks to dry in a kiln. Kiln drying reduces defects and produces a product that will not expand or contract as much as air-dried wood.

Dehumidification and Solar Kilns

Dehumidification and solar kilns are relatively new. The dehumidification method uses electricity to dry the lumber. Solar kilns use the sun's energy to produce needed heat and are the most economical. Those in current use can handle only small amounts of wood.

Unseasoned Lumber

Most lumber more than 2 in. thick is air dried. The thicker the stock, the longer it takes to achieve the desired moisture content of 19 percent. Most stock in these thicknesses is used in a green condition (more than 19 percent moisture content). It continues to dry and shrink after use, which sometimes causes problems. Moisture content can be checked on-site with a battery-operated moisture meter.

Figure 18.7 Lumber being loaded in a kiln to reduce moisture content. *(Courtesy of American Wood Dryers)*

LUMBER SIZES

After rough sawed wood has been dried, it is taken to a planing mill for surfacing and shaping into useful products, such as boards, dimension stock, timbers, molding, and trim. Softwood lumber is produced in standard sizes. Hardwoods are often **dressed** to thickness but not to any specific width or length. After the lumber is planed or shaped, it must be stored in weatherproof sheds so the moisture content will not increase.

Softwood Lumber

Softwood lumber is sold by designating its nominal size—the size of the board after it has been rough cut at a sawmill. Inch and metric sizes are shown in Tables 18.2, 18.3, and 18.4. The dressed or finished size is the actual size of the board after drying and surfacing. For example, a typical wall stud has a nominal size of 2 × 4 in. and a finished size of $1\text{-}\frac{1}{2} \times 3\text{-}\frac{1}{2}$ in. A dry, dressed metric-size stud is 38.10 × 88.90 mm. If stock

Table 18.3 Standard Sizes of Surfaced Timbers

Thickness and Width of Stock					
Customary Units (in.)			Metric Units (mm)		
Nominal	Actual Dry	Actual Green	Nomen-clature[a]	Actual Dry	Actual Green
5	—	4 1/2	114	—	114.3
6	—	5 1/2	140	—	139.7
7	—	6 1/2	165	—	165.1
8	—	7 1/2	191	—	190.5
9	—	8 1/2	216	—	215.9
10	—	9 1/2	241	—	241.3
12	—	11 1/2	292	—	292.1
14	—	13 1/2	343	—	342.9
16	—	15 1/2	394	—	393.7
18	—	17 1/2	445	—	444.5
20	—	19 1/2	495	—	495.3

[a]*Nomenclature means the size used to describe the member. Actual size may be slightly larger or smaller.*
Courtesy Canadian Wood Council and the Southern Forest Products Association

Table 18.2 Standard Sizes of Surfaced Dimension Lumber

Thickness of Stock					
Customary Units (in.)			Metric Units (mm)		
Nominal	Actual Dry	Actual Green	Nomen-clature[a]	Actual Dry	Actual Green
2	1 1/2	1 9/16	38	38.10	39.69
2 1/2	2	2 1/16	51	50.80	52.39
3	2 1/2	2 9/16	64	63.50	65.09
3 1/2	3	3 1/16	76	76.20	77.79
4	3 1/2	3 9/16	89	88.90	90.49
4 1/2	4	4 1/16	102	101.60	103.19
Width of Stock					
2	1 1/2	1 9/16	38	38.10	39.69
3	2 1/2	2 9/16	64	63.50	65.09
4	3 1/2	3 9/16	89	88.90	90.49
5	4 1/2	4 5/8	114	114.30	114.47
6	5 1/2	5 5/8	140	139.70	142.87
7	6 1/2	6 5/8	165	165.10	168.28
8	7 1/4	7 1/2	184	184.15	190.50
10	9 1/4	9 1/2	235	234.95	241.30
12	11 1/4	11 1/2	286	285.75	292.10
14	13 1/4	13 1/2	337	336.55	342.90
16	15 1/4	15 1/2	387	387.35	393.70

[a]*Nomenclature means the size used to describe the members. Actual size may be slightly larger or smaller.*
Courtesy Canadian Wood Council and the Southern Forest Products Association

Table 18.4 Metric Lengths for Softwood Lumber

Nominal Length (ft.)	Metric Length (m)
3	0.91
4	1.22
5	1.52
6	1.83
7	2.13
8	2.44
9	2.74
10	3.05
11	3.35
12	3.66
13	3.96
14	4.27
15	4.57
16	4.88
17	5.18
18	5.49
19	5.79
20	6.10
21	6.40
22	6.71
23	7.01
24	7.32

Courtesy Canadian Wood Council

is surfaced green, it is made larger so that when it dries and shrinks it will be the same size as dried stock. The size of worked stock includes the overall size, as shown in Fig. 18.8. Worked stock refers to boards that undergo some machining operation, such as moldings or tongue-and-groove flooring.

As it seasons from a green to a dry condition, lumber shrinks in direct proportion to its loss of moisture. This makes it possible to establish separate sizes for green and dry surfaced lumber used together in building to insure uniformity (size, strength, and stiffness) once the green lumber reaches its required moisture content. The dressed sizes of green and dry softwood lumber are specified in American Softwood Lumber Standard PS 20–70 and the Canadian National Lumber Grades Authority standard CSA 0141. American and Canadian standards are coordinated and accepted in both countries. They are based on dry lumber having a moisture content of 19 percent or less and green lumber having more than 19 percent. The standard sizes for finish lumber, boards, siding, and other wood products are given in Tables 18.5 and 18.6.

Softwood lumber is sold in standard lengths of two-foot multiples ranging from 6 ft. to 24 ft. Some special lengths, such as studs precut to standard ceiling heights, are cut to exact desired dimensions. Metric lumber lengths are specified in meters (m) and decimal parts of a meter, as shown in Table 18.4.

Hardwood Lumber

Hardwood lumber is sold in random widths and lengths and often is not surfaced to standard thicknesses. Since hardwoods are used in the manufacture of cabinets and furniture, boards are cut to many sizes and thicknesses. Cutting to standard widths and lengths would cause great waste. Standard rough thickness range from $\frac{3}{8}$ to 1-$\frac{1}{2}$ in. in $\frac{1}{4}$ in. increments. Hardwood lumber is also available in 2, 3, and 4 in. thicknesses.

Hardwood lumber in Canada is sized so metric units and customary units are within 1 percent of each other.

Figure 18.8 Examples of worked stock.

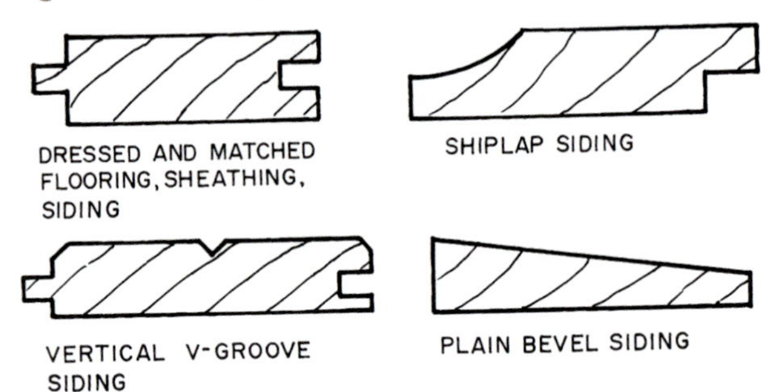

Table 18.5 Standard Sizes for Softwood Products

	Thickness (in.)		Width (in.)		
	Nominal	Worked	Nominal	Face	Overall
Bevel siding	1/2	3/16 × 7/16	4	3 1/2	3 1/2
	5/8	7/16 × 9/16	5	4 1/2	4 1/2
	3/4	3/16 × 11/16	6	5 1/2	5 1/2
	1	3/16 × 3/4	8	7 1/4	7 1/4
Drop siding	5/8	9/16	4	3 1/8	3 3/8
Rustic and drop siding (dressed and matched)	1	23/32	5	4 1/8	4 3/8
			6	5 1/8	5 3/8
			8	6 7/8	7 1/8
			10	8 7/8	9 1/8
Rustic and drop siding (shiplapped)	5/8	9/16	4	3	3 3/8
	1	23/32	5	4	4 3/8
			6	5	5 3/8
			8	6 5/8	7 1/8
			10	8 5/8	9 1/8
			12	10 5/8	11 1/8
Flooring	3/8	5/16	2	1 1/8	1 3/8
	1/2	7/16	3	2 1/8	2 3/8
	5/8	9/16	4	3 1/8	3 3/8
	1	3/4	5	4 1/8	4 3/8
	1 1/4	1	6	5 1/8	5 3/8
	1 1/2	1 1/4			
Ceiling	3/8	5/16	3	2 1/8	2 3/8
	1/2	7/16	4	3 1/8	3 3/8
	5/8	9/16	5	4 1/8	4 3/8
	3/4	11/16	6	5 1/8	5 3/8
Partition	1	23/32	3	2 1/8	2 3/8
			4	3 1/8	3 3/8
			5	4 1/8	4 3/8
			6	5 1/8	5 3/8
Paneling	1	23/32	3	2 1/8	2 3/8
			4	3 1/8	3 3/8
			5	4 1/8	4 3/8
			6	5 1/8	5 3/8
			8	6 7/8	7 1/8
			10	8 7/8	9 1/8
			12	10 7/8	11 1/8
Shiplap	1	3/4	4	3 1/8	3 1/2
			6	5 1/8	5 1/2
			8	6 7/8	7 1/4
			10	8 7/8	9 1/4
			12	10 7/8	11 1/4
Dressed and matched	1	3/4	4	3 1/8	3 3/8
	1 1/4	1	5	4 1/8	4 3/8
	1 1/2	1 1/4	6	5 1/8	5 3/8
			8	6 7/8	7 1/8
			10	8 7/8	9 1/8
			12	10 7/8	11 1/8

Courtesy Southern Forest Products Association

The standard thicknesses of rough lumber are 15 mm to 60 mm in 5 mm increments and 70 mm to 100 mm in 10 mm increments. Surfaced lumber thicknesses are 5 mm less than the rough size for green or air-dried lumber and 5 to 8 mm for kiln-dried lumber. Standard lengths

Table 18.6 Metric Sizes for Surfaced Boards, Shiplap Siding, and Centre Match Siding

Actual Thickness (mm) Dry, Dressed	Actual Width (mm) Dry, Dressed
17	38
19	64
25	89
32	114
	140
	165
	184
	210
	235

Courtesy Canadian Wood Council

are 1.2 m through 4.8 m in increments of 30 cm. Hardwood lumber is sold by volume in cubic meters (m^3).

Buying Lumber

Most lumber in the United States is sold by the board foot. A **board foot** is equal to a piece of lumber with an actual size of 1 in. thick, 12 in. wide, and 1 ft. long. For example, a board 1 in. thick, 12 in. wide, and 8 ft. long contains 8 board feet. These are nominal sizes. The actual size is $\frac{3}{4}$ in. × 11 $\frac{1}{4}$ in. × 8 in. Usually, stock under 1 in. thick is figured as 1 in. To calculate board feet multiply thickness in inches by width in inches by length in feet and divide by 12 (which converts the width to feet) (Fig. 18.9).

Metric lumber is sold by the cubic meter (m^3). The volume is based on the actual size. This gives the actual volume of the piece in cubic meters. It is computed by the formula:

Volume = thickness (mm) × width (mm) × length (m). Thickness and width are given in millimeters (mm), and the length is in meters (m). An example is shown in Fig. 18.9. The Canadian Wood Council provides tables for converting metric volume to board feet.

Other materials, such as molding and trim, are sold by the linear foot. Posts and pilings are sold by the piece. Wood shingles are sold by the square (a square covers 100 ft.^{2}).

LUMBER MANUFACTURING

The production of all types of lumber begins in the forest where trees are felled (Fig. 18.10). After branches are cut away, the log is moved to a loading site. Logs may be skidded to the site by a tractor or hauled down a steep hillside with a long cable called a choker line. Once they reach the loading site, they are generally loaded on trucks. The trucks operate on dirt roads built into the forest especially for the logging operation (Fig. 18.11). At the mill, logs are stacked into large piles and often sprayed with water to keep them from drying out and splitting. Most mills use a mechanical peeler to

Figure 18.9 Examples representing one board foot and metric volume of lumber.

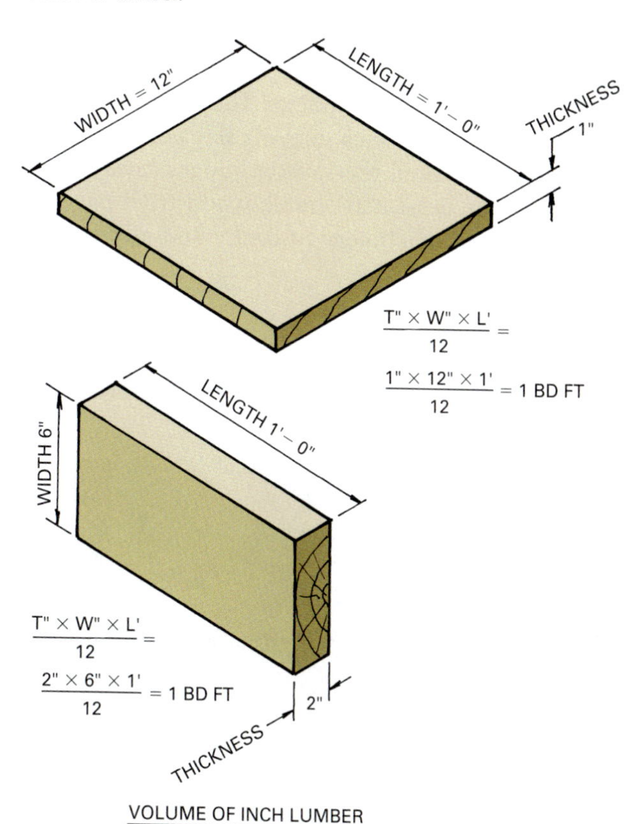

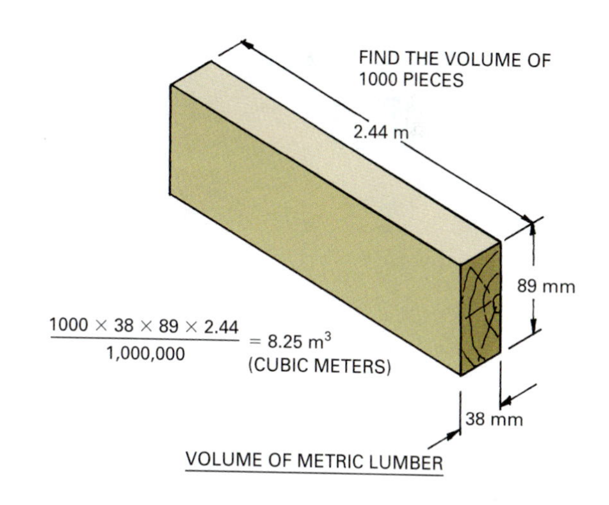

grind off the bark. After peeling, a log is ready to be cut into lumber of various types.

Logs are hauled onto the carriage of a large band saw machine called a heading. Small mills use circular saw blades 3 to 4 ft. in diameter. The carriage moves a log into the blade, which cuts off slabs of wood. The saw operator, called the sawyer, judges how to efficiently get the most marketable wood from each log and adjusts the machinery to rotate and advance the log into the saw. Modern mills use computer controls to assist with this process (Fig. 18.12). Slabs cut from the log fall on a conveyor and move to an edger that squares edges and cuts the slabs to desired widths. Pieces then move to a trim saw, which cuts the rough, square-edged wood to standard lengths. Hardwoods are not edged or cut to length.

Most lumber used for construction purposes is plain-sawed. This method produces the maximum yield and the widest stock from the log (Fig. 18.13). Plain sawed lumber produces boards that have a broad grain running their length. Another common sawing method, **quarter sawing,** produces boards with the edges of the annual rings showing on the face. Flooring is often quarter sawed because the annual rings are hard and withstand wear better than the wide areas of springwood exposed on plain-sawed boards. Examples of commonly used softwoods showing plain sawed, quarter sawed, and end grain are shown in Fig. 18.14.

Figure 18.10 A logger felling mature trees. *(Image copyright Mikhail Olykainen, 2009. Used under license from Shutterstock.com)*

Figure 18.11 Southern pine logs being moved to a mill. *(Image copyright Sally Scott, 2009. Used under license from Shutterstock.com)*

Figure 18.12 A log being processed by a precision computer-controlled laser-guided system.

Figure 18.13 Typical cutting approach for plain-sawed and quarter-sawed lumber.

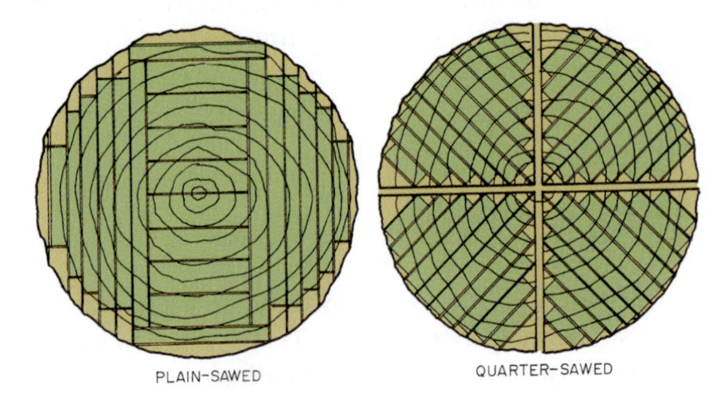

Once cut to size, the lumber moves on a conveyor past a **lumber grader**. The grader marks each board to indicate its grade as it passes from the trim saw. This is a highly technical job because of the extensive specifications for each of the many grades and types of lumber. Finally, green lumber goes to an air drying area where it is stacked with sticks between each layer. After a number of weeks, the lumber is moved, still stacked, to a kiln for final moisture reduction.

After the lumber has been dried, it is dressed to finished size in a planing mill. All dimensional lumber (up to but not including 5 in. or 114 mm), if surfaced dry, must be marked S-Dry (surfaced dry). If it was surfaced green (moisture content more than 19 percent), it must be grade stamped S-GRN. If the lumber is surfaced at 15 percent moisture content it is often stamped MC15. Hardwood lumber is generally shipped dried but unsurfaced to manufacturers that use it for furniture, cabinets, and other products.

LUMBER GRADING AND TESTING

Lumber Grades

The grades of softwood lumber are based on American Softwood Lumber Standard PS 20–70, published by the U.S. Department of Commerce. Under the provisions of this standard, a National Grading Rule Committee (NGRC) was established to develop uniform nationwide grade requirements for dimensional lumber. The functions of the committee were to "establish, maintain, and make fully and fairly available grade strength ratios, nomenclature, and descriptions of grades of dimension lumber conforming to American Lumber Standards." The specifications developed are known as the National Grading Rule for Dimensional Lumber (NGRDL). They form a part of the grading rules of all wood association grading rules, such as those of the Western Wood Products Association, the Southern Pine Inspection Bureau, West Coast Lumber Inspection Bureau, and the Redwood Inspection Service. The grading rules of these individual associations pertain to those species produced in their geographic areas and contain the National Grading Rules for Dimensional Lumber specifications plus additional specs unique to that association (Table 18.7).

Grades of Canadian lumber are identical with those used in the United States (or the same as the requirements of American Softwood Lumber Standard PS 20–70). The grades are published by the Canadian National Lumber Grades Authority (NLGA) and appear in the publication Standard Grading Rules for Canadian Lumber (Table 18.7). The mills that manufacture lumber according to these standards place a grade stamp on each piece (Fig. 18.15).

Softwood lumber grading standards cover the nomenclature and bending strength ratios for structural light framing, light framing, structural joists and planks, appearance framing, and stud grades. The bending stress ratios indicate the relationships between basic stresses of clear wood free of strength-reducing defects (knots, splits) and the stresses developed in a particular grade of lumber of the same species. Some examples using customary units are given in Table 18.8. Data for Canadian species are given in kilograms per square meter (kg/m^2) and are available from the Canadian Wood Council. A description of the most commonly used lumber grades for softwoods, as used by the Southern Pine Inspection Bureau, is shown in Table 18.9. Other U.S. regional and Canadian associations have established similar grades for species harvested in their regions. The higher the grade, the higher the allowable stresses. Lower grades of lumber are more economical.

Lumber Classifications as to Manufacture

Rough lumber has not been dressed and shows saw marks on all four surfaces.

Blanked lumber is dressed to a size larger than standard dressed sizes but smaller than the nominal size. It may be surfaced on one surface (S1S), one edge (S1E), or surfaced on all four sides (S4S).

Dressed lumber has one or more surfaces smoothed in any combination of surfaces and edges, such as S1E, S2E, S1S1E, S1S2E, S4S, and so on. For example, S1E means surface one edge, S4S means surface four sides, and S1S1E means surface one side and one edge.

Worked lumber has been dressed and matched, shiplapped, or patterned. Matched lumber has a tongue on one edge and a groove on the other, providing a tongue-and-groove joint. Shiplapped lumber has been worked or rabbeted on both edges of each piece to provide a lapped joint when the pieces fit together. Patterned lumber has been worked to a shaped or molded form in addition to being dressed, matched, or shiplapped.

Re-sawn lumber is produced by re-sawing any thickness of lumber into thinner lumber. Ripped lumber is produced by sawing any width lumber in narrower pieces.

Size Classifications

The following size classifications, summarized in Table 18.10, are used by the Southern Pine Inspection Bureau.

Figure 18.14 Common woods used in construction. *(Courtesy Forest Products Laboratory, USDA Forest Service, Madison, WI)*

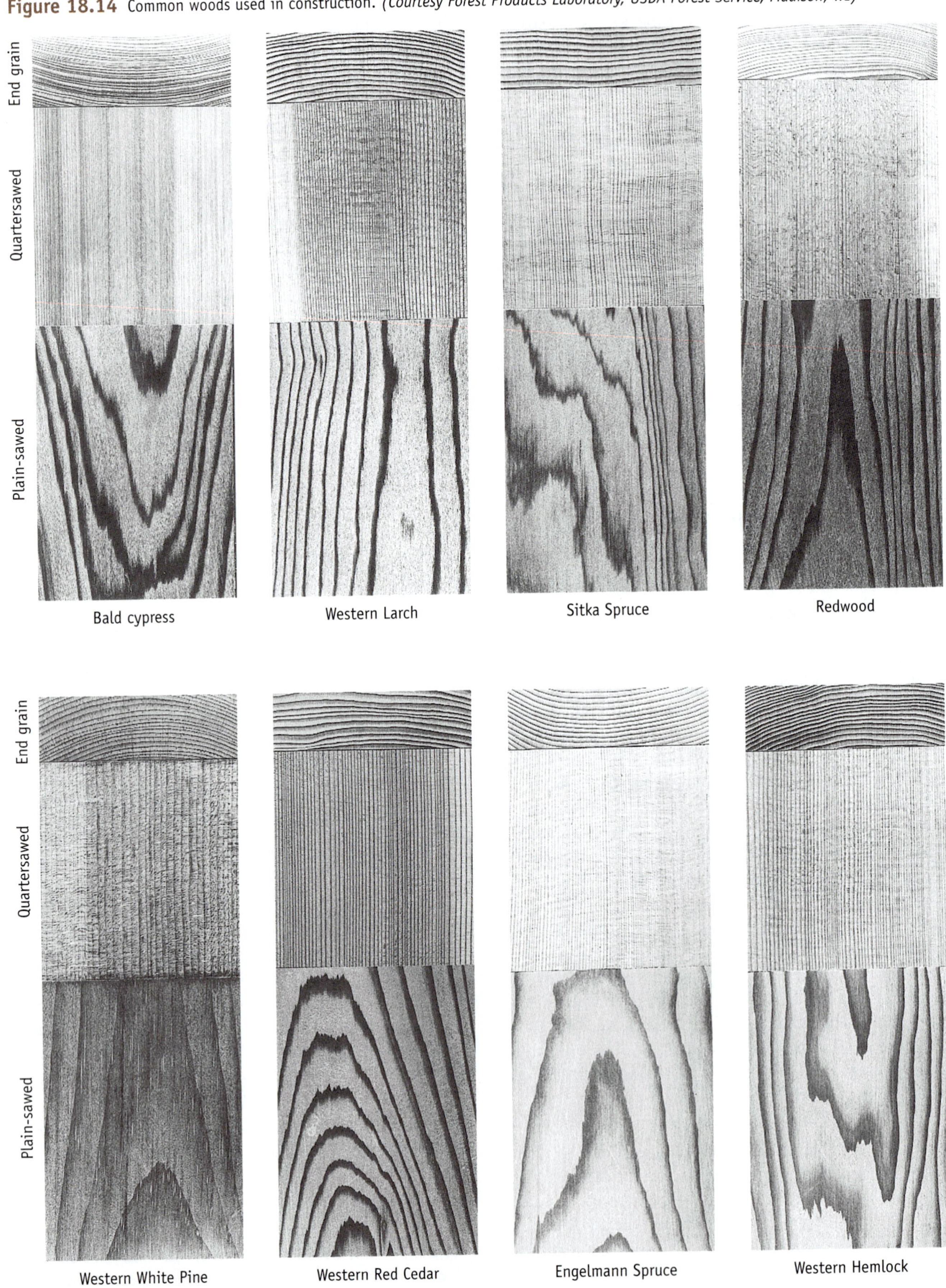

Figure 18.14 *Continued*

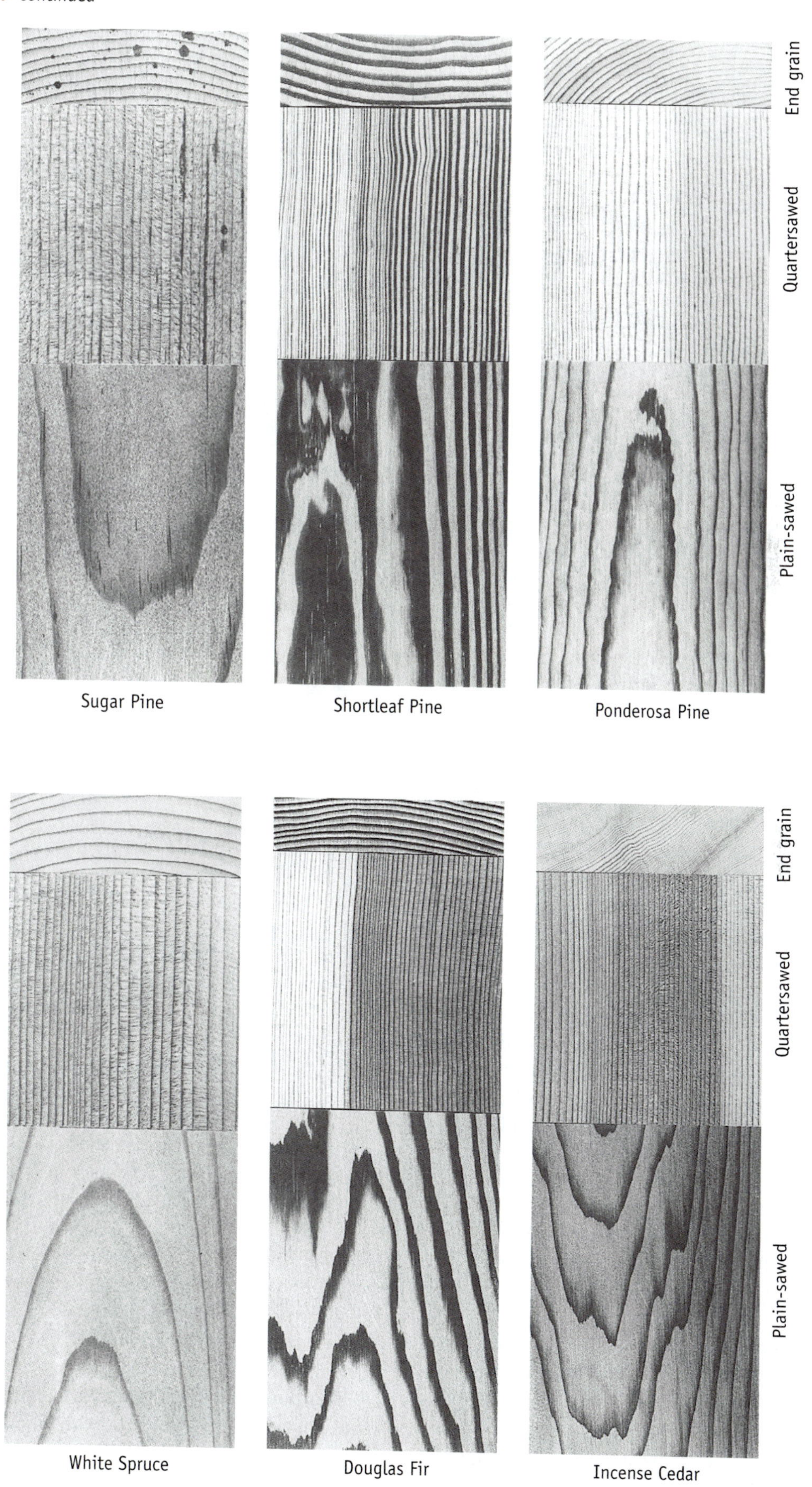

Table 18.7 U.S. and Canadian Softwood Grading Authorities by Region

United States			
Softwood Region	**Species**	**Lumber Association**	**Grading Authority**
Western wood region	Western red cedar Ponderosa pine Douglas fir White fir Western hemlock Englemann spruce Western larch Sitka spruce Lodgepole pine Idaho white pine Sugar pine	Western Wood Products Association	Western Wood Products Association West Coast Lumber Inspection Bureau
Southern pine region	Shortleaf pine Longleaf pine	Southern Forest Products Association	Southern Pine Inspection Bureau
Redwood region	Redwood Douglas fir	California Redwood Association	Redwood Inspection Service
Canada			
	Spruce Pine, several types Fir, several types Larch Douglas fir Hemlock Western red cedar Aspen Poplar	Canadian Wood Council and 12 lumber grading authorities	Canadian National Lumber Grades Authority

Boards are less than 2 in. in nominal thickness and 1 in. or more in width. If they are less than 6 in. wide, they are classified as strips. **Dimension lumber** is from 2 in. up to but not including 5 in. thick and 2 in. or more in width. It is subdivided into five classes: structural light framing, light framing, structural joists and planks, appearance framing, and studs. **Timbers** are 5 in. or more in their smallest dimension and are subdivided into classes, such as beams, posts, and girders.

Figure 18.15 Typical grade stamp for dimension lumber used by various grading authorities in the United States and Canada. *(Courtesy of Western Wood Products Association)*

Stress-Rated Lumber

Each piece of lumber is evaluated at a mill by mechanical stress-rating equipment that subjects each piece to bending stress. The modulus of elasticity (E), a measure of stiffness, is measured by instruments on the machine. The stress grade is electronically calculated and includes consideration of the effects of grain slope, knots, growth rate, moisture content, and density. The machine then puts the grade stamp on the piece.

The grade stamp on machine stress-rated lumber indicates that the stress-rating system used to make the test meets certification requirements. The grade stamp shows the agency trademark, the mill name or number, the phrase "Machine Rated," the species, and the modulus of elasticity rating in millions of pounds per square inch (Fig. 18.16).

The Southern Pine Inspection Bureau has established fifteen categories of stress-graded lumber 2 in. or less in thickness. These categories can be used for most structural purposes, such as for trussed rafters.

Table 18.8 Working Stresses for Selected Softwood Species

Species and Commercial Grade	Size Classification	Extreme Fiber in Bending "F_b"[b] Single-Member	Repetitive Member	Tension Parallel to Grain "F_t"	Horizontal Shear "F_v"	Compression Perpendicular to Grain "$F_{c\perp}$"	Compression Parallel to Grain "$F_{c//}$"	Modulus of Elasticity "E"
		Allowable Unit Stresses[a] in lb/in.²						
Douglas Fir-Larch (Surfaced dry or surfaced green. Used at 19% max. m.c.)								
Select Structural		2,100	2,400	1,200	95	385	625	1,800,000
No. 1		1,750	2,050	1,050	95	385	625	1,800,000
No. 2	2" to 4" thick	1,450	1,650	850	95	385	625	1,700,000
No. 3	2" to 4" wide	800	925	475	95	385	625	1,500,000
Appearance		1,750	2,050	1,050	95	385	625	1,800,000
Hem-Fir (Surfaced dry or surfaced green. Used at 19% max. m.c.)								
Select Structural		1,650	1,900	975	75	405	1,300	1,500,000
No. 1		1,400	1,600	825	75	405	1,050	1,500,000
No. 2	2" to 4" thick	1,150	1,350	675	75	405	825	1,400,000
No. 3	2" to 4" wide	650	725	375	75	405	500	1,200,000
Appearance		1,400	1,600	825	75	405	1,250	1,500,000
Southern Pine (Surfaced at 15% maximum moisture content, K.D. 15. Used at 15% max. m.c.)								
Select Structural		2,150	2,500	1,250	105	565	1,800	1,800,000
No. 1	2" to 4" thick	1,850	2,100	1,050	105	565	1,450	1,800,000
No. 2	2" to 4" wide	1,550	1,750	900	95	565	1,150	1,600,000
No. 3		850	975	500	95	565	675	1,500,000

[a]*Values shown are for normal loading conditions at maximum 19% m.c., as in most covered structures.*
[b]*Values in bending for "repetitive member uses" are intended for design of members spaced not over 24" o.c., in groups of not less than three members and when joined by floor or roof sheathing elements.*

There are five classes with lower allowable bending stresses in relation to the modulus of elasticity. The classes work when lower bending stress can be used, such as in floor joists.

The Southern Pine Inspection Bureau also has two grades of stress-rated timbers 5 in. x 5 in. and larger: Select Structural (SR) and Dense Select Structural (DSR). They are divided into No. 1 SR, No. 1 Dense SR, and No. 2 SR, No. 2 Dense SR. There are also a series of grades for stress-rated industrial lumber. Other regional associations in the United States and Canada have similar grading procedures.

In-Grade Testing Program

The In-Grade Testing Program is the result of a twelve-year research program conducted by the U.S. Department of Agriculture Forest Products Laboratory in cooperation with U.S. and Canadian lumber industry associations. The program was initiated to verify softwood lumber design values for visually graded lumber and to provide a scientific basis for wood engineering similar to that used for steel and concrete. Thousands of lumber specimens of many species, grades, and sizes were tested and design values were applied.

The In-Grade Testing Program developed six new species groups for Western lumber species and Eastern lumber species (Table 18.11). The Canadian species groupings are shown in Table 18.12. Southern pine is a separate grouping for woods in the south and southeastern United States. Their design values are available in the Supplement to the National Design Specifications for Wood Construction. The various lumber associations also have publications pertaining to species in their areas.

Base Values The design data are presented in the form of Base Values for the various species groupings. Base Values can be adjusted for a particular application in which the structural lumber is to be used. The Southern Forest Products Association refers to these as empirical values. Base Values are assigned to six Basic Properties of wood. The six Basic Properties are: (1) extreme fiber stress in bending (Fb) (bending strength); (2) tension

Table 18.9 Southern Pine Softwood Lumber Grades

Product	Grade	Character of Grade and Typical Uses
Finish	B&B	Highest recognized grade of finish. Generally clear, although a limited number of pin knots permitted. Finest quality for natural or stain finish.
	C	Excellent for painted or natural finish where requirements are less exacting. Reasonably clear but permits limited number of surface checks and small tight knots.
	C&Btr	Combination of B&B and C grades; satisfies requirements for high-quality finish.
	D	Economical, serviceable grade for natural or painted finish.
Boards S4S	No. 1	High quality with good appearance characteristics. Generally sound and tight-knotted. Largest hole permitted is $^1/_{16}$". A superior product suitable for wide range of uses, including shelving, form, and crating lumber.
	No. 2	High-quality sheathing material, characterized by tight knots. Generally free of holes.
	No. 3	Good, serviceable sheathing, usable for many applications without waste.
	No. 4	Admit pieces below No. 3 which can be used without waste or contain usable portions at least 24" in length.
Dimension Structural light framing 2" to 4" thick 2" to 4" wide	Select Structural Dense Select Structural	High quality, relatively free of characteristics that impair strength or stiffness. Recommended for uses where high strength, stiffness, and good appearance are required.
	No. 1 No. 1 Dense	Provide high strength; recommended for general utility and construction purposes. Good appearance, especially suitable where exposed because of the knot limitations.
	No. 2 No. 2 Dense	Although less restricted than No. 1, suitable for all types of construction. Tight knots.
	No. 3	Assigned design values meet wide range of design requirements. Recommended for general construction purposes where appearance is not a controlling factor. Many pieces included in this grade would qualify as No. 2 except for single limiting characteristic. Provides high-quality, low-cost construction.
Studs 2" to 4" thick 2" to 6" wide 10¹ and shorter	Stud	Stringent requirements as to straightness, strength, and stiffness adapt this grade to all stud uses, including load-bearing walls. Crook restricted in 2" × 4"–8' to $^1/_4$", with wane restricted to $^1/_3$ of thickness.
Structural joists and planks 2" to 4" thick 5" and wider	Select Structural Dense Select Structural	High quality, relatively free of characteristics that impair strength or stiffness. Recommended for uses where high strength, stiffness, and good appearance are required.
	No. 1 No. 1 Dense	Provide high strength; recommended for general utility and construction purposes. Good appearance; especially suitable where exposed because of the knot limitations.
	No. 2 No. 2 Dense	Although less restricted than No. 1, suitable for all types of construction. Tight knots.
	No. 3 No. 3 Dense	Assigned stress values meet wide range of design requirements. Recommended for general construction purposes where appearance is not a controlling factor. Many pieces included in this grade would qualify as No. 2 except for single limiting characteristic. Provides high-quality, low-cost construction.
Light framing 2" to 4" thick 2" to 4" wide	Construction	Recommended for general framing purposes. Good appearance, strong, and serviceable.
	Standard	Recommended for same uses as Construction grade, but allows larger defects.
	Utility	Recommended where combination of strength and economy is desired. Excellent for blocking, plates, and bracing.
	Economy	Usable lengths suitable for bracing, blocking, bulkheading, and other utility purposes where strength and appearance not controlling factors.
Appearance framing 2" to 4" thick 2" and wider	Appearance	Designed for uses such as exposed-beam roof systems. Combines strength characteristics of No. 1 with appearance of C&Btr.
Timbers 5" × 5" and larger	No. 1 SR No. 1 Dense SR No. 2 SR No. 2 Dense SR	No. 1 and No. 2 are similar in appearance to corresponding grades of 2" dimension. Recommended for general construction uses. SR in grade name indicates Stress Rated.
Structural lumber	Dense Str. 86 Dense Str. 72 Dense Str. 65	Premier structural grades from 2" through and including timber sizes. Provides some of the highest design values in any softwood species with good appearance.

Courtesy Southern Forest Products Association

Table 18.10 Classifications, Sizes, and Grades of Softwood Lumber

Standard Units			
Classification	**Thickness Nominal (in.)**	**Width Nominal (in.)**	**Grades**
Finish	$^3/_8$–4	2–16	B&B, C, C&Btr, D
Boards	$1^1/_2$	2–12+	No. 1, No. 2, No. 3, No. 4
Structural light framing	2–4	2–4	Select Structural, No. 1, No. 2, No. 3
Light framing	2–4	2–4	Construction, Standard, Utility
Studs	2–4	2–6	Stud
Structural joists and planks	2–4	5 and wider	Select Structural, No. 1, No. 2, No. 3
Appearance framing	2–4	2 and wider	Appearance
Timbers, nonstress	5 and larger	5 and larger	Square-edge and sound, No. 1, No. 2, No. 3
Timbers, stress-rated	5 and larger	5 and larger	Sel, Str, SR, DNS, Sel Str SR, No. 1 SR, No. 1 DNS SR, No. 2 SR, No. 2 DNS SR

Metric Units			
Classification	**Thickness Nomenclature[a] (mm)**	**Width Nomenclature[a] (mm)**	**Grades**
Finish	8–64	38–286	B&B, C, C&Btr, D
Boards	17–32	38–387	No. 1, No. 2, No. 3, No. 4
Structural light framing	38–102	38–387	Sel Str, No. 1, No. 2, No. 3
Light framing	38–102	38–387	Construction, Standard, Utility
Studs	38–89	38–140	Stud
Structural joists and planks	38–89	114–387	Sel Str, No. 1, No. 2, No. 3
Appearance framing	38–89	38 and wider	Appearance
Timbers, nonstress	114 and larger	114 and larger	Square edge and sound, No. 1, No. 2, No. 3
Timbers, stress-rated	114 and larger	114 and larger	Sel Str SR, DNS Sel Str SR, No. 1 SR, No. 2 DNS SR, No. 3 DNS SR

[a]Nomenclature means the size used to describe the members. Actual size may be slightly larger or smaller.
Courtesy Southern Forest Products Association (standard) and Canadian Wood Council (metric)

parallel to the grain (Ft); (3) horizontal shear (Fv); (4) compression parallel to the grain (Fc//); (5) compression perpendicular to the grain (Fc') (side-grain crushing); and (6) the modulus of elasticity (E or MOE) (stiffness) (Table 18.13).

Adjustment Factors Base Values are adjusted for various Conditions of Use. Conditions of Use and their application to Base Values are shown in Table 18.14.

Figure 18.16 Grade stamp for machine stress rated lumber. *(Courtesy Western Wood Products Association)*

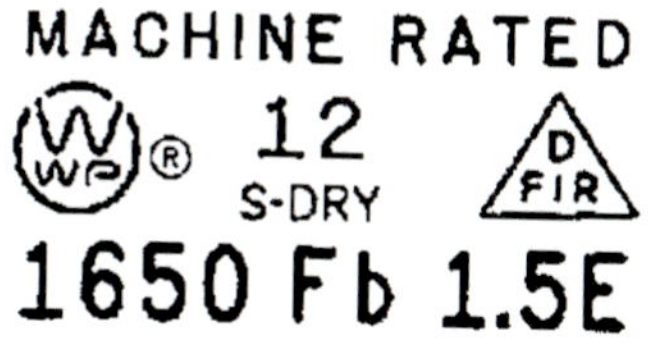

MACHINE RATED
WWP® 12 S-DRY D FIR
1650 Fb 1.5E

The seven Conditions of Use are:

Size Factors (Cf)—Applied to dimension base values

Repetitive Member Factors (Cr)—Applied to size-adjusted Fb (bending stress)

Duration of Load Adjustment (Cd)—Applied to size-adjusted values

Horizontal Shear Adjustments (Ch)—Applied to Fv (horizontal shear) values

Flat Use Factors (Cfu)—Applied to size-adjusted Fb (bending stress)

Adjustments for Compression Perpendicular to Grain (Cc') (compression perpendicular) values

Wet Use Factors (Cm)—Applied to size-adjusted values

How to Apply Adjustment Factors How to find the adjusted bending stress, Fb, for a Select Structural 2 × 6 in.

Table 18.11 Western Lumber Species Groups

Douglas Fir-Larch Douglas fir Western larch	Spruce-Pine-Fir (South) Engelmann spruce Sitka spruce Lodgepole pine
Douglas Fir-South Douglas fir grown in AZ, CO, NV, NM, & UT	Western Woods Ponderosa pine Sugar pine Idaho white pine Mountain hemlock
Hem-Fir Western hemlock Noble fir California red fir Grand fir Pacific silver fir White fir	Western Cedars Incense cedar Western red cedar Port orford cedar Alaska cedar

Table 18.12 Canadian Lumber Species Groups

Douglas Fir-Larch Douglas fir Western larch
Hem-Fir Western hemlock Amabilis fir
Spruce-Pine-Fir White spruce Red spruce Black spruce Engelmann spruce Lodgepole pine Jack pine Alpine fir Balsam fir
Northern Species All species graded in accordance with the NLGA standard grading rules for Canadian lumber

member from the Douglas Fir-Larch group: The Base Value for DF-L in SS grade is 1450 psi. This is multiplied by the Size Value, which is 1.3: 1450 × 1.3 = 1885 psi (size adjusted Fb). This can now be adjusted for other conditions of use, such as for a Repetitive Member Factor, which is 1.15: 1885 × 1.15 = 2167 psi (adjusted Fb).

Various lumber associations have published new span tables for members, such as joists and rafters. These tables include some applications of conditions of use. Before using the span tables it is necessary to note which conditions of use have been applied.

PHYSICAL AND CHEMICAL COMPOSITION OF WOOD

Since wood is a naturally occurring material, it has considerable variation in its physical properties, including color, density, weight, and strength. The physical and chemical composition of wood determines its properties and, therefore, its uses.

Porosity of Wood

Wood is a cellular material, as shown in Figs. 18.17 and 18.18. Softwood cellular structure contains large longitudinal cells called tracheids and smaller radial cells called rays, both of which store and transfer nutrients. The annual rings are also cellular. The structure of hardwoods is more complex, having two different types of longitudinal cells, small-diameter fibers and larger diameter vessels or pores, which transport the sap of the tree. They also have a higher percentage of rays than is found in softwoods.

Table 18.13 Base Values for Western Dimension Lumber[a]

Species or Group	Grade	Extreme Fiber Stress in Bending "F_b" Single	Tension Parallel to Grain "F_t"	Horizontal Shear "F_v"	Compression Perpendicular "F_c"	Compression Parallel to Grain "$F_{c//}$"	Modulus of Elasticity "E"
Douglas Fir/Larch	Select Structural	1,450	1,000	95	625	1,700	1,900,000
	No. 1 & Btr.	1,150	775	95	625	1,500	1,800,000
	No. 1	1,000	675	95	625	1,450	1,700,000
	No. 2	875	575	95	625	1,300	1,600,000
	No. 3	500	325	95	625	750	1,400,000
	Construction	1,000	650	95	625	1,600	1,500,000
	Standard	550	375	95	625	1,350	1,400,000
	Utility	275	175	95	625	875	1,300,000
	Stud	675	450	95	625	825	1,400,000

[a]Sizes: 2" to 4" thick by 2" and wider
Courtesy Western Wood Products Association

Table 18.14 Adjustment Factors for Base Values for Western Dimension Lumber

Size Factors (C_F) (Apply to Dimension Lumber Base Values)

Grades	Nominal Width (depth)	F_b 2″ & 3″ thick nominal	F_b 4″ thick nominal	F_t	$F_{c//}$	Other Properties
Select Structural, No. 1 & Btr., No. 1, No. 2 & No. 3	2″, 3″ & 4″	1.5	1.5	1.5	1.0	1.0
		1.15	1.0			
	5″	1.4	1.4	1.4	1.1	1.0
	6″	1.3	1.3	1.3	1.1	1.0
	8″	1.2	1.3	1.2	1.05	1.0
	10″	1.1	1.2	1.1	1.0	1.0
	12″	1.0	1.1	1.0	1.0	1.0
	14″ & wider	0.9	1.0	0.9	0.9	1.0
Construction & Standard	2″, 3″, & 4″	1.0	1.0	1.0	1.0	1.0
Utility	2″ & 3″	0.4	—	0.4	0.6	1.0
	4″	1.0	1.0	1.0	1.0	1.0
Stud	2″, 3″, & 4″	1.1	1.1	1.1	1.05	1.0
	5″ & 6″	1.0	1.0	1.0	1.0	1.0

Repetitive Member Factor (C_r) (Apply to Size-Adjusted F_b)

Where 2″- to 4″-thick lumber is used repetitively, such as for joists, studs, rafters, and decking, the pieces side by side share the load and the strength of the entire assembly is enhanced. Therefore, where three or more members are adjacent or are not more than 24″ apart and are joined by floor, roof, or other load distributing elements, the F_b value can be increased 1.15 for repetitive member use.

Repetitive Member Use
$F_b \times 1.15$

Duration of Load Adjustment (C_d) (Apply to Size-Adjusted Values)

Wood has the property of carrying substantially greater maximum loads for short durations than for long durations of loading. Tabulated design values apply to normal load duration. (Factors do not apply to MOE or F_c).

Load Duration	Factor
Permanent	0.9
Ten years (normal load)	1.0
Two months (snow load)	1.15
Seven day	1.25
One day	1.33
Ten minutes (wind and earthquake loads)	1.6
Impact	2.0

Confirm load requirements with local codes. Refer to Model Building Codes or the National Design Specification for high-temperature or fire-retardant treated adjustment factors.

Flat Use Factors (C_{fu}) (Apply to Size-Adjusted F_b)

Nominal Width	Nominal Thickness 2″ & 3″	4″
2″ & 3″	1.00	—
4″	1.10	1.00
5″	1.10	1.05
6″	1.15	1.05
8″	1.15	1.05
10″ & wider	1.20	1.10

Horizontal Shear Adjustment (C_H) (Apply to F_v Values)

Horizontal shear values are based upon the maximum degree of shake, check or split that might develop in a piece. When the actual size of these characteristics is known, the following adjustments may be taken.

2″ Thick Lumber

For convenience, the table below may be used to determine horizontal shear values for any grade of 2″-thick lumber in any species when the length of split or check is known and any increase in them is not anticipated.

When Length of Split on Wide Face is:	Multiply Tabulated F_v Value by:
No split	2.00
1/2 of wide face	1.67
3/4 of wide face	1.50
1 of wide face	1.33
1 1/2 of wide face or more	1.00

3″ and Thicker Lumber

Horizontal shear values for 3″ and thicker lumber also are established as if a piece were split full length. When specific lengths of splits are known and any increase in them is not anticipated, the following adjustments may be applied.

When Length of Split on Wide Face is:	Multiply Tabulated F_v Value by:
No split	2.00
1/2 of narrow face	1.67
1 of narrow face	1.33
1 1/2 of narrow or more	1.00

Adjustments for Compression Perpendicular to Grain (C_c) (For Deformation Basis of 0.02″ Apply to F_c Values)

Design values for compression perpendicular to grain (F_c) are established in accordance with the procedures set forth in ASTM Standards D 2555 and D 245. ASTM procedures consider deformation under bearing loads as a service ability limit state comparable to bending deflection because bearing loads rarely cause structural failures. Therefore, ASTM procedures for determining compression perpendicular to grain values are based on a deformation of 0.04″ and are considered adequate for most classes of structures. Where more stringent measures need to be taken in design, the following formula permits the designer to adjust design values to a more conservative deformation basis of 0.02″:

$$Y_{02} = 0.73\ Y_{04} + 5.60$$

Example: Douglas Fir-Larch: $Y_{04} = 625$ psi
$Y_{02} = 0.73\ (625) + 5.60 = 462$ psi

Wet Use Factors (C_M) (Apply to Size-Adjusted Values)

The design values shown in the accompanying tables are for routine construction applications where the moisture content of the wood does not exceed 19%. When use conditions are such that the moisture content of dimension lumber will exceed 19%, the Wet Use Adjustment Factors below are recommended:

	Property	Adjustment Factor
F_b	Extreme fiber stress in bending	0.85*
F_t	Tension parallel to grain	1.0
F_c	Compression parallel to grain	0.8**
F_v	Horizontal shear	0.97
F_C	Compression perpendicular to grain	0.67
E	Modulus of elasticity	0.9

**Fiber Stress in Bending Wet Use Factor 1.0 for size-adjusted F_b not exceeding 1,150 psi.*
***Compression Parallel to Grain in Wet Use Factor 1.0 for size-adjusted F_c not exceeding 750 psi.*
Courtesy Western Wood Products Association

Figure 18.17 A greatly enlarged example of wood cellular structure. Tracheids (TR) are vertical cells that make up the major part of the structure. Rays (WR) are cells that run radially from the center to the outside of the tree. Annual rings (AR) are made up of small hard summerwood cells (SM) and larger, softer springwood cells (S). Resin is in horizontal resin ducts (HRD) and vertical resin ducts (VRD) centered in fusiform wood rays (FWR). Simple pits (SP) allow sap to pass back and forth between ray cells and longitudinal cells. Border pits (BP) transfer sap between longitudinal cells. Face RR indicates a radial cut through the wood, and TG indicates a tangential cut. *(Courtesy Forest Products Laboratory, USDA Forest Service, Madison, WI)*

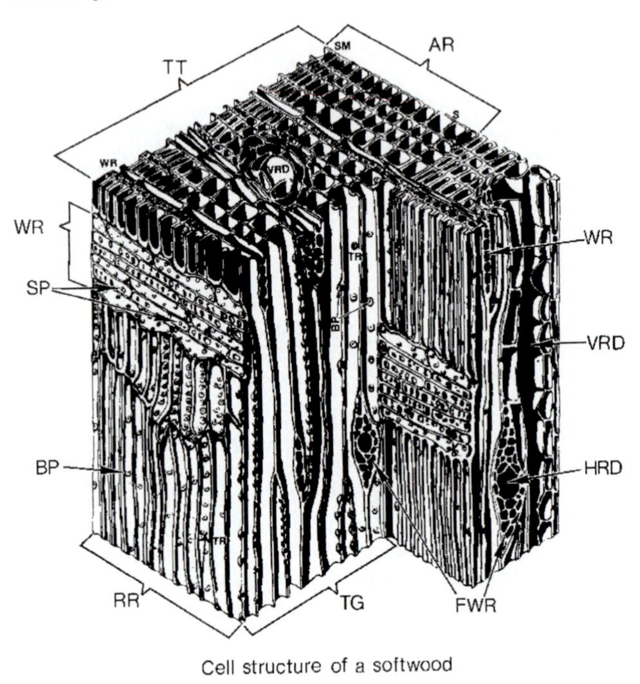

Figure 18.18 Hardwood trees have a more complex structure. The rays (WR) make up the major mass and produce the grain features. Large vertical cells with pores (P) move the sap. Wood fibers (F) give the tree structural strength. Pits (K) transfer sap from one cell to another, and rays (WR) run radially from the center of the tree. A radial cut is indicated by RR and a tangential cut by TG. An annual ring is shown by AR. *(Courtesy Forest Products Laboratory, USDA Forest Service, Madison, WI)*

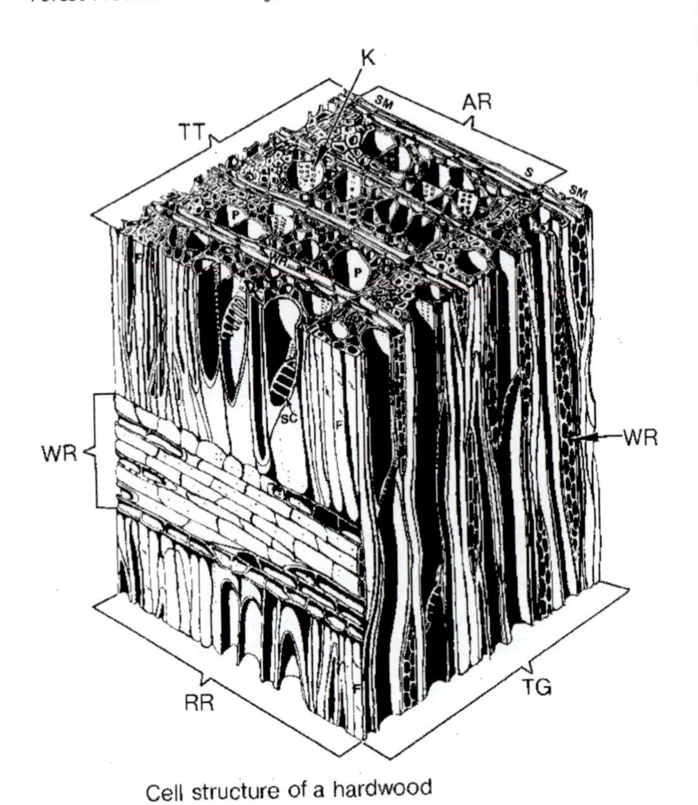

The surface area of these cells is very large and gives wood several important properties. First, it makes it possible for wood, when dry, to absorb toxic chemicals needed to prevent decay and insect attack. Second, it can absorb moisture repellents to minimize moisture exchange and therefore control shrinkage. Third, the cells are air pockets that provide insulating qualities. Fourth, it enables wood to shrink and swell as moisture content varies. Fifth, it contributes to the ease of adherence of paint, adhesives, and other synthetic resins used on wood surfaces.

Composition of Wood

The cells are made of mainly cellulose and hemicellulose fibers, that are bonded together with an organic substance called **lignin.** Cellulose and hemicellulose are complex glucose compounds. Glucose is a sugar useful to fungi and insects as a food. Although the exact amounts of cellulose, hemicellulose, and lignin vary with different species of wood, the general composition in kiln-dried woods is shown in Table 18.15. These elements are what give wood its strength, susceptibility to decay, and hygroscopic properties. **Hygroscopic** refers to the property of wood that permits it to absorb and retain moisture. Cellulose provides strength in tension, toughness, and elasticity. Lignin, because it bonds the fibers together in fiber bundles, gives wood its compressive strength. The remaining substances in wood do not contribute to its structure but give each species its unique color, odor, taste, density, resistance to decay, and flammability.

Table 18.15 Composition of Kiln-Dried Wood

Material	Softwoods Percent	Hardwoods Percent
Cellulose	40–50	40–50
Hemicellulose	20	15–35
Lignin	23–33	16–25
Extraneous materials	5–10	5–10

Paper products are made using the cellulose and hemicellulose in wood. The wood is cut into chips and cooked to separate the fibers from the lignin. The fibers form the pulp used to make paper and cardboard products. Hemicellulose plays an important part in fiber-to-fiber bonding in the papermaking process. See Chapter 23 for additional information on paper products.

Cellulose is also used in some textiles, plastics, and other products requiring cellulose derivatives. Extractive materials derived from wood include tannins, coloring matters, essential oils, fats, resins, waxes, gums, starch, and simple metabolic intermediates.

HYGROSCOPIC PROPERTIES OF WOOD

Wood is hygroscopic, expanding when it absorbs moisture and shrinking when it loses it. During its life a tree contains considerable water. In this condition, wood is called green. To be useful in construction, furniture, and other products, it must be in a dry condition. The moisture content of wood must be reduced to a level acceptable for its intended purpose.

Moisture Content

The **moisture content** of wood is the weight of water in wood divided by the weight of the wood, when oven dry, expressed as a percentage. Green samples are weighed, oven dried, and weighed again to ascertain moisture content. For example, if a sample weighs 4 oz. green and 3 oz. dry, it has a moisture content of 1 oz. of water (or 33 percent). On the job, moisture content is checked with a moisture meter. When a probe is placed in contact with the wood, this battery-operated unit shows the moisture content on a meter.

The moisture content of lumber from newly cut trees varies by species but ranges from 30 percent to as much as 200 percent. The moisture content of sapwood is higher than that of heartwood. For example, in the various species of pine, the heartwood moisture content ranges from 30 to 98 percent, and the sapwood ranges from 100 to 220 percent. In general, hardwoods have lower moisture contents when freshly cut than softwoods.

Fiber Saturation Point

In a living tree or freshly cut wood (green), moisture is present in the cell fibers as absorbed water and in the cell cavities as free water. As the wood begins to dry, the water in the cell cavities begins to disappear. Once all free water in the cell cavities is gone, the cell fibers making up the cell walls contain any remaining water. This is the **fiber saturation point**. Most softwoods have a moisture content of about 30 percent at this point. As moisture is removed from the cell walls, they begin to shrink. Therefore, the fiber saturation point is the point at which shrinkage begins (Fig. 18.19).

The total shrinkage in lumber occurs between the fiber saturation point, 30 percent, and the desired moisture content, such as the 15 percent required for most lumber. Shrinkage is approximately proportional to the amount of moisture loss. Wood shrinks about $\frac{1}{30}$ of the total shrinkage with each 1 percent reduction in moisture below the fiber saturation point. The reduction mentioned above, 30 percent to 15 percent, is about half the possible shrinkage.

Equilibrium Moisture Content

Green lumber gives off moisture to the air and, if exposed long enough, will lose moisture until it has the same moisture content as the surrounding air. It will continue to take on and give off moisture as the air's moisture content varies. When this point is reached, the wood has reached its **equilibrium moisture content**.

Since wood reacts to constantly varying changes in humidity and temperature, it is constantly seeking to reach equilibrium with the surrounding air. Variations tend to be seasonal, and there is less variation inside heated buildings than on the exterior.

As wood seeks equilibrium, it gradually shrinks or expands. Excessive expansion or contraction can cause problems in a building, such as sticking doors and windows. Cracks occur in the interior wall finish if studs and plates experience excessive shrinkage. To minimize the amount of expansion and contraction, wood should be installed at a moisture content as close as possible to the equilibrium moisture content it will have after installation. The recommended moisture content for interior wood, such as flooring, trim, cabinets, windows, and doors, is between 6 and 12 percent. This varies by region, from the dry Southwest to the more moisture-laden areas of the Northwest and Southeast (Fig. 18.20). Since wood shipped to a job can have its moisture content increased or decreased during shipping or storage, it is recommended that wood elements be stored in the area where they will be used for several days so they can approach equilibrium before installation.

The moisture content for exterior wood products is usually 12 to 15 percent for most areas of the

Figure 18.19 As the wood dries, the cells lose free water. Then the cell wall fibers dry and the cells shrink.

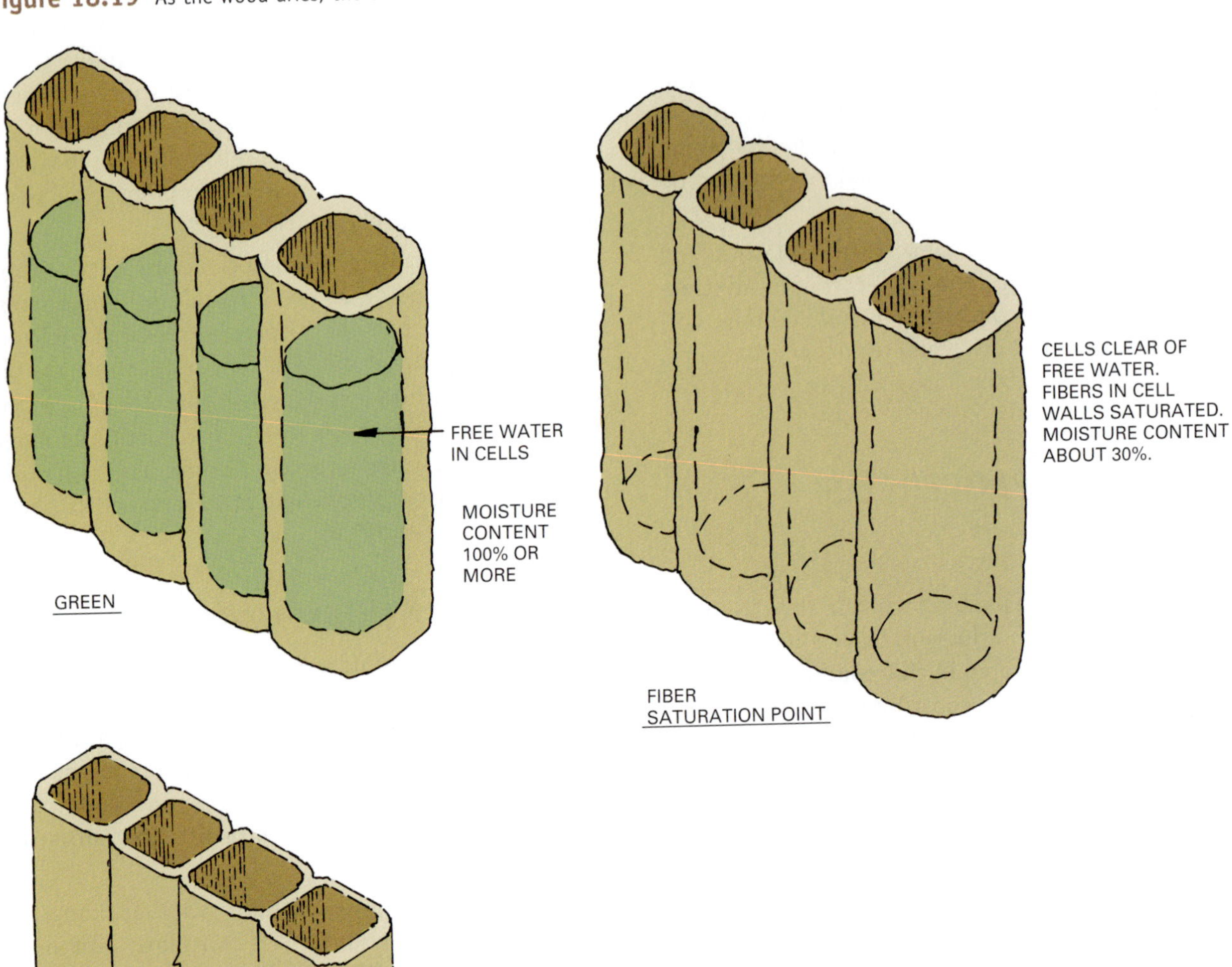

country. Products such as wood siding are exposed to greater variations in humidity and temperature and reach equilibrium at a higher moisture content. For example, surfaced framing lumber having a moisture content at maximum 15 percent is stamped MC15. If it is surfaced and has a moisture content at a maximum of 19 percent, it is stamped S-DRY (surfaced dry). If it has more than 19 percent moisture, it is stamped S-GRN (surfaced green).

To ensure that lumber does not exceed the moisture content stamped on it, mills sometimes dry it several percentages below the specified level. In dry regions, framing lumber should have a moisture content of not more than 15 percent when interior finishes are installed. In most of the country, a maximum moisture content of 19 percent is satisfactory.

How Moisture Affects Wood Properties

The moisture content of wood affects its size, dimensional stability, strength, stiffness, decay resistance, glue bonding, and paintability.

Construction Materials

Sustainable Forestry

Sustainable forestry describes the management of forests according to the principles of sustainable development. The negative environmental impacts of commercial forestry activities include the clear cutting and destruction of forests, resulting in loss of wildlife habitat, soil erosion, and water and air pollution. Sustainable forest management seeks to avoid these impacts by integrating environmental goals with the continuing need for forest products.

A working definition of sustainable forest management was developed by the Ministerial Conference on the Protection of Forests in Europe (MCPFE). It defines sustainable forest management as:

> "the stewardship and use of forests and forest lands in a way, and at a rate, that maintains their biodiversity, productivity, regeneration capacity, vitality, and their potential to fulfill, now and in the future, relevant ecological, economic, and social functions, at local, national, and global levels, and that does not cause damage to other ecosystems."

Sustainable forestry is concerned with maintaining a balance between society's increasing requirements for forest products and the preservation of forest health and biodiversity. The elements of sustainable forestry include sustainable harvesting techniques, preserving habitat for the biodiversity of plants and animals, minimizing the use of toxic chemicals, and protecting soil and water quality.

Forest managers must assess and integrate a wide array of sometimes conflicting factors, including commercial values, environmental considerations, and community, to produce sound forest management plans. In most cases, forest managers develop their plans in consultation with citizens, businesses, government organizations and other interested parties in and around the forest tract being managed.

Growing environmental awareness and consumer demand for more socially responsible businesses helped third-party forest certification emerge as a credible tool for communicating the environmental and social performance of forest operations. With forest certification, an independent organization develops standards of good forest management, and independent auditors issue certificates to forest operations that comply with those standards. This certification verifies that forests are well managed, as defined by a particular standard, and ensures that certain wood and paper products come from responsibly managed forests. Products that are certified bear the stamp of the organization.

Third-party forest certification is an important tool for those seeking to ensure that the paper and wood products they purchase and use come from forests that are well managed and legally harvested. Incorporating third-party certification into forest product procurement practices can be a centerpiece for comprehensive wood and paper policies that include factors such as the protection of sensitive forest values, thoughtful material selection, and efficient use of products. Certifying agencies include the Forest Stewardship Council (FSC), the Sustainable Forestry Initiative (SFI), and Canada's National Sustainable Forest Management Standard (CSA).

Sustainable forestry uses third-party certification to ensure non-destructive forest management policies. *(Image copyright Stephen Aaron Rees, 2009. Used under license from Shutterstock.com)*

Figure 18.20 The equilibrium moisture content varies in different sections of the country. *(Courtesy Forest Products Laboratory, USDA Forest Service, Madison, WI)*

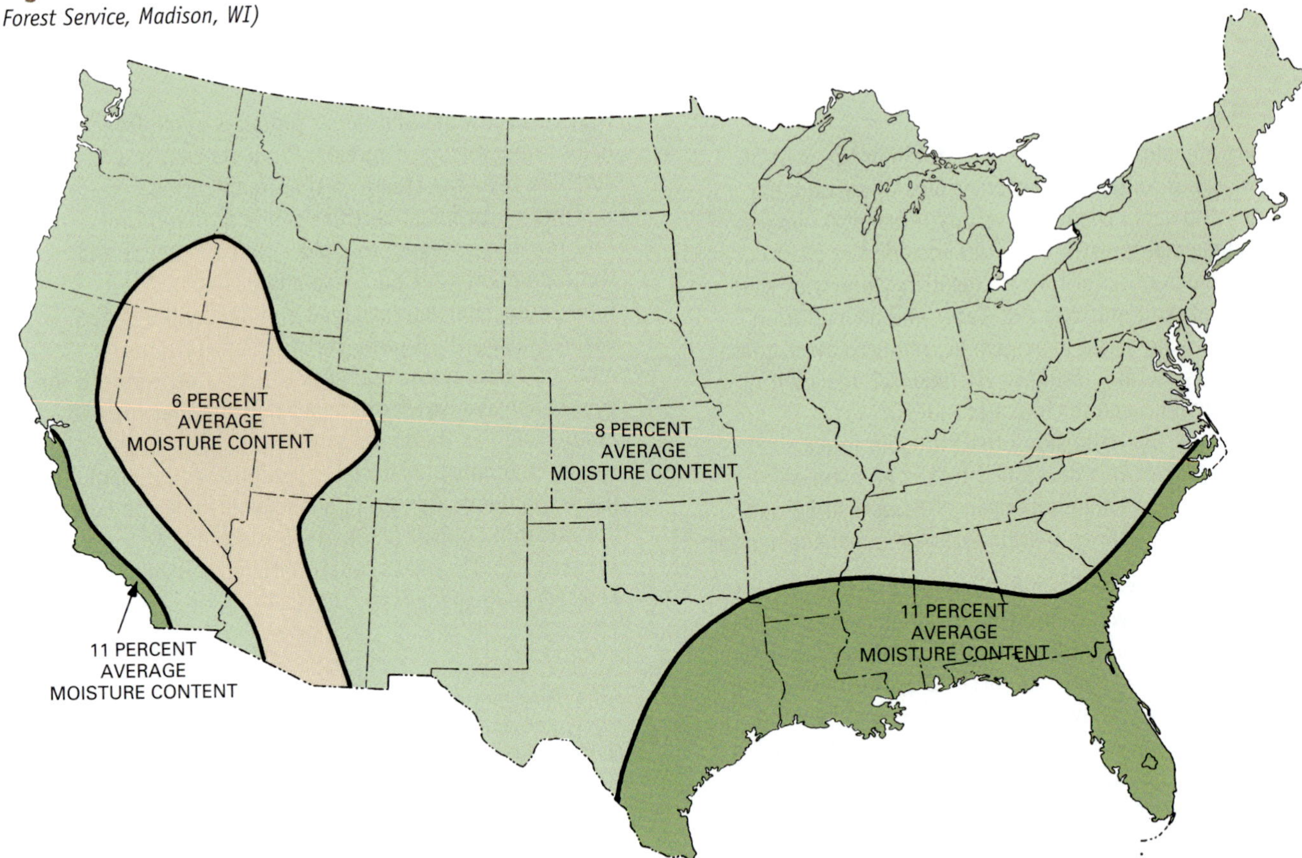

Size and dimensional stability are directly affected by the moisture content of wood. Shrinkage and swelling occur only after the moisture content falls below the fiber saturation point. Most woods shrink and swell very little parallel with the grain (longitudinally). This shrinkage has very little influence on construction uses. Wood shrinks and swells a great deal in thickness and width across the grain. Shrinkage is greatest in the direction parallel to the annual growth rings (tangential) and about twice as much as the shrinkage across the rings (radial) (Fig. 18.21).

Figure 18.22 shows the combined effects of tangential and radial shrinkage as wood dries from its green condition. Notice that the shrinkage is affected by the direction of the growth rings. Tangential shrinkage is about twice as great as radial shrinkage. Tangential shrinkage is that which is tangent to the circumference of the tree. Radial shrinkage is that which occurs parallel to a line passed through the center of the tree. This produces distortions in the lumber. The nature of the distortion depends on the location the piece of lumber occupied in the tree. These distortions due to shrinkage are greater in plain-sawed lumber.

Most hardwoods shrink more than softwoods, and heavier species shrink more than lighter species. Stock with a large cross-sectional area, such as 6 in. × 6 in. timbers, do not shrink as much proportionately because the inside does not dry at the same rate as the outside. The outer layers dry faster, become set, and keep the inner area from shrinking normally as it dries.

Softwood lumber shrinks about $\frac{1}{32}$ in. per inch of face width while drying from a green condition to about 19 percent moisture content. This means a rough cut 2 in. × 10 in. board shrinks about $\frac{1}{4}$ in. in width as it dries to 19 percent moisture content and about $\frac{5}{16}$ to $\frac{3}{8}$ in. when it reaches 15 percent moisture content.

As the moisture content of wood decreases, its strength increases. This is caused by the stiffening and strengthening of the fibers in the cell walls and the fact that as the wood shrinks it becomes a denser material. Crushing strength and bending strength increase a great deal in dry woods, while stiffness increases only moderately. Shock resistance depends on the pliability of the material and is lower for dry wood than for the original green stock. Green wood also bends farther before it breaks than dry wood.

Wood that maintains a moisture content below 20 percent will be free from decay. This can be accomplished in any of several ways, including painting or treating with wood-preserving chemicals.

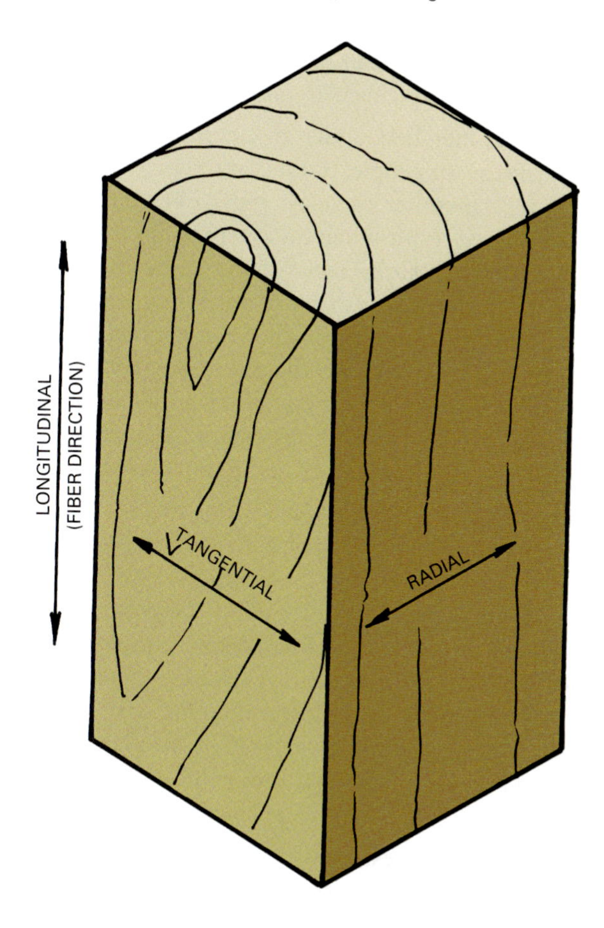

Figure 18.21 Wood shrinks in three directions in relationship to the direction of the grain and growth rings.

The recommended moisture content for wood destined for painting is 16 to 20 percent. Green wood should not be painted, nor should wood that has surface moisture, from rain, for example.

Reducing the moisture content of wood also improves the strength of glue bonds. Wood used for interior purposes is glued at 5 to 6 percent moisture content. Exterior wood can be glued satisfactorily at 10 to 12 percent moisture content. Veneers, such as those used to make plywood, usually have 2 to 6 percent moisture content. Woods glue satisfactorily up to about 15 percent moisture content, but they should be in equilibrium with the air in which they will be used. Interior wood at 12 to 15 percent moisture can be glued satisfactorily, but the glued members will check and warp as the wood loses moisture and seeks equilibrium with the air.

Specific Gravity

Specific gravity is a ratio of the weight of a certain volume of a material to the weight of an equal volume of water at 39.2°F (4°C). The specific gravity of wood is found by weighing a piece of wood of known volume, such as 6 in.[3], and dividing this by the weight of an equal volume of water. Water has a specific gravity of 1. If a material has a specific gravity less than this, it will float; if it is greater than 1, it will sink. You can see some examples for selected woods in Table 18.16.

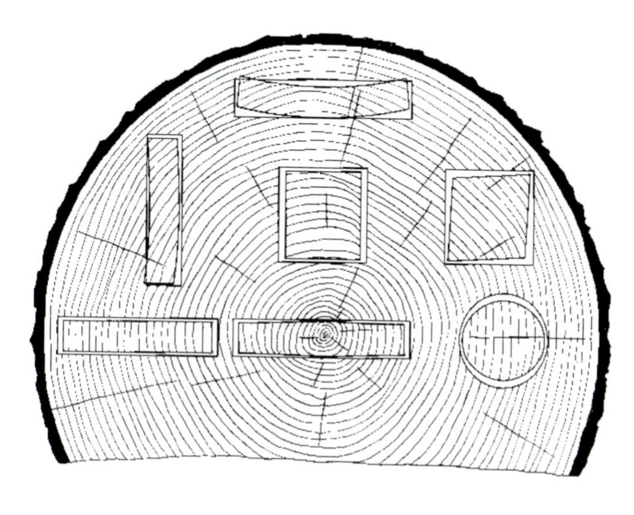

Figure 18.22 Typical shrinkage and distortion of various wood shapes in relation to the direction of growth rings. *(Courtesy Forest Products Laboratory, USDA Forest Service, Madison, WI)*

Table 18.16 Specific Gravity and Weight of Woods Commonly Used in Construction[a]

Softwoods	Specific Gravity	Density[b]
Cedar, western	0.32	20.0
Cypress	0.46	28.7
Douglas fir, coast	0.48	30.0
Larch, western	0.52	32.4
Pine, western white	0.38	23.7
Pine, lodgepole	0.41	25.5
Pine, ponderosa	0.40	25.0
Pine, shortleaf	0.51	31.6
Redwood	0.40	25.0
Spruce, Englemann	0.35	21.6
Spruce, Sitka	0.40	25.0
Hardwoods		
Birch, yellow	0.62	38.7
Cherry, black	0.50	31.2
Maple, sugar	0.63	39.3
Oak, red	0.61	37.9
Oak, white	0.64	39.9

[a]*Oven-dry samples.*
[b]*Density in points per cubic foot based on specific gravity as shown and 0% moisture.*
Courtesy Forest Products Laboratory, USDA Forest Service, Madison, Wis.

The specific gravity of wood fibers forming the cells is about 1.5, so the fibers would sink. However, most woods float, because much of the volume is made up of air-filled cells. The size of the cells and the thickness of the cell walls influence the specific gravity. Specific gravity is an approximate indicator of the density of wood and an indication of strength. In general, woods that are dense weigh more, have a higher specific gravity, and can handle higher stresses.

STRUCTURAL PROPERTIES OF WOOD

Wood is an orthotropic material. Orthotropic pertains to a mode of growth that is more or less vertical. Wood has unique and independent mechanical properties in the directions of three mutually perpendicular axes: longitudinal, radial, and tangential. The longitudinal axis is parallel to the fibers (grain). The radial axis is normal to the growth rings (perpendicular to the grain in the radial direction). The tangential axis is perpendicular to the grain but tangent to the growth rings.

Wood is a fibrous material. The fibers are bonded together with lignin forming the walls of the cells making up the material. The fibers in hardwoods are about $\frac{1}{25}$ in. long and from $\frac{1}{8}$ to $\frac{1}{3}$ in. in softwoods. The strength of wood does not depend on the length of the fibers but on the thickness of the cell walls and the direction of the fibers in relation to applied loads. Most fibers are oriented with their lengths parallel with the vertical dimension of the tree. Strength of wood parallel with the fibers (parallel with the grain) is greater than the strength perpendicular to the fibers (perpendicular with the grain).

Structural members under an external load, such as wind, furniture, or people, produce internal forces called stresses in a member to resist these external forces. Tensile stresses result when an external force tends to stretch a member. Compressive stresses occur when a member is under a squeezing force. When a member, such as a beam, is loaded so that the applied force is acting approximately perpendicular to the member, the load produces a bending stress. When a member is under bending stresses, it develops compressive stresses in the upper part and tensile stresses in the lower part.

The strength of wood under various stresses is found by testing samples in a laboratory. Tests include bending, shear, stiffness, tension, and compression. Samples free of any defects are tested to find the basic stresses for each species used in construction. Samples of the various grades in each species are tested to find realistic stresses for each. These are lower than the basic stresses and are called working stresses. The working stress takes into account things that lower the load a member can carry, such as knots, pitch pockets, or checks.

The working stresses for several species of wood used in construction are shown in Table 18.17.

Table 18.17 Working Stresses for Selected Species of Structural Light Framing Lumber[a]

Species	Grade	Extreme Fiber in Bending "F_b"	Tension Parallel to the Grain "F_t"	Compression Parallel to the Grain "F_c"	Horizontal Shear "F_v"	Compression Perpendicular to the Grain "F_c"	Modulus of Elasticity "E"
Southern Pine	Dense Select Structural	2,500	1,500	2,100	105	475	1,900,000
	Select Structural	2,150	1,250	1,800	105	405	1,800,000
	No. 1 Dense	2,150	1,250	1,700	105	475	1,900,000
	No. 1	1,850	1,050	1,450	105	405	1,800,000
	No. 2 Dense	1,800	1,050	1,350	95	475	1,700,000
	No. 2	1,550	900	1,150	95	405	1,600,000
	No. 3 Dense	1,000	575	800	95	475	1,500,000
	No. 3	850	500	675	95	405	1,500,000
	Stud	850	500	675	95	405	1,500,000
Hem-Fir	Select Structural	1,650	975	1,300	75	405	1,500,000
	No. 1/Appearance	1,400	825	1,050/1,250	75	405	1,500,000
	No. 2	1,150	675	825	75	405	1,400,000
	No. 3	650	375	500	75	405	1,200,000
Douglas Fir-Larch	Select Structural	2,100	1,200	1600	95	625	1,800,000
	No. 1/Appearance	1,750	1,050	1,250/1,500	95	625	1,800,000
	No. 2	1,450	850	1000	95	625	1,700,000
	No. 3	800	475	600	95	625	1,500,000

[a]*Structural light framing 2" to 4" thick and 2" to 4" wide MC 15%*

These are for normal loading conditions, which include dead loads (weight of structural and finish materials) and live loads (weight of occupants or furniture). They are based on the normal duration of loading, which assumes a fully stressed member under the full maximum design load for 10 years and the application of 90 percent of this maximum normal load continuously through the remainder of the life of the structure.

Working stresses are given in pounds per square inch and include working stresses for bending (Fb), **horizontal shear** (Fv), **vertical shear** (V), modulus of elasticity (E), compression parallel to the grain (Fc), fiber stress in tension (Ft), and compression perpendicular to the grain (Fc').

Bending

Wood beams under load deflect, producing bending stresses in the fibers. Wood has high fiber strengths in bending. However, if a beam is loaded enough to produce stresses greater than the fiber strength of the wood, the beam will fail. The stresses developed in the fibers are greater the farther they are from the central axis of the beam. Fibers located twice as far from the central axis of the beam as other fibers have twice the stress. Therefore, when a beam bends, the maximum stresses are developed in the outer fibers of the top and bottom of the beam. This is called the extreme fiber in bending, Fb (Fig. 18.23).

Shear

A beam is subject to vertical and horizontal shear (Figs. 18.24 and 18.25). Vertical shear refers to the tendency for one part of a beam to move vertically in relation to an adjacent part, allowing the beam to slip down between supports. Horizontal shear refers to the tendency of the sapwood fibers to move horizontally in relation to the bottom fibers. Horizontal shear is more of a factor in beam failure than vertical shear. The fiber stress in horizontal shear must be kept below the working stress values indicated by the symbol Fv.

Modulus of Elasticity

The modulus of elasticity (E) is a measure of a beam's resistance to deflection, or its stiffness. Stiffness is important when selecting columns, beams, rafters, joists, and other structural members (Fig. 18.26).

Tension

Wood under tension has good tensile strength. The allowable unit stress in tension parallel to the grain (Ft) is applied to the entire cross section of a wood member. Knots, checks, holes, and splits reduce the allowable unit stress in a member. Wood is weak in tension perpendicular to the grain (Fig. 18.27).

Figure 18.23 Forces designated as extreme fiber stresses in bending occur along the faces of the beam.

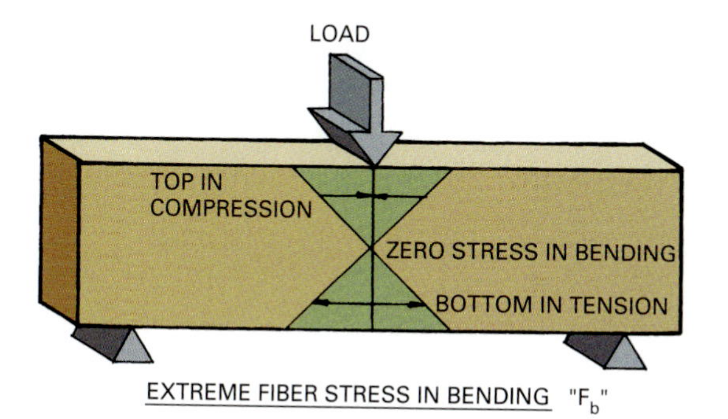

Figure 18.24 Horizontal shear stress tends to cause fibers to slice horizontally, much the same as when you bend a deck of cards.

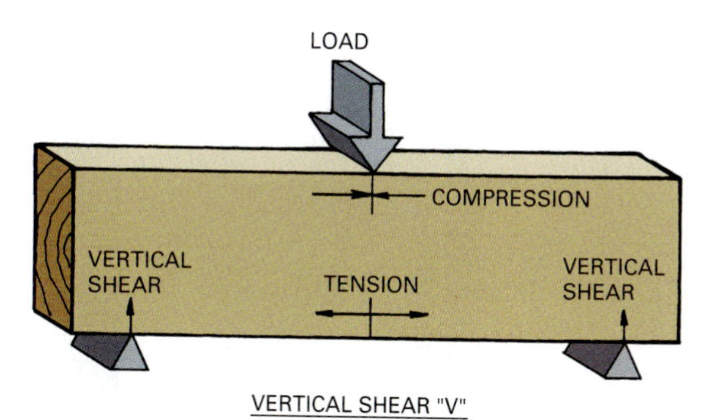

Figure 18.25 Vertical shear tends to cause fibers in one part of a beam to move vertically in relation to those adjacent to them.

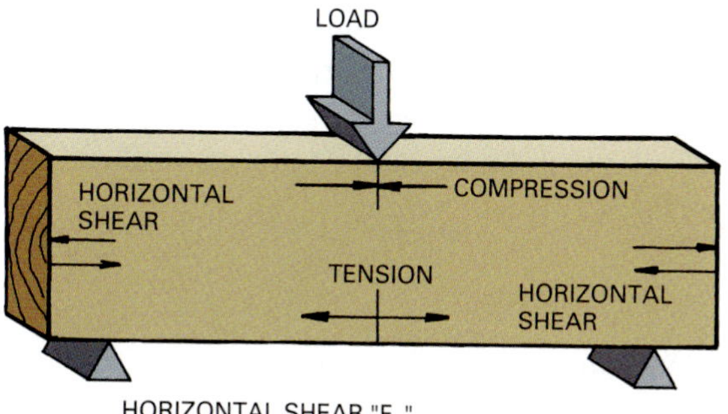

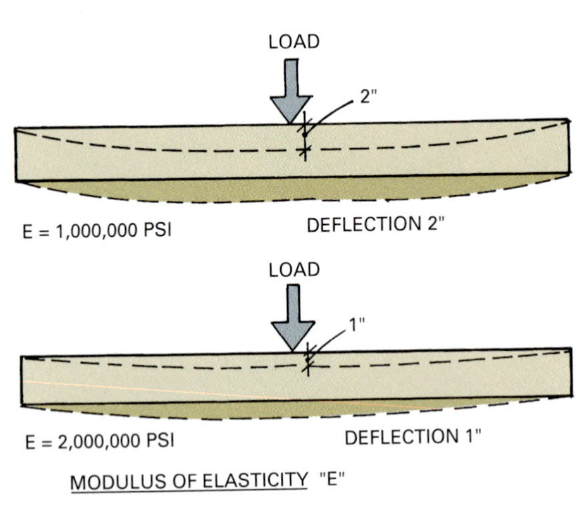

Figure 18.26 The modulus of elasticity is a measure of the stiffness of a beam.

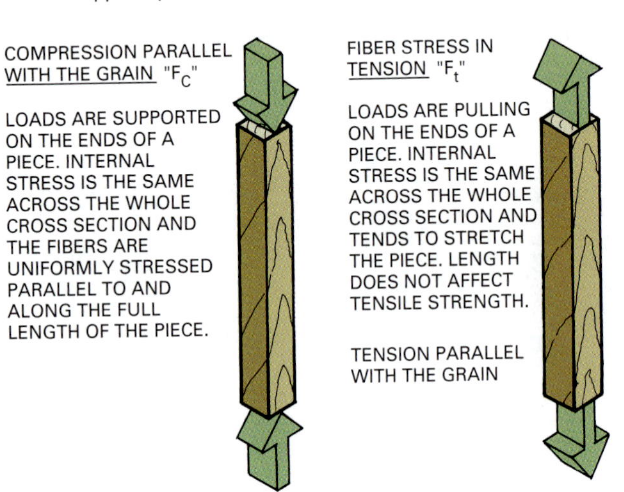

Figure 18.27 These are typical tension and compression stresses applied parallel to the grain of a member.

Compression

The unit of stress in compression parallel to the grain is several times greater than that perpendicular to the grain (Fig. 18.28). Selected values for compressive strength perpendicular to grain (Fc) and parallel to grain (Fc) are given in Table 18.17.

Gluing Properties

Glue bonding is improved as moisture content decreases, with 10 to 12 percent moisture effective for exterior products. In addition, some species of wood bond better than others (Table 18.18). A satisfactory glue joint is one that is as strong as the wood itself.

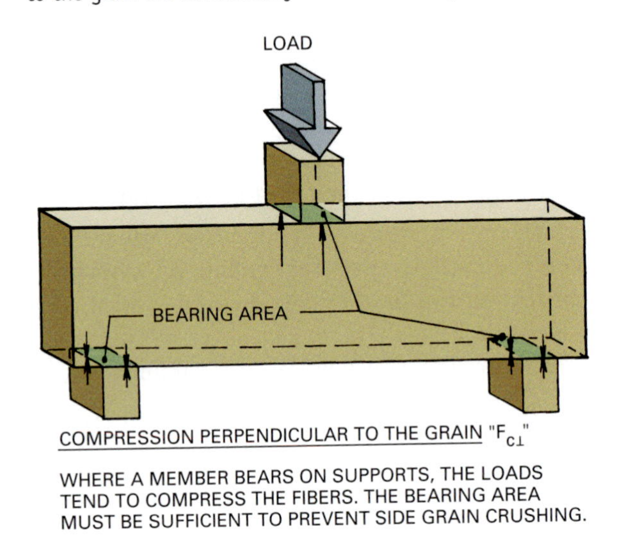

Figure 18.28 The working compression stresses perpendicular to the grain are considerably less than those parallel to the grain.

Table 18.18 Gluing Properties of Selected Woods

Bond easily	Alder Aspen Fir: white, grand, noble, pacific Pine: eastern, white, western white Red cedar, western Redwood Spruce, Sitka
Bond well	Elm, American rock Maple, soft Sycamore Walnut, black Yellow poplar Douglas fir Larch, western Pine: sugar, ponderosa Red Cedar, eastern Mahogany: African, American
Bond satisfactorily	Ash, white Birch: sweet, yellow Cherry Hickory: pecan, true Oak: red, white Pine, southern
Bond with difficulty	Persimmon Teak Rosewood

Courtesy Forest Products Laboratory, USDA Forest Service, Madison, Wis.

Properly made joints are often stronger than the wood, with the wood breaking before the joint does. The density and structure of the wood also influence the strength of the bond. In general, heartwood does not bond as well as sapwood, and hardwoods do not bond as well as softwoods.

OTHER PROPERTIES OF WOOD

Wood has other properties, some unique to it as a material, that need to be considered. These include thermal properties, decorative features, decay resistance, and insect damage.

Thermal Properties

Softwoods have a thermal conductivity of approximately 1 Btu/in. of thickness. Therefore, wood is a fairly good insulator but not as good as other materials used specifically for insulation. As the moisture content of wood increases, it becomes less efficient as an insulator, because the moisture increases thermal conductivity. Lighter (less dense) woods are better insulators because of their larger cell structure.

Wood also experiences thermal expansion. It expands when heated and contracts when cooled. For most construction purposes this is not a factor because the amount of change is so small. Expansion due to moisture increases is much greater and is the factor to consider.

Decorative Features

Wood has unique decorative features that make it a prized material. Various species have unique colors, including the mature heartwood and the new growth sapwood. Species vary in density, producing wood with smooth, closed grain to rough, very open grain, which can facilitate beautiful decorative appearances. Wood can also be stained or bleached to alter its color.

Wood takes on decorative colors and texture when it is permitted to weather with no protective coating. Some woods, such as cedar and cypress, are decay resistant and weather to a silver gray. Other woods develop a rough, checked surface when weathered, while others do not. All woods tend to warp as they weather, some more than others. Decay resistant woods, such as redwood, cedar, and cypress, warp less than most. Narrow boards, such as 1 × 6 and 1 × 8 siding, warp less than wider boards.

Decay Resistance

Some species of wood have natural decay-resistant substances in their cell structures that enable the heartwood to resist attack by decay or insects, such as termites. Decay-resistant woods include cedars, redwood, cypress, black walnut, and black locust. Insect-resistant woods include cypress, some cedars, and redwood. These species may be used for all above-ground uses that are exposed to weather. To be most effective they should be 100 percent heartwood. Wood used in the ground as foundations for permanent structures should be pressure treated with a wood preservative.

Wood can be attacked by fungi (microscopic plants) that cause decay, molds, and stains. Fungi develop in wood when the moisture content is above 20 percent, temperatures are mild (40° to 100°F), there is sufficient oxygen, and the wood provides an adequate food supply. Wood in the ground is exposed to moisture, oxygen, and a rather uniform mild temperature, and therefore ideally situated for fungi attack (decay). Wood submerged in water lacks an adequate supply of oxygen.

Often parts of a building under renovation are dry but found to be heavily decayed. Although this condition is often called "dry rot," it occurs because the wood over the years has been intermittently wet (as from a leaking roof) and dry.

A harmless type of fungi called white pocket is found in living softwood trees. When the tree is cut into lumber the fungi ceases to develop. It will not spread in the cut lumber or transfer to other lumber stored on it. It is an acceptable defect and permitted in most grades of lumber, excepting select grades.

Molds and stains are also caused by fungi but do not damage the wood. The wood may be discolored, but its strength is not impaired. If the wood is used where appearance is important, staining may be undesirable. Molds can often be removed by surface brushing. Stains appear as blue, blue-black, or brown specks, as streaks or as spots.

Insect Damage

Insects bore holes in living trees and cut lumber. Some holes are very small, called pinholes, and others are larger, called grub holes. Lumber graders watch for this damage because it can influence the strength of the stock. When structural integrity is important, insect damaged lumber is usually rejected.

Completed buildings are also subject to attack by insects, particularly termites. Most parts of the country have termites, but they occur in larger numbers in milder climates. Most termites are subterranean, living in nests in the ground and building tunnels through the earth to get to material containing cellulose, which is their food. This type of termite must have a moist atmosphere to exist. When they attack wood in a building, they must build mud tunnels up the foundation to reach the floor joists. The tunnels protect them from the atmosphere and keep them moist.

Termites will also enter a building through wood that has direct contact with the earth or through very small cracks in a foundation or concrete floor slab. They digest the insides of wood members, leaving a thin outer layer to protect them from drying.

Following are things a builder can do to reduce termite attack.

1. Install a metal termite shield on top of the foundation. Termites have been known to build tunnels around these shields.
2. Remove all wood scraps, paper, and cardboard from the construction site. Any scraps buried, as when backfilling, will provide a food supply and attract a colony of termites.
3. Chemically treat the soil to form a barrier to repel termites. Many cities require this as part of the permitting process.
4. Use chemically treated, pressure impregnated wood for sills, posts, and other parts near the soil.
5. Use poured concrete foundations rather than concrete block. They are less likely to crack and have no open cells through which termites can build tunnels sight unseen inside the foundation.
6. If a concrete block foundation is used, it should be capped with a 4 in. solid concrete cap.

A few areas have a non-subterranean type of termite. They can live in damp or dry wood and do not require contact with the ground. They tend to leave fine wood particles (sawdust) in openings where they entered the wood.

PRESSURE TREATED WOOD AND PLYWOOD

Pressure treatment is a process that forces preservatives deep into the cellular structure of wood. When properly treated, the wood resists attacks by insects and fungal decay for many years. The most commonly used approved wood preservatives for general use are Alkaline Copper Quat (ACQ), Copper Azole (CA), and Bardac 22C50. Detailed current information on these and other wood preservatives can be found at *www.epa.gov. oppad001/reregistration/cca.*

Before specifying a type of pressure treated wood, the local building code and a product's manufacturer should be consulted. The Environmental Protection Agency (EPA) regularly reviews requirements and sometimes approves the use of a particular treatment for a set number of years. The American Wood Preservers Association (AWPA) is involved in setting the standards for pressure treated wood.

Metal fasteners and connections of pressure treated components should be stainless steel or hot dipped galvanized steel containing 1.85 oz. of zinc.

Safety Information

There are certain hazards when working with pressure treated wood. Workers should always wear gloves, goggles, hats, and a high-quality dust mask or respirator.

Pressure treated wood should never be allowed to come into contact with drinking water.

It must be properly disposed of and never burned.

Chromated Copper Arsenate (CCA)

CCA is no longer approved as the primary wood preservative used in residential, recreational, and general consumer construction. It has been replaced by arsenic-free alternatives such as Alkaline Copper Quest (ACQ), Copper Azole (CA), and Baddac. CCA is approved for a variety of commercial and industrial applications, such as poles, marine piling, highway guard rails, highway noise barriers, and some agricultural uses.

Alkaline Copper Quat (ACQ)

Alkaline Copper Quat was developed as an alternative to CCA. It is an EPA-approved, nonarsenic, nonchromium, water-based wood preservative that provides a durable lumber and plywood product used for structural applications where protection from decay and insects is required. It is approved for full exposure to above ground, ground contact, and freshwater applications. Typical uses include building framing, playground equipment, posts, and decks. It is suitable for use in sensitive aquatic environments. It can be painted and stained and should be sealed with a water repellant wood sealer.

Three formulations of ACQ are available. ACQ-B is used mainly to treat Western wood species such as Douglas fir. ACQ-C is used in all parts of the United States. ACQ-D is used in all parts of the country except on the West Coast (Fig. 18.29).

Bardac-Treated Wood

Bardac-treatment can be applied to framing lumber and plywood using a pressure treatment process. It is used on wood framing that is continuously protected from the weather. It is not suitable for exposure to the

Figure 18.29 Alkaline Copper Quat–treated wood is widely used for posts and decks.

ground. It protects against wood-destroying insects and decaying fungi. Framing members are sawed and assembled using fastener systems typically employed with untreated lumber.

Copper Azole (CA)

Copper Azole is a water-based wood preservative that prevents decay from fungi and insects. There are two types: CBA-A and CA-B. CBA-A is primarily used for treating softwood species, including Southern Pine and hemlock-fir. CA-B is primarily used to treat the wide range of softwood species.

Copper Azole–treated wood is approved for full exposure to above ground, ground contact, and freshwater applications. Typical applications include millwork, shingles, shakes, siding, plywood, structural lumber, and wood used on above ground applications, such as decks and playground equipment. It should be sealed with water repellant sealer and reapplied regularly to keep surfaces from cracking and splitting.

Pentachlorophenol

Pentachlorophenol was used for many years as a wood preservative. It is not EPA-registered for residential uses as a preservative.

Methyl Isothiocyanate

Methyl Isothiocyanate is registered by the EPA as a wood preservative pesticide to control wood rot and decay-causing fungi. It is used on large structural timbers, such as utility poles, pilings, and bridge timbers. It is a restricted-use product for use only by specially trained certified applicators. It is reviewed regularly by the EPA.

Applying Preservatives

Wood products are treated by pressure and non-pressure processes.

Pressure Treatment Pressure treatment involves impregnating wood with a desired preservative by placing the wood in a closed vessel and raising the pressure considerably above atmospheric pressure (Fig. 18.30). The pressure forces the preservative into the wood until the desired amount has been absorbed. Considerable preservative is absorbed with relatively deep penetration. There are several different pressure processes in use, but the basic principles are the same. Pressure processes provide a closer control over preservation retentions and penetrations and generally better protection than non-pressure methods.

Non-pressure Processes A wide range of non-pressure processes are used. These differ widely in penetration and retention; therefore, the degree of protection afforded varies. Application by brushing is the easiest but least effective. Dipping for a few minutes in a preservative gives greater assurance that all faces and checks have been coated. This is often used for assembled window sashes, frames, and other exterior millwork. A third method is cold soaking. Well-seasoned wood is soaked for anywhere from several hours to several days in low-viscosity oil preservatives. Some woods, such as pine, are successfully treated this way and have a long life when in contact with the ground. The diffusion process is used with green and dry wood. It uses a water-borne preservative that diffuses out of the treating solution or paste into the wood. The double diffusion process is more effective. It involves steeping wood first in one chemical and then another. Other diffusion processes include injecting a preservative at the ground line of a

Figure 18.30 Logs being moved into a pressure chamber for treatment by ACQ Preserve®. *(Courtesy Chemical Specialties, Inc.)*

pole, applying a paste to the surface, and pouring the chemical into holes bored in the pole at ground line.

Another non-pressure process is a thermal process. In a hot bath, the air in the wood expands and some is forced out. In the cooling bath that follows, the air contracts, which creates a partial vacuum that forces liquid into the wood. It is used for fence posts and lumber. One last non-pressure process is the vacuum process. It utilizes a quick, low initial vacuum in a chamber, followed by a brief immersion in the preservative, followed by a final high vacuum. It is used to treat millwork with water-repellent preservatives and construction lumber with water-borne and water-repellent preservatives.

Effect of Treatment on Strength

Chemicals in water-borne salt preservatives, such as chromium, copper, arsenic, and ammonia, are reactive with wood. They are potentially damaging to mechanical properties and can cause corrosion in mechanical fasteners. At retention levels required for ground contact, mechanical properties are essentially unchanged, but maximum load in bending, impact bending, and toughness are reduced somewhat. Heavy salt loadings for marine use may reduce bending strength by 10 percent and work properties by 50 percent.

Quality Certification

The American Wood Preservers Association (AWPA) is a professional society responsible for establishing consensus standards for the wood preserving industry. The American Lumber Standards Committee certifies wood preservation inspection agencies to provide quality control services to individual wood preserving plants. The accrediting agencies permit the use of their quality stamp, which indicates to the consumer that a product meets established standards.

The National Wood Window and Door Association (NWWDA) has a testing, plant inspection, and certification program for non-pressure-treated products. They provide standards for water repellent preservatives, wood window units, wood flush doors, wood sliding patio doors, wood skylight and roof windows, and wood swinging doors.

FIRE-RETARDANT TREATMENTS

In certain applications, wood must be treated with fire-retardant chemicals. The two general methods for applying fire-retardant chemicals are pressure-impregnating the wood with water-borne or organic solvent-borne chemicals or applying fire-retardant chemical coatings to the wood's surface. Pressure treating chemicals include inorganic salts and more complex chemicals. Salts are most commonly used. They react to temperatures below the ignition point of wood, causing the combustible vapors generated in the wood to break down into nonflammable water and carbon dioxide. After treatment, the wood should be dried to its original required moisture content.

Coatings have low surface flammability, and when they are exposed to fire they form an expanded low-density film. The film insulates the surface from high temperatures. Most fire-retardant chemicals do not resist exposure to weather, so it is necessary to use leach-resistant types for exterior use, such as on wood shakes. Fire-retardant treatment results in some slight reduction in the strength properties of wood, so design values for allowable stresses are reduced. Most chemicals cause fasteners to corrode. Designers must select a combination of chemicals and metal fasteners that can coexist without corrosion. Crystal salts in wood have an abrasive effect on cutting tools. Carbide-tipped cutting tools should be used when working with fire-retardant wood. Gluing is also a problem. Special resorcinol-resin adhesives have proven acceptable. Fire-retardant wood can be painted if the moisture content has been reduced sufficiently.

Review Questions

1. How can you tell a coniferous tree from a deciduous tree?
2. What natural defects are found in wood?
3. What are the two ways wood is seasoned?
4. What is the recommended moisture content for lumber used for framing?
5. Why is surfaced green softwood lumber produced more than softwood lumber that is surfaced when dry?
6. What standard is followed in the production of softwood lumber?

7. How many board feet are in a 2 × 10 × 12 in. piece of stock?
8. What is the volume of 250 pieces 38 mm × 140 mm × 3.66 m?
9. How does a contractor know lumber received on a job has been dried to a 15 percent moisture content?
10. How can you differentiate between dimension lumber and timbers?
11. What is stress-rated lumber?
12. What is meant by the equilibrium moisture content?
13. In what ways does moisture affect wood?
14. What will be the size of a 5 in. × 8 in. rough cut timber after it dries?
15. What is the recommended moisture content for wood to be painted?
16. Explain why wood floats even though the wood fibers have a specific gravity of 1.5.
17. What is a good moisture content for producing a satisfactory glue bond?
18. Which species of wood have natural resistance to decay?
19. What types of preservatives are used on wood?
20. What are the two processes used to apply fire-retardant chemicals to wood?

Key Terms

air drying
beam
boards
board foot
dimension lumber
dressed lumber
equilibrium moisture content
fiber saturation point
grade
green lumber
hardwood
heartwood
horizontal shear
hygroscopic
kiln dried
lignin
lumber grader
moisture content
preservatives
quarter sawing
lumber
sapwood
seasoning
softwood
stress-rated lumber
timber
vertical shear
warp

Activities

1. Collect a variety of samples of wood species used in construction and conduct various tests, such as:
   a. Tension and compression tests.
   b. Moisture-content tests. Place samples outdoors for several weeks and test again. What happened to the moisture content?
   c. Drive common nails into the samples and devise a way to measure the pounds of pull required to remove them. Which woods had the best holding characteristics?
2. Prepare a visual aid showing actual samples of species used in construction. Label each and list its properties and characteristics. Try to find samples having commonly found defects and identify them.
3. Collect samples of freshly cut (green) woods used in construction. Machine them to a carefully selected thickness, width, and length. Measure the moisture content. In an oven, or by some other method (air drying), let the samples dry until they reach a moisture content of 15 percent, then measure the size of the samples. Record your findings in a written report.
4. Visit a local lumber yard and examine and make a list of the sizes and species available. Pay special attention to the type of treatment used on pressure-treated rot-resistant woods.
5. Invite a local exterminator to address the class and cite the chemicals and applications used to treat a building for termites and other insects. Ask questions about hazards and safety precautions taken.

Additional Resources

Timber Construction Manual, Herzog, Natterer, Schweizer, Volz, Winter, Birkhauser Edition Detail.

Timber Construction Manual, American Institute of Timber Construction, Englewood, CO.

Wood Building Technology, Canadian Wood Council, Ottawa, Ontario, Canada.

Wood Handbook: Wood as an Engineering Material, Forest Products Laboratory, Madison, WI.

Other resources include:

Many publications available from U.S. Department of Housing and Urban Development, HUD USER, Rockville, MD.

Numerous publications from the Southern Forest Products Association, Kenner, LA.

Pressure-treated wood information: Environmental Protection Agency; ASRC Aerospace Corp.; Ariel Rios Building, 1200 Pennsylvania Ave. NW, Washington, DC 20460; *www.epa.gov/oppad001/rereregistration/cca*.

Forestry Stewardship Council, United States, *www.fscus.org/green_building*.

19 CHAPTER

Products Manufactured from Wood

LEARNING OBJECTIVES

Upon completion of this chapter, the student should be able to:

- Use technical information on industrial plywood to make construction product decisions.
- Select the appropriate reconstituted wood-panel products for various applications.
- Identify and select hardwood plywood for construction applications.
- Identify and select manufactured wood structural components used for building construction.
- Be aware of the types, sizes, and properties of many products manufactured from wood.

Build Your Knowledge

For further study on these materials and methods, please refer to:

Chapter 30 Doors, Windows, Entrances, and Storefronts

Chapter 36 Interior Walls, Partitions, and Ceilings

Topic: Wood Product Wall Finishes

Chapter 37 Flooring

A variety of products and components are made from wood. High-quality solid wood is used to make doors, windows, and a variety of millwork. Engineered wood is manufactured by bonding together wood strands, veneers, lumber, or other forms of wood fibers to produce larger composite units that are stronger and stiffer than the sum of their parts. A variety of wood products are combined under high pressure with various resins, glues, and adhesives to produce solid wood substitutes. Engineered wood products provide higher strength and stiffness in smaller cross sections than solid lumber. These products also utilize otherwise undervalued species, as well as wood chips and other forestry waste products, resulting in an environmentally superior product. This chapter presents information on a variety of engineered wood products.

PLYWOOD AND OTHER PANEL PRODUCTS

A range of veneered and manufactured panels are made from wood. These include **plywood**, **composite panels**, **waferboard**, **oriented strand board**, **particleboard**, and **hardboard.**

Plywood Panel Construction

One of the most widely used wood products is plywood, which has existed for nearly a century. Plywood panels are made by bonding together thin layers, or plies, of wood. The grain in each ply layer is arranged perpendicular to one another. Panels always have an odd number of plys, such as three, five, or seven. Each ply may be a single thickness veneer or two veneers glued together.

Plywood types commonly used in building construction are veneer core, lumber core, particleboard core, and medium-density fiberboard core (Fig. 19.1). The veneer core panel has from three to nine plies. The lumber core panel consists of strips of solid wood glued together as the core and two veneers glued to each side. Particleboard core and medium-density fiberboard core panels contain a single sheet of one of these materials as

Figure 19.1 Gluing and assembling plywood veneers into panels. *(Courtesy of APA—The Engineered Wood association)*

a core and a single ply glued to each side. Notice that each type has the same number of plies on each side of the core to ensure balanced construction.

Specifications for Plywood Panels

Some grades of veneered plywood panels are manufactured under specifications or performance testing standards of U.S. Product Standard PS 1-83, Construction and Industrial Plywood. This manufacturing specification was developed cooperatively by members of the plywood industry and the Office of Product Standards Policy of the National Bureau of Standards. Other veneered panels, including a number of performance-rated composite and non-veneered panels, are manufactured under provisions of APA—The Engineered Wood Association performance standards. APA-rated panels that meet PS 1-83 requirements have the designation "PS 1-83" in the APA trademark.

In Canada, there are three standards for softwood plywood, all of which are in metric terms: CSA 0121-M Douglas Fir Plywood, CSA 0151-M Canadian Softwood Plywood, and CSA 0153-M Poplar Plywood. Allowable stresses and section properties for Douglas fir plywood are laid out in CAN3-086-M.

Construction and Industrial Plywood

Construction and industrial plywood are manufactured according to the specifications in U.S. Product Standard PS 1-83. Canadian plywood is made following the standards just mentioned.

Grades The outer veneers of construction and industrial plywood are classified in five appearance groups: N, A, B, C, and D. N is the best grade and D the poorest (Fig. 19.2).

Species Construction and industrial plywood is made using about seventy species of wood. These may be mixed within a panel and can include hardwoods and softwoods. Species used are divided into five groups, depending on their strength and stiffness. Group 1 includes the strongest and stiffest species. Group 5 includes those with the lowest properties (Table 19.1). The inner plies in Groups 1, 2, 3, and 4 may be any species in Groups 1, 2, 3, and 4. Inner plies in Group 5 may be any of the species listed. Front and back plies are of the same species and must derive from that group number.

Sizes of Panels In customary units, standard nominal thicknesses of sanded construction and industrial plywood panels are $\frac{1}{8}$ through 1-$\frac{1}{4}$ in., in $\frac{1}{8}$ in. increments. Unsanded panels are available in thicknesses of $\frac{5}{6}$ to 1-$\frac{1}{4}$ in. Panel widths are 36, 48, and 60 in., and lengths are available from 60 to 144 in., in 12 in. increments (Table 19.2). Soft converted metric thicknesses are also shown in Table 19.2. Panels sized 1200 × 2400 mm are designed for use in hard metric designed buildings while panels 1219 × 2438 mm are soft conversions and can be used in buildings designed using customary units.

APA Performance-Rated Panels

APA Performance-Rated Panels are manufactured to performance standards of APA—The Engineered Wood Association, which establishes performance criteria for

Figure 19.2 Plywood veneer grades. *(Courtesy APA—The Engineered Wood Association)*

A Smooth, paintable. Not more than 18 neatly made repairs, boat, sled or router type, and parallel to grain permitted. May be used for natural finish in less demanding applications. Synthetic repairs permitted.

B Solid surface. Shims, circular repair plugs, and tight knots up to 1 in. across grain permitted. Some minor splits permitted. Synthetic repairs permitted.

C Plugged Improved C veneer with splits limited to $\frac{1}{8}$ in. width and knotholes and borer holes limited to $\frac{1}{4} \times \frac{1}{2}$ in. Admits some broken grain. Synthetic repairs permitted.

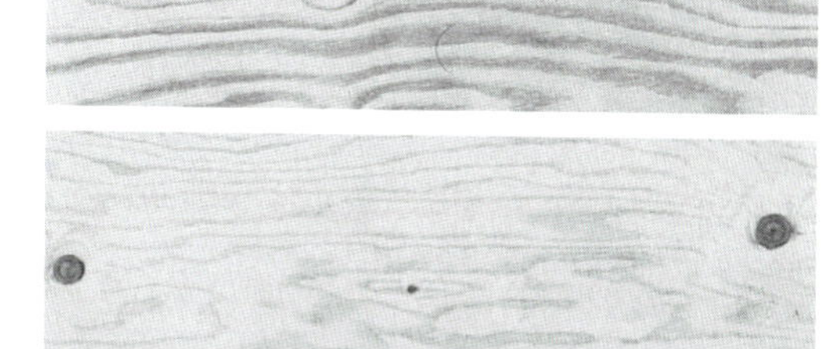

C Tight knots up to $1\frac{1}{2}$ in. wide. Knotholes to 1 in. across grain and some to $1\frac{1}{2}$ in. if total width of knots and knotholes is within specified limits. Synthetic or wood repairs. Discoloration and sanding defects that do not impair strength permitted. Limited splits allowed. Stitching permitted.

D Knots and knotholes up to $2\frac{1}{2}$ in. wide across grain and $\frac{1}{2}$ in. larger within specified limits. Limited splits allowed. Stitching permitted. Limited to Interior, Exposure 1, and Exposure 2 panels

Table 19.1 Plywood Outer Veneer Group Numbers

Group 1	Group 2		Group 3	Group 4	Group 5
Apitong	Cedar, Port Orford	Maple, black	Alder, red	Aspen	Basswood
Beech, American	Cypress	Mengkulang	Birch, paper	Bigtooth	Poplar, balsam
Birch	Douglas fir 2	Meranti, red	Cedar, Alaska	Quaking	
Sweet	Fir	Mersawa	Fir, subalpine	Cativo	
Yellow	California red	Pine	Hemlock, eastern	Cedar	
Douglas fir 1	Grand	Pond	Maple, bigleaf	Incense	
Kapur	Noble	Red	Pine	Western red	
Keruing	Pacific silver	Virginia	Jack	Cottonwood	
Larch, western	White	Western white	Lodgepole	Eastern	
Maple, sugar	Hemlock, western	Spruce	Ponderosa	Black (western poplar)	
Pine	Lauan	Red	Spruce	Pine	
Caribbean	Almon	Sitka	Redwood	Eastern white	
Ocote	Bagtikan	Sweetgum	Spruce	Sugar	
Pine, southern	Mayapis	Tamarack	Engelmann		
Loblolly	Red	Yellow poplar	White		
Longleaf	Tangile				
Shortleaf	White				
Slash					
Tanoak					

Courtesy APA—The Engineered Wood Association

Table 19.2 Plywood Panel Thicknesses and Dimensions

Nominal Thickness	
In.	mm[a]
1/4	6.4
5/16	7.9
11/32	8.7
3/8	9.5
7/16	11.1
15/32	11.9
1/2	12.7
19/32	15.1
5/8	15.9
23/32	18.3
3/4	19.1
7/8	22.2
1	25.4
1 3/32	27.8
1 1/8	28.6

Nominal Dimensions (width × length)		
ft.	mm	m[a]
4 × 8	1,219 × 2,438	1.22 × 2.44
4 × 9	1,219 × 2,743	1.22 × 2.74
4 × 10	1,219 × 3,048	1.22 × 3.05

[a]*Soft converted metric sizes*
Courtesy APA—The Engineered Wood Association

designated construction applications. The four panels specified by APA are plywood, oriented strand board, composite panels (Com-Ply®), and APA Rated Siding (Fig. 19.3).

Plywood panels are made using all veneer plies, producing the strongest type of panel. **Oriented strand board** (OSB) is manufactured from strands or wafers arranged in one general direction. These layers of oriented strands are bonded together at right angles to each other. Usually three to five layers are bonded to form a panel. Most OSB panels are textured on one side to produce a non-slick surface.

Composite or Com-Ply® panels have a core of reconstituted wood bonded between solid wood veneers. This produces a panel that allows for efficient use of wood materials and has a wood grain surface on the front and back.

APA Rated Siding is made by bonding wood veneers in the same manner as plywood. It is available in a variety of surface textures and designs. Panels are available in 4 × 8, 4 × 9, and 4 × 10 ft. dimensions. A guide to APA Performance-Rated Panels is given in Fig. 19.4.

APA Rated Sheathing is used for subfloors, walls, and roof sheathing where strength and stiffness are required. Nominal panel thicknesses range from $\frac{5}{16}$ to $\frac{3}{4}$ in. Structural I is a type of rated sheathing used where cross-panel strength and increased resistance to wracking are needed. It is used on structural diaphragms and panelized roofs. APA Rated Sturd-I-Floor® is a single-layer flooring for use under carpets. It may eliminate the need for installing additional underlayment. Panels are available with square and tongue-and-groove edges, and in thicknesses from $\frac{9}{32}$ to 1-$\frac{1}{8}$ in.

Exposure Durability Classifications APA Performance-Rated Panels are manufactured in four exposure durability classifications: Exterior, Exposure 1, Exposure 2, and Interior. These classifications signify how well the bonding agent used to join the veneers can resist exposure to moisture.

Exterior panels use a waterproof adhesive and are designed for use on applications subject to permanent exposure to moisture and weather. Exposure 1 panels use the same waterproof adhesive as Exterior panels and are designed to withstand exposure to moisture and other weather for long periods before being permanently protected. They are also used where panels may be subjected to occasional moisture after they are in service. They differ from Exterior panels in some compositional aspects and are not recommended for permanent exposure to the weather.

Exposure 2 panels are used in applications in which they will be briefly exposed to the elements during construction before being permanently protected. They are actually an interior panel with an intermediate adhesive.

Interior panels are manufactured with interior glue intended for indoor use only. They are identified by the abbreviation "INT-APA" and the omission of glueline information on their trademark. Veneer grades used on APA Performance-Rated Panels are A, B, C, C plugged, and D (Fig. 19.2).

Product Identification APA trademarks on the types of plywood used in construction are shown in Fig. 19.5. The sanded panels with B-grade or better veneer are used for various construction applications. The specialty grades include panels designed for a specific use, such as Plyform® for concrete forms and underlayment. Most of the information on the trademark is self explanatory save for span rating. A span rating of $\frac{32}{16}$ indicates the panel can be used as roof sheathing with rafters spaced up to 32 in. on-center and as subflooring with floor joists spaced up to 16 in. on-center.

Figure 19.3 APA Performance-Rated Panels. *(Courtesy APA—The Engineered Wood Association)*

Plywood
All-veneer panels consisting of an odd number of cross-laminated layers, each layer consisting of one or more piles. Many such panels meet all of the prescriptive or performance provisions of U.S. Product Standard PS 1-83/ANSI A199.1 for Construction and Industrial Plywood.

APA Rated Siding
All-veneer panels constructed in the same manner as plywood panels. Available in a variety of panel sizes and thicknesses. Various surface textures and designs available.

Composite (Comply)®
Panels of reconstituted wood cores bonded between veneer face and back plies.

Oriented Strand Board
Panels of compressed strand-like particles arranged in layers (usually three to five) oriented at right angles to one another.

Specialty Plywood

Many plywood manufacturers make various specialty plywood products that are not included in the national standards for plywood manufacture. The user must refer to information supplied by a manufacturer when selecting these products because they meet no standard. Several of the most frequently used are **overlaid plywood**, siding panels, and interior paneling. The APA does have standards and trademarks for several specialty panels.

Overlaid plywood is a high-grade exterior-type panel that has a resin impregnated fiber ply bonded to one or both sides. It is made in two types: high density and medium density. High density panels have a hard, smooth chemically resistant surface. They are made in several colors and require no additional finish. Medium density panels have a smooth, opaque, non-glossy surface that hides the grain of the veneers below it. They are used when a high-quality paint finish is needed.

Siding panels are used as finished exterior siding and are available in a variety of surface finishes, such as rough sawed, V-grooved, and reverse batten grooves.

Paneling is used on interior walls as a finish material. It is usually prefinished and only requires installation. It is available in a wide variety of wood species and surface fixtures.

Hardwood Plywood

Hardwood plywood is made using various species of hardwood veneers for the face and back surfaces of the panel. They are bonded to plywood or particleboard panels that form the core. It is manufactured following standards established by the American Society for Testing and Materials and the American National Standards Institute. The Hardwood Plywood and Veneer Association provides certification services. Canadian hardwood plywoods are made according to metric standard CSA 0115-M, Hardwood and Decorative Plywood.

Species Species used in hardwood plywood are divided into four categories. These categories reflect the modulus of elasticity (stiffness) of each species and its specific gravity. Category A is the stiffest and D has the lowest rating.

Figure 19.4 APA—The Engineered Wood Association trademarks and uses of APA Performance-Rated panels. *(Courtesy APA—The Engineered Wood Association)*

GUIDE TO APA PERFORMANCE RATED PANELS[a][b]
FOR APPLICATION RECOMMENDATIONS, SEE FOLLOWING PAGES.

APA RATED SHEATHING Typical Trademark	Specially designed for subflooring and wall and roof sheathing. Also good for a broad range of other construction and industrial applications. Can be manufactured as OSB, plywood, or other wood-based panel. BOND CLASSIFICATIONS: Exterior, Exposure 1. COMMON THICKNESSES: 5/16, 3/8, 7/16, 15/32, 1/2, 19/32, 5/8, 23/32, 3/4.
APA STRUCTURAL I RATED SHEATHING[c] Typical Trademark	Unsanded grade for use where shear and cross-panel strength properties are of maximum importance, such as panelized roofs and diaphragms. Can be manufactured as OSB, plywood, or other wood-based panel. BOND CLASSIFICATIONS: Exterior, Exposure 1. COMMON THICKNESSES: 5/16, 3/8, 7/16, 15/32, 1/2, 19/32, 5/8, 23/32, 3/4.
APA RATED STURD-I-FLOOR Typical Trademark	Specially designed as combination subfloor-underlayment. Provides smooth surface for application of carpet and pad and possesses high concentrated and impact load resistance. Can be manufactured as OSB, plywood, or other wood-based panel. Available square edge or tongue-and-groove. BOND CLASSIFICATIONS: Exterior, Exposure 1. COMMON THICKNESSES: 19/32, 5/8, 23/32, 3/4, 1, 1-1/8.
APA RATED SIDING Typical Trademark	For exterior siding, fencing, etc. Can be manufactured as plywood, as other wood-based panel or as an overlaid OSB. Both panel and lap siding available. Special surface treatment such as V-groove, channel groove, deep groove (such as APA Texture 1-11), brushed, rough sawn and overlaid (MDO) with smooth- or texture-embossed face. Span Rating (stud spacing for siding qualified for APA Sturd-I-Wall applications) and face grade classification (for veneer-faced siding) indicated in trademark. BOND CLASSIFICATION: Exterior. COMMON THICKNESSES: 11/32, 3/8, 7/16, 15/32, 1/2, 19/32, 5/8.

(a) Specific grades, thicknesses and bond classifications may be in limited supply in some areas. Check with your supplier before specifying.

(b) Specify Performance Rated Panels by thickness and Span Rating. Span Ratings are based on panel strength and stiffness. Since these properties are a function of panel composition and configuration as well as thickness, the same Span Rating may appear on panels of different thickness. Conversely, panels of the same thickness may be marked with different Span Ratings.

(c) All plies in Structural I plywood panels are special improved grades and panels marked PS 1 are limited to Group 1 species. Other panels marked Structural I Rated qualify through special performance testing.

Grades of Veneers Hardwood plywood offers six grades of hardwood veneers. There are also specific requirements for softwoods used in hardwood plywood panels.

A grade (A) is the best. The face is made of hardwood veneers carefully matched by color and grain.

B grade (B) is suitable for a natural finish, but the face veneers are not as carefully matched as on the A grade.

Sound grade (2) provides a smooth face. All defects have been repaired. It is used as a base for a painted finish.

Industrial-grade (3) face veneers can have surface defects. This grade permits knotholes up to 1 in. (25 mm) in diameter, small open joints, and small areas of rough grain.

Backing grade (4) uses unselected veneers having knotholes up to 3 in. in diameter and certain types of splits. No defect is permitted that affects the strength of the panel.

Figure 19.5 A trademark of the APA—The Engineered Wood Association. *(Courtesy APA - The Engineered Wood Association)*

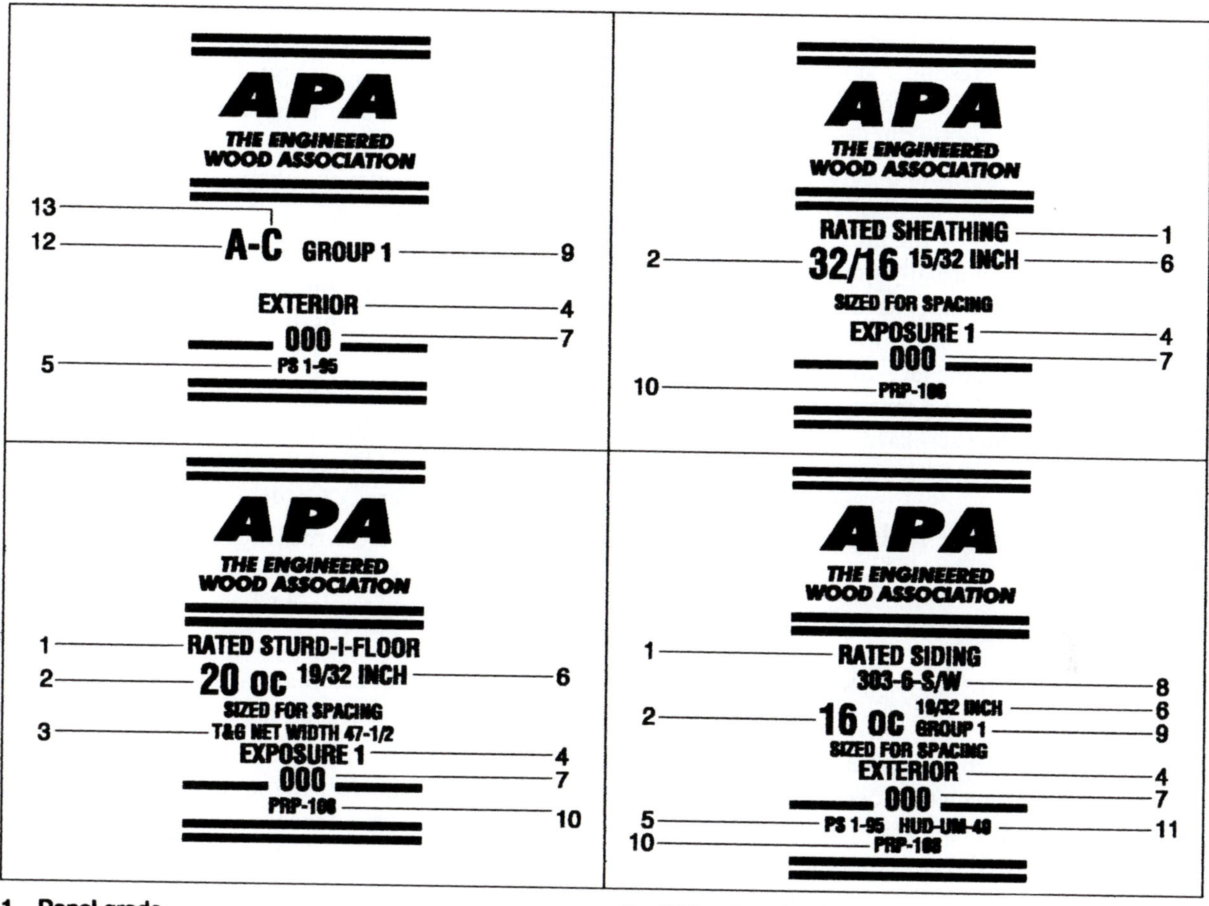

1. Panel grade
2. Span rating
3. Tongue and groove
4. Exposure durability classification
5. Product standard governing manufacture
6. Thickness
7. Mill Number
8. Siding face grade
9. Species group number
10. APA's Performance-Rated Panel Standard
11. FHA Use of Materials Bulletin number
12. Grade of face veneer
13. Grade of back veneer

Specialty grade (SP) includes veneers having characteristics unlike any of those in the other grades. The characteristics are agreed upon between manufacturer and purchaser. For example, species such as wormy chestnut or bird's-eye maple are considered specialty grade.

Construction of Hardwood Plywood The construction of hardwood plywood is described by identifying the core. The commonly available constructions are:

Hardwood and Softwood veneer core: Have an odd number of plies, such as 3-ply, 5-ply, and so on.

Hardwood and Softwood lumber core: Used in 3-ply, 5-ply, and 7-ply constructions.

Particleboard core: Used in 3-ply and 5-ply constructions.

Medium-density fiberboard core: Used in 3-ply construction.

Hardboard core: Used in 3-ply construction.

Special cores: Used in 3-ply construction. Special cores are made of materials not listed above.

Sizes and Thicknesses of Panels Hardwood plywood is available in panels 48 in. wide and 96 and 120 in. long. Standard thicknesses are $\frac{1}{4}$, $\frac{3}{8}$, $\frac{1}{2}$, and $\frac{3}{4}$ in. Lumber core panels and particleboard core panels are available in only the $\frac{3}{4}$ in. thickness.

Canadian hardwood plywood is available in incremental thicknesses from 4.8 to 32 mm.

Product Identification The backstamp certified by the Hardwood Plywood and Veneer Association (HPVA) is stamped on the back of decorative panels (Fig. 19.6). The flame spread rating, smoke generated rating, formaldehyde emissions, and bond types are established by the American Society for Testing and Materials (ASTM) and the American National Standards Institute (ANSI). Since these products are generally used on interiors, it is important they meet the minimum standards. Stock panels usually do not carry the HPVA certification backstamp but are still graded by the mill. The information on these panel stamps is normally simple, in text format without any HPVA logo, and is usually stamped on the edge of the panel and contains the thickness, face-veneer grade, back-veneer grade, the face-cutting method (such as plain cut or rotary cut), the species of wood, and the designation of the product standard.

Reconstituted Wood Products

In addition to the APA Performance-Rated Panels discussed, a number of reconstituted wood panels are used, primarily in cabinet and furniture construction. These include hardboard, particleboard, fiberboard, and waferboard.

Hardboard is made from wood chips converted into fibers and bonded into panels under heat and pressure and manufactured following standards developed by the American Hardboard Association. It is available in thicknesses from $\frac{1}{12}$ up to 1-$\frac{1}{8}$ in. The most frequent panel size is 4 × 8 ft., but other sizes are available through special order.

Metric thicknesses for standard hardboard are 2.1, 2.5, 3.2, 4.8, 6.4, 7.9, and 9.5 mm. Panels are 1200 × 2400 mm.

Hardboard is available in five classes, which are defined as follows:

Class 1: Tempered. Tempered class is impregnated with siccative material and stabilized by heat and special additives to impart substantially improved properties of stiffness, strength, hardness, and resistance to water and abrasion, as compared with the Standard Class.

Class 2: Standard. Standard class remains in the form in which it comes from the press—it receives no further treatment. It has high strength and water resistance.

Class 3: Service-tempered. Service-tempered class is impregnated with siccative material and stabilized by heat and additives. Its properties are substantially better than Service Grade.

Class 4: Service Grade. Service Grade remains in the form in which it comes from a press but has less strength than Standard Class.

Class 5: Industrialite. Industrialite class is a medium-density hardboard that has moderate strength and lower unit weight than the other classes.

Hardboard is manufactured in the United States according to the following standards: ANSI/AHA A135.4-1988, Basic Hardboard; ANSI/AHA A135.5-1988, PF Hardboard Paneling; and ANSI/AHA A135.6-1990, Hardboard Siding. These are promoted by the American Hardboard Association. Panels manufactured to these standards have an AHA grade stamp.

Canadian hardboard products are manufactured following CGSB 11-GP-3M, Hardboard, and CGSB 11-GP-5M, Hardboard for Exterior Cladding.

Figure 19.6 The HPVA backstamp used to certify the properties of prefinished plywood panels. *(Courtesy Hardwood Plywood and Veneer Association)*

HARDWOOD PLYWOOD & VENEER ASSOCIATION		
FORMALDEHYDE EMISSION 0.20 PPM CONFORMS TO HUD REQUIREMENTS ASTM E 1333	SIMULATED DECORATIVE FINISH ON PLYWOOD HARDWOOD PLYWOOD AND VENEER ASSOCIATION hpva ®	CLASS C FLAME SPREAD 200 OR LESS SMOKE DEVELOPED 450 OR LESS ASTM E84
LAY UP 16 3.6 MM THICK HP-SG-96	MILL 000 SPECIALTY GRADE	BOND LINE TYPE II ANSI/HPVA HP-1-2004

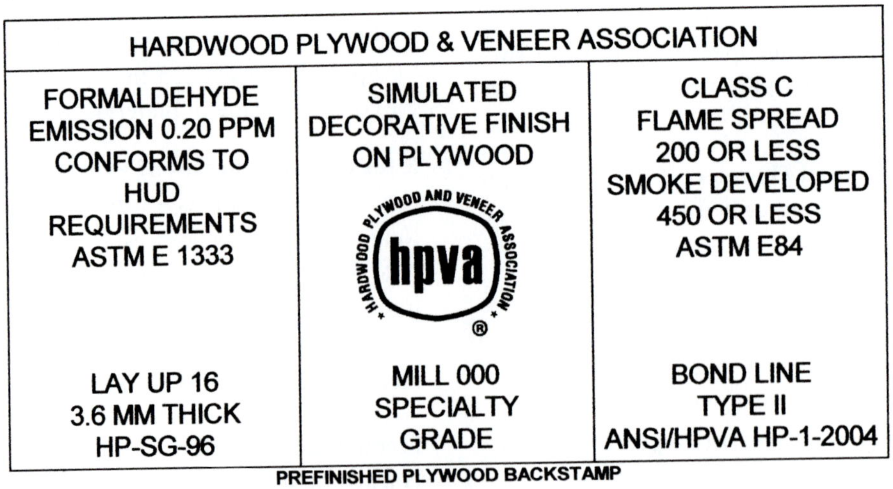

PREFINISHED PLYWOOD BACKSTAMP

Figure 19.7 Other panels composed of reconstituted wood that are designed to carry known loads over known distances.

All-veneer panels consisting of an odd number of cross-laminated layers, each layer consisting of one or more piles. Many such panels meet all of the prescriptive or performance provisions of U.S. Product Standard PS 1-83/ANSI A199.1 for Construction and Industrial Plywood.

A APA Rated Siding
All-veneer panels constructed in the same manner as plywood panels. Available in a variety of panel sizes and thicknesses. Various surface textures and designs available.

B Composite (Comply)®
Panels of reconstituted wood cores bonded between veneer face and back plies.

C Oriented Strand Board
Panels of compressed strand-like particles arranged in layers (usually three to five) oriented at right angles to one another.

Particleboard is made from wood chips, water, and a synthetic resin binder bonded with heat and pressure. It is manufactured to standards developed by the National Particleboard Association. It is mainly used for furniture and cabinet construction (Fig. 19.7).

The general uses and grades are shown in Table 19.3. The grades are identified by a letter designation followed by a hyphen and a digit or letter. The second digit or letter designation indicates the grade identification within a particular density or product description. For example, "M-2" indicates a medium-density particleboard, Grade 2.

If there is a third designation it denotes a special characteristic, such as panel M-3 Exterior Glue. This is a medium-density panel, Grade 3, with exterior glue.

Grades 1, 2, and 3 relate to the relative magnitude of the mechanical properties. For example, Grade 3 has the highest modulus of elasticity, modulus of rupture, and hardness. Grade 1 has the lowest of these properties. Canadian particleboard is made following ANSI 208.1, Particleboard Standards.

Waferboard is made by bonding large wood flakes $1\text{-}\frac{1}{2}$ in. or longer into panels having the same thicknesses and sizes as particleboard.

In the United States, fiberboard consists of two products, softboard and hardboard. In Canada, these are listed separately, and softboard panels are identified as fiberboard. Softboard panels are manufactured from loosely bound paper pulp and other

Table 19.3 Grades and Uses of Particleboard

Type	Grade	Use
High density	H-1, H-2, H-3 H-1, H-2, H-3 Exterior Glue	High-density industrial High-density exterior industrial
Medium density	M-1 M-2, M-3 M-1, M-2, M-3 Exterior Glue	Commercial Industrial Exterior construction Exterior industrial
Medium density—specialty grade	M-S	Commercial
Low density	LD-1, LD-2	Door core
Underlayment	PBU	Underlayment
Manufactured home decking	D-2, D-3	Flooring in manufactured homes

Courtesy National Particleboard Association

types of fibers into panels with insulating properties. They are used as rigid insulation on walls and roofs.

In Canada, softboard panels have no standard thickness. This is established by a panel's manufacturer. They are cut into 1200 × 2400 mm and 4 × 8 ft. panels. Canadian fiberboard is manufactured according to CSA Standard A247-M, Insulating Fiberboard. In the United States, $\frac{3}{4}$, 1, 1-$\frac{1}{2}$, and 2 in. thicknesses are available, and sheet sizes are 24 × 48 in., 48 × 48 in., and 4 × 8 ft.

STRUCTURAL BUILDING COMPONENTS

Trusses are used for floor and roof construction. They are carefully engineered to carry known loads over specified distances. Most trusses are made from 2 × 4 in. and 2 × 6 in. lumber joined with wood or metal gusset plates (Fig. 19.8).

Glued laminated wood members, or **glulams**, are formed of solid sawn lumber glued end to end and then face bonded in laminations. Three-quarter-inch thick wood strips are used for curved members and 1-$\frac{1}{2}$ in. thick wood strips for straight members. Laminations in Canadian products are 19 and 38 mm. Members of any size can be made, but manufacturers produce a wide range of standard sizes. Widths range from 3-$\frac{1}{8}$ to 10-$\frac{3}{4}$ in. and depths range from 6 in. to 81 in. Standard Canadian widths range from 80 to 365 mm and depths from 114 to 2128 mm.

Figure 19.8 Wood trusses are made from 2 × stock joined with metal gusset plates. *(Image copyright Melissa Dockstader, 2009. Used under license from Shutterstock.com)*

Wood can be laminated into various shapes and curves, resulting in members that are stronger than solid wood the same size. Glued laminated wood members are available in both structural and architectural grades. They are manufactured in the United States according to ANSI/AITC A190.1-1983, Structural Glued Laminated Timber. In Canada, the standards include CSA 0177-M, Qualification Code for Manufacturers of Structural Glued-Laminated Timber; CSA 0122, Structural Glued-Laminated Timber; and CSA3-086-M, Code for Engineering Design in Wood.

A typical glulam member is shown in Fig. 19.9. The transverse sections can be variable, achieving aesthetically pleasing and structurally appropriate elements. Its bigger sections make it more fire resistant than steel; the elements burn slowly, producing a protective char—where steel collapses, glulam beams will generally withstand fires.

Laminated veneer lumber (LVL) is produced by bonding thin wood veneers together into a large billet with the grain of all veneers parallel to the long direction. The billet is then sawn to desired dimensions, depending on the construction application. LVL members display almost no shrinking, checking, twisting, or splitting.

Figure 19.9 Glued laminated beams can span long distances and carry heavy loads. *(Image copyright robcocquyt, 2009. Used under license from Shutterstock.com)*

Figure 19.10 Laminated veneer lumber is made by bonding together layers of wood veneer with an exterior adhesive.

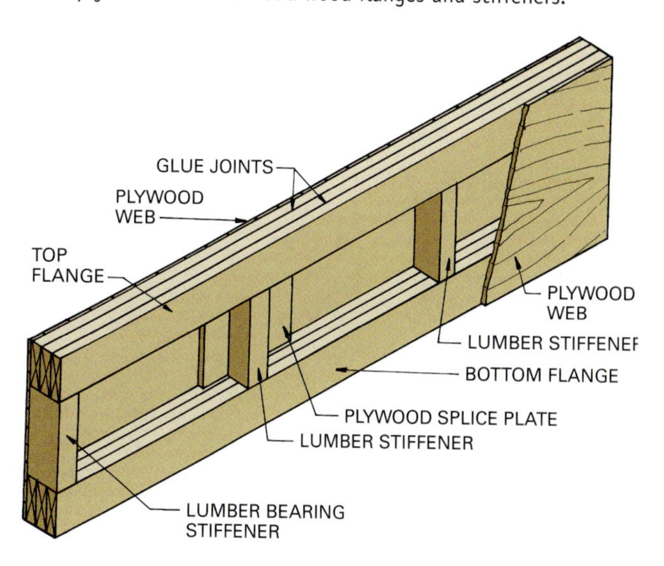

Figure 19.11 A typical plywood-lumber beam (box beam) with plywood webs and solid wood flanges and stiffeners.

They have more load-bearing capacity per pound than sawed lumber with a modulus of elasticity of about 2.0 and an allowable bending stress of 3100 psi. LVL is available in depths from 9-$\frac{1}{4}$ to 18 in. and thicknesses of 1-$\frac{1}{2}$ and 1-$\frac{3}{4}$ in. (Fig. 19.10).

A comparison of the physical properties of **glued laminated lumber**, laminated veneer lumber, and solid lumber is shown in Table 19.4. Although actual data vary for species, grades, and manufacturing requirements, this comparison shows that glued manufactured wood structural members exhibit better properties than those of most grades and species of solid woods.

Plywood-lumber beams (box beams) are made according to carefully engineered specifications as to the size of members and nailing requirements. They are glued and built in factories under controlled conditions. One example is shown in Fig. 19.11.

Stressed skin panels are prefabricated panels using solid lumber stringers and headers and plywood skins. They can be produced in a variety of shapes for use in floors, walls, and roofs. They facilitate a building's rapid erection by covering large areas and spanning long distances (Fig. 19.12).

A type of load-bearing structural joist, the **I joist**, is made of softwood veneer bonded top and bottom flanges connected with a composite panel core (Fig. 19.13). Their "I" configuration provides high bending strength and stiffness characteristics. I joists are used extensively in residential construction and their light weight precludes the need for costly handling equipment.

Parallel strand lumber (PSL) is a relatively new product that utilizes almost all the wood from a log. It is manufactured under the registered name Parallam® and made from Douglas fir or southern pine. The logs are peeled to produce a veneer that is dried and screened to remove strength-reducing defects. Then the sheets of veneer are clipped into strands up to 8 ft. in length and $\frac{1}{10}$ and $\frac{3}{8}$ in. thick. Small defects are removed, and the strands are coated with a waterproof adhesive. The

Table 19.4 Properties of Laminated Veneer, Glued Laminated Members, and Solid Lumber

Property	LVL[a]	Glued Laminated[b]	Solid Lumber[c]
Extreme fiber in bending (F_b) psi	3,000	1,600–2,400	760–2,710
Horizontal shear (H_v) psi	290	140–650	95–120
Compression perpendicular to grain ($F_{c\perp}$) psi	3,180	375–650	565–660
Compression parallel with grain (F_c) psi	—	1,350–2,300	625–1,100
Tension parallel with grain (F_t) psi	2,300	1,050–1,450	450–1,200
Modulus of elasticity (E) million psi	2.0	1.2–1.8	1.3–1.9

[a]*Limited to one size member*
[b]*Data from one manufacturer*
[c]*Data for one species of lumber*

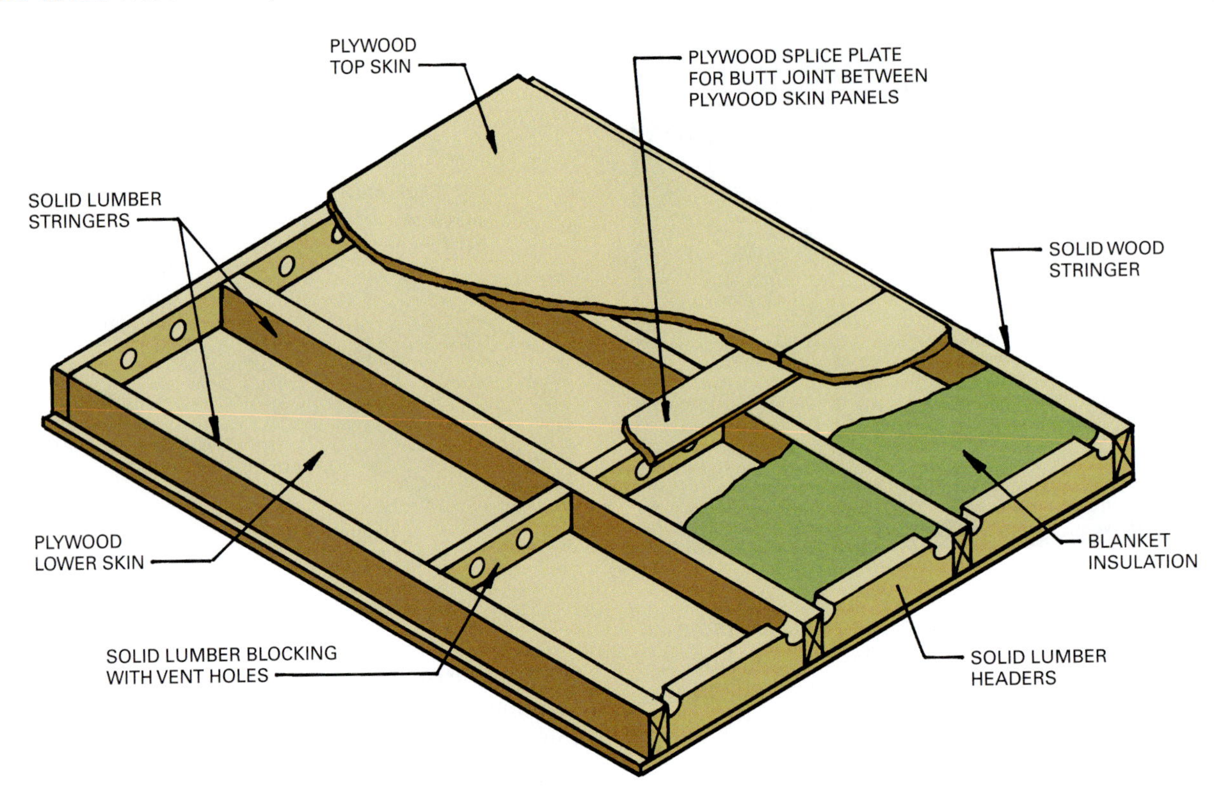

Figure 19.12 A stressed skin panel.

long oriented strands are fed into a rotary belt press and cured under pressure using microwave energy. This produces a PSL billet that can be cut to standard sizes in lengths up to 66 ft. (Fig. 19.14).

Parallam products have been tested for fire resistance, fastener holding ability, moisture response, long-term loading, flexural strength, stiffness, and internal bond and have proved to exceed the performance of competing wood products.

Parallam structural products include beams, headers, posts, and columns. Beam and column sizes are given in Table 19.5. They are cut, drilled, and installed in the same manner as solid wood and laminated wood structural members.

Structural Insulated Panels (SIPS) are a composite building material consisting of an insulating layer of foam sandwiched between two layers of structural board. The board is usually oriented strand board and

Figure 19.13 This I joist is made with flanges made of solid lumber and a composite wood web. *(Courtesy Louisiana Pacific Corporation)*

Figure 19.14 Parallam® parallel strand lumber is made by bonding thin wood strands into structurally sound members. *(Courtesy of iLevel by Weyerhauser)*

Table 19.5 Stock Sizes of Parallam® Parallel Strand Lumber

Parallam Beams and Headers	
Thickness (in.)	**Depth (in.)**
$1\frac{3}{4}$, $3\frac{1}{2}$, $5\frac{1}{4}$, 7	$7\frac{1}{4}$, $9\frac{1}{4}$, $9\frac{1}{2}$, $11\frac{1}{4}$ $11\frac{1}{2}$, $11\frac{7}{8}$, 12, $12\frac{1}{2}$ 14, 16, 18
Parallam Columns	
Dimensions (in.)	
$3\frac{1}{2} \times 3\frac{1}{2}$, $3\frac{1}{2} \times 5\frac{1}{4}$, $3\frac{1}{2} \times 7$, $5\frac{1}{4} \times 5\frac{1}{4}$, $5\frac{1}{4} \times 7$, 7×7	

Courtesy MacMillan Bloedel Limited

the foam either expanded polystyrene foam, extruded polystyrene foam, or polyurethane foam. SIPS panels perform structurally similar to beams, with the rigid insulation core acting as the web and the OSB sheathing as the flanges. SIPS replace several components of conventional building, such as studs and joists, insulation, vapor barrier, and air barrier. The panels are used for many different applications, such as exterior wall, roof, floor, and foundation construction.

In the United States, SIPS panels are available in sizes from 4 ft. (1.22 m) to 24 ft. (7.32 m) in width. Elsewhere, typical product dimensions are 300, 600, or 1200 mm wide and 2.4, 2.7, and 3 m long, with roof panels up to 6 m long. Structures using SIPS panels have tighter building envelopes and higher insulation value, resulting in fewer drafts and lower operating costs for maintaining comfortable interior environments. The standardization of SIPS also leads to reduced construction time and requires fewer trades for system integration. Several examples of SIPS construction are shown in Figs. 19.15 and 19.16.

Figure 19.15 A structural insulated panel being readied for installation.

Figure 19.16 Structural insulated panels can be used for floors, walls, and roofs.

American Wood Systems

American Wood Systems (AWS) was created by APA—The Engineered Wood Association to serve the needs of the engineered wood systems industry. The AWS created a new trademark, APA-EWS (Engineered Wood Systems). This trademark appears on glued laminated beams and other engineered products, such as wood I beams, structural composite lumber, and other engineered wood products. These products receive the same technical and promotional services APA provides to the manufacturers of structural wood panels. A typical APA-EWS trademark is shown in Fig. 19.17.

WOOD SHINGLES AND SHAKES

Wood shingles and shakes are popular as a finished roof covering and are also used as exterior siding on wood-framed buildings (Fig. 19.18). Red cedar shingles are smooth sawed on both faces and uniform in thickness at the butt end. They are tapered from thick butt end to thin top edge. They are available in grades No. 1 (blue label), No. 2 (red label), No. 3 (black label), and in an under coursing grade (green label). They come in lengths of 16, 18, and 24 in. Widths range from a 3 in. minimum to a 14 in. maximum.

Red cedar shakes are available in two types: Certi-Sawn® and Certi-Split®. Certi-Sawn shakes are sawed on both faces, as are shingles, but they are not as precisely manufactured. They are thicker on the butt end and vary more in thickness than shingles, which produces a rougher, more textured appearance. They are available in 18 and 24 in. lengths and widths from 4 to 14 in. Certi-Split shakes have a rough, split face and a

Case Study

Aldo Leopold Legacy Center

Architect: The Kubala Washatko Architects	Building type: Visitor and Interpretive Center	Completed April 2007
Contractor: The Bolt Company	Size: 11,900 sq. ft. (1,100 sq. m)	Rating: U.S. Green Building Council
Location: Baraboo, Wisconsin	Project scope: 3 1-story buildings	Level: Platinum (61 points)

Overview

Aldo Leopold, one of the founders of the modern conservation movement, advocated a holistic stewardship of the land, including careful management of natural resources and the protection of biodiversity. The headquarters of the Aldo Leopold Legacy Center consists of a grouping of buildings facing an exterior courtyard. In addition to meeting spaces, the center includes offices, the Leopold archive, and an interpretive hall. The building is located on the site of the Leopold Reserve, a property where the Leopold family planted thousands of trees as part of their restoration efforts during the 1930s and 1940s. The goal for the project was to provide a model of how sensitive design and construction techniques can integrate the built environment with natural systems. Part of the design concept was to restore and enhance the previously damaged site.

Design

The design process incorporated an integrated team, with architect, contractor, and environmental and other consultants working closely together from the onset of the project. The team sought to integrate simple, low-tech strategies with new, cutting edge technologies to create ecologically sensitive buildings that fit well into their rural setting.

Working with the U.S. Fish and Wildlife Service, the project was designed to manage all storm water on site. Native landscaping and pervious paving surfaces for parking and drives allows all water to naturally filter back into the ground.

Energy

The building utilizes high levels of thermal insulation to reduce heat loss across the exterior envelope. A heavy timber frame is clad in structural insulated panels, resulting in insulation values twice what is required by local building codes. Photovoltaic arrays on the rooftop produce more than the required energy for building operations, allowing surpluses to be sold back to the local utility. A ground source heat pump provides radiant heating and cooling within the building floor slab. The resulting buildings use 70 percent less energy than a conventional project of the same size.

Materials

The design team chose to harvest trees planted by the Leopold family directly from the building site in the construction of the building. The existing 1,500-acre forest reserve had been poorly managed, resulting in overcrowding and a declining growth rate. Almost 100,000 board feet of lumber was harvested from the site and used for structural beams and trusses, siding, flooring, doors, windows, and other interior finishes. The lumber was debarked, air dried, and milled on site, reducing both costs and transportation energy in securing materials. The surrounding forest health was improved through the careful thinning and harvesting of the timbers. The lumber used for the project was certified through the Forest Stewardship council. Other materials used were chosen based on their high recycled content and low emission potential, including composite wood products, native stone, recycled content cellulose insulation, and zero VOC interior paints.

The project incorporates other sustainable strategies, including natural daylighting to reduce electric consumption, passive solar design, waterless urinals, and water-efficient plumbing fixtures. Ongoing learning programs coupled with building energy and occupancy sensors utilize the buildings themselves in the education of visitors. The Aldo Leopold Legacy Center demonstrates how human activities, when carefully planned and executed, can work in harmony with ecological systems for the benefit of both.

UNF 19.1 *(Courtesy The Kubala Washatko Architects, Inc.)*

The Aldo Leopold Center utilized wood for both the building structure and finishes that was harvested and processed directly on the project site.

UNF 19.2 a–d *(Courtesy The Kubala Washatko Architects, Inc.)*

(1) Indicates structural use:
B - Simple span bending member.
C - Compression member.
T - Tension member.
CB - Continuous or cantilevered span bending member.
(2) Mill number.
(3) Identification of ANSI Standard A190.1, Structural Glued Laminated Timber.
(4) Applicable laminating specification.
(5) Applicable combination number.
(6) Species of lumber used.
(7) Designates appearance grade, INDUSTRIAL, ARCHITECTURAL, PREMIUM.

Figure 19.17 A typical APA-EWS trademark. 117-93 is the number of the manufacturing standards and specification for glulam beams published by the American Institute of Timber Construction. 24F means the allowable bending stress in fiber is 24,000 psi, and V4 is the lamination layup for Douglas fir. *(Courtesy APA—The Engineered Wood Association)*

smooth sawed back. They are tapered from a thick butt end. Thickness varies with the grade. Certi-Sawn cedar shakes are available in grades 1, 2, and 3, as described for shingles, but these shakes grades have different specifications. Certi-Split shakes are available in grade No. 1 and premium grade, which is the best. They come in 18 and 24 in. lengths and widths from 4 to 14 in.

Cedar shakes and shingles can be impregnated with fire-retardant polymers that penetrate the innermost cells of the wood. These meet the requirements for class C and B shake and shingle roof systems and class A shake roof systems. They have undergone the intermittent-flame test, spread-of-flame test, burning brand test, flying brand test, rain test, and weathering test. Southern pine taper-sawed shakes are available in grade No. 1 (the best) and No. 2. Special hip and ridge units are made from No. 1 shakes. The shakes are $\frac{3}{16}$ in. thick and 18 or 24 in. long.

WOOD WINDOWS, DOORS, AND CABINETS

Wood Windows

Windows made of wood are available in a wide range of sizes, styles, and quality. Some are covered with vinyl or aluminum to eliminate the need for painting. There have been recent improvements in window

Figure 19.18 Cedar shingles and hand split shakes are widely used for finished roof coverings. (a) Image copyright Tom Grundy, 2009. Used under license from Shutterstock.com.(b) Image copyright Michael Shake, 2009. Used under license from Shutterstock.com.

Figure 19.19 Windows in a variety of types and sizes are fully assembled in millwork plants and ready for installation.

design and construction, with new models that are more energy efficient and easy to operate. Figure 19.19 shows some of the more frequently used types. The National Wood Window and Door Association standards IS-2 and IS-3 established specifications for wood windows and sliding glass doors. NWWDA also has standard IS-4, which pertains to water-repellent preservative treatment.

Wood Doors

A variety of types and styles of doors are made from both solid and hollow core wood construction (Fig. 19.20). They are available in solid wood and with wood veneer, hardboard, and plastic laminate faces. Other doors have wood stiles and rails covered with a fiberglass skin. Wood and plastic doors can have glass or louvered openings.

Many companies manufacture doors to the specifications of the National Wood Window and Door Association. These standards establish minimum requirements for material, design, construction, and pressure treatment. Doors meeting these standards are recognized by the NWWDA seal.

Cabinets

Cabinets are generally made from wood and wood products, although some steel cabinets exist. They come in a wide variety of styles, such as colonial, contemporary, and provincial, and use a wide range of woods, such as birch, cherry, and walnut (Fig. 19.21). Some are covered with a plastic laminate over a particleboard substrate.

Figure 19.20 A decorative wood door. *(Image copyright Mette Brandt, 2009. Used under license from Shutterstock.com)*

Cabinet manufacturers build their units to meet the standards of the Kitchen Cabinet Manufacturers Association set forth in ANSI A161.1-1980. Cabinets manufactured to these standards have the KCMA seal. More information on cabinets can be found in chapter 22.

Figure 19.21 Kitchen cabinets are manufactured in a variety of styles and wood finishes. *(Image copyright Jorge Salcedo, 2009. Used under license from Shutterstock.com)*

WOOD FLOORING

Both hardwoods and softwoods are used for finish flooring. Wood flooring is popular because of the warmth and beauty of its grain and color. The most commonly used hardwoods are oak, beech, birch, pecan, and maple. Softwoods include southern pine, western larch, bald cypress, eastern Englemann, eastern spruce, red pine, ponderosa pine, eastern hemlock, and Douglas fir.

Grading rules for wood flooring are shown in Table 19.6. They specify requirements for kiln drying, grading, and control of moisture and establish standard sizes. Many manufacturers produce additional sizes for special applications, such as very thick flooring for industrial use.

Wood flooring is available both quarter-sawed and plain-sawed. It is produced in four basic types: strips, planks, parquet (thin wood blocks), and solid end grain blocks (Fig. 19.22). The construction of these products varies widely and manufacturers should be consulted.

Oak flooring is made from both plain-sawed and quarter-sawed wood. Maple, beech, and birch flooring are usually plain sawed because these woods are hard, dense, and close-grained. Southern pine flooring is graded under the rules of the Southern Pine Inspection Bureau. It is available in both flat grain and edge grain. Douglas fir flooring is available in both flat grain and edge grain.

Strip Flooring

Strip flooring is available in soft and hard woods. It is manufactured with square edges, side matched (tongue and groove), and side and end matched (tongue and groove on sides and ends). It usually has a small hollow relief area cut on the bottom and is available in solid wood and laminated wood strips. Strip flooring is supplied either unfinished or prefinished. Unfinished flooring usually has square, tight-fitting edge joints, which produces a smooth floor. Prefinished strip flooring usually has a V-joint on the edges and ends of each strip. A wide range of woods and finishes are available. Sizes of strip flooring are listed in Table 19.7.

Planks

Planks may be solid wood or laminated. Solid wood planks are usually $\frac{3}{4}$ in. (18 mm) thick and from 3-$\frac{1}{2}$ to 8 in. (87 to 204 mm) wide. Laminated planks are usually

Table 19.6 Grades of Unfinished Hardwood and Softwood Flooring

Unfinished Oak[a]	Prefinished Oak[a]
Clear Plain or Clear Quartered (best appearance)	Prime (excellent)
Select Plain or Select Quartered (excellent appearance)	Standard and Better (mix of Standard and Prime)
Select and Better (mix of Clear and Select)	Standard (variegated)
No. 1 Common (variegated)	Tavern and Better (mix of Prime, Standard, and Tavern)
No. 2 Common (rustic)	Tavern (rustic)
Beech, Birch, and Hard Maple[a]	**Pecan[a]**
First Grade White Hard Maple (face all bright sapwood)	First Grade Red (face all heartwood)
First Grade Red Beech (face all red heartwood)	First Grade White (face all bright sapwood)
First Grade (best)	First Grade (excellent)
Second and Better (excellent)	Second Grade Red (face all heartwood)
Second (variegated)	Second Grade (variegated)
Third and Better (mix of First, Second, and Third)	Third Grade (rustic)
Maple[b]	**Softwood[c]**
First (highest quality)	B and Btr (best quality)
Second and Better (mix of First and Second)	C (good quality, some defects)
Second (good quality, some imperfections)	C and Btr (mix of B & Btr and C)
Third and Better (mix of First, Second, and Third)	D (good economy flooring)
Third (good economy flooring)	No. 2 (defects but serviceable)

[a]*Hardwood flooring grades of the National Oak Flooring Manufacturers Association.*
[b]*Maple flooring grades of the Maple Flooring Manufacturers Association.*
[c]*Softwood flooring grades of the Southern Pine Inspection Bureau.*

Figure 19.22 Basic types of wood flooring products.

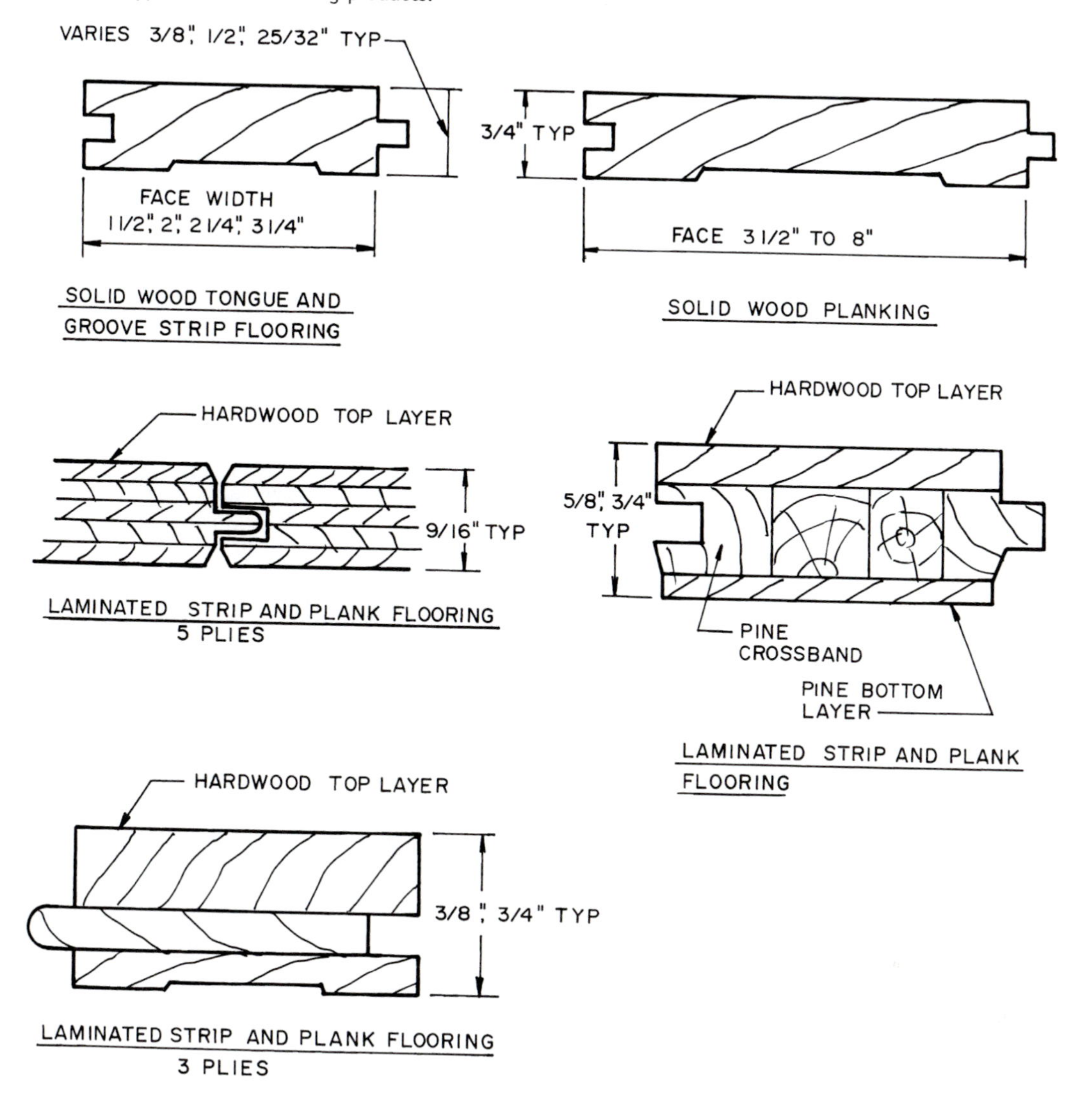

thinner, ranging from $\frac{7}{16}$ to $\frac{3}{8}$ in. thick. They are used to produce floors representative of the wide flooring used in America's early days. Some are secured with screws covered with wood plugs.

Parquet Flooring

Parquet floors consist of thin blocks made from solid or laminated wood. These blocks are joined in patterns to form a parquet floor. They are available unfinished and prefinished. Generally, the blocks are square, but other units are available from various manufacturers (Fig. 19.23). The American Parquet Association and NBS-PS 27-70, Mosaic Parquet Hardwood Flat Flooring, grading rules are used for the manufacture of the flooring. Laminated hardwood block flooring is produced following ANSI/HPMA LHF 1982.

Solid block flooring is cut so the face exposes the end grain. It is used in areas where heavy traffic occurs, a factory floor, for example, and where easy maintenance is required.

Canadian hardwood strip flooring is manufactured in the customary sizes compatible with those used in the United States and in metric sizes. Strip and parquet flooring are measured by surface area in square meters (m^2). The two Canadian industry standards for strip and hardwood flooring are published by the Canadian Lumbermen's Association.

Table 19.7 Stock Sizes for Wood Strip Flooring

Southern Pine Flooring[a]				
	Actual Thickness		Actual Width[b]	
Species	in.	mm	in.	mm
Longleaf pine, slash pine, shortleaf pine, loblolly pine	$^{5}/_{16}$	8	$1^{3}/_{8}$	35
	$^{7}/_{16}$	11	$2^{3}/_{8}$	60
	$^{9}/_{16}$	14	$3^{3}/_{8}$	86
	$^{3}/_{4}$	76	$4^{3}/_{8}$	101
	1	101	$5^{3}/_{8}$	136
	$1^{1}/_{4}$	107		

[a]*Courtesy Southern Pine Inspection Bureau*
[b]*Widths available in all thicknesses.*

Hardwood Strip Tongue-and-Groove Flooring[c]				
	Actual Thickness		Actual Width	
Species	in.	mm	in.	mm
Oak, beech, birch, hard maple, hickory	$^{3}/_{4}$	19	$1^{1}/_{2}$, 2, $2^{1}/_{4}$, $3^{1}/_{4}$	38, 51, 57, 82.5
	$^{11}/_{32}$	8.7	$1^{1}/_{2}$, 2	38, 51
	$^{15}/_{32}$	12	2, $2^{1}/_{4}$	38, 51
	$^{33}/_{32}$	26	$3^{1}/_{4}$	51, 57, 82.5

[c]*Courtesy National Oak Flooring Manufacturers Association*

Maple Strip Tongue-and-Groove Flooring			
Actual Thickness		Actual Width	
in.	mm	in.	mm
$^{25}/_{32}$, $^{33}/_{32}$	20, 26	$1^{1}/_{2}$, $2^{1}/_{4}$, $3^{1}/_{4}$	38, 57, 82
$^{41}/_{32}$	32.5	$2^{1}/_{4}$, $3^{1}/_{4}$	57, 82

Figure 19.23 Typical parquet flooring blocks and heavy wood block flooring. (a) Image copyright Kruchankova Maya, 2009. Used under license from Shutterstock.com. (b) Image copyright Petinovs, 2009. Used under license from Shutterstock.com.

Review Questions

1. What are the differences between plywood, oriented strand board, and composite panels?
2. What are the plywood exposure durability classifications? Give an example of where each could be used.
3. How does hardwood plywood differ from industrial plywood?
4. What are the differences between laminated veneer lumber and glued laminated members?
5. How do wood shingles and shakes differ?
6. What types of wood flooring are available?
7. What are the four plywood constructions commonly used in building construction?
8. What appearance groups are established for the outer veneers of construction plywood?
9. How is the strength of the woods used in plywood indicated?
10. What is a performance-rated wood panel?
11. What are the grades of outer veneers for hardwood plywood?
12. What are the classes of hardboard?

Key Terms

APA Performance-Rated Panels
composite panels
glued laminated lumber (glulam)
hardboard
hardwood plywood
I joist
laminated veneer lumber
oriented strand board
overlaid plywood
Parallel Strand Lumber
particleboard
plywood
Structural Insulated Panels
trusses
waferboard

Activities

1. Collect samples of as many kinds of plywood products as you can find. Prepare a display. Identify each one with as much technical information as you can gather.
2. Immerse samples of interior and exterior plywood in water. Continue until you delamination occurs. Keep a record of the length of time it takes each to fail. Did some types never fail?
3. Cut 2-inch-wide strips of $\frac{1}{2}$-inch-thick fir plywood 4 ft. long. Support a strip between two chairs with the flat side facing down. Add weights incrementally and note the amount of deflection until the strip breaks or slides off a chair. Then stand the strip on its edge and repeat the test. In which position did it carry the greatest load? Why?
4. Build scale models of the commonly used trusses.
5. Collect samples of the various types of wood flooring. Your building supply dealer may have manufacturers' samples for you to use. Label each.

Additional Resources

APA Engineered Wood Construction Guide, and numerous other related technical publications, APA—The Engineered Wood Association, Tacoma, WA.

Plywood Handbook and Plywood Design Fundamentals, CANPLY, Canadian Plywood Association, North Vancouver, BC, Canada.

Williamson, T. G., APA Engineered Wood Handbook, McGraw-Hill Publishing Co.

Wood Engineering Handbook, Forest Products Laboratory, Madison, WI 53705.

Wood Reference Handbook, Canadian Wood Council, Ottawa, Ontario, Canada.

See Appendix B for addresses of professional and trade organizations and other sources of technical information.

CHAPTER 20

Wood and Metal Light Frame Construction

LEARNING OBJECTIVES

Upon completion of this chapter, the student should be able to:

- Develop an understanding of the methods used to construct light frame buildings.
- Understand the differences in framing when using the various materials and products available.
- Be aware of the influence codes and ordinances have on the design of light frame buildings.

Build Your Knowledge

For further study on these materials and methods, please refer to:

Chapter 22 Finishing the Exterior and Interior of Light Wood Frame Buildings

Chapter 36 Interior Walls, Partitions, and Ceilings
Topic: Wood Product Wall Finishes

Wood light frame construction is the most widely used system for the construction of residences in the United States and Canada. Developed in the 1830s, the method has remained largely unchanged due to its inherent flexibility and economy despite enormous technical advances in other fields during the same time period. Light frame structures can be erected quickly and easily, requiring little in the way of expensive tools.

Wood light frame construction is a very flexible system that permits an almost unlimited range of design possibilities. Architects can produce designs that are classic (Fig. 20.1) or contemporary (Fig. 20.2), simple or complex, low cost or expensive, and can accommodate almost any electrical, heating, air conditioning, plumbing, and security system desired. With the vast array of products on the market, a building can be insulated, sealed, and waterproofed to facilitate long life and low maintenance. Building can occur in almost any climate and on any site that will accept an adequate foundation.

The system has evolved from employing basically a solid wood structure to the use of a variety of reconstituted wood products, as discussed in Chapter 19. Many problems that developed with early wood light frame construction—wood decay, swelling and shrinking of members, sticking doors and windows, and creaking floors—have been overcome by improved construction techniques and materials. Considerable effort has been made to utilize factory-assembled panels and modules, thereby reducing on-site labor costs associated with what is often referred to as "stick built" construction.

ORDINANCES AND CODES

Building Codes

Building codes play an important role in the design and construction of wood and metal light frame construction. Codes contain extensive requirements for the construction of foundations, walls, and roofs in both wood and metal framing. Allowable design loads on various building components are carefully detailed in the codes, and structural components must meet minimum deflection allowances.

Fire-resistance ratings for various assemblies of buildings are specified. The fire code stipulates requirements for fire ratings and approved noncombustible materials. For example, a wall between two adjacent family dwellings should have at least a one-hour fire rating. Typically these requirements are met by using

Figure 20.1 Light wood frame systems are used widely for residential construction. *(Image copyright Rafael Ramirez Lee, 2009. Used under license from Shutterstock.com)*

Figure 20.2 A contemporary wood frame structure. *(Photo taken by Bobak Ha'Eri. September 2, 2006. http://en.wikipedia.org/wiki/Image:09-02-06-ThorncrownChapel1.jpg)*

additional gypsum wallboard on interior walls and, in some cases, adding fire-resistant sheathing to exterior walls. In some areas, sprinkler systems are required by local codes. Controls are placed on materials regulating flame-spread and smoke-developed ratings.

Construction in areas where basic wind speeds equal or exceed 110 miles per hour must meet special design criteria. Areas subject to earthquakes have seismic design requirements. Codes also regulate the design of interior spaces and such things as light, ventilation, glazing, and sanitation.

Specific details are available in the International Building Code published by the International Code Council. The code classifies buildings into five types of construction.

Types I and II are those structures constructed of noncombustible materials. Type III buildings have exterior walls of a noncombustible material and interior building elements of any material permitted by the code. Type IV construction (often referred to as heavy timber) has exterior walls of noncombustible materials and interior building elements of solid or laminated wood. Type V construction permits the use of structural elements, exterior walls, and interior walls made from any of the materials permitted by the code. Light frame construction typically falls under this classification. Additional information on building codes is given in Chapter 2.

PREPARING THE SITE

Once a building has been designed, bids accepted, and building permits issued, the site must be readied for construction. Unwanted brush and trees in the way of the building, side walks, or drives are removed. In some areas, tree removal is carefully regulated, and each tree must be marked and its removal approved before the site is cleared. When possible, mature trees and other plantings should be preserved.

The surveyor stakes the corners of the foundation, following the location on the site plan. The builder sets up batter boards on each corner, as shown in Fig. 20.3. Batter boards are usually 1 × 6 lumber held by 2 × 4 posts driven into the ground. A chalk line is pulled across the boards on each side of the proposed building based on the locations of the corner stakes. Once lines are located, they are tied to a nail driven into the batter board or run in a saw cut in the board. Fig. 20.4 illustrates the typical corner layout. The corners are then checked for squareness using the 3-4-5 right triangle method. When the base of a right triangle measures 3 ft. and the vertical side measures 4 ft. the hypotenuse will measure 5 ft. If it does not, the corner is not square. Some prefer to use longer measurements in multiples of three, such as 9, 12, and 15 ft.

Figure 20.3 Batter boards and chalk lines are used to locate the corners of a foundation.

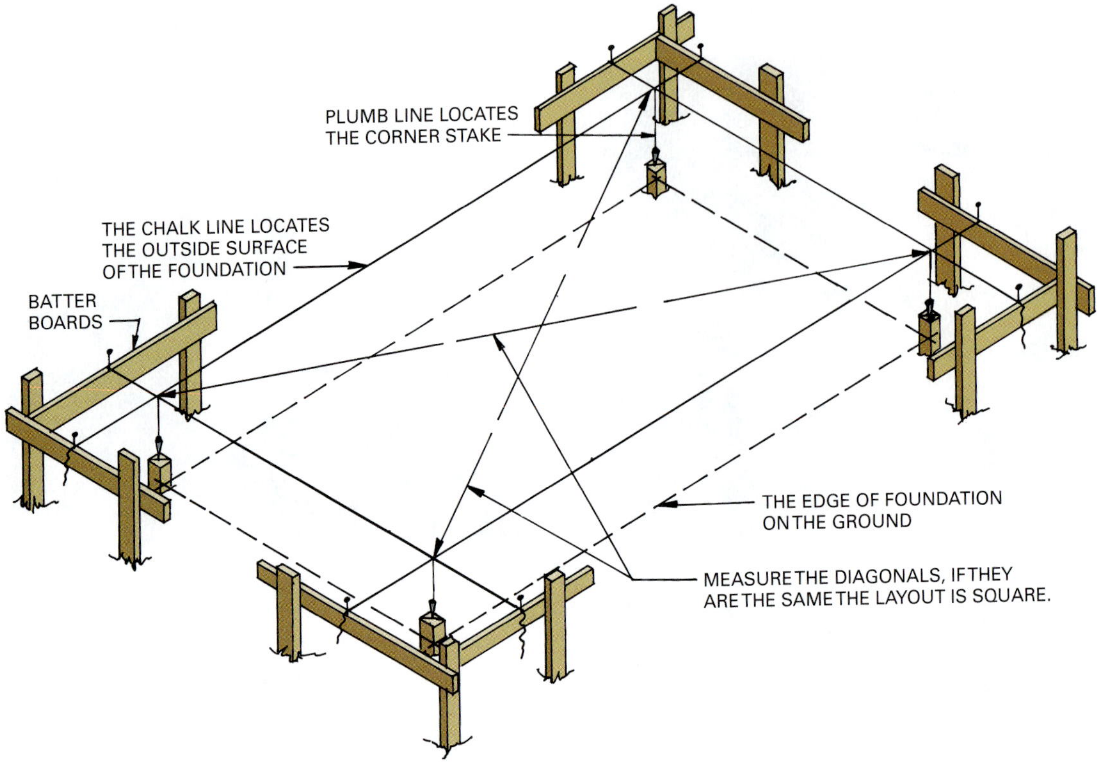

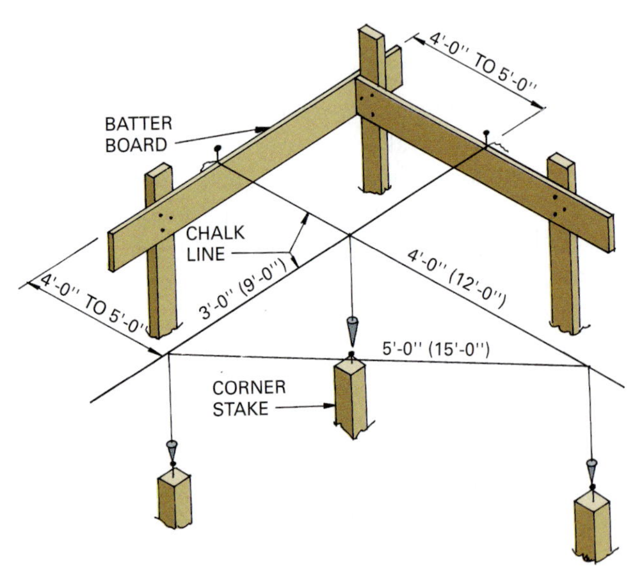

Figure 20.4 The squareness of a corner marked by a chalk line can be checked by the 3-4-5 method. Notice the batter boards are set back about 4 ft. from the corner stake.

Excavation

Excavation for footings and basements are usually accomplished with a bulldozer or backhoe. A plumb line is dropped from the batter board chalk lines to establish the corner of the foundation and excavation (Fig. 20.5). Excavations are then dug to the required width and depth.

Footings must rest on undisturbed soil so care against over excavating must be exercised. If soil crumbles on the sides of the footing excavation, wood forms must be installed (Fig. 20.6). Additional footings for columns, piers, and fireplaces are located and dug. Specified reinforcing is placed in the footing excavation and concrete poured. Excavations for footings for buildings with crawl spaces or slab floors follow a similar procedure.

FOUNDATIONS

Footings for light frame construction are made of cast-in-place reinforced concrete. The foundation stem wall may be concrete block, brick, cast-in-place concrete, or pressure treated wood. Detailed information on foundations is given in Chapter 6. The following discussion will review common foundations for wood light framed buildings, including basements, crawl spaces, and concrete slab on grade floors.

Basements are used predominantly in colder climate regions where the depth of the frostline requires deep excavations. They provide considerable space for a small cost, especially considering that a crawl space requires a footing and some foundation wall and provides no living space.

Figure 20.5 The basement is excavated to the specified depth and footings are dug. Footing forms are used if the sides of the footing excavation crumble.

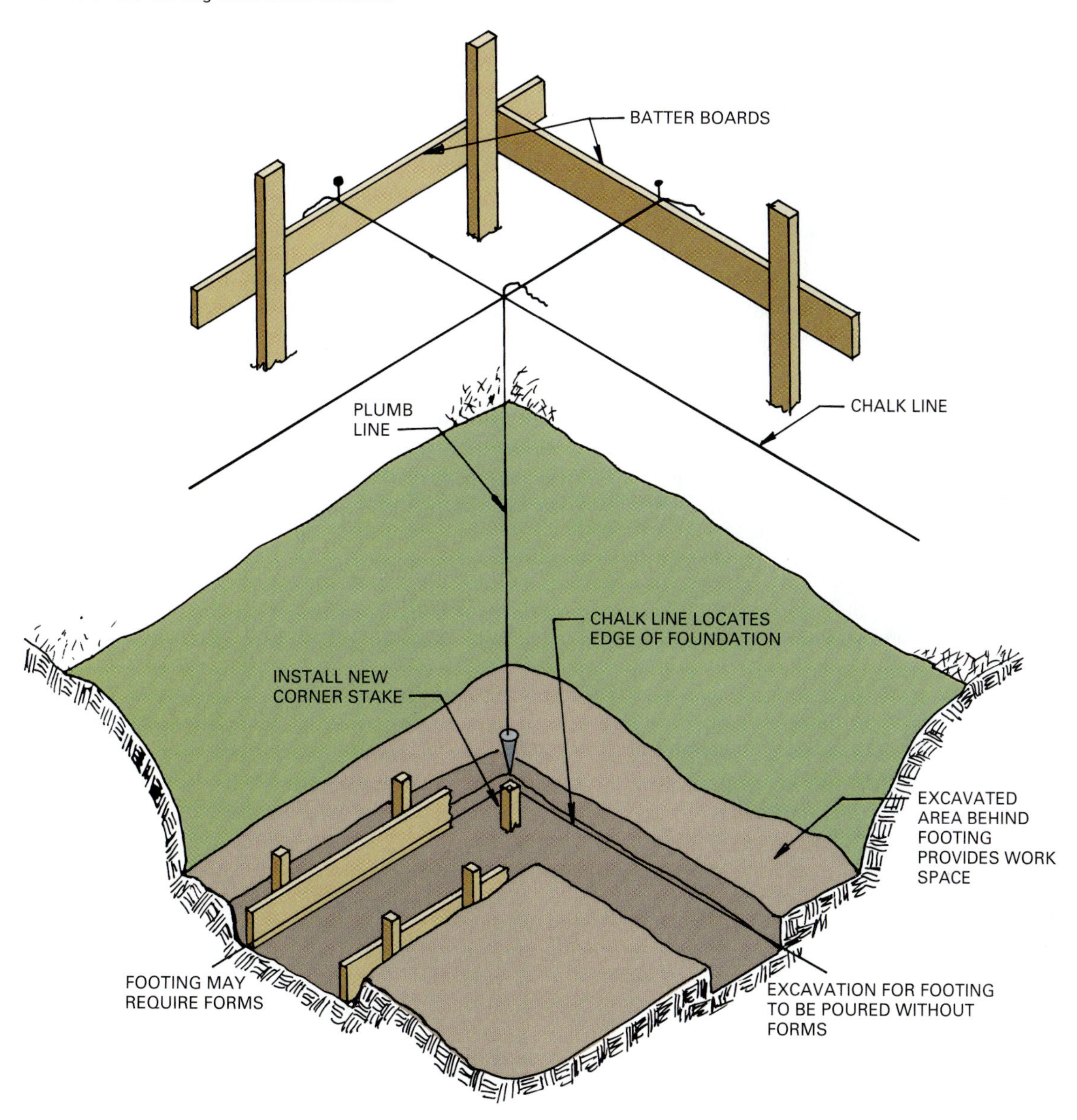

Basement walls can be cast-in-place reinforced concrete or concrete block. Two typical basement details can be found in Fig. 20.7. Footings are poured first. When they have reached adequate strength, the concrete block wall can be laid, or the forms for a poured concrete foundation installed. In both cases, vertical steel reinforcing acts to tie the footing to the stem wall.

Figure 20.6 A footing form with reinforcing steel is ready to be poured. *(Courtesy Eva Kultermann)*

The foundation wall is usually topped with a metal flashing termite shield that projects to the interior and prevents termites from reaching the wood framing. A treated wood plate attached with anchor bolts caps the foundation walls. After the basement foundation has been built, the soil should not be backfilled until the floor platform is in place to provide protection against horizontal stresses created by the weight of the soil. The foundation wall must also be insulated, waterproofed, and provided with a perimeter drain set in gravel.

The foundations in Fig. 20.8 are typical for a building with a crawl space A brick veneer exterior requires that a foundation have a brick ledge that is usually near grade. Building codes limit the minimum height between the ground and the bottom of the floor joists. Since plumbing, heating, air-conditioning, and electrical service are often run in the space below the joists, this height is often set at well above the minimum.

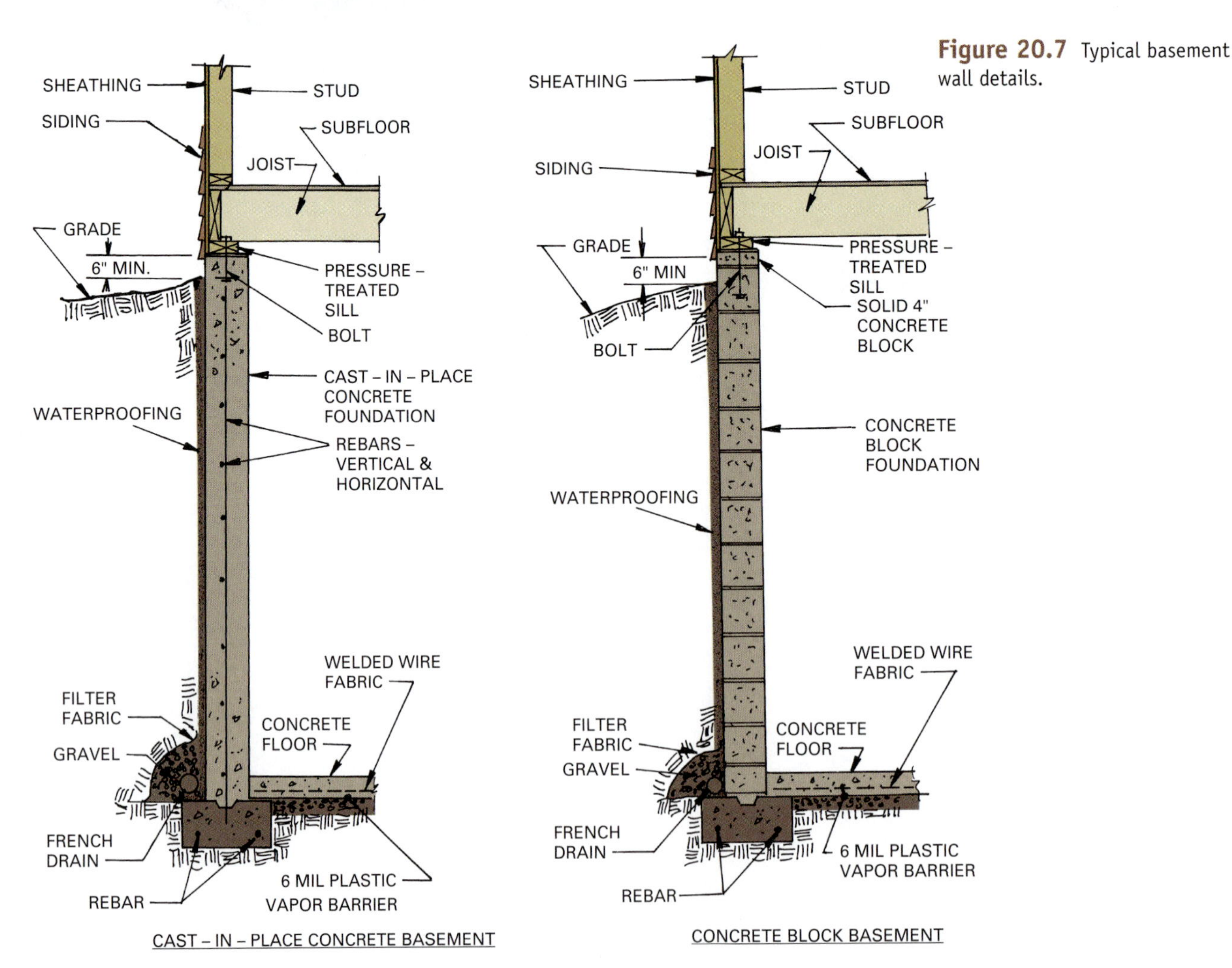

Figure 20.7 Typical basement wall details.

Figure 20.8 Typical details used on buildings having a crawl space.

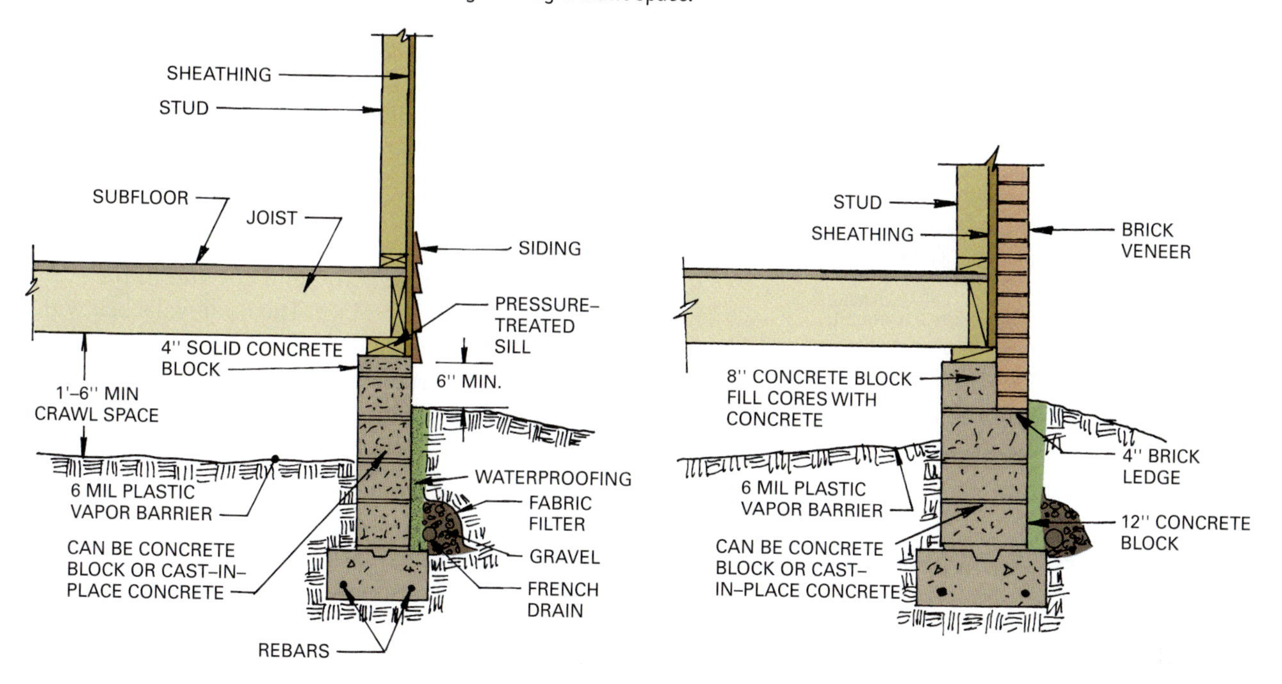

Anything below a 2 ft. (610 mm) high crawl space is claustrophobic and difficult to work in when servicing utilities.

Since it is important to keep the ground in the crawl space dry, the exterior of the foundation below grade should be waterproofed, and gravel and a French drain should be placed at the footing, draining to a drywell or daylight. The ground in the crawl space should be covered with plastic sheet material to retard moisture passage from soil to crawl space. The perimeter of the crawl space should be insulated in colder climate regions.

Several foundations for houses with concrete slab on grade floors are illustrated in Fig. 20.9. A monolithic poured foundation and floor is used in areas where freezing does not occur. Others are used where footings reach deep below grade. Anchor bolts for the connection of the bottom plate are set in concrete before it hardens.

Wood Foundations

Wood foundations can be used for buildings with basements or crawl spaces. The materials are stress graded to insure they can withstand lateral soil and subsurface water pressures. Wood materials must have the stamp of the American Wood Preservers Association (AWPA). Foundation panels are often shop-built and then shipped to site. The studs forming the panel face the inside of the building and are insulated in the same manner as an exterior above-grade wall.

A typical basement foundation panel is shown in Fig. 20.10. Panels must be carefully assembled using bronze, copper, silicon, or stainless steel nails that resist corrosion. Notice that the bottom plate is set on a bed of gravel rather than on a concrete footing. The wooden floor joists rest on the double top plate of the foundation wall.

Piers and Columns

In addition to perimeter foundation walls, footings for piers or columns to support beams that will carry the floor joists and other point loads must also be considered. Concrete blocks compose a typical pier used in wood light frame buildings with crawl spaces (Fig. 20.11). The size of the pier depends upon the loads to be carried.

In buildings with basements, steel columns and beams can support floor loads at the interior. Fig. 20.12 **shows an example**. Steel baseplates are used to bolt the column into the footing or slab. A finished foundation, ready for carpenters to install the first floor platform, is shown in Fig. 20.13.

Figure 20.9 Types of concrete slab floor construction used with wood light frame construction.

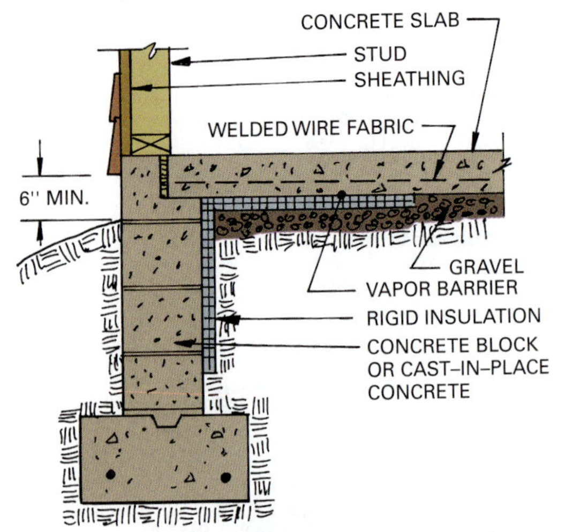

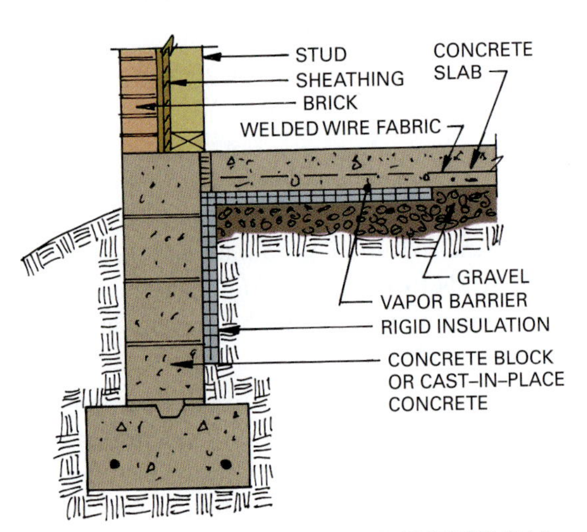

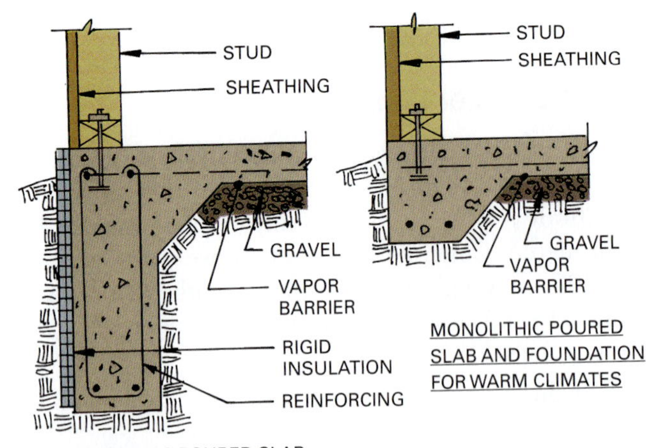

EVOLUTION OF WOOD LIGHT FRAME CONSTRUCTION

Early wood frame construction in the U.S. colonies involved building a structural frame of heavy timbers that were fitted together with mortise and tenon joints. The spaces inside this frame were infilled with smaller wood members or masonry. This construction required a large group of workers to hand cut and fit the joints and lift the timbers into place. The infill walls below the horizontal timbers were non-load-bearing. This type of heavy timber construction has been largely replaced by wood light frame construction.

The forerunner of the current widely used **platform framing** system of wood light frame construction was developed in the early 1800s. It involved using smaller vertical members called studs to carry the roof and second floor and eliminated the heavy timber frame. The system was called the **balloon frame** because it seemed light enough to float in comparison to heavy timber construction. The studs not only carried the loads but also served as the enclosure of the building (like the infill walls did on the heavy timber frame). Balloon frames used wall studs running the full two stories with floors hung from inlet ledger boards. The detail in Fig. 20.14 shows studs resting on the sill on top of the foundation and running continuously to the double top plate. The floor joists rested on a ribbon let into the stud and were nailed to the stud. Balloon framing has been replaced by platform framing and is seldom used anymore.

PLATFORM FRAMING

Platform framing involves building the first floor deck on top of the foundation (Fig. 20.15). Upon this, walls for the first floors are assembled, erected, and braced (Fig. 20.16). Joists for the second floor are laid on top of the double plate of the exterior first-floor walls and supported by load-bearing interior walls. Second floor joists are covered with subflooring, forming the second floor platform, and walls for the second floor are assembled, erected, and braced. The ceiling joists are laid on the double plate of the second-floor exterior walls and supported by interior load-bearing walls. Finally, the rafters, which rest on the double plate of the second-floor walls, are erected (Fig. 20.17). The walls and roof are sheathed, providing solid surfaces for siding and finished roofing materials. Sheathing greatly strengthens the roof framing and the rigidity of the entire structure.

Figure 20.10 Permanent wood foundations are installed using pressure-treated (PT) lumber over a gravel bed for the footing. *(Courtesy APA - The Engineered Wood Association)*

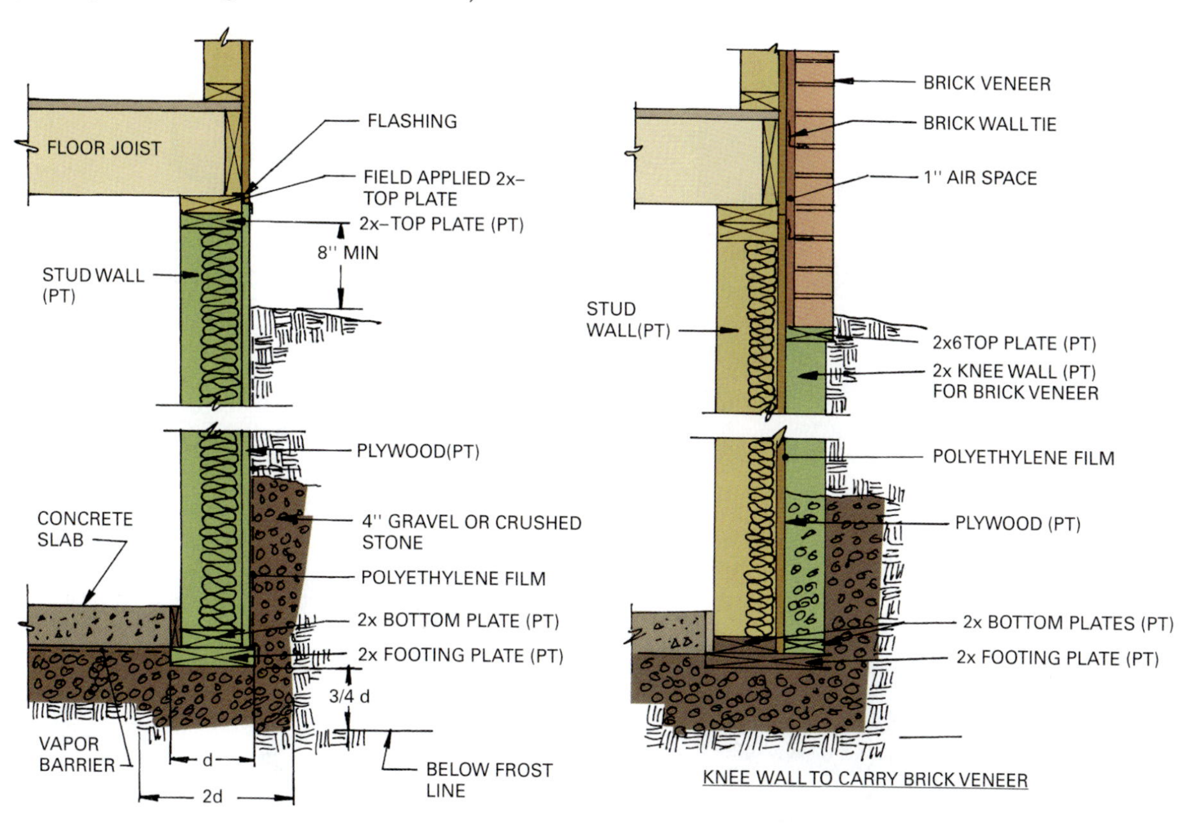

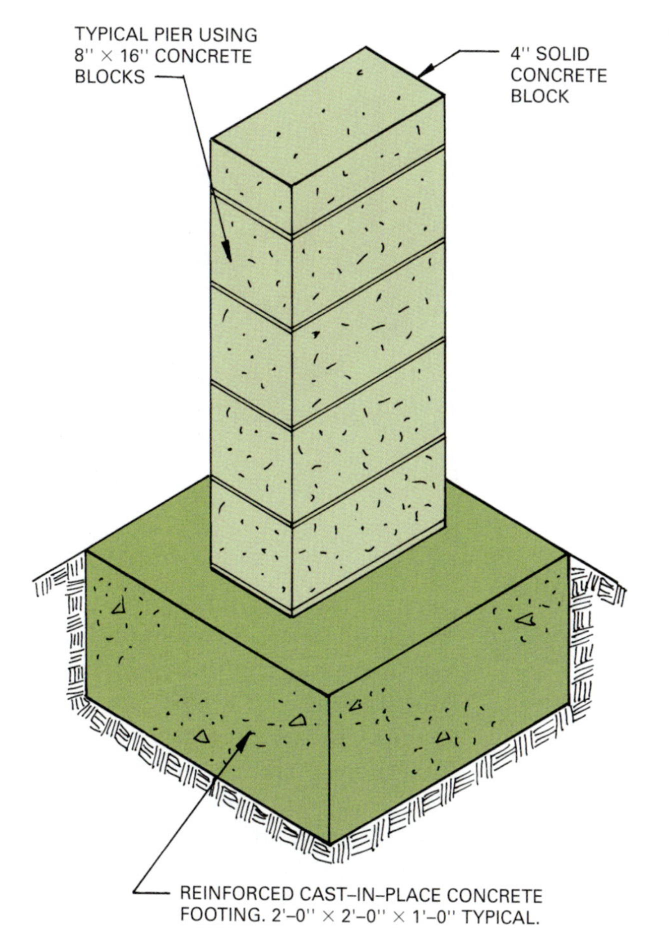

Figure 20.11 A typical concrete block pier used in residential construction.

Figure 20.12 Steel columns and beams are used to support wood floor joists. *(Image copyright Christina Richards, 2009. Used under license from Shutterstock.com)*

Figure 20.13 The foundation ready for the beams and floor joists to be installed. *(Image copyright V. J. Matthew, 2009. Used under license from Shutterstock.com)*

Figure 20.14 A typical framing detail for balloon framing.

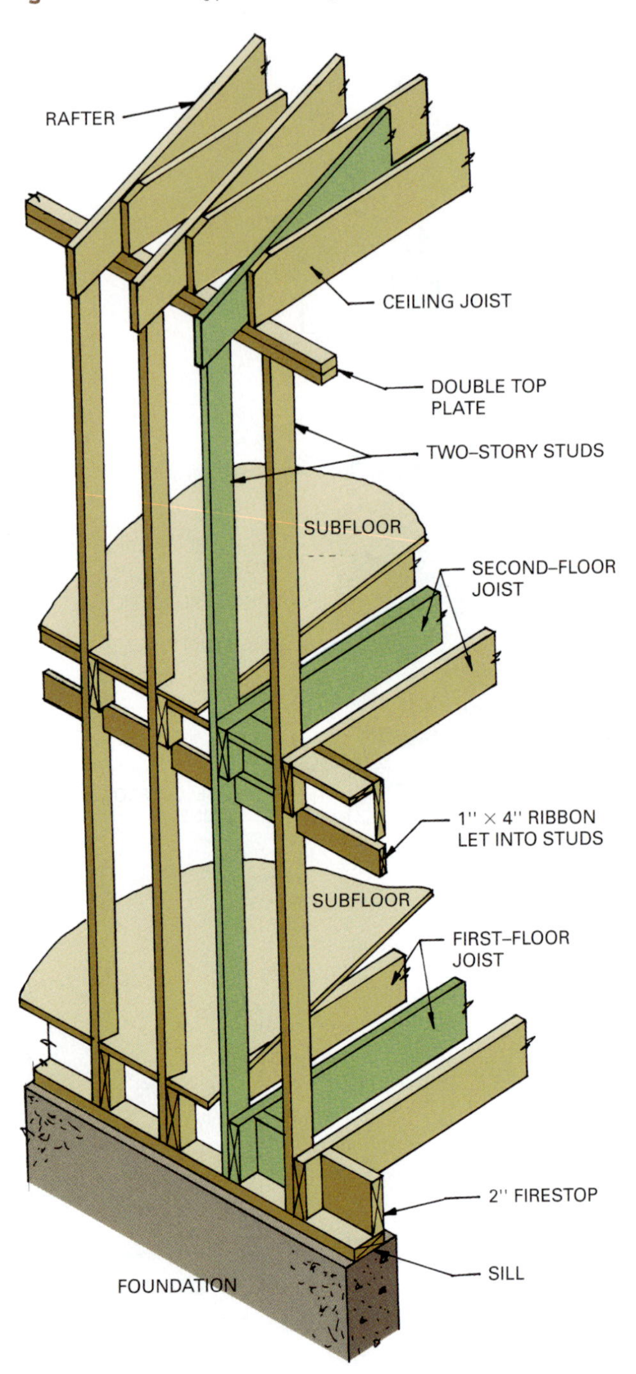

If roof trusses are used, there is no need for ceiling joists because the bottom chord of the truss serves as one. Trusses are erected bearing on the double top plate of the second-floor walls. Trusses are discussed in more detail later in this chapter.

Floor Framing

Floor joists are the horizontal members of a frame that transfer loads to sills and girders. Floor joists sometimes cannot span the entire length or width of a foundation, so a beam is required to support the floor at the mid-span. The beam may be built of steel, wood, or a manufactured wood product. It is supported at intervals by piers or columns (Fig. 20.18).

In Fig. 20.19, the first floor has been built on the foundation using box sill construction. Box sill construction has floor joists resting on a treated lumber sill that is bolted to the foundation. The joists butt a band or **rim joist**, which, in effect, forms the floor framing into a box. The rim joist is end nailed to the joist and the joists are toe-nailed to the sill. Sometimes metal connectors are used. Floor joists may consist of dimensional lumber or engineered wood I joists and are usually spaced 16 or 24 in. on center (o.c.) (Fig. 20.20).

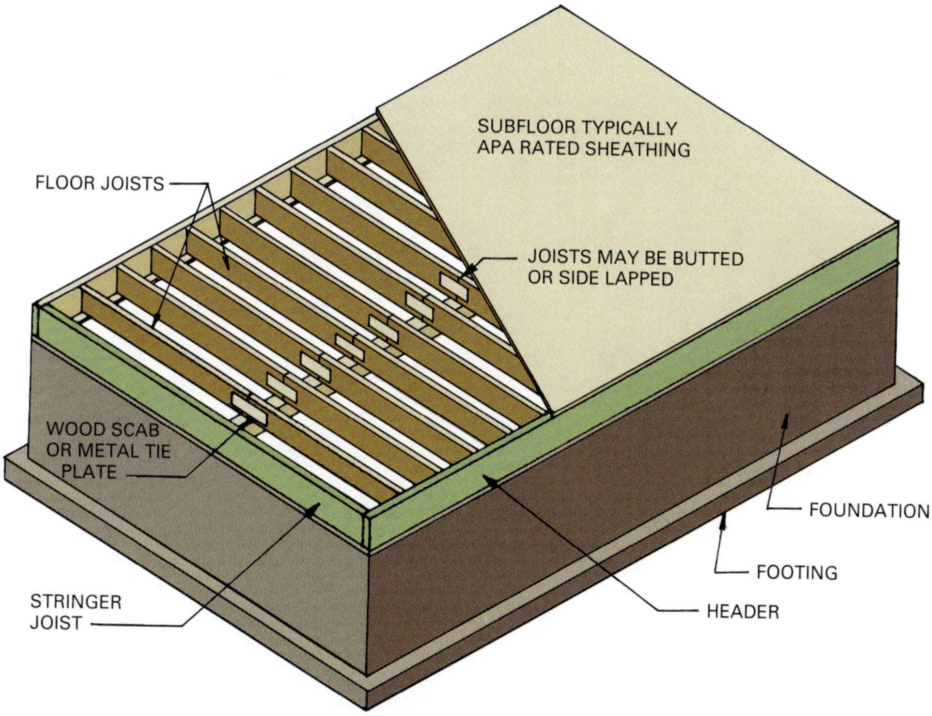

Figure 20.15 The framing for a floor with wood joists.

After the first floor joists are in place, the subfloor is glued and nailed to them (Fig. 20.21). Subfloors provide the structural floor upon which finish materials are installed. Usually a tongue and groove $\frac{3}{4}$ in. plywood is affixed with construction adhesive and then nailed in place.

Wall Framing

Figure 20.22 illustrates the framing for a typical exterior wall with an opening. Studs are typically 2 × 4 stock, but 2 × 6 studs provide space for additional insulation. The studs are generally spaced 16 in. (406 mm) on center, however, other spacings, such as 12 and 24 in. on center (305 and 610 mm) are also used. The load-carrying capacity of a wall and the size of its studs determine the spacing to be used. On-center spacings are always set in four foot modules to accommodate panels of sheathing that are 4 × 8 ft. (1.2 × 2.4 m). This ensures the panel edges will rest on a stud.

Wall construction begins with the careful layout in pencil of all wall components on the top and bottom plates. Walls use a single plate at the bottom and a double plate at the top. The double top plate locks intersecting walls together and helps transfer ceiling loads more effectively. The layout of studs in a wall follows the layout of the joists beneath to ensure that load bearing is transferred directly to the foundation. All components of the wall are laid in place on the floor deck and then nailed together, either by hand or with pneumatic nailers (Fig. 20.23). The wall is then lifted into position, straightened between corners, nailed to the floor framing, plumbed, and temporarily braced.

Framing an opening in a load-bearing wall requires that a header large enough to carry the imposed loads span the opening and transfer loads to adjacent studs. Figure 20.24 shows how an opening is framed. Figure 20.25 shows a variety of details for how headers can be built. Openings in interior non-load-bearing partitions can be framed with a single flat member. Framing details for a door opening in a load-bearing wall are shown in Fig. 20.26. Notice the rough opening door size includes the size of the door, plus space for the door frame, and a little room to allow a carpenter to plumb the door frame.

The corners of exterior walls must be built to allow for side walls to butt front or rear walls and leave a nailing surface for interior finish materials. Figure 20.27 shows the preferred framing method that leaves an opening for the insertion of insulation later in the construction process. Where interior partitions meet exterior walls, the top plate overlaps the lower plate of the double top plate, as shown in Fig. 20.28. Where interior partitions butt an outside wall or another partition, they can be joined by installing blocking or extra studs, as shown in Fig. 20.29.

When the walls are in place sheathing can be installed. Sheathing is an exterior covering placed on the studs that serves as a base for exterior cladding. Sheathing may be plywood, oriented strand board, asphalt-impregnated fiber, or rigid foam insulation

Figure 20.16 The second floor platform is built on the first floor walls and the second floor walls are assembled and raised.

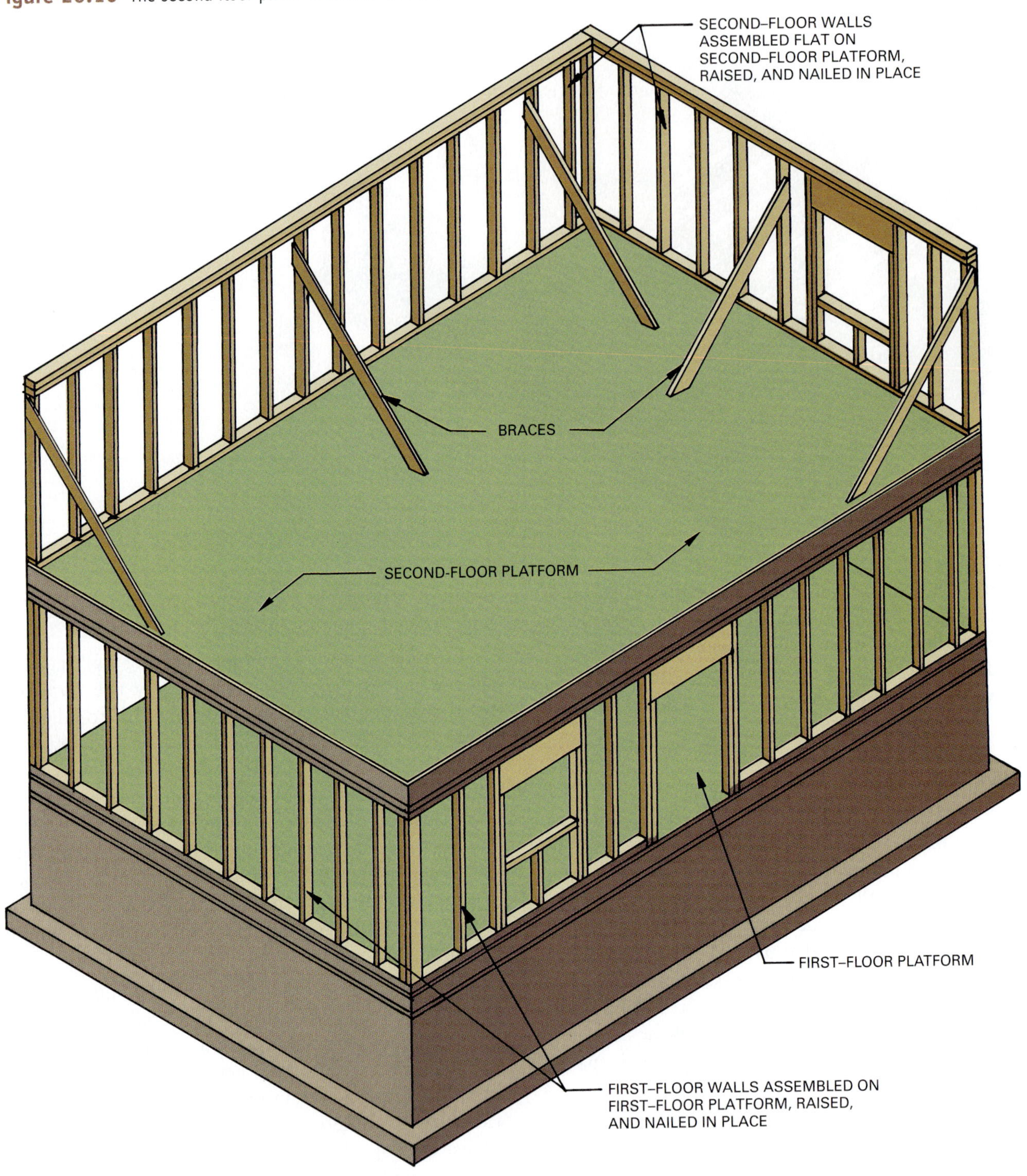

Figure 20.17 Typical platform framing details for a two-story building.

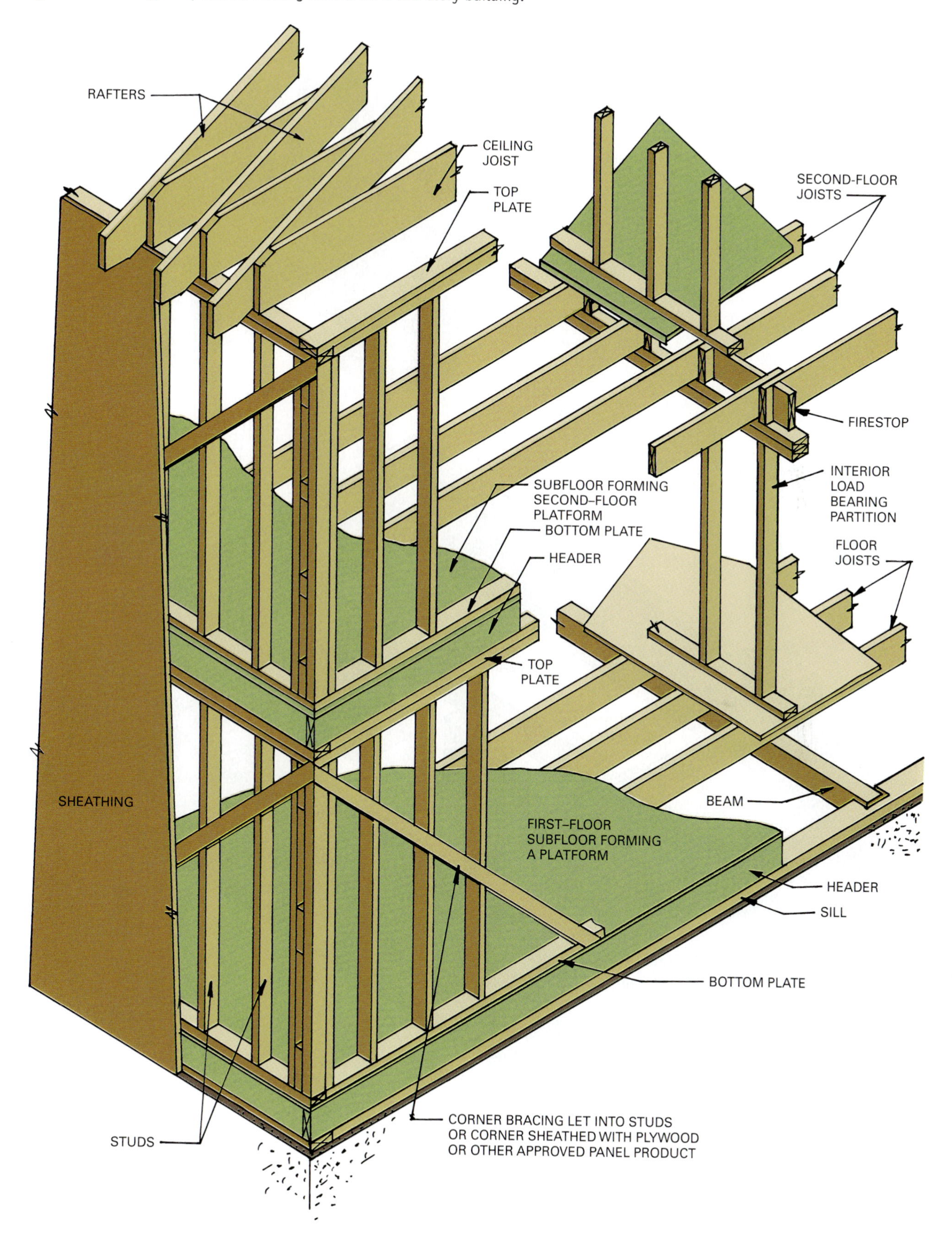

Figure 20.18 Generally, joists will not span the required distance and a beam is needed to shorten the span.

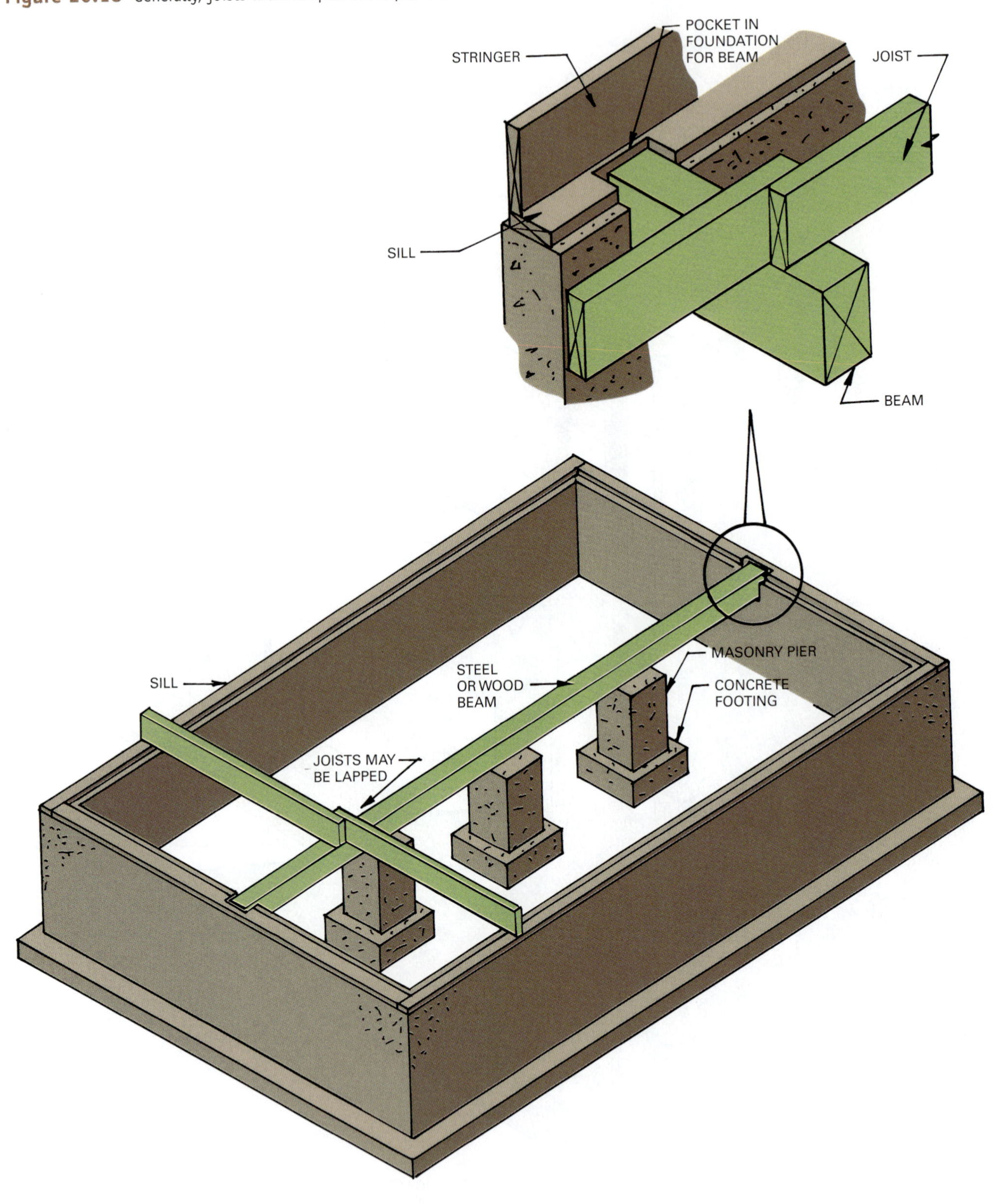

Figure 20.19 Typical rim joist construction.

Figure 20.20 Typical wood I-Joist installation. *(Courtesy Boise Cascade)*

Figure 20.21 Installation of APA-rated panel subfloor.

Figure 20.22 Framing details for a typical exterior wall.

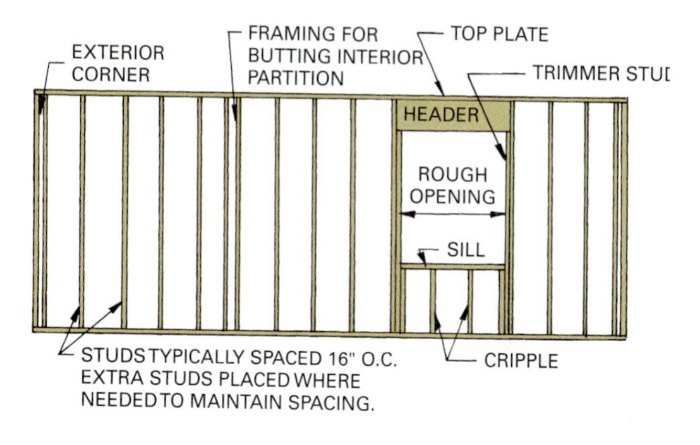

sheets. When plywood or oriented strand board are used, they provide the required lateral bracing for the wall. When rigid foam plastic sheets are used, the wall is braced with let-in diagonal bracing (1 × 4 boards notched into studs diagonally and then nailed when the wall is plumb). Or sheets of plywood or oriented strand board are nailed at each corner and at intervals along the wall, as specified by an architect. Figure 20.30 shows rigid panels on the corner and nonstructural asphalt-impregnated sheets on the remainder of the wall.

Ceiling Framing

Ceiling joists are supported on exterior walls and load-bearing interior partitions. They carry the finish ceiling materials and insulation. If they are subject to other loads, such as a second floor storage area, the architect will size them accordingly. When they become joists for the second floor of a two-story building, their framing is identical to the framing of the first floor (Fig. 20.31). The first ceiling joist is set in from the outer edge of the plate on the gable end to permit construction of the framing for the gable end wall. Ceiling joists can be trimmed, as shown in Fig. 20.32, to follow the rafter shape.

Framing the Roof

The commonly used types of roofs are shown in Fig. 20.33. Roof framing is one of the most difficult parts of wood light frame construction and requires skilled carpentry. The following discussion is limited

Figure 20.23 Walls are assembled on the floor deck and then lifted into place. (b) Image copyright Orange Line Media, 2009. Used under license from Shutterstock.com.

Figure 20.24 This opening in a load-bearing wall is framed with a solid wood header. *(© Rick Barrentine/Corbis)*

to the basic stick-built gable and hip roofs. The various components of roof framing are identified in Fig. 20.34. Rafters rest on top of a wall's double plate and are nailed to it. In areas with high winds, metal straps are used to provide a stronger connection to the plate (Fig. 20.35). Overhangs on the gable end can be built using lookouts supported on the gable end wall and nailed to the nearest rafter beyond (Fig. 20.36). The gable end can be site framed with 2 × 4 in. studs or built with a premanufactured gable end truss. Information on finishing cornices can be found in Chapter 22.

WOOD TRUSSES

Wood Roof Trusses

Wood roof trusses are widely used for framing wood light frame buildings. A roof truss is a triangulated structural unit made by assembling structural wood members into a rigid frame. A typical example is shown in Fig. 20.37.

The advantages of wood trusses are that they speed on-site erection time and can span the width of most buildings without interior load-bearing walls. Since they are carefully engineered and factory built, they have a consistent, reliable quality. One disadvantage is that they make it difficult to use an attic for storage. However, trusses are available that allow the center of a building to be open for storage or second floor rooms. A few of the many types of wood trusses available are shown in Fig. 20.38.

Small trusses can be hung between exterior walls with the ridge point down and then raised by workers standing on the platform using wood poles to spin the point into an upright position. Workers on the top plate set it in position, secure it to the top plate, and brace it. Generally, trusses are set by a crane that raises them over the building and lowers them to the top plate, and workers on the plate secure and brace them, as shown in Fig. 20.39.

Figure 20.25 Various types of header details.

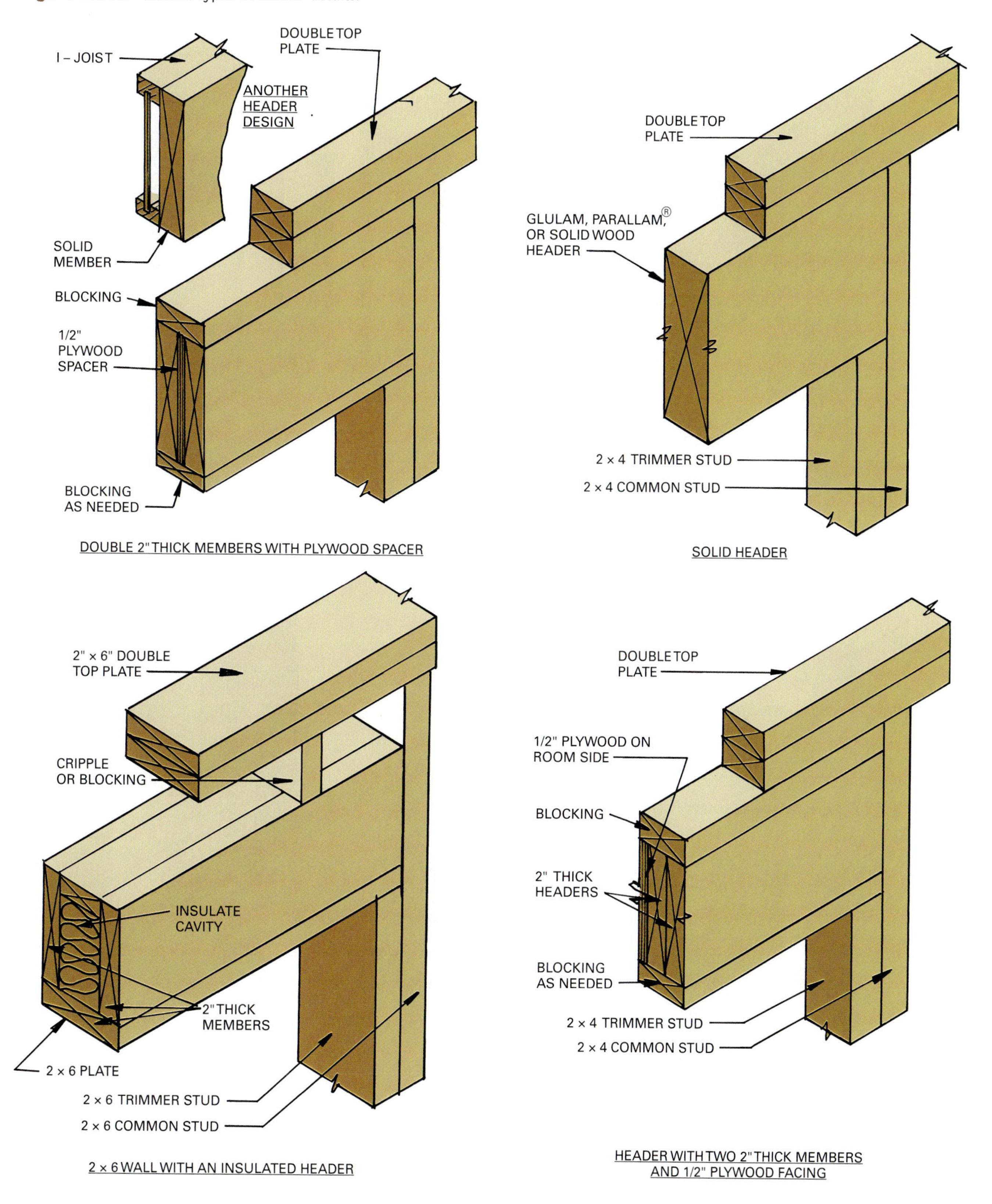

Figure 20.26 A section through the framing of a door opening in an exterior wall.

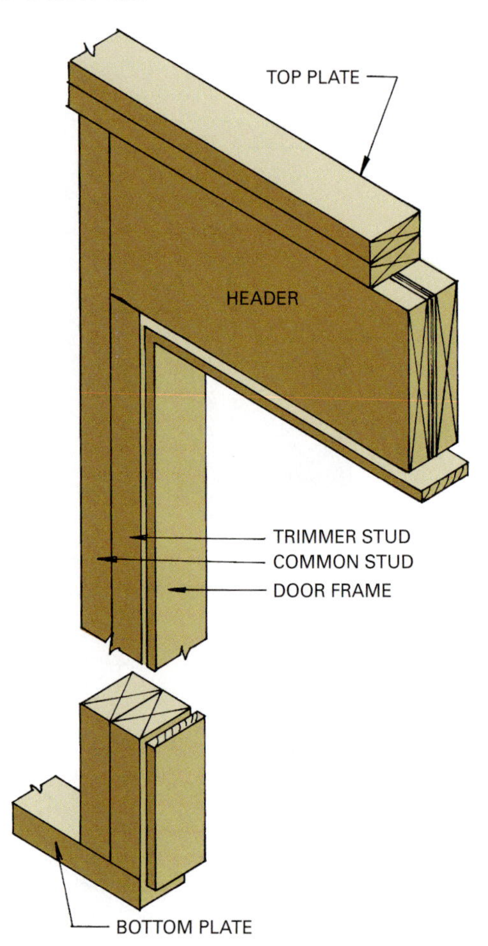

Figure 20.27 An exterior wall corner framing detail.

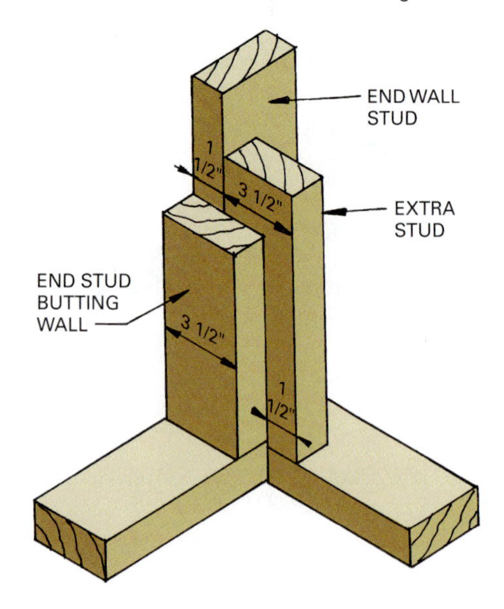

Figure 20.28 When exterior wall partitions butt exterior walls, the top plate overlaps to provide a strong connection.

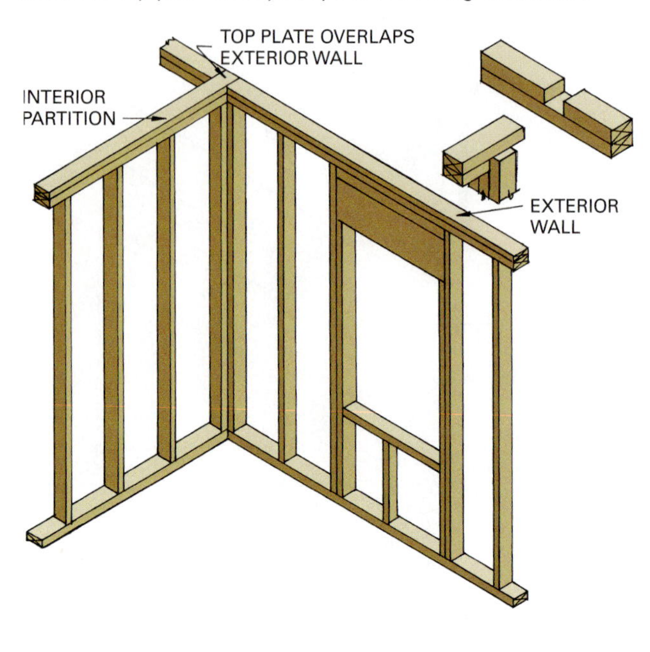

Figure 20.29 A connection detail between the exterior wall and an interior partition.

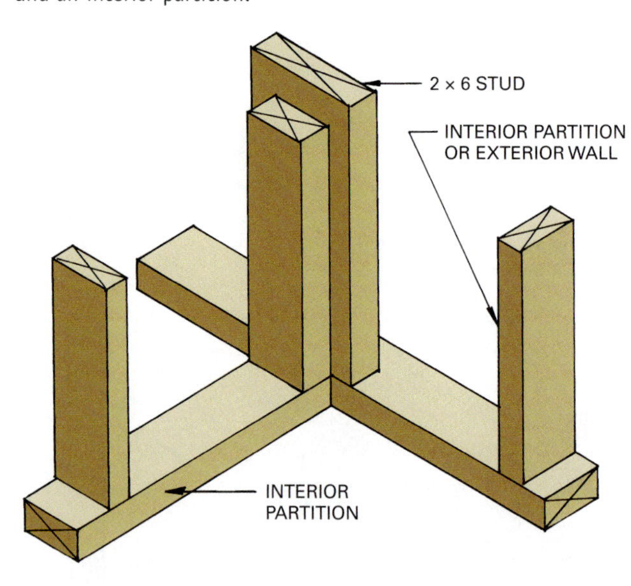

Figure 20.30 Exterior walls can be braced with panels of rigid sheathing, such as plywood. *(Image copyright Breadmaker, 2009. Used under license from Shutterstock.com)*

Figure 20.31 Completed second floor framing using a PSL rim joist.

As trusses are erected, they are secured by braces to the ground and platform, and braces are nailed between them until the sheathing is installed. Once placed and braced, roof decking is installed. Roof decking can be plywood or oriented strand board and is installed similar to floor decking (Fig. 20.40).

Wood Floor Trusses

Wood floor trusses are frequently used instead of solid or engineered wood joists. They provide long spans and often eliminate the need for one or more beams to support a long-spanned floor. Since they have open webs, it is easy to run electrical, mechanical, and plumbing systems through them. Wood floor trusses may be all wood or utilize metal webs between wood top and bottom chords (Fig. 20.41). Wood floor trusses must be installed following the manufacturer's instructions.

Panelized Frame Construction

Panelized construction uses a series of standard factory-assembled wall, floor, and roof units that are constructed on the job. In the United States, the units are made in 16 in. units based on a 4 in. module. Metric units are based on a 100 mm module in 400 mm units. Some typical wall panels are shown in Fig. 20.42. They include provisions for doors and windows. The actual length of the units available depends on the manufacturer, but they are typically 16, 32, 64, 80, 96, and 144 in., based on a 4 in. module. Comparable metric units are 400, 800, 1600, 2000, 2400, and 3600 mm, based on a 100 mm module.

When the units are erected on the subfloor, they are tied together with a top plate. Manufacturers install the sheathing, and sometimes windows, in the factory. The design of the panels must include ways to join exterior and interior corners. This is accomplished by adding extra studs and providing an overlapping flange of sheathing, as shown in Fig. 20.43. Slots are provided for the connection of interior partitions to exterior panels.

Roofs for panelized construction are often framed with factory-assembled gable ends and trusses. They are set in place with a crane and braced until sheathing is applied. The gable end comes sheathed and has the siding applied. The floor can be built with conventional joists and plywood subflooring. However, large factory-assembled floor panels can be used, and widely spaced beams can replace the floor joists.

Construction Methods

Optimum Value Engineering

Optimum Value Engineering (OVE), also referred to as advanced or in-line framing, refers to a series of wood framing techniques designed to utilize less lumber in the construction of a structure. Advanced framing construction detailing can reduce the amount of lumber used in a building while still providing adequate structural strength. Many OVE strategies have the additional benefit of reducing the amount of labor required in the construction process.

The most commonly used framing procedure employed entails the use of 2 × 4 studs spaced at 16 inches (406 mm) on center. This practice has become the nationwide standard over the last 100 years. The on-center spacing can be enlarged to 24 inches (610 mm) with no other necessary changes or detrimental effects to structural integrity. In-line framing aligns studs with the floor and roof joists. For example, 2 × 6 studs align with 2 × 10 engineered joists above, transferring their load directly "in line" to the foundations. An enlargement of the on-center spacing of structural members provides benefits beyond the conservation of material resources. A lower percentage of wall studs has the additional benefit of increasing the area available for insulation and reducing the amount of thermal bridging inherent in wall wood studding.

Headers are the structural members over door and window openings that transfer ceiling and roof loads to vertical "trimmers" and the foundations. The standard header is built from two two-inch-thick framing members with a piece of one-half inch plywood sandwiched between them, which equals the required 3–½ in. wall width of the standard 2 × 4 wall. In order to expedite the construction process, headers are commonly built of 2 × 12 (38 x 305 mm wide) material. A header of this size can be fastened directly to the top plate of a typical eight foot wall to provide the standard six foot eight inch window and door head height. Although full 2 × 12 headers are not required in interior non-load-bearing partitions, most builders put them in every opening to ensure safety. By carefully sizing headers to their anticipated load, tremendous material savings result.

Engineered lumber, such as I-joists and rafters, combines smaller dimensional lumber and particleboard and reduces the amount of cut off waste. Using OVE techniques can result in reduction of cut-off waste from standard-sized building materials; reduction in the number of top plates needed; and reduction of the number of studs in building corners and exterior bearing walls. Higher energy efficiency resulting from decreased thermal bridging across structural members can have a measurable impact on heating and cooling costs.

STEEL FRAMING

Several manufacturers offer a complete **lightweight steel framing** system for walls, floors, and roofs. Because quality solid wood framing is getting more expensive, steel framing will find increasing use.

Steel members are perforated to lighten them and permit the passage of utility lines, such as plumbing and electrical wiring. They are noncombustible and various types of exterior sheathing and siding can be applied; they are easy to insulate; and gypsum and other interior finish materials can be secured to them. Steel members can be preassembled into wall panels to speed erection. Steel framing is widely used for interior and exterior wall construction in commercial and, more recently, residential construction.

The lightweight steel framing system includes studs, tracks, joists, bracing, hangers, and other accessories needed to assemble the unit. Bearing walls can be single or double studs to increase the load-bearing capacity of the wall. The studs sit in a metal track that is screwed to the subfloor (Fig. 20.44). The studs are connected to the track with special power driven screws. Some connections, such as bracing or joining beams to the wall's top track, must be welded following the manufacturer's specifications. Posts are made by assembling studs inside tracks. Headers over door and window openings are built up using joists secured inside track material.

Assembled walls require metal furring for strength that are screwed and spaced as specified (Fig. 20.45). Floors and ceilings are assembled using lightweight steel joists. Bridging may consist of straps or sections of joists forming solid bridging. In a multistory building, the second-floor joists rest directly above a first-floor stud that may be single or double. Short sections of joist are welded in on the end of each floor joist to stiffen it and help carry the load of the second-floor

Figure 20.32 Ceiling joists commonly span from the exterior wall to a load-bearing interior partition.

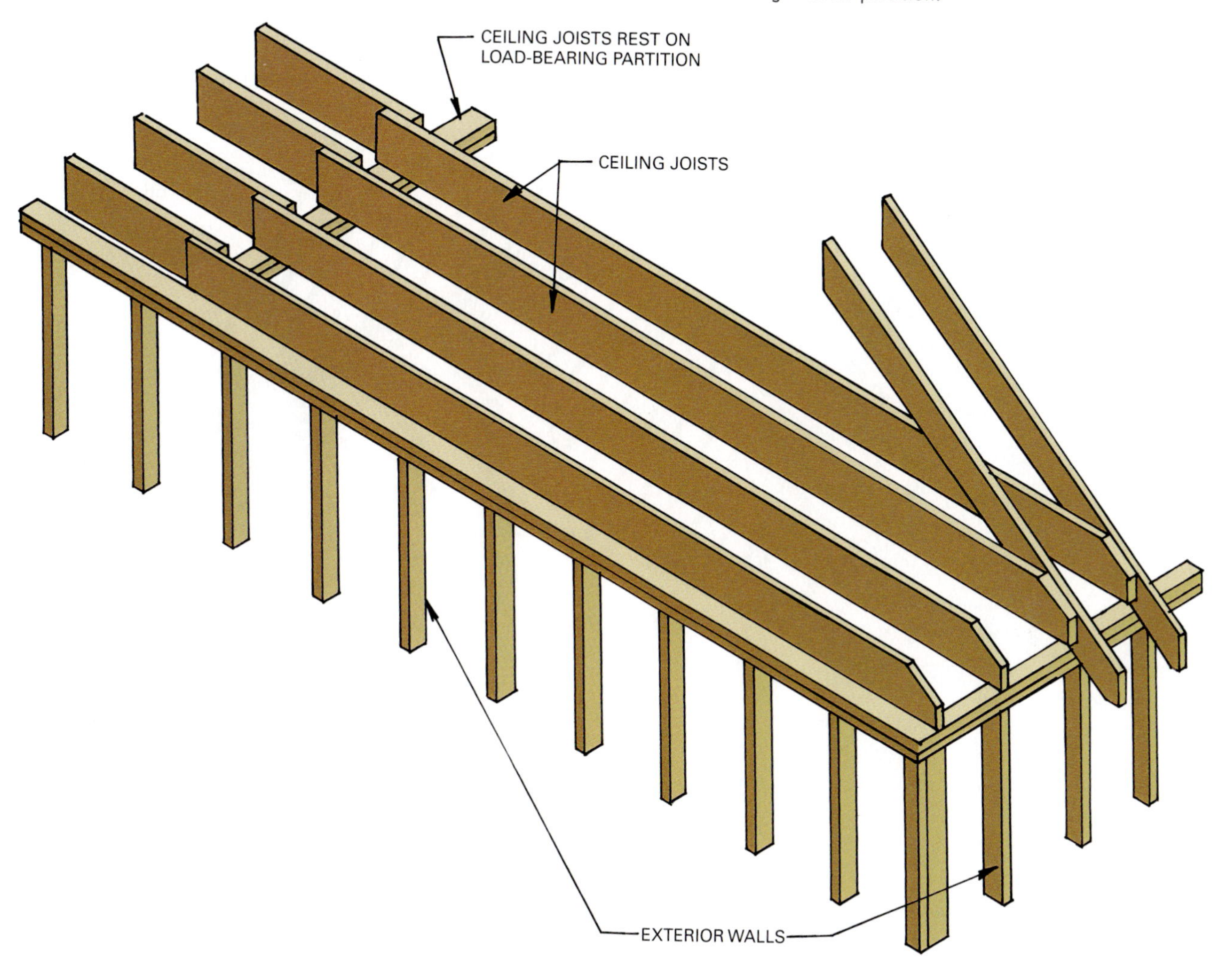

Figure 20.33 The most commonly used roof forms in light frame construction.

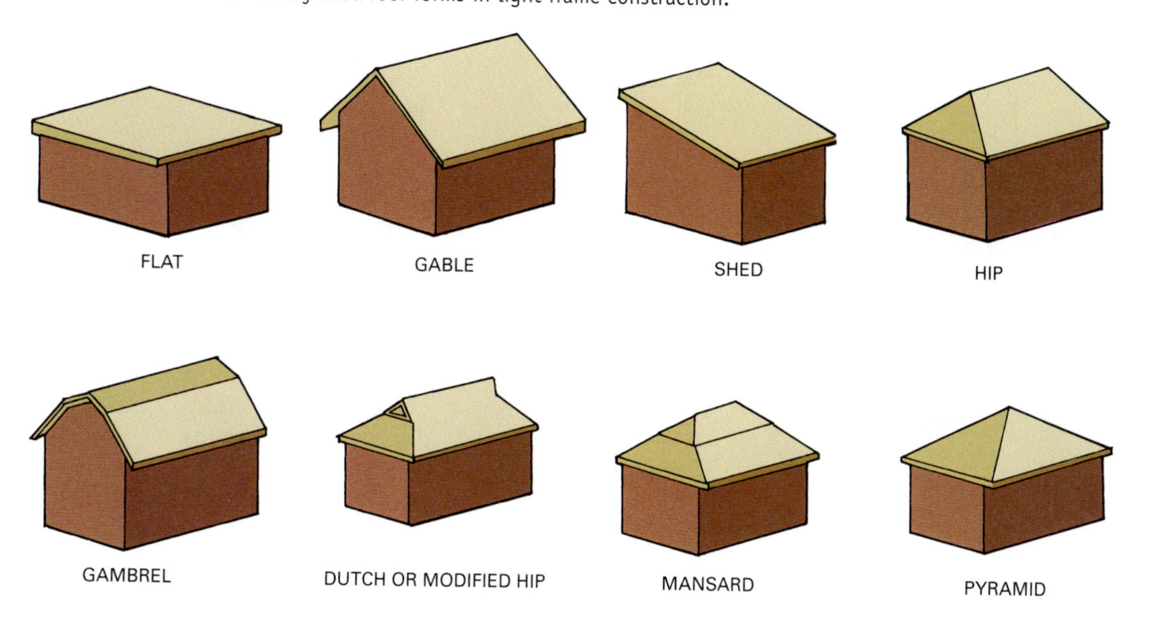

Figure 20.34 The framing members of a typical stick built roof.

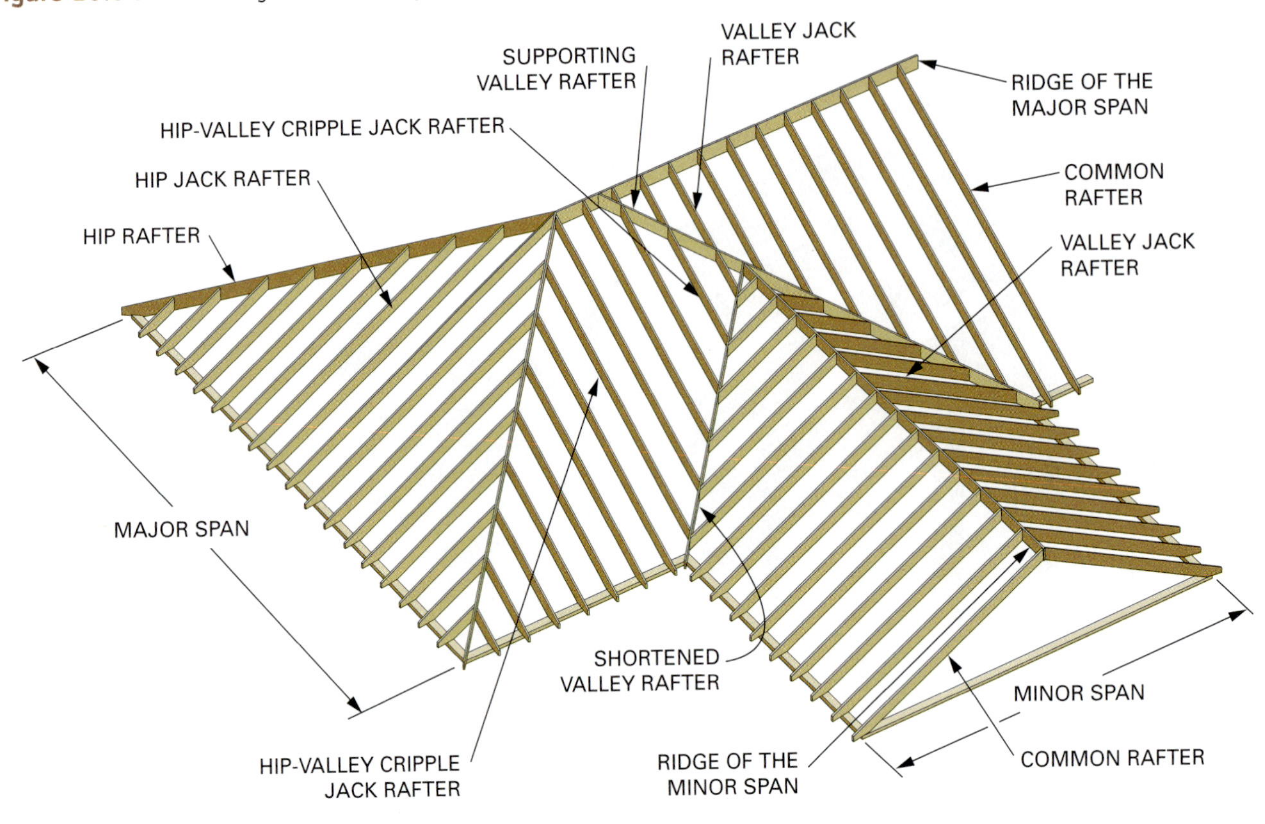

Figure 20.35 Rafters, trusses, and roof joists can be toenailed to a top plate or use metal connectors for increased strength.

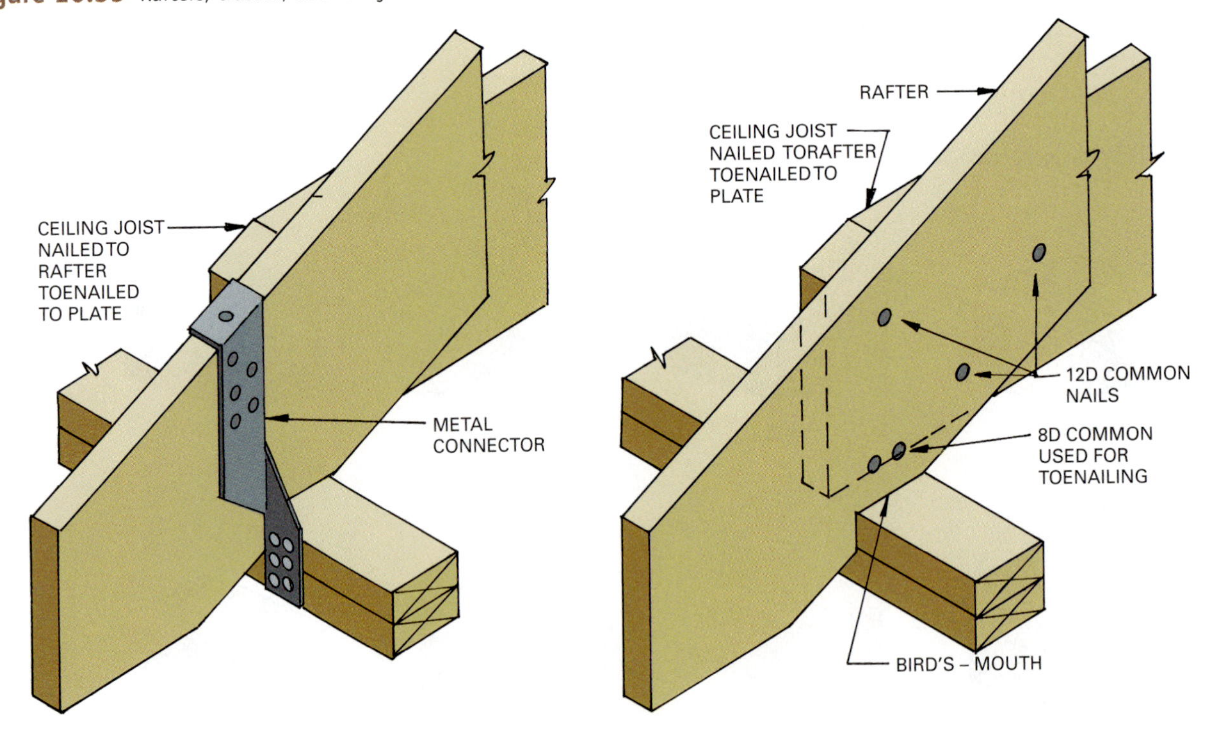

Figure 20.36 Gable end roof overhangs are framed using lookouts.

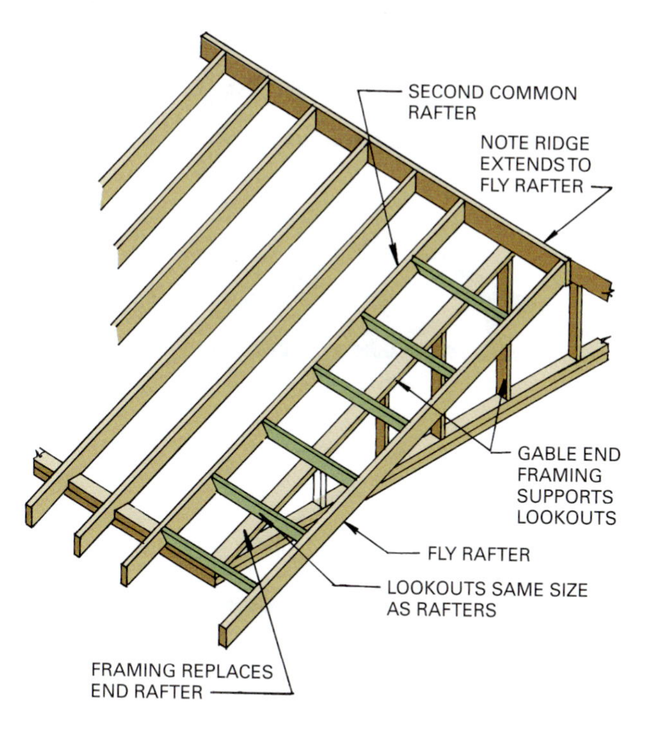

wall. Roofs are built using trusses assembled from metal channels. Refer to Chapter 17 for additional details.

Steel studs designed for nailing to wood top and bottom plates provide rapid assembly of walls and partitions. Walls are assembled by laying out studs on the subfloor and nailing them to the wood plates in the same manner as a wood stud. The wall is then lifted into place. Door and window openings are often lined with wood to allow for easier fastening of finish materials (Fig. 20.46). Additional lightweight steel framing details for commercial construction can be found in Chapter 17.

Figure 20.37 A wood truss is held together by metal gusset plates.

Figure 20.38 Some commonly used wood trusses.

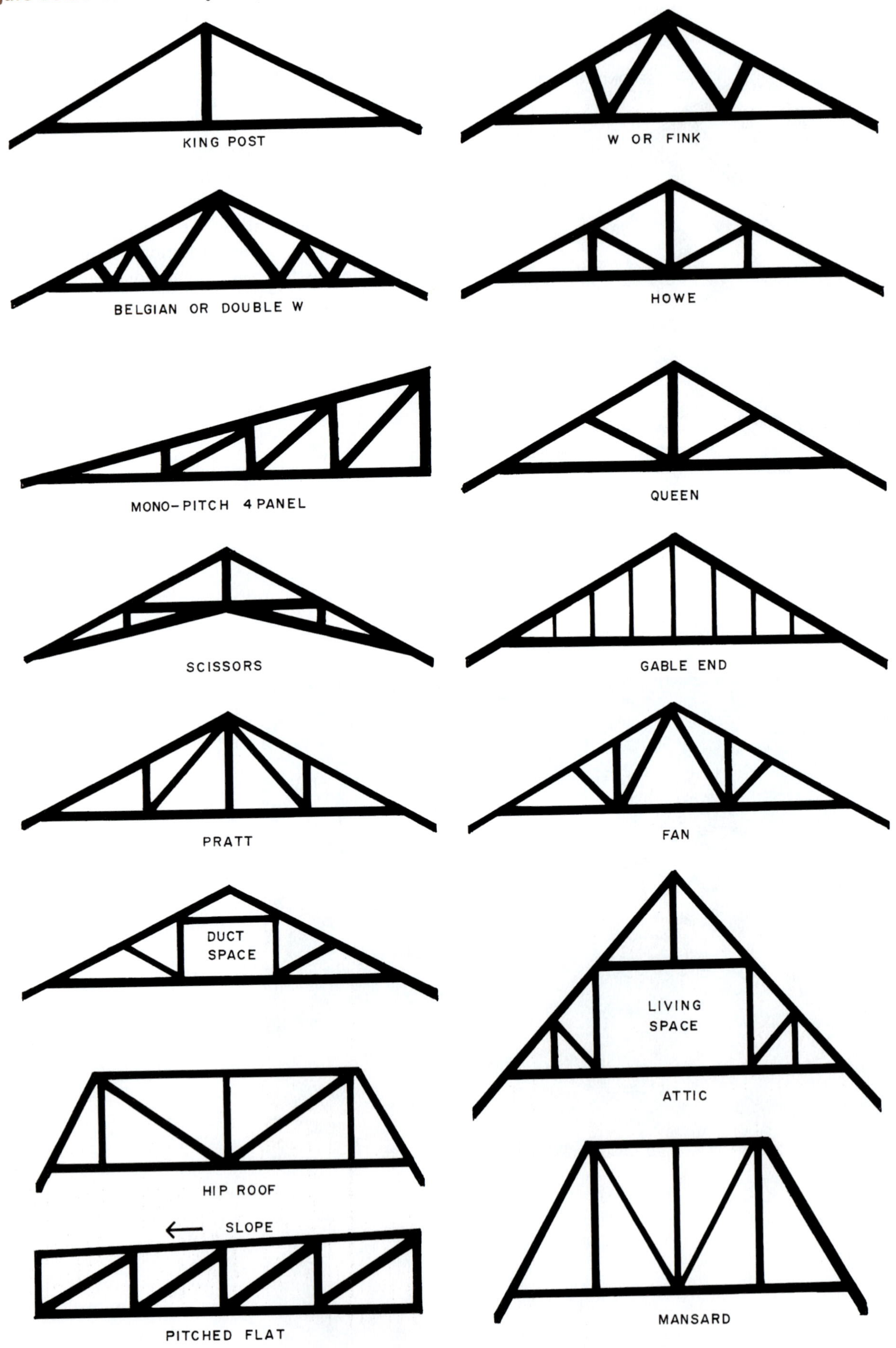

Figure 20.39 Roof trusses are set and temporarily braced against each other before the roof deck is installed.

Figure 20.40 Decking a trussed roof with plywood.

Figure 20.41 Wood floor trusses allow space for mechanical systems to be run inside the floor sandwich. *(Courtesy Joel Sartore/Getty Images)*

Figure 20.42 Examples of prefabricated wall panels.

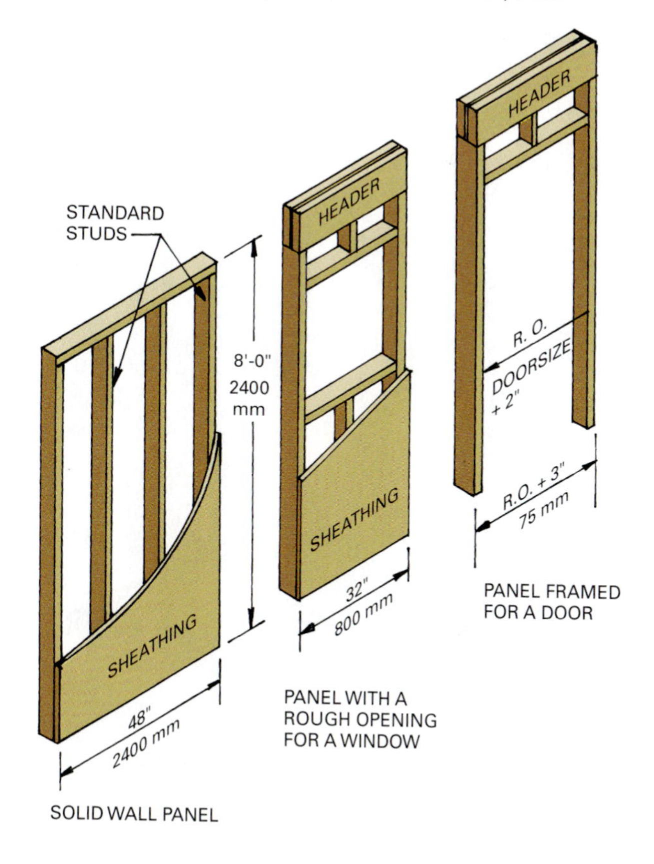

Figure 20.43 Connection details for structural insulated panels.

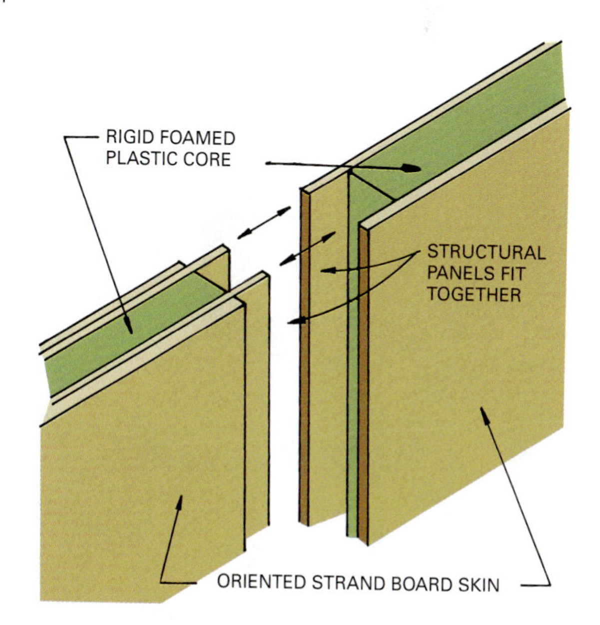

Figure 20.44 Steel studs come in several widths, lengths, and thicknesses.

Figure 20.45 Furring channels (hat track) are used in both ceiling and wall installations.

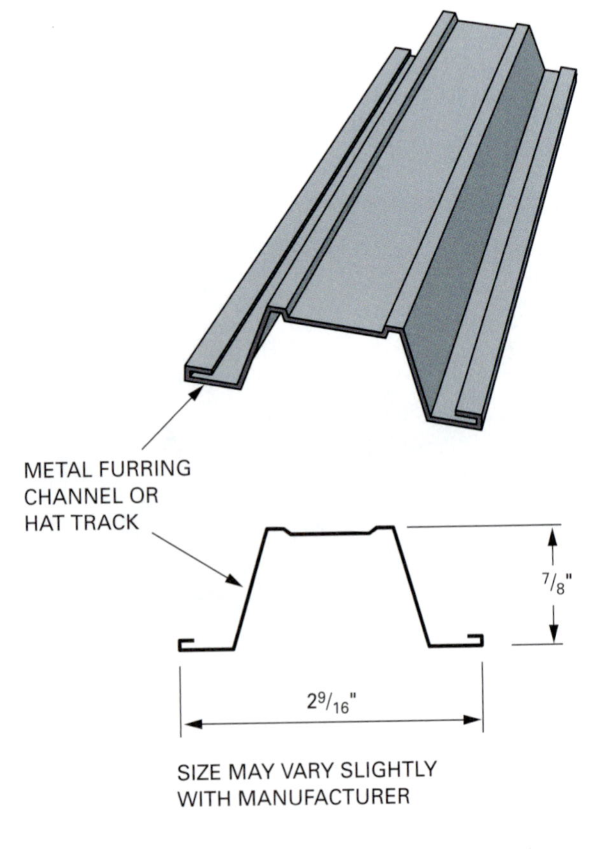

Figure 20.46 Window and door openings may be cased in wood for easier fastening of finish materials.

Review Questions

1. What are the commonly used framing methods used in light frame construction?
2. What is box sill construction?
3. When is a beam required under floor joists?
4. How do the studs in balloon framing differ from those in platform framing?
5. How are second-floor joists supported when using balloon frame construction?
6. What components make up an assembled truss unit used for truss framed construction?
7. What is meant by panelized construction?
8. What members are used to make a wood I joist?
9. What are the advantages of using wood floor and roof joists?
10. What are the members used to construct a wall with lightweight steel members?
11. How are posts built using lightweight steel members?

Key Terms

balloon framing
lightweight steel framing
panelized construction
platform framing
rim joist

Activities

1. Invite a local building official to speak to the class about how plans are checked to see if they conform to building codes.
2. Lay out the corners of a foundation for a small building and set the batter boards.
3. Visit as many construction sites as possible and observe the procedures used to lay out and construct light frame buildings.
4. Construct a two-story scale model of a small building using platform framing.

Additional Resources

AISC Manual of Steel Construction, American Institute of Steel Construction, Chicago, IL 60601.

JLC Field Guide to Residential Construction, Journal of Light Construction, 186 Allen Brook Lane, Williston, VT 05495.

Spence, W.P., *Carpentry and Building Construction*, Sterling Publishing Co., New York, 1999.

Structural/Seismic Design Manual (Volume 2), International Code Council, Falls Church, VA.

Wood Building Technology, Canadian Wood Council, Ottawa, Ontario, Canada.

Wood Engineering and Construction Handbook, prepared by APA—The Engineered Wood Association, published by McGraw-Hill Corporation, New York.

Wood Frame House Construction, Metric Edition, Canada Mortgage and Housing Corp., Ottawa, Ontario, Canada.

Other resources include:

Publications of the Construction Institute of the American Society of Civil Engineers, Reston, VA.

See Appendix B for addresses of professional and trade organizations and other sources of technical information.

CHAPTER 21

Heavy Timber Construction

LEARNING OBJECTIVES

Upon completion of this chapter, the student should be able to:

- Understand the requirements of the building code as it pertains to heavy timber construction.
- Become familiar with the materials and methods for framing heavy timber buildings and the types of connections used.

Build Your Knowledge

For further study on these materials and methods, please refer to:

Chapter 18 Wood, Plastics, and Composites

Chapter 19 Products Manufactured from Wood

Topic: Structural Building Components

Heavy timber construction is a method of creating framed structures of heavy timbers joined together with wood joinery, pegs, or metal connectors. Diagonal bracing is used to strengthen structures against racking. Timber braced wall construction developed during the middle ages and was eventually brought to North America by British carpenters. To provide enclosure for these early structures, the spaces between the timbers were infilled with brick, rubble, or wattle and daub (Fig. 21.1). Timber frame buildings tend toward an aesthetic of strength and craftsmanship. Since the frames require no interior load-bearing walls, interior spaces are open and flexible.

Wood products commonly used in heavy timber construction include solid timbers, glue laminated members, parallel strand lumber, and laminated veneer lumber. The selection of material is governed by required load-bearing capacities, appearance, and availability of product. Modern complex structures and timber trusses often incorporate steel joinery for both structural and architectural purposes. Sheathing and decking consists of thick wood members capable of spanning distances between wood structural members (Fig. 21.2). The structural frame can be enclosed in an envelope of insulated panels for high levels of energy efficiency.

Figure 21.1 Half timbering utilized braced frames infilled with brickwork or wattle and daub. *(Image copyright Torsten Lorenz, 2009. Used under license from Shutterstock.com)*

Figure 21.2 This heavy timber frame building uses tongue-and-groove decking on the roof and standard stud construction on the exterior wall areas between the columns. *(Reproduced with permission from The Building Systems Integration Handbook, Richard Rush, ed., Butterworth-Heinman Publisher, Newton, Mass., 1986)*
Some of the glued laminated structural timber members available from various manufacturers.

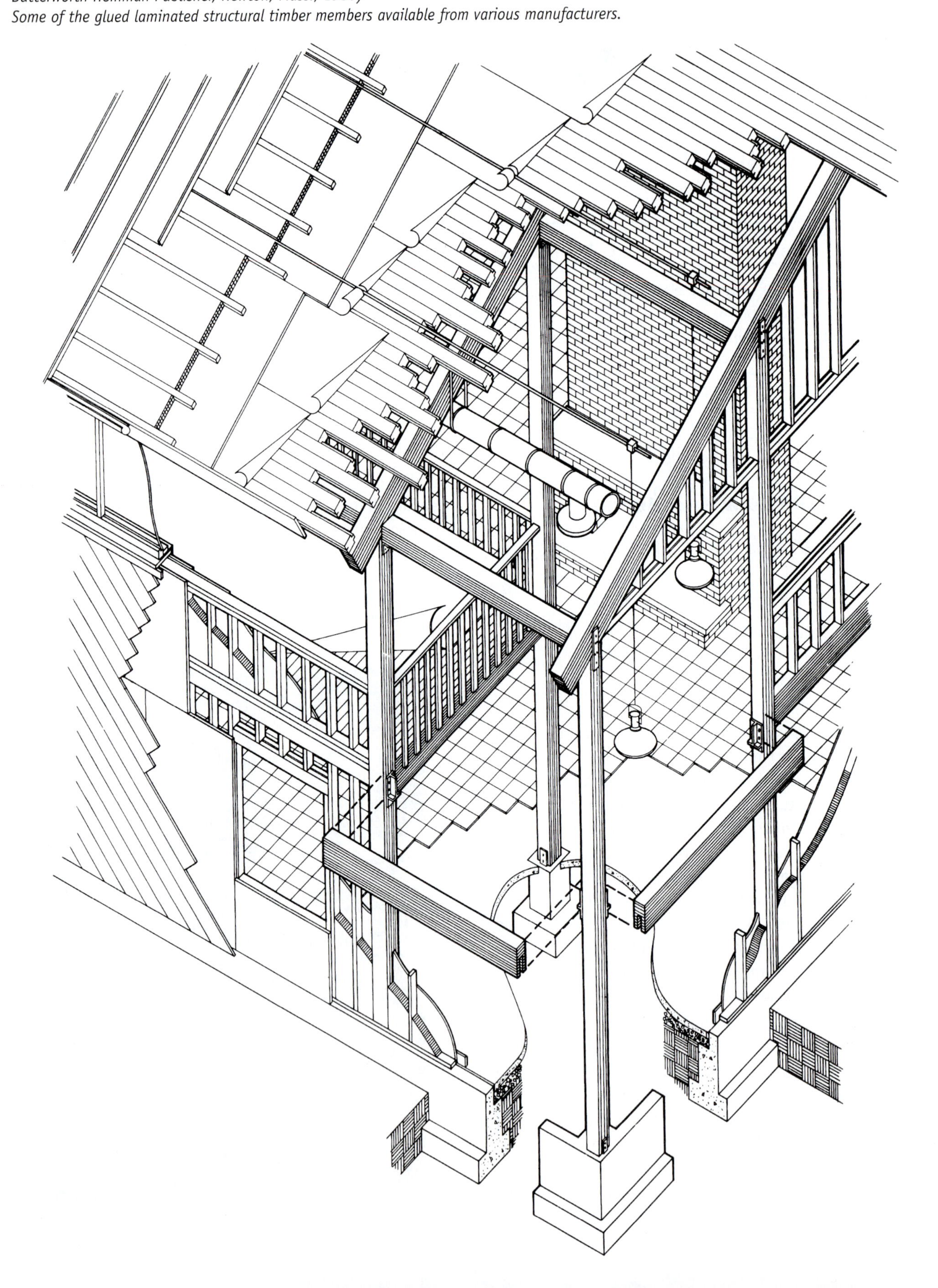

BUILDING CODES

Requirements for heavy timber construction are outlined in the building codes. They typically specify the size and type of approved structural members for columns, floor framing, roof framing, floors, and roof decks. Approved types of connections are detailed, and the fire resistance rating of the structural elements is given. Codes also require seismic analysis and design in earthquake-prone areas. Diaphragms, shear panels, and selection and spacing of fasteners may be specified. The building codes restrict the height of heavy timber buildings to three to five floors, depending on design and occupancy.

The International Building Code classifies buildings into five types. Heavy timber construction typically falls into Type IV and, in some cases, Type V. Under Type IV, exterior walls are constructed of approved noncombustible materials and internal members may be of solid or laminated wood without concealed spaces or approved noncombustible materials. Under Type V, exterior walls, load-bearing walls, partitions, floors, and roofs may be any approved material that includes heavy timber members. Type V construction utilizes heavy timber in conjunction with light wood frame walls for smaller buildings. Refer to Chapter 2 for more information on building codes.

FIRE RESISTANCE

Heavy timber frame structural systems are able to absorb heat and flames by producing a char when exposed to fire. The char produces a layer of fire protection that shields the wood from sustained fire damage and prevents strength loss. Much of their structural strength is retained until the cross sectional area becomes so small that it will not carry the load.

SOLID HEAVY TIMBER CONSTRUCTION

Solid timber construction is used for residential, commercial, and religious buildings. Timber-framed structures differ from light wood framed buildings in a number of ways. Timber framing uses fewer, larger members, commonly with dimensions in the range of 6 to 12 in. (15 to 30 cm), as opposed to light wood framing, which uses many more timbers with small cross areas. The methods of fastening frame members also differ: in conventional framing, members are joined with nails, while timber structures use complex joints that are usually fastened using only wooden pegs.

In the design of timber structures, the choice of material, the types of connections used, and the selection of a structural system all play an important role. An architect and engineer must work together from the onset to define a project's objectives. Timber frames may be supported on concrete foundations or masonry walls. Wood, as a hygroscopic material, can experience significant amounts of expansion and contraction with moisture content, especially perpendicular to the grain. Timber frames are engineered to manage this movement through careful detailing.

Traditional heavy timber construction uses a system of structural frames called bents. Precut and dressed timbers are laid out on the floor for assembly. The bents are constructed with mortise and tenon connections between column, beam, and bracing elements to produce a rigid, braced frame. The bents are raised and braced horizontally to one another. A wide range of materials can be used to serve as exterior sheathing between timber members. Figure 21.3 shows a traditional timber frame building using a series of bents as the structural frame.

Timber Joinery

Timber joinery is the traditional method of connecting heavy timber structural members. It uses interlocking wood joints, such as tenons, dados, or halving joints, that are cut into the wood members. Once hand worked by experienced carpenters, these details are today

Figure 21.3 This heavy timber framing uses mortised connections, producing an exposed interior framing free from metal connectors. *(Courtesy The Murus Company, Inc.)*

produced by modern machinery. Some common wood-to-wood connection details are shown in Fig. 21.4. Although these connections take longer to make, the finished joint demonstrates hand craftsmanship and the natural beauty of the wood. Experienced timber framers still use timber joinery, although metal connectors have largely replaced their use.

The structural framing in Fig. 21.5 shows typical beam and column construction and a heavy timber frame roof. Figure 21.6 illustrates a detail for timber beam floors supported on masonry walls. The beam is set into a steel pocket anchored into the masonry bearing wall. A space is left between wood and masonry to prevent moisture migration through the wall from reaching the wood. The end of the beam is fire cut to allow it to fall during a prolonged fire without damage to the wall.

Beam-to-column connections may be made with metal connectors or wood bearing blocks bolted to a column with **split ring connectors** to control vertical forces (Fig. 21.7). Beam-to-girder or beam- or purlin-to-beam connections generally use some type of metal connector. However, wood ledgers with metal lateral ties are used in some cases. An engineered connection using metal connectors and wood angled supports bolted to the connector is shown in Fig. 21.8. The work of a structural engineer is critical when designing structural systems and details to carry imposed loads.

Figure 21.4 Typical joints used in joining solid wood framing members.

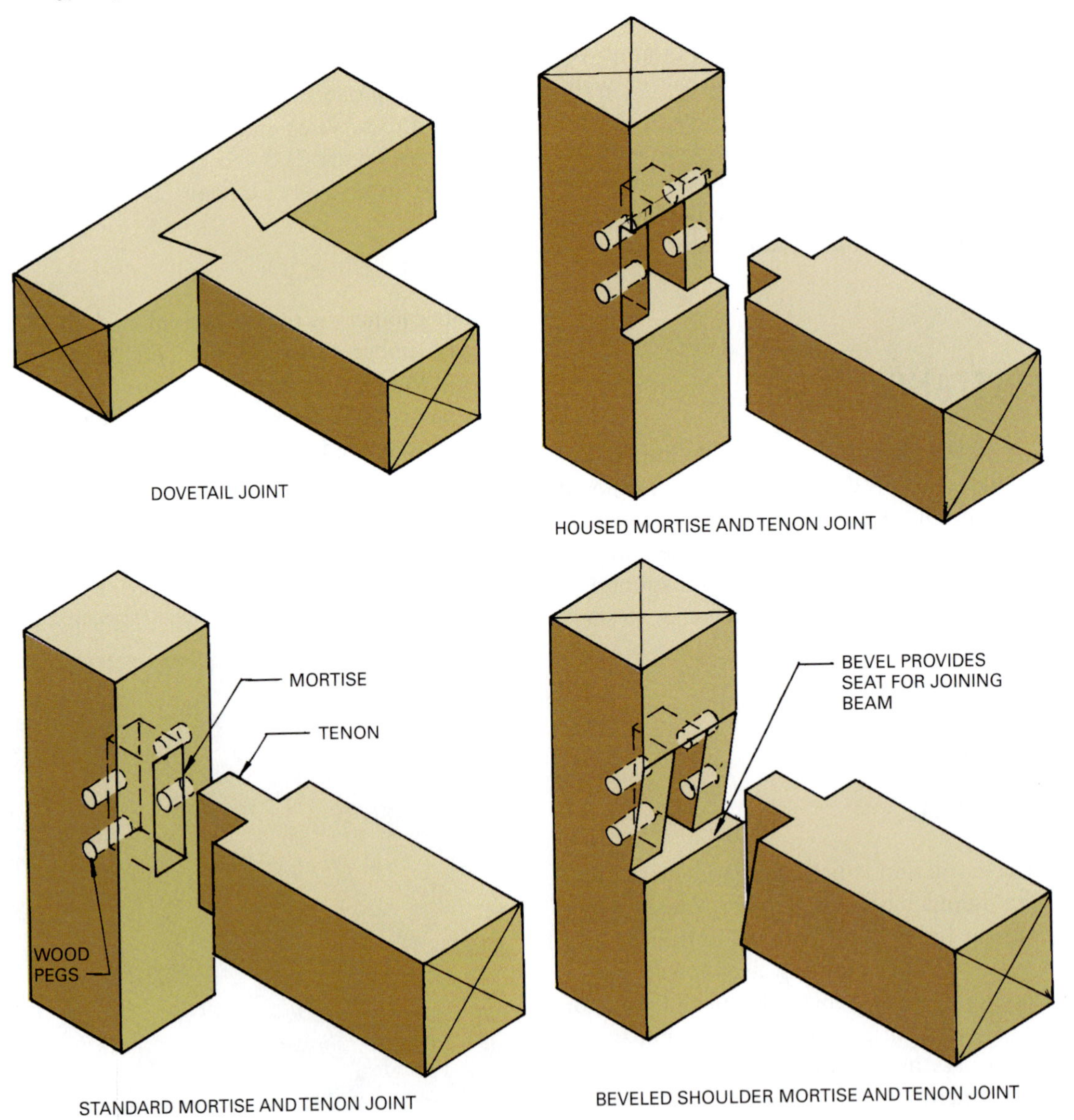

Figure 21.5 A typical beam, column, and rafter connection. *(Image copyright Troy, 2009. Used under license from Shutterstock.com)*

Figure 21.6 Wood beams are set in metal anchors that are tied to the foundation or a masonry wall. *(Courtesy American Forest & Paper Association, Washington, D.C.)*

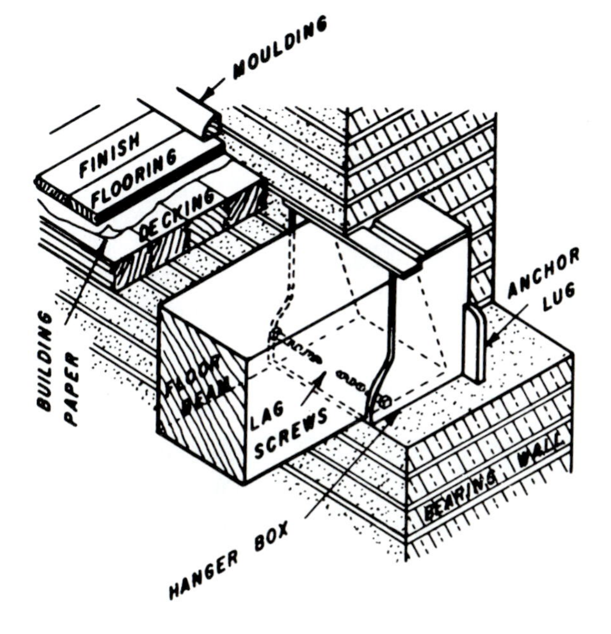

Figure 21.7 Ways to tie solid wood beams to columns. *(Courtesy American Forest & Paper Association, Washington, D.C.)*

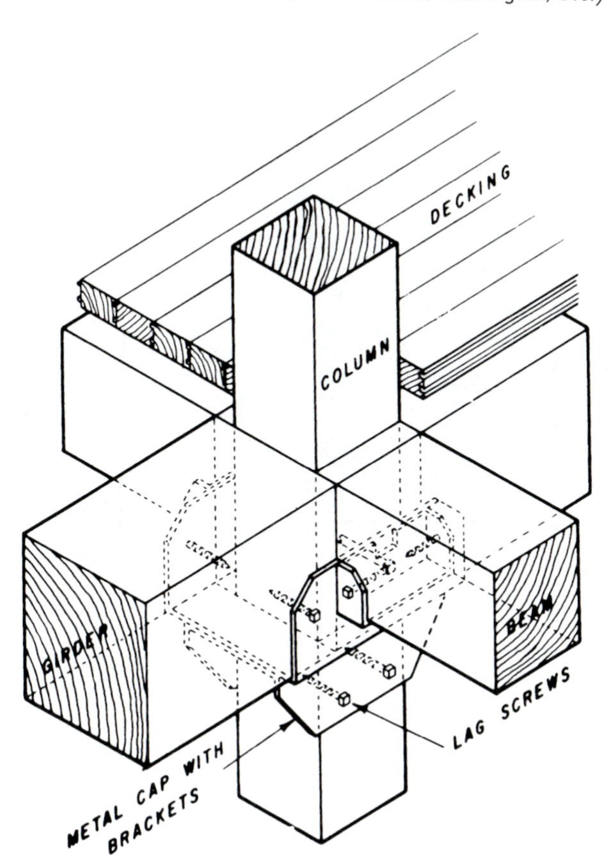

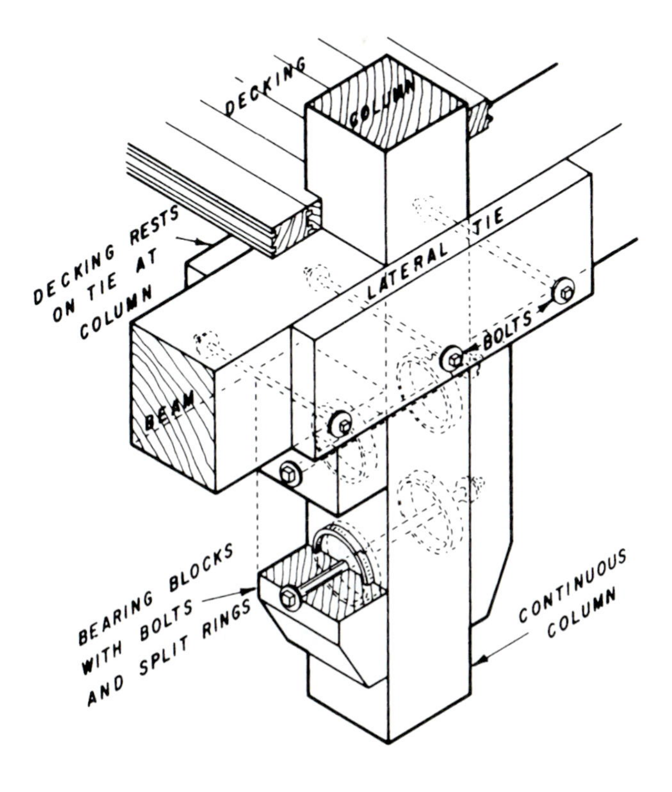

Figure 21.8 This connection has angled wood members tied to overhead beams with metal connectors. *(© Keith Hunter/Arcaid/Corbis)*

GLUED LAMINATED CONSTRUCTION

The use of engineered wood allows designers to employ larger members with better structural properties than solid wood. Columns, beams, joists, rigid frames, arches, domes, and decking can be made from glued laminated or other engineered wood products. Information about these products is available from the American Institute of Timber Construction (AITC) and in Chapter 19.

Glued laminated members are engineered, stress-rated members made by laminating soft wood members with adhesives and resins. Standard glulam timber members are shown in Fig. 21.9. The actual sizes of commonly available rectangular glued laminated members are shown in Table 21.1. Glulams are formed using adjustable presses that can produce a wide variety

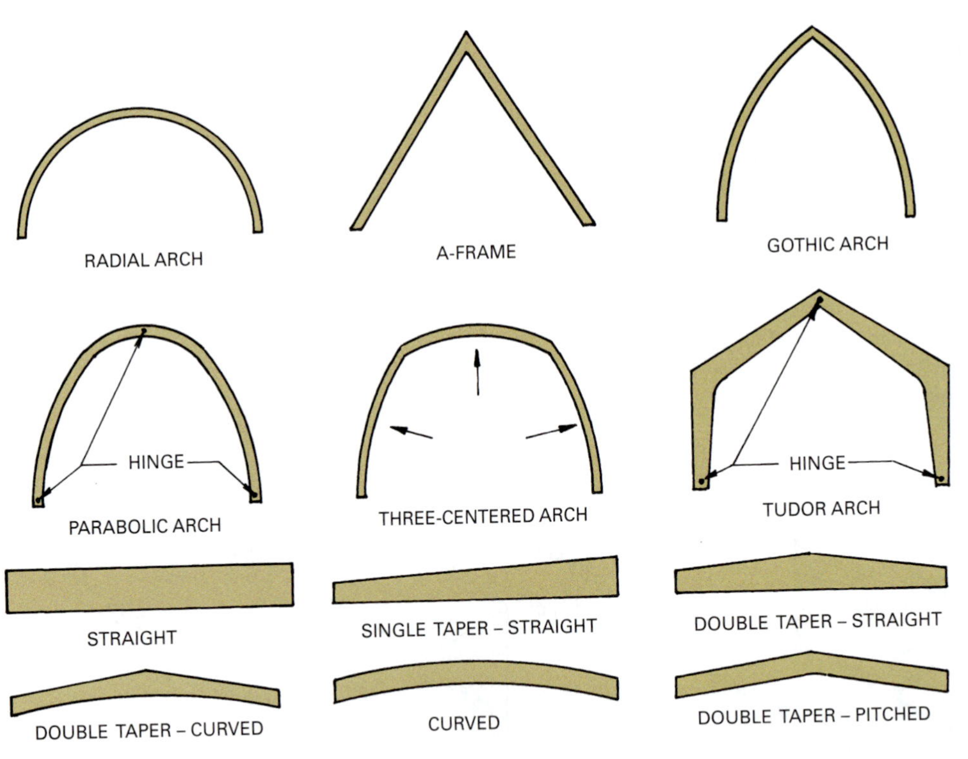

Figure 21.9 Commonly used forms for structural glue laminated members.

Table 21.1 Actual Sizes of Commonly Available Rectangular Glued Laminated Members

Laminations $1^{1}/_{2}$ in. (38 mm) Thick				Laminations $1^{3}/_{8}$ in. (35 mm) Thick			
Width (in.)	depth (in.)	width (mm)	depth (mm)	width (in.)	depth (in.)	width (mm)	depth (mm)
$3^{1}/_{8}$	$7^{1}/_{2}$	79.3	190.5	3	$6^{7}/_{8}$	76.2	174.6
$5^{1}/_{8}$	6	130.1	152.4	5	$6^{7}/_{8}$	127	174.6
$5^{1}/_{8}$	9	130.1	228.6	5	$8^{1}/_{4}$	127	209.5
$5^{1}/_{8}$	$10^{1}/_{2}$	130.1	266.7	5	11	127	279.4
$6^{3}/_{4}$	9	171.4	228.6	$6^{3}/_{4}$	$8^{1}/_{4}$	171.4	209.5

of forms, their length limited only by transportation restrictions. In addition to beams, rectangular-, tapered-, and varying-shaped columns are also manufactured.

Arches and Domes

Glued laminated arches and domes may be two-hinged, with hinges at each base, or three-hinged, with an additional hinge at the crown (Fig. 21.10). A **hinge joint** is any joint that permits movement but in which there is no appreciable separation of adjacent members. These members produce considerable horizontal thrust at their base, which is controlled by the foundation and tie rods.

Laminated arches can span long distances, up to 70 ft. (21.4 m) or more. Span capabilities depend on the type of arch, imposed loads, roof pitch, and species of wood. Arches can be left exposed inside a building, forming an open ceiling (Fig. 21.11).

Typical construction details for arches are shown in Figs. 21.12 through 21.14. Figure 21.12 shows details for connecting parabolic arches and domes to a buttress foundation. The shoe plate is anchored to the foundation and bolted to the metal connection on the end of the arch. Space is left between the metal and wood and weep holes allow for the removal of water. A bridge pin connects the base shoe imbedded in concrete to the arch bracket. Several types of crown connections are shown in Fig. 21.13. Long-span arches may require sections added on-site because the assembled arch would be too large to transport.

Glued laminated wood domes are designed as a radial arch or a triangulated system. The triangulated arch can span greater distances than the radial arch. The arch shown in Fig. 21.14 acts as both roof and exterior wall. Another type of dome construction, shown in Fig. 21.15, produces a dramatic curved ceiling. Arched laminated wood structural members are also used for a variety of other projects, such as pedestrian bridges, towers, and exterior pavillions.

Columns, Beams, and Trusses

Several ways to secure both solid and glued laminated beams to foundations are shown in Fig. 21.16. Heavy steel clips and bolts resist both vertical and horizontal

Figure 21.11 The glue laminated arches of Boston's Back Bay are stabilized by metal tie rods. (© ART on FILE/CORBIS)

Figure 21.10 Arches and domes may be two-hinged or three-hinged. The horizontal thrust at the base is controlled with metal tie rods.

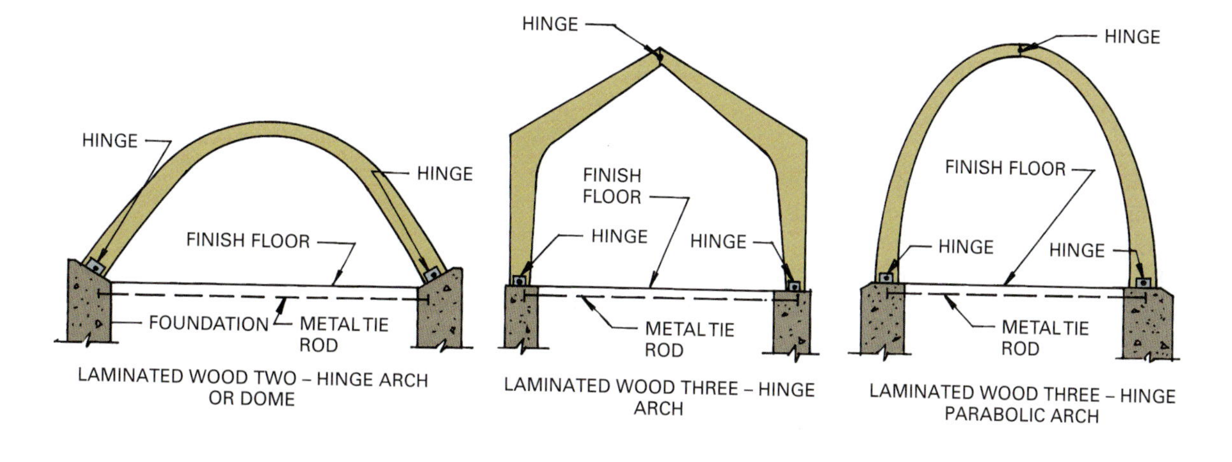

Figure 21.12 Dome-type arches are anchored to concrete foundations with steel anchor plates secured to them. *(Courtesy American Institute of Timber Construction, 7012 S. Revere Parkway, Suite 140, Englewood, CO 80112)*

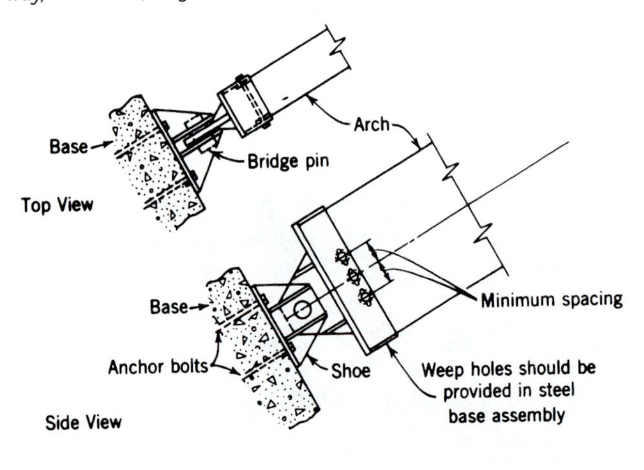

TRUE HINGE ANCHORAGE FOR ARCHES.

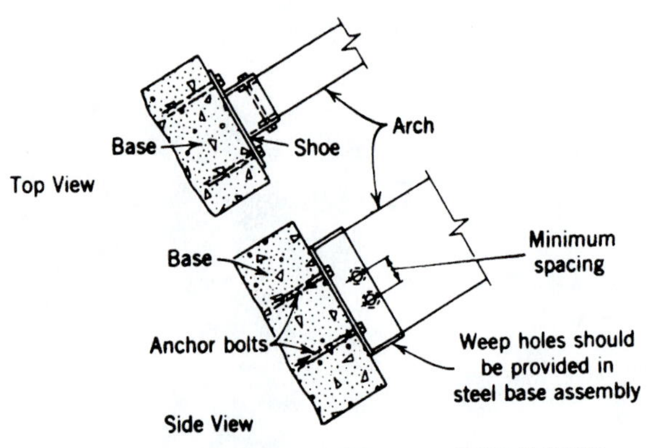

ARCH ANCHORAGE WHERE TRUE HINGE IS NOT REQUIRED.

Figure 21.13 Typical connections for the crown of arches.

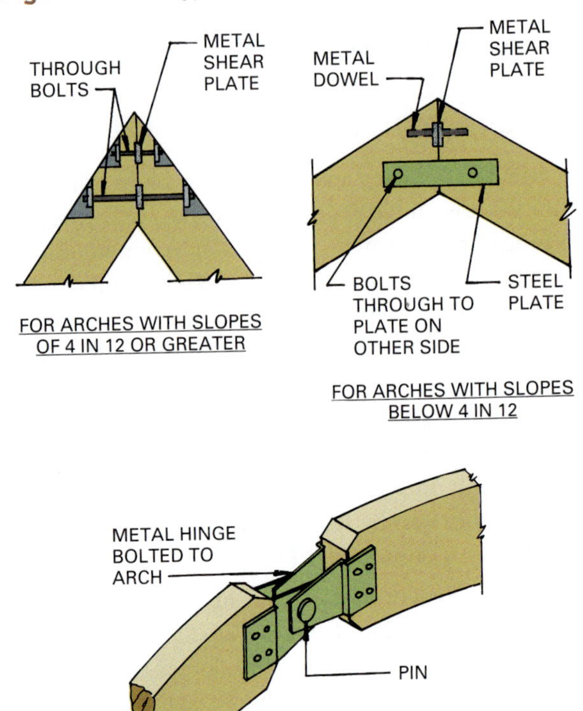

forces. It is recommended that beams rest on a metal plate. Column-to-foundation connections are shown in Fig. 21.17. The column should rest on a metal plate and be at least 3 in. (76 mm) above grade or the finished floor. Large hold-down anchors are used to meet structural tie-down and tilt-over requirements specified by building codes. Steel connectors are used to secure beams and girders to the columns (Fig. 21.18). The design engineer must indicate the type of fastener to be used and the size and number of bolts. Several beam-to-beam and beam-to-girder connections are shown in Fig. 21.19.

Heavy timber trusses are used to span the width of a building and support purlins that carry the roof decking. The bolted, multimember truss in Fig. 21.20 is supported on corbelled masonry pilasters that are an integral part of the masonry wall. The bolted connections have split rings.

Connectors

Two connectors used to provide additional strength to a bolted joint are split ring connectors and shear plate connectors.

Split ring connectors are ring-shaped metal inserts placed in grooves cut into mating wood pieces that are secured with bolts (Fig. 21.21). They are used on wood-to-wood connections to distribute lateral forces over a larger bearing area than the bolts alone provide. Split ring connectors greatly increase the load-bearing capacity of a joint. They are available in $2\frac{1}{2}$ in. (60 mm) and 4 in. (102 mm) diameters. The $2\frac{1}{2}$ in. split ring is used with $\frac{1}{2}$ in. (12 mm) bolts, and the 4 in. split ring requires a $\frac{3}{4}$ in. (19 mm) bolt. They are also used with lag screws instead of bolts.

Another type of connector used in heavy timber construction is a shear plate (Fig. 21.22). **Shear plates connectors** are used for metal-to-wood connections. The example shows a shear plate and bolts securing a steel plate forming a connection between two wood members. Shear plates are available in $2\text{-}\frac{5}{8}$ in. (52 mm) and 4 in. (102 mm) diameters. A $\frac{3}{4}$ in. (19 mm) bolt is used with the $2\text{-}\frac{5}{8}$ in. shear plate and a $\frac{3}{4}$ or $\frac{7}{8}$ in.

Figure 21.14 Triangulated glue laminated wooden domes can achieve spans up to 500 feet.

Figure 21.15 The glulam framing of this wooden dome becomes an architectural feature. *(© John Edward Linden/Arcaid/Corbis)*

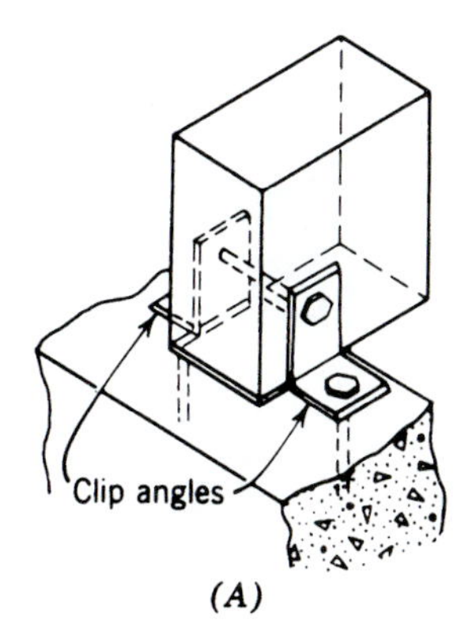

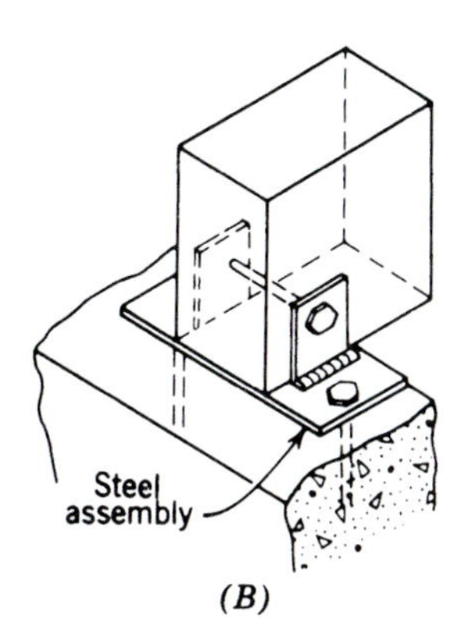

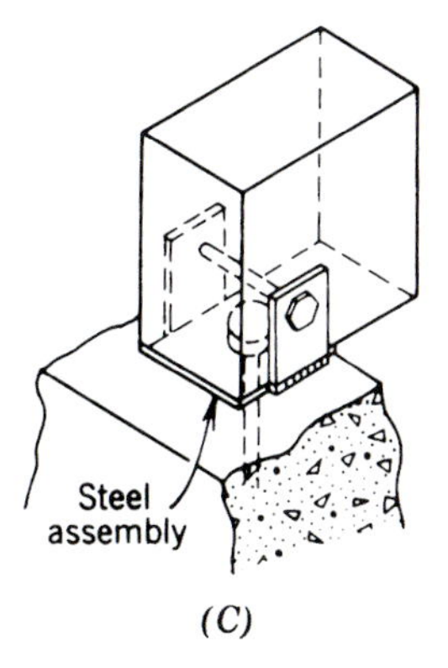

Figure 21.16 Typical glued laminated connections used to secure a beam to the foundation. *(Courtesy American Institute of Timber Construction, 7012 S. Revere Parkway, Suite 140, Englewood, CO 80112)*

Figure 21.17 Examples of glued laminated column-to-foundation connections. *(Courtesy American Institute of Timber Construction, 7012 S. Revere Parkway, Suite 140, Englewood, CO 80112)*

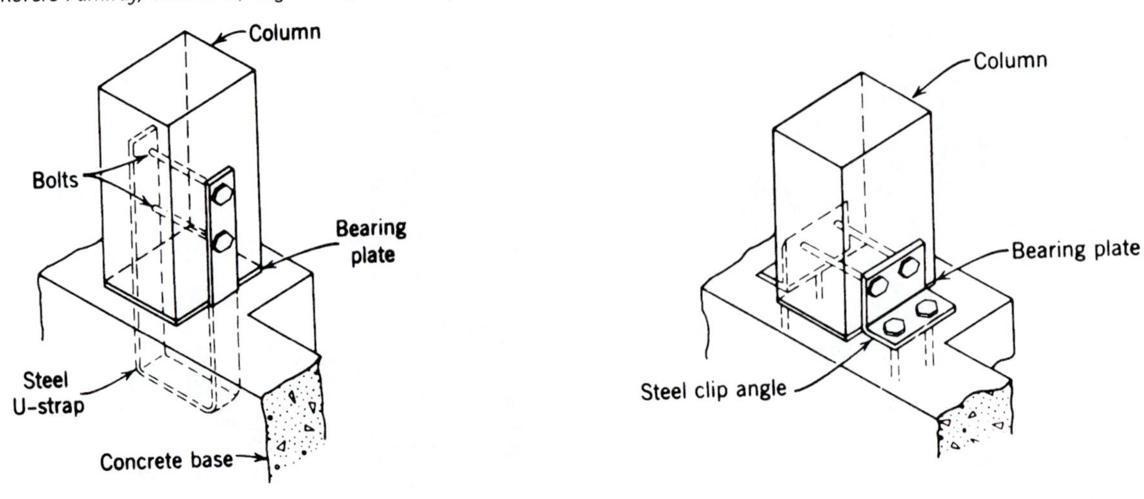

Figure 21.18 Typical beam-to-column and girder-to-column connections. *(a) Courtesy American Institute of Timber Construction, 7012 S. Revere Parkway, Suite 140, Englewood, CO 80112; (b) Courtesy Canadian Wood Council.*

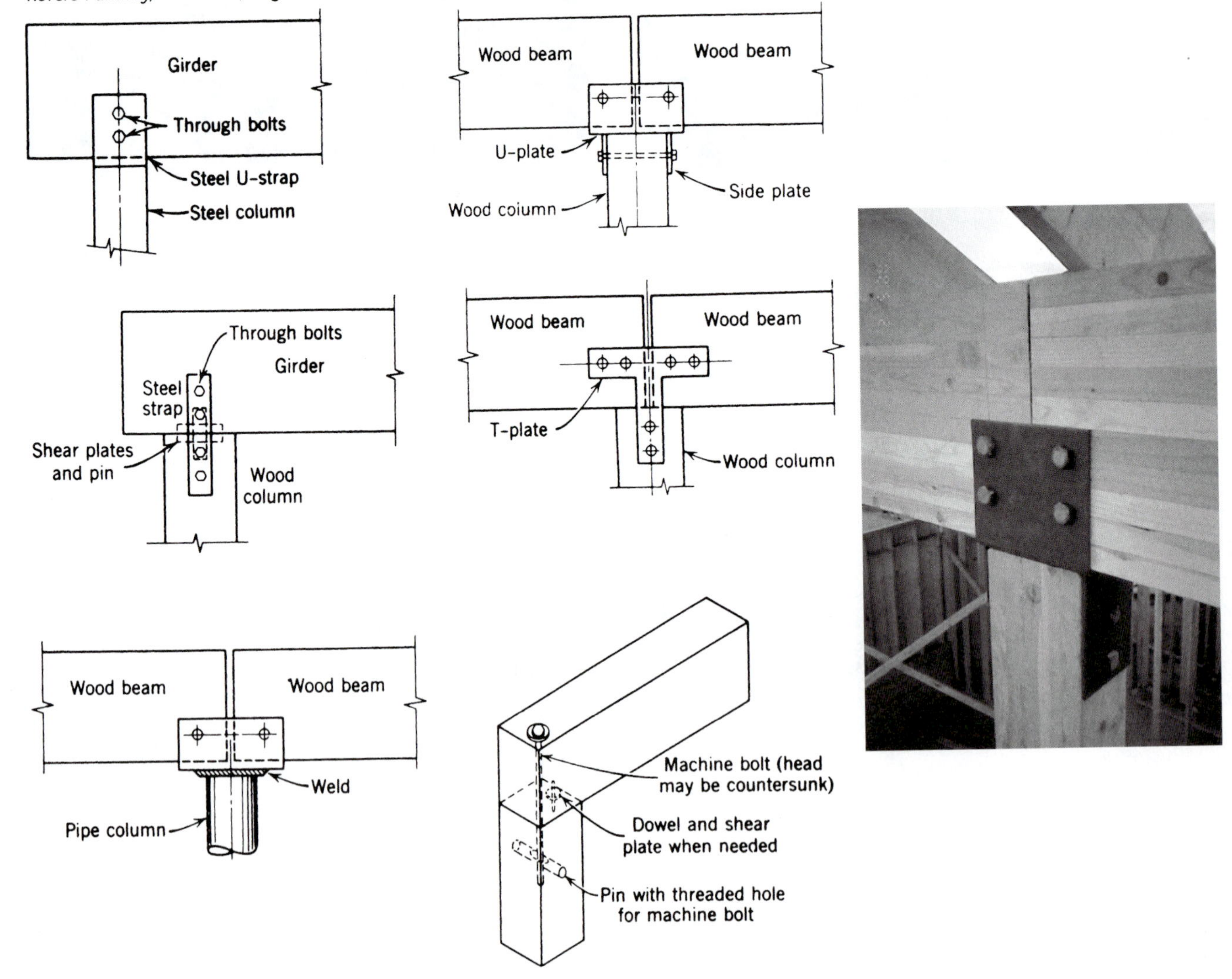

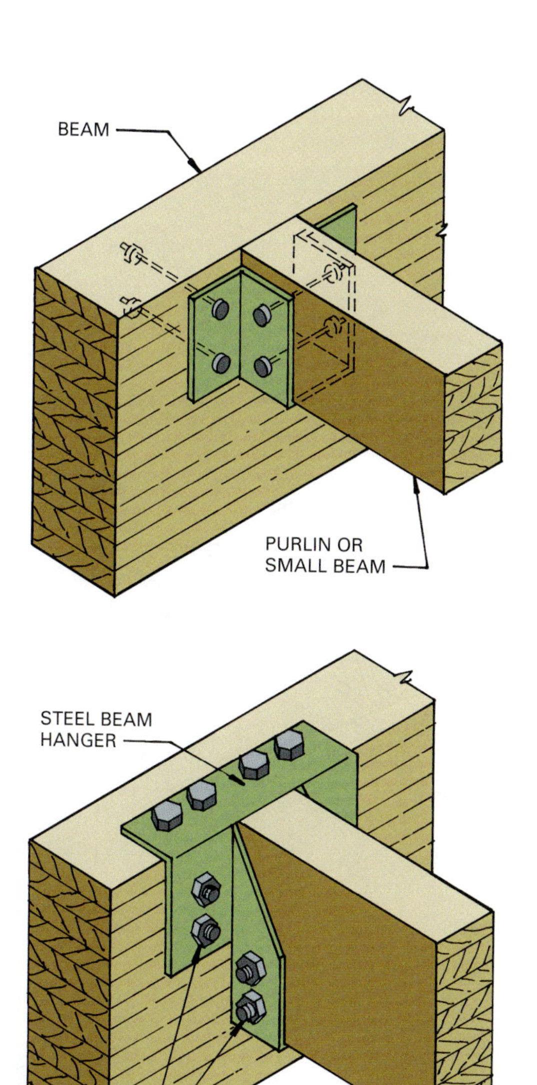

Figure 21.19 Typical beam-to-girder connections.

(22 mm) bolt is used with the 4 in. shear plate. Lag screws are also sometimes used with shear plates instead of bolts.

Decking

Decking systems for heavy timber construction include laminated and solid wood decking, stressed skin panels, and plywood panels, as shown in Fig. 21.23. Typical glulam tongue-and-groove decking products are shown in Fig. 21.24. Laminated decking is stronger than solid wood and is used for longer spans. The sizes of laminated decking vary depending on the species of wood used. Typical actual sizes include $2\text{-}\frac{7}{8} \times 5\text{-}\frac{3}{8}$ in. and $3 \times 7\text{-}\frac{1}{8}$ in. Solid wood sizes are shown in Fig. 21.25. Decking is usually installed in random lengths with joints centered on supporting timbers. Joints are staggered for better structural strength.

Sheathing

The walls of timber frames may be sheathed in solid wood similar to the kind used for roof decking. Other panelized systems can also be utilized. Structural Insulated Panels consist of two rigid wood panel materials with a foamed insulating material secured between them, either by gluing billets, as in EPS (Expanded Polystyrene), or with polyurethane foamed and formed in

Figure 21.20 These multimember heavy wood trusses carry purlins upon which the wood decking is secured. *(Courtesy American Forest & Paper Association, Washington, D.C.)*

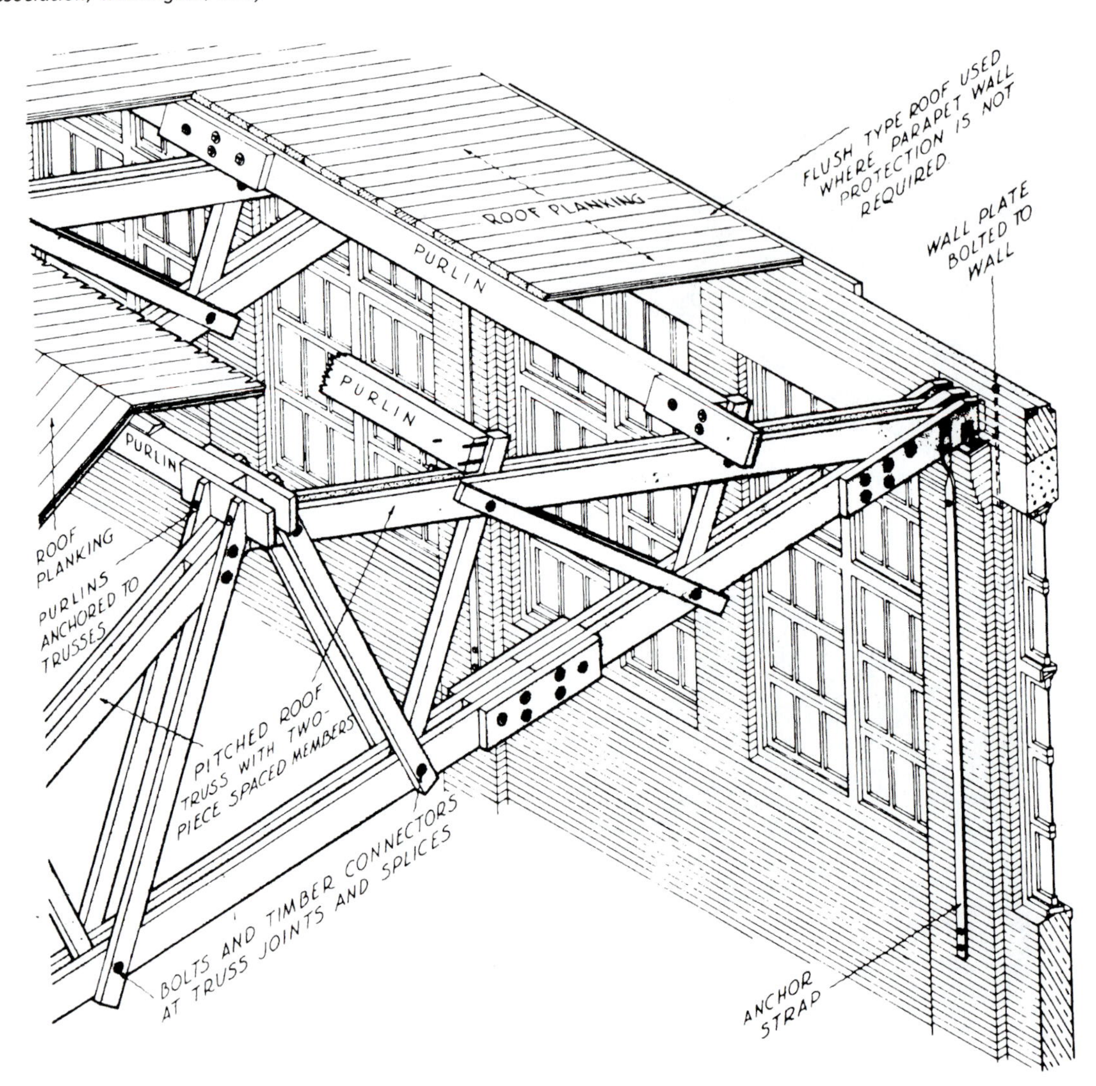

Figure 21.21 Split ring connectors are inserted in recesses cut in the joining members.

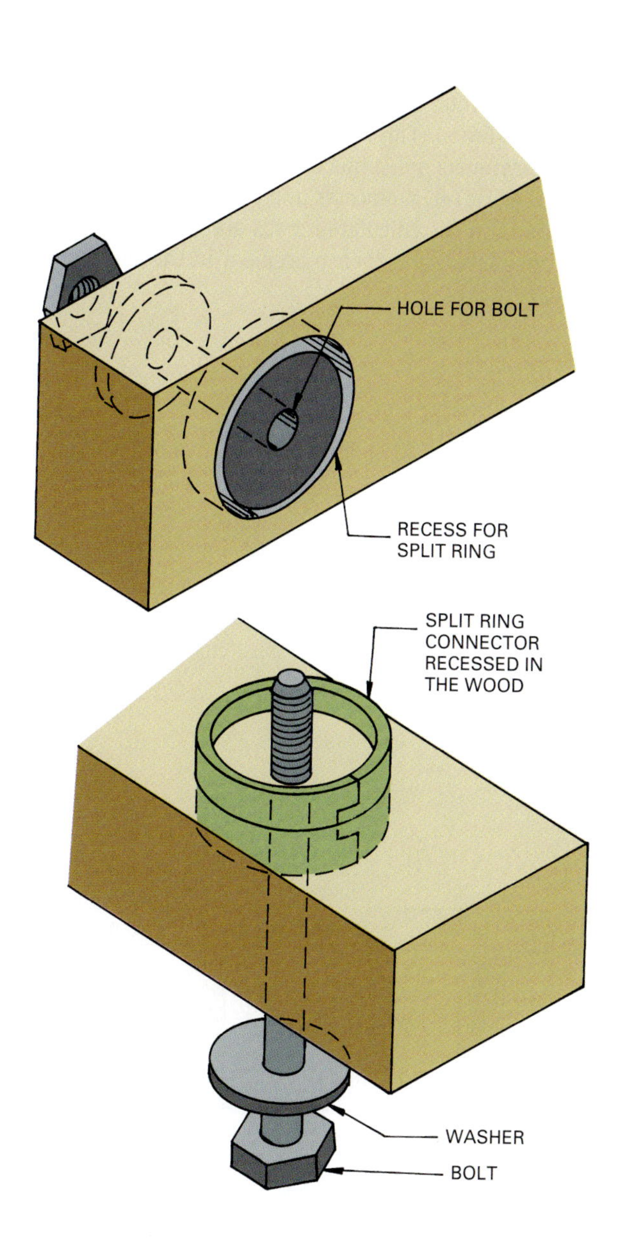

Figure 21.22 Shear plates distribute lateral forces over a larger bearing area in metal-to-wood connections.

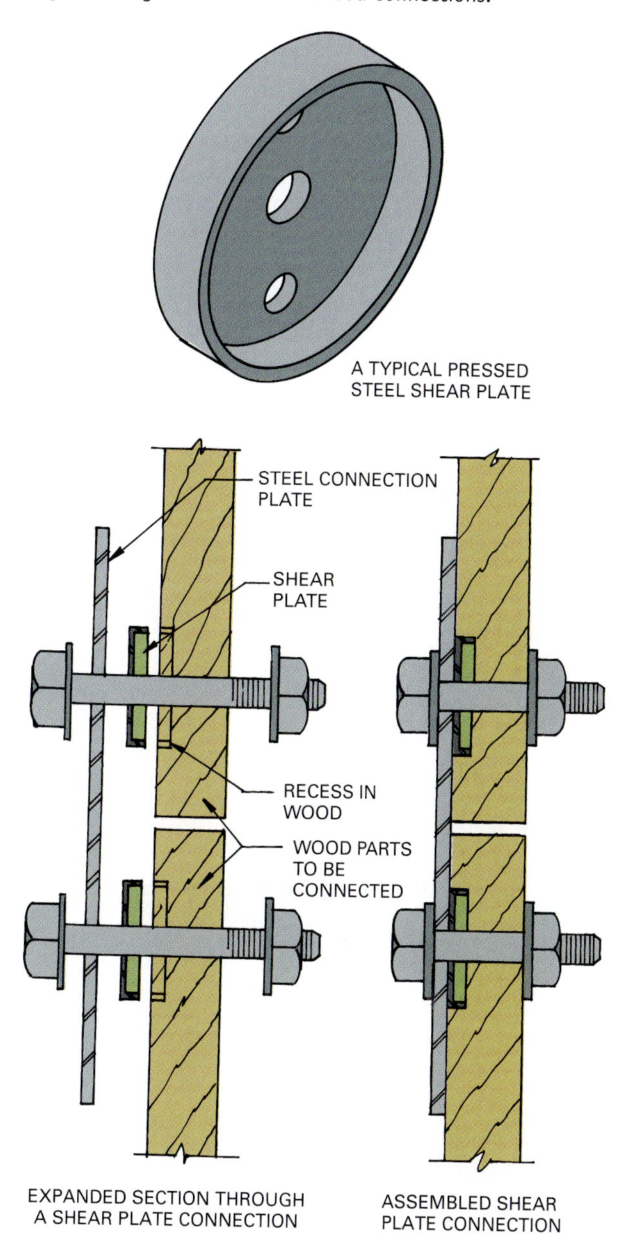

Construction Techniques

Metal Fasteners

There are many types of metal fasteners used to join wood members. These range from nails and staples, used for light frame construction, to bolts, side plates, and other types of hardware. The effectiveness of metal fasteners depends on their being large enough to carry loads and transfer them over large areas so the wood fiber in contact with the fastener is not deformed. The spacing between metal fasteners and between the ends and edges of wood members and fasteners is critical to a successful union (Fig. A).

Metal fasteners are subject to various types of loads, as shown in Fig. B. Architects and engineers design connections so the calculated loads are sustained by the connectors and their placement in the wood members.

Figure A

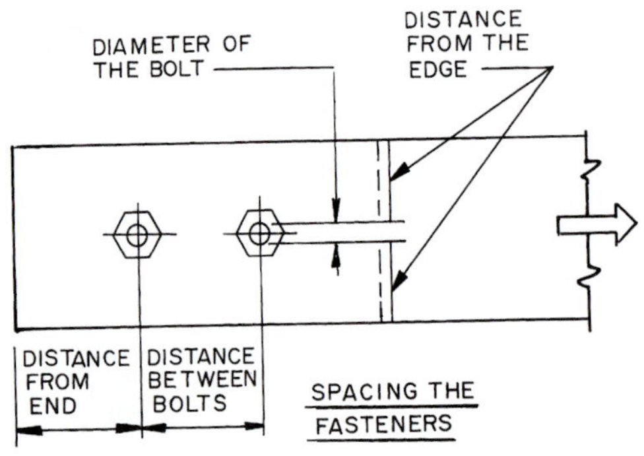

Figure B

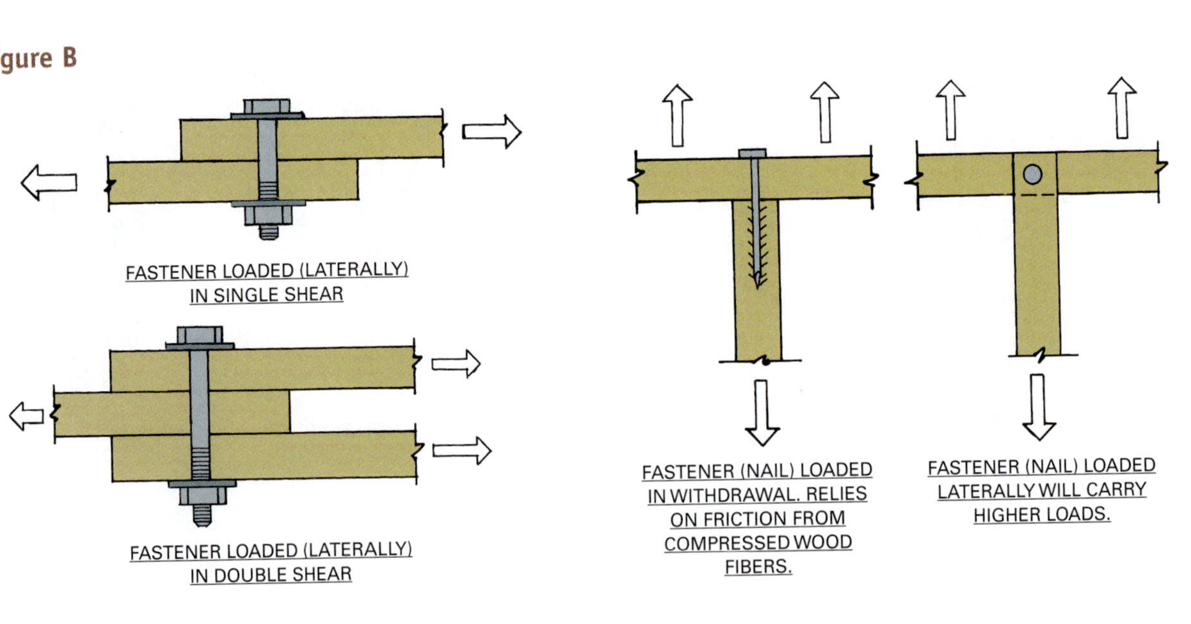

Figure 21.23 Types of decking used with glued laminated framing. *(Courtesy American Institute of Timber Construction, 7012 S. Revere Parkway, Englewood, CO 80112)*

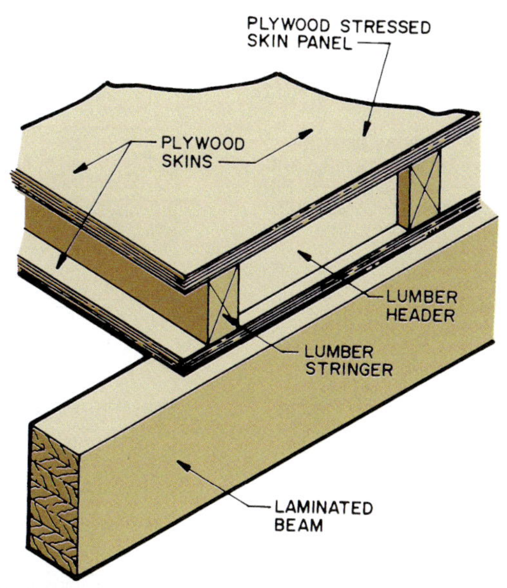

STRESSED SKIN PANELS ON LAMINATED BEAM SYSTEM.

Stressed skin panels, which have a practical span range of 32 ft, are fastened directly to the main laminated timber beams by lag screws or gutter spikes.

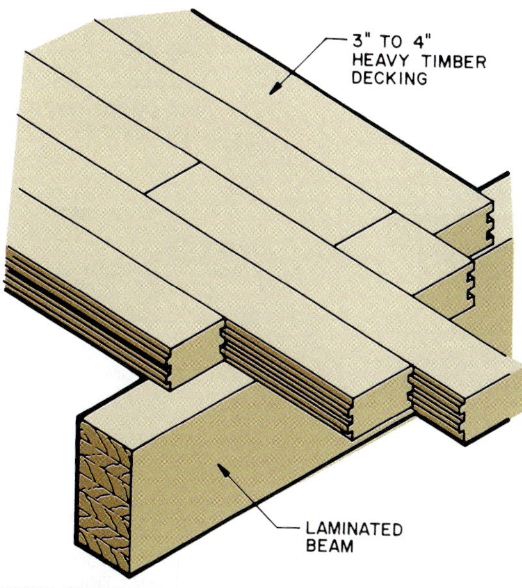

HEAVY TIMBER DECKING ON A LAMINATED BEAM SYSTEM.

Heavy timber decking either laminated or solid 3 or 4 in. nominal thickness is nailed directly to the main laminated beams. The economical span range for the heavy timber decking is 8 to 20 ft depending upon the thickness and loading conditions.

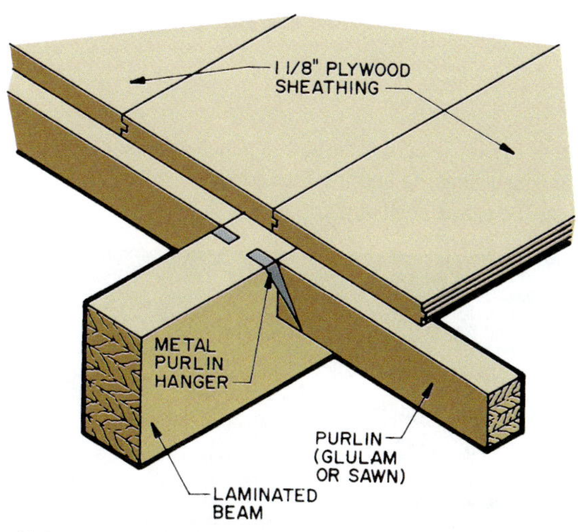

ONE AND ONE-EIGHTH-INCH PLYWOOD ON A LAMINATED BEAM AND PURLIN SYSTEM.

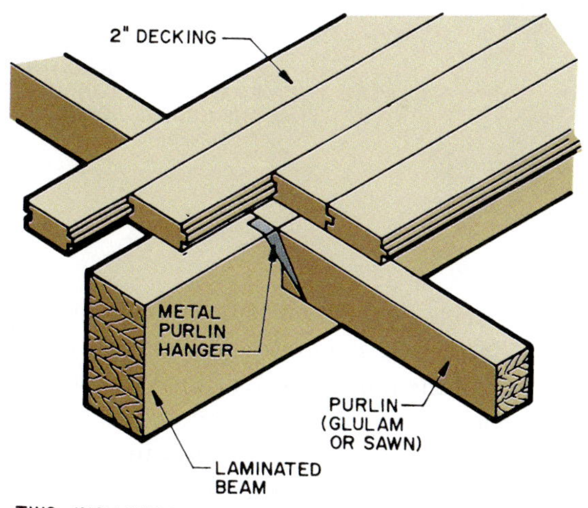

TWO-INCH DECKING ON A LAMINATED BEAM AND PURLIN SYSTEM.

Two-inch nominal thickness decking with an economical span range of 6 to 12 ft is nailed directly to glulam or sawed wood roof purlins, typically on 8 ft centers. Purlins are connected to the main laminated timber beams by metal purlin hangers.

Figure 21.24 Examples of the types of glued laminated decking available.

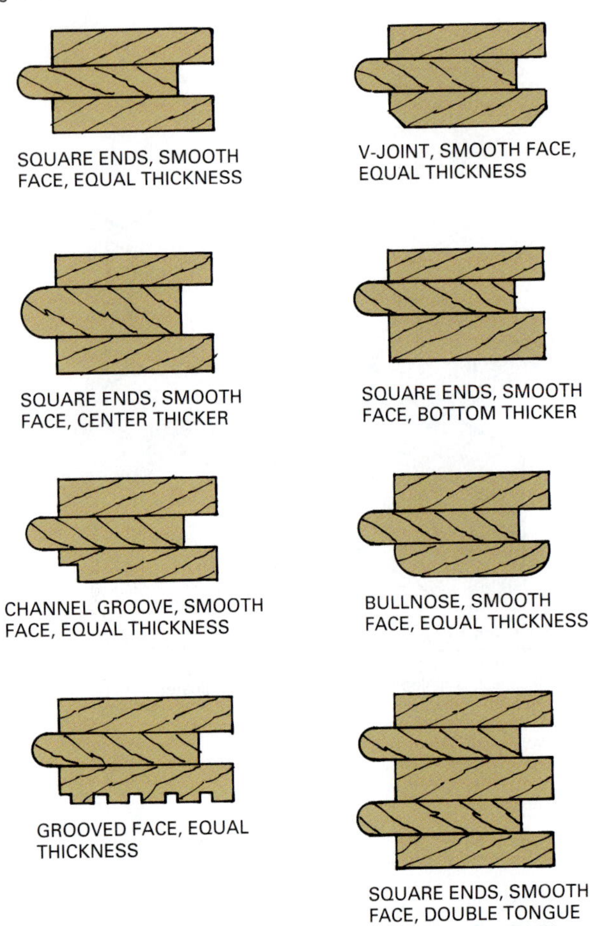

Figure 21.25 Sizes of glue laminated wood decking made from Douglas fir, larch, and southern pine.
Base connections for glued laminated arches use some type of steel shoe. Notice the tie rods shown. *(Courtesy American Institute of Timber Construction, 7012 S. Revere Parkway, Suite 140, Englewood, CO 80112)*

ACTUAL SIZES OF DOUGLAS FIR, LARCH, AND SOUTHERN PINE GLUED LAMINATED DECKING

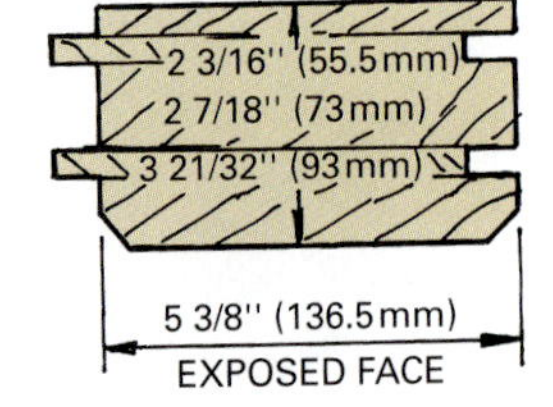

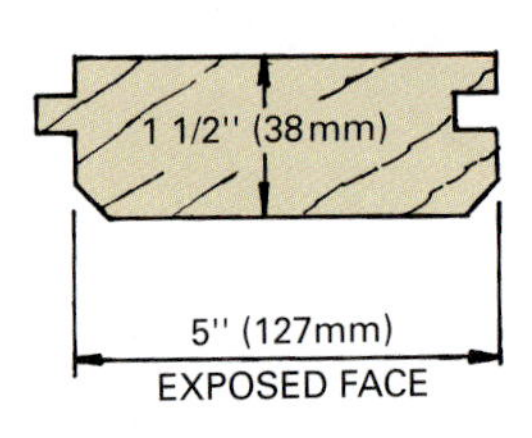

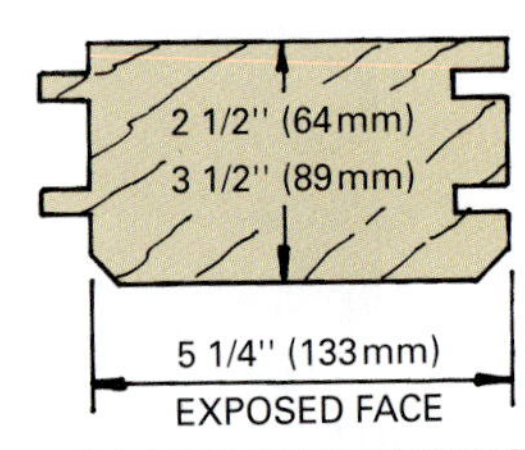

ACTUAL SIZES OF DOUGLAS FIR AND LARCH SOLID DECKING.

NOTE: EXPOSED FACE MAY BE SMOOTH, BRUSHED, GROOVED, OR STRIATED.

place. These panels can be used as both sheathing and roof decking, providing a complete insulated enclosure at the exterior of the frame (Fig. 21.26). An added advantage of panels is a decreased dependency on bracing and auxiliary members, since the panels can span considerable distances and serve to increase the stiffness of the timber frame itself. Using this method of enclosure ensures that timbers can only be seen from inside the building. The use of structural insulated panels results in a building with efficient heat insulation properties.

Figure 21.26 Structural insulated panels can be used to simultaneously sheathe and insulate a heavy timber frame. *(Courtesy The Murus Company, Inc.)*

Review Questions

1. What are the types of wood products used to frame heavy timber buildings?
2. What is the difference in reaction to fire between timber structural systems and steel systems?
3. What is meant by a hinged arch?
4. What two types of design are used for glued laminated dome construction?
5. Why do codes require steel hold downs?
6. What are the types of span details used with glued laminated wood decking?
7. Why are split ring connectors used?
8. When are shear plates used?

Key Terms

heavy timber construction
hinge joint
shear plate connector
split ring connector
timber joinery

Activities

1. Build a scale model of the structural frame of a heavy timber building.
2. Make sketches of the various types of metal connectors.
3. Copy from the local building code the requirements that apply specifically to heavy timber construction.

Additional Resources

Post-Frame Building Design Manual, National Frame Builders Association, Lawrence, KS.

Timber Construction Manual, American Institute of Timber Construction, Englewood, CO.

Wood Building Technology, Canadian Wood Council, Ottawa, Ontario, Canada.

Wood Handbook, Forest Products Laboratory, Madison, WI.

Wood Reference Handbook, Canadian Wood Council, Ottawa, Ontario, Canada.

See Appendix B for addresses of professional and trade organizations and other sources of technical information.

CHAPTER 22

Finishing the Exterior and Interior of Light Wood Frame Buildings

LEARNING OBJECTIVES

Upon completion of this chapter, the student should be able to:

- Have an understanding of the many materials and design possibilities available for finishing the exterior of light wood frame buildings.
- Be familiar with the many materials and processes and some of the installation procedures required to finish light wood frame building interiors.

Build Your Knowledge

For further study on these materials and methods, please refer to:

Chapter 25 Thermal Insulation and Vapor Barriers

Chapter 33 Interior Finishes

Chapter 36 Interior Walls, Partitions, and Ceilings

Chapter 37 Flooring

Chapter 38 Carpeting

Once the structural frame of a building is complete and sheathing, decking, and roof drainage plane are applied, exterior finish work can begin. Roof eaves and rakes must be framed and finished. Roofing contractors arrive at the site to install finished roofing and seal the building off from the weather. Doors, windows, and exterior siding are installed to complete the enclosure. Only then can interior finish work begin. While the interior finish work proceeds, final grading and drive and landscape work can occur simultaneously. Detailed information on the various materials used to enclose light frame buildings is given in subsequent chapters. This chapter outlines the basic construction procedures in finishing light frame buildings in the order that they occur.

FINISHING THE EXTERIOR

Exterior finish work includes completing eaves; installing roofing, siding, and gutters; installing windows and exterior doors; and doing the exterior painting or staining. With the installation of the exterior enclosure, the building is weather-tight and ready for interior work.

Framing the Eaves and Rake

The eaves and rake are framed after the roof is sheathed. **Eaves** and **rakes** are designed to protect walls from excessive wear, provide shade, and complement the architectural style of the house. Common details include flush rakes and overhangs. Typical designs for sloped roofs are shown in Figs. 22.1 and 22.2, and a flat roof design is shown in Fig. 22.3. The finished board covering the ends of the rafters is called the **facia**. A rough fascia, usually 2 × material, provides spacing support for the ends of the rafters and a means of securing gutters and the finished fascia. The finished fascia, commonly 1 in. (38 mm) thick, can be painted composite wood or wood covered with aluminum or vinyl. A **soffit**, the underside of an eave, can be finished in vinyl, aluminum, or plywood, hardboard, or another exterior reconstituted wood

Figure 22.1 A closed cornice provides no overhang and uses a frieze board at the intersection of the roof and wall sheathing.

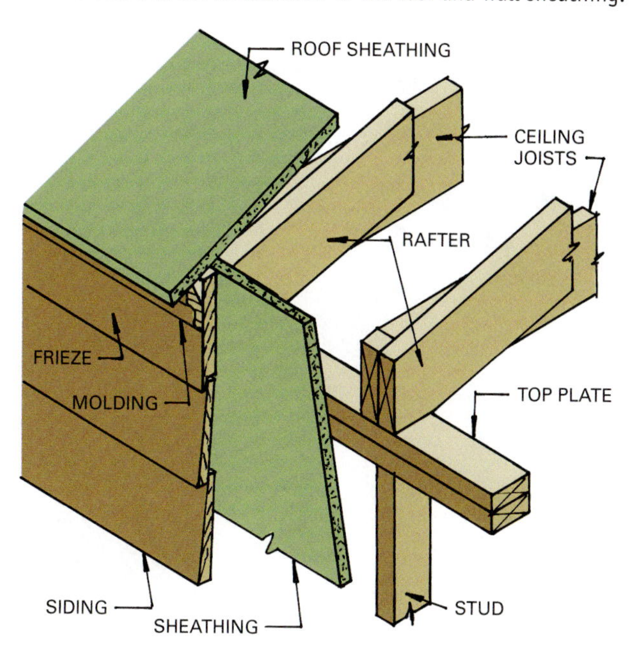

Figure 22.2 This wide boxed cornice has a continuous attic vent strip.

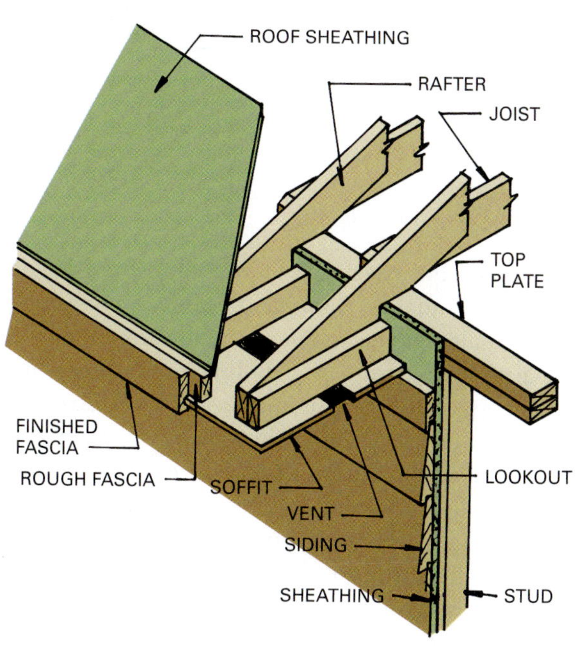

Figure 22.3 A detail for framing an overhanging cornice on a flat roof.

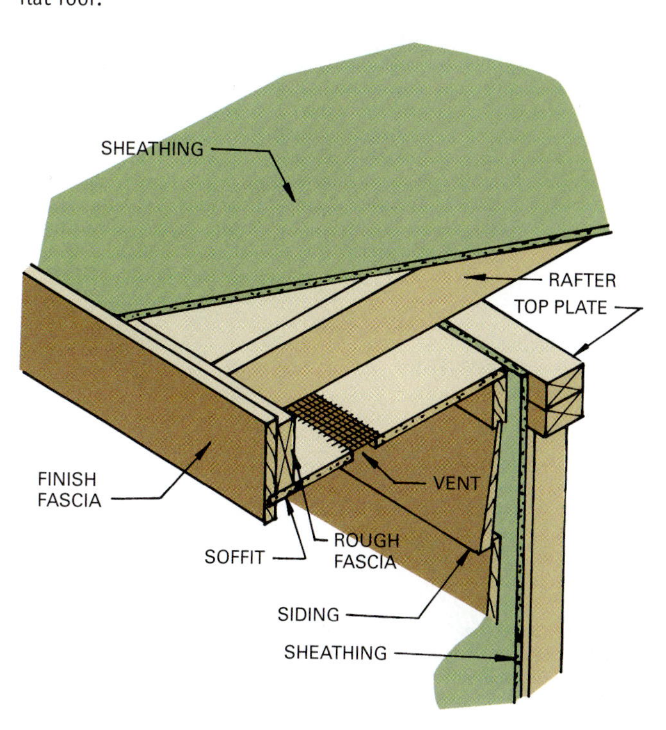

Figure 22.4 This vinyl soffit is perforated to provide attic ventilation. The fascia is also vinyl-clad.

product. Vinyl and aluminum soffits are perforated, providing for air flow into the attic (Fig. 22.4). Solid soffits require that metal vent strips be installed (Fig. 22.5).

Rakes are finished in a fashion similar to eaves. Rake fascia may be set flush over siding or extend from a gable end, as shown in Figs. 22.6 and 22.7.

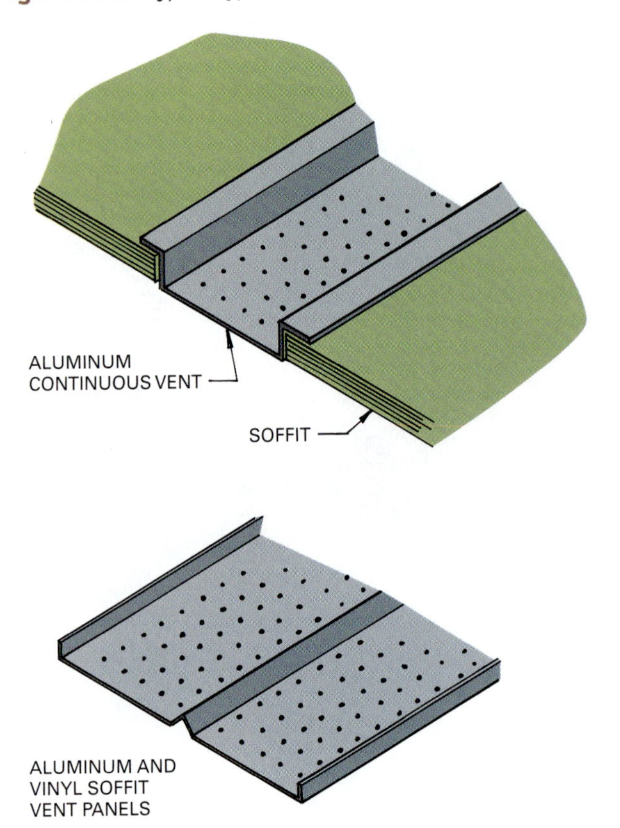

Figure 22.5 Typical types of soffit vents.

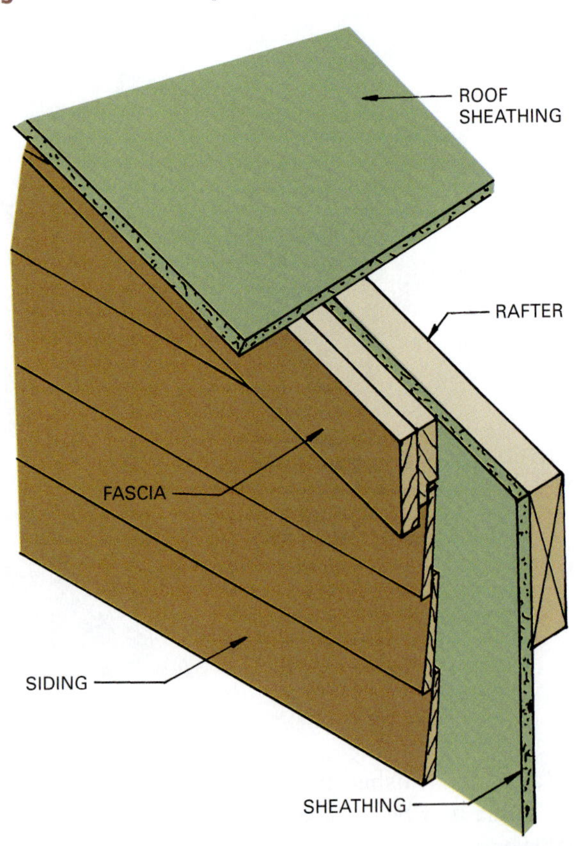

Figure 22.6 A framing detail for a flush fascia on the rake.

Figure 22.7 Rake fascias are generally extended beyond the end wall with a soffit returning to the wall.

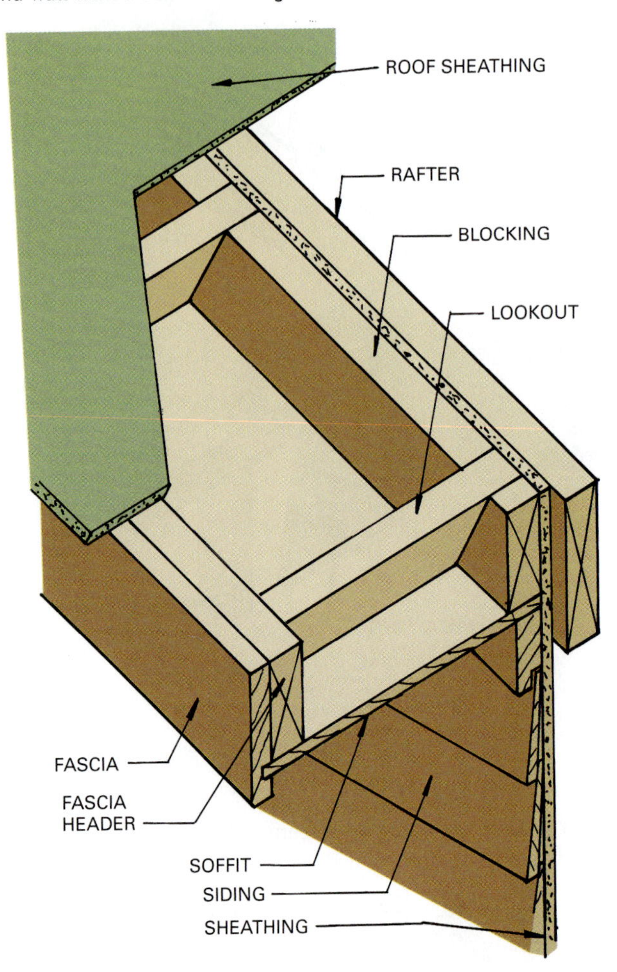

Ventilating the Attic

Ventilation is the exchange of air to allow for drying and improved air quality. A building envelope must control unwanted air leakage and ventilate unwanted moisture. With airtight construction detailing and proper ventilation, energy costs of a building are reduced. Attic ventilation is required to remove hot air in the summer and prevent the accumulation of moisture in both summer and winter. In the winter, ventilation is necessary to keep an attic's air temperature congruent with the outside air. In an attic with insufficient insulation, heated interior air causes the air temperature in the attic to rise and snow on the roof to melt. An **ice dam** forms when the melted snow refreezes as it runs to an uninsulated eave (Fig. 22.8). This snow melt can back up under the shingles and possibly leak into the exterior wall and ceiling insulation. In cold climates, extra layers of builder's felt are laid from the eave up the roof to provide additional waterproofing.

Figure 22.8 Ice dams form when melted snow flows over the cold eave and refreezes.

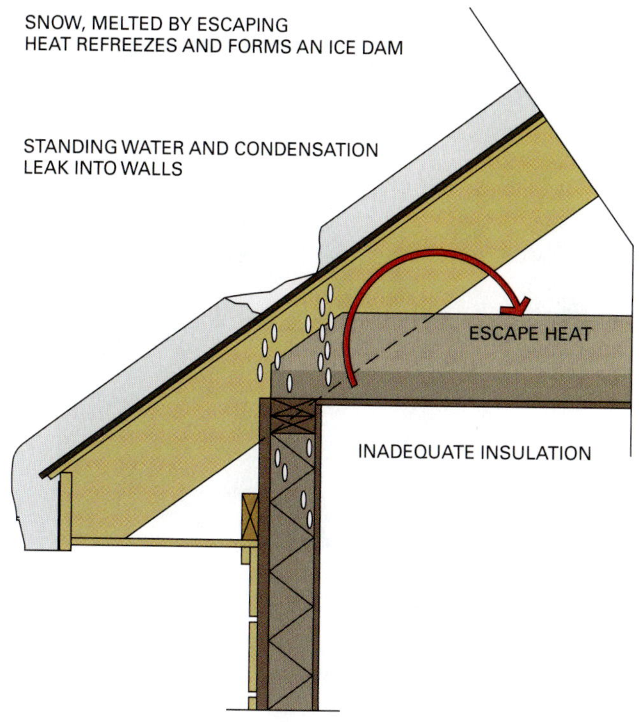

Figure 22.9 Sufficient insulation and soffit and roof vents prevents snow from melting.

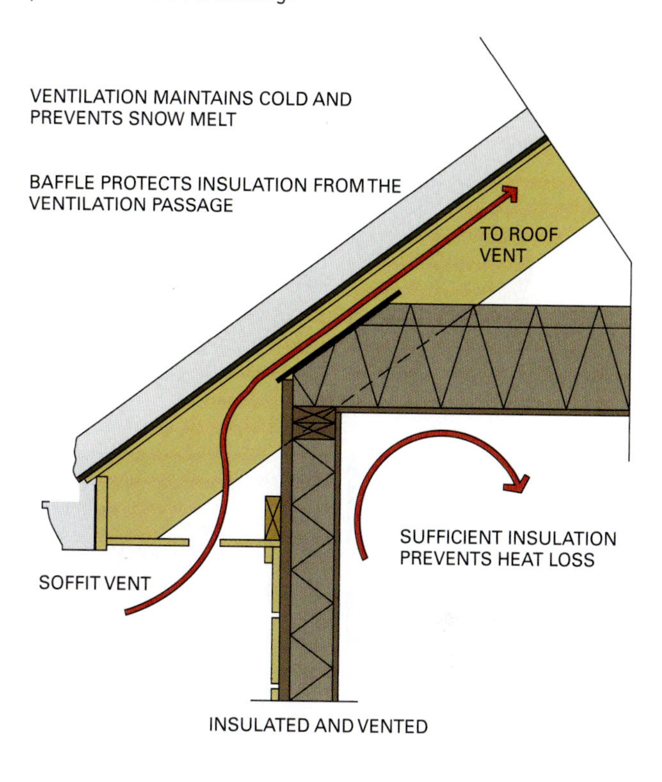

When heated air comes into contact with colder attic air, its relative humidity increases, which can cause condensation and moisture damage. With a well-insulated ceiling and adequate ventilation, attic temperatures are lowered and excessive moisture is removed (Fig. 22.9). The use of both soffit and ridge vents is an effective way to vent an attic. Each rafter cavity is vented from soffit to ridge. (Refer to Figs. 22.2, 22.3, and 22.4.) It is necessary to install some type of baffle at the exterior wall so the ceiling insulation does not block the flow of air from the soffit into the attic (Fig. 22.9). Vents near the ridge are used to allow the flow of air to exit the attic (Fig. 22.10). Roof sheathing is cut back an inch from the ridge at each side and a vent assembly is nailed over it. There are many types of additional ventilators, including roof mounted exhausts, power fans, and gable end vents (Fig. 22.11). On roofs where the finished ceiling is attached to the rafters, insulation is installed between rafters with a minimum of 1 in. airspace between the insulation and the decking for ventilation.

Figure 22.10 Pre-manufactured ridge vents are widely used for attic ventilation.

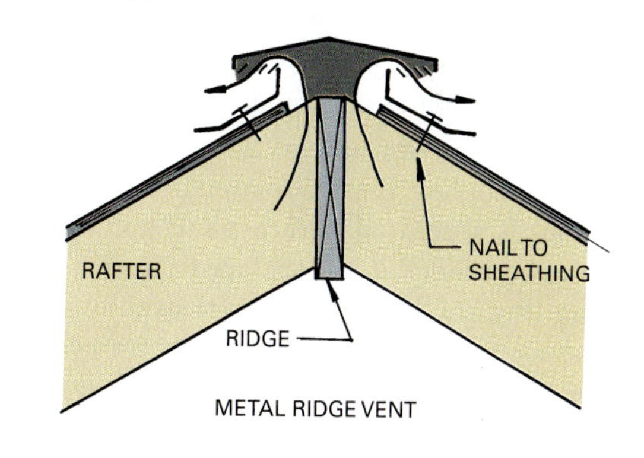

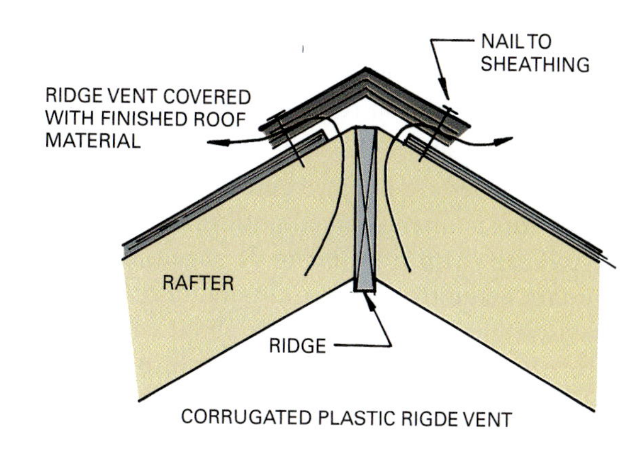

The Finished Roof Material

After roof sheathing is covered with a water shield or builder's felt, the finished roofing can be installed. Common roofing materials for light frame buildings

Figure 22.11 Attic ventilation can be achieved with gable end, dormer, cupola, or roof-mounted exhaust vents. *((a) Image copyright Kevin Norris, 2009. Used under license from Shutterstock.com (b) Image copyright Steve Fellers, 2009. Used under license from Shutterstock.com (c) Image copyright Les Palenik, 2009. Used under license from Shutterstock.com (d) Image copyright C. Kurt Holter, 2009. Used under license from Shutterstock.com)*

include asphalt and wood shingles, clay, concrete or slate tiles, and sheet metal. Roofing materials are covered in detail in Chapters 27 and 28. The common residential roofing shingle is composed of a fiberglass base that is saturated with asphalt and has mineral granules embedded in the surface to form a protective coating (Fig. 22.12). Shingles are usually applied by roofing contractors whose crews are trained to properly install both flashing and shingles. It is important they be installed according to their manufacturer's recommendations.

Clay, perlite, and concrete tiles (Fig. 22.13); wood shingles and shakes (Fig. 22.14); and metal roofing materials are also used. Flat roofs have a built-up or single-ply sheet membrane.

After shingles are in place gutters may be installed, although this is often delayed until the exterior is finished to keep them from being damaged. Gutters are commonly made from either aluminum or vinyl. The finished roofing material extends about $\frac{3}{4}$ in. over the edge of the fascia and directs water into the gutter. The gutter also serves as a decorative finish on the fascia (Fig. 22.15).

Figure 22.12 Fiberglass asphalt shingles are widely used as a finish roofing material. *(Image copyright Christina Richards, 2009. Used under license from Shutterstock.com)*

Figure 22.13 These fire-resistant roofing tiles are made from clay masonry. *(Image copyright viki2win, 2009. Used under license from Shutterstock.com)*

Figure 22.14 Wood shingles and shakes produce a rustic roof with attractive shadow lines.

Figure 22.15 Gutters and downspouts are usually installed after all exterior finish work is complete. *(Image copyright David William Taylor, 2009. Used under license from Shutterstock.com)*

Figure 22.16 The building is wrapped in a watertight, vapor permeable plastic sheeting prior to exterior finish installation.

Installing the Weather Shield

Prior to the installation of windows and siding, the exterior sheathing is covered with asphalt-impregnated builder's paper or a watertight, vapor-permeable plastic sheet made from synthetic fibers. House wrap is a thin, plastic material commonly known by the names Typar or Tyvec (Fig. 22.16). It provides a waterproof layer that permits water vapor in the wall cavity to pass through to the outside, thus preventing it from being trapped inside the wall cavity. A weathershield covers corners, window and door openings, plates and sills and is designed to survive prolonged periods of exposure to the weather. Some types of sheathing have a cover of perforated aluminum foil that serves the same purpose.

Installing Windows

Wood, steel, aluminum, and aluminum- and vinyl-clad wood windows are commonly used in light frame construction. These are described in Chapters 19 and 30. Windows are usually fully assembled and ready for installation when delivered on-site.

Installation occurs from the outside of a building after sheathing is in place. Most windows have a metal or plastic nailing flange secured to the unit. The window is placed in the opening, checked for plumb, and nailed to the studs and header through the flange (Fig. 22.17). In the second type of installation, the window unit is nailed to the studs and header through the casing (sometimes called the brick molding).

The installation procedure begins by checking the opening to see that the sill is square and level. The house wrap is carefully cut to ensure proper drainage around the window. A bead of caulk is applied around

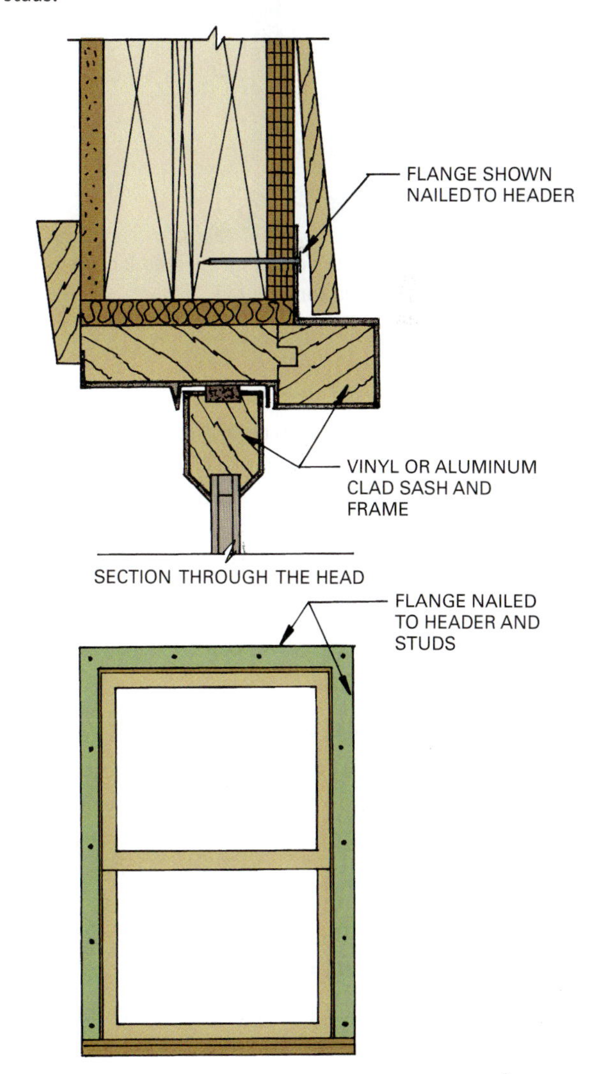

Figure 22.17 Windows made with flanges on the frame are installed by nailing through the flanges into the sheathing and studs.

Figure 22.18 The installation of a manufactured window unit.

1. The house wrap is cut and pulled into the opening.

2. A bead of caulk is applied to adhere and seal the unit to the opening.

3. The unit is placed in the opening, checked for level and plumb, and nailed through the flange.

4. The window is sealed with an adhesive tape placed to ensure the drainage of moisture.

the opening to provide an airtight seal. The window is set into the opening, shimmed to level, and checked for square before being nailed securely through the nailing flange. The nailing flanges are covered with an asphalt adhesive tape for both air and water proofing (Fig. 22.18).

Exterior Door Installation

A variety of wood, metal, and fiberglass exterior doors and frames are available. These are described in Chapters 19 and 30. Like windows, most exterior doors come in an assembled, pre-hung unit. Typical exterior door frame construction at the sill is shown in Fig. 22.19. Careful installation of the unit is essential to ensure proper operation of the door and protection against the weather.

To install an exterior door, the door frame is set in the opening and checked for plumb and level. The housewrap and caulking are treated as during window installation. Wedge-shaped shims between studs and door jamb are used to position the door in the rough opening. When the door is properly positioned the jamb is nailed through the shims to the framing. The door is checked and adjusted for proper operation before the exterior casings are installed. Units that come pre-hung with casings can be nailed through the casing first and then shimmed for final fastening.

Installing the Exterior Finish

Exterior siding may be wood, plywood, wood shingles, hardboard, plastic, vinyl, stucco, metal, brick, or stone. Details for installing wood siding are shown in Figs. 22.20 and 22.21. Siding is usually face nailed through the sheathing into the studs. When installing lap siding, the nails in the lower end of the board do not penetrate the piece of siding directly below. This allows each

Figure 22.19 Exterior door frames are plumbed with shims and nailed to the studs through the jamb and wedges. A brick casing is nailed through the sheathing into the frame.

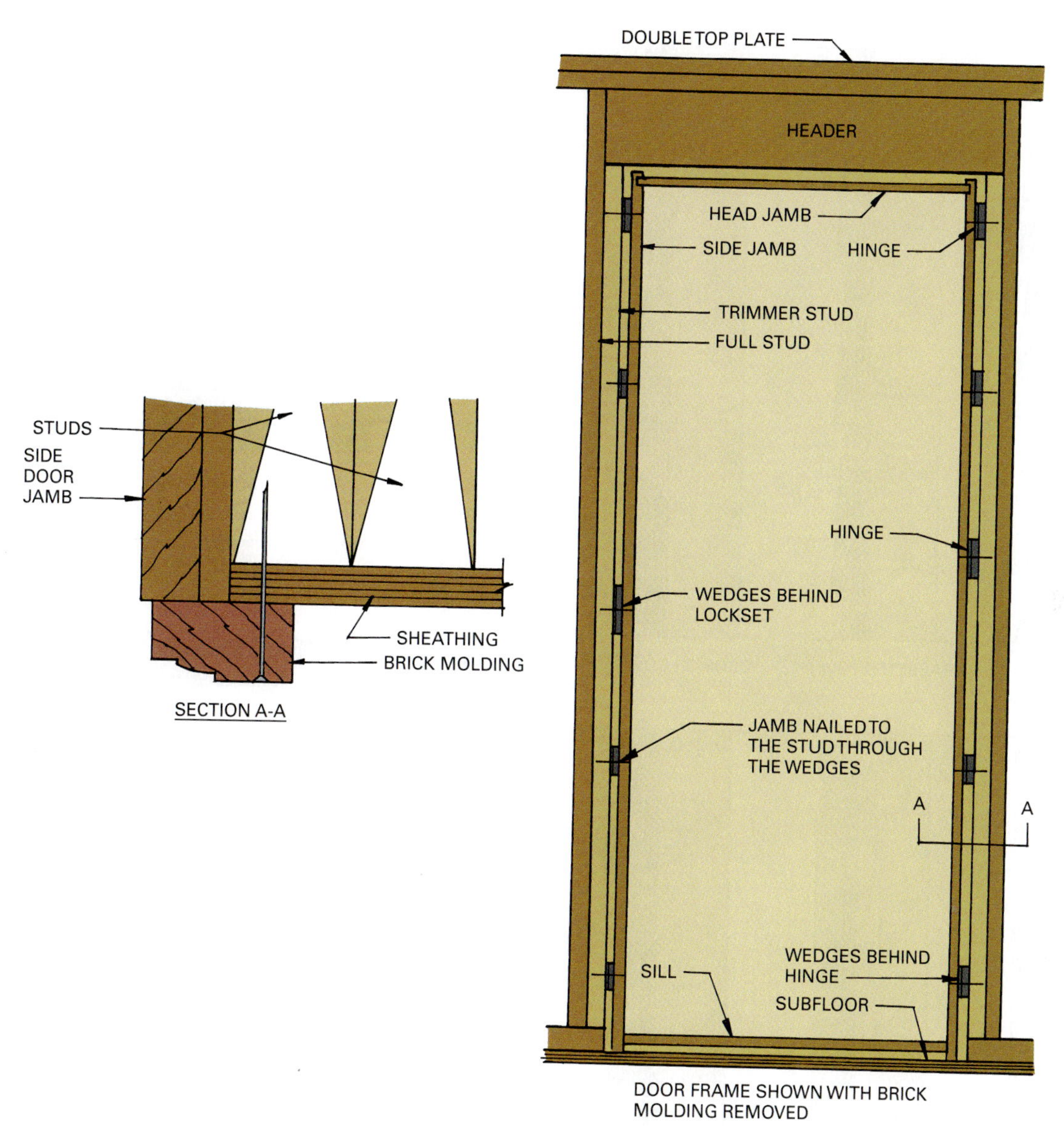

piece to expand and contract independently of the other. Hardboard siding is available in lap siding and panel siding. Lap siding installation is shown in Fig. 22.22. Hardboard panel siding is available in 4 × 8 ft. (1220 × 2440 mm) and 4 × 9 ft. (1220 × 2745 mm) sheets. It typically has batten strips nailed over it but may also be installed with a reveal (a slight offset between panels) (Fig. 22.23).

Aluminum and vinyl siding are available in a variety of colors and surface textures (Fig. 22.24). They must be nailed to nailable sheathing or directly into the studs because they have little structural strength. Specially designed channels are placed at internal and external corners and where the siding butts a surface, such as the wood frame of a door or window. A typical installation detail is shown in Fig. 22.25. Wood shingles used as siding are applied over nailable sheathing, such as plywood, and may be installed in a single or double course (Fig. 22.26).

Stucco exterior surfaces may be constructed using a finish coat of a Portland cement, lime, and sand mixture (Fig. 22.27). The stucco is troweled over a wire mesh that has been nailed to the sheathing (Fig. 22.28). Another exterior finish similar to stucco that is finding increasing use is composed of a glass-fiber-reinforced Portland cement mixture containing special bonding agents. It is applied over a glass fiber mesh bonded to the sheathing. This system has no means of draining

Figure 22.20 Types of horizontal wood siding with nailing patterns.

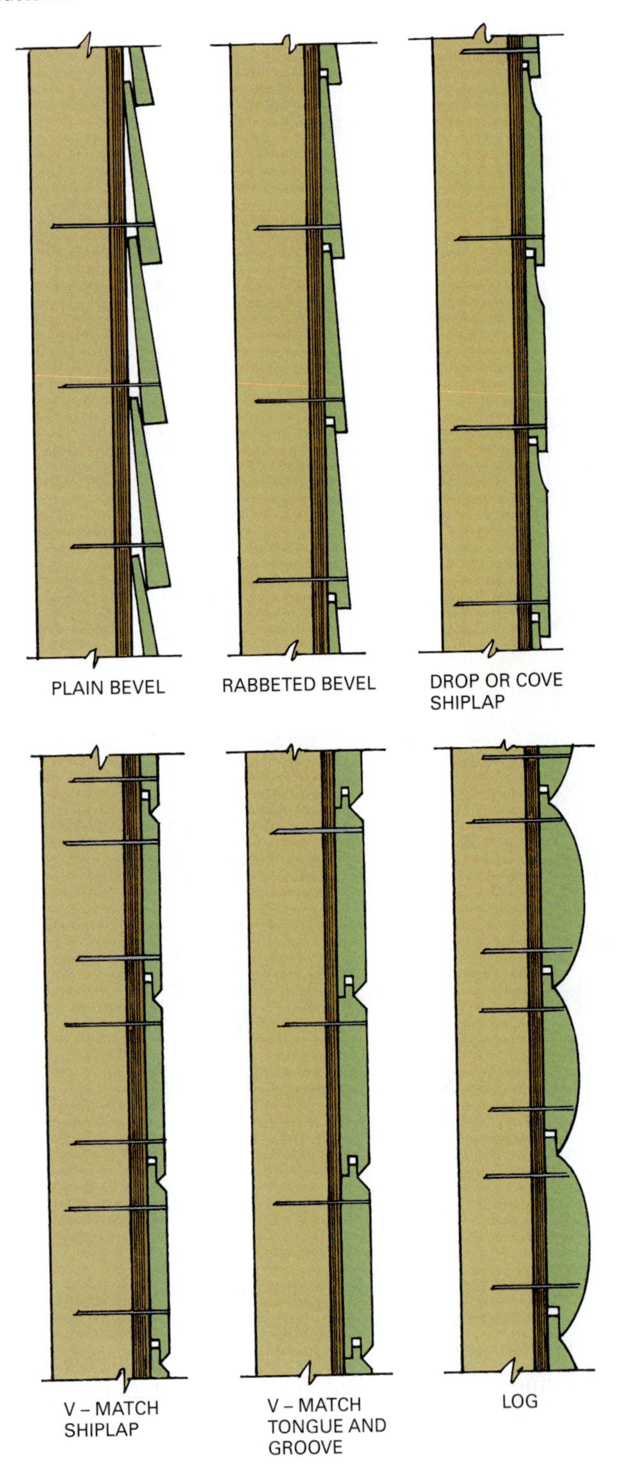

Figure 22.21 Types of vertical wood siding with nailing patterns.

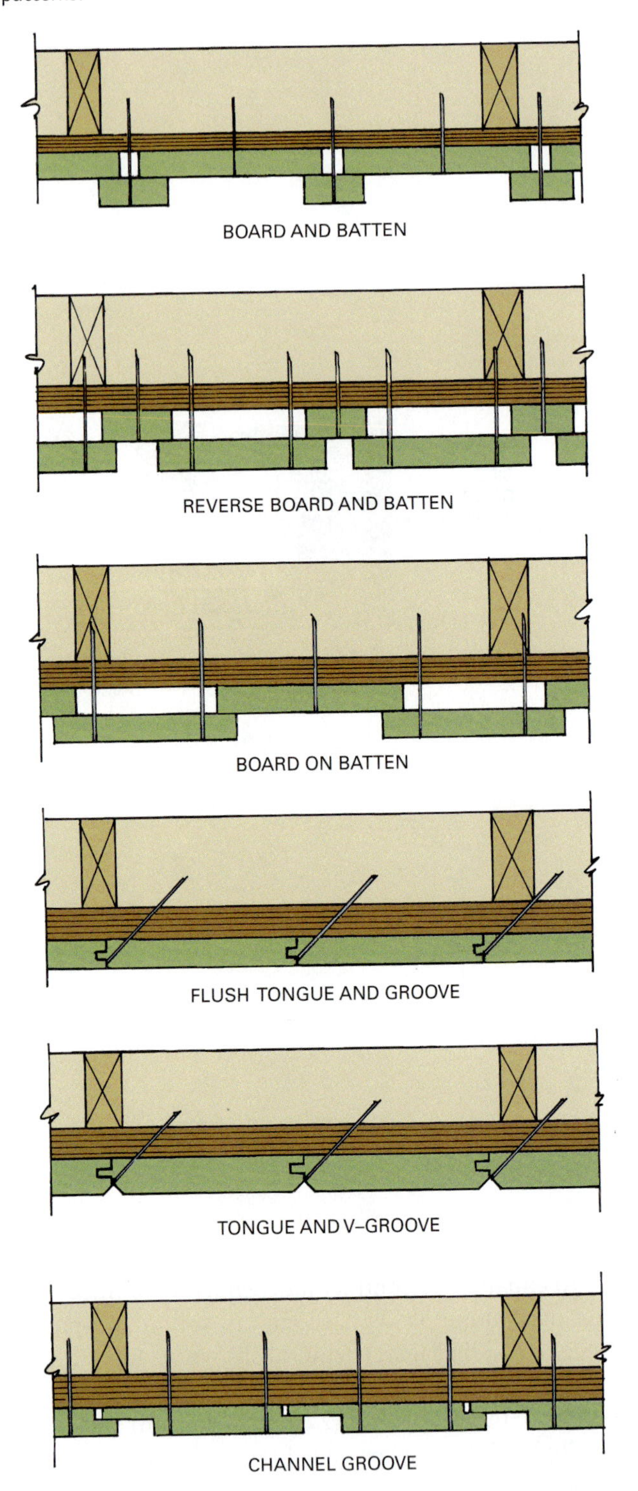

Figure 22.22 Techniques for installing hardboard siding.

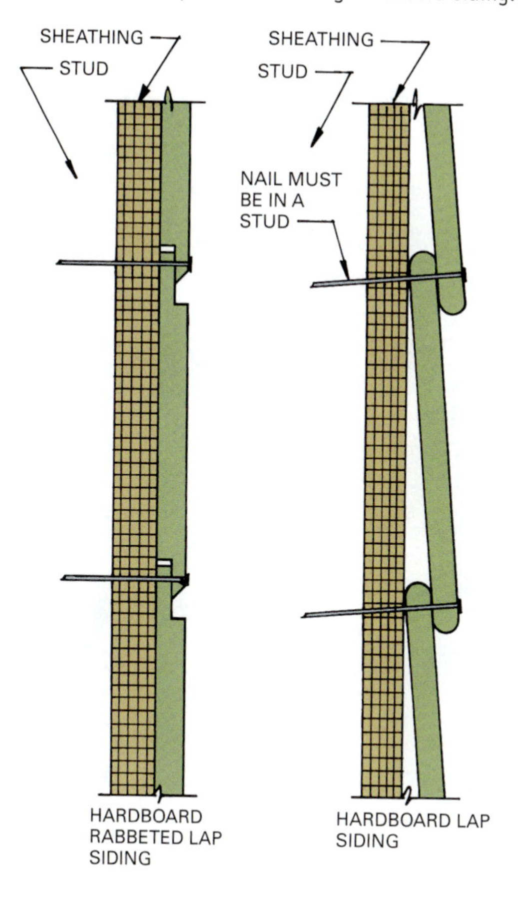

Figure 22.23 Hardboard panel siding is nailed directly into the studs.

Figure 22.24 Vinyl siding, soffits, and fascias provide a low maintenance exterior finish. *(Image copyright Kondrachov Vladimir, 2009. Used under license from Shutterstock.com)*

Figure 22.25 A typical installation for vinyl and aluminum siding uses a starter strip and interlocking units that are nailed directly into the studs. *(Image copyright Wendy Kaveney Photography, 2009. Used under license from Shutterstock.com)*

Figure 22.26 Wood shingles and shakes can be used as siding and are applied in a lapped pattern.

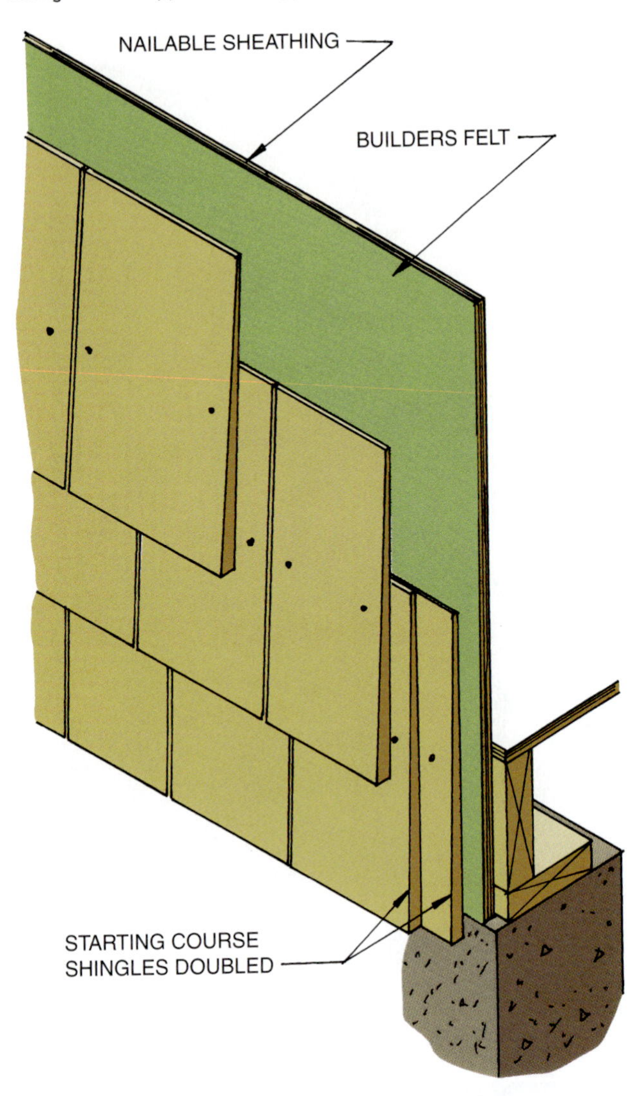

water that may penetrate the wall assembly from around poorly flashed and caulked windows and other openings. Moisture such as this can cause the sheathing and wood wall framing to deteriorate. Figure 22.29 shows the USG Corporation's Water-Management Finish System: a wall assembly that has the same exterior appearance but is flashed and permits water to weep from the wall.

Brick and stone veneers are also used as exterior siding on light frame construction. The masonry rests on a brick ledge constructed as part of the foundation. Metal ties are nailed to the studs and inserted in the mortar joints between masonry units. Wall ties are generally spaced every 16 in. (406 mm) vertically (Fig. 22.30). Masonry and stone materials are covered in Chapters 11 through 14.

Figure 22.27 Stucco siding is applied using a finish coat of Portland cement, lime, and a sand mixture. *(Image copyright Danny Ortega, 2009. Used under license from Shutterstock.com)*

Figure 22.28 Typical details for a Portland cement stucco installation.

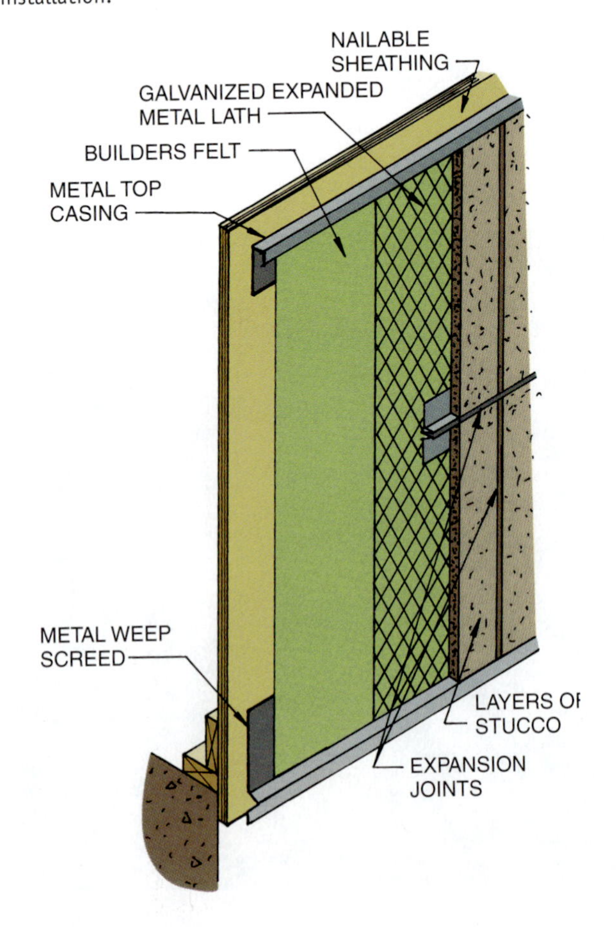

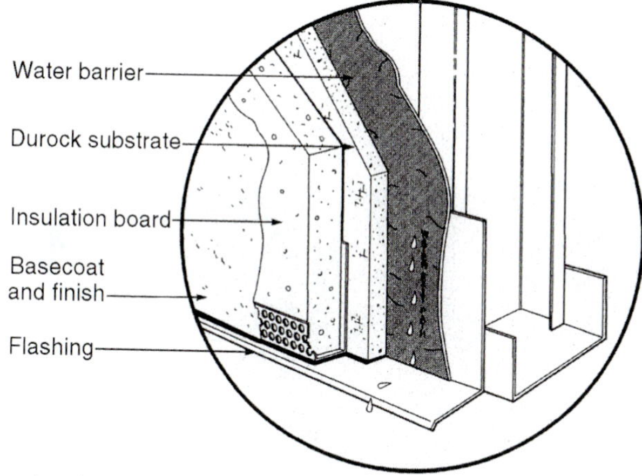

Figure 22.29 USG's Water-Management Finishing System uses a cement board (Durock) as the substrate for the insulation over which the synthetic stucco is applied. Note the use of flashing and the provision for the wall to weep at the bottom. *(Courtesy USG Corporation)*

Figure 22.30 Metal ties are used to adhere masonry veneers to wood fame walls.

Figure 22.31 Electrical wiring and service panels are installed after the building is weathertight. *(Image copyright Christina Richards, 2009. Used under license from Shutterstock.com)*

Figure 22.32 Rough-in plumbing lines are installed inside the framed walls. *(Image copyright Lisa F. Young, 2009. Used under license from Shutterstock.com)*

FINISHING THE INTERIOR

Once the building is weathertight, electricians; plumbers; and heating, ventilation, and air conditioning (HVAC) contractors begin their rough-in work. Electricians drill holes in wood studs or use existing cavities to run electrical wiring and mount the required lighting and outlet boxes (Fig. 22.31). Plumbers run water, waste disposal, and vent lines inside framed walls (Fig. 22.32). Where pipes are close to the surface of wood studs, metal plates are used to prevent puncturing water lines during the installation of interior finish materials. The heating contractor installs sheet metal air ducts or runs hot water pipes to register locations. Once all rough-in work is completed, the local building inspector checks for code compliance. The finish work and installation of fixtures, switches, outlets, registers, and other equipment will proceed after interior walls and ceilings are completed. The plumber usually sets shower and tub fixtures to allow finish materials to be installed around them. The HVAC contractor temporarily covers duct openings to prevent construction dirt and dust from entering the system.

Installing Insulation

Once the interior is protected from the weather and the mechanical systems are installed and approved by a building inspector, insulation can be placed in the walls, floors, and ceilings. Insulation materials create a space between the exterior and interior environments, breaking contact and preventing heat loss. Insulation is placed wherever the interior is exposed to exterior temperatures. A typical example is shown in Fig. 22.33. Among the materials used for insulating are glass fibers, mineral fibers (rock), organic fibers (paper, cotton), and plastics. Aluminum films are also used to reflect heat. These are described in Chapter 25.

Flexible insulation is manufactured in rolls of either blanket or batt form. Blankets and batts may have an asphalt laminated paper covering that creates a vapor barrier when placed facing the warm side of a building. The paper provides tabs used to staple the insulation to the studs (Fig. 22.34). Blankets should fit snugly against top and bottom plates and are cut to fit tightly around plumbing and electrical boxes. Unfaced batts are held in place by friction until the interior finish is applied. Proper installation of batt insulation is essential to providing an uninterrupted insulation enclosure.

Figure 22.33 Unfaced fiberglass insulation is often covered with a plastic vapor barrier. (Image copyright Sue Smith, 2009. Used under license from Shutterstock.com)

Various loose-fill and spray insulations are available today that overcome batt insulation installation problems. These are sprayed into a wall in successive layers and provide superior air-leakage control. For ceiling applications a plastic sheeting is used to contain the insulation before the finish material is installed (Fig. 22.35).

Figure 22.34 Faced insulation has a water-resistant paper barrier that is stapled to the studs.

Figure 22.35 Loose fill insulation is blown into the attic over a plastic vapor barrier.

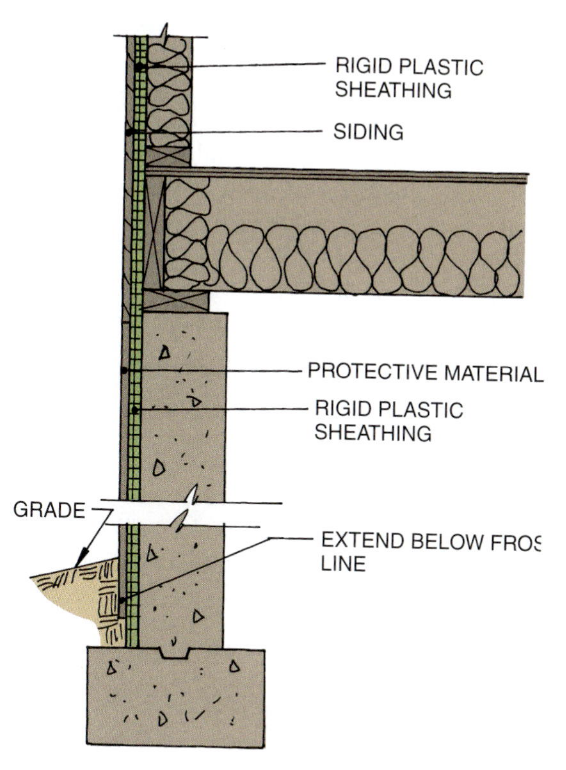

Figure 22.36 Rigid foam insulation can be used to insulate both walls and the foundation.

Figure 22.37 Gypsum wallboard is a widely used material for finishing interior walls. *(© Jim Zuckerman/Corbis)*

The insulation properties of a wall can be improved by using rigid plastic foam insulated sheathing at the exterior of the sheathing and below the house wrap (Fig. 22.36). This can be installed to extend past the framing, down over the foundation into the ground. It must be protected by a hard material where it is exposed to the weather. Some sheets are made with a protective coating already applied. Insulation may also be used to reduce the passage of sound through walls, floors, and ceilings. Various types of sound-control batts are available.

Interior Wall and Ceiling Finishes

Interior walls and ceilings in light frame construction are generally covered with gypsum board after the insulation has been installed. Other finishes include plaster, wood, and hardboard paneling. Gypsum and plaster products are discussed in Chapter 34 and wood in Chapter 19. Gypsum board and plaster finishes provide a hard, durable surface that can be covered with a wide range of decorative materials. In addition, they provide excellent fire ratings for the covered areas. The installation of these materials is illustrated in Chapter 36. Gypsum panels are usually installed by subcontractors specializing in this work (Fig. 22.37). Plaster finishes require the services of highly skilled plasterers, who may not be available in some areas. Wood, plywood, and hardboard paneling are installed by finish carpenters. Usually, wood paneling is applied over a substrate of gypsum wallboard.

Ceilings are typically finished with gypsum board, plaster, some form of composition tile, or panels set in a suspended metal frame. See Chapter 36 for additional details.

INTERIOR FINISH CARPENTRY

Interior finish carpentry, sometimes referred to as **millwork**, involves the application of trim around doors and windows; the intersection of walls, floors and ceilings; and other interior moldings. Moldings are lengths of material, shaped in a variety of patterns, for use in a particular location. Wood is used to make most moldings, although plastic, composite wood products, and metal are sometimes used. Other millwork includes the installation of wainscoting, cabinets, built-in shelving, and stair finish. Finish carpentry requires the efforts of the most skilled craftspeople and cabinetmakers. Interior trim and millwork are among the final materials installed and care must be taken to not mar finishes during construction.

Installing Interior Doors

One task performed by a finish carpenter is the installation of interior doors and jambs. Interior doors in a variety of styles and finishes are manufactured in pre-hung units. Manufacturers supply detailed installation instructions with each unit. Additional information about doors is given in Chapters 19 and 30.

Most interior swinging doors arrive pre-hung on a solid wood door jamb. The carpenter places the jamb into the rough opening and plumbs the sides by driving wood shims between the frame and the wall stud. Once the door is set, it is nailed through the side jamb and wedges into the stud. A long level is used to check for plumb (Fig. 22.38). The door is closed and the spacing between it and the jambs is checked for uniformity and to be certain the door does not stick on the jamb. Several types of split door jambs are also available. They are made with the casing attached to each half of a split jamb. The frames are slid into the rough opening from opposite sides and nailed to the studs.

Figure 22.38 An interior door frame is set plumb using wood shims to position it in the rough opening.

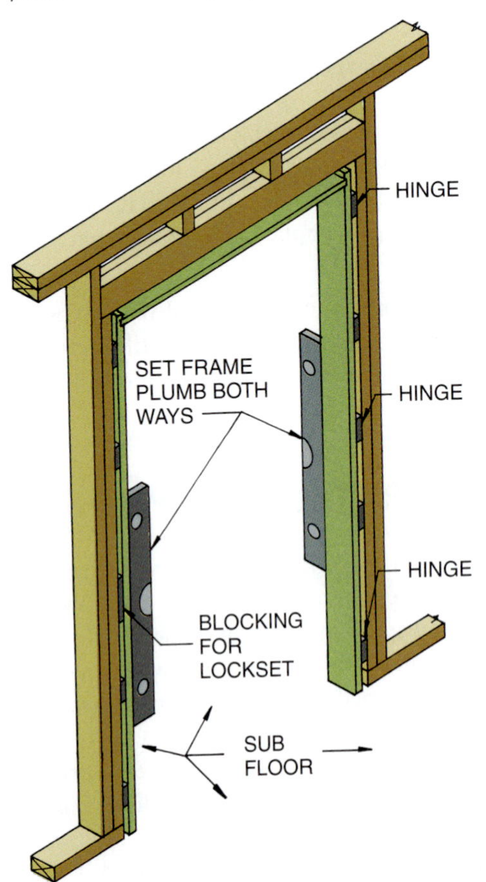

Installing Casings and Base Molding

After door jambs are in place and gypsum wallboard installed, the finish carpenter can apply the casings and baseboards. **Casings** are the moldings used to trim around doors, windows, and other openings and are available in a wide range of sizes and patterns. An architect specifies the type to use. The installation of interior door casing is shown in Fig. 22.39. The casing is usually set $\frac{1}{4}$ in. back from the edge of the jamb and nailed into the jamb and wall stud.

Traditional window casings and sills are commonly installed as shown in Fig. 22.40. Here a wooden apron supports the extension of the window sill, and the side casings die into the sill. Other designs utilize a gypsum board return, butting finished wall board directly into the window frame.

Baseboard and molding is used where a wall meets the floor. It is usually mitered at outside corners and coped at inside corners. Coping refers to cutting the ends of the base molding to the shape of the piece it will butt. If the floor will be carpeted, the base is usually placed a little above the subfloor. If hardwood flooring will be used, the base is installed after the flooring is

Figure 22.39 Interior door openings are trimmed on both sides with a casing after the finish wall materials are installed. *(© Helen King/Corbis)*

Figure 22.40 A window with an apron-supported sill and casings on three sides.

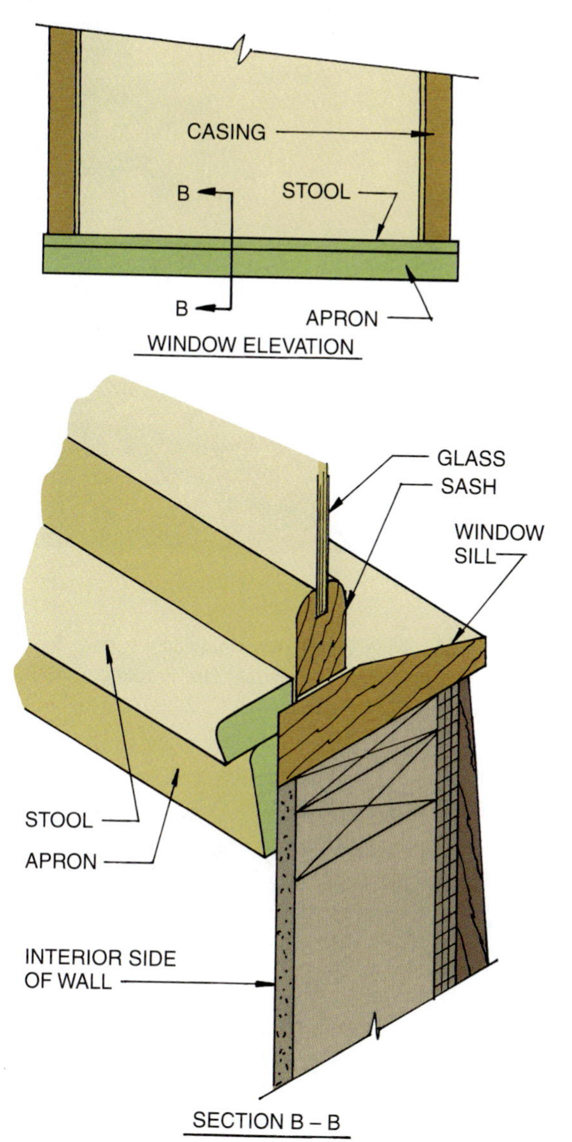

in place, and a small $\frac{3}{4}$ round molding covers the crack between the flooring and the base board (Fig. 22.41). Flooring is discussed in Chapters 19 and 37.

Other types of molding are often used in high-end architectural applications. For example, a variety of crown moldings are used where walls and ceiling meet (Fig. 22.42). Chair rails are used to protect walls from damage, especially in formal dining rooms, and to add a decorative feature. Often the wall below a chair rail is paneled or at least finished in a different way from the wall above it. Moldings are often mounted on the wall to form decorative panels (Fig. 22.43).

Installing Cabinets

Manufactured kitchen and bath cabinets come in a wide variety of styles, materials, and finishes. Most cabinets used today are made in a factory and shipped to sites ready for installation. Figure 22.44 shows standard

Figure 22.42 Crown moldings are used at the intersection of walls, beams, and ceilings. *(Image copyright John Wollwerth, 2009. Used under license from Shutterstock.com)*

Figure 22.41 Baseboard and base shoe installation details for wood and carpeted floors.

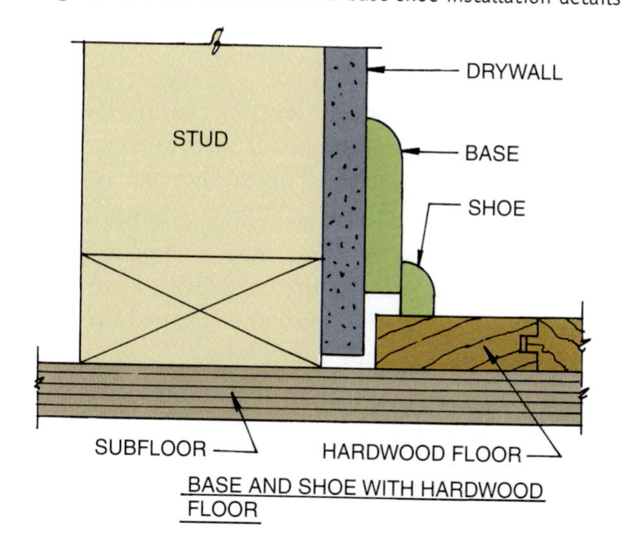

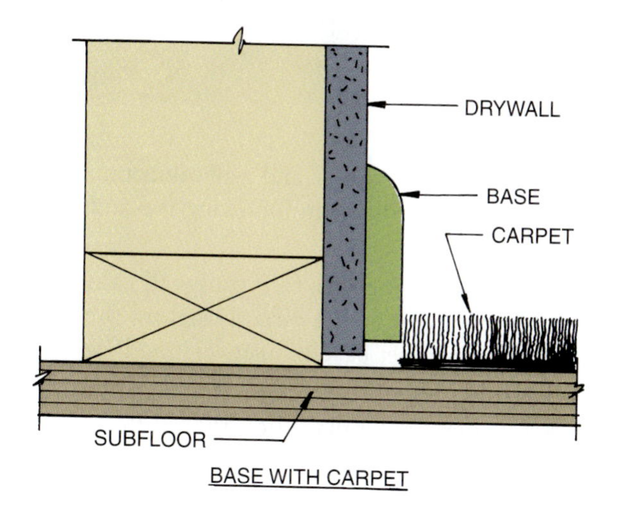

Figure 22.43 A wide variety of manufactured and site-built millwork is installed during interior finishing. *(Image copyright John Wollwerth, 2009. Used under license from Shutterstock.com)*

Figure 22.44 Manufactured kitchen cabinets are made in a factory and shipped to the site ready for installation. *(Image copyright Kristin Smith, 2009. Used under license from Shutterstock.com)*

modular sizes of both base and wall cabinet units. A finish carpenter installs them following the architectural drawings.

Some installers prefer to install wall cabinets first because they can stand directly below them while working. Others set base cabinets first and use them as a support to hold the wall cabinets as they are installed.

Base cabinets are placed on the subfloor and leveled by shimming them at the floor. They are screwed through the back rail to the wall studs. Some cabinets have adjustable metal legs for leveling purposes. Walls must be checked for straightness. If there is a bow, the base cabinets must be shimmed out so they are straight.

Figure 22.45 Base cabinets are shimmed for level and screwed to the wall studs.

Figure 22.46 Wood sub-counters provide a surface for finish countertop material. *(Image copyright Lisa F. Young, 2009. Used under license from Shutterstock.com)*

Wall cabinets are installed a specified distance above the base cabinets. They are often supported on the base cabinets and adjusted until they are level. Then they are shimmed to get them plumb and screwed through the back rails into the studs (Fig. 22.45).

Countertop materials include plastic laminate, solid plastics, granite and other stones, sheet metal, and ceramic tile, among others (Fig. 22.46). Thin materials utilize a particle board underlayment for strength.

Figure 22.47 A typical detail for a straight run wood frame stair.

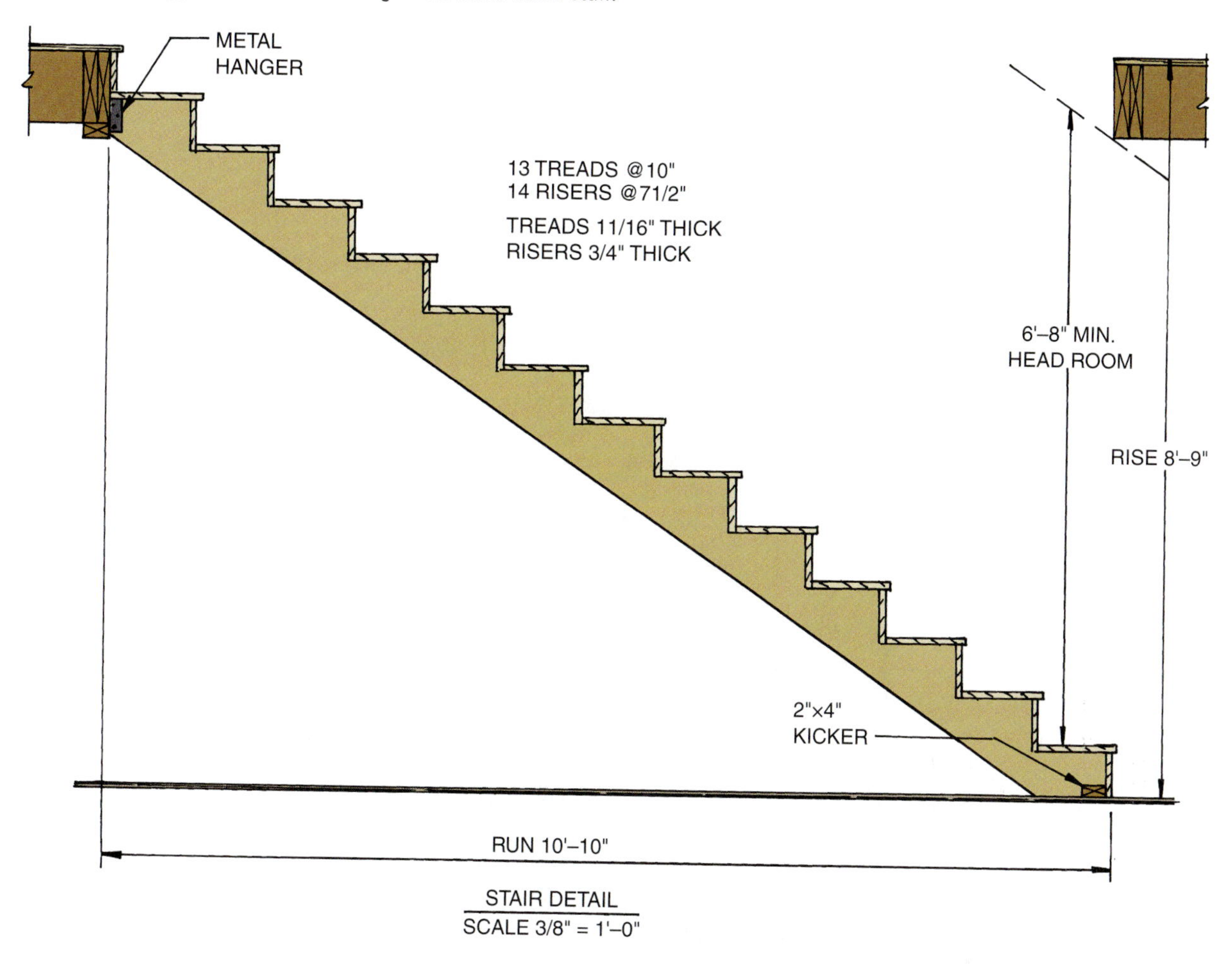

A carpenter cuts openings needed for sinks and lavatories. A plumber installs these when the building nears completion.

Installing Stairs

Stair design is shown on architectural drawings; they specify the type and quality of the treads, handrail, newel posts, balusters, and other parts. Most wood stairs are built by carpenters on-site. Stringers are constructed of solid or engineered wood, while railings and finish materials may be pre-manufactured. Completely manufactured stairs can be specified and assembled on-site. In the design of stairs, an architect must observe the requirements of local building codes that regulate riser height, tread width, stair width, and handrail specifications.

The design of a typical wood-frame stair is illustrated in Fig. 22.47. The load is carried by the carriage or stringer. Generally, a stair has three or more carriages. The rough framing for a stair with a landing is shown in Fig. 22.48. A riser is a board covering the vertical distance between treads. Wood treads are commonly 1-$\frac{1}{4}$ in. in thickness. Some stairs have the treads and risers set in housed carriages. They are held in place with glue and wedges. There are many factory-manufactured stair components that provide interesting and attractive appearances (Fig. 22.49).

Other Finish Items

Finish carpenters complete a number of additional tasks as a building nears completion. These include installing hardware, door stops, bath accessories, and fireplace mantels. Some items, such as mirrors and closet shelving, may be installed by the subcontractor supplying the item. Finally, the building is thoroughly cleaned, the subfloor scraped clean and sanded if necessary, and the carpet; composition floor covering; or tile, marble, or slate floor covering is installed (Fig. 22.50). If pre-finished hardwood floors are installed, they need to be cleaned and waxed. Unfinished floors are sanded and finished as specified.

Figure 22.48 Framing details for a stair with landing.

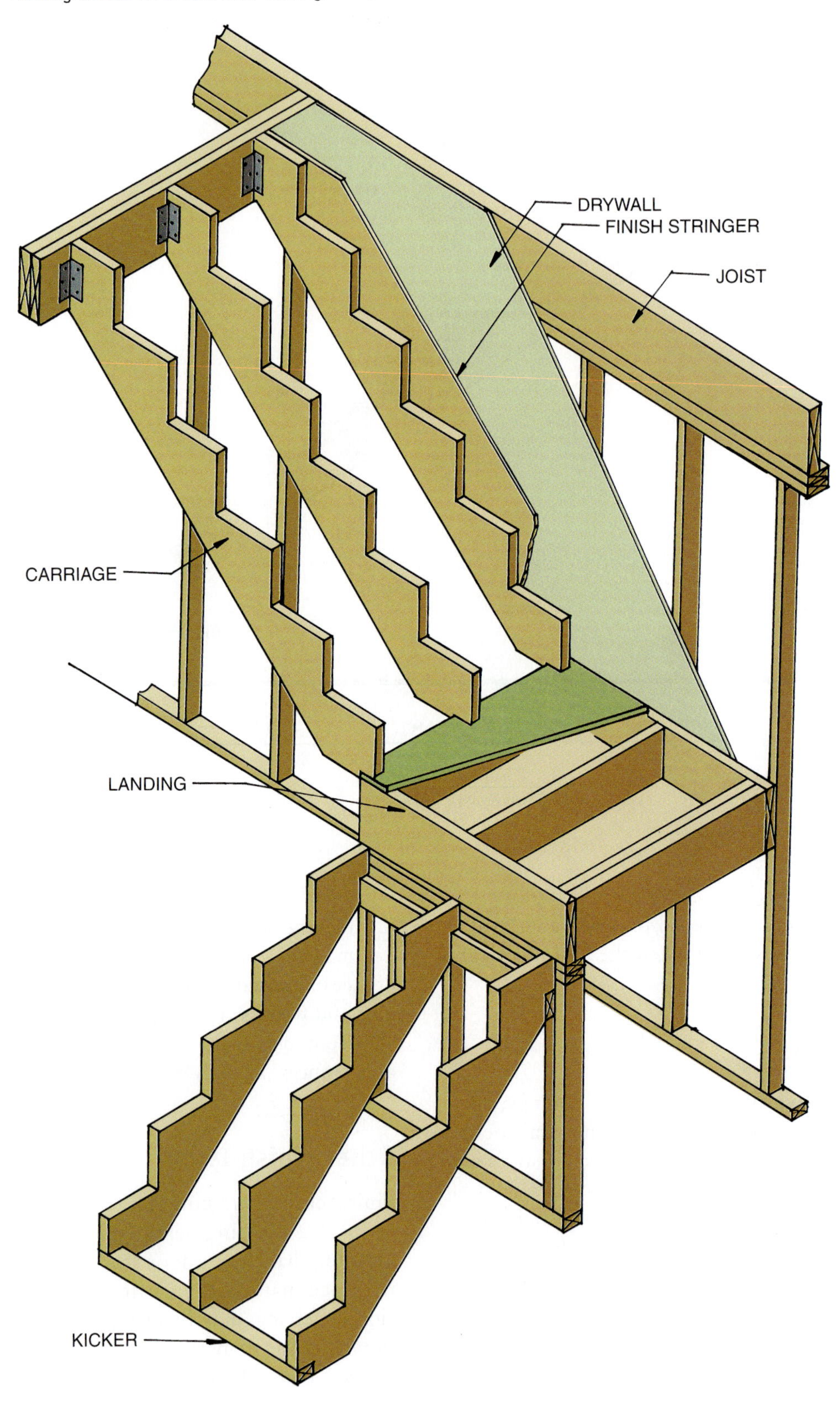

Figure 22.49 A staircase using commercially available components. *(Image copyright Chad McDermott, 2009. Used under license from Shutterstock.com)*

Figure 22.50 The interior is finished after painting and the installation of the finished flooring material. *(Image copyright Sergey Shlyaev, 2009. Used under license from Shutterstock.com)*

Other trades have to phase in their work at the proper time. This includes painting and the installation of wallpaper and other wall coverings. Window shades and curtains are installed after carpet is down and all interior finishing is complete.

Review Questions

1. What materials can be used for the finished soffit and fascia?
2. Why is attic ventilation important?
3. What are some of the ways a roof can be vented?
4. What types of exterior siding are commonly used?

Key Terms

casings
eaves
facia
millwork
rakes
soffit

Activities

1. Visit a construction site regularly and follow the building phases, from finishing the exterior through the work of the trades involved in interior finishing. Keep a log indicating when each part started, thereby producing a construction schedule for the job.
2. Sketch several ways used for framing eaves and rakes.
3. Carefully examine the construction of a staircase in a building and see if it meets recommended construction standards. If it does not, prepare a report that includes sketches citing what you would do to improve the construction.

Additional Resources

Gypsum Construction Handbook, United States Gypsum Company, Chicago, IL.

Spence, W. P., *Installing and Finishing Flooring*; *Roofing Materials and Installation*; *Interior Trim-Making, Installing Finishing*; *Installing and Finishing Drywall*; *Carpentry and Building Construction*; and *Staircases, Balustrades and Landings*, Sterling Publishing Co., New York.

Wood Building Technology, Canadian Wood Council, Ottawa, Ontario, Canada.

See Appendix B for addresses of professional and trade organizations and other sources of technical information.

23 CHAPTER

Paper and Paper Pulp Products

LEARNING OBJECTIVES

Upon completion of this chapter, the student should be able to:

- Understand the technical processes used to produce paper and paper pulp.
- Be able to use the knowledge of the properties of paper and paper pulp products when making material selections.
- Specify the required chemical treatments of paper and paper pulp products to meet the needs of a construction application.
- Be aware of the many paper and paper pulp products used in construction.

The word paper derives from the Greek term for the ancient Egyptian writing material called papyrus, which was formed from beaten strips of papyrus plants. A wide variety of paper and paper pulp products are used in the construction industry. New products appear regularly as paper and paper pulp products are combined with other materials, such as synthetic resins, plastic materials, and metal foils.

MANUFACTURE OF PAPER

Paper is manufactured from paper pulp, which is produced using fibers from materials containing large amounts of cellulose. The main source of pulp is wood from coniferous (softwood) and deciduous (hardwood) trees. Coniferous trees supply the major source of **cellulose**. Other pulp materials include straw, cornstalks, **jute**, waste-paper, rags, and **bagasse** fibers. Jute is a strong, coarse fiber taken from two types of East Indian tiliaceous plants. Bagasse fibers are derived from refuse left over from the crushing of sugarcane or beet refuse from sugar making. Wood pulp fibers are used to produce paper and fiber panels, such as wallboard, acoustical tile, fiberboard, paneling, and roof decking.

In addition to cellulose, paper pulp may include **carbohydrates** and lignin from wood and coloring materials, such as dyes, vegetable colors, and mineral pigments. A wide range of chemicals are used to produce the type of pulp needed for the finished products. Rosin, glues, and synthetic resins also are added to provide sizing, and a number of filler and coating materials, such as gypsum, clay, talc, and chalk, may be added to the pulp.

Manufacturers of paper pulp begin by removing bark from logs. The logs then go to a chipper where they are ground to a pulp or into small pieces. When ground into pulp, the wood goes directly into the papermaking process. If ground into chips, the wood is fed into a chemical process where it is digested into a soft wood-pulp mass (Fig. 23.1).

After the pulp is digested, it goes to a bleacher where, using chemicals, it is purified and the chlorine neutralized. The pulp is rinsed between stages. Some products, such as brown wrapping paper and pulp for paperboard, are not bleached to increase the brightness of the paper.

Once the pulp is out of the bleacher, it proceeds to a beater, where the fiber and added materials are beaten into a mass of the proper consistency. The pulp leaves the beater and flows to the papermaking machine. Here it passes over a wire mesh screen where some of the water drains away and a sheet of fibers forms. The sheet still

Figure 23.1 The manufacturing process used to produce paper products.

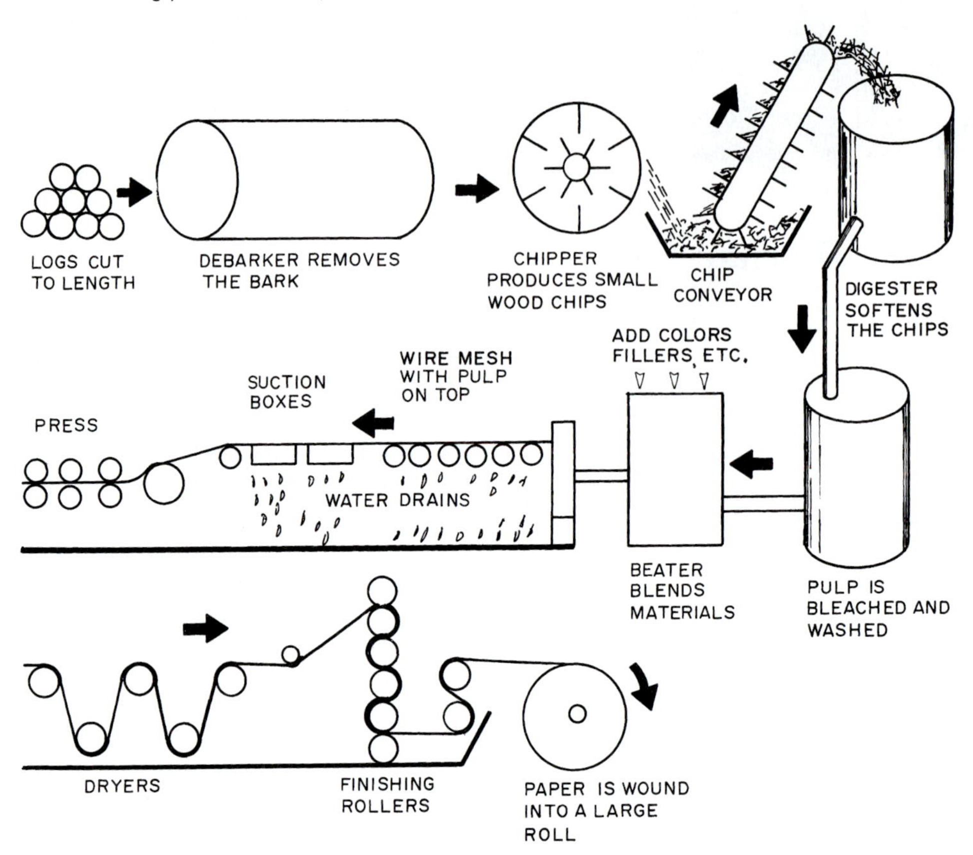

contains considerable water, so it is passed over a series of suction boxes and through smooth rollers that remove additional moisture. The sheet then goes to a series of rollers in a dryer, where the moisture is reduced to 5 to 10 percent. The dry sheet passes through a set of finishing rollers, which gives it a final surface finish. The finished paper is then wound in large rolls and ready for processing into useful products.

Paper Recycling

Industrialized paper making has an effect on the environment in terms of both the acquisition of raw materials and waste-disposal impacts. Recycling paper reduces this impact. Paper recycling is the process of recovering waste paper and remaking it into new paper products. There are three categories of paper that can be used as raw material for making recycled paper: mill broke, pre-consumer waste, and post-consumer waste. Mill broke are paper trimmings and other scraps generated during the manufacture of paper that are recycled internally in a paper mill. Pre-consumer waste is material that was discarded before it was ready for consumer use. Post-consumer waste is material discarded after consumer use, including newspaper and magazines, telephone directories, and residential mixed paper. Paper suitable for recycling is called "scrap paper."

PROPERTIES OF PAPER PRODUCTS

The properties of paper and paper pulp products vary with the composition of the pulp, the cellulose fibers used, the manufacturing process, and the finishing operations involved. Water absorption is an important factor in determining where a paper pulp product can be used. Although some forms of pulp product are waterproof, most require an environment containing little moisture. Fire resistance properties influence where they can be used in a building.

Tensile strength is important when paper products are under a load, such as in sheathing paper or as backing for insulation batts. Porosity influences acoustical and insulation properties of paper pulp products as well as their water absorption characteristics. Chemical resistance properties have a direct relationship to specific uses, including possible reactions with materials touching the paper product.

CHEMICAL TREATMENTS

Paper products can be treated with chemicals that enable them to resist fire, water and water vapors, bacteria, mold, fungi, and insects.

Papers that are **flame-retardant** are not totally fireproof, but will not burst into flames, which would release deadly combustible gases. **Fire-resistant** paper is difficult to ignite, will not support combustion, and will self-extinguish after a heat source is removed. Table 23.1 lists chemicals commonly used for providing flame-retardant and fire-resistant properties in paper and paper pulp products. Some fire-resistant and flame-retardant chemicals are destroyed by wetting and should not be used on products installed in areas exposed to moisture.

Paper and paper pulp products are waterproofed by mixing synthetic resins with pulp. Some types of wall coverings are moisture resistant because they have a plastic film applied on the exposed surface. Waterproof membranes used on walls and floors below the water level in the soil are coated with tar, asphalt, or a synthetic resin. Waterproof paper with a laminated metal foil is used for concealed flashing. Papers used for vapor barriers are also water resistant.

Table 23.1 Flame-Retarding and Fire-Resisting Chemicals Commonly Used on Paper Products

Material	Effectiveness	Moisture Resistance
Antimony trioxide	Excellent	Are moisture resistant
Zinc borate	Good	
Antimony oxide	Excellent	
Ammonium sulfate	Excellent	Are not moisture resistant
Ammonium chloride	Good	
Sodium phosphate	Good	
Boric acid-sodium borate	Good	

PAPER PRODUCTS IN CONSTRUCTION

A summary of selected paper products used in construction is provided in Table 23.2. Various types of building papers and felts are used for sheathing, reduction of dust and air filtration, vapor barriers, and sound-deadening purposes (Fig. 23.2). Some have a pulp made from cotton, old rags, and paper. Others use a kraft paper base or jute pulp. Papers such as vellum are used for the preparation of architectural drawings, reproduction of drawings, correspondence, billing, and many other of the business operations involved in construction (Fig. 23.3).

Figure 23.2 Asphalt impregnated paper is used for roofing applications. *(Image copyright Robert Pernell, 2009. Used under license from Shutterstock.com)*

Table 23.2 Selected Paper Products Used in Construction

Paper Product	Uses	Coating
Asphalt felt	Sheathing paper, built-up roofing	Cotton and rag pulp paper saturated with asphalt
Asphalt sheathing paper	Between sheathing and siding	Asphalt saturated
Floor lining paper	Between subfloor and finish floor	Asphalt or tar saturated and coated
Carpet felt	Carpet underlayment	No special coating
Copper foil laminated	Concealed flashing, vapor barrier, waterproofing	Copper foil applied to one side of kraft paper
Deadening felt	Sound deadening in walls and floors	Made from rags and paper with no special surface coating
Plastic-coated paper	Vapor barriers	Laminated kraft paper coated with plastic on one side
Building paper reinforced with glass fibers	Vapor barriers	Laminated paper reinforced with glass fibers
Rolled roofing	Final (exposed) roof layer	Felts saturated with asphalt and coated with mineral aggregate
Sheathing paper	Protection against dust and air infiltration in walls and floors	No special coating
Wallpaper	Cover finished interior walls	Clay or casein coating
Resin-coated wallpaper	Washable finished interior wall coating	Synthetic waterproof coating, such as plastic film
Kraft paper	Bags, protective wrapping, concrete forms, plastic laminate base material, concealed flashing, cover gypsum sheets	Wood pulp sized with resin
Vellum	Drafting paper	100% rag impregnated with synthetic resin
Mineral-fiber papers	Where high temperatures exist, for electrical insulation	Glass, silica, and ceramic fibers in the pulp

Figure 23.3 Papers such as vellum are used for the preparation of architectural drawings. *(Image copyright Trutta, 2009. Used under license from Shutterstock.com)*

Other paper products are used to make bags, such as protective wrappings of products, and to encase materials, such as fiber insulation batts. It is important when selecting one of these products to follow the recommendations of its manufacturer pertaining to its best use, relationship to building codes, and installation techniques.

Cardboard

Cardboard is a pulp product that varies widely in its thickness and properties. It is a relatively stiff, rigid product. In construction it is used for boxes and cartons in which products are shipped. Corrugated board is probably the most widely used. Corrugated board consists of outer plies of a coarse, stiff brown paper bonded to the face and back of a corrugated core (Fig. 23.4). Corrugated paper honeycomb cores are also used in the construction of some wall panels and hollow-core doors.

Figure 23.4 Corrugated board has coarse outer paper plies bonded to a corrugated paper core. *(Image copyright c., 2009. Used under license from Shutterstock.com)*

Paper Pulp Sheets

Paper pulp sheets are available with a variety of surface treatments, structural characteristics, chemical properties, and colors. The simplest form is a single homogeneous ply of wood pulp with no special additives or surface coatings. Other sheets have the ply treated with resin. Some sheets are made by laminating sheets of paper and impregnating them with resin. Other sheets have an aluminum foil or plastic film bonded to the surface. The pulp forming the sheet may be reinforced with glass fibers or have a thin core of plastic foam or corrugated paper. Sheets may be impregnated with asphalt or synthetic resins and can be waterproof. Some are rigid, durable, and strong. They are available with plain, perforated, and corrugated surfaces. The basic sheet is usually brown, but various surface finishes and colored finished surfaces are available. Paper pulp sheets are available in 4 × 8 ft. (1219 × 2438 mm) sheets in thicknesses of $\frac{1}{8}$, $\frac{3}{16}$ and $\frac{1}{4}$ in. (3, 5, and 6 mm).

Laminated paper sheet material, made by laminating sheets of paper, is used for finished surfaces on interior walls. It is available in 4 ft. (1220 mm) widths and lengths up to 16 ft. (4880 mm).

Fiberboard Wood Pulp Panel Products

Fiberboard wood pulp products are made from pulpwood. A wide variety of products are manufactured. They differ from pulp sheets in that they are thicker. They are available as laminated layers of paper pulp board or panels and as a single homogeneous layer of wood pulp fibers.

Single layer fiberboard (often called insulation board) is a high-density asphalt-impregnated panel used for wall sheathing. One-inch fiberboard has an R-value of about 2.6 and is available in 4 × 8 ft. (1219 × 2438 mm) sheets and thicknesses of $\frac{1}{2}$, 1, $1\frac{1}{2}$, and 2 in. (12, 25, 38, and 51 mm). Longer sizes are also available. Standard fiberboard will not hold nails, but a nail base fiberboard is available. Fiberboard is also used as backer boards behind shingles and some types of siding. Fiberboard sheets with a plastic foam or reflective foil bonded to them can be used for roof insulation.

Fiberboard finds some use as interior finished wall panels with one surface prefinished. Acoustical ceiling tiles are made with a variety of surface treatments, such as plain, patterned, textured, and perforated. Typical tiles are usually 12 × 12 in. (305 × 305 mm) or 16 × 16 in. (406 × 406 mm) square, although larger size panels are available from some manufacturers.

Mineral fiberboard is made of mats of rock wool or fiberglass with stiff paper faces on both sides to give rigidity. Fiberboards are used widely for insulation.

Fiberboard panels are installed by nailing or gluing them to studs or other surfaces. Joints are not laid tight to allow for expansion. Acoustical ceiling tiles are installed with suspended tracks or by stapling or gluing them to another surface or to wood furring. The joints between the panels on decorative interior walls can be covered with metal strips. Fiberboard panels are not recommended for areas with high moisture potential. Under these conditions, a plastic-coated panel, sealed on all edges with a waterproof caulking, should be considered.

Wood Pulp and Fiber Decking

A variety of roof decking products are available made from layers of pulp board or fiberboard bonded with a waterproof adhesive. These products come with many surface treatments, including coatings of fibrous cement, plastic film, plywood, metal foil, asphalt, roofing felt, paper, or plastic foam sheets. The coatings increase the panel's water resistance, insulation value, and strength. Coated surfaces can be exposed to view and become finished ceilings. Panels can be treated to provide acoustical dampening. Surfaces may be painted, covered with a textured fibrous sheet, or have a pre-finished plywood veneer. Most contain a vapor barrier within the assembled deck panel (Fig. 23.5).

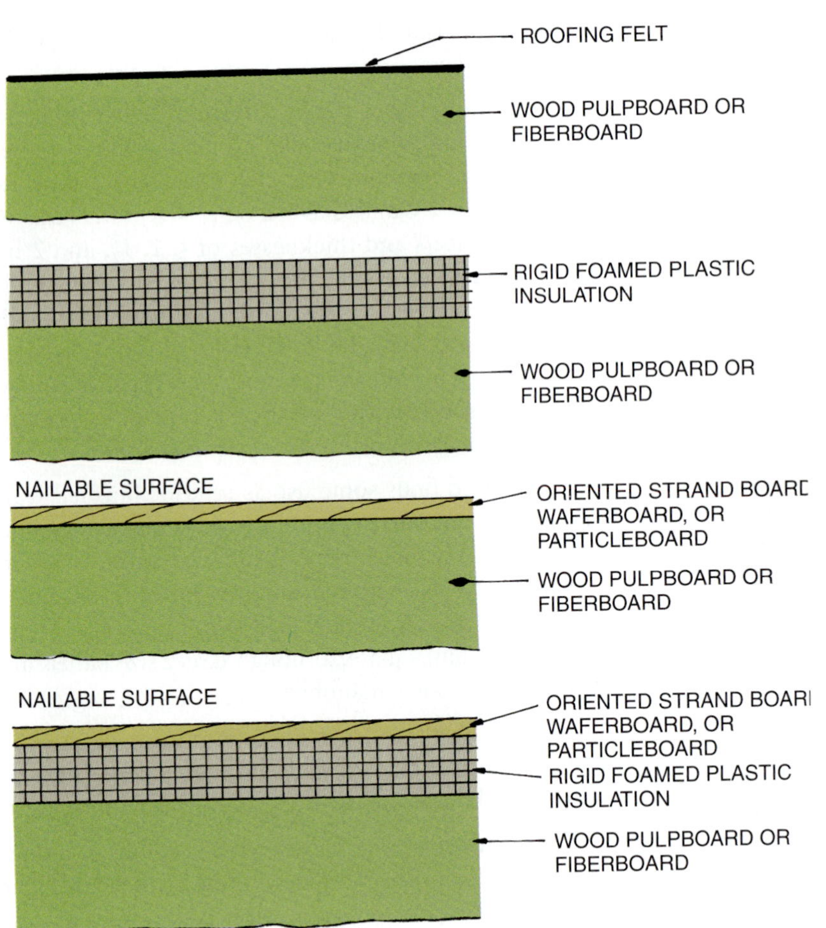

Figure 23.5 Examples of wood pulp and fiber roof decking.

When choosing a decking product, it is important to know the distance it will span under a known loading. The span varies with the applied load and thickness. Typical panel thicknesses range from 2 to 4 in. (51 to 102 mm), widths from 20 to 48 in. (508 to 1219 mm), and lengths from 48 to 144 in. (1219 to 3657 mm). Spans in general range from 16 to 48 in. (406 to 12,192 mm). Manufacturers' catalogs must be consulted for specific data.

Decking is manufactured with various edge treatments. Common edge treatments include all four edges square, all four edges tongue and groove, and tongue and groove on two edges and square on the ends. Those used with Metal T brackets have rabbeted edges.

The atmospheric conditions in which wood fiber decking is used should be considered and manufacturer's recommendations followed. Some products require that the temperature be kept 40°F (4.4°C) or higher and that relative humidity be kept below 50 percent. The type of finished roofing material to be applied must also be taken into account. If the structure is to have a built-up roof on which hot asphalt or tar is applied, the joints between panels must be sealed to keep the asphalt or tar from leaking through and damaging the finished ceiling below. If the finished roofing material, such as shingles, requires nailing or stapling, the pulp roof decking must have a plywood, oriented strand board, particleboard, or wafer board veneer to hold the nails. It must also be thick enough so the nails do not break through the finished ceiling below.

Following is a typical example of the design information provided by wood pulp and fiber decking manufacturers:

Diaphragm resistance—can be used in seismic regions and areas of severe wind loads

Meets U.L. wind uplift resistance requirements

Flame spread	20
Smoke developed	5
Roof and ceiling assembly	$\frac{11}{2}$ hr. fire rating

Linear expansion	0.2%
Noise reduction coefficient	2 in. ≤ NRC 5 .65
	3 in. ≤ NRC 5 .80
Light reflectance	60%
Sound transmission class	$\frac{21}{2}$ in. thick is STC 41
R value	2 in. ≤ 4, 3 in. ≤ 6, 4 in. ≤ 8
Compressive strength	300 psi

Fastener withdrawal loads with oriented strand board

- #14 all-purpose screw - 380 lb.
- #12 steel decks - 330 lb.
- #10 sheet metal screw - 180 lb.
- #11 gauge roofing nail - 51 lb.

Review Questions

1. What is the main source of wood pulp for papermaking?
2. In addition to wood pulp, what other materials are used in papermaking?
3. What types of construction products are made from wood pulp fibers?
4. What sizing materials are added during the papermaking process?
5. What filler materials are added during the papermaking process?
6. What is the main difference between the two methods of papermaking?
7. How does bleaching affect the paper?
8. What properties of paper products directly influence where they are used?
9. How are paper products waterproofed?

Key Terms

bagasse
carbohydrates
cellulose
fire-resistant
flame-retardant
jute

Activities

1. Collect samples of the various paper and paper pulp products used in construction and identify their properties.
2. Visit a building supply dealer and make a list of all the paper and paper pulp products you can observe.

Additional Resources

See Appendix B for addresses of professional and trade organizations and other sources of technical information.

CHAPTER 24

Plastics

LEARNING OBJECTIVES

Upon completion of this chapter, the student should be able to:

- Learn the properties and characteristics of plastic materials used in construction.
- Select the proper plastic material for various construction applications.

Build Your Knowledge

For further study on these materials and methods, please refer to:

Chapter 25 Thermal Insulation and Vapor Barriers
Chapter 30 Doors, Windows, Entrances, and Storefronts
Chapter 45 Plumbing Systems
Topics: Piping, Tubing, and Fittings

Although the use of plastics is relatively new in the construction field, plastic products have rapidly become a major material and are being used for structural and nonstructural purposes **(Fig. 24.1)**. Because extensive research in plastic development has resulted in a variety of plastics that have properties not available in conventional materials, plastic is replacing materials such as wood and metal, in many aspects of building. Plastics resist corrosion and moisture; they are tough, lightweight, and easily formed into useful products. Since they are chemically derived, a wide range of plastics with special properties can be developed. This makes them particularly useful as a construction material because of the extensive range of applications that exist.

Vinyl siding is available in a variety of colors, surface textures, and sizes and provides a tough, maintenance-free exterior. Wood windows are clad in vinyl, eliminating the need for painting, and some windows are framed entirely from solid extruded plastic. Other types of plastics are replacing glass glazing because the plastic is lighter and more resistant to shattering. Plastic lavatories, showers, and bathtubs have largely replaced those of ceramic-coated metal with cast iron fixtures. A high percentage of items used in electrical systems, such as boxes and wiring insulation, are plastic. Plastic film is used for vapor barriers and to reduce air infiltration, and plastic foams are used for insulation.

Since there are so many different kinds of plastics, each with varying properties, designers and constructors must be careful to use individual plastics for their designed purposes. For example, using plastic pipe designed to carry cold water for hot water lines will eventually cause problems.

Many families of plastic materials exist, and within each, properties can vary widely. Plastics from several families may serve the same purpose, but others exhibit unique, special characteristics.

The term **plastic** is used today to describe manmade **polymers** that contain carbon atoms covalently bonded with other elements. Plastic is obtained by breaking down materials found in nature, such as petroleum, coal, and natural gas **(Fig. 24.2)**. Plastics are **synthetic materials**, resulting from chemical manipulations of natural materials. While plastics are composed of **organic materials**, they are manmade and not found in nature. When produced, they are soft and exhibit **plastic behavior**, meaning they can be formed into desired shapes. Most are made up of molecules built around a carbon atom. An exception to this is a plastic composed of a group of materials built around the silicon atom.

Figure 24.1 Plastic products are widely used in the construction industry. *(a) Image copyright Serg64, 2009. Used under license from Shutterstock.com (b) Image copyright Lezh, 2009. Used under license from Shutterstock.com (c) Image copyright Liveshot, 2009. Used under license from Shutterstock.com (d) Image copyright foto.fritz, 2009. Used under license from Shutterstock.com*

a

b

c

d

MOLECULAR STRUCTURE OF PLASTIC

The properties of plastics can be made to vary by manipulating the molecular structure of the material. Most materials classified in the plastics family are based on the carbon atom. The carbon atom bonds with other atoms through covalent bonding. **Covalent bonding** is a process in which small numbers of atoms are bound into molecules. A single molecule is known as a **monomer**. When several monomers link together to form a chain they form a polymer. This chain formation is called polymerization. Plastic polymers are called macromolecules because they are composed of many smaller molecules ("macro" means large). Since the small molecules are joined together in a chainlike condition, they are often called chains. Plastics are made up of these polymers. Note that many plastics are described by the prefix "poly," such as polyvinyl and polystyrene. Plastic is a polymer. Polymers are moldable materials, sold in the form of granules, powder, flakes, liquids, or pellets for processing into useful products.

The molecular bonding process can be shown by the following illustration. The carbon atom has a **valence** of four, which means there are four points at which other atoms can attach themselves to the carbon atom to form covalent bonds. Other elements have different valences. For example, oxygen atoms have a valance of two. Two oxygen atoms can bond to the four carbon valence points creating carbon dioxide (CO_2) (one carbon atom and two oxygen atoms).

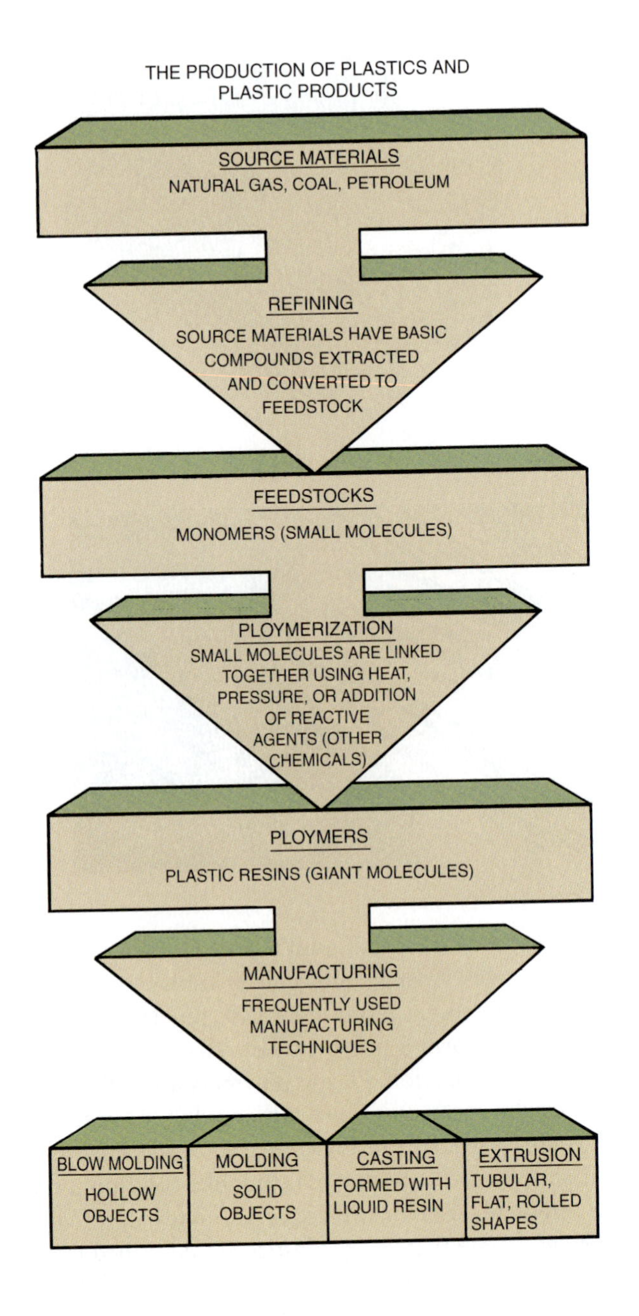

Figure 24.2 The production of plastics and plastic products. *(Courtesy the Society of the Plastics Industry, Inc.)*

CLASSIFICATIONS

Plastics can be divided into two basic classifications: thermoplastics and thermosetting materials. A new classification of bioplastics is finding wider use.

Thermoplastics

Thermoplastics are plastic materials that can be softened or re-melted by the application of heat and reformed. This quality allows for the recycling and reuse of products made with thermoplastics, although some thermoplastic materials will experience contamination and chemical degradation if they are reheated frequently.

Thermoplastic materials are composed of long chain-like molecules that are unattached to each other. These molecules can slide past one another and change shape. At normal temperatures, 70°F (21°C), the material retains its shape because the movement of molecules is slight. However, when heat is applied, the bonding between molecules weakens, and the material expands and becomes soft. At high temperatures (which vary according to composition) the plastic softens and has sufficient flow to permit molding. As the molded part cools, the plastic holds the new shape.

Thermosetting Plastics

Thermosetting plastics, also called **thermosets**, are plastics that cannot be reheated and reformed once they have been softened, constituted, and cured. They are formed with a chemical process that produces a strong bond between molecules and prevents their sliding by each other. The final form of the material is irreversible.

Bioplastics

Bioplastics, or organic plastics, are a form derived from renewable biomass sources, such as vegetable oil or corn starch, rather than from petroleum, as are fossil fuel plastics. The production and use of bioplastics is generally regarded as more sustainable when compared with petroplastics (plastic production from petroleum), because it relies less on fossil fuel carbon sources resulting in fewer greenhouse emissions. Because the material biodegrades, it significantly reduces hazardous waste caused by oil-derived plastics, which remain solid for hundreds of years. Bioplastics represent a new era in manufacturing technology.

ADDITIVES

Most plastic resins in pure form do not have the properties needed for particular applications. Therefore, **additives** are mixed with the resin to modify their properties. Frequently, more than one additive is needed to obtain the desired properties. The commonly used additives include plasticizers, fillers, stabilizers, and colorants.

Plasticizers

Plasticizers are added to plastic resin to reduce brittleness; increase flexibility, resiliency, and moldability; and, in some cases, to improve impact resistance. A plasticizer lessens the bond between the chainlike molecules that make up a plastic. Since the molecules are able to slide past each other, the plastic can bend and flex. In doing this the strength, heat resistance, and dimensional stability are lessened (Fig. 24.3).

Fillers

Some **fillers** are used to reduce a plastic's cost by providing bulk. Others are used to improve a specific property. Fillers used to increase bulk or ease of molding include finely ground hardwood and nutshells. Hardness is improved by adding mineral oxides and mineral powders. Heat resistance is enhanced by adding inorganic fillers, such as clay, silica, ground limestone, or asbestos. Quartz or mica increase electrical resistance. Toughness is improved by adding fibers, such as hemp, cotton, sisal, rayon, polyester, or nylon.

Figure 24.3 Some plastic products, such as this vinyl floor covering, have plasticizers added to the resin to reduce brittleness, increase flexibility and resiliency, and improve impact resistance. *(Image copyright ukrphoto, 2009. Used under license from Shutterstock.com)*

Mechanical properties are strengthened with the addition of glass or metal fibers.

Stabilizers

Stabilizers, or lead compounds, are used to stabilize plastic by helping it resist heat, loss of strength, and the effects of radiation on bonds between chains.

Colorants

Colorants added to the resin include organic dyes and inorganic pigments (metal-based). Some of these also have a stabilizing influence. Inorganic pigments disperse throughout the resin, rather than dissolve, as do the dyes, so they reduce the transparency of the material. They also withstand high temperatures without fading.

PROPERTIES

The important properties of plastics are illustrated in Tables 24.1 through 24.5. Notice the variances within the two classifications, thermosets and thermoplastics. The limitations on these materials for certain uses in construction can be judged by comparing them with other materials on tensile strength and other properties. Following is a general discussion of the major properties of plastics. The composition of plastics can vary considerably, so the specifics of their properties do also. To make proper use of plastic materials, the properties of a specific formulation of resin and additives must be known.

Mechanical Properties

Mechanical properties that are generally most important to consider for plastics used in construction are tensile strength, stiffness, toughness (or impact strength), hardness, and creep. As shown in Table 24.1, the tensile strength of most plastics is lower than steel and more closely resembles wood. Plastic laminates and fiberglass-reinforced plastics have tensile strengths approaching steel.

Stiffness is measured by the modulus of elasticity (E). The greater the modulus of elasticity, the stiffer the material. Overall, plastics are not as stiff as steel, aluminum alloys, and concrete (Table 24.2), so they deflect more under load. One advantage plastic structural members have is that they are lighter than conventional materials.

Impact strength (toughness) is the ability of a material to resist impact from an object striking it. In general, thermoplastics exhibit better impact strength than thermosets. Polyurethane is particularly resistant and is used for automobile bumpers. Polyethylene and polyvinyl chloride (PVC) have a wide range of impact strength and provide an array of uses, from bottles to piping. Compared to other construction materials, plastics are quite high in impact strength. This is why acrylic and polycarbonate sheet material is used for glazing in many situations.

Hardness is a measure of abrasion resistance. In general, plastics are not as good as glass or steel in resisting abrasion, as evidenced by visible scratches on plastic glazing resulting from improper cleaning.

Creep is the term used to describe the tendency of a material to flow in normal temperatures and to have a permanent change in size and shape when under continued stress. Although all materials creep, it is insignificant in materials such as steel. Creep is a significant factor for many plastics, even at room temperature. It is most apparent in thermoplastics, since they contain unattached molecules that can begin to slip past each other when under stress. When the duration of stress

Table 24.1 A Comparison of the Tensile Strengths of Plastics with Other Commonly Used Construction Materials

Material	Tensile Strength, psi
Thermoset plastics	175–35,000
Thermoplastics	1000–20,000
Reinforced plastics	800–50,000
Plastic laminates	800–50,000
Glass (annealed)	3,000–6,000
Cast iron	20,000–100,000
Hot-rolled carbon steel	43,000–130,000
Southern pine	500–1,250
Wrought aluminum	5,000–82,000

Table 24.2 A Comparison of the Modulus of Elasticity of Plastics with Other Commonly Used Construction Materials

Material	Modulus of Elasticity (E), 10^6 psi
Thermoset plastics	0–2.4
Thermoplastics	0–1.8
Reinforced plastics	0–5.00
Cast iron	15–25
Wood	0.5–2.6
Aluminum alloys	9–11
Glass	9–11
Concrete	1–5
Steel, carbon, high strength, and stainless	29.0

Table 24.3 A Comparison of the Service Temperatures of Plastics with Other Commonly Used Construction Materials

Material	Service Temperature, °F
Plastics	80–500
Plastic foam (polystyrene)	140–150
Carbon steel	900–1350
Wood	225–400
Glass	350–1600
Concrete	450–1350
Aluminum alloys	300–575

Table 24.4 A Comparison of the Coefficient of Thermal Expansion of Plastics with Other Commonly Used Construction Materials

Material	Coefficient of Thermal Expansion 10^{-6} in./in./°F
Thermoset plastics	7–167
Thermoplastics	19–195
Reinforced plastics	7–38
Cast iron	5.9–6.6
Carbon steel	6.5
Wood	7–38
Glass	1
Aluminum alloys	9
Concrete	7

Table 24.5 A Comparison of the Thermal Conductivity of Plastics with Other Commonly Used Construction Materials

Material	Thermal Conductivity BTU/hr./ft.^{2}/in./°F
Thermoset plastics	1–2
Thermoplastics	0.5–5.0
Plastic foams	0.15–0.26
Wood	0.5–1.5
Glass	5–6
Brick	3–6
Concrete	5.5–9.5
Steel alloys	300–500
Aluminum alloys	500–1500
Copper	440–2550

is short, the plastic may return to its original condition. But long term stress may produce permanent deformation.

Laminates, fiberglass reinforced plastics, and some thermosets resist creep better than thermoplastics, but they do not resist stress as effectively as steel and concrete.

Electrical Properties

Plastics have excellent electrical insulating properties and have enabled the development of greatly improved electrical equipment. In addition, they have good heat resistance, a requirement for the use of a material in many electrical devices. Silicones and fluorocarbons have these properties and are widely used. Epoxies are employed to encapsulate electric components subjected to temperatures as high as 300°F (150°C), and phenolics are used as housings for electrical parts, switches, and receptacles. In other words, plastics have excellent **dielectric strength** (the maximum voltage a dielectric can withstand without rupture). A dielectric is a nonconductor of electric current. Since electrical components are subject to damage by arcing, arc resistance is important. **Arc resistance** is measured by the total elapsed time in seconds an electric current must arc to cause a part to fail. Plastic parts carbonize and become conductors, burn, develop thin lines between electrodes, or become glowing hot. Plastics used for applications in which arcing is a possibility have an arc resistance time from 100 to 300 seconds.

Thermal Properties

Two major factors to consider when examining thermal properties are the influence of temperature on strength and expansion and contraction of the material. Many plastics lose strength at comparatively low temperatures. The **service temperature** of plastics, the maximum temperature at which a plastic can be used without affecting its properties, is low when compared with other construction materials (Table 24.3). Some plastics soften below 200°F (94°C), so they soften if placed in boiling water. In general, thermosets have higher service temperatures than thermoplastic materials. Polyamide thermosetting resins resist intermittent heat up to 930°F (500°C) under low stress and up to 480°F (250°C) under continuous stress.

All building materials expand and contract, a property that must be taken into consideration as a designer selects materials and prepares design drawings. Most plastics have a higher coefficient of thermal expansion than other construction materials (Table 24.4). In general, thermosets have a lower rate of expansion than thermoplastics.

Plastics are poor conductors of heat. Unmodified plastics compare favorably with brick, concrete, and glass (Table 24.5). Notice that foamed plastics have the lowest thermal conductivity and are therefore excellent insulators. They are widely used for all types of insulation. When comparing plastics in general with other materials it can be seen that plastics have the lowest coefficient of thermal conductivity.

Flammability is the ability of a material to resist burning. Since plastics are organic materials, they will burn. The range of flammability, however, is great. Cellulosics are highly flammable and often banned by building codes. Polyurethane and polyvinyls give off toxic fumes when burning, and building codes specify protective measures for many uses. Vinyl siding, for example, begins to burn at 698°F (370°C), but wood ignites at about 400°F (206°C). In general, thermoplastics are more flammable than thermosets, though these properties can be changed with additives. Some plastics can develop a char layer that serves as a shield to deter burning. The National Aeronautics and Space Administration uses nylon and phenolic resins on the exterior of space craft that are exposed to the searing heat of reentry.

Although plastics are no more combustible than wood, they can produce toxic fumes, which are more likely than flames to cause death in a fire. Most building codes have a separate chapter devoted to the use of plastics in building construction.

Chemical Properties

Plastics do not corrode like metals, but they can deteriorate and be damaged by chemical attack. When a metal corrodes, its surface is damaged to the extent that it loses weight. Deteriorated plastics gain weight because attacking chemicals combine with the plastic resin. This causes swelling, crazing, and discoloration. In addition, there is loss of impact, flexural, and tensile strengths. Usually the chemical resistance of a plastic decreases as the temperatures increase.

The **weatherability** of a plastic involves consideration of moisture, ultraviolet light, heat, and chemicals found in the air as ozone and hydrochloric acid. High-density polyethylene (HDPE) has great resistance to acids, water absorption, and weathering. It is used for electrical wire insulation. Acrylics also have great resistance to weathering and excellent optical qualities, so they are widely used for glazing.

Density

The density of plastic materials is in general lower than other commonly used construction materials. Glass reinforced plastics (GRP) are lighter than steel and aluminum.

Specific Gravity

The specific gravity of plastics varies considerably, ranging from 0.06 for foams to 2.0 for fluorocarbons. Specific gravity is the ratio of the mass of a volume of material to the mass of an equal volume of water at a standard temperature. The specific gravity of water is 1, so some plastics will float (lighter than water) and others will sink. The specific gravity for softwoods is 0.5, about 2.7 for aluminum, and about 8.0 for steel.

Optical Properties

Some plastics have optical properties equal to that of glass. Acrylics are as transparent as fine optical glass and have a light transmission of 92 percent. Polystyrene, polypropylene, and polycarbonates also exhibit transmission qualities of 90 percent or better.

PLASTIC CONSTRUCTION MATERIALS

The following discussion gives a brief description of the plastics most commonly used in construction and some of their major uses. They are divided into thermoplastics and thermosets.

Since plastic materials are frequently identified in technical literature and professional magazines by approved abbreviations, designers and constructors must be familiar with them. A list of these abbreviations, as compiled by the American Society for Testing and Materials, is provided in Table 24.6.

Thermoplastics

The majority of products used in construction are made from thermoplastic materials rather than thermosets. Following are commonly used thermoplastic materials and products using them.

Acrylonitrile-Butadiene-Styrene Acrylonitrile-butadiene-styrene (ABS) plastics are a combination of high-impact polystyrene, which is very tough, and acrylonitrile, which improves rigidity, tensile strength, and chemical resistance. ABS plastics are widely used for pipe and pipe fittings for water lines with water up to 180°F (83°C) (non-pressure), gas supply lines, waste, drain, and sewage vent systems (Fig. 24.4). They are also used for hardware, such as handles and knobs.

Table 24.6 ASTM[a] Abbreviations for Plastics

Thermoplastics	
Type	**Abbreviation**
Acrylonitrile-butadiene-styrene	ABS
Acrylic:	
Polymethyl methacrylate	PMMA
Cellulosics:	
Cellulose acetate	CA
Cellulose acetate-butyrate	CAB
Cellulose acetate-propionate	CAP
Cellulose nitrate	CN
Ethyl cellulose	EC
Fluorocarbons:	
Polytetrafluoroethylene	PTFE
Nylons:	
Polyamide	PA
Polyethylene	PE
Polypropylene	PP
Polycarbonates	PC
Styrene:	
Polystyrene	PS
Styrene-acrylonitrile	SAN
Styrene-butadiene plastics	SBP
Vinyl:	
Polyvinyl acetate	PVAc
Polyvinyl butyral	PVB
Polyvinyl chloride	PVC
Thermoset Plastics	
Type	**Abbreviation**
Epoxy, epoxide	EP
Fiberglass reinforced plastics	FRP
Melamine-formaldehyde	MF
Phenolic:	
Phenol-formaldehyde	PF
Polyester	—
Polyurethane:	
Urethane plastics	UP
Silicone plastics	SI
Urea-formaldehyde	UF

[a]*American Society for Testing and Materials*

Acrylics The most widely used acrylic is polymethyl methacrylate (PMMA). It has excellent optical clarity and is used for glazing. Typical uses include door and window lights and roof domes and skylights. It is also used to make lighting fixtures, but it will soften at 200°F (94°C). It finds use in outdoor signs, corrugated roofing, and molded hardware. Acrylics are also dispersed as fine particles in a liquid producing a latex used to make latex paints.

Cellulosics Two common plastics based on the cellulose molecule are cellulose acetate (CA) and cellulose acetate-butyrate (CAB). Cellulose acetate is not used for construction purposes, but CAB can be made resistant to weathering and is used in piping for gas and chemicals. Cellulosics are also used in coating compounds and adhesives.

Fluorocarbons The most widely used fluorocarbon is polytetrafluoroethylene (PTFE). It has high resistance to chemical degradation and a service temperature from 2450°F (2234°C) to 1500°F (262°C). PTFE has a low coefficient of friction and is marketed under the trade name Teflon. It is used for pipe that must handle corrosive liquids at high temperatures and for parts that require easy sliding surfaces.

Figure 24.4 ABS plastics are used to make plumbing pipe and fittings. *(Image copyright AYAKOVLEVdotCOM, 2009. Used under license from Shutterstock.com)*

Ethylene tetrafluoroethylene (ETFE) is a fluorocarbon-based polymer designed to have high corrosion resistance and strength over a wide temperature range. In addition, it has a high melting temperature and does not emit toxic fumes when ignited. It is also recyclable. Compared to glass, ETFE film is 1 percent the weight, transmits more light, and costs considerably less to install. ETFE is finding wider use in the construction industry as a cladding material (Fig. 24.5).

Nylons Polyamides (PA), also known as nylon, are tough, high in strength, and have good chemical resistance. Since they resist abrasion well, they are used for molded parts, such as locks, rollers, gears, and cams. They do not weather well.

Polycarbonates Polycarbonates (PC) have high impact strength and good heat resistance. They are dimensionally stable and transparent. Polycarbonates find use in light fixtures, molded parts, and signs. They are also used in place of glass in areas where damage is likely, such as in skylights.

Polycarbonate plastic sheets, such as Lucite and Lexan, have high impact strength, reaching up to 250 times the strength of glass and 30 times that of acrylic. They are suitable for glazing openings in areas where high security is needed and can be laminated to produce bullet-resistant panels. Light transmission varies with thickness but reaches 75 percent for 0.50 in. (12 mm) panels to 86 percent for 0.125 in. (3 mm) panels. Some manufacturers produce sheets in transparent solar tints, translucent white, and a variety of other colors.

Figure 24.5 The Beijing National Aquatics Centre is the world's largest structure made of ETFE laminate. *(Image copyright iPhotos, 2009. Used under license from Shutterstock.com)*

Double-skinned units with internal ribs and dead air spaces are available for use on vertical and sloped glazing (Fig. 24.6).

Polyethylene Polyethylene (PE) is light, strong, and flexible, even at low temperatures. It has good water resistance and low vapor transmission. Its major use in construction is as a vapor barrier on walls, floors, and ceilings. It is also used on basement walls as part of a waterproofing application. The wide use of polyethylene makes for an important environmental issue. Polyethylene is not biodegradable and takes several centuries to efficiently degrade. Though it can be recycled, most commercial polyethylene ends up in landfills.

Polypropylene Polypropylene (PP) is much like polyethylene but is more heat resistant and stiffer. It is used for hot water pipes and waste disposal systems and makes strong fibers for carpeting.

Polystyrene Polystyrene (PS) is a water-resistant, dimensionally stable, transparent plastic that maintains its properties at low temperatures but begins to soften around 200°F (94°C). It is brittle and has poor weathering qualities. When produced in a foamed condition, it is widely used as insulation. The foam is also used as a core for insulated doors and sandwich panels and for roof insulation (Fig. 24.7).

Vinyls The term vinyl describes a large group of plastics developed from the ethylene molecule. Those used in construction are polyvinyl chloride (PVC) and polyvinyl butyral (PVB).

Polyvinyl chloride is the most widely used vinyl in the manufacture of products used in construction. It is dimensionally stable and has high impact resistance, high abrasion resistance, and good aging qualities. A wide variety of products are made from PVC, including siding, gutters, floor tile, pipe, and window frames (Fig. 24.8). It can be bonded to other materials, such as plywood panels, to provide a protective skin. PVC is available as a rigid or flexible foam and used as a core material in panel construction. It is copolymerized with other plastics to produce a binder for terrazzo floors and a number of adhesives.

Polyvinyl butyral (PVB) is used as an inner layer in safety glass and as a protective coating on fabrics. Another type of vinyl, polyvinyl acetate (PVAc), is used in mortars, paints, and adhesives.

Figure 24.6 Polycarbonate panels used as glazing panels have good heat resistance and light transmission qualities. *(© Edifice/CORBIS)*

Figure 24.7 Polystyrene is widely used for basement insulation panels. *(Image copyright prism_68, 2009. Used under license from Shutterstock.com)*

Thermosets

Thermoset materials are used for products requiring greater heat resistance and stiffness than that afforded by thermoplastics. Thermosets find limited use because they are brittle and harder to form.

Epoxies Epoxies have good chemical and moisture resistance but are mainly used because of their excellent adhesive qualities. They are widely employed in coating compounds and as adhesives. Epoxies bond to almost any material and are used in the assembly of panels, the bonding of veneers and overlays, and as protective coatings. They are used as mortars for bonding concrete block and in patching material for damaged concrete. Epoxies are also used to produce some types of fiberglass reinforced plastic.

Formaldehyde Formaldehyde plastics are incapable of plastic deformation and are hard, strong, heat resistant, and brittle. There are three types that find some use in construction.

Phenol-formaldehyde (PF), generally referred to as phenolics, has fillers, such as glass fibers, added to improve impact resistance and strength. It has good electrical and thermal properties. Phenolic plastics are the most widely used of the thermosets. They are molded to form hardware and electric parts, such as switches, boxes, and circuit breakers. An important use involves the coating of kraft paper that forms the base of high-pressure plastic laminates used for countertops and other surfaces. The cardboard interior structure of hollow-core doors and panels is impregnated with phenolic resin. It is also used to make some types of adhesives, protective coating materials, and foamed insulation (Fig. 24.9).

Melamine-formaldehyde (MF), generally referred to as melamines, are hard, scratch-resistant plastics that withstand chemical attack. They are used in the production of high-pressure plastic laminates, such as those used for countertops. They are used as an adhesive in the production of plywood and to mold hardware and electrical fixtures.

Urea-formaldehyde (UF) enjoys the same uses as melamine-formaldehyde. Urea-formaldehyde is not as hard and does not have the same heat-resisting properties as melamine-formaldehyde.

Formaldehyde can be toxic, allergenic, and carcinogenic, and its use in many construction materials makes it one of the more common indoor air pollutants. At concentrations above 0.1 ppm (parts per million) in air, formaldehyde can irritate the eyes and mucous membranes. Formaldehyde inhaled at this concentration may cause headaches, a burning sensation in the throat, and difficulty breathing, as well as triggering or aggravating asthma symptoms.

Polyesters A large number of plastics fall under the generic name polyester. Polyesters include Mylar, from which drafting film is made; alkyds, used for paints and enamels; and fiberglass reinforced plastics. Reinforced plastic interior moldings are available made from fiberglass reinforced polyester. These are easy to cut and install and come primed and ready for painting. They are available in small and very large, complex-profile

Figure 24.8 Polyvinyl chloride is used to produce exterior products, such as siding, gutters, soffits, and fascias. *(Courtesy Wikipedia, http://en.wikipedia.org/wiki/Image:Vinyl_siding.jpg)*

Figure 24.9 Formaldehyde finds use in sprayed insulation. Phenolics are used to mold electrical parts.

single pieces, which makes installation easy. Complex wood moldings require that several individual pieces of molding be cut and fitted together. Other plastic moldings and trim for interior and exterior use are available made from cellular vinyl, polyvinyl chloride (PVC), and polymers. Saturated and unsaturated polyesters are produced. The most important group is the unsaturated polyesters. They can be linked to a monomer, such as styrene, forming a strong, hard thermosetting plastic. Their major use is in the production of reinforced plastics to which glass fibers are added. This produces a stable, tough, impact-resistant, and high-strength material. Major uses include molded bathtubs, showers, sheets for roofing and partitions, curtain wall exterior laminates, and window frames and sashes. Reinforced polyesters have ultimate strengths up to 50,000 psi (344,500 kPa) and E values as high as 3 x 10^6 psi. The strength of the plastic is proportional to the amount of glass reinforcement used because the glass fiber provides a tensile strength approaching 150,000 psi (1,033,500 kPa).

Polyurethanes (UP) Polyurethanes are used to produce low-density foams that can be varied from soft, open cell types that are flexible to a tough, closed-cell, rigid material. These foams have very low thermal conductivity and make excellent insulation. They have good chemical and heat resistance and good tensile strength. Polyurethane foams are used to fill wall, ceiling, and floor spaces by spraying them on surfaces or in cavities, after which they solidify. They are also used to insulate pipes, ducts, and wall panels. Rigid insulation sheets are widely used for many purposes, such as insulating flat roofs (over which a hot tar and gravel roof is applied). Polyurethanes also find use in the manufacture of **elastomers** (synthetic rubbers), caulking, adhesives, and glazing materials.

Silicones (SI) Silicone plastics are not based on the carbon atom, as all those previously discussed, but on the silicon atom. They have good corrosion resistance and are efficient electrical insulators. Heat-resistance properties enable them to have a service temperature ranging from 280°F (227°C) to 500°F (260°C). An important property involves their ability to withstand exposure to the elements.

Since silicons are very water repellent, they are applied in liquid form to exterior wall and masonry materials to provide a water-resistant coating (Fig. 24.10). This stops water from penetrating a wall, such as a brick exterior wall, yet has the permeability to allow internally developed moisture to pass through as a vapor. Silicone rubber is soft, heat resistant, and does not harden at low temperatures. It is used in sealants and gaskets where watertightness is required.

Figure 24.10 Silicone sealants and caulks are in common use to seal gaps, joints, and crevices in buildings. *(Courtesy Wikipedia, http://en.wikipedia.org/wiki/Image:Caulking.jpg)*

MANUFACTURING PROCESSES

Plastic resins in the form of liquid, powder, or beads are produced by chemical companies and sold to manufacturers of plastic products. A manufacturer combines these polymers with additives to modify the properties of the polymer to meet the needs of the product to be produced. The modified polymer is then processed by equipment of various types to produce the products.

Common manufacturing processes include blow molding, calendaring, compression molding, casting, extrusion, expandable bead molding, form molding, injection molding, laminating, rotational casting, transfer molding, and thermoforming.

Blow molding involves placing a heated, preformed plastic tube called a paison in a forming die. Air pressure is raised inside the plastic tube, forcing it to conform to the shape of the die. When it cools, the die is opened and the part is removed. Most bottles are formed this way.

Calendaring involves moving a plastic material in a liquid state through a series of rollers, which creates a thin plastic film as the material solidifies. Vapor barrier and floor covering materials are typical calendared products.

Compression molding involves placing plastic resin in powder form in a heated mold to which heat and pressure are applied. The resin melts and fills the mold.

Casting involves pouring plastic in a liquid state into a cavity in a mold. The liquid plastic fills the cavity and hardens. Extrusion is a process in which a semi-liquid plastic is forced under pressure through an opening in a die. The shape of the die opening determines the shape of the extruded member (Fig. 24.11).

Figure 24.11 The extrusion process involves forcing molten plastic through an opening in a die.

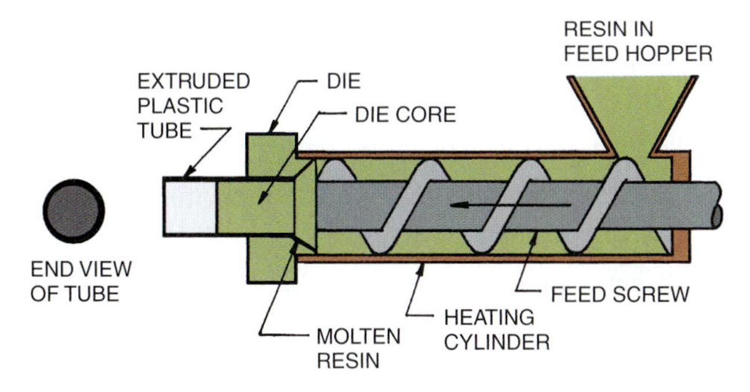

Expandable bead molding is a process in which small granules of resin, such as polystyrene, are mixed with an expanding agent and placed in a steam-heated rolling drum. When the granules have expanded, they are cooled and transferred to a mold in which they are heated until they fuse together (Fig. 24.12).

Form molding requires that an expanding agent be mixed with plastic granules or powder and injected into a mold, where it is heated, melting the resin and forming a gas that expands the resin to fill the mold cavity.

Injection molding uses granules or powder resin. This is fed into the heated cylinder of the injection molding machine and forced by a ram into a cold mold where it solidifies (Fig. 24.13).

Laminating is a process in which several layers of material are bonded together to form a single sheet. An example is laminating a colored plastic veneer to a plywood backing for use as wall paneling. The materials to be laminated are impregnated with the plastic resin, placed together, and bonded by the application of heat and pressure.

Rotational molding forms hollow one-piece items from polyethylene powders. The resin is placed inside the mold, which is heated as it rotates about two axes. The resin is distributed to the surfaces of the mold by centrifugal force and fused by the heat.

Transfer molding is a combination of compression and injection molding. The resin is made liquid in the transfer chamber outside the mold and then injected into the mold, where it fills the cavity and solidifies.

Thermoforming involves two commonly used procedures: vacuum forming and pressure forming. Vacuum forming involves placing a heated sheet of plastic over a mold cavity and pulling a vacuum below it. Atmospheric pressure forces the sheet to the shape of the mold (Fig. 24.14). Pressure forming involves placing a heated sheet of plastic over a mold cavity and increasing the pressure behind the sheet, forcing it into the cavity of the mold.

Figure 24.12 Expanded plastic beads are fused together to form a solid using the expandable bead molding process.

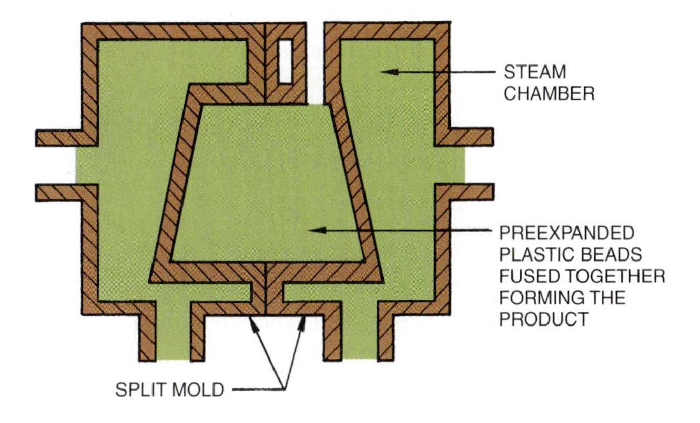

Figure 24.13 Injection-molded parts are formed by forcing a molten plastic resin into a die cavity.

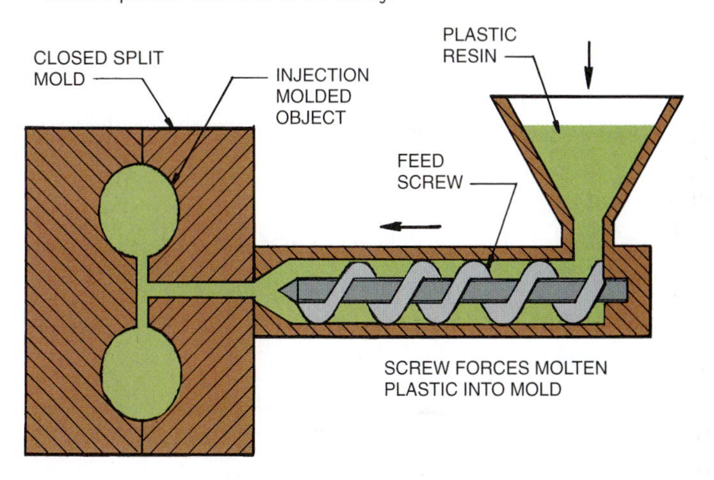

Figure 24.14 Pressure forming involves pulling a vacuum in the mold cavity, permitting atmospheric pressure to force the heated plastic sheet to the surface of the cavity.

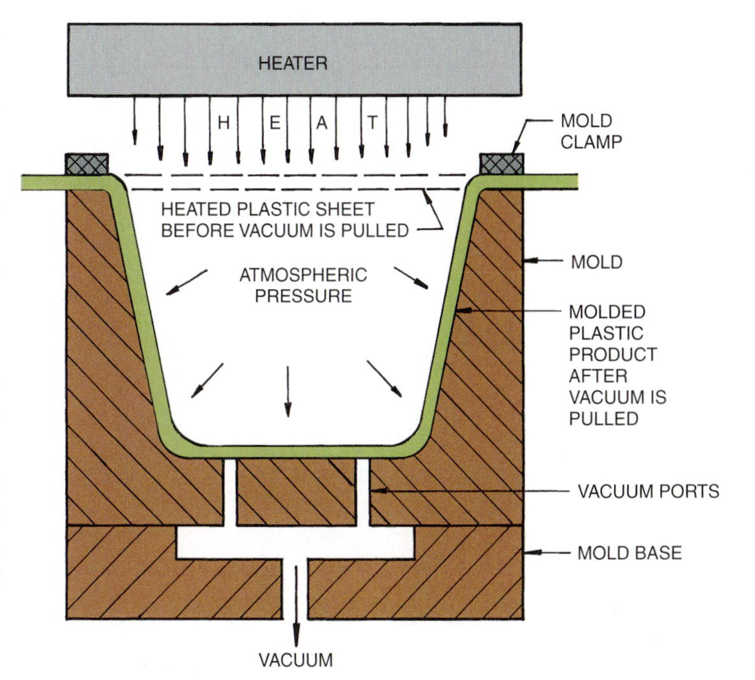

Review Questions

1. What two atoms form the nucleus for the derivation of the material plastics?
2. What is the process called that enables carbon atoms to bond with other atoms?
3. What is a single molecule called?
4. What is formed when several monomers link together?
5. What is the valence of the carbon atom?
6. What are the two major classifications of plastics?
7. Why is it possible to heat and reform thermoplastic materials?
8. Why is it that thermoset plastics cannot be reheated and reformed into new shapes?
9. What are the major groups of additives used in formulating plastics?
10. How do plasticizers change the properties of plastics?
11. What is a common way to reduce the cost of plastics?
12. What is added to plastic to increase its strength?
13. What colorants are used to produce desired colors?
14. What are the most important mechanical properties of plastics used in construction?
15. What are the two major considerations when considering the thermal properties of plastics?
16. Which plastics have the lowest thermal conductivity?
17. What is the major danger produced by burning plastics?
18. When plastics deteriorate due to chemical action, what type of damage occurs?
19. Which plastics have good optical properties?
20. Identify the following plastics by their standard abbreviation: ABS, PMMA, PA, PE, PVC.
21. What is the major use for ABS plastics?
22. What type of plastic would you choose for door and window glazing?
23. What type of plastic is used in the manufacture of latex paints?
24. The vapor barriers used in walls and floors most likely are made from what type of plastic?
25. What type of plastic weathers well and is used for exterior purposes, such as siding, gutters, and window frames?
26. What thermoset plastic is known for its use as a strong adhesive?
27. What plastics are used in the manufacture of high-pressure plastic laminates?
28. What type of plastic is used in the manufacture of showers and bathtubs?
29. What thermoset plastic is widely used to produce foamed insulation commonly sprayed on roofs and in wall cavities?
30. What uses are made of silicones in building construction?

Key Terms

additive
arc resistance
covalent bonding
creep
dielectric strength
elastomers
fillers
flammability
hardness
monomer
organic material
plastic
plastic behavior
plasticizers
polymers
service temperature
stabilizers
synthetic materials
thermoplastics
thermosets
valence
weatherability

Activities

1. Collect samples of various plastic products used in construction. Try to identify the type of plastic.
2. Develop a series of tests you can perform with the facilities available and test plastics samples. For example, see which will ignite the quickest when exposed to a continuous flame and which will burn after the flame is removed. Run tensile tests, expose samples to acids, hot water, etc., and record what happens to each.

Additional Resources

Plastic Pipe and Building Products, Volume 08.04, American Society for Testing and Materials, West Conshohocken, PA.

Structural Plastic Selection Manual, American Society of Civil Engineers, New York.

See Appendix B for addresses of professional and trade organizations and other sources of technical information.

Thermal and Moisture Protection CSI MasterFormat™

- 071000 Dampproofing and Waterproofing
- 072000 Thermal Protection
- 072500 Weather Barriers
- 073000 Steep Slope Roofing
- 074000 Roofing and Siding Panels
- 075000 Membrane Roofing
- 076000 Flashing and Sheet Metal
- 077000 Roof and Wall Specialties and Accessories
- 078000 Fire and Smoke Protection
- 079000 Joint Protection

Image copyright Christina Richards, 2009. Used under license from Shutterstock.com

CHAPTER 25

Thermal Insulation and Vapor Barriers

LEARNING OBJECTIVES

Upon completion of this chapter, the student should be able to:

- Select appropriate types of insulation for various applications in the design of a building.
- Understand heat transfer and how insulation can be used to control it.
- Be aware of the problems caused by moisture penetrating the insulation and learn how to design an assembly of materials to reduce moisture transmission.

Build Your Knowledge

For further study on these materials and methods, please refer to:

Chapter 23 Paper and Paper Pulp Products

Chapter 24 Plastics

In this period of increasing energy costs, potential shortages, and climate change resulting from the burning of fossil fuels, energy conservation is of critical importance. Structures are exposed to weather and below-ground water. New and improved products for controlling moisture inside and outside of buildings appear frequently on the market. Energy conservation and moisture control are necessary to both protect a building and provide comfort for its occupants.

Thermal insulation in buildings is an important factor in achieving both energy efficiency and interior thermal comfort. Insulation reduces unwanted heat loss and gain, thereby decreasing demands on heating and cooling systems. Considerable attention is now given to the design and construction of energy efficient buildings. Manufacturers have responded with an array of products for use in almost any environment or situation. The type and amount of insulation for a building is determined by the building design, the local climate, budget restrictions, and building code requirements.

Any exterior surface that separates conditioned spaces from outside air must be insulated. Insulation is placed around the foundations, under slabs, and in floors, exterior walls, ceilings, and floors. Insulation must fill the gaps between enclosing doors and windows. Collectively, these surfaces are called the **thermal envelope** (Fig. 25.1).

Great care must be taken when installing insulation. When improperly installed, insulation materials can be rendered useless, wasting both time and money. Materials must be made to conform to and fill all irregularities and obstructions. Loose-fill insulation must have its fluffiness maintained without bunching or squeezing. Voids in insulation of only 5 percent can reduce a structure's overall insulation effectiveness by 25 percent.

THERMAL INSULATION

Thermal insulation is manufactured from a variety of materials. Metallic insulation comes in the form of metallic foils, such as aluminum or copper, or as an organic insulation material with a metallic laminate. Organic fibrous insulation materials include cane, cotton, wood, cellulose, and synthetic fibers. Organic cellular materials include polyurethane, polystyrene, cork, cotton, recycled materials, and foamed rubber. Mineral cellular insulation materials include perlite, vermiculite, and foamed glass. Mineral fibrous materials include rock, glass, slag, and asbestos melted and spun into fibers.

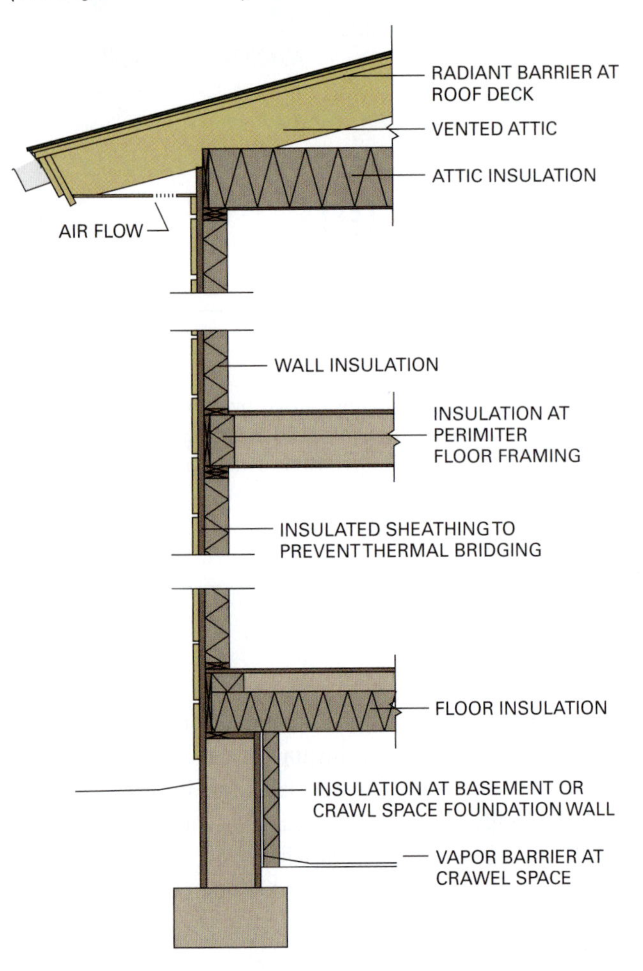

Figure 25.1 Any interior surface that separates conditioned air from outside air is insulated to form the thermal envelope. *(Courtesy Eva Kultermann)*

Thermal insulation is available in granular or fibrous loose-fill forms, flexible wool-like blankets and batts, rigid sheet material, liquid spray that uses a mineral fiber or insulating concrete, cast-in-place insulating concrete, and polyurethane foams (as well as other foamed-in-place types).

METHODS OF HEAT TRANSFER

One key to energy efficient design is an understanding of the fundamentals of heat transfer. Heat transfer is the movement of thermal energy from a hotter item to a cooler item. When an object or fluid resides at a different temperature than its surroundings or another object, transfer of thermal energy, also known as heat transfer or heat exchange, occurs in such a way that the body and its surroundings reach thermal equilibrium. Heat transfer always occurs from a hot body to a cold one according to the second law of thermodynamics. To avoid heat loss in winter and heat gains in summer, insulation is used to break this transfer.

The amount of transferable heat depends on the characteristics of a material, its thickness, and the quality of construction detailing. Porous and fibrous materials that contain air pockets, such as wood, transfer less heat than do solid materials, such as brick. A thick section of material provides better protection than a thin section of the same material.

Heat is transferred by radiation, convection, or conduction. **Radiation** involves the transmission of heat by electromagnetic waves. Heat energy passes through the air between a source and another body without heating the intervening air. Shiny materials tend to reject radiant energy while dark materials will absorb it. A white shingle roof reflects more radiant energy than a brown or black roof, thus increasing the energy efficiency of a building.

Heat transfer by **convection** involves the transfer of heat by the circulation or movement of heated liquids or gases. Heat travels by natural or forced (fanned) currents of air that absorb the heat brought to a space (as with a hot water heating system convector, for example). As warm air passes over a cool surface, it transfers some of its heat to the cool surface. This convection current is a means for heat transfer and occurs in every building.

Conduction is the transfer of thermal energy from a region of higher temperature to a region of lower temperature through direct molecular movement within a material or between materials in direct physical contact. For example, when solar radiation heats a masonry wall, the heat is transferred to the other side of the wall by conduction. When a material or assembly is uninsulated from exterior to interior, it creates a thermal bridge, allowing for conductive heat transfer from the warm side to the cold.

Thermal insulators are materials specifically designed to reduce the flow of heat by limiting conduction, convection, or both. Radiant barriers are materials that reflect radiation and reduce the flow of heat from radiation sources. Good insulators are not necessarily good radiant barriers and vice versa. Metal, for instance, is an excellent reflector and poor insulator.

DESIGNATING THERMAL PROPERTIES

Heat quantity is measured in **British thermal units.** One Btu is the amount of heat needed to raise the temperature of 1 lb. of water by 1°F. A Btu is approximately equal to the heat generated by a wooden match. In the metric system, heat quantity is measured in calories. One calorie is the heat required to raise the temperature of 1 gram of water 1°C. Calories can be converted to joules (J) by multiplying them by 4.18. A **joule** is a derived metric unit used to describe work, energy, and quantity of heat. The terms used to describe the thermal properties of materials include thermal conductivity (k), thermal resistance (R), thermal conductance (C), and thermal transmittance (U).

Thermal conductivity (k) is a measure used to indicate the amount of heat that will be conducted through a square foot of area of a material per a specified unit of thickness. A lower k value indicates better insulating qualities. In the United States' customary system of measurement, thermal conductivity is indicated as Btu · in./ft.2^2 · hr. · °F (or Btu per inch of thickness per square foot per hour per degree Fahrenheit). In the metric system it is W/m · K (or watts per meter-kelvin).

Thermal resistance (R) is a measure used to indicate the ability of a material to resist the flow of heat through it. The larger the R-value, the greater the thermal resistance and the better the insulating value. Thermal resistance, R, is the reciprocal of conductance, C, or R = 1/C. In the United States' customary system of measurement, thermal resistance is indicated as R = (hr · ft.^{2} · °F)/Btu (or hours per square foot per degree Fahrenheit per Btu). In the metric system, thermal resistance is specified as R SI = (K · m2)/W (or kelvin-meter squared per watt). Total thermal resistance, ΣR, is the resistance to heat flow through an assembly of materials.

Thermal conductance (C) is a measure used to indicate the amount of heat that will pass through a specified thickness of material. It is the reciprocal of thermal resistance: C = 1/R. In the United States' customary system of measurement, thermal conductance is indicated as C = Btu (hr. · ft.^{2} · °F) (or Btu per hour per square foot per degree Fahrenheit). In the metric system, thermal conductance is denoted as W/(m^2 · °C) (or watts per square meter-degree Celsius).

Thermal transmittance (U) is a measure of the amount of heat that will pass through an assembly of various materials, such as an exterior wall. It is the reciprocal of the total resistance (R) of the assembly: U = 1/ΣR. A smaller U-value indicates greater resistance to the transmission of heat.

INSULATION MATERIALS

Insulation is manufactured in a variety of types and forms. The decision of which type to use depends on its location in a building, the environment to which it will be exposed, its effectiveness in resisting heat transfer, and its cost relative to the savings expected from improving the energy efficiency of the building. Minimum required levels of thermal insulation for various construction assemblies are specified by building and energy codes.

The major classifications of insulation materials include loose-fill, batts and blankets, rigid, reflective, foamed-in-place, and sprayed. Insulation materials are rated by their ability to resist heat flow, which is indicated by their R-value. Typical R-values of various materials are given in Table 25.1.

Loose-Fill Insulation

Loose-fill insulation is usually composed of materials in bulk form. It can be installed (by blowing, pouring, or hand packing) into attics, finished wall cavities, and hard-to-reach areas. It is an ideal insulator because it conforms to spaces, thoroughly filling nooks and crannies. It can also be sprayed in place with water-based adhesives. Some loose-fill insulations are made of recycled materials and are relatively inexpensive.

Loose-fill insulation is available as a granular or loose fibrous material. Granular insulation is generally poured while the fibrous variety is machine-blown into areas to be insulated Granular insulation includes perlite (expanded volcanic rock), vermiculite (expanded mica), cork, and expanded polystyrene. These are available in different densities and R-values (Table 25.2). It is used when filling vertical cavities, such as cores in concrete block and wall cavities. Granular insulation materials are also used to produce a lightweight concrete roofing base material that is pumped and leveled on a roof deck. Expanded polystyrene roof insulation boards are placed on the concrete and bonded to the deck. The roof base material is spread over the insulation board, forming a base for the application of the finished roofing material.

Fibrous loose insulation is made by blowing a jet of air through molten glass, slag, or rock to form thin fibers that compose a wool-like substance when gathered. Fibrous loose insulation is difficult to work into tight spaces. It is widely used in large areas, such as insulating ceilings, where a sprayed uniform layer can be applied (Fig. 25.2).

Cellulose insulation is made from recycled newsprint and other paper stock that has been treated for fire and pest resistance (Fig. 25.3). Cellulose is increasingly regarded as a superior insulating material for three main reasons. First, it provides higher levels of acoustical insulation and fire protection. Secondly, the material is less subject to a decline in performance over time.

Table 25.1 R-Values of Insulation Materials[a]

Fiberglass Board Panels			
Thickness (in.)	Thickness (mm)	R-Value (customary units)[b]	R = Value (SI units)[c]
1	25	4.30	30
1.5	38	6.50	45
2	50	8.70	60
2.5	63	10.90	75
3	75	13.00	90
4	100	16.50	114
Extruded Polystyrene Rigid Panels			
Thickness (in.)	Thickness (mm)	R-Value (customary units)[b]	R = Value (SI units)
0.75	18	3.80	26
1	25	5.00	35
1.5	38	7.50	52
2	50	10.00	69
2.6	66	16.67	115
3.1	78	20.00	138
Polyisocyanurate Rigid Panels			
Thickness (in.)	Thickness (mm)	R-value (customary units)[b]	R = Value (SI units)
1.2	30	7.14	49
1.6	40	10.00	69
2.0	50	12.50	86
2.6	66	16.67	115
3.1	79	20.00	138
Polyurethane Rigid Panels			
Thickness (in.)	Thickness (mm)	R-Value (customary units)[b]	R = Value (SI units)
1.0	25	6.25	43
1.5	38	10.00	69
2.0	50	14.30	98
2.7	68	20.00	138
Wood Fiber Rigid Panels			
Thickness (in.)	Thickness (mm)	R-Value (customary units)[b]	R = Value (SI units)
0.5	12	1.40	10
0.75	18	2.10	14
1	25	2.80	19
1.5	38	4.20	29
2	50	5.60	38
Granular Insulation			
	Thickness (mm)	R-Value (customary units)[b]	R = Value (SI units)
Vermiculite per inch	per 25 mm	2.1 to 2.3	14 to 16
Perlite per inch	per 25 mm	2.6 to 3.5	18 to 24
Aluminum Reflective Insulation			
	Thickness (mm)	R-Value (customary units)[b]	R = Value (SI units)
Multilayer foil batts with 4 three quarter inch air spaces	18 mm air space	summer 17 winter 13	117 89
Multilayer foil batts with 6 one inch air spaces	25 mm air space	summer 30 winter 18	207 124

[a]*Data obtained from various manufacturer catalogs. Consult manufacturer's specification sheets for specific data.*
[b]*R-value (customary units) hr. · ft.^{2} · °F/Btu*
[c]*R SI (metric units) m^2 · K/W*

Table 25.2 Properties of Loose-Fill Insulation

Material	Density (lb./ft.^{3})	R-Value (hr. · ft.^{2} · °F/Btu)	Water Vapor Permeability
Cellulose	2.2–3.0	3.1–3.7	High
Expanded polystyrene	0.9–1.8	3.8–5.0	High
Fiberglass	0.6–1.0	2.9–3.7	High
Mineral wool	1.5–2.5	2.9–3.7	High
Perlite	2–11	2.7–2.9	High
Vermiculite	4–10	2.1–2.3	High

Figure 25.2 Fibrous insulation is machine-blown into the areas to be insulated. *(Courtesy Bruce Forster/Getty Images)*

Finally, cellulose insulation is very effective in creating an airtight building envelope, thereby protecting against unwanted infiltration heat losses.

Cotton, wool, hemp, corn cobs, strawdust, and other harvested natural materials are also beginning to find increased use as loose-fill insulation.

Figure 25.3 Cellulose insulation, made from recycled newsprint, is sprayed into wall cavities for an airtight enclosure.

Batts and Blankets

Batts and blankets are flexible insulation mats made from fiberglass, mineral wool, cotton and wood fibers, and recycled plastic materials (Fig. 25.4). Batts are available in 48 in. (1,219 mm) lengths, and blankets come in rolls up to 8 ft. (2,438 mm). Some thinner blankets are available in longer rolls. Common thicknesses are $3\frac{1}{2}$, $6\frac{1}{2}$, $9\frac{1}{2}$, and 12 in. (89, 158, 241, 305 mm). Common widths include 15 and 23 in. (380 and 584 mm). Other sizes are available for special applications, such as noise barrier insulation and for furred-out masonry walls. Batts and blankets work in a variety of applications, such as low-density insulation used in residential construction; extra low-flame spread types used in commercial construction where insulation remains exposed;

Figure 25.4 Fibrous loose insulation is made by blowing a jet of air through molten glass, slag, or rock and works by entrapping air within the fibers of batts. *(Image copyright BrunoSINNAH, 2009. Used under license from Shutterstock.com)*

special batts for use over panels in suspended ceilings; and as sound transmission barriers for walls, floors, or ceilings.

Batts and blankets are available unfaced, faced on one side (with moisture resistant kraft paper that forms a vapor barrier), and faced with aluminum foil (that forms a fire-resistant surface). Some batts have facing on both sides and are used in both vertical and horizontal applications, such as for walls, floors, and ceilings in commercial buildings. Blankets can be used to wrap items needing insulation, such as water heaters. Most batt insulations are stapled to the face or side of a stud. Some types are designed for placement between studs or joists and are held by friction. The vapor barrier (faced) side always faces the warm side of a building.

Rigid Insulation

Rigid insulation board is made using organic fibers, such as wood or cane; mineral wool; glass; corkboard; several forms of expanded plastics (such as expanded and extruded polystyrene and polyisocyanurate foam); and some forms of cellular hard rubber. These products are used in all parts of a building, including roof, wall, and floor insulation (Fig. 25.5).

Wood and cane fiberboard are asphalt impregnated and commonly used for exterior sheathing, roof insulation, and shingle backer boards. Rigid insulation sheets made from granulated cork are used for roof, wall, and floor insulation and are available in thicknesses from 2 to 12 in. (50 to 305 mm) and sheet sizes of 24 x 36 in. (610 x 915 mm). Mineral wool wall panels have an insulation mat bonded to a rigid back sheet and are commonly used for roof insulation.

Figure 25.5 Rigid insulation sheets are used in all parts of a building, including roof, wall, and floor insulation. *(Image copyright robcocquyt, 2009. Used under license from Shutterstock.com)*

Expanded and extruded foamed plastic rigid insulation is used on walls, floors, roofs, and foundations (Fig. 25.6). Expanded Polystyrene is available in widths up to 4 ft. and lengths up to 12 feet. If expanded polystyrene comes into contact with water it will absorb moisture, which lessens its insulating value.

Extruded polystyrene is a closed-cell foam that resists moisture absorption. It can be installed in wet locations and is often used to insulate foundations and roof decks. It is manufactured by reacting benzene with ethylene and uses an HCFC blowing agent (the only remaining ozone-depleting blowing agent for plastic foams).

Another type of board is bonded to a particleboard sheet that provides a nailing surface for exterior finish materials (Fig. 25.7). Rigid polystyrene, or polyisocynurate foam, is usually made with a facing of foil or building paper. Polyisocynurate has a superior R-value of up to 6.2 per in.

Rigid insulation panels can be used as an underlay on roof decks to be covered with insulating concrete. The concrete has a cellular structure that reduces weight and increases insulation properties. Sheets for roof insulation can be tapered to provide slope for drainage on flat roof applications. The variation in panel thicknesses allows the roof to slope to drainage sources (Fig. 25.8).

Figure 25.6 Rigid insulation board can be used on foundations below grade. *(Image copyright prism_68, 2009. Used under license from Shutterstock.com)*

Figure 25.7 This wall insulation is made by laminating rigid foamed insulation sheets to particleboard, which provides a nailing surface. *(Image copyright Vitezslav Halamka, 2009. Used under license from Shutterstock.com)*

Figure 25.9 Tongue and groove panels containing a fiber insulation core being installed on a sloped roof. *(Courtesy Ralph Orlowski/Staff/Getty Images)*

Flat panels containing a fiber insulation core may be unfaced or faced with an aluminum foil vapor barrier (Fig. 25.9). Common thicknesses include 1, $1\frac{1}{2}$, 2, $2\frac{1}{2}$, and 3 in. (25, 38, 50, 63, and 75 mm). Standard fiber-core panel sizes are 24 x 48 in. (610 2 1220 mm). Extruded and expanded plastic panels typically come in 16 and 24 in. (406 and 610 mm) widths and 48 to 96 in. (1,220 to 2,438 mm) lengths. Thicknesses from 1 to 4 in. (25 to 100 mm) are available. Other sizes are procurable and manufacturers should be consulted for sizes and R-values.

Most rigid insulation panels are manufactured from flammable materials. Building codes require they be covered with a fire-resistant material, such as gypsum board. Polystyrene products deteriorate when exposed to ultraviolet rays from the sun and must be covered to avoid this. Molded polystyrene and expanded perlite board must also be protected from water.

Figure 25.8 Polystyrene insulating board is used to insulate concrete roofs and to provide a sloping base on which insulating concrete can be poured.

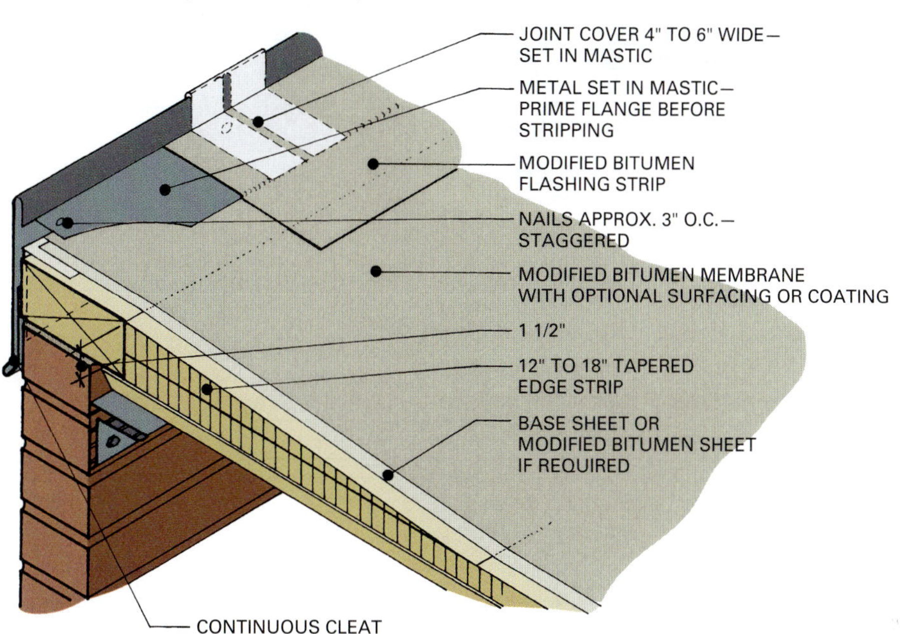

Reflective Barriers

Radiant, or reflective, barriers inhibit heat transfer by thermal radiation. Since thermal energy may also be transferred via conduction or convection, radiant barriers are typically used in addition to other insulation. On a sunny summer day, solar energy is absorbed by a roof, heating the roof sheathing and causing the underside of the sheathing and the roof framing to radiate heat downward toward the interior. When a radiant barrier is placed directly underneath the roofing material incorporating an air gap, much of the heat radiated from the hot roof is reflected back with little radiant heat emitted downwards.

Reflective barriers are usually made from aluminum or copper foil in sheets or rolls (Fig. 25.10). Rolls are typically 24 and 48 in. (610 and 1,220 mm) wide and 500 ft. (152.5 m) long. They are available in single thickness layers or in a multilayer batt with dead air spaces between the layers. The foil utilizes reflective properties to reject the passage of heat and increase the effectiveness of the dead air spaces (Fig. 25.11). The foil may be bonded onto a heavy kraft paper, insulation board, or gypsum lath (Fig. 25.12). Some fiber insulation batts have reflective foil bonded to one side that serves as a vapor barrier. Reflective insulation is used in residential and commercial construction on walls, floors, ceilings, and roofs. It can decrease heat flow by as much as 25 percent, reducing the amount of conventional insulation needed and, in some cases, reducing the size of air-conditioning units required.

Foamed-in-Place and Sprayed Insulation

Foamed-in-place insulations include open- and closed-cell polyurethane, icynene, and soy-based products. They are generally based on polyurethane or phenol-based compounds that provide excellent insulation qualities.

Figure 25.10 Reflective insulation is available in rolls. *(Courtesy Advanced Foil Systems)*

Figure 25.11 Some forms of reflective insulation have multi-layers of foil with dead air spaces between them.

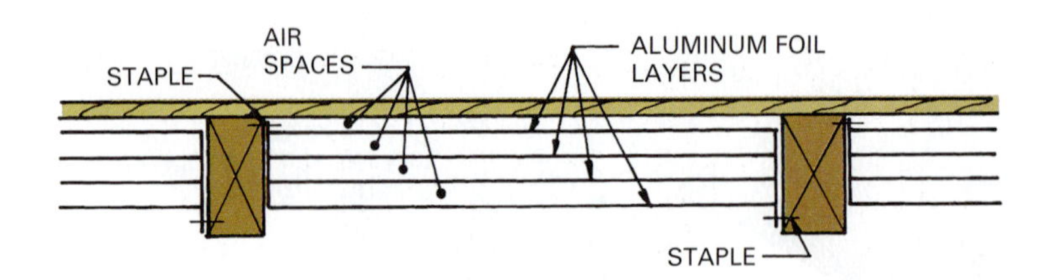

Figure 25.12 Reflective foil may be bonded onto rigid insulation for exterior sheathing applications. *(© Dennis Whitehead/Corbis)*

The ingredients are carefully measured and mixed by special equipment that maintains isocyanate and polyol levels at a one-to-one ratio. These are pumped as separate materials into a proportioning unit, which heats each and pumps them into separate heated hoses. Once mixed they are pumped through hoses into building cavities and sprayed in layers on flat and sloping surfaces, such as roof decks. The two components are mixed in a spray gun and applied to the substrate. The mixture also can be applied by power or hand rollers. Since the material expands as it hardens, the amount injected into a cavity must be carefully measured. If a wall cavity is overfilled, the pressure could damage the wall finish. After the sprayed polyurethane foam insulation is in place it must be protected from fire, exposure to moisture, and ultraviolet radiation. Commonly used protective coatings include acrylics, butyls, chlorinated synthetic rubber, modified asphalts, silicones, and urethanes.

Sprayed-on insulations are used on ceilings, walls, tanks, and other items. Although polyurethane is widely used, vermiculite or perlite aggregate combined with a gypsum or Portland cement binder are mixed with an inorganic binder and used as insulation. Sprayed-on insulation is also used to increase fire- and moisture-resistance and to improve acoustical properties. It has the advantage of being able to bond to irregularly shaped and sloped surfaces. Manufacturers should be consulted for R-values, fire resistance, flame spread, smoke development, and compressive and impact strengths.

Spray foams are also available in aerosol cans that are used to insulate small gaps between wall frames and door and window units and to seal around mechanical penetrations.

While spray foams provide excellent insulation characteristics, they should be used sparingly. Most, including Polyurethane and Isocyanate insulation, contain hazardous chemicals, such as benzene and toluene. These are a potential hazard and environmental concern during raw material production, transport, manufacture, and installation. Although CFCs are no longer used as blowing agents, many spray insulations use HCFCs or HFCs. Both are potent greenhouse gases, and HCFCs have some ozone-depletion potential. New products are now in development, including natural soy-based foam insulations.

Other Insulating Materials

Aerogel is a low-density material derived from gel in which the liquid component of the gel has been replaced with gas. The result is an extremely low-density solid with several remarkable thermal insulation properties. Since it is 99.8 percent air, it appears semi-transparent, making it ideally suitable as a thermal insulation material for windows and skylights.

Various insulating materials are now being manufactured from recycled materials. One example is an insulation made from recycled blue jeans (Fig. 25.13).

Figure 25.13 Insulation made from recycled denim and cotton fibers. *(Courtesy Thomas Northcut/Getty Images)*

Construction Materials

Superinsulation

Superinsulation (SI) is an approach to building design that provides higher-than-normal levels of thermal insulation. Superinsulated houses are intended to be heated predominantly by internal sources, including waste heat generated by lighting, appliances, and the body temperature of occupants. They typically provide no conventional heating system, instead relying on just a small backup heater. Superinsulated homes have been demonstrated to perform well even in very cold climates, but their construction requires close attention to air tightness and insulation.

There is no established definition for superinsulation, but superinsulated buildings typically include the following aspects: they utilize very thick insulation with R-values typically in the area of R-40 for walls and R-60 for roofs; careful attention is paid to detailing insulation where walls meet roofs, foundations, and other walls; airtight construction is achieved through a thorough sealing of all penetrations, especially around doors and windows; windows are generally smaller and of high quality, often with a triple layer of glazing to prevent heat loss; and, because of the hermetic nature of the envelope, superinsulated construction must incorporate a ventilation system to control air quality, sometimes in the form of a heat-recovery ventilator that provides fresh air.

A variety of insulation types are used in tandem to produce superinsulated buildings. Cavity walls may be insulated with loose-fill batts, then covered both inside and out with rigid insulation sheets, and finally finished with a radiant barrier before exterior finishes are applied. Straw-bale construction is a building method that uses straw bales as structural elements, insulation, or both. It is, by nature, "superinsulation," and easier to air seal, particularly in conjunction with a slab on grade and plastered exterior surfaces.

Superinsulated buildings conserve energy without impacting an occupant's lifestyle. They tend to produce more comfortable interiors, with no drafts, cold spots, or temperature stratification. Because of thicker walls and better windows, less outside noise penetrates their interiors. Superinsulated construction typically costs 5 to 7 percent more compared to conventional construction. Most of the cost accrues from better windows and doors, more insulation, and the additional labor involved with air sealing. While initial costs are higher, the return on the investment is quick, since SI buildings typically save 75 percent of heating and cooling costs.

SI building practices are included in Canadian building codes and are standard practice in many parts of Europe. In the United States, changes to the Model Energy Code have been implemented by the Department of Energy and the Department of Housing and Urban Development. Monetary incentives, such as Energy-Rated Mortgages, which increase the borrowing power of buyers of energy-efficient homes, are now available. Programs like the Energy Star Homes Program, sponsored by the Environmental Protection Agency and the Department of Energy, encourage builders to erect SI homes and provide a rating system to evaluate energy performance.

VAPOR RETARDERS

Vapor retarders are typically a plastic or foil sheet that resists diffusion of moisture through the wall, ceiling, and floor assemblies of a building. Water vapor moves into building cavities primarily through two mechanisms: diffusion through building materials and by air transport (infiltration). Other wetting mechanisms, such as wind-driven rain and capillary wicking of ground moisture, can be equally important. Vapor retarders slow the rate of vapor diffusion into the thermal envelope of a structure. A vapor retarder is also used to keep water vapor generated inside a building, such as by cooking and bathing, from penetrating walls and condensing as moisture on insulation (Fig. 25.14).

Various types of insulation are made with vapor retarders applied as part of a sheet or blanket. Kraft paper coated with wax or asphalt is commonly used on one side of fibrous insulation batts and blankets. Aluminum foil is a more expensive coating that has heat-reflective values as well. Polyethylene film is sometimes stapled to studs on the warm side of a building and placed under the subfloor and on the ceiling. The film is also used to cover the ground in crawl spaces to retard moisture transfer from ground to crawl space air. It is laid below welded wire fabric before concrete slabs are poured. It is an excellent vapor barrier and also assists in reducing air infiltration. Polyethylene film is available in thicknesses of 2, 3, 4, and 6 mm and rolls 3 to 20 ft. (0.9 to 6 m) wide.

The permeability of vapor retarders is rated in perms, a measure of the rate of transfer of water vapor through a material. Vapor retarders have permeability ratings of 1.0 or lower. Materials with a perm rate of ($\leq$ 1 perm) are considered impermeable, (> 1 to 10 perms) semi-permeable, and (> 10 perms) permeable.

Figure 25.14 Polyethylene vapor barriers reflect moisture generated inside a building and keep the moisture from penetrating the insulation.

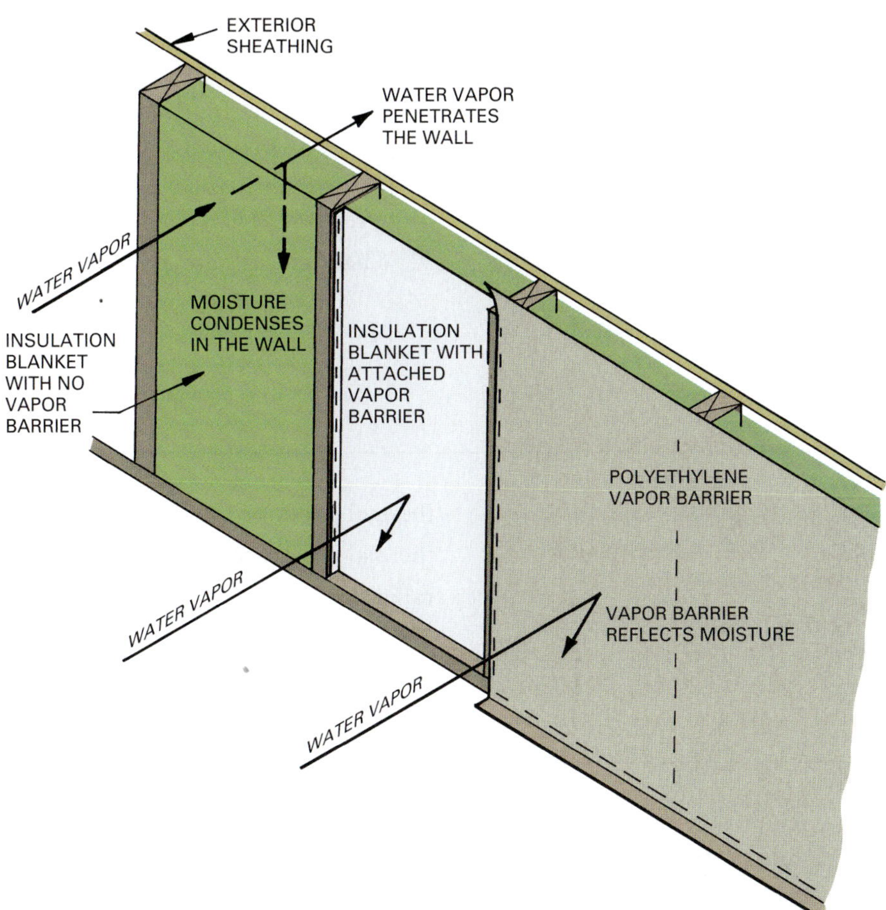

Figure 25.15 This building is being wrapped with DuPont Tyvek® Housewrap to reduce air infiltration. *(Courtesy Eva Kultermann)*

Air infiltration from the exterior of a building is also reduced by applying a plastic barrier on the exterior over the sheathing (Fig. 25.15). This material is a spunbound olefin formed into a sheet of very fine high-density polyethylene fibers. It resists tearing, punctures, and will not rot. It has a perm rating of 94, so moisture vapors that get into the wall from inside the building can pass through to the exterior. In all cases, vapor barriers and air infiltration wrap must be carefully installed, overlapped at joints, and sealed to reduce penetration. Various liquid materials can also serve as vapor barriers, including enamels, primers, latex paints, and oil-based paints.

Review Questions

1. What types of organic fibers are used for insulation?
2. What kinds of organic cellular materials are used for insulation?
3. What kinds of mineral cellular materials are used for insulation?
4. What kinds of mineral fibrous materials are used for insulation?
5. In what forms is thermal insulation available?
6. In what three ways can heat be transferred?
7. What are the four terms used to describe thermal properties of a material?
8. How are insulation materials rated on their ability to resist heat flow?
9. What materials are used for vapor barriers?

Key Terms

British thermal unit (Btu)
conduction
convection
joule
thermal envelope
thermal resistance
radiation
Thermal transmittance (U)

Activities

1. Try to set up an experiment to get a rough evaluation of the relative effectiveness of various types of insulation. For example, you might build a box from $\frac{3}{4}$; in. plywood, leaving the top open. Place a heat source in the box, such as a number of high-wattage incandescent lamps. Cover the open top with an insulation product and measure the temperature of the exterior surface after a set number of hours. Repeat with various types and thicknesses of insulation. Be alert for the possibility of fire and do not leave the experiment unattended.
2. Collect samples of various insulation products and label them, identifying their composition and properties.

Additional Resources

Insulation Manual, National Association of Home Builders, Washington, DC.

Moisture Control Handbook, U.S. Department of Energy, Oak Ridge, TN.

Thermal Design Guide for Exterior Walls, American Iron and Steel Institute, Washington, DC.

Wood Building Technology, Canadian Wood Council, Ottawa, Ontario, Canada.

Other resources include:

Publications in whole-building design and energy-simulation software, Sustainable Buildings Industry Council, Washington, DC.

See Appendix B for addresses of professional and trade organizations and other sources of technical information.

26 CHAPTER

Bonding Agents, Sealers, and Sealants

LEARNING OBJECTIVES

Upon completion of this chapter, the student shall be able to:

- Select the proper bonding agents for various construction assemblies.
- Understand how the various bonding agents cause adherence.
- Select appropriate sealers for use on exterior applications.
- Decide the best way to protect joints in exterior construction.
- Select waterproofing materials best suited for various applications.

Build Your Knowledge

For further study on these materials and methods, please refer to:

Chapter 24 Plastics

Topics: Silicones, Thermoplastic Resins, Thermoset Resins

Chapter 27 Bituminous Materials

Bonding agents, sealers, and sealants made from natural and manmade substances are used widely for both interior and exterior purposes. Bitumens and various synthetic resins are used to produce many adhesive products.

BONDING AGENTS

A **bonding agent** is a compound that joins materials together by bonding to their contacting surfaces. In many building assemblies and products, bonding agents are used to permanently fasten both structural and decorative elements. Bonding agents used in the fabrication of construction products and for on-site applications typically join materials by **mechanical action** or by **specific adhesion**. In mechanical action, the bonding agent enters the pores of a porous material, hardens, and forms a mechanical link. Bonding agents that join by specific adhesion are used to bond dense materials without pores, such as glass and metal. The bonding is caused by the attraction of unlike electrical charges. The positive and negative charges in the bonding agent are attracted by the electrical charges on the surface of the material to be bonded. This molecular attraction provides a strong holding force.

Bonding agents can have their properties varied for specific conditions and materials. Some are a combination of two or more types, such as a phenol and resorcinal resin combination or a urea resin blended with a melamine resin. They are available as powders, solids, liquids, and pastes. Some require the addition of a catalyst.

A summary of commonly used bonding agents, their uses, and the materials they will join is given in Table 26.1.

Curing of Bonding Agents

The curing of bonding agents is accomplished by loss of solvents, anaerobic environments, catalysts in two-part mixtures, and cooling of hot melts. Loss-of-solvent bonding agents cure by the loss of volatile liquids, water, or organic solvents used to dissolve the base material. The solvents evaporate or soak into porous materials to be bonded. Sometimes a solvent may damage the materials on which it is applied, so manufacturer's instructions should always be followed. **Anaerobic bonding agents** maintain a fluid condition when exposed to oxygen but

Table 26.1 Commonly Used Adhesives and Their Applications

Adhesive	Bonded Material	Typical Uses
Acrylic	Plastics to metal, plastics to plastics, rubber to metal	Curtain walls
Casein	Wood to paper, wood to wood	All interior wood-joining needs
Cyanoacrylate (anaerobic)	Acrylics, phenolic, rubber, glass, polycarbonates, ceramics, steel, copper, aluminum	Any use (known as "super glue"); electronic and electrical devices
Epoxy	Almost any material except a few plastics and silicones	Interior and exterior uses, panels, glass to metal, curtain walls
Melamine formaldehyde	Paper, textiles, hardwood, interior plywood	Interior uses, plywood manufacturing
Natural rubber	Leather, paper, cork, foam rubber	Pressure-sensitive tape
Neoprene rubber (contact cement)	Many plastics, ceramics, aluminum	Plastic laminates, other interior uses
Nitrile rubber	Many plastics, ceramics, glass, aluminum	General uses
Phenol	Wood, cardboard, cork	Exterior plywood, any exterior use
Polyvinyl acetate	Porous materials (wood)	Various interior and exterior applications
Resorcinol	Rubber, paper, cork, asbestos, wood	Furniture, wood beams, columns
Silicone	Glass, ceramics, aluminum, polyester, acrylics, phenolic, rubber, steel, textiles	Sealant, gasket material
Polyurethane	Many plastics, glass, copper, aluminum, ceramics	Bonding dissimilar materials (e.g., on steel and glass sun roofs)
Urea	Many plastics, glass, copper, aluminum, ceramics	Particleboard, furniture, cabinets
Vinyl butyral	Glass	Laminating glass

set hard when oxygen is omitted, as occurs when two components are clamped together. Two-part mixtures supply the resin and catalyst in separate containers. When mixed, the catalyst causes cross-linking of the resin. **Hot-melt adhesives** come in solid form, become liquid when heated, and set rapidly when heat is removed.

Types of Bonding Agents

Bonding agents may be divided into three major classes: adhesives, glues, and cements. Following are some of the more commonly used types.

Adhesives Adhesive bonding agents are made from synthetic materials. They fall into two types: thermoplastic and thermosetting.

Thermoplastic Adhesives Thermoplastic polymer adhesives generally have less resistance to heat and moisture and less long-term resistance to loads than do thermosetting polymers. They are moisture-resistant but not used where they will be exposed to moist conditions.

Aliphatic resins are a type of polyvinyl resin and are stronger and more heat-resistant than other polyvinyls. They are generally yellow in color, but some polyvinyls are white. Aliphatic resins are used for furniture and general carpentry work (Fig. 26.1).

Figure 26.1 Polyvinyl acetate (white glue) and aliphatic resin types of polyvinyl resin (yellow or carpenter's glue) are used for furniture and general carpentry. *(Image copyright Gillian Mowbray, 2009. Used under license from Shutterstock.com)*

Alpha-cyanoacrylate, often called "superglue," is used to bond metals, plastics, and other dense materials. It is not recommended for use with porous materials, such as wood.

Hot-melts are a mixture of polymers sold in solid form, such as rods, pellets, ribbons, or films. They are placed in an electric hot-melt applicator that melts the

Figure 26.2 Hot-melts are applied with a type of electric hot-melt gun. *(Image copyright roadk, 2009. Used under license from Shutterstock.com)*

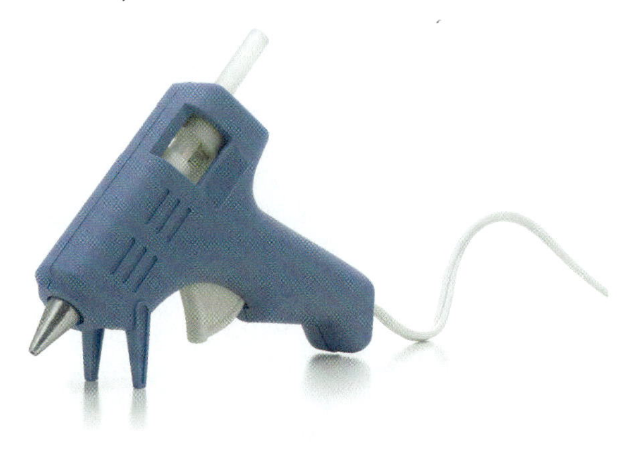

Figure 26.3 This is a waterproof polyvinyl acetate adhesive suitable for use in bonding wood used on building exteriors.

adhesive as it is applied to a surface (Fig. 26.2). There are various types for bonding plastics, particleboard, softwoods, and hardwoods. Hot-melts set up fast but are not very strong and are used primarily for interior purposes where little stress is expected.

Polyvinyl acetate adhesives, generally referred to as "white glue," have moderate moisture-resistance but high dry-strength and are recommended for interior use only. They are used to join paper, wood, and vinyl plastics (sometimes with metal) and are one of the most widely used adhesives. Typical applications include furniture assembly, bonding plastic laminates, and flush doors. A waterproof-type for bonding exterior wood is also available (Fig. 26.3).

Figure 26.4 Epoxies are two-part bonding agents with the resin and catalyst in separate tubes. *(Image copyright 1125089601, 2009. Used under license from Shutterstock.com)*

Thermoset Adhesives Thermosetting polymers are widely used as a structural adhesive because they undergo an irreversible chemical change and, if subjected to heat after use, will not soften and flow. They resist moisture and many chemicals and are very strong. Thermosetting adhesives are available in a variety of forms. Epoxy adhesives are produced in a two-part liquid form. An epoxy resin and an epoxy hardener are mixed immediately before use (Fig. 26.4). Epoxy resins will bond to almost any material and produce a strong joint. Epoxies are used in the construction of curtain walls and for bonding steel and concrete in applications such as bridge construction. Epoxies are being used for more and more structural applications and are available in a variety of types for special uses.

Melamine adhesives are produced as a powder with a catalyst to be added when used. They are cured by applying heat of about 300°F (150°C) and provide a good bond to paper and wood. Melamine adhesives are used as fortifiers in urea resins for hardwood plywood, gluing lumber, and scarf joining softwood plywood.

Phenol resin adhesives have good bonding qualities to paper and wood, good shear strength, and are resistant to moisture and temperature. They are cured in a hot press at about 300°F (150°C). Phenol resins are widely used in the manufacture of plywood and particleboard.

Resorcinol resin adhesives are used when a waterproof bond in wood products is required. The two-part adhesive consists of a resin and a catalyst. The mixture must be used within eight hours of mixing. One important use is in the manufacture of glued laminated wood structural members. Some adhesives are a mixture of resorcinol and phenol resins.

Urea resin adhesives are available in powder form and are mixed to proper consistency with water. They are moisture-resistant but not waterproof. They bond well to paper and wood and have good resistance to heat and cold. Normal clamp time is about sixteen hours but can be reduced to minutes or seconds using a radio frequency glue-drying machine. Typical uses include interior hardwood plywood, interior particleboard, flush doors, and furniture.

Polyurethane adhesives are available as low-viscosity liquid or high-viscosity mastic. Both provide superior strength and stiffness to an adhesive bonded assembly for wood, stone, metal, ceramics, plastics, and other materials. They are capable of withstanding exposure to moisture and some types are considered waterproof.

Glues **Glues** are bonding agents made from animal and vegetable products, such as bones, blood albumin, hides, fish, and milk.

Animal glue is produced as flakes or dry powder that is mixed with water and heated. The heated mixture is applied to wood surfaces. It is primarily used for furniture manufacture. Animal glue has excellent bonding and shear strength but has been largely replaced by newer products. Another form, liquid hide glue, a ready-mixed form of animal glue, is more widely used. It is not moisture-resistant and is used only on interior products, such as furniture, paper, and textiles.

Blood albumin glue is a form of animal glue used to bond paper products. It finds limited use on some forms of interior plywood. It has moderate bonding power with wood, poor resistance to moisture, and only fair resistance to heat and cold. Fish glues are used for sealing cardboard boxes and on packing tape.

Casein glue is made from dried milk curds in powdered form mixed with water. It is water-resistant and used for interior wood and on exterior products that will be sheltered and not be directly subjected to the weather, such as laminated timbers. Used on wood and paper it provides a strong joint, resists heat well, but has only moderate resistance to cold.

Vegetable glues are made from soybeans, starch, and dextrin and are used for bonding interior plywood, paper, and wallpaper.

Cements **Cements** are made from synthetic rubber, such as neoprene, nitrile, and polysulfide, suspended in a liquid.

Cellulose cements, such as cellulose acetate, cellulose nitrate, and ethyl cellulose, are used for interior purposes and mainly for bonding plastics and glass and for porous materials, such as wood and paper. Cellulose nitrate is more commonly known as a general household cement and is sold in tubes for ease of application. Cellulose cements have moderate resistance to temperature change but good moisture resistance.

Figure 26.5 Mastic is an elastomeric construction cement applied with a caulking gun.

Contact cements are neoprene cements that stick immediately upon contact, requiring no clamping time. They are used to bond plastic laminates on countertops, plastics to plastics, plastics to wood, and wood to wood. Although they have very high bonding strength and good resistance to moisture, they are generally used for interior purposes only.

Mastic is an elastomeric construction cement. Sometimes referred to as liquid nails, it is sold in tubes and applied with a caulking gun. Typical uses include bonding plywood subfloors to joists; bonding wall paneling to studs; laminating gypsum wall-board, styrene, and similar materials; and assembling wall, floor, and roof panels. Mastic is water-resistant and develops full strength after several weeks (Fig. 26.5).

Buna N resins are a form of acrylonitrile butadiene rubber used for general purpose interior cements. They are liquid cements that have good moisture resistance and strength. A summary of some of the properties of these bonding agents is given in Table 26.2.

SEALERS FOR EXTERIOR MATERIALS

Sealers are applied to the surface of a material to protect against penetration by water and moisture. Brick walls are often sealed to keep moisture from infiltrating mortar joints and leaking into the inside of the wall. Sealers have adhesive properties and are closely related to bonding agents. In addition to bonding to surfaces, they form an unbroken film over them and fill any minute pores, cracks, or other openings. Large cracks or defects must first be packed with some form of sealant before a sealer is applied because sealers will not span most cracks.

Table 26.2 Characteristics of Bonding Agents

	Form	Moisture Resistance	Application Temperature	Clamp Time
Thermoplastic Adhesives				
Aliphatic	Liquid	Low	45°F (7°C)	1 hr.
Alpha-cyanoacrylate	Liquid	Low	75°F (21°C)	Minutes
Hot melts	Solid	Low	Electronic welder 300°F (150°C)	None
Polyvinyl acetate	Liquid	Low	60°F (16°C)	1 hr.
Thermosetting Adhesives				
Epoxy	Liquid	Water resistant	60°F (16°C)	None
Melamine resin	Liquid	Water resistant	Hot press 250°F (122°C)	Minutes
Phenol resin	Liquid and powder	Waterproof	300°F (150°C)	Minutes
Resorcinol resin	Liquid	Waterproof	70°F (21°C)	16 hr.
Urea resin	Powder and liquid	Water resistant	70°F (21°C)	1 to 3 hr., or seconds with high-frequency glue curing machine
Glues				
Animal	Flakes and powder	Low	70°F (21°C)	2 to 3 hr.
Liquid hide	Liquid	Low	70°F (21°C)	2 to 3 hr.
Blood albumin	Liquid	Moderate	70°F (21°C)	2 hr.
Vegetable or fish	Liquid	Low	70°F (21°C)	2 hr.
Casein	Powder	Water resistant	32°F (0°C)	2 hr.
Cements				
Contact cement	Liquid to thick	Water resistant	70°F (21°C)	None
Mastic	Paste in tubes	Water resistant	70°F (21°C)	15 min.
Cellulose	Liquid	Low	70°F (21°C)	Minutes
Buna N	Liquid	Water resistant	70°F (21°C)	Minutes

Acrylic sealer is a high-solid-content clear acrylic sealer. It maintains the original appearance of masonry or concrete while protecting it from moisture, airborne dirt, and other pollutants. It is used on products such as exposed aggregate panels, brick, and stone. Acrylic sealer is virtually unaffected by prolonged exposure to moisture, common acids, ultraviolet rays, oils, and aliphatic solvents. It reduces damage from the freeze-thaw cycle, prevents efflorescence, and stains. It is available in a number of variations for different applications, such as methyl methacrylate acrylic polymer.

Applied by brush or roller, asphalt driveway sealer is a thin, quick-drying sealer that produces a black protective coating. Coal tar coatings are used to seal surfaces and provide protection against corrosive conditions encountered in industrial plants, water and sewage systems, and other such industries. They resist moisture and most acids, alkalies, corrosive vapors, and atmospheric corrosion and can be applied to metal, concrete, and masonry surfaces.

Silicone sealers come in liquid form for use on brick, concrete block, stucco, cement plaster, and concrete. They produce a water-repellent coating that protects surfaces but does not prevent water vapor from escaping. They seal surfaces, minimize efflorescence, and protect from absorbing staining materials, such as dirt and soot. Since protection against moisture penetration is provided, damage due to freeze-thaw cycles is also reduced.

Epoxies are used to provide a waterproof coating. They are often employed to seal floors in areas where moisture or chemicals exist, such as in food processing and industrial applications. Other uses include floors around swimming pools, decks, showers, restrooms, loading docks, and on stair treads. When mixed with an aggregate, such as emery, epoxies provide a textured, slip-resistant surface available in a variety of formulations.

Polyurethane sealers are used to minimize concrete problems such as scaling, spalling, chloride penetration, and damage due to freeze-thaw cycles. Some sealers provide ultraviolet light stabilization. They do not inhibit the transmission of vapor out of the concrete.

Wood Sealers

Sealers on wood products are used to keep the various layers of finish, such as stains and fillers, from bleeding through the final topcoat. A sealer also provides the required base for the finished topcoat. The sealer used depends on the recommendations of the manufacturer of the topcoat. If the wood is to be stained and varnished, shellac is often used as a sealer over the stain. Then wood filler can be applied and another sealer, often more shellac, follows. Then the varnish topcoat is applied. Synthetic resin sealers are also used. Lacquer topcoats require a lacquer sanding sealer. After careful sanding, the lacquer topcoats go on. Other finishes, such as polyesters and polyurethanes, may or may not require a sealer.

For exterior and interior paints, a manufacturer will have specific recommendations for sealers. Following is an example for one manufacturer:

Drywall sealer – latex primer

Plaster sealer – wall and wood primer

Wood sealer – alkyd enamel undercoat

Concrete block sealer – block filler

Masonry sealer – latex primer

Ferrous metal sealer – water-based acrylic paint

Walls to receive wallpaper or fabric coverings also must have sealers; usually some form of casein glue is used.

SEALANTS

A **sealant** is a material used to seal joints between construction members and protect materials against the penetration of moisture, air, corrosive substances, and foreign objects. Examples include expansion joints in large masonry walls, spacers between glass and frames, and openings between exterior siding and door and window units. The sealing of joints between these and other parts of a building is essential to ensure the integrity of the entire structure.

Joints in exterior walls are needed to allow materials to expand and contract. These are called expansion joints, and they must allow for this movement. If a foreign object, such as a rock, gets into a joint, contraction may be blocked, and the object may cause the materials on each side to spall or crack (Fig. 26.6).

The two basic methods for protecting joints involve the use of sealants and prefabricated covers. A sealant is a flexible adhesive material that is worked into a joint, bonds to the sides, and sets into a firm but rubbery plastic material. Prefabricated covers are typically metal and made to allow for anticipated movement. The sealant filling a joint is a substitute for the material that was removed. As such, it must meet the performance requirements of the adjoining material. It must maintain the integrity of the assembly of materials while still allowing for movement between them.

Figure 26.6 Joints between masonry materials to be sealed must be free of foreign objects.

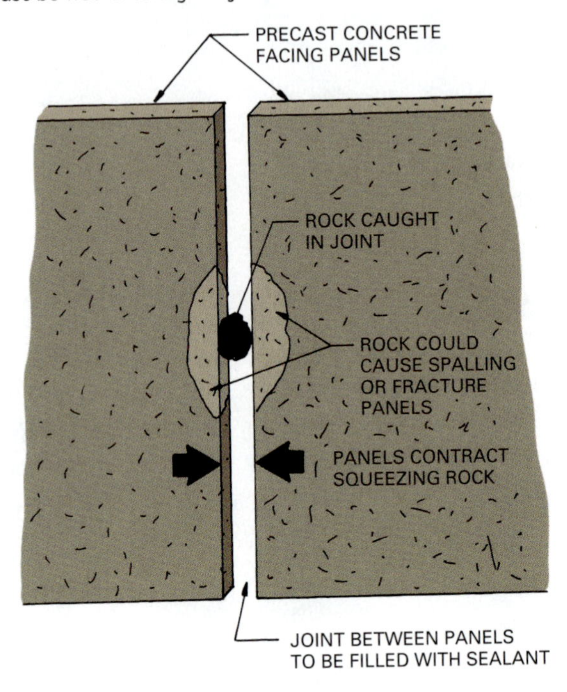

Moldable sealants may be deformable or elastic/elastomeric. The deformed sealant is installed in its natural stretched shape. It stretches as the joint widens and deforms as the joint width is reduced. The elastic/elastomeric sealant stretches as the joint widens and shrinks back to its normal size as the joint width contracts (Fig. 26.7).

Sealant Performance Considerations

A key factor in sealant design is the percent of elongation the sealant can safely stretch and still give expected protection. High-performance sealants have a 50 percent elongation, while intermediate types are usually rated up to 25 percent. A sealant must have excellent adhesion to the material with which it is expected to bond. It must be flexible and have minimum internal shrinkage through years of use; resist staining the material around it; and have a tough non-tacky elastic surface skin such that dirt and solid objects do not stick to it.

Figure 26.7 Elastic/elastomeric and deformable sealants allow for expansion and contraction at each joint.

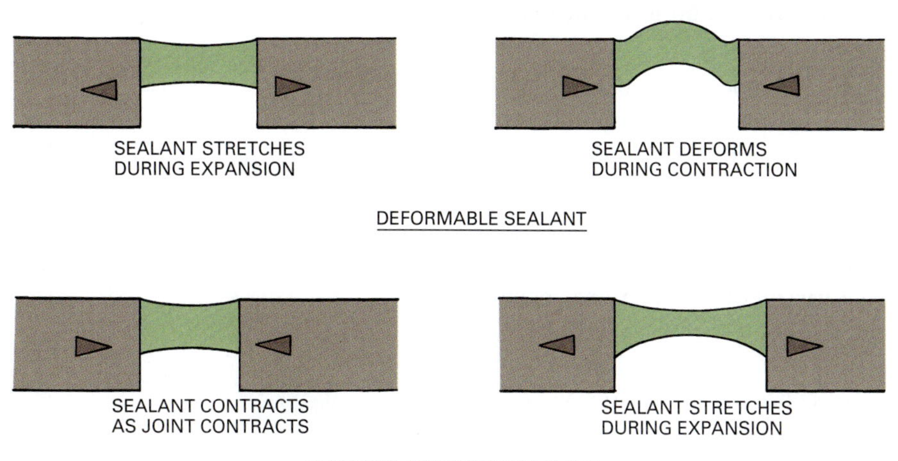

Figure 26.8 Preformed sealants are designed for specific applications.

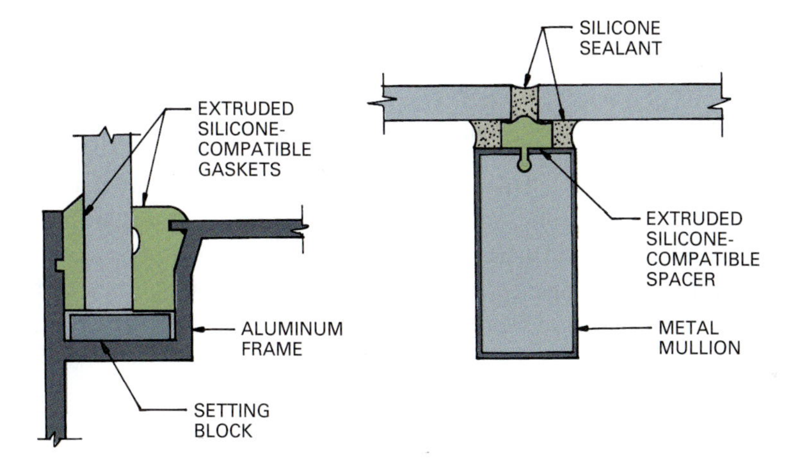

Types of Sealants

Sealants are available in two basic types: a flexible, moldable adhesive compound and a solid, flexible preformed shape. Preformed sealants are available in a variety of materials and shapes (Fig. 26.8).

Moldable sealants are manufactured in three performance levels. Low performance sealants have a life of four to seven years and are used where limited joint movement is expected. Intermediate performance sealants are more expensive but last between seven to fourteen years. They are used in joints with the greatest movement. High performance sealants have a life expectancy of twenty to thirty years and are the most expensive. They are used in joints where the anticipated movement is the greatest. Manufacturers' recommendations on joint size and the allowable percent of elongation must be observed.

Moldable bulk sealants are available in pourable form, in knife grade (such as glazing compounds), gunable form for manual or pneumatic caulking guns, and as preformed tapes that may or may not be cured.

The actual makeup of the following commonly used sealants can vary considerably and many special-purpose sealants are available.

Polyurethane sealants are general purpose sealants used for areas such as precast concrete and masonry joints, glazing and sealing around door and window openings and swimming pools. Some are designed for exterior concrete joints in roads and sidewalks. This type is resistant to damage from fuels and oils. Some are formulated for application by gun, and others are liquid and poured.

Epoxy crack filler is available for use on very small fractures that occur in concrete, concrete block, and brick walls.

Silicone rubber sealants are available in a variety of compounds. Some are for interior use on nonporous surfaces where high humidity and temperature extremes exist, such as around bathtubs. Another type is used to seal exterior building joints and will bond to nonporous materials, such as glass, ceramics, and most metals and plastics. It can be gunned in below freezing temperatures, is available in a variety of colors, and has a life expectancy of up to thirty years.

Latex sealant adheres to wood, metal, concrete, masonry, marble, porcelain, ceramic tile, glass, and many types of plastics. Some types permit joint movement up to 25 percent.

Polysulfide polymer sealants are elastomeric and bond to all masonry, concrete, wood, glass, and metal surfaces. Available in a variety of formulations, including those with one and two components, they are widely used in applications such as sealing joints between curtain wall panels, precast concrete, and various window assemblies. Polysulfide polymer sealants cure in about twenty-four hours to a rubber-like material with excellent stretch capabilities.

A urethane bitumen sealant is a two-part catalyzed 100 percent solid polyurethane-coal tar elastomeric compound. It bonds to concrete, stone, brick, glass, wood, metal, cement asbestos, concrete block, and most plastics, with the exception of polyethylene. It offers excellent resistance to acids and commonly used solvents.

Sealants and Joint Design

The width of a joint between two members, such as precast concrete facing panels, must be determined according to the anticipated structural and thermal movement. For example, one type of sealant available tolerates a joint movement of 1100 to 250 percent. The manufacturer requires the joint width be two times the expected joint movement and at least $\frac{1}{4}$ in. (6 mm) wide. A backup rod is used to control the depth of the sealant (Fig. 26.9). Other types of sealer joints are shown in Fig. 26.10.

Figure 26.9 A backup rod is used to control the depth of the sealant.

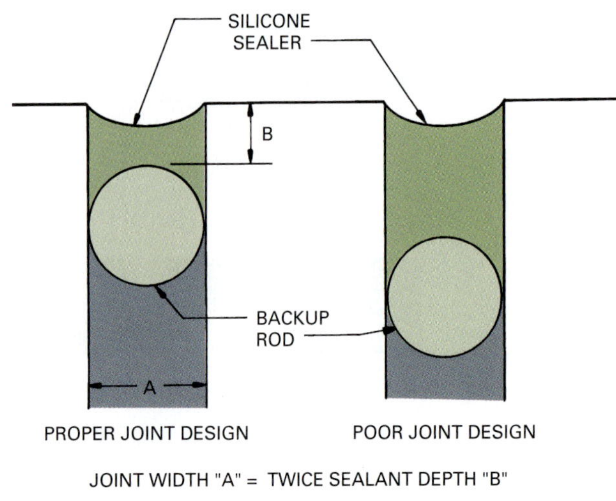

Figure 26.10 Examples of other types of joints requiring backup rods or the use of bond breaker tape. *(Courtesy Karnak Corporation)*

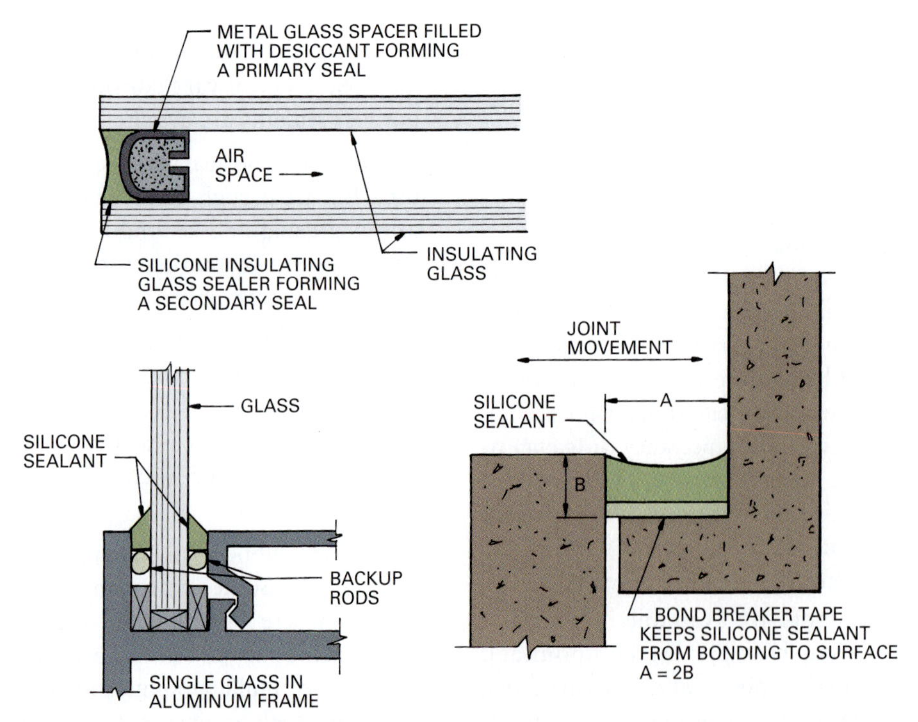

Backup Materials

Proper installation of sealants depends on the use of proper backup materials installed to control the depth of the sealant, as shown in Fig. 26.11. The backup material not only limits the depth of the sealant but also serves as a bond breaker to keep the sealant from bonding to the back of the joint. Some backers are installed in the factory as part of a unit, as in a manufactured window. Others are applied to joints in the field, such as in control joints in a concrete floor slab.

Backup materials vary depending on location and use. Rods or tubing made from polyethylene, butyl, urethane, and neoprene that control the depth of a sealant are most frequently used. Other rod types form a primary water seal to keep out moisture during construction until the sealant can be installed. Typical uses for rods and tubes are in joints between precast curtain wall panels and expansion joints in long masonry walls.

Figure 26.11 Typical applications of the use of backup rods and expansion strips.

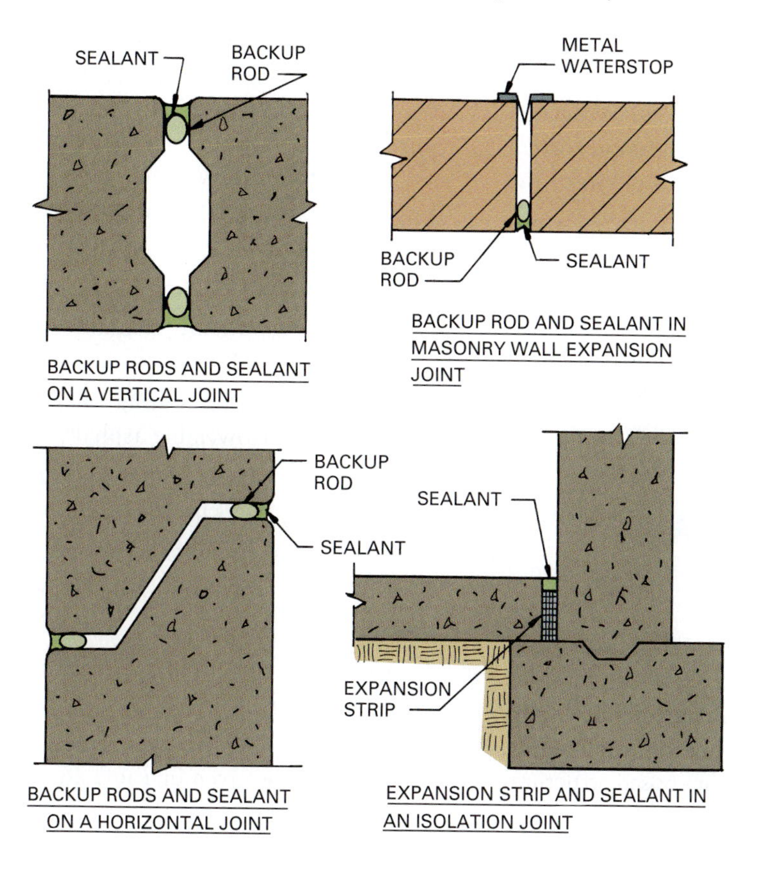

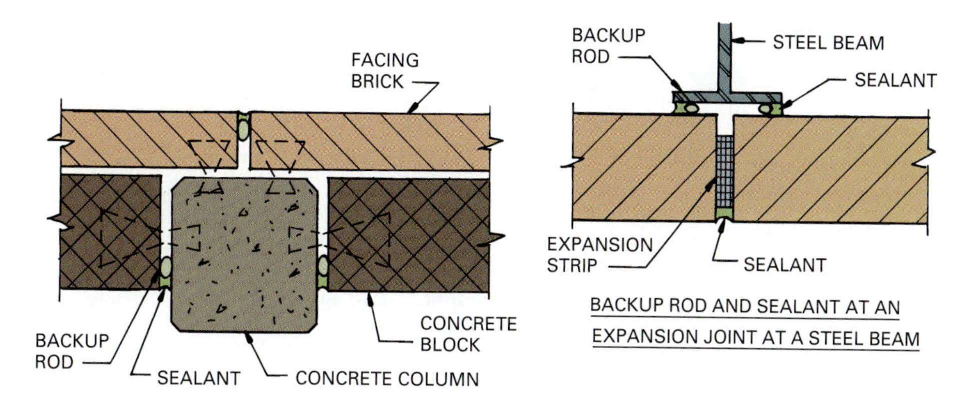

Tapes and polyethylene film are another form of backup material. They act as a bond breaker, keeping a sealant from adhering to joint surfaces where bonding is not desired.

Other materials, such as resin-impregnated fiberboard, corkboard, and dense plastic foamed strips, are used in places such as isolation and control joints in concrete and masonry construction. Rods and tubes can also be used for these purposes. Some examples are shown in Fig. 26.11.

Caulking and Glazing

Caulking is a procedure for sealing joints, cracks, or other small openings with caulking compound (sealer). **Caulking compound** is a resilient mastic material. Glazing compounds are a form of sealer used to set glass in place in frames. They serve to seal out water and air and form a cushion allowing for expansion and contraction of the glass and frames.

Caulking and glazing compounds are commonly made from silicone, acrylic, butyl, polysulfide, or polyurethane. These five types are considered gunnable, meaning their consistency allows for application with a caulk gun. Other types are referred to as "knife grade," because they are applied with a putty knife.

WATERPROOFING MEMBRANES AND COATINGS

Waterproofing involves applying a material on the surface of an assembly of materials, such as a foundation, to make it impervious to water. Many parts of a building require waterproofing, including foundation walls, roofs, exterior wood or masonry walls, and exposed structural components, including steel. Waterproofing must be able to resist the forces that tend to force water through the assembly of materials. These forces include:

Gravity, which forces water through horizontal areas, such as a roof or deck

Hydrostatic pressure on one side of a horizontal or vertical assembly (most commonly due to subsoil water)

A difference in air pressure on one side of a horizontal or vertical assembly

For surfaces not subjected to hydrostatic, gravity, or air pressure differences, waterproofing can be a light-duty coating, such as silicone or coal tar pitch. These provide a moisture-resistant membrane that dampproofs the assembly. Surfaces under pressure require a heavy duty membrane, such as tar and felt, or a synthetic membrane. Waterproofing is most effective if applied on the surface directly facing the source of moisture.

Waterproofing can be accomplished by:

- Applying a built-up bituminous membrane of felt and hot or cold tar pitch.
- Applying a heavy coating, such as Portland cement plaster or a trowelable asphalt.
- Bonding an elastomeric membrane to a wall.
- Applying a thin film or coating to the exterior of the wall, such as liquid silicone or coal tar pitch.
- Adding waterproofing admixtures to concrete as it is mixed.
- Applying a dry coating that emulsifies in place, such as bentonite clay.

A summary of the most frequently used types of waterproofing is given in Table 26.3.

Table 26.3 Types of Waterproofing

		Built-up Membranes			
Sheet membranes	Composite membranes	Hot applied	Cold applied	Liquid membrane	Applied coating
Butyl Ethyene propylene Neoprene Polyethylene Polyvinyl chloride	Elastomeric, backed Polyethylene and rubberized bitumen Polyvinyl chloride backed Saturated felts and bitumen coated	Asphalt, type I, II, III Coal tar pitch, type B Felts, saturated and coated	Bitumen emulsion Bitumen, fiberated cement Felts, coated Bentonite clay Fabric, saturated Glass fiber mesh, saturated Cementitious membrane	Butyl Urethane Polychlorene (neoprene) Polyurethane, coal tar	Acrylic, silicone Asphalt emulsions, cut backs Cementitious with admixtures Epoxy, bitumen Urethane, bitumen Bitumen, rubberized

Construction Techniques

Waterproofing Tips

There are a number of ways to waterproof a foundation wall. The manufacturer of a system usually requires the contractor to employ a certified applicator if the manufacturer's guarantee is to be valid. Common systems include liquid membranes, sheet membranes, cementitious coating, built-up systems, and bentonite.

Liquid membranes are applied with a roller, trowel, or spray. The liquid solidifies into a rubbery coating. Different materials are available, such as polymer-modified asphalt and various polyurethane liquid membranes.

Sheet membranes tend to be self-adhering rubberized asphalt coverings, typically an assembly of multiple layers of bitumen and reinforcing materials. Some companies manufacture PVC and rubber butyl sheet membranes.

Cementitious products are mixed on-site and applied with a brush. Some have an acrylic additive available that improves bonding and makes the cementitious coating more durable. One disadvantage is that these coatings will not stretch if the foundation cracks, thus opening leakage possibilities.

The widely used hot tar and felt membrane is an example of a built-up system. Alternate layers of hot tar and at least three layers of felt are bonded to a foundation. Bentonite is a clay material that expands when wet. It is available in sheets that are adhered to foundations. As groundwater penetrates the clay, it swells many times its original volume, providing a permanent seal against water penetration.

Surface Preparation

Regardless of the type of waterproofing system used, it is important to prepare a surface before application. This includes (1) drying the wall and footings; (2) removing the concrete form ties, making certain they break out inside the foundation to prevent penetration of the waterproof membrane; (3) clearing the wall of all dirt or other loose material; and (4) sweeping the wall free of dust and mud film residue. The residue left when wet mud is wiped off and left to dry on a foundation can inhibit bonding. Finally, any openings around pipes or other items that penetrate the wall must be grouted.

Safety

Waterproofing presents some hazards that must be controlled. First, possible cave-ins of soil around excavations could bury workers. Normal shoring procedures should be observed. Many materials used are flammable and solvent-based, presenting a potential fire hazard. Workers should not smoke or use tools that might cause ignition. Solvent fumes can be very harmful so workers must wear respirators. Fumes are usually heavier than air and settle around a foundation in the excavated area. The solvents, asphalt, and other materials used may cause skin problems, so protective clothing, including gloves and eye protection, is required with many products. Manufacturer recommendations should be followed closely.

Waterproofing coatings and membranes are not self supporting but must be bonded to the surface to be treated. In addition, waterproofing must be able to adjust to the stresses caused by movements of the assembly and any cracks or deterioration without losing its waterproofing capabilities. It should be noted that waterproofing coatings and membranes can be damaged during installation and construction. Roof membranes can be punctured by construction traffic. Protective pads are available for installation atop roof membranes to protect areas where workers walk. Damage frequently occurs during backfilling of foundation walls. A protective material can be placed over the waterproofing to keep rocks from piercing it.

Bituminous Coatings

Hot-applied and emulsified coal tar pitch, hot-applied and cold-applied asphalt, and emulsified asphalt can be administered to a foundation by brush, roller, or spray. These are effective only for situations in which hydrostatic pressure is not a factor.

A waterproofing system that will resist hydrostatic pressure consists of alternate layers of hot-mopped coal tar pitch, asphalt, or cold-mopped emulsified asphalt over layers of mineral or glass fiber felts (applied in much the same way as for laying a built-up roof). The number of layers of felt and asphalt depends on the hydrostatic conditions. Manufacturers of these systems have established specifications for various conditions.

Another form of light-duty waterproofing is a bituminous binder with asbestos fibers forming a trowelable mix (Fig. 26.12). It is hand-troweled on concrete foundations and often used to bond foamed plastic insulation boards to a foundation. Bituminous binders are only good for dampproofing a wall not subject to hydrostatic pressure. A thinner version can be sprayed using a mastic pump. Refer to Chapter 27 for additional information.

Figure 26.12 This trowelable dampproofing material has a non-volatile bituminous binder dispersed in water by means of selected mineral colloids.

There are a variety of solvent-based asphalt dampproofing compounds. They are available in thick, semisolid, and spray mastics and give dampproofing properties to interior and exterior above- and below-grade surfaces. They are also used on metal to prevent corrosion.

Liquid Coatings

An acrylic copolymer waterproof coating is available in a variety of colors. It may be obtained with fillers and texturing aggregates that are fused onto a concrete or masonry surface. This offers waterproofing protection and an attractive finish coating. Important uses are on above-grade exterior concrete walls, columns, and spandrels. The coatings are also used on Portland cement plaster and stucco walls, giving a textured, sandlike finish in color that minimizes surface defects. They are applied by brush, roller, or spray.

Clear silicone dampproof liquids are widely used on exterior concrete, masonry, and wood surfaces. They are not a surface coating but penetrate a material, carrying solids into its pores. They do not color a wall but do reduce efflorescence. They can be applied with a brush, roller, or spray.

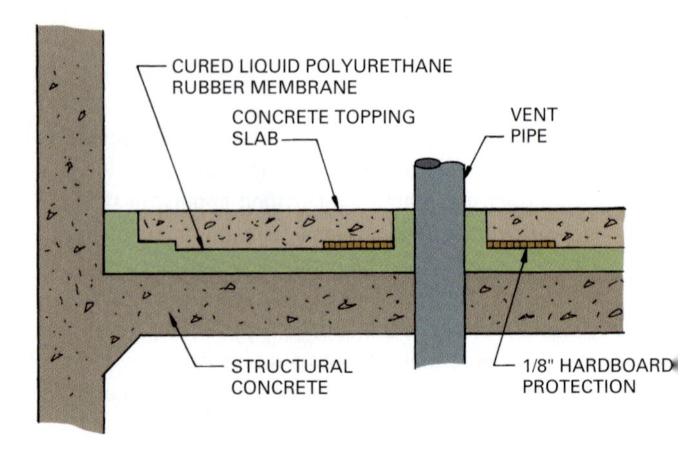

Figure 26.13 Self-curing polyurethane rubber forms a seamless waterproof membrane.

A liquid waterproofing material that is self-curing is available in a polyurethane rubber with a coal tar additive. It is applied to decks, roof slabs, and floors with a brush, roller, or squeegee. It will cover small cracks and is flexible enough to cover surfaces of irregular shapes. It can be covered with a concrete slab, hardboard, or mineral surface roll roofing (Fig. 26.13).

Another form of liquid waterproofing uses a silicate liquid gel that reacts with soluble calciums in concrete to form a glass gel in the microscopic pores. The gel will penetrate 1.5 to 2.0 in. (38 to 50 mm), providing a strong moisture-proof barrier that also reduces efflorescence.

Synthetic Sheet Membranes

Synthetic membranes are available made from neoprene, polyvinyl chloride, polyethylene, butyl, and ethylene propylene. These sheet materials are bonded to foundation walls using adhesives recommended by the system manufacturer. Since they are flexible membranes, they tend to adjust to settling and compaction of the soil and are not likely to rupture. They are available as single materials or composite membranes. Following are examples of a few of these products.

Composite Membranes Composite membranes are sheet products made by laminating two or more waterproofing materials. One frequently used composite membrane is made of a rubberized asphalt layer with a polyethylene film bonded to the outer surface. It remains stable below ground and below water. When the sheets are lapped, the rubberized asphalt back bonds to the polyethylene face of the sheet beside it. Under most conditions it will bridge gaps up to $\frac{1}{4}$ in. (6 mm). It is available in rolls 48 in. (121 mm) wide and 60 ft. (18.3 m) long. Since the back surface is very tacky, it is covered with a peel-off paper covering (Fig. 26.14).

Figure 26.14 Silicate liquid gel forms a waterproof coating by reacting with the soluble calciums in the concrete.

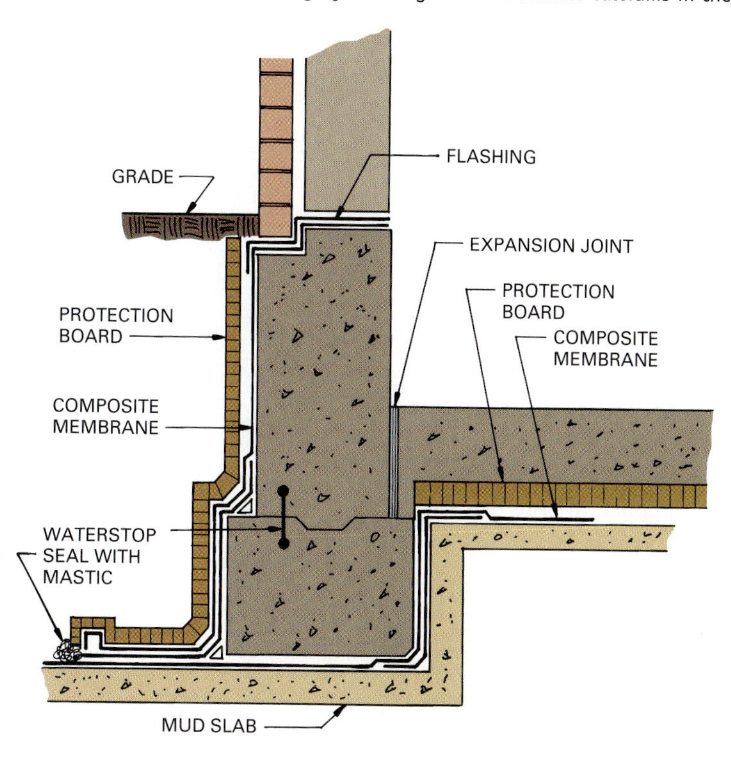

Another composite membrane is made by laminating a chlorinated polyethylene (CPE) film to a non-woven polyester fabric. This composite is bonded with water-based acrylic adhesives, avoiding problems that occur from fumes of solvent-based adhesives. It can be installed on surfaces that are damp at the time of installation.

Sheet Membranes Sheet membranes are waterproofing products composed basically of one major waterproofing material. One such product is made from polyvinyl chloride alloyed with high-density polymer resins. It is not affected by aging, mildew, or corrosion and remains flexible at low temperatures with good abrasion and tear resistance. It is bonded, and laps are sealed with a special adhesive. Some forms of PVC membrane can be formulated to be resistant to gasoline and oil. The seams can be welded with hot air or bonded with an adhesive specified for that purpose.

Another type of sheet membrane is a chlorinated polyethylene (CPE) product. It is available in a range of thicknesses with and without integral reinforcement. It is suitable for use above and below grade and on horizontal and vertical surfaces. Laps can be joined by chemical or thermal fusion. It will bridge cracks up to $\frac{1}{4}$ in. (6 mm).

Cement-Based Waterproofing

A number of cement-based heavy duty waterproof coatings are available. These have carefully graded aggregates that produce a high-density and high-strength waterproof coating. Some types are applied with a trowel while others require special procedures. They are used to waterproof concrete masonry and stone in interior and exterior locations. They are employed on water reservoirs, swimming pools, basements, parapet walls, and other heavily exposed surfaces.

Lead Waterproofing

Lead waterproofing sheets are frequently used for projects in which waterproofing must be of uncompromised security. Examples include areas of grass, fountains, and reflecting pools that may be built over underground facilities, such as parking or retail, or on the roof of a building. The gauge of the lead sheets varies for different purposes but must be at least 6 lb. lead ($\frac{3}{32}$ in. or 2.3 mm) thick to allow for the burning (welding) necessary to join sheets. If lead comes in contact with cementitious materials it must be protected with a bituminous coating. Lead wool is also used to waterproof joints in above-ground installations.

Bentonite Clay Waterproofing

Bentonite is a clay formed from decomposed volcanic ash, with a high content of the mineral montmorillonite. It can absorb large amounts of water, causing it to swell many times its original volume and form a waterproof barrier. The dried, finely ground particles are usually applied as a waterproofing membrane in three ways.

Bentonite Panels Panels, usually 4 x 4 ft. (1,200 x 1,200 mm), consist of a biodegradable paper covering over bentonite clay particles. One type is $\frac{3}{16}$ in. (2.5 mm) thick with a corrugated kraft board core. It is used on vertical walls and under structural slabs. Another type is $\frac{5}{8}$ in. (16 mm) thick, composed of layers of corrugated kraft board with the center layer holding the bentonite clay. The hollow outer layers allow space for the expansion of the clay and reduce upward pressure against a thin nonstructural slab.

The panels are ready for installation when received on-site and can be applied at all temperatures and over moist substrates. They can be nailed to green concrete walls. The panels are overlapped $1\frac{1}{2}$ in. (38 mm). A hydrated sodium bentonite gel is used to fill gaps around pipes and fittings. If the backfill contains rocks that may pierce the panels, they should be covered with a protective material (Fig. 26.15).

Figure 26.15 Bentonite panels can be nailed to concrete foundations and are overlapped at the joints between panels.

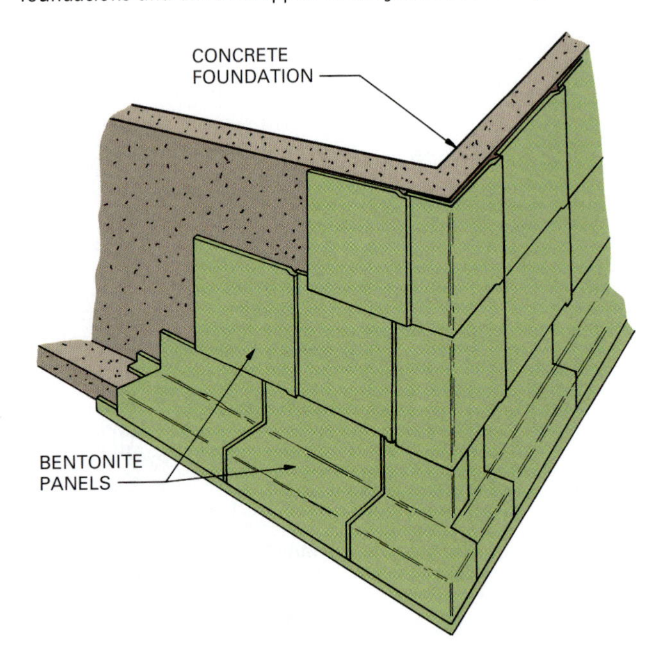

Sprayed Bentonite Bentonite clay mixed with a modified asphalt that serves as an adhesive to bond the clay to a surface is applied by spraying. A $\frac{3}{8}$ in. (10 mm) membrane is built up for normal applications, but thicker layers can be used for situations in which hydrostatic pressure is severe.

Bentonite-Sand Mixture Bentonite clay mixed with sand is used to produce a waterproof barrier under concrete slabs. The mix is carefully measured and spread over the area to be covered with concrete. It is then covered with a polyethylene sheet to protect it from the moisture in the concrete. The reinforcing is placed on top and the slab is poured. If too much bentonite is used, the forces of expansion can crack the slab.

Review Questions

1. What are the two ways bonding agents hold materials together?
2. In what ways do various bonding agents cure?
3. What are the three major classifications of bonding agents?
4. What are the two classes of adhesives?
5. What is the source of materials for making glues?
6. From what materials are cements made?
7. What is the difference between a sealer and a sealant?
8. What is a working joint?
9. What are the two basic methods of protecting exterior joints?
10. What are the two types of sealants?
11. What are the performance levels of sealants?
12. What purpose does a sealant backup rod serve?
13. Why are tapes sometimes used with sealants in joints?
14. What forces tend to force water through an assembly of materials?
15. What is bentonite?

Key Terms

adhesive
adhesive bonding agents
anaerobic bonding agents
bonding agent
caulking compound
cements
glues
hot-melt adhesives
mechanical action
sealants
sealers
specific adhesion

Activities

1. Bond identical wood samples with various types of adhesives recommended for use on porous materials. Cure following the manufacturers' recommendations. Test each sample for bond strength with a tensile testing machine. Soak other samples in water for an identical number of days and test these for tensile strength. Write up your findings.
2. Bond identical metal samples with various types of adhesives. After the recommended curing time, test for strength of bond with a tensile testing machine.
3. Visit several commercial buildings and note how various expansion joints and other areas needing to be sealed are protected. Observe any evidence of sealant failure. What would you recommend be done to repair any failures?

Additional Resources

Joint Sealers; Structural Sealant Glazing Systems and *Fenestration Sealants Guide Manual*, American Architectural Manufacturers Association, Schaumburg, IL.

The NRCA Roofing and Waterproofing Manual, National Roofing Contractors Association, Rosemont, IL.

A Professional's Handbook on Grouting, Concrete Repair, and Waterproofing, available from Five Star Products, 425 Stillson Road, Fairfield, CT 06324.

Specifications for Weatherstrips and Sealants, American Architectural Manufacturers Association, Schaumburg, IL.

Other resources include:

Chemprobe Coating Systems, TSE Inc., PO Box 23559, Columbia, SC 29224.

Publications from the Gorilla Group, Santa Barbara, CA 93103, 800-966-3458.

Publications from Architectural Products Outerwear, LLC, PO Box 347, Wood-Ridge, NJ 07075.

See Appendix B for addresses of professional and trade organizations and other sources of technical information.

CHAPTER 27

Bituminous Materials

LEARNING OBJECTIVES

Upon completion of this chapter, the student should be able to:

- Understand the properties of bitumen and how these properties influence decisions as to how bituminous materials might be used.
- Interpret the results of laboratory tests and use these for decision-making.
- Be aware of the array of bituminous products available and the best applications for each.

Build Your Knowledge

For further study on these materials and methods, please refer to:

Chapter 4 The Building Site
Topic: Paving

Chapter 26 Bonding Agents, Sealers, and Sealants
Topics: Waterproofing, Membranes and Coatings

Chapter 28 Roofing Systems

Bitumen is a mixture of organic liquids that are highly viscous, black, sticky, complex hydrocarbons that occur naturally or are heat-produced from materials such as coal and wood. Bitumen may be in gaseous, liquid, semisolid, or solid states.

Bitumen can now be made from non-petroleum-based renewable resources such as sugar, molasses and rice, corn and potato starches. Bitumen can also be made from waste material through fractional distillation of used motor oils, which are sometimes disposed of by burning or dumping into landfills. Non-petroleum-based bitumen binders can be made light-colored. Roads made with lighter-colored pitch absorb less heat from solar radiation and become less hot than darker surfaces, reducing their contribution to the urban heat island effect.

Asphalt, tar, and coal tar pitch are the most commonly used bituminous materials in construction. Asphalt is found in natural deposits or produced from petroleum. Tar is produced by the distillation of wood and coal. Fractional distillation of tar produces coal tar pitch.

PROPERTIES OF BITUMENS

Possibly the most important property of bitumens is their excellent water resistance. A variety of different bitumen products are produced as waterproof materials. Bitumens bond well to dry solid surfaces because they exist in the semi-fluid state needed by adhesives to bond. They will not bond to wet surfaces.

Bitumens are flammable and will ignite when heated to their flash point. This influences the temperatures used to heat bituminous materials for various applications.

Another important property of bitumens is their softening point. Although this varies with the composition of a product, it is crucial to consider in relation to temperatures experienced in various applications. For example, the softening point for roofing asphalts ranges from 200 to about 220°F (95 to 104°C), while coating-grade asphalts, such as those used for waterproofing, have a softening point of 50 to 55°F (11 to 13°C).

Bitumens exhibit cold-flow properties: they tend to flow, spread, or lose their shape. This is more pronounced at higher temperatures and offers the advantage of causing a check or crack on a roof membrane to seal or heal when the sun heats up the asphalt.

The **viscosity** of asphalt is an important property when considering applications. Viscosity describes the ability of a material to stay in place when subjected to heat. Asphalts have good viscosity properties up to 135°F (58°C), with some able to withstand temperatures up to 275°F (136°C).

Asphalts rate high in their ductility properties. Their molecules hold fast even when they are extended by heat and pressure. Asphalts can expand and still remain bonded to materials upon which they are placed.

ASPHALT

Asphalt is a dark brown to black cementitious material in semisolid or solid form consisting of bitumen found in deposits of natural asphalt. A manmade asphalt is manufactured from residues from distillation of petroleum. Petroleum provides the raw material for the manufacturing of most asphalt used today (Fig. 27.1).

Asphalt is an ingredient in many construction products, including roofing, siding, and paving material. It is used in some paints, adhesives, acid- and

Figure 27.1 The process for producing various asphalt products from petroleum. *(Courtesy The Asphalt Institute)*

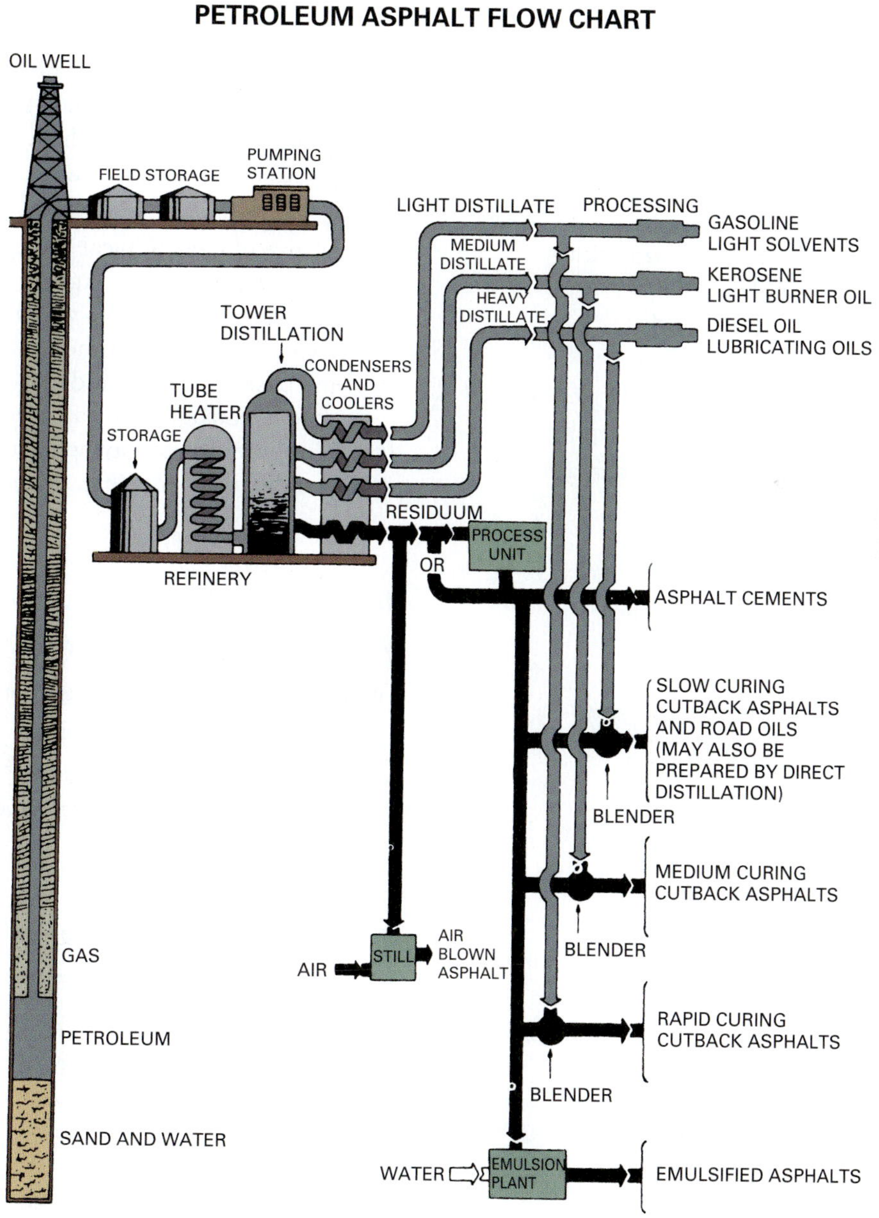

alkali-resistant coatings, and dampproofing and waterproofing solutions. Asphalt is used to coat organic fiber, fibrous plastic panels, and papers of various types. It can be found in adhesives, cementitious materials, and in the manufacture of drain and sewer pipe. Some types, such as waterproof coatings, are applied on-site, while others, such as coating for fibrous panels, are applied in a factory (Table 27.1).

The grades and types of asphalt used for coatings, cements, paints, and adhesives are detailed in Tables 27.2 and 27.3. The grade refers to the liquefaction of the asphalt. Grade 0 is the thinnest, and grade 5, which is like a paste, has the thickest consistency. Steam-refined or oxidized asphalt having a petroleum solvent is classified as slow curing (SC), medium curing (MC), and rapid curing (RC) (Table 27.4). Asphalt emulsified in chemically treated water is graded as slow setting (SS), medium setting (MS), and rapid setting (RS).

Roofing asphalt is available in four types:

Type I – Dead level

Type II – Flat

Type III – Steep

Type IV – Special steep

Physical characteristics for these four types of roofing asphalts are given in Table 27.5.

Asphalt cements are binders used to produce high-quality asphalt pavements. They are highly viscous and made in several grades based on consistency. Asphalt cements are tested for viscosity at 140°F (60°C) and are semisolid at normal ambient temperatures.

Asphalt cements are graded from the softest, AR 1000, to the hardest, AR 16000. AR 4000 is a general purpose grade, and AR 8000 is used in hot climates. The grade chosen depends on climate and the quality of the aggregate.

Asphalt cements are blended with aggregates graded into a range of sizes. Typical aggregates include crushed stone, gravel, and sand. Aggregates compose about 90 percent of a paving mix's weight. The aggregates give the mix its strength, and the asphalt acts as a binder. In some cases, tars are added because they increase resistance to damage from gasoline spilled on paved surfaces. Asphalt cement paving is used for paving roads, drives, and parking lots (Fig. 27.2).

Table 27.1 Major Classes and Uses of Bituminous Mixtures

Class	Use
Liquid	Alleviate dust (spraying) Waterproofing Impregnation or saturation of materials
Medium consistency	Sealing compound Road surface binder Adhesive compound for roofing Expansion joint caulking
Solid	Electrical insulation compound Molded products

Table 27.2 Grades of Asphalt Used for Paints, Coatings, Adhesives, and Cements

Grades of Liquefaction
0 (Thinnest)
1
3
4
5 (Thickest)

Table 27.3 Types of Asphalt Used for Paints, Coatings, Adhesives, and Cements

Type	
Steam Refined with Petroleum Solvent	**Emulsified in Chemically Treated Water**
Slow curing (SC)	Slow setting (SS)
Medium curing (MC)	Medium setting (MS)
Rapid curing (RC)	Rapid setting (RS)

Table 27.4 Liquid Asphalt Products and Their Solvents

Classification	Solvent
Rapid curing (RC)	Gasoline or naphtha
Medium curing (MC)	Kerosene
Slow curing (SC)	Slowly volatile or nonvolatile oils
Asphalt emulsions	Water and emulsifiers

Table 27.5 Physical Characteristics of Roofing Asphalts

Type	Softening Point	Flash Point
Type I, Dead level	135–151°F 58–67°C	475°F 248°C
Type II, Flat	158–176°F 70–80°C	475°F 248°C
Type III, Steep	185–205°F 85–97°C	475°F 248°C
Type IV, Special steep	210–225°F 100–108°C	475°F 248°C

Figure 27.2 Asphalt paving cement is used to pave roads, driveways, and parking lots. *(Image copyright Alex Kosev, 2009. Used under license from Shutterstock.com)*

Cutback asphalt is a broad classification of residual asphalt materials left after petroleum has been processed to produce gasoline, kerosene, diesel oil, and lubricating oils. The residual asphalt is then blended with various solvents to produce cutback asphalt in three classifications: rapid curing (RC), medium curing (MC), and slow curing (SC). Cutback asphalt is often mixed with granular soil to stabilize a roadbed before paving or to serve as a binder for a finished road surface for light traffic—on a secondary road, for example.

An emulsion group of asphalts consists of emulsified asphalt cement mixed with water and is used in road construction. The emulsion is made by adding heated asphaltic cement (in liquid form) to water mixed with an emulsifying agent, such as soap or bentonite clay. A protein stabilization agent is also added to prevent particles from blending within the mix. The suspended particles of asphalt blend with the aggregate or soil particles as the water drains away and evaporates.

Emulsified asphalts are either anionic or cationic, depending on the emulsifying agent used. Anionic emulsions are those in which the asphalt globules in the mix are negatively charged. Cationic emulsions have positively charged asphalt globules. Each comes in three grades (Table 27.6) that are used for the same purposes as cutback asphalts.

Emulsified asphalts have the advantage of being applied cold, eliminating the need to heat the asphalt to bring it to a fluid condition. They may be applied to damp aggregates, as in the base for a road, because water is already a part of the mix. Water is not volatile, unlike the solvents in cutback asphalt, so danger from fire and the production of hazardous fumes is not present. Emulsified asphalts work best when weather conditions ensure that water evaporation will not occur, such as in a rainy season.

Table 27.6 Grades of Emulsified Asphalts

Anionic
RS asphalt—rapid setting
MS asphalt—medium setting
SS asphalt—slow setting
Cationic
CRS asphalt—rapid setting
CMS asphalt—medium setting
CSS asphalt—slow setting

Plastic asphalt cements are made using asbestos fibers and asphalts with good plasticity and elastic properties. This black cement is used to bond flashing and to make roof repairs. It does not flow at high temperatures or become brittle at cold temperatures, allowing for expansion and contraction of materials without breaking.

Quick-setting asphalt cement is much like plastic asphalt cement but has greater adhesive properties and sets up rapidly. It is used to cement the free tabs on shingles and the laps between layers of roll roofing.

Asphalt roofing tape is a porous fabric strip saturated with asphalt. It is available in rolls 4 to 36 in. (100 to 914 mm) wide and 50 yd. (45.5 m) long. It is used with plastic asphalt cements to patch holes in roofs, to patch seams, and to seal flashing.

Asphalt Paving

Asphalt is widely used for paving because it is a cementitious substance and will bind other ingredients, such as aggregates, together. It withstands exposure to hot and cold weather, enjoys resistance to salt, acids, and alkalis, and experiences little cracking due to its flexible, plastic, and ductile nature.

A bituminous binder is used to hold and protect the mineral aggregate used on paving surfaces. Asphalt cements are used as binders for quality paving. Cutback asphalt is used to stabilize a roadbed before paving and as a binder on the surfaces of light duty roads. Emulsion asphalts may be applied cold to damp aggregates. Asphalt paving mix designs are tested by a variety of methods. The purpose of these tests is to produce a blend of asphalt and aggregates of various sizes that will withstand traffic and still be easy to place.

Laboratory Tests of Asphalt

Since asphalts are used where temperatures may vary considerably, standard tests ascertain their flow properties. Various testing procedures have been developed by the Asphalt Institute, the American Association of State Highway and Transportation Officials, and the American Society for Testing and Materials.

Viscosity Test Viscosity tests determine the flow properties of asphalts at application and service temperatures. One testing method uses a capillary tube viscometer. The viscometer is mounted in a constant temperature bath, and preheated asphalt is poured into the large side of the viscometer until it reaches a fill line. The viscometer is left in the bath for a prescribed length of time. A partial vacuum is pulled on the small tube, and the time it takes the asphalt to move between two timing marks is recorded. This is an indication of flow or viscosity.

Flash Point Test A flash point test determines the temperature to which asphalt may be heated without the danger of an instantaneous flash occurring when exposed to an open flame. The flash point temperature is well below the temperature at which the asphalt will burn. The burn temperature is called the "fire point."

The test is made by placing a sample in a brass cup and heating it at a prescribed rate. A small flame is played over the surface periodically. The temperature at which the vapor produces a flash is the flash point temperature (Fig. 27.3).

Figure 27.3 This apparatus is used to find the flash point of asphalt samples. *(Courtesy The Asphalt Institute)*

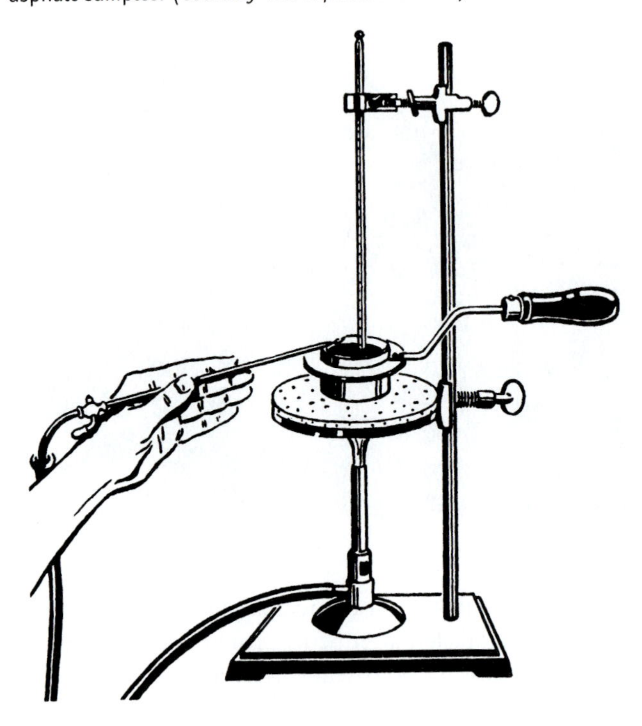

Thin Film Oven Test This test subjects a sample of asphalt to hardening conditions similar to those that occur in a hot-mix plant operation. A 50 ml asphalt sample is placed in a cylindrical flat bottom pan 5.5 in. (140 mm) in diameter and $\frac{3}{8}$ in. (9.5 mm) deep. An asphalt layer about $\frac{1}{8}$ in. (3 mm) deep is placed in the pan. This is heated at 325°F (163°C) in an oven with a revolving shelf for five hours. The sample is then subjected to a penetration test.

Ductility Test The ductility test is used to determine how much an asphalt sample will stretch at various temperatures below its softening point. A standard briquette is molded, brought to the specified test temperature, and pulled at a specified rate until the sample completely separates. The elongation in centimeters, at which the final thread of asphalt breaks, is used to indicate ductility (Fig. 27.4).

Figure 27.4 The ductility test is made to find out how much an asphalt sample will stretch at various temperatures. *(Courtesy The Asphalt Institute)*

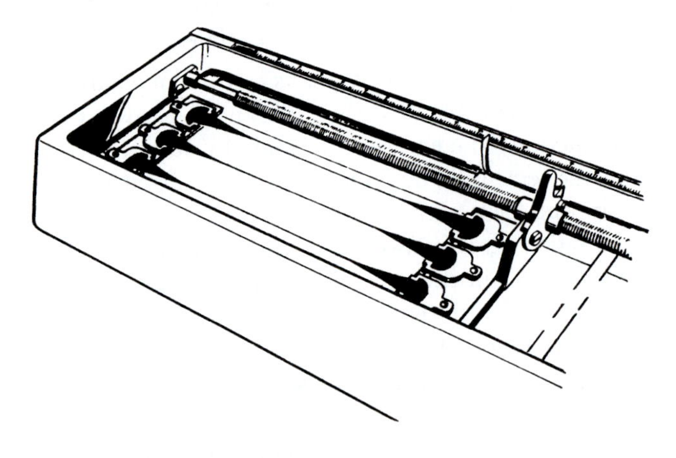

Solubility Test This test is used to ascertain the purity of asphalt cement. A predetermined amount of asphalt cement is dissolved in trichloroethylene. The soluble portion represents the active cementing constituents. Inert matter is filtered out and measured. The result is given as a percent of the soluble content.

Specific Gravity Test The specific gravity of a material is the ratio of the weight of a given volume of the material to the weight of an equal volume of water. Water has a specific gravity of 1. The specific gravity of a material will vary with the temperature because the volume of a material changes with temperature fluctuations. Asphalts with a specific gravity of 1.1 are 1.1 times as heavy as water.

Softening Point Test Various grades of asphalt soften at different temperatures. The softening point is found by the ring and ball test. The heated asphalt is poured into a brass ring. The sample is suspended in a water bath, and a steel ball of specified weight and diameter is placed in the center of the sample. The bath is heated at a controlled rate until the ball reaches the bottom of the glass bath container. The temperature of the water at this point is the softening point.

Distillation Test The distillation test is used to find the proportions of asphalt and diluent present in a sample. It is also used to find how much diluent distills off the sample at various temperatures. This information is useful to ascertain an asphalt's evaporation characteristics, which influence the rate at which road construction asphalt will cure after it is applied.

OTHER PRODUCTS MADE WITH BITUMINOUS MATERIALS

In addition to asphalt, a wide range of products are manufactured using bituminous materials as their major components. Most common among these are coal tar pitch, felts, stabilizers, and surfacing materials.

Coal Tar Pitch

Coal tar pitch is a dark brown to black hydrocarbon obtained through the distillation of coke-oven tar. It is available in several grades and is used as the basis for a number of paints, roofing products, and waterproofing materials. It has a softening point near 150°F (65°C).

Coal tar enamel is made from coal tar pitch with added mineral fillers. It is used to protect pipe in pipeline work. Cold-applied coal tar products have a solvent added to liquefy them. Hot-applied coal tar coatings give better protection than cold coal tar coatings.

Felts

Felt is a sheet material made from the cellulose fibers of organic materials such as wood, paper, rags, glass fibers, and asbestos.

Saturated felts, sometimes called tar paper, are made with an organic mat saturated with coal tar pitch or asphalt and coated with a layer of thin asphalt. Tar paper is used as an underlayment for shingles, as sheathing paper, and as laminations in built-up roof construction. Tar paper is also used to produce roll roofing and shingles.

Saturated felt is available in three weights: Types 15, 20, and 30, according to the weight per square of felt. For example, Type 15 weighs 15 pounds per square (100 sq. ft.).

Saturated felt is available in rolls 36 in. (914 mm) wide and up to 144 ft. (44 m) long.

Ice and Water Shield

Ice and water shield is a roofing membrane composed of two waterproofing materials bonded into one layer. Comprised of a rubberized asphalt adhesive backed by a layer of polyethylene, it comes in 36 in. x 75 foot rolls. The rubberized asphalt surface is backed by a release paper to protect the sticky side. The material is used in trouble spots on roofs, such as along eaves, in valleys, and in other areas where leaks are more likely.

Fiberglass Sheet Material

Fiberglass mats can be impregnated with asphalt but are not "saturated" because the glass fibers will not absorb the asphalt. The asphalt forms a coating on the surface and fills the spaces between the fibers. Fiberglass mats are available in rolls 36 in. (914 mm) wide and 108 ft. (11 m) long.

Fireproofing Paper

Fireproofing paper is made using asbestos fibers either in a pressed mat-like felt or in woven sheets. The various sheet products are used as underlayment for finished roofing materials, as vapor barriers in walls and floors, and for other similar applications. They should not be exposed to the weather because coal tar pitch oxidizes rapidly when subjected to the sun's ultraviolet rays.

Waterproof Coatings

Asphalt waterproofing is used on masonry walls above and below grade. Below grade it must resist the pressure of subsurface water and prevent it from passing through the foundation. Since it is not subjected to high temperatures below grade, asphalts with lower softening points can be used. Above grade it resists passage of water through walls or roof decking. Where it will be exposed to sunlight, asphalts with a higher softening point should be utilized.

A bituminous waterproofing coating is applied in one or more coats mopped on either hot or cold. Cold-applied coats can be reinforced by the addition of glass, plastic, or asbestos fibers. Cutbacks and emulsions are used extensively for this purpose. They may be covered with a plastic or felt membrane.

A dampproofing board product is made with an asphalt core covered on both sides by layers of asphalt-impregnated paper or felt treated with a weather-resisting coating. Additional information on waterproof coatings can be found in Chapter 26.

ROOF COVERINGS

Building codes classify roof covering requirements according to building type, size, occupancy, and location. The following topics discuss the various types of bituminous roof covering materials in common use. It should be noted that coal tar pitch and asphalt are not compatible and should not be used where they will come in contact with each other (Fig. 27.5).

Fire Ratings

The fire rating classification of roofing products is an important factor in choosing asphalt roof coverings. The following fire ratings of roofing materials are based on specifications of the Underwriters Laboratory, Inc. (UL). UL labels indicating fire ratings appear on packaged asphalt roofing products (Fig. 27.6).

Class A: Highest rating. The covering is effective against severe fire exposure, is not readily flammable, and offers a high degree of fire protection for the roof deck.

Class B: Moderate protection against fire exposure. The roof covering is not readily flammable and offers a moderate degree of fire protection for the roof deck.

Figure 27.5 A variety of roof covering materials use asphalt. *(Used by permission of Georgia-Pacific Corporation. All rights reserved.)*

Class C: Minimal protection. The roofing provides protection against light fire exposure, is not readily flammable, and offers a measurable degree of fire protection for the roof deck.

Roll Roofing

Roll roofing uses either organic felt or fiberglass mats as a base material. A viscous bituminous coating is applied to this base, forming the exposed surface. Roll roofing is made in four types: smooth-surfaced, mineral-surfaced, mineral-surfaced selvage-edged, and pattern-edged (Fig. 27.7). Smooth-surfaced roll roofing has both sides covered with a fine talc or mica to keep the surfaces from sticking as it is made into rolls.

Mineral-surfaced roll roofing has mineral granules in a wide range of colors rolled into the surface, producing a surface that is attractive and protects the bitumen from the sun's ultraviolet rays. The minerals also increase the fire resistance of the product. Mineral-surfaced roll roofing is available in rolls 36 in. (915 mm) wide and from 36 to 72 ft. (8 to 22 m) long. Smooth-surfaced rolls come in weights of 50 to 65 lb. per square, mineral-surfaced in rolls of 90 lb. per square, and fiberglass-reinforced mineral fiber in rolls of 75 lb. per square.

Mineral-surfaced selvage-edge roofing is of the same construction as mineral-surfaced roll roofing except that only 17 in. of the 36 in. wide surface is covered with granules. The 19 in. selvage edge is used for lapping with an adjoining layer, forming a two-ply covering.

Pattern-edged roll roofing is a mineral-surfaced product that has a 4 in. uncoated band in its center. The roll is semi-cut along this strip to form two 18 in. wide patterned roofing strips that are thin lapped 2 in. over the layer below. Roll roofing can be installed in a single or double thickness. The double thickness provides increased protection over a longer period of time (Fig. 27.8).

Figure 27.6 Manufacturers of shingles whose products meet the safety standards of Underwriters Laboratories, Inc. (UL) may be authorized by UL to use these labels, which include the UL Mark on those products. *(Reproduced with permission of Underwriters Laboratories, Inc.)*

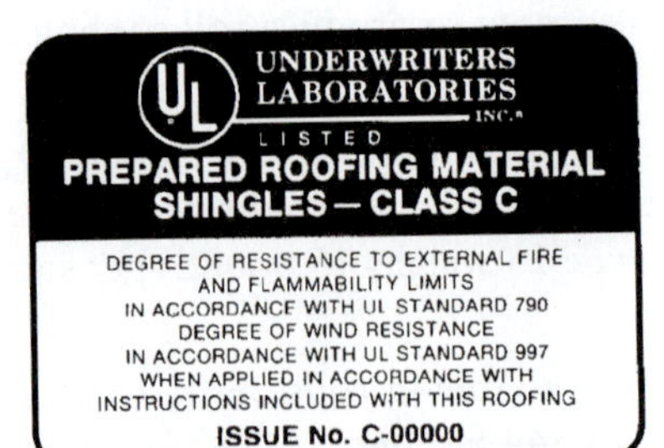

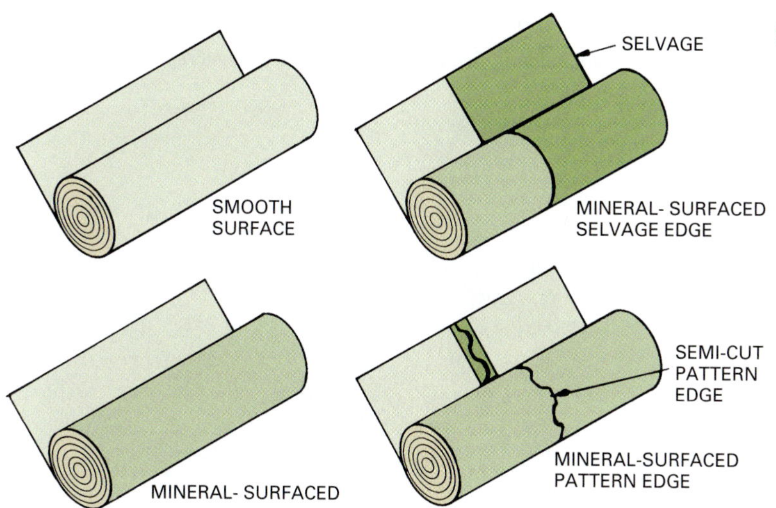

Figure 27.7 Roll roofing is available in four types.

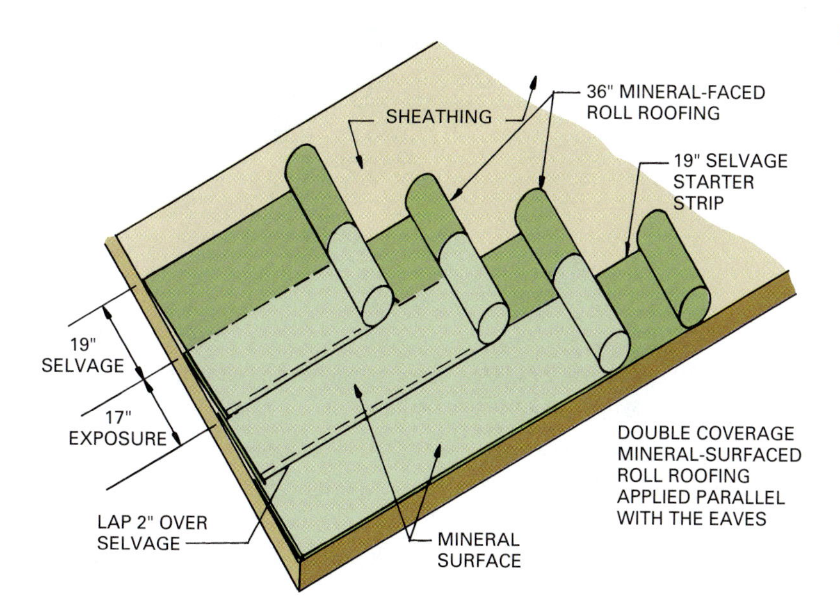

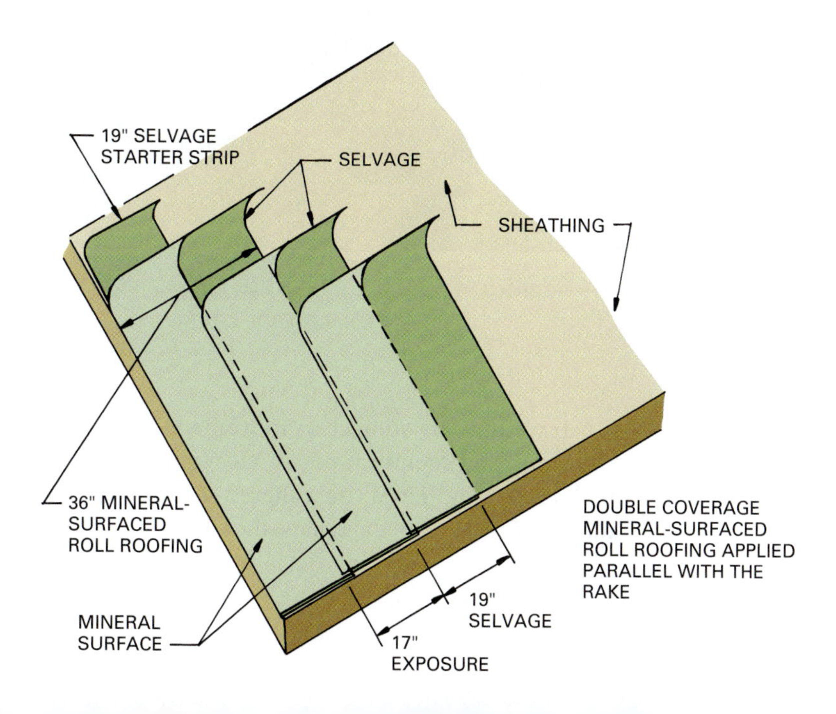

Figure 27.8 Two ways to apply double coverage mineral surfaced roll roofing.

Conventional Hot Built-Up Roof Membranes

A built-up roof consists of alternate plies of organic or fiberglass roofing felt with a hot bitumen coating mopped over each layer (Fig. 27.9). The design of the roof varies by situation but generally consists of three or more layers of felt with a bitumen layer over each and a bitumen topcoat with aggregate rolled on top. A conventional hot built-up roof system is shown in Fig. 27.10.

Felts provide the needed reinforcement to keep bitumens in each layer from alligatoring. Alligatoring refers to surface cracking caused by oxidation and shrinkage stresses, which can result in a repetitive mounding of the asphalt surface similar to an alligator's hide.

The aggregate surface forming the top of the roof is typically derived from gravel, marble chips, or slag. Gravel and marble chips are usually applied at 400 pounds per 100 sq. ft. and slag at 300 pounds per 100 sq. ft. Conventional built-up roofs can be designed for slopes up to 3 in. (76 mm) per 12 in. (305 mm). Slopes above $\frac{1}{2}$ in. (12 mm) per 12 in. (305 mm) usually require the use of steep asphalt.

Figure 27.9 Tar can be applied by hot mopping. *(Used by permission of Georgia-Pacific Corporation. All rights reserved)*

Modified Asphalt Roofing Systems

Modified asphalt roll roofing is composed of polymer-modified bitumen reinforced with one or more plies of fabric, such as polyester glass fiber (Fig. 27.11). These membranes are of uniform thickness and have consistent physical properties throughout the membrane area.

A variety of modifiers and types of reinforcing plies are designed for use on almost every type of construction assembly, including new roofing, re-roofing, domes, and spires. Modified membranes are also used below grade for waterproofing canals, water reservoirs, and landfills.

Most modified bitumen membranes are made using either styrene-butadiene-styrene (SBS) or atactic poly-propylene (APP). APP-modified membranes are generally applied using a propane torch to heat and soften the underside of the membrane. This surface becomes a molten adhesive that is placed on the substrate, rolled for adhesion, and bonds when it cools.

SBS modifies the bitumen by forming a polymer lattice within the bitumen. When this polymer lattice cools, the membrane acts somewhat like rubber. SBS membranes are more flexible than APP membranes and are used where flexibility is needed, such as when the substrate may be subject to movement or deflection. SBS-modified membranes are either mopped in hot asphalt, self-adhered, or adhered with cold-process adhesives. Some types can have their joints heat-welded.

Figure 27.10 A conventional hot asphalt built-up roof system on a nailable deck.

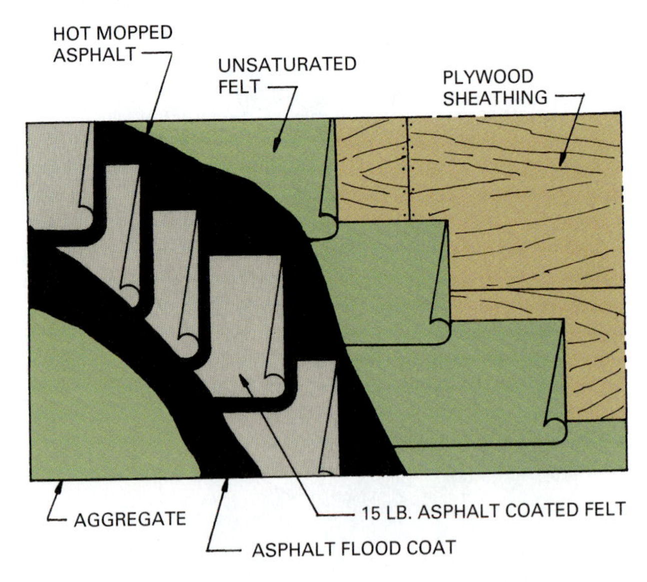

Figure 27.11 A form of a modified asphalt roll roofing installation. *(Courtesy Asphalt Roofing Manufacturers Association)*

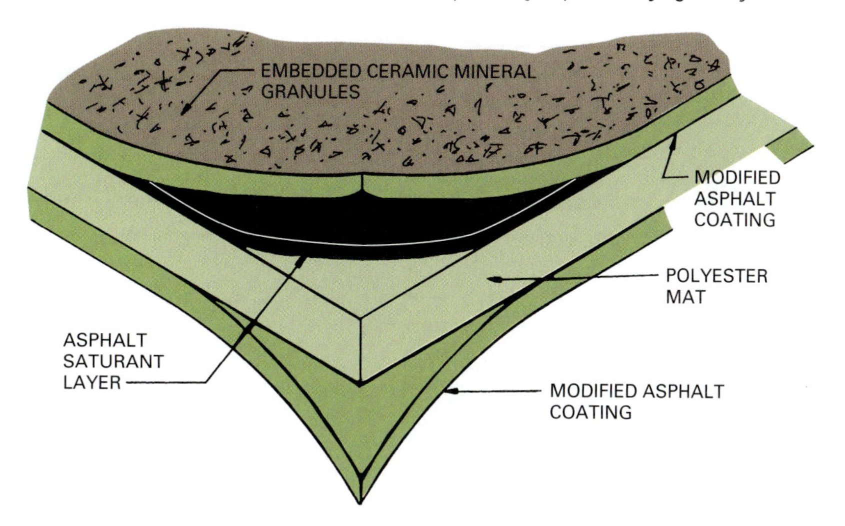

Some have factory-applied mineral aggregate surfaces to protect them from ultraviolet damage. Those without this covering need some form of ultraviolet protective coating.

Cold-Applied Asphalt Roofing Systems

Cold-applied systems use some form of coated base sheet, fabric, or similar reinforcement over which the principal waterproofing agent, which is a liquid, is applied at ambient temperatures. The selection of the cold-process application depends on the levels of maintenance, repair, and service expected and on the compatibility between the cold-applied materials and any existing substrate.

The reinforcing felts and fabrics include organic-coated base sheets, organic mineral-surfaced cap sheets, organic felt sheets, fiberglass ply sheets, fiberglass base sheets, fiberglass mineral-surfaced cap sheets, single-ply smooth surfaced sheets, single-ply mineral-surfaced sheets, fiberglass fabrics, polyester fabrics, cotton fabric, and jute burlap fabric. The base sheet, cap sheet, and single-ply roofing are bonded with cold-applied adhesives and cold-applied surface coatings to form a roof membrane.

The coatings and adhesives are designed to be brushed or sprayed at normal room temperatures. These include filled and non-filled asphalt cutbacks, asphalt emulsions, coal tar coatings, and aluminum pigmented asphalt. Toppings include gravel and aluminum chips placed in the topcoat while it is still wet. They block ultraviolet light, can reflect heat, and are decorative.

Asphalt and Fiberglass Shingles

Asphalt shingles are made using an organic felt base saturated with asphalt and coated with a mineral-stabilized coating of asphalt on both sides. The exposed top side is coated with mineral granules. They are available in a wide range of colors. The bottom is relatively smooth and coated with talc to prevent it from sticking to the shingle below when bundled for shipping (Fig. 27.12).

Fiberglass shingles are made using a fiberglass mat as the base. The mat does not have to be saturated with asphalt before it receives the asphalt top coating. Mineral aggregates are rolled into the top coating. These shingles are lighter in weight than asphalt shingles.

Figure 27.12 An example of asphalt/fiberglass shingles. *(Image copyright Brad Remy, 2009. Used under license from Shutterstock.com)*

Figure 27.13 Commonly available forms of asphalt and fiberglass shingles.

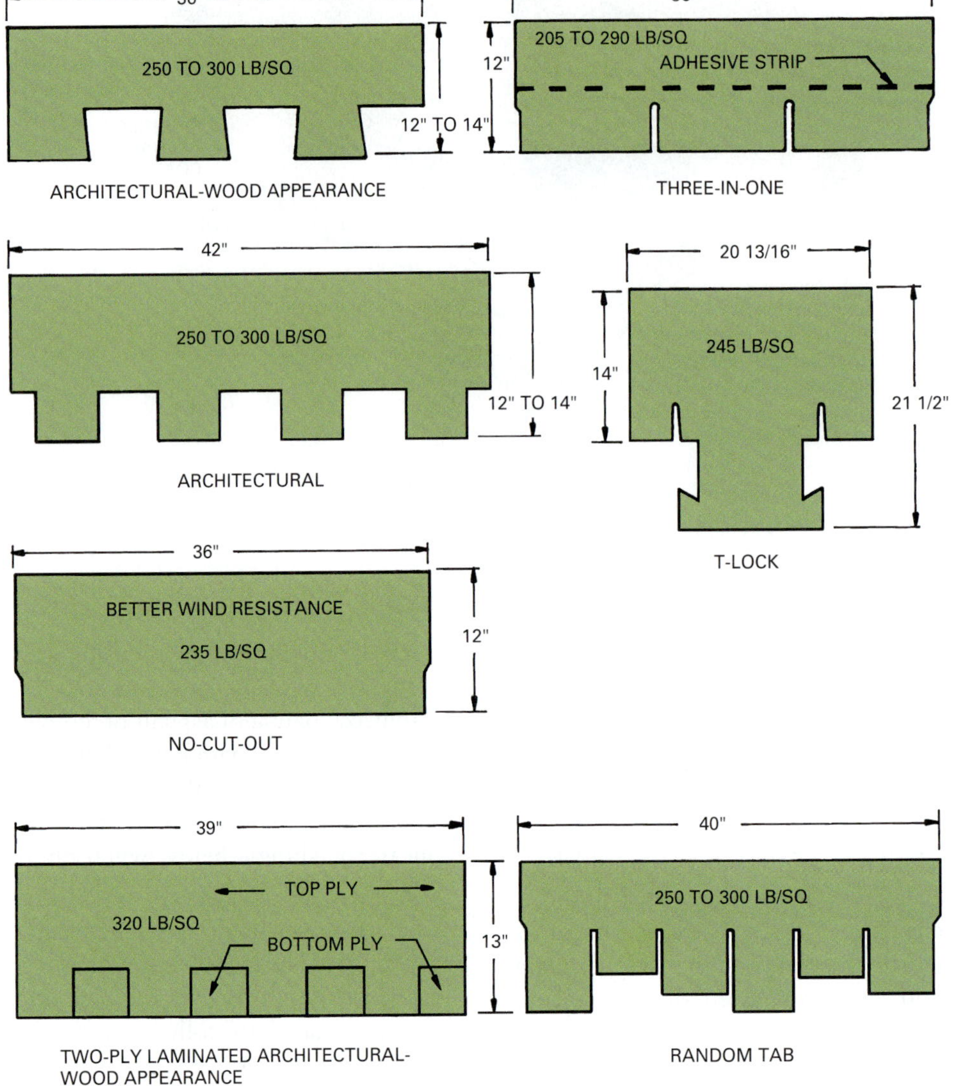

Shingles are made in a variety of styles. Most common are strips 12 x 36 in. (305 x 915 mm) in size with the exposed surface cut to resemble three smaller shingles. They are available in weights of 100 to 300 lb. per square (45 to 135 kg per square). Commonly available forms of asphalt and fiberglass shingles are shown in Fig. 27.13.

Review Questions

1. What will prevent bitumen from adhering to a concrete wall?
2. What is the point at which bitumen will burn?
3. What property of bitumen allows it to spread after it is in place?
4. What binder is used with aggregate to produce quality asphalt pavements?
5. From what source are most asphalts obtained?
6. How do cutback asphalts differ from asphalt emulsions?
7. What are the two major uses for cutback asphalts?
8. What can happen that will cause asphalt emulsions on a roadbed not to harden?
9. What asphalt product is used to make roof repairs?

10. What are the big differences between saturated felts and fiberglass felts?
11. Why are fiberglass felt mats not saturated with asphalt?
12. What types of fiber are used in making fireproofing paper?
13. What are the fire rating classes used to rate asphalt roofing materials?
14. Which fire rating class indicates the highest fire protection?
15. How does a built-up roof differ from one covered with roll roofing?
16. What are the various materials used to make fiberglass shingles?

Key Terms

asphalt
bitumen
coal tar pitch
ductility
felt
viscosity

Activities

1. Collect samples of as many bituminous products as possible.
2. Visit construction sites when the finished roofing is being applied and observe the installation techniques and safety protection used.
3. Visit a road paving project and observe the procedures for preparing, delivering, and laying the asphalt final coat. Try to find out what was done to prepare the base. If possible, find the local or state requirements for street and parking area paving.

Additional Resources

CertainTeed Shingle Applications Manual, CertainTeed Corporation, Roofing Products Group, PO Box 860, Valley Forge, PA 19482.

The NRCA Roofing and Waterproofing Manual, National Roofing Contractors Association, Rosemont, IL 60018.

Residential Asphalt Roofing Manual, Asphalt Roofing Manufacturers Association, Washington, DC 20005.

Spence, W. P., *Roofing Materials and Installation*, Sterling Publishing Co., 387 Park Avenue South, New York, NY 10016-8810.

Other resources include:

CD-ROMs: *Installation Tips & Techniques and Avoiding Common Roof Installation Mistakes*, GAF Materials Corporation, 1361 Alps Road, Wayne, NJ 07470.

See Appendix B for addresses of professional and trade organizations and other sources of technical information.

CHAPTER 28

Roofing Systems

LEARNING OBJECTIVES

Upon completion of this chapter, the student should be able to:

- Understand the vast array of roofing systems available.
- Select the proper roofing system for various applications.
- Understand the proper installation of roofing systems.

Build Your Knowledge

For further study on these materials and methods, please refer to:

Chapter 11 Clay Brick and Tile
Chapter 16 Nonferrous Metals
Chapter 19 Products Manufactured from Wood
Topic: Wood Shingles and Shakes
Chapter 25 Thermal Insulation and Vapor Barriers
Chapter 27 Bituminous Materials
Topic: Roof Coverings

A roof system consists of a number of interacting materials that when properly combined provide a weather-resistant cover to a building. Manufacturers of a variety of deck, insulation, and finished roofing materials provide technical information and installation instructions. Roofing systems are designed to serve a variety of purposes, including fire resistance, water penetration, and wind and snow load resistance. In addition, they can provide acoustical and thermal insulation and contribute to energy conservation.

Roof systems are generally divided into two categories: low-slope and steep-slope. **Low-slope roofs** are flat (or nearly flat) and provide slow rain runoff. **Steep-slope roofs** permit the rapid runoff of rain, reducing the likelihood of penetration through their water-resistant surfaces.

Low-slope roofs are generally more economical to build and can be extended over large buildings, while the limited spans possible with structural materials can restrict steep-sloped roof applications. Water tends to pond on flat roofs even when drains are provided. Finished membranes may crack or pull loose from parapets and drains as temperatures change or if there is deck movement. Expansion joints are used to diminish this problem. If water vapor enters below a finished membrane, solar heat can cause blistering in some roofing materials.

Steep-slope roofs can be finished with metal, asphalt or fiberglass shingles, clay tiles, or metal roofing. They may have an attic space below the roof surface that, when properly ventilated, reduces possible damage from vapor or high temperatures under roofing materials. The finished roofing material is highly visible on a steep-slope roof and forms an important part of the overall design of a building.

Roofing materials are tabulated by the square. A square is the amount of roofing required to cover 100 square feet of roof area.

BUILDING CODES

Building codes govern the materials, design, construction, and quality of a roof structure and covering. This includes the ability to resist rain and wind, durability specifications, compatibility of materials, physical characteristics, and fire protection. Fire protection classifications are specified by ASTM E108.

There are four classifications of roof covering. Class A roof coverings are effective against severe fire test exposure. They include clay tile, mineral fiber, concrete, slate, some fiberglass asphalt shingles, and other materials that have been so certified by an approved testing agency. Class A coverings may be used on buildings of all types.

Class B roof coverings are effective against moderate fire test exposure. They include metal sheets and shingles, some composition shingles, and other materials certified by an approved testing agency.

Class C roof coverings are effective against light fire test exposure. They include materials certified as Type C by an approved testing agency.

Non-classified roof coverings are not permitted on any buildings covered by most codes. In some cases these may be approved on some types of storage buildings.

ROOFING SYSTEM MATERIALS

Roofing systems are composed of decks, vapor retarders, insulation, and finished roofing materials. Materials for most of these components have been discussed in earlier chapters and are referenced below.

Roof Decks

Materials commonly used for roof decks are detailed in other chapters related to materials and types of construction. A properly functioning roof depends on a structurally sound deck that is compatible with the roofing system. Following are the frequently used decking materials.

Cement-wood-fiber panels are made by bonding treated wood fibers with Portland cement or some other binder and compressing them into structural panels. Concrete decks are made with either cast-in-place forms or precast concrete structural members. (Review Chapters 8 and 9.)

Gypsum concrete decks are produced by mixing gypsum, wood fibers, or mineral aggregates with water and casting it on form boards. Precast gypsum planks are also used.

Lightweight insulating concrete is made by mixing insulating aggregates, such as perlite, Portland cement, and water, and casting it on top of metal or bulb-tee and formboard decking systems.

Manufactured wood panels are produced by bonding wood veneers or wood chips. Plywood and oriented strand board are the two commonly used materials. Wood planks consist of solid wood decking or glued laminated members made by bonding solid dimensional lumber into a structural decking member. Steel decks are produced by cold-rolling sheet steel into various structural shapes.

Vapor Retarders

The flow of water vapor from inside a building into its roofing system must be carefully controlled. Water vapor can cause decking to deteriorate and potentially damage some types of finished roofing surfaces. Moisture can also damage insulation in an exposed roofing assembly. Changes in temperature and humidity create a difference in water vapor pressure between the inside and outside of a building, causing water vapor to migrate into insulation (Fig. 28.1). The vapor retarder must resist the passage of water vapor. The protected roof membrane also serves as a vapor retarder. Vapor retarders should be applied to the warm side of the insulation in roof decks in cold climates. In warm, humid climates, winter condensation in the roof assembly is not a problem, and vapor retarders are often not used. Air conditioning in warm climates may actually cause a reverse vapor migration. Vapor retarders must be used on top of concrete and poured gypsum decks, and any tears or holes in the material must be sealed.

The performance of a vapor retarder is indicated by its **water-vapor-transmission rate (WVTR)**, which is specified in perms. Perm is a measure of the porosity of a material to the passage of water vapor. A perm rating indicates the number of grains of water vapor that will pass through one square foot of a material per hour when the vapor-pressure differential between the two sides equals 1 in. of mercury (0.49 psi). The metric equivalent is expressed in terms of grams/m^2/hr./mm of mercury pressure difference between the two faces. Materials used as vapor retarders must have a perm rating of 0.00 to 0.50 perms.

Some frequently used vapor retarders are:

1. Bituminous materials, such as layers of asphalt roofing felt covered with hot asphalt. Kraft paper layers bonded with asphaltic adhesive and glass-fiber

Figure 28.1 Vapor retarders are placed on the warm side of the roof assembly below the insulation.

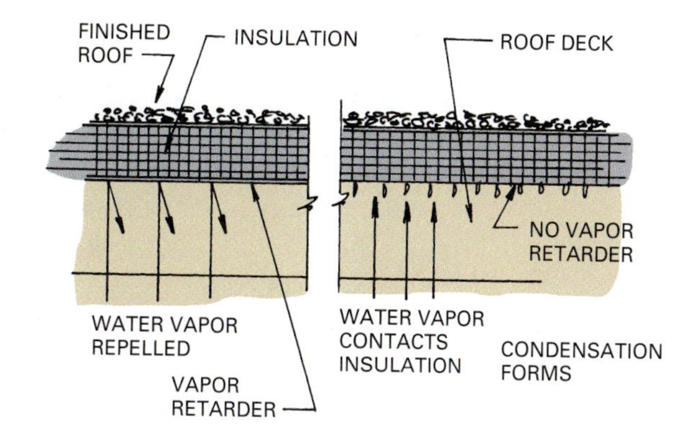

reinforcement. They are bonded to decking with cold-applied asphalt adhesive.

2. Polyethylene sheet material laid loose on a deck or attached with mechanical fasteners. The sheets are over lapped and taped or joined with an adhesive.
3. Aluminum foil bonded to the face of insulation batts or rigid insulation. It provides a vapor barrier and serves simultaneously as a reflective insulation. The joints between insulation panels should be covered with aluminum tape.

Other vapor retarders contain a combination of materials. Examples of products available include combinations such as a polypropylene scrim, kraft paper sheet, vinyl scrim (an open-weave fabric), aluminum foil, or polypropylene scrim and aluminum foil.

Insulation

The location of roof insulation varies with the design of the roof. In a typical steep slope roof with an attic, the ceiling is insulated and no roof insulation is required. A low-slope roof must have some form of insulation and ventilation. These are illustrated in Chapter 22. All of the types of insulation can find some use in the design of the roof system. Insulation materials are detailed in Chapters 22 and 25.

Finished Roofing

A wide range of finished roofing materials are available, providing the designer with considerable choices. Table 28.1 describes many of these materials. The specific properties of products provided by various manufacturers must be known and the manufacturer's installation recommendations must be followed.

Table 28.1 Materials Used for Finished Roofing

Material	Type of Roof	Descriptive Factors	Weight per Square (lb./100 ft.^{2})	Weight per Square Meter (kg/m^2)
Aluminum (sheet, shingles)	Steep-slope	Fire resistant, long life, range of colors	5–90	2.44–4.39
Asphalt (built-up)	Steep-slope, low-slope	Granular topping applied influences fire class, life 20–30 years	100–600	4.88–29.3
Asphalt shingles (fiberglass, asphalt)	Steep-slope	Fire resistance varies with product, range of colors, life 20–30 years	235–325	11.47–15.86
Cement-fiber tile	Steep-slope	Fire resistance, long life, heavy, use in warm climates	950	46.4
Clay tile	Steep-slope	Fire resistant, long life, heavy	800–1,600	39–78
Copper (sheet)	Steep-slope, low-slope	Fire resistant, long life, can be soldered	0.019" thick 160 0.040" thick 320	7.8 15.6
Lead, copper coated	Steep-slope, low-slope	Fire resistant, long life	1/32" thick 200 1/16" thick 400	9.76 19.52
Monel (Ni-Cu)	Steep-slope	Fire resistant, long life	22 gauge 1,424 26 gauge 827	69.5 40.4
Perlite-portland cement	Steep-slope	High fire rating, lightweight, long life	900–1,000	43.9–48.8
Plastic (single-ply membrane)	Low-slope	Long life, requires careful installation, limited fire classification, several types available	Loose laid Ballasted, 1,000–1,200 Fully adhered, 30–55	48.8–58.6 1.5–2.7
Plastic (liquid applied)	Low-slope	Limited fire classification follow manufacturer's directions	20–50	0.98–2.4
Slate	Steep-slope	Fire resistant, heavy, long life	3/8" thick 800 1/4" thick 900 3/8" thick 1,100 1/2" thick 1,700 3/4" thick 2,600	39.0 43.9 53.7 83.0 126.9

Table 28.1 Materials Used for Finished Roofing *(Continued)*

Material	Type of Roof	Descriptive Factors	Weight per Square (lb./100 ft.^{2})	Weight per Square Meter (kg/m^2)
Stainless steel, terne coated	Steep-slope	Fire resistant, long life	90	3.89
Steel (sheet, shingles)	Steep-slope, low-slope	Fire resistant, long life, durable colors	Copper coated 130 Galvanized 130	6.3 6.3
Wood (shingles, shakes)	Steep-slope	No fire resistance unless treated, limited life	200–450	9.8–22.0
Zinc	Steep-slope	Fire resistant, long life, can be painted	9 gauge 670 12 gauge 1050	32.7 51.2
Terneplate copper-bearing sheet	Steep-slope	Fire resistant, long life	30 gauge 540 26 gauge 780	26.4 38.1
Mineral-surfaced cap sheet	Low-slope	Limited fire resistance, limited life	55–60	2.68–2.9
Modified bitumen	Low-slope	Fire resistance, 10 years or more	100	4.9

LOW-SLOPE ROOF ASSEMBLIES

In contrast to the sloped form of roof, a low-slope roof is horizontal (or nearly so). Materials that cover flat roofs should allow water to run off freely on a very slight inclination to drainage outlets. The actual slope permitted varies with the type of roofing system used and the manufacturer's recommendations. Typical slopes fall in a range of $\frac{1}{4}$ in. rise per 12 in. run (6.4 mm per 305 mm) to 2 in. rise per 12 in. run (50.8 mm per 305 mm). Special provisions can be made for roofs with a 3 in. per 12 in. (76.2 mm per 305 mm) slope.

Low-slope roofing systems include built-up, thermoset single-ply coverings, thermoplastic single-ply coverings, modified bitumen coverings, spray-applied polyurethane foam coverings, liquid-applied coatings, and metal sheet coverings. An important feature in the design of low-slope roofs is the provision of adequate insulation.

Insulation for Low-Slope Roofs

Roof insulation for low-slope roofs is usually applied to the top side of the decking. Insulating a roof this way increases the possibility of condensation occurring within the roof system, so a vapor retarder is required. Since the insulation increases the roof temperature on hot days, the roof materials age faster and the expansion and contraction stresses on the membrane are greater. The insulation must meet building code and fire-resistance requirements.

Rigid roof insulation includes cellular glass, composite boards, glass fiber, perlitic boards, polyisocyanurate foam boards, polystyrene boards, polyurethane foam boards, and wood fiber boards.

Roof insulation may be placed as a single or double layer. The double layer is placed with offset joints, eliminating any leakage of heating or cooling energy that may occur between joints in the first layer. The first layer is mechanically joined to the deck when steel decking is used. The second layer is then bonded with hot bitumen or an adhesive (Fig. 28.2). The first layer on concrete decks is bonded directly to the deck with hot bitumen (Fig. 28.3).

Figure 28.2 The first layer of insulation is fastened to the deck with mechanical fasteners. The second layer is bonded with hot asphalt. *(Courtesy Schuller International, Inc.)*

Figure 28.3 Rigid insulation is placed over hot asphalt on a concrete deck. *(Courtesy National Roofing Contractors Association)*

Built-Up Roofing

The traditional **built-up roofing** (BUR) system used on low-slope roofs consists of bitumen (asphalt or coal tar)—usually applied hot over felts, which may be glass fiber, organic, or polyester—and a finished top surface, such as an aggregate (gravel or slag), or a cap sheet. Different manufacturers offer a variety of compositions of felts and roofing asphalts with varying recommendations of installation methods. The following examples are typical of those in use.

Typical built-up roof construction on uninsulated, nailable roof decks is shown in Fig. 28.4. The deck is covered with one ply of sheathing paper (a vapor retarder) nailed to the deck. Nailable decks include wood, plywood, structural wood fiber panels, lightweight insulating concrete, and precast and poured gypsum. Next, three to five layers of an asphalt-coated base felt are applied, bonded with coatings of hot-mopped bitumen. The top coating is covered with roofing asphalt and gravel or slag. Usually 400 lb. of gravel or 300 lb. of slag are applied per 100 sq. ft. of surface area. Manufacturers recommend that dead flat roofs be given a slight slope, such as $\frac{1}{2}$ in. per linear foot.

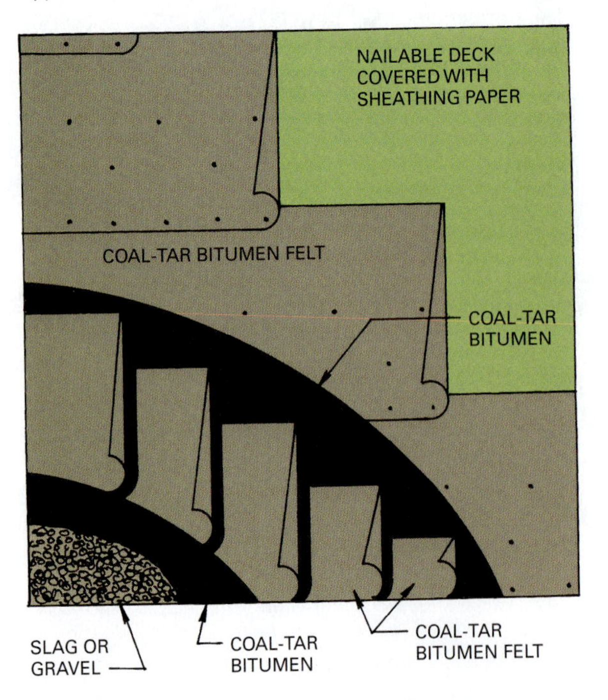

Figure 28.4 One of several ways built-up roof constructions are applied over a nailable roof deck.

Figure 28.5 Overlapping sheets of asphalt-saturated felt are bonded to rigid insulation with hot roofing asphalt. *(Courtesy National Roofing Contractors Association)*

Non-nailable decks, such as steel, precast concrete, and poured concrete, have the insulation bonded with hot bitumen or an approved adhesive. This is followed by layers of asphalt-saturated roofing felt and hot roofing asphalt. The layers of felt are laid in a full bed of hot asphalt and broomed in place (Fig. 28.5). Roofing asphalt is brought to the site in a tank truck and heated in an asphalt kettle (Fig. 28.6). The heated asphalt is pumped to a tank on the roof and moved to needed areas (Fig. 28.7).

Figure 28.6 Roofing asphalt is heated in an asphalt kettle and pumped to the roof. *(Courtesy National Roofing Contractors Association)*

Roof penetrations must be flashed (Fig. 28.8), as must roof drains, parapets, and other places where the roof butts against a wall (Fig. 28.9). Finally, gravel forming the top protective coating is lifted to the roof and spread in a bed of hot roofing asphalt (Fig. 28.10). A typical construction detail is shown in Fig. 28.11.

Figure 28.7 Hot roofing asphalt is pumped up to a tank on the roof. *(Courtesy National Roofing Contractors Association)*

Figure 28.8 Roof penetrations are flashed with hot roofing asphalt and asphalt-saturated organic felt. *(Courtesy National Roofing Contractors Association)*

Figure 28.9 Layers of asphalt-saturated organic felt are mopped with hot roofing asphalt to build up a multilayer membrane along the edge of the roof. *(Courtesy National Roofing Contractors Association)*

Figure 28.10 Stone ballast is spread in a bed of hot roofing asphalt to help hold the membrane in place and protect it. *(Courtesy National Roofing Contractors Association)*

Figure 28.11 A construction detail for an insulated low-slope roof deck with built-up roofing.

Another type of built-up roofing is an assembly using an asphalt glass-fiber roof membrane covered with a mineral-surfaced inorganic cap sheet. A mineral-surfaced roof material consists of a base felt coated on one or both sides with asphalt and surfaced with mineral granules. A typical assembly is shown in Fig. 28.12. An asphalt glass-fiber membrane is nailed to the nailable deck and additional layers are bonded with hot asphalt. The mineral-surfaced inorganic cap sheet is bonded to the asphalt glass-fiber base with hot asphalt. It is recommended that this system be used on roofs with a slope of $\frac{1}{4}$:12 in. (6.25:305 mm) or greater.

Modified Bitumen Membranes

Modified bitumen membranes combine polymer-modified asphalt and a polyester or fiberglass mat, resulting in a product of exceptional strength. Two common membranes available are SBS (styrene-butadiene-styrene) and APP (atactic polypropylene). SBS sheets have a reinforcement mat coated with an elastomeric blend of asphalt and SBS rubber. APP membranes have a reinforcement mat coated with a blend of asphalt and APP plastic. While fiberglass reinforcement is most frequently used, membranes are available from various manufacturers with several reinforcements.

The major difference between SBS and APP products is the blended asphalt used. The blend creates a product that has greater elongation, strength, and flexibility than traditional roofing asphalts. Recovery after elongation, flexibility, and cold weather performance of SBS membranes are superior to those of APP membranes.

SBS products are generally installed using hot asphalt as the bonding material. They are applied as cap sheets over a base of hot asphalt and roofing felts, as shown in Fig. 28.13. The cap sheet (SBS membrane) may have a ceramic granule surfacing to protect it from ultraviolet light, or it may be unsurfaced. The unsurfaced type must be coated with asphalt and gravel to give it ultraviolet protection (Fig. 28.14).

Figure 28.12 One installation detail for using mineral-surfaced cap sheets.

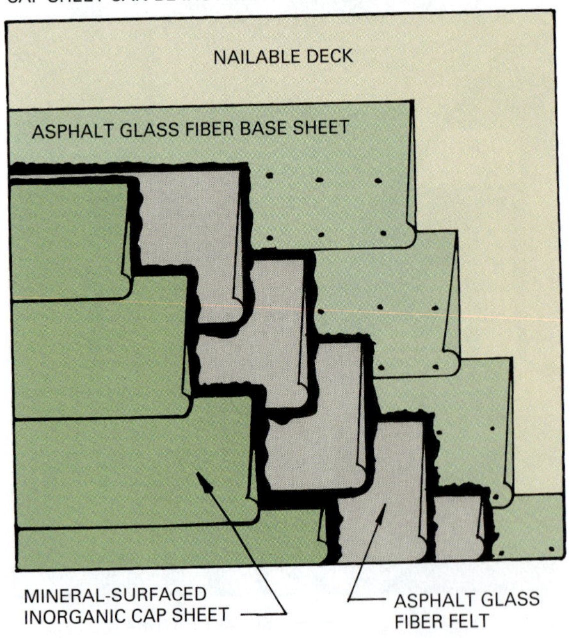

Figure 28.13 These modified SBS bitumen membranes can be laid in hot roofing asphalt. *(Courtesy Schuller International, Inc.)*

Figure 28.14 A typical assembly for a partially adhered single-ply modified bitumen roof system.

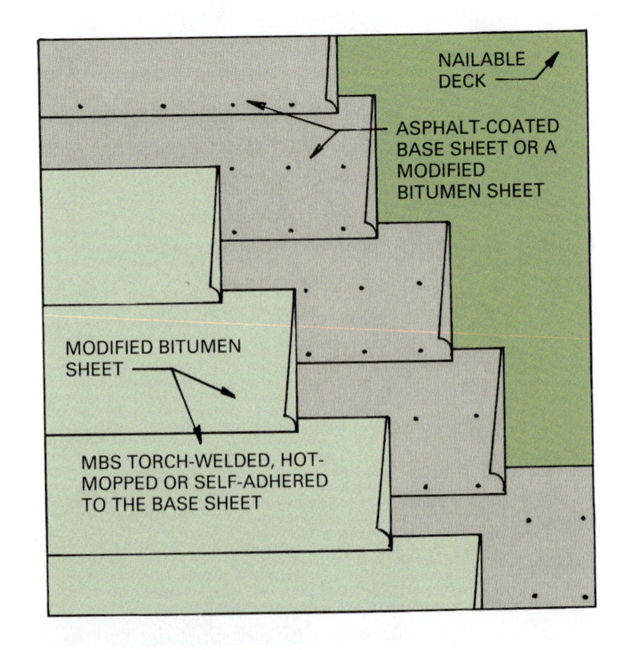

APA products are applied by a method called "torching," made possible by the unique properties of the modified bitumen. The back coating of modified asphalt is heated with a propane torch to the point at which the sheet can be bonded to the substrate. APA products cannot be installed with hot-mopped asphalt.

Single-Ply Roofing Systems

Single-ply roofing systems can be applied over almost any commonly used roof deck, as well as over existing asphalt or built-up roofs when re-roofing a building. Roofing insulation is required over a roof deck except when it's composed of very smooth concrete, plywood, or splinter-free solid wood decking. The roof must drain freely and have sufficient outlets to carry away water.

Single-ply roofing membranes are available in a number of materials, and manufacturer's specifications and installation instructions should be carefully observed. Single-ply membranes are fashioned from either thermoset or thermoplastic materials. Thermoset materials cure during manufacture and can only be bonded to themselves with an adhesive. Thermoplastic materials do not completely cure during manufacturing and can be welded together, usually with high-temperature air.

Single-ply membranes may be secured to a deck with ballast. The membrane is laid loose and covered with ballast to prevent it from being lifted by wind forces. The ballast is usually large, smooth aggregate or concrete pavers (Fig. 28.15). A variation of the ballasted roof membrane is shown in Fig. 28.16. The membrane is covered with insulation, a protective mat, and ballast. In both systems, the abutting edges of the membrane are spliced, and all edges, pipes, and other roof penetrations are flashed, as specified by the manufacturer.

Single-ply membranes can also be mechanically fastened to a deck. This can be accomplished using metal batten bars placed at intervals on top of the membrane and screwed to the deck. The bars are then covered with a batten cover strip (Fig. 28.17). Another method uses screws and metal discs placed over the edge of a layer of membrane and screwed into the deck (Fig. 28.18). The adjoining membrane is lapped over this attachment. A third method involves placing large-diameter discs on top of the membrane, spaced as required, and screwing them to the deck. A waterproof cover is placed over the discs.

Following are some of the commonly used single-ply roofing membranes.

Chlorosulfated polyethylene (CSPE) is a thermoset membrane that completes its cure after installation. It has excellent weathering qualities and is resistant to ozone, sunlight, and most chemicals.

Figure 28.15 Single-ply membranes are sometimes covered with aggregate, which serves as ballast to hold it to the deck.

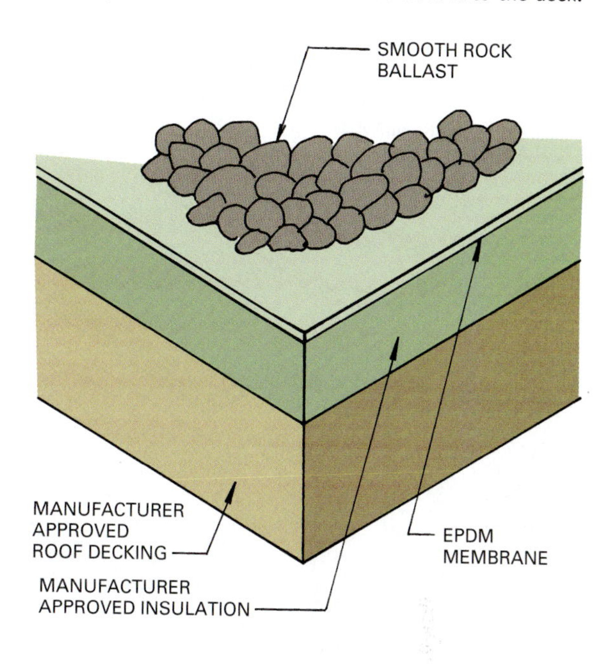

Figure 28.16 A single-ply protected membrane is covered with a layer of rigid insulation and ballast.

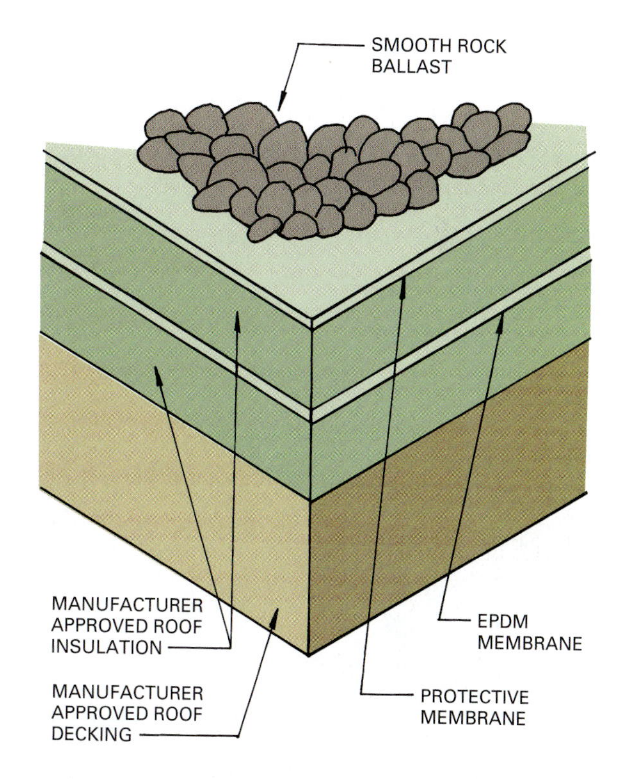

Ethylene propylene diene monomer (EPDM) is a thermoset elastomeric compound produced from propylene, ethylene, and diene monomer. It has good resistance to weathering, ultraviolet rays, abrasion, and ozone.

Polyvinyl chloride (PVC) is a thermoplastic membrane produced by the polymerization of vinyl chloride

Figure 28.17 This single-ply membrane is mechanically fastened to the deck. The metal batten strip is covered with a plastic batten cover.

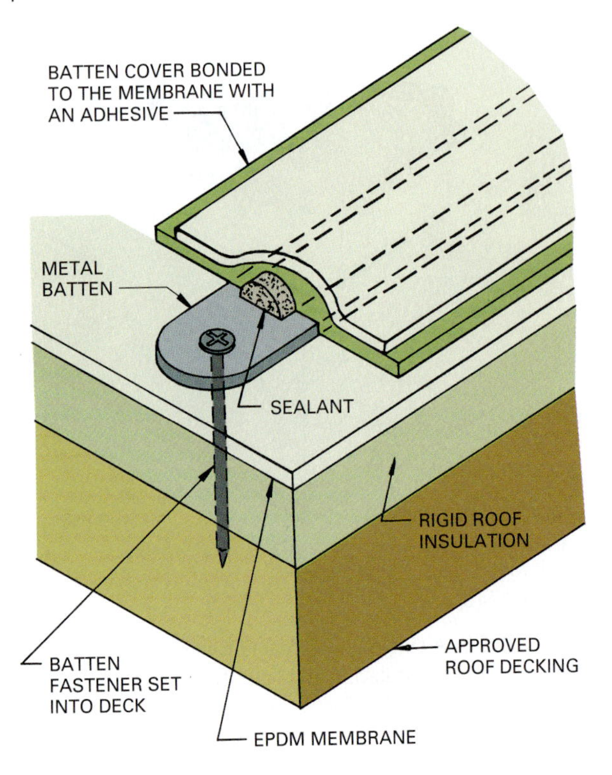

Figure 28.18 Some membranes are fastened to the deck with metal discs and sealed with a heat-welded lap joint.

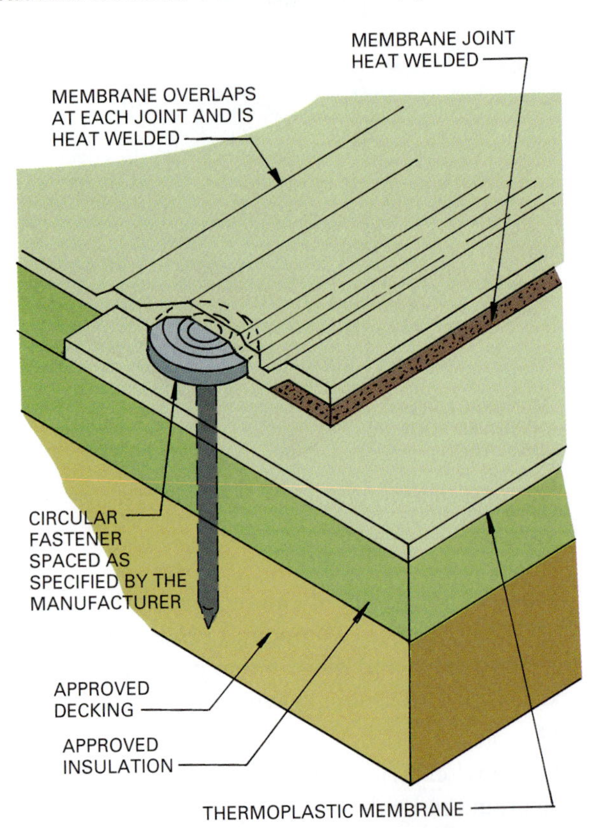

monomer, stabilizers, and plasticizers. PVC membranes are resistant to weather and chemical atmospheres, have good fire resistance, and are easy to bond. They do not work well with bituminous materials (Fig. 28.19).

Styrene-butadiene-styrene (SBS) is a thermoplastic membrane made by blending SBS with high-quality asphalt over a fiberglass mat. It has good fire resistance, and can be torched or applied with hot or cold asphalt. The method used depends on the specific composition of the membrane.

Spray-Applied Roof Coatings

Spray-applied roof coatings consist of a polyurethane foam insulation layer applied to a deck and then topped with a protective coating. The protective coating is usually acrylic, polyurethane, or silicone. Sometimes mineral granules or aggregate are applied to the wet top coating and used to provide additional protection.

Liquid-Applied Roof Systems

Liquid-applied coverings are available from various manufacturers as one- or two-component elastomeric materials. They are applied by spraying, brushing, or

Figure 28.19 The polyvinyl chloride (PVC) thermoplastic membrane is adhered on the deck. *(© Mark E. Gibson/CORBIS)*

Figure 28.20 This metal roofing system is designed for use on low-slope roofs. *(Image copyright Zibedik, 2009. Used under license from Shutterstock.com)*

rolling over the roof decking, which is typically plywood or concrete. Any joints require special attention as recommended by a manufacturer.

Metal Roof Assemblies

Various manufacturers have metal roofing systems designed for low-slope roof assemblies (Fig. 28.20). Numerous panel profiles and seaming methods have been developed to produce a watertight roof. Panels are available in a wide range of colors and may be galvanized or Galvalume steel, aluminum, copper, or terne coated stainless steel. Galvalume is a trade name for a patented steel sheet coated with a corrosion-resistant aluminum-zinc alloy applied by a continuous hot-dip process. Terne-coated stainless steel panels are made from nickel-chrome stainless steel coated on both sides with an alloy of 80 percent lead and 20 percent tin. They are recommended for use in severe chemical and marine environments. Metal roofing panels are available embossed to create a stucco-like finish. Various types of other coatings are available, such as siliconized polyester and epoxy-based coatings.

Low-slope standing-seam metal roofing systems require a minimum slope of $\frac{1}{4}$:12 for positive drainage. This type of roofing is generally held to a structure with a metal clip. The clip is joined to the bar joist or Z-purlin of the structure with metal fasteners (Fig. 28.21). The hold-down clip hooks onto the leading edge of the panel. The clips and edges of the butting panels are interlocked during the seaming process. The standing seam is formed on the roof with a portable seaming machine that runs along the seam and folds the members into a watertight joint. The clip fits into a slot in the clip base that permits the clip to move with the roof during expansion and contraction. Additionally, metal roofing panels are used extensively for re-roofing flat conventional roofs and steep-slope roofs. This requires the installation of a metal subframe that is secured through the old deck to the structural frame and supports the new metal roof deck (Fig. 28.22).

Figure 28.21 This metal roofing panel is held to the structural frame with a metal clip that is rolled into a standing seam. *(Courtesy CECO Building Systems)*

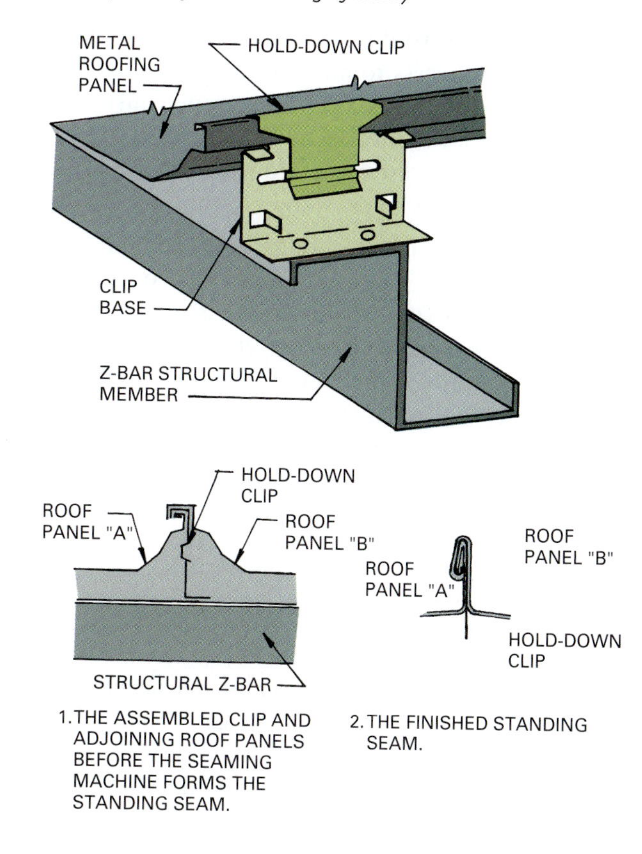

Figure 28.22 Typical construction for re-roofing a building with metal standing-seam roofing panels. *(Courtesy CECO Building Systems)*

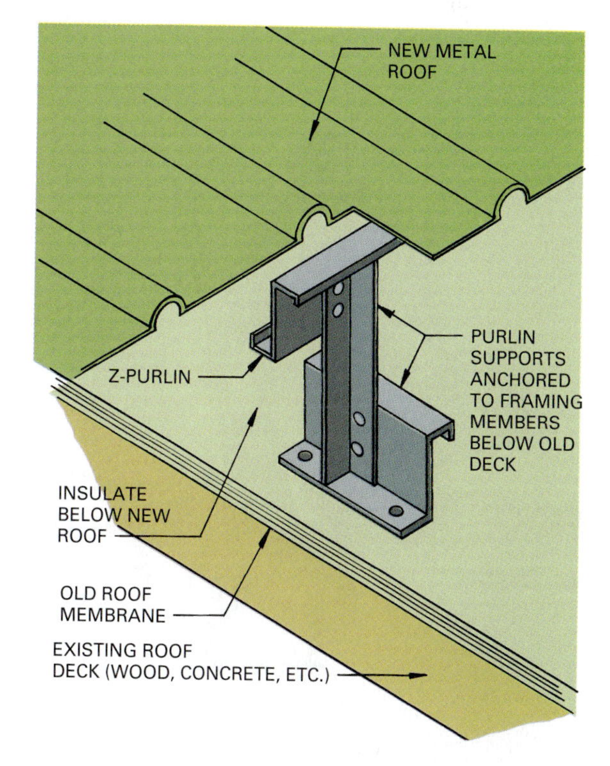

ROOFTOP WALKWAYS

Rooftop walkways are placed over standing-seam metal roofs and other flat roof membranes to provide maintenance workers access to mechanical units on a roof and to other areas (Fig. 28.23). One type consists of lightweight perforated steel planks designed to fit over standing seams. They are anchored with clips and supports attached directly to the standing seam.

Walkways used on aggregate-ballasted roof systems use concrete pavers. The pavers may rest on plastic or concrete blocks or a bed of gravel is laid over the membrane and pavers or slabs are set on it with cracks left open between the units. In both types, drainage below and between the pavers is provided. In no situation should the roof membrane be pierced to support a walkway.

FLASHING

Flashing is made of a thin material impervious to water. It is placed where needed to prevent water penetration and provide for water drainage. Typical flashing locations in roof construction include joints between roofs and walls, at expansion joints, and around objects, such as pipes, that penetrate the roof membrane.

Figure 28.23 Steel plank walkways provide access to various parts of the roof, protecting it from damage. *(Courtesy Unistrut Corporation)*

Materials used for flashing include copper, galvanized steel, lead, aluminum, stainless steel, bituminous sheet material, and plastics. In some cases, a combination of materials are used, such as galvanized steel covered with bitumen, which prevents corrosion of the steel when in contact with mortar.

Flashing materials have different thermal expansion and contraction characteristics from those of the roofing membrane. These differences can cause cracking and tears in a roof membrane unless allowance is made for this movement.

Wood nailers are used at roof edges and wherever roof insulation ends. They provide a surface to which the roof membrane and flashing can be nailed. Wood nailers must be securely fastened to the deck to prevent wind damage to flashing. The wood nailers are treated with a wood preservative. Care must be taken that the preservative does not damage roofing materials.

As the size of a roof increases, the danger of tears from expansion and contraction of the membrane also increases. The designer locates expansion joints to provide allowances for movement. An expansion joint for a built-up roof is shown in Fig. 28.24. Also included in the design are area dividers located between expansion joints. They must not restrict the flow of water. Spacing varies, depending on the climate, but they are generally located every 150 to 200 ft. (Fig. 28.25). The details in Figs. 28.26 through 28.30 show built-up roof membranes. Those for EPDM, PVC, and modified bitumens are similar. Additional details are available from the National Roofing Contractors Association.

STEEP-SLOPE ROOF ASSEMBLIES

Steep-slope roof assemblies have sufficient slope to permit water to drain to the ground via gutters and downspouts used to capture and control the water flow. A variety of roofing materials are available for use on steep slopes. Choices depend on architectural appearance and slope.

Slope is a common term used to express the steepness of a roof in terms of unit run and unit rise. The unit run is the number used as a base for the roof angle and specified as 12 in. (304 mm). The unit rise is the number of inches the roof will rise vertically for every unit of run. For example, if the unit rise is 6 in., then the roof will rise 6 in. for every 12 in. it covers horizontally, and the slope will be expressed as 6:12.

Steep-sloped roofs generally have a slope of 3:12 or higher. If a sloped roof is below the minimum acceptable slope for various types of shingles, it must be roofed

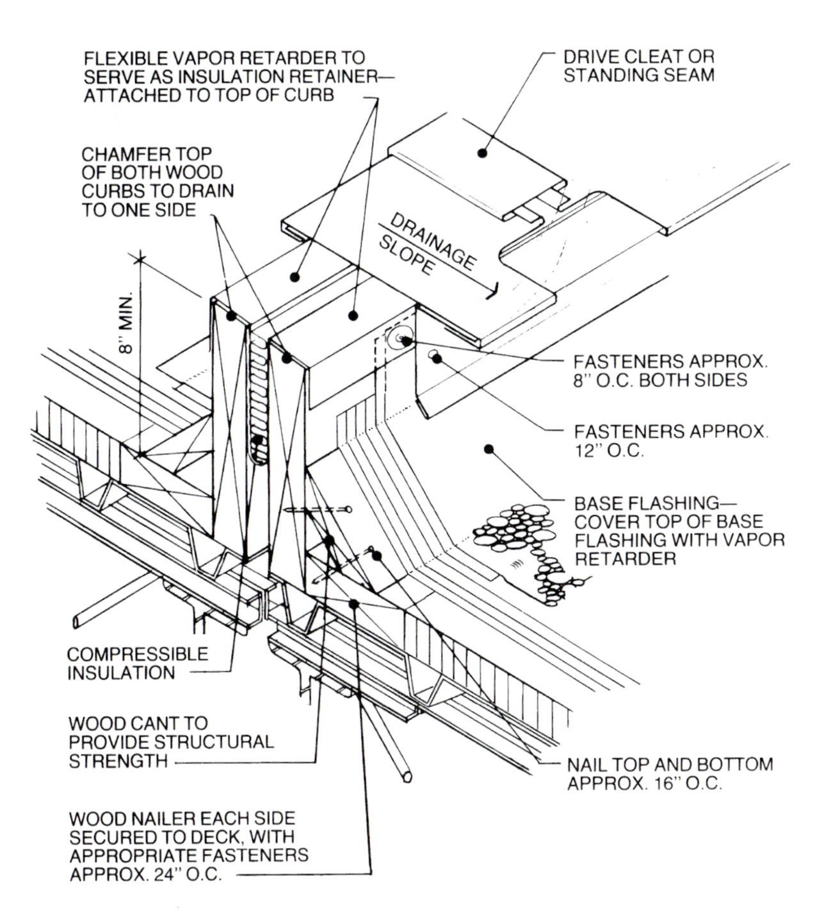

Figure 28.24 The expansion joints allow for expansion and contraction in the roof. *(Courtesy National Roofing Contractors Association)*

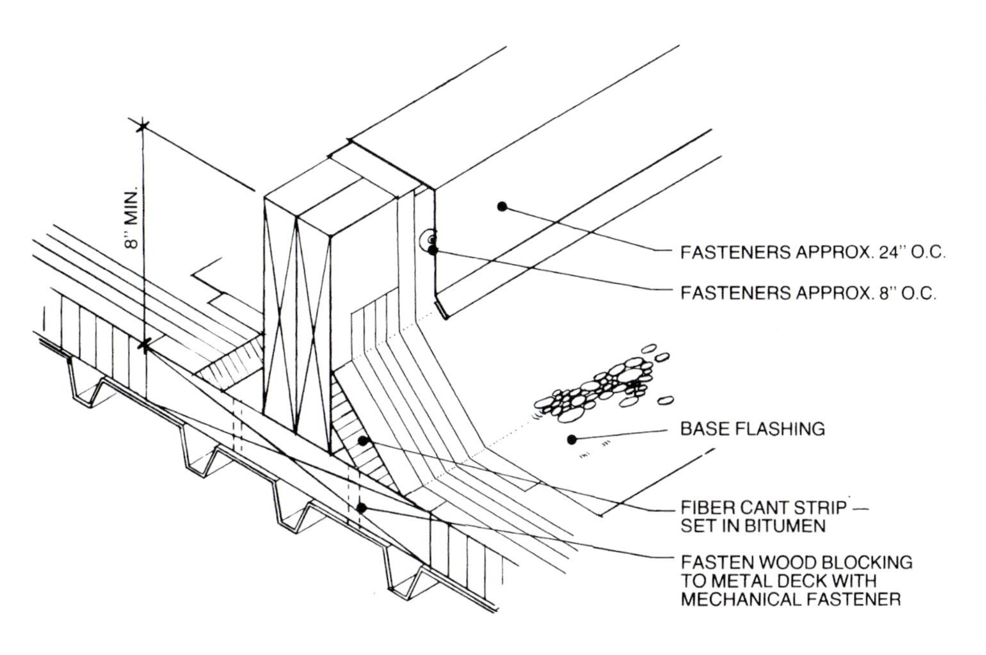

Figure 28.25 An area divider is a raised wood member attached to the deck and flashed. *(Courtesy National Roofing Contractors Association)*

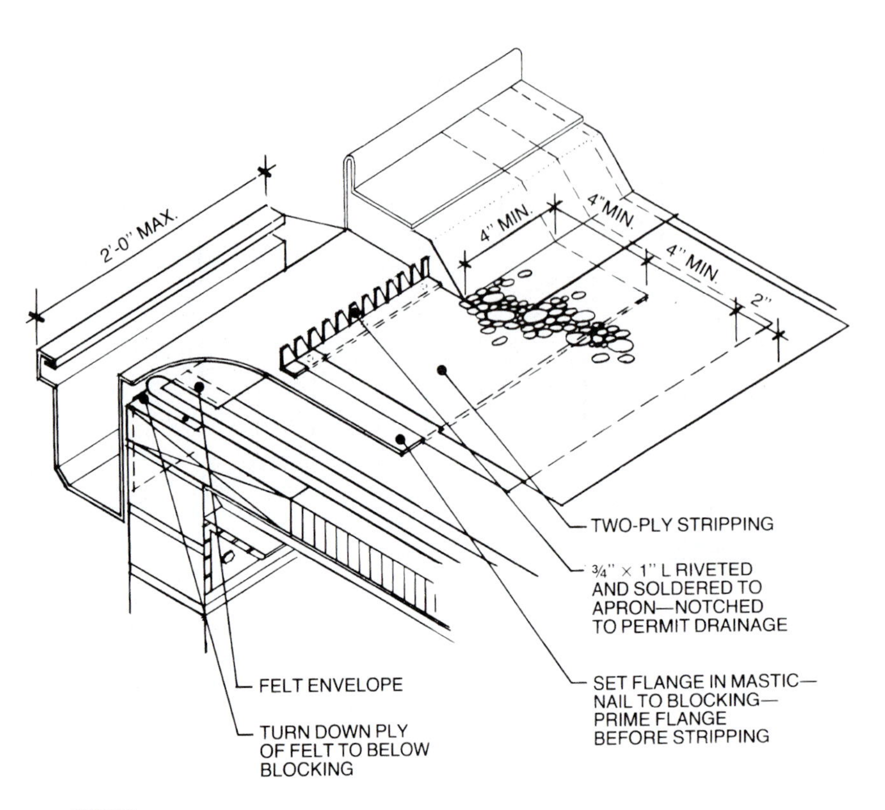

Figure 28.26 A scupper is an opening in the edge of the roof that permits water to drain to the ground, usually through a downspout. Notice the layers of felt laid over the sheet metal from the scupper that is secured to the deck. *(Courtesy National Roofing Contractors Association)*

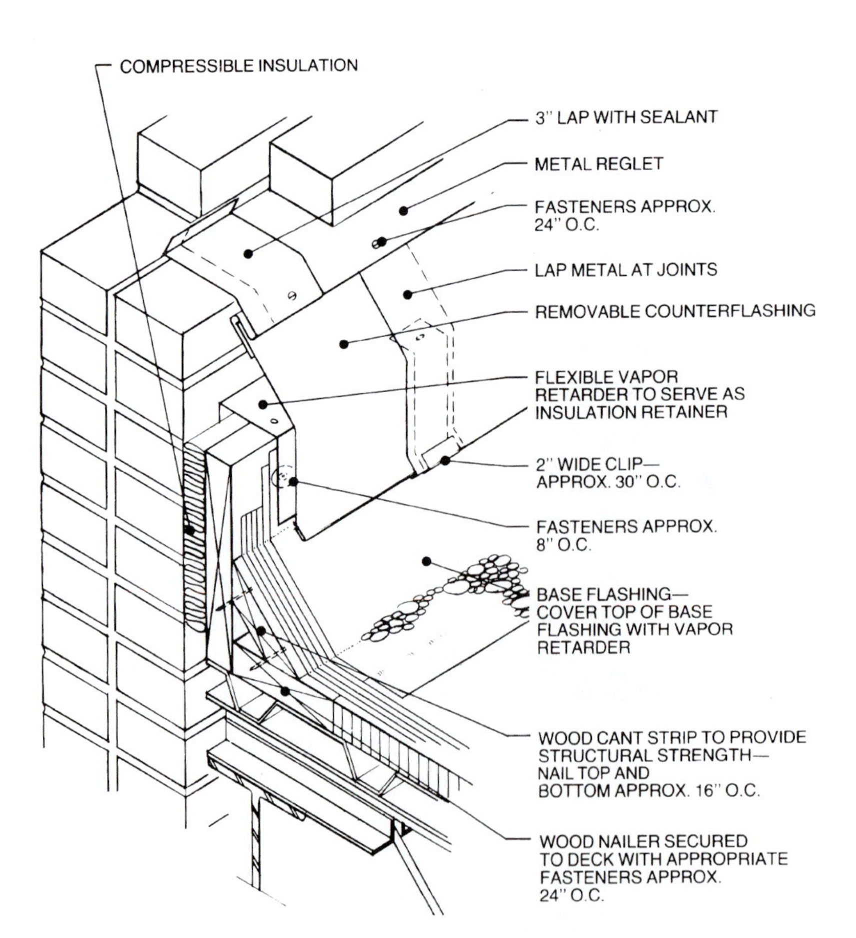

Figure 28.27 This detail shows a roof butting a masonry wall. A wood ledger protects the edge of the insulation. A wood header butts the vertical wall and has a layer of insulation between it and the wall. The corner is sloped with a wood cant strip. Metal flashing is set into a mortar joint and overlaps the felt laid up over the cant strip. This assembly allows movement between the roof and the masonry wall. *(Courtesy National Roofing Contractors Association)*

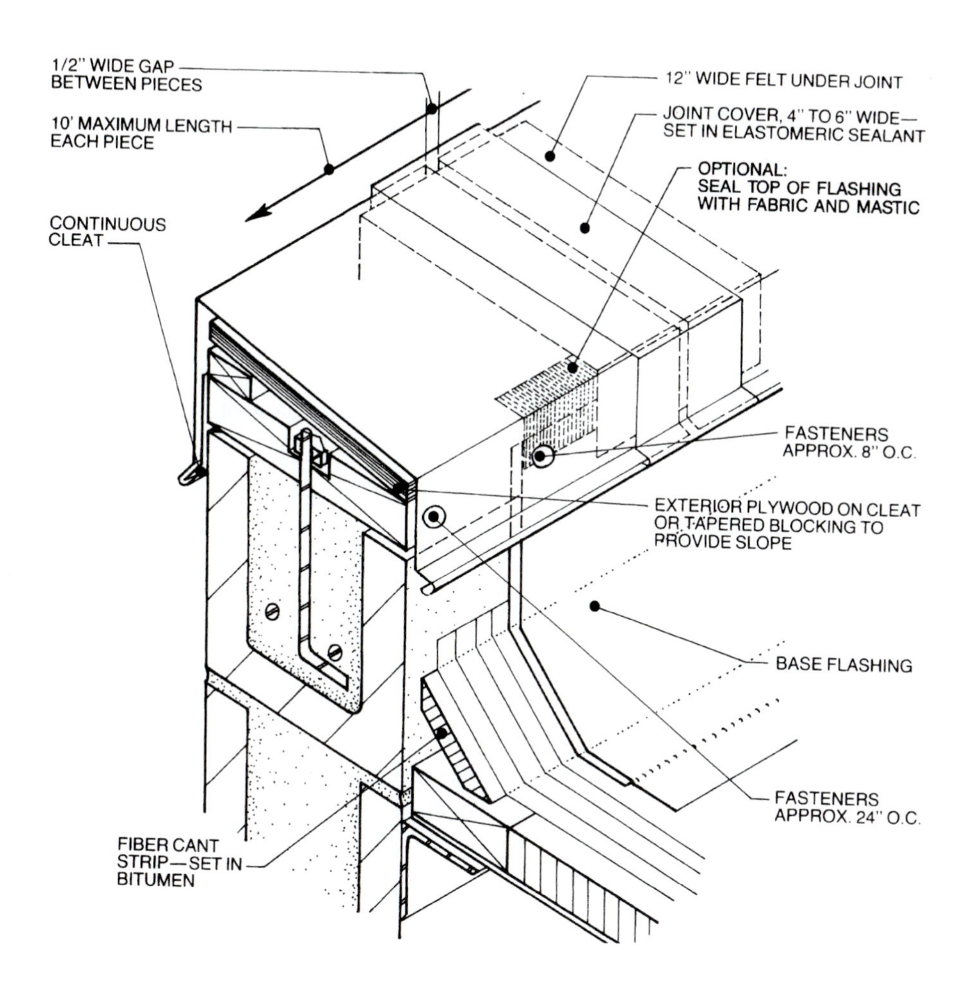

Figure 28.28 This light-metal parapet cap uses a wood nailer and cant strip. The base flashing is overlapped with a metal parapet cap. This construction is used only when the deck is supported by the wall. *(Courtesy National Roofing Contractors Association)*

Figure 28.29 Lead flashing used to seal a pipe penetrating the roof membrane. Several layers of roofing felt are laid over the lead flashing and sealed to the pipe with mastic. The built-up roof membrane is laid over this assembly. *(Courtesy National Roofing Contractors Association)*

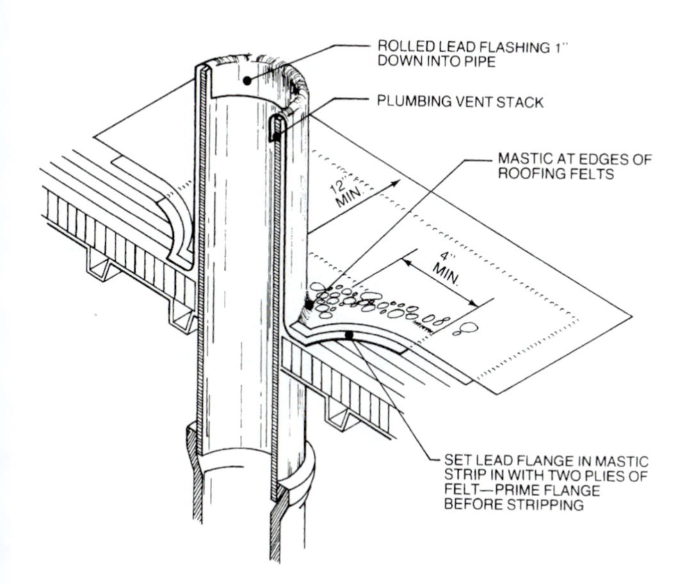

Figure 28.30 Roof drains are located in various parts of the roof and carry away water through pipes running down inside the building. A variety of products are manufactured. Installation details are provided by the manufacturer. *(Courtesy National Roofing Contractors Association)*

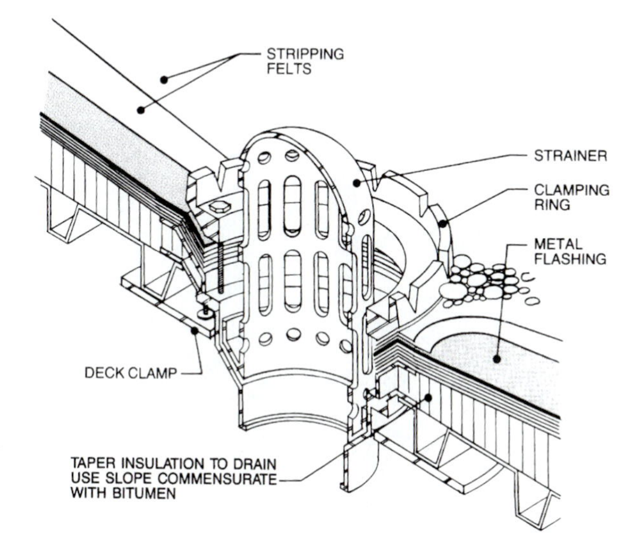

Table 28.2 Typical Minimum Slopes for Roofing on Steep-Slope Roofs[a]

Material	Allowable Slope
Asphalt shingles	4:12 if one layer of asphalt-saturated felt underlayment is used. 2:12 if two layers of asphalt-saturated felt underlayment are used.
Clay and concrete tile	4:12 if interlocking tile are used with one layer of 4 lb. cap sheet underlayment. Less than 3:12 when noninterlocking tile are used with two layers of No. 40 asphalt-saturated felt underlayment.
Slate	4:12 with one layer of 30 lb. asphalt-saturated felt underlayment. 2:12 with double underlayment.
Wood shingles and shakes	3:12 for shingles and 4:12 for shakes with 30 lb. felt underlayment and interlayment with shakes or shingles.
Metal shingles and shakes	3:12 and 4:12 depending on the material and style. Requires 30 lb. asphalt-saturated felt underlayment.

[a] *Consult local building codes and observe the manufacturer's recommendations on acceptable slope and installation details.*

with a built-up membrane or metal standing-seam roof, as described for low-slope roofs. Generally accepted slopes for steep-slope roofs are listed in Table 28.2.

Asphalt Shingles

Asphalt shingles are the most commonly used roofing for residential and light commercial roof construction. They are available in a variety of styles and colors and provide protection from the weather for 20 to 30 years. Asphalt shingles have either an organic or fiberglass base. Organic asphalt shingles consist of a wood fiber base that is saturated with asphalt and coated with colored mineral granules. Fiberglass asphalt shingles have a fiberglass mat that has top and bottom layers of asphalt, also coated with mineral granules. A self-healing adhesive seals the shingles to one another and prevents wind uplift. Asphalt shingles are die cut from the mat, typically in three tabs with square edges. Other shapes available are illustrated in Chapter 27.

A typical installation starts with the underlayment. The material is rolled out on the deck starting at the bottom of the roof and overlapped in successive layers. The underlayment is doubled at the eave, as shown in Fig. 28.31. In cold climates this double layer should extend over the roof until it is 24 in. (610 mm) inside the exterior wall.

A corrosion-resistant metal drip edge is placed along the edge of the roof at the eave and rake to facilitate water dripping away from the roof edge. The underlayment goes below the drip edge on the rake and on top at the eave (Fig. 28.32). On most roofs, shingles are

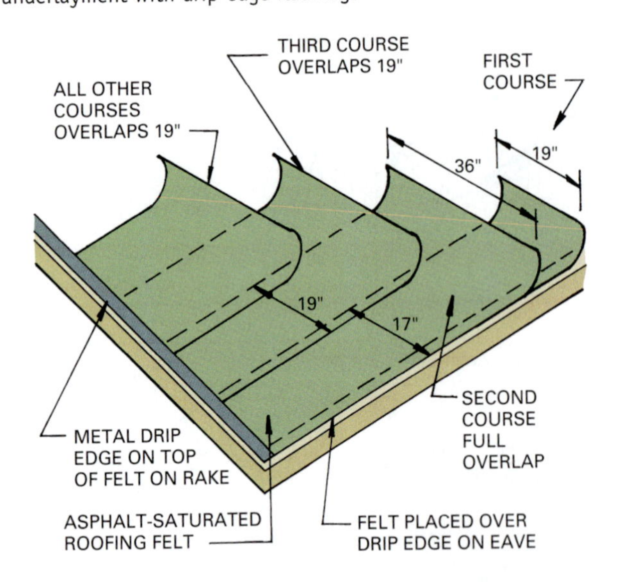

Figure 28.31 A double layer asphalt-saturated roofing felt underlayment with drip-edge flashing.

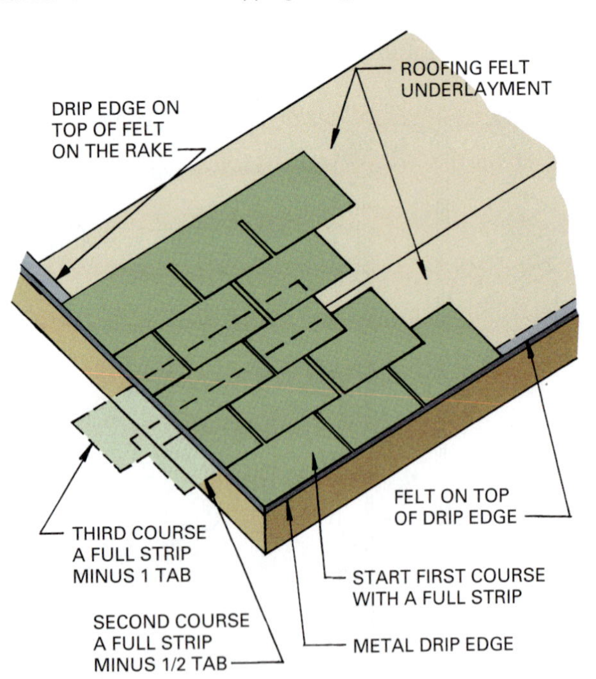

Figure 28.32 Asphalt shingles are spaced so the slots between the tabs on overlapping shingles do not line up.

applied starting from the rake. A starter course is laid with tabs facing up to fill spaces between tabs on the first regular course of shingles. Additional courses are overlaid in a lapped configuration. The maximum exposure of the shingles depends on weather conditions and ranges between 4 to 6 in. A variety of patterns are possible, depending on the shingle arrangement. One method involves keeping the cutouts centered over the tabs, as shown in Fig. 28.32.

Asphalt and fiberglass shingles are fastened using nails or staples. The fasteners are placed at the midpoint of each shingle, where the next course of shingles will cover them (Fig. 28.33). The fastener length should be sufficient to penetrate the deck at least $\frac{3}{5}$ in. or through approved panel sheathing. Common nails used are 11 or 12 gauge hot-dipped galvanized roofing nails with shanks $\frac{7}{8}$ to 1 in. (25 mm) long. Required staple sizes and spacings are specified in building codes.

To prevent water from entering a building, flashing is used in various locations on roofs susceptible to leakage. Roof valleys are particularly vulnerable to leaking. Shingle valleys may be either open or closed. Shingles cover valley centerlines in closed valleys, while open valleys are fashioned with no shingles or roofing material installed in valley centers (Figs. 28.34 and 28.35).

When a shingle roof meets a wall, step flashing is used. Folded at a 90 degree angle, each piece is nailed to the wall and rests on top of each shingle. The wall siding material is installed over the flashing (Fig. 28.36).

Figure 28.33 Asphalt shingles can be nailed by hand or fastened with a pneumatic roofing nailer.

Figure 28.34 The shingles are woven over roll roofing valley flashing, forming the valley.

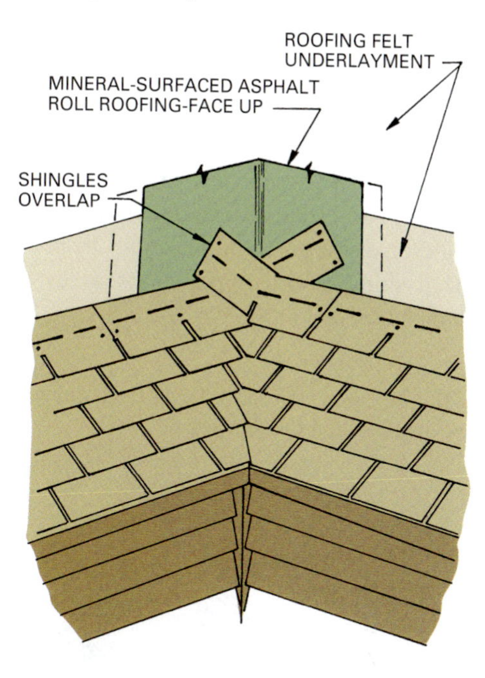

Figure 28.35 Metal valley flashing is placed over asphalt felt that is bonded to the felt on the roof with plastic asphalt cement.

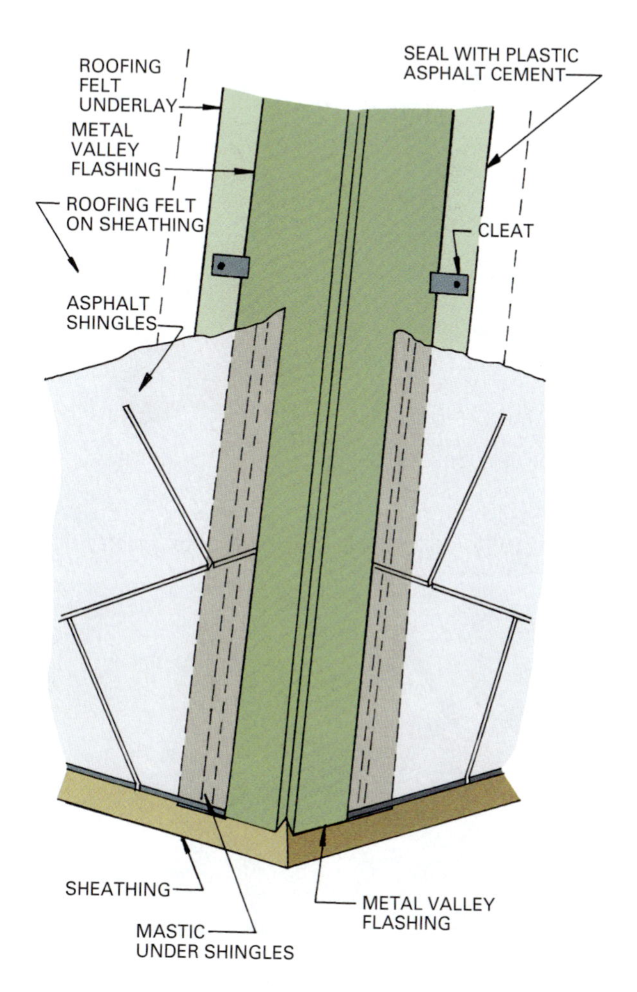

When the roof meets a brick wall, flashing is set into the mortar joint. Finally, hips and ridges are capped with single tabs overlapping each other, as shown in Fig. 28.37.

Wood Shingles and Shakes

Wood shingles and shakes are made from cedar, redwood, southern pine, and other woods. Shakes are hand-hewn and shingles are sawed. Building codes in many areas require that shingles and shakes be treated with fire-retardant chemicals.

Figure 28.36 The siding is placed over the flashing and kept 1 in. above the surface of the shingles.

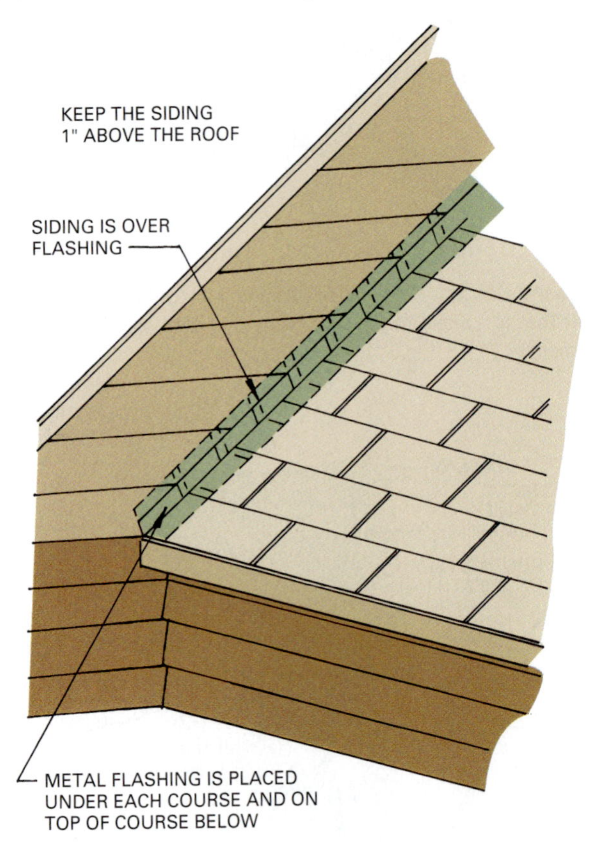

Figure 28.37 Applying ridge cap shingles to a vented ridge.

Usually wood shingles are applied over spaced 1 × 4 in. or 1 × 6 in. wood boards called purlins. The size used depends on the amount of shingle to be exposed to the weather. Wood shakes are placed over 1 × 6 in. spaced wood decking. Unspaced decking is used at the eaves and runs upwards until it is 24 in. (610 mm) inside exterior walls. The deck is covered with roofing felt (Fig. 28.38).

Shingles are doubled or tripled at eaves while shakes are usually doubled. A layer of felt lays over the top edge of each row of shakes. Hips and ridges are finished with cap shingles that may be hand-cut or purchased ready-made (Fig. 28.39). Valleys use metal flashing placed over roofing.

Shakes can be applied to roofs with slopes less than 4:12 if special construction is used. In this application, each wood shingle or shake should be applied with two corrosion-resistant nails, such as aluminum, stainless steel, or hot-dipped zinc-coated steel. Staples should be stainless steel or aluminum (Fig. 28.40).

Slate Shingles

Slate shingles are available in a variety of colors originating out of the rock from which it was quarried. They are split to thickness and trimmed to size from larger slabs of slate. Holes for fasteners are either drilled or punched. Additional details are discussed in Chapter 13.

Figure 28.38 Typical installation detail for wood shakes.

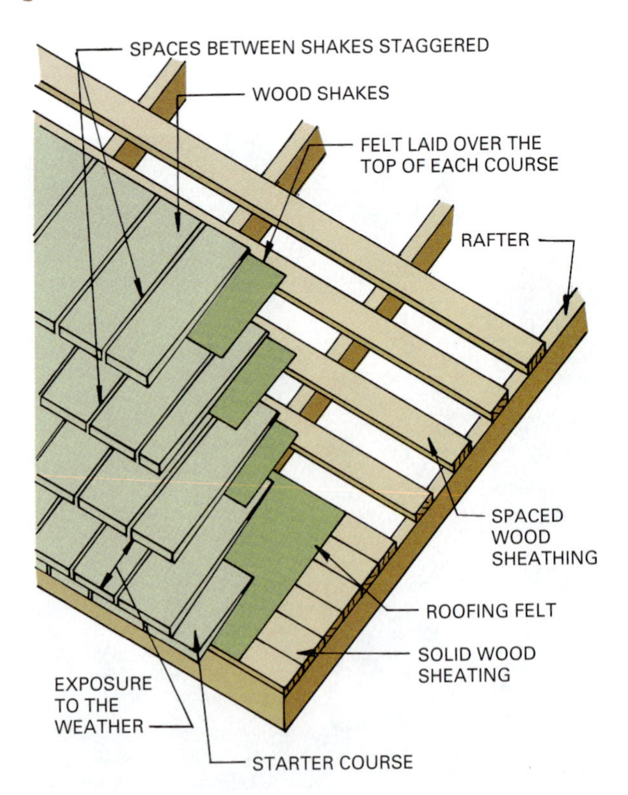

Figure 28.39 Preassembled wood hip and ridge caps are used to seal the shingles over hips and ridges.

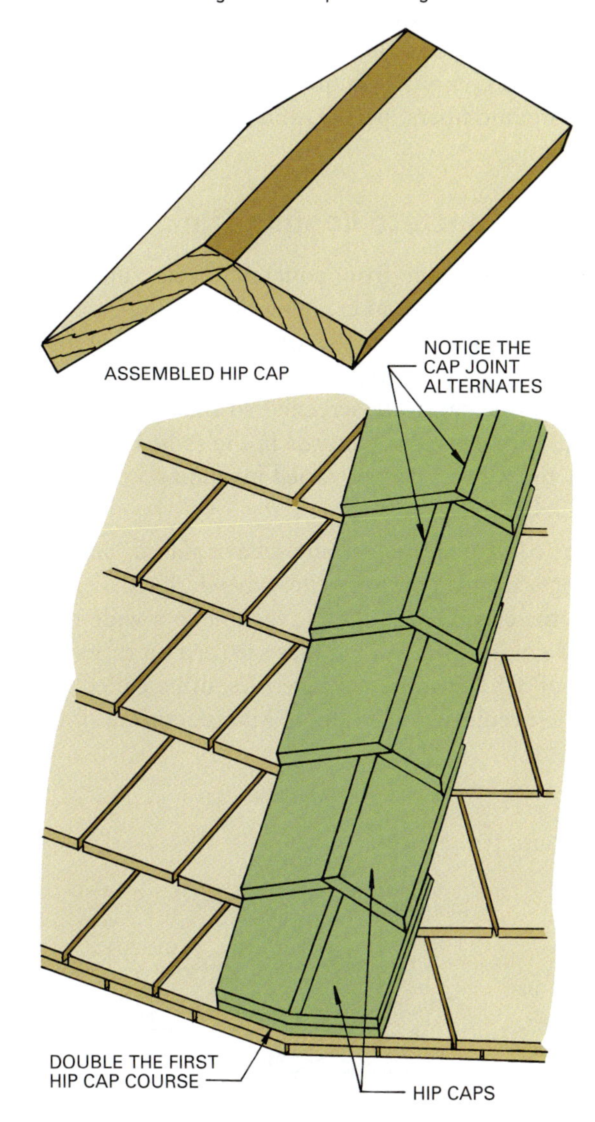

Slate is a heavy material, and the roof structure must be designed to carry the weight. Slate shingles are fire-resistant and have a long life but are more costly than most roofing materials. They can produce a roof that is durable and enhances the architectural beauty of a building (Fig. 28.41).

Slate roofing is divided into two classifications: textural (or random) and standard commercial. Textural slate is delivered to the job site in a range of thicknesses and sizes and must be sorted by roofers. Standard commercial slate is graded at the quarry by thickness, length, and width, with thicknesses of usually $\frac{1}{4}$ in. (6 mm).

ASTM C406 Standard Specification for Slate Roofing addresses material characteristics, physical requirements, and sampling procedures for the selection of roofing slate. Roofing slate is available in three grades: grade S1 has an expected life of 75 years; S2 40 to 75 years; and S3 20 to 40 years.

Slate installation details are shown in Fig. 28.42. It is usually applied over solid sheathing covered with an asphalt-saturated roofing felt underlayment. Codes require double underlayment for roofs with less than specified slopes. Fastened with large-head copper nails, each course of slate covers the joints in the course below it. Plastic cement is applied over joints to be covered. Each slate must have at least two fasteners.

Slates are laid in the same manner as other shingles. The exposed ends may be in even rows or laid in a random edge. Hips and ridges cannot be nailed and are held in place with a waterproof elastic slater's cement that is colored to match the slate.

Figure 28.40 Wood shakes are fastened with corrosion-resistant nails or staples. *(© Raymond Gehman/CORBIS)*

Figure 28.41 Slate shingles produce an attractive, durable, fire-resistant roof covering. *(Image copyright Kevin Britland, 2009. Used under license from Shutterstock.com)*

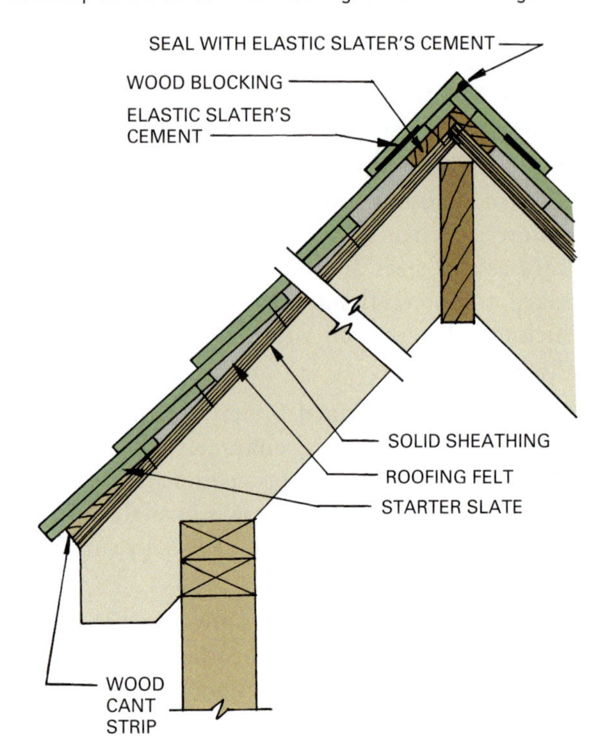

Figure 28.42 Slate shingle installation is started with a wood cant strip at the eave. Wood blocking is used at the ridge.

Cement-Fiber Shingles

Cement-fiber shingles are made by combining Portland cement and various fibrous materials. One common product is a composition of Portland cement, organic and inorganic fibers, perlite, and iron oxide pigments for color.

Clay and Concrete Roofing Tile

Tile roofing is made from concrete or clay units that overlap and interlock to create a durable, fire-resistant, and heavy roof. Some forms of concrete tile are made using lightweight aggregates. It is important that the aggregates be carefully screened so the sizes are controlled. They are manufactured in the same basic types as clay tile. Colors can be varied by adding iron oxides (Fig. 28.43).

Clay roofing tile may be unglazed or glazed. Unglazed tile ranges from an orange-yellow to a dark red, depending on the clay. Glazed tile is available in a wide range of solid colors. Interlocking tiles are used on roofs with slopes of 3:12 or more. Shingle tiles, often called Norman tiles, require a 5:12 slope. Manufacturing of clay

Figure 28.43 The shapes of concrete roofing tile available from one manufacturer. *(Courtesy Monier)*

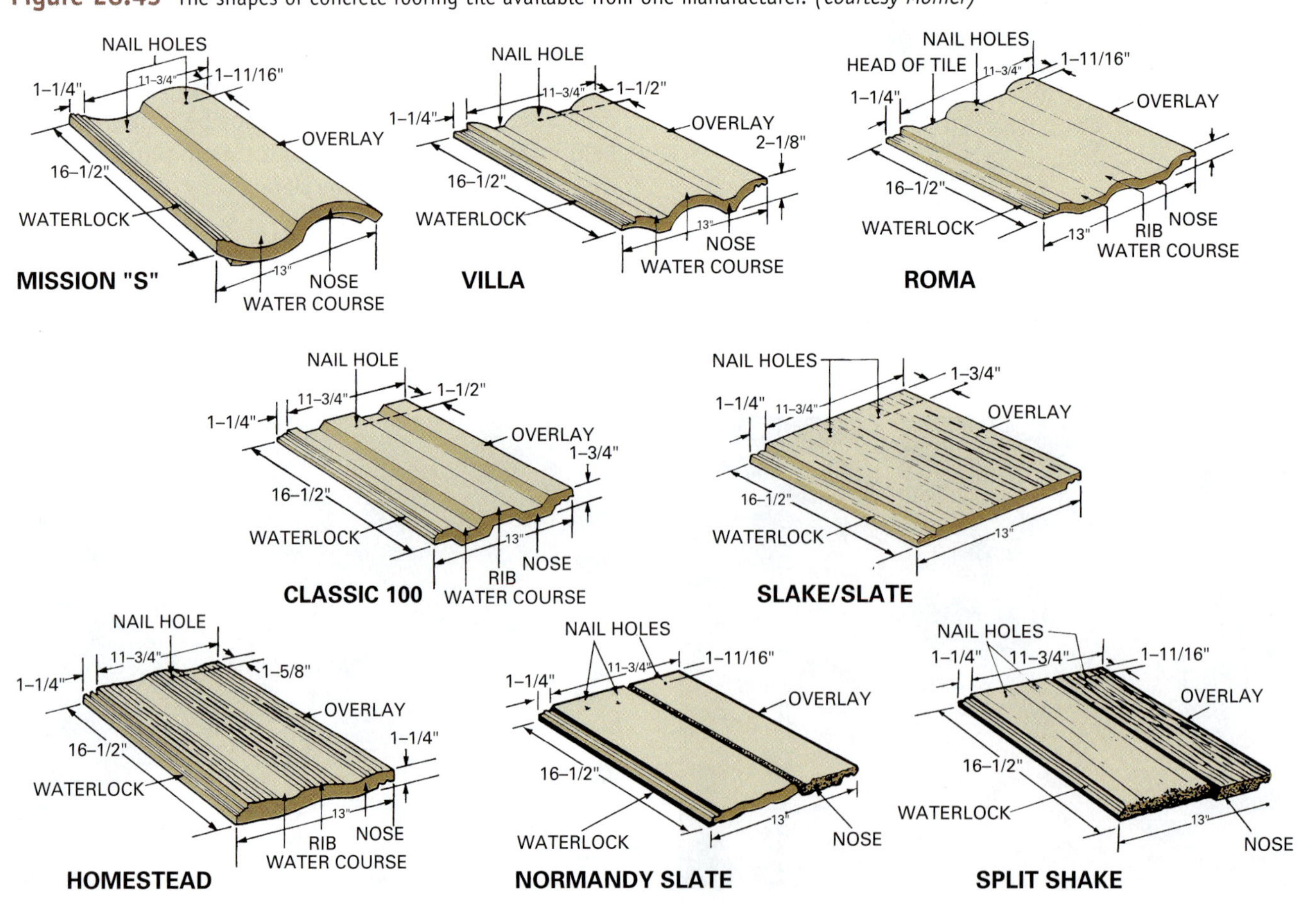

tile is discussed in Chapter 11. The specific installation details for clay and concrete tile vary by tile manufacturer. The following are typical examples.

Flat interlocking clay tile installation is shown in Fig. 28.44. The tiles rest on solid sheathing covered with asphalt-saturated roofing felt. A flat under-eave tile or an under-eave tile with an apron is used along the eaves. The gable rake is covered with special right and left rake tiles. Ridges and hips are covered with V-shaped tiles set in a cement mortar and nailed.

Flat clay tile are installed much like slate shingles. Spaced sheathing can be used for roofs with over 4:12 slopes with roofing felt laid between layers.

A number of curved profile tiles are available. Two common ones are shown in Fig. 28.45. These include a gable rake tile and curved ridge and hip tiles. In addition to being nailed to sheathing, they are set in a cement mortar. Valleys are flashed as described for other roofing products. A typical installation is shown in Fig. 28.46.

Figure 28.44 Typical installation detail for flat clay tile using special eave, rake, and ridge tile.

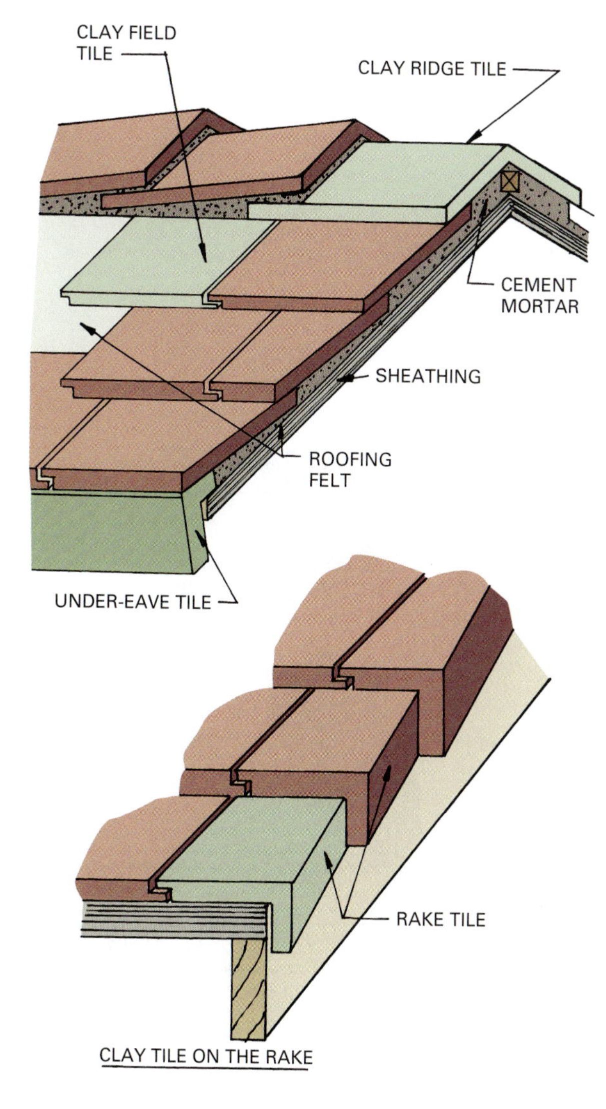

Metal Roof Systems

Low-slope standing-seam metal roofs use either standing, capped, or batten seams. Some metal roofing is installed over solid decking, and other types have sufficient strength to span between properly spaced purlins. Examples of purlin-supported metal roofing are shown in the low-slope section of this chapter (see Fig. 28.22).

A typical batten seam is shown in Fig. 28.47. The roofing panels have an L-flange that butts a cleat. The cleat fastens to the deck and folds over the flanges. A metal batten cap snaps over this assembly.

A capped seam is shown in Fig. 28.48. A T-shaped cleat fastens to the deck and the roofing flanges butt it. The cleat is folded over the flanges and a metal cap snaps over the assembly.

Another type of standing seam is shown in Fig. 28.49. A cleat fastens to the deck and folds around one roof panel flange. Then the butting panel is placed next to the cleat and folded over the assembly, and the entire assembly is folded over again. In all cases, it is

Figure 28.45 Two of the commonly manufactured tile profiles. *(a) Image copyright IBI, 2009. Used under license from Shutterstock.com. (b) Image copyright ZTS, 2009. Used under license from Shutterstock.com.*

Figure 28.46 A typical installation of concrete roofing tiles. *(Courtesy Monier)*

Figure 28.47 A batten seam is constructed by snapping a metal batten cap over the cleats that secure the edge of the metal roofing to the deck.

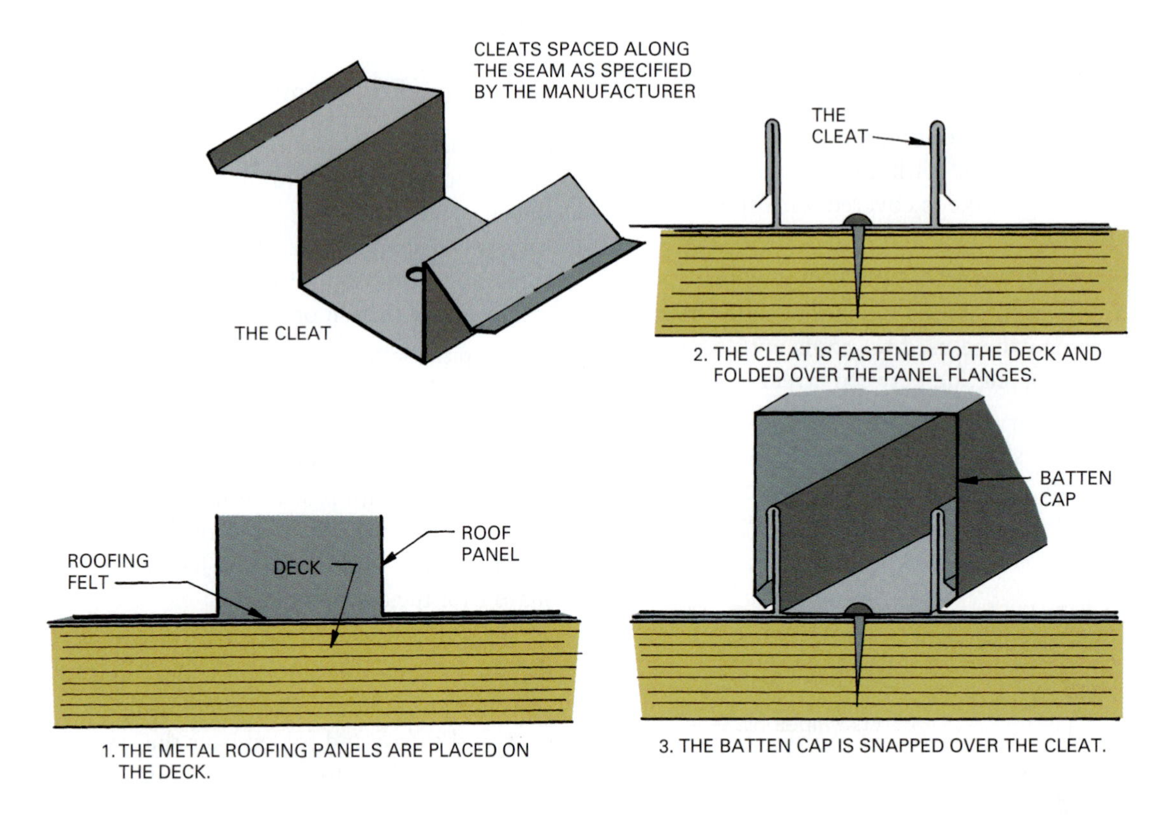

Figure 28.48 A capped seam provides a watertight seal over the union of metal roofing panels.

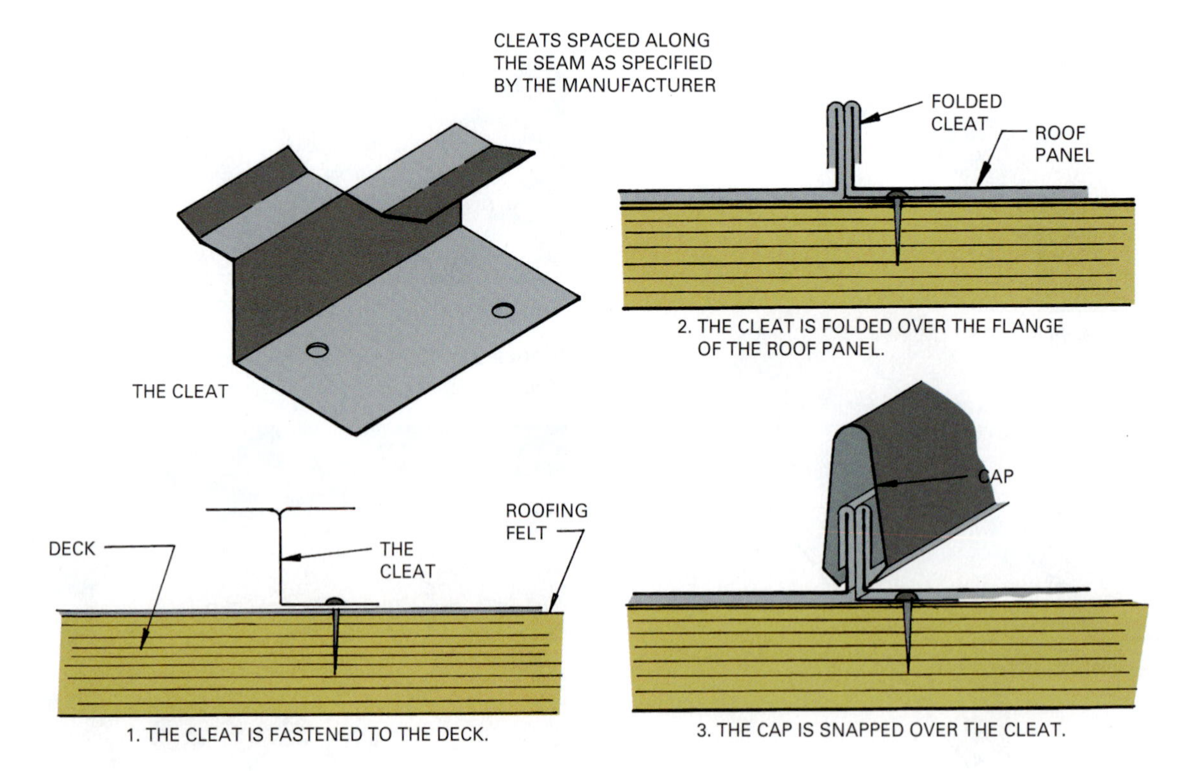

Figure 28.49 This standing seam is formed by rolling over the edges of each metal roof panel and the cleat that secures them to the deck.

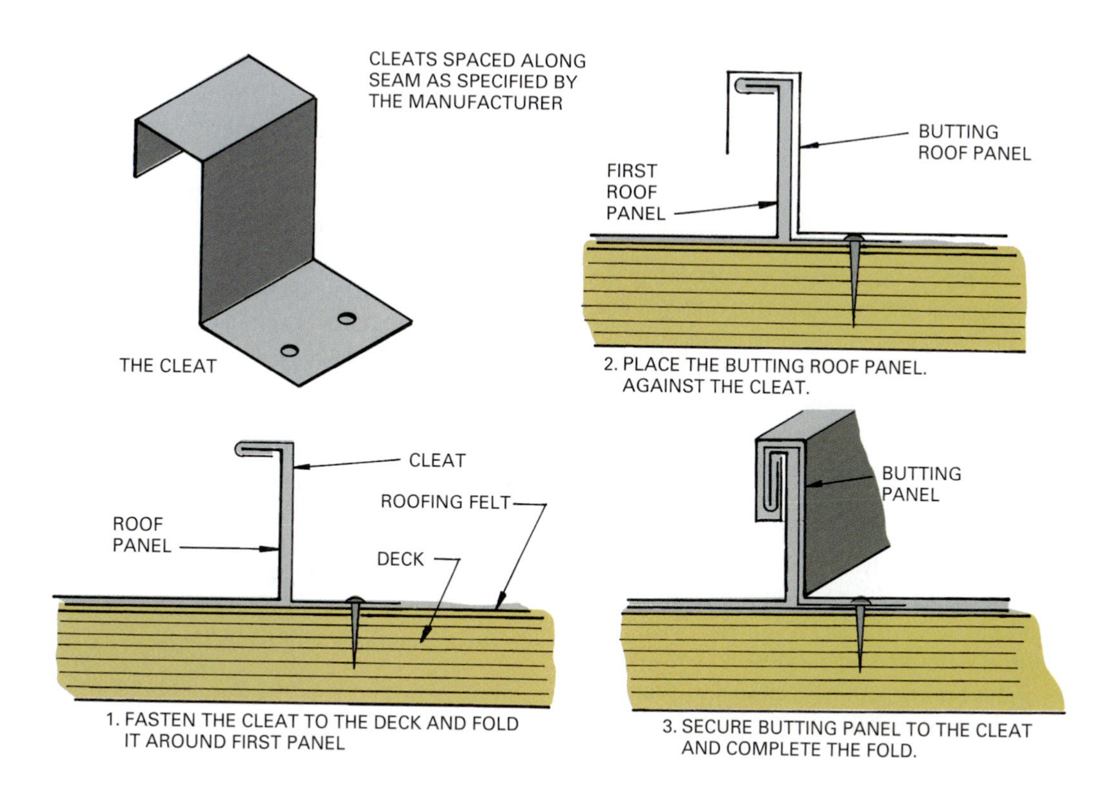

important that the cleats and fasteners provided by the manufacturer of the roofing panels be used. A finished standing seam metal roof is shown in Fig. 28.50.

Metal shingles and shakes are available with embossed surfaces in a range of patterns and textures. The shingles lock together and are secured to a deck with metal clips. The fastening system used varies with the manufacturer of the shingles.

Cool Roofing

A cool roof is defined as a roof surface that stays relatively cool when compared to the surrounding temperature. Materials with highly reflective, light color surfaces help roofs to absorb less heat and stay considerably cooler than conventional roofing materials. The temperature of a surface depends on the surface's reflectance and emittance. The Solar Reflectance Index (SRI) incorporates both of these properties and varies from 100 for a white surface to zero for a black surface. Materials with the highest SRI are the coolest and the most effective in reducing a building's cooling loads and mitigating the heat island effect. Cool roofing systems utilize reflective white membranes and ballast, white metal, or paints and coatings to both reflect and emit heat.

Figure 28.50 This durable standing-seam roof will provide years of service. *(Image copyright ppl, 2009. Used under license from Shutterstock.com)*

Construction **Methods**

Green Roofs

With increased interest in sustainable building technologies, the practice of applying vegetation directly to a roofing system is on the rise. Green roofs are classified as either intensive or extensive. An intensive green roof is installed on an adequately supported roof structure and creates actual roof gardens encompassing trees, bushes, and terraced surfaces. Extensive green roofs are used for smaller applications in both low-slope and pitched roofs. They are characterized by a shallow growing medium and utilize low maintenance sedums, grasses, and mosses.

The basic components of a green roof include four essential layers:

A waterproofing layer may consist of a liquid-applied membrane, a specially designed single-ply roofing, or multiple layers. Correct and meticulous application of the waterproofing membrane is essential to insure the water tightness of a green roof. A thorough water flood test is usually conducted to check for membrane leaks before other layers are applied.

Green roofs have a drain layer that prevents water from accumulating on the roof or in the substrate. Pre-manufactured drainage systems—consisting of a high-strength plastic core with two layers of geo-synthetic fabric attached to top and bottom—are available. The geo-synthetic fabric acts as a root barrier that prevents plant roots from affecting the membrane and drainage systems.

The third layer consists of the growing medium or soil. Because natural soils are heavy, particularly when wet, green roofs often involve the use of lightweight engineered soil mixes. Soil cannot contain any silt that might clog the filter fabric and should be water permeable and resistant to rot, heat, frost, and shrinkage.

Vegetation is the final and most visible layer of the green roof. Plants add aesthetic value and ultimately determine the success or failure of the roof. Compatibility of plant selection with the local environment is essential. Characteristics of plants typically used on green roofs include: a shallow root system, good regenerative qualities, resistance to direct sun, drought, frost, and wind tolerance, and compatibility with local ranges of temperature, humidity, rainfall, and sun.

Green roofs exhibit a number of positive attributes when compared to conventional roofing systems. The soil layers and plants protect the roofing membrane from ultraviolet radiation damage, which extends the life of the roof. Roofs can reduce and slow the amount of storm-water run off. Many municipalities now offer tax breaks and other incentives to builders in an effort to encourage the use of green roofs.

Review Questions

1. List the four fire classifications for roof coverings.
2. What types of roof decking are used for low-slope roofs?
3. Why are vapor retarders an important part of a roof assembly?
4. What materials are used for vapor retarders?
5. Describe the process for installing a built-up roof covering.
6. What is a modified bitumen?
7. How are single-ply roofing membranes secured to a roof deck?
8. What types of single-ply roofing membranes are available?
9. Describe the makeup of a spray-applied roof coating.
10. How are standing-seam metal roof panels joined to form a watertight joint?

11. How are standing seam metal roof panels secured to a deck?
12. Why are rooftop walkways needed?
13. List several locations where flashing is required.
14. What roof covering materials are used on steep-slope roofs?

Key Terms

built-up roofing
flashing
low-slope roofs
modified bitumens
single-ply roofing
steep-slope roofs
water-vapor transmission rate (WVTR)

Activities

1. Collect samples of as many roofing products as possible. Label each, listing its properties, fire rating, and advantages and disadvantages.
2. Set up a partial roof deck outside the classroom and install as many types of roofing systems as possible.
3. Ask a roofing contractor to visit the classroom and discuss bidding procedures, relative costs of various systems, and safety requirements that must be met during installation.
4. Visit various construction sites and observe the procedures being used for the installation of roofing systems.

Additional Resources

Schunk, Oster, Barthel, Kiessl, *Roof Construction Manual, Birkhauser Edition Detail, Munich.*

Concrete and Clay Roof Installation Manual, Roof Tile Institute, Eugene, OR.

NRCA four-volume publication, *Roofing and Waterproofing Manual*, National Roofing Contractors Association, Rosemont IL.

Residential Asphalt Roof Manual, the Asphalt Roofing Manufacturers Association, Calverton, MD.

Spence, W. P., *Roofing Materials and Installation*, Sterling Publishing Co., 387 Park Avenue South, New York 10016.

Steep-Slope Roofing Materials Guide, National Roofing Contractors Association, Rosemont, IL 20018.

Other resources include:

CD-ROMs: *Installation Tips & Techniques and Avoiding Common Roof Installation Mistakes*, GAF Materials Corporation, 1361 Alps Road, Wayne, NJ 07470.

See Appendix B for addresses of professional and trade organizations and other sources of technical information.

DIVISION 08

Openings
CSI MasterFormat™

081000 Doors and Frames
083000 Specialty Doors and Frames
084000 Entrances, Storefronts, and Curtain Walls
085000 Windows
086000 Roof Windows and Skylights
087000 Hardware
088000 Glazing
089000 Louvers and Vents

Image (Courtesy Eva Kultermann)

CHAPTER 29

Glass

LEARNING OBJECTIVES

Upon completion of this chapter, the student should be able to:

- Recognize the various types of glass products used in building construction.
- Select appropriate glass products for various applications.

Build Your Knowledge

For further study on these materials and methods, please refer to:

Chapter 30 Doors, Windows, Entrances, and Storefronts
Chapter 31 Cladding Systems

The selection of glazing systems for a building is a process of ever-growing complexity. In the recent past, the range of glazing products available to designers was much smaller than what is available today. Windows and glazed walls have historically been regarded as net energy drains on a building. Recent developments and the introduction of insulated windows, glass tinting, and selective film coatings are beginning to change this perception. With the current concern with energy efficiency, choosing windows that will enhance the energy performance of a building becomes more crucial (Fig. 29.1). Only by gaining a comprehensive understanding of the key physical parameters used to characterize glazing systems can an appropriate choice be made.

Glass is a major component in bringing daylight into a building. As lighting is often the single largest consumer of electricity in buildings, all efforts should be made to integrate the use of daylight in building design. The challenge is to find a balance between the utilization of useful daylight and minimizing the solar gains that accompany it. By utilizing daylight, the need for electrical lighting will be reduced, as will the heat gains produced by artificial lighting, thereby reducing cooling loads. Shading and diffusing devices can be used to offset the possibility of glare, which accompanies natural light.

Glass is an inorganic mixture that has been fused at a high temperature and cooled without crystallization. It has an unusual internal structure because mechanically it is rigid and has the characteristics of a solid, yet the atoms in glass are arranged in a random order similar to those in a liquid. Glass is technically a supercooled liquid.

TYPES OF GLASS

The six basic types of glass are classified by their ability to resist heat (Table 29.1). Soda-lime-silica is the type commonly used for door and window glazing, as well as consumer bottles. About 90 percent of all glass produced is of this type. Its general composition includes 74 percent silica, 15 percent soda, 10 percent lime, and 1 percent alumina. Soda-lime-silica glass is easy to form and cut, has fair chemical resistance, and does not resist high temperatures or rapid thermal changes. It is easily shattered into small, sharp pieces.

Lead-alkali-silica glass has similar properties to soda-lime but is more expensive.

Fused silica glass is composed of about 99 percent silicon dioxide and is the most expensive. It has the highest resistance to heat, managing temperatures ranging from 1,650 to 2,190°F (900 to 1,200°C). It also has the highest corrosion resistance and affords excellent transmission of ultraviolet rays.

Figure 29.1 Glazing can provide daylight and views while enhancing the energy performance of a building. *(Courtesy Eva Kultermann)*

Table 29.1 Thermal Properties of Commonly Used Glass

Categories of Glass	Thermal Expansion	Heat Resistance
Soda-lime-silica	High	Low
Lead-alkali-silica	High	Low
Fused silica	Low	High
96% silica	Low	High
Borosilicate (Pyrex)	Medium	Medium
Aluminosilicate	Medium	Medium

Ninety-six percent silica is used when high thermal hardness and the ability to withstand the thermal shock of going rapidly from hot to cold is required. Borosilicate glass is a thermally hard glass that has been used for many years. It is best recognized by its Corning trade name, Pyrex®. Aluminosilicate glass is more costly than borosilicate but can withstand high service temperatures and is similar in its ability to withstand thermal shock.

This chapter is largely devoted to the uses of soda-lime-silica glass, because it is the most widely used glass for construction applications.

THE MANUFACTURE OF GLASS

Although several glassmaking processes have been used for many years, most glass today is produced worldwide by the **float process**. The first production of glass by this method was in 1959 by the English firm Pilkington Brothers, Ltd.

Float Glass

The float process involves producing molten glass in a furnace from which it is conveyed to a float bath. Here the molten glass floats across a bath of molten tin (Fig. 29.2). The molten tin provides a very flat surface that supports the glass as it is polished by the application of heat from above. The heat melts out any irregularities. The ribbon of glass moves to a cooling zone, which permits the glass to solidify enough to be conveyed on to the annealing lehr. After the glass has been annealed, it is cut into lengths, inspected, and packed. The sheets of glass produced by this method have parallel surfaces, a smooth, clear finish, and high optical clarity.

Float glass is a flat glass available as regular float glass or heavy float glass. Thicknesses range from $\frac{3}{32}$ to $\frac{1}{2}$ in. (2.5 to 12 mm). Regular float glass is made in three forms with specific qualities: silvering, which is used in selected high-quality pieces for optical uses and mirrors; mirror glazing, for general-purpose mirrors; and glazing for door and window glazing. Float glass provides clarity and visual transparency with a minimum of distortion. Float glass is used for many products, such as reflective glass, mirrors, tinted glass, laminated glass, and insulating glass.

Sheet Glass

Sheet glass is a type of flat glass that is more economical than float glass. It is made using older methods that involve drawing a ribbon of molten glass along a series of rollers where its thickness is established and it is annealed, cooled, and cut to size. Sheet glass has more distortion than float glass and is not as widely used. It is available in single strength, $\frac{3}{32}$ in. (2.3 mm) thick; double strength, $\frac{1}{8}$ in. (3.1 mm) thick; and heavy sheet, $\frac{3}{16}$ in. (4.7 mm) and $\frac{7}{32}$ in. (5.6 mm) thick. Picture glass is a thinner version—$\frac{3}{64}$, $\frac{1}{16}$, and $\frac{5}{64}$ in. (1.2, 1.6, and 2.0 mm) thick—that is used for covering pictures and for other purposes where strength is not a factor. Sheet glass is available in three grades—AA (best), A (good), and B (general glazed)—and as clear, tinted, reflective, tempered, or heat-treated products.

Figure 29.2 The float glass manufacturing process.

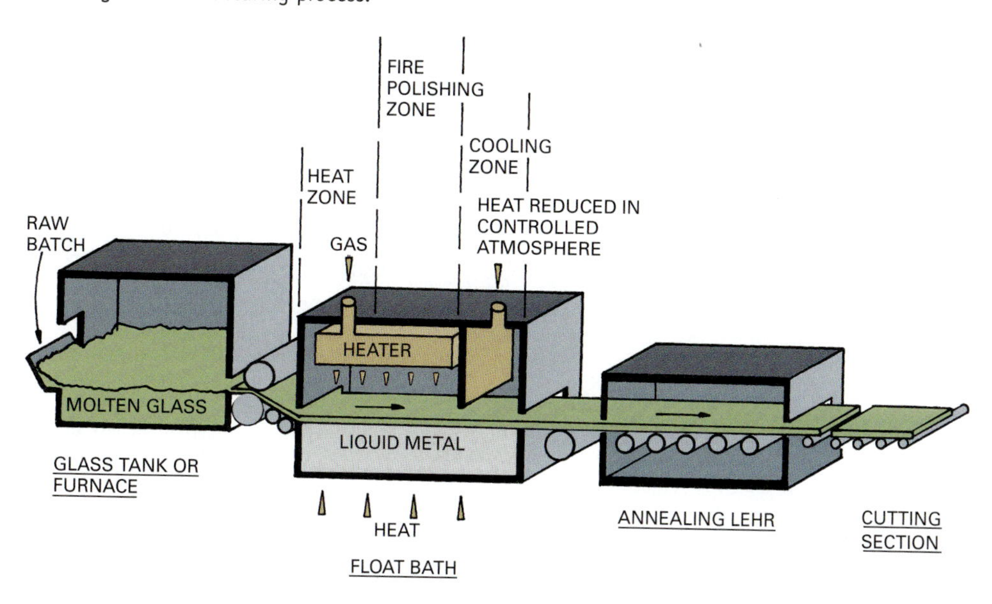

PROPERTIES OF SODA-LIME-SILICA GLASS

The following discussion presents the general properties of soda-lime-silica glass. Glass properties can be varied by altering the composition of its ingredients.

Mechanical Properties

When deciding whether or not to use glass in a specific application, its ability to withstand breakage is a major consideration. Glass is brittle yet remains elastic up to its ultimate tensile strength. This means that glass can be bent up to a breaking point and, if released, will return to its original position. Glass breaks by bending or stretching; therefore, tensile strength is a major mechanical property to be considered.

Glass does not have a clearly defined tensile strength because its actual strength depends on the surface condition of the glass. A small scratch or nick is sufficient to cause tensile stresses to concentrate at that weaker point. Therefore, actual tensile strength is less than the theoretical tensile strength. Mechanical strength can be increased by chemical and heat treatments.

Thermal Properties

Heat can be transferred by conduction, convection, and radiation. The thermal conductivity of glass is high but lower than most metals. The U-value for a single-strength sheet of glass is about 1.04 to 1.10. Double glazing with a $\frac{1}{2}$ in. (12 mm) air space reduces the U-value to about 0.49 to 0.56. Typical thermal properties for selected flat glass products are given in Table 29.2.

For single glazed glass, the majority of thermal resistance is borne at the outdoor and indoor surfaces. Indoors, about two-thirds of heat flows by radiation to room surfaces and one-third flows by convection. Heat transfer of the interior surface of glazing can be reduced a great deal by adding low-emissivity metallic films (referred to as Low-E) to the glass. This also reduces the U-value for glazing units with air spaces. A typical double glazed opening with suspended Low-E film has a U-value of 0.31 to 0.32. Other glass products, such as tinted glass, reflective glass, and insulating glass, also reduce heat gain and loss.

The thermal expansion of glass must be considered as glazing units containing glass are designed. The coefficient of expansion of soda-lime glass is about 4.5×10^{-6}, while the coefficient for aluminum is about 13×10^{-6}. An aluminum window frame contracting under low temperatures imposes stress on glass edges unless sufficient clearance is allowed for the difference in thermal movement.

Chemical Properties

Glass used in typical applications in building construction is very durable and more resistant to corrosion than many other materials. Since it is not porous, it will not absorb moisture or chemical elements in the ground or atmosphere. There are a few exceptions that may occur in isolated industrial applications. Hot concentrated alkali solutions and superheated water can cause soda-lime-silica glass to dissolve, while hydrofluoric acids will cause corrosion.

Table 29.2 Thermal Properties of Flat Glass

Glass	Thickness (in.)	Winter Nighttime U-value/R-value	Summer Daytime U-value/R-value
Clear single	1/4	1.3/0.78	1.04/0.96
Clear double	1/8	0.49/2.04	0.52/1.92
Clear double	1/4	0.49/2.04	0.56/1.79
Clear double with Low-E film	1/4	0.31/3.32	0.32/3.03
Light brown single	1/4	1.13/0.88	1.10/0.91
Dark brown single	1/4	0.89/0.88	0.89/1.88
Double glass light brown/clear	1/4	0.31/2.04	0.33/1.75
Double glass dark brown/clear	1/4	0.41/2.04	0.47/1.72
Light green single	1/4	1.13/0.88	1.10/0.91
Dark green single	1/4	0.95/1.14	0.98/1.12
Double light green/clear	1/4	0.50/2.04	0.59/1.75
Double dark green/clear	1/4	0.42/2.50	0.50/2.13
Clear insulating glass with suspended Low-E film	1/4	0.23/4.30	0.37/2.70

Electrical Properties

Glass is a good electrical insulator and is widely used for applications where this and other properties make it useful. Light fixtures often consist of glass, as do light bulb envelopes, because it is strong, heat resistant, and an electrical insulator.

Optical Properties

Glazing products are not completely transparent to incoming radiant energy. A fraction of solar radiation is reflected, another absorbed, and a third transmitted. Clear sheet glass permits the passage of about 86 to 89 percent of visible light. Double glazed lights (window panes) permit about 80 to 82 percent light passage. Typical daylight transmittance figures are shown in Table 29.3.

Table 29.3 Daylight Transmittance for Selected Glass Units

Glass	Percent Daylight Transmittance
Clear single	86–89
Clear double	80–82
Clear double with Low-E film	72
Clear triple	72–74
Clear insulating with Low-E film	37–69
Light brown single	52
Light green single	75
Light gray single	41

When glass is treated to reflect or absorb light, the percent of light transmitted is greatly affected. Tinted glass frequently transmits less than 50 percent of the solar radiation striking it. Glass with reflective coatings transmits 6 to 50 percent of visible light. Optical quality is also influenced by lack of distortion. If the front and back surfaces of a sheet of glass are not parallel, images viewed through the sheet will form some angles, appearing wavy or distorted. With the increased use of float glass, distortion has been reduced and optical properties improved.

HEAT TREATING GLASS

The properties of glass can be improved by various heat-treating methods. These include annealing, tempering, and heat strengthening.

Annealing

Annealing is part of the normal production of glass. After a glass ribbon has been formed it is passed through an annealing lehr where temperatures are carefully controlled. The temperature of the glass is raised high enough to relieve strains developed during forming in the float bath. Then the temperature is slowly lowered, allowing all parts of the glass to cool uniformly.

Tempering

Tempering increases strength. **Tempered glass** is used when strength beyond that of standard annealed glass is required.

Tempering involves raising the temperature of glass just shy of its softening point and then blowing jets of cold air on both sides, quickly cooling it. Surfaces harden and shrink while the interior is still fluid. As the interior cools and shrinks, the exterior remains unchanged in size, resulting in compression forces along the surfaces and edges and tension in the interior. The opposing compressive and tension forces balance each other, resulting in a stronger glass.

Tempered glass is three to five times as resistant to damage as annealed glass. When the thin tempered skin on the glass is broken, the entire sheet disintegrates into small pebble-like particles instead of sharp slivers. Tempered glass is made in accordance with Federal Specification DD-G-1403B, which requires minimum compression values of 10,000 lb. per sq. in. (surface) or 9,700 lb. per sq. in. (edge). Fully tempered glass can meet safety standards required by ANSI Z97.1-1975 and Federal Standard CPSC 16 CFR 1201, Safety Standard for Architectural Glazing Materials.

Heat Strengthening

Heat-strengthened glass is heated and cooled much like tempered glass. Heat-strengthened glass has a compression range of 3,000 to 10,000 lb. per sq. in. (surface) or 5,500 to 9,700 lb. per sq. in. (edge), making it about twice as strong as annealed glass.

Chemical Strengthening

Glass can be **chemically strengthened** via immersion in a molten salt bath, causing larger potassium ions in the salt to replace the smaller sodium ions already in the glass. This crowds the surface, causing compressive stresses.

FINISHES

A variety of finishes are available for glass depending on its end use. Although most glass produced is clear and transparent, finishes make it useful for other applications.

Etching

Etching glass produces a surface that can range from almost opaque to smooth but translucent. The degree of opacity depends on the length of time the glass is exposed to hydrofluoric acid or other etching compounds. The glass may be dipped or sprayed with the etching fluid. The finished surface is opaque but remains smooth to the touch.

Sandblasting

Glass is sandblasted by blasting its surface with coarse-grained sand particles blown by compressed air. This produces a translucent finish that is generally rougher than etched glass. Etching and sandblasting abrade the surface, greatly lowering the strength of the sheet.

Patterned Finish

Patterned glass has a texture or pattern rolled into the surface as the glass is drawn through the furnace. It may be colorless or tinted and is available in a wide variety of patterns. It may be relatively transparent from one side and translucent from the other.

Silvering

Mirrors are produced by spraying silver nitrate and tin chloride on the surface of glass as it moves on a conveyor. For additional protection, this layer can be covered with shellac, varnish, or paint, or a layer of copper can be electroplated over it.

Considered the highest quality, float glass mirrors are available in silvering, mirror glazing, and glazing quality, with silvering being the best. Sheet glass used for window glazing can be used for low-quality mirrors. These come in two grades: A and B.

Ceramic Frit

Glass surfaces can be coated with a ceramic frit that is fired at high temperatures. Ceramic frit is composed of small powdered particles produced by quenching a molten glossy material. This produces a colored surface layer that is highly weather resistant. Both translucent and opaque colored layers are available. These are often used for finished exterior surfaces on wall spandrels (Fig. 29.3).

Tinting

Glass can be infused with color by tinting. One function of tinted glass is to minimize the amount of solar gains entering a building. Tinted glass is also known as heat absorbing glass, since the glass itself absorbs much of the incoming energy. Tinting is achieved by adding color-producing ingredients to the glass at the beginning of its production process. The color is technically not a surface finish but rather infused in the glass.

Tinted glass is often used to reduce glare. Blue-green glass achieves a moderate reduction in glare and brightness. Bronze tints are used primarily to reduce heat gains. Gray tints give a wide range of light transmittance values. Translucent glass reduces the entry of direct sunlight yet permits considerable transmission of diffused light.

Figure 29.3 Glass surfaces, such as those used on building exteriors, may be coated with a colored ceramic frit.

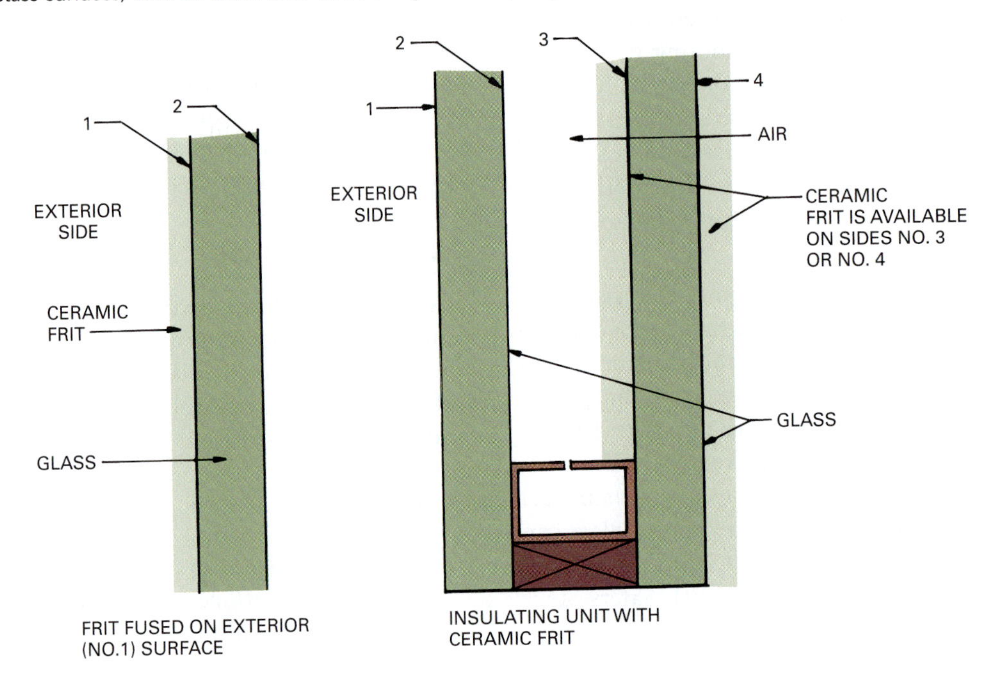

Physical Parameters of Glazing Systems

A number of physical characteristics are used to describe the performance of glazing systems. Manufacturers supply these performance parameters directly with commercially available glazing units.

Total Solar Energy Transmittance

The total solar energy transmittance (g) of a glazing unit is its solar heat gain factor. It is the energy entering the interior through a glazing unit directly, either as direct or diffuse beam radiation, with a smaller amount absorbed by the glass being transferred inside via infrared radiation or convection. The concept is important in determining the amount of heat gain that will enter a space through a glazing unit. Values for "g" are given on a scale from zero to one, with "one" representing 100% of solar heat being transmitted and "zero" indicating no heat being transmitted.

U-Value (U)

The U-value is a measure of the overall heat transfer of an assembly—or the rate of conductive, convective, and radiative heat that occurs due to the difference in temperature between the inside and outside. This can be either heat loss or heat gain, depending on exterior climate conditions. For a window unit, the U-value represents the sum of heat transfer from the centre of the glass, the edge of glass, and the frame. The smaller the U-value, the better the thermal performance of the window. Heat transfer can be minimized through the use of coatings that reduce radiative heat losses and specialized frames that limit conductive losses.

Visible Light Transmittance (Tv)

Visible light transmittance is that portion of incoming solar radiation that occurs in visible wavelengths—or electro magnetic radiation to which the human eye is sensitive: a small amount from 380-760 nanometers. Only this visible transmittance is useful for daylighting purposes. The visible component is about one half of the total incoming radiation; the rest is mostly in the infrared, with small amounts in the ultra-violet. In general, a high value (.6 or above) of Tv is desirable, as this will reduce the need for artificial lighting.

Shading Coefficient (Sc)

The solar heat gain of a window is related to the total energy transmittance with a shading coefficient. The **shading coefficient** is defined as the ratio of solar heat gain of a glazing unit to the solar heat gain of clear float glass. The lower the shading coefficient, the more efficient the glazing unit is at reducing solar gains.

Emissivity (e)

Another important characteristic of glazing materials is their emissivity, which is the ability of an object to radiate energy. A material that radiates energy perfectly will have an emissivity of one, while one that radiates no heat will have a value of 0. Materials with smooth and reflective surfaces will have a lower emissivity than those with rougher, more textured surfaces. Clear float glass has an emissivity of one. The emissivity of a glazing unit can be altered by the introduction of coatings on its glass surfaces. Low-emissivity (low-e) coatings reduce the radiative heat transfer of a window unit. The positioning of the coating within the unit can produce either enhanced solar control or solar gains.

GLASS PRODUCTS

A vast variety of glass products are available for specific purposes. Specifications vary by manufacturer.

Low-E Glass

Low-E glazing units have special coatings applied to the glass panes that are designed to reduce a window unit's heat transfer. Solar radiation in the form of short wavelengths has the ability to pass through the glass of a window unit. Once admitted through the glass into a space, this radiation is absorbed by objects and re-radiated as long-wave radiation, or heat energy.

The properties of low-E coatings are such that the transmission of solar radiation in the short wavelengths is high but drops off dramatically in the long wavelengths. The net result is that solar radiation is allowed to enter a space but is prevented from re-radiating back though the glazing unit. This allows the window unit to act as a collector of solar energy within the space.

Low-E glass insulating units have a thin metallic coating on the inside glass surface, so the coating is fully protected. This coating selectively reflects the ultraviolet and infrared wavelengths of the energy spectrum. Low-E coatings permit the use of natural daylighting techniques, because they are virtually invisible and have a high visible light transmission (Fig. 29.4).

Laminated Glass

Laminated glass is used in areas where the glazing is subject to possible breakage or security of glazed openings is required. Typical applications from a safety standpoint include residential and commercial door glazing and windows in high-impact areas, such as gymnasiums. Special security applications may be required in prisons and banks.

Laminated glass is made by bonding layers of float glass with intermediate layers of plasticized polyvinyl butyral (PVB) resin or polycarbonate (PC) resin. The glass is chemically strengthened by immersing it in a molten salt bath. Laminated glass is made to meet ANSI Z97.1-1975 and Consumer Products Safety Commission Regulation 16 CFR1201 Categories 1 and 11.

Many variations and thicknesses are available from various manufacturers. Laminated glass is available with ultraviolet filtering laminates and wire glass laminates. The widths of the various products vary by manufacturer, but thicknesses from $\frac{1}{4}$ to 3 in. (6 to 75 mm) are common. Typical lamination designs are shown in Fig. 29.5.

Figure 29.4 Low-E glass helps keep heat in a building in the winter and out of the building during summer.

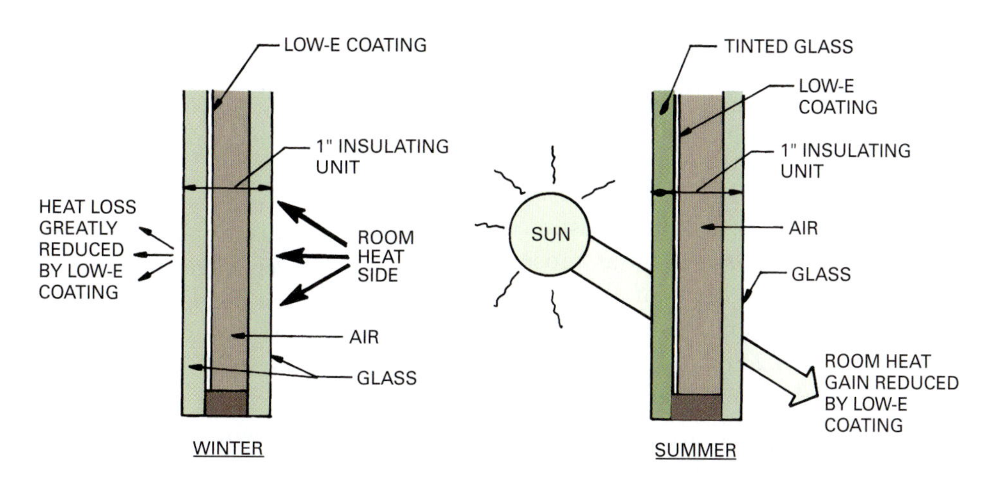

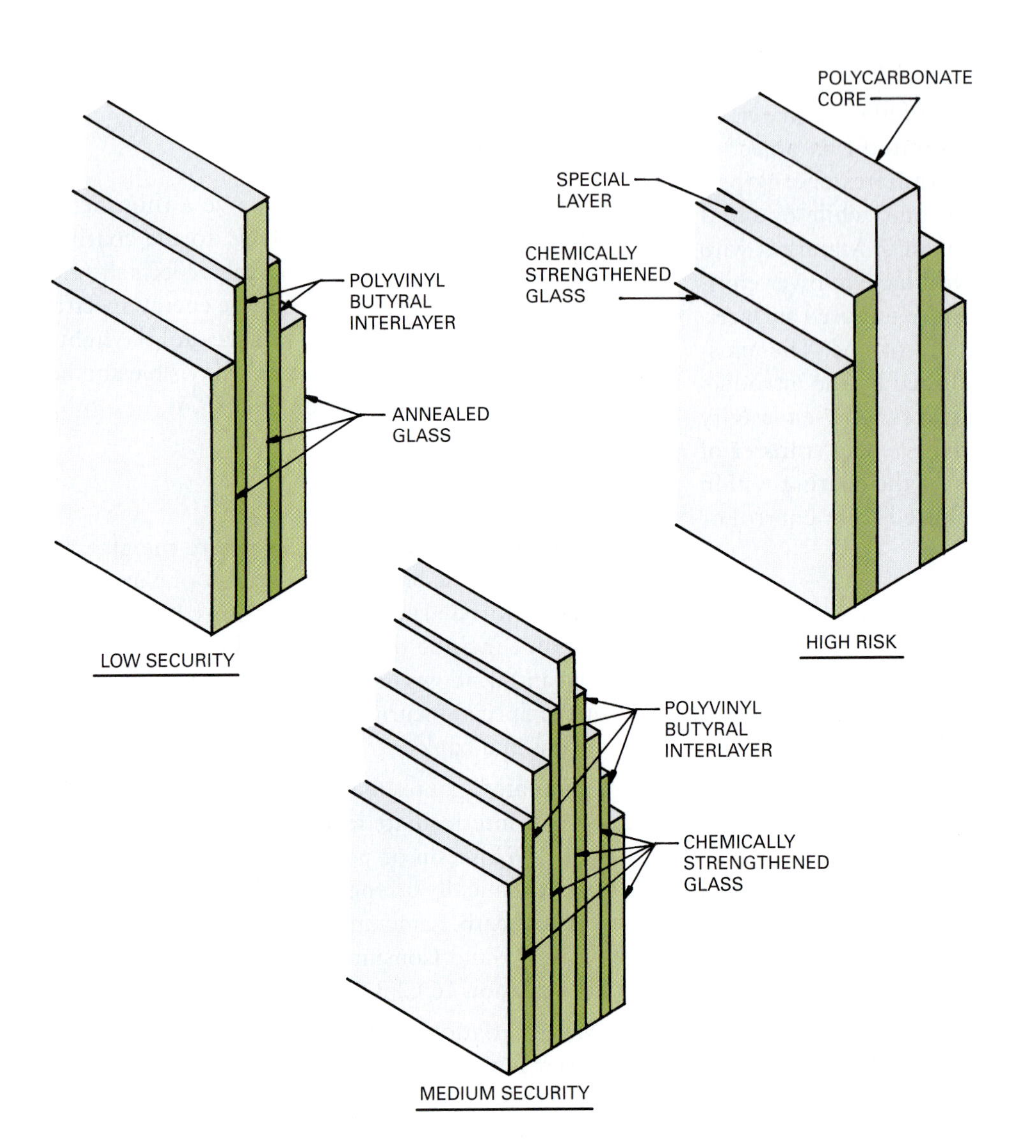

Figure 29.5 Several types of laminated glass used to increase the security of glazed openings.

Laminated glass for use in high-traffic, security areas is composed of three sheets of annealed glass and two intermediate layers. It will withstand impact by heavy wood and other small, lightweight objects. When more security is needed, a five-glass panel with four plastic intermediate layers is used to provide protection against attack by bricks, bottles, and chairs. A laminate for high-risk areas has a polycarbonate core from $\frac{1}{4}$ to $\frac{3}{8}$ in. (6 to 9 mm) thick, with special intermediate-layer materials plus chemically strengthened glass. It is made with multiple cores for greater strength, can withstand pounding by sledge hammers, and is used where maximum security is needed. Bullet-resistant laminated glass has a similar construction as those just mentioned but with additional layers of glass and polyvinyl butyral intermediate layers. It is available in thicknesses ranging from $1\frac{3}{16}$ to $2\frac{1}{2}$ in. (30 to 63 mm). Greater thicknesses are available on special order.

Another type of laminated glass is **acoustical glass.** It is manufactured with a sound-absorbing plastic sandwiched between two or more pieces of glass. The sound reduction is achieved because this soft interlayer allows the glass unit to bend in response to pressure from sound waves. It is also available in insulated constructions, thus reducing heat loss and gain as well as sound transmission. Acoustical glass comes in thicknesses of $\frac{1}{4}$, $\frac{1}{2}$, and $\frac{3}{4}$ in. (6, 12, and 18 mm), with sound class ratings of 36, 40, and 41. It is commonly used for office partitions and in radio, television, and recording studios.

Insulating Glass

Insulating glass is a manufactured glazing unit composed of two layers of glass with an airtight, dehydrated air space between them. The insulated, or double-pane window, utilizes two panes of glass separated by a spacer with a hermetically sealed air gap between. A wide range of systems are designed for use on commercial and residential buildings. Examples of these are shown in Fig. 29.6. Additional designs, including

Figure 29.6 Typical details for insulated glass units.

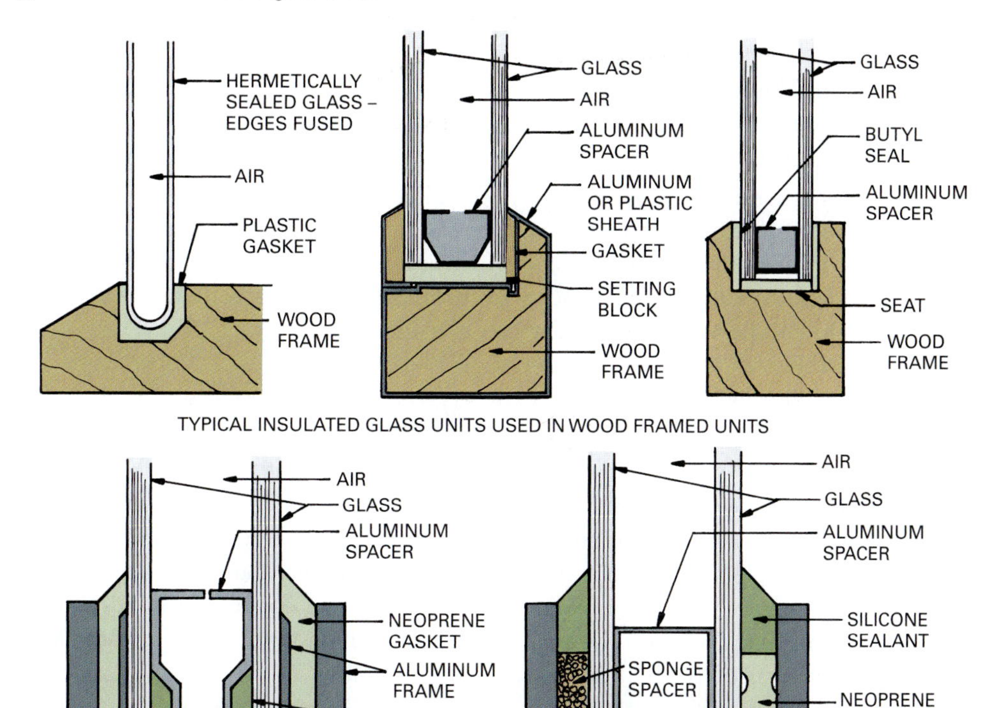

triple-pane units, are available from various manufacturers. Insulating glass lowers heating and cooling costs by reducing air-to-air heat transfer.

A better rating of thermal resistance can be achieved through the introduction of materials, such as argon or krypton, to replace the air inside the glazing cavity. The introduction of gases, which have a lower thermal conductivity than air, will reduce the conductive heat loss within the cavity, thereby reducing the overall heat transfer of the window unit.

When glass must be installed where it will be subjected to unusual thermal stresses or high wind loads, heat-strengthened or tempered glass is used. Insulating glass units also use tinted, reflective, clear laminated, and Low-E glass. The sizes of typical units are shown in Table 29.4.

Table 29.4 Sizes of Insulating (Float) Glass

Maximum size	90 × 144 in. (2286 × 36576 mm)				
Minimum size	15 × 20 in. (380 × 508 mm)				
Glass Thicknesses		**Air Space**		**Unit Thickness**	
in.	**mm**	**in.**	**mm**	**in.**	**mm**
3/32	2.5	1/4	6.0	7/16	11.0
1/8	3.0	1/4	6.0	1/2	12.7
5/32	4.0	1/4	6.0	9/16	14.2
3/16	5.0	1/2	12.0	7/8	22.2
1/4	6.0	1/2	12.0	1	25.4

Reflective Glass

Adhered reflective metal coatings reduce the heat buildup of the glazing associated with tinted glass. Here the majority of incoming radiation is simply reflected directly from the exterior surface. To understand how reflective glass reduces the amount of solar energy transmitted through a glass opening, consider the following discussion.

When solar energy strikes a glass surface it can be (1) transmitted through the glass, (2) reflected, or (3) absorbed in the glass. Of that portion that is absorbed in the glass, part of the energy will be re-radiated and convected (transmitted) inward and part of it will be re-radiated and convected outward. The solar heat gain consists of that portion that is transmitted through the glass plus the portion of the absorbed energy that is re-radiated and convected to the inside. The solar heat rejected is that portion of the original solar incidence that is returned to the exterior. This includes the energy reflected from the glass surface plus any absorbed energy that is re-radiated and convected outward.

Another source of energy transfer through a glass window results from convection. Convection is the transfer of heat by currents of air resulting from differences in air temperature and density in a heated space. The total heat gain of a window is the sum of solar heat gain and convective heat gain (Fig. 29.7). It is these factors designers consider when choosing a type of glazed opening.

Reflective glass has one surface covered with thin, transparent layers of metallic film. Several metals and mineral oxides are used, producing a variety of colors and heat reflection ratings. On single-sheet glazing, the metallic film is applied to the surface facing the inside of a building. On insulating units, the film is on the outside face of the glass facing the inside of the building (Fig. 29.8). Reflective glass is available in colors such as silver, green, blue, bronze, copper, and gold. Variations of these colors are available from different manufacturers (Fig. 29.9).

Visible light transmittance varies by coating but ranges in general from 8 to 45 percent. Coatings are available on clear or tinted heat-absorbing glass. Most reflective glasses are $\frac{1}{4}$ in. (6 mm) thick, but thicker sheets are available. A summary of the solar-optical properties of clear, tinted, and reflective float glass is given in Table 29.5.

Wired Glass

Wired glass is safety glass with a mesh of small diameter wires rolled into sheets of molten glass. When it breaks, the wires hold the glass shards together for increased safety. It also maintains its integrity as a fire barrier and is used in fire doors and window glazing. Tempered and laminated glazing has replaced wire glass in many applications.

Figure 29.7 Reflective glass reduces the amount of solar energy transmitted through a glazed opening.

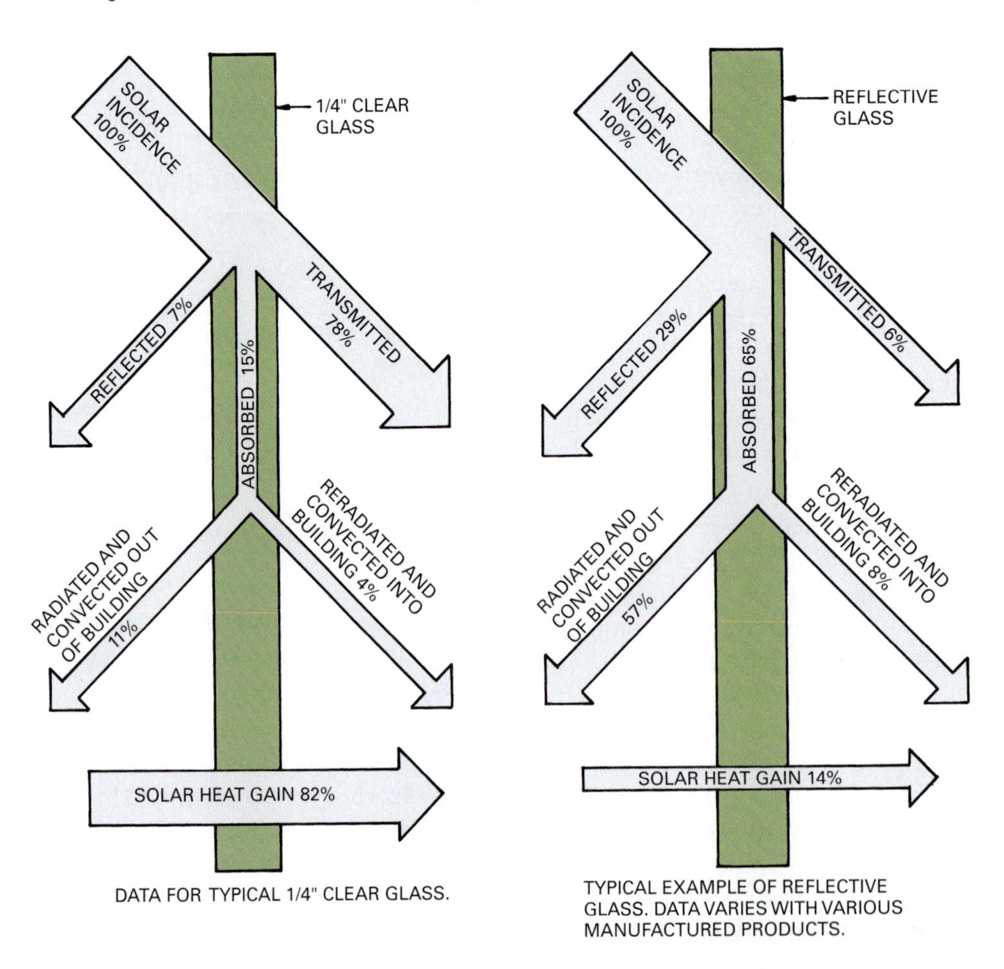

Figure 29.8 Insulating glass units use a transparent layer of metallic film to reduce energy transfer through the glass by convection.

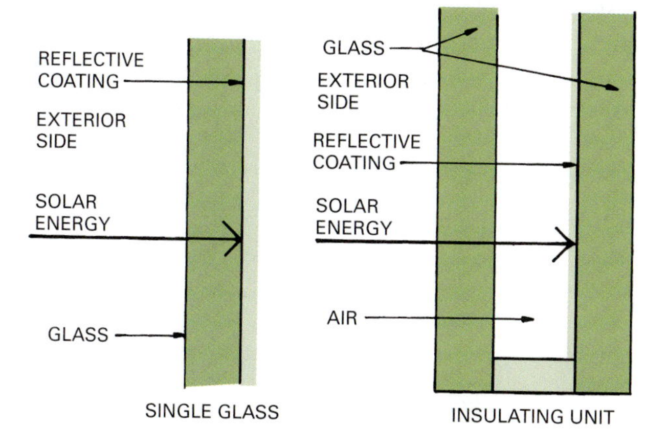

Figure 29.9 The facade is covered with high-performance reflective glass that lets daylight enter the building but reflects the rays of the sun. *(Image copyright Sailorr, 2009. Used under license from Shutterstock.com)*

Spandrel Glass

Ceramic-coated spandrel glass is an effective way to clad exterior wall areas. Spandrel panels eliminate the problem of corrosion, greatly reducing deterioration of a building (Fig. 29.10). Glass spandrel panels use $\frac{1}{4}$ in. (6 mm) heat-strengthened glass with an opaque ceramic frit fired on its surface. Some companies fire the frit on the exposed exterior surface, while others favor the protected interior surface. Spandrels are insulated, usually with 1 to 2 in. (25 to 50 mm) fiberglass or another acceptable insulation with a 1 to 2 in. (12 mm) air space and an aluminum foil vapor barrier facing the panel's interior side. Panel sizes vary, but typical sizes available include minimum panels of 18 × 33 in. (457 × 838 mm) and maximum panels of 84 × 144 in. (2130 × 3660 mm).

Bent Glass

Bent glass provides designers the opportunity to produce components with unique and dramatic looks (Fig. 29.11). It is available in clear float, tinted float, obscure, wire, Pyrex, patterned, and other types. Some forms of reflective glass can also be bent. Laminated safety glass and double-glazed thermal insulating units can be fabricated. Thickness and size specifications must be developed with the help of a manufacturer.

Architectural Beveled Glass

Architectural beveled glass provides a unique transition from outside to inside and is an outstanding architectural detail. The glass is made in a wide range of standard sizes and designs as well as custom designs. The pieces of beveled glass are assembled into panels with lead caming (Fig. 29.12).

Glass Blocks

Glass blocks are made as solid and cavity units and are laid in a mortar bed similar to masonry units. Various designs permit light to be diffused, reduced, or reflected. The translucence or transparency varies by manufacturer (Fig. 29.13).

Glass blocks are made by fusing two halves of pressed glass together, creating a space with a partial vacuum. This produces an insulating value equal to a 12 in. (305 mm) thick concrete wall (U-value 0.51, R-value 1.96). Glass blocks are very strong and provide excellent security.

Glass blocks serve well as solar glazing in walled areas facing in a southerly direction. Blocks are available with a variety of surface patterns that help provide privacy, brightness, and control light transmission. Blocks also can have a highly reflective thermally bonded oxide surface coating that reduces heat gain and transmitted light. Selected properties of glass block are given in Table 29.6.

Table 29.5 Solar Optical Properties of Glass

	Light (%)		Solar Energy (%)		U-value Btu/hr./ft.^{2}/°F	
	Transmitted	Reflected	Transmitted	Reflected	Winter	Summer
Single Glazed						
Sheet clear	91	8	86–89	7	1.13	1.02–1.03
Float clear	79–90	8	58–86	7	1.00–1.13	.98–1.03
Float blue-green	74–83	7	48–64	6	1.10–1.13	1.08–1.09
Float bronze	26–68	5–6	24–65	6	1.08–1.13	1.08–1.09
Float gray	19–62	4–6	22–63	5	1.06–1.13	1.08–1.09
Float reflective coating	8–34	6–44	6–37	14–35	0.90–1.11	0.89–1.12
Double Glazed						
Clear	78–82	14–15	60–71	14	0.49–0.57	0.55–0.62
Tinted	37–70	6–12	34–55	6–11	0.49–0.57	0.57–0.64
Reflective coating	7–30	6–44	5–29	14–35	0.41–0.49	0.47–0.58
Triple Glazed						
Clear	70–73	20	46–60	19–20	0.31–0.38	0.40–0.46
Tinted	11–56	8–17	25–46	8–16	0.31–0.38	0.40–0.46
Reflective coating	16–22	8–44	5–22	8–44	0.28–0.31	0.34–0.41

Figure 29.10 Glass spandrel panels with an opaque ceramic frit fired on the surface are used to conceal structural elements on fully glazed facades. *(Image copyright Roman Sigaev, 2009. Used under license from Shutterstock.com)*

Figure 29.11 Bent laminated glass is used in a variety of applications. *(Image copyright iofoto, 2009. Used under license from Shutterstock.com)*

Figure 29.12 Beveled glass provides a unique decorative appearance. *(Image copyright Maureen Rigdon, 2009. Used under license from Shutterstock.com)*

Figure 29.13 Glass blocks serve well as solar glazing in walled areas faci ng in a southerly direction. *(Image copyright Carlos Neto, 2009. Used under license from Shutterstock.com)*

Table 29.6 Properties of Single Cavity Glass Block

	Nominal Sizes		Compressive Strength (psi)	Light Transmission (%)	U-value
	3″ thick	4″ thick			
Clear	3 × 6 6 × 6 4 × 8 8 × 8 4 × 12 6 × 8 12 × 12	6 × 6 8 × 8 12 × 12	400–600	75	.52
Clear with reflective coating		8 × 8 12 × 12	400–600	5–21	.56
Light-diffusing pattern		8 × 8 12 × 12	400–600	39	.60
Decorative pattern	3 × 6 6 × 6 4 × 8 4 × 12 8 × 8 6 × 8 12 × 12		400–600	20–75	.65

Smart Glazing

The past decades have seen the development of a number of new, dynamic glazing products that have the ability to respond to changing environmental conditions.

Chromogenic glazing materials display an ability to change their optical properties in response to fluctuations in solar radiation intensity, changes in temperature, or through the introduction of electric charges (photo chromic, thermo chromic, electro chromic). The transmittance, reflectance, and absorption characteristics of the glazing are altered according to changing environmental conditions. These products offer a new way to regulate solar gains over the range of daily and yearly environmental conditions.

Figure 29.14 Photo-voltaic glazing provides enclosure and light for a building while simultaneously generating electricity.

Studies conducted to test control strategies find that electrochromic windows have excellent optical clarity; no coating aberrations; uniform density of color across their entire surface during and after switching; smooth, gradual transitions when switched; and excellent synchronization (or color-matching) between groups of windows during and after switching.

Photo-Voltaic Glazing

Research is now concentrating on the development of building integrated photo-voltaic (BIPV) systems. This new technology will enable glazing to not only provide enclosure and lighting for a building but also to simultaneously generate electricity.

A typical PV glazing unit uses two panes of clear or tinted glazing with the photo-voltaic modules adhered to the front of the window's inside pane (Fig. 29.14). Most PV glazing is available only by special order. The PV cells can be spaced in a variety of patterns and densities to aid in blocking sunlight from interior spaces.

Review Questions

1. What are the basic types of glass?
2. What type of glass is used in most construction applications?
3. What are the two types of float glass?
4. What are the grades of sheet glass?
5. How does a small scratch or nick in glass reduce the tensile strength of a sheet?
6. How does the thermal conductivity of glass compare with that of metal?
7. Why is glass resistant to corrosion?
8. How does glass rate on optical clarity?
9. Which types of soda-lime-silica glass are annealed?
10. List the types of heat-treated glass from weakest to strongest.
11. In addition to heat treating glass, how can it be strengthened?
12. List the types of surface finishes commonly used on glass.
13. What type of glass helps reduce heat transfer into a building in the summer and heat loss to the outside during winter?
14. How does tinting affect the performance of glass?
15. What are the main reasons for using laminated glass?
16. How does insulating glass differ in construction from laminated glass?
17. What three things occur to produce solar heat gain when solar energy strikes a glass surface?
18. How much visible light is transmitted through reflective glass?
19. What are the two important features of wired glass?
20. What type of finish is used on spandrel glass?

Key Terms

acoustical glass
annealing
chemical strengthening
chromogenic glazing
float process
insulating glass
laminated glass
low-E glass
reflective glass
shading coefficient (Sc)
tempered glass

Activity

1. Collect samples of glass products used for glazing, identify each, and describe their properties. Glass manufacturers often have small sample pieces available.

Additional Resources

Schittich et al., *Glass Construction Manual*, Birkhauser Edition Detail.

Glass Block Installation Manual, Pittsburgh Corning Corp., 800 Presque Isle Drive, Pittsburgh, PA 15239.

Merritt, F. and Rickets, J., *Building Design and Construction Handbook*, McGraw-Hill Publishing Co., New York.

Spence, W. P., *Encyclopedia of Construction Methods and Materials*, Sterling Publishing Co., 387 Park Avenue South, New York 10016.

Spence, W. P., *Windows & Skylights*, Sterling Publishing Co., 387 Park Avenue South, New York 10016.

See Appendix B for addresses of professional and trade organizations and other sources of technical information.

CHAPTER 30

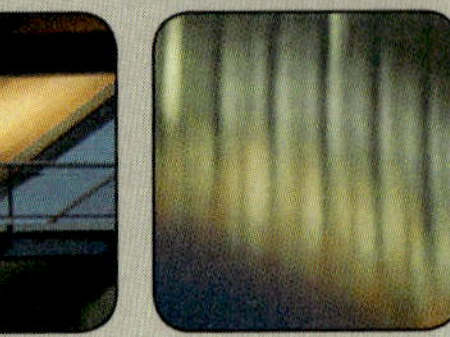

Doors, Windows, Entrances, and Storefronts

LEARNING OBJECTIVES

Upon completion of this chapter, the student should be able to:

- Know the types of doors, windows, entrances, and storefronts available.
- Apply this information to the decision making processes as these units are selected for a building.
- Understand the properties of various units as they relate to fire, security, privacy, and operation.

Build Your Knowledge

For further study on these materials and methods, please refer to:

Chapter 19 Products Manufactured from Wood
Topic: Wood Doors and Windows

Chapter 29 Glass
Topic: Glass Products

DOORS

Doors are used to provide security, privacy, thermal, and fire protection to access openings in interior and exterior walls. The choice of a door depends on the volume of traffic, security requirements, and desired appearance. In residential buildings, interior doors provide privacy and sound control while exterior doors provide security, protection from the weather, and an important part of the exterior appearance. In commercial buildings, doors must permit ingress and egress as required by building codes. They are subject to considerable use and must be durable with hardware able to withstand continued use. Other doors provide access to large equipment, such as vehicles and aircraft. They are usually mechanically opened and closed due to their size and weight. Doors are also used to divide large interior spaces into smaller rooms.

Doors are available manufactured from wood, hardboard, plastic, metal, and glass. Typical residential and commercial door types are shown in Figs. 30.1 and 30.2.

Hollow-Core Metal Doors

Hollow-core metal doors are available in steel and aluminum. They are used on interior and exterior openings and are available in a wide range of designs (Fig. 30.3). Exterior surfaces may be flush or panelized, integrate windows or louvers, and finished with a baked enamel finish on a smooth or textured surface. Decorative doors may be covered with plastic veneers of wood grain or a variety of colors and patterns.

The exact construction of hollow metal doors varies by manufacturer, but two types of designs are common. One uses a tubular framework over which the face panels are welded. Insulation and sound-deadening material are placed in the hollow core. Another type uses a core material, such as kraft paper honey-comb, polyurethane, polystyrene, a steel grid, mineral fiberboard, or vertical steel stiffeners, over which metal panels are bonded. The door may be seamless, full-flush, flush-panel, or industrial-tube construction (Fig. 30.4). Seamless doors have the edge seams hidden by welding and grinding them smooth. Full-flush doors have visible seams on the edges but no seams on their faces. Flush-panel doors are either of a stile-and-panel or a stile-and-rail construction. Face panels may be slightly recessed. Industrial tube doors are made using tubular steel stiles and rails with recessed panels. Metal doors are classified by grades and models, as shown in Table 30.1.

Figure 30.1 Common door types and styles.

Figure 30.2 Common types of door operation.

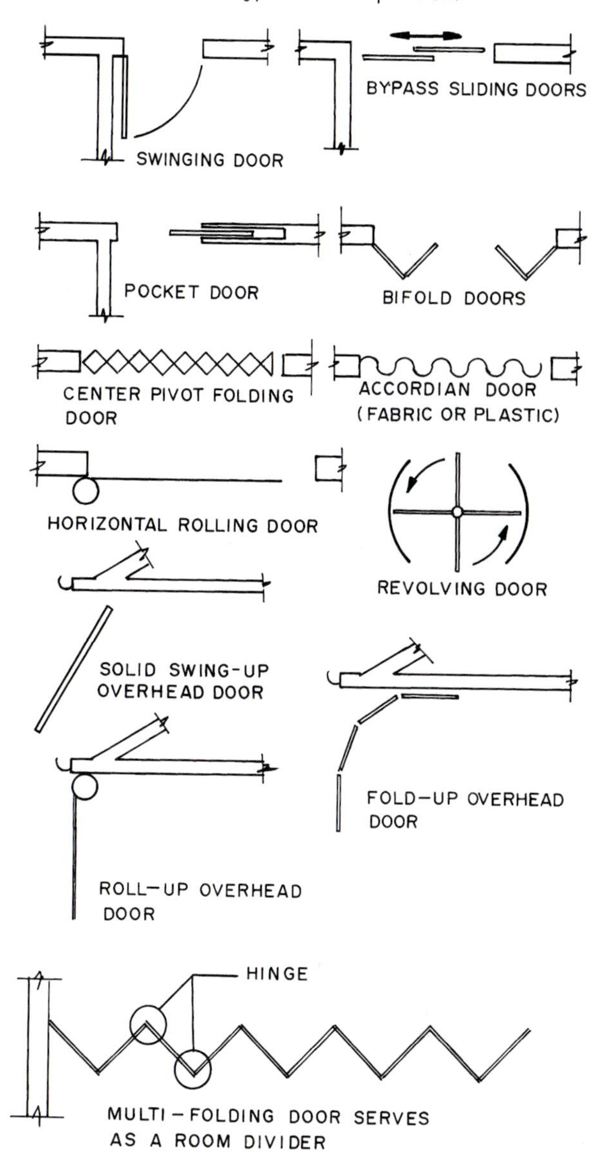

The higher grades utilize a thicker-gauge metal. Hollow steel door types are designated by letter symbols, such as F, for flush; L, for louvered; VL, for vision lite; and FG, for full glass. Some typical construction details for the top and bottom edges of these doors are shown in Fig. 30.5. Some of the commonly used meeting stiles for double hollow metal doors are shown in Fig. 30.6.

Hollow Metal Frames

Hollow metal frames are available in steel or aluminum and are supplied knocked down or preassembled. A wide range of designs can accommodate many types of interior and exterior wall constructions (Fig. 30.7). Hollow metal frames are anchored to masonry walls with a metal strap or wire loop (Fig. 30.8). Usually three anchors are required on each jamb. Several strap designs are used to secure hollow metal door frames to wood or metal studs. Brackets secure door bases to floors (Fig. 30.9).

The frame and the door are prepared to receive a lockset and strike plate. The holes, recesses, and screw holes are precut into the frame (Fig. 30.10). Assembled door frames frequently have doors installed before a unit ships to a job. The required steel gauge for door frames is related in standards to the door grade and

Figure 30.3 A typical insulated core metal door. *(Image copyright Beth Van Trees, 2009. Used under license from Shutterstock.com)*

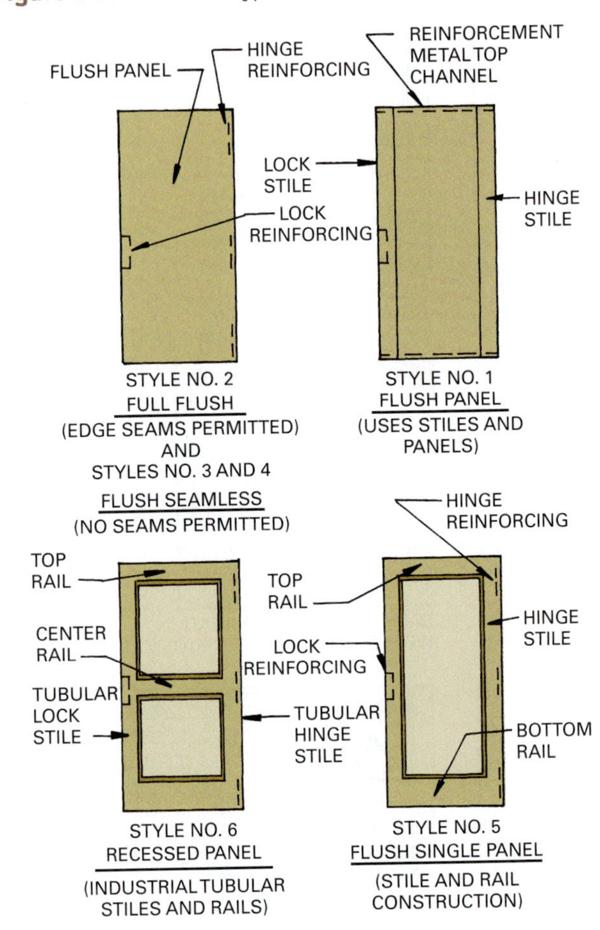

Figure 30.4 Standard types of steel doors.

model. The higher the door grade the thicker the steel used for the door frame.

Metal Fire Doors and Frames

Fire doors are an assembly of a door and frame manufactured and installed to give protection against the passage of fire through an opening in a wall. The design, location, and installation are controlled by national standards and local building codes. Approved fire door assemblies are required to meet the test requirements specified in ASTM E152, Methods of Fire Tests of Door Assemblies. Standards for fire doors and windows are outlined in ANSI/NFPA 80, a publication prepared by a committee of the National Fire Protection Association and approved by the American National Standards Institute.

All fire door assemblies must bear the label of an approved agency and include the name of the manufacturer, the fire protection rating, and the maximum transmitted temperature end point. Some assemblies are automatic, keeping the door open and closing it automatically in case a fire raises the temperature or exposure to smoke.

Table 30.1 Grades and Models of Standard Steel Doors

Grade	Model
Grade I—Standard duty, $1^3/_8$" and $1^3/_4$" thick	Model 1—Full flush design, hollow metal and composite Model 2—Seamless design, hollow metal and composite
Grade II—Heavy duty, $1^3/_4$" thick	Model 1—Full flush design, hollow metal and composite Model 2—Seamless design, hollow metal and composite
Grade III—Extra heavy duty, $1^3/_4$" thick	Model 1 and 1A[a]—Full flush design, hollow metal and composite Model 2 and 2A[a]—Full flush design, hollow metal and composite Model 3—Stile and rail, flush panel

[a]1A and 2A are made with a heavier gauge metal than models 1 and 2.
Courtesy The Steel Door Institute

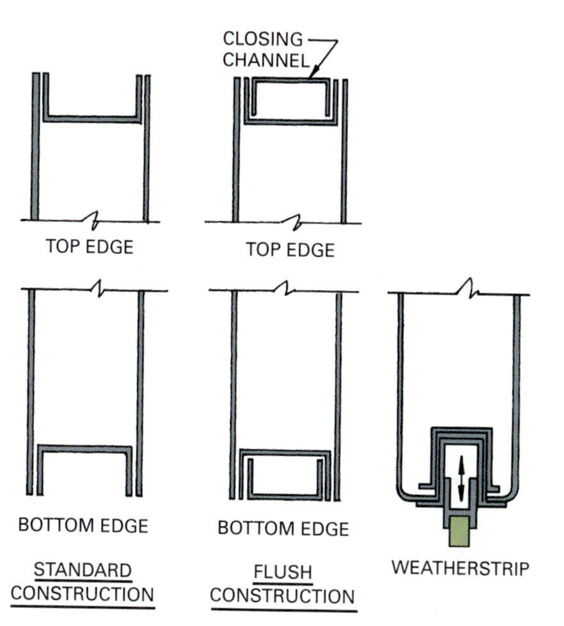

Figure 30.5 Some commonly used top and bottom construction details for hollow metal doors.

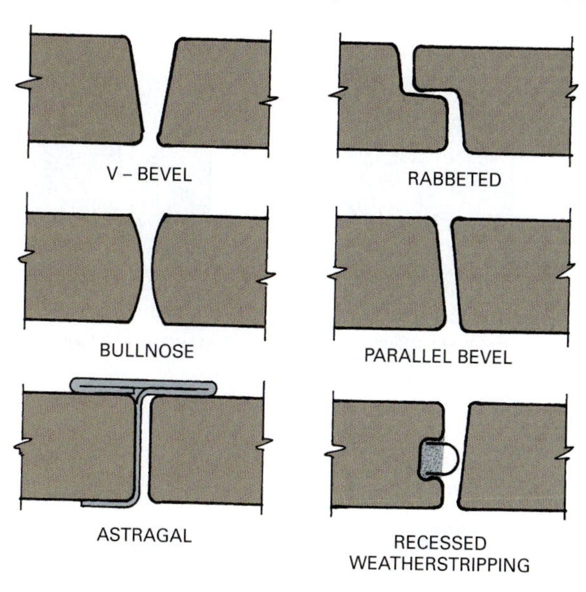

Figure 30.6 Typical meeting stile profiles used on metal doors.

Figure 30.7 Some typical metal door frame details.

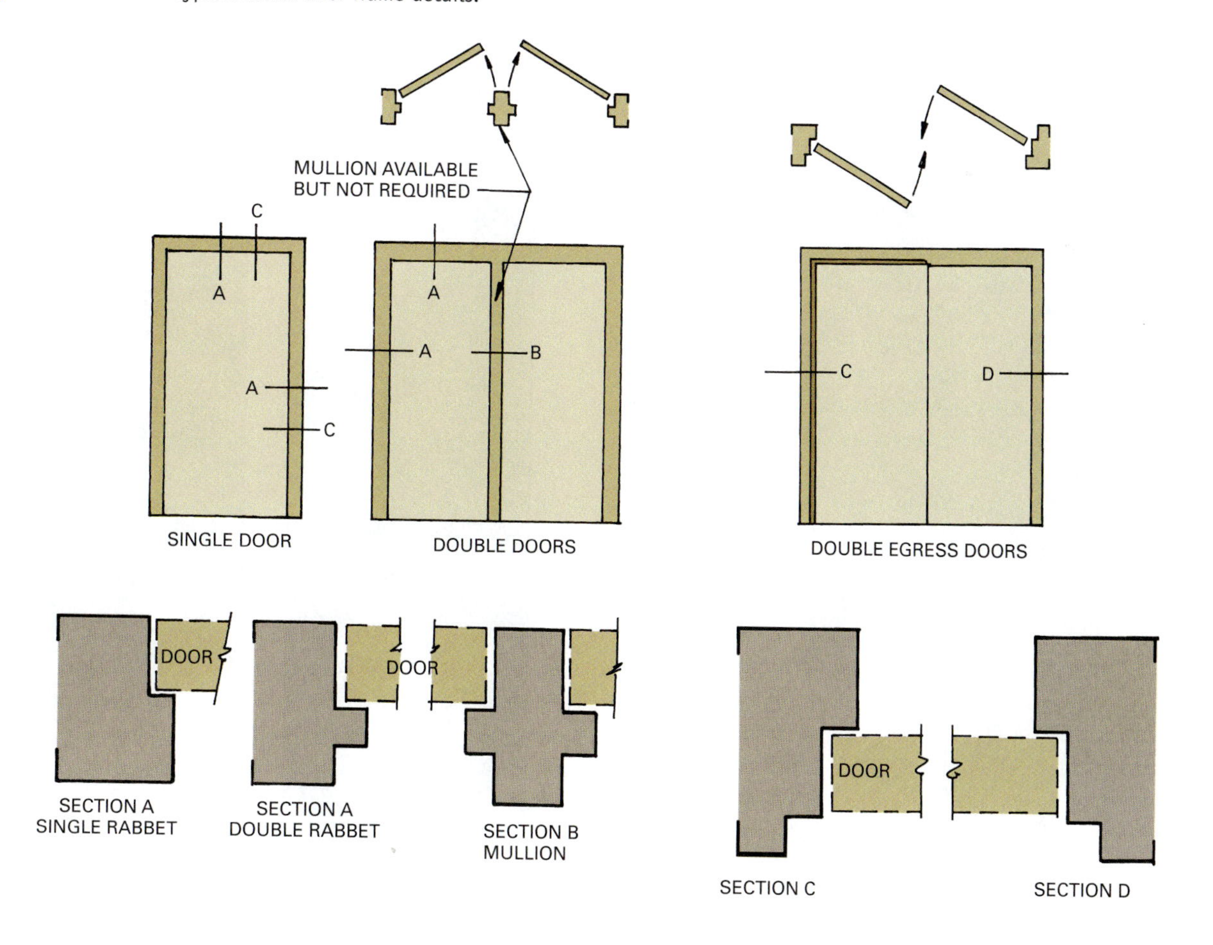

Figure 30.8 Some of the types of anchors used to secure steel door frames to walls.

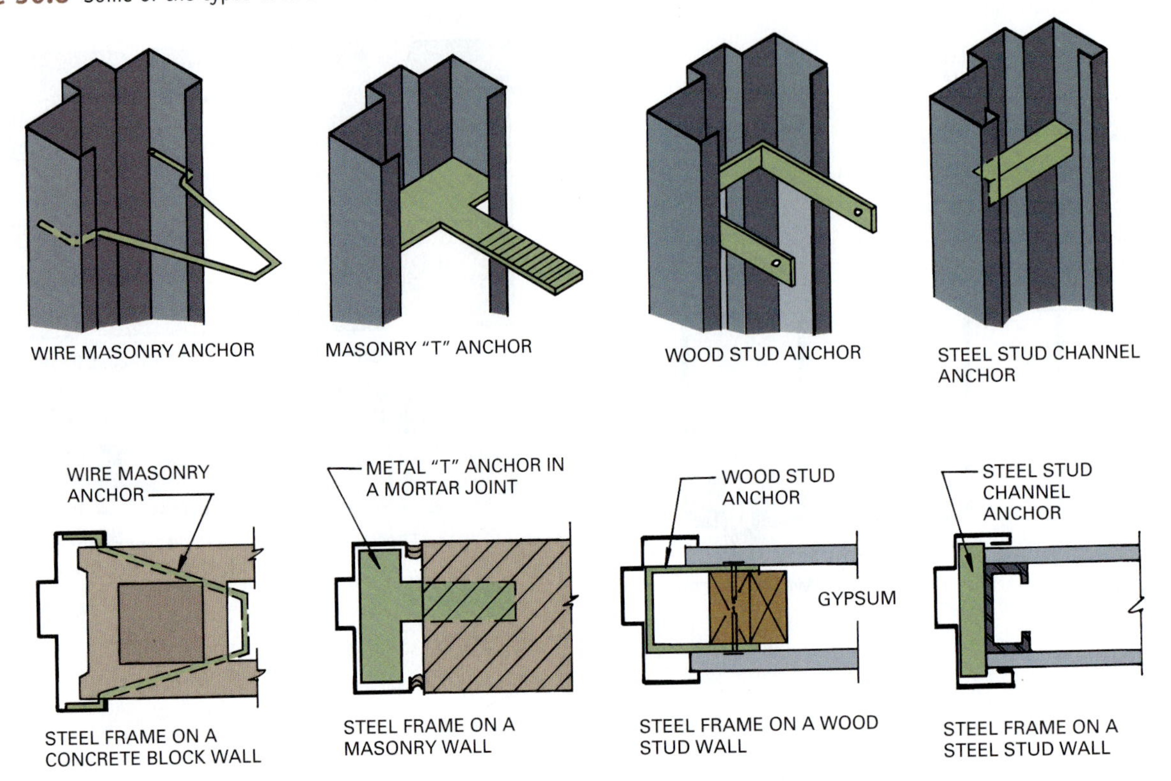

Figure 30.9 Metal door frames are anchored to a floor with different types of metal angles.

Figure 30.10 The metal door frame has recessed hinge and lockset back plates to receive the hinges and strike plate.

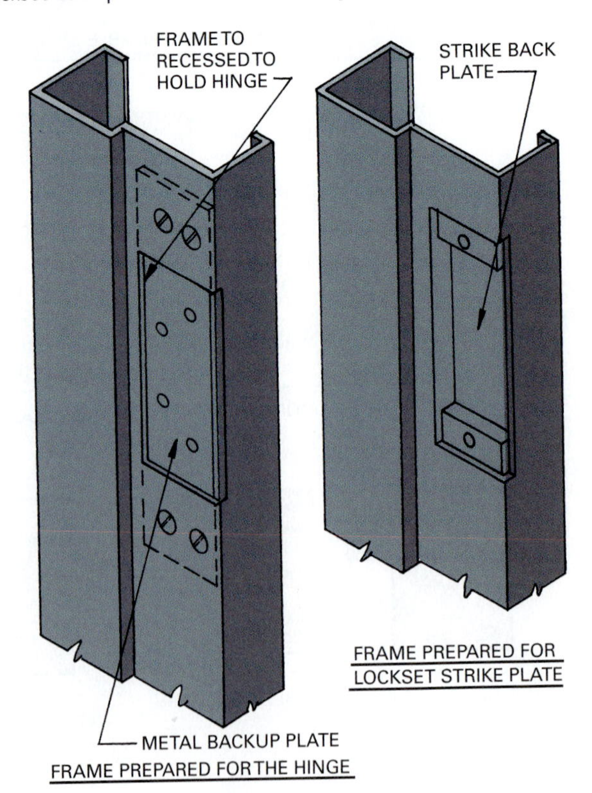

Fire doors are classified by the following designation systems:

1. Hourly rating designation.
2. Alphabetical letter designation.
3. A combination of hourly and letter designations.
4. Horizontal access doors use a special listing indicating the fire rated floor, floor-ceiling, or roof-ceiling assembly for which the door will be used.

The hourly designation indicates the duration of the fire test exposure in hours and is called the fire protection rating. The alphabetical letter designation in use follows:

Class A—Openings in fire walls and in walls dividing a single building into fire areas

Class B—Openings in enclosures of vertical communications through buildings and in two-hour rated partitions providing horizontal fire separations

Class C—Openings in walls or partitions between rooms and corridors having a fire resistance rating of one hour or less

Class D—Openings in exterior walls subject to severe fire exposure from outside a building

Class E—Openings in exterior walls subject to moderate or light fire exposure from outside a building

Special listings are tested in accordance with NFPA 251, Standard Methods of Fire Tests of Building Construction and Materials. It indicates the fire-rated assembly and its hourly rating. Fire door rating details are shown in Table 30.2. Labels on doors specify the class of door, which indicates the time interval the door will meet. For example, a Class A door has a 3 hr. time interval. Metal door frames have fire ratings of $\frac{3}{4}$, $1\frac{1}{2}$, and 3 hours. The hardware used must be fire rated to maintain the fire rating of the door and frame.

Typical types of materials used for fire doors include:

1. Wood core fire doors have wood, hardboard, or plastic laminate faces bonded to a solid wood core or a wood particleboard core.
2. Hollow metal fire doors may be of flush or panel design with a steel face of 20 gauge or thicker.
3. Metal clad fire doors have wood cores or stiles and rails with insulated panels. The face is 24 gauge steel or lighter.
4. Sheet metal fire doors are made using 22 gauge or lighter steel.
5. Tin clad fire doors have a solid wood core covered with a 24 to 30 gauge terneplate or galvanized steel facing.
6. Composite fire doors consist of some combination of wood, steel, or plastic laminate bonded to a solid core material.
7. Rolling steel fire doors are steel doors that move on an overhead barrel enclosed in a hood. They may have an automatic closing mechanism (Fig. 30.11).
8. Curtain-type fire doors have interlocking steel blades forming a steel curtain in a steel frame.

Special-purpose fire doors include a fire door and frame assembly. Acoustical fire doors, security fire doors, armored attack resistant fire doors, radiation shielding fire doors, and pressure resistant fire doors are examples. Also, a variety of automatic fire vents are manufactured (Fig. 30.12).

Wood and Plastic Doors

Commonly available types of wood doors are illustrated in Fig. 30.13. Flush doors have wood veneers or hardboard face panels bonded to a core of solid wood strips (called a solid-core door) or a core made of wood or kraft paper forming a hollow, honeycomb interior (called a **hollow-core door**). Wood doors can have openings cut for windows and louvers. Stile and rail doors have solid wood vertical and horizontal members enclosing panels made of wood, glass, or louvered sections. Some are made from pressed wood fibers, forming a paneled hardboard door. Fire-rated doors use a mineral composition core. Acoustical doors use sound-dampening cores or sheets of lead over the core. Solid wood core doors reduce sound transmission and retard fire better than do hollow-core doors.

Table 30.2 Fire Door Ratings

Class	Hour	Glazing	Location
A	3	No glazing	In fire wall and walls that divide a building into fire areas
B	$1^1/_2$	100 sq.in. of glazing	In enclosures of vertical communication through buildings
C	$^3/_4$	1,296 sq.in. of glazing per light	Openings in walls requiring a fire resistance rating of 1 hr. or less
D	$1^1/_2$	No glazing	Openings in exterior walls subject to fire exposure from the outside of the building
E	$^3/_4$	720 sq.in. glazing	Openings in exterior walls subject to moderate or light fire exposure from outside the building

Figure 30.11 A rolling steel fire door. *(Image copyright Joe Gough, 2009. Used under license from Shutterstock.com)*

Figure 30.12 This automatic fire door opens during a fire to let the rising smoke and gases clear from the building. *(Courtesy The Bilco Company)*

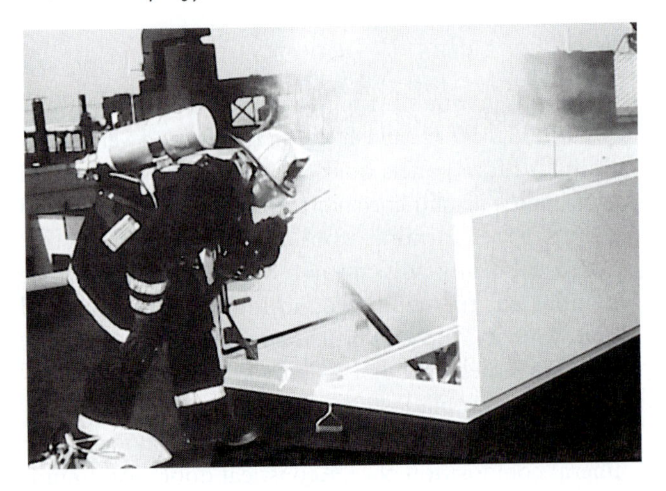

Glass Doors

Glass doors may be frameless with metal channels to hold a pivot hinging apparatus or have a very narrow metal frame on all sides or a wider frame that provides a stronger door for use in areas where heavy traffic is expected (Fig. 30.14). The glass must meet building code safety requirements.

Special Doors

Many doors are manufactured to meet special needs other than providing occupant access within a building.

Figure 30.13 Common types of wood doors.

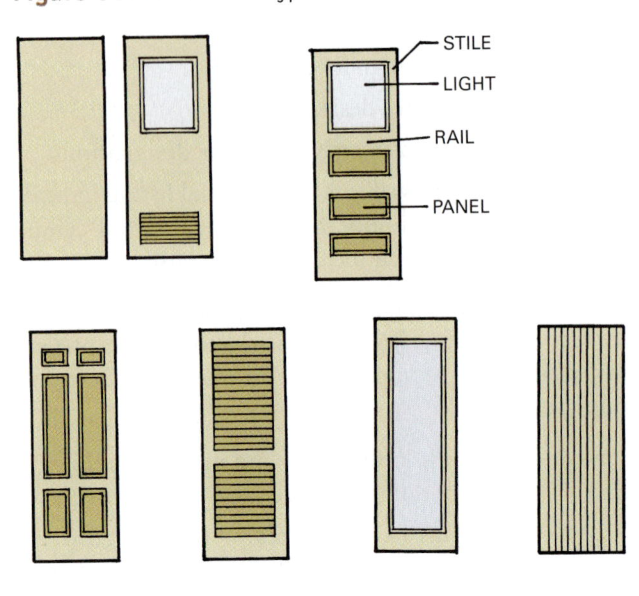

These include units that open to an area such as a roof (Fig. 30.15), various types of sliding doors, doors to resist blast impacts, and doors that provide an airtight or a watertight closure. Special doors and glazing are used for security purposes. Various types of folding and accordion doors are widely used in residential and commercial construction. There are many variations of rolling doors, such as those used on aircraft hangers and openings in manufacturing plants. Many are power activated. Various types of rolling grille doors are used to provide security for stores opening onto a public mall. Required overhead clearance issues need to be addressed to allow for motors and door and roller space.

WINDOWS

The basic types of windows used in residential construction are illustrated in Chapter 19. Various windows are manufactured primarily for use in commercial buildings. Windows are available made from solid wood, wood clad with plastic or aluminum, solid plastic, steel, stainless steel, aluminum, bronze, and composite materials.

A major consideration in the selection of windows is energy savings. Various types of energy efficient glazing are available, including low-E glass, double glazing, and glass with louver blinds set between the panes of glass for shading. Some double-glazed units have a gas inserted in the space between the panes. See Chapters 29 and 31 for additional information on glass and glazing. Windows must have airtight unions between both fixed and moving parts to reduce air

Figure 30.14 Some of the glass doors used in commercial construction.

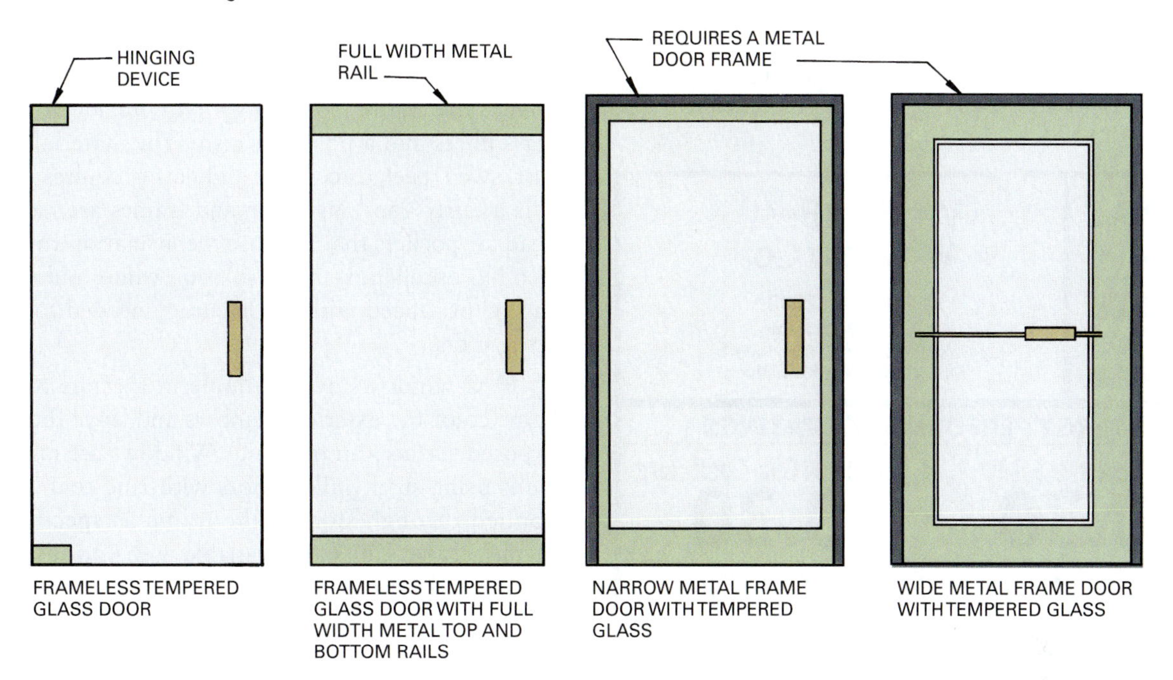

Figure 30.15 A roof scuttle provides safe and easy access to the roof. *(Courtesy The Bilco Company)*

infiltration. Metal units transmit heat and cold rapidly through their material, so designs must provide insulation, called thermal breaks, between adjacent interior and exterior parts.

Windows are also a major source of light to the inside space. From a design standpoint, they are a major feature of the appearance of an exterior. Windows are also a major source of natural ventilation.

Quality windows are tested and certified against ASTM standards for air leakage, water infiltration, uniform wind load structural requirements, and uniform load deflection.

The energy performance of fenestration (windows and other exterior openings) products is rated by the National Fenestration Rating Council (NFRC). The NFRC combines U-value, solar heat gain factors, optical properties, air infiltration, condensation resistance, and other characteristics into a uniform rating system to reflect annual energy performance. Products are certified and labeled, providing a means for comparing products manufactured by various companies. The NFRC certification label is shown in Fig. 30.16.

A wide range of locking and operating devices are available and should be considered as window units are selected. These include manual and electric operating devices, hinges and other hardware, and safety devices, such as a keyed lock or safety bar that limits how much a window can be opened.

Local fire codes also influence the choice of windows. These codes typically control the sizes of windows, height of sills above floors, and the ability to open a window from the inside without the necessity of special equipment. Windows used in fire walls must be fire-rated and labeled to restrict the spread of fire and

Figure 30.16 A generic example of the certification label of fenestration products rated by the National Fenestration Rating Council. The blank space is used to indicate solar heat gain, air infiltration, long-term energy performance, and other energy performance attributes. *(Courtesy The National Fenestration Rating Council)*

smoke. Security windows are tested to meet the standards required to provide resistance to forced entry. Finally, the material and finish must fit with the surrounding area and be able to withstand any corrosive elements to which it will be exposed.

Wood Windows

Wood frame windows are available with fixed and operable sashes. They may be clad with plastic or aluminum or left natural and painted. Wood windows provide better insulation than do metal and plastic windows. However, they swell and shrink as the moisture content changes. The cladding of wood window exteriors has reduced this problem considerably.

Wood windows can be installed in wood frame walls by nailing into a metal or plastic flange, through the sheathing, and into the wood studs and headers in the rough opening (Fig. 30.17). Some are designed with a wood molding that is nailed to the sides of the wood frame rough opening (Fig. 30.18). An example of another installation method involves the use of metal clips to secure a window in a wall framed with metal studs, as shown in Fig. 30.19. Windows come with various types of glazing and insect screens.

Plastic Windows

Plastic windows are available in the same basic types as described for wood windows. Structural and casement frames are made from polyvinyl chloride (PVC) and glass fibers and a polyester resin. The material will not rust, swell, peel, or corrode and never requires painting. It is a fairly good insulator, and frames are made with dead air pockets that increase the insulation value. PVC also has excellent sound insulation value, and members can be produced with the accuracy needed to provide airtight fits.

PVC windows are available in a range of colors. Some color the exterior members and leave the interior exposed surfaces an off white. Windows are fastened to walls using steel wall anchors with zinc-coated screws applied through slots in the frame as specified by a manufacturer. Units are available with various types of glazing and insect screens.

Metal Windows

Metal windows used in residential construction are most often aluminum. Although some types of steel windows are designed for use on residential buildings, they are mainly used for commercial, public, and industrial buildings.

Steel Windows

Steel windows are made from structural grade steel. Residential grade windows are made from lighter gauge steel than windows designed for use in industrial and commercial buildings, which are classified as intermediate grade. Heavy duty intermediate grade windows are used where the size of a window requires additional strength or must meet other conditions, such as resisting high winds. Manufacturers' catalogs list the gauges and recommended uses for their products.

Steel window units are rigid and strong and require narrower members than other materials, producing a smaller sight line. They are available primed or galvanized and ready for painting. Some types have a polyvinyl chloride or urethane plastic coating to provide permanent protection. The units are weather stripped and available with single and double glazing. Manufacturers provide a range of stock sizes but will produce custom-designed units of any size.

Steel windows used in residential construction are typically casement type, but other designs are available, as shown in Fig. 30.20. A section through a typical steel window is shown in Fig. 30.21.

Figure 30.17 An installation detail for a clad casement window secured in a wood frame wall by nailing through the plastic flange. *(Courtesy Pella Windows and Doors)*

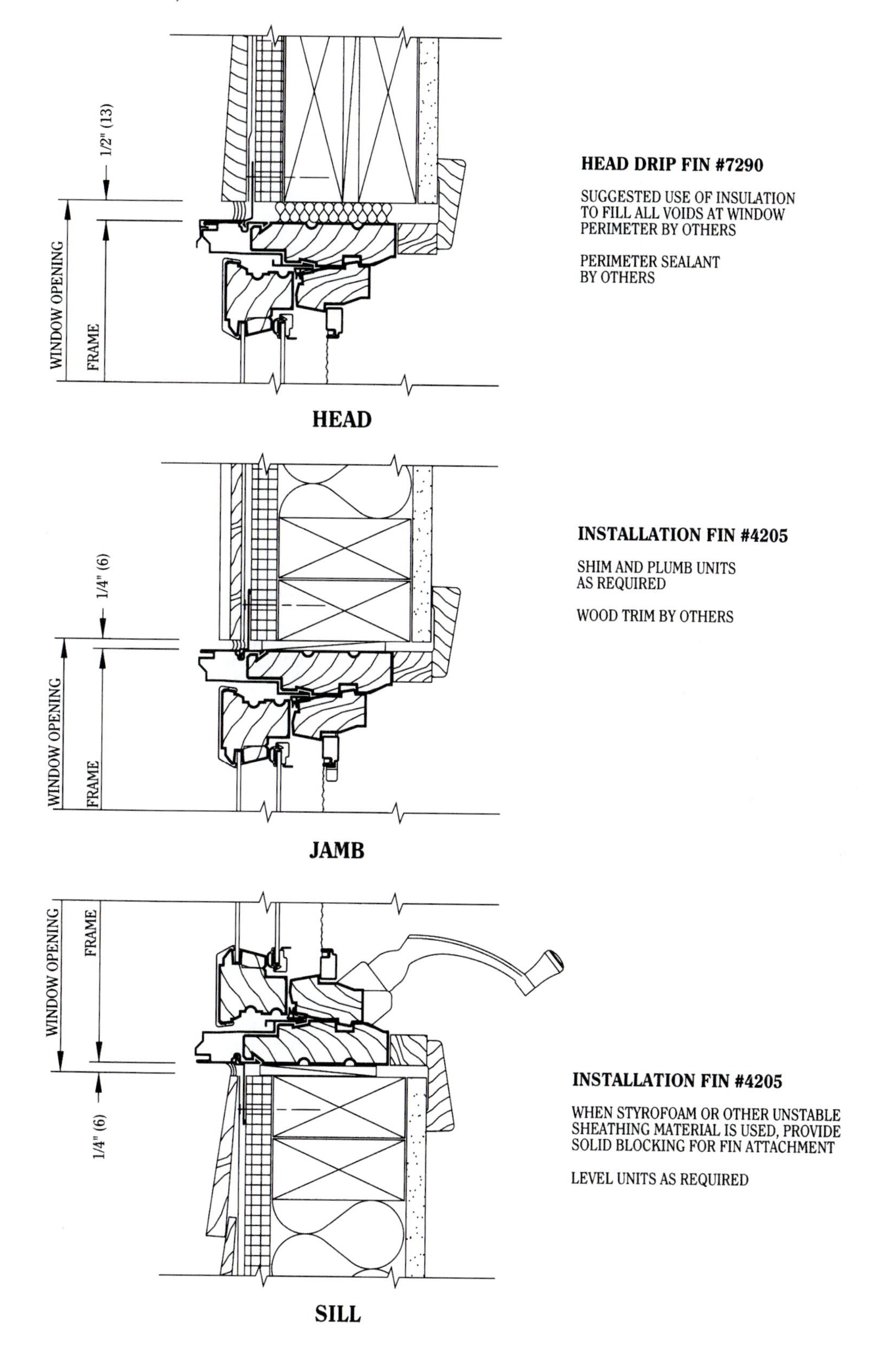

Figure 30.18 An installation detail for a wood frame double-hung window in a wood frame wall installed by nailing through the exterior molding. *(Courtesy Pella Windows and Doors)*

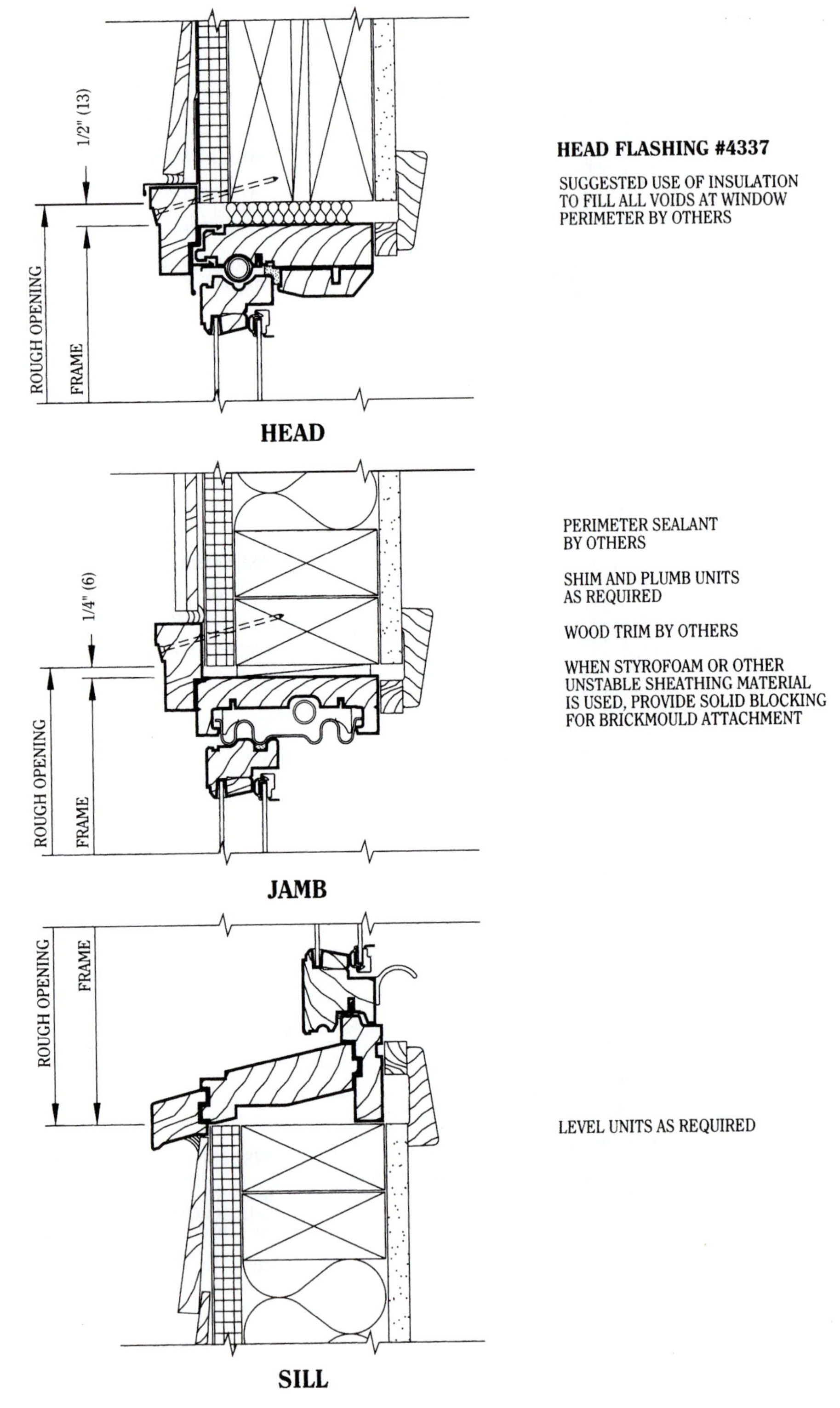

Figure 30.19 An installation detail for a clad double-hung window installed in a steel stud frame wall with clips that are secured to the metal studs. *(Courtesy Pella Windows and Doors)*

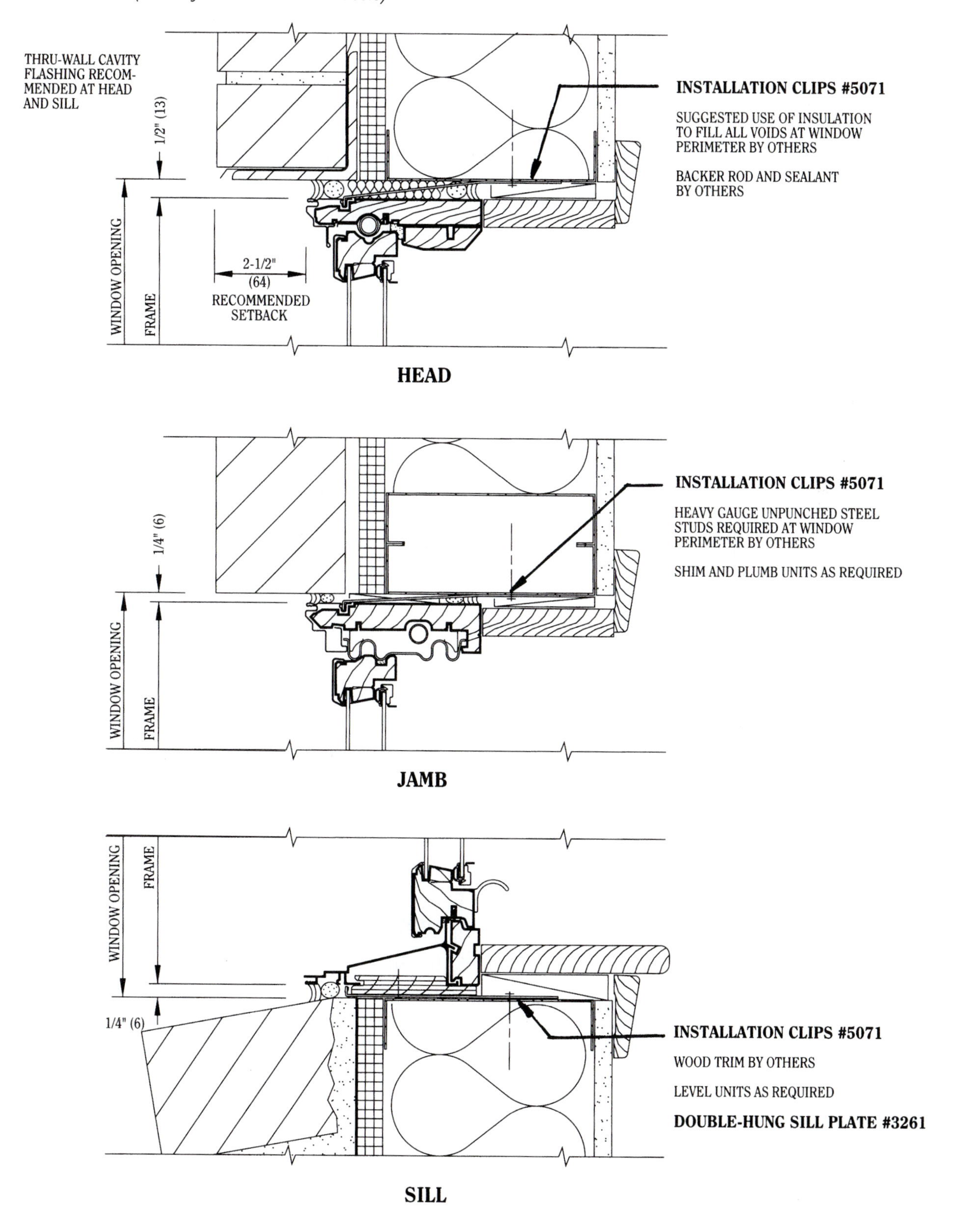

Figure 30.20 Typical types of steel windows available.

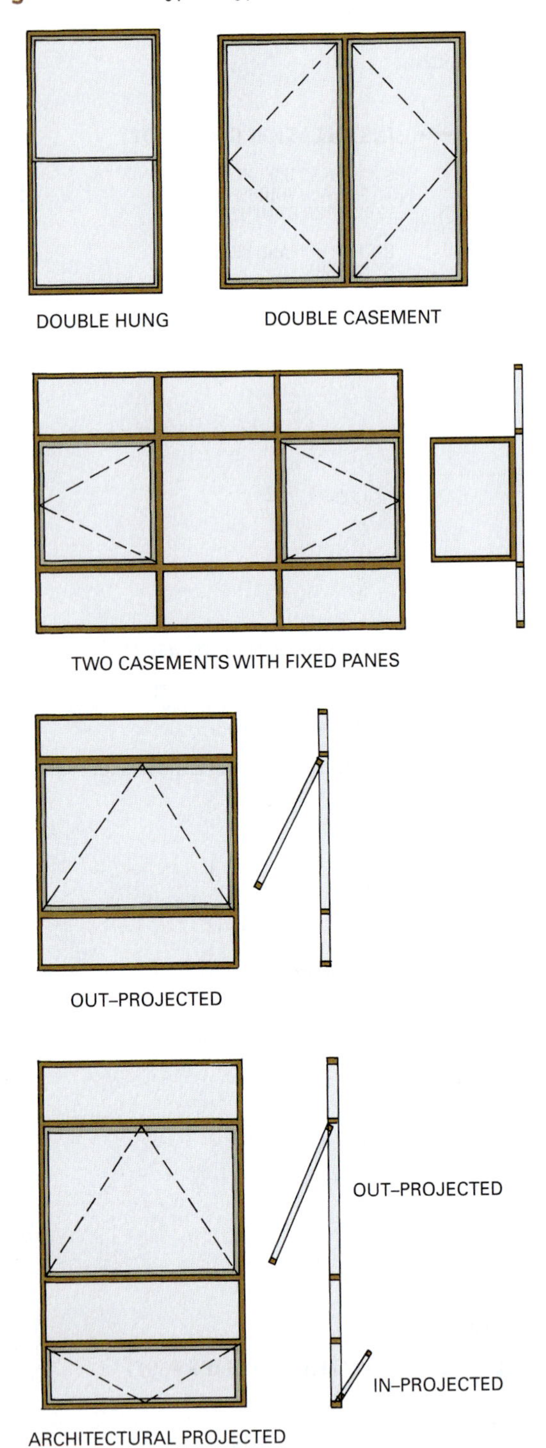

Figure 30.21 A section through a typical steel window.

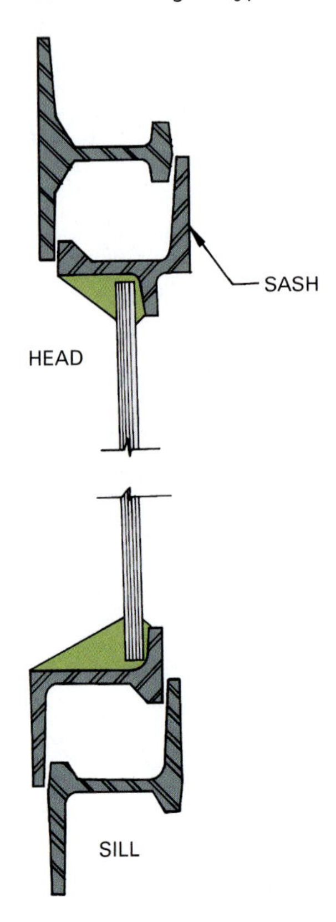

Stainless Steel Windows

Stainless steel windows are usually more expensive than steel or aluminum windows but provide long life and may save money on replacement costs. They are made from various types of stainless steel formed into structural members that are welded to form a unit. The sizes and types of window vary by manufacturer and many are custom built to an architect's design. Typical types available include folding, awning, casement, and various hinged types.

Aluminum Windows

Aluminum windows are made from extruded aluminum structural members. Since aluminum is easy to work, complex unit designs are readily produced. Aluminum has good structural properties and resists corrosion and damage from weather. Various organic protective coatings can be applied in areas where damage may occur. It can also be colored using anodized finishes. Additional information on aluminum is given in Chapter 16.

Aluminum windows are assembled using both mechanical connections and welding. Any connectors, fasteners, hardware, or anchors must be aluminum or a material (such as stainless steel) that is compatible with aluminum so corrosion does not occur from galvanic action.

Several aluminum alloys are used in the manufacture of aluminum windows. These affect the properties of a product and vary with the purpose for which a window will be used. Standards for the quality of aluminum windows are established by the Architectural Aluminum Manufacturers Association. These vary with the type of window and its intended use. These windows are of three types: residential, commercial, and monumental. They are available in the same operating types as described for plastic and steel windows. Some manufacturers have other special designs and will produce custom-designed units for special applications. A section through a typical aluminum window is shown in Fig. 30.22. Additional window details are discussed in Chapter 31.

Figure 30.22 A section through the sill of an aluminum frame window showing the thermal barriers.

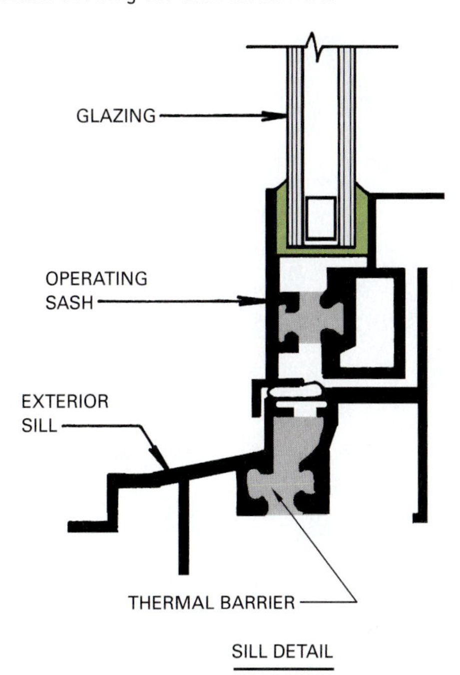

Composite Materials

Increasing use is being made of composite materials in construction. Composites are formed by combining several different materials. One such product is shown in Fig. 30.23. This window is made entirely of composites (a cellular PVC material) and wood byproducts, producing a material that looks like wood but will not split, warp, or rot. It has twice the insulation value of wood and is available in several colors.

Figure 30.23 These windows are made from a composite material composed of wood by-products and a cellular PVC material. *(Courtesy Jeld-Wen, Inc.)*

Special Windows

A number of special windows are manufactured for special applications, such as skylights, security, pass, and storm windows. Skylights and roof windows perform the same basic function as other windows, but they are located in a roof. Skylights can have fixed or operable glazing, or the sash on roof windows can be opened to provide ventilation. They may be installed as single units, as is common in residential construction, or in strings or clusters, admitting considerable light (Fig. 30.24). Skylights are glazed with glass or plastic lights. Some operable units open manually with a crank, and other types are electrically opened. They may be designed to blow out the glazing if an explosion occurs or to open automatically if the temperature increases, serving as a fire vent.

Most skylights used in commercial, industrial, and monumental work come with aluminum frames. The small units used in residential applications have wood frames. Small-domed and barrel skylights may have single or double glazing.

Skylights are available for enclosing large spaces, such as malls, building entries, or pool areas (Fig. 30.25). They also find use in enclosed walkways, protecting interiors from the weather and providing abundant light (Fig. 30.26). They are glazed with glass and plastic materials.

Security windows (also called detention windows) are tested to classify their resistance to forced entry. This includes testing the locking device, the impact resistance

Figure 30.24 Skylights are used to provide natural daylight from above. *(Courtesy Eva Kultermann)*

Figure 30.25 Commercial and public buildings make extensive use of skylights. *(Image copyright Olexa, 2009. Used under license from Shutterstock.com)*

Figure 30.26 Enclosed walkways use full glazing to provide ample daylight. *(Image copyright Sailorr, 2009. Used under license from Shutterstock.com)*

of the sash and frame, the resistance of security bars, and the resistance of the glazing. The degree of security is specified as one of four classes:

Class 1: Minimal

Class 2: Moderate

Class 3: Medium

Class 4: Relatively high

In some occupancies, window security is specified in building codes that require a high-grade window in certain areas of a wall. If windows are protected by devices, such as bars, screens, or shutters, provisions for meeting exit requirements in fire codes must be met. Manufacturers indicate the ANSI/ASTM security standards their windows meet. In addition to a secure window, the window frame, locking mechanism, hinges, and glazing must also be secure. Laminated glass used on security windows is discussed in Chapter 29.

ENTRANCES AND STOREFRONTS

Entrances and storefronts are typically designed with extruded aluminum members. The aluminum frame is typically filled with flat architectural glass products, such as tempered and heat-strengthened glass, laminated glass, and insulating glass. Storefront systems differ from curtain walls in that they are designed to span one or two floors only.

Typically, tempered glass doors are available with door closers, top and bottom pivots, locks, and push-pulls. Units may include one or more doors, sliding glass panels, and transoms (Fig. 30.27). The doors may be manually or automatically opened. The metal frame, typically

Figure 30.27 Entrances and storefronts are typically designed with extruded aluminum members. *(Image copyright Khafizov Ivan Harisovich, 2009. Used under license from Shutterstock.com)*

aluminum, is usually anodized, and a variety of colors are offered. Stainless steel frames are also available. Glazing is available in clear, obscure, and several colors.

Entrances may be custom-built for a particular opening using stock door sizes and custom-built frames. The frames are mechanically joined and then welded. Figure 30.28 shows a high-traffic entrance system for shopping malls, schools, concert halls, and convention centers. The doors, frames, and hardware are all reinforced heavy-duty assemblies. The door walls are heavier, and the hinges and pivots are directly mounted into the reinforced frame.

Storefronts use various types of entrance units plus glazing and solid panels of various types. When designing a glazed storefront, it is essential to include wind-load calculations. Many window manufacturers have this information for their products.

Figure 30.28 Storefront entrances are often custom built for a project. *(Image copyright Irina Fischer, 2009. Used under license from Shutterstock.com)*

Entrances and storefronts are subject to a variety of requirements in local building codes. Among these are considerations of fire resistance; wind and snow loads; hazards, such as human impact loads; the means for supporting the doors and related glazing; hardware; locking arrangements; requirements for power-operated doors; and the size and number of doors required.

Review Questions

1. Describe the construction of various types of hollow-core metal doors.
2. Explain how hollow metal door frames are secured to walls and floors.
3. Identify each of the letter designations used on metal fire doors.
4. List some of the specialty doors available.
5. What are some of the types of energy-efficient glazing available?
6. What association has established energy performance rating for fenestration products?
7. What two methods are commonly used to secure wood windows to wall framing?
8. What are the grades of steel windows?
9. How do windows made of composite materials compare with solid wood windows?
10. What are the classes of security windows?
11. What are typical building code requirements placed upon entrances and storefronts?

Key Terms

hollow-core door

fenestration

Activity

1. Walk through a commercial development and list the various doors, windows, entrances, and storefronts used. Record the materials and glazing used. Check your local codes to see if those you observed meet the requirements specified.

Additional Resources

How to Install Interior and Exterior Door Frames, Wood Molding & Millwork Products Association, Wood-land, CA.

GANA Glazing Manual, *GANA Fully Tempered Heavy Glass Doors and Entrance Systems Design Guide*, Glass Association of North America, Topeka, KS.

Performance Specifications and Standards for Doors and Windows, American Architectural Manufacturers Association, Schaumburg, IL.

Spence, W. P., *Doors & Entryways*, Sterling Publishing Co., 387 Park Avenue South, New York 10016.

Spence, W. P., *Encyclopedia of Construction Methods and Materials*, Sterling Publishing Co., 387 Park Avenue South, New York 10016.

Spence, W. P., *Windows & Skylights*, Sterling Publishing Co., 387 Park Avenue South, New York 10016.

See Appendix B for addresses of professional and trade organizations and other sources of technical information.

31 CHAPTER

Cladding Systems

LEARNING OBJECTIVES

Upon completion of this chapter, the student should be able to:

- Understand the forces that must be considered when designing cladding systems.
- Know the building code requirements pertaining to cladding design and installation.
- Distinguish between various types of cladding systems.

Build Your Knowledge

For further study on these materials and methods, please refer to:

Chapter 9 Precast Concrete
Chapter 12 Concrete Masonry
Chapter 26 Bonding Agents, Sealers, and Sealants
Topic: Sealants
Chapter 29 Glass

Cladding is a non-load-bearing exterior wall enclosing a building. It may be brick, aluminum, steel, bronze, plastic, glass, stone, or other acceptable material. Cladding is exposed to the weather and must resist forces generated by wind, rain, earthquakes, and temperature changes. It must repel penetration by rain, condensation, and high winds. It must control the transfer of heat into and out of a building and meet building code requirements pertaining to weather, fire resistance and structural needs. Cladding therefore facilitates indoor environment control year round to maintain conditions required by the occupants of a building. A building may be clad with precast concrete panels, masonry panels, or some type of lightweight curtain wall **(Fig. 31.1)**.

CLADDING DESIGN CONSIDERATIONS

The design and construction details of cladding systems require careful engineering analysis and the consideration of a wide range of factors. The various elements and the final assembled materials must be tested to ensure they will meet the design requirements of the building codes. Following are some of the critical factors to be considered when selecting and designing cladding systems (Fig. 31.2).

Structural Performance

Panels can run continuously one or more stories in height and may be made up of spandrels and vision glass or continuous glass panels. Structural requirements include factors such as the ability of cladding to support its own weight over span distances between connectors, accommodating forces of cyclical expansion and contraction, resisting wind loads, and meeting seismic requirements. The method of connecting cladding to a structural frame is also subject to analysis. One consideration is whether a designer should rely on local and national wind-load standards or conduct wind tunnel tests for a building. Some cities require a wind analysis of buildings above a specified height.

Figure 31.1 Some of the commonly used cladding systems.

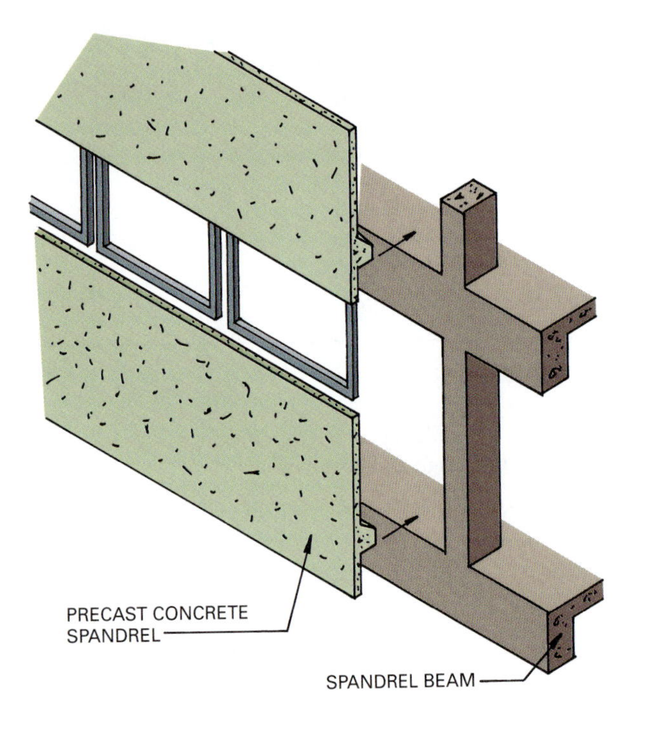

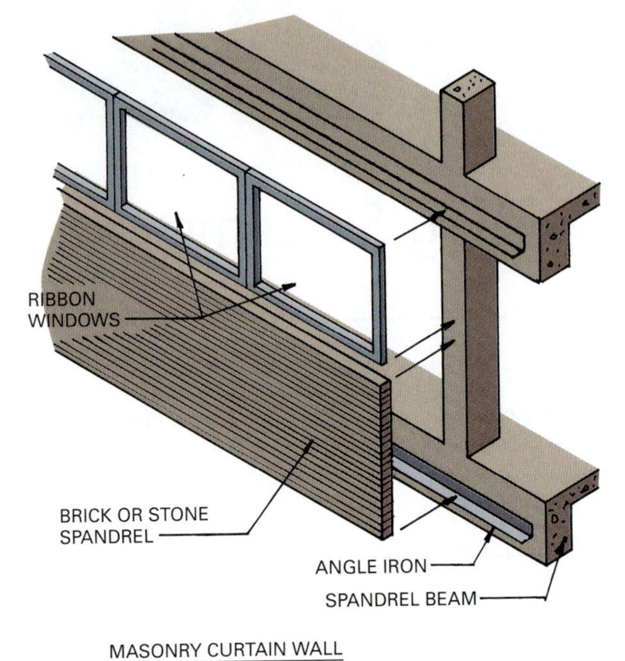

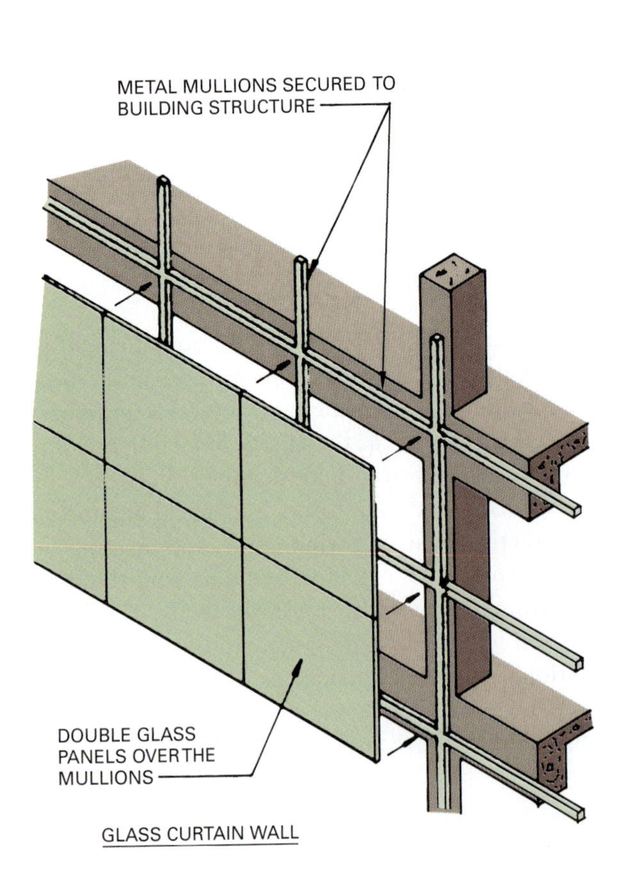

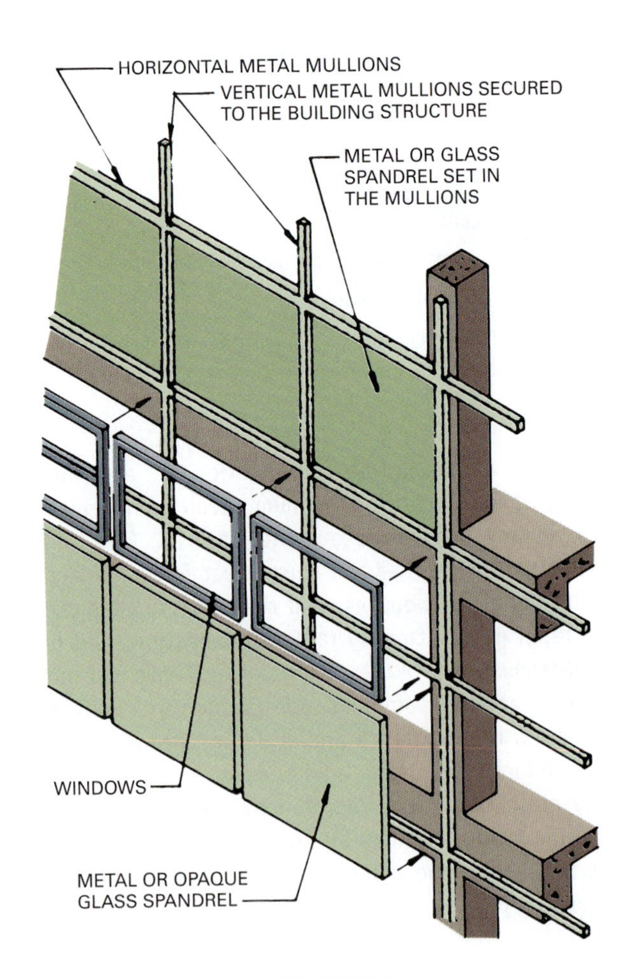

Figure 31.2 Curtain wall cladding systems may be made up of vision glass with spandrel panels or continuous glass panels. *(Image copyright Bryan Busovicki, 2009. Used under license from Shutterstock.com)*

Stiffness and Weight

Transportation to a site, lifting into place, and an ability to support its own weight are factors that must be accounted for in a panel's structural design. Since cladding is non-load-bearing, it should not carry additional loads. And cladding stiffness requirements are a major structural factor in the resistance of wind loads.

Connections

The connections securing a panel to a structural frame must support the panel's weight and resist other forces acting upon it, such as wind loads. Connections are generally designed and tested for structural requirements by a panel's manufacturer. Examples of various connection details are shown later in this chapter.

Wind Forces

Cladding systems must have sufficient structural strength to resist forces produced by winds. These forces may create positive pressure against a panel that cause an inward deflection. Wind blowing around a building is likely to create a negative pressure. Negative pressure (suction) tends to pull the panel from the structure. On high-rises, these negative pressures are usually greatest near the corners of the building. Indoor air pressure, due to mechanical ventilation systems, may be higher than outdoor air pressure, creating additional stress on the curtain wall from inside the building. If a panel lacks adequate stiffness or connections, it can blow off (Fig. 31.3).

The higher the floor on a low- or high-rise building, the greater the wind velocity and forces it is subjected to. In an area with surrounding multistory buildings, wind pressures are influenced by the direction of prevailing winds, the shape of a building and those around it, site positioning of a building, and the topography of the surrounding area.

Movement

Since a building is constantly subject to various forces, its cladding system must allow for movement within the structural system. For example, wind and earthquake

Figure 31.3 Wind forces subject curtain walls to positive and negative forces.

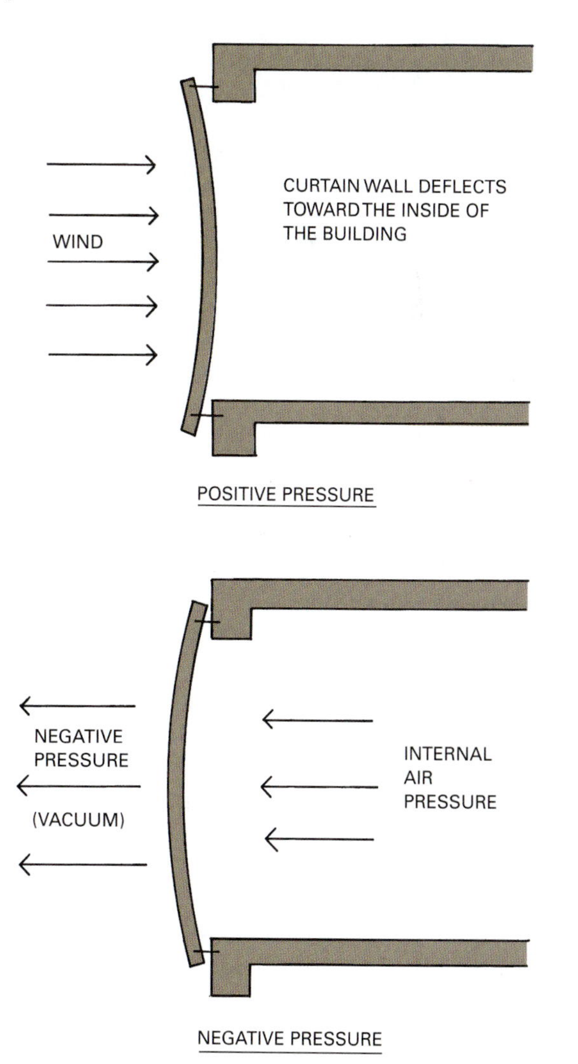

Figure 31.4 Earthquakes cause wracking and twisting of the structural frame. Wind causes movement in the frame. Both put stress on the curtain wall panels, windows, and connections, which can lead to possible failure.

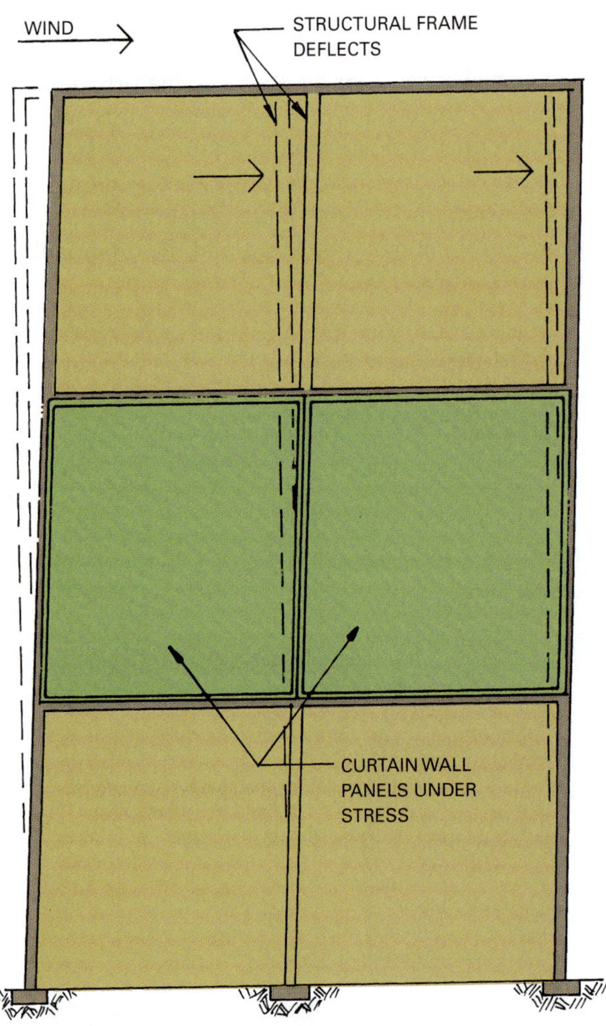

forces may cause the structural frame to deflect or twist, putting the cladding panels, windows, and connections under stress (Fig. 31.4). In addition, cladding may be subject to forces created by gravity, thermal expansion and contraction, moisture penetration, and condensation (Fig. 31.5). The weight of construction materials and differential (uneven) settling or heaving (lifting) of a foundation can cause beams to deflect. Creep that may occur over a long period of time can cause beams, columns, and girders to produce additional stress on the cladding. Moisture on the curtain wall can cause movement due to swelling and drying. A designer must produce a cladding system that serves to shelter a building's interior while remaining structurally intact.

Cladding panels must incorporate expansion joints that permit a sliding overlap (or another method for accommodating movement). Large glass panes must be glazed into frames that provide a watertight seal yet permit movement between the glass and the frame. Should the design prove inadequate, windows could break, cladding attachments could pull loose, panels could rupture, and components could even separate from the building.

Air Infiltration

The methods for controlling air infiltration closely resemble those for containing water penetration. Air leaks permit unconditioned air to enter a building, which requires additional energy to heat or cool the unwanted air. Water vapor can enter cladding panels through air leaks and condense inside them. Air leaks also permit pollution and unwanted noise to enter.

Water Penetration

A typical curtain wall is an assembly of different materials with a variety of joints between similar and dissimilar materials. The watertightness of exterior cladding is critical to the success of the installation. Sealing materials must bond to joint surfaces; withstand temperature changes; and repel rain, wind, and stress due to expansion, contraction, and wind loads. All of these directly influence the watertightness of cladding systems.

Since the cladding is directly exposed to the weather, it is subjected to wind-driven rain, snow, and hail. The upper levels of high-rise buildings are subject to the greatest possible penetration because of higher wind velocities at these levels. In some areas, water penetration seals must withstand chemical attack by polluted air. Temperature variations also place stress on seals.

In multistory structures, large amounts of wind-driven rain tend to cascade down a building's face. Provisions must be made to carry away this water and maintain the integrity of the water-penetration seals. The design of joints and proper installation of seals is the major factor in controlling water penetration. In some types of panels, such as aluminum-faced, provision is made for internal drainage of water that may penetrate a joint. These provisions can consist of flashing, drainage channels, and weep holes.

Water penetration of cladding may be caused by wind-driven rain and other forces, as shown in Fig. 31.6. Gravity works constantly to force water downward into improperly designed joints. Wind-driven rain possesses sufficient kinetic energy to carry through openings in joints and wall surfaces. **Kinetic energy** is the energy in

Figure 31.5 Many forces acting on a structural frame can put excessive stress on the curtain wall, possibly causing failure.

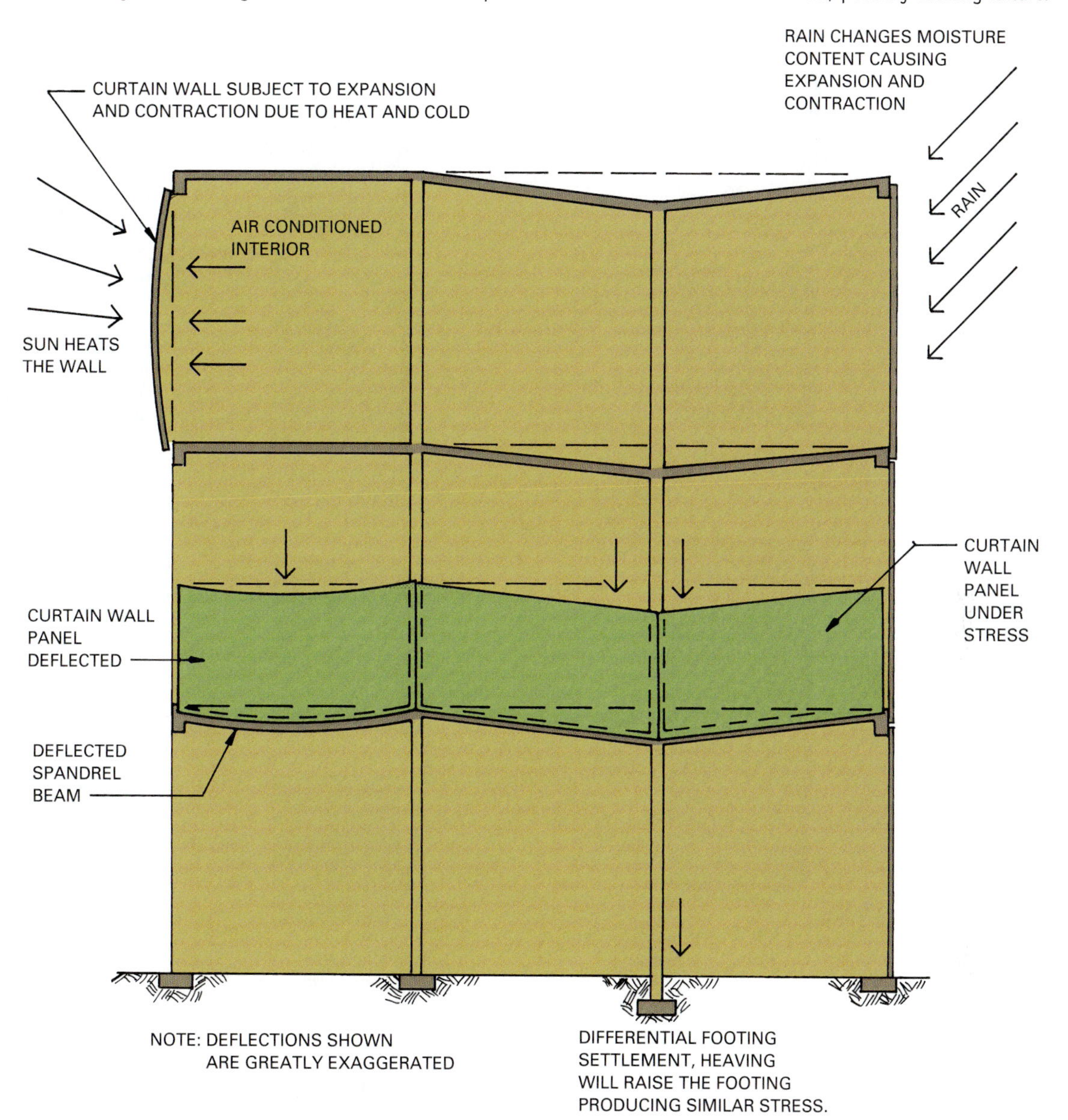

a body caused by motion. Covering joints with some type of batten or internal baffle can mitigate penetration by kinetic energy–driven water.

Water also can penetrate poorly designed joints through surface tension, which allows it to cling to and flow along soffit areas. This can be prevented by adding an outer edge drip. Another force to consider when designing joints is **capillary action**. This can occur when the butting surfaces of a joint provide a very small opening along which water clings and travels. One way to control this is to provide an air gap in the joint, which breaks the capillary path. These four forces can be easily controlled by proper joint design (Fig. 31.7).

Wind forces may also cause a difference in pressure between a cladding panel's outside face and its interior. When air pressure is greater at the exterior of a panel, water can be forced through a joint into the panel. This type of penetration can be controlled by the rainscreen principle.

The **rainscreen principle** is a pressure equalization method for cladding systems. The rainscreen is composed of a wall's exposed outer surface backed by an air space. Joints are designed to prevent water from penetrating but are not sealed. This permits air pressure on the panel's face to equalize with the air pressure inside the panel.

Figure 31.6 These forces can cause water penetration in joints and other openings in exterior walls.

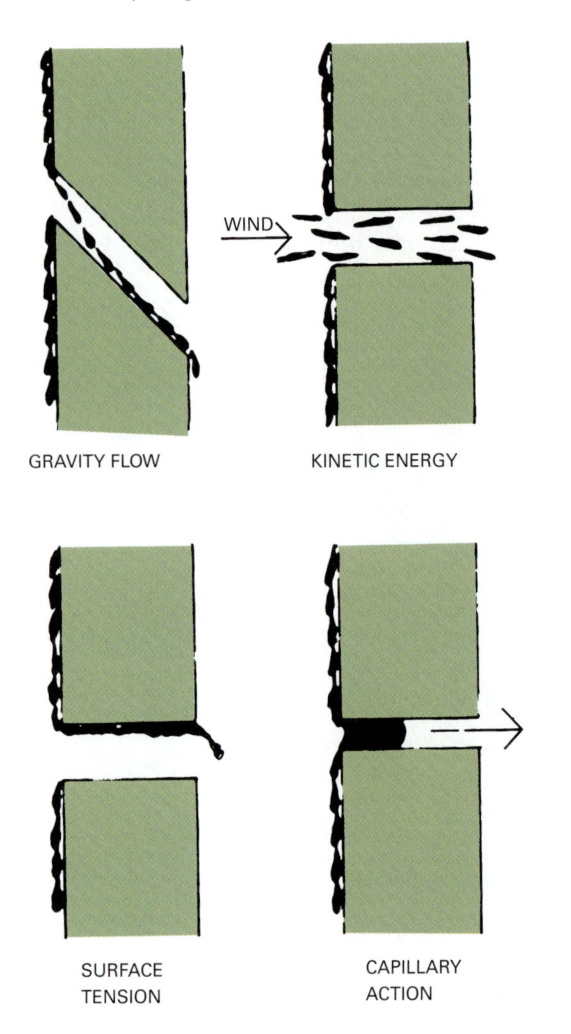

Figure 31.7 Techniques for controlling water penetration at joints. *(Courtesy American Architectural Manufacturers Association, from Curtain Wall Manual CW 1-9 The Rainscreen Principle)*

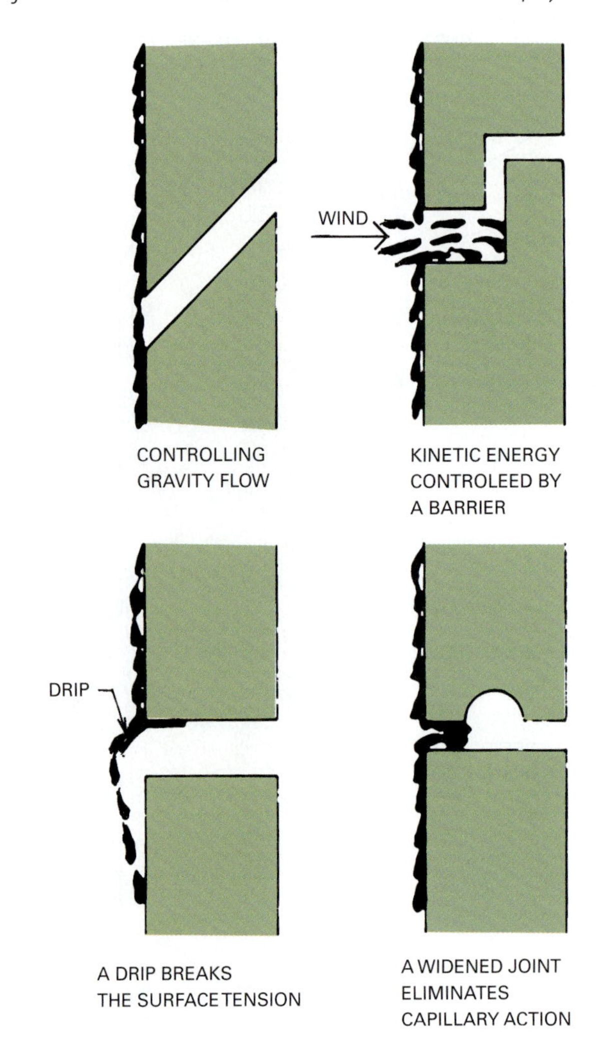

The size of the air space behind the rainscreen is limited by compartmentalizing it into relatively small subdivisions, with each area having openings to the exterior. A waterproof air barrier must be provided on the interior side of the panel. This is necessary to prevent leakage to the interior of the building. Use of the rainscreen principle with precast concrete curtain wall panels is detailed in Fig. 31.8.

Pressure equalization techniques can also be used with glazing details. Glazing units may have a water-resistant seal, a pressure equalization chamber along the edge of the glass, and an inner vapor seal (Fig. 31.9). Additionally, the rainscreen principle can be used to resist water penetration by air pressure differences in traditional wood and masonry walls.

ASTM tests for water penetration and wind infiltration are made on curtain wall units with glazing sealants or gaskets installed. Actual conditions can vary during a storm and may, under certain circumstances, exceed those normally specified.

PERFORMANCE FACTORS

Cladding systems are designed to last for long periods of time. They must control condensation, radiation and conduction, noise and pollution, and the amount of natural light entering a building. Cladding also must meet building code requirements for fire resistance.

Durability and Maintenance

Cladding should last over the expected life span of a building. Material components used—including connections, seals, and joints—should have a proven record of durability. Maintenance requirements over the life of a building will be greater if reduced durability is accepted. A designer must decide between more expensive, more durable materials with higher initial costs or cheaper, less durable materials that incur higher maintenance costs over time.

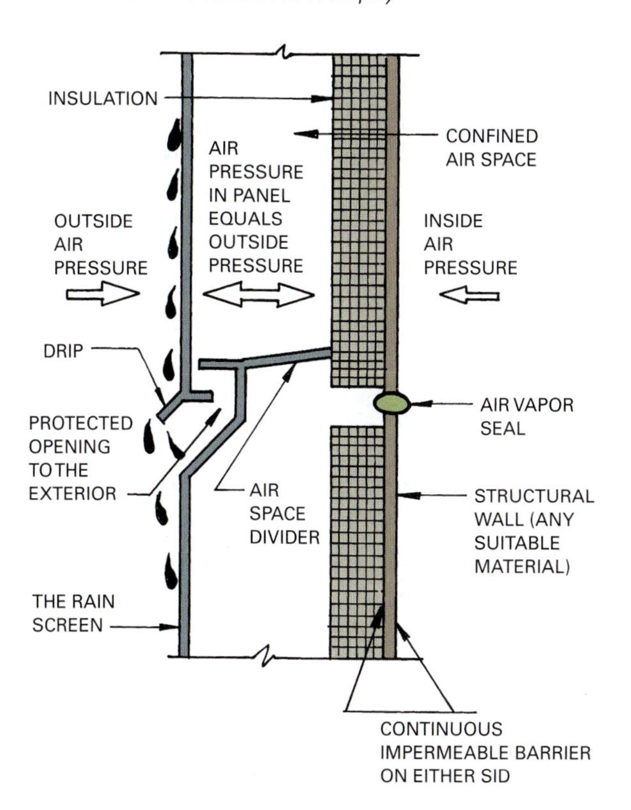

Figure 31.8 This illustrates the rainscreen principle. This aluminum curtain wall panel has divided air spaces that are vented with protected openings to the exterior. This equalizes the air pressure in the panel with the outside air. *(Courtesy American Architectural Manufacturers Association, from Curtain Wall Manual CW 1–9 The Rainscreen Principle)*

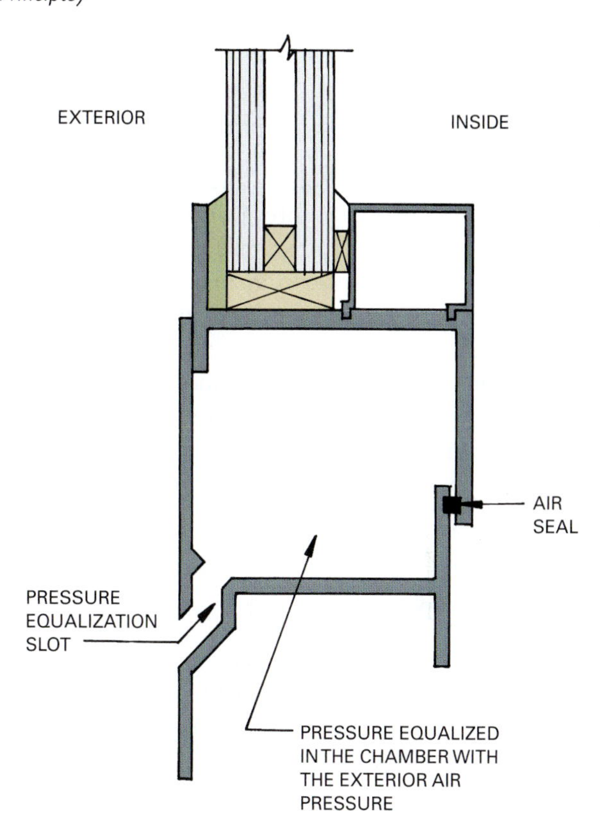

Figure 31.9 A typical detail for using the rainscreen principle with glazed units. *(Courtesy American Architectural Manufacturers Association, from Curtain Wall Manual CW 1–9 The Rainscreen Principle)*

Generally, masonry coverings require little maintenance and tend to show weather streaking less than impervious skins. The exterior of a building can be routinely cleaned using vertical ladders or a scaffold (or cradle) hanging from gantry arms on a trolley. Some windows are designed so their exteriors can be cleaned from inside a building. But in many buildings, windows are fixed and must be cleaned from outside.

Condensation Control

As is true with any type of wall construction, building cladding must have a vapor barrier on the inside wall facing to prevent water vapor from passing into the wall assembly. Water vapor will condense and run down the surface of masonry and concrete cladding. If it penetrates a hollow assembly, such as an aluminum panel, it will condense, reduce the value of the insulation, and cause damage from freezing and thawing. Some form of drainage is provided to remove any condensation that may occur. Provision for the escape of moisture vapors to a building's exterior must be provided. To reduce condensation, interior surfaces of panels should be insulated such that they are warmer than the dew point of air within the panel.

Radiation and Conduction Control

Cladding must control the influx into a building of solar radiant heat that can cause physical discomfort and increased air-conditioning costs. Radiant heat is actually radiant energy; heat is induced when the radiant energy strikes something. Thermal energy always moves from a hot body to a cold body. For example, a person sitting near a cold wall radiates heat to the wall, making the person feel cold. Radiant energy from the sun passes through windows and heats up people and materials it strikes inside a building.

Heat is also transferred by thermal conduction. Cladding must prevent the conduction of heat into and out of a building through walls. Thermal conduction is the process of heat transfer through a material. Cladding systems use insulation and thermal breaks in materials that are good thermal conductors. A **thermal break** is an insulating material or gap placed between thermal conducting materials in contact to retard the passage of heat or cold. The use of multiple glazing and energy-efficient glass also offers radiant controls, as discussed in Chapter 29.

Sound and Pollution Control

The importance of controlling the influx of exterior sounds and air pollution into a building varies with the type of occupancy. For example, a hospital requires greater control than other occupancies. In other cases, it may be necessary to keep noise or pollution inside a building, such as in the case of manufacturing or chemical processing operations. These cladding systems must be airtight, well insulated, and, in many cases, have sound-deadening material as part of their assembly. Generally, glazed wall areas admit more sound than spandrel panels.

Natural Light

Natural light is used to provide illumination and a connection to the exterior environment. Sunlight must be controlled to prevent distracting glare and unwanted radiant energy (heat). Although some control can be had using various types of solar screening and energy-efficient glass, the amount of glazing used is a major control factor.

Windows are a part of exterior cladding systems. They are exposed to the same factors as the actual panels, including wind load, thermal movement, air infiltration, and water penetration. Their design must enable them to meet structural requirements as well as have adequate fastening systems and durability.

Fire Resistance

Cladding is subject to the requirements of building and fire codes. Cladding fire resistance ratings are determined by assembly tests following ASTM procedures. These involve the combustibility of materials in the cladding, the fire resistance ratings of the assembled panels and spandrels, and the nearness of other buildings. The required fire resistance of non-load-bearing cladding varies with a building's occupancy and fire separation distance. Fire separation specifies the distance between adjacent buildings. It is used to reduce the risk of fire spreading from one building to another. Distances specified in the code are usually rendered in feet and measure between a building and the closest lot line, from the centerline of the street, or from a halfway distance between two buildings on the same site.

Cladding connectors must hold cladding in place during a fire for the time specified by the fire resistance rating of the assembly. Exterior trim and other wall finish materials are also subject to established fire resistance ratings. In addition, codes contain requirements for fire protection of openings in walls.

Firestopping uses approved materials to prevent the movement of flame and gases through small vertical and horizontal concealed openings in building components (Fig. 31.10). Materials commonly approved for use as firestops include masonry set in mortar; concrete, mortar, or plaster on metal lath; plasterboard; sheet metal of approved gauge; asbestos–cement board; mineral, slag, or rock wool, if solidly compacted into a confined space. These materials, commonly referred to as **safing**, must be permanently fastened in place.

Figure 31.10 One example of firestopping a curtain wall at the spandrel beam.

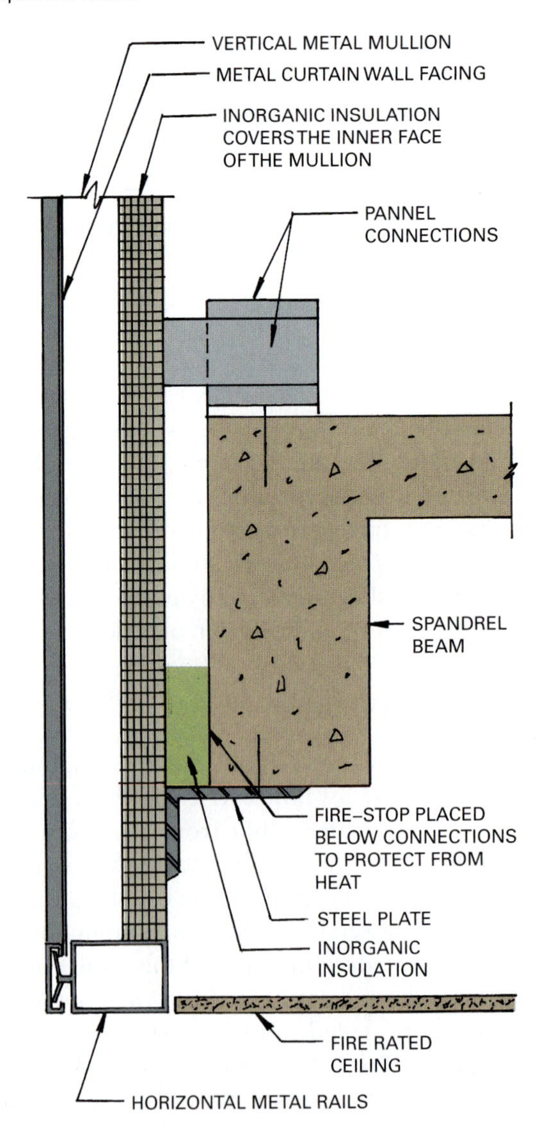

Case Study

GSW Administration Building

Architect: Sauerbruch Hutton Architekten | Building type: Commercial office
Location: Berlin, Germany | Completed 1999

Overview

The high-rise slab and lower wing of the GSW Administration building houses the corporate offices of a large real estate company. The complex of buildings incorporates an existing 1961 office tower. The most striking aspect of the project is a west-facing double-skin glass façade that incorporates red and orange sun-shading devices.

A double-skin façade consists of two glass skins separated by an air space ranging in width from a few inches to several feet. The main layer of glass, usually insulating, serves as part of the conventional structural wall or a curtain wall, while the additional layer, usually single glazing, is placed either in front of or behind the main glazing. The air space acts as an insulating barrier against temperature extremes, noise, and wind. As demonstrated in the GSW Administration building, sun shading-devices can be integrated between the two skins.

Double-skin facades provide a number of environmental benefits, including increased day-lighting, acoustical and pollution protection, weather protection of the facade and shading devices, and wind protection for manually operated windows in high-rise applications, allowing for natural ventilation that helps eliminate sick building syndrome (SBS). The façade also provides solar pre-heating of ventilation air, which serves to reduce heating and cooling demands.

Day Light and Ventilation

The narrow floor plate and fully glazed west wall of the GSW building provides ample daylight for all offices. The building further supports natural air flow through added ventilation elements integrated into the design of the east façade (Figure A). Fresh air enters from the east side of the tower through specially designed louvered vents. The air is drawn through the building to the western façade where it is exhausted by convection through operable glazing units in the outer layer of the façade. The natural thermal lift that occurs in the gap between the two facades draws the air through the narrow floor plate.

Sun Shading

In order to prevent excessive heat gains from the fully glazed west façade, adjustable shading devices are integrated into the double layer of the façade. Vertical louvers can be automatically operated by thermostatically controlled sensors or manually adjusted by occupants (Figure B).

Energy Conservation

The double-skin construction of both the east and west facades provides an insulated air space, minimizing heat loss by conduction. The concrete structure of the floor slabs act to moderate temperature extremes in both summer and winter. In summer, the thermal mass of the slabs serves as a heat sink, absorbing heat during the day that is flushed by ventilation at night. In winter, the slabs absorb solar radiation in the day time and re-radiate the heat at night.

The operations of the ventilation and thermal storage systems allow the complete omission of mechanical air conditioning. Central mechanical heating is used only as a backup system for extreme cold weather. Double-skin facades are expensive, with additional costs generated by the engineering of these systems and the amount of additional glass required.

CURTAIN WALL CLADDING

A curtain wall is defined as a non-load-bearing exterior building wall or skin, supported by the building framework. A metal curtain wall is an exterior non-load-bearing building wall consisting of metal, glass, or other surfacing materials supported by a metal framework. A window wall is a type of metal curtain wall composed of metal framing members containing operable sash, fixed lights, ventilators, or opaque glass panels. All of these systems are the result of years of engineering development and testing. Various manufacturers supply patented systems engineered for a wide range of applications (Fig. 31.11).

A curtain wall system typically includes panels that form the exposed exterior walls, glazing, mullions, connection components, gaskets, and sealants. Curtain walls are lightweight and non-load-bearing; anchored to columns, spandrel beams, and floors; and generally supported at their bottom edge. Facing panels are made from metal, glass, modified stucco, molded glass, fiber-reinforced polyester, or other code-approved materials. Since curtain wall panels are non-load-bearing, they can be thin and lightweight, thus reducing the loads on a building's foundation. This can produce significant savings in both the weight and construction costs of high-rise buildings. Curtain walls range from a simple single-thickness material, such as metal siding, to multilayer sandwich panels that incorporate insulation. The choice of system depends on desired appearance, environmental conditions, and requirements for the interior environment. Following are general examples of details for commonly used curtain walls.

Masonry Curtain Walls

Masonry curtain walls are usually supported by steel shelf angles secured to the structural frame of a building. The shelf may be anchored to a concrete spandrel beam or bolted or welded to a structural steel spandrel beam

Figure 31.11 This building is clad with several types of curtain walls. On the right is an all-glass wall with exposed mullions and rails. *(Image copyright MalibuBooks, 2009. Used under license from Shutterstock.com)*

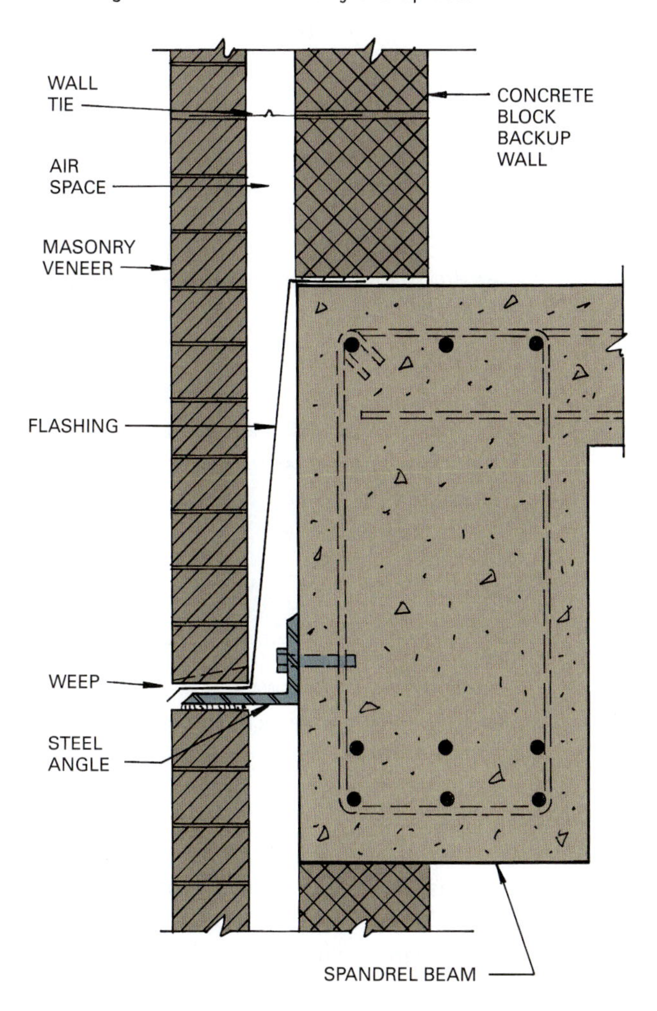

Figure 31.12 A typical detail for a brick masonry curtain wall placed over a concrete structural system. The brick is mounted on a steel angle and tied to a masonry backup wall.

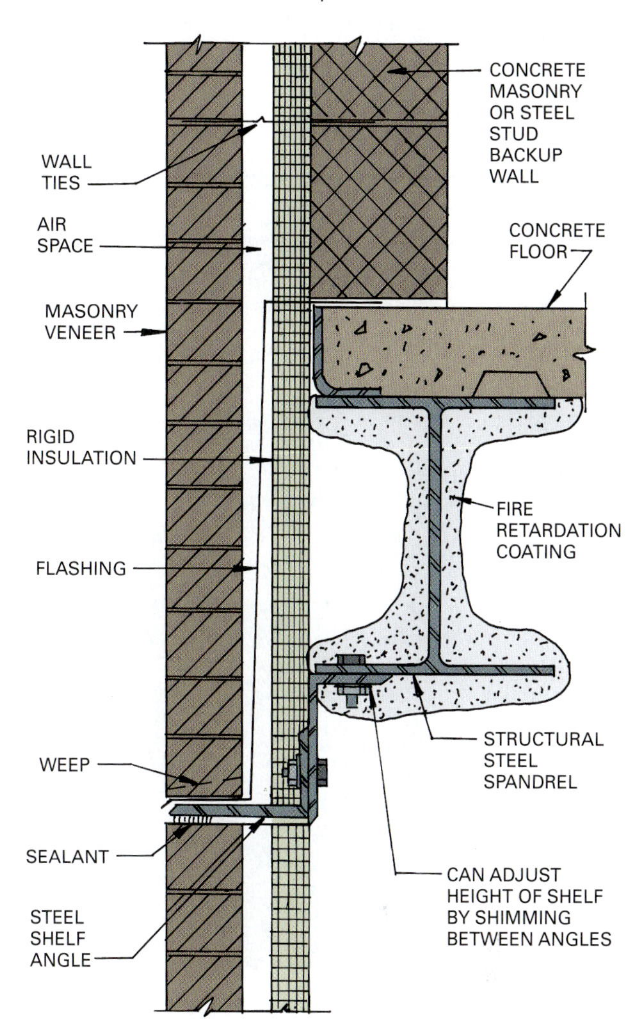

Figure 31.13 A typical detail for a brick masonry curtain wall placed over a steel frame building. The brick is mounted on a steel shelf and is tied to a backup wall.

(Figs. 31.12 and 31.13). Walls can be laid up a brick at a time, as is done with low-rise masonry construction. Provisions are made for horizontal and vertical expansion joints to permit panels to expand and contract as temperatures vary. Masonry curtain wall panels can also be prefabricated on the ground and lifted into place with a crane. In either case, to secure them, high-bond mortars are used for their excellent bonding characteristics and high compressive and tensile strength.

Masonry curtain walls typically have a backup wall, often concrete masonry units or steel studs (Fig. 31.14). The back up wall is non-load-bearing and supports the sheathing, insulation, and air and vapor barriers, as well as the interior finish materials. The curtain wall is affixed to the backup wall with masonry ties. Ties are stiff but can flex enough to accommodate differential movements between the veneer and the backup wall.

Most single-width masonry curtain walls experience some penetration by driving rain. This moisture is handled by the rainscreen principle, which provides a cavity behind the masonry and weep holes to drain the moisture. The cavity acts as a pressure equalization chamber and provides some degree of pressure equalization between the exterior and interior surfaces of the masonry. Some architectural designs permit the structural system to be visible on the exterior of the finished building, with cladding systems filling the spaces between.

A typical detail showing a ribbon window installation on top of a masonry curtain wall is shown in Fig. 31.15. The aluminum frame has a sill extending into the room and a slot to hold the top of the interior wall finish material.

Masonry curtain walls are often constructed using prefabricated panels that are hoisted into place. Two methods are commonly used. One uses a plastic gasket form liner inside a form into which bricks are placed. This spaces the bricks and holds them in place. Reinforcement is laid over the bricks, and

Figure 31.14 A typical detail for a brick masonry curtain wall placed over a structural concrete frame building. It is tied to a steel stud backup wall. The weep hole permits drainage of moisture and equalizes the air pressure (the rainscreen principle) on both sides of the masonry.

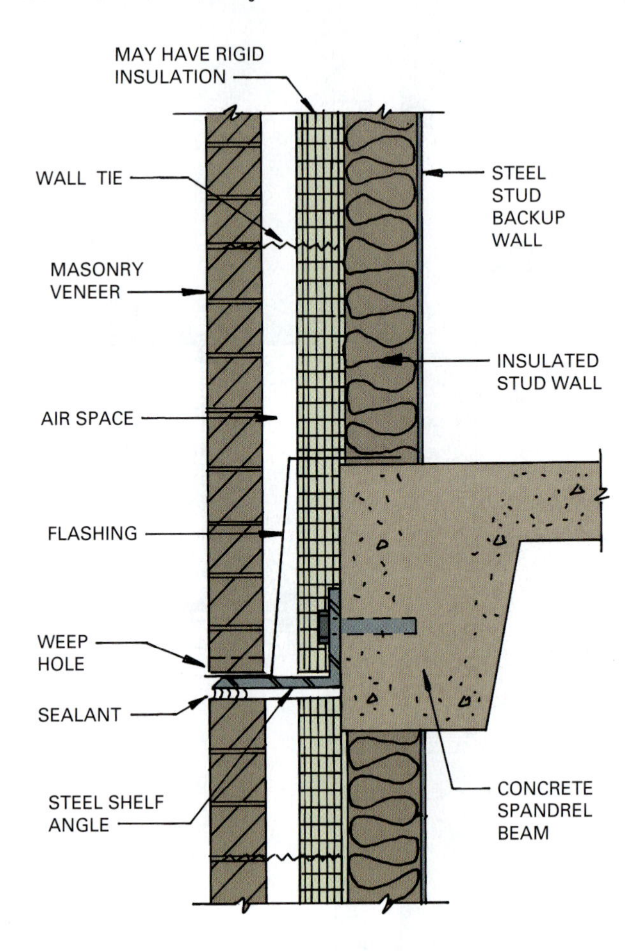

Figure 31.15 A typical installation of a metal window on a masonry curtain wall panel.

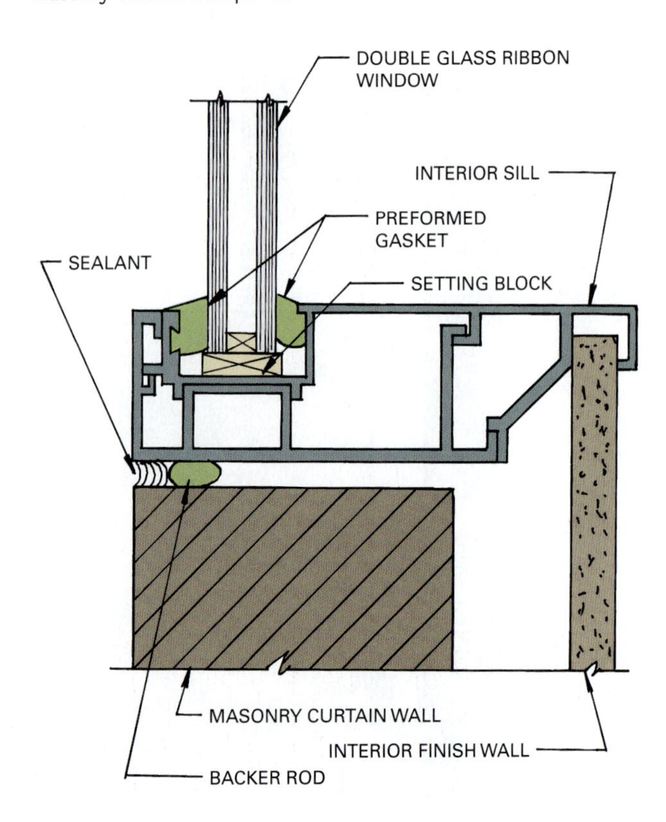

a concrete panel is poured over the assembly. Once cured, the liner is removed and the panel lifted into place (Fig. 31.16).

Stone Curtain Walls

Stone is a widely used material for facing low-rise, mid-rise, and high-rise buildings. Commonly used stones include granite, slate, marble, and lime-stone, among others. The material is cut into panels of various thicknesses—ranging from 1 $\frac{1}{4}$ in. to 4 in. (32 to102 mm), depending on the material and the size of the panel—that are secured to the structural frame of a building.

An architect must consider the specifications related to the physical characteristics of the material. For example, granite characteristics relating to material density, compressive strength, and modulus of rupture are established by ASTM standards. Stone panels are available in a variety of finishes, ranging from smooth and highly polished to sand-blasted rough surfaces. Some manufacturers produce panels with the rough-sawed surface exposed. Joints between panels are typically $\frac{1}{4}$ in. (6 mm) in width but vary widely by product. Stone products are discussed in Chapter 13.

Manufacturers often supply or recommend engineering-approved connections and installation details. A typical way to support stone panels is with a metal subframe (Fig. 31.17). This frame can be mounted to the structural concrete frame or to steel or masonry backup walls. A number of connection systems are shown in Figs. 31.18 and 31.19. One method of cladding a soffit at a window opening is shown in Fig. 31.20, and a typical parapet construction is shown in Fig. 31.21.

Preassembled stone cladding systems are made by mounting stone panels on a steel frame. The assembly is lifted into place and secured to the structural frame of a building. Other systems join stone panels to a reinforced concrete backing with steel anchors. Preassembled panels reduce the time required for plumbing and leveling the cladding because multiple panels are installed in one operation. A building with stone veneer cladding is shown in Fig. 31.22.

Figure 31.16 A preassembled curtain wall panel is hoisted into place. *(Courtesy American Brick Company)*

Figure 31.17

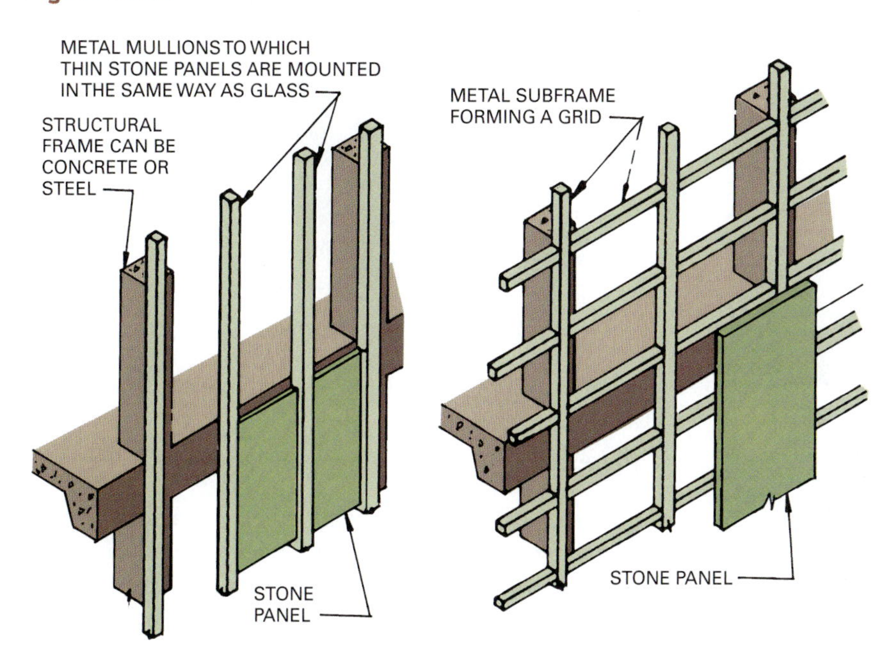

Typical types of steel curtain wall subframes used with stone panels. The subframes are secured to steel or concrete structural systems.

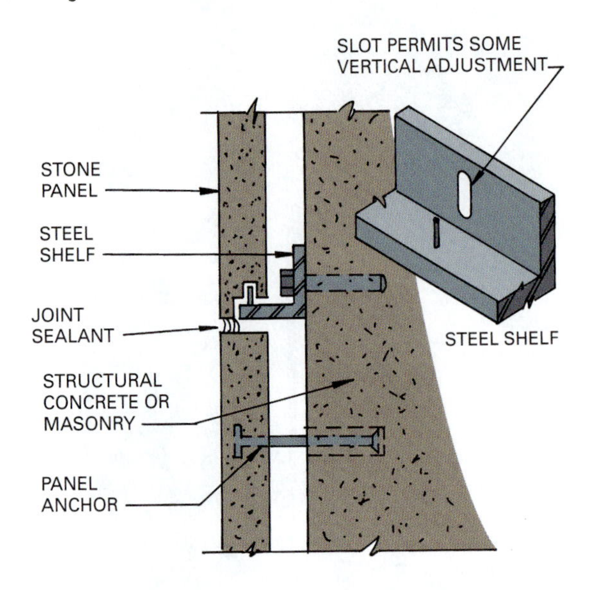

Figure 31.18 A stone curtain wall panel is mounted on a metal angle seat and tied to the masonry backup wall.

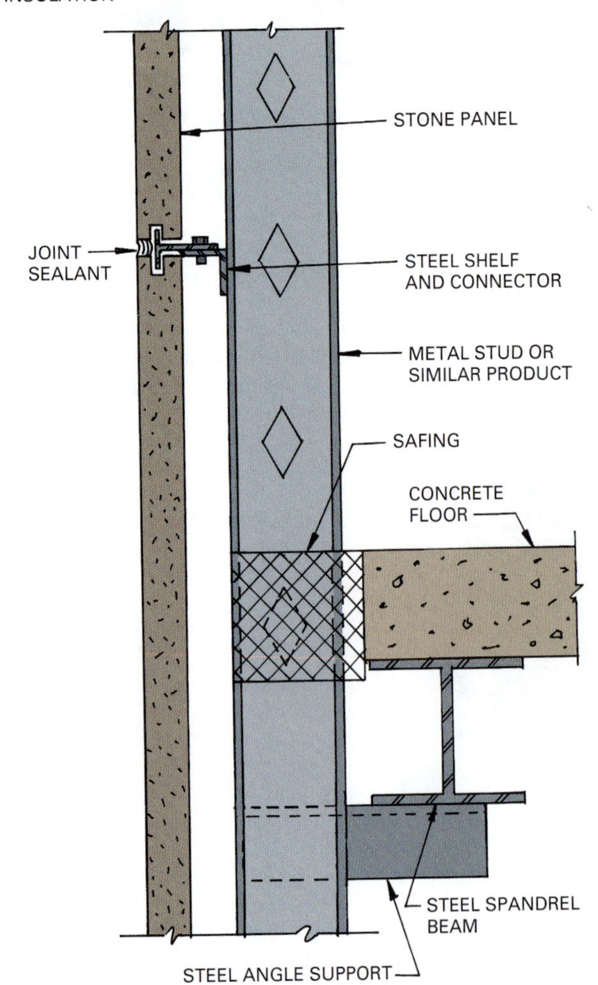

Figure 31.19 A stone curtain wall panel mounted over a steel structural system and steel studs.

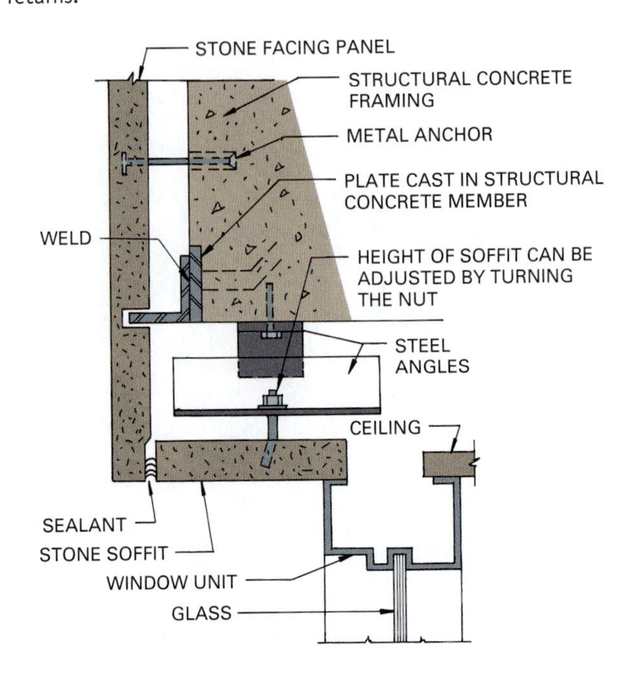

Figure 31.20 A typical stone soffit detail as used at window returns.

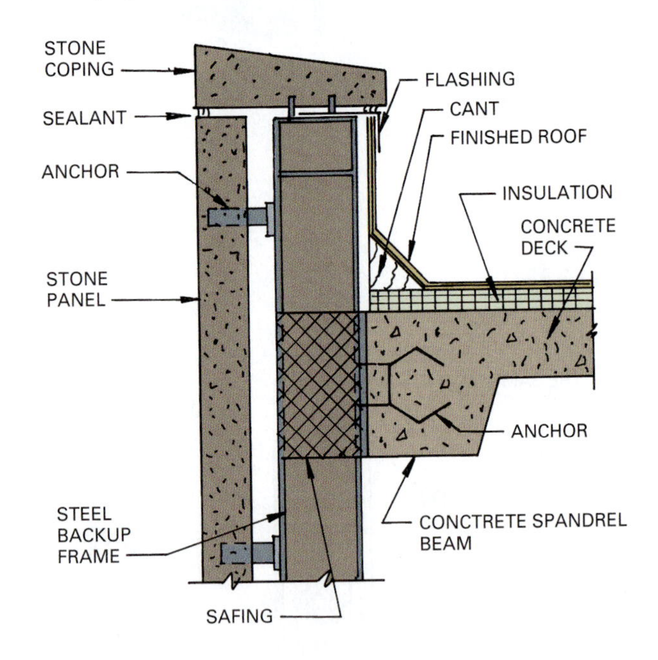

Figure 31.21 One design for a stone-faced parapet with a stone coping.

Precast Concrete Curtain Walls

Precast concrete curtain wall panels are available with a variety of surface finishes and thicknesses. Surface finishes can be produced by the texture of a casting form or after casting and prior to hardening. Methods such as brooming, stippling, floating, troweling, or exposing the aggregate can be used in the latter case. Surfaces

Figure 31.22 This high-rise building is clad with stone panels. *(Image copyright Brad Sauter, 2009. Used under license from Shutterstock.com)*

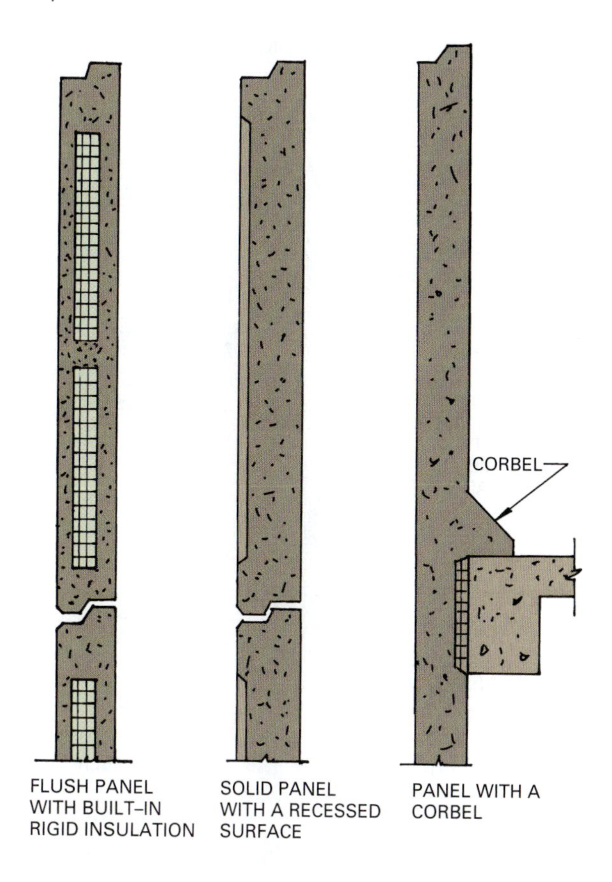

Figure 31.23 Some of the types of precast concrete curtain wall panels available.

may be recessed or panelized. The design of a reinforced precast concrete curtain wall panel can vary depending on exterior appearances specified by an architect and an engineer's structural design. Some typical panel types are shown in Fig. 31.23. They can be made using conventional reinforcing or as pre-stressed units. The use of glass fiber reinforcing in addition to steel enables the production of lighter and thinner panels. The glass fibers reduce the need for the secondary steel reinforcing used on heavier panels to resist cracking and help control thermal expansion and contraction.

Precast panels are installed with anchors cast into them and bolted to angles on a structural frame. Panels can be insulated by bonding rigid insulation to their inside surfaces or sandwiching rigid insulation in the center of a panel. Allowances must be made between panels for expansion and contraction. An adequate sealant is required to provide a waterproof exterior surface. Typical details are shown in Figs. 31.24 and 31.25. Panels are installed over steel and structural concrete frames. Typical parapet and window opening details are shown in Fig. 31.26.

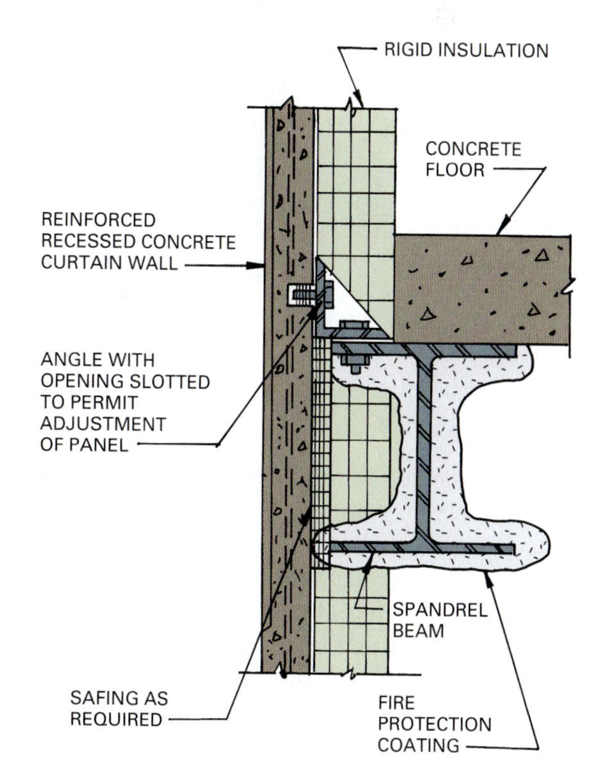

Figure 31.24 A typical precast concrete curtain wall panel over a steel structure frame.

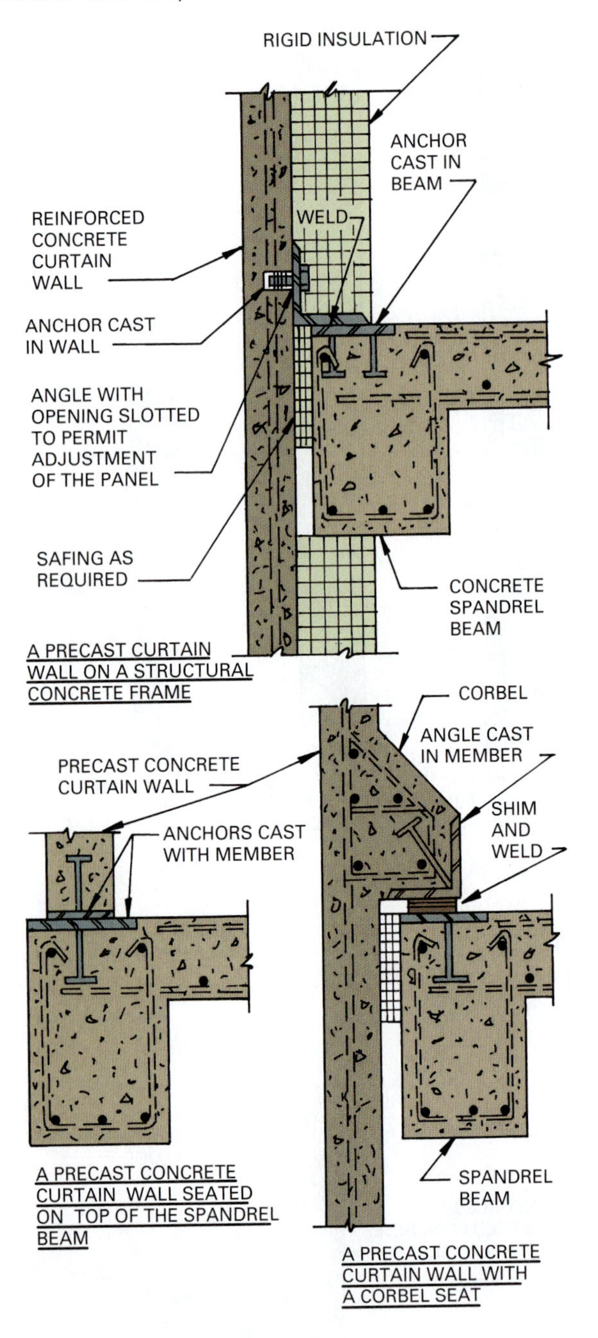

Figure 31.25 Several ways precast concrete curtain walls are secured to concrete spandrel beams.

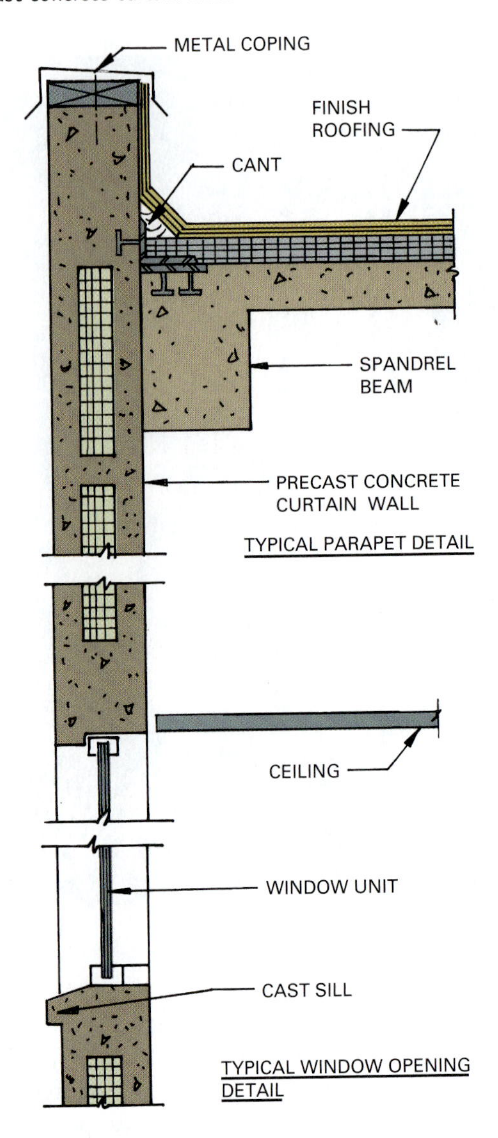

Figure 31.26 Typical parapet and window opening details for precast concrete curtain walls.

Modified Stucco Curtain Walls

Modified stucco curtain wall panels are an assembly of materials providing both insulation and a weathertight rain-screen. The system is referred to as an exterior insulation and finish system (EIFS). EIFS systems begin with a sheathing material over which expanded polystyrene (EPS) insulation board is secured with special mechanical fasteners. The substrate often uses a board with inorganic glass mat facings, a water-resistant, silicone-treated gypsum core, and an alkali-resistant surface coating. A base coat is troweled over the insulation board; then a reinforcing mesh is embedded into it. The synthetic plaster finish coat is troweled over the base coat (Fig. 31.27).

Curtain wall panels one or more stories high can be constructed of steel studs covered with EIFS materials. The assembled panels are lifted into place and bolted or welded to connectors on the structural frame of a building in the same manner as described for other types of curtain wall construction. A finished commercial building is shown in Fig. 31.28.

A number of similar panel systems are manufactured for lightweight exterior wall cladding. For example, a composite panel similar to the one just described

Figure 31.27 The Dryvit ® Infinity EIFS provides a watertight membrane and an insulation board with drainage channels to capture, control, and discharge any incident moisture that may enter the system. *(Copyrighted and used with the permission of Dryvit Systems, Inc.)*

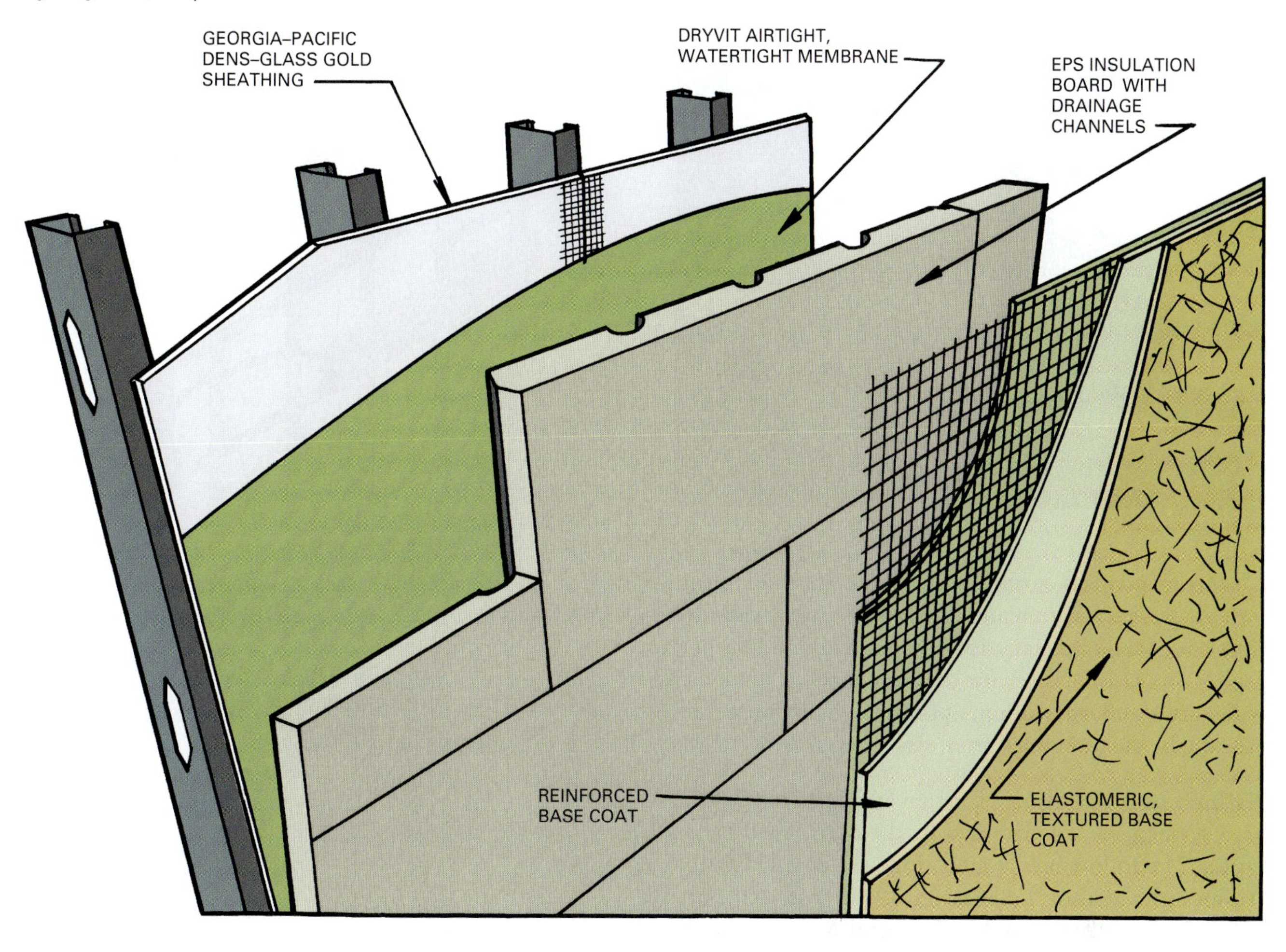

Figure 31.28 Modified stucco curtain wall panels can be several stories high. *(Image copyright Vladislav Gurfinkel, 2009. Used under license from Shutterstock.com)*

is sometimes used, but it has a hard polymer exterior finish. Another type provides a stucco, exposed aggregate or simulated brick or stone exterior surface.

Metal and Glass Curtain Wall Systems

Glass and metal curtain walls are widely used because they provide a lightweight exterior wall that can be rapidly erected (Fig. 31.29). The metal framing is usually aluminum, and various types of glass panels are used. The glass is typically insulating glass, laminated glass, heat-reflective glass, and spandrel glass. Spandrel glass is heat-strengthened with a ceramic frit, a molten glossy material, fired on it. It is recommended that rigid insulation or some other rigid backing material be bonded to the back of the glass spandrel panel to hold the glass in place in case of fracturing due to high internal temperatures or storm damage.

Figure 31.29 A glass-to-the-front, stick fabrication framing system has the advantage of fast field erection and significant savings. *(Image copyright Cynthia Farmer, 2009. Used under license from Shutterstock.com)*

Metal and glass curtain wall systems are available in two basic types: custom and standard. Custom walls are those designed especially for a particular building and built by a manufacturer for that application. Standard walls are standardized units produced by manufacturers and can be assembled from stock parts.

Methods of Installation Custom and standard curtain wall systems can be classified by the method of installation. The following classifications are detailed in the Aluminum Curtain Wall Design Guide Manual published by the American Architectural Manufacturers Association.

In the stick system, vertical mullions are installed followed by horizontal rails into which glazing and spandrel panels are placed (Fig. 31.30).

The unit system uses large preassembled wall panels. The vertical edges of the units are joined, serving as a mullion, and the bottom of one panel joins the top of the one below it, forming a horizontal rail (Fig. 31.31).

The unit-and-mullion system uses mullions secured to a building structure with preassembled wall panels installed between them (Fig. 31.32).

The panel system uses homogenous wall panels that can be precast concrete, molded plastic, or stamped from sheet metal. They may or may not have openings for windows. They are connected to a structural frame as shown in Fig. 31.33. A typical building is shown in Fig. 31.34.

The column-cover-and-spandrel system uses long spandrel panels that span column covers. Glazing may be installed above the spandrel, and columns are enclosed with a hollow cover (Fig. 31.35).

Figure 31.30 The stick curtain wall system: (1) anchors; (2) mullion; (3) horizontal rail; (4) spandrel panel; (5) horizontal rail; (6) vision glass; (7) interior mullion trim. *(Courtesy American Architectural Manufacturers Association, from Aluminum Curtain Wall Design Guide Manual)*

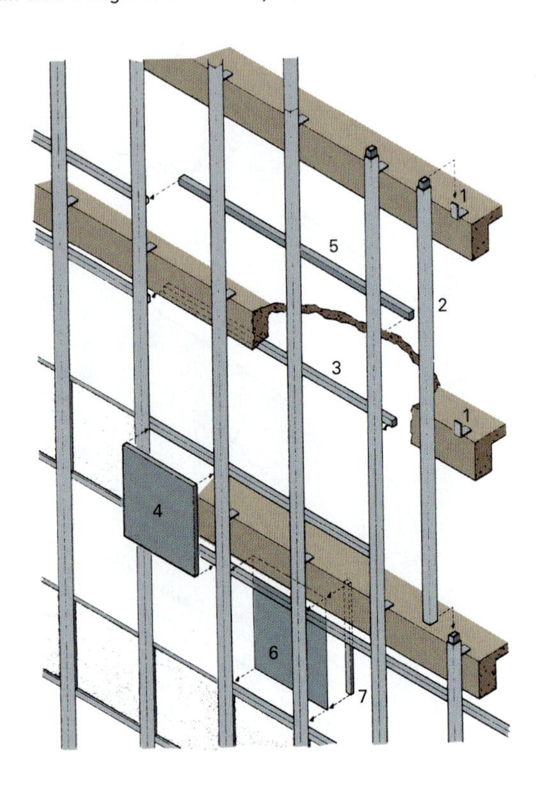

Figure 31.31 The unit curtain wall system: (1) anchor; (2) preassembled framed curtain wall unit. *(Courtesy American Architectural Manufacturers Association, from Aluminum Curtain Wall Design Guide Manual)*

Figure 31.32 The unit-and-mullion curtain wall system: (1) anchors; (2) mullion; (3) preassembled curtain wall unit lowered into place behind the mullion from the floor above; (4) interior mullion trim. *(Courtesy American Architectural Manufacturers Association, from Aluminum Curtain Wall Design Guide Manual)*

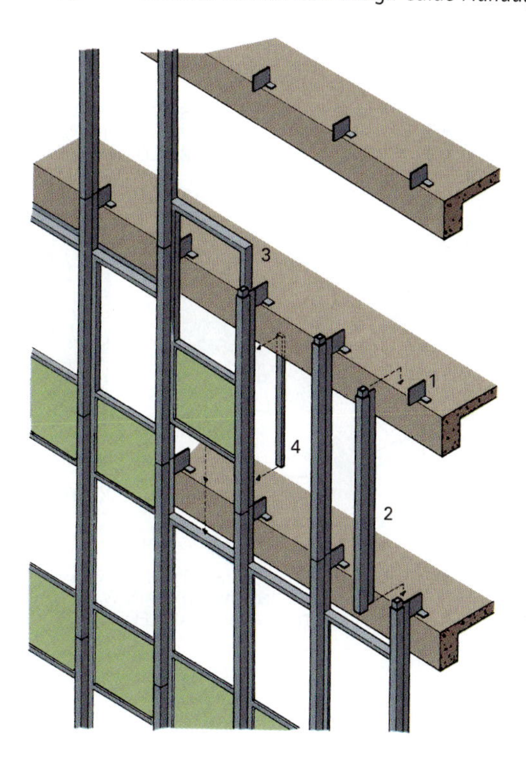

Figure 31.33 A panel curtain wall system: (1) anchor; (2) preassembled curtain wall panel. *(Courtesy American Architectural Manufacturers Association, from Aluminum Curtain Wall Design Guide Manual)*

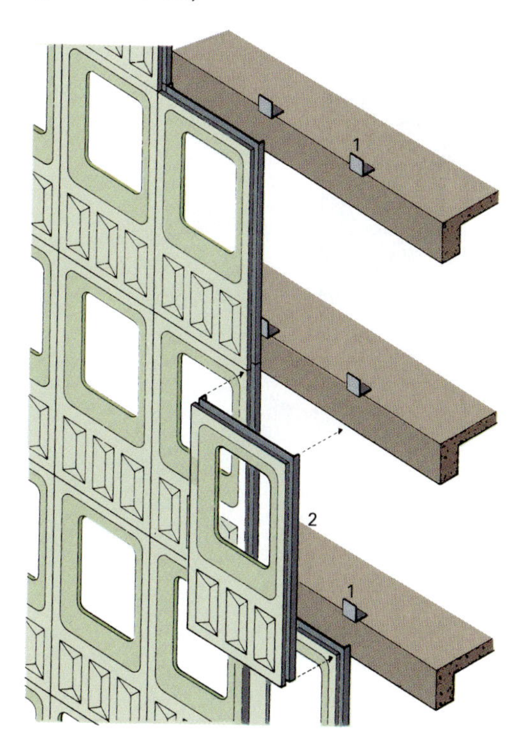

Figure 31.34 This curtain wall system uses precast concrete wall panels. *(Image copyright Aaron Roe, 2009. Used under license from Shutterstock.com)*

Figure 31.35 The column-cover-and-spandrel curtain wall system: (1) column cover; (2) spandrel panel; (3) glazing unit. *(Courtesy American Architectural Manufacturers Association, from Aluminum Curtain Wall Design Guide Manual)*

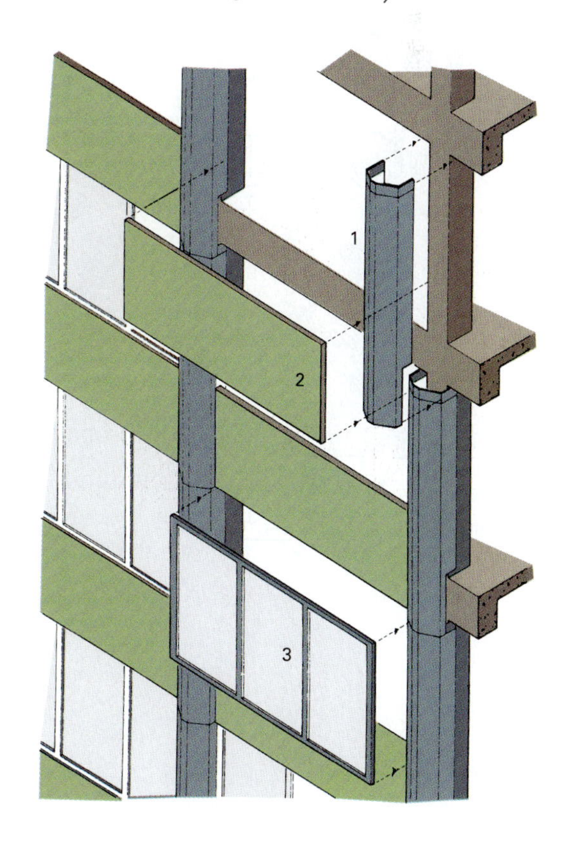

Figure 31.36 These are generalized details for installing a double-glazed ribbon window curtain wall.

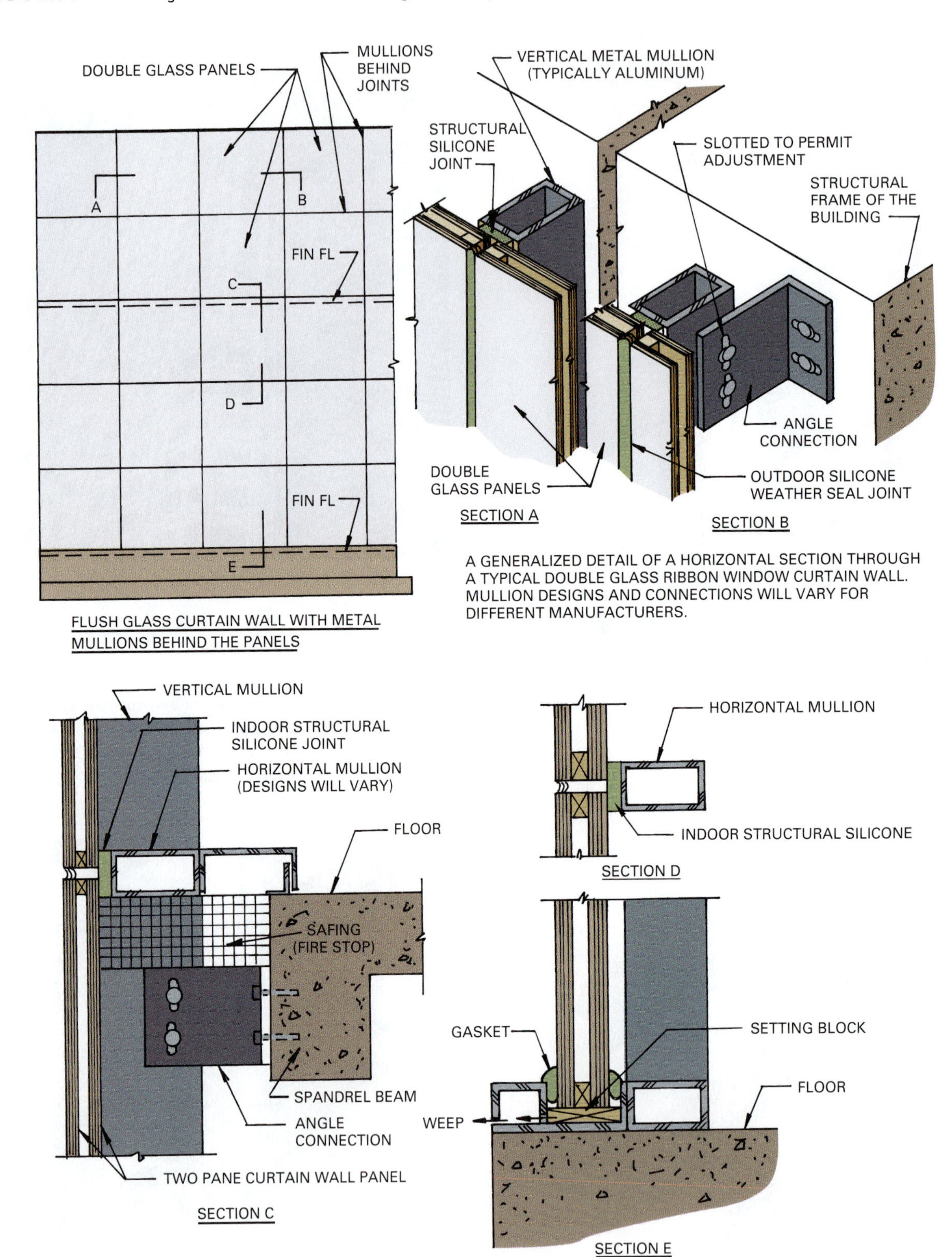

Many other system designs are being developed, including some with little or no exposed framing.

Design Factors There are many factors to consider when designing glass curtain walls and glass spandrels. For example, when glass is exposed to sunlight, its temperature rises. If glass panels are glazed directly to a material that easily absorbs heat, such as concrete, glass edges are cooler than the center of a glass panel. The hotter center expands more, causing increased stresses at the edges. This may fracture the panel, and proper glazing is required to reduce this danger. Interior heat traps, as mentioned for spandrel panels, can cause temperature differences in a panel, leading to fracture.

Exterior sun shading of glass curtain walls also can produce uneven temperatures in a panel because often only a part of the glass is shaded. If most of a panel is shaded, the stresses are minimal. A design must allow for air circulation. This is also a consideration when interior shades are used. They must be hung several inches from the curtain wall to permit air circulation between shade and glass.

Large glass panels used in curtain wall construction are heavy. They must have the strength to span a distance equal to the width of the panel and to resist wind loads. They must be isolated from the frame so expansion, contraction, and possible bending from wind loads will not fracture the glass. The curtain wall system is designed so the frame provides enough of a grip around the edge of the glass to hold it in place when under stress. Frames and mullions are designed to withstand the combined stresses of weight and wind loads.

Manufacturers frequently specify the minimum compressive pressure of gaskets on a given glass surface. Typically, a pressure of at least 4 lb. per linear in. (700 N/m) is needed to provide glass edge support and a watertight seal. Excessive pressures of more than 10 lb. per linear in. (1750 N/m) can increase mechanical stress and may contribute to glass breakage.

Design Examples Construction details for a glass curtain wall using structural silicone sealant to attach panels to metal mullions hidden behind the glass are shown in Fig. 31.36. This design clads the entire wall with glass curtain wall panels that admit natural light and permit occupants to see out. They are referred to as vision-type glass ribbon windows. The glass may be any of several types, such as reflecting or heat resisting. A building with a total glass curtain wall is shown in Fig. 31.37. In Fig. 31.38, a similar construction is used, but opaque glass spandrel panels cover interior structural members and other construction components, such as ceiling panels, heating, plumbing, and electrical runs.

Figure 31.37 This glass curtain wall system has concealed mullions and capped horizontal rails. *(Image copyright Tatiana Grozetskaya, 2009. Used under license from Shutterstock.com)*

Neopariés are another type of curtain wall made with crystallized glass. Neopariés are unique glass ceramic products used for interior and exterior wall cladding. With a marble-like appearance, they are harder and lighter than natural stone, will not absorb water, and are available in flat and curved panels. A flat panel installation is shown in Fig. 31.39.

Possibly the most commonly used metal curtain wall has exposed mullions and rails. In the stick system, the panels are held by metal frames, as shown in Fig. 31.40. The mullions are often larger than the rails, but this varies considerably. This system is shipped disassembled, and mullions are usually installed first, followed by the horizontal rails. Then the glazing and spandrel panels are installed. This system requires more on-site time to install than other factory-prepared panel systems.

Spandrel panels may be made of materials other than glass. Manufactured spandrel panels with aluminum, stainless steel, stone aggregate set in a plastic matrix, ceramic tile panels, and fiberglass-reinforced plastic sheet materials are available.

Aluminum spandrel panels are the most commonly used. They may be left natural or finished with an organic coating (enamel, lacquer), vinyl or other plastic coatings, clear and color anodic finishes, and porcelain enamel. These spandrel panels are installed in aluminum frames in much the same manner as described for

Figure 31.38 A generalized detail for a double-glazed ribbon window with glass spandrels. Actual details vary depending on the manufacturer of the system.

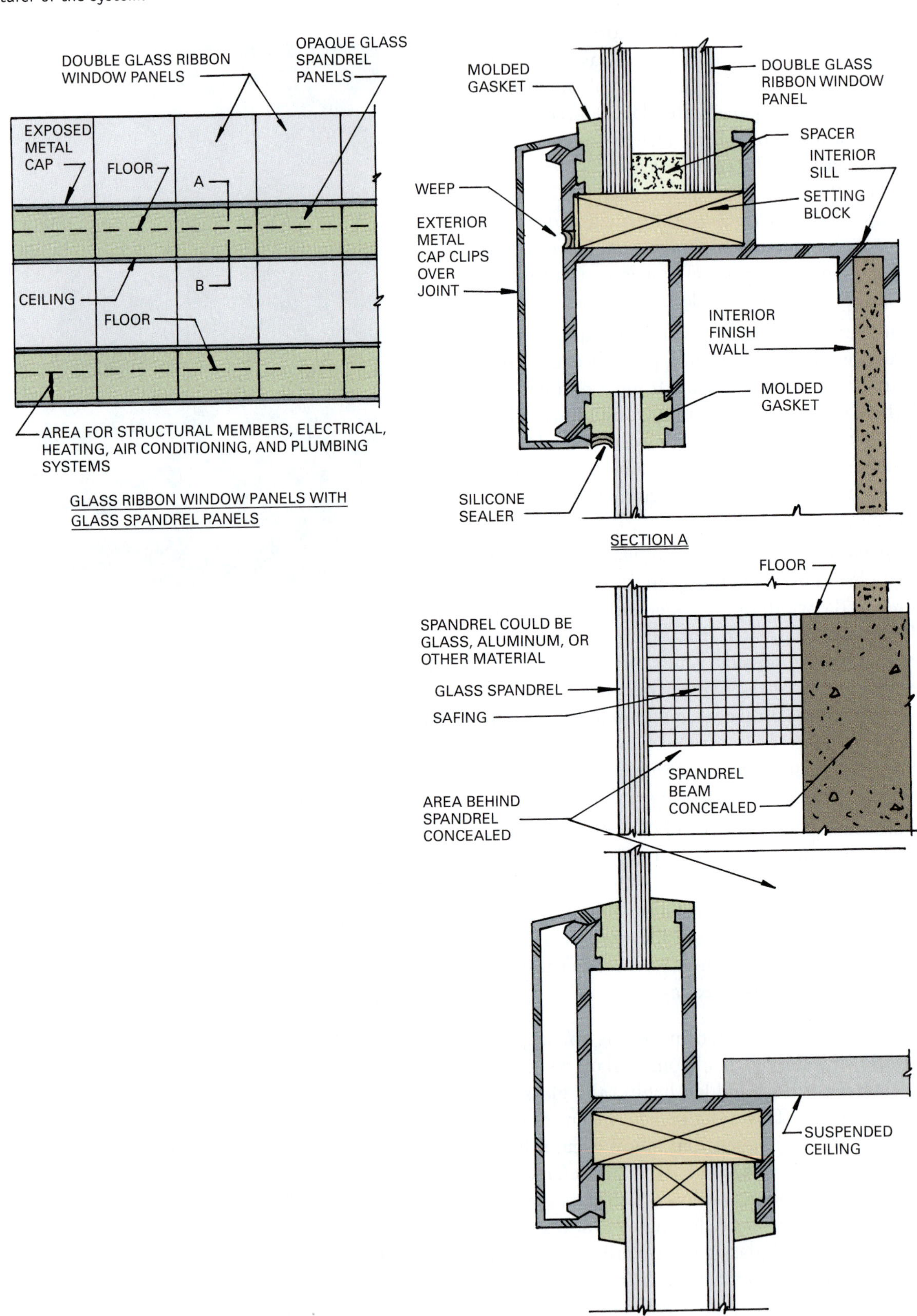

glass spandrel panels. Aluminum panels are also used to cover an entire wall when windows are not wanted. Panel construction varies by manufacturer.

A number of manufacturers provide wall systems for sloped glazing. The sloped glazing serves as the roof and is integrated with the vertical curtain wall. It produces a naturally lighted area and a dramatic setting.

Figure 31.39 These crystallized glass curtain wall panels are secured to the structural frame with metal connectors. Soffits are set with an adhesive and a metal channel. *(Courtesy N.E.G. America, Inc.)*

The building in Fig. 31.41 illustrates the freedom of architectural design a curtain wall system can provide. Some of the structural system is left exposed, and other parts are covered by the glass curtain wall. The exposed columns and spandrel beams create an interesting architectural feature.

A point supported curtain wall is shown in Fig. 31.42. Point supported connections that carry the glass through drilled holes are available in a variety of patented versions (Fig. 31.43). Glass panes supported by point connections require the use of thicker glazing due to the higher stresses encountered. Cast aluminum clamps or brackets that are secured to structural frame elements can also be used to support glass without drilled holes. The weight of the glass is transferred with setting blocks.

Curtain Wall Joint Sealants

The choice of joint sealant depends on the materials to be sealed, the design of the joints, and possible changes that may occur in the joints after the sealant

Figure 31.40 Glass curtain walls may have exposed mullions and rails.

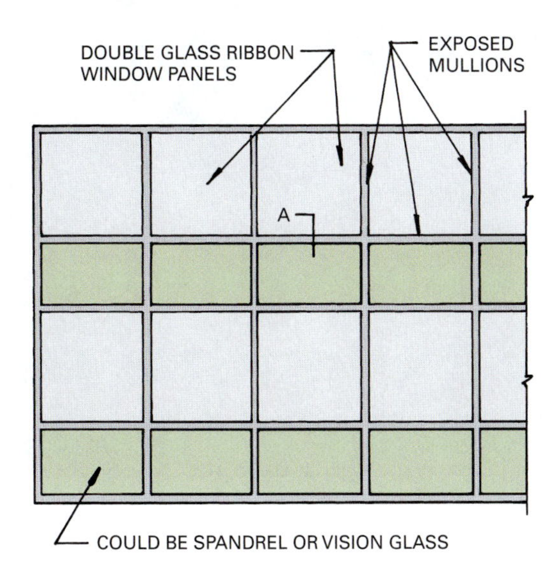

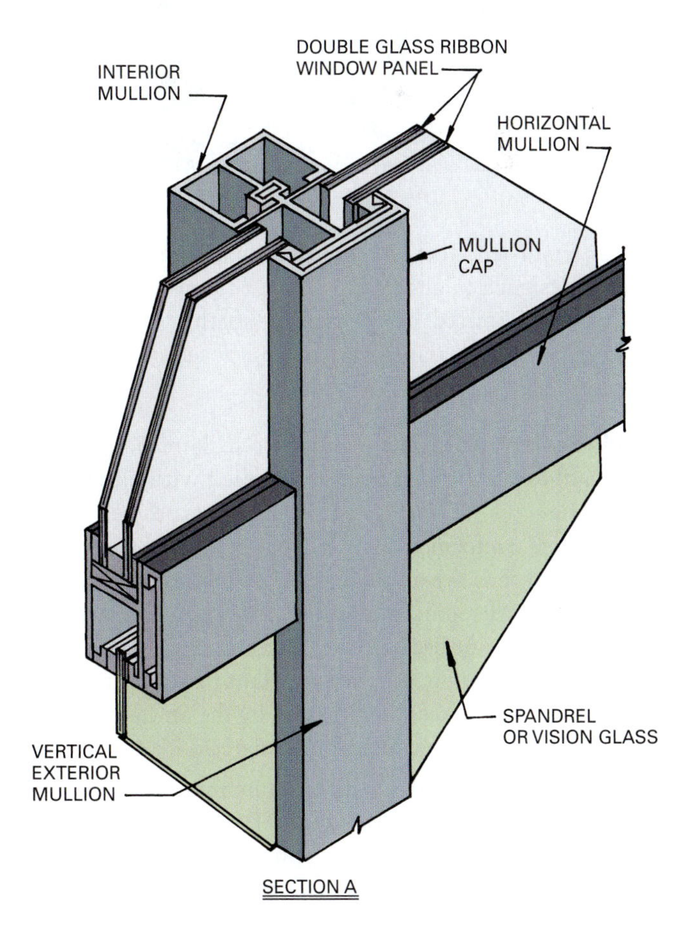

Figure 31.41 The high-rise building has a fixed glass curtain wall with stone facings on the structural members exposed to view.

has been installed. Joints are either working joints, which are designed to allow movement, or non-working joints that are joined by a fastener so they do not move (Fig. 31.44). Sealants may be composed of flowable compounds or solid materials.

Flowable Sealant Materials Flowable sealant materials have adhesive qualities and are applied with either a sealant gun or a knife (Fig. 31.45). Some types come in an extruded preformed strip. They adhere to the surfaces on which they are applied and cure into a rubbery material, sealing the joint. They allow for expansion and contraction in a joint because they can stretch and contract without fracturing. Flowable sealant materials are grouped into three classes, determined by the amount of change in joint size they can accommodate.

Low-performance sealants have minimum movement capacity and are used in stable joints. Typical materials include oil-and-resin-based compounds, bituminous-based caulks and mastics, and polybutene compounds.

Figure 31.42 A point supported curtain wall structured with cable struts. *(Courtesy Eva Kultermann)*

Figure 31.43 Point supported curtain walls use thick glass with drilled holes to support the glazing. *(Courtesy Eva Kultermann)*

Medium-performance sealants can accommodate elongations in the range of 65 to 612.5 percent. Typical materials in this group include acrylics, butyls, and neoprene.

High-performance sealants are used when cyclic movement is expected to be between 612.5 to 625 percent. Typical materials in this group include polymercaptans, polysulfides, polyurethanes, silicones, and some solvents that release acrylics.

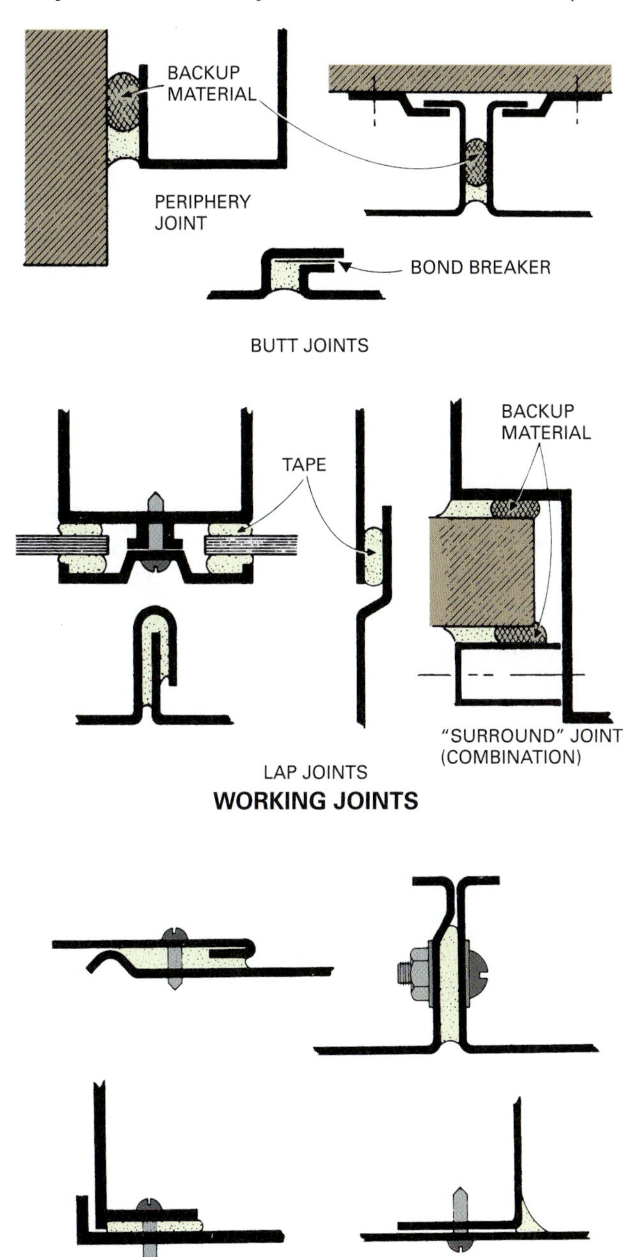

Figure 31.44 Some of the frequently found shapes used as edge joints of joining members. *(Courtesy American Architectural Manufacturers Association, from AAMA Joint Sealants Manual)*

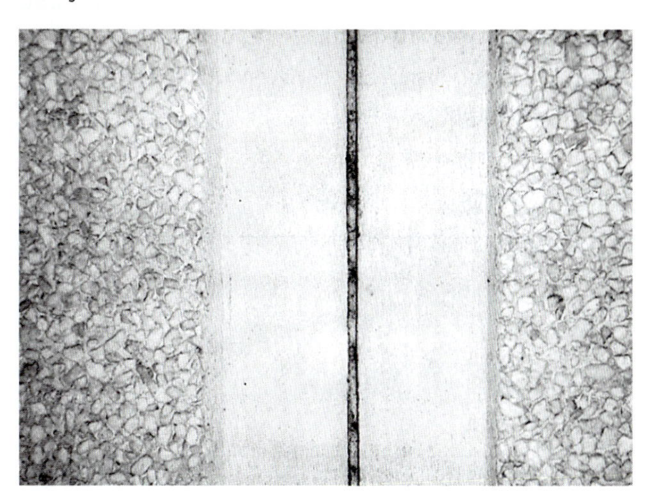

Figure 31.45 These precast concrete curtain wall panels have their joints filled with a flowable sealant.

Detailed information on sealants and their applications is given in Chapter 26.

Solid Sealant Materials Solid sealant materials include tapes and gaskets. Preformed solid tapes are made from polybutene or polyiso-butylene and have adhesive on one or both sides. They are available in rectangular, square, circular, and wedge shapes. They are completely cured when manufactured, seal by being compressed, and are mainly used on lap joints.

Preformed cellular tapes are made from polyurethane, neoprene, vinyl nitrile, and polyvinyl chloride. They contain a chemical solvent and are delivered in a tightly compressed condition. When installed, they expand, filling the joint and bonding to contacting surfaces as they cure.

Gaskets are preformed solid elastomeric materials designed to fit the configuration of components and panels to be sealed. They are compressed and forced into a joint, forming a watertight seal.

Review Questions

1. What types of cladding are typically used for multistory buildings?
2. What are the major considerations when designing a cladding system?
3. What problems do wind forces cause for cladding systems?
4. What forces can cause movement within a cladding system?
5. What is the rainscreen principle?
6. What forces may cause water to penetrate the joints in cladding?
7. What materials are used as firestopping?

8. Define a curtain wall.
9. Describe each of the systems used for metal and glass curtain wall design.
10. What is the difference between working joints and non-working joints?
11. Describe the classes of flowable sealants.
12. Describe the types of solid sealant materials used in curtain wall construction.

Key Terms

capillary action
cladding
kinetic energy
rainscreen principle
safing
thermal break

Activities

1. Try to find buildings with various cladding materials and examine the finished installation. Prepare a report detailing how the architect handled the various factors, such as meeting fire resistance requirements, controlling condensation, and resisting water penetration.
2. Examine manufacturers' brochures (as in Sweet's Catalogs). Sketch the designs proposed and indicate the materials to be used, such as glazing materials, sealants, and thermal breaks. Show and label construction details, such as how units are installed, insulated, and vented, and how condensation drainage is provided.
3. Invite a local architect to address your class on how decisions are made regarding various cladding systems and relate personal experiences with the actual performance of systems.

Additional Resources

Aluminum Curtain Wall Design Guide Manual, Architectural Manufacturers Association, Palatine, IL.

Exterior Wall Construction in High-Rise Buildings, Canada Mortgage and Housing Corp., Ottawa, Ontario.

Howard, P. J., Editor, *Precast Concrete Cladding*, Halsted Press, New York.

The Metal Curtain Wall Manual, American Architectural Manufacturers Association, Palatine, IL.

Reed, R. D., Editor, *The Building Systems Integration Handbook*, the American Institute of Architects, published by Butterworth-Heinman, 80 Montvale Avenue, Stoneham, MA 02180.

See Appendix B for addresses of professional and trade organizations and other sources of technical information.

DIVISION

09

Finishes
CSI MasterFormat™

092000 Plaster and Gypsum Board
093000 Tiling
095000 Ceilings
096000 Flooring
097000 Wall Finishes
098000 Acoustic Treatment
099000 Painting and Coating

(Image copyright Zastol`skiy Victor Leonidovich, 2009. Used under license from Shutterstock.com)

CHAPTER 32

Interior Finishes

LEARNING OBJECTIVES

Upon completion of this chapter, the student should be able to:

- Understand the integration of various interior finishing processes and other aspects of a building's construction.
- Select interior finish materials based on knowledge of interior finish processes.

Build Your Knowledge

For further study on these materials and methods, please refer to:

Chapter 22 Finishing the Exterior and Interior of Light Frame Buildings
Topic: Interior Finish carpentry

Chapter 33 Protective and Decorative Coatings
Topic: Clear Coatings, Opaque Coatings

Chapter 34 Gypsum, Lime, and Plaster
Topic: Gypsum Plaster Products, Gypsum Board Products, Cement Board

Chapter 35 Acoustical Materials
Topic: Sound Control

Interior finishes refers to wall, ceiling, and floor finishes; materials applied to them, such as paneling and wainscoting; and materials used for acoustical treatment, decoration, and other features **(Fig. 32.1)**. Interior finish installation is carefully regulated by building codes, which vary depending on the proposed occupancy of a building.

Interior finishes include a wide range of materials, as shown by the MasterFormat outline for Division 9. Details on these systems and materials are described in other chapters.

In addition to meeting code requirements, the selection of interior finish materials also depends on factors such as cost, appearance, durability, health issues, acoustical considerations, and fire requirements. An architect and owner work to select materials that meet the requirements of both design intent and building code.

Interior finishing begins after a building has been enclosed, protecting the interior from the weather. The roof must be finished, cladding installed, and doors and windows set. The electrical and mechanical trades can then begin installing electrical, communications, telephone, and computer systems, as well as waste and potable water lines, automatic sprinkler systems, and heating, cooling, and ventilation systems. Installation of transportation systems, such as elevators, escalators, and moving walks and ramps, can also start.

ELECTRICAL AND MECHANICAL SYSTEMS INTEGRATION

The installation of electrical and mechanical systems is discussed in Chapters 44 through 47. A building must be designed to facilitate the installation of these systems. Horizontal runs of electrical, plumbing, and ventilation ducts are often placed below the floor and concealed with a suspended ceiling. Some buildings feature cellular floor construction using metal or precast hollow-core concrete decking through which wires and plumbing can be run (Fig. 32.2). Some designs use a raised floor system that provides access below the finished floor for flexibility. Restrooms usually have a plumbing wall installed so the plumbing can be placed out of sight yet accessible for servicing when needed. The restrooms may

Figure 32.1 Interior finishes refers to materials that create finished wall, ceiling, and floor surfaces. *(a) Image copyright Jerome Scholler, 2009. Used under license from Shutterstock.com (b) Image copyright Gualtiero Boffi, 2009. Used under license from Shutterstock.com (c) Image copyright Invisible, 2009. Used under license from Shutterstock.com (d) Image copyright Galtiero Bffi, 2009. Used under license from Shutterstock.com*

Figure 32.2 Electrical and mechanical systems may be run horizontally below floors or through cellular metal decking or hollow-core concrete decking. Raised flooring systems also provide space for horizontal runs. *(Reproduced with permission from* The Building Systems Integration Handbook, *Richard D. Rush, ed., Butterworth-Heinmann Publishers, Newton, Mass., 1986.)*

be part of a central **service core**, which stacks plumbing walls on each floor directly above the one below. This facilitates the vertical run of plumbing lines.

In multistory buildings, areas to house the major electrical service entrance, sewer connections to the central sewage system, potable water, elevator operating equipment, and heating, cooling, and ventilation units are generally placed in a basement or subbasement. These systems are run up through the building to each floor through a vertical central service core (Fig. 32.3) from which they are distributed as required to each floor. The service core may also house stairs, elevators, restrooms, electrical and mechanical chases, fire protection equipment storage areas, and ventilation air shafts. Each floor will have an area allotted to the services provided by the central service core. A typical example is shown in Fig. 32.4. Areas such as electrical and mechanical rooms are enclosed on each floor with a floor and ceiling. Areas such as air and elevator shafts are often open for the entire height of the building. The enclosed rooms must meet fire codes and be designed to contain noise generated within them. Mechanical rooms may contain air handling fans and pumps capable of transmitting sound vibrations through the floor to occupied spaces outside the core. Adjacent interior finishes must control these and other disturbances.

Interior finishes on walls of air and elevator shafts are carefully controlled by codes. These shafts can serve as a passage for flames, smoke, and gases from a fire on one floor to all floors above. An interior finish is critical to the design of the service core.

It should be noted that in high multistory buildings, major electrical and mechanical equipment is usually located on a number of intermediate floors. For example, major equipment on the tenth floor could service the core areas on floors ten through nineteen, and another major equipment area on floor twenty could service core areas on floors twenty through twenty-nine.

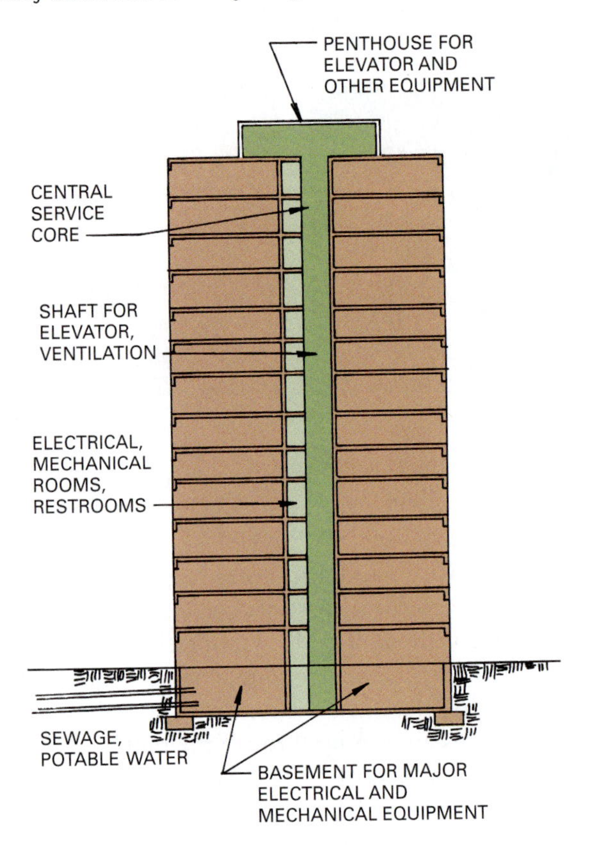

Figure 32.3 The services required for multistory buildings are usually distributed vertically using a central service core.

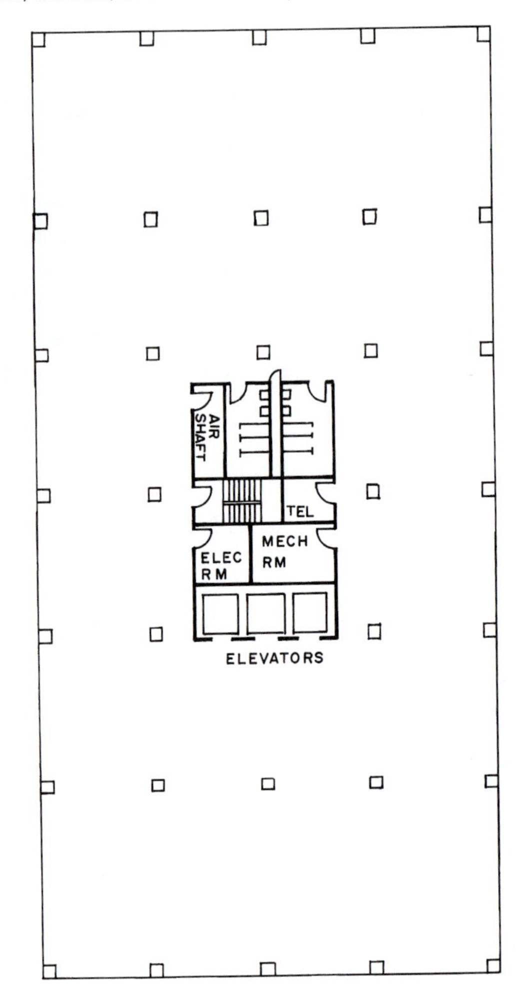

Figure 32.4 The central service core provides space for services, utilities, and vertical transportation on each floor.

Similarly, banks of elevators typically serve only the first ten floors, while a second bank might serve floors eleven through twenty.

ENVIRONMENTAL CONCERNS

Environmental characteristics of products should be considered when selecting interior finish materials. Studies show that we spend 80 percent of our time indoors, in an environment where hundreds of pollutants have been identified in building finish materials. Products that minimize the use of volatile organic compounds (VOCs) and other toxins that reduce indoor air quality should be avoided. Generally speaking, products not made from petroleum-based resources create the healthiest interior environment.

Products are now available that are certified by independent organizations to meet stringent indoor air quality standards. The GREENGUARD Environmental Institute is an example of a not-for-profit organization that oversees the GREENGUARD Certification Program. As an ANSI Accredited Standards Developer, GEI establishes acceptable indoor air standards for indoor products, environments, and buildings. GEI's mission is to improve public health and quality of life through programs that improve indoor air.

COST

The selection of interior finish materials involves consideration of material costs plus the labor to install them, including a contractor's overhead and profit. Other cost factors include the expected life of materials before replacement is required, regular maintenance, minor repairs, and the possibility of increased replacement costs years later because of inflation and increased labor costs. Using life cycle costing, calculations can reveal if it would be less costly over the long run to use higher quality, more expensive materials. This is frequently the case if an owner plans to retain title to a building over a long time period.

APPEARANCE

Interior finishes generally hide the structural, mechanical, and electrical systems and provide attractive interior surfaces. In some cases, portions of the structural system are left exposed and become part of the interior design (Fig. 32.5). Materials used are directly influenced by desired appearances of interior spaces and their proposed use. For example, a hotel lobby may have darker textured walls, carpeted floors and stairs, and elaborate wall hangings. The lighting system can provide soft general illumination yet spotlight features such as paneled walls and paintings. The interior of a health and fitness center will be in sharp contrast to this by having light, durable wall surfaces, acoustical ceilings, considerable natural light, and floor coverings to suit the activity in a given area. The halls and lobby may have durable clay tile floors, the aerobics room a soft, resilient floor covered with carpet, and the basketball area a composition floor.

Figure 32.5 These exposed wooden beams provide a major architectural detail to the interior of the room. The ceiling is exposed finished wood decking. *(Image copyright Photoroller, 2009. Used under license from Shutterstock.com)*

The color of an interior finish establishes the mood and reactions to an area. Warm colors, such as those ranging from reds through orange and yellow, create a feeling of friendliness and warmth. Cool colors, such as those ranging from greens through blues and purples, create a more distant feeling. Lighter colors increase illumination levels, while darker colors absorb light, reducing the amount of reflective light. Generally, interior finishes use a range of both warm and soft colors. Textures of finished walls and ceiling surfaces influence not only appearance, but also other qualities, such as acoustical properties and light reflection.

Coatings

Protective and decorative coatings are applied over many types of interior walls, ceilings, and floors. These are discussed in detail in Chapter 33. Coatings are layers of material in liquid form applied to decorate, preserve, protect, seal, or smooth a surface. When a coat solidifies, it may leave a flat, a semi-gloss, or high gloss finish. Coatings may be transparent, semitransparent, or opaque. Coatings are made from a wide range of materials, which determine the color, hardness, and opacity of a coating. The type of coating used is regulated by building codes. Fire resistance and the degree to which a material supports combustion are important. Codes specify the flame spread, smoke, and fuel contributed ratings of coatings. See Chapter 33 for more information.

Other surface coverings include paper, plastic, fabric, or thin wood veneers. They are subject to the same code restrictions as liquid coatings.

DURABILITY

Durability of interior finishes includes the ability of exposed material (and any protective or decorative coating) to resist damage from abrasion or dirt. Hard-surfaced materials provide considerable protection and are used in areas where wear and tear is expected. In some areas, such as commercial kitchens and restrooms, the ability of a material to withstand water and high humidity is important. Floor coverings are especially vulnerable to wear and damage. A lobby of a busy building will require a very durable floor covering while other areas may be able to use something less robust. The ability to easily clean materials with minimum damage to them is also a factor. Finally, the expected life of a product and replacement costs must factor in when determining degrees of durability.

ACOUSTICAL CONSIDERATIONS

The acoustical properties of finish materials on walls, ceilings, and floors are another major consideration. The type of material, texture of a surface, and any coatings influence the acoustical properties. Under consideration is the ability of a finish to control sound created within an area as well as restrict its flow through walls, ceiling, and floor to adjoining areas. Sound transmitted through materials may occur because of vibrations in caused by the actions of occupants or by machinery operating in a building. Assemblies can be tested to ascertain their impact noise rating. Impact noise can be reduced by insulating floors and covering them with soft materials, such as underlayments, carpet padding, and carpet. This must be considered as a finish is planned. Acoustical design and materials are discussed in Chapter 35.

FIRE CONSIDERATIONS

Interior finish and trim materials are subject to a wide range of fire code requirements. The combustibility of an interior finish material is rated by testing the flame spread of the surface of the material. Flame spread is the rate at which combustion will move across a material. The **flame spread rating** is a single number that designates the ability of a material to resist flaming combustion over its surface. The rate of flame travel is measured under ASTM testing procedures. Noncombustible cement-asbestos board has a rating of 0. Untreated specified species of wood have a designated rating of 100. The lower the rating, the slower flame will spread across the surface of a material. Flame spread ratings are classified as: Class A flame spread, 0–25; Class B flame spread, 26–75; Class C flame spread, 76–200. The acceptable flame spread ratings for various types of buildings and occupancies are specified by building codes (Table 32.1).

Table 32.1 Maximum Flame-Spread Class *(Courtesy International Code Council)*

INTERIOR WALL AND CEILING FINISH REQUIREMENTS BY OCCUPANCY[k]

GROUP	SPRINKLERED[l]			NONSPRINKLERED		
	Vertical exits and exit passageways[a, b]	Exit access corridors and other exitways	Rooms and enclosed spaces[c]	Vertical exits and exit passageways[a, b]	Exit access corridors and other exitways	Rooms and enclosed spaces[c]
A-1 & A-2	B	B	C	A	A[d]	B[e]
A-3[f], A-4, A-5	B	B	C	A	A[d]	C
B, E, M, R-1, R-4	B	C	C	A	B	C
F	C	C	C	B	C	C
H	B	B	C[g]	A	A	B
I-1	B	C	C	A	B	B
I-2	B	B	B[h, i]	A	A	B
I-3	A	A[j]	C	A	A	B
I-4	B	B	B[h, i]	A	A	B
R-2	C	C	C	B	B	C
R-3	C	C	C	C	C	C
S	C	C	C	B	B	C
U	No restrictions			No restrictions		

For SI: 1 inch = 25.4 mm, 1 square foot = 0.0929 m^2.

a. Class C interior finish materials shall be permitted for wainscotting or paneling of not more than 1,000 square feet of applied surface area in the grade lobby where applied directly to a noncombustible base or over furring strips applied to a noncombustible base and fireblocked as required by Section 803.4.1.
b. In vertical exits of buildings less than three stories in height of other than Group I-3, Class B interior finish for nonsprinklered buildings and Class C interior finish for sprinklered buildings shall be permitted.
c. Requirements for rooms and enclosed spaces shall be based upon spaces enclosed by partitions. Where a fire-resistance rating is required for structural elements, the enclosing partitions shall extend from the floor to the ceiling. Partitions that do not comply with this shall be considered enclosing spaces and the rooms or spaces on both sides shall be considered one. In determining the applicable requirements for rooms and enclosed spaces, the specific occupancy thereof shall be the governing factor regardless of the group classification of the building or structure.
d. Lobby areas in Group A-1, A-2 and A-3 occupancies shall not be less than Class B materials.
e. Class C interior finish materials shall be permitted in places of assembly with an occupant load of 300 persons or less.
f. For churches and places of worship, wood used for ornamental purposes, trusses, paneling or chancel furnishing shall be permitted.
g. Class B material is required where the building exceeds two stories.
h. Class C interior finish materials shall be permitted in administrative spaces.
i. Class C interior finish materials shall be permitted in rooms with a capacity of four persons or less.
j. Class B materials shall be permitted as wainscotting extending not more than 48 inches above the finished floor in exit access corridors.
k. Finish materials as provided for in other sections of this code.
l. Applies when the vertical exits, exit passageways, exit access corridors or exitways, or rooms and spaces are protected by a sprinkler system installed in accordance with Section 903.3.1.1 or 903.3.1.2.

Smoke development ratings classify materials by the amount of smoke they will give off as they burn. Most codes prohibit materials with a rating of 450 or more for use inside a building. Materials are tested using ASTM test procedures. Class A, Class B, and Class C ratings permit smoke amounts to range from 0 to 450.

The **fuel contributed rating** is a measure of the amount of combustible substances in a given material. The **fire resistance rating** indicates a material's capacity to withstand fire for a specified time and under conditions of standard intensity such that it will not fail structurally nor permit the non-fire side to become hotter than a stipulated temperature in hours.

Codes also specify regulations pertaining to flame retardance. Curtains, draperies, and other decorative materials are required by code to be fire-retardant.

Floor finish materials, such as wood, composition tile, and carpet, are also regulated by building codes as are doors and windows. Typical fire doors are discussed in Chapter 30.

Review Questions

1. What materials are included in Division 9 of the MasterFormat?
2. What factors are considered as an interior finish is planned?
3. How are electrical and mechanical services distributed in multistory buildings?
4. How do flame spread, smoke contributed, and fuel contributed considerations influence interior finish?
5. What acoustical measurements are considered when selecting interior finish materials?
6. What purpose do coatings serve on interior finishes?

Key Terms

fire resistance rating
flame spread rating
fuel contributed rating
service core
smoke developed ratings

Additional Resources

The GREENGUARD Environmental Institute, *www.greenguard.org*.

CHAPTER 33

Painting and Coatings

LEARNING OBJECTIVES

Upon completion of this chapter, the student should be able to:

- Recognize the range of protective and decorative coatings available and the properties of each.
- Select appropriate protective and decorative coatings for interior and exterior work.

Paints and coatings are layers of materials in liquid form applied to surfaces to decorate, preserve, protect, and seal them. When a liquid coat solidifies it leaves a thin layer over the **substrate** to which it is applied.

Coatings are used to protect materials from heat, soiling, solar radiation, moisture, chemicals in the air, and corrosion, as well as to provide **abrasion resistance**. They also provide decorative surfaces. Many coatings are applied in three layers; a primer coat, an intermediate coat (sometimes called an undercoat), and a finish coat, also referred to as a topcoat.

The **primer coat** is applied directly to the substrate (such as steel or wood). It must have good **adhesion** to the surface, appropriate flexibility, permit the intermediate coat to bond to it, retard corrosion of the substrate, and resist weathering long enough to permit the application of the intermediate and finish coats.

The **intermediate coat** is applied over the primer coat. It must provide an adequate film thickness and structural strength, bond to the primer coat, permit the finish coat to bond to it, and create a barrier to chemicals and other environmental contaminates.

The **topcoat** (finish coat) is applied over the intermediate coat. It provides the final finish surface, the desired aesthetic, and may serve other functions, such as providing a nonskid surface or resisting mildew.

The total coating system must be one that is compatible with the substrate, and each coating must be compatible with every other. If this does not occur, the coating will deteriorate and have to be completely removed before the substrate can be recoated.

Coatings may be clear, semitransparent, or opaque. **Clear coatings** let most of the natural color and texture of the substrate show through. They are used when the natural qualities of materials are desired, such as the color and grain of wood or the color and texture of colored concrete.

Semitransparent coatings allow some color and texture of an original surface to show but obscure much of it. They are used when the appearance of a substrate needs augmenting, for example by adding some color yet permitting the existing material to show through.

Opaque coatings completely obscure the color and much of the texture of a substrate. They are used when a solid, uniformly colored surface is desired. The material composing the substrate is often not identifiable.

FEDERAL REGULATIONS

Federal regulations that affect the use of paint and other coatings include the 1992 Residential Lead-Based Paint Hazard Reduction Act, which provides guidelines for the evaluation, reduction, and notification of any known lead-based paint hazards in residential occupancies. The 1970 Clean Air Act is legislation that enforces regulations to protect the public from exposure to airborne contaminants that are known to be hazardous to human health, including paints and coatings.

Lead in Coatings

Lead is now prohibited from being used in paints. An ingestion of loose paint dust or chips can cause lead poisoning. Typical hazards include harm to workers on remodeling projects where paint containing lead was used in previous years and harm to children living in older houses and apartments who may eat loose paint chips pulled from the trim and cabinets. See the Construction techniques section of this chapter for additional details.

Air Quality Regulations

Information related to air quality regulations can be found in the Code of Federal Regulations. The regulation for finishing materials is called the National Volatile Organic Compound (VOC) Emissions Standards for Architectural Coatings, citation i40CFE 59.400. The standard regulates emissions from many sources, including products used as finishing materials in construction that emit **volatile organic compounds**, or VOCs.

Emissions from solvents in coatings are one of the largest sources of air pollution. We are surrounded by paints, stains, and varnishes both at home and in other buildings. Although paints are becoming less toxic, many are made from petro chemicals, particularly those for exterior use.

VOCs include any solvent, propellant, or other substance, except water, that evaporates from a coating as it dries. VOC is calculated as the combined weight of all solvents in a given volume of coating, excluding certain exempted materials. VOC is specified in grams per liter (g/l), which can be converted to pounds per gallon (lb./gal.) by dividing g/l by 119.8.

Architectural coatings are limited to about 250 to 700 grams per liter (2.1 to 5.8 pounds per gallon). A few examples are given in Table 33.1. These limits are the amount of VOC per liter or gallon of coating thinned to the manufacturer's maximum recommendations, excluding the volume of water, exempt compounds, or colorant base material. Contractors must observe VOC ratings of coatings they use or determine if exempt products are available. Manufacturers include VOC ratings of products on containers and in sales literature.

Table 33.1 Limits on VOC Emissions for Selected Finishing Materials

Coating Category	Grams VOC per liter	Pounds VOC per gallon
Concrete protective coating	400	3.3
Clear fire-retardant coatings	850	7.1
Flat exterior coatings	250	2.1
Flat interior coatings	250	2.1
Lacquers	680	5.7
Nonflat exterior coatings	380	3.2
Nonflat interior coatings	380	3.2
Enamels	450	3.8
Rust preventative coatings	400	3.3
Interior sealers	400	3.3
Clear shellac	730	6.1
Opaque stains	350	2.9
Varnishes	450	3.8
Clear wood sealer	550	4.6

For more information, see the National Volatile Organic Compound Emission Standards for Architectural Coatings, Code of Federal Regulations, Environmental Protection Agency.

Figure 33.1 A variety of pigments are used to add color and opacity to paints. (Image copyright Adam Radosavljevic, 2009. Used under license from Shutterstock.com)

COMPOSITION OF COATINGS

Many different kinds of coatings are available, and their composition varies widely. However, they all have the same basic composition. This includes a binder, a solvent, and pigments. The **binder** is a non-volatile natural or synthetic **resin** that forms the base of the hardened coating. The **solvent** is the volatile part of the coating in which the binder is dispersed. The mixture of the binder in the solvent forms the **vehicle** of the coating. **Pigments** are insoluble particles that are suspended in the vehicle to add color and opacity (Fig. 33.1). The binder plus the pigments form the solids of a coating and produce the layer remaining after the solvent has evaporated.

Clear coatings are made of binder and solvent but contain no pigments. Semitransparent coatings have a small amount of pigment to provide some opacity.

Opaque coatings contain considerable pigment and totally obscure the face of a substrate. The binder bonds itself to the substrate. It also must have the properties needed to meet the requirements of the coating. For example, it must have plasticizers to make it flexible, stabilizers to enable it to resist solar radiation or sources of heat, and driers to control the rate of curing.

Coatings are either solvent-based or water-based. Solvent-based coatings are volatile, while **water-based coatings** have the binders and pigments dissolved in water. Solvent-based coatings require the use of a solvent, such as paint thinner, to thin the mix and clean brushes and spray guns. The solvent used depends on the composition of a coating. Water-based coatings are thinned, and the tools used to apply them are cleaned with water.

Coatings are specified for specific uses. Manufacturer's recommendations clearly indicate whether a coating can be used for exterior or interior purposes, or both. They specify the type of substrate to which a coating can be applied and whether it resists attack by chemicals, heat, and ultraviolet rays. In addition, the final appearance is indicated as clear, semitransparent, or opaque and high gloss, satin gloss, or flat (no gloss). Coating specifications may include wet film thickness and/or dry film thickness.

Wet Film Thickness

To ensure optimum performance, the proper thickness of paint must be applied. This is specified by the minimum wet film (MWF) thickness and stated as "applied at the rate of ×× mils per coat minimum wet film thickness (MWF)." A mil is a unit of measure equal to 0.001 in. (0.025 mm).

Wet film thicknesses vary depending on **spreading rate** per gallon. The wet film thickness of a coating can be determined by measuring the wet film thickness with special gauges designed for that purpose or by using an industry-developed formula that relies on the spreading rate. Table 33.2 shows some typical wet film thicknesses found using the formula. The minimum wet film thickness for various coatings can also be found by noting the recommended spreading rates indicated by the manufacturer of a coating material.

Table 33.2 Wet Film Thickness Based on the Spreading Rate

Spreading Rate (sq. ft./gal.)	Wet Film Thickness (mils)
1,600	1.0
1,000	1.6
700	2.3
400	4.0
200	8.0
160	10.0

Dry Film Thickness

Specifications state the minimum dry film (MDF) thickness per coat for a coating material. Dry thickness determines the amount of protection a coating will provide. Sufficient wet mils must be applied to get a specified dry film thickness.

The dry film thickness is found by multiplying the wet film thickness in mils by the percent of solids in the paint. This gives a theoretical dry film thickness. The percent of solids in a coating is determined by the manufacturer of a material.

FACTORS AFFECTING COATINGS

After a coating has been applied to a substrate and has hardened, it is subject to a number of external factors, and these must be considered before choosing a coating.

Water resistance is one of the most common factors affecting coatings. Rain on external surfaces may cause a thermal shock if the coating is very hot. It may penetrate the coating through checks and cracks and freeze or become very hot, causing blisters. Water vapor may penetrate a coating from the outside, resulting in damage to the substrate that causes the coating to peel and crack. Water vapor may penetrate the substrate from behind the coating, again causing blisters and peeling. Vapor barriers used on the warm side of exterior walls are designed to prevent this. Substrates, such as wood, may contain moisture before a coating is applied. The moisture content of a substrate must be within specified limits for successful coating. Excess moisture in a substrate will cause poor adhesion of a coating.

Solar radiation bombards coatings with ultraviolet radiation, which can cause pigment fading and other chemical reactions in the solvents and binders that lead to coating deterioration. This permits ultraviolet radiation to reach the substrate, possible causing further damage. Solar radiation can increase the temperature of both the coating and substrate, causing both to expand. A coating must be flexible enough to expand and contract without cracking. This includes withstanding expansion of a substrate, which may exceed the normal expansion of a coating at the same temperature. Likewise, freezing temperatures cause damage, especially if moisture has penetrated a coating.

Coatings are subject to the effects of dust and dirt. The extent of possible damage depends on their location. A ceiling, for example, usually has minimum exposure, while a baseboard or door has greater exposure. Location also dictates exposure to abrasion. Impact due to natural causes (such as hail), normal wear (as on doors), and vandalism must be considered.

In some situations, coatings are affected by chemical fumes, solutions, and reactions. Chemical fumes are generated by many sources, including power plants and automobile emissions. Sea water, oils, solvents, and other chemical solutions impact heavily on coatings. Soluble alkaline salts in mortar and concrete can dissolve and crystallize on the surface, damaging the coating. This is called **efflorescence**. Sealants may react unfavorably, and rust may streak a coating surface. Wood knots heavy with resin may bleed through a coating. These and other possible reactions must be carefully considered before choosing a coating.

The absorption of the surface of a substrate will affect a coating. Some surfaces are hard and smooth and absorb none of the coating. Such surfaces may require roughing by sanding, sandblasting, or etching for better adhesion. Various porous surfaces have different rates of absorption, providing different levels of adhesion, which may cause some cracking of a coating.

New coatings applied over existing ones must be compatible. The surface of the old coating must be stable and have good adhesion to the substrate or the new coating will fail. Old coatings chalk as they age. If this is the only deterioration, a new coating can be applied over it. If there are cracks and loose peeling sections, the old coating must be removed before recoating.

Following is a discussion of commonly used field-applied coatings. There are many other products finished in the factory that use automated application and drying techniques.

CLEAR COATINGS

In addition to traditional clear coatings, such as shellac, varnishes from natural resins, and lacquer, a number of products using synthetic resins are available. Clear coatings used in exterior locations in general do not have the durability of opaque coatings because they lack the pigments that protect against ultraviolet damage from the sun. They let the natural color and texture of a substrate show as well as protect it from moisture, abrasion, and other forms of damage.

Natural Resin Varnishes

Varnishes are one of the oldest finishes used to coat wood surfaces. Several types are available, and their properties and composition vary considerably. Varnishes fall into two broad classifications: natural-resin varnishes and synthetic varnishes.

There are three basic types of natural resin varnish: linseed oil varnishes, tung oil varnishes, and spirit varnishes. Natural varnishes are made from either fossil resins or resins obtained from a variety of trees in tropical countries. The vehicle is some form of drying oil into which the resin is dissolved. The oil-to-resin ratio determines the classification of a varnish and is expressed as the number of gallons of oil that are mixed per 100 lb. (45 kg) of resin. Varnishes made with natural resins and oils are called oleoresinous varnishes.

Turpentine is a common solvent for varnish, although mineral spirits, naphtha, and benzene are also used. The solvent evaporates, causing the varnish to harden. The drying oils cure by oxidation and polymerization following the loss of the solvent. This is why varnish is a slow-drying coating.

Linseed oil varnishes are available in three types: long-oil, medium-oil, and short-oil. Long-oil varnish is sold under the name "spar varnish." It is used on exterior surfaces exposed to intermittent wetting. Synthetic resin varnishes are better suited for very moist conditions. Medium-oil varnishes dry faster than long-oil types and have a harder film but are not as water resistant. Synthetic binders, alkyd and phenolic, are used to produce a modified spar varnish. Short-oil varnishes contain the least oil, dry rapidly, are brittle, and do not resist abrasion very well.

Tung oil varnishes are used in areas where heavy use is expected, such as on school furniture. They are usually factory applied.

Spirit varnish, also known as shellac, uses a resin obtained from the exudation of the lac insect, which is found in Southeast Asia and India. The resin is dissolved in denatured alcohol and is orange in color (referred to as orange shellac). It can be bleached, producing white shellac (Fig. 33.2).

Shellac is available in various grades depending on the amount of resin dissolved in a gallon of denatured alcohol. The grades are referred to as "cuts," such as a 4 lb. (1.8 kg) cut, which denotes 4 lb. of resin dissolved in one gallon of denatured alcohol. Shellac dries rapidly but does not resist moisture. It is mainly used in construction to seal knots and other resinous places in wood over which water-resistant coatings are applied. Shellac will not withstand exposure to sunlight.

Synthetic Resin Varnishes

Synthetic resins utilize plastic materials suspended in a solvent. The most commonly used resins are **alkyds**, polyurethane, silicone, epoxy, acrylics, and phenolics. The vehicle may consist of the same drying oils used for natural resins, although synthetic materials have been developed. A listing of some clear synthetic resin varnishes and their solvents are shown in Table 33.3.

Acrylic resin varnishes produce a thermoplastic film that is resistant to age-induced yellowing, ultraviolet rays, and oxidation. They have good gloss retention, are almost colorless, and are used on metal, wood, plastics, textiles, and paper. Acrylic resin coatings use either a solvent base or a water base. The solvent-based formulation has a high gloss and is impermeable to water vapor. The water-based formulation has a semi-gloss finish and is permeable to water.

Figure 33.2 Shellac is a natural finish material made from a sticky substance given off by the lac insect, which thrives in India and Burma. *(Image copyright Thomas M Perkins, 2009. Used under license from Shutterstock.com)*

Table 33.3 Clear Finishes

Binder	Base	Uses
Acrylic	Solvent or water	Waterproofing, sealing surface against dirt Used on concrete, masonry, stucco
Alkyd (spar varnish)	Solvent	Used on interior and exterior protected surfaces
Phenolic (spar varnish)	Solvent	Exterior wood exposed to moisure, marine, applications
Silcone	Solvent	Waterproofing, sealing surface against dirt Used on concrete, masonry, stucco, wood
Polyurethane	Solvent	Resists chemical attack, abrasion, heavy foot traffic

Alkyd resin varnishes are made by a chemical reaction between an alcohol and an organic acid. They are used for waterproofing and for reducing dirt retention on concrete, masonry, and stucco walls. Some types are used on exterior metal surfaces.

Phenolic resin varnishes include phenol formaldehyde and modified phenolic types.

Phenol formaldehyde varnishes are excellent for materials exposed to the weather and can be used on wood marine products. They are resistant to caustic substances and acids but tend to yellow and darken with age and lose gloss with exposure to the sun. Modified phenolic varnishes do not resist weathering as well but have an abrasion-resistant quality that makes them useful for interior applications, such as furniture and floors.

Epoxy varnishes have excellent resistance to caustic materials, adhesion qualities, and resistance to abrasion. Epoxy coatings come in a variety of forms, each with special characteristics. Some are highly flexible. They can be used on a variety of materials, such as concrete, metal, and wood. Epoxies are often used as primers. Some are formulated to resist solvents, heat, chemicals, and salt water.

Polyurethane resin varnishes protect against abrasion resistance better than natural varnishes and offer resistance to solvents, chemicals, and oxidation. Polyurethane coatings have a wide range of applications on wood products and wood floors (Fig. 33.3) and are also used as penetrating **sealers** to control dust on concrete floors. They provide a fast drying, durable, mar-resistant surface. Three types are one-component coatings, and two types have two components. Type 1 cures by exposure to oxygen in the air. Type 2 requires 30 percent humidity in the air to cure. Types 1 and 2 can be applied on-site. Type 3 cures with heat. Types 3, 4, and 5 are factory applied.

Silicone varnishes have excellent heat resistance, water resistance, and resistance to corrosive atmospheric conditions. A silicone-acrylic coating has a high gloss and high resistance to blistering and crazing. Silicones are also used to modify alkyds by improving their durability. Silicone-polyester coatings will resist damage by heat up to 550°F (288°C) if the mix contains 75 percent silicone.

Silicones bond well to wood, but steel must be cleaned and bonderized to achieve proper adhesion. Silicone solutions are formulated for application to concrete and

Figure 33.3 A fast-drying, clear polyurethane top coat used on interior surfaces, such as flooring and trim. *(Image copyright Harry Hu, 2009. Used under license from Shutterstock.com)*

masonry materials; they provide a water-repellent coating by penetrating pores, leaving no film on a surface.

Lacquer

Lacquer is made from synthetic materials and is quick-drying due to the rapid evaporation of the solvent. Lacquer has a nitrocellulose base used in combination with various resins and plasticizers. Drying oils are added to improve adhesion and elasticity. A variety of natural and synthetic resins are used to improve adhesion and hardness and give a desired gloss. Those used depend on the end use of the lacquer. Commonly used resins are alkyds, epoxy, acrylic resins, and cellulose acetate.

Solvents commonly used in lacquer include diethylene glycol, acetone, and amyl, ethyl, butyl, and isopropyl acetate. Formulating lacquer requires several different solvents to dissolve the synthetic and natural materials used. This blending of solvents affects the gloss, ease of flow, setting time, and amount of bubbling experienced when applied.

Thinners are sometimes added to lacquer before application. This is especially important when using sprayed lacquers. Thinners adjust consistency and rate of drying. In general, lacquers dry to the touch in five to ten minutes and form a firm film in thirty minutes to three or four hours, depending on their formulation. Some sprayed lacquers can have a second coat applied just fifteen to twenty minutes after the first. Commonly used lacquer thinners are toluol, xylol, benzol, and ethyl, amyl, butyl, and isopropyl alcohol. Lacquer thinner spilled on a hardened lacquer finish will dissolve the finish.

Most lacquers are sprayed on. A brushing lacquer can be applied with a pure bristle brush but dries slower. Wood surfaces require a wash coat of shellac or lacquer sealer before a finish coat is applied. Several finish coats are required because lacquer film is very thin. Lacquer is used mainly on interior products, such as cabinets and furniture. Pigments are added to provide color and opacity.

OPAQUE COATINGS

Opaque coatings obscure the natural color of a surface, provide protection, and add a decorative feature through a wide range of colors. Type of coating material used depends on the surface to be coated and its intended use. Exposure to weather, soil, chemical fumes, salt water, and other conditions must be considered when selecting a type of coating. Some of the frequently used coatings for various materials are listed in Table 33.4.

Before a surface is coated it must first be cleaned and prepared to receive a primer, if required. After priming, the requisite number of finish coats are applied.

Primers for Opaque Coatings

A primer is a coating applied to a surface to seal and prepare it for a finish topcoat. Primers help hide discoloration and stains and seal areas that may bleed through a finish topcoat. Resin in wood presents a common bleeding problem. Primers are also used to smooth porous surfaces, such as those on concrete blocks. The use of the proper primer is vital to the successful application of a finish topcoat, and manufacturer's recommendations must always be followed. Table 33.4 gives examples of some commonly recommended primers and associated topcoats.

Cleaning Metal Surfaces for Priming The preparation of steel surfaces depends on which primer and finish topcoat will be used. In some cases cleaning with a detergent or solvent is adequate. Wire brushes, sandpaper, and sandblasting are also used to clean surfaces and roughen them. Galvanized metal can be cleaned with a dilute acetic acid, steel with phosphoric acid. After a surface is washed and dried, the proper primer can be applied. Some metals are allowed to weather for six to nine months prior to the application of a primer.

Aluminum painted on-site should be allowed to weather for at least a month. It can then be wiped with mineral spirits or other solutions as recommended by the coating manufacturer. Most aluminum products arrive on-site with a factory-applied finish.

Copper, bronze, and their alloys can be cleaned by wiping them with mineral spirits or a dilute solution of hydrochloric or acetic acid, which must then be completely washed away. Corrosion and material stuck on a surface can be removed by brushing or light sanding.

Table 33.4 Commonly Used Primers and Topcoats

Material	Primer	Topcoat	Remarks
Aluminum	Vinyl red lead	Vinyl	Exposure to weather
	Zinc chromate	Chlorinated rubber	Exposure to rain, salt water spray
	Zinc chromate	Alkyd or acrylic	Used on trim, flashing
	Self-priming	Epoxy ester	Exposure to fumes
Ferrous metal	Self-priming	Phenolic	Exposure to weather, high humidity
	Zinc silicate	Silicate, alkyd	Exposure to weather
	Zinc silicate	Silicone, aluminum pigmented	Exposure to weather
	Self-priming	Vinyl	Exposure to rain, salt-water spray
	Red lead	Acrylic	Will not resist abrasion
	Self-priming	Urethane	Exposure to corrosion, chemicals, abrasion
	Self-priming	Epoxy	Exposure to acids, alkalis, chemicals
	Self-priming	Coal tar	Apply hot to metal to be below ground
	Chlorinated rubber with red lead or zinc chromate	Chlorinated rubber	Exposure to rain, salt-water spray, chemicals
	Red lead	Oil-based paints	Normal exterior conditions
	Red lead and alkyd-based red lead	Alkyd	Not abrasion resistant, exposure to severe weather
	Zinc-polystyrene	Polystyrene	Chemical fumes, exposure to fresh and salt water
Ferrous metal (galvanized)	Zinc dust or zinc chromate-zinc dust	Alkyd	Does not require topcoat
	Zinc dust or zinc oxide	Chlorinated rubber	Exposure to rain, salt-water spray
Ferrous metal in ground	Self-priming	Coal-tar-epoxy	Used on pipelines, buried structural steel
Gypsum wallboard	Vinyl	Alkyd	Light duty
	Self-priming	Acrylic	Heavy duty
Gypsum plaster	Self-priming	Acrylic	Plaster must be dry
Concrete and concrete masonry (dry), brick masonry	Self-priming	Acrylic	Interior locations, scrubbable
	Self-priming	Vinyl	Dry locations
	Self-priming	Epoxy esters	Exterior use, resists fumes, scrubbing
	Self-priming	Polychloroprene	Resists water, solvents, impact, exterior uses
	Self-priming	Urethane	Washable, interior locations
Concrete floors, no moisture exposure	Self-priming	Urethane	Light to moderate traffic
	Self-priming	Epoxy	Moderate to high traffic
Concrete, heavy moisture	Self-priming	Alkali-resistant chlorinated rubber	Water reservoirs, swimming pools
Portland cement plaster	Self-priming	Vinyl	Dry locations
	Styrene-butadiene	Alkyd	Dry locations
Wood, interior	Self-priming	Vinyl	Walls and floors
	Self-priming	Alkyd	Doors, paneling, trim light-duty floors
	Self-priming	Urethane	Surfaces subject to impact, scrubbing, heavy-duty floors
	Self-priming	Acrylic	Surfaces subject to impact, scrubbing
Wood, exterior	Self-priming	Urethane	Porch decking, exterior stairs
	Self-priming	Alkyd	Siding, plywood, cedar shakes, trim
	Self-priming	Acrylic	Siding, plywood
	Self-priming	Phenolic	Siding, plywood, trim
	Oil-based primer	Oil-based vehicle	Wood siding, exterior trim, plywood siding
	Self-priming	Epoxy	Any exterior wood

An etching primer is available for cleaning ferrous and nonferrous metals and some alloys. It cleans and chemically etches a surface, leaving a thin protective film. Generally, a standard prime coat is applied over this material.

Preparing Wood Surfaces for Priming Exterior wood to be painted should have a moisture content of 12 to 15 percent. Interior wood should have a moisture content of about 6 percent. Wood with knots that might bleed should have a coat of shellac or other sealer applied before priming. Mold, fungus, and other stains must be removed.

Preparing Concrete Surfaces for Priming It is advisable to let concrete completely cure before priming. Concrete aged less than thirty days is generally considered not suitable for painting. Paint manufacturers' recommendations on curing time should be observed. Concrete surfaces should be free of dirt, oil, and other substances. Loose material can be removed by brushing or sanding.

Priming Coats A primer is a first coat applied to a substrate. It seals and fills the pores of a surface, inhibits rust in ferrous metals, and improves the adhesion of subsequent coats of paint (Fig. 33.4). The following discussion covers some of the frequently used primers.

Gypsum plaster can be primed with latex (acrylic), alkyd, or oil-based primer after a thirty day curing period. Latex flat interior finish may be used as a primer if the topcoat is a gloss or semigloss coating. Damp new plaster requires an **alkali**-resistant primer.

Gypsum wallboard and other paper-faced products are primed with polyvinyl acetate if an alkyd topcoating is used. Latex (acrylic) topcoatings are self-priming. Portland cement plaster often is painted with a vinyl topcoating, which requires no primer.

A wide range of primers and topcoatings are used on concrete, concrete masonry, and brick masonry. Most of these topcoats are self priming. Surfaces of concrete and masonry are usually rough and porous. They can be coated with a latex-Portland cement grout or latex blockfiller to produce a smoother surface. This results in a better, watertight coverage because the topcoat will not have to bridge pores in the surface. A thinned coat of catalyzed-epoxy coating material can also be used. Concrete walls can be primed with a latex primer-sealer, after which a topcoat of latex or alkyd paint can be applied.

Aluminum is often primed with zinc chromate. Bare ferrous metals use a variety of primers depending on which topcoat will be used. Since ferrous metals rust rapidly, they must be coated with a rust-inhibiting primer. Red lead is the oldest of the primers and is almost exclusively used as a corrosion-inhibiting metal primer, typically on bridges and structural steel. Zinc silicate and zinc chromate are used when there will be considerable exposure to the weather. Coatings designed for special purposes often have a primer developed specifically for use with that product. Etching primers that clean and seal surfaces also are effective.

Figure 33.4 Gypsum wall board is primed prior to the application of a final topcoat with latex paint. *(© Fancy/Veer/Corbis)*

Galvanized metal can be chemically etched before painting to provide needed adhesion. Etching can be omitted if a latex metal primer designed for galvanized metal is used. Another primer is a varnish-based material pigmented with zinc dust, zinc oxide, or Portland cement.

Types of Opaque Coatings

Opaque coatings are used to hide the color of previous coatings or substrate and obscure much of the grain or texture of a substrate (Fig. 33.5). Commonly used opaque coatings are listed in Table 33.5.

Alkyd Coatings Alkyd coatings have been a major type of organic coating. However, because of requirements relating to the amount of volatile organic compounds (VOCs) that alkyd coatings release to the environment, water-based products are largely replacing them.

Alkyd emulsions are formulated using a variety of synthetic alkyd resins in different coating formulations. Alkyd coatings have only mild alkali resistance but provide excellent water resistance and weather well. This makes them useful for exterior applications—as enamels for exterior stairs and porches, for example. Alkyd coatings retain their color well and permit the formulation of a range of light colors. With some reformulation, alkyd emulsions are used in making baking enamels, like those used on kitchen appliances.

Alkyd resins are added to other coatings to improve both adhesion and durability. For example, when modified

Figure 33.5 Opaque coatings include rapid drying, water-based latex used on interior and exterior surfaces. *(Image copyright Hannamariah, 2009. Used under license from Shutterstock.com)*

with phenolic resins, water resistance and alkali resistance are improved and a coating will penetrate rusted surfaces. When alkyd resins are formulated with rust-inhibiting pigments, such as red lead, iron oxide, or zinc chromate, they are used as rust-inhibiting primers. A gloss enamel resistant to chalking is produced by adding vinyl chloride acetate to alkyd resins. Silicone aids in color and gloss retention.

Chlorinated Rubber Chlorinated rubber–based coatings are solvent thinned and have excellent resistance to alkalis and acids and some resistance to salt water and salt-air exposure. They exhibit superior water and water vapor resistance and are used around swimming pools and on basement walls. They have good abrasion resistance and can be used on masonry, plastic, concrete, and metal surfaces. Chlorinated rubber coatings are not recommended for use on wood because they are not permeable and blistering can occur, but they do resist attack by microorganisms. However, they do bond to metal, and, through separating dissimilar metals, they are used to stop corrosion caused by galvanic action.

Enamel Coatings Enamels are a form of pigmented coating that use varnish as a vehicle. They form a gloss or semigloss film that is hard and durable. Oil-based and resin-based paints form similar coatings and can be classified as enamels.

Baked enamels are formulated to be applied in a factory by spraying. They are generally thermosetting materials that cure to a hard coating at 200 to 300°F (93 to 149°C). The coating becomes insoluble in the solvent used in its formulation. Enamels are available in a wide range of colors, are hard and washable, and resist alkalis and acids.

Epoxy

Epoxy coatings are available in a variety of formulations. Epoxy-ester is an epoxy resin reacted with a drying oil. It has properties similar to phenolic varnishes and alkyd resins but offers better resistance to chemical fumes and exposure to water. Epoxy-polyamide offers excellent resistance to chemical fumes, oils, atmospheric acids, and alkalies. It resists abrasion; adheres to concrete, metal, or wood; and will cure even when wet. Epoxy-bitumens may be formulated with coal tar or asphalt. They are used on items buried in soil, such as tanks and piping. Epoxy-polyester coatings are heavy-bodied two-part systems used to protect masonry and concrete. They have a high solids vinyl filler that is applied to a concrete surface followed by a high solids epoxy-polyester pigmented topcoat.

Table 33.5 Fire-Retardant Coatings[a]
Surface Burning Characteristics (Based on 100 for Untreated Red Oak)

Coating Type[b]	DS - Clear	PR - Clear	PR - White		"DS 11" - Clear	
Surface	Douglas Fir	Douglas Fir	Douglas Fir	Cellulose Board	Douglas Fir	Cellulose Board
Flame spread	5	5	5	5	10	10
Smoke developed	0	0	0	0	30	20
Number of preliminary coats	None	None	None	None	None	None
Number of fire-retardant coats	2	2	2	2	2	2
Rate per coat (sq. ft./gal.)	200	200	200	200	200	200
Number of overcoats	None	None	None	None	None	None

[a]*Tests conducted in accordance with ASTM E/84 (UL 723 and ULC-S-102).*
[b]*Coatings tested are Exolit Fire Retardant Coatings manufactured by American Vamag Company, Inc.*
Courtesy American Vamag Company, Inc.

Epoxy resins are chemically setting and solventless. They harden with the application of heat rather than by the evaporation of solvent. When exposed to weather, they may fade and chalk, but the film remains undamaged.

Latex Coatings Latex is a term applied to emulsions containing synthetic resins that are thinned with water. Since they contain no flammable solvent, they present no fire hazard when stored or applied. They release only minimum amounts of volatile organic compounds into the atmosphere. Latex emulsion coatings dry rapidly. The common types available are acrylic, polyvinyl acetate, and styrene-butadiene. They are used for coating interior and exterior vertical surfaces (Fig. 33.6).

Acrylic Coatings Acrylic coatings are thermoplastic resins that have a range of properties varying from hard coatings to softer finishes. They are water-based and available in clear and pigmented coatings. They offer excellent protection to concrete against weathering. They are also used as factory-applied coatings on aluminum and steel wall panels because they have good durability and resistance to salt spray and chemicals. Acrylic coatings remain flexible and do not lose their color as they age. Some acrylic coatings are solvent-based and are therefore not latex emulsions, using solvents such as xylol or toluol. One-part acrylic emulsions (latex) are used on interior and exterior vertical surfaces, such as wood, masonry, gypsum wallboard, plaster, and metal. They are permeable to water vapor.

Two-part epoxy modified acrylic coatings are water-based and used for interior and exterior vertical surfaces. They are tough coatings that resist stains and will withstand scrubbing.

Styrene-Butadiene Coatings These water-based coatings are formulated using styrene and butadiene, producing a rubber-like film. They resist alkali well and are permeable to water vapor. Styrene-butadiene coatings are used as exterior fillers over porous concrete surfaces. Some formulations are used on interior masonry, plaster, and gypsum wallboard, producing a washable film that resists abrasion. They generally are not used on wood.

Figure 33.6 Latex paint can be applied by brush and roller or by pneumatic paint spray equipment. *(Image copyright Anthony Hall, 2009. Used under license from Shutterstock.com)*

Vinyl Coatings Vinyl and polyvinyl acetate emulsions are available as water-based coatings. One formulation is used for interior application and another for exterior use. Polyvinyl chloride copolymerized with polyvinyl acetate is an opaque solvent-based coating that has poor adhering properties and requires a special primer. It has good durability and resistance to oils, alkalies, acids, and salt water.

Oil-Based Coatings Oil-based coatings are formulated by combining a body, pigment if needed, and a vehicle (drying oil, a thinner, and a drier). The paint body is a solid, fine material that provides superior hiding power. White lead, lithopone, titanium white, and zinc oxide are used for bodies of white oil-based paint. Both natural and synthetic pigments are added to give the coating color. Natural pigments are obtained from minerals, as well as animal and vegetable products. For example, red lead is a typical red pigment. Synthetic pigments are generally derived from coal tar.

Oil based paints use a nonvolatile fluid vehicle that suspends the body particles in solution. It is composed primarily of drying oil with small amounts of thinner and dryer. Thinners are volatile and evaporate. They are used to regulate the flow of the paint. Turpentine, a product formed by distilling gum from pine trees, is a high-quality thinner. Naphtha and benzene are used in some formulations. Driers accelerate the oxidation and hardening processes. Organic salts of iron, zinc, cobalt, and manganese are commonly used driers.

Oil-based paints are permeable to water vapor and thus minimize blistering over porous surfaces, such as wood. They should not be used in corrosive or alkaline conditions. Since they have excellent wetting properties, they are widely used as primers.

Phenolic Coatings

Phenolic coatings are made by polymerization of phenol and a formaldehyde reactant. They are solvent-based and dry by the evaporation of the solvent, leaving a strong, flexible coating. Phenolic coatings work on exterior concrete, plaster, wood, metal, and gypsum wallboard. They are used where resistance to acids, alkalies, and some solvents is required, as well as immersion in hot distilled water. Two-part phenolic coatings have a catalyst added on-site and harden via a chemical reaction. This coating protects against harsh conditions, including chemical fumes.

Construction Techniques

Testing for Lead Paint

Lead paint found in older buildings can poison construction workers doing remodeling or additions. Common symptoms of lead poisoning include stomach cramps, nausea, loss of appetite, headache, and joint and muscle aches. The use of lead pigments in paint was banned in 1978, but the practice continued until 1980. Most houses built before 1980 contain some lead paint.

The Residential Lead-Based Paint Hazard Reduction Act of 1992 sets forth detailed instructions on what is required of a contractor before beginning work on projects where lead paint exists and how workers and occupants in a building must be protected. For example, workers must wear respiratory filter masks and protective clothes, and work site facilities for their cleanup and clothes changing must be provided. Those who deal regularly with lead paints are required to undergo periodic blood testing. OSHA follows meticulous regulations when inspecting work where lead-pigmented paints are known to exist.

Before accepting work where lead paints may be encountered, a contractor should have tests conducted on painted surfaces. Three tests are used: x-ray fluorescence, laboratory analysis, and chemical spot tests.

A portable x-ray fluorescence analyzer measures the amount of lead on a painted surface (expressed in milligrams per square centimeter). The analyzer can read through multiple layers of paint but cannot indicate which of the layers contain lead pigment.

Laboratory analysis is the most accurate test method for lead. It is known as atomic absorption spectrophotometry and gives results as a percentage of lead by weight. Paint samples are removed from a surface, making certain all layers of paint down to the substrate in each chip will be tested. No portion of the substrate should be adhered to the backs of the samples. The chips are then sent to laboratories for testing.

Spot tests for lead use a chemical kit containing a swab or dropper used to apply a chemical reagent that reacts to lead by changing color if a paint contains 0.5 percent or more lead by weight. These kits are sold by many retail paint stores. Since the test kit only checks top layers of paint, it is necessary to scrape down and test each layer. Generally, this test is used for preliminary analysis because it is inexpensive and fast. If it appears to indicate the presence of lead, paint chip samples should then be sent to a laboratory for a more accurate determination.

Urethane or Polyurethane Coatings

Urethane resin coatings are available as one-part and two-part formulations. One-part coatings are moisture-cured (in reaction to atmospheric moisture) and are generally clear (no pigment). They have better abrasion-resistance than alkyd enamels. An oil-modified one-part urethane coating is available. Although it has better gloss retention, it has lower resistance to chemical attack.

Two-part formulations range from hard to rubber-like surface films with good resistance to abrasion, water, and solvents. Adhesion to steel and concrete is poor, and surfaces must be carefully prepared. Both one- and two-part types are used for heavy duty wall coatings and surfaces subjected to heavy traffic, such as a gymnasium floor. They resist scrubbing, abrasion, and impact and have better abrasion-resistance than regular varnishes.

SPECIAL PURPOSE COATINGS

Special purpose coatings are formulated to meet a specific need and are therefore not suitable for most coating situations. Several widely used types include bituminous, asphalt, reflective, and fire-retardant coatings.

Bituminous Coatings

Bituminous coatings are formulated by dissolving natural bitumens, such as coal tar or asphaltic products, in an organic solvent. They are most effective when used below ground because they do not react well when exposed to sunlight. One special weather-resistant type is a solvent-based asphalt roof repair coating. Fillers are added to provide body and prevent it from running. Another type uses coal tar emulsified in water, which enables it to be used on damp roofs. It has good resistance to sunlight. Coal tar paints are made with a coal tar–produced solvent, such as coal tar naphtha or xylol. They are often mixed with other synthetic resins to produce high-quality water resistant coatings for metal.

Asphalt Coatings

Asphalt coatings are produced from petroleum and found in a variety of emulsions, cold-applied paints, and enamels. They have good moisture resistance.

Reflective Coatings

Reflective coatings absorb the ultraviolet band of solar radiation and reflect it as visible light. The life expectancy of reflective coatings exposed to sunlight is about one year.

Pigmented Fire-Retardant Coatings

A variety of pigmented fire-retardant coatings are available from various manufacturers. Products are rated for surface-burning characteristics on combustible and noncombustible substrates. Class 1 flame spread (0–25) is required by codes for numerous applications. Many pigmented fire-retardant paints meet this requirement. Although these coatings retard the spread of flame on a surface, they do not protect a substrate from fire or heat. If substrate protection is required, an intumescent coating must be used.

Intumescent Fire-Retardant Coatings

When exposed to flame, **intumescent coatings** develop a thick, rigid foam protective layer that insulates a substrate and prevents the spread of fire. They are applied to wood, hardboard, cellulose board, and other wood-based products. They are noncombustible and produce no toxic fumes. A common type uses a urea-formaldehyde resin with an intumescent agent. It is a water-based material that dries rapidly via water evaporation. When exposed to fire or heat of about 350°F (178°C), it expands and develops a thick, insulating mat hundreds of times thicker than the original paint film. Usually one or two coats are sufficient. The important factor is not the number of coats but the amount of coating applied per square foot. The insulating layer delays contact between fire and the combustible material below it. This impedes flame spread on a surface, holds down smoke, and, since it retards heat transfer, delays the ignition of a substrate, giving occupants additional time to evacuate a building (Table 33.5). These coatings should not be applied over other covering materials and are genrally not used for exterior applications due to their water sensitivity.

STAINS

Stains are used on exterior wood to provide color and weather protection. Those designed for interior wood provide color, and protection is accomplished with a transparent topcoat, such as varnish or lacquer.

Exterior Stains

Exterior stains are blends of oil, driers, resins, a coloring pigment, a wood preservative (such as creosote or pentachlorophenol), a **mildewcide**, and a water repellent. Low-pigmented stains are penetrating types made to soak into wood. Heavily pigmented stains have the same formulation but contain more pigment. Heavily pigmented stains are frequently used for staining cedar shakes and shingles. They usually do not have the durability of paints. Both types are used on a variety of exterior wood applications, including siding, plywood, fencing, decks, and trim.

Exterior wood stains are available in solid and semitransparent formulations. Solid stains hide wood color but allow its texture to show through. Semitransparent stains retain the natural wood color (Fig. 33.7).

Stains that are oil- and water-based are available for application on wood. Oil-based alkyd stains are solvent-thinned and available in opaque and semitransparent types. Acrylic latex stains are water soluble.

Interior Stains

Interior stains are used on doors, trim, cabinets, and other wood products that require a furniture-quality finish. Many types are available (Fig. 33.8). A summary of them is shown in Table 33.6. Note that a variety of vehicles are used and that some are penetrating stains while others are pigmented stains. Penetrating stains are made with dyes and do not obscure wood grain. Pigments in pigmented stains stay on the surface of the wood and

Figure 33.7 Semi-transparent exterior stains and sealers provide excellent water-repellent properties. *(Image copyright Jeffrey Sheldon, 2009. Used under license from Shutterstock.com)*

Figure 33.8 This interior stain penetrates into the wood, colors it, and leaves a protective seal coat. *(Image copyright Liz Van Steenburgh, 2009. Used under license from Shutterstock.com)*

Table 33.6 Wood Stains Commonly Used for Interior Application

Type of Stain	Vehicle Solvent	Staining Action	Remarks
Alcohol	Alcohol	Penetrating	Dries quickly but fades easily Sold in powder from Will raise grain
Gelled wood stain	Mineral spirits or turpentine	Pigmenting	Slow drying and does not fade Sold in gelled form
Latex stain	Water	Pigmenting	Slow drying and does not fade Sold in liquid form
Non-grain-raising stain	Alcohol Glycol	Penetrating	Dries quickly and fades easily but does not raise the grain Sold in liquid form
Oil stain (penetrating)	Mineral spirits or turpentine	Penetrating	Dries quickly and fades easily but does not raise the grain Sold in liquid form
Oil stain (pigmenting)	Mineral spirits or turpentine	Pigmenting	Slow drying and does not fade Sold in liquid form
Penetrating resin stain	Mineral spirits	Penetrating	Sold in liquid form Contains stain and protective coating in one coating
Water	Water	Penetrating	Dries quickly but fades easily Sold in powder form Will raise the grain

partially obscure the grain. Some stains are penetrating compounds that contain aliphatic hydrocarbons. When applied, their color penetrates the wood and a protective coat forms over it. They are combustible and good ventilation is needed or a NIOSH-approved respirator should be worn as it is applied.

Fillers

Woods fall into two general groups: close-grained, such as maple, and open-grained, such as oak and walnut. When finishing these woods, a filler is applied to achieve a smoother final surface. Close-grained woods are smooth and use a liquid filler to seal surface pores. Some varnishes act as liquid fillers. Open-grained woods have visible open pores that must be filled with a paste wood filler to achieve a smooth surface. Paste wood fillers contain a vehicle such as linseed oil, a solvent such as mineral spirits, silex (ground quartz), and a drier. Silex mixed with the linseed oil makes a paste. It is forced into pores and the excess is wiped off before it hardens. Silex is a neutral colored material, and color pigment can be added so the filled pores match the wood.

FLAME SPREAD, SMOKE, AND FUEL CONTRIBUTED RATINGS

The speed at which flames spread over a surface is indicated by a material's **flame spread rating.** Coating manufacturers have their products tested and provide results for consumers. Flame spread is generally specified for different situations by building codes.

It is measured by following ASTM tests that give noncombustible cement-asbestos board a value of 0 and untreated red oak a value of 100. Flame spread ratings are specified as Class I (0–25), Class II (26–75), and Class III (76–200).

Other considerations when selecting a coating are the fuel contributed rating, which is an indication of the level of combustible material in a coating, and the smoke developed rating, which classifies the coating by the amount of smoke produced as it burns. These ratings are on a scale of 0 to 100. The nearer the rating is to 0 the less smoke is developed. Fuel contributed and smoke developed ratings use the same scale as flame spread.

Review Questions

1. What is the purpose of regulations related to volatile organic compounds in coatings?
2. What are the VOC limits for architectural and industrial maintenance coatings?
3. What are the major components of a coating?
4. What are the two types of coating bases?
5. What do MWF and MDF mean when used in coating specifications?

6. What are the major elements that affect the life of coatings?
7. What are major natural resin varnishes?
8. What are major synthetic resin varnishes?
9. What is a primer?
10. What methods are used to clean metal surfaces for painting?
11. What is the recommended moisture content for wood to be painted?
12. What are the major types of opaque coatings?
13. What are some special purpose coatings?
14. What is meant by a flame spread rating?
15. What are flame spread classifications?
16. What does a smoke developed rating indicate?
17. What does a fuel contributed rating indicate?
18. How do semitransparent exterior stains affect wood siding?

Key Terms

abrasion resistance
adhesion
alkali
alkyd
binder
bituminous coatings
clear coating
coating
durability
efflorescence
enamel
epoxy varnish
flame spread rate
fuel contributed rating
hiding power
intermediate coat
intumescent coatings
lacquer
latex
mildewcide
oil-based paints
opaque coatings
pigments
primer coat
resin
sealer
semitransparent coatings
smoke developed rating
solvent
spreading rate
stain
substrate
topcoat
varnish
vehicle
volatile organic compounds (VOCs)
water-based coatings

Activities

1. Visit a paint store and collect color charts and samples. Prepare a display citing the properties and some suggested uses for each type of material collected.
2. Secure several lead sampling kits and test finishes in several old local residences and campus buildings for lead. Write up your findings and submit a report to the class.

Additional Resources

Lead Paint Safety, U.S. Department of Housing and Urban Development National Lead Information Center, 1-800-424-5323.

Merritt, F., and Rickets, J. *Building Design and Construction Handbook, (5th ed.)*, McGraw-Hill Publishing Co., New York.

Paint Tests for Chemical, Physical, and Optical Properties: Appearance, Volume 06.01, American Society for Testing and Materials, West Conshohocken, PA.

Performance Specifications and Standards for Coatings and Finishes, American Architectural Manufacturers Association, Schaumburg, IL.

Small Entity Compliance Guide: National Volatile Organic Compound Emissions Standards for Architectural Coatings, U.S. Environmental Protection Agency, Research Triangle Park, NC 27711.

Williams, R., and Knaebe, M., *Finishes for Exterior Wood*, U.S. Department of Agriculture, Forest Service, Forest Products Laboratory, Madison, WI.

Other resources include:

Many publications available from U.S. Department of Housing and Urban Development, HUD USER, Rockville, MD.

See Appendix B for addresses of professional and trade organizations and other sources of technical information.

CHAPTER 34

Gypsum, Lime, and Plaster

LEARNING OBJECTIVES

Upon completion of this chapter, the student should be able to:

- Learn the properties and applications of gypsum, lime, and plaster products.
- Make appropriate choices when selecting interior finish materials and designing to meet codes.

Build Your Knowledge

For further information on these materials and methods, please refer to:

Chapter 36 Interior Walls, Partitions, and Ceilings

Topics: Interior Gypsum Wall Systems; Plaster Wall Finishes of Mortar

Gypsum is a hydrated calcium sulfate mineral found in natural rock formations. It is identified by the chemical formula $CaSO_4 \cdot \frac{1}{2} H_2O$, denoting a compound of lime, sulfur, and water. The mined rock is crushed, ground, and calcined (heated). Calcining evaporates most of the chemically combined water. The **calcined gypsum**, called **plaster of Paris**, is then ground into a fine powder that is used to produce products such as wall board and gypsum plaster. Various materials are added to produce properties needed for the different gypsum products.

When calcined gypsum is mixed with water, the plaster absorbs the water and solidifies. This bonds the gypsum and any additives into a solid, hardened mass.

GYPSUM PLASTER PRODUCTS

A plaster finish consists of a supporting base, either gypsum or metal lath, and one or more coats of plaster. The main ingredient in plaster is usually gypsum, but some types use Portland cement. Gypsum plasters meet the requirements of ASTM C28, Standard Specification for Gypsum Plasters, or ASTM C587, Standard Specification for Gypsum Veneer Plaster.

Gypsum plasters include gauging, wood-fiber, neat, and ready-mixed plasters. Following are plaster products used for interior finish applications.

Plaster of Paris is a calcined gypsum mixed with water to form a thick paste. It sets in fifteen to twenty minutes and is used in construction for ornamental plastering and repairing plaster walls. If mixed with lime putty, it can be used as a **finish coat** on plaster walls. Plaster of Paris hardens rapidly and experiences little shrinkage.

Unfibered gypsum is a neat gypsum with no aggregate or filler added. It is applied in a series of layers, including an initial **scratch coat**, followed by a **brown coat** and additional leveling coats. Three common types are used. Regular is used with sand aggregate and applied by hand. LW is used with lightweight aggregates and is also manually applied. Machine-applied gypsum is used with either sand or lightweight aggregates.

Gypsum neat plaster is a calcined gypsum plaster mixed at a mill with other ingredients to control its working quality and setting time. Sand, perlite, and vermiculite are used as aggregates.

Fibered gypsum is a neat gypsum with cattle hair or organic fibers, such as sisal, added to hold the plaster together. It is generally used for base coats and applied by machine.

Wood-fibered gypsum is a manufactured product in powder form that requires the addition of water on-site.

There are two types: regular, for use over gypsum lath, and masonry, for use over masonry surfaces.

Bond coat gypsum is used as a base coat on monolithic concrete surfaces that are smooth and dense but lack sufficient suction for standard plaster base coats. Finish plaster is applied over bond coat.

Gauging plaster is a coarsely ground gypsum plaster composed of screened particles that regulate setting time. It is available in quick set (30 to 40 min.) and slow set (50 to 70 min.) mixtures. When mixed with slaked lime it produces a finishing coat.

Keene's cement is a double-calcined gypsum with almost all water removed. Two types are available: regular (slow setting) and quick-setting. It is high in strength, quite resistant to moisture, and used with lime and sand for floated or sprayed finishes.

Casting plaster is made from plaster of Paris that has been ground down to a finer powder. It is used for ornamental applications, such as moldings and cornices. Lime putty is added to increase plasticity. Casting plaster sets up slower than regular plaster of Paris, allowing more time to form ornamental profiles.

Finish plaster is used to cover a brown (second) coat with a hard, smooth finish or a putty coat that can be painted or papered. It is mixed with hydrated lime putty and water in proportions of one part plaster to three parts lime putty. Prepared finish plaster has no lime and requires only the addition of water. It does not dry as white as finish plaster, but it does dry faster.

Acoustical plaster is a calcined gypsum mixed with a lightweight mineral aggregate. Several compositions of acoustical plaster exist, incorporating either gypsum, lime, or Keene's cement with aggregates such as rock wool, pumice, perlite, or vermiculite. Some mixtures include a foaming agent that forms small voids, producing a porous plaster.

Texture plaster is a type of finish plaster used to produce a rough, textured finished surface. It is applied in two coats over a plaster base coat or gypsum lath. The final coat can be applied by sponge, brush, or trowel, depending on the desired texture.

Texture spray is a gypsum-based material mixed with water and applied with a pneumatic spray machine as a final coat. The texture of the finished surface can be varied by adjusting the air pressure or spray orifice size, or by varying the amount of water, thus changing the consistency of the mix by the addition of aggregate (Fig. 34.1). It is used to provide a textured finish on gypsum wallboard plaster and monolithic concrete walls and ceilings.

Veneer plasters are specially formulated high-strength gypsum plasters. They can be a thin, monolithic base-coat plaster over which a finish is applied or formulated for application as a single-coat finish plaster.

Figure 34.1 Textures typically used to finish interior drywall ceilings and walls. *(Courtesy Gold Bond Building Products, National Gypsum Company)*

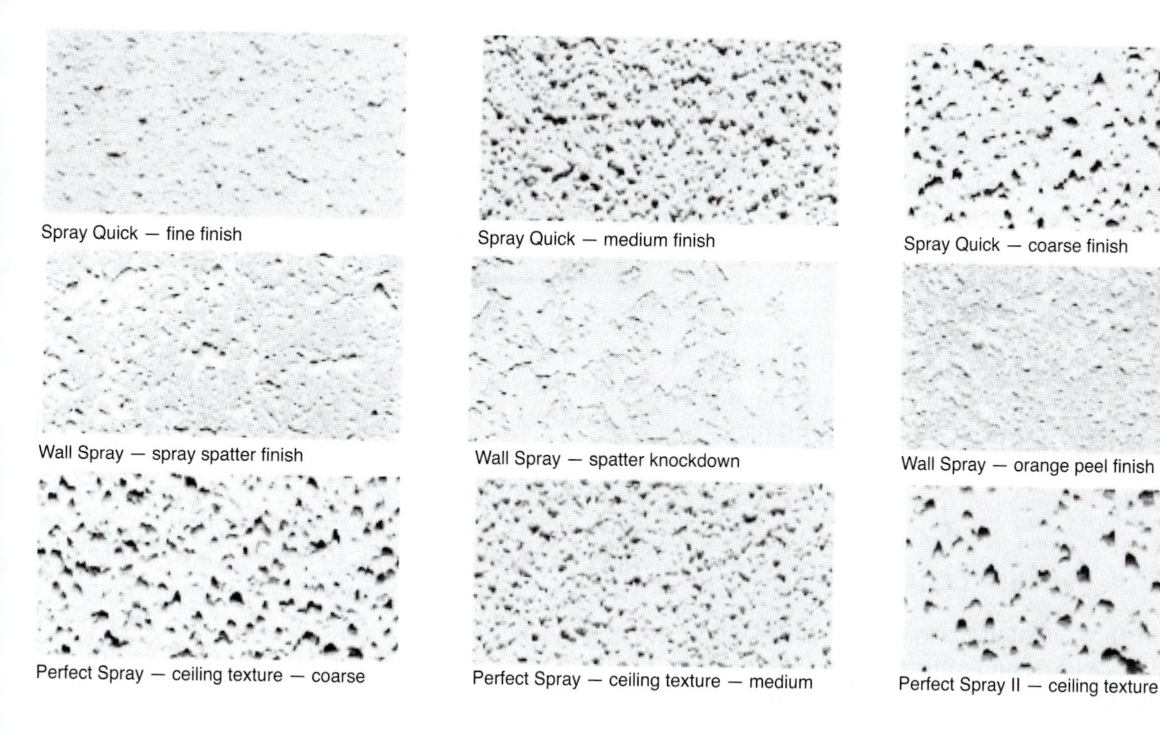

Figure 34.2 Drywall joint compound is available ready-mixed for immediate use. *(Image copyright R. MACKAY PHOTOGRAPHY, 2009. Used under license from Shutterstock.com)*

Setting time and compressive strength are controlled by a manufacturer via plaster composition. Veneer plasters can be applied over gypsum plaster base, masonry, or concrete surfaces.

Fireproofing plasters contain inorganic binders, or lightweight aggregates. They are sprayed directly onto bare steel shapes, providing fire protection as specified by codes. These coatings are easily damaged and not intended to be exposed finish coats.

Joint compound is used to cover nail heads and joints between sheets of gypsum wallboard. It is sold ready-mixed for immediate use (Fig. 34.2). It is available in various setting times ranging from thirty minutes to six hours. The rapid chemical hardening and low shrinkage permit same-day finishing and next-day decorating. Detailed information on the installation and finishing of plaster is covered in Chapter 36.

GYPSUM BOARD PRODUCTS

Gypsum board, often called drywall or wallboard, is the generic name for a series of panel products having a noncombustible core of calcined gypsum with paper surfacing on faces, backs, and long edges. These products are manufactured to ASTM specifications.

Advantages of Gypsum Board

Gypsum board is fire-resistant material that finds common use on walls, ceilings, and other parts of buildings. Type X is specifically formulated to meet increased fire code requirements. When exposed to fire, the chemically combined water is slowly released, retarding the transfer of heat through the panel. In addition, Type X gypsum board has a low flame-spread and smoke-density index. Type X panels meet the requirements of ASTM C36 and tests made by the Underwriters Laboratories.

Gypsum board is also an effective barrier to the transfer of sound when used for partitions and floor and ceiling constructions. It is a durable material that produces high quality walls and ceilings while maintaining excellent dimensional stability. Gypsum products are inexpensive, easy to install, and accept a wide range of finishes.

Manufacturing Gypsum Board

Gypsum board products are made by mixing calcined gypsum with water and additives to form a slurry. The slurry is fed between continuous layers of paper on a board machine. As the board moves down a conveyer, the calcium sulfate re-crystallizes or re-hydrates, forming a solid core that bonds to the paper. Finally, the board is cut to size and passed through dryers to remove excess moisture.

Types of Gypsum Board Products

Regular gypsum wallboard is used as the finished surface of walls and ceilings. It is produced with smooth, off-white paper on the finish side and gray paper on the back side of the gypsum core. Special boards are available with an aluminum foil back covering that acts as a vapor barrier. Standard sheet size is 4 × 8 ft, but lengths in two foot increments up to 14 ft are available. Metric sizes are approximately 1200 × 2400 mm. Sheets are available in thicknesses of $\frac{1}{4}$, $\frac{5}{16}$, $\frac{3}{8}$, $\frac{1}{2}$, and $\frac{5}{8}$ in. (6.4, 8, 9.5, 12.7, and 16 mm). A variety of edge profiles are manufactured, including tapered, square, beveled, rounded, or tongue and groove (Fig. 34.3).

Fire-resistant gypsum board (Type X) has improved fire-resistance qualities due to the addition of special core materials. It is used in assemblies that must meet fire code ratings.

Pre-decorated gypsum board has finish surfaces covered with a printed, textured, painted, coated, or vinyl film. It provides a finished surface that requires no further treatment after installation.

Water-resistant gypsum board has a water-resistant core and is faced with a green water-repelling paper. It is used as a base for the installation of ceramic tile and plastic panels in bath, shower, kitchen, and laundry areas. Available with a regular or fire-resistant core , it comes in $\frac{1}{2}$ in. (12 mm) and $\frac{5}{8}$ in. (16 mm) thicknesses. It is not recommended for use on ceilings with joist spacing greater than 12 in. because tile weight could exceed panel strength.

Figure 34.3 Regular gypsum wallboard is available with a variety of edge shapes.

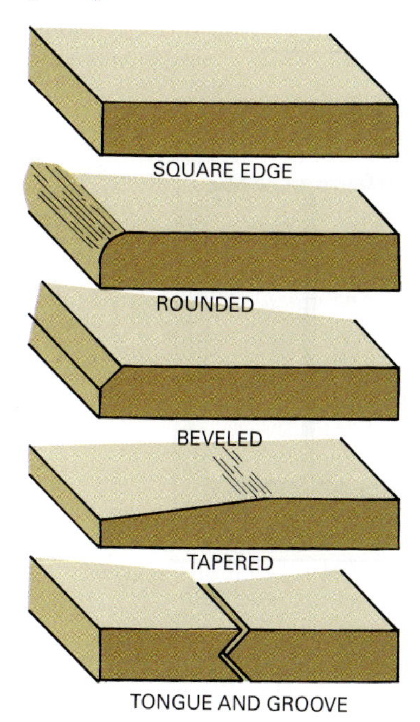

Figure 34.4 Gypsum panels are used with suspended ceiling systems. *(Courtesy Chicago Metallic Corporation)*

Backer board is used as the base ply or plies in assemblies that require more than one layer of gypsum board. It is also used as a backing for acoustical tile, plywood paneling, and other decorative wall paneling. It is available in regular and fire-resistant panels.

Gypsum coreboard is a 1 in. (25 mm) thick panel used in shaft walls and laminated gypsum panels.

Gypsum liner board has a special fire-resistant core enclosed in a moisture-resistant paper. It is used as a liner panel on shaft walls, stairwells, chaseways, corridor ceilings, and area separation walls. It is available in $\frac{3}{4}$ in. (18 mm) and 1 in. (25 mm) thicknesses and 48 in. (1220 mm) widths.

Exterior gypsum soffit board is available with regular and fire-resistant cores in $\frac{1}{2}$ in. (12 mm) and $\frac{5}{8}$ in. (16 mm) thicknesses. It is used in exterior areas where it will experience only indirect exposure to the weather, such as soffits, canopies, and carport ceilings.

Gypsum sheathing panels have fire- and water-resistant gypsum cores faced front and back (and on long edges) with specially treated water-repelling paper. They are used on wood and steel framing in residential and commercial construction. Panels are available with beveled, square, and V-shaped tongue and groove edges and in $\frac{1}{2}$ in. (12 mm) and $\frac{5}{8}$ in. (16 mm) thicknesses.

Sound deadening board is a $\frac{1}{4}$ in. (6 mm) gypsum panel used in connection with fire-resistant gypsum panels to meet the dual requirements of sound and fire resistance.

Lay-in ceiling panels are made from $\frac{1}{2}$ in. (12 mm) fire-resistant gypsum wallboard. They are cut into 2 × 2 ft. (610 × 610 mm) and 2 × 4 ft. (610 × 1220 mm) panels for easy installation in suspended ceiling grids (Fig. 34.4).

Gypsum lath is used as a base to receive layers of hand- or machine-laid plaster. It is available in 16 in. (406 mm) and 24 in. (610 mm) panels that are 48 in. (1220 mm) long, and in 16 in. (9 mm) and 24 in. (12 mm) thicknesses. It may have an aluminum foil backing to provide heat-reflecting qualities.

Gypsum fiberboard is a composite material of gypsum, cellulose fibers, and fibers from recycled newspapers. It sometimes includes perlite to reduce panel weight. The cellulose fiber reinforces the surrounding gypsum, making a product that works more like particleboard than typical gypsum wallboard. It is a solid material and does not have the paper facing used on conventional wallboard. It can be installed with a pneumatic stapler, nail gun, or drywall screws in wood stud applications. It is strong enough to permit anchoring of mirrors and towel racks to it (rather than to studs). More stable than conventional gypsum wallboard, it resists moisture, fire, impact, mildew, and sound. Unlike conventional gypsum wallboard, its seams are not taped but filled and smoothed with two layers of a specially formulated joint compound. Lightweight panels and a heavier, high-density type are also available. They come 4 ft. wide and in stock lengths of 8, 10, and 12 ft.

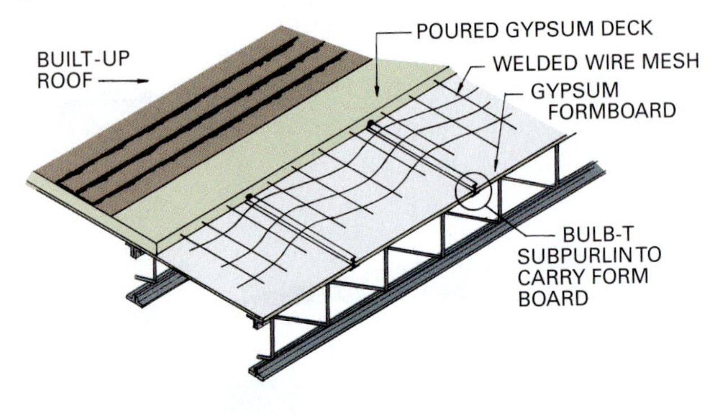

Figure 34.5 Gypsum formboard is supported on structural members and provides a deck upon which a gypsum roof deck is poured.

(2440, 3050, and 3660 mm). Widths to 8 ft. (2440 mm) and lengths to 20 ft. (6100 mm) are available by special order. Panels are made in thicknesses of $\frac{3}{8}$, $\frac{7}{16}$, $\frac{1}{2}$, and $\frac{5}{8}$ in. (10, 11, 13, and 16 mm). They are available with four square edges or two or four tapered edges. Panels can be used as interior wall finish, floor underlayment, tile backing, and gypsum sheathing.

Poured gypsum roof decks are formed by pouring gypsum concrete over permanently installed gypsum form boards. The form boards are supported by structural steel roof framing. The gypsum comes in a thick, creamy condition and is pumped through a hose to the deck. A construction detail for a poured gypsum roof deck is shown in Fig. 34.5.

Moisture-Resistant Drywall

Over the years, manufacturers have worked to improve drywall resistance to mold growth. To grow, mold needs mold spores, moisture, heat, and food. Food can be any organic material, and with drywall, it is the paper covering. For years, a product referred to as goodboard and other similar paper facings were used with some success. These paper facings have wax-impregnated gypsum cores and chemically treated paper. Newer products use inorganic facings made from gypsum cellulose or fiberglass. They provide no food for mold and are quite effective in preventing its growth.

APPLICATIONS

Various plaster products are used for finish wall and ceiling construction and cast ornamental features in residential and commercial building. Plaster is used for sound attenuation and fire protection for structural members, such as steel beams and columns (Fig. 34.6). It is applied over gypsum plaster base panels and metal and wire lath that has been secured to structural framing.

Figure 34.6 Gypsum board is used to provide fire protection for structural members.

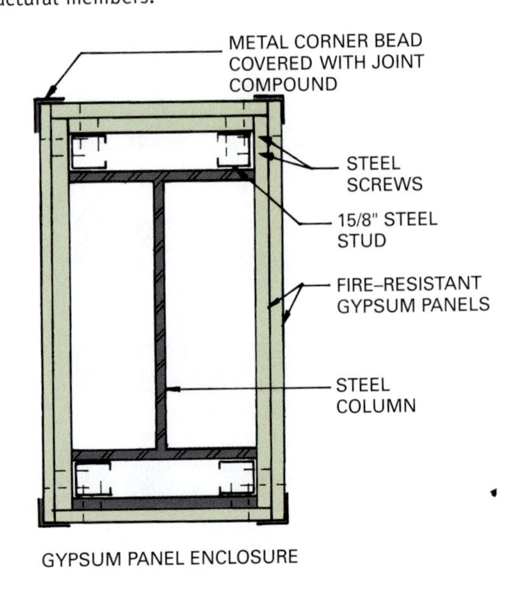

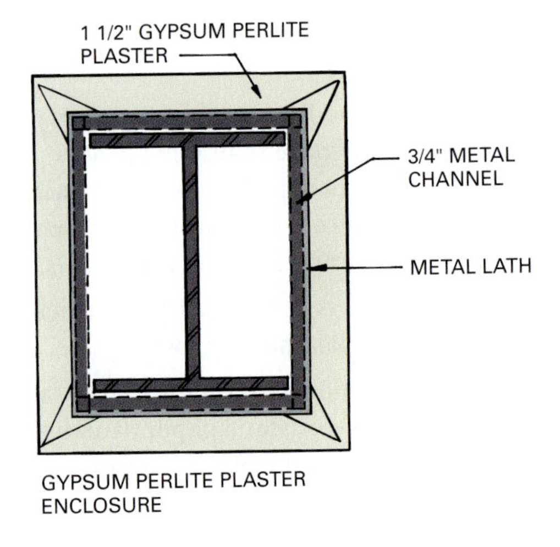

Paperless interior gypsum panels have an inorganic glass fiber mat on both front and back surfaces and provide protection from all interior-developed moisture. These panels have moisture-resistant, noncombustible gypsum cores that are reinforced with inorganic glass fibers, which increases panel strength. They are used on interior walls and ceilings and installed with nails or screws. Mold thrives on moist surfaces of paper-faced gypsum panels. The inorganic glass-faced panels remove the conditions needed for mold to thrive.

Gypsum wallboard products are used for finishing interior walls and ceilings (Figs. 34.7 and 34.8), providing sound attenuation, wall and ceiling fire resistance, and fire protection for structural members, such as steel beams and columns. Some types have water-resistant

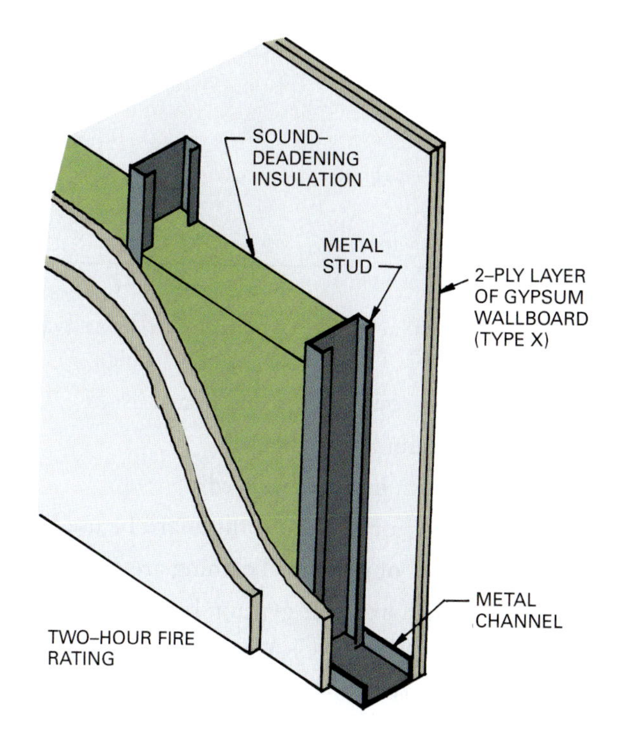

Figure 34.7 A typical interior wall with a two-ply gypsum wallboard finish.

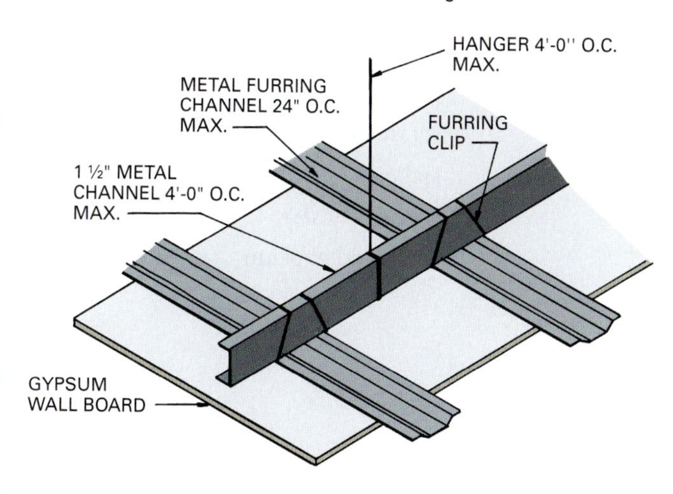

Figure 34.8 Gypsum wallboard can be secured to suspended steel channels to form a fire-resistant ceiling.

coverings and are used on walls to be tiled. Other types are used for exterior sheathing. Gypsum wallboard is also used over subfloors to increase the fire resistance of an assembly. Installation details and applications on interior walls and ceilings are illustrated in Chapter 36.

LIME

Lime is produced by burning limestone, marble, coral, or shells in a kiln at 2,000°F (1,100°C) to remove carbonic acid gases. The purest lime (97 percent) is used in plaster and constitutes a finishing lime. The lime produced by a kiln is called quicklime. It has the capacity to **slake** or **hydrate** when allowed to soak up to two or three times its weight in water. Slaking or hydration is a chemical reaction that causes the temperature of a mix to rise rapidly. This mixture is allowed to cool and sit for three weeks before it can be used. Contractors buy lime already slaked to avoid the waiting period.

Slaked quicklime is transformed by chemical reaction into **hydroxide of lime,** a fine, dry powder. This is the lime product used in plaster. There are two types: Type N (normal) and Type S (special). Type N must be soaked twelve to sixteen hours before it can be added to plaster. Type S can be added to plaster as soon as it is mixed with water.

CEMENT BOARD

Cement board is a product made with an aggregated Portland cement–core reinforced with polymer-coated glass fiber mesh embedded in both surfaces (Fig. 34.9). Although it is not a gypsum product, it is used for similar applications. Cement board is a durable, fire- and water-resistant panel used on both interior and exterior load-bearing and non-load-bearing walls. It can be attached to wood or steel framing with special screws for each application. Holes must be drilled to accept fasteners.

Galvanized roofing nails can secure cement board to wood framing. Joints can be covered with an open glass-fiber mesh tape. Panels can be finished with a Portland cement mortar containing dry latex polymers.

Figure 34.9 Cement board is used where a wall or ceiling will be exposed to water and high humidity.

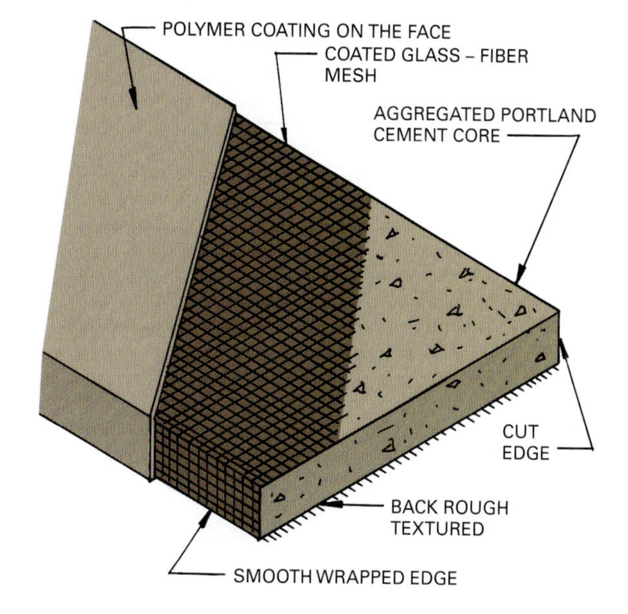

These materials are supplied by manufacturers, and their recommendations must always be carefully followed.

Cement board may be applied in areas with high humidity, such as baths, kitchens, and pools. It also works for soffits, fences, and chimney enclosures. It is used as a substrate for the application of ceramic tile, slate, and quarry tile on interior surfaces. And it can serve as a heat shield when installed behind stoves and heaters.

Review Questions

1. What use is made of casting plaster?
2. Where is perlite gypsum plaster used?
3. What gypsum plaster product is used as a base coat on dense, smooth concrete surfaces?
4. What gypsum product would you use in areas subject to high moisture?
5. What material is used to cover the heads of nails or screws that hold gypsum wallboard to studs?
6. How does gypsum wallboard retard the spread of fire?
7. What are the advantages of gypsum wallboard?
8. What sizes of regular gypsum wallboard are available?
9. Where is water-resistant gypsum board used?
10. What is a gypsum backer board?
11. Where is gypsum liner board used?
12. Where can exterior gypsum soffit board be used?
13. What two types of gypsum sheathing are available?
14. What is the major use for gypsum board substrate?
15. What are the differences between the uses for gypsum base and gypsum lath?
16. How are poured gypsum roof decks constructed?

Key Terms

acoustical plaster
brown coat
calcined gypsum
cement board
fibered gypsum
finish coat
finish plaster
fire-resistant gypsum
gypsum
gypsum board
gypsum neat plaster
hydrate
hydroxide of lime
lime
plaster of Paris
scratch coat
slake
texture plaster
unfibered gypsum

Activities

1. Install gypsum panels on a mock stud wall and practice taping the seams. Use various joint compounds and note drying times and sanding properties.
2. Build a stud wall and cover it with gypsum lath and metal lath. Apply a three coat plaster finish on each.
3. Visit a building under construction and observe the drywall crew installing and taping gypsum panels.

Additional Resources

Guide to Portland Cement-Based Plastering, American Concrete Institute, Farmington Hills, MI 48333.

Gypsum Construction Handbook, and many other technical publications, United States Gypsum Company, Chicago, IL 60680.

Melander, J., and Isberner, A., "Portland Cement Plaster," *Journal of Light Construction*, Publishers, Richmond, VT 05495.

Spence, W., *Installing and Finishing Drywall*, Sterling Publishing Co., 387 Park Avenue South, New York 10016.

Other resources include:

Many publications from the Gypsum Association, Washington, DC 20002.

See Appendix B for addresses of professional and trade organizations and other sources of technical information.

CHAPTER 35

Acoustical Materials

LEARNING OBJECTIVES

Upon completion of this chapter, the student should be able to:

- Understand the factors involved when considering the specifications and recommendations for the acoustical treatment of a building.
- Select appropriate materials to provide acoustical control of various interior spaces.

Build Your Knowledge

For further information on these materials and methods, please refer to:

Chapter 12 Concrete Masonry
Chapter 29 Glass
Topic: Glass Products
Chapter 34 Gypsum, Lime, and Plaster
Topic: Gypsum Plaster Products

Acoustics is the science of sound, including the generation, transmission, and effects of sound waves. Acoustical materials are used to reduce the levels of sound within an area, such as a room, by absorption and to control sound transmission between adjacent areas caused by sound vibrations in a building structure.

SOUND

Sound is the sensation produced by human organs hearing vibrations transmitted through the air. This includes vibrations traveling in the air at a speed of about 1130 ft./sec. (345 m/sec.) at sea level and mechanical vibrations transmitted through an elastic medium, such as steel. Sound is the movement of air molecules traveling in a wavelike motion. A sound wave produces changes in the atmospheric pressure above and below the existing static pressure. This deviation in atmospheric pressure is called sound pressure.

Sound Waves

Sound waves move in a spherical direction from a source in all three dimensions (Fig. 35.1). There is a delay between the time a sound is created and the time it is heard. Since it takes audio sound about a second to travel 1,130 ft. (345 m) in the air, a person 2,260 ft. (690 m) away will hear the sound two seconds after it is generated. When lightning strikes in the distance, thunder is not heard for several seconds due to the differing speeds of sound and light. Sound travels at different speeds in various materials. In wood it travels about 11,000 ft. /sec. (3,355 m/sec.) and in steel about 16,000 ft./sec. (4,880 m/sec.).

Frequency of Sound Waves

The **frequency** of sound is the number of cycles of like waveforms per second. Sound travels in sine waves, as shown in Fig. 35.2. Wavelength equals the velocity of a sound (1,130 ft./sec. in air), divided by the frequency (number of cycles). For example, a sound having a frequency of 15 cps would have a wavelength of 1,130/15 or 75 ft. The frequency, cycles per second, is expressed in units of hertz (Hz). A hertz is one cycle per second (cps), so, in the above example, the frequency is expressed as 15 hertz (15 cps = 15 Hz).

The frequency of sound determines the pitch. Low, deep sounds have low frequencies, and high-pitched sounds have high frequencies. The human ear can receive sounds ranging from a low of about 20 Hz to a high of about 20,000 Hz (or 20 KHz).

Figure 35.1 Sound travels in all directions at once.

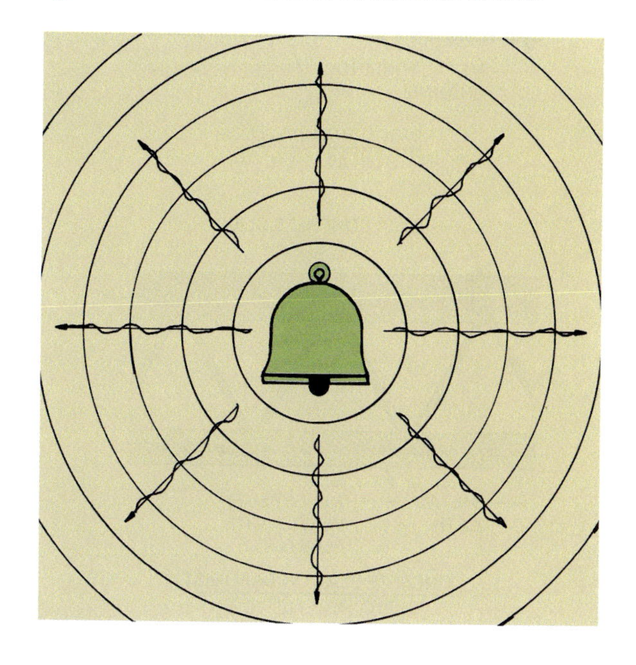

Figure 35.2 The sine wave audio frequency.

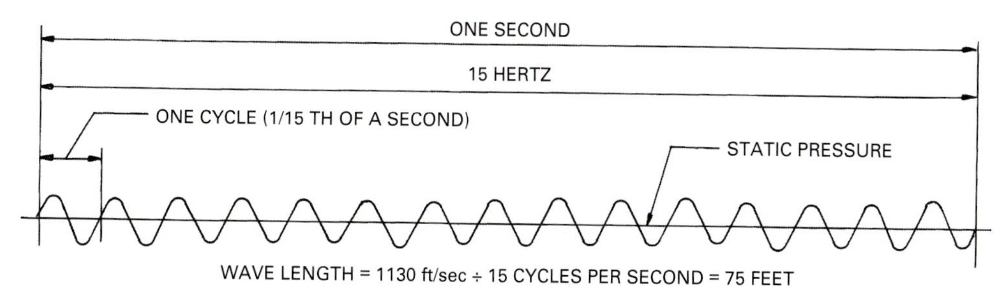

The frequency of sound to be considered varies, often rapidly and constantly. For example, the sound produced by music from a radio often ranges from low to high frequencies rapidly and often. Frequency from other sources, such as a dishwasher, is more limited in range.

Sound Intensity

Sound intensity depends on the strength of the force that sets off the sound vibrations. Sound intensity is measured in **decibels** (dB). The decibel scale for normal applications ranges from 0 dB, which is just below the lowest audible sound (about 20 Hz), to 120 dB, which can produce a feeling of vibration in the ear. Intensities above this can cause actual damage to human hearing and structural damage to buildings. Representative sound levels in decibels for selected situations are given in Table 35.1. The intensity of sound varies in much the same way as described for frequency. Music can produce very high decibel levels, but a dishwasher produces lower, steady decibel intensity.

SOUND CONTROL

Sound generated within a space is controlled by acoustical materials with sound-absorbing properties and by materials that reduce sound transmission through assemblies of materials, such as a wall or floor.

Table 35.1 Sound Pressure Levels of Selected Sounds

Sound Levels (decibels)	Source of Sound	Sensation
140	Near a jet aircraft	Deafening
130	Artillery fire	Threshold of pain
120	Elevated train, rock band, siren	Threshold of feeling
110	Riveting, air-hammering	Just below threshold of feeling
80–100	Power mower, thunder close by, symphony orchestra, noisy industrial plant	Very loud
60–80	Noisy office, average radio or TV, loud conversation	Loud
40–60	Conversation, quiet radio or TV, average office	Moderately loud
20–40	Average residence, private office, quiet conversation	Quiet
0–20	Whisper, normal breathing, threshold of audibility	Very faint

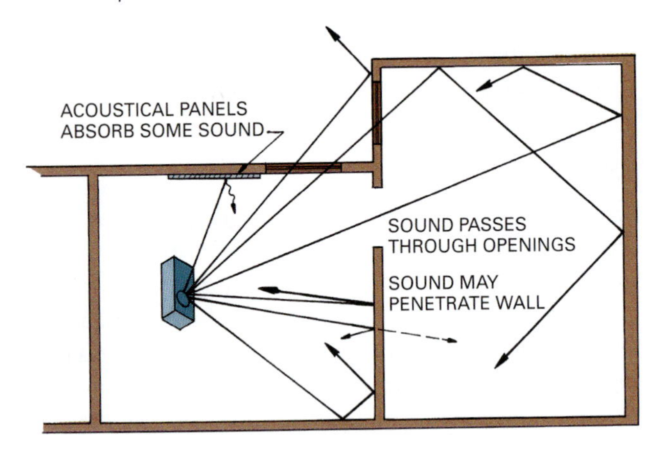

Figure 35.3 Sound is reflected when it strikes hard, non-absorptive surfaces.

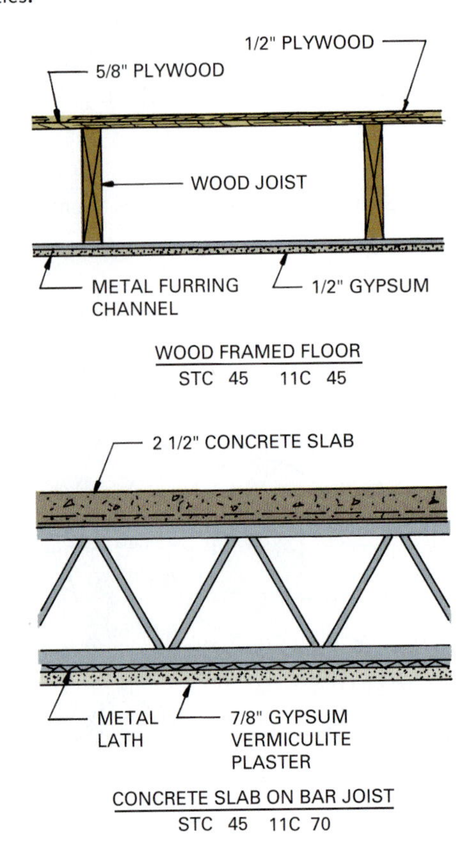

Figure 35.4 STC and IIC ratings for typical floor-ceiling assemblies.

Sound-absorbing materials soak up some sound waves that strike them and reflect the rest into an area. Acoustical engineers select materials to control sound in a desired manner. Some materials, such as plaster walls, reflect most of the sound striking them into a room. As the sound hits other walls, it is reflected again (Fig. 35.3). Sound-absorbing materials are porous and have openings in and out of which vibrating air particles move. This movement causes friction, which generates heat. Some of the sound energy is lost as heat, which may be reflected into a room or transmitted through the material.

Sound Transmission Class

Sound transmission class (STC) is a single number rating that indicates the effectiveness of a material or an assembly of materials in reducing the transmission of airborne sound. The larger the STC number, the more effective the material is as a sound transmission barrier. Materials are tested following the specifications in ASTM E90-70.

Typical building code STC requirements for residences, apartments, and hotels are as follows:

1. All separating walls and floor-to-ceiling assemblies must provide an STC of 50.
2. All penetrations in assemblies, such as openings for piping and electrical wiring, must be sealed to maintain the STC 50 rating.
3. Entrance doors and their seals must have an STC rating of 26 or more.

Manufacturers of materials used to reduce sound transmission indicate STC ratings for each material and its various thicknesses. Figure 35.4 shows STC ratings for several assemblies of materials.

Impact Isolation Class

Impact isolation class (IIC) is a single number giving an approximate measure of the effectiveness of floor construction to provide isolation against sound transmission from impacts. Impacts include walking, skidding, or dropping items on the floor, all of which set up vibrations that radiate to areas below. Improved performance for floors with low IIC ratings can be realized by applying carpet or other sound-absorbing materials to them. Several design suggestions are shown in Figure 35.5. IIC ratings of 45 to 65 are common for floor-ceiling assemblies in multifamily dwellings. Building codes typically require that floor-ceiling assemblies have an IIC rating of 50.

Noise Reduction Coefficient

The **noise reduction coefficient** (NRC) is an indication of the amount of airborne sound energy absorbed by a material. The single number rating is the average of the sound absorption coefficients of an acoustical material at frequencies of 250, 500, 1,000, and 2,000 Hz. The larger the NRC number, the greater the efficiency of a material to absorb sound. Some typical examples are given in Table 35.2.

Figure 35.5 STC and IIC ratings for typical floor construction.

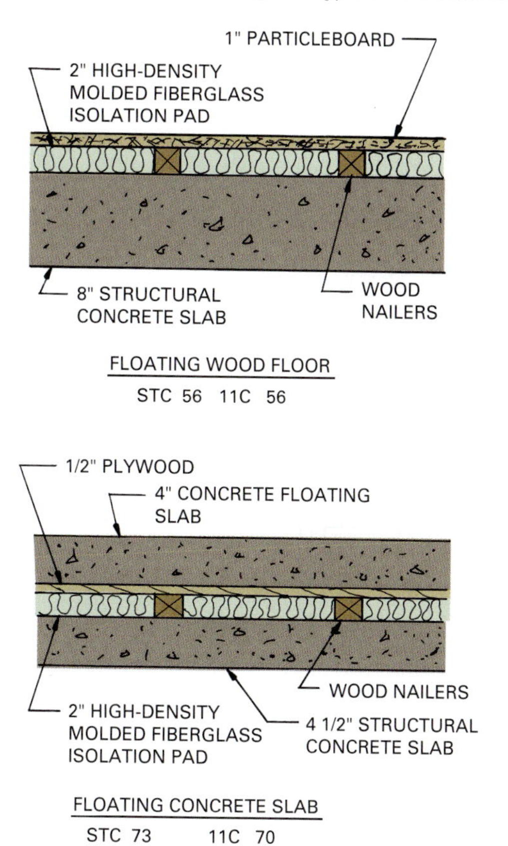

Table 35.2 Noise Reduction Coefficients for Selected Materials

Material	NRC
Unpainted brick wall	0.02–0.05
Painted brick wall	0.01–0.02
Glazed clay tile	0.01–0.02
Concrete wall	0.01–0.02
Lightweight concrete block	0.45
Heavyweight concrete block	0.27
Standard plaster wall	0.01–0.04
Gypsum wall	0.01–0.04
Acoustical plaster wall	0.21–0.75
Glass	0.02–0.03
Fiberglass	0.50–0.95
Wood panel	0.10–0.25
Mineral wool	0.45–0.85
Acoustical tile	0.55–0.85
Carpeting	0.45–0.75
Vinyl floor covering	0.01–0.05

ACOUSTICAL MATERIALS

The control of sound transmission and the absorption of sound can be accomplished using a wide variety of materials. Some standard construction materials, such as brick or concrete block, are used to control sound transmission. Other materials are especially designed for acoustical purposes. Many of these materials are discussed in detail in other chapters.

Floor Coverings

The installation of carpet and vinyl composition floor covering reduce impact noise and dampen airborne noise. Carpeting is much more effective than vinyl floor covering and its effectiveness is increased when installed over a cushion. See Chapter 38 for additional information on carpets.

Acoustical Plaster

Acoustical plaster is composed of a plaster made with perlite or vermiculite aggregate. It may be applied with a brush but is usually sprayed on surfaces. It has the advantage of uniformly covering curved and irregular shapes. The plaster is applied in several layers, with a finished thickness of about $\frac{1}{2}$ in. (12 mm). It has an NRC of about 0.21 to 0.75. See Chapter 34 for more information about plaster.

Another form (though not a plaster) of spray-applied acoustical wall and ceiling covering is composed of cellulosic fibers in a bonding agent. It will bond to almost any surface and provides acoustical control and thermal insulation.

Acoustical Ceiling Tile and Wall Panels

A variety of acoustical ceiling tiles and wall panels are available. They are made from various materials, such as wood fibers, sugarcane fibers, mineral wool, gypsum, fiberglass, aluminum, and steel. Most ceiling tiles have some form of perforation in their surface. Drilled or punched holes in a variety of patterns are common. Some tiles have slots, fissures, or striations (Fig. 35.6). Other types use a sculptured, irregular molded surface. Following are brief descriptions of some tiles available.

Wood fiber and sugarcane fiber ceiling tiles are available in thicknesses ranging from $\frac{5}{8}$ in. (16 mm) to $1\frac{1}{2}$ in. (38 mm). Tile sizes commonly available include 12 × 12 in. (305 × 305 mm), 12 × 24 in. (305 × 610 mm), 24 × 24 in. (610 × 610 mm), and 24 × 48 in. (610 × 1220 mm).

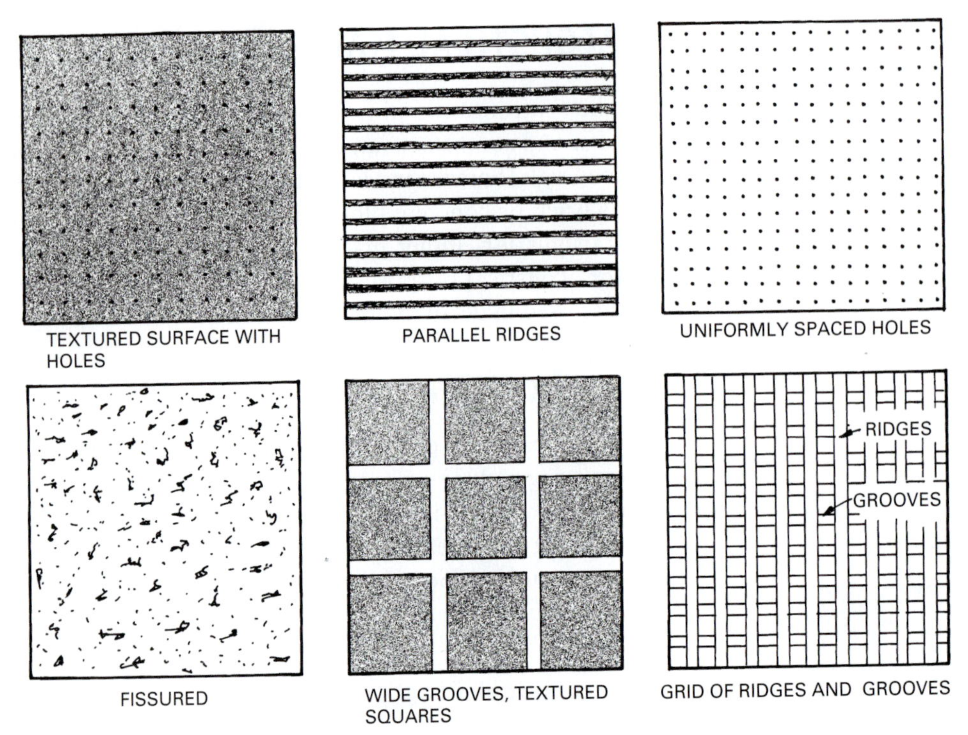

Figure 35.6 Acoustical surface treatments used on ceiling tile.

Ceiling tiles and wall panels are also made from molded mineral fibers and are cast with a wide choice of surface textures. Some manufacturers use mineral fiber or fiberglass sound-deadening panels covered with fabric. This provides an attractive finished ceiling or wall panel. This type has STC ratings of 35 to 40, NRC ratings of 0.60 to 0.80, and flame spread ratings of 0 to 25.

Some ceiling tile and wall panels can resist damage from moisture and bumps. One common type is a ceramic ceiling tile made from mineral fibers in a ceramic bond. It is fire resistant and used in areas of high humidity. It has STC ratings of 40 to 44, NRC ratings of 0.50 to 0.60, and flame spread rating 0. Another type uses perforated aluminum or steel tiles. Still others apply a vinyl coating over a perforated aluminum tile that is backed by a mineral fiber substrate. The sound passes through the holes and is absorbed by the mineral fiber. Metal tiles have STC ratings of 35 to 45, NRC ratings of 0.60 to 0.70, and flame spread rating 0. Another type that resists damage is made by bonding wood fibers into a porous panel resembling particleboard that resists moisture and heavy blows.

Lightweight panels are made utilizing a vinyl covering over a laminate of high-density molded glass fiber bonded to a core of 1 or 2 in. (25 to 50 mm) sound absorbing fiberglass. Wall panels made from these materials are applied in hallways, restaurants, gymnasiums, offices, and other places where sound control is necessary (Fig. 35.7). One type uses a perforated zinc-coated steel or aluminum panel bonded to a 2 in. (50 mm) thick fiberglass panel. These panels are mounted several inches off a wall (Fig. 35.8).

Figure 35.7 Acoustical wall panels of molded glass fiber can be used to control sound reverberation.

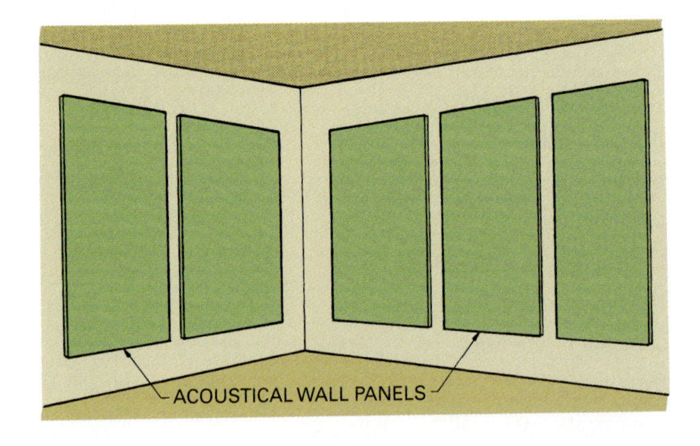

Figure 35.8 A perforated metal acoustical panel.

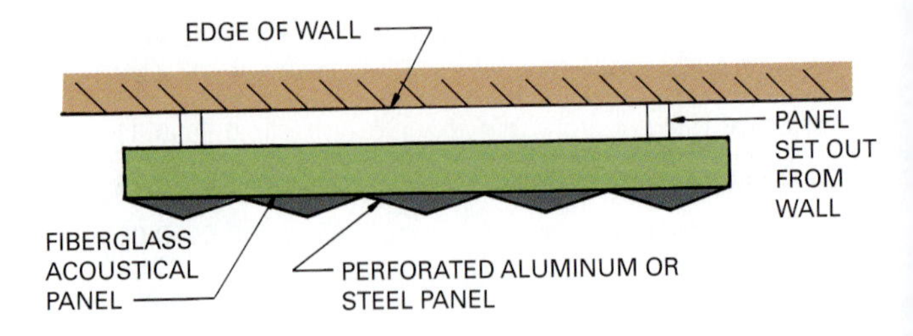

Another type of acoustical treatment involves the use of ceiling baffles. Baffles are acoustical panels hung from a ceiling to reduce airborne sound in large spaces, such as factories, restaurants, or auditoriums (Fig. 35.9). Some baffles use a lightweight panel of vinyl-covered fiberglass. Panels are 1 in. (25 mm) thick, 10 in. (254 mm) high, and from 2 to 4 ft. (610 to 1220 mm) long. Another type uses a flame-resistant polyethylene cover over a $1\frac{1}{2}$ in. (38 mm) fiberglass core. It is flexible, like a blanket, and hung from a ceiling (Fig. 35.10). Another type of baffle uses rigid wood-fiber sound-deadening panels. These baffles can be joined to form an attractive grid on a ceiling.

Sculptured acoustical wall units made from a high-density molded fiberglass layer bonded to a sound-absorbing glass fiber blanket are used to attenuate sound and provide a decorative feature. They are covered with a wide range of fabrics, typically in round, octagonal, and triangular shapes.

Figure 35.9 Acoustical materials are used on ceiling baffles to reduce sound reverberation. *(© Emeraldlight/Corbis)*

Figure 35.10 Fabric baffles at the ceiling used to control noise in a hallway. *(Paul Prescott)*

Some ceiling tiles are designed for gluing to ceiling substrates while others can be nailed or stapled to a ceiling. Tile installed with bonding agents has nut-size daubs of adhesive placed on its back. The tile is pressed against a ceiling substrate and slid into place.

If ceiling tiles are to be installed to the bottom of floor joists without a substrate, wood 1 × 3 in. (25 × 75 mm) strips are nailed perpendicular to the joists at 12 in. (305 mm) on center spacing. The ceiling tiles are then nailed or stapled through their tongues to the wood strips (Fig. 35.11).

Ceiling tiles and panels are made with a variety of edge shapes, which produce different appearances (Fig. 35.12). Suspended ceiling systems are widely used in commercial construction and find some use in residential work. They consist of a grid of metal runners, hung from overhead on wires, into which acoustical ceiling panels fit. Light fixtures are also designed to fit into this grid. A typical system is shown in Fig. 35.13. The space between the metal grid and the overhead structure can be used to run mechanical and electrical systems.

A number of grid systems are available, but basically they fall into three types: exposed grid, semi-exposed grid, and concealed grid. Manufacturers produce acoustical tiles with edges designed to fit their grid, resulting in the type of exposure desired. Exposed systems have the main runner and cross runner exposed, producing a square or rectangular grid appearance on the ceiling. The acoustical units rest on the flange of the T-shaped runner. Semi-exposed systems have the main runner exposed and the cross runner concealed, resulting in a series of long parallel lines in the ceiling. The concealed system has no metal runners showing (Fig. 35.14).

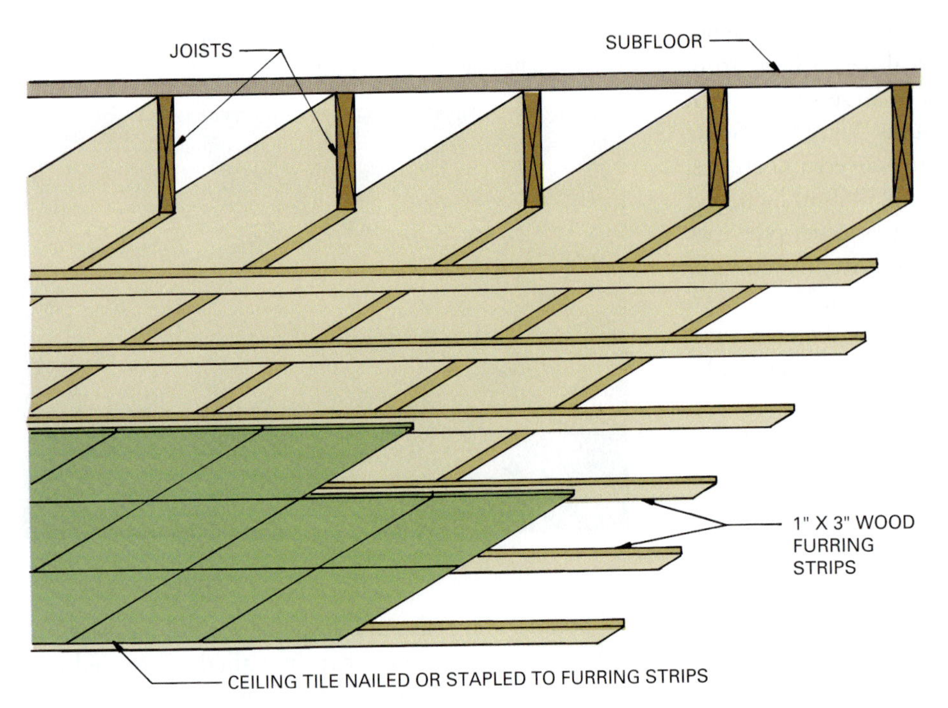

Figure 35.11 Ceiling tile can be installed by nailing or stapling it to wood furring strips.

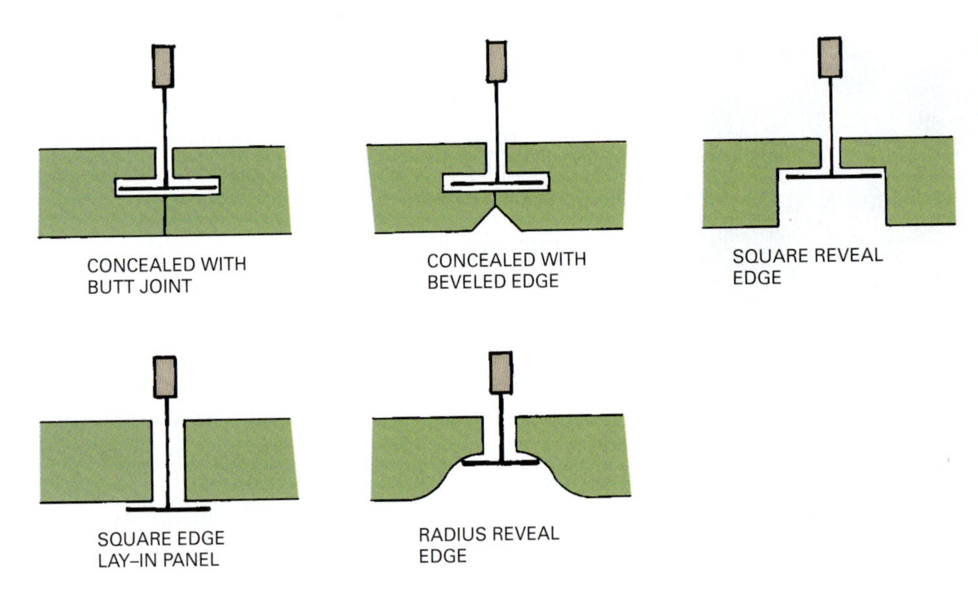

Figure 35.12 Typical panel edges for drop-in ceiling tiles and panels.

Sound Barriers

A wide range of materials can be used to block the transmission of sound through an assembly, such as a wall. Many of these, such as brick, concrete, concrete block, and gypsum, are discussed in earlier chapters. These materials control sound by providing mass, which blocks its transmission. For example, a 4 in. (100 mm) brick or concrete masonry unit has an STC of about 40. A 4 in. (100 mm) concrete floor has an STC of 44.

Lead sheet material is used in commercial buildings to block sound transmission. For example, a lead sheet can be laid on top of a suspended ceiling or hung from the bottom of the floor above, blocking the transmission of sound that penetrates the ceiling. Lead is available in sheets and foils as thin as 0.0005 in. (0.013 mm). It is used in walls, doors, and other areas where a thin but effective sound barrier is needed.

Acoustical sealant is available as a pumpable material and is applied to all openings through which sound

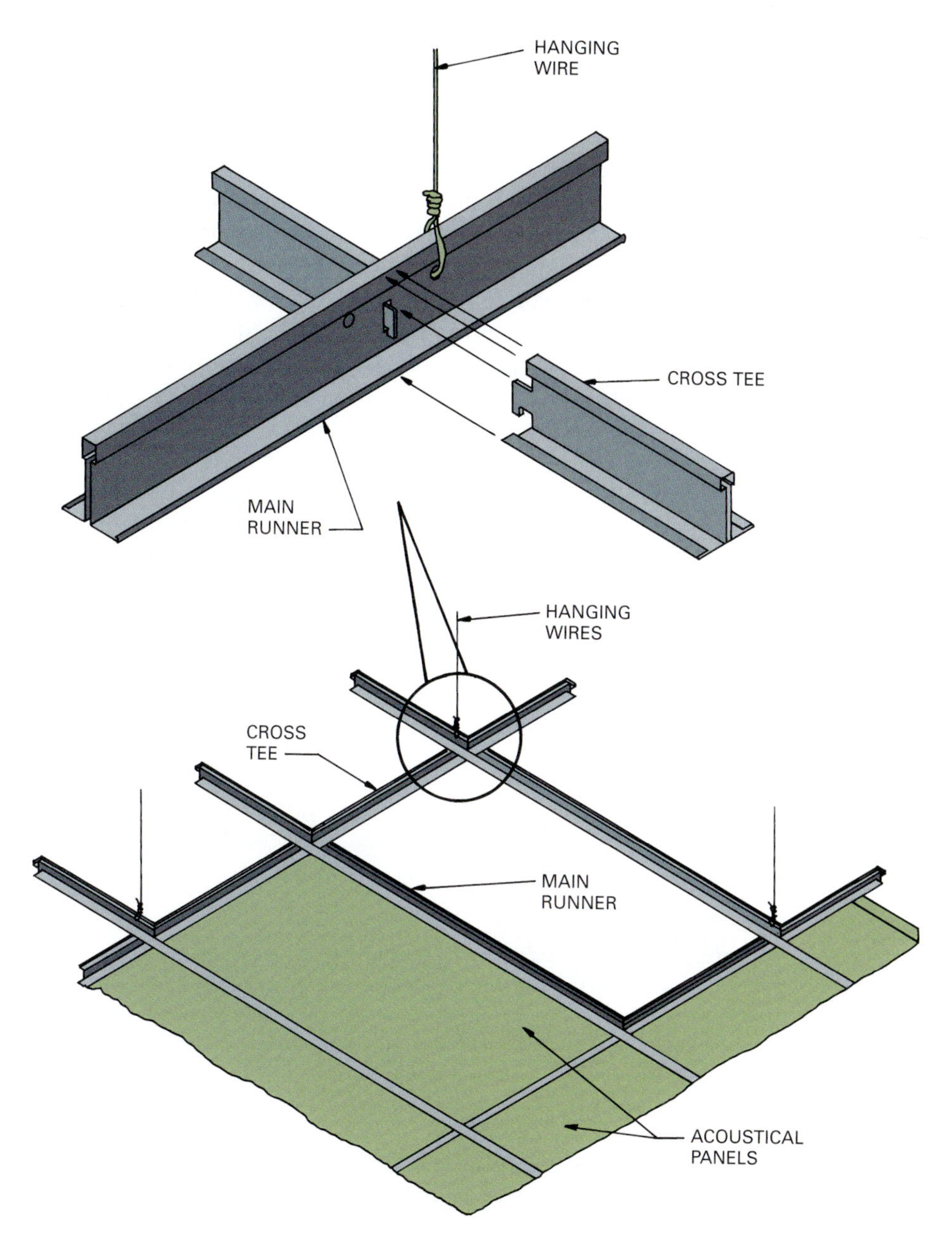

Figure 35.13 An exploded view of a suspended metal grid with acoustical tile that rests on the grid flanges.

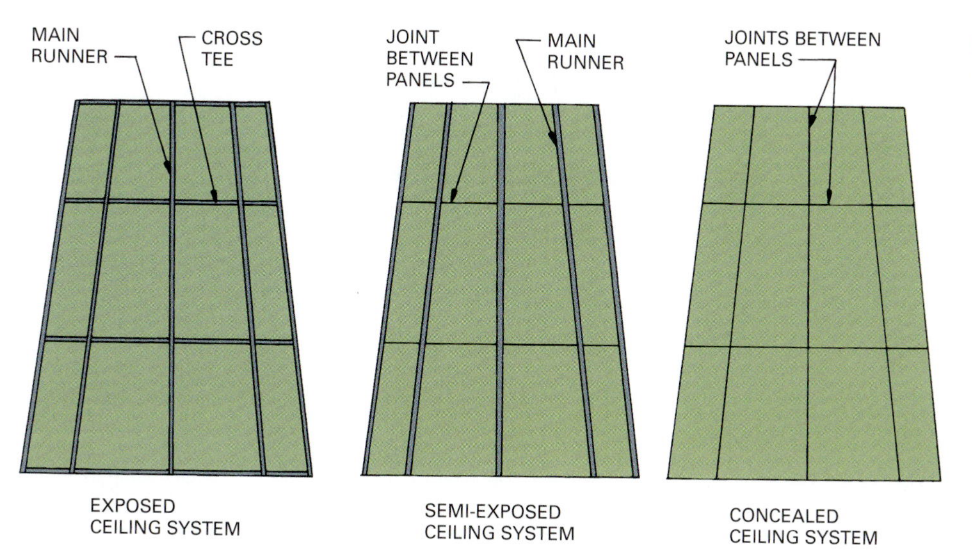

Figure 35.14 Types of suspended ceiling systems.

may penetrate. For example, space around electrical boxes and pipes that pierce a wall or floor must be sealed. Openings, even very small ones, reduce the effectiveness of an otherwise efficient sound barrier.

CONSTRUCTION TECHNIQUES

The following discussion outlines common ways to construct walls, floors, and ceilings to reduce sound transmission.

Frequently used methods for increasing the STC rating of wood and metal frame walls are shown in Fig. 35.15. All of these could have sound-deadening insulating batts installed to increase the STC value.

Several floor-ceiling assemblies appear in Figs. 35.4 and 35.5. The floating floor technique shown is used in commercial construction. The same type of construction can be used with residential wood frame floors.

CONTROLLING SOUND FROM VIBRATIONS

Mechanical equipment mounted on the structural frames of buildings produce vibration sounds that can travel for great distances to other parts of a building. Vibrating water and sewer pipes also create sound that needs to be controlled. Much of the sound can be dampened by mounting the mechanical equipment on various types of isolator pads. Rubber, neoprene, cork, and fiberglass pads are used for small pieces of equipment or larger pieces located in basement areas (Fig. 35.16). For heavier-duty units mounted within a building or on its roof, a steel spring mounting is used. The base of the spring pad must itself be isolated from the structure with an isolator pad to prevent the transmission of audible high-frequency vibration through the spring to the structure.

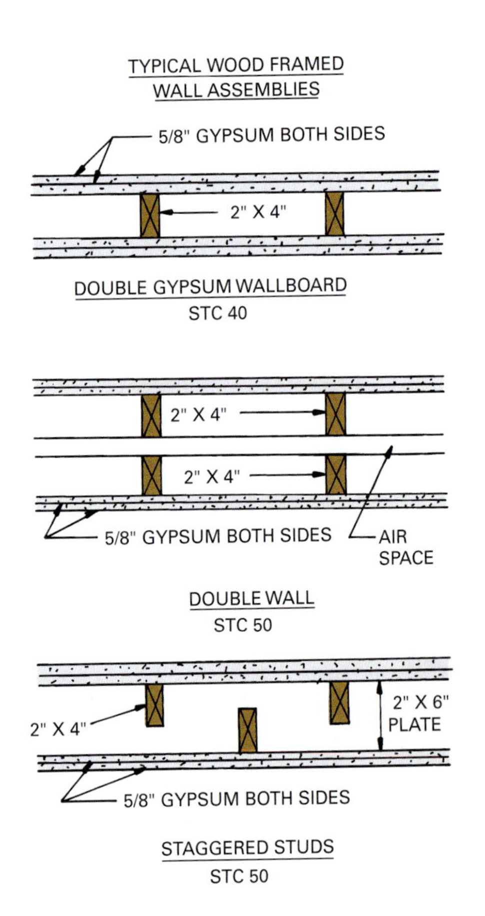

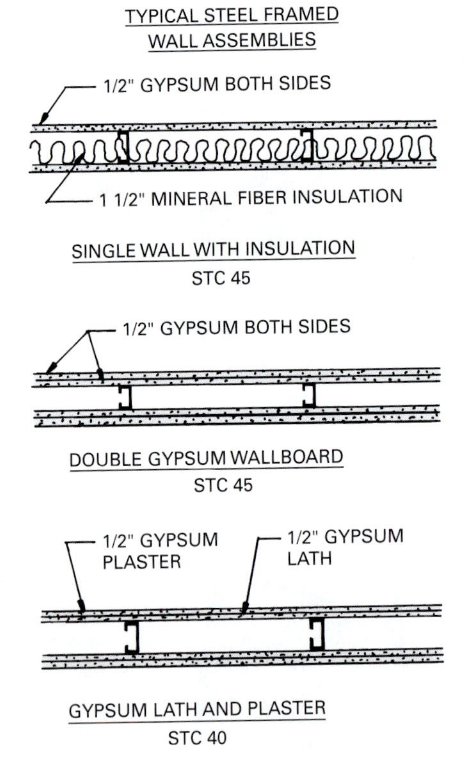

Figure 35.15 Examples of STC ratings for selected wall assemblies.

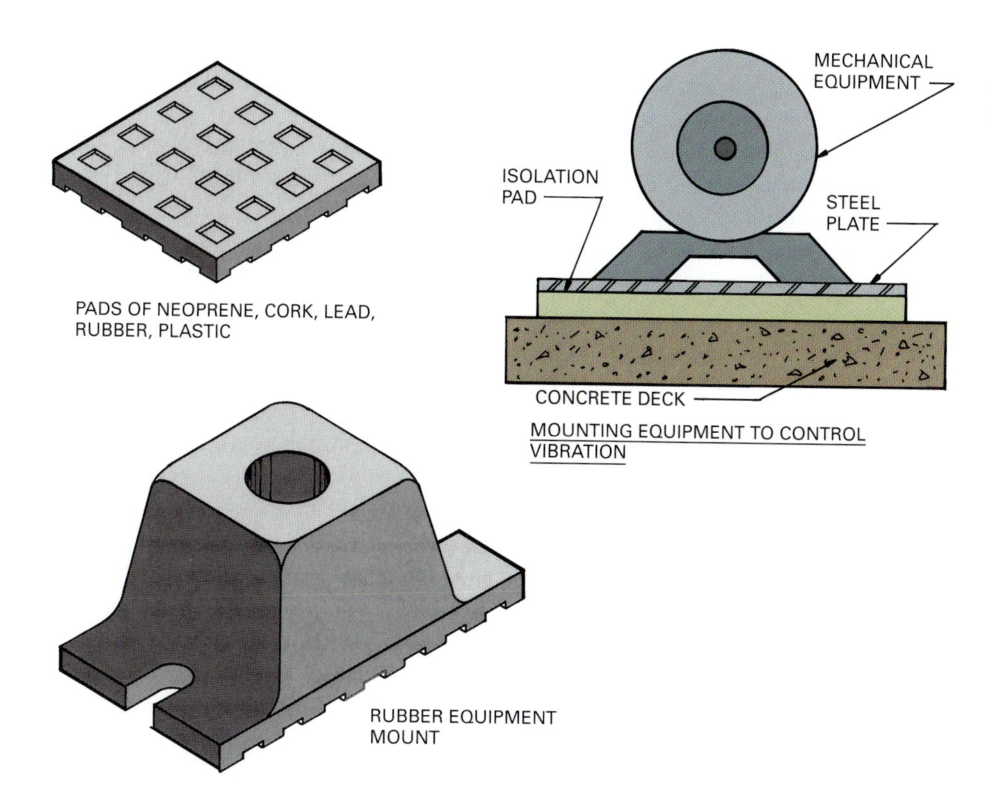

Figure 35.16 Types of pads used to damp the sound produced by vibrations of mechanical equipment.

Review Questions

1. In what two ways do acoustical materials control sound?
2. What is sound pressure?
3. In what directions do sound waves move?
4. How long will it take a sound to travel one mile from its source?
5. What is the wavelength of a sound having a frequency of 25 cps?
6. What is the frequency in hertz of a sound having a frequency of 35 cps?
7. How does sound frequency influence the pitch of a sound?
8. What is the normal range of sound that can be heard by the human ear?
9. How many decibels of sound will damage a person's hearing?
10. What is the difference between sound absorbing materials and sound reflective materials?
11. What are typical STC and IIC requirements for apartment construction?
12. How does acoustical plaster differ from standard gypsum plaster?
13. What acoustic product is used to reduce airborne sound at the ceiling in a large area, such as an auditorium?
14. What are the three types of grids used on suspended ceilings?
15. How is sound transmission of vibrations from mechanical machinery controlled?

Key Terms

acoustics
decibel (dB)
frequency
impact isolation class (IIC)
noise reduction coefficient (NRC)
sound
sound transmission class (STC)

Activities

1. Visit rooms used for various types of activities, such as a classroom, music room, auditorium, and mechanical room in your school. List the materials on the ceiling, walls, and floor, plus any special acoustical treatments. Write a report explaining what might be done to improve the acoustics of each space.
2. Ask the director of the physical plant of your school to show you the devices used on mechanical equipment that reduce the vibration transmissions into the structure. Write a report and make sketches of the various devices.

Additional Resources

See Appendix B for addresses of professional and trade organizations and other sources of technical information.

36 CHAPTER

Interior Walls, Partitions, and Ceilings

LEARNING OBJECTIVES

Upon completion of this chapter, the student should be able to:

- Be aware of the code requirements for various interior walls, partitions, and ceilings.
- Select appropriate designs and materials for interior walls, partitions, and ceilings.
- Understand the installation of the various types of finish walls and ceilings.

Build Your Knowledge

For further study on these materials and methods, please refer to:

Chapter 11 Brick and Tile
Chapter 19 Products Manufactured from Wood
Topic: Plywood and Other Panel Products
Chapter 32 Interior Finishes
Chapter 34 Gypsum, Lime, and Plaster
Topic: Gypsum Board Products

Interior walls and partitions may be load-bearing or non-load-bearing and constructed with wood or steel studs or masonry. Various types of finish material can be applied over a partition structure. Walls and partitions divide a building into areas having various types of occupancy and conceal electrical and mechanical systems that may run inside wall cavities. They also provide privacy, security, acoustical control, insulation, and fire and smoke protection.

Ceilings are a major component of the architecture of a space. Ceilings make available space for electrical and mechanical systems while simultaneously providing sound attenuation and fire protection.

Interior finish materials used on partitions and ceilings include gypsum board products, plaster, wood, hardboard panels, fiberboard, and a variety of metals. Detailed information of the characteristics of these materials is given in other chapters.

INTERIOR WALLS AND PARTITIONS

In many commercial construction projects, interior walls and partitions are non-load-bearing and serve to divide an open space into usable areas; they also enclose openings such as stairs, shafts, and exits. A building's structural system carries the loads of roof and floors, interior walls, and furnishings. The requirements an interior wall or partition must meet are specified by building code. Table 36.1 lists minimum fire resistance requirements for various building elements, as specified in the International Building Code.

Non-load-bearing non–fire-resistant partitions are not used as fire separation walls. They must meet code requirements specifying the combustibility of materials for the type of construction in which they are used. Codes specify fire ratings for fire barrier assemblies and fire walls.

Fire separation walls or fire partitions are installed to enclose shafts, floor openings, and exits and to subdivide usable building area. Codes specify the selection and assembly of materials for types of construction and occupancy. Walls are used to control the spread of fire between areas on a floor. The fire separation wall or fire partition should have a fire-resistance rating equal to code requirements specified for the occupancy of the separated fire areas (Table 36.2). They must extend from the top of a fire-resistant floor to the bottom of a fire-resistant ceiling (or roof slab above) and be securely

Table 36.1 Fire-Resistance Rating Requirements for Building Elements (hours)

Building Element	Type I A	Type I B	Type II A	Type II B	Type III A	Type III B	Type IV HT	Type V A[d]	Type V B
Structural frame[a] Including columns, girders, trusses	3	2	1	0	1	0	HT	1	0
Bearing walls Exterior[f]	3	2	1	0	2	2	2	1	0
Bearing walls Interior	3	2	1	0	1	0	1/HT	1	0
Nonbearing walls and partitions Exterior	See Table 602								
Nonbearing walls and partitions Interior[e]	0	0	0	0	0	0	See Section 602.4.6	0	0
Floor construction Including supporting beams and joists	2	2	1	0	1	0	HT	1	0
Roof construction Including supporting beams and joists	$1\frac{1}{2}$	1	1	0	1	0	HT	1	0

For additional information, see Chapter in the International Building Code published by the International Code Council.
Courtesy the International Code Council.

Table 36.2 Fire-Resistance Rating Requirements for Fire Barrier Assemblies between Fire Areas

Occupancy Group	Fire-Resistance Rating (hours)
High Hazard Groups 1 and 2	4
Factory Group 1	3
High Hazard Group 3	3
Storage Group 1	3
Assembly	2
Business	2
Educational	2
Factory Group 2	2
High Hazard Groups 4 and 5	2
Institutional	2
Mercantile	2
Residential	2
Storage Group 2	2
Utility and Miscellaneous	1

For additional information, see Chapter 7 in the International Building Code published by the International Code Council, Falls Church, VA.
Courtesy International Code Council.

fastened to them. Opening sizes in fire separation walls are restricted by code, and openings must be protected by fire doors, windows, or shutters.

Fire walls are **fire-rated partitions** that restrict the spread of fire and extend continuously from a structure's foundation to or through its roof. If the fire wall stops at the roof, the roof assembly must consist of noncombustible material. If it extends above the roof, the height of the extension is specified by code. The fire wall should be able to remain intact even if construction on either side collapses due to fire damage. Fire-resistance ratings of fire walls are shown in Table 36.3. Fire walls must be smoke tight where they meet exterior walls. Opening sizes in fire walls are restricted by code, and openings require protection by fire doors or wire glass.

Party walls are fire walls used jointly by two parties under an easement agreement. They are erected on an interior lot line dividing two parcels of land that are each separate real estate entities. A party wall is actually a common wall between two buildings that meet on a lot line.

Smoke barriers are fire-resistant continuous membranes used to restrict the movement of smoke. They have a fire-resistance rating specified by building code. They form a continuous membrane from one outside wall to another and from floor slab to roof (or floor slab above). Doors in smoke barriers must meet special code requirements and have automatic closers that activate when detectors sense smoke. They typically have a one-hour minimum fire-resistance requirement.

Interior walls, partitions, and ceilings are also subject to specifications relating to sound transfer between adjacent areas and the control of reverberating sound within rooms. Acoustical principles and construction details are discussed in Chapter 35.

Table 36.3 Fire Wall Fire-Resistance Ratings

Occupancy Group	Fire-Resistance Rating (hours)
Assembly	3
Business	3
Educational	3
High Hazard Group 4	3
Institutional	3
Residential Groups 1 and 2	3
Utility	3
Factory and Industrial Group 1	3
High Hazard Groups 3 and 5	3
Mercantile	3
Storage Group 1	3
High Hazard Groups 1 and 2	4
Factory and Industrial Groups 2	2
Storage Group 2	2
Residential Group 3 and 4	2

For additional information, see Chapter 7 in the International Building Code published by the International Code Council, Falls Church, VA.
Courtesy International Code Council.

INTERIOR GYPSUM WALL SYSTEMS

Metal Stud Non-Bearing Interior Walls

Partition walls in larger buildings often use light-gauge metal framing. Metal studs are formed from light-gauge steel and come in a range of standard sizes. For exterior and load-bearing walls, a heavier-gauge steel is used. Metal studs for non-load-bearing interior walls and partitions are typically made from 18-gauge steel sheet with either solid or perforated webs (Fig. 36.1). Perforations in the web facilitate installation of bridging, pipe, and electrical conduit.

Another type of metal stud uses round rod bent on a diagonal truss design between vertical cords (Fig. 36.2). These studs are used primarily for walls covered with gypsum lath or metal lath. The lath is joined to the studs with clips. Studs are secured to floor and ceiling by runner tracks. The solid and perforated web studs are secured to runners with self-drilling, self-tapping screws that are made of hardened steel and driven with an electric screwdriver. The tracks are secured to concrete floors and overhead materials with power-driven fasteners. Guns use a gunpowder cartridge to drive a steel fastener through a track and into a concrete floor (Fig. 36.3). Gypsum wallboard sheets are carefully positioned and screwed to the metal studs (Fig. 36.4).

Figure 36.1 Light-gauge steel studs are used for interior walls and partitions.

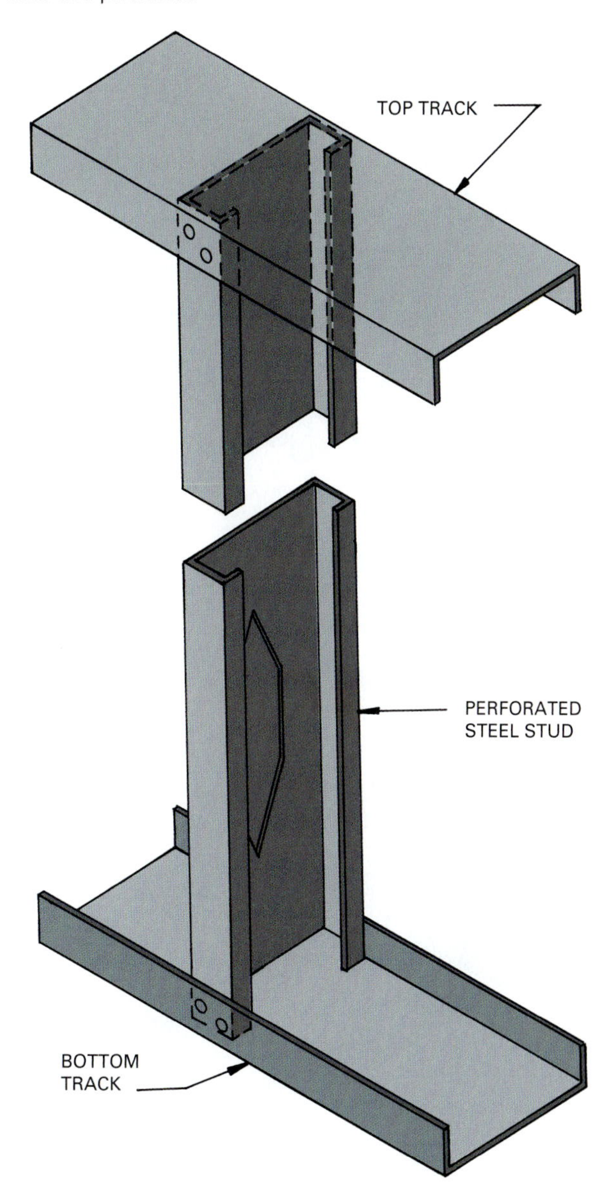

Typical Metal Stud Wall Assemblies

The actual assembly of gypsum wallboard walls depends on desired sound transmission levels, thermal insulation, and building code fire-resistance specifications. Fig. 36.5 shows a typical assembly for a cavity wall that, depending on the circumstances, could serve as a fire, separation, or party wall. This is a continuous, vertical non-load-bearing wall assembly with metal studs and furring, sound attenuation, fire blankets, gypsum liner panels, and water- and fire-resistant gypsum facing panels. The studs used are C-H steel studs with steel E-studs at wall ends. A wall with this construction can achieve a two-hour fire rating.

Figure 36.2 A truss stud is set in snap-in runners and secured to the top runner with a stud shoe.

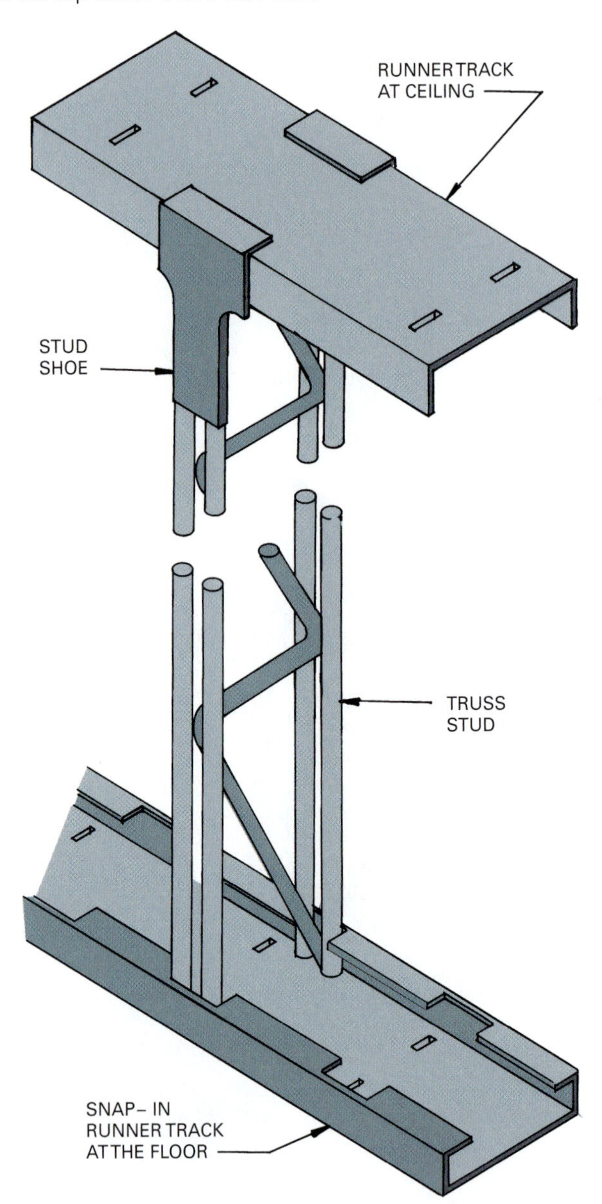

Figure 36.3 The metal runner is fastened to the concrete floor with steel fasteners driven by a gun that uses a small gunpowder charge. (Courtesy USG Corporation)

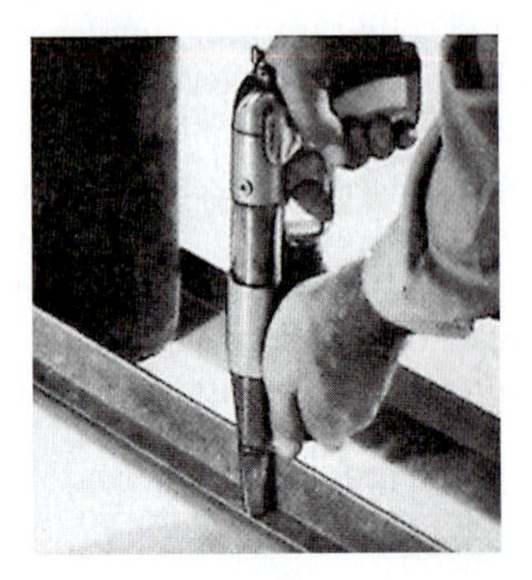

Figure 36.4 Gypsum panels cover a large wall area and are screwed to the metal studs. *(Courtesy Georgia-Pacific Corporation)*

Figure 36.5 A cavity-type separation wall utilizing a gypsum liner panel and a sound-attenuation fire blanket.

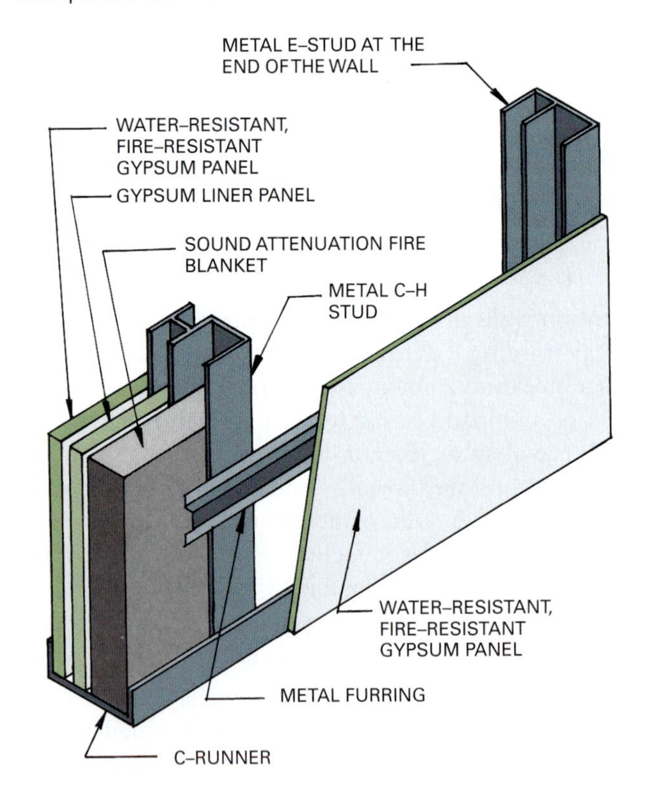

A typical assembly for a fire-rated partition designed to offer effective sound control and a high fire rating is shown in Fig. 36.6. It uses steel studs, gypsum lath on steel furring, a gypsum finish coat, and a sound attenuation blanket in the wall cavity.

Fire-resistant drywall partitions are used to enclose shafts in multistory buildings, including those for elevators, mechanical equipment, stairwells, and air returns.

Figure 36.6 A sound-attenuation steel-frame wall with fire-resistant properties.

A typical assembly for a non-load-bearing fire-resistant gypsum board partition for enclosing shafts, air ducts, and stairwells is shown in Fig. 36.7. It uses a C-H stud system, a gypsum panel liner, and a single layer of fire-resistant gypsum board on each side and achieves a two-hour fire rating.

Gypsum fire-resistance wall assemblies are considerably lighter than masonry fire-resistant walls. Fire-resistance ratings of metal partition walls can be increased by adding layers of wall board. Figure 36.8 shows several steel frame partition systems with gypsum board, lath and plaster, and veneer plaster cladding. These contain no fire-resistant insulation or gypsum liner panels. The manufacturers of gypsum products have a wide range of wall designs for various fire ratings.

Figure 36.7 Typical construction of a single-layer fire-resistant non-load-bearing cavity shaft wall, as used on elevator and mechanical shafts, air ducts, and stairwells.

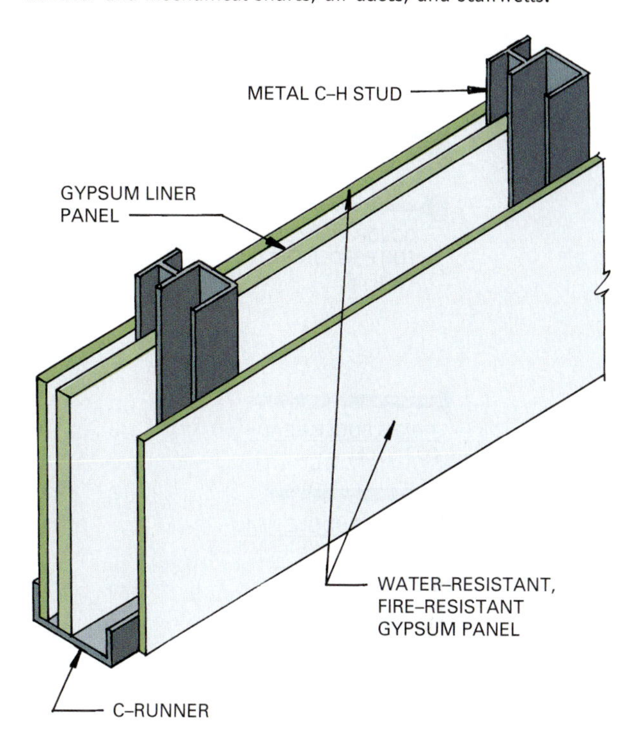

Figure 36.8 Typical fire ratings for various types of metal-frame gypsum wallboard and plaster partitions. Consult a manufacturer for specific data on recommended stud sizes and panel thicknesses.

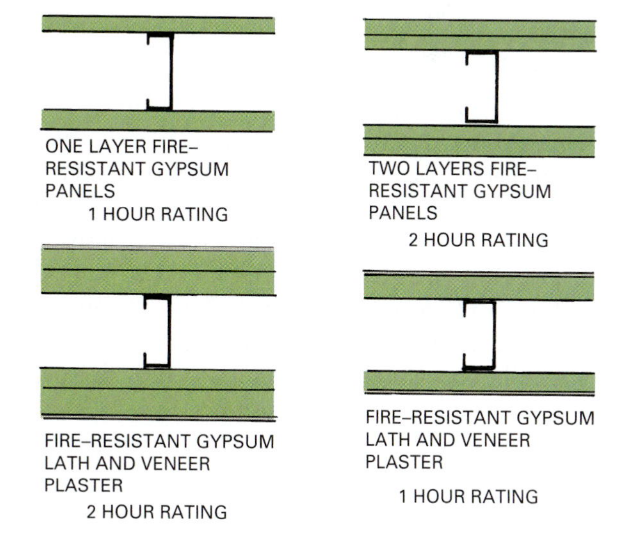

Installing Gypsum Wallboard

Gypsum wallboard is fastened to wood studs and ceiling joists with special nails or power driven screws and to metal studs and metal furring with self-drilling, self-tapping screws (Fig. 36.9). Panels can be installed with

Figure 36.9 Nails and screws are used to secure gypsum wallboard to wood and metal studs.

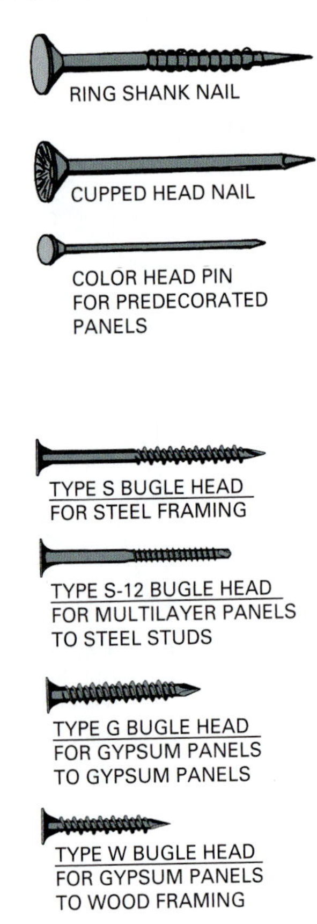

the long edge perpendicular to or parallel to the studs. Perpendicular application is usually used because it reduces the amount of joints that need taping (Fig. 36.10). Perpendicular application also places the strongest dimension of the panel across the studs.

Ceiling panels are installed first and run perpendicular or parallel to joists, depending on which method produces the fewest joints. Panels may be single or double nailed. Nails on single-nailed ceilings are usually spaced 7 in. (178 mm) O.C. and 8 in. (203 mm) O.C. on side walls (Fig. 36.11). Screws are spaced 12 in. (305 mm) on a ceiling and 16 in. (406 mm) on side walls. To reduce the possibility of nail popping after a wall is finished, panels can be double nailed, as shown in Fig. 36.12. Fasteners are driven until their heads are set in shallow dimples without breaking the paper covering. Nails are driven with a drywall hammer with the required shape and diameter (Fig. 36.13).

Gypsum panels are often installed in a double layer. The first layer is nailed or screwed to the framing, and the second layer is bonded to it with an adhesive (plus a few nails or screws). The adhesive is applied to the backs of panels with a trowel. This produces a wall with greater sound attenuation and fire resistance and reduces the possibility of nail popping.

Figure 36.10 Gypsum wallboard panels are often installed with the long edge perpendicular to the studs. *(© John Wilkes Studio/Corbis)*

Figure 36.11 Suggested single-nailing pattern for gypsum wallboard panels.

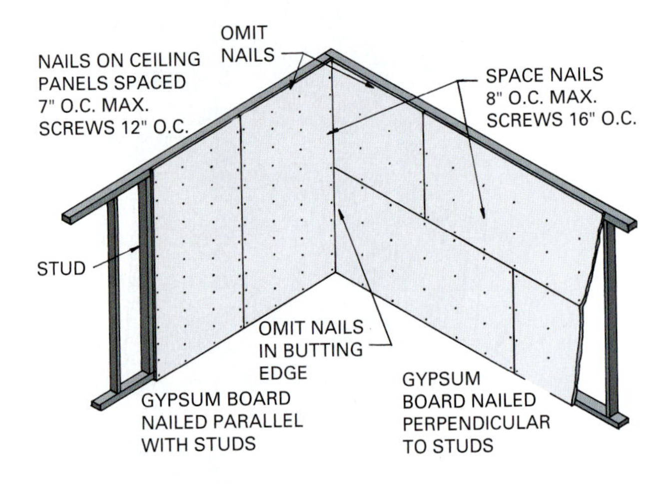

Figure 36.12 Suggested double-nailing pattern for gypsum wallboard panels.

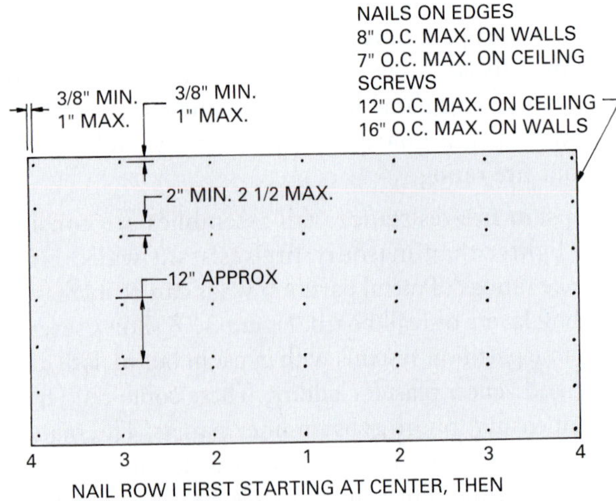

Figure 36.13 The face on this drywall hammer is crowned to produce a shallow dimple around nail heads. *(Courtesy Georgia-Pacific Corporation)*

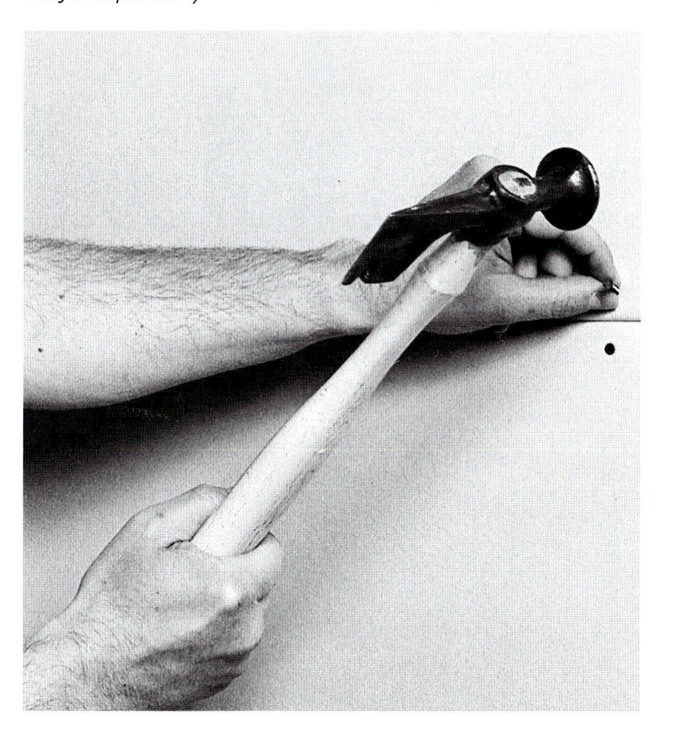

Gypsum panels are cut by scoring their front surface with a utility knife along the edge of a metal T-square. A panel is then bent to break the core, and the paper is scored on the back to separate the panel parts.

Joints between panels are covered with layers of **joint compound** and tape. The long edges of panels are made with a taper, allowing compound and tape to fill it flush to the surface. Short ends of panels are not tapered and produce a slight bulge when taped, so end joints are avoided during installation whenever possible (Fig. 36.14). Both types of joints are covered with a thin layer of compound and the tape is pressed into it with a finishing knife (Fig. 36.15). A thin skim coat is applied over the tape and the compound is allowed to dry. Additional layers of compound are applied, allowed to dry, and sanded. Each layer is wider than the one before and is feathered out with taping knives (Fig. 36.16). The tape and joint compound can be mechanically applied to wall and ceiling joints with a variety of machines. The compound may be hand sanded, but a pole sanding unit with a vacuum greatly reduces dust. Nail and screw heads are covered with three layers of joint compound.

Internal corners are finished with a folded paper or fiberglass tape bonded in joint compound and finished in the same manner described for joints. External corners are finished with metal or plastic corners nailed or screwed to the studs and covered with joint compound

Figure 36.14 The joints between panels are covered with layers of joint compound and tape.

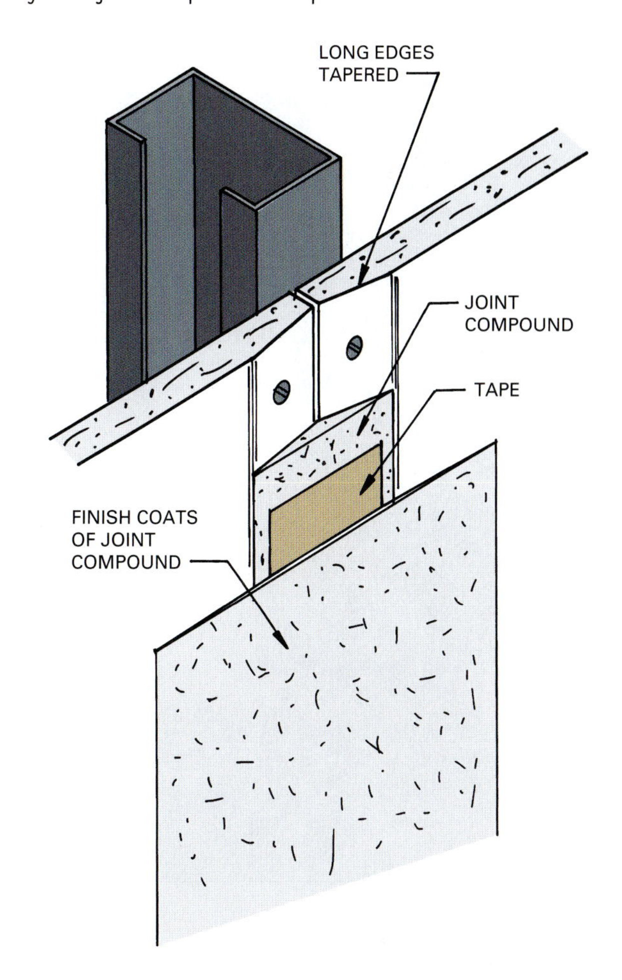

Figure 36.15 The tape is pressed into the first layer of joint compound. *(Courtesy Georgia-Pacific Corporation)*

Figure 36.16 This taping tool applies the tape to the joint after the first coat of joint compound is in place. *(a. Courtesy USG Corporation; b. Image copyright Feverpitch, 2009. Used under license from Shutterstock.com)*

(Fig. 36.17). Other finishing accessories include edge trim, control joints, and trim for round corners, such as those found on arches (Figs. 36.18 and 36.19).

Door frames on walls with metal or wood studs are installed as shown in Fig. 36.20. These details show the installation of metal and wood door jambs.

Figure 36.17 External corners are covered with a metal or plastic corner bead and joint compound.

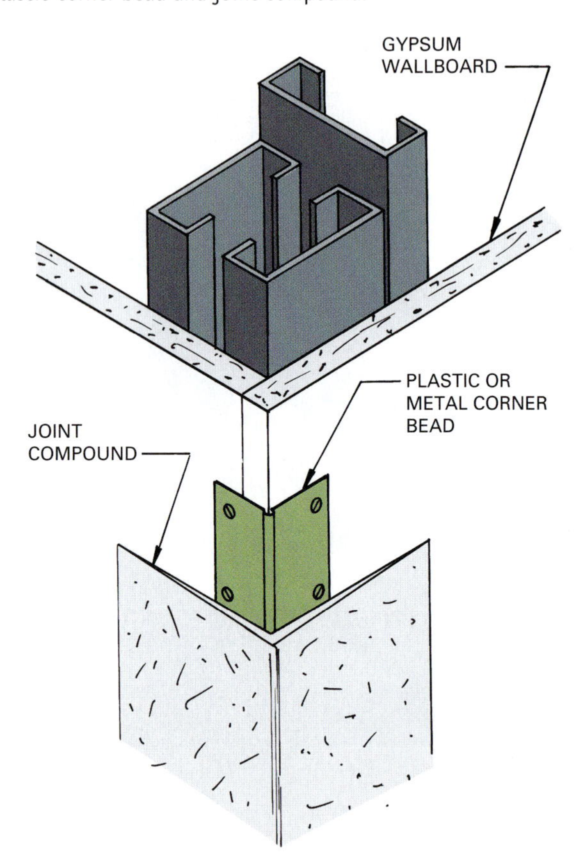

PLASTER WALL FINISHES

The types of plaster and gypsum base materials used to finish wall surfaces are detailed in Chapter 34. **Plaster** is applied over expanded metal lath, gypsum lath, or veneer **plaster base. Metal lath** is made by slitting sheets of thin metal alloy and stretching it to form a diamond-shaped mesh or by punching and forming the openings in the metal sheet. Typical types are rib-lath, diamond-mesh lath, sheet lath, and wire lath (Fig. 36.21). The lath is usually wired to steel studs or secured with self-tapping screws.

Gypsum lath provides a rigid fire-resistant base to which gypsum plasters are applied. Gypsum lath is available in several thicknesses and as two different products. A standard gypsum lath is nailed or stapled to wood studs or screwed to metal studs and furring. A second type includes a fire-resistant core.

Veneer gypsum base has a gypsum core and is faced with paper. It is used as the base for **veneer plaster**.

Plaster over expanded metal lath is applied in a three-coat system (Fig. 36.22). The first coat (scratch coat)

Figure 36.18 Drywall finishing accessories are available in metal and plastic. *(Courtesy USG Corporation)*

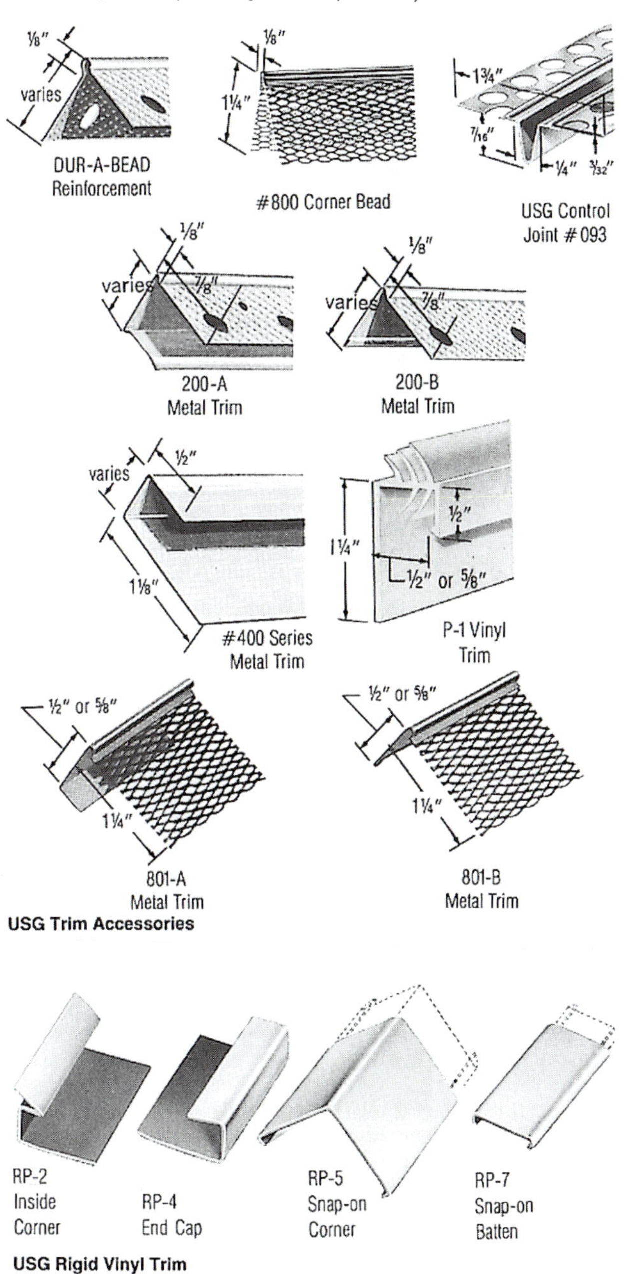

is applied to the metal lath with enough pressure to force it into the mesh and form a good bond. It is cross-raked, leaving a flat but textured surface. After the scratch coat has hardened, the second coat (brown coat) is applied over it. This coat is leveled and allowed to harden. The third coat (finish coat) goes over the brown coat and is finished to the desired texture to produce the final surface.

Figure 36.19 Applications of drywall edge trim.

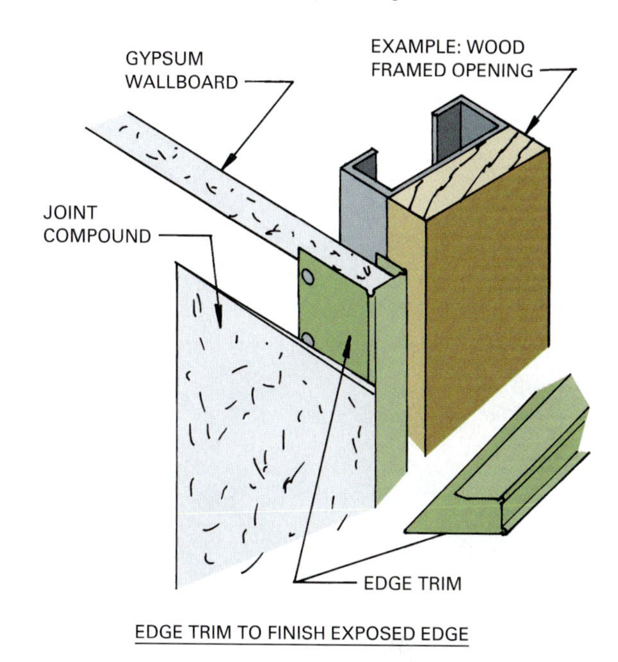

Figure 36.20 Typical installation details for metal and wood door jambs.

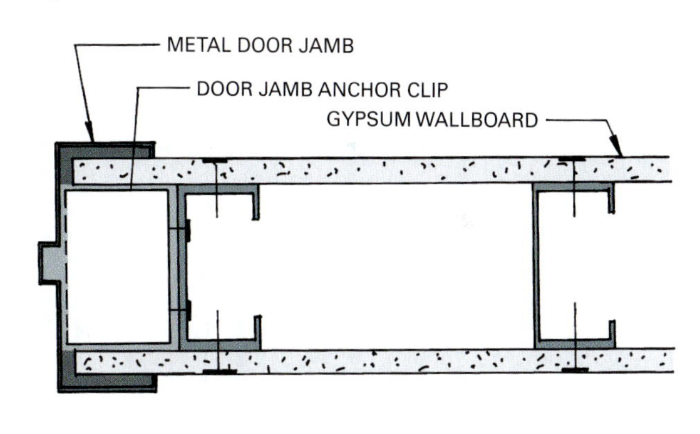

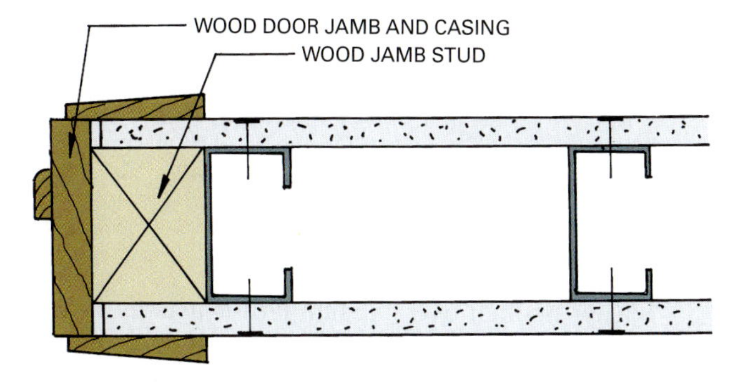

Figure 36.21 Types of metal lath. *(Courtesy USG Corporation)*

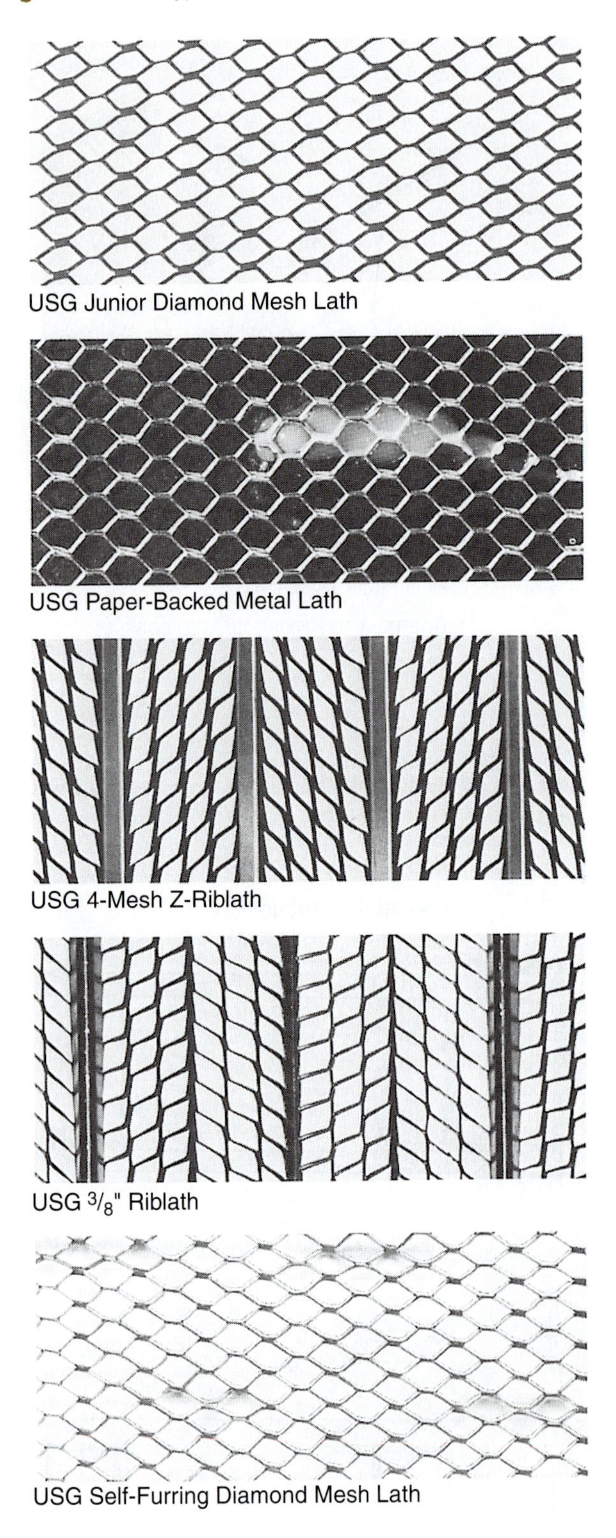

Figure 36.22 A three-coat plaster wall finish over metal lath.

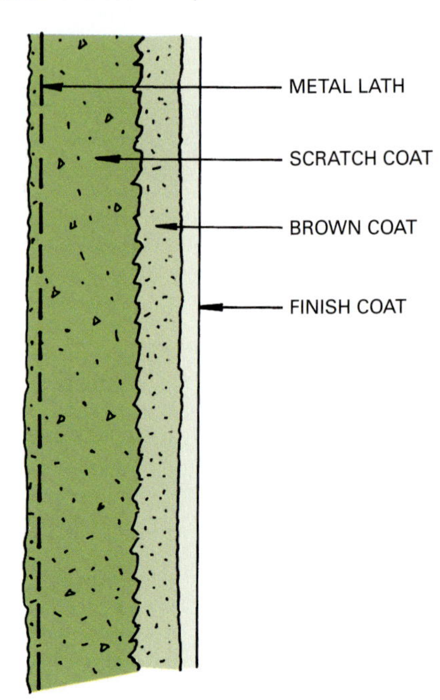

Figure 36.23 A two-coat plaster wall finish over gypsum lath.

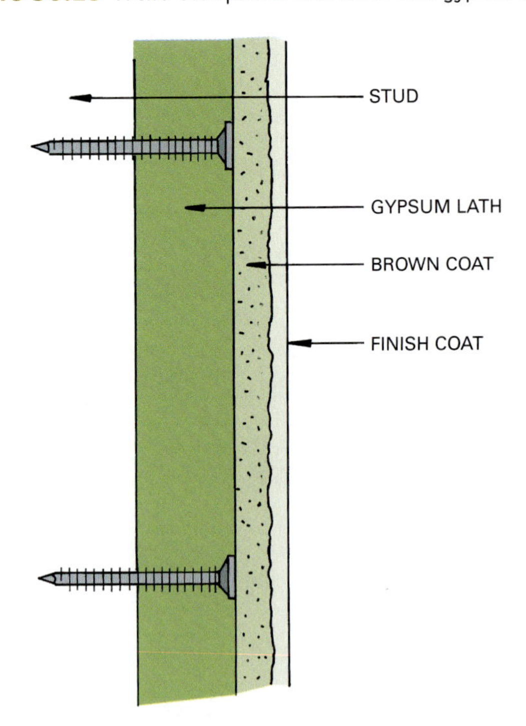

Plaster applied over gypsum lath is usually a two-coat system, but three coats may be used (Fig. 36.23). The first coat is applied with pressure so it bonds to the lath and provides a level coat with a rough surface. The second coat goes over the first coat and is finished to the desired texture.

Veneer gypsum base is secured to studs and covered with one or two coats of veneer plaster (Fig. 36.24). The first coat is a thin layer that is covered by a second thin coat before the first one has hardened. The second coat is troweled to the desired texture. Veneer plaster usually hardens completely within twenty-four hours.

Figure 36.24 Veneer plaster is a single-coat plaster applied over a gypsum veneer plaster base.

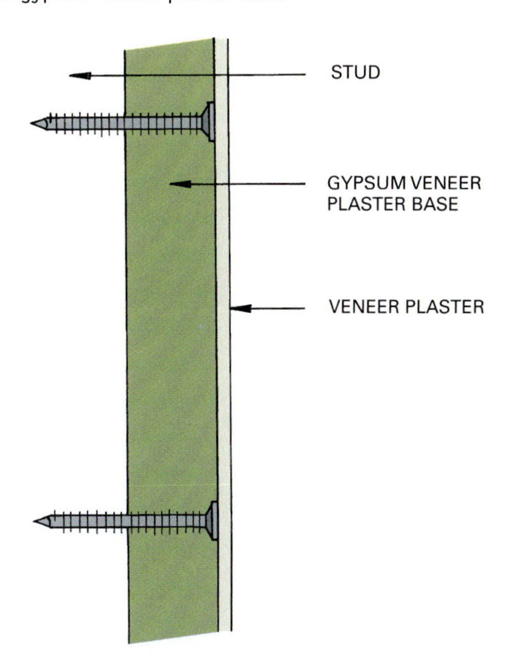

Plaster over Masonry and Concrete Walls

Plaster may be applied directly to masonry, concrete, and clay tile walls. Concrete block walls are porous and provide a satisfactory base. Clay tile and brick walls may be used as a base for plaster walls, but they must be sufficiently porous to provide suction for plaster and also scored to increase mechanical bonding. Smooth-surface glazed or semi-glazed tile cannot be covered. These require that the wall be furred and a metal or gypsum plaster base be installed. Monolithic concrete walls are porous but usually require an application of an agent to produce the adhesive bond necessary for direct application of gypsum plasters. Generally, masonry walls in contact with the exterior are not plastered directly to the masonry because of the likelihood of water seepage and condensation that can saturate the plaster.

Masonry concrete and clay tile walls can have gypsum lath installed by securing metal furring to their surface. On exterior walls, this provides an air space, keeping the plaster base and plaster away from the masonry wall. Furring can be shimmed to help true a wall with minor surface irregularities. Furring may be placed horizontally or vertically (Fig. 36.25).

Plastering

Plaster is applied to base material with a hand trowel, as shown in Fig. 36.26. In one hand, the plasterer holds a square, flat tool called a hawk. The hawk holds a supply of

Figure 36.25 Masonry walls can be covered directly with a plaster finish or use furred out walls.

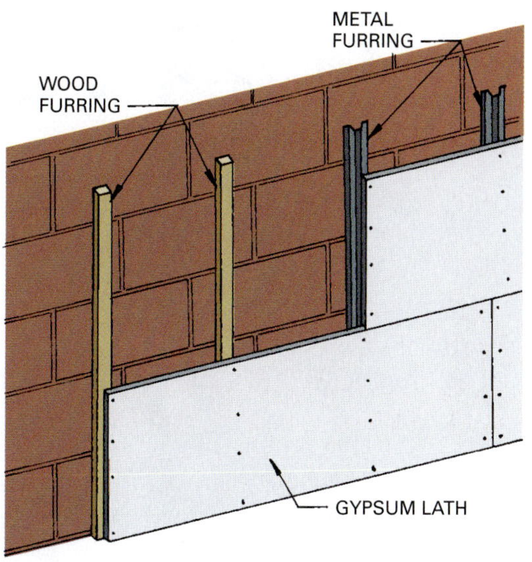

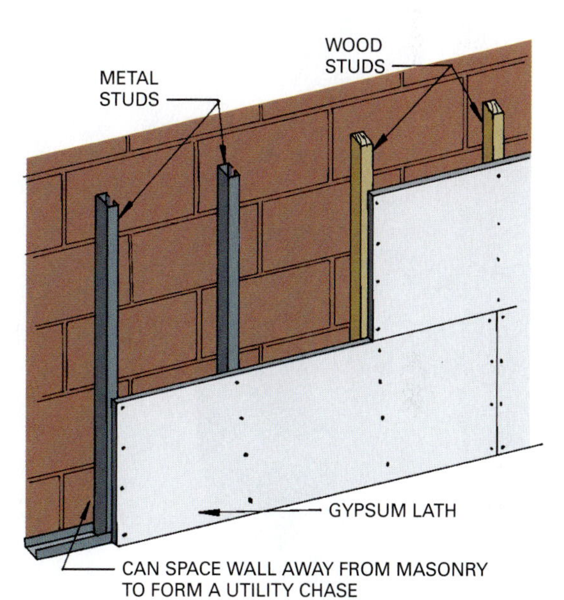

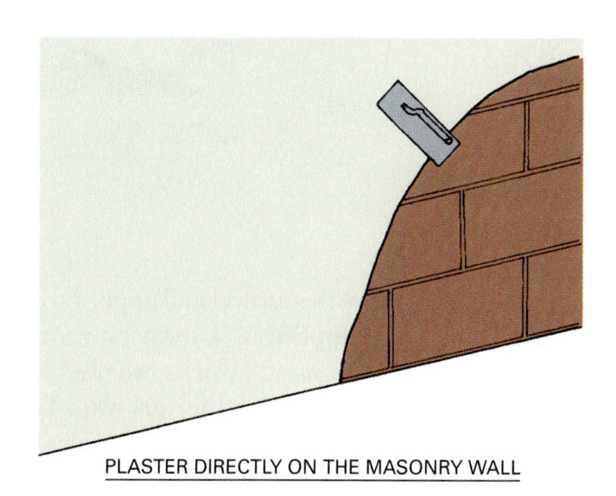

Figure 36.26 A plasterer hand-troweling the scratch coat to a gypsum lath base that is secured to metal studs with wire clips. *(Courtesy USG Corporation)*

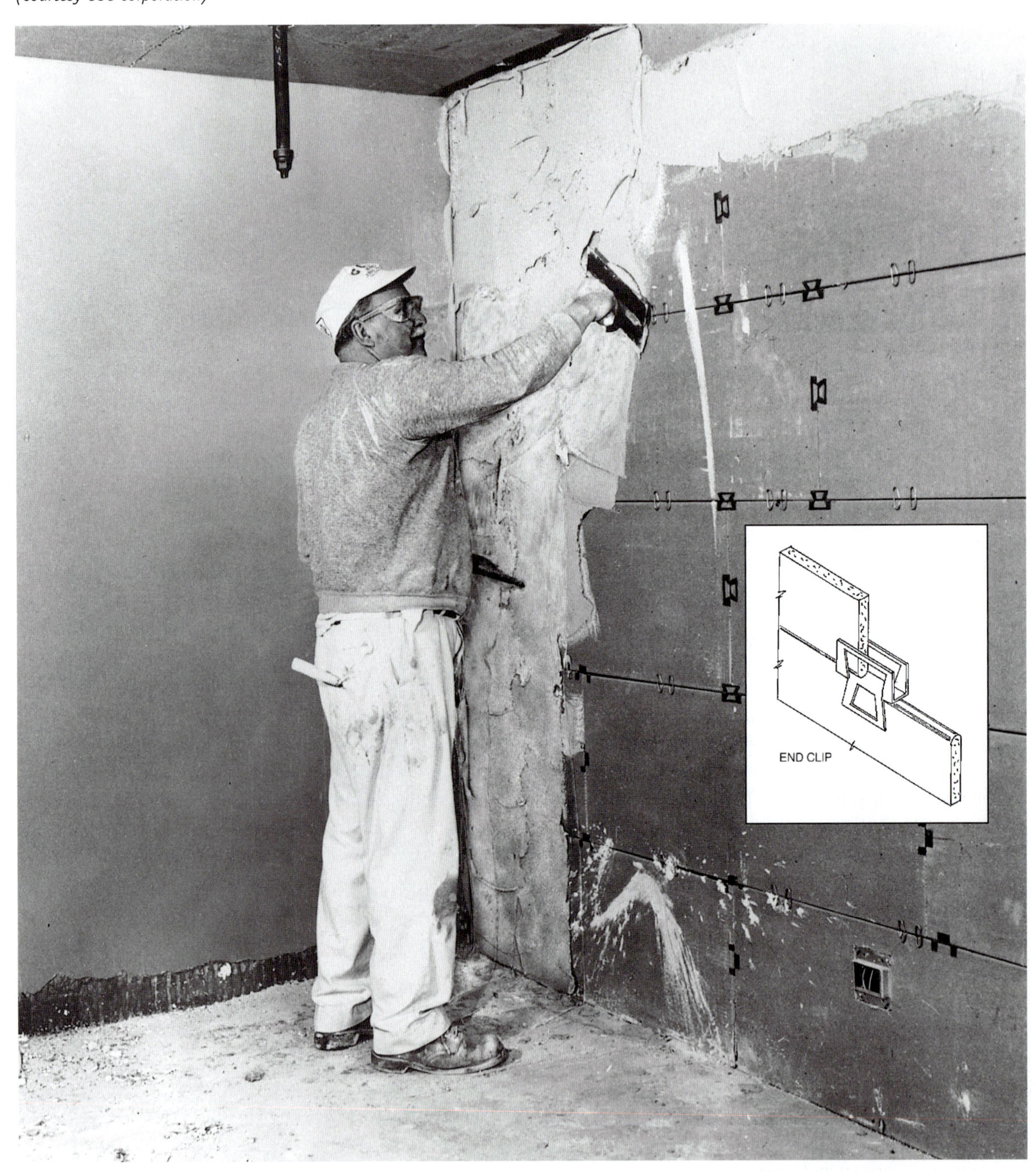

plaster, which is removed with a trowel and applied to the wall. Plaster may also be applied by a spraying process, as shown in Fig. 36.27. Various accessories are used with plaster finishes, much as is described for drywall. Typical among these are casing beads, external corner beads, flexible corner beads, and expansion joints (Fig. 36.28).

SOLID GYPSUM PARTITIONS

Solid gypsum partition systems may be studless or use channel studs. The solid, studless partition has a vertical core of metal mesh lath (Fig. 36.29). Specifications for the lath vary with a wall's height. The lath is connected

Figure 36.27 The plasterer on the right is applying a scratch coat by spraying it on the base. The plasterer in the center is hand-troweling the brown coat, while the person on the left is leveling the coat by pulling a darby across it. After it has been leveled, it may be troweled again to smooth the surface. *(Courtesy USG Corporation)*

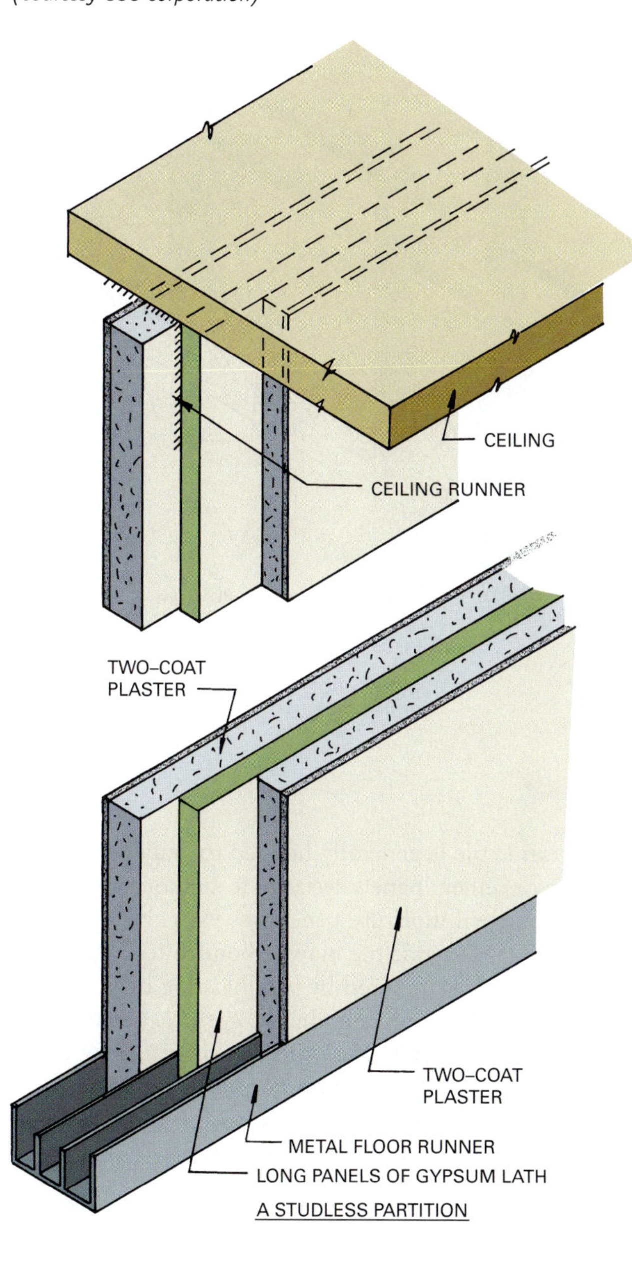

Figure 36.28 Accessories used with plaster finishes.

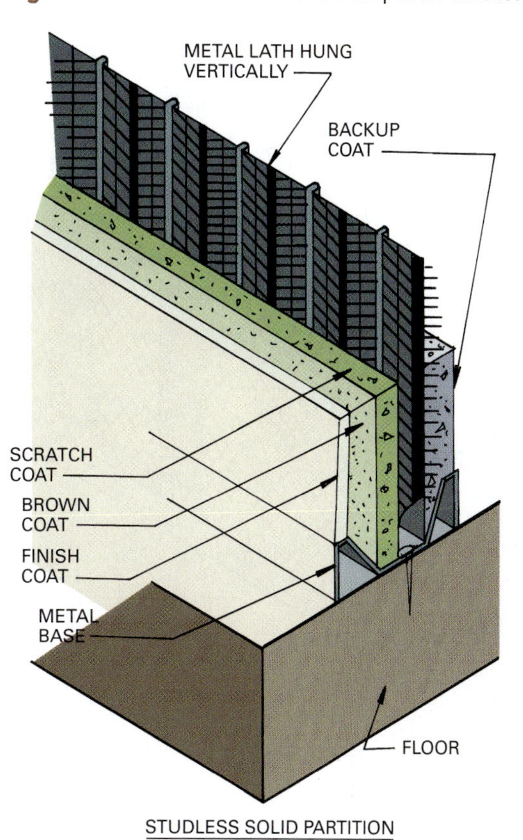

Figure 36.29 Typical construction of a studless solid gypsum partition built around vertical metal lath.

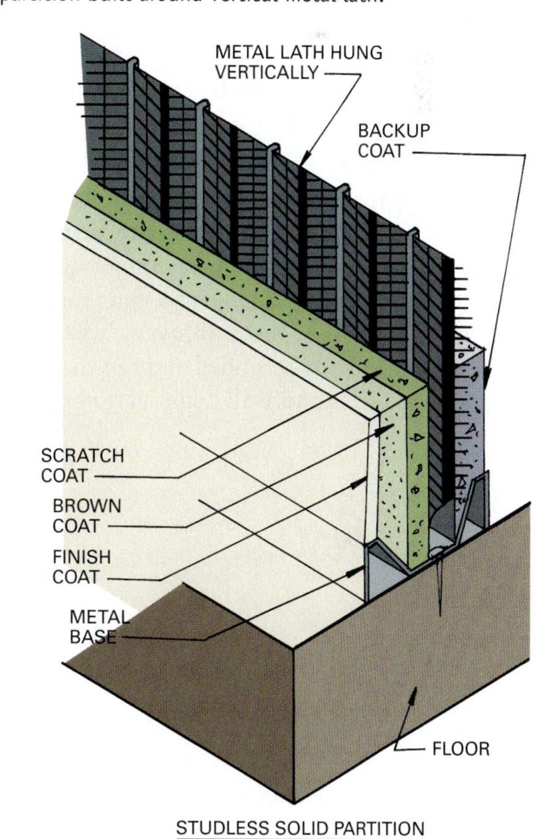

to floor and ceiling with attached runners (Fig. 36.30). Plaster is applied in successive scratch, brown, and finish coats.

Another partition construction uses a double-channel-stud hollow partition. Vertical studs are secured to ceiling and floor with metal runners. Wire mesh is tied to each row of studs, and conventional plastering procedures are used (Fig. 36.31). The studs are stiffened and braced with horizontal channels. This type of partition is useful for running electrical and mechanical system components.

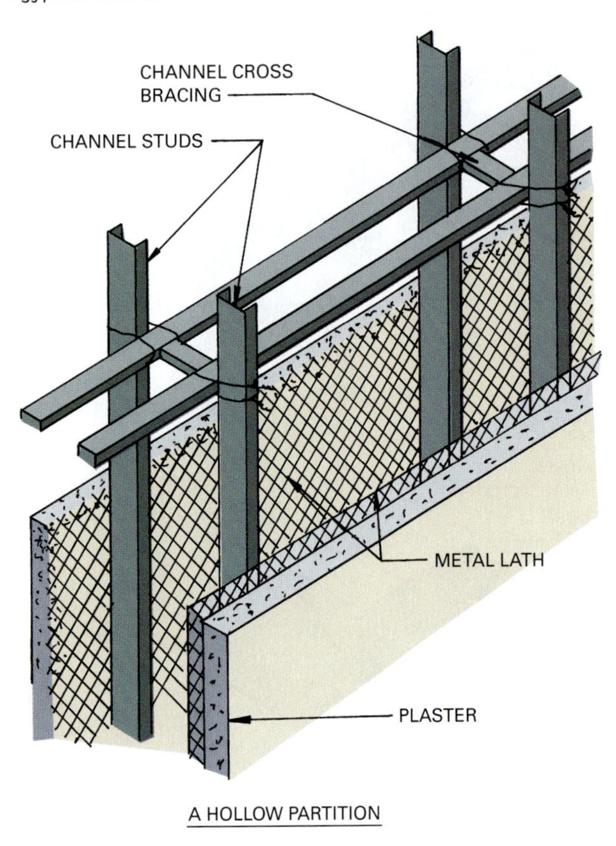

Figure 36.30 A studless solid partition built around a gypsum lath core set in metal tracks.

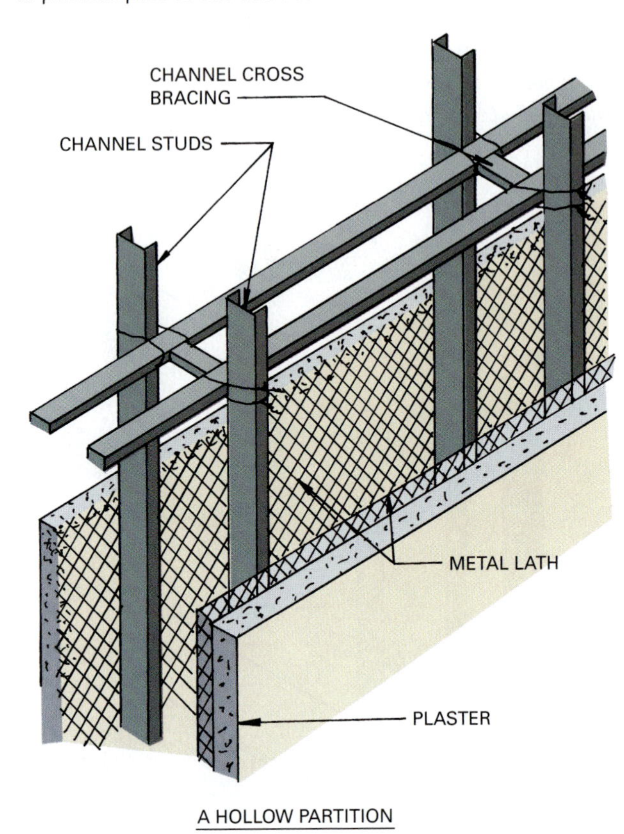

Figure 36.31 A double-channel hollow partition can be used to provide space to run utilities.

STRUCTURAL CLAY TILE PARTITIONS

Structural clay tile partitions are made with hollow clay masonry units that are glazed on one or both sides. The glazes provide a smooth surface that is impervious to penetration by water and is available in a wide range of colors (Fig. 36.32). The tiles resist abrasion, wear, and mildew and can be routinely scrubbed and sanitized. Information about tile sizes and wall construction can be found in Chapter 11.

CERAMIC TILE WALL FINISHES

Ceramic tile used to finish walls and partitions is available in a wide range of sizes and colors. Tiles provide a hard, water-resistant facing and are durable and easily cleaned (Fig. 36.33). Information about tile sizes and installation can be found in Chapter 11.

Ceramic tile is generally bonded to walls faced with gypsum or cement panels designed to support the tile and resist damage if moisture penetrates joints between tiles (Fig. 36.34). Ceramic tile may be bonded to concrete or masonry that is prepared by sandblasting or scarifying to provide a true uncontaminated surface. Ceramic tile may also be applied to wall surfaces consisting of metal lath covered with a cement mortar bed. Portland cement mortar is used for setting tile in a thick bed. Other mortars and adhesives are used for thin beds.

WOOD PRODUCT WALL FINISHES

An interior wall finish may consist of many types of solid wood paneling or reconstituted wood panels (such as plywood or hardboard). They are available prefinished, including some with vinyl wall coverings. Details about these products are given in Chapter 19. The use of wood interior wall finishes may be severely limited or prohibited by code regulations for certain building types and occupancies.

Figure 36.32 Partitions and walls are constructed with glazed structural ceramic tile units, providing a fire-resistant wall that is easily cleaned and resists damage. *(Courtesy Stark Ceramics, Inc.)*

Figure 36.33 Ceramic tile wall facing provides a fire-resistant, durable surface that is easily cleaned. *(Image copyright Christina Richards, 2009. Used under license from Shutterstock.com)*

CEILING CONSTRUCTION

A ceiling serves a variety of functions. It provides a finished appearance to overhead areas, contributes to the acoustical treatment of a room, and provides a light-reflecting surface to enhance illumination. Some types provide a space below a floor structure above in which electrical and mechanical systems can be run out of view. A ceiling may support lights, provide a measure of fire protection to the floor or roof above, and permit the use of sprinkler heads in a fire-control system. Ceilings may be flat or curved and constructed on any of a wide variety of sloping surfaces that provide for the architectural enhancement of an area (Fig. 36.35).

Ceiling finish materials may be plastic, metal, plaster, gypsum panels, wood, or one of many fibrous panels available. Floor or roof decking may be left exposed and used as a finished ceiling. Types of ceiling constructions include suspended, furred, and contact.

Suspended Ceilings

Suspended ceilings are possibly the most widely used in commercial construction. They offer considerable advantages by allowing electrical and mechanical systems to be hidden while providing a variety of finished surface materials.

Suspended ceilings are hung from a floor or roof structure by a series of wires that support a metal grid that holds the finished ceiling (Fig. 36.36). Wire lengths can vary, so a flat horizontal ceiling can be installed below a structural floor or roof that has members of

Figure 36.34 Some of the assemblies used when building a wall destined to have a ceramic tile face.

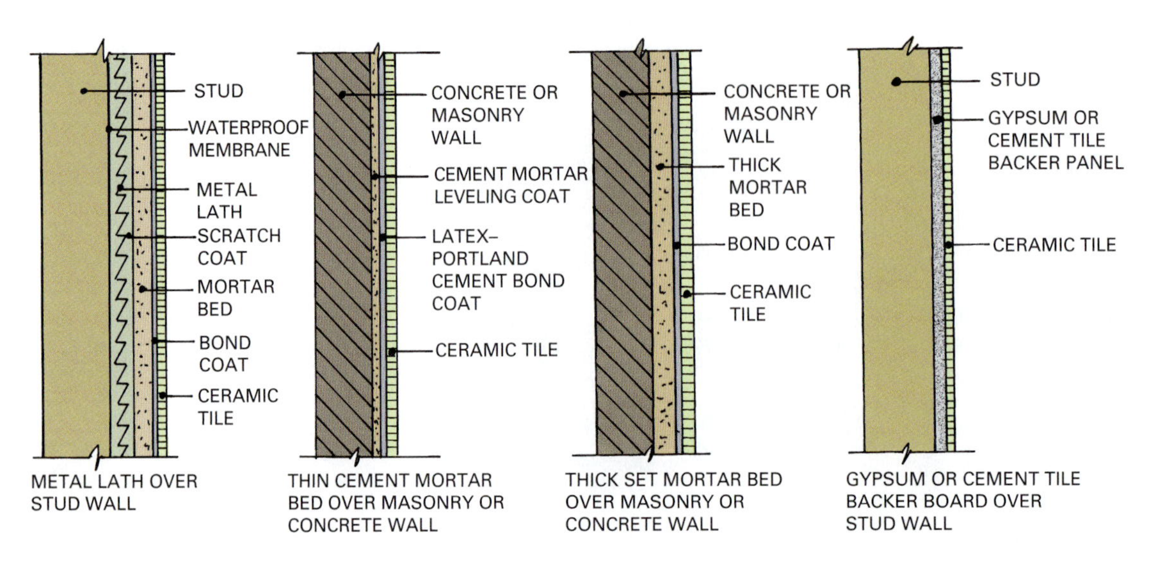

Figure 36.35 This flat aluminum ceiling accommodates lights and heating-system vents. *(Image copyright Bidouze Stéphane, 2009. Used under license from Shutterstock.com)*

Figure 36.37 This fire-resistant acoustical ceiling has recessed incandescent and fluorescent light fixtures. *(Image copyright Matt Niebuhr, 2009. Used under license from Shutterstock.com)*

Figure 36.36 A suspended ceiling is hung from overhead by wires. The wire carries the main runners, which are joined by cross tees to form a metal grid. In this design, ceiling panels are laid into the grid, forming the ceiling.

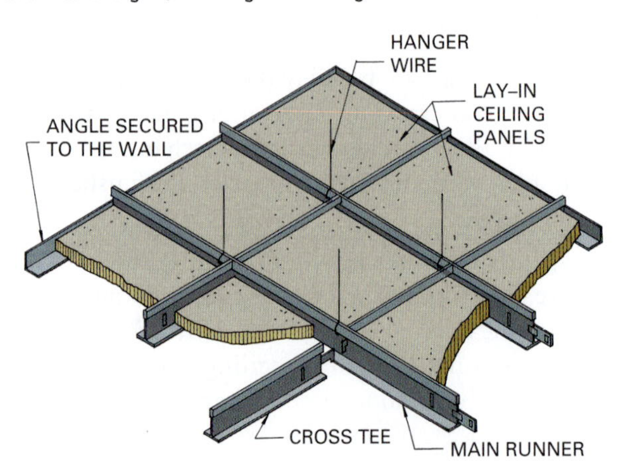

various sizes. Typically, a lighting system is set in the grid and is flush with the ceiling (Fig. 36.37). Some types of panels have acoustical properties and others provide a degree of fire protection below the floor or roof above.

Suspended plaster ceilings are constructed by hanging channel runners to which furring channels are tied with wire. The wire or metal lath plaster base is wired to the furring. This construction produces a hard, durable, highly fire-resistant ceiling. Typical construction details are shown in Fig. 36.38.

Various types of suspended acoustical systems are available that use a suspended metal grid into which lightweight panels and lights are placed. Grids

Figure 36.38 A suspended plaster ceiling has the main channel hung from overhead by wires, and metal furring channels are wired to the main channel. The plaster coats are the same as described for finished walls.

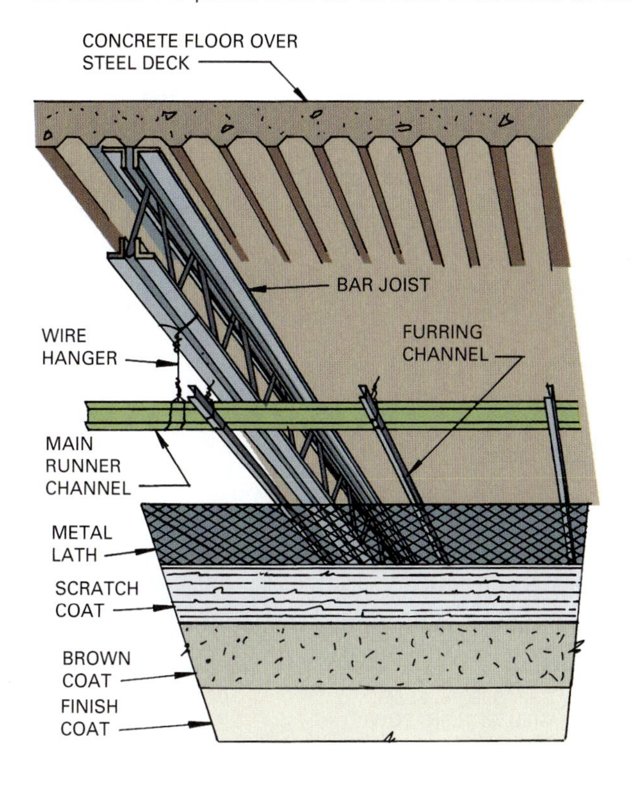

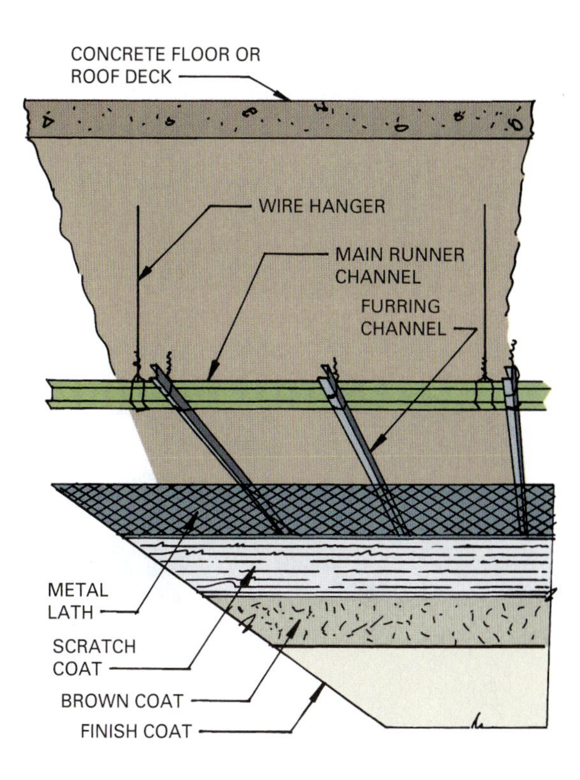

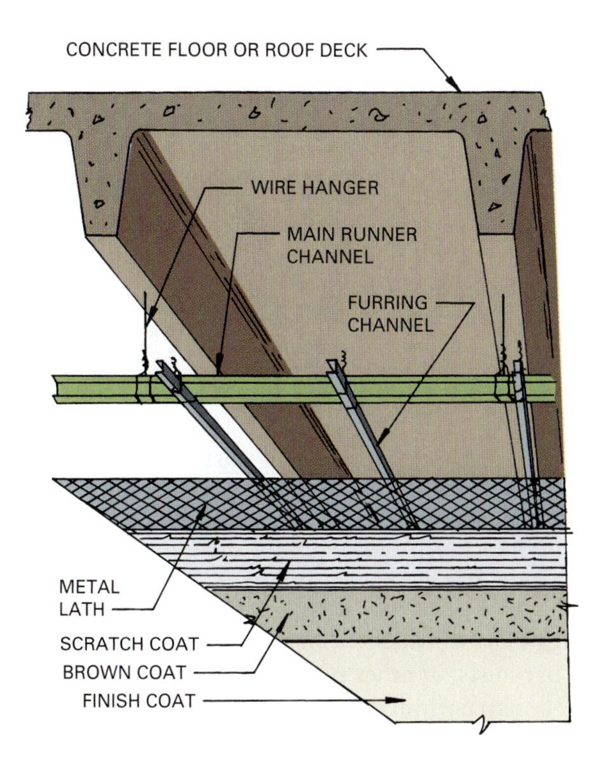

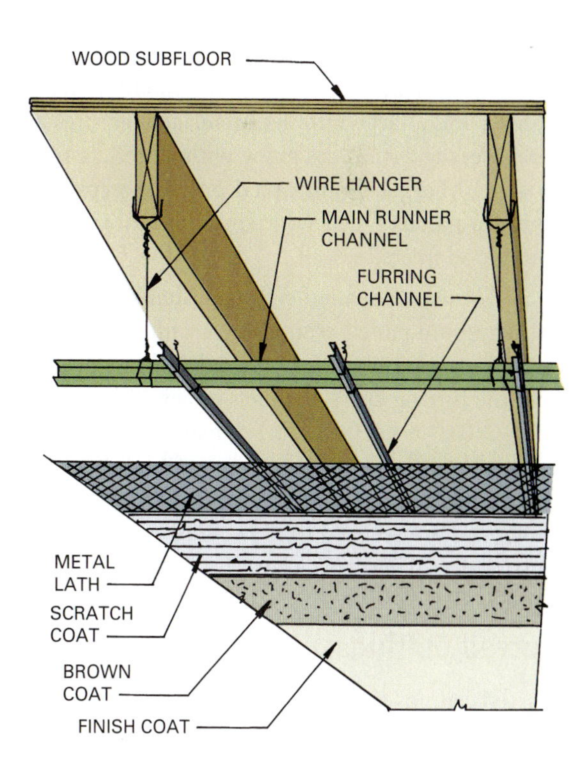

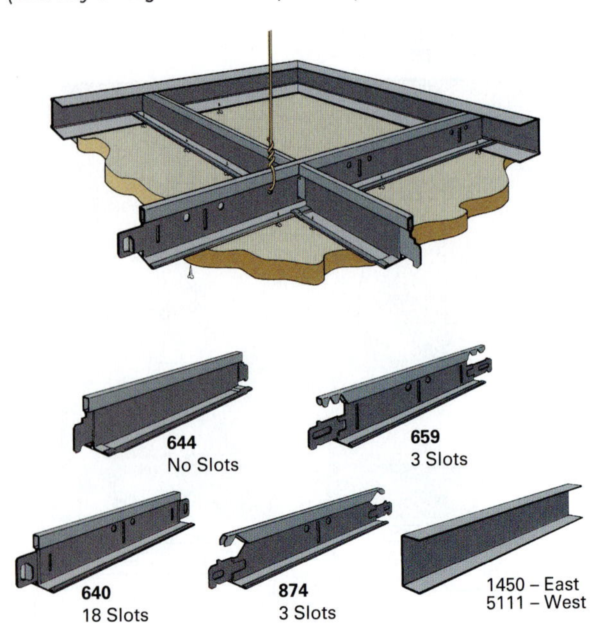

Figure 36.39 This suspended ceiling system has a variety of grid members. It is designed to support a gypsum ceiling. *(Courtesy Chicago Metallic Corporation)*

are assembled with some form of locking connection (Fig. 36.39). Drop-in panels may have an exposed grid or a concealed grid (Fig. 36.40). These systems also can accommodate heat ducts, sprinkler heads, and other required penetrations. Some grid systems have recessed panels, and a large number of surface textures are available. Ceiling panels are made with materials such as aluminum, fiberglass-reinforced polymer gypsum cement, gypsum panels faced with vinyl and fabrics, fiberglass, and wood fibers.

Another suspended ceiling system uses a metal grid with gypsum panels secured to it with screws. In this system, the main runners are spaced 48 in. (1219 mm) O.C. and the furring cross channels either 16 in. (406 mm) or 24 in. (610 mm) O.C. The gypsum drywall panels are attached to the grid with conventional drywall screws. The ceiling grid is secured to walls with a wall track, and partitions are secured to the ceiling by screwing them to a drywall furring cross-channel or main runner.

Furred Ceilings

Furred ceilings are constructed using metal furring strips secured to a metal or wood structural system. Concrete overhead floors and roofs often have a C-channel hanger hung below them, and the furring strips are wired to the hanger. The metal lath is wired to the furring strip. The spacing and size of the hangers depends on the design of the system (Fig. 36.41).

Figure 36.40 Drop-in ceiling panels may have an exposed, recessed, or concealed grid.

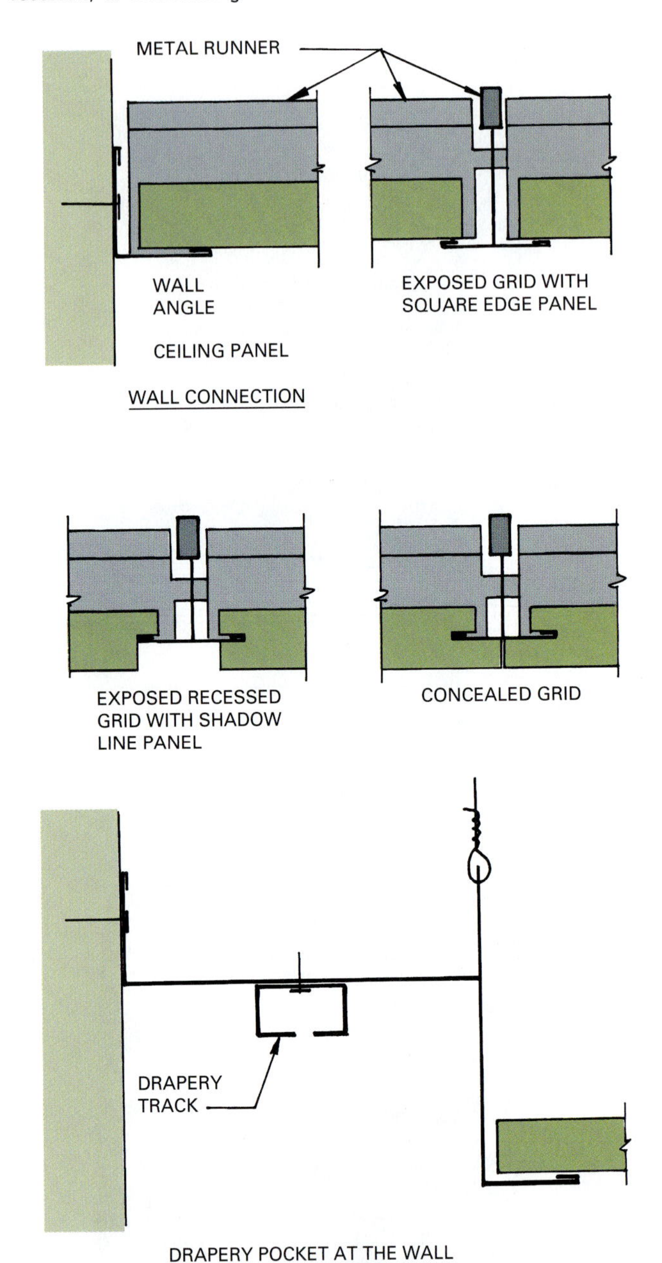

Contact Ceilings

Ceilings can be attached directly to floor or ceiling joists, bar joists, or other structural members. In wood-frame construction, gypsum wallboard panels are screwed directly to the joists, as shown in Chapter 20. Plaster ceilings can have the metal lath secured directly to steel, wood, or concrete structural systems, as shown in Fig. 36.42. The ceilings are then finished with several layers of plaster, in the same manner as described for plaster wall construction.

Figure 36.41 Furred ceilings have the furring channels secured to the floor or roof structure.

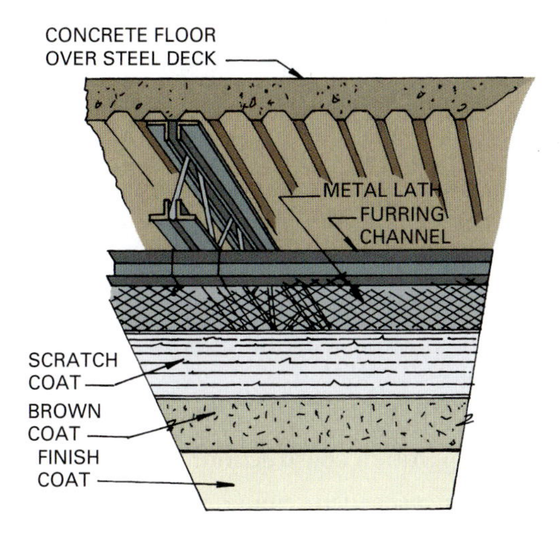

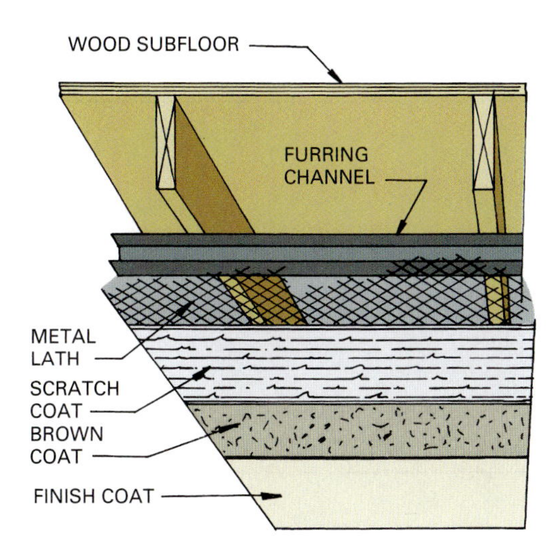

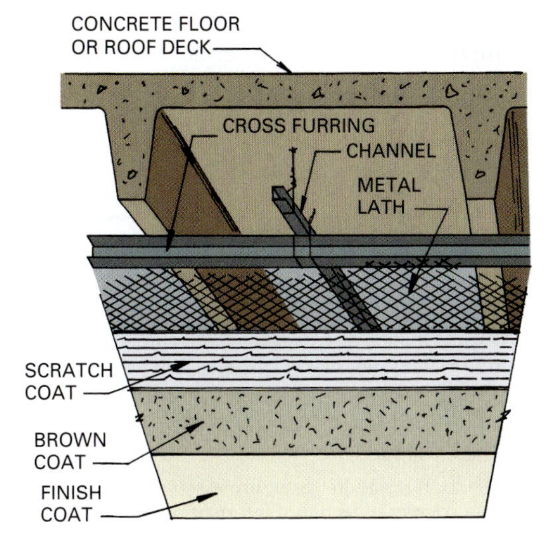

Figure 36.42 Contact ceilings have the finished ceiling material secured directly to the overhead floor or roof structural system.

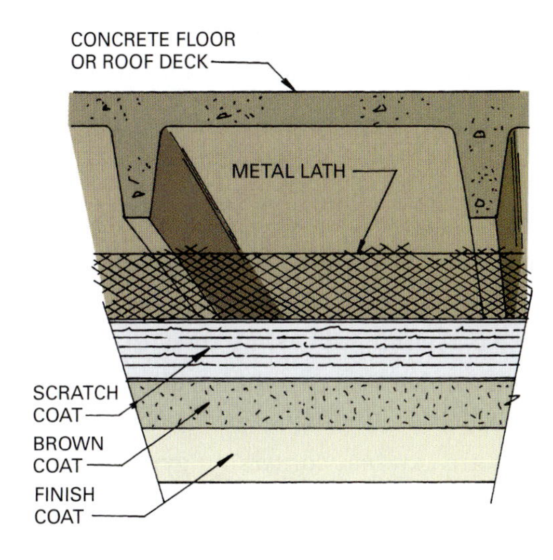

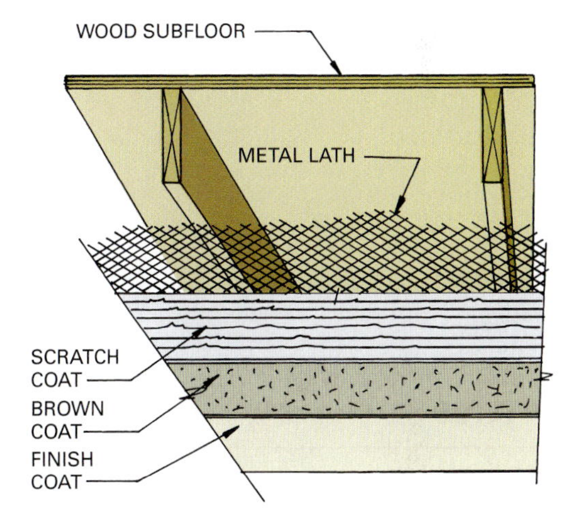

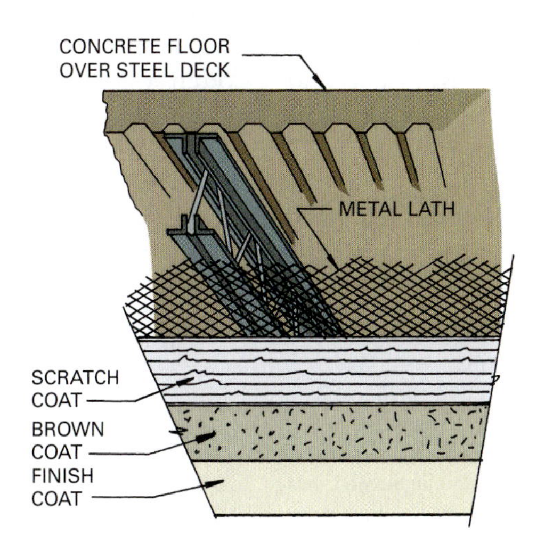

Review Questions

1. What purposes do interior partitions serve?
2. What is the difference between a fire partition and a fire wall?
3. How can electrical wiring be run inside partitions having metal studs?
4. What assembly of materials is often used in a fire-rated partition that also requires effective sound control?
5. How is gypsum wallboard secured to metal studs?
6. What can be done to reduce nail popping in gypsum wallboard installation?
7. How are external corners on gypsum wallboard finished?
8. What base materials are used when constructing a plaster interior finish?
9. What types of plaster coats are applied over metal lath?
10. Why can concrete block walls have plaster applied directly to them?
11. How can you plaster over a glazed tile wall?
12. Describe the construction of typical solid gypsum partitions.
13. What are the merits of using glazed hollow clay tiles for partition construction?
14. What is a suspended ceiling?

Key Terms

Fire-rated partition
fire wall
gypsum lath
joint compound
metal lath
plaster
plaster base
smoke barriers
veneer gypsum base
veneer plaster

Activities

1. Examine catalogs of companies that manufacture interior walls, partitions, and ceiling systems. For each, list materials used, fire rating, sound transmission class, and sketch a section of the assembly.
2. Tour various buildings on your campus and record the location of different types of interior wall assemblies, partitions, and ceilings. Do those in older buildings meet the current codes? What should be done to upgrade each system found lacking?

Additional Resources

Gorman, J. R., Jaffe, S., Pruter, W. F., and Rose, J. J., *Plaster and Drywall Systems*, BNI Building News, 1612 Clementine Street, Anaheim, CA 92802.

Gypsum Construction Guide, National Gypsum Company, Gold Bond Building Products, 2001 Rexford Road, Charlotte, NC 28211.

Gypsum Construction Handbook, United States Gypsum Company, 125 S. Franklin Street, Chicago, IL 60606.

Handbook for Ceramic Tile Installation, Tile Council of North America, Anderson, SC.

Melander, J., and Isberner, A., "Portland Cement Plaster," *Journal of Light Construction Publishers*, Richmond, VT.

Remodelers Guide to Suspended Ceilings, Chicago Metallic Corp., 4849 S. Austin Avenue, Chicago, IL 60638.

Spence, W., *Installing and Finishing Drywall*, Sterling Publishing Co., 387 Park Avenue South, New York, 10016.

Other resources include:

Many publications from the Gypsum Association, 810 First Street NE, Washington, DC 20002.

See Appendix B for addresses of professional and trade organizations and other sources of technical information.

CHAPTER 37

Flooring

LEARNING OBJECTIVES

Upon completion of this chapter, the student should be able to:

- Understand the importance of building codes in the selection of flooring.
- Select the appropriate flooring materials for various applications.
- Understand the types of flooring available and their sizes and properties.

Build Your Knowledge

For further information on these materials and methods, please refer to:

Chapter 8 Cast-in-Place Concrete

Chapter 11 Clay Brick and Tile

Chapter 13 Stone

Chapter 19 Products Manufactured from Wood
Topic: Wood Flooring

Chapter 24 Plastics

Chapter 38 Carpeting

Finish flooring materials are installed in a variety of applications over substrates of various kinds. A designer must consider the characteristics of a substrate, expected traffic loads on a floor, its required maintenance, fire resistance, building code requirements, and appearance. The finish flooring must meet the prescribed conditions at an acceptable cost. Some applications may require that special conditions be met, such as a need for water resistance, sanitation, or other unique situations like those found in hospitals or laboratories.

BUILDING CODES

Building codes detail specific requirements for floor finish materials. Codes related to enclosed spaces, vertical exits and passageways, and corridors that provide access to exits are given. Requirements vary depending on the classification and occupancy of a building. The most commonly used finish flooring materials, such as wood, vinyl, terrazzo, and clay tile, do not present unusual hazards and are generally not subject to building code interior finish requirements. Carpeting may present indoor air quality hazards and must meet requirements specified by U.S. Department of Commerce standard DOC FF1 and the National Fire Protection Association Standard NFPA 253. Carpeting is discussed in Chapter 38.

WOOD FLOORING

Both hardwoods and softwoods are used to form strip flooring and various types of parquet and heavy wood block flooring (Fig. 37.1). Some types have a factory-applied finish, and others are sanded and finished after installation. Most are nailed to a wooden sub floor, but they can also be bonded with a mastic adhesive. Composite wood floorings are available in a very thin veneer that is prefinished.

Standard wood strip flooring is manufactured in several thicknesses. Typical installation details are shown in Fig. 37.2. Manufacturers supply detailed recommendations for the types of fasteners and adhesives that should be utilized. Parquet flooring is generally installed with adhesives. It is available in a wide variety of wood species and patterns (Fig. 37.3).

Figure 37.1 Hardwood strip flooring protected with a durable coating provides a beautiful finished floor. *(Image copyright Elena Elisseeva, 2009. Used under license from Shutterstock.com)*

Other wood flooring products include a durable hardwood plank and parquet flooring that is acrylic impregnated. This prefinished laminated product comes in $\frac{3}{8}$ in. (915 mm) thickness and meets slip resistance standards set forth by the Americans with Disabilities Act of 1990. It has a Class B flame spread (ASTM E-84).

ENGINEERED AND LAMINATE FLOORING

Engineered and laminate flooring are manufactured products that use a multi-ply construction to create a stable product that contracts less than many other flooring materials. Engineered flooring is a laminate of three layers of solid wood with tongue-and-groove edges and ends. The top veneer is visible and available in many wood species (Fig. 37.4).

Figure 37.2 Several installation details for wood flooring.

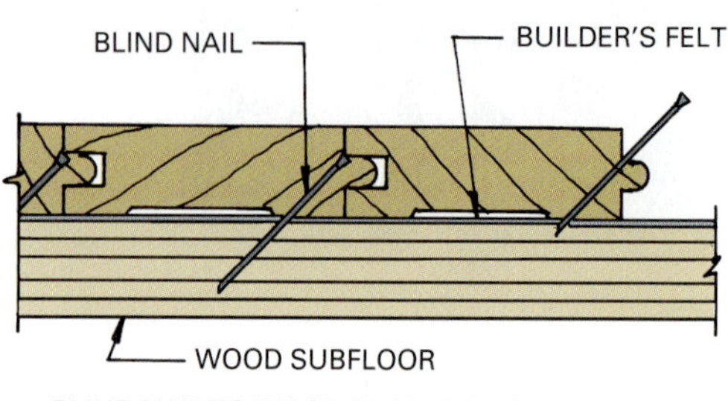

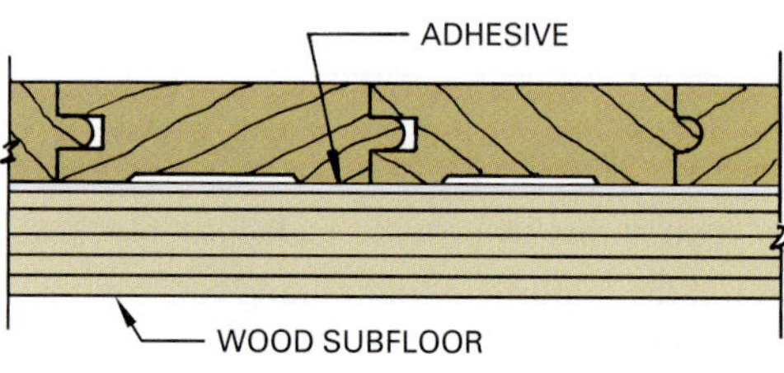

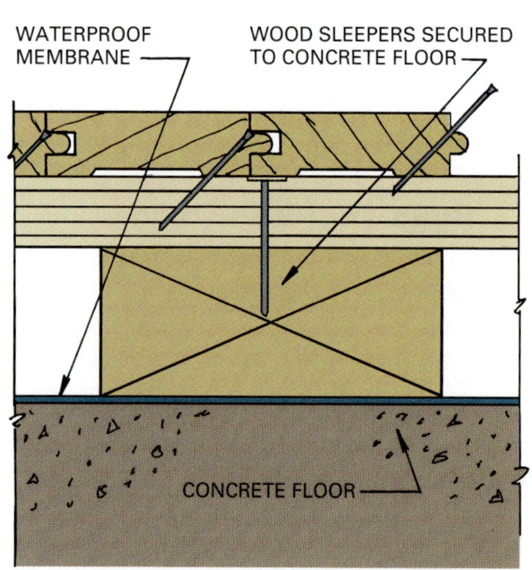

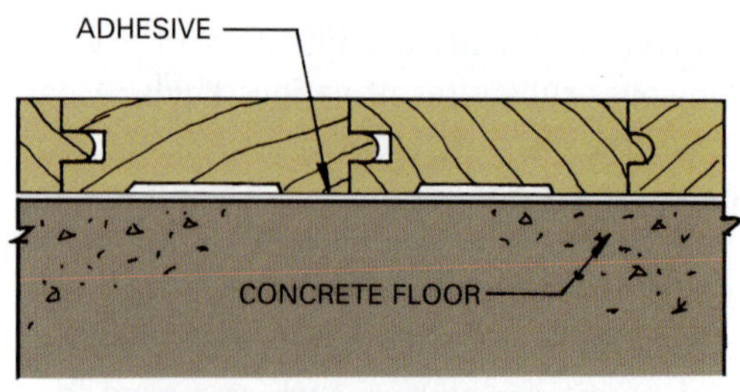

Laminate flooring is a rigid floor covering that has a top layer consisting of one or more thin sheets of fibrous material, often a paper product impregnated with aminoplastic thermosetting resins, such as a melamine. There are

Figure 37.3 Parquet flooring has been widely used for many years and provides a variety of patterns. *(Image copyright Semjonow Juri, 2009. Used under license from Shutterstock.com)*

two types of laminate flooring: direct-pressure and high-pressure (Fig. 37.5). Direct-pressure laminates fuse the top wear layer to a core material using pressure between 400 and 500 pounds per square inch. High-pressure laminates fuse the multilayers at about 1,000 pounds per square inch. The laminate and backer layer are bonded to a water-resistant, high-density fiberboard core with a urea-based adhesive.

RESILIENT FLOORING

Resilient floor covering in a polyvinyl chloride material is available in sheets and individual tiles. Sheet vinyl has a top vinyl layer with a composition backing. Vinyl composition tiles are composed of vinyl resins, plasticizers, stabilizers, fibers, and pigments and are available in 9, 12, 18, and 36 in. (22.8, 30, 45.7, and 91.4 mm) squares and

Figure 37.4 Engineered flooring is a laminate of three layers of solid wood that has a tongue-and-groove edge and end. *(Image copyright Sharon Meredith, 2009. Used under license from Shutterstock.com)*

some rectangular shapes. Available thicknesses are $\frac{1}{8}$ and $\frac{3}{32}$ in. (2.4 and 3.2 mm). Sheet vinyl flooring is available in rolls 6, 9, and 12 ft. (1.8, 2.7, and 3.7 m) wide and up to 50 ft. (15.2 m) long, in a wide range of thicknesses from about 0.069 to 0.224 in. (1.75 to 5.69 mm). Also available are solid vinyl, commercial vinyl, and PVC resilient flooring. Solid vinyl floor covering offers the maximum wear resistance. PVC flooring is used when a heat-welded surface is required.

Vinyl composition floor coverings (Fig. 37.6) are tested for fire spread and smoke production. They can be installed over wood or concrete subfloors with a mastic, as recommended by a manufacturer.

Figure 37.5 The components of direct- and high-pressure laminate flooring.

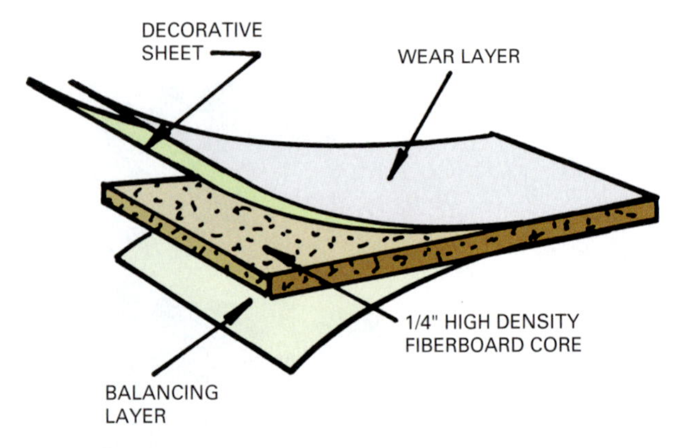

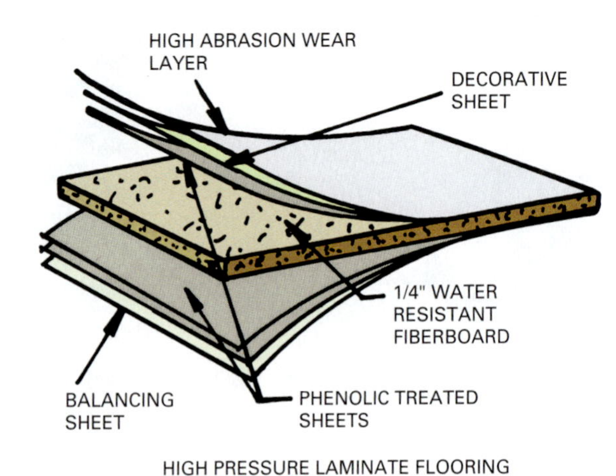

Asphalt resilient flooring is available in tiles usually 9 or 12 in. (2,745 or 3,660 mm) square. They are a composition of an asphalt binder for standard tile or a resinous binder for greaseproof tile, with inert fillers, fibers, and pigments for color. They come in grades A, B, C, and D, based on color, and are usually $\frac{1}{8}$ or $\frac{3}{16}$ in. (3 or 4.8 mm) thick. Bonded to wood or concrete subfloors with special mastics, they are durable and fire resistant.

Rubber resilient flooring is available in tiles and sheets (Fig. 37.7). Tile sizes vary by manufacturer, but 12 and 36 in. (305 and 915 mm) square tiles are common. Sheets range from 36 to 50 in. (915 to 1,270 mm) in width. Flooring is typically $\frac{1}{8}$ and $\frac{3}{16}$ in. (3 and 4.7 mm) thick. Rubber flooring is comfortable for walking, wears well, and resists damage from oils, solvents, alkalis, acids, and other chemicals. Most types have some form of gridded surface that aids traction and a roughened or recessed back grid. A number of new flooring products are available that utilize rubber from recycled automobile tires. They are available in a wide range of patters and colors (Fig. 37.8).

Installation techniques depend on given situations, and manufacturer's recommendations should always be observed. Typically, rubber flooring is bonded to concrete or wood subfloors with an approved adhesive. On-grade slabs require the use of a moisture-resistant adhesive. Manufacturers also produce rubber flooring products for use as stair tread, entrance mats recessed in the floor, and long runners that are not bonded to the subfloor.

Figure 37.6 Resilient vinyl composition tile, available in a variety of textures, finishes, and colors, provides a durable, easy to clean floor for business and residential applications. *(Image copyright Kimberly Hall, 2009. Used under license from Shutterstock.com)*

Figure 37.7 Rubber finish flooring is available in tiles and sheets. *(Image copyright Whaldener Endo, 2009. Used under license from Shutterstock.com)*

Figure 37.8 Rubber flooring made of post-consumer recycled materials can be used where a slip-proof resilient surface is required. *(Image copyright Invisible, 2009. Used under license from Shutterstock.com)*

Figure 37.9 Quarry tile provides a clean, durable finish floor. *(Image copyright Paul-André Belle-Isle, 2009. Used under license from Shutterstock.com)*

CLAY TILE

Clay tile products used for flooring include quarry tile, paver tile, and ceramic tile. Quarry tile is unglazed and made from shales and fire clays. It is durable, cleans easily, resists damage from freezing, thawing, and abrasion, and is used on floors subject to heavy and extra heavy use (Fig. 37.9). Some types are made with an abrasive grain surface to provide slip resistance. Although various manufacturers produce a variety of sizes and shapes, square, rectangular, and hexagonal tiles are most common. Various trim units, such as stair nosing and cove base units, are also available.

Paver tiles may be glazed or unglazed and are much like ceramic tiles but larger. Typical sizes are 4 x 4 in. (102 × 102 mm), 4 × 8 in. (102 × 203 mm), and 8 in. (203 mm) hexagons. They are weather- and abrasion-resistant. Typical uses include residential and moderate duty commercial floors, such as in shopping centers and restaurants, and heavy duty floors, such as those found in a commercial kitchen. A variety of trim materials are available.

Ceramic mosaic tile is available glazed or unglazed in square, rectangular, and hexagonal shapes. It comes in a wide range of sizes, but tiles are typically small, 1 or 2 in. (25.4 or 50.8 mm) square or hexagonal shapes, or 1 × 2 in. (25.4 × 50.8 mm) rectangles. Ceramic tile is used for interior floors, countertops, walls, swimming pools, and exterior floors and walls with special installation procedures. Tile with a variety of slip-resistant surfaces is available (Fig. 37.10), as is base, bullnose and many other trim tiles.

Figure 37.10 Paver tiles are weather- and abrasion-resistant. *(Image copyright Pchemyan Georgiy, 2009. Used under license from Shutterstock.com)*

Glazed ceramic floor tile is available with smooth and textured surfaces. Textured surfaces provide greater slip resistance. Tiles are available in 4 and 8 in. (101 and

202 mm) squares, 8 in. (202 mm) hexagons, and 4 × 8 in. (101 × 202 mm) rectangles. They are used for interior residential and moderate-duty commercial floors, such as in restaurants and shopping malls (Fig. 37.11).

Mortars and Adhesives for Tile Floors

Clay floor tile can be installed using a variety of bonding agents. Portland cement mortar is recommended for a thick-setting bed. Epoxies and furans that do not contain Portland cement can be used but are more expensive.

Portland cement mortar is a mixture of Portland cement, hydrated lime, sand, and water. It can be used to provide a leveling bed up to 1-$\frac{1}{4}$ in. (32 mm) thick. Tiles are set into this bed while it is still plastic. After curing, tiles may be bonded to the leveling bed with a thin or dry set coat or latex Portland cement mortar.

Dry-set mortar is a mixture of Portland cement, resinous materials, sand, and water. It is a thin-set mortar (about $\frac{3}{32}$ in. or 2.4 mm thick). It may be applied over the hardened thick-set Portland cement mortar bed and used to bond tiles in place.

Epoxy mortar contains no Portland cement but is a mix of epoxy resin and a hardener. It is a thin-coat adhesive with high bond strength and resistance to impact and chemical attack.

Latex Portland cement mortar is a mixture of Portland cement, a latex additive, and sand. It is a thin-coat application, usually about $\frac{1}{8}$ in. (3 mm) thick.

Organic adhesives harden by evaporation. They are applied with a notched trowel (Fig. 37.12), leaving a thickness of about $\frac{1}{16}$ in. (1.6 mm), and are not used for exterior applications and interior uses exposed to considerable water.

Furan mortar is a mixture of furan resin and a hardener. It has high resistance to chemicals.

Figure 37.11 Ceramic mosaic tile floors can be laid in a variety of colors and geometric patterns. *(Image copyright Kutlayev Dmitry, 2009. Used under license from Shutterstock.com)*

Figure 37.12 Cement mortars for tile products are applied with a notched trowel. Temporary plastic spacers maintain consistent grout widths between tiles. *(Image copyright Titi Matei, 2009. Used under license from Shutterstock.com)*

Tile Grouts

Grouts typically are a mixture of Portland cement, fine sand, and lime. They generally come premixed, requiring the addition of water. Latex, furan, epoxy, or silicone rubber can be added to influence properties. A manufacturer should be consulted to ensure choosing the most desirable grout for a situation.

BRICK FLOORING AND BRICK AND CONCRETE PAVERS

Brick flooring and brick pavers are used to produce durable finish flooring. Brick pavers are thinner than standard brick, ranging from $\frac{1}{2}$ to 2-$\frac{1}{2}$ in. (12 to 57 mm) thick, and are preferred to standard brick for flooring applications. Pavers are also available in a variety of concrete and asphalt units (Fig. 37.13). Glazed brick is also used for finished floors. Brick flooring can be utilized for both interior and exterior floors. Hard-backed types are suitable for heavy use on industrial floors.

Bricks can be laid with either the large flat face exposed or on edge, exposing the narrow face. Many techniques and patterns are used to lay brick pavers (Fig. 37.14). Exterior brick patios are often laid on a bed of sand with no mortar between units. When a floor will be subject to heavy loads, pavers are laid over a concrete slab.

Figure 37.13 Brick pavers provide an attractive, durable finish flooring for both interior and exterior applications. *(a. Image copyright Amy Walters, 2009. Used under license from Shutterstock.com; b. Image copyright Xtuv Photography, 2009. Used under license from Shutterstock.com; c. Image copyright Péter Gudella, 2009. Used under license from Shutterstock.com)*

Figure 37.14 Examples of typical floor details using clay brick, clay pavers, or concrete pavers.

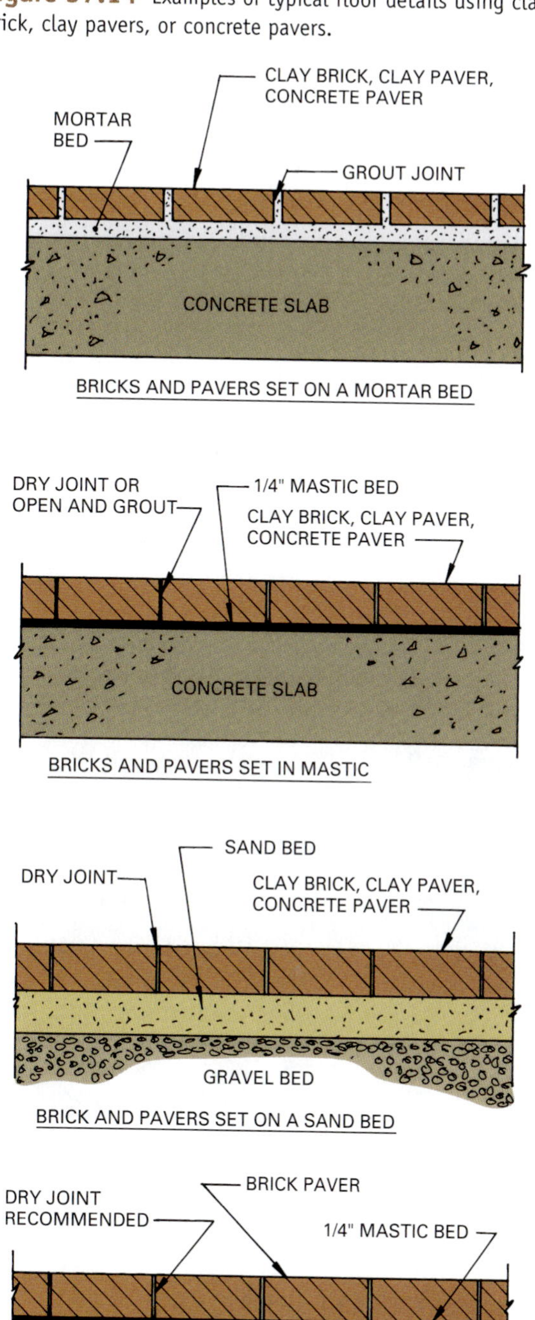

CONCRETE FLOORS

Finished concrete floors are widely used and provide a durable and economic solution to many flooring situations. Properties of concrete and placing techniques are discussed in Chapters 7 and 8.

A concrete floor surface may be finished in several ways. A slab can be poured, leveled, and screeded and its surface smoothed with a wood float. A smoother surface can be produced by smoothing the wood float surface with a steel trowel before the concrete firmly sets. Various textured surfaces can be achieved by techniques such as broom brushing a surface after troweling but before it sets fully. An exposed aggregate finish is made by leveling a slab, embedding aggregates with a float to get a level surface. After the concrete begins to set, it is sprayed with water, leaving the aggregate exposed and forming the finished surface (Fig. 37.15). A high-quality smooth finished surface can be produced by covering the base concrete with a 1 in. (25 mm) thick specially formulated concrete layer. This is applied before the base concrete hardens.

Colored surfaces are produced by adding color pigments, usually metallic oxides, to a concrete mix as a batch is prepared. Another technique involves pouring and finishing a floor, then spreading a dry shake coloring material prepared especially for it. The dry shake is spread evenly over the surface before it hardens, then floated to smooth the surface and even out the color.

Other finishes include painting, staining, applying an abrasive aggregate to a surface to produce a non-slip finish, adding metallic aggregate to a surface to produce a more wear-resistant finish, and the use of various chemical coatings to harden a surface (Fig. 37.16).

Figure 37.15 Exposed aggregate concrete floors provide a durable and attractive finished surface. *(Image copyright LianeM, 2009. Used under license from Shutterstock.com)*

Figure 37.16 Structural concrete slabs can be stained, sealed, and polished to produce an attractive finish floor. *(Image copyright oksana.perkins, 2009. Used under license from Shutterstock.com)*

Heavy duty concrete floors are used in many industrial applications. They are usually constructed as a two-layer floor. The reinforced floor base is poured and topped with a durable concrete layer made with special abrasion-resistant aggregate, such as emery or iron filings.

TERRAZZO

Terrazzo is a matrix consisting of marble or granite chips, Portland cement, and water or a synthetic resin. It is placed as a top layer over a concrete substrate, steel decking, or a wood subfloor.

Portland cement terrazzo is divided into sections by metal or plastic dividing strips, which reduce the possibility of cracking. The three methods for casting Portland cement terrazzo are monolithic, bonded, and sand cushion (Fig. 37.17). Monolithic terrazzo uses a $\frac{5}{8}$ in. (16 mm) thick layer placed as a topping on a green concrete slab.

Bonded terrazzo has a topping 1-$\frac{3}{4}$ in. deep (44.5 mm) or thicker. The cured concrete slab is cleaned and coated with a neat Portland cement, and a concrete underbed of about 1 in. (25 mm) is laid. Divider strips are placed on the underbed and pushed into it. The terrazzo topping is leveled over the underbed.

Sand cushion terrazzo has a $\frac{1}{2}$ in. (12.7 mm) bed of dry sand laid over a concrete slab. The sand is covered with a waterproof membrane, and a concrete underbed is laid over it. Dividing strips are placed on the underbed, and the terrazzo topping is poured and leveled. The sand cushion isolates the slab from the floor slab, protecting it from damage caused by movement of the floor slab.

Synthetic resin matrix terrazzo should be installed as per a resin manufacturer's directions. Synthetic resins are applied in thin coats ranging from $\frac{1}{8}$ to $\frac{1}{4}$ in. (3 to 6 mm) thick. The stone chips used in synthetic resin terrazzo are much smaller than those used in Portland cement terrazzo. Dividing strips are bonded to a wood, metal, or concrete underbed with an adhesive.

After the terrazzo topping has hardened, it is ground smooth and polished. The polishing produces a gloss and brings out the color of the chips. Sometimes a less glossy non-slip surface is ground. A protective sealer is applied over the finished floor (Fig. 37.18).

Figure 37.17 Typical Portland cement terrazzo floor installations over various types of floor decks.

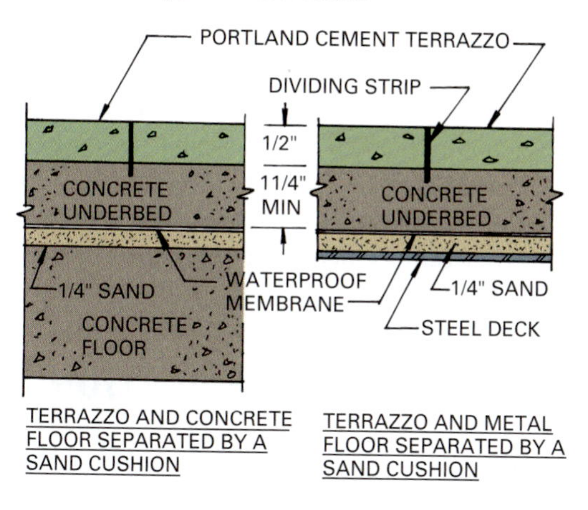

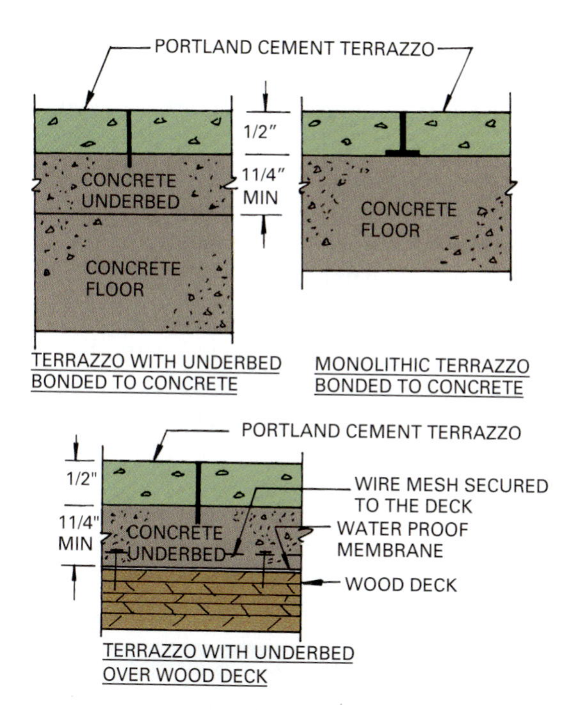

Figure 37.18 A close-up view of a finished terrazzo floor. *(Image copyright cheyennezj, 2009. Used under license from Shutterstock.com)*

Terrazzo precast units are available for a variety of specialty applications. Synthetic resin terrazzo is cast in square and rectangular shapes in thicknesses from $\frac{1}{8}$ to $\frac{1}{4}$ in. (3 to 6 mm). Precast stair treads, risers, window stools, wall base, and other shapes are available precast in Portland cement terrazzo.

STONE FLOOR COVERING

Stone finish flooring is typically slate, marble, or granite (Fig. 37.19). When used on a building's interior, stones are laid over a wood or concrete slab in the same manner as

Figure 37.19 A variety of stone products, including marble, granite, and slate, are used for finished flooring. *(a. Image copyright Andrew Kerr, 2009. Used under license from Shutterstock.com; b. Image copyright Feng Yu, 2009. Used under license from Shutterstock.com)*

Figure 37.20 Various ways stone flooring may be laid for interior and exterior applications.

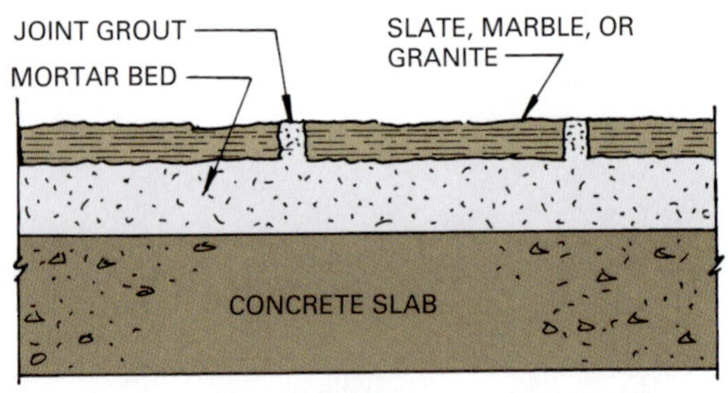

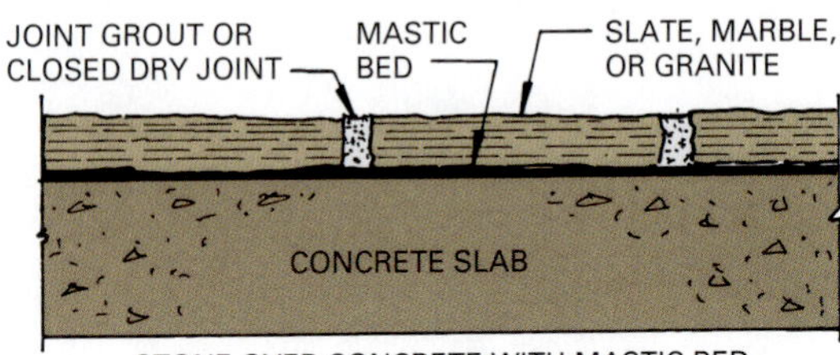

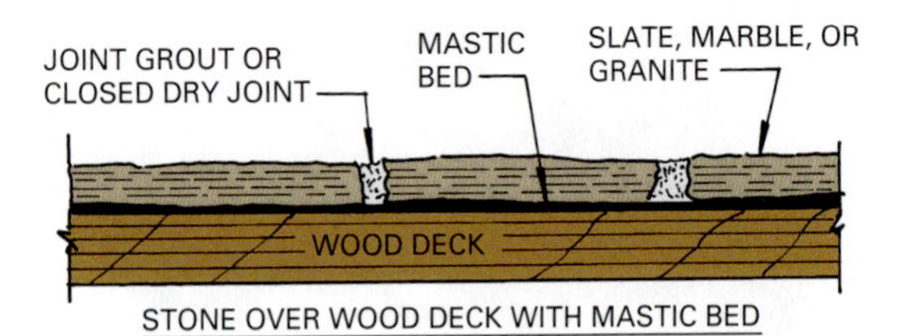

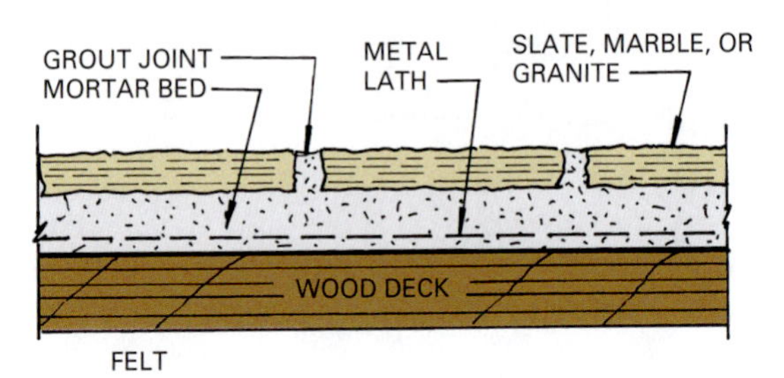

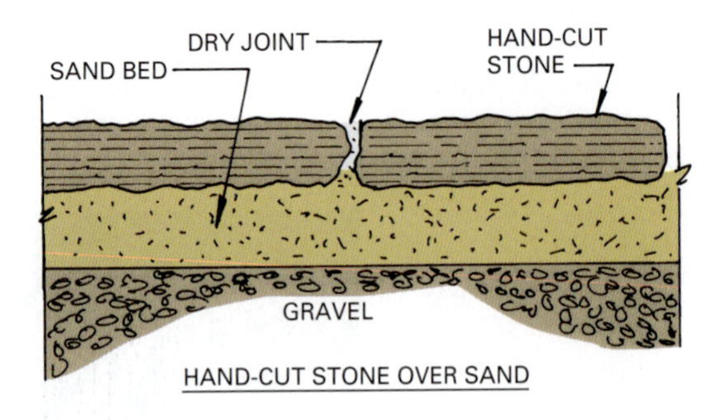

brick pavers. Sand-laid stone is usually used for exterior work, and the joints are left open. Interior stone floors may have joints filled with grout or be set tight without it (Fig. 37.20).

Another stone floor covering material is a cast stone tile flooring formed by impregnating and encapsulating stone aggregate with high-technology resins that create a very dense surface that meets Underwriters Laboratory requirements for slip resistance.

OTHER FLOORING APPLICATIONS

Industrial buildings are generally built with a concrete slab. Finish flooring mainly protects the slab from corrosive materials and the movement of heavy traffic. The concrete can be treated with chemicals to harden its surface (Fig. 37.21) or covered with a flooring, such as wood blocks standing on end such that the end grain forms the surface (Fig. 37.22). Asphalt mastic floors also can be used. An entire floor, or at least certain work areas, may need treatment for slip resistance, which can be incorporated in a concrete top coating (or accomplished via rubber mats).

Some floors must have a low electrical resistance to prevent sparks from static electricity. For example, non-sparking conductive flooring is used in areas where volatile vapors are present. Other flooring

Figure 37.21 This concrete garage floor has been finished with an epoxy primer and a single coat of urethane that cures to produce a durable, weather-resistant coating. *(Image copyright Losevsky Pavel, 2009. Used under license from Shutterstock.com)*

Figure 37.22 End grain wood-block floors are used in industrial areas where heavy wear is expected.

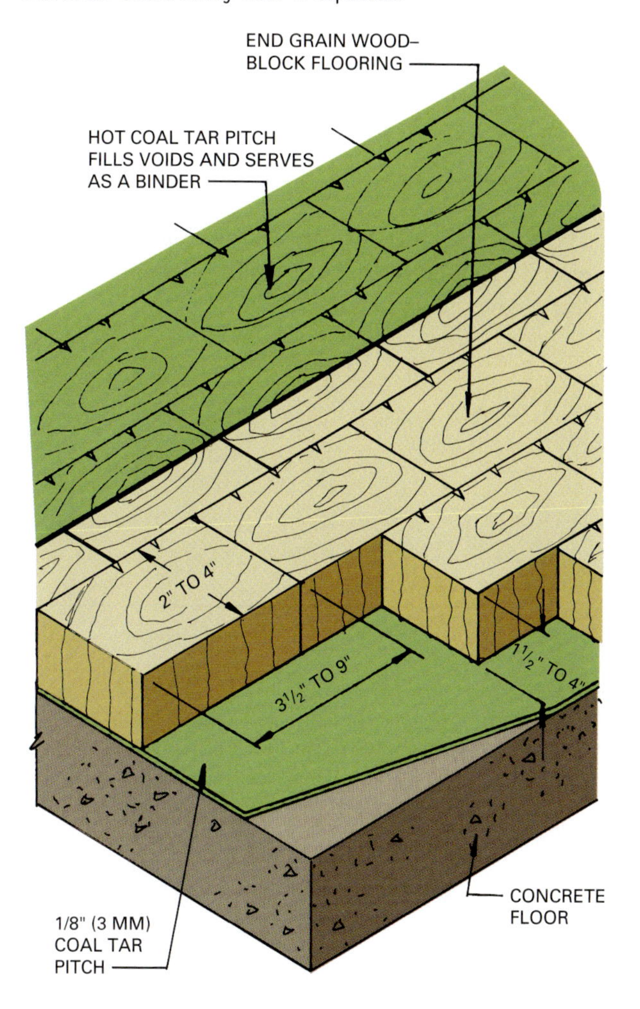

materials must be grease resistant, slip resistant, fire resistant, acid and alkali resistant, static discharging, or x-ray protective.

Various types of poured seamless floor coverings are available. Some are formed from plastic, such as acrylics, latex epoxy, resins, methyl methacrylate, and polyacrylate resins, and are spread over a concrete or wood subfloor. A wide range of flooring for athletic use is available. This includes coatings for showers and saunas; basketball, tennis, and volleyball courts; tracks; and locker rooms. Materials used include elastomers, rubbers, polymers, and urethane products (Fig. 37.23).

Figure 37.23 Typical poured-in-place floors are made from a two-component polyurethane elastomer.

Review Questions

1. What types of finish flooring are in common use?
2. What type of finish flooring presents the greatest hazard in a fire?
3. What techniques are used to secure wood flooring to a substrate?
4. How can a designer find the fire spread and smoke production data for various finish flooring products?
5. What type of floor tile is used for areas subject to heavy use?
6. How is the surface of rubber flooring designed to provide slip resistance?
7. What types of mortars and adhesives are used to install tile floors?
8. Why is grout used to finish the installation of ceramic tile floors?
9. What substrate should be used under brick floors that will be subject to heavy use?
10. How is an exposed aggregate concrete floor surface produced?
11. How are concrete floors given a colored surface?
12. What substrates can be used when constructing a terrazzo finished floor?
13. What are the commonly used types of terrazzo flooring?
14. Where are precast terrazzo products used?
15. What materials are used to form seamless floors?

Key Terms

clay tile

resilient floor covering

terrazzo

Activities

1. Visit floor covering dealers to obtain samples of as many different products as possible. Arrange the samples in a display and label each, citing the brand, material makeup, and places where its use would be appropriate.
2. Visit buildings under construction when floor finishing crews are at work. Observe the tools and procedures used.
3. Have the managers of various flooring supply dealers visit class and explain how they help a client select a covering material.

Additional Resources

Guide for Flooring Covering Installers, Mannington, Mills, Inc., P.O. Box 30, Salem, NJ 08079.

Installing Hardwood Flooring and other publications, National Oak Flooring Manufacturers Association.

Sheet Flooring Installation Guide, Armstrong Floor Products, P.O. Box 3001, Lancaster, PA 17604.

Spence, W. P., *Finish Carpentry*, Sterling Publishing Co., 387 Park Avenue South, New York 10016.

Spence, W. P., *Installing and Finishing Flooring*, Sterling Publishing Co., 387 Park Avenue South, New York 10016.

See Appendix B for addresses of professional and trade organizations and other sources of technical information.

38 CHAPTER

Carpeting

LEARNING OBJECTIVES

Upon completion of this chapter, the student should be able to:

- Understand various types of carpet construction.
- Choose carpet for various applications.
- Understand installation procedures for carpeting.

Carpet is widely used as floor covering in both residential and commercial construction. It competes well with other finish floor products in terms of cost, durability, and maintenance. In addition, it is available in a wide range of fibers, textures, and colors. It is manufactured in widths of 6, 8, 12, 25 ft. and greater, resulting in a material that produces few seams and rapidly covers large floor areas. Although carpeting is often selected for its warmth and comfort, it also provides sound-deadening properties by reducing impact noise and sound reflection within a room (Fig. 38.1).

The use of carpet affects indoor air quality in a variety of ways. All carpet types trap dust, moisture, and pollutants to varying degrees, depending on the depth of the pile, the carpet density, and the type of carpet. Synthetic Fibers are traditionally made from petroleum and can off-gas. Seam sealants, carpet padding, and carpet treatments often contain volatile organic compounds, formaldehyde, and other pollutants. In response to these issues, the Carpet and Rug Institute (CRI) has developed the "Green Label" indoor air quality (IAQ) testing and labeling program for carpet and the adhesives used with them. CRI tests each carpet line four times a year for four categories of emissions. Those that contain no hazards to building occupants are certified by the Green Label.

Over 3 million tons of carpeting is produced every year and about 2 million tons are discarded. Carpet accounts for a considerable amount of municipal solid waste. The carpet industry has begun voluntarily to take responsibility for its products once their useful life is over. Discarded carpet is collected by manufacturers and reused in the production of new carpet and other products.

Figure 38.1 Carpeting provides a soft, beautiful finished floor. *(Image copyright Lein de León Yong, 2009. Used under license from Shutterstock.com)*

PILE FIBERS

Wool, a **natural fiber**, has been a major component in carpet construction for many years. The development of **synthetic fibers**, however, has enabled carpet manufacturers to produce products with a wider range of characteristics. Carpet piles are often a blend of natural and synthetic fibers. For example, a wool carpet may have a blend of 10 to 30 percent nylon fibers to increase its toughness. Acrylic, modacrylic, nylon, polyester, and polypropylene (olefin) are commonly used synthetic fibers.

Wool Fibers

Wool used in carpets is a natural fiber imported into the United States from wool producing countries such as Australia, New Zealand, and Scotland. Different breeds of sheep make for a widely varied product. Wool fibers range in size from $3\frac{1}{2}$ to 7 in. (89 to 178 mm) long. The wool pile is made by spinning the fibers into yarn. Wool has good resistance to abrasion and aging, resists damage from mildew and sunlight, and is relatively unaffected by weak alkalis and organic solvents. Generally, wool is more expensive than synthetic fibers.

Acrylic Fibers

Synthetic acrylic fiber is similar to wool in texture and abrasion resistance. It is composed of more than 85 percent acrylonitrile by weight. Manufactured in a wide range of colors, it has good resistance to mildew, aging, sunlight, moths, and chemicals.

Modacrylic Fibers

This synthetic fiber is a member of the acrylic group. It contains at least 35 percent (but less than 85 percent) acrylonitrile. It is soft, resilient, quick drying, abrasion- and flame-resistant, and resists damage from acids, alkalis, sunlight, and mildew.

Nylon Fibers

Nylon is a synthetic petrochemical product produced as a continuous filament but often cut into short lengths, varying from $1\frac{1}{2}$ to 6 in. (38 to 152 mm). The short fibers are spun into yarn much the same as wool fibers. Nylon is a strong fiber that has anti-stain properties and is easily dyed. It is resilient, low in moisture absorbency, and resistant to mildew, aging, and abrasion.

Polyester Fibers

Polyester fibers are synthetic fibers that have high tensile strength, good resistance to mineral acids, and excellent mildew, aging, and abrasion resistance. Prolonged exposure to sunlight may cause some loss of strength. They are not as durable as nylon fibers.

Polypropylene Fibers (Olefin)

Polypropylene is a synthetic fiber that is a class of olefin consisting of 85 percent propylene by weight. It has the lowest moisture absorption rate of all carpet fibers in use. It resists mildew, abrasion, aging, sunlight, and common solvents. While not as resilient as nylon, it is less expensive. Olefin is very lightweight, strong, and soil resistant.

CARPET CONSTRUCTION

Although the construction of carpets varies somewhat, Figure 38.2 illustrates the terms commonly used to describe their makeup. Pile yarns form the top surface that is exposed to wear and abrasion. Pile yarns are made from wool or one or more synthetic fibers. Backing yarns consist of weft yarns (running the width of the carpet) and warp yarns (running the length of the carpet). Stuffer yarns run lengthwise and give strength and stability to the carpet. The warp yarns pass over the weft yarns, pulling the pile yarns into place.

WOVEN CARPETS

Woven carpets include Axminster, loomed, velvet, and Wilton construction.

Axminster Construction

Axminster carpet is woven on a loom that inserts each tuft of pile yarn individually, allowing the color of each tuft to vary. This provides great flexibility for design patterns. Axminster is usually a cut pile even in

Figure 38.2 Terms used to identify the parts of carpeting.

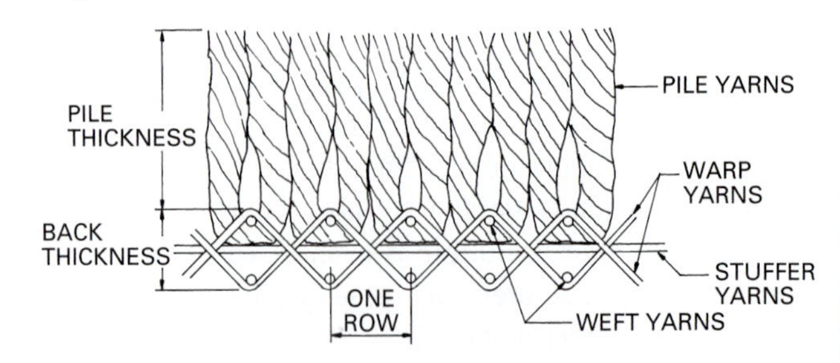

height. The backing is heavily ridged and sized, making it so stiff that it must be rolled lengthwise. A typical construction detail is shown in Fig. 38.3.

Loomed Construction

Loomed carpet has the pile yarn bonded to a $\frac{3}{16}$ in. (5 mm) thick rubber cushion in a low-loop single-level pile. The pile back has a waterproof coating to which a rubber cushion is bonded. A typical construction detail is shown in Fig. 38.4.

Velvet Construction

A **velvet** weave is shown in Fig. 38.5. The pile, stuffer, and weft yarns are held together with double warp yarns. The setting of the loom establishes the height of the piles, which begin as loops before being cut by a knife-edged wire to a desired height. A textured surface can be produced by setting wires to form and cut loops of different heights. Patterned designs cannot be produced by this loom, but tweed effects can be created by varying yarn colors.

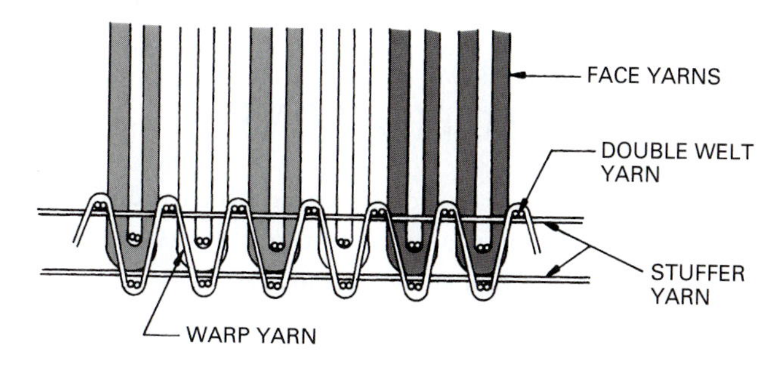

Figure 38.3 Construction details of Axminster carpet.

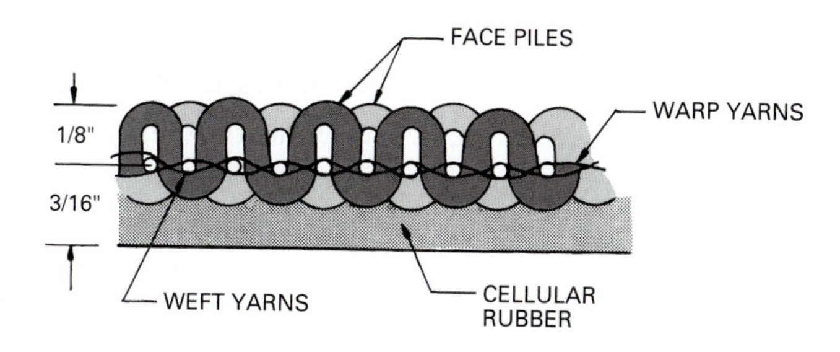

Figure 38.4 Loomed carpet construction bonds the pile yarn to a rubber cushion.

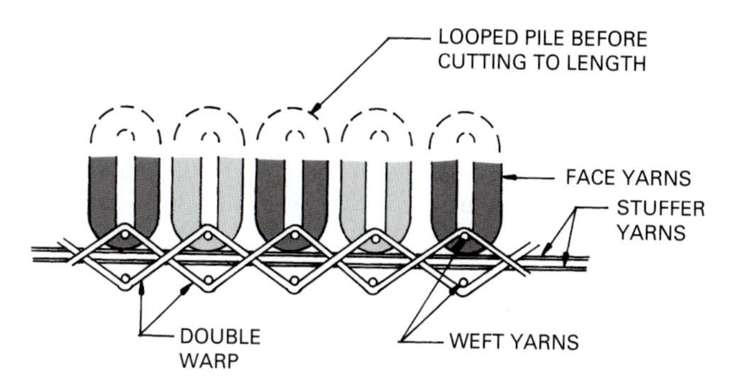

Figure 38.5 Velvet construction has double warp yarns.

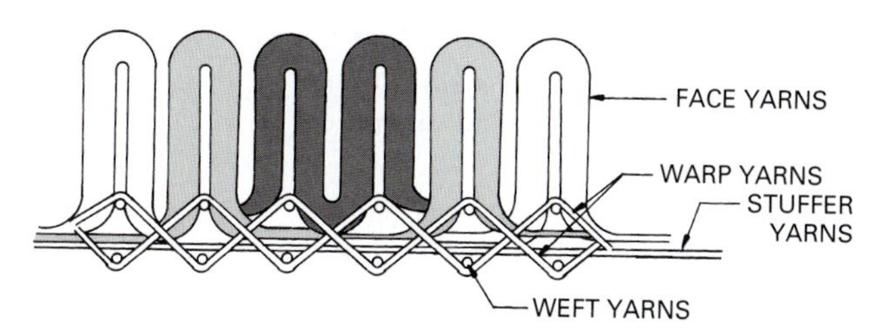

Figure 38.6 Wilton construction is used to produce carpets containing multicolored yarns.

Wilton Construction

Wilton construction is produced on a loom similar to the one used for velvet weaves, but a Wilton loom has a mechanism that feeds yarns of various colors. Different color yarns are selected by a series of punched cards. The cards regulate which color yarn is looped over the loop-forming wires. These become the visible pile tufts, and the other colored yarns are buried in the body of the carpet, producing a stiff-backed, multicolored product (Fig. 38.6).

Wilton looms can produce sculptured and embossed textures by varying the pile height and effecting a combination of cut and uncut loops. Uncut loop Wilton weaves are also available.

TUFTED CONSTRUCTION

Tufted construction involves stitching pile yarn through a backing material, much like sewing on a power sewing machine. The back is coated with a latex bonding agent that holds the tufts to the base and bonds a second backing over it. The first backing is usually made of cotton, polypropylene olefin, or jute. The second backing is usually a coarse jute fabric (Fig. 38.7). Tufted carpets are available in level loop, multilevel, cut, uncut, and a mixture of cut and uncut piles. Carved and textured surfaces can also be produced.

KNITTED CONSTRUCTION

Knitted carpet is made by looping together the pile yarn, stitching, and backing in a single operation. A coat of latex spread on the back bonds it all together and stiffens the carpet (Fig. 38.8). Knitted carpets are available in solid colors and tweeds and in single or multilevel uncut pile.

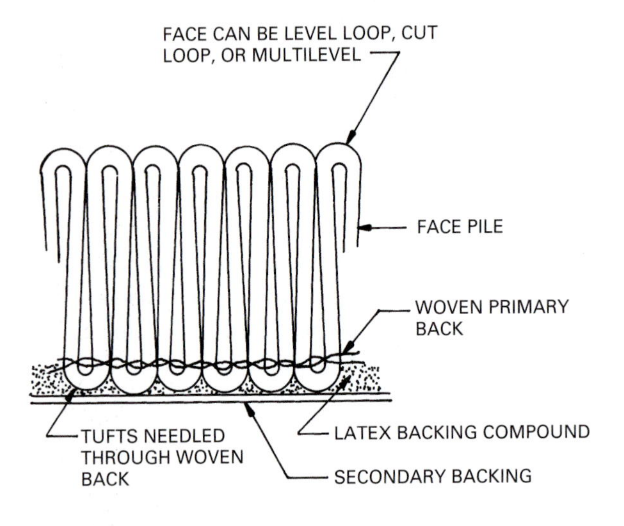

Figure 38.7 Tufted carpet is made by stitching pile yarns through a backing material and bonding two layers of backing over it.

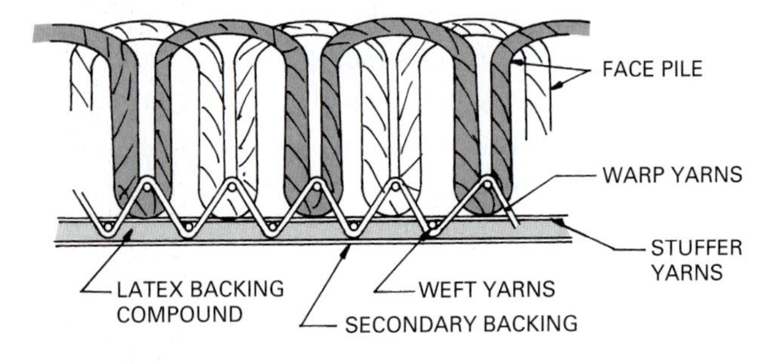

Figure 38.8 Knitted carpet construction involves looping the pile yarns, stitching, and backing in a single operation.

FLOCKED CONSTRUCTION

Flocked construction involves electrostatically spraying short strands of pile yarn onto an adhesive-coated backing sheet. The pile strands become vertically imbedded in the adhesive, a second backing is laminated to the first, and the unit is cured (Fig. 38.9).

FUSION BONDED CONSTRUCTION

Fusion bonded carpet is made by bonding pile yarn between two parallel sheets of backing. These are coated with vinyl adhesive, making a sandwich product with the pile yarn in the center. After the adhesive hardens, the pile yarn is cut in the middle to form two pieces of cut pile carpet. A secondary backing is often bonded to the first backing sheet (Fig. 38.10).

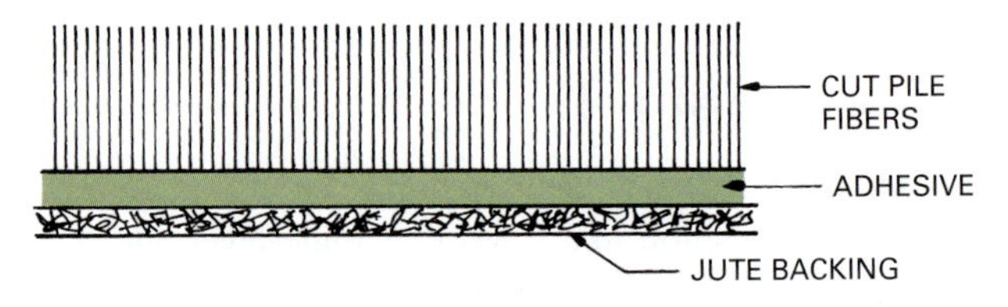

Figure 38.9 Flocked carpet construction involves electrostatically spraying short strands of pile yarn onto an adhesive-coated backing.

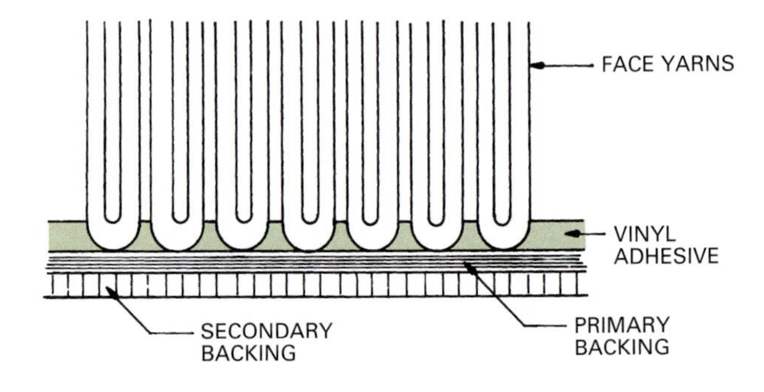

Figure 38.10 Fusion bonded carpet construction involves bonding pile yarn to backing sheets coated with a vinyl adhesive.

QUALITY SPECIFICATIONS

Carpet is made in various qualities suitable for different locations. Following is a discussion of the basic parameters to consider when choosing carpet.

The Federal Housing Administration carpet quality specifications include pile yarn weight, pile density, and pile thickness. ASTM tests indicate how to measure these qualities. The General Services Administration has also issued carpet specifications.

Pile yarn weight (W) is the average weight of pile yarn in ounces per square yard, excluding the backing.

Pile thickness (t) is the height of pile tufts above the backing in inches.

Pile density (D) is the weight of pile yarn per unit of volume in the carpet, stated in ounces per cubic yard. It is 36 times the finished pile weight divided by the average pile thickness:

$$D = 36W/t = \text{oz./cu.yd.}$$

The minimum acceptable pile density is 2,800 oz./cu.yd.

Minimum requirements for pile weight-density (WD) and pile yarn weight (W) are given in Table 38.1. Pile weight density equals pile weight (W) multiplied by the minimum pile density, 2,800 oz./cu.yd.

Carpet can be specified by the number of tufts per square inch; rows, wires, and stitches per linear inch; and pitch, gauge, and number of needles. Needles

Table 38.1 Minimum Physical Requirements for Carpeting in Moderate Traffic Areas[a]

Pile Fiber	Minimum Pile Yarn Weight (W) oz./yd.^{2}	Minimum Weight-Density Factor (WD)[b]
Acrylic	25	70,000 (25 × 2,800)
Modacrylic		
Wool		
Nylon (staple and filament)	20	56,000 (20 × 2,800)
Polypropylene (olefin)		

[a]*Based on FHA Use of Materials Bulletin UM-44A*
[b]*Pile density (D) = 2,800*

describe the spacing of the tufts. Yarn size and the number of piles can also be specified. Piles are the number of strands twisted together to form the pile yarn.

Tufts per square inch is an indication of the overall density of a carpet. The greater the number of tufts per square inch, the denser the carpet.

Rows are a way of indicating the number of tufts per inch. Wiltons and velvets are described by the number of wires per inch (Fig. 38.11). This spacing of tufts indicates the pile density lengthwise. The more rows or wires per linear inch, the denser the carpet. Since tufted carpets have no weft yarns, their density is indicated by the number of tufts per inch in the backing. These are referred to as stitches per inch.

Pitch, gauge, and needles are used to describe the number of tufts across the width of a carpet. Pitch is the number of tufts in a 27 in. (686 mm) width of carpet. The higher the pitch number, the tighter the weave, and the denser the carpet (Fig. 38.12). When no continuous warp exists, the term "gauge" is used to indicate the spacing of tufts across a width of carpet in inches. For example, "$\frac{1}{8}$ gauge" means 8 tufts per inch, and "$\frac{2}{16}$ gauge" means 16 tufts per 2 inches. The term "needles" is sometimes used instead of gauge to indicate number of tufts per inch.

Physical specifications for carpeting in moderate use areas are given in Table 38.1. Physical specifications by the GSA for heavy traffic areas are listed in Table 38.2.

Building codes specify flammability requirements for carpets in various locations and occupancies. The flame resistance of carpets is most commonly ascertained by the Flooring Radiant Panel Test. A carpet specimen is mounted on the floor of a test chamber and exposed to intense radiant heat from above, and the rate of flame spread is measured.

Figure 38.11 Rows are used to identify number of tufts per inch.

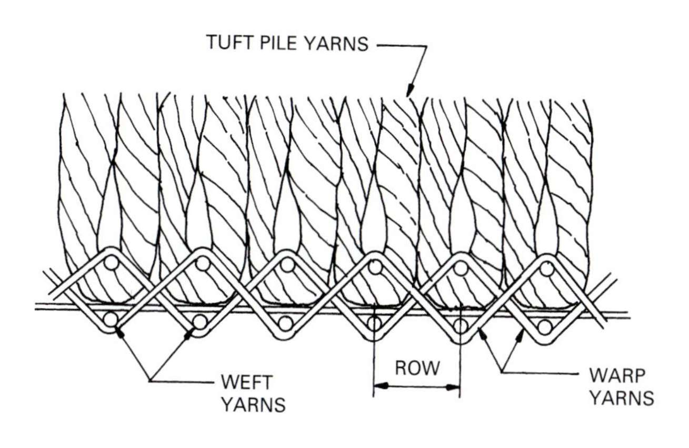

Figure 38.12 Pitch indicates number of tufts in a 27 in. (686 mm) width of carpet.

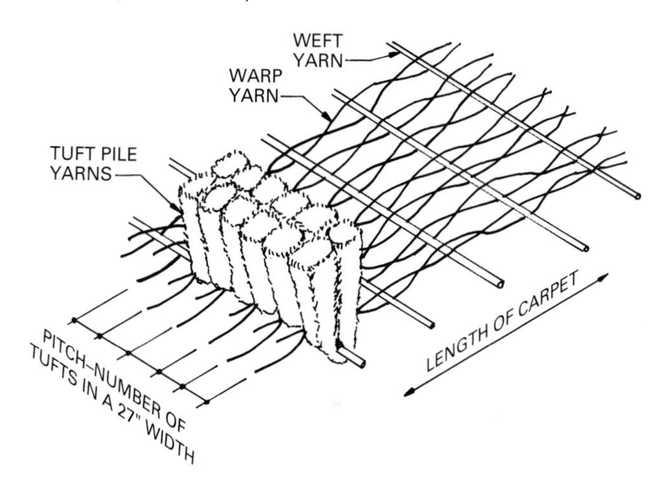

CARPET CUSHIONS

Carpet is usually installed over a **cushion** because this increases its life, absorbs considerable traffic noise, has insulation value, and provides a soft, resilient floor covering. There are two common types of carpet cushion installations: carpet can be installed over a separate cushion or carpet can be made with a foam cushion attached to its back. Carpet can also be laid without a cushion (Fig. 38.13).

Carpet cushions are available in three basic materials: urethane, fiber, and rubber (Table 38.3). They are manufactured in two types: Class 1 (light and moderate traffic) and Class 2 (heavy-duty traffic). These recommendations meet U.S. government requirements for cushion used in FHA-financed housing. Rubber cushion is more expensive, but it lasts longer, resists mildew and mold, and is non-allergenic.

Table 38.2 Minimum Physical Requirements for Carpeting in Heavy-Traffic Areas[a]

	Velvet					
	Wilton Wool or Acrylic	Wool or Acrylic	Nylon	Wool or Acrylic	Tuffed Wool or Acrylic	Knitted Wool or Acrylic
Pile Description	Single Level Loop Woven Thru Back	Single Level Loop Pile Woven Thru Back	Single Level Loop Pile Woven Thru Back	Multilevel Loop Woven Thru Back	Single Level Loop	Single Level Loop
Tufts/sq.in.	52	60	32	34	60	26
Frames	3	–	–	–	–	–
Shots/wire	2	2	2	2	–	–
Weight (oz./sq.yd.) Pile	48	42	29	44	42	37
Total	72	60	50	64	–	58
Pile height	Min. 0.250 Max 0.320	Min. 0.210 Max. 0.310	Min. 0.200 Max. 0.290	Min. 0.190 Max. 0.370	Min. 0.250 Max. 0.320	Min. 0.230 Max. 0.290
Chain	Cotton and/ or rayon	Cotton and/ or rayon	Cotton and/ or rayon	Cotton and/ or rayon	–	Cotton, rayon or nylon
Filling	Cotton and/or rayon, or jute	Cotton and/or rayon, or jute	Cotton and/or rayon, or jute	Cotton and/or rayon, or jute	–	Jute or kraftcord
Stuffer	Cotton, jute or kraftcord	Cotton, jute or kraftcord	Cotton, jute or kraftcord	Cotton, jute or kraftcord	–	–
Back coating (oz./sq.yd.)	–	8	6	8	–	–
Tuft bind (oz.)	80	80	80	80	100	14
Ply twist turns (per inch)	Min. 1.5 Max. 3.5	Min. 1.5 Max. 3.5	Min. 1.0 Max. 3.0	Min. 1.5 Max. 3.5	Min. 2.5 Max. 4.5	Min. 1.5 Max. 3.5

[a]*Based on General Services Administrative Federal Specifications DD-C-95.*

Figure 38.13 Types of carpet cushions. *(Courtesy Cushion Carpet Council)*

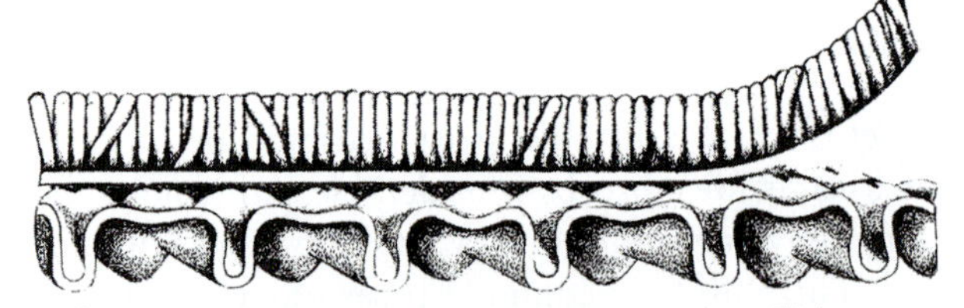

1. Carpet installed over separate cushion. This can be done by stretching the carpet in over tack strip, or by glueing the cushion to the floor and the carpet to the cushion.

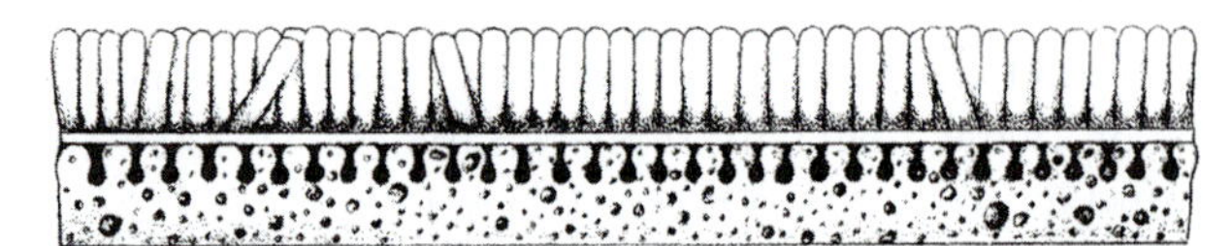

2. Carpet with attached cushion.

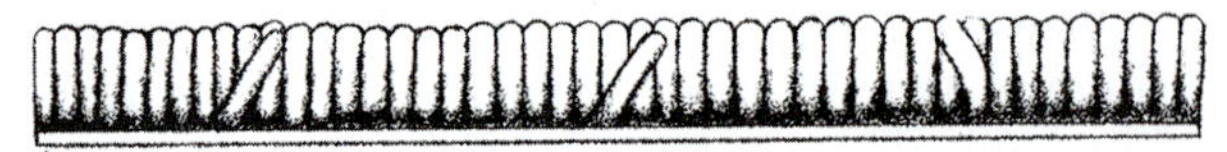

3. Direct glue down: carpet cemented directly to the floor without any cushion.

INSTALLING CUSHION AND CARPET

Carpet is installed over concrete, plywood, particleboard, and other firm, smooth substrates in several ways. The most frequently used method uses tackless strips. These are thin wood strips with tack points sticking through them. Metal strips with hooked points are also used. They are nailed to a particleboard, plywood, or concrete subfloor (Fig. 38.14). The carpet is laid over the cushion (Fig. 38.15), and any seams are secured.

Seams between sections of carpet are bonded with a hot melt tape (Fig. 38.16) applied beneath the seam. An electric seaming iron is placed below the carpet on one edge of the tape to melt the adhesive. This edge of the carpet is pressed into the hot adhesive. Then the other edge of the tape is heated and the butting carpet pressed against the tape (Fig. 38.17). The seam is then pressed with a roller, closing the gap and firming the bond.

After the seam adhesive hardens, the carpet is stretched against one wall. A knee kicker (Fig. 38.18) is used to stretch the carpet into place, after which it

Table 38.3 Minimum Cushion Recommendations for Residential Applications

Type	Key Characteristics	Class 1	Class 2
Urethane			
A. Prime	Density lbs./cu. ft. min. Thickness, in. min.	2.2 0.375	Not Recommended for class 2
B. Grafted prime[a]	Density lbs./cu. ft. min. Thickness, in. min.	2.7 0.375	2.7 0.25
C. Densified prime	Density lbs./cu. ft. min. Thickness, in. min.	2.2 0.313	2.7 0.25
D. Bonded[b]	Density lbs./cu. ft. min. Thickness, in. min.	5.0 0.375	6.5 0.375
E. Mechanically frothed[a]	Density lbs./cu. ft. min. Thickness, in. min.	10.0 0.250	12.0 0.250
Fiber			
A. Rubberized hair jute	Weight, oz./sq. yd. min. Thickness, in. min. Density lbs./cu. ft. min.	40.0 0.27 12.3	50.0 0.375 11.1
B. Rubberized jute	Weight, oz./sq. yd. min. Thickness, in. min. Density lbs./cu. ft. min.	32.0 0.3125 8.5	40.0 0.375 8.9
C. Synthetic fibers[a]	Weight, oz./sq. yd. min. Thickness, in. min. Density lbs./cu. ft. min.	22.0 0.250 6.5	28.0 0.300 6.5
D. Resinated recycled textile fiber[a]	Weight, oz./sq. yd. min. Thickness, in. min. Density lbs./cu. ft. min.	24.0 0.25 7.3	30.0 0.30 7.3
Rubber			
A. Flat rubber	Weight, oz./sq. yd. min. Thickness, in. min. Density lbs./cu. ft. min.	56.0 0.22 18.0	64.0 0.22 21.0
B. Rippled rubber	Weight, oz./sq. yd. min. Thickness, in. min. Density lbs./cu. ft. min.	48.0 0.285 14.0	64.0 0.330 16.0

Maximum thickness for any product is 0.5 in.

Class 1: *Light and moderate traffic (such as living rooms, dining rooms, bedrooms, recreational rooms, and corridors. Class 2 cushion may be used in Class 1 applications.)*

Class 2: *Heavy duty traffic (for heavy traffic use at all levels, but specifically for public areas such as lobbies and corridors in multi-family facilities. The Carpet Cushion Council also recommends Class 2 for stairs and hallways.)*

[a]*The Carpet Cushion Council has endorsed HUD UM72 as its recommended minimum standards. Products noted above are Carpet Cushion Council recommendations for inclusion in UM72a, HUD's proposed revision of UM72.*

The UM72 standards include other technical characteristics for each type of carpet cushion. Please contact the Carpet Cushion Council for further details.

[b]*Bonded 8 lb. density can qualify for Class 2 with minimum thickness of .25–5%.*

This chart illustrates products that meet ***minimum*** *guidelines. Better grades of carpet cushion than the minimum suggested arc always recommended when possible to provide more support and cushion for carpet. In areas where heavy use is expected, the Carpet Cushion Council suggests using firmer grades of cushion. These areas include stairways, halls, and areas where heavy furniture is used (such as living rooms and dining rooms). Softer cushion may be used in bedrooms and lounge areas where use is lighter and a plusher "feel" is desired.*

Figure 38.14 Some carpet is installed using wood tackless strips around the edge of a room to hold the carpet tightly in place.

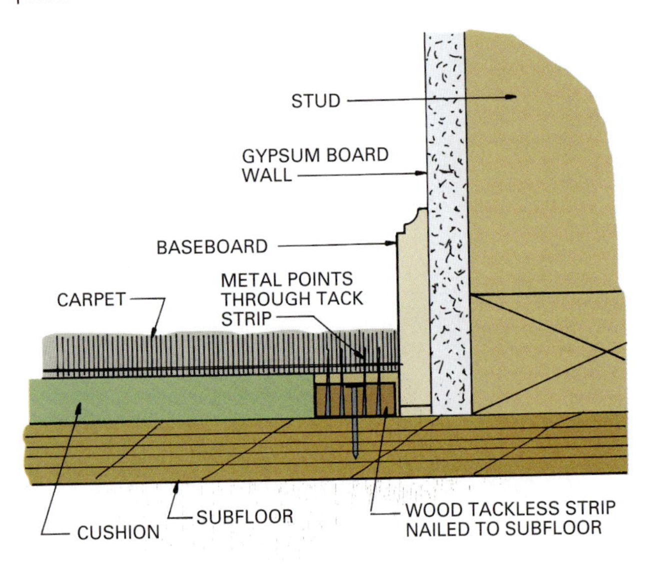

Figure 38.15 A carpet cushion provides an underlayment for the finish carpet. *(© John Madere/CORBIS)*

Figure 38.16 Hot melt tape is used to join butting edges of carpet and hold seams tightly closed.

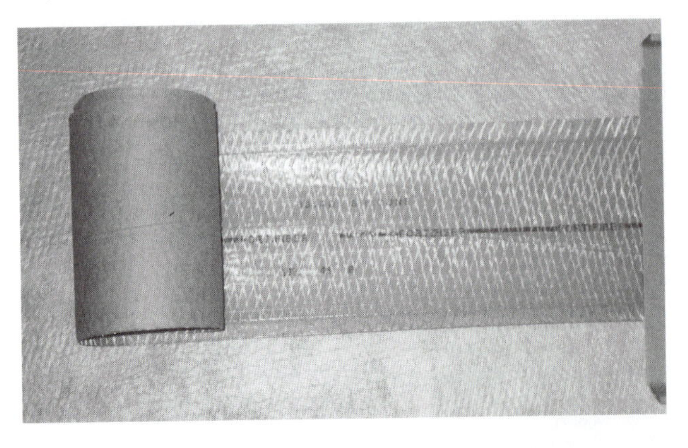

Figure 38.17 After the first edge of a seam is bonded to the tape, the tape under the butting carpet is heated, and the carpet is pressed against the tape.

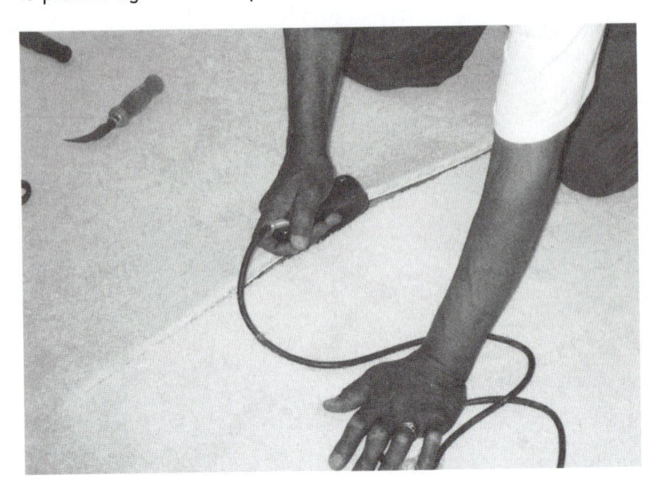

Figure 38.18 The knee kicker pad is placed about one inch from the tackless strip. The pins are pressed into the carpet and the kicker is driven with blows by the installer's knee.

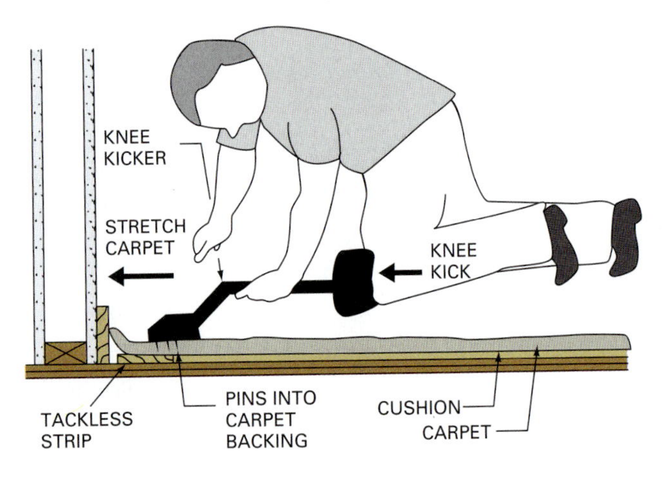

is pressed on to the tackless strip. Then the carpet is stretched toward the other side of the room, as shown in Fig. 38.19. At the wall edge, the carpet is pressed into the space between the baseboard and the tackless strip with a stair tool (Fig. 38.20).

Carpet can be glued directly to a wood or concrete floor without a cushion. But if placed in a frequently used area, un-cushioned carpet will wear out faster. When placing carpet over concrete, the base must be absolutely dry and not sweat with changes in temperature. The application of a coat of sealer to the concrete is recommended.

A plywood subfloor can be installed directly over concrete or on wood sleepers. Sleepers provide an air space that can be ventilated along the walls.

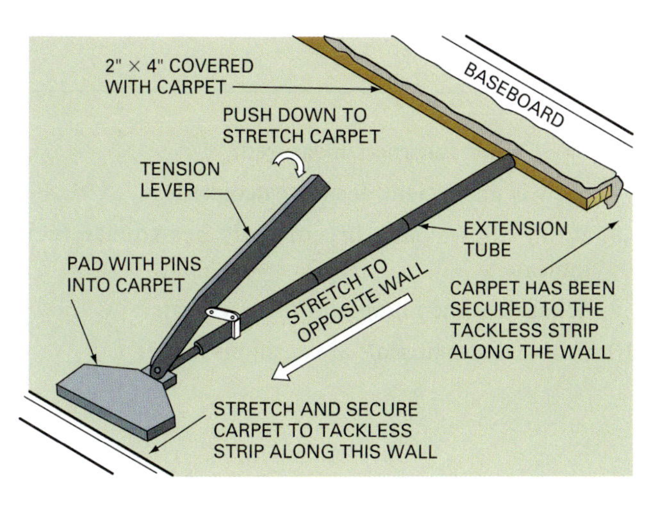

Figure 38.19 A stretcher tool stretches the carpet to the opposite wall, where it is pressed into the pins of the tackless strip.

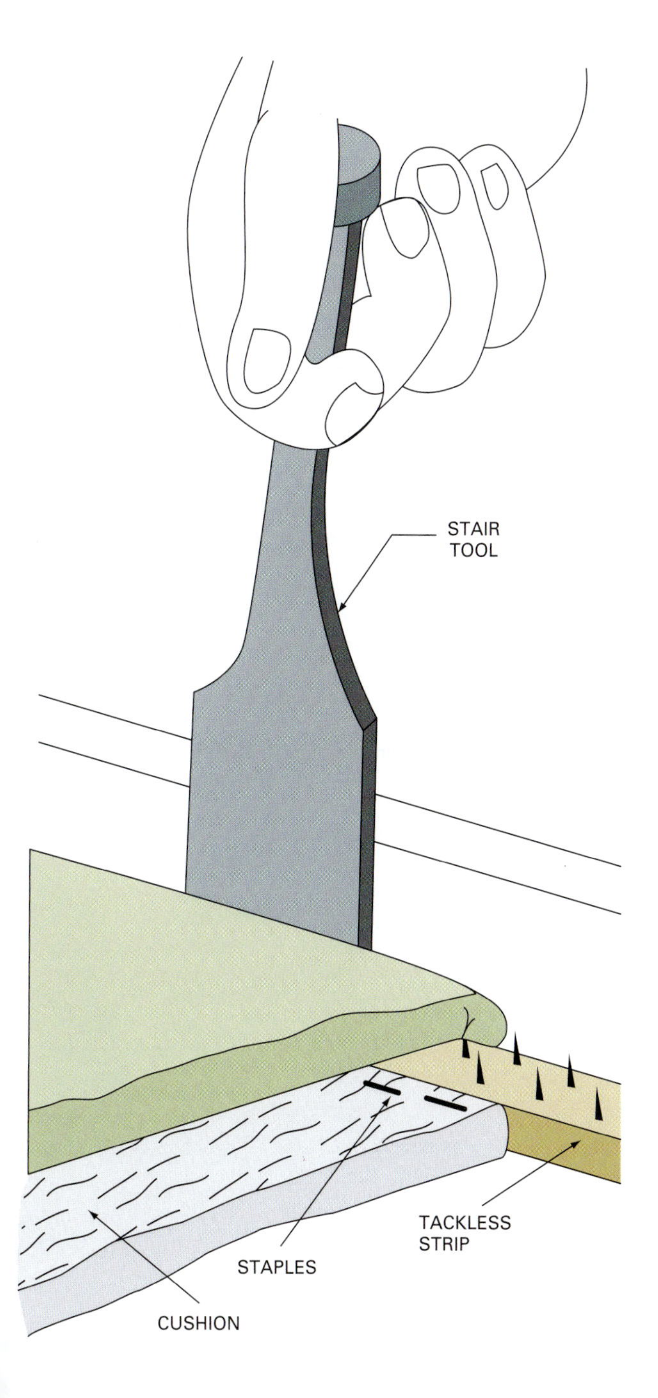

Figure 38.20 A stair tool is used to press carpet at the edge of a wall into the space between the baseboard and the tackless strip. *(Image copyright Thorsten Rust, 2009. Used under license from Shutterstock.com)*

Review Questions

1. In what widths is carpet generally available?
2. What is the major natural fiber used in carpet construction?
3. What are the commonly used synthetic fibers used in carpet construction?
4. What are the major types of carpet construction?
5. What is meant by pile yarn weight?
6. How is pile thickness measured?
7. How is pile weight density calculated?
8. What does a measure of tufts per square inch indicate?
9. What does the number of rows indicate?
10. From what materials are cushions made?

Key Terms

Axminster construction
cushion
flocked construction
fusion bonded carpet
knitted carpet
loomed carpet
natural fibers
pitch
rows
synthetic fibers
tufted construction
velvet construction
Wilton construction

Activities

1. Collect carpet samples from manufacturers and local carpet retail outlets. Arrange a display and label the samples, detailing the fibers, methods of construction, and suitable applications.
2. Review the local building code pertaining to carpeting and prepare a brief summary.

Additional Resources

Spence, W., *Installing and Finishing Flooring*, Sterling Publishing Co., 387 Park Avenue South, New York 10016.

Standard for Installation of Commercial Carpet, and many other publications, the Carpet and Rug Institute, Dalton, GA. Carpet and Rug Institute (*http://www.carpet-rug.com*)

Other resources include:

Publications of the Carpet Cushion Council, Riverside, CT 06878.

See Appendix B for addresses of professional and trade organizations and other sources of technical information.

DIVISION

10

Specialties

CSI MasterFormat™

101000 Information Specialties
102000 Interior Specialties
103000 Fireplaces and Stoves
104000 Safety Specialties
105000 Storage Specialties
107000 Exterior Specialties
108000 Other Specialties

(Image copyright Roger Bruce, 2009. Used under license from Shutterstock.com)

CHAPTER 39

Specialties

LEARNING OBJECTIVES

Upon completion of this chapter, the student should be able to:

- Identify some specialty products as classified by the Construction Specifications Institute.
- Make informed decisions when deciding on and specifying specialty items.

Commercial and industrial buildings have widely varying requirements when it comes to items needed to prepare a facility for use by occupants. Products listed as specialties encompass a vast spectrum of items used to provide a limited special function. Following are brief descriptions of some of these. The entire CSI outline of specialties is given on the page introducing Division 10 of the MasterFormat.

VISUAL DISPLAY BOARDS

Chalkboards are available with the exposed surface made from slate, porcelain enamel on steel, aluminum, or glass (Fig. 39.1). The glass units use tempered float or plate glass coated with a colored, vitreous glaze that contains a fine abrasive. Chalkboards may be permanently mounted on a wall or they can be portable. Some have panels that slide horizontally or vertically. They are available in sizes up to 6 × 12 ft. (1,830 × 3,660 mm). Natural slate chalkboards are typically from $\frac{1}{4}$ to $\frac{3}{8}$ in. (6 to 10 mm) thick, in widths up to 4 ft. (1,220 mm), and in random lengths up to 6 ft. (1,830 mm). Other chalkboards are available in widths up to 4 ft. (1,220 mm) and lengths up to 10 ft. (3,050 mm).

Marker boards may have a vinyl sheet surface or one of porcelain enamel on steel. The substrate can be particleboard, fiberboard, Sheetrock, hardboard, plywood, or a plastic honeycomb. Special ink markers are used to write on a board's surface. Some are a form of watercolor and can be wiped off the marker board with a wet cloth. Another type has semi-permanent ink and requires a special liquid cleaner.

Figure 39.1 A chalkboard is one of the more common visual display devices. *(Image copyright Ronny Lambotte, 2009. Used under license from Shutterstock.com)*

Pegboard panels are usually $\frac{1}{4}$ in. (6 mm) thick hardboard with holes spaced 1 in. (25 mm) on center. They are typically manufactured with an aluminum frame and special wire hanging clips.

Bulletin boards, or **tackboards**, typically have a cork surface laminated to a fiberboard substrate. Some have a woven fabric facing and can hold paper with pins or hook-and-loop fasteners that attach to the fabric. Various types of manual and motorized projection screens are available. They may be integrated into a larger unit with chalkboards, tack boards, and marker boards.

COMPARTMENTS AND CUBICLES

These products include enclosures, such as metal, plastic laminate, and stone toilet compartments; shower and dressing **compartments**; and various **cubicles**. Also included are dividers, screens and curtains for showers, hospital cubicles, toilets, and urinals. Toilet enclosures may be supported by a floor, ceiling, and wall (Fig. 39.2). Toilet enclosures hung from a ceiling make floor cleaning easier. Ceiling-hung units require a structural steel ceiling support.

Metal toilet compartments may be stainless steel or steel with a baked enamel finish. Plastic laminate compartments have a $\frac{1}{8}$ in. (1.6 mm) plastic laminate sheet adhered to particleboard or high-pressure plastic laminate core. Stone compartments are typically made of marble.

Cubicle enclosures are hung from an overhead track whose track path can be varied to suit the conditions. It can be mounted directly to a ceiling or suspended.

Figure 39.2 Toilet enclosures may be supported by a ceiling, wall, and floor. *(Image copyright R, 2009. Used under license from Shutterstock.com)*

LOUVERS, VENTS, GRILLES, AND SCREENS

A variety of **grilles**, screens, louvers, and vents are considered specialty items. These include louvers and other devices for ventilation that are not an integral part of a building's mechanical system, as well as operable, stationary, and motorized metal wall louvers; louvered equipment enclosures; door louvers; and various types of wall and soffit vents. For example, a curtain wall louver is a specialty product. Interior and exterior grilles and screens of any material are used for air distribution, sun screen, and other functions in addition to ventilation purposes (Fig. 39.3).

Grilles are open gratings or barriers used to cover, conceal, protect, or decorate an opening. They can be steel, aluminum, cast iron, or wood (Fig. 39.4). Various types of grating are used to cover trench frames that control surface water flow. These are typically cast iron (Fig. 39.5).

Figure 39.3 These wooden louvers are part of the building's architecture and are used to provide sun shading. *(Image copyright Arunkumar. T.D, 2009. Used under license from Shutterstock.com)*

Figure 39.4 These cast metal grilles provide security and are a major architectural feature. *(Image copyright Miledy, 2009. Used under license from Shutterstock.com)*

Figure 39.5 This cast-iron trench drain is used to control surface water. *(Image copyright M. Ali Khan, 2009. Used under license from Shutterstock.com)*

Street and parking lot construction must be designed to carry away surface water from rain and melting snow. Curb inlets are used as intakes for this purpose. Tree grates are another form of grating used both inside and outside buildings. They provide protection while permitting a tree to receive water (Fig. 39.6).

SERVICE WALL SYSTEMS

Service wall systems incorporate services such as clocks, fire hoses, drinking fountains, fire extinguisher cabinets, telephones, and waste receptacles (Fig. 39.7).

Figure 39.6 A cast-iron tree guard protects tree roots from surface damage yet permits rain to reach them. *(Image copyright Darla Hallmark, 2009. Used under license from Shutterstock.com)*

Figure 39.7 Drinking fountains are a part of the service wall system. *(Image copyright Adriana Muzyliwsky, 2009. Used under license from Shutterstock.com)*

WALL AND CORNER GUARDS

Wall and corner guards are protective devices made from metal, plastic, rubber, and other materials that can withstand impacts and, for exterior use, exposure to the weather. Figure 39.8 shows cast iron wheel guards used to protect the sides of a large opening. Rubber bumpers are secured to walls, at a truck loading dock, for example, to protect them and trucks from damage. Other types of guards, typically constructed of molded rubber or plastic, are used to protect wall corners that might be damaged by moving carts, forklifts, or other passing traffic.

Figure 39.8 These cast iron wheel guards will withstand severe impact and abrasion as they protect the sides of garage door openings. *(Courtesy Neenah Foundry Company)*

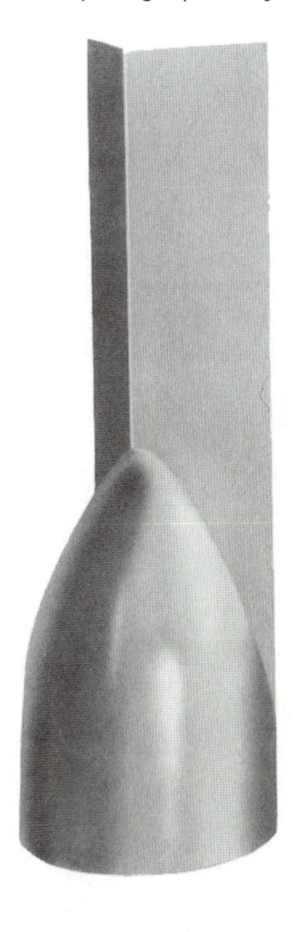

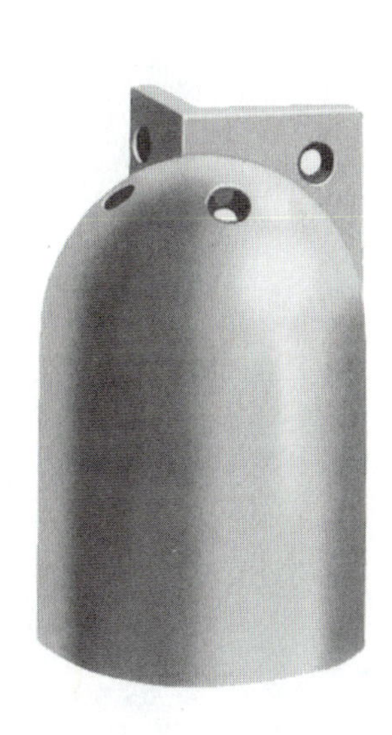

ACCESS FLOORING

Access flooring is freestanding flooring made in modular units and raised above a basic floor system to provide space to run mechanical and electrical services. The systems commercially available have some type of adjustable metal pedestals on which the modular floor panels rest. The panels typically are covered with vinyl composition floor covering (Fig. 39.9), although carpet can be used.

PEST CONTROL

Various mechanical, chemical, and electrical pest-repellent systems and protective devices are available. These include devices that control rodents, birds, and insects.

FIREPLACES AND STOVES

Manufactured fireplaces and stoves, including dampers, metal chimneys, and fireplace screens and doors, are specialties. Manufactured fireplaces typically are metal-framed firebox and vent units designed to accommodate

Figure 39.9 An access floor permits computer and electrical runs to be located below the floor and easily accessed for changes as needed. *(Reproduced with permission from* The Building Systems Integration Handbook, *Richard D. Rush, Editor, Butterworth-Heinmann, 313 Washington St., Newton, Mass. 02158–1626)*

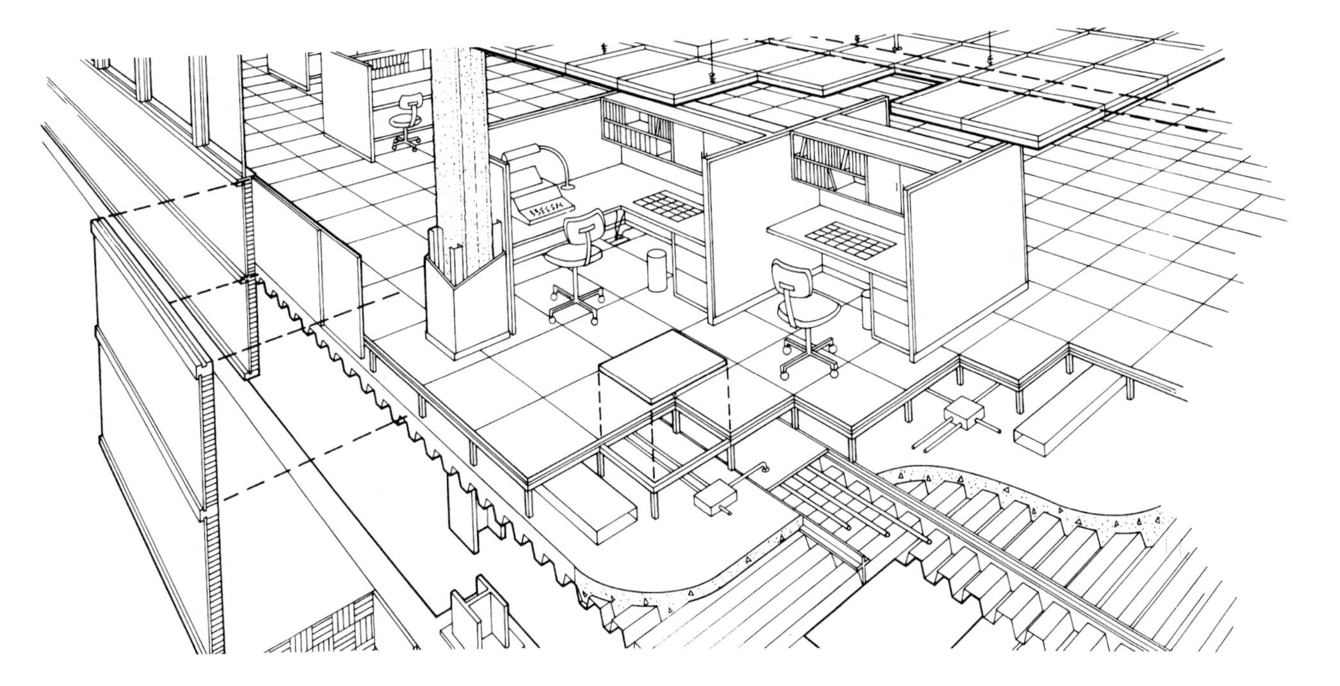

Figure 39.10 A manufactured direct vent gas fireplace is quickly installed and has high efficiency and heat output. *(Image copyright John Wollwerth, 2009. Used under license from Shutterstock.com)*

Figure 39.11 This freestanding gas-fired stove provides an attractive focal point in a room. *(Image copyright Nicola Gavin, 2009. Used under license from Shutterstock.com)*

gas-fired logs. Some may be vented vertically through a metal pipe or horizontally through a wall, while others do not require venting. They are delivered completely assembled (Fig. 39.10). A wide range of gas and wood-burning stoves are available. Some are used not only as a heat source but also as part of a room's decor (Fig. 39.11). Others are large units with ducts running to several rooms.

MANUFACTURED EXTERIOR SPECIALTIES

These items include stock and custom designed steeples, spires, cupolas, and weather vanes. A **steeple** is an ornamental construction usually ending in a spire (Fig. 39.12). It is usually erected on a roof or tower and may hold a clock or bells. A **spire** is a tall pointed pyramidal roof built on a tower or steeple. A **cupola** is a light-roofed structure mounted on a roof. Cupolas typically are circular or polygonal. In addition to adding architectural beauty, a cupola often serves as a roof vent (Fig. 39.13).

FLAGPOLES

This specialty area includes complete flagpole assemblies, including required hardware and accessories. Flagpoles are typically made from wood, metal, or fiberglass. They may be anchored in the ground or wall mounted (Fig. 39.14).

IDENTIFYING DEVICES

Identifying devices include directories, direction signs, bulletin boards, and various letters, signs, and plaques used to communicate or identify. These may be fixed devices, or they may be lighted, computerized, or some other type of electronic device.

Figure 39.12 This steeple contains a bell tower and is crowned with a tall spire. *(Image copyright Ethan Boisvert, 2009. Used under license from Shutterstock.com)*

Figure 39.13 A cupola with louvered sides provides attic ventilation. *(Image copyright Mark Winfrey, 2009. Used under license from Shutterstock.com)*

Figure 39.14 A flagpole flies the Stars and Stripes. *(Image copyright Melody Mulligan, 2009. Used under license from Shutterstock.com)*

Directories usually have replaceable letters and a locked glass door (Fig. 39.15). They are available as wall-mounted or freestanding units. Bulletin boards may have a cork face or a cork face covered with nylon or vinyl fabric. The substrate is typically a fiberboard panel.

Various types of accessibility signs have become standardized for use across the country. Many building codes require accessible elements to be identified by the international symbol of accessibility (Fig. 39.16). The international symbol of accessibility is posted to indicate all elements and spaces of accessible facilities. This includes facilities such as parking spaces reserved for individuals with disabilities, accessible passenger-loading zones, accessible entrances, and accessible toilet and bathing facilities. In addition, signs are used to identify volume-control telephones, text telephones, and assistive listening systems (Fig. 39.17).

Figure 39.15 This building directory gives room locations for offices in the courthouse. *(Courtesy Claridge Products and Equipment, Inc.)*

Figure 39.16 The International Symbol of Accessibility.

Figure 39.17 Symbols used to identify telephones and assistive listening devices for those with hearing impairments.

Examples of code regulated signs include exit identifiers and signs indicating the direction of a means of egress. Illuminated exit signs are used to denote stairs, elevators, or doors exiting directly from a building to the outside (Fig. 39.18). Signs also are used to identify areas of refuge and the direction to move to reach them. An area of refuge is a designated space that provides protection from smoke and fire. The maximum occupant capacity for assembly rooms, restraints, and other areas where numbers of people gather is indicated by posted signs. Control of parking and identification of facilities for the handicapped are other major parts of any sign system related to a building. These and many other signs are required by building codes.

Figure 39.18 Lighted exit signs are one of the more important identifying devices in a building. *(Image copyright Warren Chan, 2009. Used under license from Shutterstock.com)*

Other symbol signs are widely used to present a graphic message and to help communicate with people using various languages. These are found extensively in public buildings, such as airports, hotels, and hospitals, and are used as highway traffic control signs (Fig. 39.19).

PEDESTRIAN CONTROL DEVICES

Products used for pedestrian control include portable posts and railings, rotary gates, turnstiles, electronic detection and counting systems, and items to limit access (Figs. 39.20 and 39.21). These are strong, durable products that can withstand exposure to the weather.

LOCKERS AND SHELVING

Lockers provide temporary storage in areas such as airports and athletic facilities (Fig. 39.22). All types of open manufactured metal and wood shelving used for general storage are specialty items. This includes open shelving, various manufactured shelving systems, and mobile storage systems.

FIRE PROTECTION SPECIALTIES

This specialty includes all portable firefighting devices and their storage facilities but excludes items directly connected to an installed fire protection system. Typical

Figure 39.19 Some of the symbols used to communicate graphically.

Figure 39.20 A non-penetrable security turnstile provides total passageway security. *(Image copyright Manfred Steinbach, 2009. Used under license from Shutterstock.com)*

Figure 39.21 This self-closing turnstile permits electronic card access. *(Image copyright bery45, 2009. Used under license from Shutterstock.com)*

items include fire extinguishers, extinguisher cabinets, fire blankets and their storage cabinets, and any fire extinguishing units on wheels (Fig. 39.23).

PROTECTIVE COVERS

Various types of awnings, canopies, marquees, covered walkways, and sheltered bus and car stop structures are available as standard manufactured units or as custom designed and built structures. They are made from any of the commonly used materials that withstand weather exposure.

Figure 39.22 Double-tier lockers are widely used in schools and athletic facilities. *(Image copyright Ken Schulze, 2009. Used under license from Shutterstock.com)*

POSTAL SPECIALTIES

Postal specialty products include all types of postal service facilities and devices, such as view windows, letter slots, letter boxes, collection boxes, and chutes. In large commercial buildings these can become an extensive system used to speed mail collection (Fig. 39.24).

DEMOUNTABLE PARTITIONS

Demountable (movable) partitions include various room dividers, screens, and enclosures. They may be freestanding post-and-panel units that hang from a wall or ceiling, or they can be secured to a floor and ceiling. Some types are solid, offering privacy and a measure of sound control (Fig. 39.25). Others are made of open mesh or screen panels, giving an area security but permitting visual examination. Certain types incorporate folding gates or doors (Fig. 39.26). They are made of a variety of materials, including wood, fabrics, plastic laminates, and metal.

Figure 39.23 A typical fire extinguisher stored in a protective cabinet. *(Image copyright Scott David Patterson, 2009. Used under license from Shutterstock.com)*

Figure 39.24 Post office boxes typically used in multi-unit residential buildings. *(Image copyright Andrejs Pidjass, 2009. Used under license from Shutterstock.com)*

Figure 39.25 Low demountable partitions provide privacy while using the general illumination and heating and air-conditioning systems. *(Image copyright Chad McDermott, 2009. Used under license from Shutterstock.com)*

Figure 39.26 These floor-to-ceiling prefabricated wall panels are used to enclose an area and provide privacy and security yet allow visual oversight when required. *(Image copyright Aura Castro, 2009. Used under license from Shutterstock.com)*

OPERABLE PARTITIONS

Operable partitions are manually or power operated, allowing a moving partition to enclose or divide a larger area. They are suspended from a ceiling track and supported by a steel floor track. Some are top hung and have a rubber seal at the floor. They include folding panel partitions, accordion folding partitions, and sliding and coiling partitions. Folding and accordion doors are outside of this classification.

Figure 39.27 These perforated metal louvers are used to provide partial sun screening. *(Courtesy Eva Kultermann)*

EXTERIOR PROTECTION DEVICES FOR OPENINGS

An extensive range of protection devices are manufactured, including products such as sun control screens, shutters, louvers, screens for security, and panels that provide insulation and protection against storms. These may be fixed or controlled manually or electrically (Fig. 39.27).

TELEPHONE SPECIALTIES

Telephone specialties include manufactured telephone enclosures that may be completely enclosed, supported on a floor, or wall hung (Fig. 39.28). Telephone directory and shelving units are usually wall hung; they are available in many sizes. Outdoor telephone enclosures are weathertight and equipped with integral lighting for night use (Fig. 39.29). Some companies will manufacture enclosures to an architect's design.

TOILET AND BATH ACCESSORIES

Toilet and bath accessories include all items manufactured for use in connection with bathroom facilities. These encompass products used in most commercial restrooms (Fig. 39.30) plus special units designed for use in hospitals and areas where accessibility and security are important. Items in this specialty area include mirrors, air fresheners, electric dryers, grab bars (Fig. 39.31), waste receptacles, and dispensers for soap, toilet tissue, towels, razor blades, lotion, and paper cups. Hospital products include special cabinets and bedpan storage units. Security items include mirrors that resist theft and breakage and specially designed soap and tissue dispensers.

Figure 39.28 This row of individual floor-mounted telephone kiosks offers a degree of privacy. *(Image copyright SVLumagraphica, 2009. Used under license from Shutterstock.com)*

Figure 39.29 Exterior Telephone units are equipped with integral lighting. *(Image copyright palms, 2009. Used under license from Shutterstock.com)*

Figure 39.30 This paper towel dispenser has a waste disposal container. *(Image copyright Carolyn Brule, 2009. Used under license from Shutterstock.com)*

Figure 39.31 This wall unit provides the required grab bar for the handicapped plus a toilet paper and seat cover supply. *(Image copyright Carolyn Brule, 2009. Used under license from Shutterstock.com)*

SCALES

Any type of weighing device is considered a specialty item.

WARDROBE AND CLOSET SPECIALTIES

These specialties are units used to store clothing, such as hat and coat racks. Among them are wardrobes that provide hanging space as well as several drawers. Other units typically used in hospitals and dormitories combine a wardrobe with a vanity area and lavatory. Wardrobes often are constructed with a plastic laminate over a substrate such as particleboard. Racks are usually steel.

A wide variety of racks, wall mounted and floor supported, are used to store coats, hats, and packages. Automatic coat and hat checking systems are used in restaurants, hotels, theaters, and other places where large numbers of people gather. They are activated by hand and foot remote controls or by automatic selection devices using a customer's check number to locate and retrieve garments.

Review Questions

1. What finished surfaces are available when selecting chalkboards?
2. What is a marker board?
3. What are the typical ways toilet enclosures are hung?
4. What materials are used for panels on toilet enclosures?
5. What purposes do grilles serve?
6. What services are available with various service wall systems?
7. Why are access floors used?
8. How are gas-fired fireplaces vented?
9. What is the difference between a steeple, a spire, and a cupola?
10. What types of identifying devices are commonly used?
11. How do building codes govern the design of identifying devices?
12. What are some frequently used pedestrian control devices?
13. Where are protective covers used?
14. What are the ways various partition systems may be installed?
15. List some exterior protection devices.

Key Terms

access flooring
chalkboard
compartment
cubicle
cupola
grille
marker board
spire
steeple
tackboard

Activity

1. Take a walk around your school or a major building and record the location and use of any specialty items you observe. Decide if the proper choice was made and suggest replacements for those you consider improperly used.

Additional Resources

Sweets Catalog File, Architects, Engineers, and Construction Edition, Section 10, McGraw-Hill Construction, New York.

See Appendix B for addresses of professional and trade organizations and other sources of technical information.

DIVISION 11

Equipment
CSI MasterFormat™

111000 Vehicle and Pedestrian Equipment
111500 Security, Detention, and Banking Equipment
112000 Commercial Equipment
113000 Residential Equipment
114000 Foodservice Equipment
115000 Educational and Scientific Equipment
116000 Entertainment Equipment
116500 Athletic and Recreational Equipment
117000 Healthcare Equipment
118000 Collection and Disposal Equipment
119000 Other Equipment

(Image copyright ariadna de raadt, 2009. Used under license from Shutterstock.com)

40 CHAPTER

Residential and Commercial Equipment

LEARNING OBJECTIVES

Upon completion of this chapter, the student should be able to:

- Understand the extensive array of residential and commercial equipment a designer and contractor must be able to specify and install.

Division 11 of the CSI MasterFormat covers equipment related to residential and commercial construction. The items included in this section are shown on the introductory page to this chapter. A few of these are discussed in the following pages.

MAINTENANCE EQUIPMENT

Maintenance equipment consists of freestanding and built-in equipment used to provide maintenance for a building. This includes devices such as window washing systems, automatic vacuum systems, floor and wall cleaning equipment, and various types of housekeeping carts. Central vacuum systems use a centrally powered vacuum unit connected to pipes run through walls to sealed outlets in various rooms. The flexible plastic hoses with various end attachments are connected to an outlet when an area needs vacuuming.

Window washing systems for high-rise buildings have hoists mounted on a building's roof. The exterior washing equipment is controlled by an apparatus that usually rides on tracks on the roof (Fig. 40.1). Some systems lower scaffolding from a roof, providing a platform from which workers can wash windows.

Figure 40.1 Window washing equipment assemblies for high-rise buildings sometimes use hoists mounted on a roof to lower scaffolding, allowing workers to reach and wash windows. *(Image copyright oksana.perkins, 2009. Used under license from Shutterstock.com)*

SECURITY AND ECCLESIASTICAL EQUIPMENT

Security and vault equipment is designed to store money or other valuables and includes vault doors (Fig. 40.2), day gates, security and emergency systems, safes, and safe deposit boxes.

Teller and service equipment is designed for use where money transactions occur. This includes service and teller windows, package transfer units, automatic banking systems (Fig. 40.3), and teller equipment systems.

Ecclesiastical equipment includes items related to churches, such as baptisteries and various chancel items (Fig. 40.4). These articles vary according to the

Figure 40.2 Security vault doors provide maximum protection for valuables. Various types are made for different levels of security. The one in this illustration rates GSA Class 5. It is rated to protect against surreptitious entry (20 hours), covert entry (30 minutes), forced entry (10 minutes), lock manipulation (20 minutes), and radiological attack. *(Courtesy Overly Manufacturing Company)*

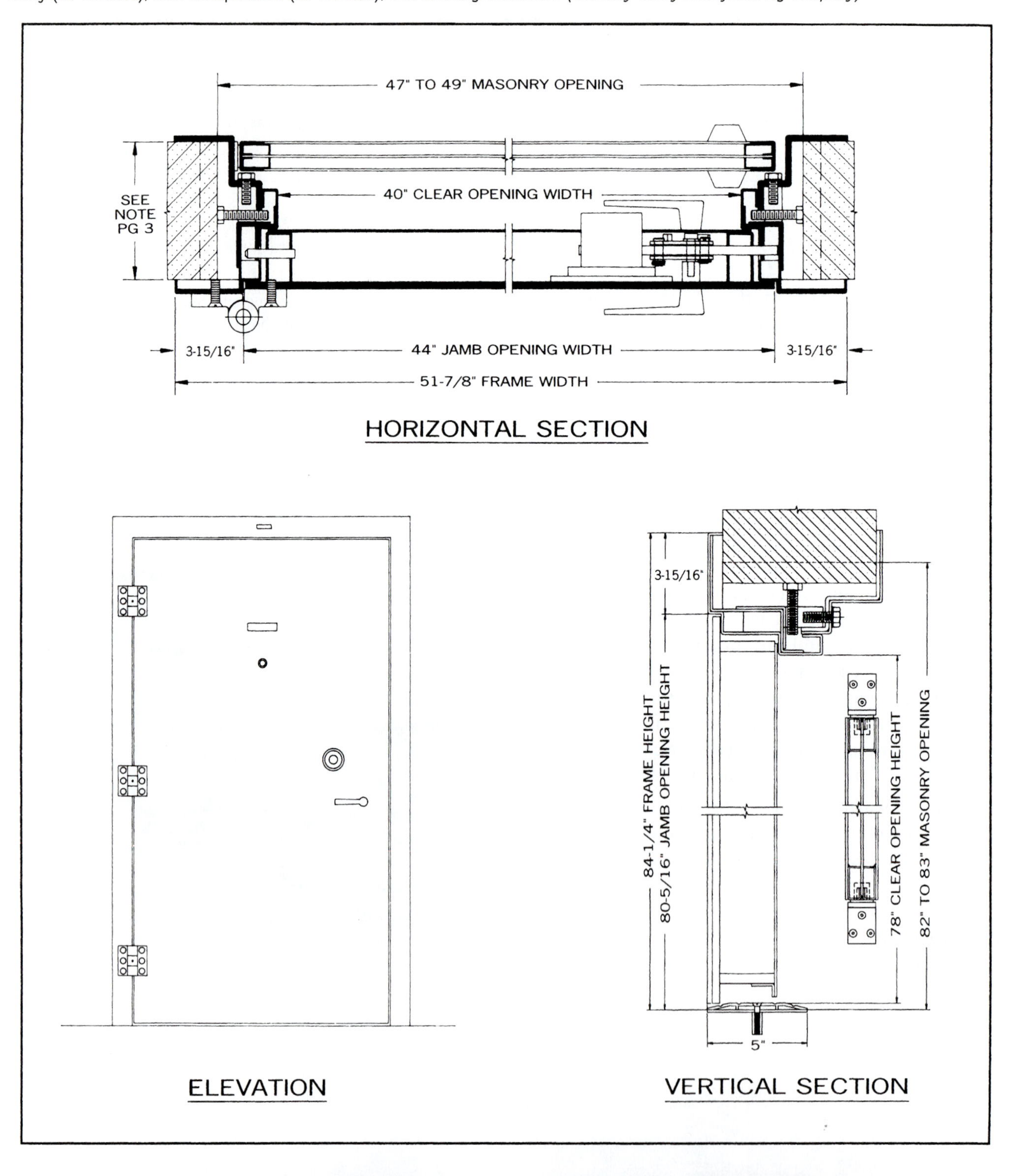

Figure 40.3 Automatic teller machines are widely used to facilitate the banking needs of clients. *(Image copyright Stephen Coburn, 2009. Used under license from Shutterstock.com)*

Figure 40.4 A custom built pulpit. *(Image copyright Jim Lopes, 2009. Used under license from Shutterstock.com)*

Figure 40.5 Retail display fixtures are designed to exhibit a wide variety of merchandise of different sizes, weights, and values. *(Image copyright Losevsky Pavel, 2009. Used under license from Shutterstock.com)*

denomination and desires of a congregation. Some items are manufactured stock, and others are custom built and can utilize all types of materials, including stone, wood, and metal.

MERCANTILE EQUIPMENT

This includes equipment used in retail and service stores, such as display cases (Fig. 40.5), cash registers, and checking and food processing machinery. Mercantile equipment also includes items in specialty establishments, such as barber and beauty shops.

SOLID WASTE HANDLING EQUIPMENT

Waste handling equipment includes units that collect, shred, compact, and incinerate solid waste. The systems manufactured include package incinerators, compactors, storage bins, pulping machines, pneumatic waste transfer systems, and various chutes and storage collectors. One such system is shown in Chapter 43.

OFFICE EQUIPMENT

Office equipment includes products such as copiers (Fig. 40.6), computers, word processors, printers, fax machines, and modems. It does not include any office furniture, such as desks, files, or chairs, which are detailed in Division 12.

Construction Materials

Security of Valuables

Security and vault equipment is a major item for banks, wholesale and retail stores, hotels, and many other business establishments. Available are both built-in and freestanding equipment especially designed for the storage of money and other valuables, such as artwork, jewelry, and stock certificates. Large vaults are available for storing sizable items. Vaults require ventilation systems and specialized security networks.

Vault doors have fire ratings up to at least six hours (Fig. A). The controls include a time lock, day guard lock, alarm activator, lighting, emergency ventilator, and a connection for a telephone handset. Manufacturers offer centralized twenty-four-hour alarm systems for installations within the United States and maintain fleets of trucks manned by trained service technicians.

Safe deposit boxes come in stackable, interlocking sections to fit various vault sizes (Fig. B). Various types of key systems are available, which greatly reduces the ability to improperly duplicate a key.

Figure A Vault doors are massive and provide maximum security. *(Image copyright Sebastian Kaulitzki, 2009. Used under license from Shutterstock.com)*

Figure B Safe deposit boxes are installed in vaults and have a two-key security system. *(Image copyright Andriy Rovenko, 2009. Used under license from Shutterstock.com)*

Office equipment is manufactured by many companies. Since some of these products must communicate with each other, care should be exercised when different brands are combined.

RESIDENTIAL EQUIPMENT AND UNIT KITCHENS

Residential equipment includes all appliances that are either built-in or freestanding, such as stoves, washers, dryers, freezers, microwaves, dishwashers, compactors, refrigerators, surface cooking units, ovens, range hoods, and indoor barbecue units. When combining units built by different manufacturers, it is important to consider possible differences in appearance (Fig. 40.7).

Unit kitchens are manufactured assemblies that combine a refrigerator, cooking element, sink, and cabinets into one unit. They are delivered totally assembled for rapid installation (Fig. 40.8). This section does not include kitchen cabinets, which are detailed in Division 12.

Figure 40.6 Equipment for offices includes items such as this rapid feed production copier and these computer network servers. *(a) Image copyright Albo, 2009. Used under license from Shutterstock.com (b) Image copyright Eimantas Buzas, 2009. Used under license from Shutterstock.com*

(a)

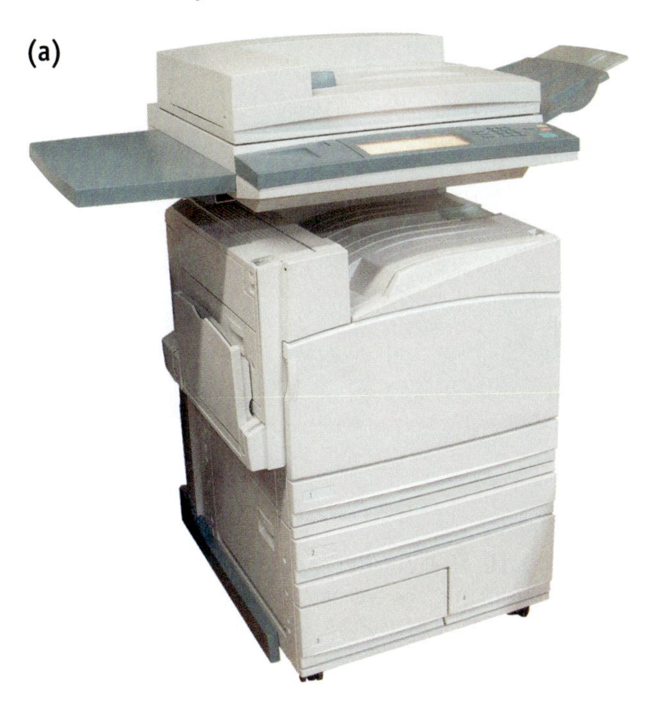

(b)

Figure 40.7 Kitchen appliances include a wide range of electric and gas units in a variety of sizes and colors. *(a) Image copyright Susan Quinland-Stringer, 2009. Used under license from Shutterstock.com (b) Image copyright Shi Yali, 2009. Used under license from Shutterstock.com*

(a)

(b)

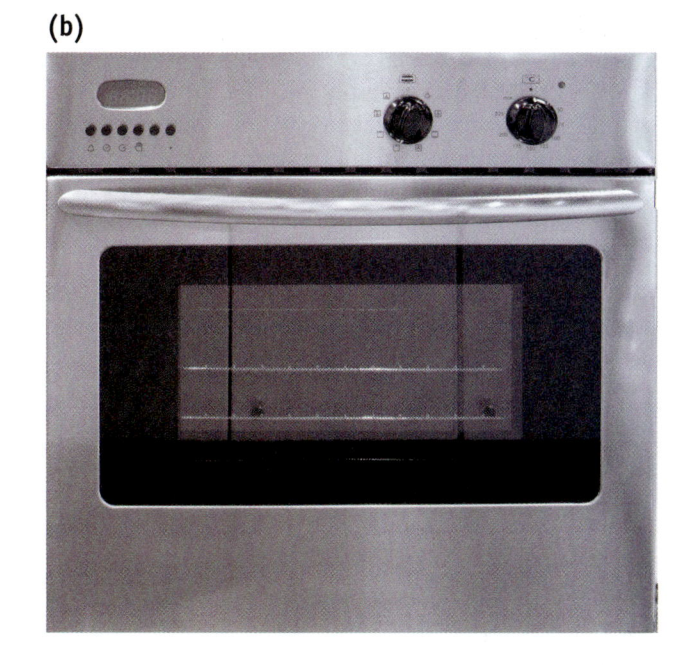

Figure 40.8 Unit kitchens are compact single arrays that fit along one wall. *(Image copyright Zastol`skiy Victor Leonidovich, 2009. Used under license from Shutterstock.com)*

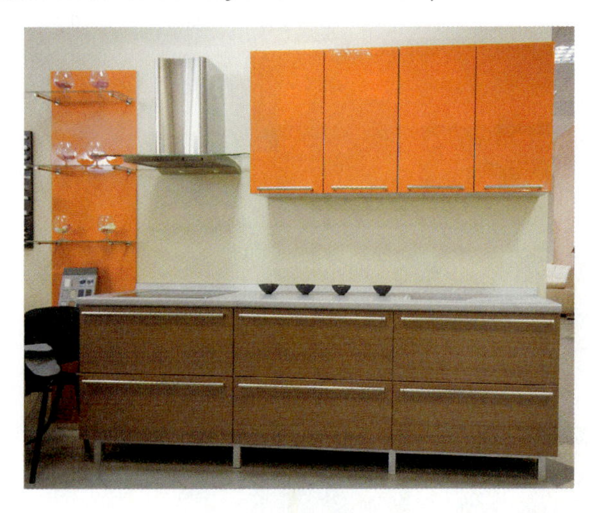

Figure 40.9 Typical equipment in a dental operatory. *(Image copyright bravajulia, 2009. Used under license from Shutterstock.com)*

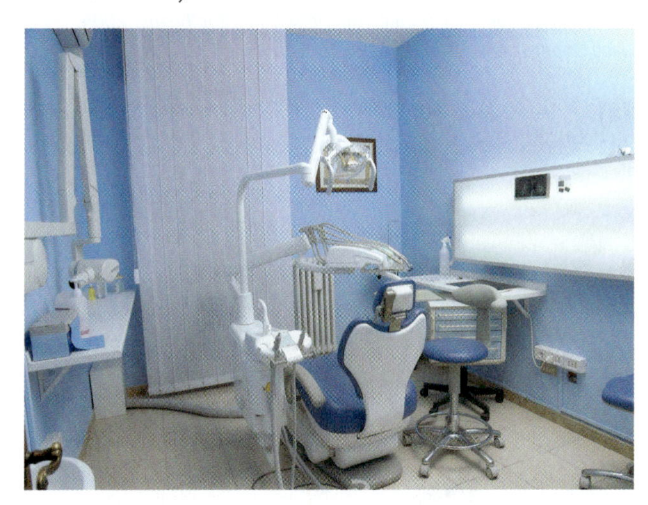

FLUID WASTE TREATMENT AND DISPOSAL EQUIPMENT

This section includes a wide range of equipment used to treat and dispose of fluid waste. A few of the products include an oil/water separator, fluid pumping stations, sewage and sludge pumps, scum removal equipment, aeration equipment, and package sewer treatment plants. Unit selection comprises a major part of the engineering design of a waste treatment system. The units must be durable and dependable.

MEDICAL EQUIPMENT

The array of medical equipment used for human and animal healthcare is extensive. It includes sterilizing equipment, examination and treatment equipment, optical and dental equipment (Fig. 40.9), and items used in operating rooms and radiology facilities (Fig. 40.10). Technical changes and improvements occur rapidly in this arena, and those planning these facilities must be constantly alert for new equipment. These items must be of the highest quality.

The many items used in laboratories for testing, research, and developing new products may be preassembled by a manufacturer or designed for assembly on-site. Typical items include fume hoods, incubators, sterilizers, refrigerators, and emergency safety appliances designed specifically for laboratory use—along with a range of special service fittings and the accessories to use them.

Figure 40.10 Medical facilities utilize a large amount of testing, treatment, and examination equipment. *(Image copyright Dalton Dingelstad, 2009. Used under license from Shutterstock.com)*

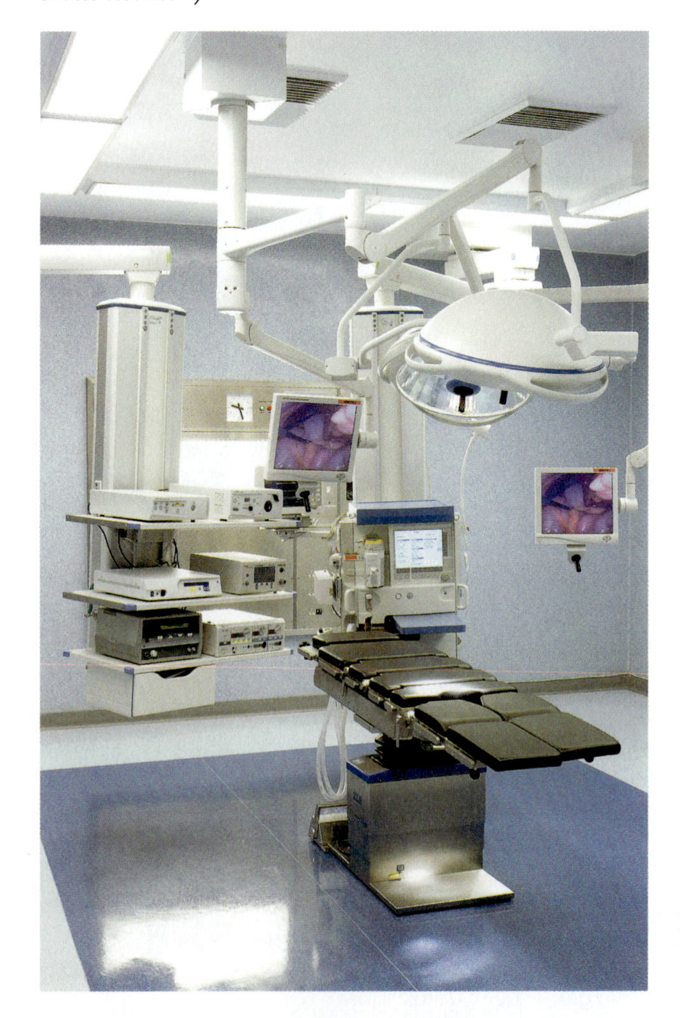

Review Questions

1. What types of equipment are used for building maintenance?
2. What types of ecclesiastical equipment can you name?
3. What operations are performed by solid waste handling equipment?
4. How does a unit kitchen differ from a typical residential kitchen?
5. What types of units are used for fluid waste treatment and disposal?

Activity

1. Visit a hospital, a large restaurant, and a school and record the types of equipment that have been installed. Discuss the merits of each with the person responsible for operation and maintenance and record that person's comments on each item. Does your interviewee feel that any items are not adequate for their assignment?

Additional Resources

Sweets Catalog File, Architects, Engineers and Contractors Edition, Division 11 Equipment, McGraw-Hill Construction, New York.

See Appendix B for addresses of professional and trade organizations and other sources of technical information.

Furnishings CSI MasterFormat™

121000 Art
122000 Window Treatments
123000 Casework
124000 Furnishings and Accessories
125000 Furniture
126000 Multiple Seating
129000 Other Furnishings

(Image copyright Gregor Kervina, 2009. Used under license from Shutterstock.com)

41 CHAPTER

Furnishings

LEARNING OBJECTIVES

Upon completion of this chapter, the student should be able to:

- Understand both the range of furnishings used in residential and commercial buildings and the places that various art forms are used.

Build Your Knowledge

For further study on these materials and methods, please refer to:

Chapter 40 Residential and Commercial Equipment

Topics: Mercantile Equipment; Residential Equipment and Unit Kitchens; Medical Equipment

Furnishings include products that are decorative in nature, such as artwork, rugs, and window treatments, and other items such as casework, furniture for commercial buildings, and multiple seating. An interior decorator plays a major role in the selection of these items and sometimes becomes involved in the design of special furnishings. The architect also is an important part of the overall design process. The CSI MasterFormat listing for furnishings is on the page introducing this section.

ARTWORK

Murals and Wall Hangings

A popular form of interior artwork is photo murals. They are available in a wide range of sizes and scenes. Murals painted on walls are much more expensive but frequently become an important art item.

Mosaic murals are formed by bonding pieces of stone—such as marble, clay, or glass—together on heavy paper to form a desired design. These sheets are then installed against a mortar setting bed on a wall. When the setting bed hardens, the paper is soaked off the face, leaving the mosaic mural exposed (Fig. 41.1).

Figure 41.1 This mosaic was formed using small ceramic tiles. *(Image copyright Regien Paassen, 2009. Used under license from Shutterstock.com)*

Many types of wall decorations are used, including paintings, prints, fabric wall hangings, and **tapestries** of various types. High-quality woven rugs are also used as wall hangings.

Sculpture

Sculpture is another form of artwork used in both interior and exterior locations. Some types are carved from stone, and others are cast in bronze, lead, or aluminum (Fig. 41.2). Sculpture made by welding various shapes of metal is called reconstructed sculpture. Additional materials, such as stone, may also be integrated into a reconstructed piece. Relief artwork is usually a panel-type piece with images raised above its surface, producing a three dimensional appearance.

Stained Glass

Stained-glass windows are mosaics of translucent glass shards cut to a desired shape. Sometimes the glass requires additional detailing, shading, or texturing. This is accomplished by painting it with special mineral pigments and firing it in a kiln. The two major types of stained glass are leaded and faceted.

Leaded glass windows have glass pieces joined with H-shaped lead strips called cames (Fig. 41.3). Large windows require round bracing bars be wired to the leaded glass. This provides support while allowing for thermal movement. Exterior leaded glass is pressed into a bed of glazing sealant or tape (Fig. 41.4).

Faceted glass is fashioned from colored glass slabs 1 in. (25 mm) thick. The glass is cut or faceted to reflect light rays in many directions. Faceted glass is made in a wide range of translucency, opacity, and transparency, which influences the color and degree of reflected light.

Faceted glass is set in a reinforced concrete or epoxy resin matrix (the process is similar to setting brick) (Fig. 41.5). Light passes through the faceted glass pieces bound together by the matrix. The panels are mounted in wood or metal frames or in a masonry wall.

Ecclesiastical Artwork

Ecclesiastical artwork includes pieces of religious significance to specific denominations. They can take a variety of forms, some of which are discussed in this chapter. Specific types can include altar pieces and religious symbols.

Figure 41.2 Cast metal sculpture has been used for many years to produce striking art objects. *(Image copyright Paul Prescott, 2009. Used under license from Shutterstock.com)*

Figure 41.3 Leaded glass is formed by joining glass pieces cut in various sizes and shapes with lead cames.

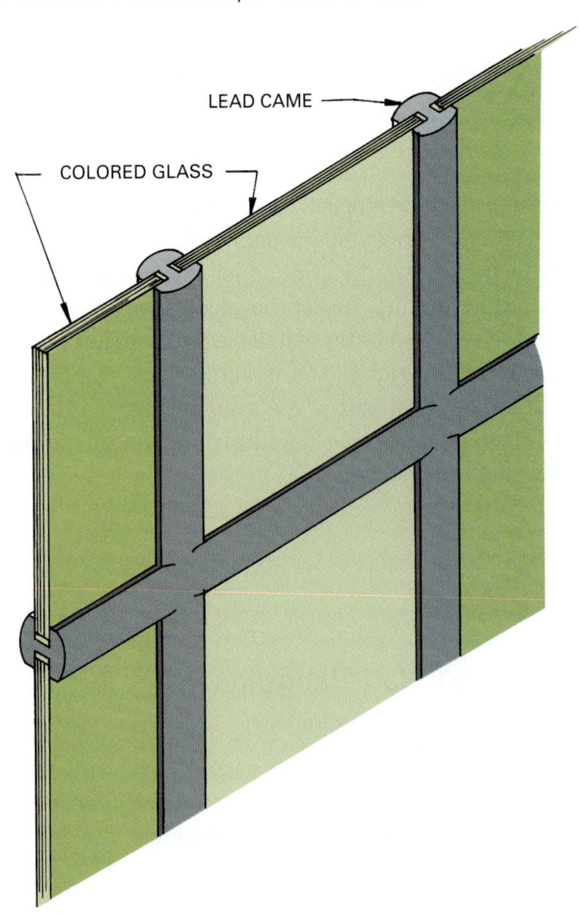

Figure 41.4 These dramatic leaded glass windows contribute to the atmosphere of the church interior. *(Image copyright Tom Cummins, 2009. Used under license from Shutterstock.com)*

Figure 41.5 Faceted glass walls are formed by bonding glass cut in various shapes and bonded with an opaque epoxy resin or reinforced concrete.

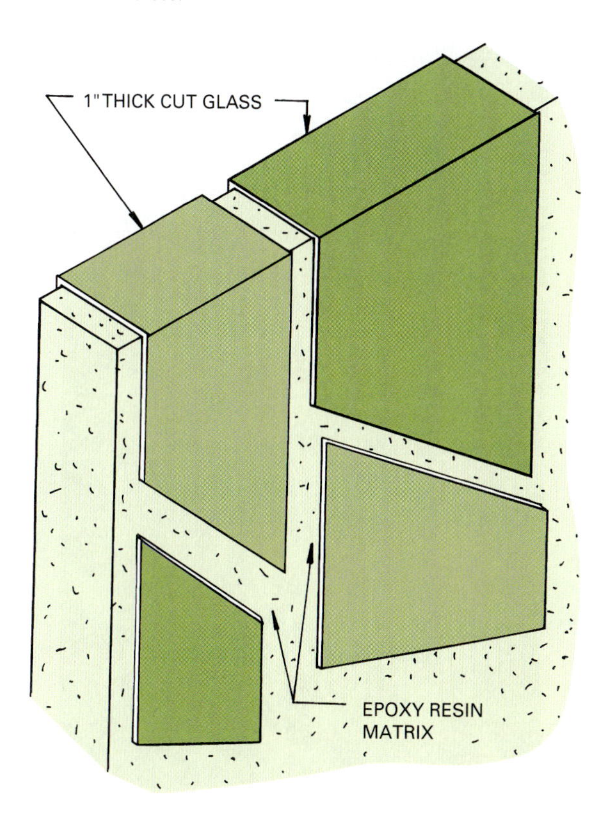

Figure 41.6 Fabrics are used extensively in interior design for upholstery, curtains, draperies, and wall hangings. *(© Dana Hoff/Beateworks/Corbis)*

FABRICS

This section relates to fabrics, leathers, and furs for upholstery, curtains, draperies, and wall hangings; this includes the various fabric treatments and fillers. Fire codes may require these products be fireproofed (Fig. 41.6).

MANUFACTURED CASEWORK

Manufactured casework includes stock cabinets fashioned from wood, steel, and plastic laminates. It includes countertops, sinks, and any accessories or fixtures mounted on countertops. For residential applications it includes both kitchen and bath casework (Fig. 41.7).

Medical applications include casework designed for dental, hospital, and optical uses, for nurses stations (Fig. 41.8), and for veterinary uses. All types of laboratory casework used in medical, research, and other laboratories are part of this section.

Figure 41.7 Residential casework includes kitchen and bath cabinets. *(Image copyright Baloncici, 2009. Used under license from Shutterstock.com)*

Figure 41.8 Casework for medical applications includes cabinetry for nurse's stations. *(Image copyright Kiselev Andrey Valerevich, 2009. Used under license from Shutterstock.com)*

Various manufacturers produce specialized casework with a limited market. Included is casework for banks, dormitories, ecclesiastical uses, hotels, motels, schools, restaurants, and various display units.

WINDOW TREATMENT

Interior window treatment can be decorative, and it can be practical, blocking the intrusion of sunlight and providing privacy. Blinds, often called Venetian blinds, are adjustable narrow wood or metal slats that can be easily positioned to block sun and provide privacy or permit a view of the outside (Fig. 41.9). They can be raised to a window's head so they are completely out of view.

Vertical blinds are made with fabric slats hanging from the top of a window. They can be set open or closed (Fig. 41.10). Shutters made of wood or plastic are also used on window interiors. Some have adjustable louvers that let in light but prohibit much of an exterior view.

A number of shades are manufactured, including insulating, lightproof, translucent, and woven wood or plastic types. They can be pulled down for privacy or rolled up for viewing. Accordion types with fiber-glass fabric are also available.

Curtains of various styles can serve the same purpose as shades (Fig. 41.11). Some drapes have metal tracks across the top that permit opening or closing over

Figure 41.9 Venetian blinds have horizontal wood or metal slats that can be adjusted to block views and sunshine or opened to permit both. *(Image copyright Lykovata, 2009. Used under license from Shutterstock.com)*

Figure 41.10 Vertical blinds are widely used in residential and commercial construction. *(Image copyright Gordon Ball LRPS, 2009. Used under license from Shutterstock.com)*

Figure 41.11 Draperies, curtains, and other wall hangings in commercial buildings must be noncombustible or maintained as flame resistant. *(Image copyright Mayer George Vladimirovich, 2009. Used under license from Shutterstock.com)*

windows. Other units remain on each side or above a window, serving a decorative purpose. Fire codes must be observed when selecting shades and curtains.

FURNITURE AND ACCESSORIES

Furniture under CSI MasterFormat Division 12 includes all types of freestanding units for commercial and residential use. The listing of open office furniture includes office partitions, storage, work surfaces, shelving (Fig. 41.12), and light fixtures. General furniture includes all types from residential through commercial areas, such as classroom, hotel, library (Fig. 41.13), medical, office, and restaurant furniture. Furniture accessories, such as clocks, ashtrays, lamps, and waste receptacles are also included (Fig. 41.14).

Figure 41.12 Libraries use large shelving units to store their inventory of publications. *(Image copyright SharonPhoto, 2009. Used under license from Shutterstock.com)*

Figure 41.13 Various types of upholstered furniture are specified under CSI Division 12 (Furnishings). *(Image copyright Caruntu, 2009. Used under license from Shutterstock.com)*

Figure 41.14 Clocks are one type of furniture accessory. *(Image copyright Francesco Ridolfi, 2009. Used under license from Shutterstock.com)*

Figure 41.15 Multiple seating includes products used in any area where large numbers of people gather. *(Image copyright Glen Jones, 2009. Used under license from Shutterstock.com)*

RUGS AND MATS

This section includes loose rugs and mats, gratings, and foot grilles (carpet covering floor areas is included in Division 9). Typical products include chair pads, floor mats, entrance tiles, and floor runners. These are made from a variety of materials, including rubber, carpet, vinyl, aluminum, and stainless steel.

MULTIPLE SEATING

Multiple seating includes seating for areas in which audiences gather, such as in schools, restaurants, theatres, churches, stadiums, and auditoriums (Fig. 41.15). It includes fixed, portable, and telescoping seating. Booths and various table and seat modules, as found in restaurants, also are included.

INTERIOR PLANTS AND PLANTERS

Extensive use of live and artificial plants is common in all types of commercial buildings. This section includes plants, plant holders, and any landscaping accessories and maintenance materials required (Fig. 41.16).

Figure 41.16 An extensive variety of live and artificial plants are used for interior decoration. *(Image copyright Jill Lang, 2009. Used under license from Shutterstock.com)*

Review Questions

1. What types of window blinds are commonly available?
2. How are mosaic murals made?
3. What types of sculpture are commonly available?
4. How are leaded stained glass pieces joined?
5. What are the differences between leaded and faceted glass?
6. What types of units are used for window treatment?
7. Where is multiple seating commonly found?

Key Terms

faceted glass
leaded glass window
mosaic
tapestry

Activity

1. Visit various commercial buildings and shopping malls and record the types of furnishings used both inside and out.

Additional Resources

DeChiara, J., Panero, J., and Zelnik, M., *Interior Design and Space Planning*, McGraw-Hill, New York.

Sweets Catalog File, Architects, Engineers and Contractors Edition, Division 12, McGraw-Hill Construction, New York.

See Appendix B for addresses of professional and trade organizations and other sources of technical information.

DIVISION 13

Special Construction CSI MasterFormat™

131000 Special Facility Components
132000 Special Purpose Rooms
133000 Special Structures
134000 Integrated Construction
135000 Special Instrumentation

(Courtesy DX Broadrec, http://en.wikipedia.org/wiki/File:Tokyo_Dome_2007-1.jpg)

42 CHAPTER

Special Construction

LEARNING OBJECTIVES

Upon completion of this chapter, the student should be to:

- Understand the wide range of special construction facilities available and how they function.

Special construction encompasses an extensive array of structures, systems, and assemblies, each designed to serve a specific purpose. Construction ranges from complex facilities and large structures to individual components of a small scale, such as special purpose rooms or an ornamental fountain.

SPECIAL FACILITY COMPONENTS

The CSI MasterFormat lists special facility components for water-related constructions, including swimming pools, fountains, aquariums, and ice rinks, among others.

Pools and Spas

Swimming pools are constructed in a variety of sizes, from large public pools to residential applications and specialty spas (Fig. 42.1). They are designed to withstand all anticipated loadings, both in an empty and full state. Small pools may be constructed of gunite, a mixture of pea stone, sand, cement, flyash, and water, sprayed on a reinforcing framework. Surfaces can be finished by painting or with a vinyl membrane lining. Other pools and tubs are made of a prefabricated fiberglass. Larger pools use heavy reinforced concrete construction clad with tile or a variety of other finish materials.

All swimming pools must be equipped with a filtration system to clarify the water. Systems normally consist of one or more filter units containing sand, diatomaceous earth, or a cartridge-type filter. A disinfectant feeder keeps the microbiological, chemical, and physical characteristics of pool water within prescribed limits. Pool equipment rooms house pumps, filters, water treatment feeders, and heaters.

Figure 42.1 Swimming pools are constructed in a variety of sizes, from large public pools to residential applications. *(Image copyright Ramzi Hachicho, 2009. Used under license from Shutterstock.com)*

Saunas and steam rooms require insulated enclosures with integral mechanical systems to supply heat and steam. Saunas are usually clad in cedar or similar woods that perform well in extreme environments (Fig. 42.2). Saunas can be divided into two basic types: conventional saunas that warm the air or infrared saunas that warm objects. Infrared saunas use various heating agents, such as charcoal or active carbon fibers.

Figure 42.2 Saunas are often finished in water- and moisture-resistant materials, such as cedar or redwood. *(Image copyright Sandra Kemppainen, 2009. Used under license from Shutterstock.com)*

Heat sources include wood, electricity, natural gas, and other methods, such as solar power. There are wet saunas, dry saunas, steam saunas, and those that work with infrared waves. There are two main types of stoves: continuous heating and heat storage-type. Continuously heating stoves have a small heat capacity and can be heated quickly on an on-demand basis, whereas a heat storage stove has a large heat (stone) capacity and can take much longer to achieve high temperatures.

SPECIAL-PURPOSE ROOMS

Special-purpose rooms are designed and constructed to meet specific performance requirements, such as fire protection, thermal control, or sound control, among others. The CSI MasterFormat lists planetariums, athletic rooms, soundproof rooms, clean rooms, cold storage rooms, insulated rooms, shelters, booths, saunas, steam rooms, and vaults under the category "special-purpose rooms."

Figure 42.3 An anechoic chamber is a shielded room designed to attenuate sound or electromagnetic energy.

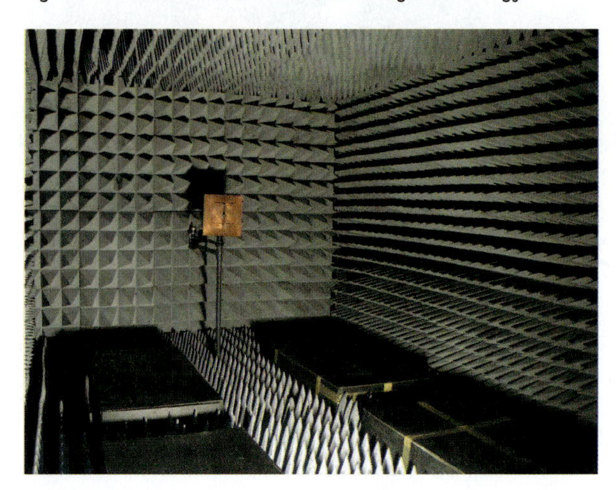

An example of one such room is an anechoic chamber, a shielded room designed to attenuate sound or electromagnetic energy (Fig. 42.3). Anechoic chambers were originally used to study acoustics caused by internal reflections of a room, but more recently they have also been used to provide shielded environments for radio frequency and microwaves.

Anechoic chambers range from small compartments to chambers as large as aircraft hangars. The size of an anechoic chamber depends on the size of the objects to be tested and the frequency range of the radio or microwave signals used. Radio frequency interference (RFI) is the unwanted reception of radio signals. Radio frequency interference sources include lightning, electrical equipment, fluorescent lighting, cell phones, and transmitting equipment from radio stations.

SPECIAL STRUCTURES

Special structures listed in CSI MasterFormat Division 13 include fabric structures, space frames, geodesic structures, fabricated engineered structures, and towers.

Fabric Structures

Fabric structures are architecturally innovative forms constructed of fibers that provide a variety of free-form building designs for both permanent and temporary applications (Fig. 42.4). Custom-made fabric structures are engineered and fabricated to meet structural, flame retardant, and weather-resistant specifications.

Most fabric structures are constructed of fabric, meshes, or films. Typically, the fabric is coated with synthetic materials for increased strength, durability,

Figure 42.4 This custom-made fabric roof structure uses a coated fabric to provide durability and weather resistance. *(Image copyright Konovalikov Andrey, 2009. Used under license from Shutterstock.com)*

and weather resistance. Among the most widely used materials are polyesters laminated with polyvinyl chloride (PVC) and woven fiberglass coated with polytetrafluoroethylene (PTFE). Some form of top coating is usually applied to the exterior to facilitate easy cleaning. Top coating provides a hard surface on the outside of the material, forming a barrier that aids in preventing dirt from sticking to the material, while allowing the fabric to be cleaned with water.

Air Supported Structures

Air-supported (pneumatic) **structures** derive their structural integrity from pressurized air used to inflate a flexible material, like structural fabric, making air the main support of the structure (Fig. 42.5). To maintain structural integrity, a structure is pressurized to produce internal pressure that equals or exceeds external pressures, such as wind forces. All access points to a structure's interior must be equipped with two sets of doors or a revolving door airlock. These lightweight structures are secured by ground anchors, or they are attached to a foundation.

Figure 42.5 Pneumatic structures use pressurized air to inflate a flexible material. *(Image copyright GeoM, 2009. Used under license from Shutterstock.com)*

Figure 42.6 This is an unreinforced air-supported structure utilizing a three-ply membrane.

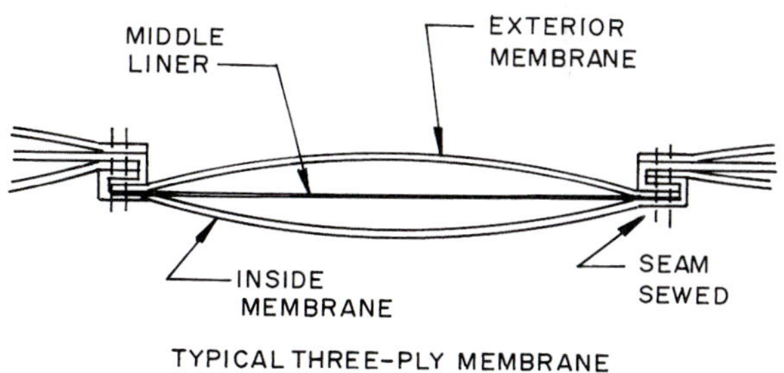

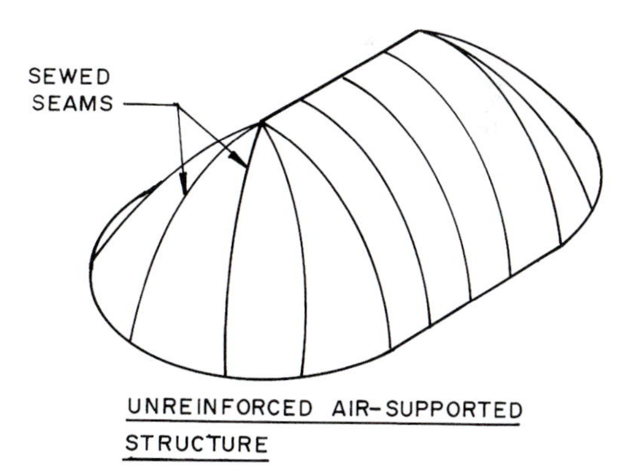

Air-supported structures use either single or multiple wall enclosures made of a flexible material, such as Teflon-coated fiberglass fabric or polyester coated with vinyl or neoprene (Fig. 42.6). Some are unreinforced membranes in which the membrane is the primary structural element. Another type uses a membrane reinforced by a network of cables or webbing. The webbing forms the primary structural system, and the membrane spans the area between the cables or webbing. Air locks are used as entrances, and emergency exits are counter-pressure balanced, self-closing, and have panic hardware. Membranes are often light admitting, and must meet local fire codes.

Geodesic Structures

A **geodesic dome** is a spherical shell structure based on a network of triangular elements that provide triangulated rigidity and distribute stresses across the entire structure.

The architect Buckminster Fuller coined the term "geodesic dome," receiving a U.S. patent in the late 1940s. Geodesic domes are very strong and provide an enclosed space free of structural supports. The infill material most widely used is glass, although other opaque materials are also utilized (Fig. 42.7). The basic structure can be erected quickly and efficiently using small, lightweight members. Domes can range in size from small enclosures to spaces as large as 150 ft. (50 m.) in diameter. A geodesic dome provides a very aerodynamic structure that is able to withstand considerable wind loading. Numerous manufactures provide frame materials and engineering for geodesic domes.

Figure 42.7 Geodesic domes provide a structurally strong, light-filled space free of internal supports. *(Image copyright Massimiliano Pieraccini, 2009. Used under license from Shutterstock.com)*

Space Frames

A **space frame** is a truss-like, lightweight rigid structure constructed by interlocking struts in a geometric pattern. Space frames are essentially three dimensional trusses that can accomplish very long spans with minimal internal supports. They derive their strength from the inherent rigidity of the triangular frame, with tension and compression loads transmitted along the length of each strut. Materials used to construct space frames include steel tubes, hollow sections, and I-beams. Space frames are an increasingly common architectural structure, especially for large roof spans in modern commercial and industrial buildings (Fig. 42.8).

Figure 42.8 This steel space frame uses three-dimensional trusses to create a long span roof canopy.

Figure 42.9 An integrated ceiling can incorporate lighting, security, sound, and sprinkler systems. *(Image copyright Shi Yali, 2009. Used under license from Shutterstock.com)*

INTEGRATED ASSEMBLIES

Integrated ceilings are a type of integrated assembly (Fig. 42.9). They are pre-engineered and prefabricated for assembly on-site. A wide range of products are available, as discussed in Chapter 36. Some integrated systems include lighting, air supply registers, ducts, air return grilles, fire sprinkler systems, and communication linkage. These systems are coordinated with CSI MasterFormat Divisions 15 and 16.

PRE-ENGINEERED STRUCTURES

All types of pre-engineered prefabricated buildings and structures are classified as special construction, even if they are erected on temporary foundations. Some of the structures included are metal building systems, glazed structures (such as a greenhouse), portable buildings, grandstands and bleachers, cable supported, fabric, and log structures. Information on some of these is given in Chapters 1 and 17.

WASTE AND UTILITIES SECTIONS

A number of special constructions occur in the area of waste treatment and systems involved with utilities. Filters under drains and media include the piping and filters used in water and fluid waste treatment. Filter media include anthracite, charcoal, diatomaceous earth, mixed, and sand media. Digester covers and appurtenances include special tank covers and assemblies used on digestion tanks.

Oxygenation systems include site-assembled piping systems and related equipment for the dissolution and mixing of gaseous oxygen in liquid waste. This includes oxygen generators, oxygen storage, and dissolution systems.

Utility control systems include the operating and monitoring systems for water supply, wastewater, and electrical power generation plants. They include metering devices, display panels, control panels, and sensing and communicating equipment.

MEASUREMENT AND CONTROL INSTRUMENTATION

Measurement and control instrumentation are covered in a section of Division 13. The types of controls vary by industry and can include mechanical, electrical, fluid, pneumatic, and computer controlling devices. These systems may be totally automated or semi-automated. They can monitor phenomena such as temperatures or pressures, regulate a system so it operates at the required speed and quality, and control safety within a manufacturing, processing, or assembly plant.

Devices listed as recording instruments are installed to measure and record such various occurrences as seismic information, stresses in structures, and meteorological information, such as solar and wind energy. Transportation control instrumentation describes systems used to monitor and control the various aspects in transportation systems, such as airport control, railway control, subway control, and transit vehicle control.

Review Questions

1. What types of covering materials are used for air-supported structures?
2. How are entrances to air-supported structures treated so air pressure within a structure is maintained as people enter and leave?
3. What is meant by special-purpose rooms?
4. What types of structures are available as pre-engineered units?
5. What kinds of materials are used to construct space frames?
6. What is the basic structural element of the geodesic dome?

Key Terms

air-supported structure
geodesic dome
space frame

Activity

1. Locate any of the special construction areas mentioned and arrange a visit. Ask the person in charge to conduct a tour and explain how the system operates.

Additional Resources

Sweets Construction Catalog, Architects, Engineers, and Contractors, Edition, Division 13, McGraw-Hill Construction, Two Penn Plaza, New York 10121.

See Appendix B for addresses of professional and trade organizations and other sources of technical information.

14

Conveying Equipment CSI MasterFormat™

(Image copyright Christophe Testi, 2009. Used under license from Shutterstock.com)

CHAPTER 43

Conveying Systems

LEARNING OBJECTIVES

Upon completion of this chapter, the student should be able to:

- Understand the codes related to the design and installation of conveying systems.
- Understand the various operating mechanisms used on elevators.
- Be familiar with additional conveying systems used in buildings.

Many different types of conveying systems are available, each designed for a special purpose. Some move materials and equipment, and others move people. They range from a simple dumbwaiter used to carry mail between floors in a building, to large, complex elevator systems carrying people or heavy materials and equipment.

ELEVATOR HOISTWAYS

The **hoistway** is a vertical, fire-resistant enclosed shaft in which an elevator moves. It has a **pit** at the bottom and openings at each floor. The openings are protected by automated doors controlled by an operating system. Some types require a penthouse on the roof above the shaft. Codes typically require that elevator shafts have a 2 hour fire rating and that the opening be protected with doors having a $1\frac{1}{2}$ hour fire rating. A metal or concrete deck is required on the top of the hoistway (Fig. 43.1).

Hoistway Doors

Building codes specify permissible door types and their opening requirements. Passenger elevators generally employ horizontal sliding doors, although swinging doors may also be used. Vertical sliding doors are used on freight elevators.

Elevator doors are closed automatically before the car leaves the landing zone. The **landing zone** is an area 18 in. (5490 mm) above or below the landing floor. The elevator car will not move unless all doors on all levels are closed and locked. Doors have locks that prohibit them from being opened by someone on the landing side. Emergency access is available for maintenance and rescue personnel.

Machine Rooms

Machine rooms are part of the hoistway and provide a fire-resistant enclosure for the required hoisting machinery, controls, pumps, and hydraulic oil storage. Since control systems are computerized, the area should be equipped with temperature control. Only equipment involved with the operation of the elevator is permitted in the machine room. The size of the machine room will vary depending on the type of elevator.

Machine rooms for traction-type elevators are generally situated in a penthouse on the roof over the hoistway (Fig. 43.2). The structural system must be designed to carry the weight of the machinery plus loads required for lifting and lowering the car. In some cases, traction elevator machine rooms can be located in the pit at the bottom of the shaft. Hydraulic elevator machine rooms are usually located to the side of the hoistway (Fig. 43.3). They contain the hydraulic equipment and controls.

Venting Hoistways

In a fire, the elevator hoistway may become a vertical flue for gases and smoke and be used to carry away these noxious materials via venting at the top. Some

Figure 43.1 The elevator hoistway is constructed with fire-resistant walls, top, and pit.

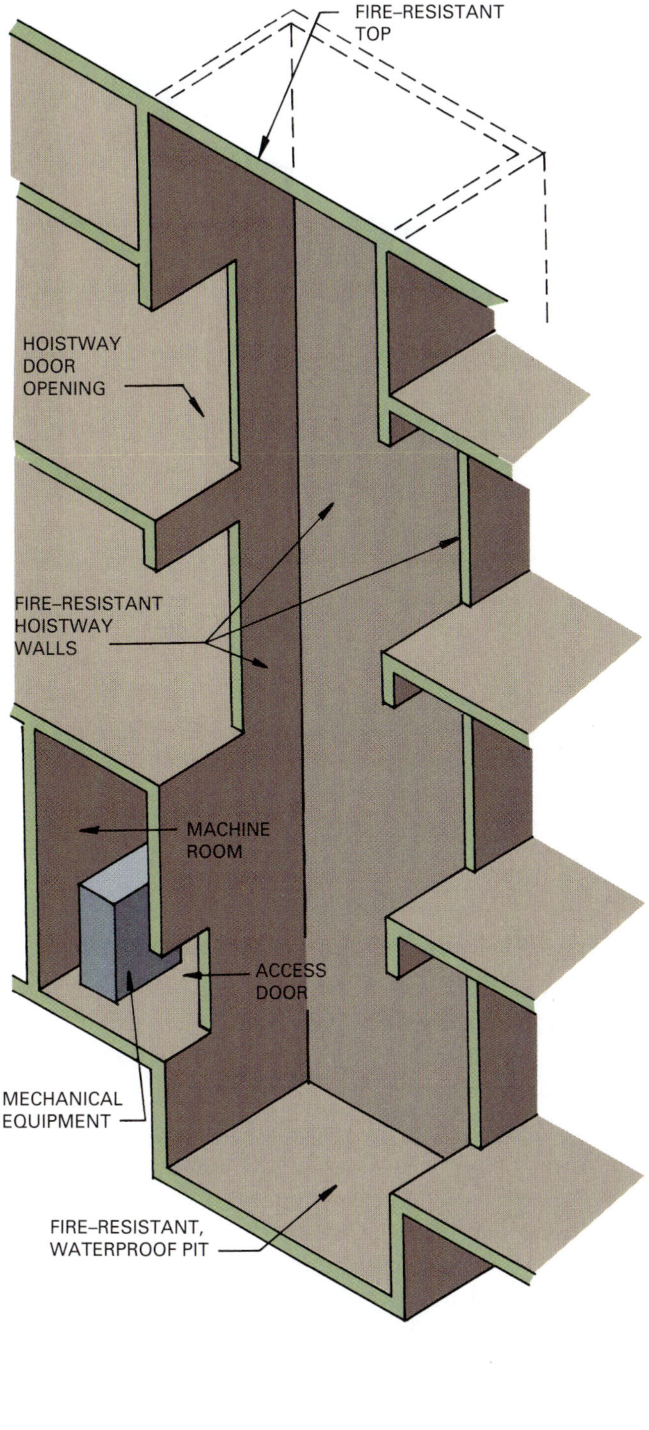

Figure 43.2 A traction electric elevator generally has the hoisting machine at the top of the hoistway. *(Courtesy Adamantios, http://en.wikipedia.org/wiki/File:Radio-frequency-anechoic-chamber-HDR-0a.jpg)*

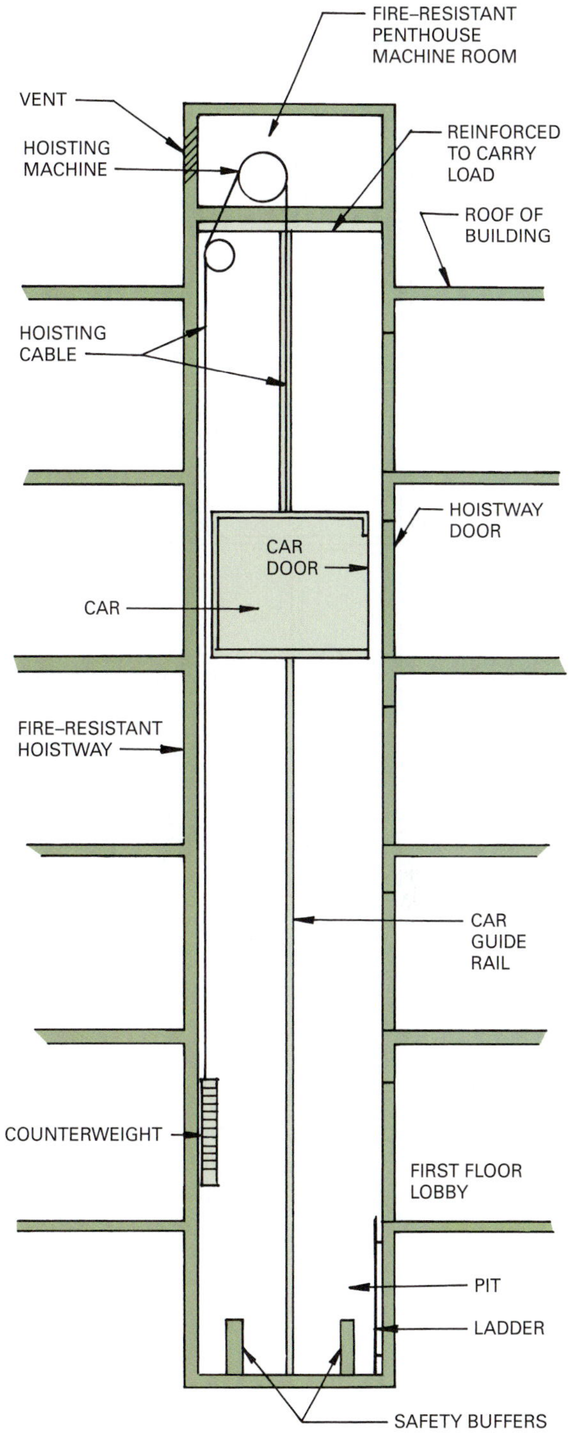

building designs use other means of containing and venting smoke and gases from a fire site to prevent them from moving to unaffected floors. Codes may require the hoistway to be vented, but other design considerations for controlling smoke must be considered as the hoistway is designed.

Hoistway Sizes

The clear inside dimensions for hoistways housing standard-size elevators are specified in the *National Elevator Industry Standard: Elevator Engineering Standard*

Figure 43.3 A hydraulic elevator typically has the machine room on the side of the hoistway.

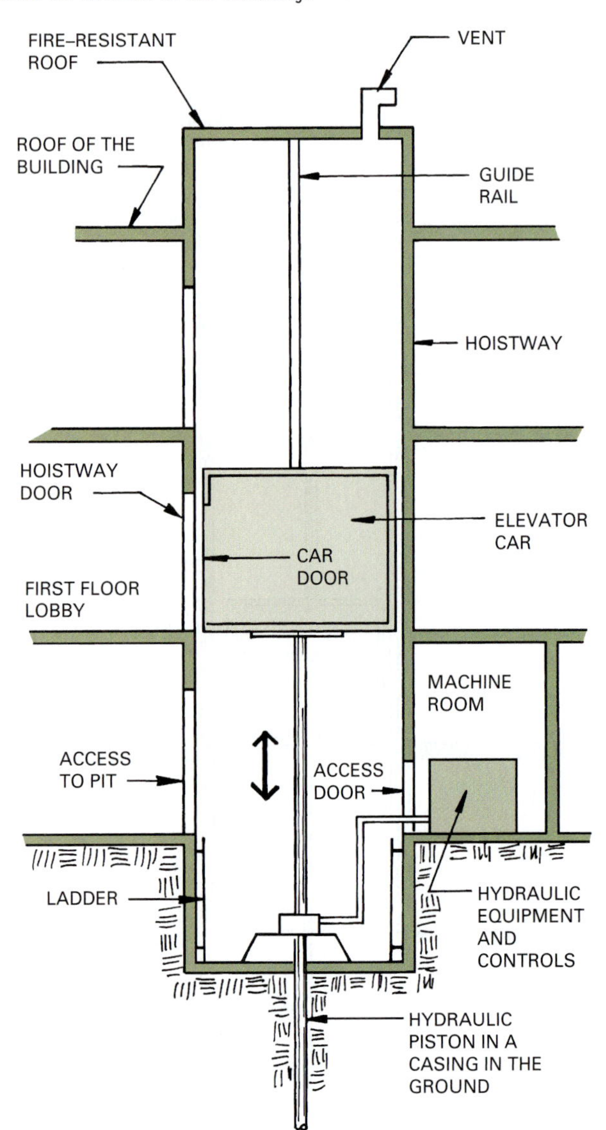

Layouts. Maximum and minimum clearances between the hoistway and cars, moving weights, and other equipment must be observed. For specially designed elevators the hoistway may accommodate larger or smaller car and moving equipment, and the required clearances become critical.

ELEVATOR CARS

The **elevator car**, a platform enclosed by walls and a roof, is designed to carry passengers, freight, and other specialty loads, such as hospital beds and stretchers. Each car contains lighting, controls, venting units, handrails, telephones, and various types of wall, ceiling, and floor finish materials. Passenger elevators are finished to be attractive and durable (Fig 43.4). Freight elevator

Figure 43.4 A typical elevator car. *(Courtesy Dover Elevator Systems, Inc.)*

cars must have durable, abrasion-resistant interior finishes. Typically, they have steel floors and walls.

The car in a hydraulic elevator is mounted on top of a piston. Cars on electric elevators have a frame to which wire hoisting ropes are connected. The design of elevator cars is detailed in ANSI/ASME A17.1.

Car doors are automatically operated and cannot be opened while a car is moving or outside a landing zone. A car will not move if the door is open. Both single- and double-entrance elevator cars are available. Double entrance cars open to the front and rear and are widely used on hospital and freight elevators (Fig. 43.5). Elevator car doors may be center opening, two-speed sliding doors, and single sliding doors (Fig. 43.6). Center-opening doors provide the quickest access and egress and are used on high-speed systems. Two-speed sliding doors provide the widest possible opening but operate slower than center-opening doors. Single sliding doors are the most economical and the slowest. They move right or left, depending on the car design, and the opening width is limited by the width of the car.

Elevator cars contain a panel with several controls. Control panels must be located low enough to accommodate a person in a wheelchair. They feature a button that sounds an alarm outside the hoistway and an

Figure 43.5 Single and double entrance doors are available.

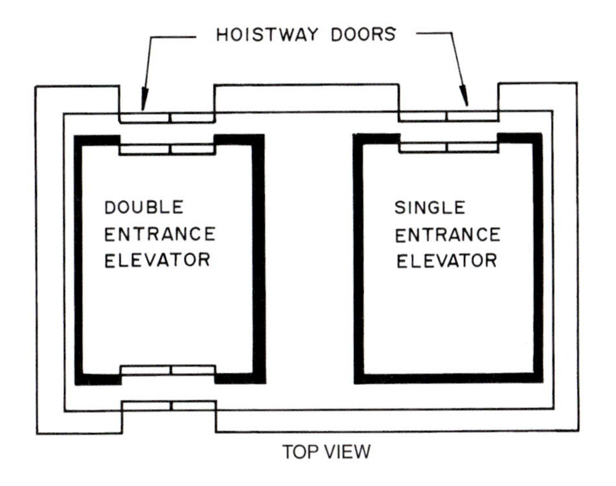

Figure 43.6 Typical types of car door action. *(Image copyright Adam Radosavljevic, 2009. Used under license from Shutterstock.com)*

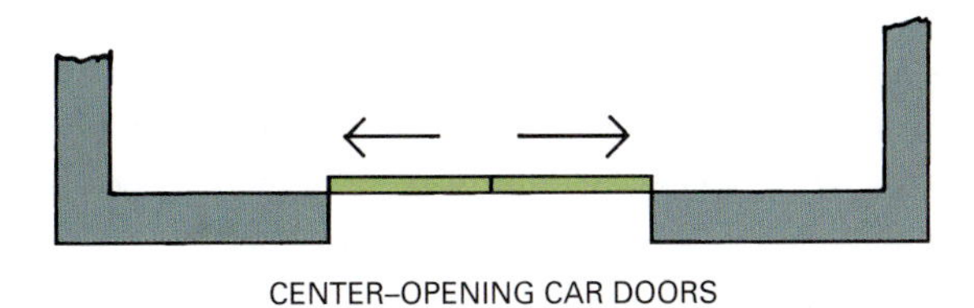

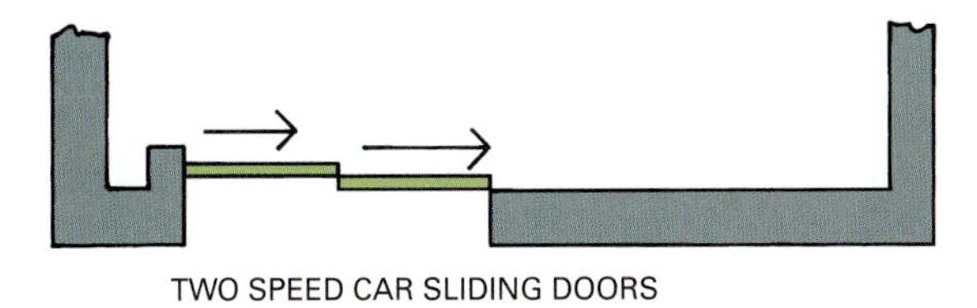

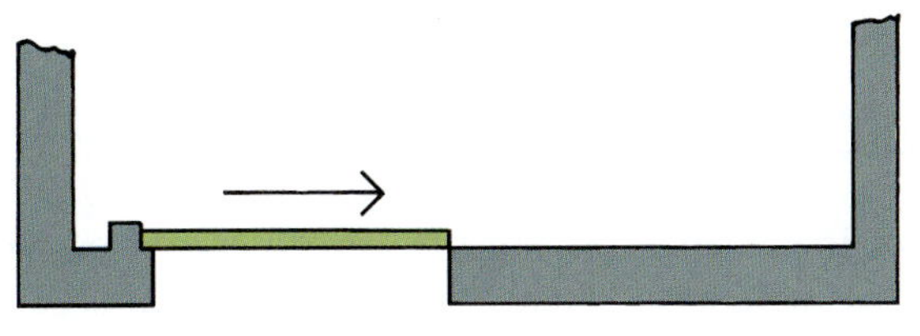

Figure 43.7 Typical elevator car controls. *(Image copyright Vladimir Mucibabic, 2009. Used under license from Shutterstock.com)*

emergency stop switch. A telephone provides a means of communication with building personnel in case of emergency. The panel also houses buttons for each floor and one that holds doors open when the computer tries to close them (Fig. 43.7).

The capacity of a car is determined by the maximum number of people it can safely carry. The capacity is the total load in pounds divided by 150 (the allotted weight per person). For example, an elevator rated at 1,500 lbs. would be marked to carry a maximum of 10 people.

Car sizes vary with the different manufacturers, but typical sizes for passenger elevators range from 6 ft. × 4 ft. (1.8 × 1.2 m) to 8 ft. × 6 ft. (2.4 × 1.8 m). Door are typically 3 ft. to 4 ft. (0.9 to 1.2 m) wide and 7 ft. to 9 ft. (2.1 to 2.7 m) high. Hospital elevators are generally 9 ft. × 7 ft. (2.7 × 2.1 m). Freight elevators range from 5 ft. × 7 ft. (1.5 × 2.1 m) to 12 ft. × 16 ft. (3.7 to 4.9 m). They have clear door widths from 5 ft. to 12 ft. (1.5 to 3.7 m).

ELEVATOR CODE STANDARDS

Elevators, escalators, moving walks, and dumbwaiters are designed and installed following the code requirements of the American National Standard Safety Code for Elevators, Dumbwaiters, Escalators, and Moving

Walks, ANSI/ASME A17.1, and local building codes. Standard elevator sizes and shapes have been developed by National Elevator Industries, Inc. (NEII). Standards such as Elevator Engineering Standard Layouts and Suggested Minimum Passenger Elevator Requirements for the Handicapped are also available. These standards establish basic rules that require specific compliance and should be coordinated with the design criteria of the complete elevator installation. Following are some examples.

Hoistway design criteria specify that hoistways be enclosed for their entire height with fire-resistant materials such as masonry, drywall, or concrete. Pits must also be constructed of noncombustible material and waterproofed to prevent groundwater entry. Hoistway tops must be enclosed by concrete or metal floors. Hoistway assembly design must prevent the accumulation of gases and smoke in case of fire. No windows are permitted, and the hoistway and machinery space must be free of any pipes or ducts.

The elevator machine room must have fire-resistant enclosures and doors. The only machinery allowed in the room is that needed for the operation of the elevator. Since regular maintenance will be necessary, permanent and easy access to the machine room is required.

Electrical equipment and wiring must conform to the National Electrical Code, ANSI/NFRA 70. All main electrical feeds are installed outside the hoistway. The only electrical equipment allowed in the hoistway are those devices directly connected with the elevator. The machine room should have permanent lighting and natural or mechanical ventilation.

Buffers are required in the pit. Buffers are energy-absorbing units located at the bottom of the hoistway that absorb any impact from a car that descends below the normal lowest level. Code requirements also include stipulations for required tests, emergency condition operation, venting, opening protection, signals, and signs.

ELEVATOR TYPES

An elevator consists of a hoisting mechanism connected to a car or platform that slides vertically on guides on the sides of a fire-resistant hoistway. It is used to move passengers and materials between floors of multistory buildings. Passenger elevators are designed to transport people between floors. Freight elevators carry materials between floors. Safety regulations permit only the operator or others needed to handle the materials being transported. Hospital elevators have cars large enough to transport patients on stretchers or beds along with their attendants. They can also serve as passenger elevators.

The Americans with Disabilities Act specifies the following for elevator requirements in new construction.

One passenger elevator should serve each level, including mezzanines, in all multistory buildings. An exception is that elevators are not required in facilities less than three stories unless the building is a shopping center, shopping mall, or the office of a health care provider.

Elevators are not considered part of the system of egress for code purposes because they might not operate as required during an emergency, such as a fire or an earthquake, and people could get stranded between floors. If power fails, an elevator will automatically return to the lowest landing and the doors will open, allowing passengers to exit. In the event of a fire, the elevator fire service will be activated by the building's smoke alarm or by the fire service key switch located in a hallway. When this happens, all car calls are canceled, cars return to the main floor, and the doors open. Since elevators are vital for use by firefighters and other emergency personnel, they may be reactivated for emergency use.

Elevators are either electric or hydraulic. The electric elevator uses an electric motor to supply power. The mechanism includes an electric motor, a brake, a driving sheave or drum, gearing, belts, and wire rope. Electric elevators are used on low-, medium-, and high-rise buildings and are faster than hydraulic elevators.

ELECTRIC PASSENGER ELEVATORS

Electric passenger elevators suspend cars from wires and use weights to counterbalance them. A car is guided by vertical guide rails. The electrically driven hoisting mechanism may be located at the top or bottom of the shaft. The wire rope runs over traction sheaves. The two types in common use are geared and gearless traction mechanisms.

Traction Driving Mechanisms

Electric elevators are powered by traction machines consisting of an electric motor connected to a driving sheave. This connection can be direct or operated through a series of gears. A wire rope runs through grooves in the face of the sheave or traction is provided by friction.

Gear-driven traction machines provide slower rising speeds and are used when slower speeds are desired. Gearing may utilize a helical gearbox or a worm gear. Gearless direct drive machines have the traction sheave connected directly to the motor. They provide high speeds and are usually used on high-rise buildings.

Geared Traction Elevators

A geared electric passenger elevator is shown in Fig. 43.8. This hoisting mechanism uses a worm gear to drive a large spur gear connected to the traction sheave. A sheave is a pulley with a grooved rim for retaining a wire rope used to transmit force to the rope. The wire rope runs over the traction sheave and moves as the sheave rotates. This type of drive is used when low speeds and high lifting capacity are required. Speeds can be changed by varying the size of the spur gear, which changes the gear ratio. Typical car rise speeds for geared-traction passenger elevators range from 350 to 500 ft. /min. (106 to 152 m/min.). The range of lifting capacity is typically from 2,000 to 4,500 lb. (900 to 2,025 kg). The hoisting mechanism is usually housed in a penthouse on the roof. Geared traction freight elevators typically have a rise from 50 to 200 ft. /min. (15 to 60 m/min.) and carry loads up to about 20,000 lb. (9,072 kg).

Gearless Traction Elevators

A **gearless traction elevator** is shown in Fig. 43.9. The traction sheave and brake are mounted directly on the motor shaft. The speed of rotation of the traction sheave (which contains the wire rope) is the same as the speed of the motor. The speed can be varied by using a DC electric motor built to run at the speed required. Car rise speeds can range from 500 to 1200 ft./min. (152 to 366 m/min.). Gearless elevators can also be operated using a motor generator drive. The range of lifting capacity is typically from 2,000 to 7,000 lb. (900 to 3,150 kg). The hoisting mechanism is normally housed in a penthouse on the roof.

Electric Elevator Control

The elevator control system regulates the starting, stopping, safety devices, speed of movement, direction of movement, acceleration, and deceleration of the car. Two types of controls are in general use: multivoltage and variable-voltage variable-frequency control (VVVF).

Figure 43.8 This is a geared electric passenger elevator that uses an electric motor to drive a gear box that regulates the speed of the traction sheave. *(Courtesy Otis Elevator Company)*

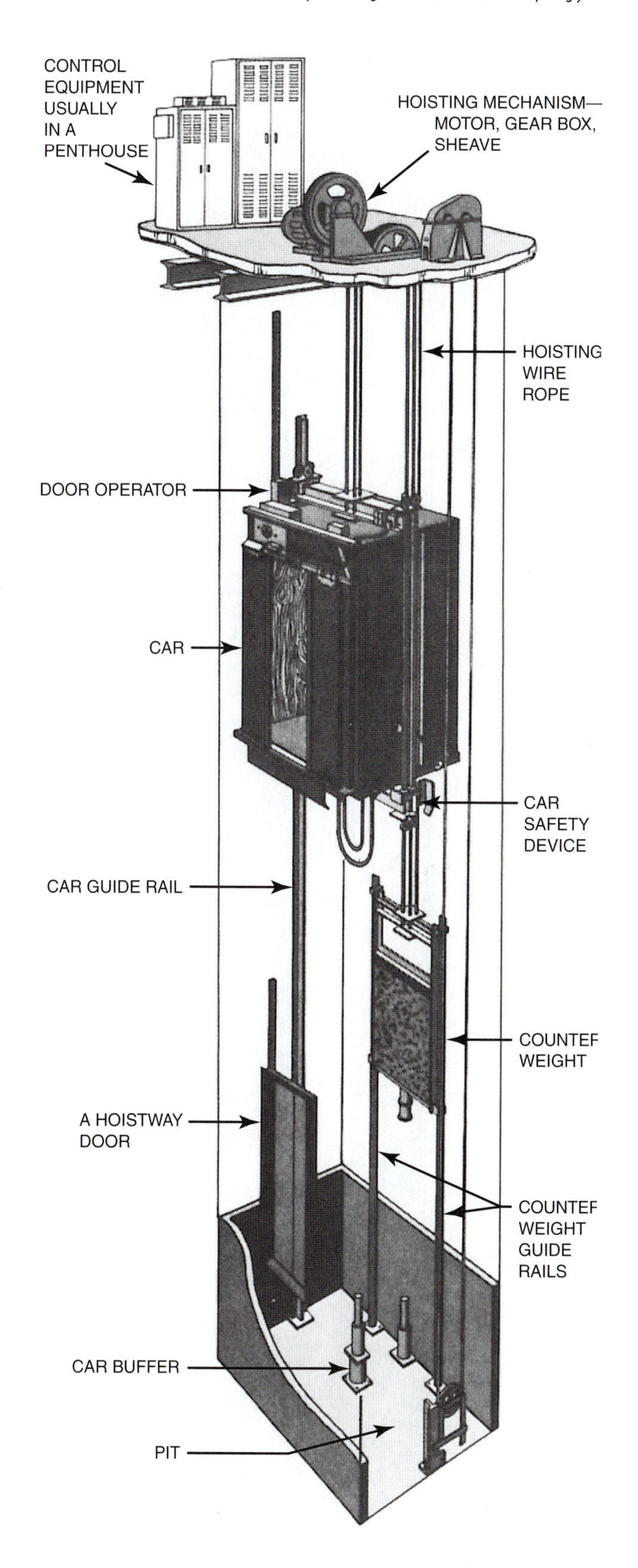

Figure 43.9 This gearless electric passenger elevator has the traction sheave and brake connected directly to the motor shaft. *(Courtesy Otis Elevator Company)*

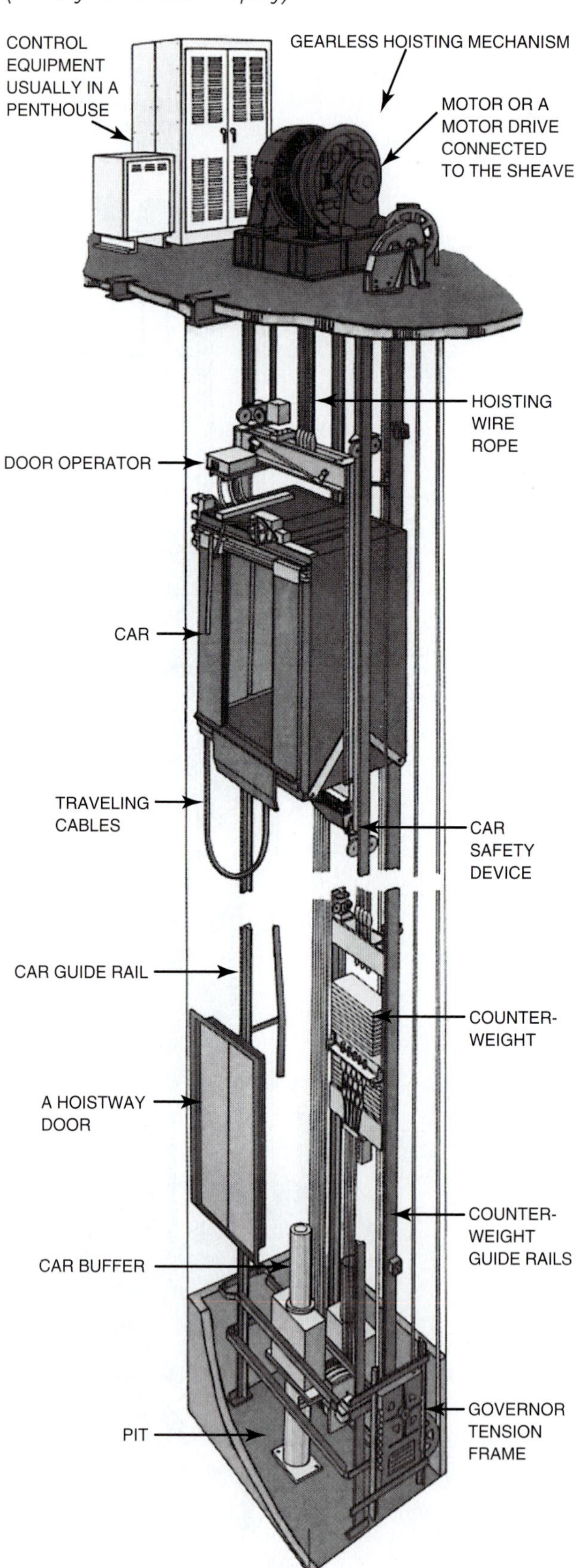

Multi-voltage control is used with machines having DC motors. It controls speed by varying the voltage supplied to the motor armature. Multi-voltage control provides a smooth regulation of speed and is widely used on passenger elevators.

VVVF control is used to control AC motors and produces a smooth overall operation of the car. It is more efficient than multi-voltage control and DC motor operation, and it is gaining popularity over DC controls and motors.

Car Safeties

Car safeties are used to stop the movement of a car and hold it in position. A device called a governor monitors speed and activates a car safety if the car exceeds safe velocities. A car safety applies brake shoes against the guide rails, stopping the car. The governor also switches off the electrical power to the motor. In addition, some types activate a brake shoe to the motor drive shaft.

Roping

Traction-type machine cars must be suspended from at least three hoisting ropes (ANSI 17.1). The wire rope consists of steel strands laid helically around a hemp core. Each steel strand is made up of steel wires wrapped helically around a steel wire core (Fig. 43.10).

The arrangement of the roping of traction-type elevators greatly affects the speed of the car and the loads on the hoisting wires. Typical roping systems are shown in Fig. 43.11.

Single-wrap roping has the wire pass over the driving sheave once before it moves on to the counterweight. The deflection sheave moves the wire clear of the moving car. This provides a 1:1 ratio, which means it will move the car 1 ft. (305 mm) for every 1 ft. (305 mm) of travel along the circumference of the sheave.

Double-wrap roping has the wire wrap twice around the driving sheave and a secondary sheave and then on to the counterweight. Double-wrap roping can be used to provide a 1:1 or 2:1 ratio. The 1:1 double-wrap ratio causes less wear on the wire but increases the load on the sheave. The 2:1 double-wrap roping is used for heavily loaded elevators. This system moves the car 1 ft. (305 mm) for every 2 ft. (610 mm) of rope travel along the circumference of the sheave. Other roping systems are also used.

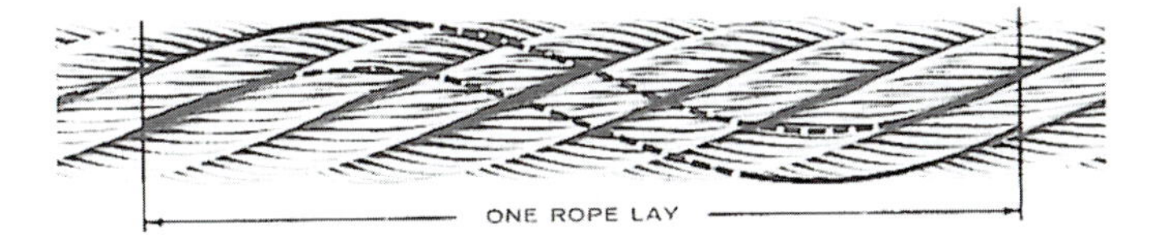

Figure 43.10 Typical types of wire rope. A lay is the length of a strand required to make a single wrap around the core. It is identified by the way the wires have been laid to form the strands and the way the strands are laid around the core. Left lay wire rope strands slant down to the left. Right lay strands slant to the right. *(Courtesy Bethlehem Steel Corporation)*

Figure 43.11 Several types of roping systems used on electric traction elevators.

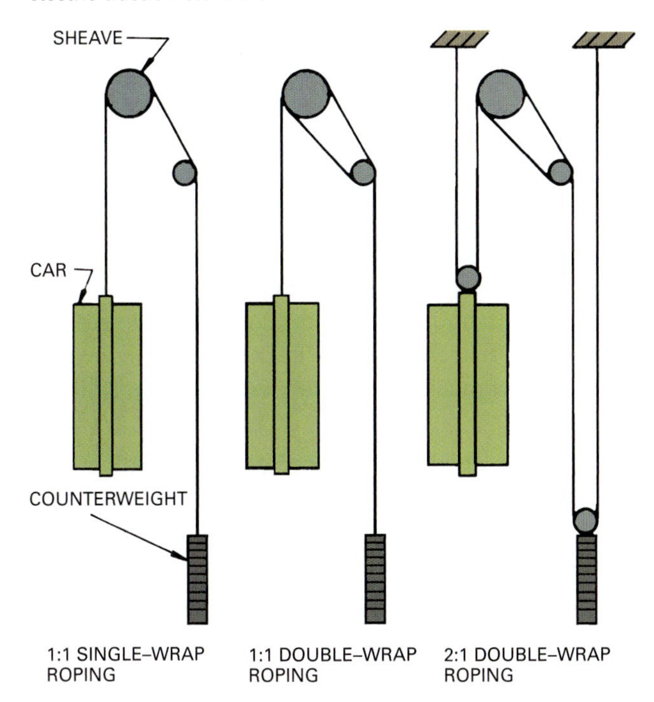

Winding Drum Machines

A winding drum hoisting mechanism has a gear driven drum. The hoisting wire is attached to the drum and winds or unwinds around it as it rotates. This type is used with dumbwaiters and some small residential elevators.

Counterweights

Counterweights are steel plates that slide on vertical steel guides. As an elevator rises, they lower, and as the elevator lowers, they rise. Counterweights generally equal the weight of the unloaded car and hoisting wires plus a percentage of the load capacity of the elevator. Since they rely on gravity, they help raise the car, thus reducing the overall power required.

Operating Systems

Elevator operating systems control the operation of an elevator. They range from simple systems to complex, automatic systems.

Typical operator-controlled systems include the car-switch and signal systems. In the car-switch system, the operator controls the direction of travel and initiates car movement. In the signal system, the operator presses buttons indicating the floor stops and then presses a start button. The car automatically stops at each floor for which a button was pressed.

Automatic operating systems do not require an operator. Signals are initiated by the passengers or by an automatic operating device. Typical automatic operating systems include the single, selective collective, and group systems.

Single automatic systems start when a passenger pushes a button to indicate the floor desired (see Fig. 43.7). The car moves automatically to that floor, stops, and the doors open. By pressing a button on the wall next to the elevator doors, anyone can call a car to any floor.

The selective collective automatic system has two call buttons, up and down (Fig 43.12). When the up button in a car or on a hall wall is pressed, the car moves up, stopping at floors where the hall wall up button has been pressed. When it reaches the top up floor, it descends, stopping as indicated by passengers using the elevator's control panel and on floors where a hall wall down button has been pressed (Fig. 43.13).

Group automatic systems control the operation of several cars that serve the same floors using a supervisory control system. The control system automatically decides which cars to send to various floors and coordinates the flow of traffic. The cars are dispatched from the first floor on predetermined intervals, and whichever car is nearest a floor where a passenger is waiting stops to make the pickup. This reduces waiting time and increases the number of passengers that may be carried.

Figure 43.12 Call buttons signal the desired direction of travel of the person waiting for the elevator. *(Image copyright Max Romeo, 2009. Used under license from Shutterstock.com)*

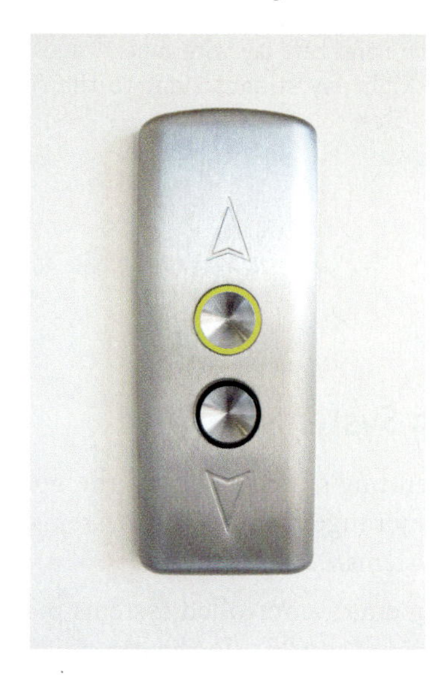

Figure 43.13 Call buttons on the wall bring the elevator to a floor. The overhead direction arrows indicate the direction the arriving elevator is moving.

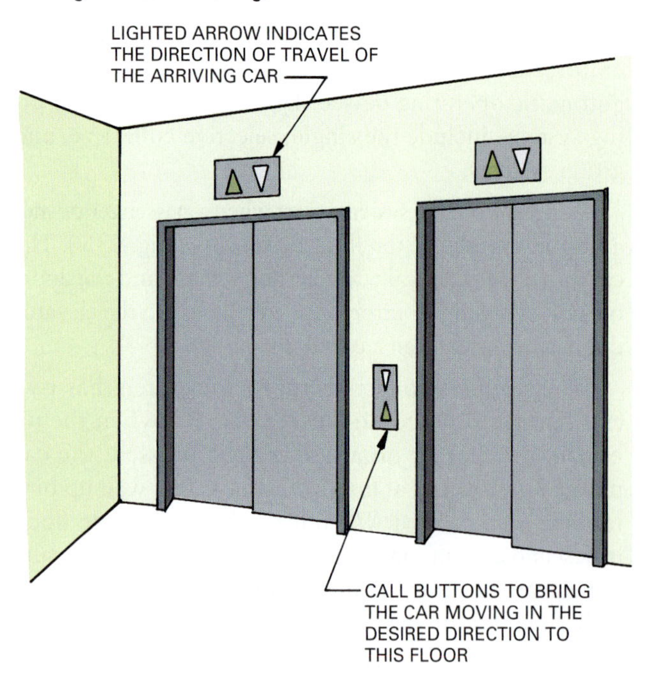

The supervisory control system is able to make adjustments in car flow and direction to handle times when requirements vary, such as late afternoon in a high-rise office building when many workers are trying to leave at the same time. The use of computers to make adjustments and control car movement has increased the efficiency of group car operations in heavily loaded situations.

Computerized systems vary, but many use a traffic information database to decide which pattern of car utilization will produce the shortest waiting time. For example, in peak up traffic times the cars may make fewer stops per round trip, returning to the lobby more often. A car may be assigned to serve only one group of floors and return quickly to the lobby. Another car can serve another group of floors. This reduces the stops for each car and facilitates quicker lobby return for more trips. The reverse can occur when the down trips become heavily loaded. The control system display in the lobby indicates which floors each elevator will serve.

HYDRAULIC ELEVATORS

Hydraulic elevators are used mainly for low-rise installations. The car is mounted on top of a piston that slides inside a hydraulic pressure cylinder. Hydraulic oil is pumped under pressure by an electric pump to the bottom of the piston, forcing it to rise and moving the car with it. When the car reaches the desired floor, the pump automatically stops and the oil under pressure holds the car at that level. The car lowers by releasing oil from the pressure cylinder into a storage tank. The installation requires that the pressure cylinder be sunk into the ground a distance equal to the length of the cylinder (Fig. 43.14).

Hydraulic elevators are more economical than electric ones, and their operating mechanism is simpler. They do not require a penthouse on the roof to house the hoisting mechanism.

The speed of rise of hydraulic passenger elevators typically falls within the range of 100 to 150 ft./min. (30 to 46 m/min.). The range of lifting capacity is usually 1500 to 4,000 lb. (675 to 1,800 kg). Hydraulic freight elevators ordinarily rise at speeds of 50 to 100 ft./min. (15 to 30 m/min.) and carry loads from 3,000 to 12,000 lb. (1,350 to 5,400 kg). Since gravity affects the system, these elevators lower faster than they rise.

FREIGHT ELEVATORS

Freight elevators are designed to carry general freight, motor vehicles, and other heavy concentrated loads. Hydraulic systems can be used for freight elevators in low-rise buildings, while electric systems are necessary in mid- and high-rise buildings. Three classes of freight elevators are specified in the American National Standard Safety Code for Elevators, Dumbwaiters, Escalators, and Moving Walks (ANSI/ASME A17.1). Class A is limited to carrying a maximum of one-quarter of the

Figure 43.14 A hydraulic elevator moves the car with a long piston controlled by a hydraulic pump and valves. *(Courtesy Otis Elevator Company)*

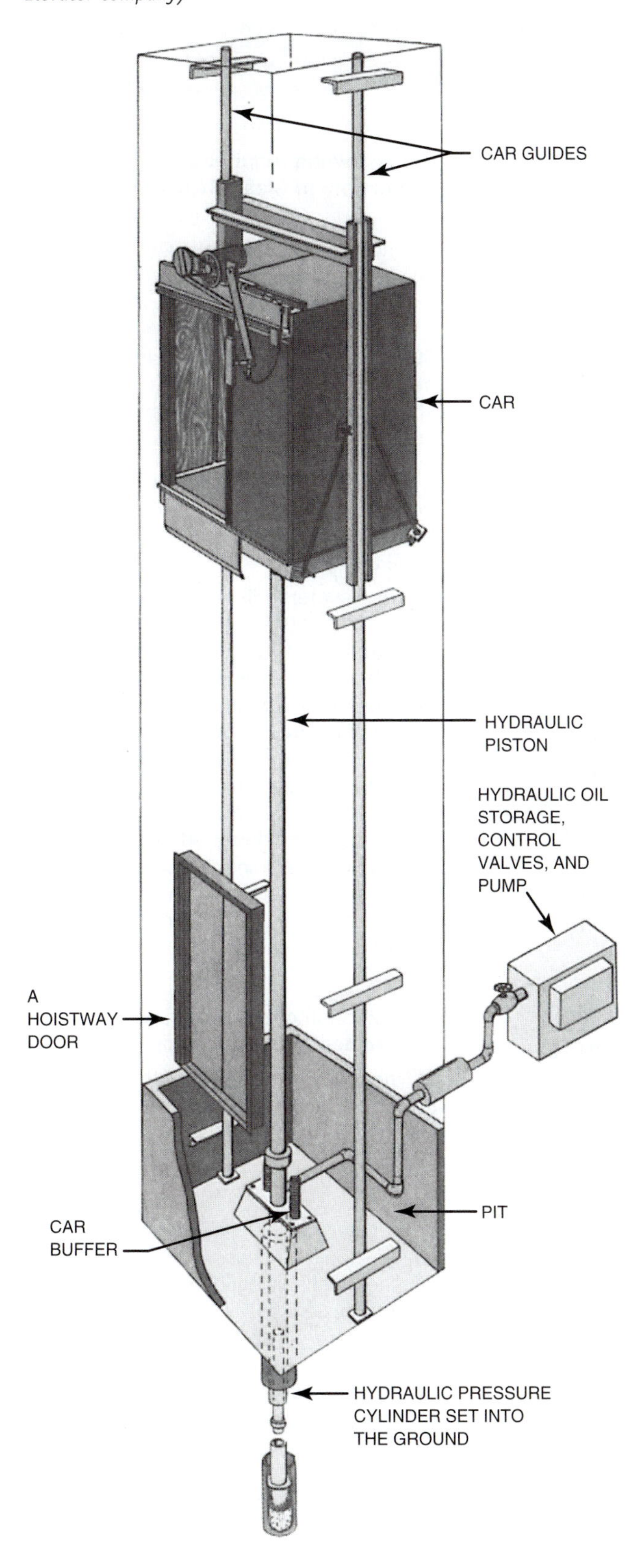

rated load. It is for general freight that is loaded and unloaded by hand (often with a lightweight hand truck). Class B handles only motor vehicles. Class C elevators are designed to handle heavy concentrated loads. A Class C1 elevator can carry materials plus industrial trucks (forklifts, pallet trucks, etc.). Class C2 elevators are used when materials are loaded by industrial trucks but the trucks are not carried with the materials. Class C3 elevators carry heavy concentrated loads other than trucks. These classifications are detailed in Fig. 43.15.

Pre-manufactured freight elevators are available, but many are custom designed. Size of car and loading capacity vary widely with industrial requirements. Platforms are engineered to carry anticipated loads and withstand extra weight and movement when being loaded. For example, a truck loading materials can cause sideways movement and shock impacts if a load is not slowly lowered to the floor. A load placed off-center in a car produces eccentric loading, the possibility of which must be considered as the system is designed.

The doors on freight elevator cars are often made of an open steel mesh and operate vertically. The door to the hoistway is typically a solid, vertical biparting door with a locking system that will not allow it to open until the car arrives at the floor (Fig. 43.16).

Generally freight elevators are automatically operated (in the same manner as described for passenger elevators). Some operator controlled systems are available.

OBSERVATION ELEVATORS

Observation elevators move up a hoistway secured to the exterior of a building. Cars are designed with glass walls to provide panoramic views of surrounding areas as the elevator rises (Fig. 43.17). The elevators may be geared, gearless, or hydraulic, and the hoisting machinery is hidden. The lower end of the hoistway (at the ground) must be enclosed with a safety barrier.

RESIDENTIAL ELEVATORS

Residential elevators are available in a range of sizes, including cars large enough to carry wheelchairs (Fig. 43.18). Designs vary, but residential elevators may use hydraulic or electric traction systems similar to those described for commercial passenger elevators. The car in an electric system is directed by channel guides, and doors have safety locking devices that prevent them from opening until the car arrives at a designated floor. The hydraulic unit uses a piston to raise the car.

Figure 43.15 Classifications, capacities, and loading requirements for freight elevators. *(Courtesy Dover Elevator Systems)*

DOVER® Capacity and Loading Requirements

All Dover freight elevators are designed and manufactured strictly in accordance with ASME A17.1 according to the following loading classifications:

CLASS A: GENERAL FREIGHT LOADING.

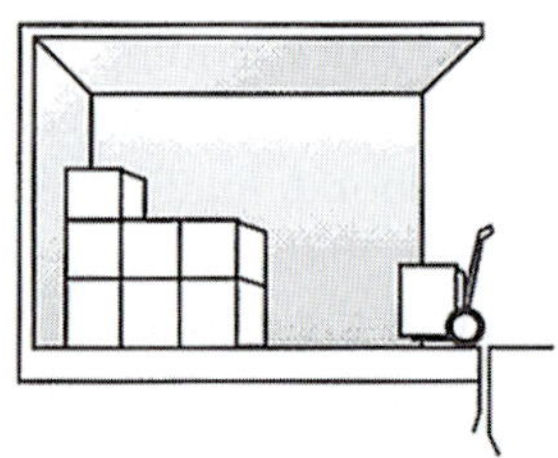

Where the load is distributed, the weight of any single piece of freight or any single hand truck and its load is not more than 1/4 the capacity of the elevator, and the load is handled on and off the car platform manually or by means of hand trucks. For this class of loading, the capacity shall be based on not less than 50 lb/ft2 (244.10 kg/m2) of inside net platform area.

CLASS B: MOTOR VEHICLE LOADING.

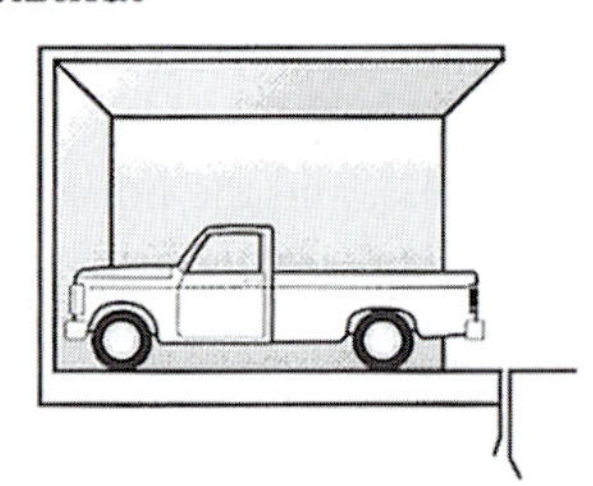

Where the elevator is used solely to carry automobile trucks or passenger automobiles up to the rated capacity of the elevator. For this class of loading, the capacity shall be based on not less than 30 lb/ft2 (146.46 kg/m2) of inside net platform area.

There are 3 Types of Class C Loading:

CLASS C1: INDUSTRIAL TRUCK LOADING.

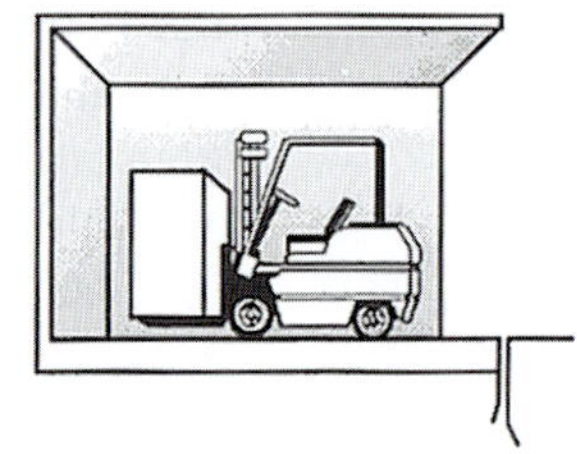

Where truck is carried by the elevator.

CLASS C2: INDUSTRIAL TRUCK LOADING.

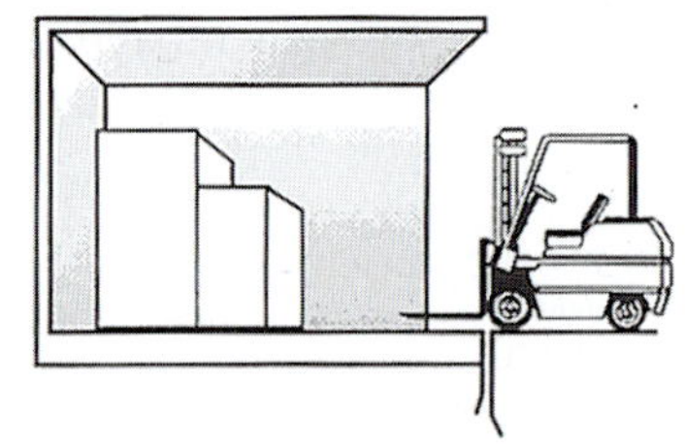

Where truck is not carried by the elevator but used only for loading and unloading.

CLASS C3: OTHER LOADING WITH HEAVY CONCENTRATIONS.

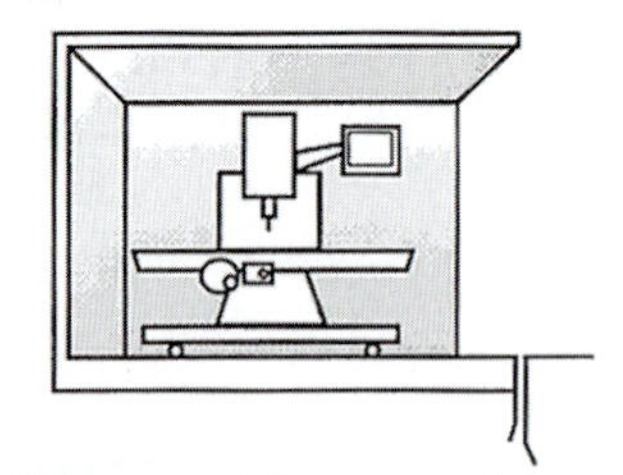

Where truck is not used. Determined on the actual loading conditions, but not less than that required for Class A loading. Consult your Dover representative to assist with the specific needs of this type of loading.

The following requirements shall apply to Class C1, C2, and C3:

The capacity of the elevator shall be not less than the load (including any truck) to be carried, and shall in no case be less than 50 lb/ft2(244.10 kg/m2) of inside net platform area. The elevator shall be provided with two-way automatic leveling.

For Class C1 and C2 the following additional requirements shall apply:

For elevators with a capacity of 20,000 lbs. (9,072 kg) or less, the car platform shall be designed for a loaded truck of weight equal to the capacity or for the actual weight of the truck to be used, whichever is greater. For elevators with a capacity exceeding 20,000 lbs. (9,072 kg), the car platform shall be designed for a loaded truck weighing 20,000 lbs. (9,072 kg) or for the actual weight of the loaded truck to be used, whichever is greater.

For C2 loading, the following requirements shall also apply:

The maximum load on the car platform during loading or unloading shall not exceed 150% of rated load.

Elevator size, capacity, and speed should be chosen to give you the most efficient and economical system possible. Your local Dover representative will be glad to work with you in designing an elevator system that will give you years of rugged use and dependable service.

Figure 43.16 Freight elevators are designed to carry anticipated loads and resist actions from forklifts and other devices used to load them. *(Courtesy Thomas Northcut/Getty Images)*

Elevator cars are steel-reinforced and available with a variety of interior wall, floor, and ceiling finishes. Car sizes range from 36 in. (10,980 mm) to 42 in. (12,810) square. A custom 36 × 48 in. car can accommodate a wheelchair. Most handle loads up to 450 lb. (202.5 kg).

Figure 43.17 Observation elevators are glass enclosed and travel outside of a hoistway or in a hoistway open on one side. *(Image copyright Carlos Neto, 2009. Used under license from Shutterstock.com)*

Figure 43.18 This elevator, used in both commercial and residential applications, is large enough to handle a wheelchair. *(Courtesy Getty Images)*

AUTOMATED TRANSFER SYSTEMS

Automated transfer systems provide for vertical and horizontal distribution of materials. They are widely used in hospitals, clinics, office buildings, hotels, and manufacturing plants. Among them are dumbwaiters, tote box and cart transfer systems, and horizontal tote box transfer systems.

Dumbwaiters

A dumbwaiter is a mechanism used to raise and lower a small car vertically within a building. Dumbwaiters are used in hospitals, restaurants, libraries, and office buildings to move mail, supplies, and materials (such as food, medicine, and books) from one floor to another. Car sizes are controlled by local and national codes.

Standard heights are 3, $3\frac{1}{2}$, and 4 ft. (915, 1,067, and 1,220 mm), with a maximum platform size of 9 sq. ft. (0.837 m2). Units range from light-duty lifts that carry 25 to 50 lb. (11.25 to 22.50 kg) to heavy-duty types that carry up to 500 lb. (225 kg). They may be manually or electrically powered. Manually operated units have an endless rope connected to a large pulley that is connected by gears to a pulley that connects to the hoisting mechanism (Fig. 43.19). Dumbwaiters have automatic braking mechanisms and are usually limited to two-story applications.

Electrically powered dumbwaiters are used on buildings of any height. They are available as drum type, traction type, or hydraulic, as explained in the section on passenger elevators. The drum type has a maximum height of rise of 40 ft. (12.2 m), but the traction type height is unlimited.

Electric dumbwaiters move 50 to 150 ft./min. (15.25 to 45.75 m/min.). The higher speeds are typically used in buildings more than 50 ft. (15 m) high. Some types permit the standard speed to be reduced to as low as 25 ft./min. (7.6 m/min.) for carrying fragile items. Cars can open from the front, front and rear, or front and side (Fig. 43.20). Standard doors are power or manually operated with biparting, slide-up, slide-down, and swinging door configurations.

Dumbwaiters may be counter loading or floor loading. Floor loading types enable carts to load directly onto the dumbwaiter (Fig. 43.21). Electric traction and drum dumbwaiters ride on vertical rails secured to each floor with brackets, the whole of which forms the vertical structure (Fig. 43.22). The lifting mechanism is typically located on the top of the shaft.

Cart and Tote Box Transfer Systems

Cart and tote box transfer systems use high-performance dumbwaiter and elevator lift equipment that may work in connection with a horizontal transfer system. The systems available will lift up to 1,000 lb. (450 kg). Carts come in various sizes but may be up to 30 in. (9,150 mm) wide, 55 in. (16,775 mm) deep, and 65 in. (19,825 mm) high. Tote boxes are about 15 in. (4,575 mm) wide, 20 in. (6,100 mm) long, and 10 in. (3,050 mm) deep. They can be carried on standard dumbwaiters. Small carts with weights not exceeding the capacity of a standard dumbwaiter can be carried by them. Tote box systems use counter-high doors.

Horizontal transfer systems move tote boxes and other items into a vertical transfer system (Fig. 43.23). Typically, items such as mail are placed in a tote box, which is placed on a conveyor. The conveyor system is installed in front of the vertical transfer system door. The boxes are placed on the conveyor and the operator

Figure 43.19 Manually operated dumbwaiters use a hand rope over a pulley to operate the gears used to lift and lower the car. *(Courtesy Vincent Whitney Co.)*

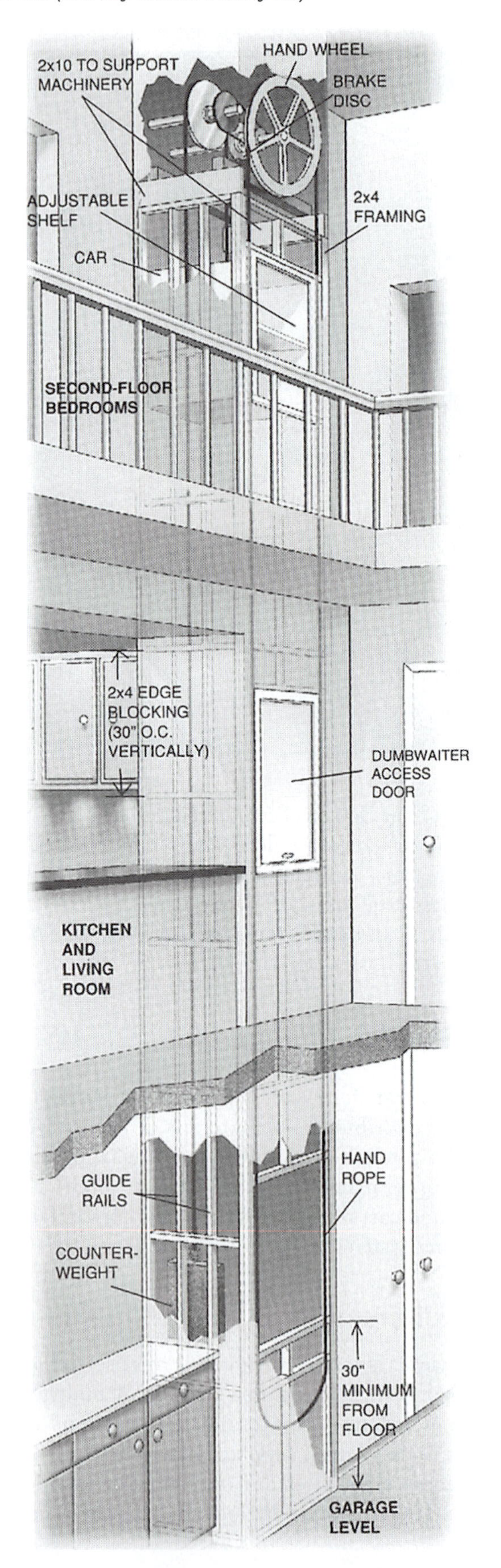

Figure 43.20 Modern dumbwaiters are available in a variety of styles and several door arrangements. *(© Goodshoot/Corbis)*

presses the desired floor destination button. When a car arrives at the receiving floor, the door opens automatically, and a car transfer device loads the tote box into the car of the tote box transfer system. The door closes and the car moves up or down to the desired floor, where the door opens and the box is automatically unloaded onto a conveyor.

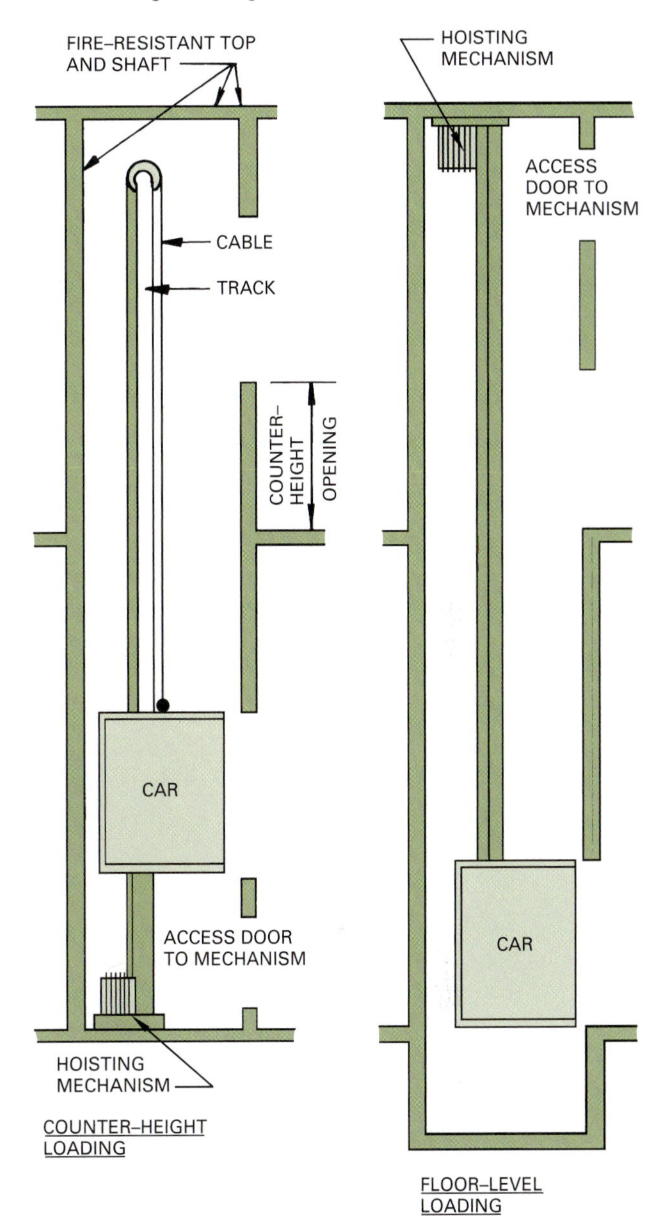

Figure 43.21 Dumbwaiters may have floor-level or counter-height loading.

WHEELCHAIR LIFTS AND STAIR LIFTS

Wheelchair lifts are used for both interior and exterior locations. They are constructed of a steel platform with steel sides and a front gate that lowers to form a ramp that helps load the wheelchair. The ramp has a rubber skid-proof surface. Wheelchair lifts are operated by an electric motor and have an automatic stop switch that activates whenever a user releases the mechanism that controls movement. Lifts will not operate until all entry and exit doors are closed.

Typical maximum lifting capacity falls in the range of 400 to 500 lb. (180 to 225 kg), and the speed of rise is generally about 8 ft./min. (2.8 m/min.). Various lifting heights are available, with 3 to 9 ft. (915 to 2,745 mm) being common The platform area is generally about 12 square feet (1.1 m^2).

Another type of wheelchair lift moves the passenger and wheelchair up a stair to another level. This system uses a platform with side enclosures to move a wheelchair up a stair on a rail system fastened to the wall or stair treads. It can travel up a multilevel straight stair. Two lifts are used when turning a 90° or 180° corner is necessary (Fig. 43.24).

Stair lifts are used to move people who do not use a wheelchair but have difficulty climbing stairs. The lift consists of a chair that runs along a track that may be mounted on the stair or along the wall of the stairwell. The chair can move around corners and across stair

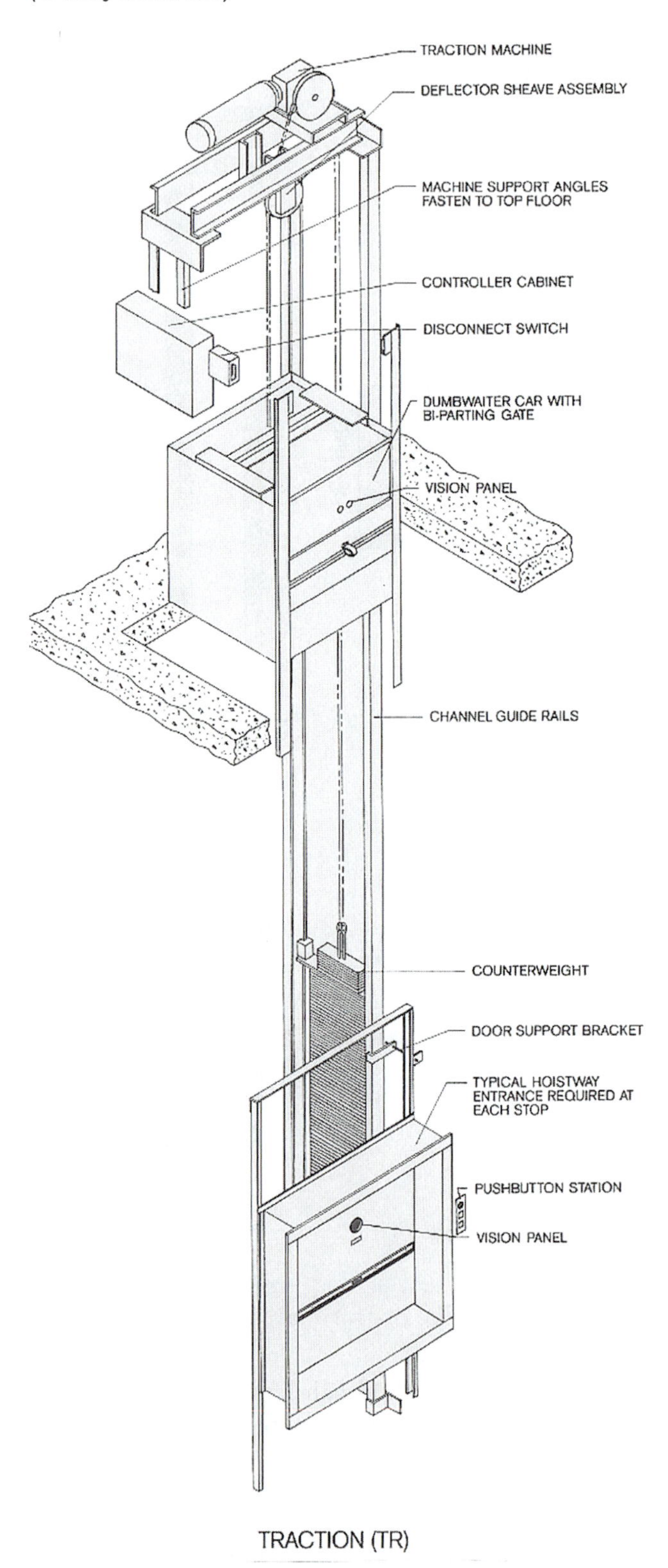

Figure 43.22 A traction electric-powered dumbwaiter. *(Courtesy Matot, Inc.)*

Figure 43.23 Horizontal transfer systems put items to be moved in tote boxes that are automatically removed from a dumbwaiter when it arrives at a specified floor; the tote boxes then load onto a conveyor. *(© FK Photo/FK PHOTO/Corbis)*

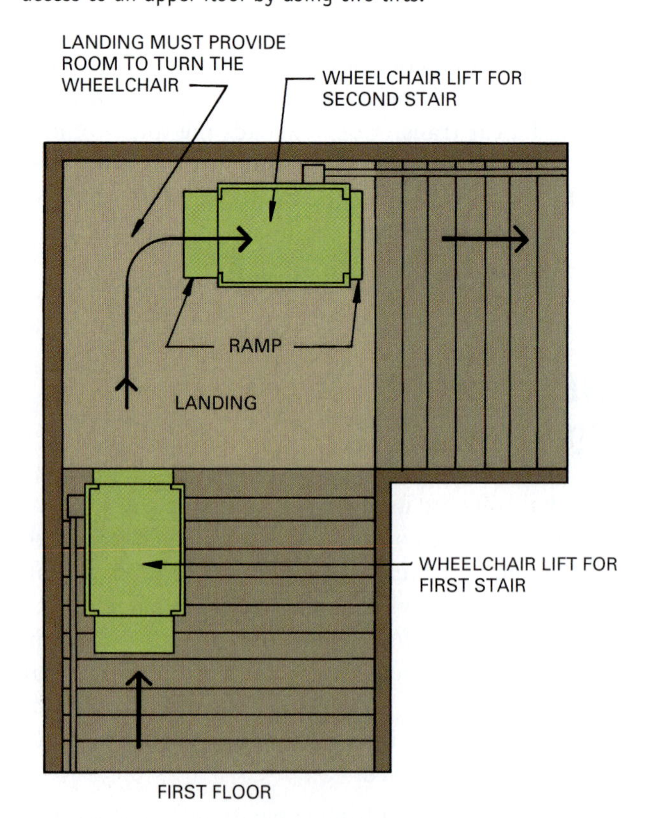

Figure 43.24 This stair-climbing wheelchair lift provides access to an upper floor by using two lifts.

landings (Fig 43.25). The rail can continue past the top of the stair, permitting a passenger to disembark a safe distance from the stairway (Fig 43.26). Lifts are electrically powered, and a series of gears and shafts provide the motion.

Figure 43.25 A chair lift moves a person up a stair while comfortably seated. *(Courtesy Wikipedia, Peter Clarke, http://en.wikipedia.org/wiki/File:Stairlift_with_remote_control.jpg)*

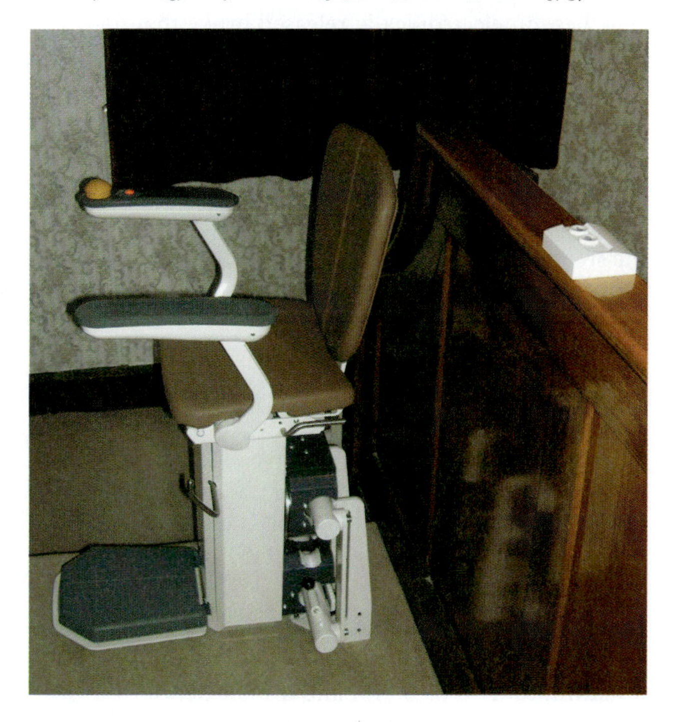

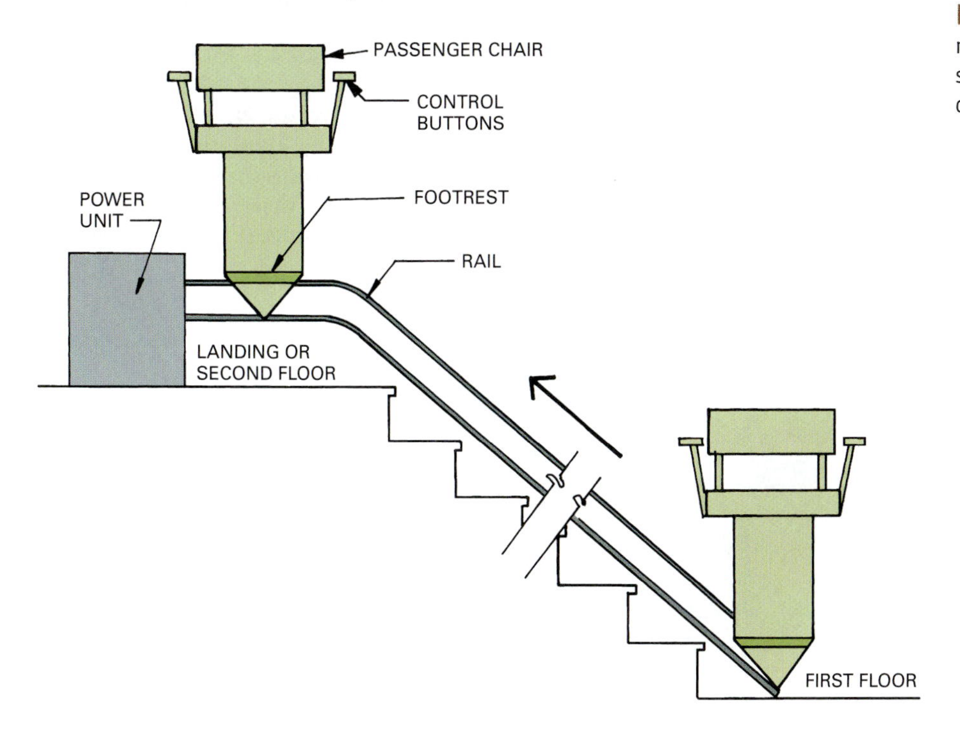

Figure 43.26 The stair lift moves the chair past the top step to provide a safe place to dismount.

ESCALATORS

Escalators are inclined, continuous, power-driven stairways used to move passengers up or down between floors. An escalator should be located where it will be most accessible to traffic. Adequate space must be allowed at each landing, because people will be concentrated in these areas (Fig. 43.27). An alternate method for moving between floors, such as stairs parallel with

Figure 43.27 This escalator has a large open area at each landing to provide needed space as people crowd on and off. *(Image copyright Christophe Testi, 2009. Used under license from Shutterstock.com)*

the escalator, is necessary when codes require it. This provides access should the escalator fail or fall under repair. Escalators may be used as a required means of exiting if they meet all requirements for an emergency egress stairway. This includes having floor openings enclosed to provide fire and smoke protection and the provision of an approved sprinkler system. Escalators not serving as a required exit should have floor openings enclosed or protected by one of the following systems: a partial enclosure that uses self-closing doors; an automatic self-closing rolling shutter; a system of high-velocity water-spray nozzles; or an automatic water curtain with an air exhaust system.

Escalators can move large numbers of people much faster than elevators. Typical speeds of up 90 to 100 ft./min. (27 to 30 m/min.) can move 2,000 to 4,000 or more people per hour. However, they are seldom used to move passengers more than five or six stories.

Escalator Components

An escalator has a welded steel truss structural frame. The stair is a series of moving steps that are cast metal and grooved to provide safe footing. The treads and risers are secured to a continuous chain that is moved by an electric-geared driving unit. Each side of the stair has a solid balustrade that covers the ends of the stair and supports a handrail. The handrail moves at the same speed as the stair. The escalator has electronic control devices for operation and emergency situations (Fig. 43.28).

Standards and Safety

Escalator standards are available in the publication, *American National Standard Safety Code for Elevators, Dumbwaiters, and Escalators*, ANSI/ASME A17.1, and the *Life Safety Code of the National Fire Protection Association*. In Canada, the requirements are given in Standard CAN 3B-44. The safety features covered by the codes and the various manufacturers include emergency stop buttons; broken step and drive chain switches; an electronically released brake that activates during power failures; a switch to prevent the exceeding of design speeds; step lights; switches to stop the escalator if a handrail breaks; switches to control handrail speed; a switch to shut down operation if a step breaks; smooth balustrades that protect sides of steps; and landing plates that protect against items getting caught as the steps flow under the floor.

Installation Examples

To provide smooth movement of passengers, escalators often are installed in pairs with one moving up and one heading down. They may be located in a parallel arrangement (Fig. 43.29) or in a crisscross pattern (Fig. 43.30).

Escalator Sizes

The stair sizes established in ANSI/ASME A17.1 require a maximum rise of $8\frac{1}{2}$ in. and a max depth of $15\frac{3}{4}$ in. (2593 mm and 4804 mm). The widths of escalator treads

Figure 43.28 Escalators are built using heavy structural components and drive mechanisms that can withstand constant year-round operation. *(Courtesy Dover Elevator Systems)*

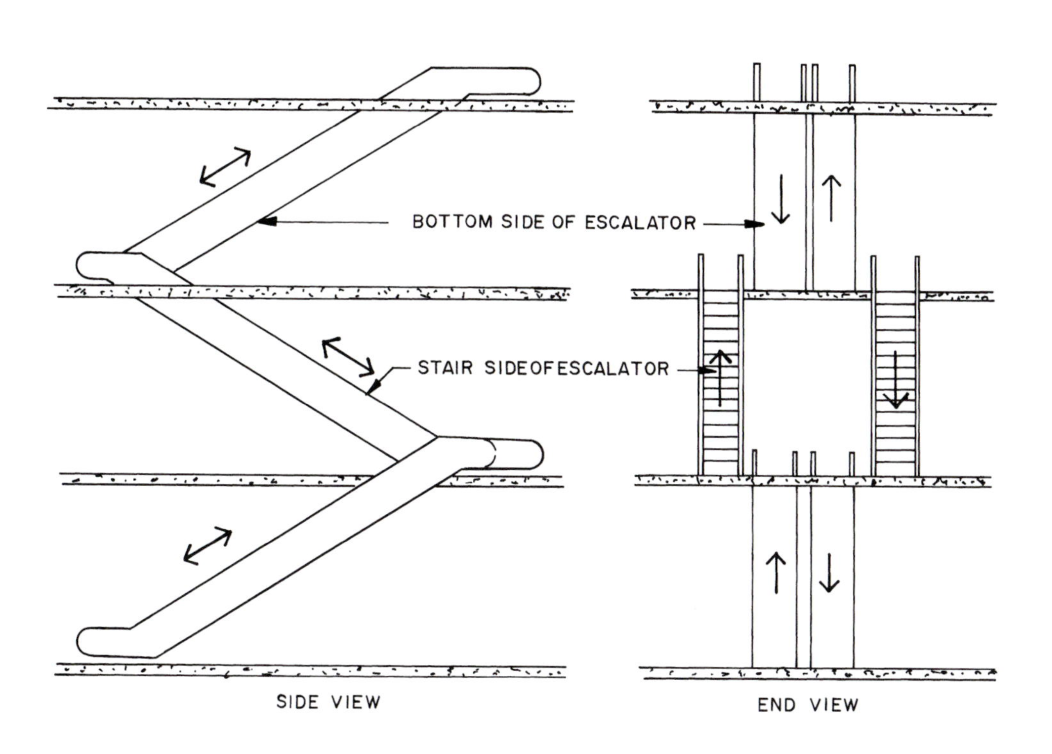

Figure 43.29 Escalators may be installed in a parallel arrangement.

Figure 43.30 This is a typical crisscross escalator installation. *(Image copyright zhu difeng, 2009. Used under license from Shutterstock.com)*

available are typically 24, 32, and 40 in. (0.6, 0.8, and 1.0 m). They are built on an angle of 30° and have various maximum rises, depending on design. Rises of 20 to 30 ft. (6.1 to 9.2 m) are common for a single unit. The installation in Fig. 43.31 shows design sizes for one style of escalator.

MOVING WALKS AND RAMPS

Moving walks are horizontal conveyor belts designed to move people (Fig. 43.32). They may have a slight rise or fall, seldom exceeding 5°. Moving ramps convey people up or down inclines (with a maximum slope of 12°) and connect with moving walks. They are used where large numbers of people need to be transported over long distances, such as in an airport terminal.

Typical widths of the moving flexible rubber-covered belt are 24, 32, and 40 in. (0.6, 0.8, and 1.0 m). The 24 in. belt can accommodate one adult, the 32 in. belt provides room for an adult and a child, and the 40 in. accommodates two adults or one adult with luggage.

The sides of the walks and ramps are enclosed with solid balustrades covered with an endless moving rubber handrail. The steel truss structural system is set in a concrete pit (Fig 43.33). The electrically driven apparatus has a mechanical and electrical safety system and controls similar to those used on escalators.

Figure 43.31 Typical escalator design dimensions. *(Courtesy Otis Elevator Company)*

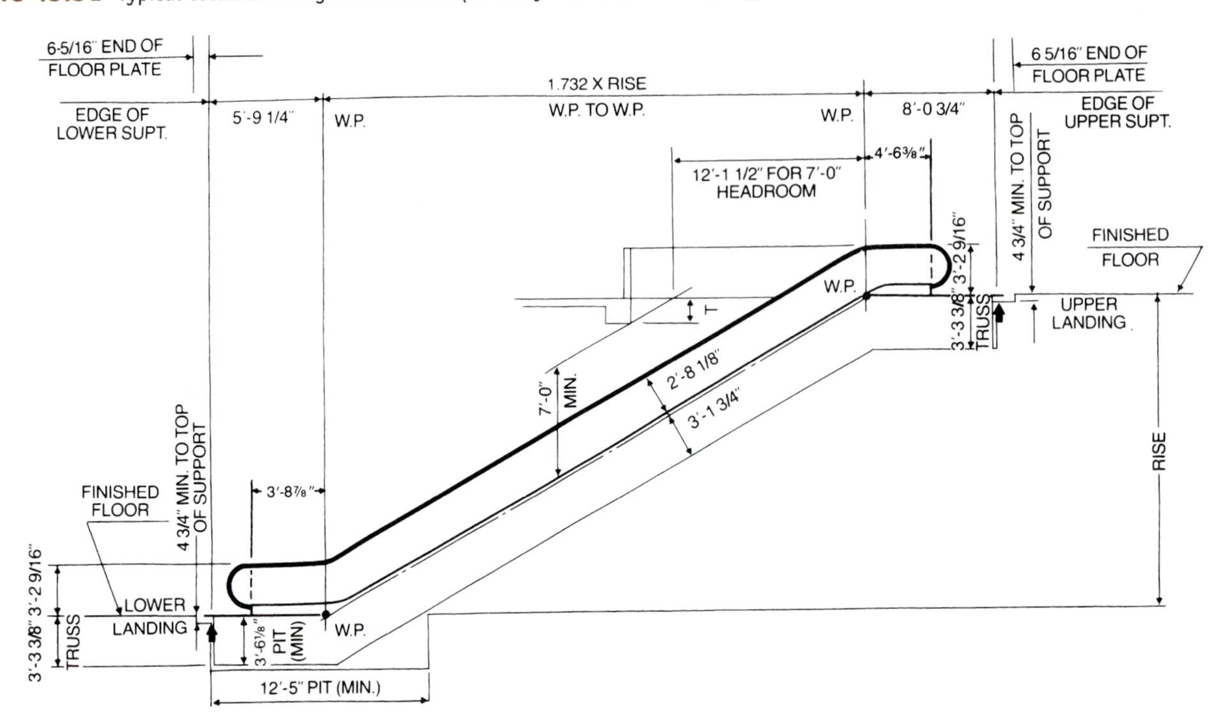

Figure 43.32 A typical moving walk used for longer distance horizontal travel within buildings. *(Image copyright Denis Babenko, 2009. Used under license from Shutterstock.com)*

Figure 43.33 A generalized illustration showing a section through a moving walk.

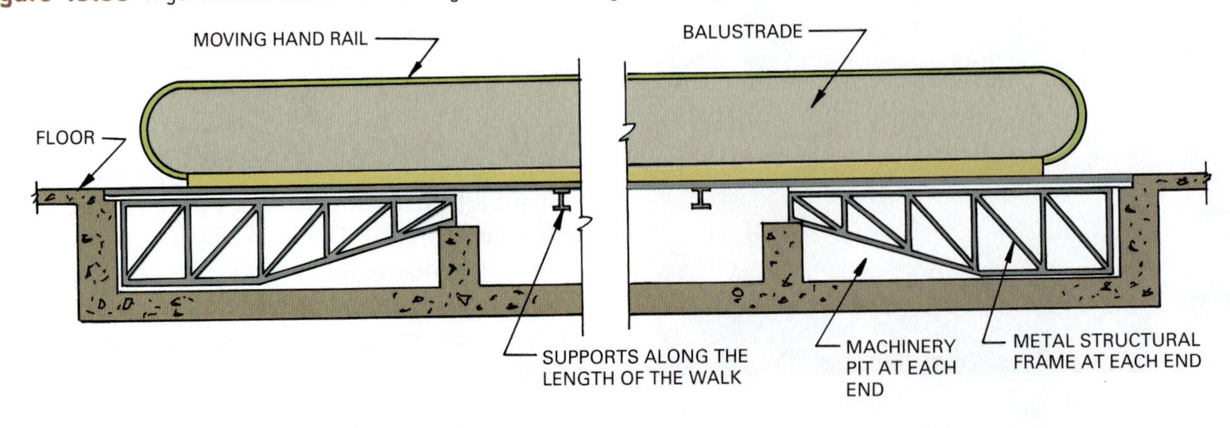

OTHER CONVEYING EQUIPMENT

In addition to the equipment already discussed in this chapter, there is a wide variety of other, less widely used conveying equipment for both people and materials. These include shuttle transit, conveyors, pneumatic tube systems, material conveyors, cranes, hoists, and special-purpose material-moving systems.

SHUTTLE TRANSIT

Shuttle transit systems provide horizontal transportation for distances beyond the practical limits of moving walks. They have many applications, including linking office or retail areas to remote parking facilities or moving passengers between terminals in large airports. Business and industrial parks sometimes use them to provide rapid internal transportation on large campuses (Fig. 43.34).

Shuttle transit systems function much like elevators. The cars operate on automatic controls and may be dispatched on a regular schedule or on call. Each boarding station has signs to indicate scheduled arrival times. The system uses standard elevator gearless traction machine drives and cable equipment. A steel cable is attached along the side of the shuttle. The steel guide rails and power rails are located adjacent to the running surface. Vertical loads are supported by a cushion of air developed between pads on the bottom of the car and the guideway running surface. The cars may be operated singly or several can be joined to increase passenger capacity.

CONVEYORS AND PNEUMATIC TUBES

Various types of conveyor and pneumatic tube systems are used to move materials within a building. An electric track vehicle conveyor system consists of self-propelled electric cars that travel over a network of tracks connecting stations within a building. Cars typically carry up to 50 lb. (22.5 kg) and move vertically and horizontally at speeds up to 120 ft./min. (36.6 m/min.). They are used to transport paperwork, tools, test samples, and other small items (Fig. 43.35). Systems can be installed in a number of ways. The simplest is a single track system in which the car moves back and forth on one track. A dual track permits cars to go both ways at

Figure 43.34 People are moved rapidly and comfortably between buildings by monorail systems. *(Image copyright Thorsten Rust, 2009. Used under license from Shutterstock.com)*

Figure 43.35 This electric track vehicle conveyor system moves containers within the building following a single or dual track. *(Courtesy Swisslog/Translogic Corporation)*

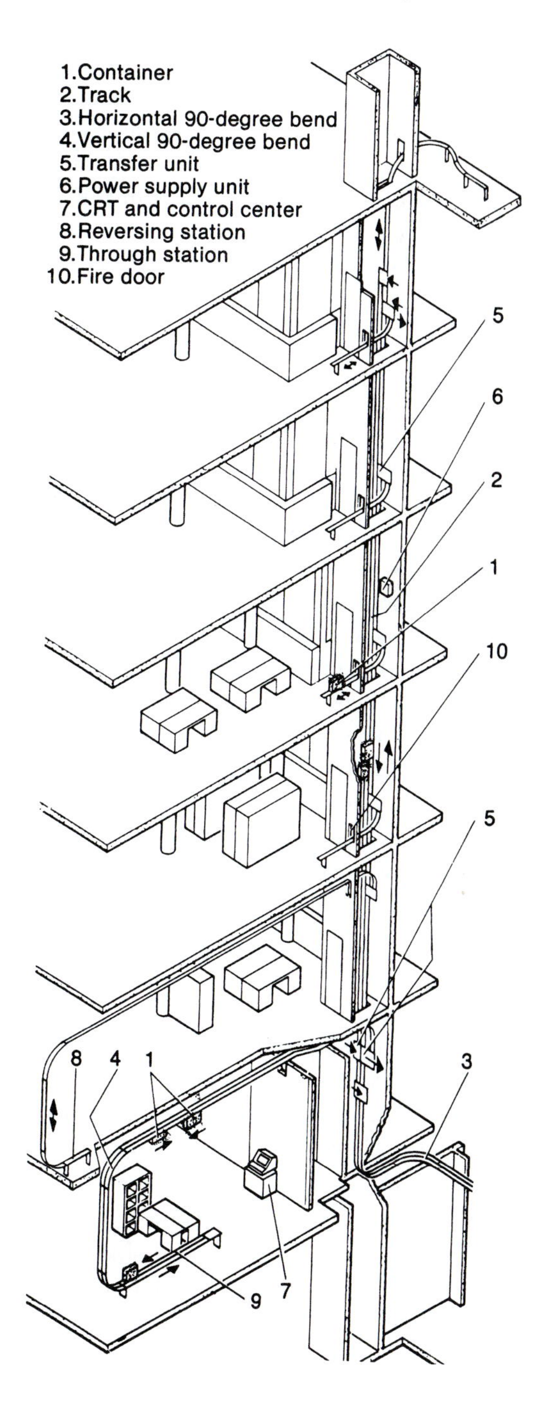

the same time, providing faster service. The loop system stores cars and sends them forward when called, after which they return to storage until called again.

Pneumatic Tube Systems

Computerized pneumatic tube systems serve a variety of purposes. The system shown in Fig. 43.36 transmits small items placed in tube-like carriers. They are conducted through a system of transmission piping and managed by a computerized control center. The system can be a simple single zone route or have multiple zones, which speeds up carrier movement.

Another pneumatic system consists of a large-diameter closed-pipe vacuum network used to move trash or linens. The linen collector system, as used in hospitals, can handle plastic- or cloth-bagged linens or linens in loose form. This system can be configured as a single- or multiple-bag installation, the latter permitting the loading of 10 to 20 bags in the collector before cycling starts. The trash collector system can handle loose trash consisting of paper, glass, plastic, and viscous liquids. A trash disposal system can feature an added shredding station.

Material Conveyors

Material conveyors are used to move items, such as packages, luggage, parts, aggregate, and concrete, within a building or on a construction site. They include belt, roller, and segmented moving surfaces. Belt conveyors may be flat or troughed and are powered by an electric motor. The flat belt conveyor is used to move items, such as packages and manufactured parts, within a building to stations where items can be loaded or unloaded from the moving belt. The system can turn corners using special power belt curves (Fig 43.37). It is able to move items horizontally or down small inclines. The troughed conveyor belt runs on rollers forming a U-shape and is used to move dry loose materials.

Roller conveyors may have solid steel rollers across a unit or use a series of individual wheels, sometimes referred to as skate wheels. Solid roller units may be gravity operated or power operated. They are used for medium and heavy duty work and may be permanently installed or movable. Skate wheel conveyors are used for light duty use, such as unloading light packages off delivery trucks. They are gravity operated and usually portable.

The segmented conveyor has a moving surface made up of flat sections joined with hinge-like connectors (Fig. 43.38). Airport luggage conveyors of this type can handle heavy loads and are power driven.

Figure 43.36 A typical computerized pneumatic tube conveying system transmits items stored in tube-shaped containers through tube transmission piping. *(Courtesy Swisslog/Translogic Corporation)*

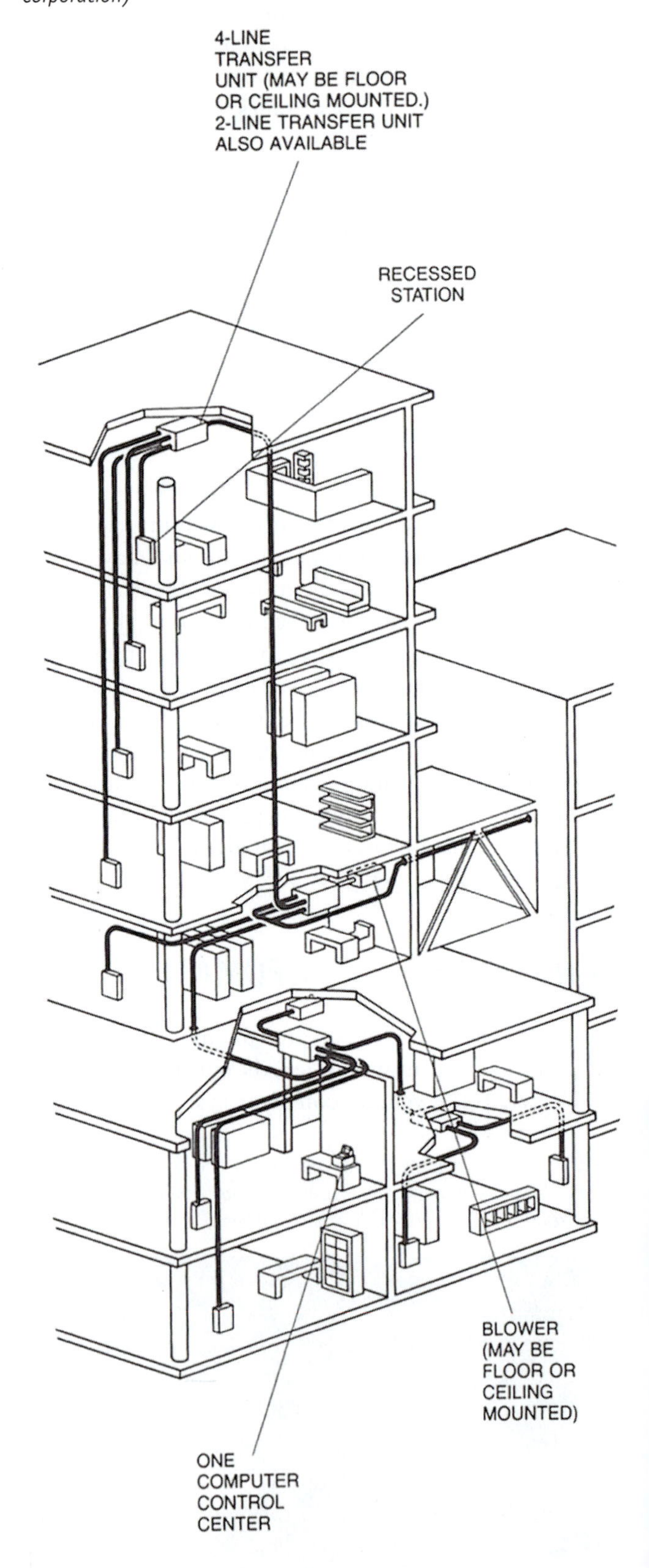

Figure 43.37 On roller conveyors, the items being moved roll directly upon a solid roller or skate wheels. *(Image copyright Vladimir V. Georgievskiy, 2009. Used under license from Shutterstock.com)*

Figure 43.38 Segmented conveyor belts can move materials around corners. *(Image copyright Carlos E. Santa Maria, 2009. Used under license from Shutterstock.com)*

CRANES AND HOISTS

Cranes and hoists are used to move materials and heavy items at outdoor locations and within buildings. The choice of system depends on the applications required and is an important design aspect of a structure. Overhead, monorail, and under-hung cranes run on rails generally supported by the structural frame of a building, so loads imposed by the crane must be carefully calculated. Some types use a structural system independent of the building structure.

Overhead cranes move along the top of fixed overhead rails. They are often used for specific jobs, such as moving steel members from storage to fabrication areas and moving finished members to storage or shipping areas (Fig. 43.39). The trolley moves horizontally along a steel track with the lifting action produced by a hoist that is part of the trolley. A monorail has the hoist slung below a single steel track in a similar arrangement. The underslung crane is much like the overhead crane except it travels along the bottom flange of a track. A variation of this is the wall-mounted jib crane shown in Fig. 43.40. These cranes are powered by electric motors and generally controlled by an operator on the floor. A control cable extends from hoist to floor. The operator can manipulate the hoist to lift or lower loads and the trolley to move along the track.

Figure 43.39 An overhead crane runs on the top of a track.

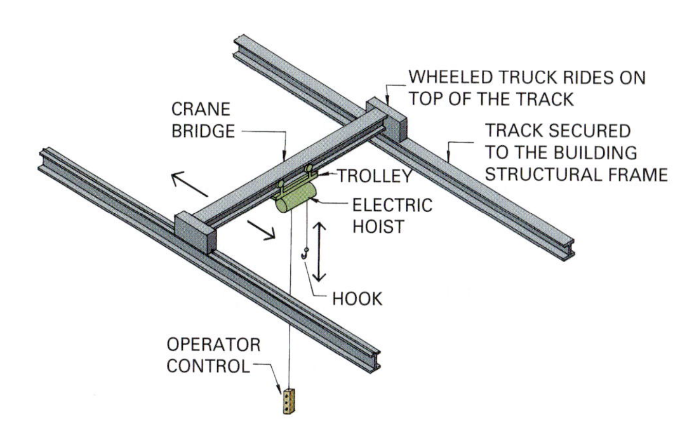

Figure 43.40 A wall-mounted jib crane rides on a horizontal track, which can rotate on a hanger pin, increasing the floor area it can serve.

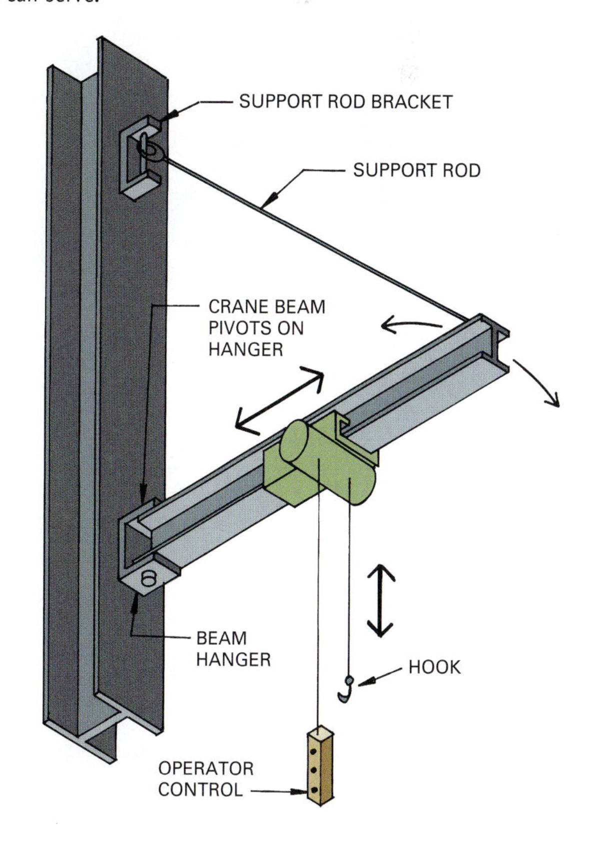

A gantry crane has an overhead track and a trolley that runs on a horizontal track supported by legs that run on rails secured to the floor. Gantry cranes are used by industries that must move very heavy loads. They may operate inside or outside a building and do not require support from its structural system (Fig. 43.41). On large cranes, the operator sits in a cab located below the bridge.

A diesel powered mobile gantry is shown in Fig. 43.42. It moves on wide-base radial tires, so it is more versatile than the track-bound gantry. It is available in a wide range of structural configurations. The cab is located to give the operator excellent visibility and is reinforced with steel guards. It also has a heater, defroster, windshield wipers, tinted glass, and a dome light. The control panel includes a load weight indicator. It is available with two-wheel and four-wheel drive.

Figure 43.41 This heavy-duty gantry crane rides on rails and can lift loads of over a thousand tons. *(© epa/Corbis)*

Figure 43.42 Mobile hoists can carry large, heavy loads.

Review Questions

1. What code is used to ascertain the safety standards of elevators and escalators?
2. Where can information on elevator size and shape standards be found?
3. What items are permitted in an elevator machine room?
4. What types of doors are used on various types of elevators?
5. What controls are normally located on an elevator car panel?
6. How is the capacity of an elevator car calculated?
7. What happens to elevator cars in case of a power failure?
8. What types of traction machines are used for hoisting elevators?
9. What are the major differences between gear-driven and gearless traction machines?
10. How is the speed of gearless traction elevator drive mechanisms varied?

11. What are the two types of elevator control systems?
12. What types of roping systems are used with traction elevators?
13. What types of elevator operating systems are available?
14. What mechanism is used for moving hydraulic elevators?
15. What are the classes of freight elevators?
16. What purposes do automated transfer systems serve?
17. What types of lifts are used to move the physically handicapped from floor to floor?
18. What escalator installation configurations are generally used?
19. What emergency features are included in an escalator system?
20. What is a moving walk?
21. Where is a shuttle transit system likely to be used?
22. What are the commonly used material conveyors?

Key Terms

buffer, elevator
car, elevator
car safeties, elevator
elevator
escalator
gearless traction elevator
hoistway, elevator
hydraulic elevator
landing zone, elevator
moving walk
pit, elevator
sheave

Activity

1. Arrange to visit buildings having a variety of conveying systems. Inspect the mechanical rooms and other behind-the-scenes aspects of the mechanism. Ask about maintenance activities and schedule a visit for each system. Detail these in a written report.

Additional Resources

American National Standard Safety Code for Elevators, Dumbwaiters, Escalators, and Moving Walks, American National Standards Institute, New York.

Other resources include:

Publications from the National Association of Elevator Contractors, Conyers, GA.

Publications from the National Elevator Industry, Teaneck, NJ.

See Appendix B for addresses of professional and trade organizations and other sources of technical information.

Fire Suppression
CSI MasterFormat™

210000 Fire Suppression
211000 Water-Based Fire-Suppression Systems
212000 Fire-Extinguishing Systems
213000 Fire Pumps
214000 Fire-Suppression Water Storage

(Image copyright Tomasz Gulla, 2009. Used under license from Shutterstock.com)

44 CHAPTER

Fire-Suppression Systems

LEARNING OBJECTIVES

Upon completion of this chapter, the student should be able to:

- Have an understanding of the fire codes and the agencies that maintain them.
- Develop a working knowledge of the various types of fire-suppression systems.

Build Your Knowledge

For further study on these materials and methods, please refer to:

Chapter 39 Specialties

Topic: Fire Suppression Specialties

Fire protection is a major concern in the planning of commercial and industrial buildings. A thorough strategy for means of egress for the occupants of a building is a part of the planning and regulated by codes. This is especially critical in large, multistory buildings. The design process considers compartmentalization of the building, smoke control, and the type of fire-control system to be used.

Passive fire protection attempts to contain fires or slow their spread through the use of fire-resistant walls, floors, and doors. Many of these materials and strategies have been outlined in previous chapters. Active fire protection is characterized by systems that require automation and equipment in their operation. As a building is designed, consideration of the fire-suppression system becomes integrally involved with the design of the plumbing, mechanical, communications, and signaling systems. The fire-suppression system must provide for early detection of a fire, give adequate warning, and be connected to a source of emergency electrical power.

FIRE CODES

Model fire codes are adopted by local governments to regulate the design, construction, and maintenance of buildings. The Uniform Fire Code is a widely used model code that is promoted by the National Fire Protection Association (NFPA). It includes considerations of building design, hazardous materials, regularity procedures, code enforcement, and other fire-related activities. In addition, NFPA offers other codes and standards related to fire protection and associated activities.

The International Code Council (ICC) offers the International Fire Code, which addresses equipment, hazardous materials, fire safety, and other fire protection requirements. In addition, the Underwriters Laboratory, Inc. (UL) tests and approves fire protection equipment and issues reports in its Fire Protection Equipment List.

FIRE-SUPPRESSION SYSTEMS

Fire-suppression systems include various types of sprinkler systems, foam and fog extinguishing systems, gas systems, and chemical systems. Manual water extinguishing systems are also used.

Automatic Water Sprinkler Systems

A typical automatic fire-suppression sprinkler system uses a water supply from the city water system. Tall buildings require a backup supply, such as a storage tank on the roof (Fig. 44.1). The inflow is controlled by a valve. When water flow is detected, an alarm valve activates (Fig. 44.2). The riser has a **Siamese connection** (Fig. 44.3) located outside the building and used by the fire department to pump additional water into the system from an outside source, such as a secondary water supply or a street hydrant. Each area on each floor protected by sprinklers has a similar system. In large buildings, separate risers to various sections are used.

Figure 44.1 A simplified illustration of an automatic water fire-suppression sprinkler system.

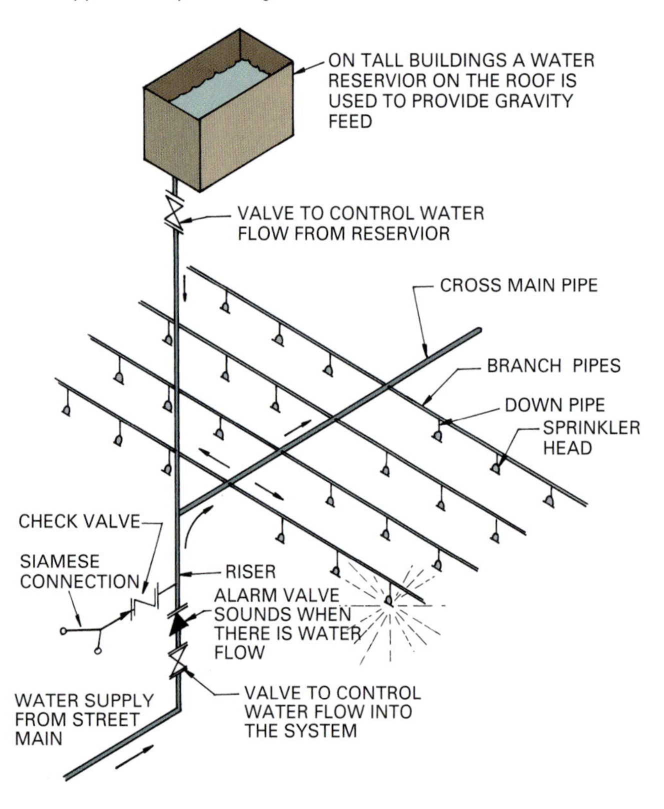

Figure 44.2 A sprinkler alarm sounds a warning when the sprinkler system is activated. *(Image copyright StudioNewmarket, 2009. Used under license from Shutterstock.com)*

Figure 44.3 A Siamese connection is used by the fire department to connect a supplemental source of water to the building's fire-suppression system. *(Image copyright EugeneF, 2009. Used under license from Shutterstock.com)*

Sprinkler heads of various designs are available, as shown in Fig. 44.4. They may be an upright or pendant types (Fig. 44.5). Automatic water sprinkler systems include wet pipe, dry pipe, deluge, pre-action, and various types of water fog and liquid foams, each of which use some type of sprinkler head or nozzle.

Wet pipe systems keep water under pressure in the system of pipes at all times. When sprinkler heads are activated by the heat from a fire, water is immediately released. This is the most widely used water system (Fig. 44.6). In order to limit water damage, only sprinkler heads over the area where fire is located activate. In areas where pipes may freeze, the system is filled with an antifreeze and water mixture.

Dry pipe systems are maintained with air or nitrogen under pressure. When a sprinkler head opens due to heat from a fire, the air pressure releases, causing the dry pipe to open. The pipes fill with water, which moves through the open sprinkler heads (Fig. 44.7). Since the pipes are dry, this system is widely used in areas subject to freezing temperatures. Normally, upright sprinkler heads are used because pendant heads may hold water and freeze.

Deluge systems are designed to deliver as much water as quickly as possible. Sprinkler heads or spray nozzles are kept open at all times and the pipes are dry. The system activates by a sensitive fire detection

Figure 44.4 Typical sprinkler heads: (a) An upright head on top of an exposed branch pipe and below the ceiling *(Image copyright dcwcreations, 2009. Used under license from Shutterstock.com)* (b) A recessed pendant that extends below the line of the ceiling *(Image copyright Photoroller, 2009. Used under license from Shutterstock.com)* (c) An adjustable concealed head with a round plate that provides a decorative appearance, (d) An adjustable pendant used in residential applications and standard commercial systems. *(Courtesy Central Sprinkler Company)*

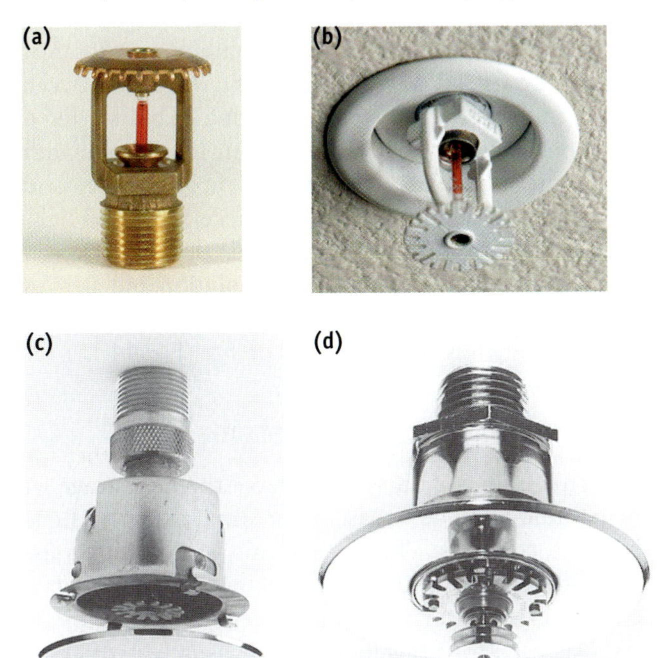

Figure 44.5 Sprinkler heads may be upright or pendant types.

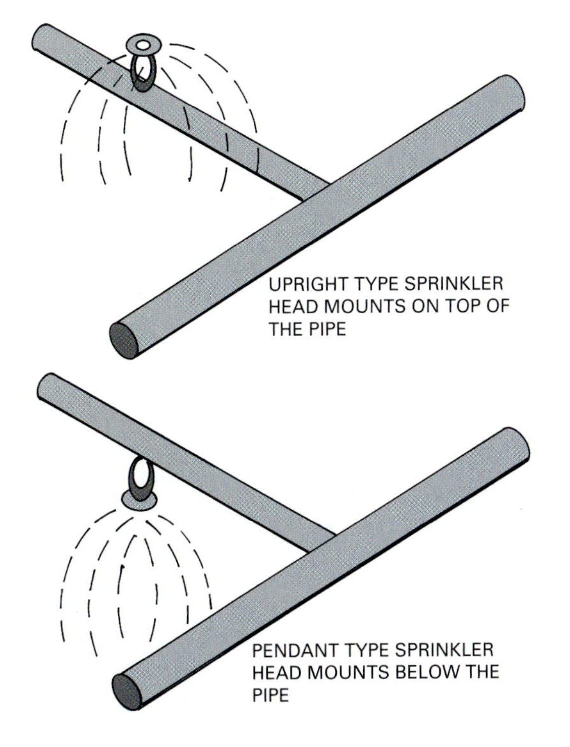

Figure 44.6 A wet pipe sprinkler system maintains water under pressure in the pipes.

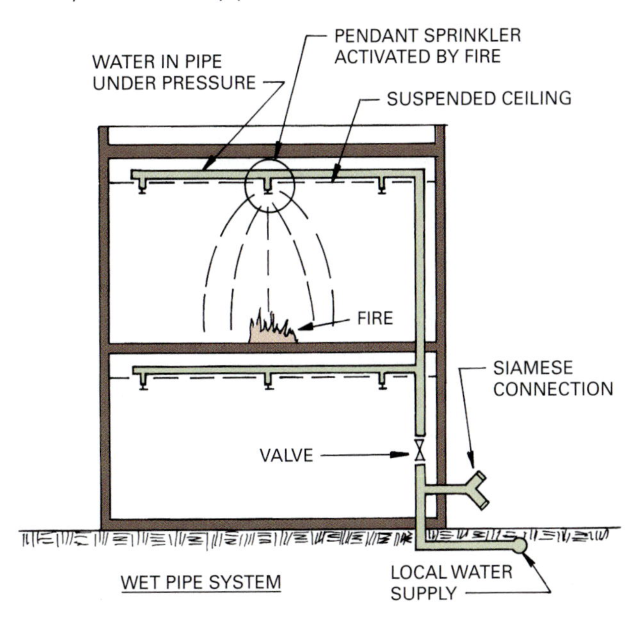

Figure 44.7 A dry pipe system maintains the sprinkler pipes dry and under air pressure.

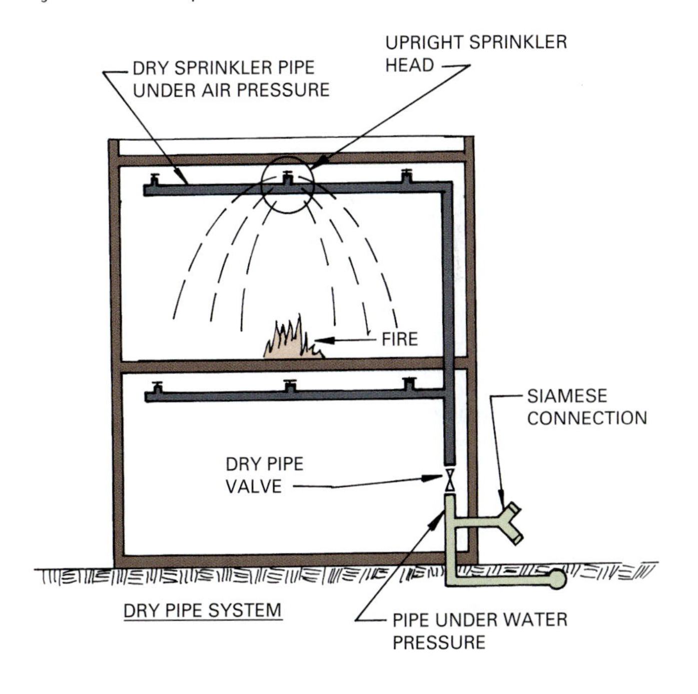

mechanism that opens the deluge valve rather than activating each sprinkler head (Fig. 44.8). The deluge system delivers water to an entire area because all sprinklers are open.

Pre-action systems operate with sprinkler heads closed and pipes dry. They use a fire detection system that is more sensitive than typical. This detection system opens the pre-active valve, allowing water to flow only to sprinkler heads opened by heat from the fire (Fig. 44.9).

Figure 44.8 A deluge pipe system activates all the sprinkler heads in an area at the same time.

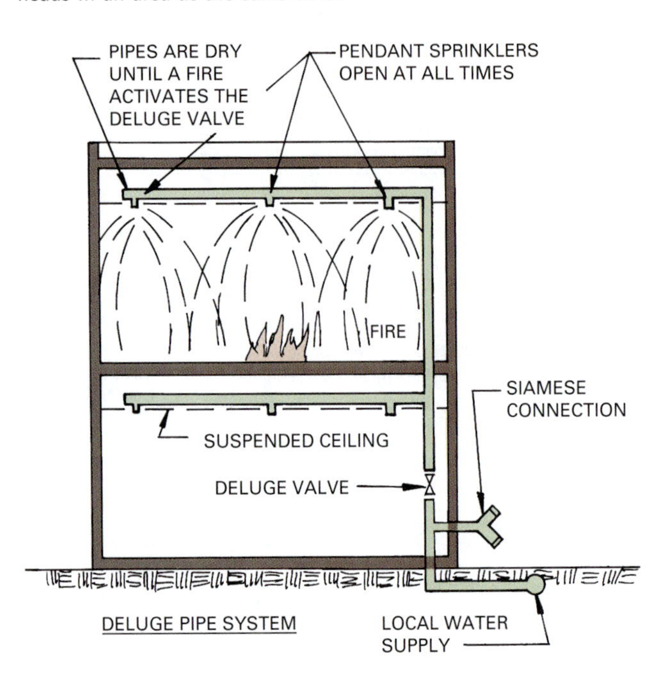

Figure 44.9 A pre-action pipe system maintains a dry pipe until the sprinkler valve is opened by a sensitive fire detector. Water is sprayed only by sprinklers activated by the fire.

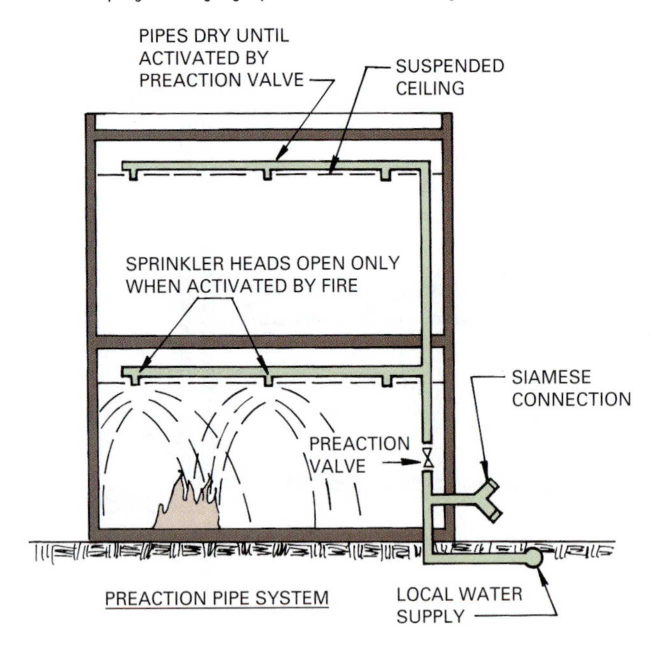

The accidental opening of a sprinkler head will not cause the system to discharge water because the pipes are kept dry. This protects areas with sensitive or costly equipment from damage due to accidental openings.

Water fog systems use a standard sprinkler piping system and have spray heads or nozzles instead of sprinkler heads. They are used in areas where highly flammable materials are stored. The fog tends to cool materials, keeping them below ignition temperature.

Foam Fire-Suppressant Systems

Foam fire-suppressant systems are used on fires involving gasoline or chemicals for which water systems are ineffective. Foams are masses of air-filled or gas-filled bubbles formed by mechanical or chemical techniques. Chemical foam forms by a reaction between water and several chemicals, resulting in bubble-filled foam produced by carbon dioxide. Air foam forms with water and a chemical. It is moved through pipes and hoses and sprayed through discharge nozzles. High-expansion foam forms by passing air through a screen constantly wetted with a chemical solution and a small amount of water. It is moved to fire areas in large ducts and generally fills compartments on fire (Fig. 44.10).

Gas Fire-Suppression Systems

Gas fire-suppression systems have the advantage of being able to flood areas, suppressing fire with little harm to contents. This is especially important in areas with sensitive equipment, such as computer rooms, commercial aircraft, telephone exchanges, and libraries. Gas used for fire suppression is usually carbon dioxide or Halon 1301. Non-ozone depleting substitutes for Halon are being developed.

Carbon dioxide covers fire with a blanket of heavy gas that reduces the oxygen content in the surrounding air so combustion is extinguished. It is used in small

Figure 44.10 Foam systems produce large amounts of air-filled or gas-filled bubbles and fill the compartment with foam from large-diameter ducts. *(Courtesy Wikipedia, http://en.wikipedia.org/wiki/File:COMPUTER_ROOM_DISCHARGE.JPG#filelink)*

compartmentalized areas, such as electrical cabinets. It is not used in areas where people may be present. Carbon dioxide is stored in liquid form under great pressure and is released as a gas that cools and smothers a fire. Fire-sensing systems activate control valves that release gas through a series of pipes with nozzles directed at the point of a potential fire.

Halon 1301 extinguishes fire by interfering with the combustion process, thus preventing it from occurring. Low concentrations of Halon 1301 extinguish flames rapidly, and higher concentrations completely diffuse into the atmosphere, preventing further burning or explosion. Halon 1301 is stored as a gas in cylinders under pressure (Fig. 44.11). It vaporizes as it enters a fire area through discharge nozzles. The vapor diffuses into the surrounding atmosphere, leaving no residue.

While Halon 1301 is an effective fire-suppression agent, it is an ozone-depleting gas. Since the early 1990s, manufacturers have successfully developed safe and effective Halon alternatives. Generally, Halon replacement agents available today fall into two broad categories: in-kind (gaseous extinguishing agents) and not-in-kind (alternative technologies). In-kind gaseous agents generally fall into two further categories: Halocarbons and Inert Gases. Not-in-kind alternatives include such options as water mist or the use of early warning smoke detection systems.

Figure 44.11 Cylinder gas canisters for use in extinguishing fire without damaging equipment. *(Courtesy Wikipedia, William Viker (c) 2006)*

Dry Chemical Systems

Dry chemical systems do not penetrate burning materials but remain on the surface and smother fire. Chemicals frequently used include sodium bicarbonate, potassium chloride, and monoammonium phosphate. They are effective on flammable liquid, electrical materials, and ordinary combustible materials. They have a high tolerance for extreme weather conditions and temperatures.

Automatic systems pipe chemicals directly to areas in which fire might occur. These tend to be limited areas, such as a piece of equipment that is composed of or uses materials that make it a constant fire hazard. Dry chemicals are also widely used in handheld fire extinguishers.

Manual Fire-Suppression Systems

Manual fire-suppression systems have fire hose stations on each floor of a building that are connected to the water piping system (Fig. 44.12). When water pressure is insufficient to supply stations on the upper floors of a building (or will not supply water in adequate quantity), standpipes are added to the system. A standpipe is a pipe or tank on the roof of a building that stores a supply of water. Standpipes provide an extra supply when normal water pressures fail. Standpipe systems can include pumps to increase water pressure (Fig. 44.13). A standpipe also provides a reserve for the potable water needed for daily use in a building. Additional information on fire detection devices and alarm control systems is provided in Chapters 39 and 42.

Figure 44.12 A typical manually operated fire hose station. *(Image copyright Angelo Gilardelli, 2009. Used under license from Shutterstock.com)*

Figure 44.13 A simplified schematic of a manual fire-suppression system utilizing standpipes. A water storage tank on the top of the building supplies water to the standpipes in addition to that from the local water supply. The standpipes supply water to the fire hose stations. The storage tank provides a short-term gravity-fed extra supply to assist until the fire department can connect to the Siamese connection.

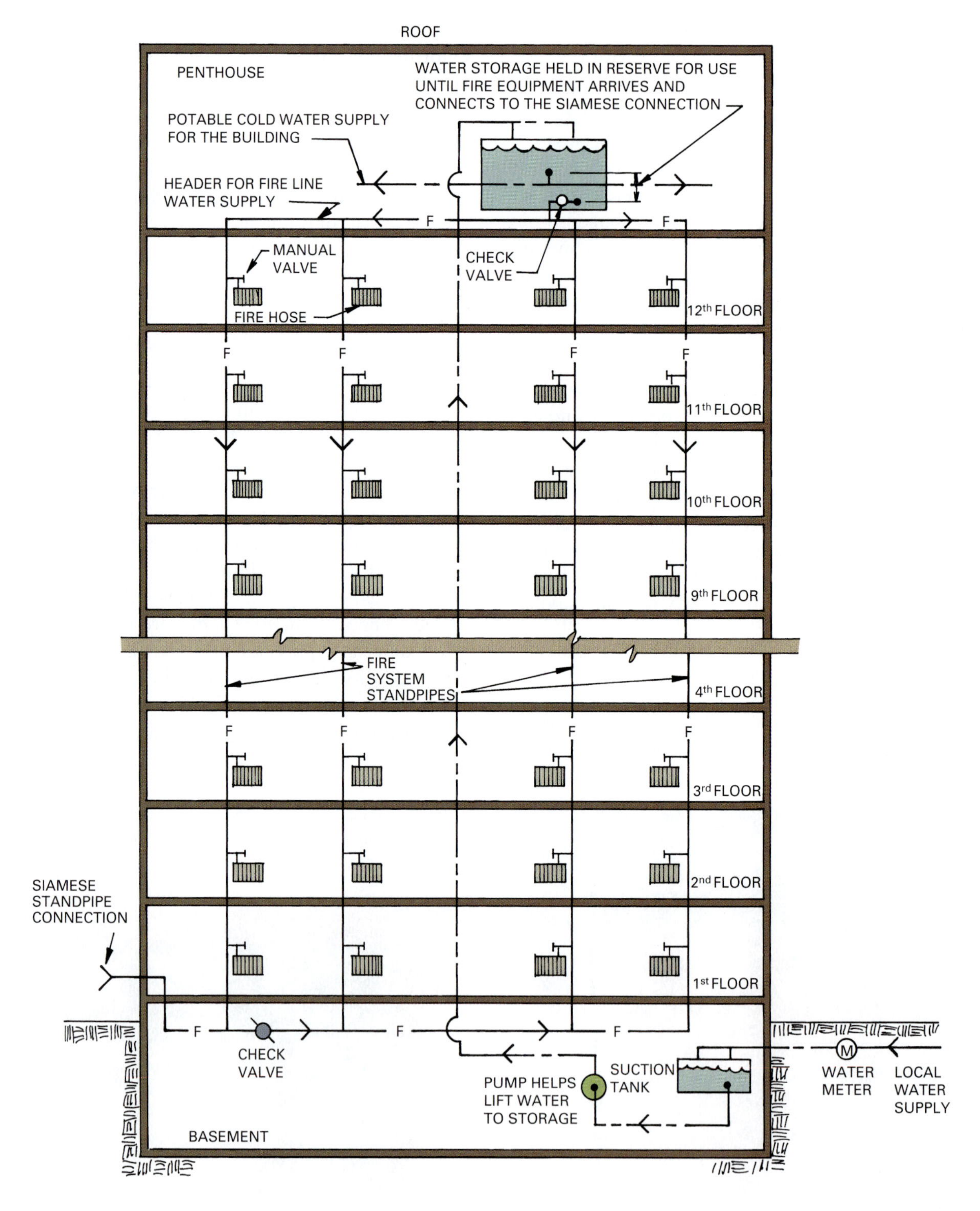

Review Questions

1. What organization publishes the major fire protection requirement manuals?
2. How is the Underwriters Laboratory involved in fire protection issues?
3. What purpose does a Siamese connection serve?
4. How does the deluge fire protection system differ from the wet pipe system?
5. What is a suitable fire protection system for areas exposed to freezing temperatures?
6. How does the pre-action system protect an area from water damage if a sprinkler head is accidentally opened?
7. What types of foam fire-suppression systems are available?
8. What is a major advantage of the gas fire-suppression systems?
9. What gases are commonly used in gas fire-suppression systems?
10. How do dry chemical systems extinguish a fire?
11. In a manual fire-suppression system, how is water supplied to individual fire hose stations?

Key Terms

Siamese connection

Activities

1. Tour your campus, local shopping mall, hospital, and other commercial buildings and record the type of fire-suppression systems found in each. Take photos of the visible components and prepare a labeled display.
2. Invite the local fire chief and building official to address your class. The fire chief can discuss the role and responsibilities of the fire department in making inspections. The building official can address fire requirements in local codes.

Additional Resources

Cote, A.E., *Fundamentals of Fire Protection*, National Fire Protection Association, Quincy, MA.

Fire and Safety Inspection Manual, National Fire Protection Association, Quincy, MA.

Gypsum Association Fire Resistance Manual, Gypsum Association, Washington, DC.

International Fire Code, International Code Council, Falls Church, VA.

National Fire Alarm Code; *National Fuel Gas Code*; *National L-P Gas Code*; National Fire Code Subscription Service, National Fire Protection Association, Quincy, MA.

UL Fire Resistance Manual, Underwriters Laboratories, Inc., Northbrook, IL.

Uniform Fire Code, National Fire Protection Association, Quincy, MA.

See Appendix B for addresses of professional and trade organizations and other sources of technical information.

DIVISION

22

Plumbing
CSI MasterFormat™

221000 Plumbing, Piping, and Pumps
223000 Plumbing Equipment
224000 Plumbing Fixtures
225000 Pool and Fountain Plumbing Systems
226000 Gas and Vacuum Systems for Laboratory and Healthcare Facilities

(Image copyright Lisa F. Young, *2009. Used under license from Shutterstock.com)*

45 CHAPTER

Plumbing Systems

LEARNING OBJECTIVES

Upon completion of this chapter, the student should be able to:

- Relate plumbing codes to plumbing system design.
- Understand the types of piping, tubing, and fittings used on plumbing systems.
- Be knowledgeable about the sources of potable water and its distribution systems.
- Understand the parts of sanitary piping systems.

Build Your Knowledge

For further study on these materials and methods, please refer to:

Chapter 15 Ferrous Metals
Topic: Steel Products

Chapter 16 Non-Ferrous Metals

Chapter 24 Plastics

Chapter 40 Residential and Commercial Equipment
Topics: Solid Waste Handling Equipment, Fluid Waste Treatment and Disposal Equipment

A basic need for a successful building is a well-functioning plumbing system. Moving potable water to required areas is a considerable task in a multistory building. Even more formidable is the task of removing liquid waste. A typical building plumbing system includes hot and cold potable water distribution systems and a sanitary disposal system. Many other plumbing systems are designed to suit a special need, such as compressed air systems, fuel oil systems, natural gas systems, and oxygen gas systems. All of these are designed and installed as specified by plumbing codes.

PLUMBING CODES

Materials, plumbing system design, and installation procedures are strictly regulated by local plumbing codes. Local governments adopt the *International Plumbing Code*, sponsored by the International Code Council. The plumbing codes sponsored for many years by various code organizations have been cooperatively joined to form the International Plumbing Code.

POTABLE WATER SUPPLY

Of the approximately 370 billion gallons of water on earth, almost 98 percent resides in the oceans; less than one percent is fresh water in the atmosphere, rivers, lakes and groundwater; and the remaining two to three percent is contained in glaciers and ice caps.

Potable (drinkable) water for public and privately owned central water systems may be obtained from rivers, lakes, ponds, surface runoff, and wells. Surface-collected water tends to contain a number of contaminates that must be removed to produce potable water. Quality problems include hardness caused by calcium and magnesium salts, discoloration caused by manganese or iron, corrosion caused by acidity in the water, pollution by sewage and organic matter, odor and taste problems caused by organic matter, and turbidity caused by silt and other suspended matter. Water treatment plants process water to correct these problems (Fig. 45.1).

Water treatment includes processes such as screening water at intake, sedimentation (allowing particles to drop in a settling basin), coagulation (removing suspended matter with a chemical, such as hydrated aluminum sulfate, in a settling basin), filtration (removing suspended particles and some bacteria through a

Figure 45.1 Water treatment plant processes include settling, screening, sedimentation and coagulation, filtration, disinfecting, aeration, and softening. *(Image copyright Wade H. Massie, 2009. Used under license from Shutterstock.com)*

filter, such as sand or diatomaceous earth, or by chlorination filtration), disinfection (removing harmful organisms using bromine, iodine, ozone, or heat treatment), softening (removing calcium and magnesium), and aeration (exposing water to air, by spraying, for example, to improve taste and color).

In areas without a central supply, water can be obtained from wells. These are usually drilled, driven, or jetted. A drilled well involves placing pipe in a deep hole in the earth dug with a steel auger. A driven well is formed by placing a steel point on the end of sections of pipe and driving them into the earth. A jetted well has a well point from which a high-pressure stream of water pumps, opening a hole in the earth for the well pipe to move into.

Well water, as a rule, does not require the extensive treatments needed for surface water. But it must be tested for purity and treated as required to reduce hardness, iron, or magnesium.

Once potable water is delivered to a building, the quality must be maintained as the water is stored and distributed through the building piping system. Potable water systems must be kept isolated from all other plumbing systems.

PIPING, TUBING, AND FITTINGS

Water is piped through buildings to a number of locations for various purposes. Potable water is delivered to sinks, lavatories, and water heaters. Typically, it is also run to toilets (water closets) and exterior hose bibs. Although these do not require potable water, a less pure and less expensive supply (known as gray water) is not generally available. Pipes for most applications are hidden in the walls, floors, or ceilings of a building structure. Pipes can be exposed in areas where their visibility is not important, but must always be protected from freezing. Plumbing systems use valves to control the flow of water into a building, within parts of large buildings, and to individual fixtures.

The types of pipe used for potable water systems are detailed in Table 45.1. Decisions on which to use are based on cost, length of service expected, and the required quality of the water. Some water contains large amounts of minerals and is very corrosive. A pipe that carries hot water must withstand temperatures of 180°F (82°C) or higher. In special applications, the effect of the liquid or gas to be carried must be considered. Gasoline, liquified petroleum gas, and similar materials have a substantial effect on pipes.

Plastic, copper, and steel pipe are manufactured with a variety of diameters and wall thicknesses. The thicker the pipe wall the greater the pressure the pipe can carry.

Table 45.1 Pipe and Tubing used for Water Supply Systems

Type	Connections	Diameters	Special Qualities
Galvanized steel	Threaded	1/8–4 in. (3–102 mm)	Strong, long life
Welded steel	Threaded	1/8–4 in. (3–102 mm)	Strong, long life
Copper tube	Soldered, brazed	1/4–6 in. (6–152 mm)	Corrosion resistant
Red brass	Threaded, brazed	1/8–6 in. (3–152 mm)	Corrosion resistant
Plastic	Solvent joined, heat fusion, serrated inserts	1/4–6 in. (6–152 mm)	Lightweight, corrosion resistant

Galvanized steel pipe is strong, long lasting, and can be used for hot and cold water. Piping and fittings are joined with threaded connections. The pipe is available in three grades: standard weight, extra strong, and double extra strong.

Gas is piped to a meter outside a building by a utility company. The building's gas piping is run inside to individual fixtures that require fuel, such as a gas-fired furnace. Codes regulate materials and methods of installation. Generally, a building's gas inlet must be above grade. Gas piping within the building is black steel pipe meeting the requirements of ASTM A53 or ASTM A106. It uses steel or malleable iron fittings. Some codes permit the use of copper or brass pipe if the gas type will not cause them to corrode. Some types of plastic pipe are permitted for underground exterior installation. They must receive a protective coating to help resist corrosion from soil elements.

Welded steel pipe is made by rolling a flat hot steel strip into a circular shape and butt-welding the edges as they are pressed together. It is available in the same grades as galvanized steel pipe. The outside diameter is the same for all three grades and the difference in the thickness of the wall is taken up on the inside of the pipe. Welded steel pipe is joined by threaded fittings.

Red brass pipe is used for water lines, especially if the water contains corrosive elements. It is manufactured with threaded fittings and plain ends that are brazed to socket-type fittings.

Copper water tubing is an excellent hot and cold water distribution material. It is resistant to corrosion and does not rust. Flexible copper tubing is easy to bend into various shapes, reducing the number of fittings needed. The joints fit together in a socket arrangement and are secured by soldering. The solder used cannot contain lead because lead will contaminate the water, causing potential health problems.

Copper water tubing has a high coefficient of expansion. When carrying heated liquids, it can expand enough in length to cause damage to the pipe unless the expansion is allowed for. Typically, a section of pipe is bent into a loop, allowing it to expand and contract (Fig. 45.2).

Different types of copper tubing are shown in Table 45.2. Copper water tubing is available in types K, L, and M. Type K has the heaviest wall and is marked by a green stripe along its length. Because of its strength, it is widely used for underground runs. Type L has a medium wall and a blue stripe, while type M has a light wall and a red stripe. All three are used for potable water, heating and air-conditioning, and fuel and fuel oil systems. Type ACR is used for air-conditioning and refrigeration systems; DWV (yellow stripe) is used for drain, waste, and ventilation systems; and several other types identified as OXY and MED are used for medical gas applications.

Figure 45.2 Ways to allow for expansion and contraction of copper tubing. *(Courtesy Copper Development Association)*

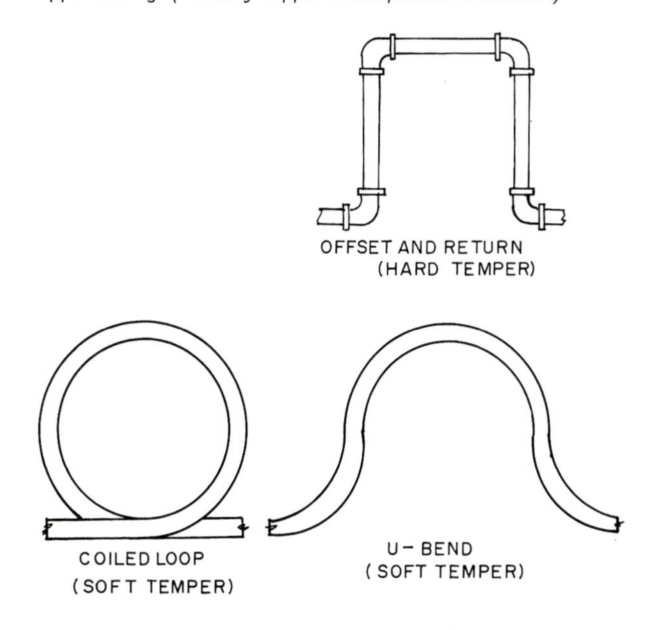

Copper tubing is manufactured in hard and soft tempers. Soft-tempered pipe bends easily, reducing the need for many connections. The hard-tempered pipe requires soldered connectors to turn corners. Copper tube is available in diameters from $\frac{1}{4}$ to 8 in. (6 to 203 mm).

Plastic pipe is manufactured in several synthetic resins. The type of resin used greatly influences the strength and use of the pipe. Table 45.3 lists the commonly used resins for producing plastic pipe. Notice that most are not acceptable for hot water piping. Plastic pipe is lightweight, flexible, and available in long lengths. Various types are joined with solvent cement, elastomeric seals for bell-end piping and fittings, serrated insert fittings secured with stainless steel clamps, heat fusion used on plastics for which there is no solvent, and threaded fittings that can connect to threaded metal pipe.

Plastic piping and sanitary piping are manufactured in diameters from $\frac{1}{2}$ in. (12.7 mm) to 48 in. (1,219 mm) and in lengths to 20 ft. (6 m). The wall thicknesses are identified by a schedule number, such as Schedule 40, 80, or 120, with the larger numbers indicating a thicker pipe wall.

Pipes made from glass, nickel, silver, and chrome are corrosion resistant and used for special applica-

Table 45.2 Copper Tubing: Types, Uses, and Sizes

Type	Color	Uses	Nominal Diameter
K	Green	Potable water, fire protection, solar, fuel/fuel oil, HVAC, snow melting	Straight lengths $^{1}/_{4}$–12 in. (6–305 mm) Coils $^{1}/_{4}$–2 in. (6–50.8 mm)
L	Blue	Potable water, fire protection, solar, fuel/fuel oil, HVAC, snow melting	Straight lengths $^{1}/_{4}$–12 in. (6–305 mm)
M	Red	Potable water, fire protection, solar, fuel/fuel oil, HVAC, snow melting	Straight lengths $^{1}/_{4}$–12 in. (6–305 mm)
DWV	Yellow	Drain, waste, vent, HVAC, solar	Straight lengths $1^{1}/_{4}$–8 in. (32–203 mm)
ACR	Blue Soft ACR not marked	Air-conditioning, refrigeration, natural gas, liquified petroleum gas	Straight lengths $^{3}/_{4}$–$4^{1}/_{8}$ in. (9.5–105 mm)
OXY, MED, OXY/MED, OXY/MED OXY/ACR,	K-Green L-Blue	Medical gas	Straight lengths $^{1}/_{4}$–8 in. (6–203 mm)
G	Yellow	Natural gas, liquified petroleum gas	Straight lengths $^{3}/_{8}$–$1^{1}/_{8}$in. (9.5–28.5 mm) Coils $^{3}/_{8}$–$^{7}/_{8}$in. (9.5–22 mm)

Table 45.3 Major Types and Uses of Plastic Pipe

			Maximum Operating Temperature		
Type	Condition	Connections	°F	°C	Typical Uses
Acrylonitrile butadiene styrene (ABS)	Rigid	Threaded, serrated fittings, solvent	100 pressure 180 nonpressure	38 82.9	Cold water, waste, vent, sewer, drain, conduit, gas
Chlorinated polyvinyl chloride (CPVC)	Rigid	Threaded, couplings, serrated fittings, solvent	180 at 100 psig (type 11)	82.9	Cold and hot water, chemical piping
Polyethylene (PE)	Flexible	Serrated fittings, fusion in socket, butt fusion	100 pressure 180 nonpressure	38 82.9	Cold water, gas, waste, chemicals
Polypropylene (PP)	Rigid	Mechanical couplings, butt fusion, socket fusion	100 pressure 180 nonpressure	38 82.9	Chemical piping, chemical drainage
Polyvinyl chloride (PVC)	Rigid	Solvent, threading, mechanical couplings, serrated fittings	100 pressure 180 nonpressure	38 82.9	Cold water, gas, waste, vents, drains, sewers, conduit
Styrene rubber plastic (SRP)	Rigid	Solvent, serrated fittings, elastomer seal	150 nonpressure (not used under pressure)	66	Sewage disposal field, storm drainage, soil drainage

Figure 45.3 No-hub pipes are joined with gaskets and stainless steel straps.

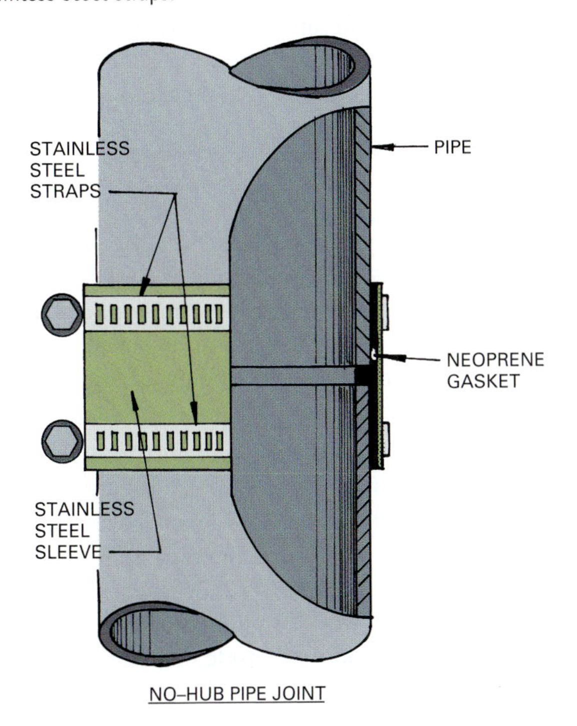

tions. Most metal pipes have threaded connections. Glass pipe is connected by neoprene gaskets secured with stainless steel straps over a stainless steel sleeve (Fig. 45.3).

Pipe Connections

The most common methods for connecting pipe are shown in Fig. 45.4. Water pipe uses butt-welded, socket, and threaded fittings. The bell and spigot connection is used on sewer lines, while the flanged connection is used in applications such as petrochemical and power generation piping.

Water Pipe Fittings

Figure 45.5 illustrates the most commonly used water pipe fittings. Couplings are used to connect two pipes end to end. Elbows change the direction of the pipe. A tee permits one pipe to intersect another. The end of a pipe can be closed by installing a coupling and screwing a plug into it with a cap that screws over the end of the pipe.

Figure 45.4 Common types of pipe and fitting connections.

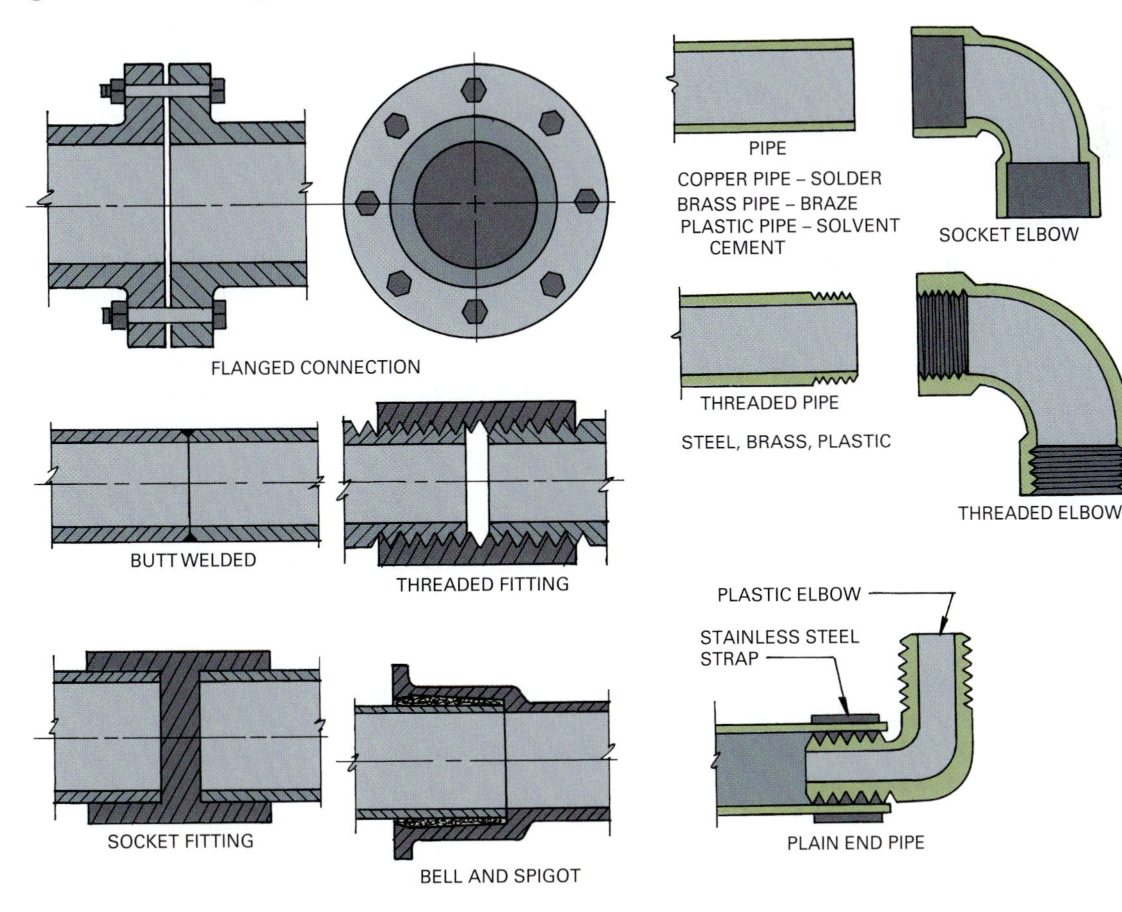

Figure 45.5 Some of the commonly used pipe fittings.

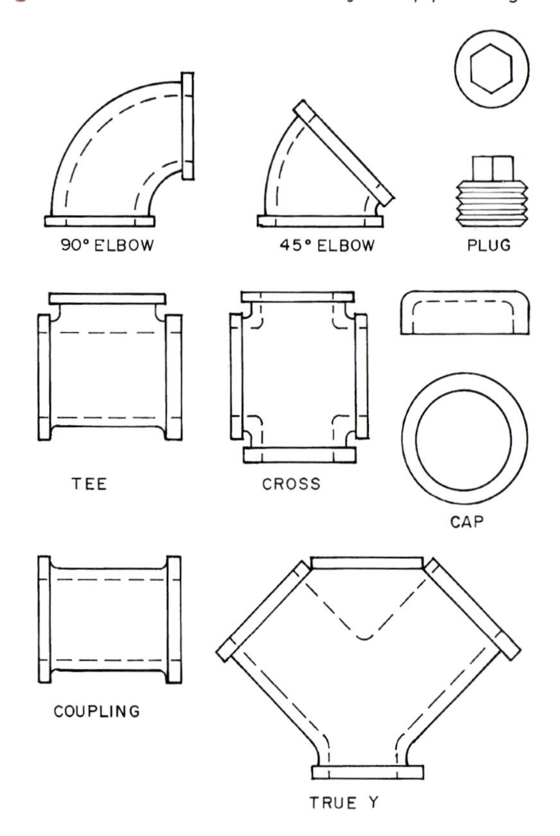

VALVES

Valves are used to control the flow of water, oil, gas, and chemicals in pipe distribution systems. Some of the frequently used types are shown in Fig. 45.6. A pressure regulator valve limits water pressure to acceptable levels, preventing damage to piping and equipment. A water hammer arrestor has a hydraulic piston that absorbs shock waves produced by sudden changes in water flow; this reduces banging in the pipes.

Backflow preventers keep water from backing up in a system, and expansion tanks mitigate excess pressure by absorbing extra water volume created when water is heated and expands. Expansion tanks utilize a rubber bladder that flexes against the water pressure. A "T and P" valve is a safety relief valve that senses a buildup of temperature or pressure and opens to release the excess pressure and allow cold water to enter and prevent an explosion.

Float valves are used to control water levels in tanks. As the water level rises, the float rises and shuts off the supply when a set capacity has been reached. Strainers typically have a 20 mesh screen that collects dirt and debris. They may be on the main line or on lines to a particular piece of equipment. Saddle valves are installed on piping that carries water under pressure. They clamp on a pipe and then pierce it.

Gate-style stop and waste valves were once commonly used to shut off water to a building on the owner's side of the meter. Now the more dependable ball valve is used for shutoffs. Globe valves are used to control the flow of hot and cold water, oil, and gas.

INSULATION

Condensation forms on the exterior of cold water pipes when they come in contact with warm humid air. Condensation can drip from a pipe into a wall cavity, ceiling, or floor, wetting insulation and penetrating drywall or other wall finishes.

To prevent damage from condensation, pipes are covered with preformed fiberglass or foam insulation that is usually $\frac{1}{2}$ to 1 in. (12 to 25 mm) thick (Fig. 45.7). The insulation is fitted around the pipe and taped. Hot water pipes are insulated to reduce heat loss to the cooler atmosphere. Most plumbing units, such as water heaters, have insulation built inside their jackets and extra batts can be taped to the outside.

Pipes carrying chilled water, brine, refrigerant, domestic hot water, commercial hot water and steam, will condensate, and require the use of minimum insulation, as shown in Table 45.4. If hot and cold piping run parallel they should be 6 to 8 in. (152 to 203 mm) apart to reduce the chance of heat and cooling exchange.

WATER PIPE SIZING

The size of the required water supply pipe depends on the available pressure, the vertical and horizontal distances involved (friction in the pipe), the number of fittings (tees, elbows), and use demand at the fixture. Experience has allowed values to be determined for various fixture flows under recommended minimum pressures. The values of these are considered as the pipe diameter is calculated. Several fixtures on one water line do not require the water flow to increase directly for each because not all of the fixtures are likely to be in use at the same time. Typical flow pressures and flow rates are shown in Table 45.5.

POTABLE WATER DISTRIBUTION SYSTEMS

Codes specify that all fixtures in a building must be supplied with potable water at required pressures. **Potable water** in residential buildings generally uses water pressure from the central water system or pump on a well

Figure 45.6 Typical valves used in water distribution systems.

(a) Ball valve *(Courtesy of Conbraco)*

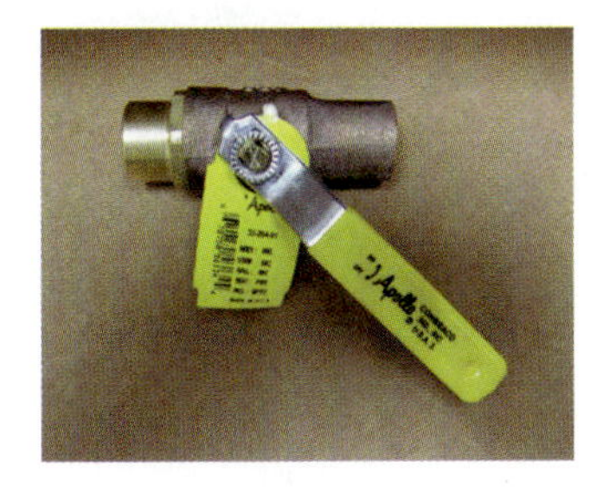

(b) Pressure-reducing valve *(Courtesy of Wilkins, a Zurn Company)*

(c) Stop and waste valve *(Courtesy of A.Y. McDonald Mfg. Co.)*

(d) Swing check valve *(Courtesy of Watts Regulator)*

(e) Strainer *(Courtesy of Conbraco)*

(f) Relief valve *(Courtesy of Conbraco)*

Figure 45.7 Hot and cold water pipes are insulated to maintain water temperature, prevent freezing, and stop condensation from dripping from the pipes. *(Image copyright rfx, 2009. Used under license from Shutterstock.com)*

to supply the fixtures (Fig. 45.8). An up-feed system is one in which the water moves up from the source into the building and to the fixtures. The piping used must consist of the proper size to carry the amount of water needed by each fixture. Circulation pipes and risers usually feed more than one fixture, so total expected flow must be calculated.

Water enters a building through a meter and a shutoff valve. In cold climates, the meter is located inside the building and the water service line below the frost line. The depth of the frost line varies in different geographical areas. The water continues to a water softener if required, then on to a water heater and the horizontal circulation lines. Water moves to fixtures in the upper stories through **risers** and to individual fixtures through small-diameter pipe connectors. Hot water runs from the water heater through a similar piping distribution system.

Low multistory buildings may also use an up-feed system if water demand is not excessive (Fig. 45.9). A series of basement pumps provide the needed pressure to raise the water to desired heights and maintain pressure as fixtures are used.

Table 45.4 Minimum Insulation for Pipes (Inches)

Fluid Design Operating Temperature Range (°F)	Insulation Conductivity		Nominal Pipe Diameter (in.)					
	Conductivity Range [BTU-in./(h-ft³-°F)]	Mean Rating Temperature (°F)	Runouts[a] up to 2	1 and less	1–1¼ to 2	2½ to 4	5 and 6	8 and up
Heating Systems (Steam, Steam Condensate, and Hot Water)								
Above 350	0.32–0.34	250	1.5	2.5	2.5	3.0	3.5	3.5
251–350	0.29–0.31	200	1.5	2.0	2.5	2.5	3.5	3.5
201–250	0.27–0.30	150	1.0	1.5	1.5	2.0	2.0	3.5
141–200	0.25–0.29	125	0.5	1.5	1.5	1.5	1.5	1.5
105–140	0.24–0.28	100	0.5	1.0	1.0	1.0	1.5	1.5
Domestic and Service Hot Water Systems[b]								
105 and greater	0.24–0.28	100	0.5	1.0	1.0	1.5	1.5	1.5
Cooling Systems (Chilled Water, Brine, and Refrigerant)[c]								
40–55	0.23–0.27	75	0.5	0.5	0.75	1.0	1.0	1.0
Below 40	0.23–0.27	75	1.0	1.0	1.5	1.5	1.5	1.5

Source: Reprinted by permission from ASHRAE/IES Standard 90.1-1989, Energy Efficient Design of New Buildings Except Low-Rise Residential Buildings, *© 1989 by the American Society of Heating, Refrigerating, and Air-Conditioning Engineers, Inc., Atlanta, Ga.*

[a] *Runouts to individual terminal units not exceeding 12 ft. in length.*

[b] *Applies to recirculating sections of service or domestic hot water systems and first 8 ft. from storage tank for nonrecirculating systems.*

[c] *The required minimum thicknesses do not consider water vapor transmission and condensation. Additional insulation, vapor retarders, or both, may be required to limit water vapor transmission and condensation.*

Table 45.5 Minimum Flow and Pressure Required by Typical Plumbing Fixtures

Fixture	Flow Pressure (psi)	Flow Pressure (kPa)	Flow Rate (gpm)	Flow Rate (L/s)
Ordinary basin faucet	8	55	2.0	0.13
Self-closing basin faucet	8	55	2.5	0.16
Sink faucet, ⅜ in. (9.5 mm)	8	55	4.5	0.28
Sink faucet, ½ in. (12.7 mm)	8	55	4.5	0.28
Bathtub faucet	8	55	6.0	0.38
Laundry tub faucet, ½ in. (12.7 mm)	8	55	5.0	0.32
Shower	8	55	5.0	0.32
Ball-cock for closet	8	55	3.0	0.19
Flush valve for closet	15	103	15–40	0.95–2.52
Flushometer valve for urinal	15	103	15.0	0.95
Garden hose (50 ft., ¾-in. sill cock) (15 m, 19 mm)	30	207	5.0	0.32
Garden hose (50 ft., ⅝ in. outlet) (15 m, 16 mm)	15	103	3.33	0.21
Drinking fountains	15	103	0.75	0.05
Fire hose 1½ in. (38 mm), ½ in. nozzle (12.7 mm)	30	207	40.0	2.52

Reproduced from Manual of Individual Water Supply Systems, *U.S. Environmental Protection Agency*

Figure 45.8 Components of a typical residential potable water distribution system.

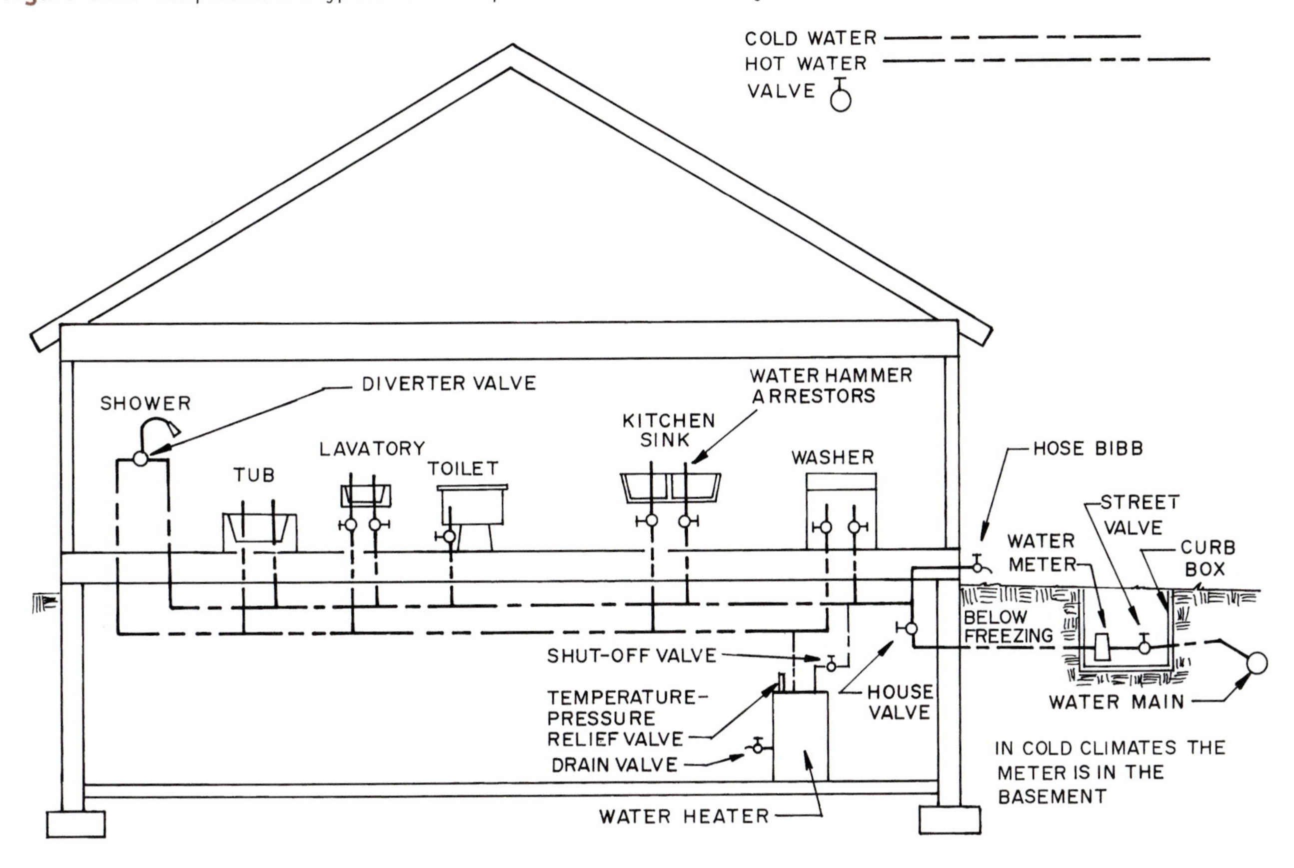

Up-feed systems are useful on multistory buildings that are not high enough to warrant the expense of a rooftop water storage system but do have height limitations. A system may require two or more pumps. As water demand increases, a second or third pump will come on line to increase flow and maintain pressure. It should be noted that the supply available from the central water system main should be large enough so that demands of the building can be met without reducing water service to neighboring buildings. Notice that this system does not have a reserve water supply.

Tall multistory buildings utilize a down-feed potable hot and cold water distribution system. Water from the central system is pumped from a street main or water storage suction tank to a roof storage tank with one or more pumps. This storage tank also holds a reserve for the fire-suppression system (see Chapter 44) and a supply of potable water for fixtures. The water flows from a header pipe to down-feed risers from which branch water lines run to fixtures on each floor (Fig. 45.10). Very tall buildings usually are divided into several zones, each having its own pumps and storage tanks (Fig. 45.11).

SANITARY PIPING

Piping used for sanitary systems may be cast iron, copper, plastic, lead, glass, and clay. These are used for various **drainage, waste, and vent installations (DWV).**

Cast-iron soil pipe and fittings are gray iron castings suitable for installation and service for storm drain, sanitary, waste, and vent piping. Hub-type cast-iron soil pipe and fitting specifications are detailed in ASTM A74-87. They are available in Extra Heavy and Service classifications in diameters shown in Table 45.6. The pipes and fittings must be coated with a material to protect their surfaces. Hubless cast-iron soil pipe and fittings are specified in ASTM A888-90. Neither type is intended to be used on pressure systems. The selection of the proper design size ensures space for the free air needed for gravity drainage. Both types are available in a wide range of fittings. A few are shown in Figs. 45.12 and 45.13. The hub-type joints may be sealed with lead and **oakum** or with a neoprene compression gasket. Hubless connections use a gasket and a stainless steel casing and retaining clamps (Fig. 45.14).

Figure 45.9 A simplified schematic of an up-feed potable water distribution system for a low-rise building.

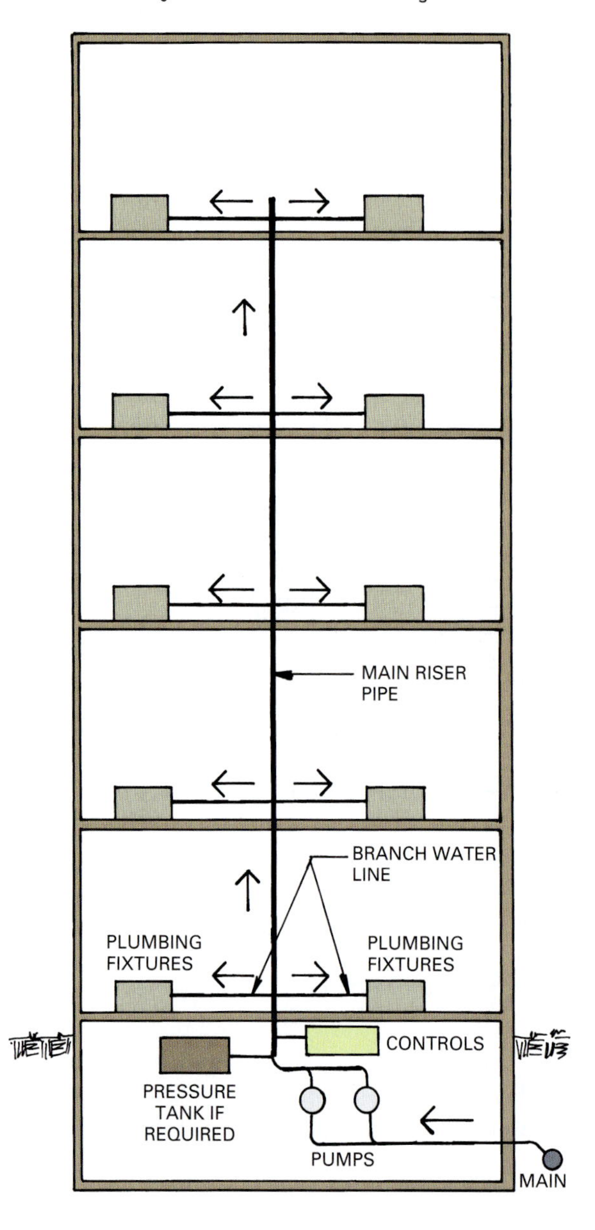

Copper tubing used for sanitary waste systems is classified as DWV (drainage, waste, vent). The available diameters are given in Table 45.2. Copper tubing is used in residential, low-rise, and high-rise buildings for all parts of drainage plumbing, including soil and vent stacks and soil, waste, and vent branches. In high-rise buildings, expansion must be considered as the system is designed. Changes of 1°F can cause DWV pipe to change up to 0.001 in. (0.025 mm) for a 10 ft. section. One way to control thermal movement is to anchor the pipe to the floor, as shown in Fig. 45.15. The number of anchors is a design decision, but a rule of thumb is to anchor at every eight floors if the temperature rise is anticipated to be up to 50°F (10°C).

Joints in copper DWV tubing are made by properly cleaning and fluxing the joining parts and allowing the solder to flow between them by capillary action. The space between the parts is sized to allow molten solder to flow into the joint. Brazed connections are used if greater strength is needed.

Plastic pipe suitable for DWV systems include acrylonitrile butadiene styrene (ABS), styrene rubber plastic (SRP), and polyvinyl chloride (PVC). All three are used for sewer systems, and ABS and PVC can be used for drain, waste, and vent systems. Local plumbing codes should be checked to verify approved uses.

Plastic pipe and fittings have identification symbols on each piece, as shown in Fig. 45.16. They are joined with solvent cement. Some plumbers prefer to clean pipe ends fitting insides by brushing on a coat of priming solvent. After about fifteen seconds, a thick coat of solvent cement is applied to the end of the pipe and the inner surface of the fitting (Fig. 45.17). The pipe is inserted and twisted a quarter turn to spread the cement. At least three minutes must pass before starting the next joint.

Lead and glass drain pipe is used for the disposal of special liquids, such as in a chemical plant or research laboratory. Lead and glass resist attack by many chemicals. Lead pipe is linked by welding the joints. Glass pipe is connected with a gasket and stainless steel clamp, as shown for hubless pipe in Fig. 45.14.

Clay pipe is used for waste disposal lines outside a building. Its use is carefully controlled by building codes. The pipe is made from clay and burned in a kiln (similar to clay brick). It is available in diameters from 4 to 36 in. (102 to 914 mm) and has a variety of fittings. It uses a hub joint filled with a packing compound. Clay pipe is impervious to acids and alkalines and does not deteriorate with age. It is not used where it might be subjected to shock or loads.

Calculating Waste Pipe Sizes

The diameter of pipes used in a sanitary piping system depends on the amount of waste they are to carry. Plumbing codes typically establish the number of drainage fixture units running through each pipe size (Table 45.7). A **fixture unit** is a measure of the probable discharge into the drainage system by the various plumbing fixtures. It is expressed in units of cubic volume per minute. The value for a particular fixture depends on the volume rate of discharge, the time duration of a single discharge, and the average time between successive discharges. A fixture unit is generally equal to $7\frac{1}{2}$ gallons of flow.

Figure 45.10 A simplified schematic of a down-feed potable hot and cold water distribution system typically used in high-rise buildings.

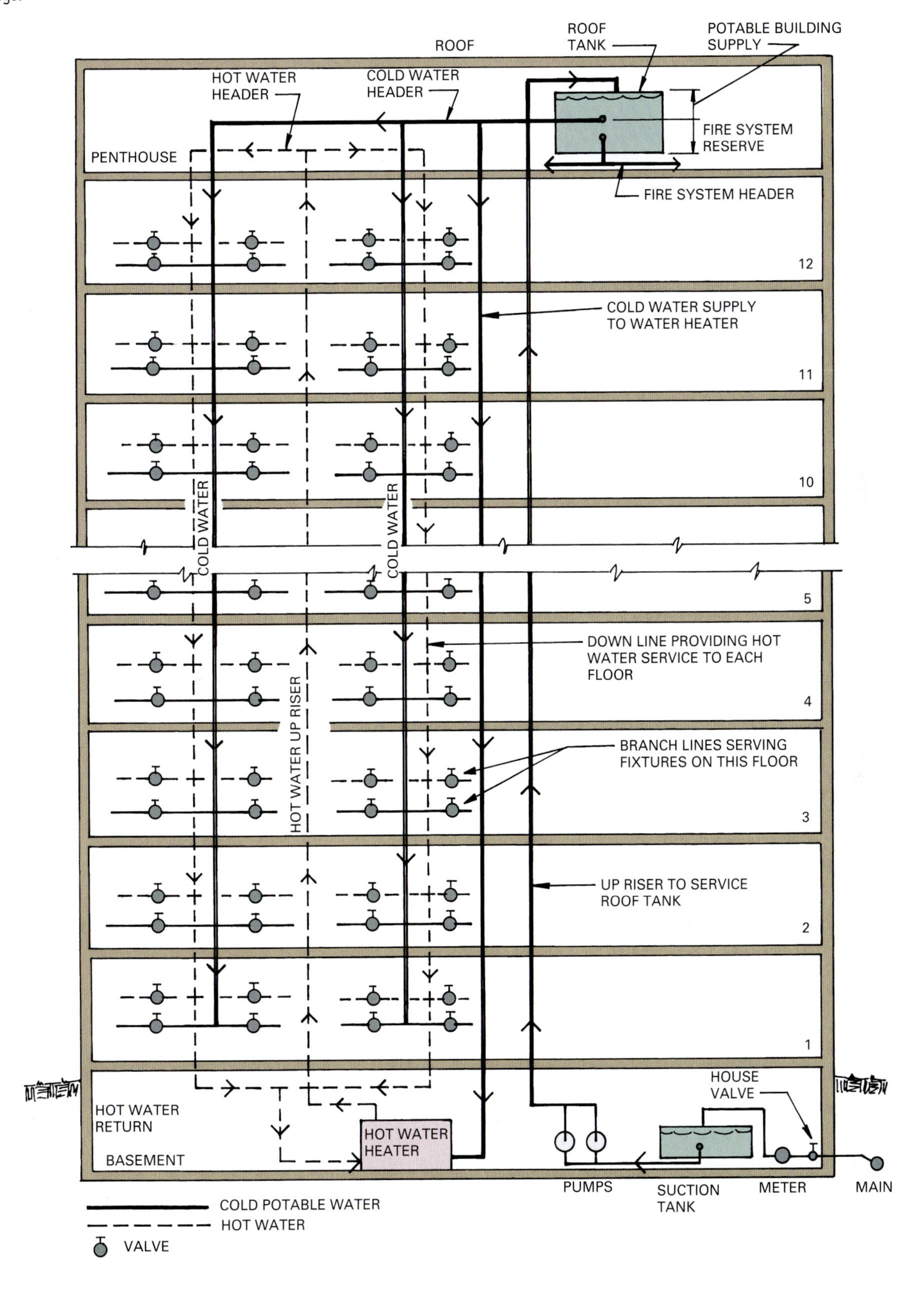

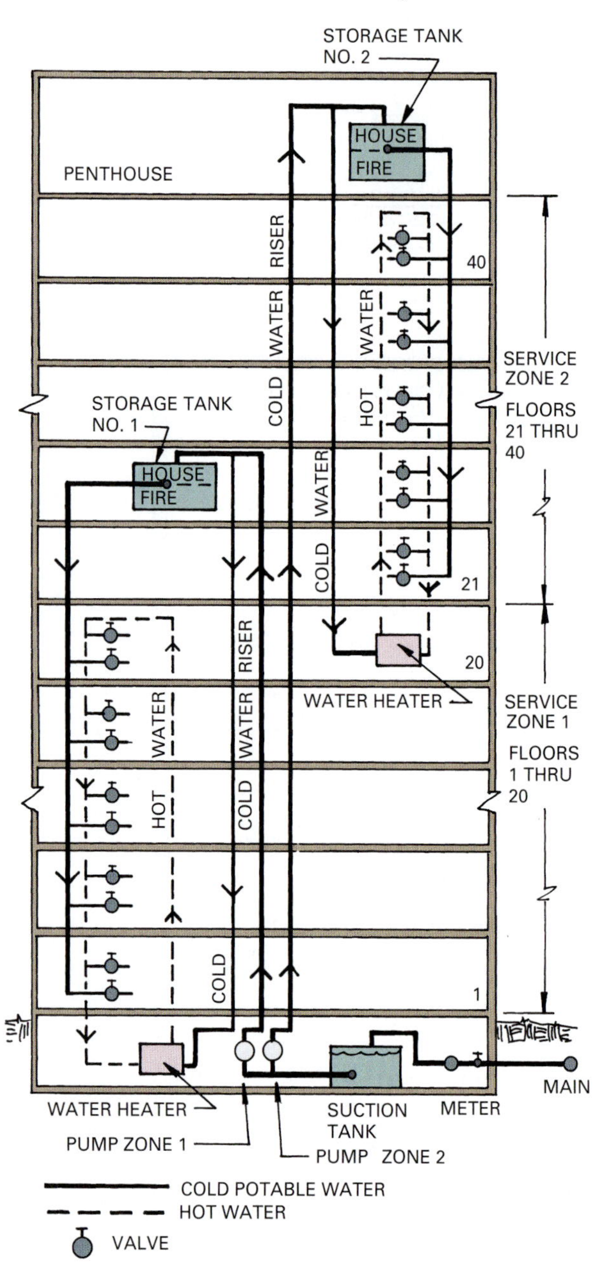

Figure 45.11 A simplified schematic of a two-zone down-feed hot and cold water distribution system.

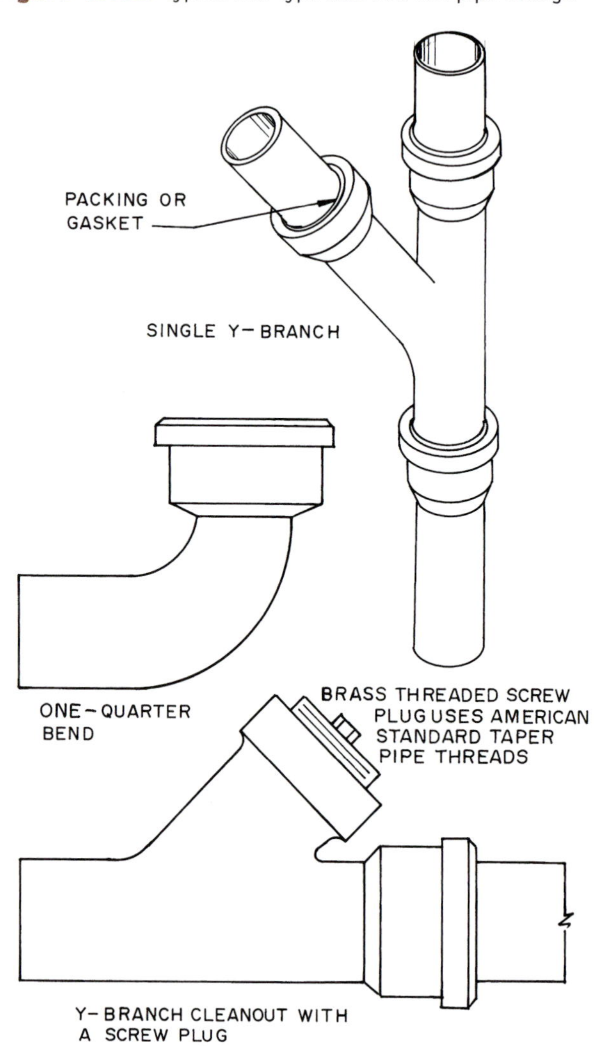

Figure 45.12 Typical hub-type cast-iron soil pipe fittings.

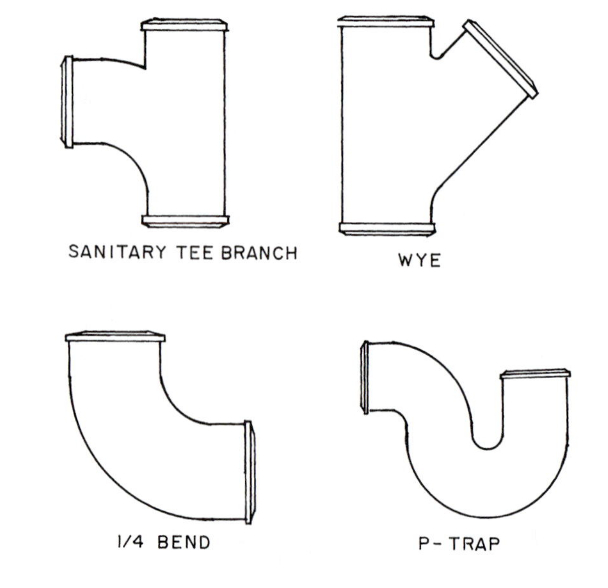

Figure 45.13 Typical hubless-type cast-iron soil pipe fittings.

Table 45.6 Cast-Iron Soil Pipe Uses and Sizes

Type	Uses	Nominal Diameter
Hubless soil pipe, standard and extra heavy	Sanitary and storm drains, waste, vents	2–15 in. (50.8–381 mm)
Hub type soil pipe, service and extra heavy	Sanitary and storm drains, waste, vents	2–15 in. (50.8–381 mm)

Figure 45.14 Joints used to join hub-type and hubless cast-iron soil pipe. *(Courtesy Cast Iron Pipe Institute)*

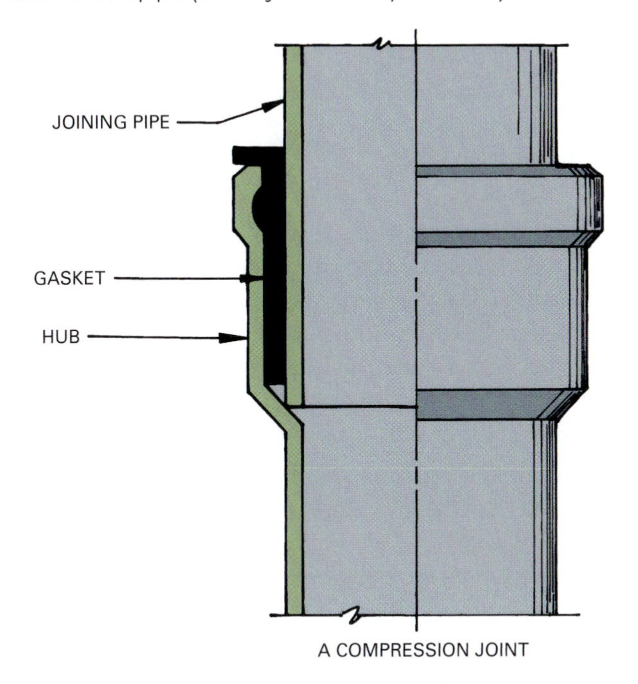

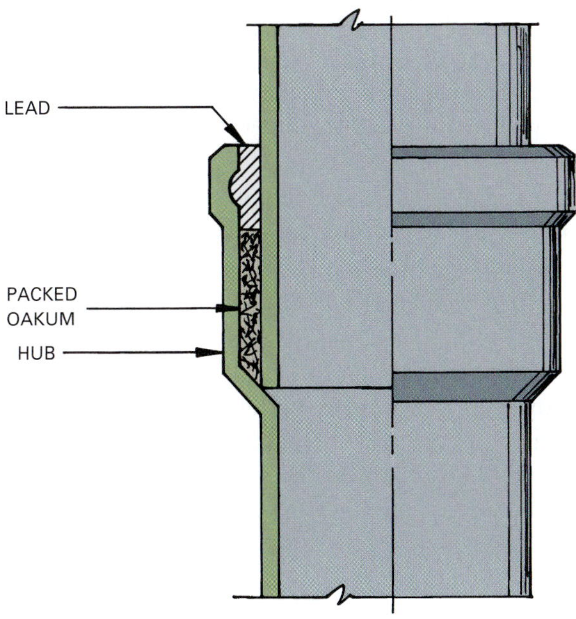

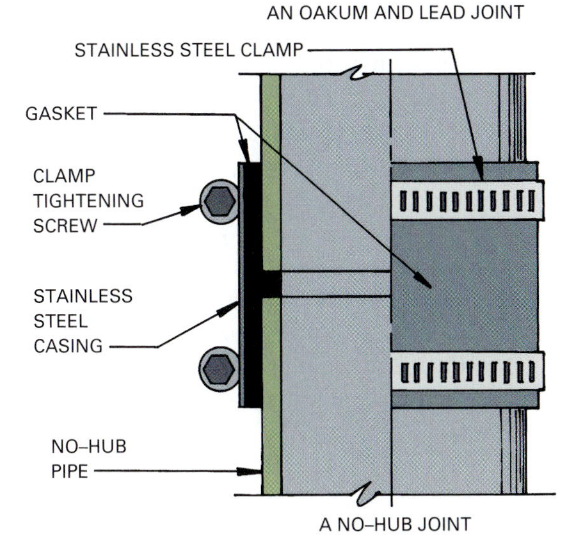

Figure 45.15 Thermal movement in long vertical runs of copper pipe can be controlled by anchoring the pipe to the floor as required by a system's design. *(Courtesy Copper Development Association)*

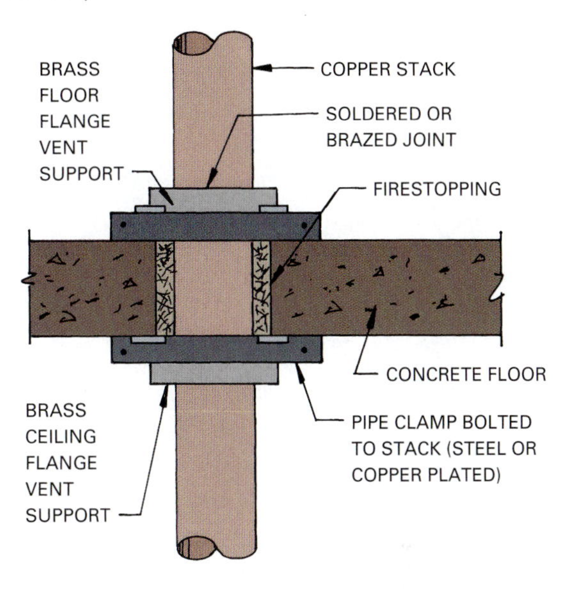

Figure 45.16 Identification symbols used on plastic pipe. *(Courtesy Plastic Pipe Institute)*

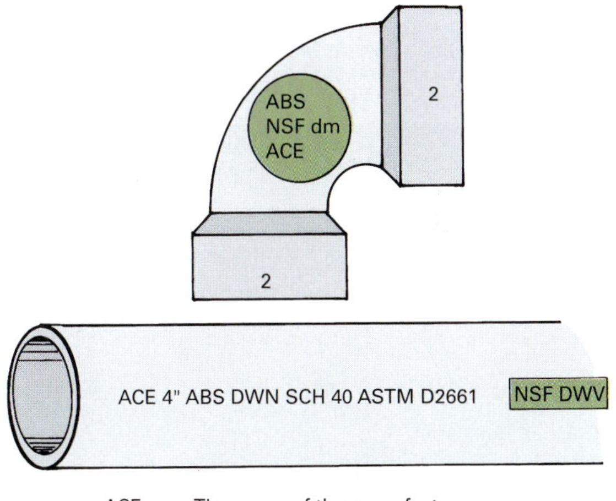

ACE	The name of the manufacturer.
4 in.	Diameter of the pipe.
ABS	Acrylonitrile-Butadiene-Styrene, the material.
DWV	Suitable for drainage waste and vent.
SCH 40	Schedule 40. This identifies the wall of thickness of the pipe.
ASTM D2661	"Standards Number" assigned by the American Society for Testing and Materials.
NSF DWV	Tested by the National Sanitation Foundation Testing Laboratory. The pipe meets or exceeds the current standards for sanitary service.

Figure 45.17 Adjoining surfaces of the plastic pipe are cleaned and primed before coating with a solvent cement. *(Image copyright Christina Richards, 2009. Used under license from Shutterstock.com)*

Table 45.7 Drainage Fixture Units for Selected Plumbing Fixtures

Plumbing Fixture	Drainage Fixture Unit Value
Automatic clothes dryer	3
Bathtub with or without overhead shower	2
Clinic sink	6
Dental unit	1
Drinking fountain	1/2
Dishwasher, domestic	2
Floor drain, 2 in. waste pipe	3
Kitchen sink, 1 1/2 in. trap	2
Kitchen sink with dishwasher	3
Lavatory, 1 1/4 in. waste pipe	1
Shower stall, residential	2
Urinal, stall, washout	4
Water closet, tank operated	4
Water closet, valve operated	6

From National Standard Plumbing Code, *National Association of Plumbing, Heating, and Cooling Contractors*

Sanitary Piping Systems

A **sanitary piping** system removes the water and other waste materials discharged at the various fixtures. The system carries the material through the building to a **building drain** from which it is discharged to the **building sewer** and on to the public **sanitary sewer** system or septic tank. The waste material flows from the building at a level below the lowest fixture and moves by gravity flow to the building sewer. If the public sewer is not below the sloped building sewer, waste will not drain and pumps must be used to lift the waste and move it to the public sewer.

Pipe diameters for sanitary piping systems are sized to carry the anticipated waste material flow rapidly without clogging pipes or creating annoying water noises. Pipe diameters also are selected to produce minimum pressure variations where fixtures connect to **waste pipes**, where waste pipes connect to **branch soil pipes**, and where branch soil pipes connect to the **soil stack**.

The system relies on gravity flow, so it is not under pressure. The effluence increases in speed as it proceeds down the piping system. Friction limits speed and must be considered when sizing pipe.

Plumbing codes specify the required slope for horizontal piping. The requirement varies with pipe diameter but is typically from $\frac{1}{8}$ to $\frac{1}{2}$ in. (3 to 12.7 mm) per foot of run. Flow velocity increases with the amount of slope, and high velocities in horizontal pipes can increase siphonage. The capacity of a pipe should be increased by using a larger diameter pipe rather than by increasing the slope.

The sanitary drainage system receives waste matter that may be difficult for private or public waste treatment plants to process. Such waste will often clog the pipes in the sanitary drainage system.

Interceptors are used to catch large items of waste material near the point of origin, intercepting it before it moves very far into the system. Manufacturers have a variety of devices available to catch grease, hair, oil, glass and plastic grindings, plaster, and other undesirable materials. The interceptor must be emptied on a periodic maintenance schedule.

Indirect Wastes

Indirect waste, such as that from food-handling equipment, dishwashers, sterilizers, and commercial laundries, are discharged with an indirect waste pipe. The waste from the indirect waste pipe is not discharged directly into the building sanitary piping system but into a fixture connected to the building sanitary system. This provides an air gap between the two systems. An **air gap** is used to prevent one system from accidentally backing up into another. Toxic and corrosive waste discharges must be automatically diluted with water or chemically neutralized before being introduced into the building sanitary system.

Case Study

Bronx Zoo Eco-Restrooms

Architect: Edelman Sulton Knox Wood Architects LLT

Contractor: Summit Construction Services Group

Location: Bronx, New York

Building type: Service Building

Completed: 2006

Overview

Founded in 1895, the Wildlife Conservation Society (WCS) was one of the first conservation organizations in the U.S. The Society began with a clear mandate: advance wildlife conservation, promote the study of zoology, and create a first-class zoo. The Bronx Zoo opened in 1899 with a clear mandate: to advance wildlife conservation and promote the study of zoology. An existing bathroom facility at one of the zoo gateway entrances had fallen into disrepair. Zoo management decided to inaugurate a new building that would showcase strategies to responsibly deal with human waste while simultaneously preventing pollution of local waterways. The aged facility had operated on a septic system that over time had leached into the adjacent Bronx River. In addition to the use of natural daylighting and other energy saving features, the new building incorporates a number of cutting edge water conservation technologies (Fig. A).

Plumbing Fixtures

The Eco-Restroom's plumbing fixtures include fourteen foam-flush toilets, four waterless urinals, and ten composting chambers. Composting foam-flush fixtures use between three to six ounces of water mixed with a drop of soap that flushes waste down to the composters below. In the composter, solid and liquid wastes are separated and treated by natural biological decomposition into a topsoil-like compost and strong liquid fertilizer (Fig. B). Even when compared to new efficient 1.6 gallon per flush toilets, the composting system has the potential of saving more than one hundred thousand gallons of water per year.

Graywater System

Water from the restroom's non-toilet fixtures is collected via conventional plumbing conduits in a tank in the basement. The graywater is distributed

Fig A: The exterior of the Bronx Zoo bathroom facility. *(Courtesy Clivus Multrum)*

Fig. B: One of the composting toilet fixtures. *(Courtesy Clivus Multrum)*

through buried pipes into the root zone of a small adjacent garden where plants use the water and nutrients contained in it. Because graywater is used for plant irrigation and toilet wastes are contained in the composters, there is no waste water discharge like there is in conventional buildings attached to sewers or septic systems.

Rainwater Harvesting

In addition to the composting toilets and graywater system, the new restroom also harvests rainwater to prevent stormwater runoff from entering the municipal sewer system. Rainwater is collected from the roof in a series of barrels, where it is stored for later use in landscape irrigation. Rainwater harvesting is especially useful in urban areas where it can increase soil moisture levels for urban greenery and supplement the ground water table through artificial recharge

The various green building technologies incorporated in the Eco-Restroom are clearly explained through signage designed to educate visitors about environmental issues. The project was recently named Eco Project of the Year by New York Construction, and plans are under way for two additional green buildings in the zoo.

RESIDENTIAL SANITARY PIPING

A typical simple sanitary piping system for a residence is shown in Fig. 45.18. The waste from the fixtures drains through waste pipes or branch soil pipes into a soil stack. Each waste pipe is sized to carry the flow from a fixture. The branch soil pipe is sized to carry the flow from all the fixtures flowing into it. The soil stock extends below the building and connects to the building drain and sewer, which in turn connects to the central sewer system or a septic tank. Each fixture has a vent pipe connected to a **vent stack** running through the roof. The **vent** pipe keeps the water in the **trap** of the fixture under atmospheric pressure, thus eliminating the chance of its being siphoned out when another fixture is used (Fig. 45.19). For example, if the waste system were a closed installation, a trap in a lavatory could have the water siphoned out when a connecting toilet is flushed. Some fixtures, such as toilets, have a trap built into the unit. The house drain will have one or more cleanouts

Figure 45.18 A simplified schematic of a typical residential sanitary piping system.

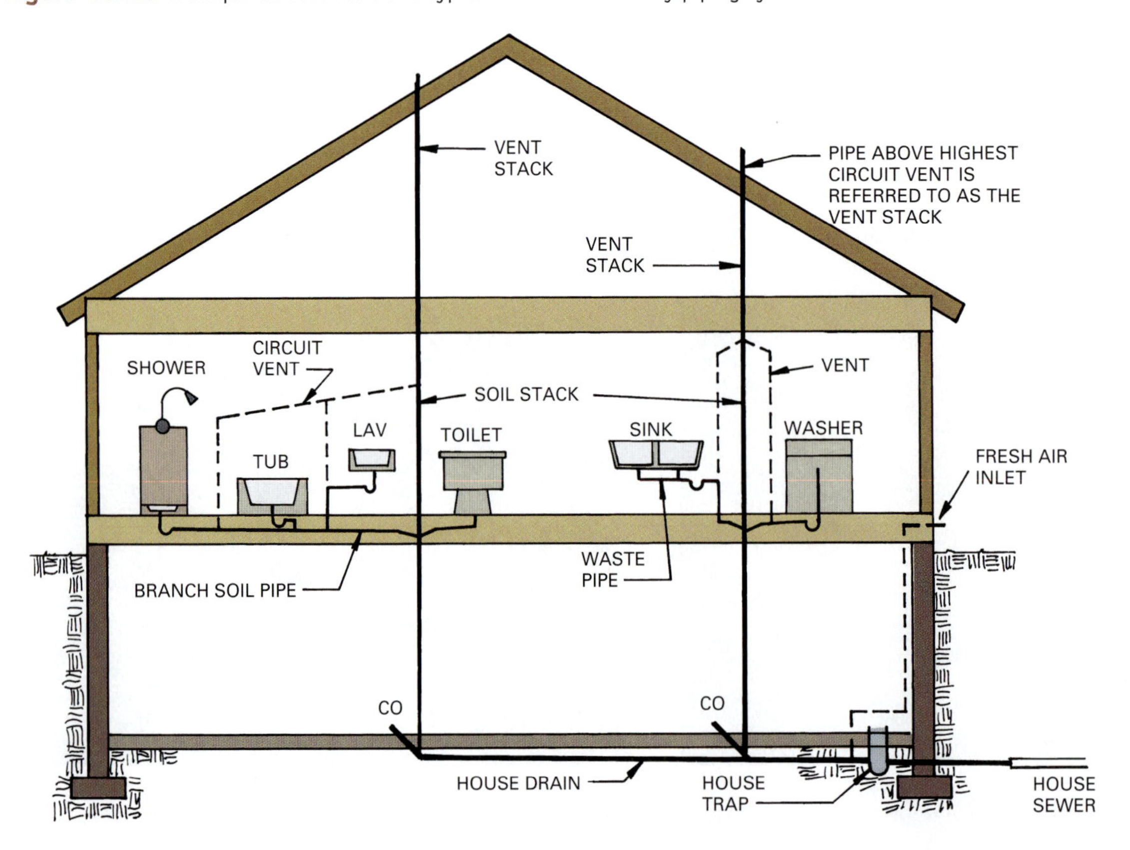

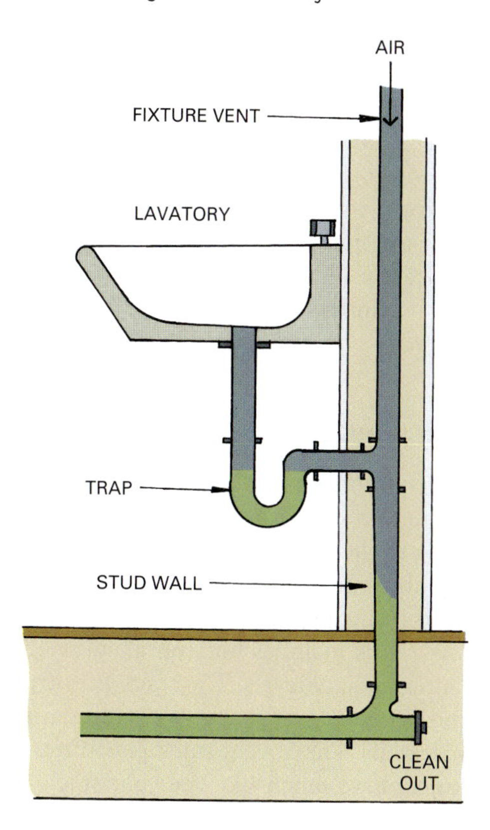

Figure 45.19 Fixtures are vented to keep water in the trap under atmospheric pressure so it is not siphoned out when a nearby fixture discharges waste in the system.

(C.O.). A **cleanout** is a pipe fitting with a removable plug through which an auger may be run to dislodge obstructions in the pipe.

Various types of **traps** are used (Fig. 45.20). Lavatories typically have a p-trap, while tubs use a drum trap. To keep plumbing costs low, designers attempt to place fixtures on the same wall. This reduces the piping required and eliminates long horizontal runs to reach out of the way fixtures (Fig. 45.21). In wood frame construction, the wall containing the plumbing uses at least 2 × 6 in. studs.

MULTISTORY BUILDING SANITARY PIPING SYSTEMS

In multistory buildings, it is common to design **plumbing chases** on each floor directly above the other to run water and waste disposal systems the height of the building (Fig. 45.22). Heating, air-conditioning, and fire suppressing systems can also use these chases for the efficient distribution of all systems. This consolidates the plumbing, reduces costs, and makes planning

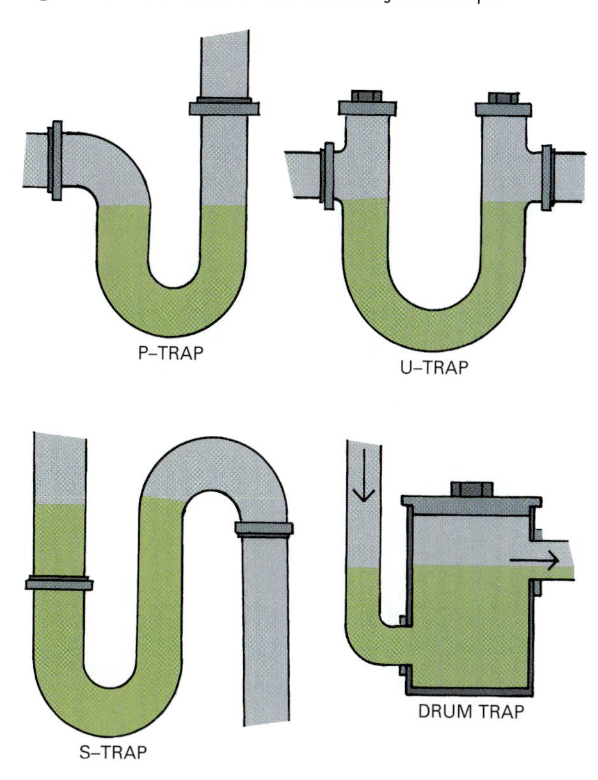

Figure 45.20 Some of the commonly used traps.

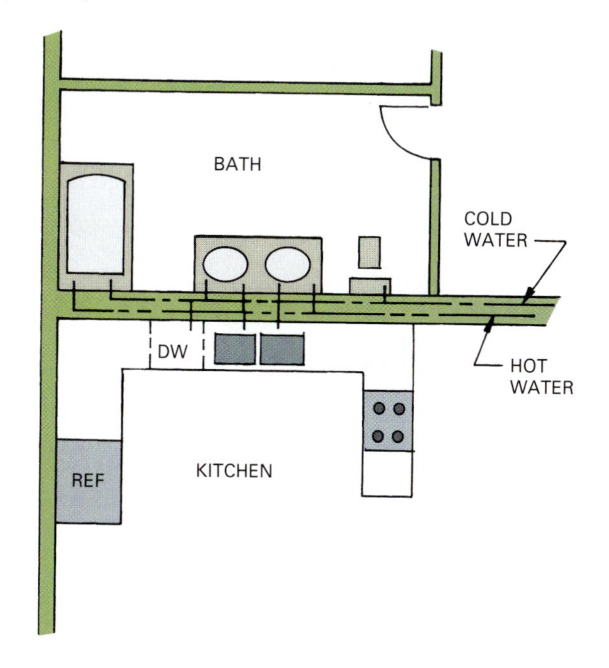

Figure 45.21 Plumbing costs can be reduced by aggregating plumbing fixtures.

of the uses of the space on each floor more flexible. Plumbing fixtures are typically located next to the sides of the vertical chase. Horizontal runs to locations within a floor area can also be made from the chase. In other cases, pipe risers can be located inside the

Figure 45.22 A plumbing chase is used to stack fixtures on floors above each other.

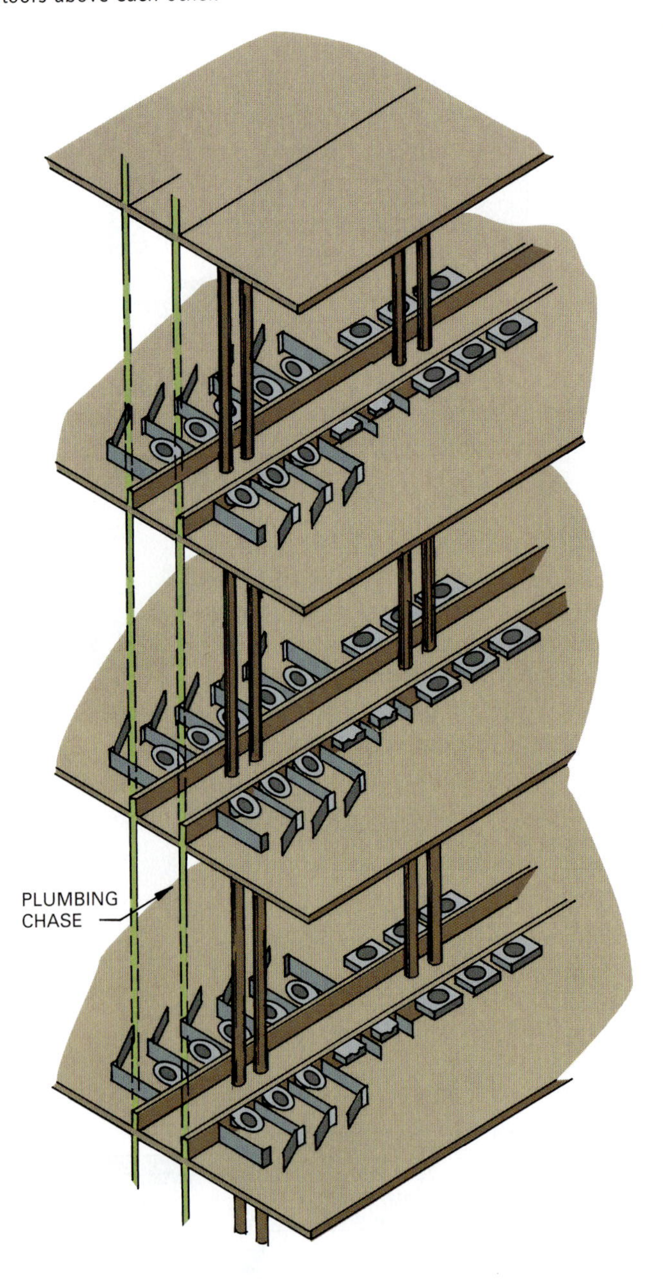

finished enclosure of a structural column. This enables placement of fixtures in various locations within the building.

A schematic of a small multistory building sanitary piping system is shown in Fig. 45.23. The fixtures for the restrooms with their related piping are detailed on the second floor with notes indicating that they are identical on all other floors. The soil stack, vent stack, and building drain are shown for the entire building. Notice the venting of multiple fixtures to horizontal circuit vents that connect to the vent stack. In some cases the fixture waste pipes connect directly to the soil stack, but in other situations they feed into a horizontal branch soil pipe that connects to the vertical soil stack. The circuit vents, waste pipes, and branch soil pipes slope toward the soil stack, providing gravity flow.

The system in Fig. 45.23 requires two sets of pipes, a vent system, and a waste disposal system. This two-pipe system is most commonly used in the United States. The venting controls the possibility of siphoning water from the traps, which would allow sewer gas to enter the building. Another system, the Sovent system, is used in Europe and Africa and has been approved for use in the United States by some national plumbing codes.

The Sovent System

The Sovent system is a vertical cast-iron drainage and waste system that conveys waste from the upper levels of a building to the base of each Sovent stack. The stack begins just above the lowest deaerator fitting and continues up to just above the highest fixture connection (Fig. 45.24). This includes horizontal stack offsets located at intermediate levels. The stack uses traditional fittings and pipe made from approved drain, waste, vent (DWV) materials. The Sovent stack penetrates the roof to the atmosphere much like the vent stack in traditional systems.

Waste flowing in a vertical pipe clings to the interior wall and moves down the pipe in a swirling motion. This leaves an open airway in the center needed for flow to continue. As the falling waste gathers speed, it meets air resistance, flattens out, and may form a complete blockage of the pipe. This blockage is eliminated by the Sovent aerator fitting (Fig. 45.25). Aerator fittings are available in many DWV materials and have no moving parts. The waste enters the offset chamber, which disrupts the waste from forming a plug and reduces its velocity. As the waste exits the offset chamber, it clings to the interior pipe surfaces, leaving an open center air space. It then enters the mixing chamber. Here the incoming waste from horizontal branches hits the baffle, drops, and is mixed with the down-falling waste. The waste flows down the stack to the lowest level that contains a deaerator fitting.

The deaerator fitting is designed to handle pressure fluctuations that occur when falling water suddenly turns horizontal (Fig. 45.26). Here, the velocity lessens while additional waste continues to pile behind, causing a wave (hydraulic jump) to form, and possibly the development of positive and negative pressures. The water strikes the nosepiece, which reduces its velocity and allows the air and waste to separate and continue to flow down the horizontal pipe. The relief line provides

Figure 45.23 A simplified schematic of a sanitary piping system for a low-rise multistory building. Underground tanks can collect rain water from the roof for distribution to the non-potable water system used for toilets and landscaping irrigation.

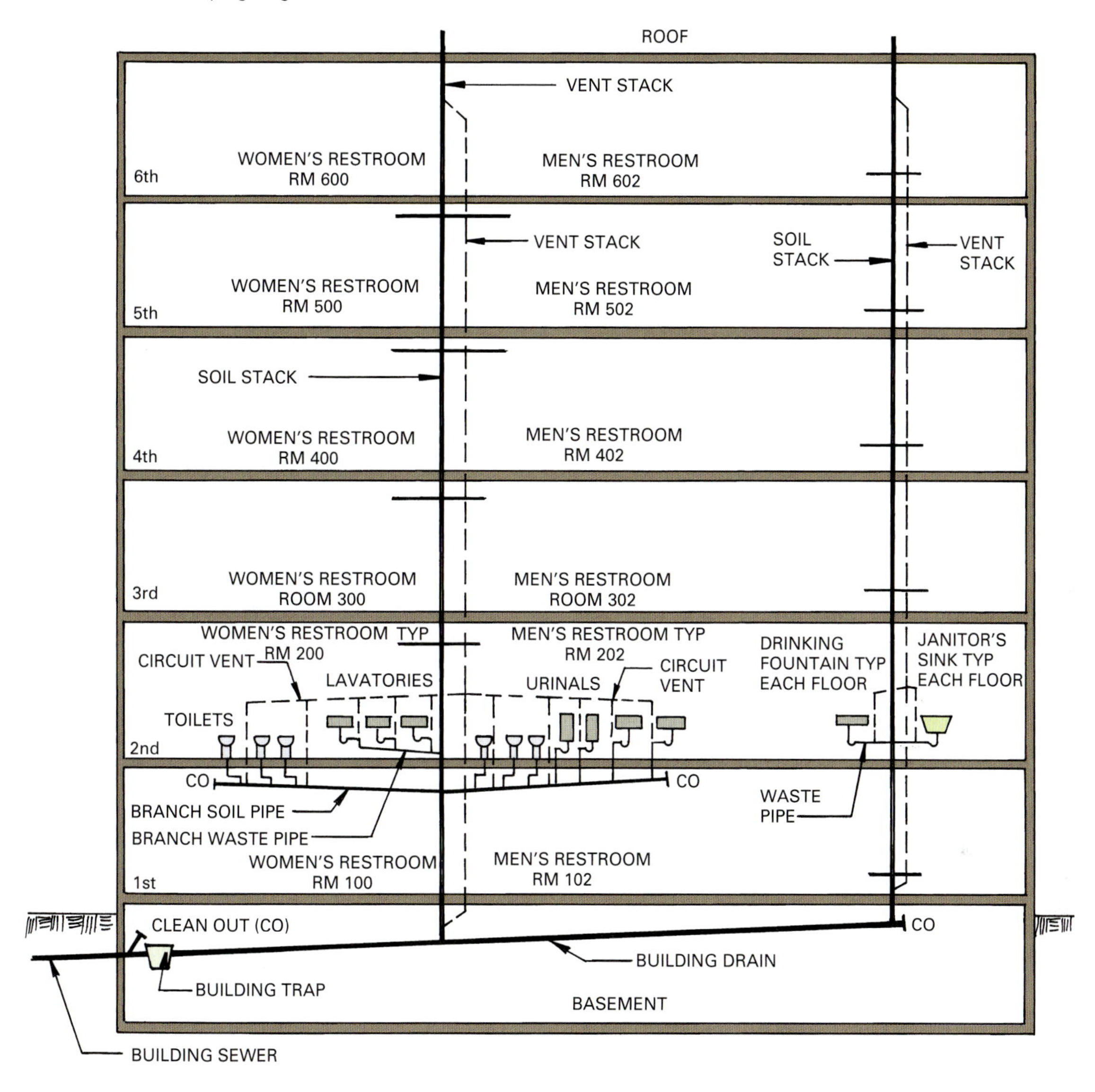

an outlet for pressure and eliminates any chance of the system's developing a vacuum, which would pull the water out of the traps at each fixture.

NON-POTABLE WATER SYSTEMS

Increasing awareness of the need for water conservation has produced a need for new technologies and alternate methods of handling water and wastewater. In buildings where large amounts of water are used, a considerable flow of wastewater is produced. Recycling this flow helps conserve the water supply and reduces the load on wastewater disposal facilities.

Graywater is non-industrial wastewater generated from domestic processes, such as dish washing, laundry, and bathing. It comprises wastewater generated from all of the house's sanitation equipment except for the septic tank (water from toilets being "blackwater"). Graywater makes up between 50 to 80 percent of residential wastewater. By designing plumbing systems to separate it from blackwater, graywater can be recycled for flushing toilets, landscape irrigation, and exterior washing.

A schematic for a wastewater treatment and recycling facility is shown in Fig. 45.27. Potable water is delivered to drinking fountains, lavatories, and other fixtures where it is needed (1). The potable supply is distributed in a completely separate system from the

Figure 45.24 A vertical waste stack using cast iron Sovent aerators and deaerators and standard code-approved drain, waste, vent (DWV) piping and fittings. *(Courtesy Conine Manufacturing Co., Inc.)*

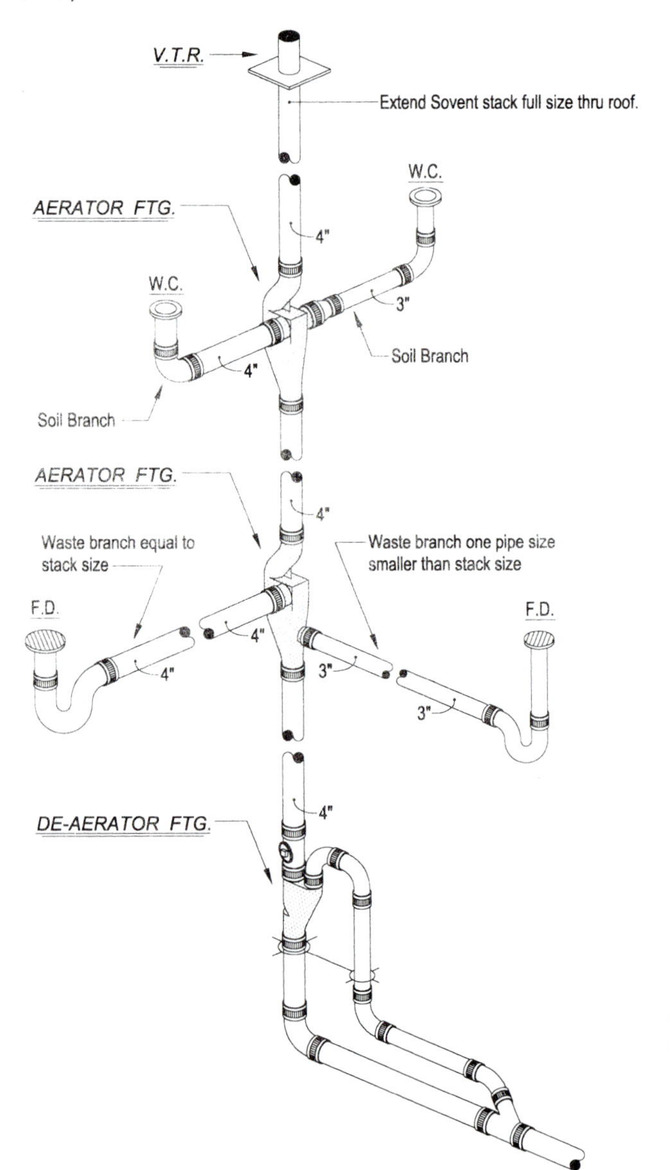

Figure 45.25 The cast-iron Sovent aerator offsets the waste flow around the horizontal branch inlets and breaks up the hydraulic plug that occurs in unrestricted vertical flow, reducing the velocity of falling waste. *(Courtesy Conine Manufacturing Co., Inc.)*

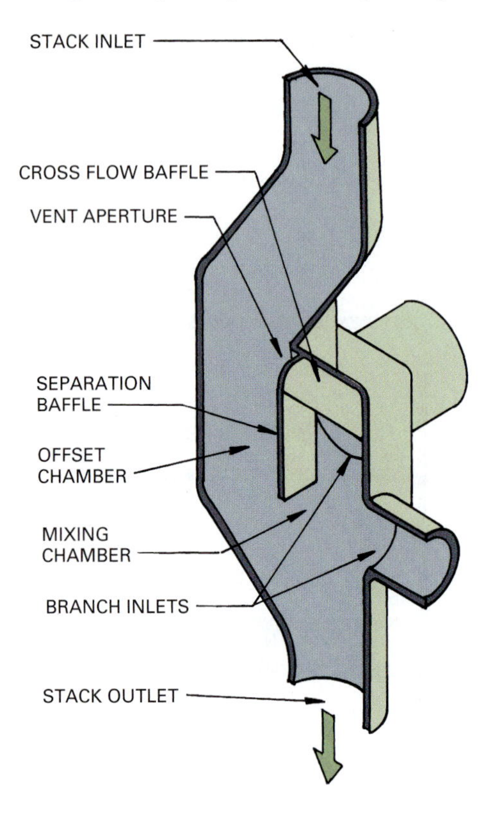

Figure 45.26 The cast-iron Sovent deaerator fittings relieve excessive pressures that result from vertical flows changing to horizontal flows. *(Courtesy Conine Manufacturing Co., Inc.)*

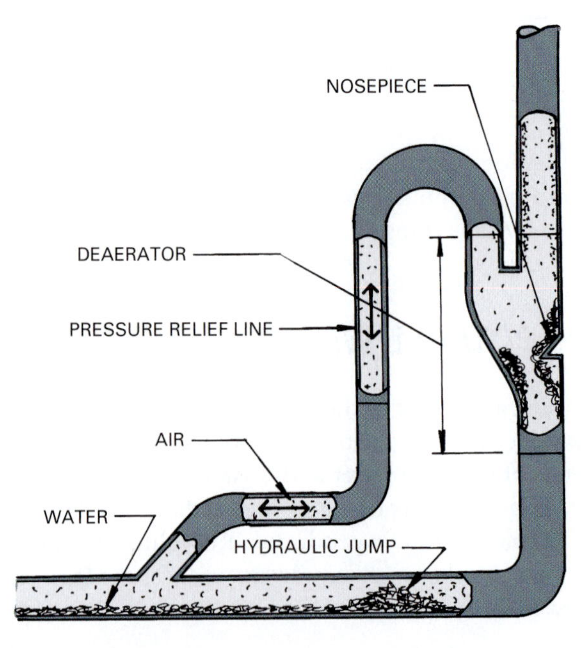

non-potable water system. The wastewater from the potable supply enters a pretreatment trash trap (2), and a sump (3) provides temporary storage in case of a system mechanical failure. Usually this will hold several days' supply of wastewater. The wastewater proceeds automatically through the various processes, including biological treatment, filtering, color removal, disinfection, and ozone treatment. This produces treated water, which is kept in a storage reservoir (12). The recycled water is moved under pressure to the piping serving the toilets (13).

Figure 45.27 A schematic for an in-building wastewater treatment and recycling facility. *(Courtesy SFA Enterprises, Inc. and John Irwin, Thetford Systems, Inc.)*

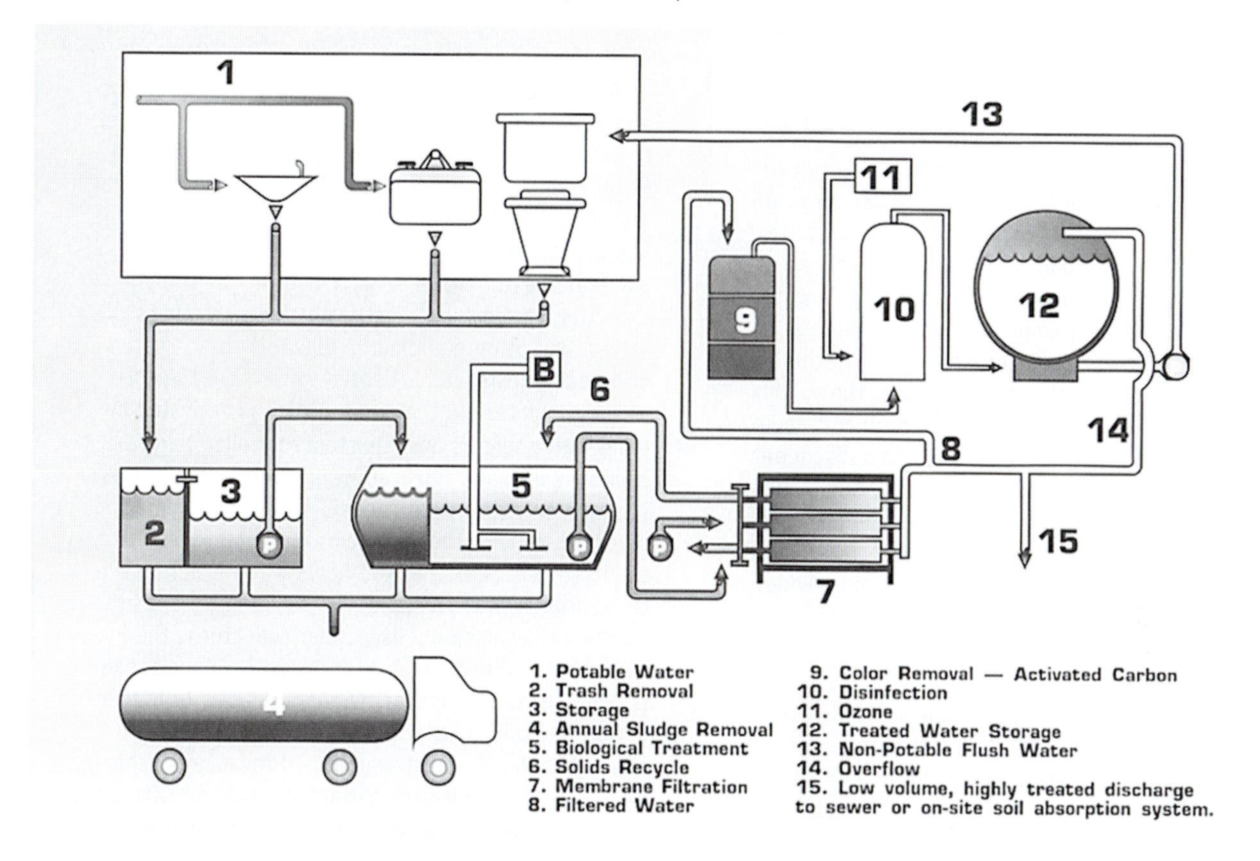

It is vital that the potable and non-potable systems are designed to be kept separate. All piping must be clearly marked "non-potable water supply" or color coded and marked with colored tape. Valves, wall outlets, and other possible places of attachment are marked with warning tags. The National Sanitation Foundation (NSF) has established a standard for recycled water quality in the publication *Certification Standard No. 1.*

ROOF DRAINAGE

Areas such as flat roofs or balconies on multistory buildings collect water during storms and require a drainage system. This system is separate from the sanitary sewer and must drain into the public storm sewer system (Fig. 45.28).

Flat roofs and other such surfaces usually slope toward interior roof drains. The roof drains are raised cast iron, plastic, or aluminum strainers set in the center of a sloped area. Water drains through downpipes, which may be hidden internally or mounted on the exterior of a building. The downpipes connect to a storm drain below the building, which connects to the storm sewer outside the building. Another technique is to slope the roof toward **scuppers**, which provide outlets in the parapet wall and drain the surface water through them into downpipes and on to the storm sewer (Fig. 45.29).

PLUMBING EQUIPMENT

Buildings use a variety of plumbing equipment that connects to distribution and drainage systems.

Water Heating

There are many ways to provide domestic hot water (DHW) needs for a building. Most water heaters utilize a storage tank that is heated either by electricity, natural gas, or heating fuel. A thermostat monitors the water temperature and controls the heating mechanism. When the water is heated to a desired temperature, the heat source automatically shuts off. When the thermostat senses that the water is below the desired temperature, the heating cycle resumes.

Construction Techniques

Solar Hot Water

Water heating accounts for a substantial portion of energy use in both residential and institutional buildings. Solar water heating can provide up to 85 percent of hot water needs, depending on climate, without fuel cost or pollution and with minimal operational and maintenance expenses. Solar water systems are generally composed of solar thermal collectors filled with a fluid system to move the heat from the collector to its point of usage. The system may use electricity for pumping the fluid and have a reservoir or tank for heat storage and subsequent use. Systems are either direct, using the sun's energy to heat water directly, or indirect, utilizing a fluid, such as antifreeze, that indirectly heats water through a heat exchanger. Indirect systems are more widely used in cold climates where freeze protection is required.

The simplest of all solar water heating systems is a batch system. A metal water tank painted with a heat absorbing black coating is placed in an insulating container with a glass cover that admits sunlight (Fig. A). Unheated water enters the solar collector and remains there until it is heated to the desired temperature. The heat is prevented from escaping by the glazing. The heated water is then stored for use as needed, with a conventional water heater providing any additional heating that might be necessary. Batch systems are inexpensive and require little maintenance.

Figure A *(Image copyright Vakhrushev Pavel, 2009. Used under license from Shutterstock.com)*

A passive thermosiphon uses a flat plate collector and an integrated storage tank that is located higher than the collector. The storage tank receives heated water coming from the top of the collector by natural convection while colder water from the bottom of the storage tank is drawn into the solar collector to replace the heated water (Fig. B).

Active Systems use electrically powered pumps, valves, and other equipment to help circulate water or a heat-transfer fluid. Heated water from the collector circulates via the pump, which is activated by temperature sensors of the system controller. Figure C shows a schematic illustration of a solar water heating system with typical control, safety, and operating features. Many of these systems incorporate a separate solar pre-heat tank that pre-heats water entering the conventional water heater.

The most commonly used solar collector is the insulated glazed flat panel. Less expensive panels, like polypropylene panels (for swimming pools) or higher-performing ones like evacuated tube collectors, are also available. The best annual performance is achieved by facing solar collectors toward the equator with a tilt up from the horizontal equal to the latitude of the site.

Figure B *(Image copyright Georgios Alexandris, 2009. Used under license from Shutterstock.com)*

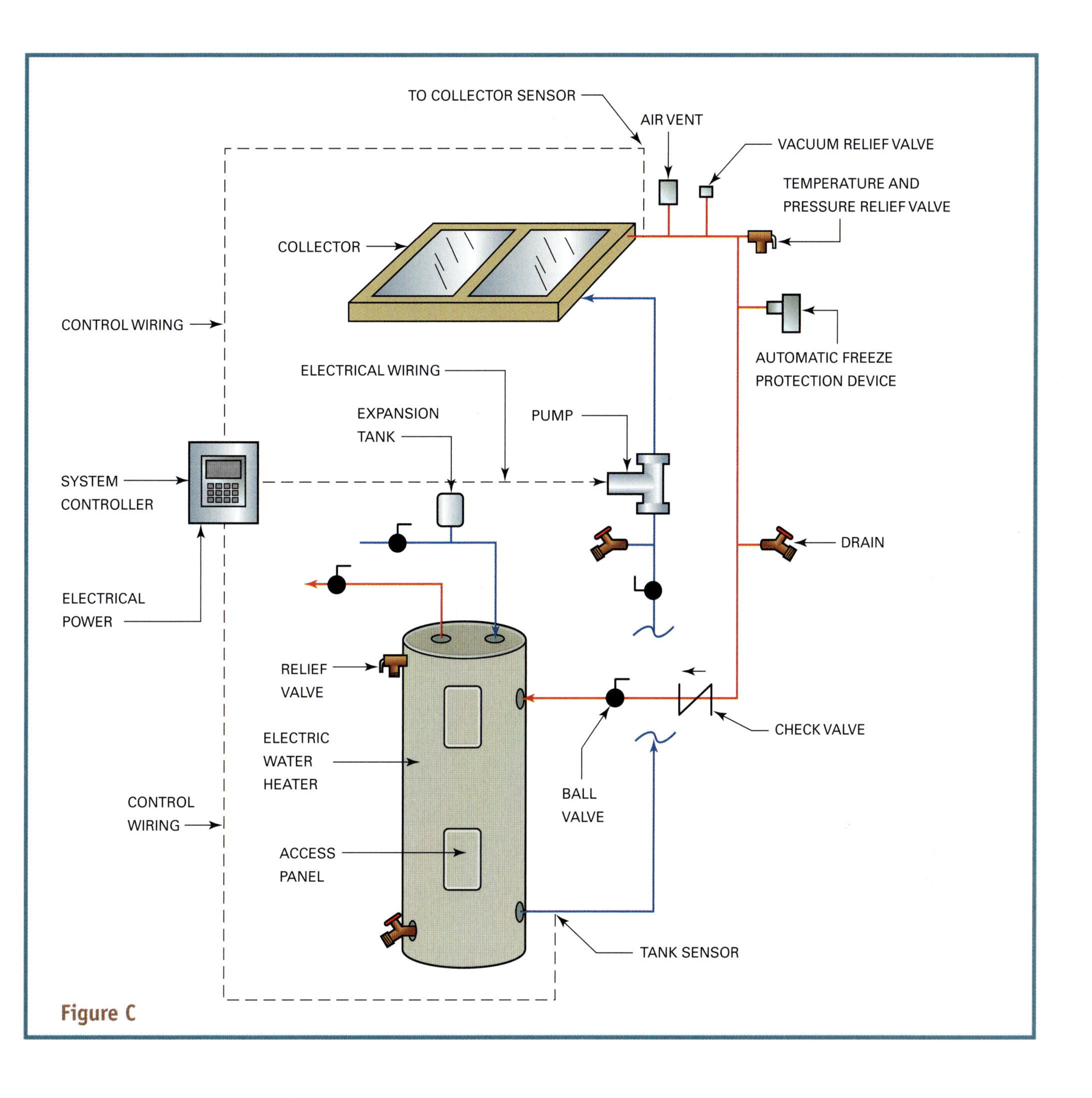

Figure C

Residential gas water heaters are rated in terms of the gallons of hot water they can produce. They are available in a variety of heights, diameters, and sizes, with the most common having a 40 or 50 gallon storage tank capacity. The exhaust fumes from gas water heaters contain carbon monoxide, so direct venting from the heater through the roof or exterior wall is required to exhaust fumes. Tankless, or instantaneous, water heaters take up considerably less space by eliminating the storage tank. In an **instantaneous water heater**, a hot water faucet opens, and water heats as it flows through the unit.

PLUMBING FIXTURES

Minimum requirements for the number of plumbing fixtures are specified in the various plumbing codes according to occupancy. An example of some selected types of building occupancy requirements as specified by the Uniform Plumbing Code is given in Table 45.8. These are minimum recommendations and known conditions may warrant the use of additional fixtures.

Plumbing fixtures require a steady supply of clean water to assist with the discharge of waste materials. They are under constant wear from water, bacteria, and

Figure 45.28 A simplified schematic showing a storm drain system for a roof on a multistory building.

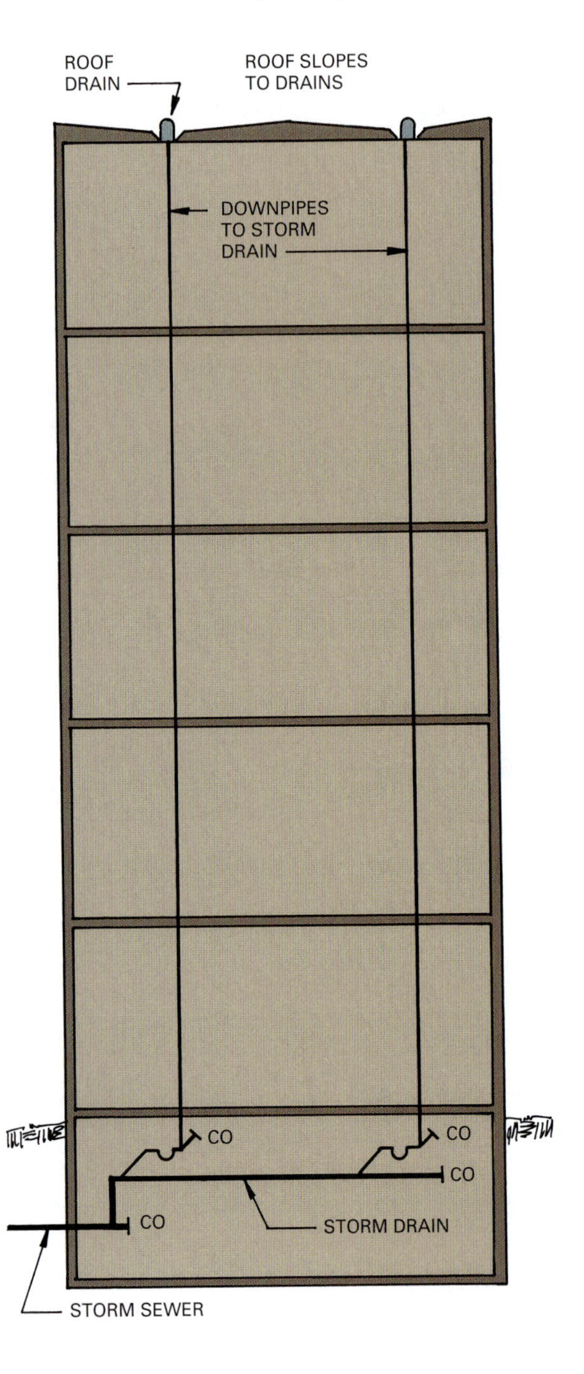

other harmful elements and must be made from durable materials having a smooth, nonporous surface. Typical materials include stainless steel, copper, brass, enameled cast iron, vitreous china, molded plastics, gel-coated fiberglass, and acrylic-faced fiberglass.

The selection of materials and fixture styles is usually made by an owner with the assistance of his or her architect.

Special fixtures for the disabled and their design specifications are available in the publication *Americans with Disabilities Act Accessibility Guidelines*. It specifies heights, grab bar requirements, and other features required for all aspects of a building's design to accommodate those with physical disabilities. Recommendations relating to fixture placement are shown in Figs. 45.30 through 45.32.

Water use in both residential and commercial occupancies accounts for more than 50 percent of all the water supplied to U.S. communities by public and private utilities.

Increasing water efficiency can preserve more water for the environment at large and reduce water supply and wastewater treatment demand. Federal legislation passed in 1992 requires all U.S. plumbing manufacturers and importers to meet or beat the following water efficiency standards for various fixtures: toilets 1.6 gpf (gallons per flush), urinals 1.0 gpf, lavatory faucets 2.5 gpm (gallons per minute), and showerheads at 2.5 gpm.

Water closets (toilets) are available as wall hung or floor mounted units. Floor mounted, tank-type toilets are common in residential buildings. Wall hung water closets are widely used in industrial and public restrooms because it is easier to keep the floor clean around them (Fig. 45.33). Residential water closets typically use a flush tank while commercial establishments use a high-pressure flushing system. Water closets have various methods for flushing. New low-flush models use as little as 1.0 to 1.6 gpf and can cut annual water use by 20 to 25 percent, with a corresponding reduction in wastewater flow. Dual flush toilets allow the user to choose either a full 1.6 gallon flush for solids or a 0.8 gallon flush for liquids (Fig. 45.34).

A **composting toilet** is any system that converts human waste into an organic compost and usable soil through the natural breakdown of organic matter into its essential minerals. Microorganisms accomplish this over time, working through various stages of oxidation and anaerobic breakdown. Self-contained composting toilets complete the composting within the unit, while central ones flush waste to a remote composting unit in a utility space below the toilet. A composting toilet with a separate composting room for the compartment is called a Clivus Multrum. These units incorporate a composting reactor that is connected to one or more toilets and an exhaust system, often fan-forced, to remove odors, carbon dioxide, water vapor, and the by-products of aerobic decomposition and provide oxygen (aeration) for the aerobic organisms in the composter. A means for draining and managing excess liquid and an access door for removal of the end product complete the installation.

Figure 45.29 A strainer protects the storm-water drain used on flat roofs. *(Courtesy National Roofing Contractors Association)*

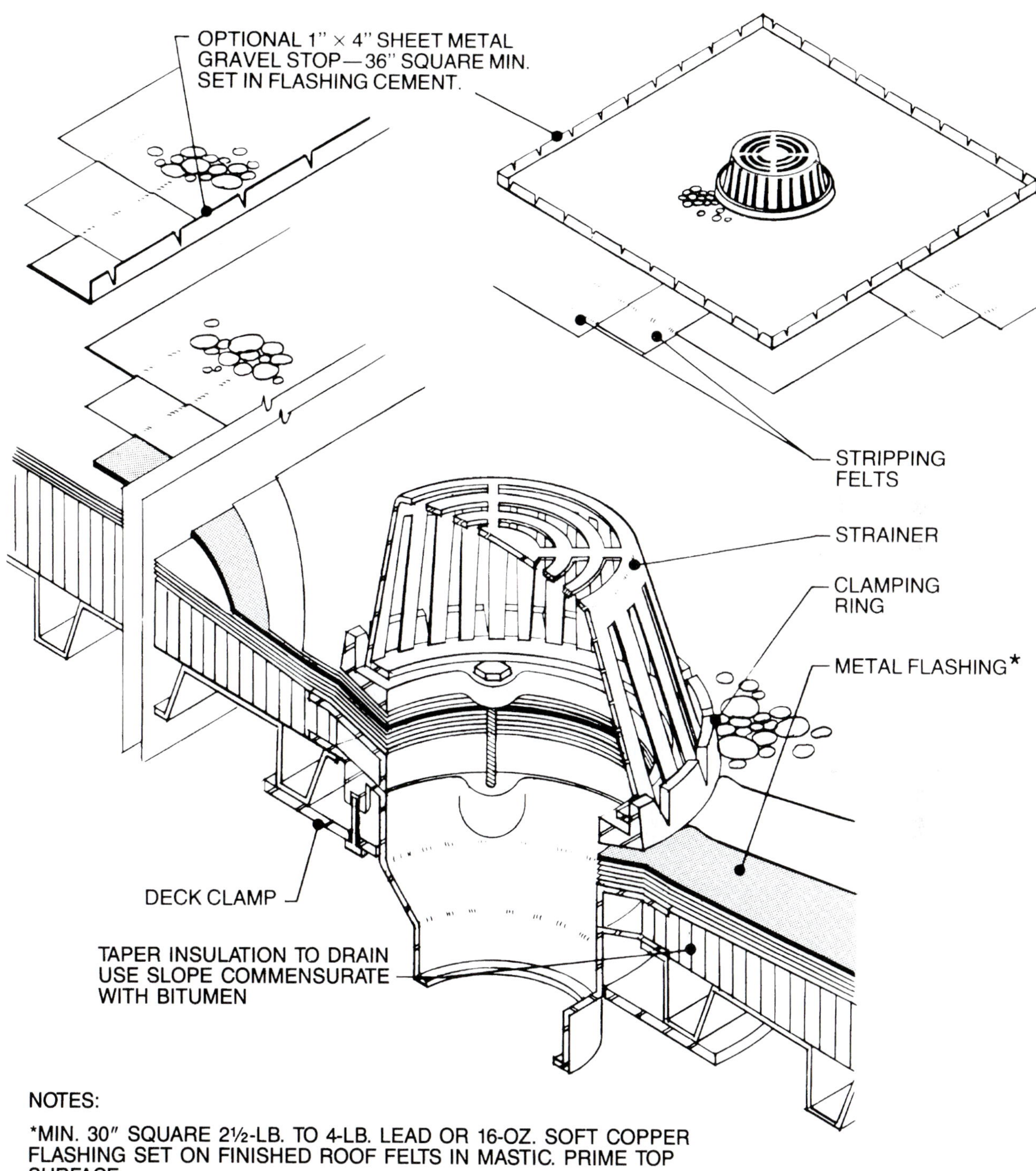

NOTES:

*MIN. 30″ SQUARE 2½-LB. TO 4-LB. LEAD OR 16-OZ. SOFT COPPER FLASHING SET ON FINISHED ROOF FELTS IN MASTIC. PRIME TOP SURFACE
BEFORE STRIPPING.

MEMBRANE PLIES, METAL FLASHING, AND FLASH-IN PLIES EXTEND UNDER
CLAMPING RING.

STRIPPING FELTS—EXTEND 4″ AND 6″ BEHOND EDGE OF FLASHING SHEET, BUT NOT BEYOND EDGE OF SUMP.

THE USE OF METAL DECK SUMP PANS IS NOT RECOMMENDED.

Table 45.8 Minimum Plumbing Facilities

Type of Building or Occupancy	Water Closets	Urinals	Lavatories	Bathtubs or Showers	Drinking Fountains
Assembly places (theaters, etc.)	1 per 1–15 2 per 16–35 3 per 36–55 Over 55, 1 per additional 40	1 per 50 males	1 per 40	—	1 per 75
Hospitals Individual room Ward room	1 per room 1 per 8 patients	— —	1 per room 1 per 10 patients	1 per room 1 per 20 patients	— —
Restaurant	1 per 1–50 2 per 51–150 3 per 151–300	1 per 1–150 males	1 per 1–150 2 per 151–200 3 per 201–400	—	—
Worship place, assembly area	1 per 300 males 1 per 150 females	1 per 300 males	1 per toilet room	—	1 per 75
Worship place, educational and activities	1 per 250 males 1 per 125 females	1 per 250 males	1 per toilet room	—	1 per 75

Reprinted from the Uniform Plumbing Code™ *with the permission of the International Association of Plumbing and Mechanical Officials ©copyright 1994.*

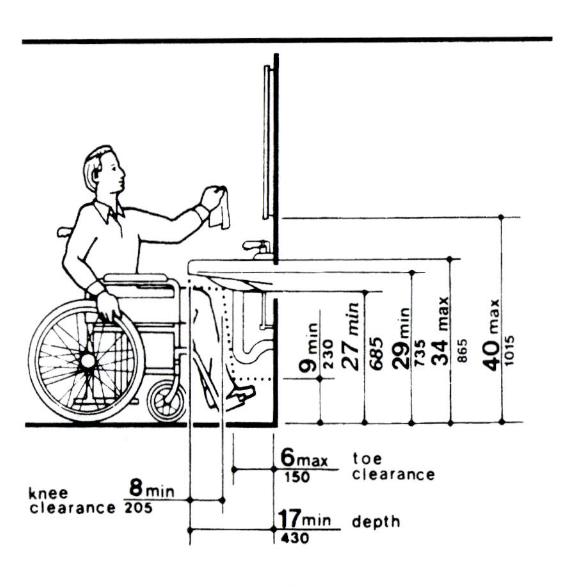

Lavatory Clearances

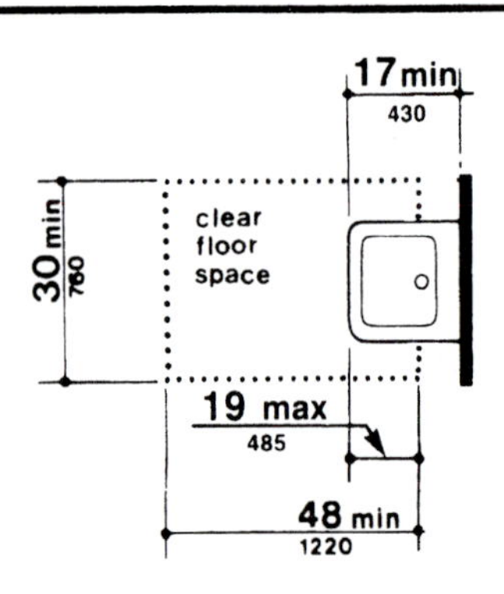

Clear Floor Space at Lavatories

Figure 45.30 Lavatories must be mounted with their top surface not more than 34 in. (865 mm) above the floor. They must be positioned so a person in a wheelchair will have the required knee room. Any exposed pipes must be located out of the way, and any sharp edges must be covered. *(From Americans with Disabilities Act, U.S. Architectural and Transportation Barriers Compliance Board, Washington, D.C.)*

Figure 45.31 Water closets not in stalls require a minimum of 48 in. (1,220 mm) clearance space from other items. The water closet must be 17 to 19 in. (430 and 485 mm) high. Grab bars to the side and rear are required. *(From Americans with Disabilities Act, U.S. Architectural and Transportation Barriers Compliance Board, Washington, D.C.)*

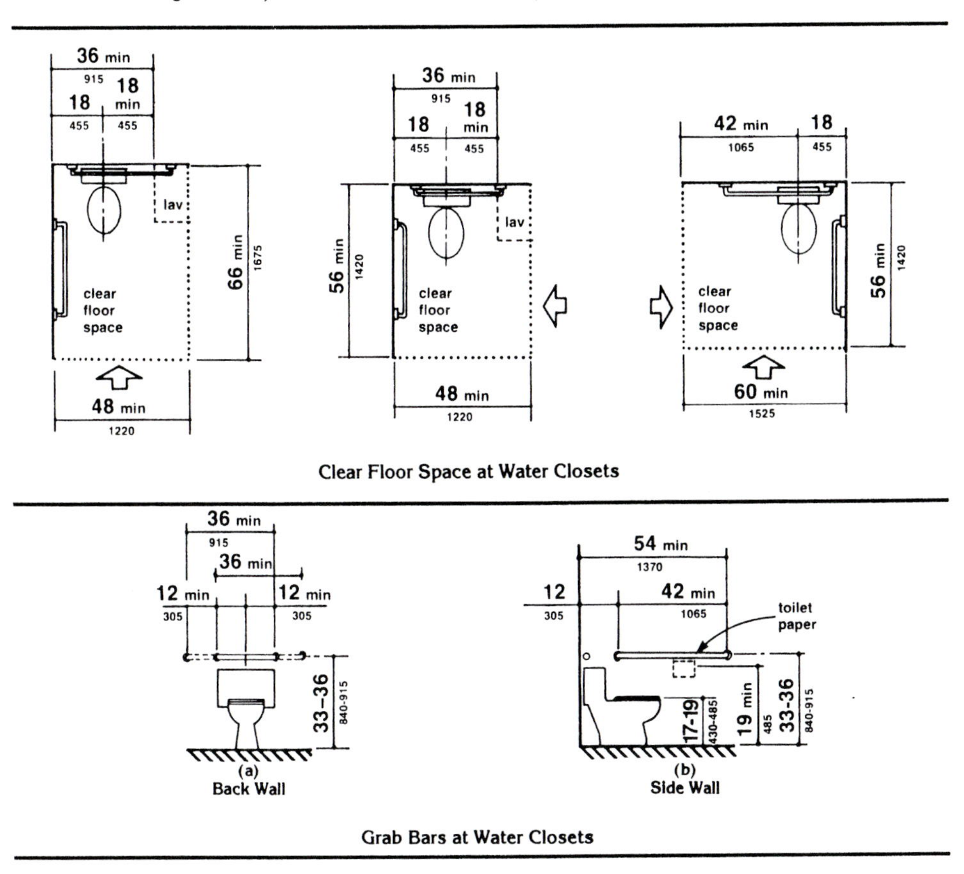

Urinals are wall-hung units used in men's restrooms in public facilities to reduce the number of required water closets. They may have a manual high-pressure flush valve or be connected to an automatic high-pressure flushing system. Sometimes one is set lower than others for use by children. A more recent innovation is urinals that do not use water at all. Waterless urinals work completely without water or flush valves (Fig. 45.35). They install to conventional waste lines and are equipped with a special cartridge that acts as a funnel, allowing liquid to flow through a sealant liquid that prevents any odors from escaping. The cartridge filters sediment, allowing the remaining liquid to pass freely down the drain. Waterless urinals can save up to 45,000 gallons of water per year per urinal.

Lavatories in residences are typically mounted in a base cabinet. The lavatory may be set into a top covered with plastic laminate or ceramic tile. Others have a molded plastic top with the lavatory bowl and top as one integral piece (Fig. 45.36). This reduces problems that occur around the stainless steel edge of those set into the top. Another popular lavatory is a pedestal type. The lavatory is actually wall hung and the pedestal covers up the plumbing below (Fig. 45.37). Wall-hung residential lavatories often have decorative metal legs instead of a pedestal.

Lavatories in public facilities are typically wall-hung units, although some set into a wall-hung countertop without a cabinet base below. These usually have self-closing faucets that prevent anyone from letting the water run after they leave. One type of faucet has an infrared control that turns on the water when hands are placed below the faucet and turns it off when hands are removed. This saves on water use and reduces the cost of hot water.

Wash fountains are used in restrooms and dressing rooms in which a large number of people need to wash up at the same time, such as in an industrial plant. Wash fountains are typically half-round wall-hung units or round freestanding units. Large units can accommodate up to eight people at one time.

Figure 45.32 Drinking fountains must have spouts no higher than 36 in. (915 mm) above the floor. The water flow must be projected so a person in a wheelchair can move below it. Controls must be front-mounted or side-mounted near the front edge. *(From Americans with Disabilities Act, U.S. Architectural and Transportation Barriers Compliance Board, Washington, D.C.)*

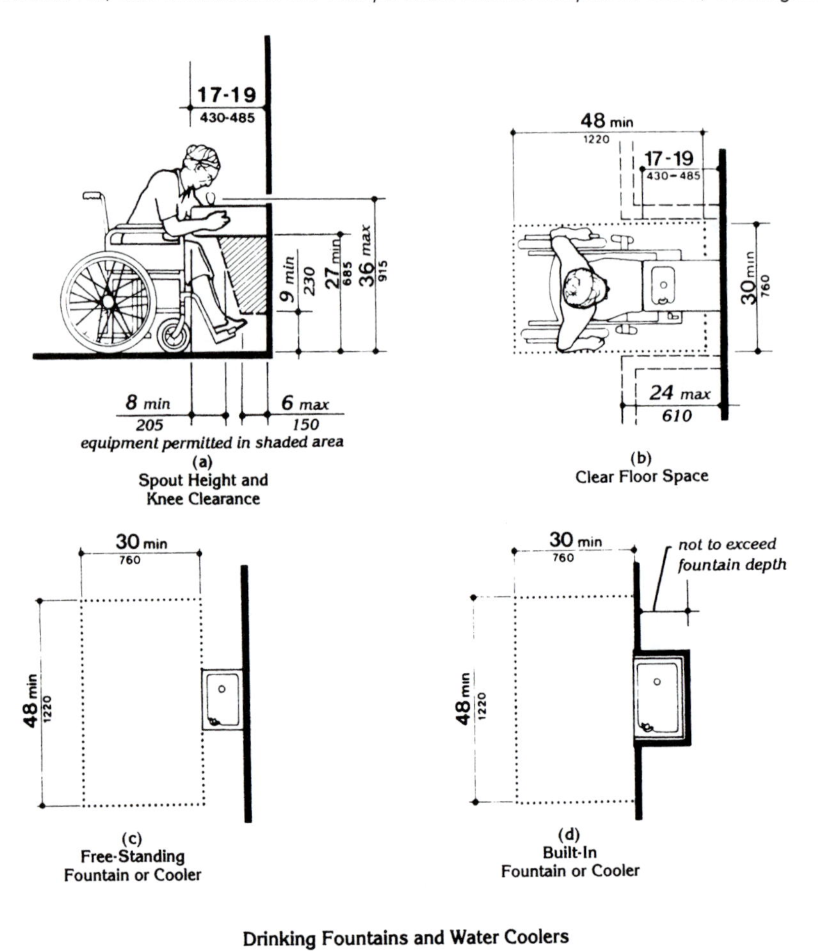

Figure 45.33 A floor-mounted high-pressure toilet used in public restrooms. *(Image copyright Laura Clay Ballard, 2009. Used under license from Shutterstock.com)*

Figure 45.34 Water-conserving dual-flush toilets allow users to choose either a full or partial flush. *(Image copyright Ljupco Smokovski, 2009. Used under license from Shutterstock.com)*

Figure 45.35 Waterless urinals install to conventional waste lines and work completely without water or flush valves. *(Courtesy Chris Goddard, http://en.wikipedia.org/wiki/File:Waterless_urinal_(Armitage_Shanks,_UK).jpg)*

Figure 45.37 A decorative pedestal lavatory conserves space. *(Image copyright Barbara Kennedy, 2009. Used under license from Shutterstock.com)*

Figure 45.36 This molded plastic bathroom lavatory has the top and bowl cast as a single unit. *(Image copyright bhowe, 2009. Used under license from Shutterstock.com)*

Kitchen sinks are available in a range of types with the most typical being some form of two-bowl unit. However, single- and triple-bowl units are available. One bowl serves for general use while the second contains the garbage disposal. Sinks are installed in plastic laminate, stone, and a variety of other countertop materials.

Service sinks are used by janitors to clean mops and for other cleaning activities. They are usually wall hung and deep. Mop service basins are floor-mounted units about 1 ft. (305 mm) high.

Bathtubs are available in a wide range of sizes and designs. Some have two closed sides and fit in a corner. Others have one closed side and fit between end walls. Whirlpool tubs have pumps that circulate the water (Fig. 45.38). Some bathtubs made from gel-coated fiberglass have the tub and wall enclosure formed as a single unit, leaving no joints to form mold, leaving the unit easy to clean.

Showers made of one piece, gel-coated fiberglass units are commonly used in residential applications. They are free from any joints, making them one solid, easy-to-clean unit. Other types are available with a cast terrazzo base. Shower walls may be covered with ceramic tile or have a prefabricated enclosure made from galvanized bonderized steel with an enamel finish or molded plastic panels. Shower stall are often installed with tempered glass doors (Fig. 45.39).

Figure 45.38 This whirlpool bathtub rests on the floor and provides a soothing circulation of hot water. The floor must be reinforced to carry the load and provision must be made to install the pump below the floor. *(Image copyright Rade Kovac, 2009. Used under license from Shutterstock.com)*

Figure 45.39 Shower enclosures are often fitted with glass doors. *(Image copyright Gordon Ball LRPS, 2009. Used under license from Shutterstock.com)*

Various types of drinking fountains are available. Some are wall hung and protrude into the room, while others are recessed or semi-recessed into a wall. Free-standing and some recessed drinking fountains have a water cooling system in their base that produces temperature-controlled water. Units that provide access for the handicapped are wall mounted and have an electrically activated push-button valve on the front that requires only a light touch. Some pedestal types are designed for outdoor use and may have a foot-operated valve that is frost-proof. Stainless steel is the major material used on the surfaces exposed to water. Drinking fountains require a sanitary drain as well as a source of potable water.

REVIEW QUESTIONS

1. What types of pipe are used to carry natural gas?
2. How are copper pipe joints secured?
3. How can the expansion of copper pipe used to carry hot water be handled to prevent breakage?
4. How are the various types of copper water tubing identified?
5. What resins are used to produce plastic pipe?
6. What methods are used to join various types of plastic pipe?
7. Where would glass pipe be used?
8. Where would you use a relief valve?
9. Why are pipes insulated?
10. What processes are used at a water purification plant to produce potable water?
11. What is meant by an up-feed potable water distribution system within a building?
12. What type of potable water distribution system is used in large multistory buildings?

13. What materials are used for pipes in sanitary piping?
14. How does a plumbing contractor know what slope is required on sanitary piping installations?
15. What are indirect wastes?
16. How does the Sovent drainage and waste system operate without the venting piping used in traditional systems?
17. What uses are made of non-potable water within a building?

KEY TERMS

air gap
branch soil pipe
building drain
building sewer
cleanout
composting toilet
DWV
fixture unit
graywater
instantaneous water heater
interceptor
non-potable water
oakum
plumbing chase
potable water
riser
sanitary piping
sanitary sewer
scupper
soil stack
trap
valve
vent
vent stack
waste pipe

ACTIVITIES

1. Arrange a tour of a multistory building in which the plumbing system can be inspected. Attempt to view the hidden aspects, such as plumbing chases and vertical distribution systems.
2. Visit the local water treatment plant and sewage disposal plant. Request the process be explained in the order treatment occurs. Obtain information about regulations within which each facility must operate.

ADDITIONAL RESOURCES

Americans with Disabilities Act Accessibility Guidelines, U.S. Architectural and Transportation Barriers Compliance Board, 1331 F Street, NW, Suite 1000, Washington, DC 20004.

Applications of Water Pipe Sizing, International Code Council, Falls Church, VA.

International Plumbing Code, International Code Council, Falls Church, VA.

International Private Sewage Disposal Code, International Code Council, Falls Church, VA.

LP-Gas Code; *National Fire Alarm Code*, National Fire Protection Association, Quincy, MA .

See Appendix B for addresses of professional and trade organizations and other sources of technical information.

DIVISION

23

Heating, Ventilating, and Air-Conditioning CSI MasterFormat™

Image copyright Chad McDermott, 2009. Used under license from Shutterstock.com

46 CHAPTER

Heating, Air-Conditioning, Ventilation, and Refrigeration

LEARNING OBJECTIVES

Upon completion of this chapter, the student should be able to:

- Understand the many factors to be considered when designing heating and air-conditioning systems.
- Make decisions concerning needs and types of ventilation for acceptable indoor air quality.
- Recognize the fuels used by heating, air-conditioning, and ventilation systems.
- Select appropriate air heating and cooling systems.
- Select appropriate designs for steam and hot water heating systems.
- Discuss the systems and equipment available for cooling systems in large buildings.

Heating, ventilating, and air-conditioning is based on the principles of thermodynamics, fluid mechanics, and heat transfer. Heating, venting, and air-conditioning (HVAC) equipment is used to accomplish climate control in buildings, producing comfortable humidity and temperature levels that must be closely regulated to maintain safe and healthy conditions. The aim is to provide thermal comfort, acceptable indoor air quality, and reasonable installation, operation, and maintenance costs. A solution for a small residential building may be insufficient for a medium to large office building or a skyscraper. Solutions vary depending on climate, scale, occupancy, and make-up of the exterior envelope. For example, office building design gives major consideration to human comfort, while manufacturing plant design addresses a range of other considerations.

Design factors include the type and amount of glazing; insulation and air infiltration levels; heating and cooling requirements due to machinery; industrial processes; solar load; materials used for walls, ceilings, floors, and roofs and their coefficients of thermal conductivity. Space within a building must be provided to house mechanical units and to access various parts of the structure containing ducts, pipes, and other parts of a system. Some other factors—such as a need to control humidity; removal of chemicals, noxious gases, or dust; availability and cost of fuel; cost of the various systems; expected maintenance expenses; possibility of down time; and climatic factors—illustrate a few additional things HVAC engineers have to consider as a system is designed.

HVAC CODES

The installation, control, and operation of heating, ventilating, and air-conditioning systems involves a wide range of equipment, fuels, and a host of safety considerations. These are regulated by the various HVAC codes and supporting manuals available from professional organizations. Some of the codes from the National Fire Protection Association include the *Liquefied Petroleum Gas Code*, the *National Fuel Gas Code*, and the *Boiler and Combustion Systems Hazards Code*. The International Code Council offers code publications such as the *International Mechanical Code* and the *International Fuel Gas Code*.

HEAT BALANCE

HVAC design is concerned with establishing indoor temperature and humidity at comfortable levels and maintaining them. In hot weather, a system must remove excess heat from a building's interior at the same rate that it gains heat. In cold weather, it must add interior heat at the same rate it loses heat to the cold outside environment. This provides a balance that achieves comfortable conditions without large fluctuations in temperature.

HEAT TRANSFER

Heat transfer is the movement of thermal energy from a heated item to a cooler item. When an object or fluid resides at a different temperature than its surroundings, transfer of thermal energy, also known as heat exchange, occurs in such a way that the solid or liquid body and the surroundings reach thermal equilibrium. Heat transfer always flows from a hot object to a cold one, a result of the second law of thermodynamics. This flow occurs via radiation, conduction, or convection.

Thermal radiation transfers heat energy emitted by a source, such as waves through space, which is absorbed by a material it touches. For example, the sun radiates heat energy through space to the earth where it warms anything it strikes. Radiant solar energy can be used by solar collectors that receive it and through a transfer system that moves it to a building's interior. The sun provides **radiant heat** through windows, which can help warm a room or make it uncomfortably hot. Likewise, a cold exterior room wall absorbs heat from occupants' bodies, leaving them cold (Fig. 46.1). In the summer, a hot exterior wall radiates heat, potentially making occupants uncomfortable.

Thermal conduction involves the transfer of heat energy through a material by transmitting kinetic energy from one molecule to the next. The movement is caused by a temperature change in the material. For example, when a match is held on one end of a short piece of copper wire the other end will quickly get hot. Heat gain or loss due to conduction occurs when your body has physical contact with a material. If you touch a metal object (a good thermal conductor) at room temperature, it will absorb body heat and feel cool to the touch. If you touch a wooden object (a poor thermal conductor), it will not absorb as much heat and feel warm to the touch.

Thermal convection involves the transfer of heat by the circulation of heated air, gas, or liquid. In natural convection, air surrounding a heat source receives heat, becomes less dense, and rises. The surrounding, cooler air then moves to replace it. This cooler air then heats, and the process continues, forming a convection current. Heat transfers to the ceiling, walls, and other building elements. Convection continues until thermal equilibrium is reached. Hot air furnaces supply forced heated air through ducts and use convection to heat a room. Hot water and steam-heating systems also use convection by heating air with radiators.

Figure 46.1 The HVAC engineer must consider various sources of radiant heat transfer energy and their influence on human comfort.

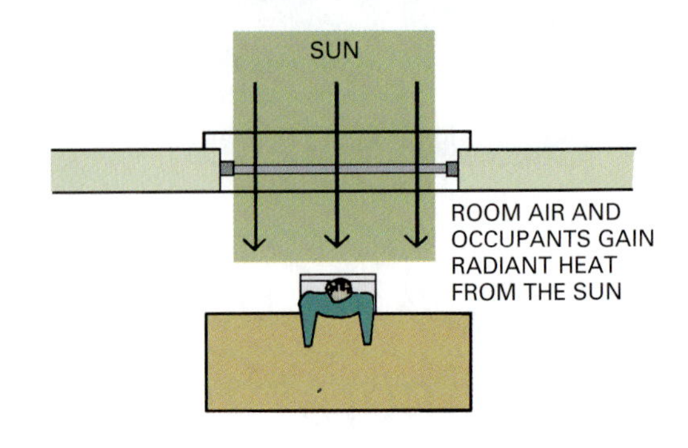

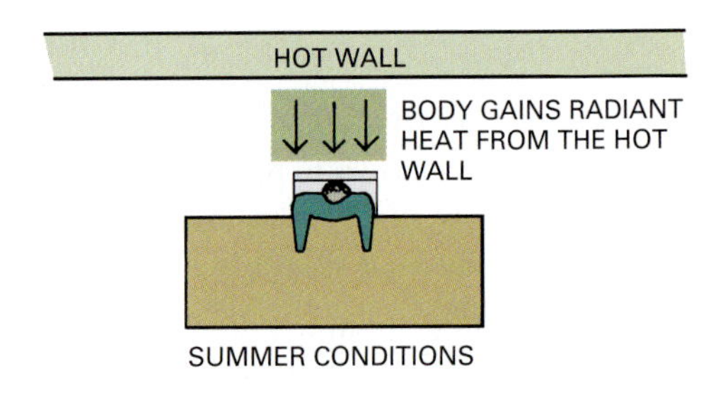

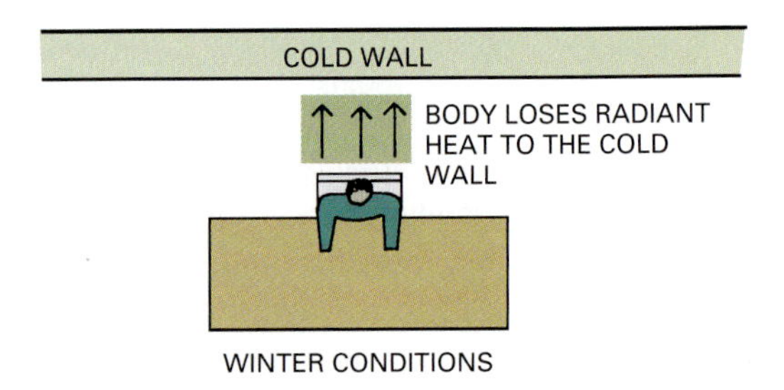

If air temperature is lower than body temperature, your body loses heat to the air, and your body cools. If air temperature is higher than body temperature, your body absorbs heat from the air and becomes warmer.

SENSIBLE AND LATENT HEAT

Sensible heat is heat that can be detected by touch when heat energy is added to or removed from a material. It is associated with a change in temperature.

Latent heat involves action that changes the state of a substance, such as when water turns to steam. This is an important factor to consider in the design of HVAC systems. For example, by adding heat to a unit of water until it reaches the boiling point, 212°F (100°C), an engineer can calculate the Btu's of heat required to reach this point. A **British thermal unit** (**Btu**) is the amount of heat required to raise the temperature of 1 lb. of water 1°F. If heat continues to be applied, the temperature of the water will not increase, but it will begin to boil and undergo a physical change into water vapor, also known as **steam.**

Table 46.1 High Ambient Temperature Tolerance Times

Air DB Temperature	Tolerance Time
180°F (82°C)	Almost 50 minutes
200°F (93°C)	33 minutes
220°F (104°C)	26 minutes
239°F (115°C)	24 minutes

Reproduced with permission from the ASHRAE Handbook of Fundamentals, American Society of Heating, Refrigerating and Air-Conditioning

HUMAN COMFORT

HVAC systems are designed to maintain recommended room temperatures. These vary by season, with acceptable winter temperatures sitting a few degrees below acceptable summer conditions. For example, residences and similar occupancies, such as offices and classrooms, are typically most comfortable if kept in the range of 73 to 79°F (22.5 to 26°C) in the summer and 68 to 74°F (20 to 23.5°C) in the winter. The air temperature for most other occupancies, such as retail shops, medical facilities, and restaurants, is also kept within this range. Some areas, such as showers, will have a higher design temperature. Despite this range of recommended temperatures, people feel comfortable at different temperatures. Differing levels of physical activity—manual labor versus working at a desk, for example—also necessitate different temperature requirements.

A space in which one area is cold while another is warm produces discomfort. Experiencing a rapid temperature change, such as exiting an air-conditioned building on a very hot day, may cause immediate discomfort. Some control can be exercised by keeping inside temperatures close to those outside, so long as the interior environment is comfortable.

Research by the American Society of Heating, Refrigerating, and Air-Conditioning Engineers (ASHRAE) has established tolerances for high ambient temperature (**Table 46.1**). (Ambient temperature is that of the surrounding air.) When a person is exposed to high temperatures beyond these limits they are subject to heat stress, which can be dangerous. Likewise, exposure to very cold situations without adequate protection loses body heat faster than it can be replaced, which results in discomfort or worse. If uncorrected, this creates a life-threatening situation.

ASHRAE Standard 55–2004, *Thermal Environmental Conditions for Human Occupancy*, establishes the indoor conditions and personal factors required for human comfort. The factors involved include clothing, air speed, temperature, thermal radiation, and occupants' activity level.

VENTILATION

Ventilation is used to control heat, condensation, humidity, odors, and contaminants in the air. Ventilating involves the process of replacing air in a space to control temperature or remove moisture, odors, smoke, heat, dust, and airborne bacteria. Ventilation includes both the exchange of air to the outside as well as circulation of air within a building. It is one of the most important factors for maintaining acceptable indoor air quality in buildings. Methods for ventilating can be divided into mechanical and natural.

Air typically contains minute particles, such as dust raised by the wind and various pollens. The air where large numbers of people assemble, such as in a theater, becomes contaminated by carbon dioxide given off by occupants. This carbon dioxide must be removed and replaced with fresh air. In some situations, mist, fog, and various gases and vapors exist, as well as a variety of bacteria and other microorganisms. These can cause serious infections in humans and are a particular problem in medical facilities. Some business and industrial plants develop odors that must be removed. Unpleasant odors can cause physical discomfort, such as headaches and nausea. Others are an indication of airborne toxic substances that could cause physical harm to occupants.

ASHRAE outdoor air ventilation specifications for various indoor occupancies can be found in ASHRAE Standard 62.1–2004, *Ventilation for Acceptable Indoor Air Quality*. Selected examples from this publication for health care facilities are listed in **Table 46.2**. The standard lists outdoor air requirements for commercial and institutional spaces in proportion to the number of persons in the space. For example, patient room ventilation requirements are based on the expectation that the maximum occupancy will be ten persons per 1,000 sq. ft. (100 m^2). Physical therapy room ventilation requirements are based on an occupancy of twenty persons per room. For residential spaces, the required outdoor air supply is specified in air changes per hour or cubic feet per minute. ASHRAE 62.1–2004 recommendations for residential facilities are listed in **Table 46.3**.

Table 46.2 Outdoor Air Requirements for Ventilation of Health Care Facilities (Hospitals, Nursing, and Convalescent Homes)

Application	Estimated Maximum** Occupancy P/1,000 ft² or 100 m²	Outdoor Air Requirements cfm/ person	L/s* person	cfm/ft²	L/s · m²	Comments
Patient rooms	10	25	13			Special requirements or codes and pressure relationships may determine minimum ventilation rates and filter efficiency. Procedures generating contaminants may require higher rates.
Medical procedure	20	15	8			
Operating rooms	20	30	15			
Recovery and ICU	20	15	8			
Autopsy rooms	20			0.50	2.50	Air shall not be recirculated into other spaces.
Physical therapy	20	15	8			

Table prescribes supply rates of acceptable outdoor air required for acceptable indoor air quality. These values have been chosen to dilute human bioeffluents and other contaminants with an adequate margin of safety and to account for health variations among people and varied activity levels.
***Net occupiable space.*
Reproduced with permission from ASHRAE Standard 62.1–2004, Ventilation for Acceptable Indoor Air Quality, © American Society of Heating, Refrigerating and Air-Conditioning Engineers, Inc. www.ashrae.org

Table 46.3 Outdoor Air Requirements for Ventilation of Residential Facilities (Private Dwellings, Single, Multiple)

Applications	Outdoor Requirements	Comments
Living areas	0.35 air changes per hour but not less than 15 cfm (7.5 L/s) per person	For calculating the air changes per hour, the volume of the living spaces shall include all areas within the conditioned space. The ventilation is normally satisfied by infiltration and natural ventilation. Dwellings with tight enclosures may require supplemental ventilation supply for fuel-burning appliances, including fireplaces and mechanically exhausted appliances. Occupant loading shall be based on the number of bedrooms as follows: first bedroom, two persons; each additional bedroom, one person. Where higher occupant loadings are known, they shall be used.
Kitchens[b]	100 cfm (50 L/s) intermittent or 25 cfm (12 L/s) continuous or openable windows	Installed mechanical exhaust capacity.[c] Climatic conditions may affect choice of the ventilation system.
Baths, Toilets[b]	50 cfm (25 L/s) intermittent or 20 cfm (10 L/s) continuous or openable windows	Installed mechanical exhaust capacity[c]
Garages: Separate for each dwelling unit	100 cfm (50 L/s) per car	Normally satisfied by infiltration or natural ventilation
Common for several units	1.5 cfm/ft² (7.5 L/s m²)	See "Parking garages"

[a]*In using this table, the outdoor air is assumed to be acceptable.*
[b]*Climatic conditions may affect choice of ventilation option chosen.*
[c]*The air exhausted from kitchens, bath, and toilet rooms may utilize air supplied through adjacent living areas to compensate for the air exhausted. The air supplied shall meet the requirements of exhaust systems as described and be of sufficient quantities to meet the requirements of this table.*
Reproduced with permission from ASHRAE Standard 62.1–2004, Ventilation for Acceptable Indoor Air Quality, © American Society of Heating, Refrigerating and Air-Conditioning Engineers, Inc. www.ashrae.org

Ventilated air is brought from outside and filtered before entering an interior air distribution system. In some cases, the interior air is recycled, having contaminants removed and odors reduced. It is not always permissible to release air from a building directly outside without some form of treatment. Exchanging outside air with inside air can also increase heating and cooling costs, but heat exchanger units are available to reduce them. Heat exchangers use the heat in outgoing air to warm the cooler incoming fresh air.

Methods of Ventilation

Ventilation may be natural or mechanical. Natural ventilation utilizes outside air without using fans or other mechanical assistance. It can be achieved with operable windows, skylights, roof ventilators, or other openings. In more complex systems, warm air in a building can be allowed to rise and exit through upper openings (stack effect), thereby drawing in cool air naturally through openings in lower areas.

Mechanical, or forced, ventilation is used to regulate indoor air quality. Excess humidity, odors, and contaminants are controlled via dilution or replacement with outside air. Mechanical ventilation uses fans or blowers to pull air into a building through louvered exterior openings; it then directs the air to needed areas via ducts (Fig. 46.2). Mechanical ventilation is more reliable than natural ventilation because the fans, openings, and ducts can be sized to meet specific needs, and mechanical systems operate continuously to provide the specified ventilation. Natural ventilation performance varies, due to uncontrollable fluctuation in temperature and exterior weather conditions.

Some ventilation systems remove smoke, gases, and heat during fires. These are designed following the National Fire Protection Association Standard NFPA 90-A, *Installation of Air-Conditioning and Ventilating Systems*. Other ventilation systems, such as those in restaurant kitchens, are specifically designed to remove fumes and heat. The air contains greasy materials that coat ducts and fans and become a fire hazard. In addition to routine maintenance, these systems must have an automatic fire-control system. This typically includes fire dampers that close ducts when a fusible link breaks and activates some fire-smothering material, such as foam.

In certain occupancies, it is not permissible to recirculate ventilated air. Areas such as restrooms, biology and chemistry laboratories, hospital areas, including operating rooms, and industrial facilities in which various flammable vapor or hazardous fumes are developed, must exhaust the air—in some cases treating it before release.

Types of Air Filters

The most commonly used air cleaners include fibrous panel filters, renewable filters, electronic filters, and combination air cleaners.

Viscous impingement fibrous panel filters are made of porous, coarse fibers coated with an adhesive substance that captures particles in the air. They are inexpensive and good for lint but do little with pollen or atmospheric dust. They are available in thickness from $\frac{1}{2}$ to 4 in. (12.7 to 102 mm) and in a variety of standard sizes. A common application is in hot air residential furnaces and air-conditioning systems. Filter materials include coated animal hair, glass, vegetable and synthetic fibers, metallic wool, expanded metals and foils, crimped screens, random matted wire, and synthetic open-cell foams. Some types, such as the glass fiber residential filter, are disposable (Fig. 46.3). Metal filters can be cleaned and reused. They are typically washed with water or steam and may be recoated with adhesive (Fig. 46.4).

Figure 46.2 A fixed-bar louvered grille used with fans or blowers to pull exterior air into a building. *(Image copyright Ken Schulze, 2009. Used under license from Shutterstock.com)*

Figure 46.3 A disposable glass fiber air filter panel. *(Image copyright NATALE MattÃ?Â©o, 2009. Used under license from Shutterstock.com)*

Figure 46.4 This metal mesh and plastic fiber mesh air filter can be washed and reused.

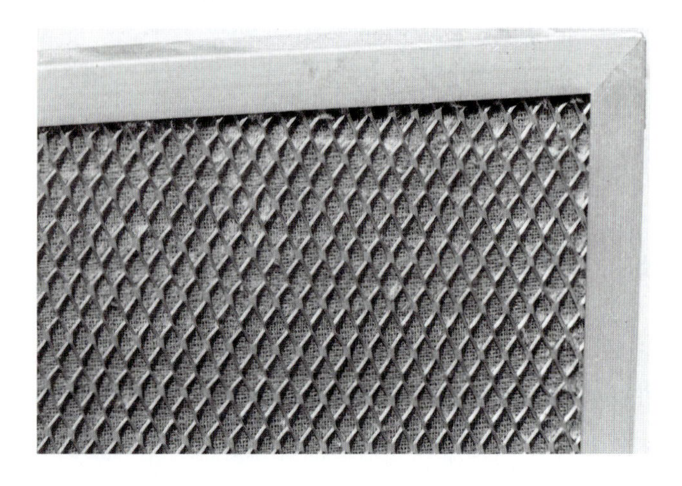

Dry-type extended surface air filters are made of random fiber mats or blankets using bonded glass fiber, cellulose fibers, synthetics, and wool felt supported by a wire frame that forms pockets or pleats. Some types require replacement of the entire filter, while others permit the replacing of mats inside the wire frame. Generally, dry-type filters are more efficient and have higher dust-holding capacities than viscous impingement fiber panel filters.

Renewable filters are viscous-impingement-type filters in roll form. They are either moved by hand or automatically unrolled across the filter area. As the clean filter material unrolls, the dirty portion rolls onto a roller at the bottom. When the entire roll is used up it is removed, disposed of, and replaced. Random dry fiber filter material is also used.

Electronic air cleaners use electrostatic precipitation to collect dust, pollen, and smoke. The filters have ionization and collection plate sections. The ionization section has small diameter wires with a positive direct current potential of 6 to 25kV DC. They are suspended between grounded plates, as shown in Fig. 46.5. The wires create an ionizing field through which particles in the air pass and receive a positive charge. The collecting plate section has a series of equally spaced parallel plates. Alternate plates have a 4 to 10kV DC positive charge. The plates not charged (they are negative) are grounded. The charged particles in the air pass into the collecting section and are forced by their electrical fields onto the negative grounded plates.

Electronic air cleaner cells are cleaned with a detergent and hot water. Large commercial installations may include an automatic washing system that cleans the cells in place. Small units, as used in residential systems, generally are removed and cleaned manually.

Figure 46.5 A cross section through an ionizing electronic air cleaner.

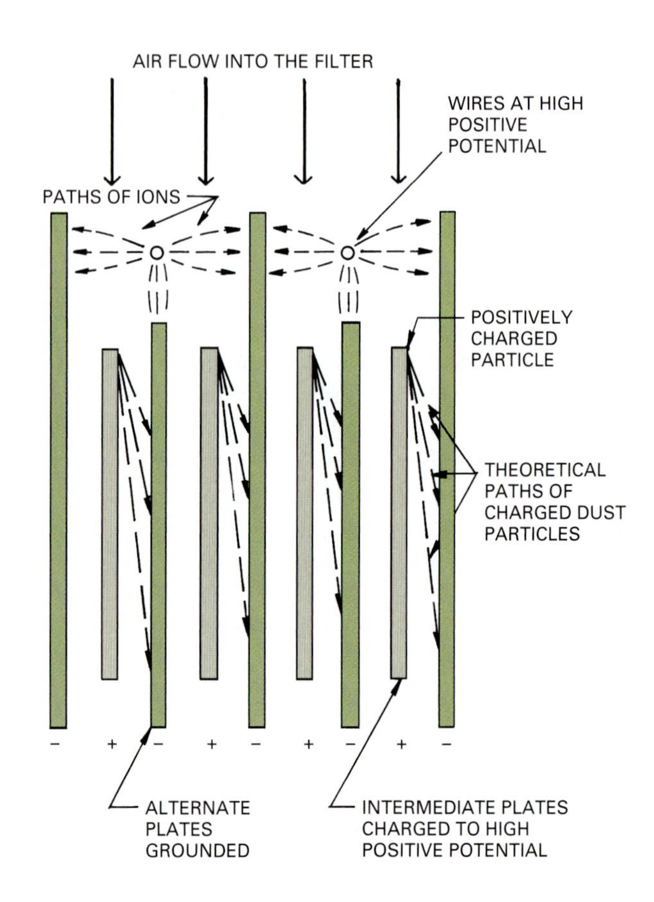

Combination air cleaners are an assembly of the systems just discussed. Typically a dry-type filter might be used in line after an electronic filter has processed the air. The dry filter can catch any particles that may escape the collector plates before they pass through the system.

FUEL AND ENERGY USED FOR HVAC

The major fuels used for mechanical systems include coal, fuel oil, natural gas, propane, butane, solid waste, and wood. All but natural gas require some form of storage. The energy sources include electricity, wind, geothermal, ground heat, and solar. Solar energy can be used directly with a means for storing heat. The others sources depend on an outside source of energy supply.

Coal

Coal is a combustible fossil fuel composed of elemental carbon, hydrocarbons, complex organic compounds, and various inorganic materials. The types commonly used for commercial and industrial heating and the production of electricity are anthracite, bituminous, lignite, semi-anthracite, and subbituminous. Coal is pulverized for burning in commercial furnaces. It is converted into liquid and gaseous products that may be used as fuels and also made into coke, which is used to produce steel. Systems that use coal as a fuel must have emissions controls to reduce air pollution and a means for disposing of the ash. Anthracite coal has a heat content of about 14,600 Btu/lb or 33,980 kJ/kg (kilojoules per kilogram).

Fuel Oil

Fuel oil is a fossil fuel derived from petroleum refining. It is a mixture of liquid hydrocarbons containing about 85 percent carbon, 12 percent hydrogen, and 3 percent other elements. The American Petroleum Institute (API) has a system for grading fuel oil. Kerosene is Grade 1, characterizing it as a very light and easily vaporized liquid. As the grade numbers increase so does the viscosity. Viscosity is a measure of the flow quality of a liquid. Grades 1 and 2 fuel oils are frequently used in residential and small commercial furnaces (Grade 3 is sometimes used for these purposes). They flow easily and can be vaporized and mixed with air as required for burning. Grades 4, 5, and 6 are heavier and used in large industrial and commercial burners. Grades 5 and 6 are very heavy and must be preheated before they can be pumped and burned. No. 2 fuel oil has a heat content of about 141,000 Btu/gal. or 39,300 kJ/L (kilojoules per liter).

Natural Gas

Natural gas is a combustible hydrocarbon gas consisting primarily of methane with small amounts of ethane and traces of butane, propane, pentane, and hexane. It is odorless in its natural form but an added odorant acts as a safety measure so leaks can be detected by smell. Natural gas has a heat content of approximately 1,050 Btu/ft.^{3} (39,100 kJ/m^3). It is obtained from wells drilled in earth strata that contains pockets of gas. Natural gas is supplied to buildings through underground utility pipelines, so on-site storage is not required. A utility company is responsible for maintaining a continuous supply.

Propane and Butane

Propane and butane are liquid petroleum products and very volatile. They are colorless, flammable gases derived from petroleum and natural gas. They come in liquified form and are transported in pressurized cylinders. Propane and butane have a heat content of about 2,500 Btu/ft.^{3} (93,150 kJ/m^3).

Solid Waste

The collection of burnable solid waste is an ever-increasing activity, and its use as a fuel is becoming more popular. It takes specialized incinerator equipment to produce a clean burn that results in the desired heat energy. Typically, such a facility is built near a large city to ensure a steady supply of waste or near a commercial or industrial plant that has contracted to use the energy. This has the advantage of helping reduce the growing problem of disposing of large amounts of waste materials while producing a usable end result.

In some systems, the waste is shredded, and electromagnets draw out ferrous materials in the recycling process. These plants burn this refuse-derived fuel (RDF) with coal as a backup fuel. The fuel produced from the waste is the main fuel used to burn the incoming waste. About 50 percent of waste materials entering the plant drive turbine generators that produce electricity. In this case, the electricity is used by the city to operate sewer and water treatment plants, street lights, and to serve other public and private customers.

Another waste-to-energy system is illustrated in Fig. 46.6. It accepts mixed waste and sorts it for recycling and composting before turning the remainder into refuse-derived fuel (RDF). In all cases, major efforts are required to make waste-to-energy systems economical to operate. It requires cooperation from the local trash collectors and local and state governments.

Wood

Wood is a renewable resource, and the harvesting of trees is accompanied by a program of replanting. Wood has become a minor source of heat energy, typically used only to heat small residences. Overall, wood has a heat content of about 7,000 Btu/lb. (16,290 kJ/kg).

A variety of wood-burning stoves are available, most of which operate more efficiently than fireplaces. Many newer-designed stoves meet the Environmental Protection Agency's Phase II emission standards. These have the EPA approval label, consume less wood, and burn 80 to 90 percent cleaner than older wood-burning stoves.

The EPA Phase II stoves are available in two types: catalytic and non-catalytic. The catalytic stove has a cored combuster in the smoke path just past the main combustion chamber. The combuster has a catalytic coating that causes combustion gases to burn completely, thus reducing emissions. After the fire in the main combustion chamber reaches 500°F (262°C), a bypass damper opens, directing gases through the catalyst rather than straight up the chimney. The combustor does deteriorate with use and must be occasionally replaced.

Figure 46.6 This waste-to-energy system accepts unsorted waste and sorts, recycles, and composts it to produce refuse-derived fuel (RDF). *(Courtesy Great Lakes Regional Biomass Energy Program, Council of Great Lakes Governors)*

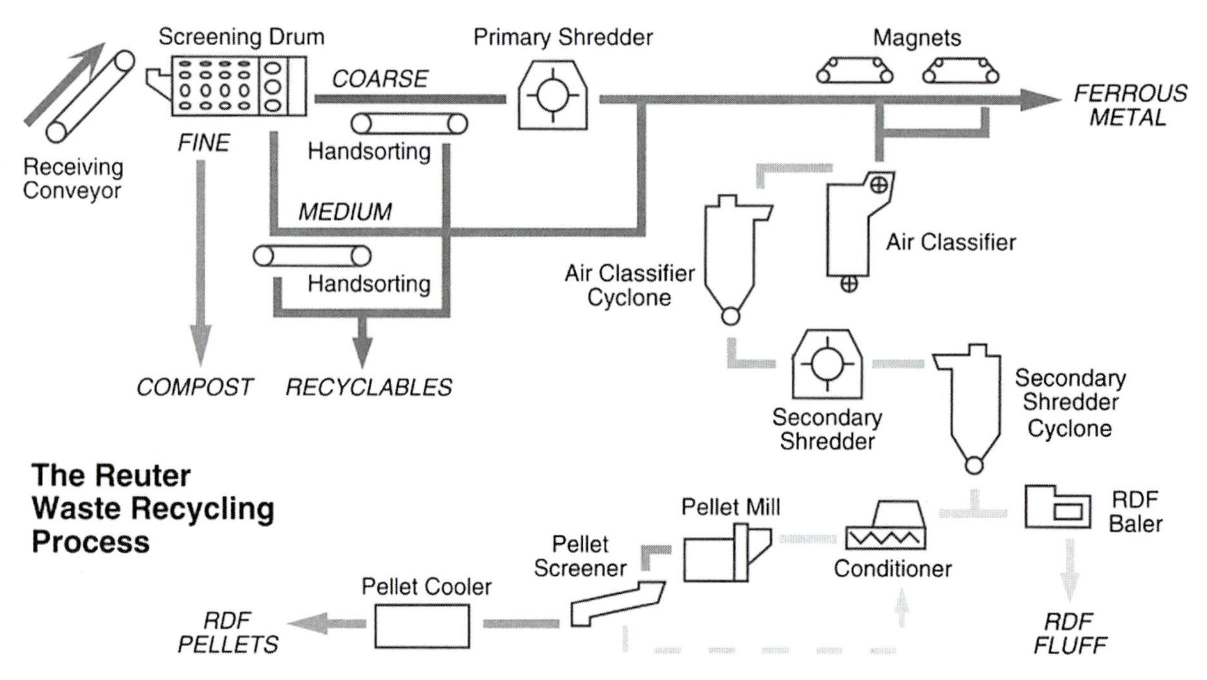

The non-catalytic stove can also achieve low emissions, but it does this by keeping high temperatures in the main combustion chamber and adding turbulent secondary air at the top of the chamber where hot gases leave the stove. This air provides additional oxygen so combustion gases burn more completely.

Pellet stoves are designed to burn a manufactured product made from wood and wood by-products. The pellets are small, about the size of large feed grain, and are sold in waterproof bags. They are a concentrated material that burns hot and clean, producing low emissions. Pellet fireplace inserts are also available. They use a motor-driven auger to feed pellets from a fuel storage hopper. The typical hopper holds enough fuel to last several days.

Some types of fireplace inserts function much like stoves. The insert is a metal liner with pyroceramic glass doors that permit generated heat to radiate into a room. They are airtight and burn hot, producing low emissions. This type of fireplace insert has emission qualities equal to the EPA Phase II woodstove. They can be installed in a wood-burning fireplace or have an approved brick or wood surrounding and are vented with a metal chimney. Some have an electric blower to increase heat distribution into a room.

Electricity

Electricity is a major energy source for operating HVAC systems. It powers the blowers, pumps, and controls and supplies heat energy through various types of resistance heating units. It usually has the lowest installation cost and is easy to install anywhere in a building. It cannot always compete with other fuels on a cost basis. It requires no on-site storage, but large installations require space for transformers and switchgear. Power is usually provided by a utility company that guarantees a steady supply, although some on-site power generation is used on large installations (as discussed in Chapter 47).

Geothermal

Geothermal heat energy derives from the earth's internal heat. The typical example is geysers that send up jets of hot water and steam from within the earth. This source of heat is limited to a geyser basin (an area containing a group of geysers). The system can be tapped to supply heat to an entire village from this central source.

Ground Heat

Ground temperatures fluctuate by season, but they remain fairly stable as depth increases. Exact temperatures vary with the geographic location. For example, in Minnesota's cold climate, the temperature is a relatively constant 50°F (10°C) at a depth of 16 to 26 ft. (5 to 8 m). Earth-sheltered houses use this stable temperature to provide a steady source of heat through exterior walls. If a 50°F (10°C) interior air temperature is maintained, very little supplemental heating is required. Ground source heat pumps provide another important use of ground heat. One type uses an earth coil that involves burying plastic pipe in either drilled holes or in a continuous ground loop. The water in the pipe absorbs ground heat and is pumped to a water source heat pump, which extracts the heat and recirculates the water. Tapping well water also works. A heat pump removes water from a well, extracts the heat, and disposes of the water in a second well, putting it back into the aquifer. Groundwater temperatures vary by geographic region, but groundwater heat pumps in the range of 55 to 60°F (13 to 16°C) are commonly used.

Solar

Solar energy derives from the sun. It is a major source of heat energy and keeps all living things alive. Solar energy can be tapped by various collectors, stored, and used to heat and cool buildings or produce electricity. It is a source of natural light. A variety of solar heating systems are available. Some require mechanical devices to gather, store, and distribute heat energy as needed; these are active solar systems. But passive solar systems do not require hardware of any type.

AIR HEATING EQUIPMENT

Warm air heating systems incorporate a heat-generating device (furnace), controls, and a distribution system of ducts. A furnace may be fired by oil, gas, electricity, and some solar systems that use warm air distribution. Dual systems include an air-conditioning unit and both heat and cool buildings.

Warm Air Furnaces

Warm air furnaces use blowers to move heated air through ducts. These include upflow, downflow, and horizontal furnaces. An upflow warm air furnace moves air out the top of the unit into a duct system. It is used in basements when ducts are run under a floor or on the floor when ducts are run in an attic (Fig. 46.7). A downflow warm air furnace moves air out the bottom of a unit. This type is typically used when ducts are run in a concrete slab floor or below the floor in a crawl space (Fig. 46.8). Horizontal warm air furnaces are mounted in an attic or hung below floor joists, as shown in Fig. 46.9. Each of these may have an air-cooling unit that uses a furnace blower and ducts to provide cool air to building spaces (Fig. 46.10).

Figure 46.11 shows a horizontal gas-fired warm air furnace with both heating and cooling capabilities. It can be installed outdoors on a concrete slab for horizontal flow or on a roof for downflow installation. It can burn natural gas or be converted to propane.

Figure 46.7 Typical configurations of upflow warm air heating systems.

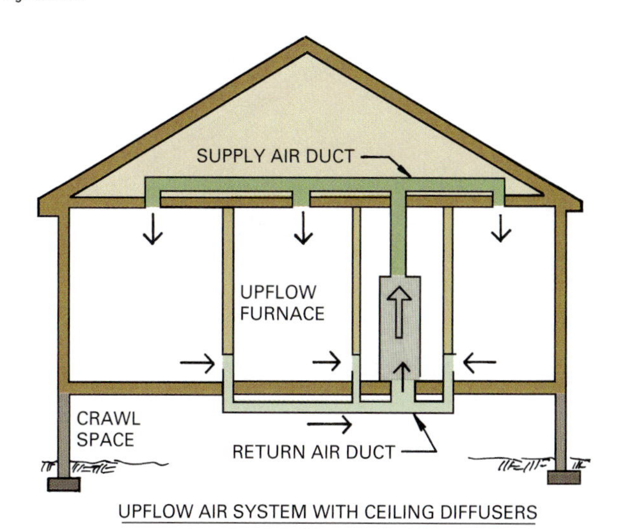

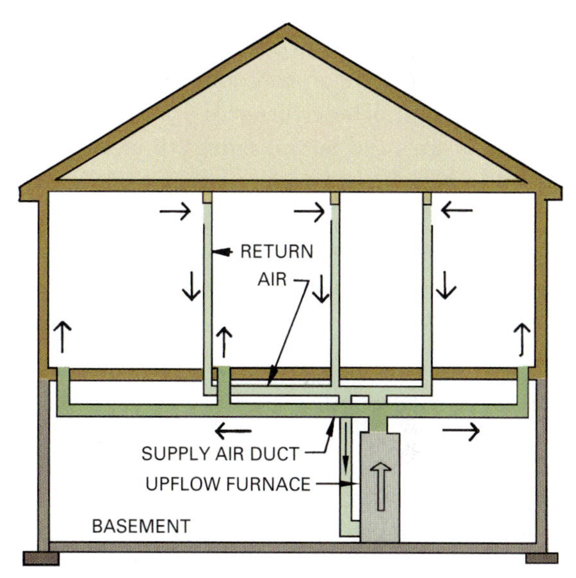

Figure 46.8 Downflow warm air heating systems are used in buildings with a crawl space or concrete floor.

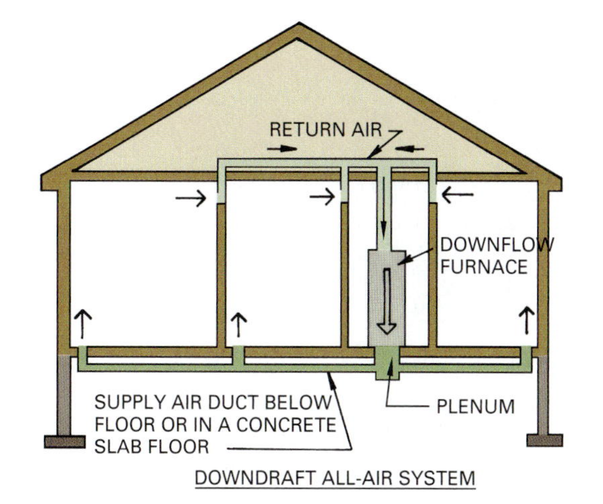

Figure 46.9 Horizontal warm air furnaces are hung below a floor or in an attic.

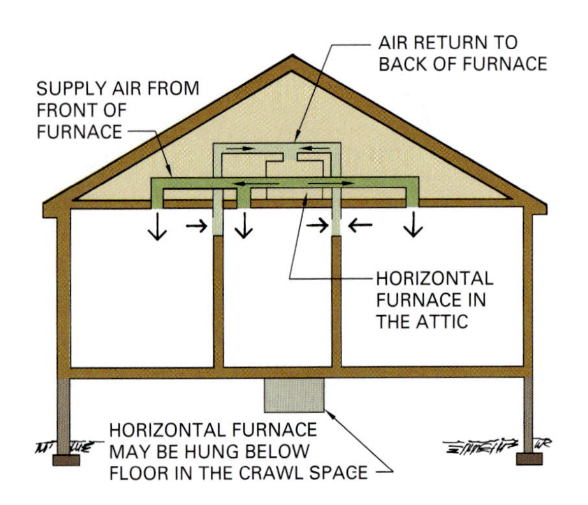

The internal construction of a gas-fired warm air furnace includes a blower that operates on two speeds and circulates cool air when an air-cooling unit is mounted on top of the unit. These furnaces are available in a wide range of sizes and output ratings (Btuh). Gas-fired furnaces can be switched on-site from an upflow/horizontal configuration to a downflow setup.

Fig. 46.12 shows a high efficiency gas-fired warm air furnace. Used in residential construction, this blower saves consumers electricity costs and controls temperature swings. The variable-speed motor tailors airflow for better moisture control and comfort. Warm air furnaces may also use electric resistance coils to heat air, which is then distributed through ducts by a blower. These types are similar to those shown for gas furnaces. One major problem when using them is the high cost of electricity in many areas (Fig. 46.13).

Oil-fired furnaces require a large exterior storage tank from which to pump fuel oil to the furnace burner. A typical installation is shown in Fig. 46.14. The location and method of placement of the unit is regulated by codes. Oil tank construction must meet Underwriters Laboratory (UL) specifications. Tanks may be placed above or underground below the frost line. Some codes permit oil storage tanks inside a building. An oil-fired furnace is much like gas-fired models, but it has an oil burner unit with a nozzle that breaks the oil into a fine spray. A high-voltage ignition system using spark electrodes ignites this vapor.

Heat Pumps

A heat pump is a machine that can heat or cool a building using forced air through ducts. It takes heat from one source, such as outside air, and transfers it to another, such as the air inside a building. The commonly available types include air-to-air, water-to-water, and air-to-water. Use of ground source heat pumps is increasing.

Figure 46.10 A horizontal gas-fired warm air furnace. *(Courtesy Bryant Air Conditioning)*

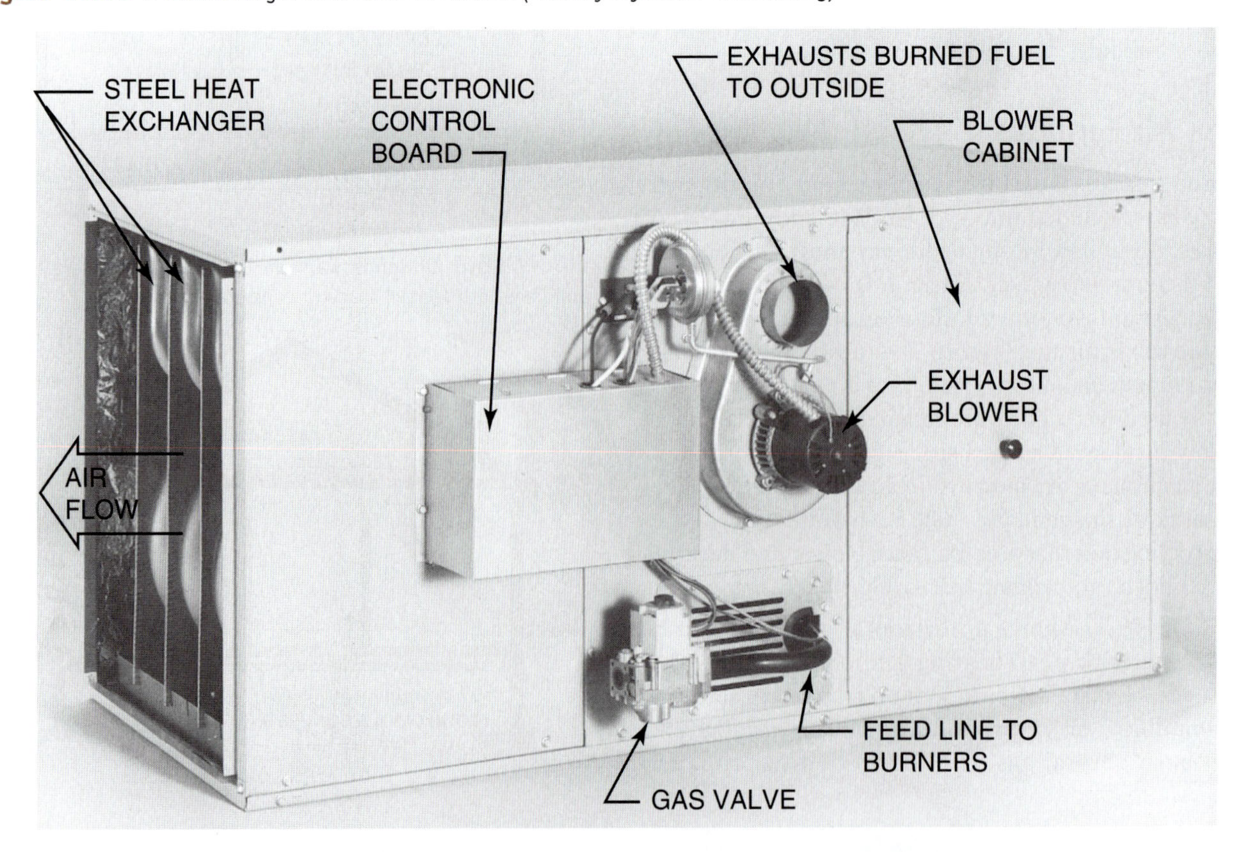

Figure 46.11 This single-package gas-fired warm air furnace has an electric cooling coil, thus providing summer cooling in addition to heating. It can be mounted on a concrete slab on the ground or on a roof. *(Courtesy The Trane Company)*

Figure 46.12 This is a high-efficiency gas-fired upflow furnace with a direct spark ignition that eliminates the pilot light. It uses pulse combustion and is available in fixed and variable speed models. *(Courtesy Lennox International Inc.)*

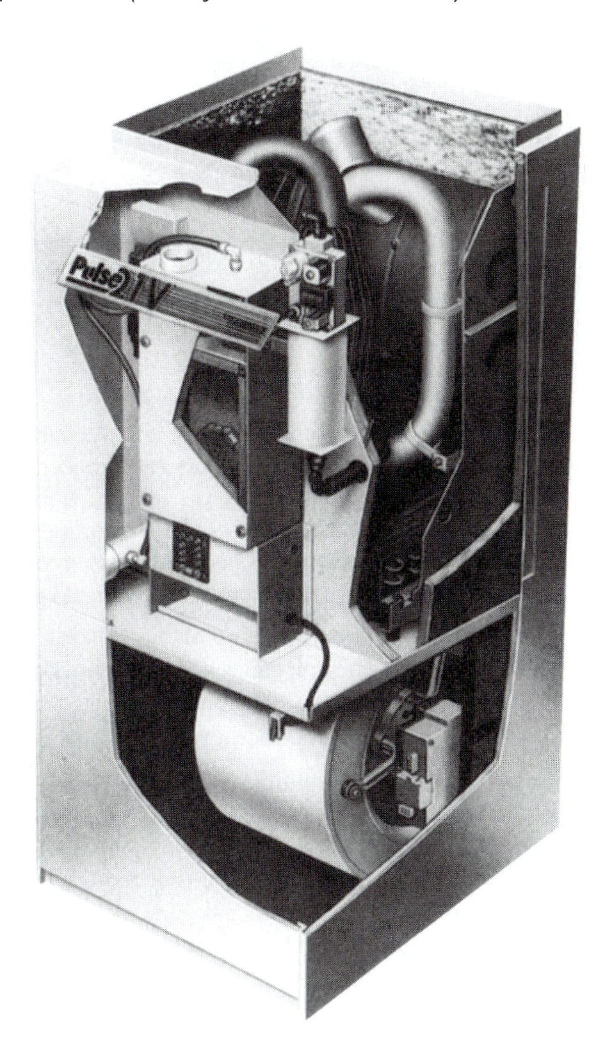

Figure 46.13 This electric central heating/cooling furnace has options of 5 to 25KW and up to 5 tons of cooling capacity. *(Courtesy Rheem Manufacturing Company)*

Figure 46.14 A typical installation for an oil-fired warm air furnace.

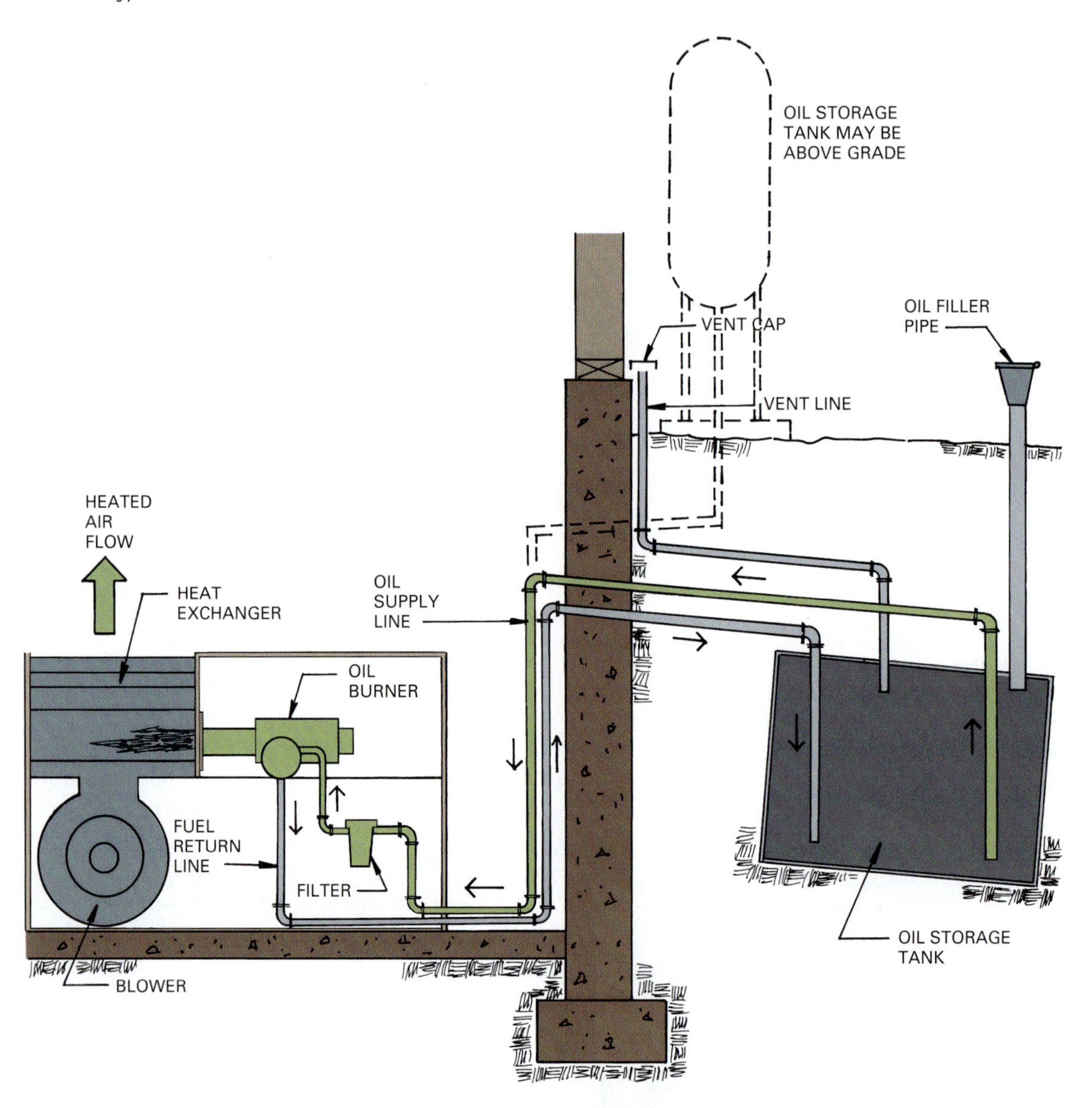

The most frequently used heat pump is an air-to-air unit. It has a **compressor** similar to those used in refrigerators. A **refrigerant** (a gas) circulates in coils. In heating mode, the coils in the exterior unit enable the refrigerant to absorb heat from outdoor air. The refrigerant is then transferred to a compressor where compression raises its temperature. It then transfers to indoor coils, a blower moves air over it, which captures the heat and distributes it through a building via ducts (Fig. 46.15). In cooling mode, the reverse happens. Heat is absorbed by the inside coils and dispersed to outside air by the outside coils.

Figure 46.16 shows an air-to-air heat pump compressor unit. The assembled outdoor unit rests on a concrete slab, and the coils are fully exposed to the air. The fan on top pulls air through the coils.

A groundwater heat pump (GWHP) is a water-to-air system. A water-to-air system pumps water from a well to the heat pump. The well water typically maintains a constant temperature, which varies with the geographic area but is typically between 55 and 65°F (13 to 18°C). The refrigerant in the heat pump coils absorbs heat in the water and discharges it into a building through an air handling unit and ducts. The unit can reverse the process and remove heat from inside a building and discharge it to the water, which is drained in a disposal well (Fig. 46.17).

Water-to-water heat pumps can be used to simultaneously heat and cool water, if necessary. For example, the pump can remove heat from a liquid source, such as a fluid in a manufacturing operation that requires chilling, and move the heat to another liquid, such as water for cleaning operations.

Figure 46.15 The heating cycle of an air-to-air electric heat pump.

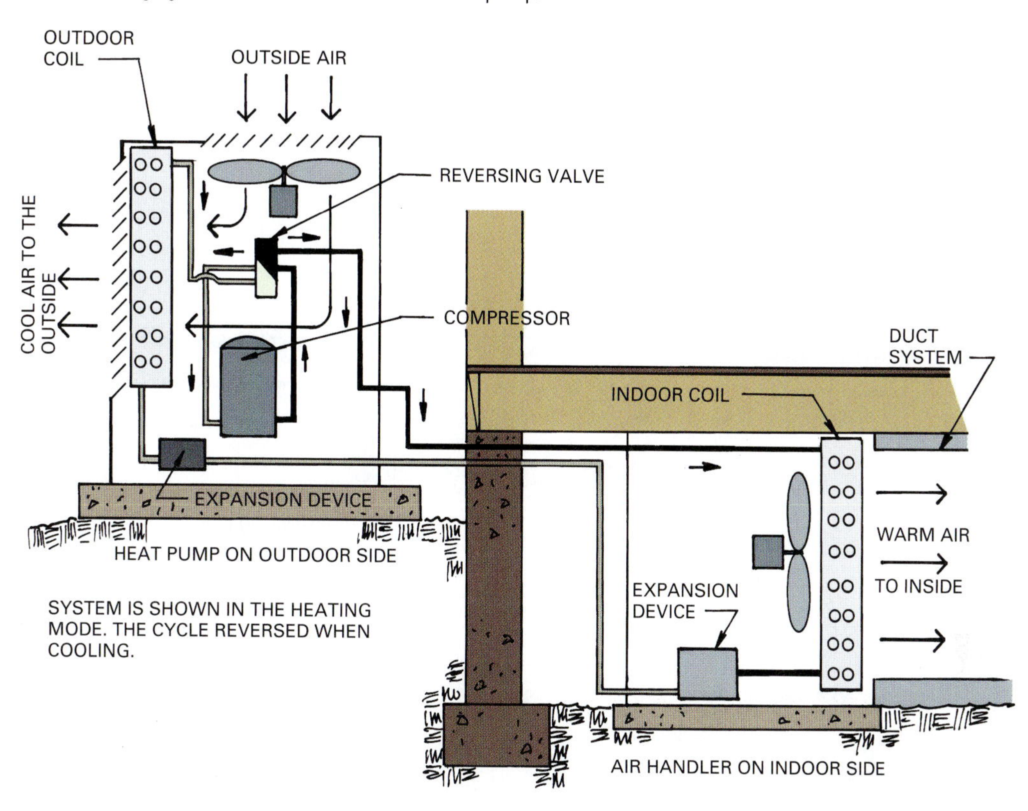

Figure 46.16 An air-to-air heat pump. *(Courtesy Lennox International, Inc.)*

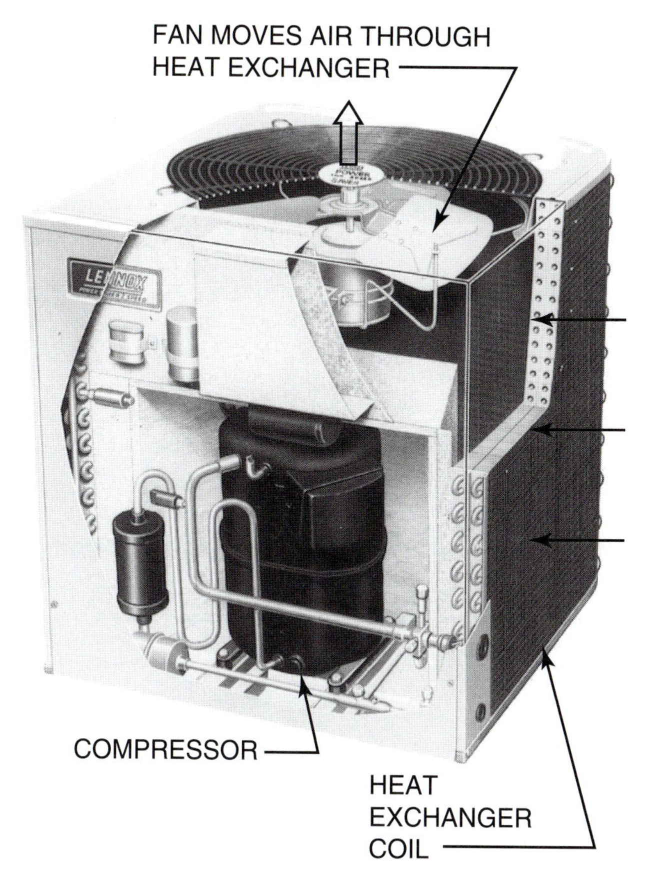

Air-to-water heat pumps can be used to cool air in a room and transfer captured heat to water or another liquid that needs heating. An apartment complex could cool air in living spaces and use the resultant heat to keep water in a swimming pool at a desired temperature.

A ground loop heat pump is used in areas where groundwater is not readily available, too costly to acquire, or of poor quality for heat pump use. The ground loop avoids the maintenance issues caused by the mineral content of water in the groundwater systems. The system may use an earth coil in a horizontal loop (Fig. 46.18) or be installed vertically in drilled holes (Fig. 46.19). The earth coil system transfers heat to or from a water/antifreeze solution circulated through plastic pipes buried in the earth. The heat from the earth is transferred to the refrigerant in a heat exchanger and compressed and distributed by a blower through ducts. In cooling mode, the reverse occurs with indoor heat being transferred to the water/antifreeze solution, which disperses it to the cooler ground.

The choice of a horizontal or vertical ground loop depends on the results of a study that analyzes surface and subsurface conditions, including soil type, moisture content, rock strata, and ground temperatures. Horizontal loops require sufficient open space free from paved surfaces, underground utilities, and trees and shrubs.

Figure 46.17 A water-to-air heat pump can be used wherever groundwater is available and wells can be drilled. The system also requires a water return system that returns the water to the earth. *(Courtesy Kansas Electric Utilities Research Program, research by Mark Hannifan and Joseph King, AIA)*

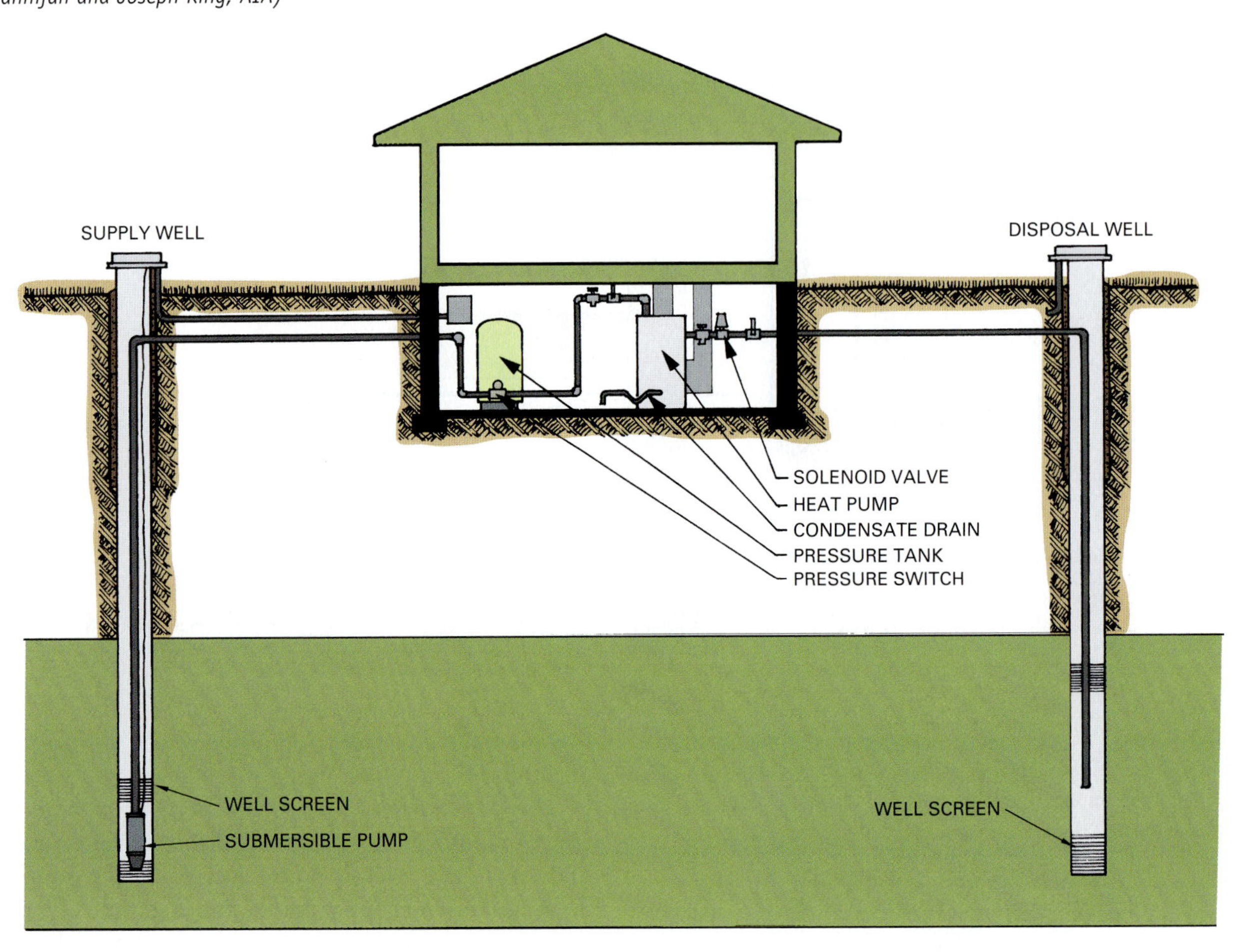

Figure 46.18 A horizontal ground loop heat pump installation. The ground loop is made of plastic pipe carrying a water/antifreeze mixture that circulates through the pipe to the water-to-refrigerant heat exchanger in the heat pump. *(Courtesy Kansas Electric Utilities Research Program, research by Mark Hannifan and Joseph King, AIA)*

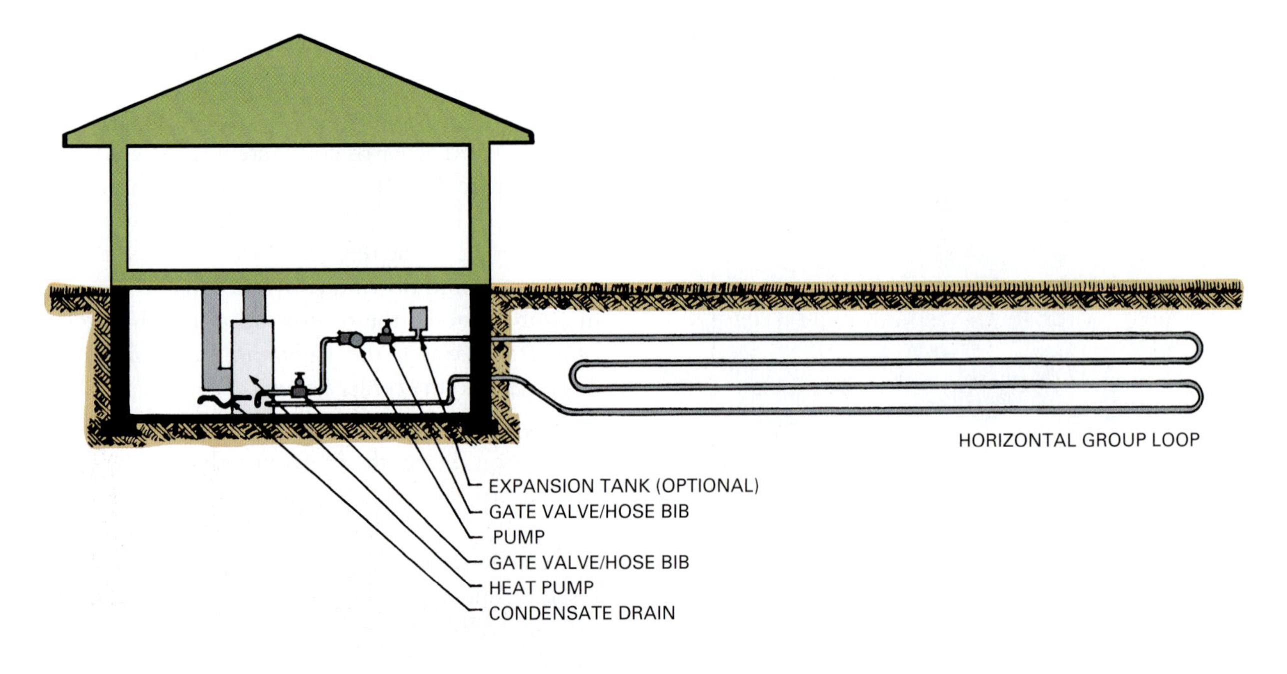

Figure 46.19 A vertical ground-loop heat pump installation. In areas with low soil moisture, vertical loops are recommended if the soil permits drilling deep enough to get adequate lengths on the vertical loops. *(Courtesy Kansas Electric Utilities Research Program, research by Mark Hannifan and Joseph King, AIA)*

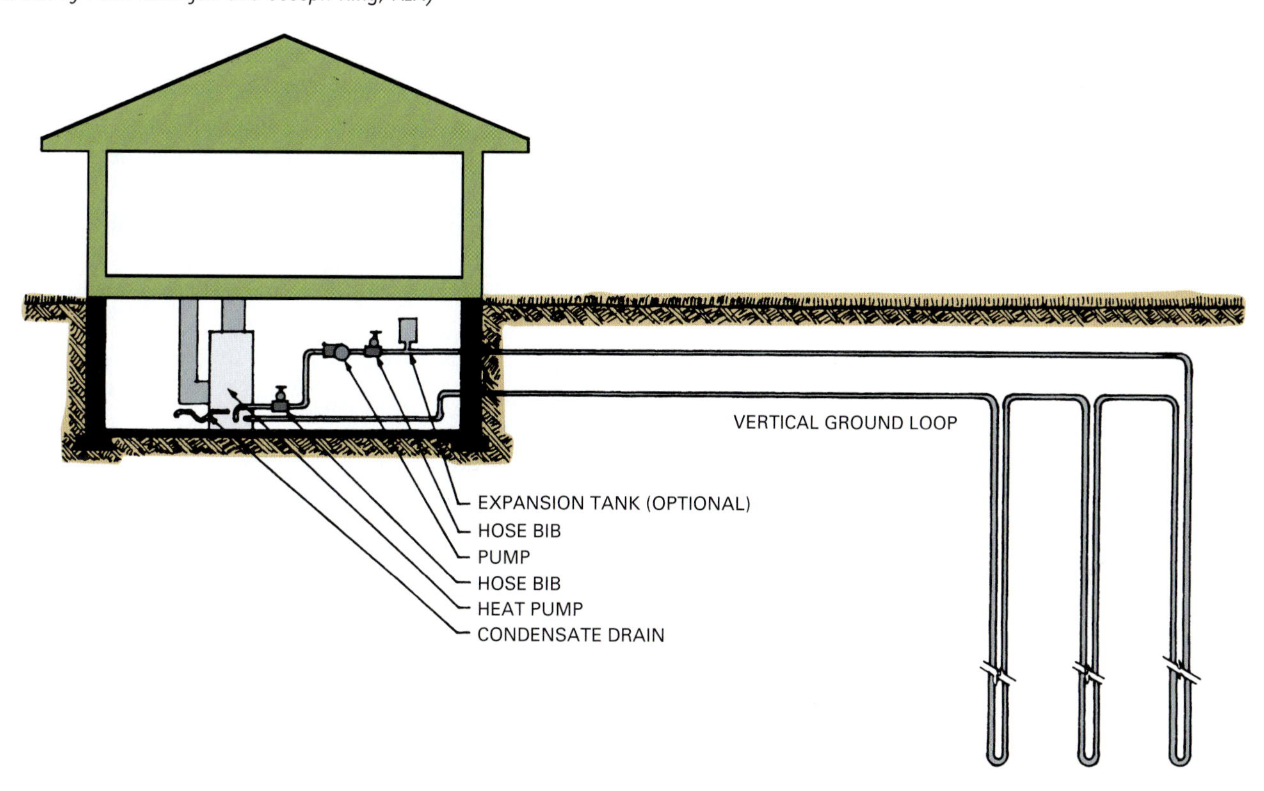

Efficient pumps designed to heat and cool one room with no ducts are manufactured as a single unit. The coil and fan are housed in a decorative interior case. The compressor and exterior coil extend through an opening in the wall and are exposed to the outside air (Fig. 46.20). Similar units are available with total electric heating and cooling capacities.

Figure 46.20 A section showing a typical through-the-wall heat pump installation.

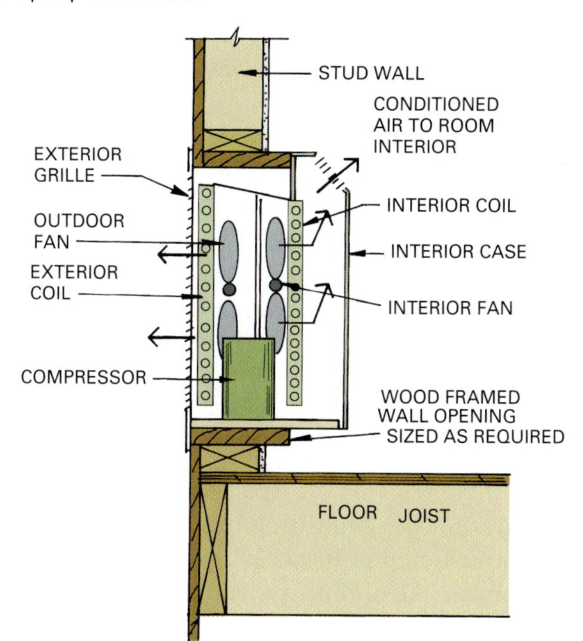

A natural gas heating and cooling unit is shown in Fig. 46.21. This unit uses natural gas as the primary energy source for both heating and cooling, operating much like the all-electric heat pump. The compressor is driven by a quiet natural gas engine, and its function

Figure 46.21 This gas-fired heating and cooling unit can be used for residential and small commercial buildings. It uses the basic heat pump cycle. However, it uses a high-efficiency natural gas engine to operate the compressor. *(Courtesy York International Corporation)*

is the same as described for electric heat pumps. The unit has a recuperator that recovers engine heat, thus increasing efficiency. An auxiliary gas heater provides supplemental heat during extremely cold weather. The system is microprocessor-controlled to provide continuous comfort. A gas engine uses pressure formed by combustion of expanding gas to produce mechanical energy by supplying power to the motor shaft, which connects to the compressor.

AIR-CONDITIONING WITH DUCT SYSTEMS

Duct systems are also used to air-condition buildings. The heat pump previously discussed provides both heating and cooling modes. When oil, gas, or electric warm air furnaces are used, a cooling coil unit is placed on the furnace and connected to an outside air conditioner, as shown in Fig. 46.22. They are connected by pipes that carry refrigerant from the compressor in the outdoor unit to the indoor coil. The furnace blower moves air over the cold coils and through the ducts into the building.

A common residential external electric air-conditioning unit is shown in Fig. 46.23. The refrigerant absorbs heat from air blown over the indoor coil. It then moves to the outdoor air-conditioning compressor, where the absorbed heat is transferred to the atmosphere. The refrigerant then moves back to the interior coil to repeat the process.

Figure 46.22 An electric air-conditioning unit is installed out of doors and supplies the chilled refrigerant to cooling coils on the furnace.

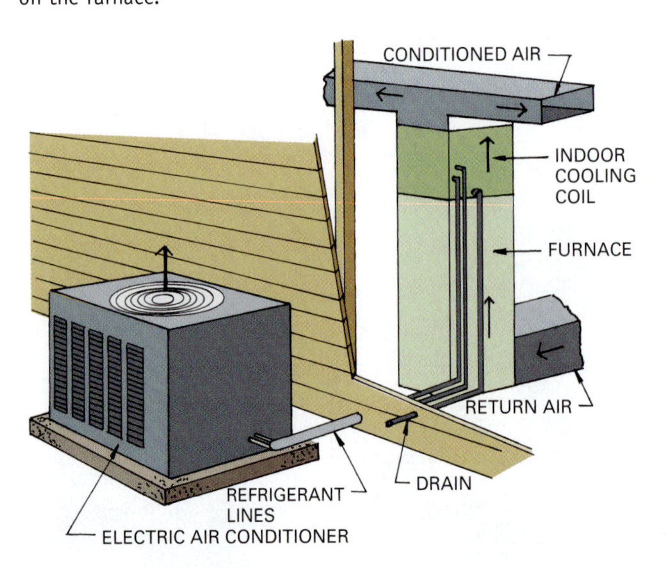

Figure 46.23 An electric air-conditioning unit that supplies cooled refrigerant to indoor cooling coils located on a furnace or in an air handler. *(Courtesy The Trane Company)*

Figure 46.24 This roof-mounted unit is designed for commercial buildings. It is available as an electric-cooling/gas-heating unit or an electric-cooling/electric-heating unit. *(Courtesy Rudd Air Conditioning)*

If a building has hot water or steam heat, a similar system can be used to air-condition it. A duct system is installed for cooling, and the indoor unit is an air handler with a blower and coil. The outdoor air-conditioning unit is the same as the one just described.

Figure 46.24 shows a roof-mounted heating/cooling unit used on light commercial construction. When combined with a duct system, as shown in Fig. 46.25, they can heat and cool zones within a building. Each zone has its own **thermostat** that operates a zone damper.

ALL-AIR DISTRIBUTION SYSTEMS

Procedures for designing all-air duct distribution systems can be found in the 1989 ASHRAE Handbook-Fundamentals. All-air duct systems may be low or high

Figure 46.25 A typical duct system for a roof-mounted heat pump or electric/gas-fired unit. The controls permit it to condition the air in various zones within the building.

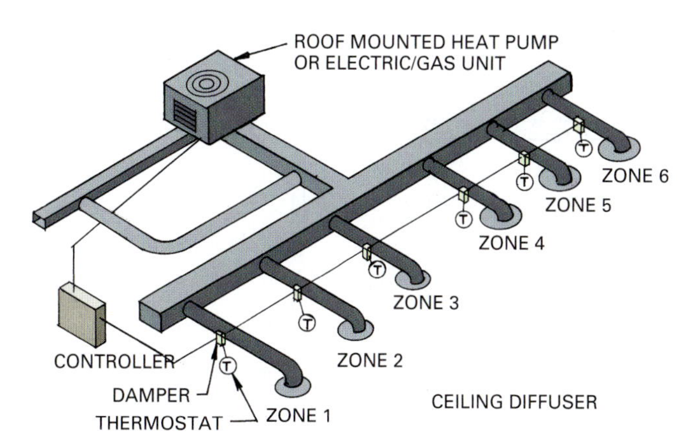

velocity, with the high-velocity design requiring smaller diameter pipes and high pressures. A designer must consider how duct size influences performance, outside air requirements, supply air temperatures, airflow, air changes, zoning humidity control, heat gain and heat loss of the space to be conditioned, control of noise in the system, possible use of heat recovery devices, and many special factors, like those in hospitals, manufacturing plants, and laboratories.

Systems in common use include single zone, multizone, reheat, variable air volume, and dual duct. Various combinations of these can be used to meet design requirements.

Single-Zone Systems

Single-zone systems use one air handling unit (AHU) to supply an entire building or a portion of a building that is considered a single zone. The furnace and air-handling unit can be installed outside of or within the space to be conditioned. A **return air** duct system is usually required. A typical application is a small residence. A large residence may have two or more heat and cooling sources using two or more single zone systems. A simple schematic is shown in Fig. 46.26. It should be noted that this system supplies air at a constant rate. Therefore, room temperature is varied by changing the air temperature.

Multizone Systems

Multizone systems produce heated and cooled air in an air-handling unit. The air for each zone is blended at the air-handling unit into a single temperature and fed to the zone by a single duct (Fig. 46.27). This enables a zone requiring the lowest air temperature to receive it or, if needed, the temperature can be raised to a higher level. Likewise, areas requiring higher temperatures can be accommodated by blending little or no cool air with the heated air sent to an area.

The flow of air in the ducts is regulated by motorized dampers controlled by thermostats. Since this system uses a constant volume of air, it passes by the heating and cooling coils even if the air is not needed.

Figure 46.26 A single-zone heating/cooling duct system for a small building that has a single source of heating and cooling.

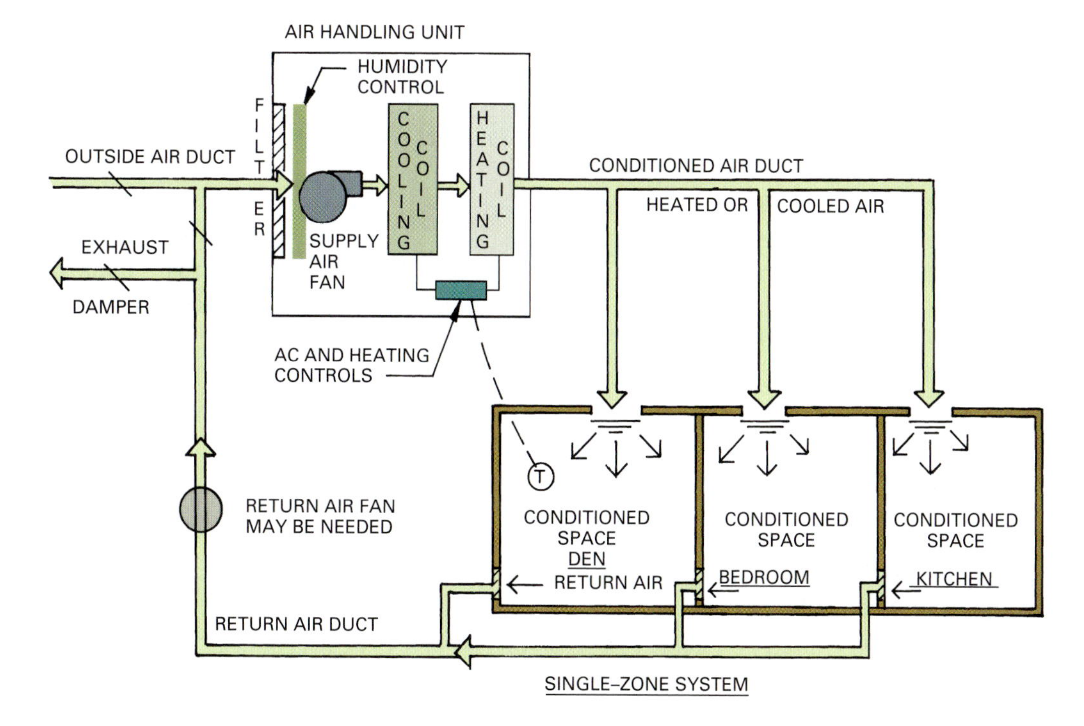

Figure 46.27 A multizone system designed to blend cool and heated air to produce various temperatures required for the various zones.

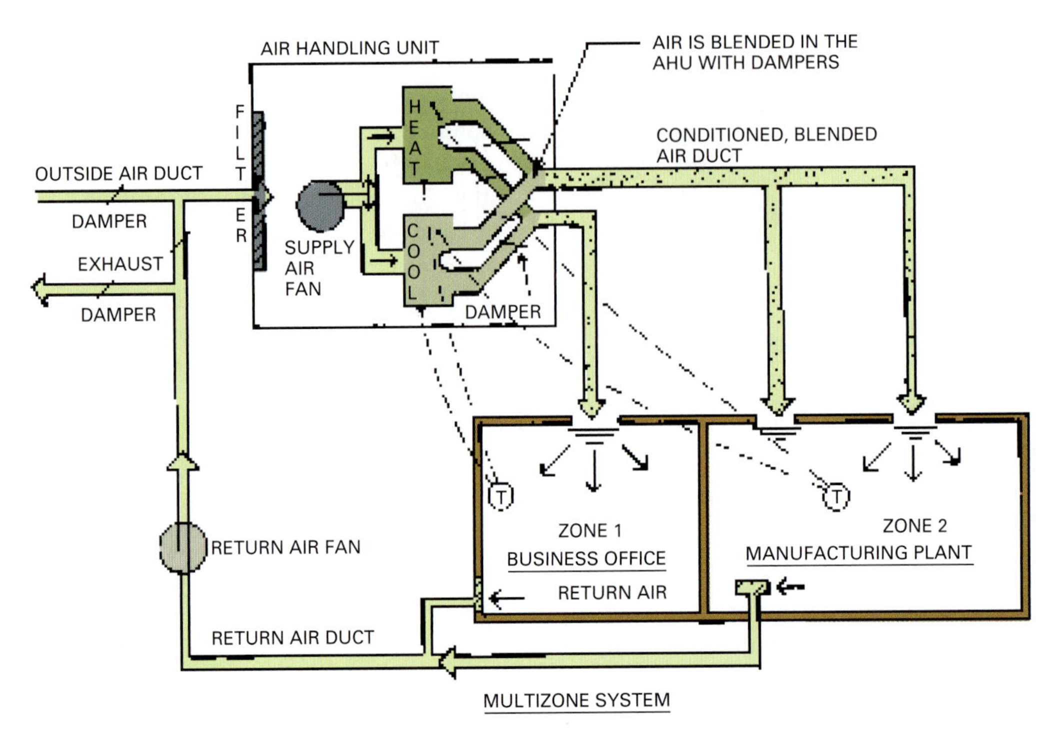

The system is not cost efficient nor is it widely used, because more efficient zoning methods are available—like reheat systems.

Reheat Systems

A reheat system is a variation of the single-duct single-zone and the multizone systems. It supplies a single source of preconditioned or recirculated air at a constant rate through ducts to several zoned areas. The air is processed through an air-handling unit, in which it can be filtered, humidified, and cooled. The temperature of the air is that required for the coolest zone. The cool air is then sent through ducts to each zone. The duct to each zone has a reheat coil that heats the incoming air to the temperature required for that zone. The reheat can be an electric resistance unit or a hot water or steam coil (Fig. 46.28). This system permits the simultaneous cooling and heating of areas with different requirements.

Variable Air Volume Systems

Variable air volume (VAV) systems control a space's environment by supplying air at a constant temperature and varying the quantity supplied rather than changing the temperature of the air (as is done in the single duct, reheat, and multiple zone systems). The air from the handling unit moves through single ducts to each zone, where a variable air volume terminal is located. This terminal varies the air supply to the space, but the air temperature is held constant (Fig. 46.29).

Variable air volume systems may use a common or separate fan system, and a reheat coil can be added at zone entrances if necessary. If reheat is used, the volume of air can be reduced, producing a more cost-efficient operation. Variable air volume systems are not useful in situations where humidity control is important. They are well suited for use in offices and other areas where temperature control for human comfort is important but precise humidity control is secondary.

Dual Duct Systems

Dual duct systems are similar to the multizone system. The difference is that the conditioned air in the air-handling unit is moved to spaces by two parallel ducts. One duct carries cold air and the other warm air (Fig. 46.30). At each space, mixing valves combine the warm and cold air in the proportion needed to meet the air temperature requirements of the space. These may be constant volume or variable volume air systems. Constant volume systems may use a reheat. Variable air volume systems may use a single fan or dual supply fans. The dual duct system is not cost efficient because it uses energy to both heat and cool air and then mix them to get desired temperatures. Additional details on these systems can be found in the 1992 ASHRAE Handbook, *HVAC Systems and Equipment*.

Figure 46.28 A reheat warm air system supplies conditioned air at the lowest required temperature and uses reheat units to raise the temperature at zones requiring higher air temperatures.

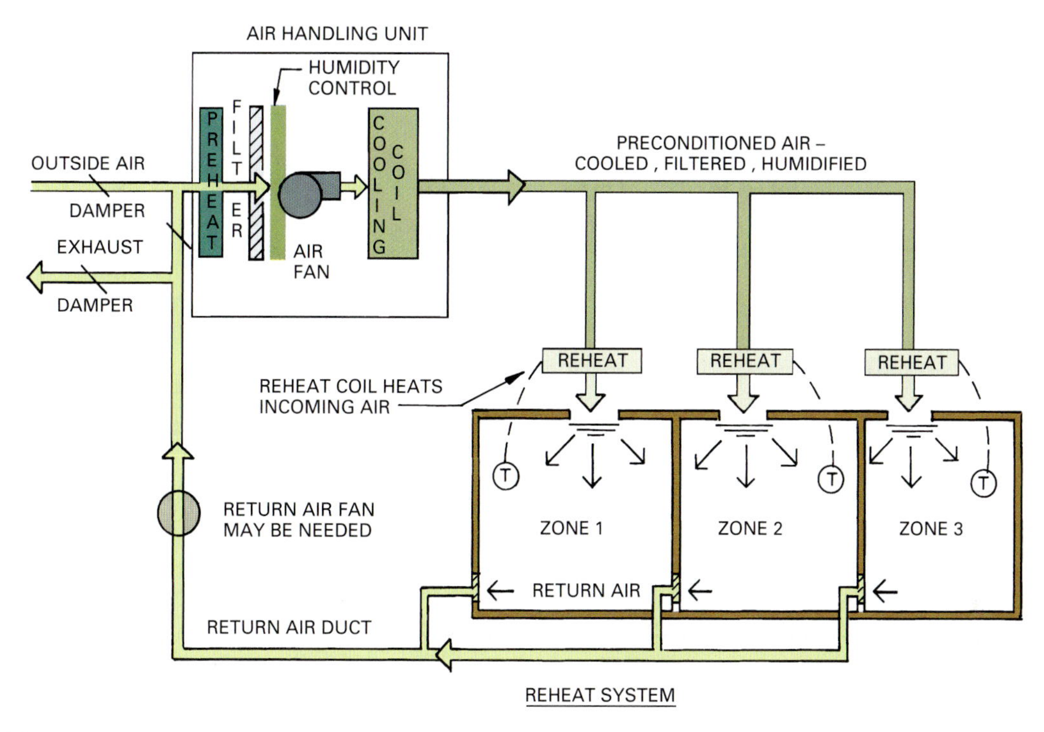

Figure 46.29 A variable-air-volume warm-air system controls air temperature in each zone by varying the amount of air flowing into that zone.

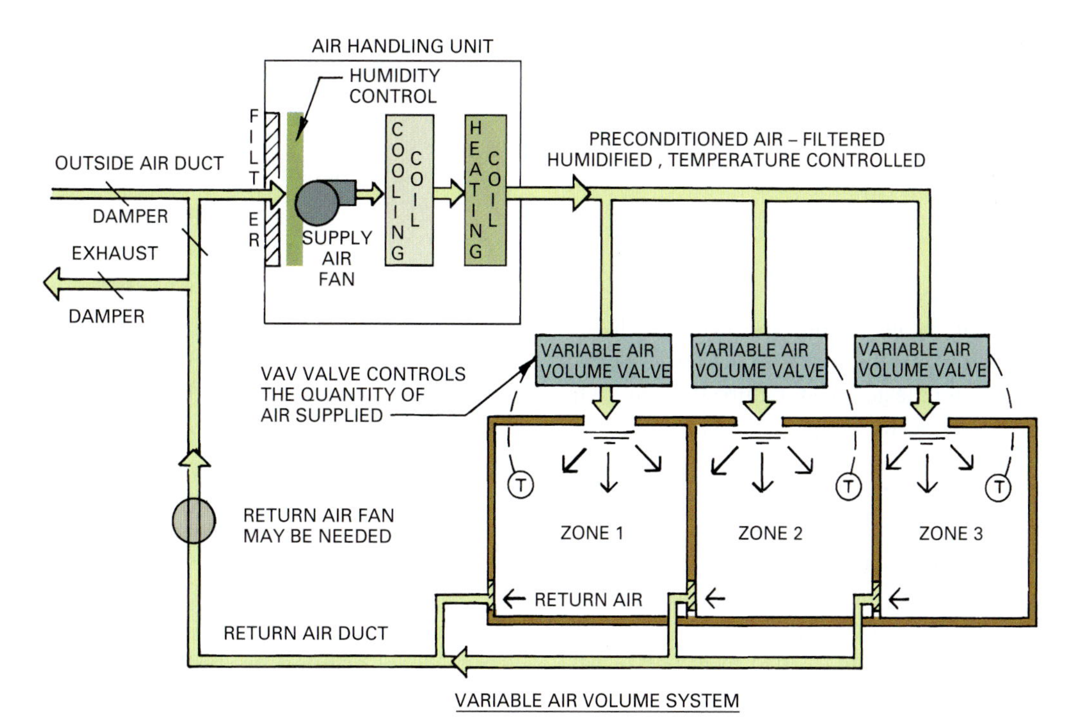

TYPES OF DUCT SYSTEMS

Air distribution systems receive heated or cooled air from a furnace. The air moves into an air-handling unit, where it may be filtered, humidified, or dehumidified. The air-handling unit has a fan, filters, humidifiers, coils, and dampers. From the air-handling unit, the conditioned air moves through ducts to diffusers in rooms needing heating or cooling. A separate return system of ducts moves air from these spaces back to the furnace for reconditioning, possibly exhausting some of the air and adding fresh air from outside the building. All of this is accomplished by a series of electrical controls.

Figure 46.30 Dual duct warm air systems use separate warm and cool air ducts that meet at a mixing valve to produce air at the temperature required.

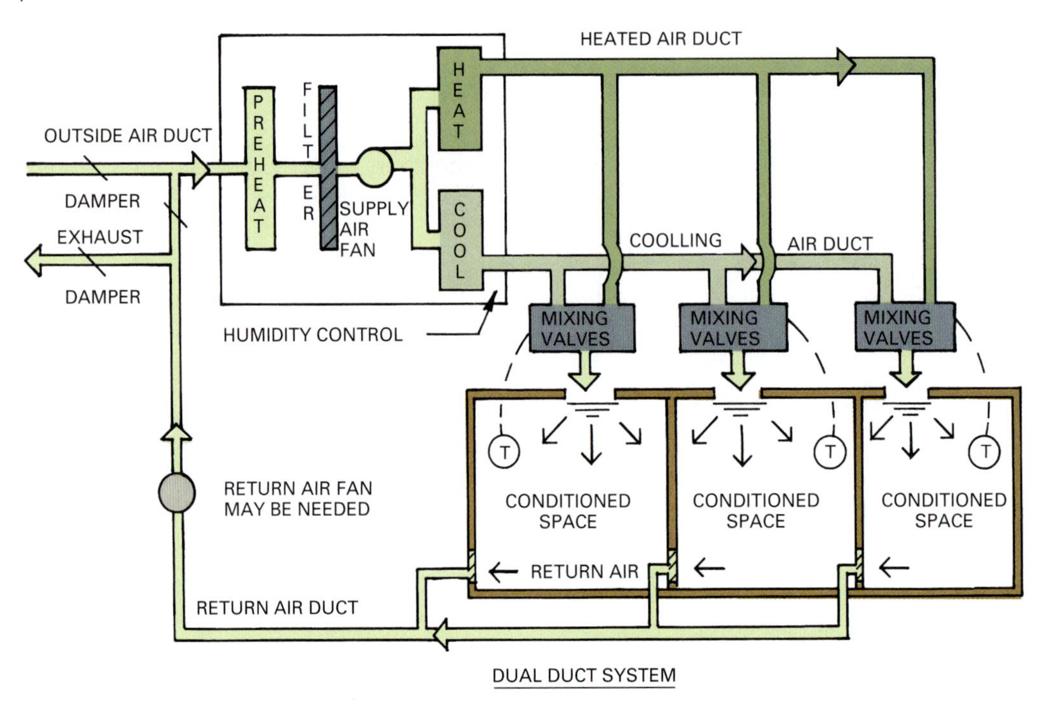

Ducts

Warm air distribution systems use ducts to move heated air from a furnace to diffusers (outlets) in various rooms. The design of the ducts and selection of materials are vital to a properly functioning system. A designer must carefully determine required sizes, ascertain proper air velocity, and calculate pressure.

Duct Pressure

The movement of air through a duct results in both velocity pressure and air pressure loads on the duct. A duct designer must calculate the actual static pressure on each section of a duct and specify the pressure classification. The pressure classifications for residential, commercial, and industrial ducts are shown in **Table 46.4**. The pressure classifications for residential, commercial, and industrial ducts are specified in inches of water. An inch of water in the inch-pound (I-P) system is a unit of head equal to a column of liquid water 1 inch high at 39.2°F. (Head is the energy per unit mass of fluid divided by gravitational acceleration.) The I-P system uses customary units (inches and pounds) rather than metric units.

A standard for duct design is published by the Air Conditioning Contractors of America (ACCA) Manual D, *Duct Design for Residential Winter and Summer Air Conditioning*. The Sheet Metal and Air Conditioning Contractors National Association (SMACNA) also has publications relating to duct design and installation.

Table 46.4 Allowable Static Pressure Classifications for Ducts in Various Applications

Application	Allowable Static Pressure
Residences	±0.5 in. of water
	±1 in. of water
Commercial systems	±0.5 in. of water
	±1 in. of water
	±2 in. of water
	±3 in. of water
	+4 in. of water
	+6 in. of water
	+10 in. of water
Industrial systems	Any pressure

Courtesy American Society of Heating, Refrigerating and Air-Conditioning Engineers, Inc.

Duct Classification

Duct systems are regulated by various laws, building codes, local ordinances, and standards. These must be considered by an engineer as a duct system is designed. Projects built for the federal government will have standards issued by various agencies, such as the General Services Administration and the Federal Construction Council.

Duct construction is classified in terms of pressure and use. Commercial duct systems include HVAC systems for applications in educational, business, general factory, and mercantile structures. Industrial duct systems include those used for industrial exhaust and air pollution control.

Model project specifications for the construction of ducts include Masterspec, produced by the American Institute of Architects (AIA), and Spectext, available from the Construction Specifications Institute (CSI). Residential ducts are specified by local building codes. An often used source for multifamily dwellings is National Fire Protection Association (NFPA) Standard 90A. Supply ducts may be galvanized steel, aluminum, or other materials rated by Underwriters Laboratory (UL) Standard 181. Rigid and flexible fiberglass supply ducts must meet the Fibrous Glass Duct Construction Standards of the SMACNA.

Commercial ducts are also usually regulated by NFPA Standard 90A and UL181. This classifies ducts into two groups:

Class 0. Zero flame spread, zero smoke spread

Class 1. 25 flame spread, 50 smoke developed

Class 0 ducts are made from iron, steel, aluminum, concrete, masonry, or clay tile. Class 1 ducts include many of the flexible and rigid fiberglass ducts manufactured. Industrial ducts are specified by NFPA Standard 91. These are used for duct systems that might convey flammable vapors or air containing various particles. Particle conveying ducts are available in four classifications:

Class 1. Nonparticulate applications (**makeup air**, general ventilation, and gaseous emission control)

Class 2. Moderately abrasive particles in the air (sanding or buffing)

Class 3. Highly abrasive material in low concentration (handling sand or abrasive cleaning)

Class 4. Highly abrasive particles in high concentration

Abrasive ratings are specified in *Round Industrial Duct Construction Standards* by the SMACNA. Industrial ducts are generally made from galvanized steel, uncoated carbon steel, or aluminum. Aluminum is not used if the air will contain abrasive particles. Those carrying corrosive vapors must have appropriate protective coatings.

Duct standards include many other specifications, such as insulation, type and spacing of hangers, welding when necessary, seismic requirements, ducts exposed to outdoor atmospheric conditions, and ducts below ground.

Ducts are available in round, rectangular, and flat-oval shapes. Specifications for these vary, depending on their use. They are often insulated to control heat loss or gain and to control condensation that occurs when moist air strikes a duct.

FORCED AIR DUCT SYSTEMS

Duct systems used in residential and small commercial buildings include the perimeter loop, perimeter radial, and extended plenum systems. The perimeter loop system is typically used with concrete slab floors and downflow furnaces or in a building with a basement or crawl space (Fig. 46.31). The perimeter duct is placed in the thickened edge of a slab. It is essential that the edge and bottom of the slab be insulated. Registers are placed along the perimeter duct as needed. The return ducts in this system are in the attic.

The perimeter radial system is also used in concrete slab construction but can be used in basements and crawl spaces, too. It uses a downflow furnace (Fig. 46.32). A variation of this system uses an upflow furnace with the radial ducts in the attic (Fig. 46.33).

The extended plenum system may have the furnace in the basement or on the first floor. It can also be used with horizontal furnaces in a crawl space or attic. The plenum is extended to provide needed airflow to feed ducts that run from it to outlets.

If the furnace is on the first floor, the plenum can be in the attic with ducts running over ceiling joists and diffusers running through the ceiling into the room (Fig. 46.34). Horizontal warm air furnaces are commonly placed below the floor or in the attic.

The ducts in residential and many small commercial systems are hidden from view, but in some large commercial and industrial buildings, they are left exposed. Typically these systems are used where large open areas exist, such as in an airport or a sports complex (Fig. 46.35).

Figure 46.31 A perimeter loop warm air duct system circulates air through a continuous duct system fed by several horizontal ducts.

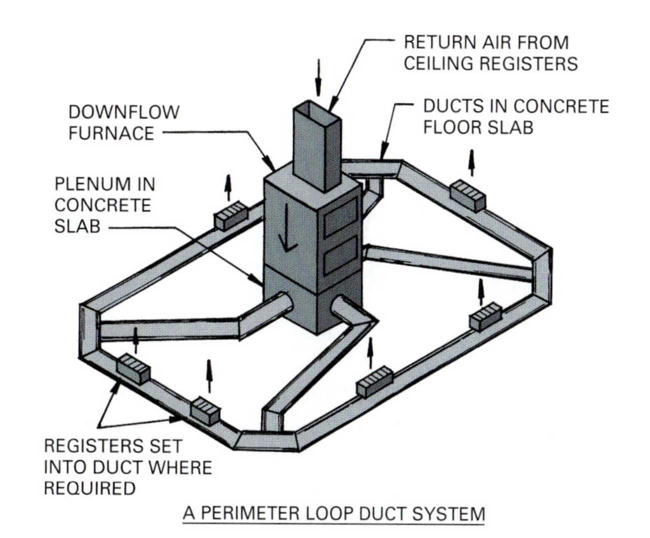

Figure 46.32 A perimeter radial duct system extends individual ducts to each space needing heating and cooling.

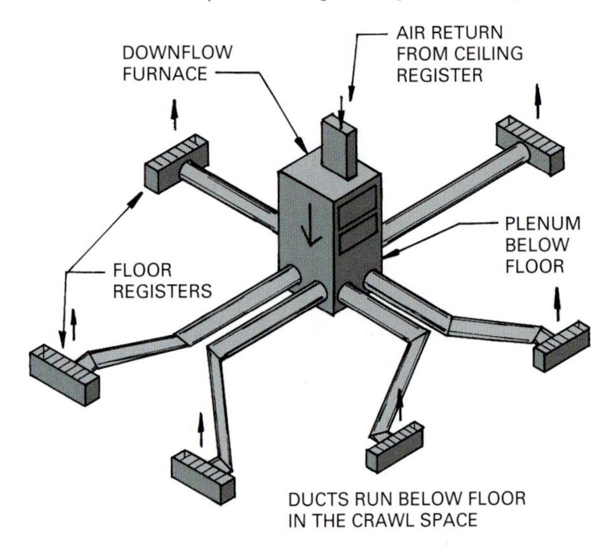

Figure 46.33 A perimeter radial duct system can place ducts in an attic and use ceiling diffusers.

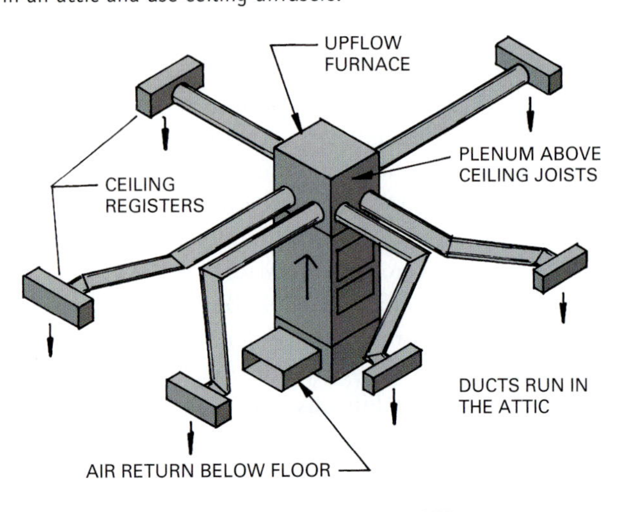

Figure 46.34 The extended plenum runs from the furnace, and individual ducts are taken off it to the rooms to be heated and cooled.

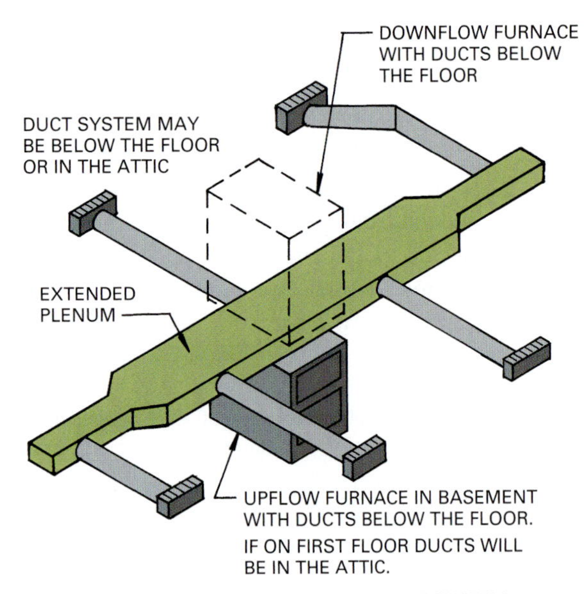

Figure 46.35 Exposed duct systems are used in buildings with large open areas. *(Image copyright Darryl Brooks, 2009. Used under license from Shutterstock.com)*

SUPPLY AIR OUTLETS AND RETURN AIR INLETS FOR WARM AIR HEATING SYSTEMS

The design and selection of **supply air** outlets and return air inlets is critical to providing adequate air conditioning in a room. Poorly located or poorly chosen outlet and inlet grilles and diffusers can reduce the effectiveness of even the best heating and cooling units. An engineer has a considerable number of factors to consider, and a decision is often one of compromise. Following are examples of some of the things to be considered.

Air velocity and air temperature in a supply duct must be greater than that permitted within a room. This air must therefore be emitted into the room so air velocity is not offensive and people are not exposed to areas of high temperature. This requires the system to mix the air supply with room air to control convection currents and uneven temperatures.

Air must be supplied to a room with a termination device (a grille or diffuser) that provides the required air dispersion pattern needed for human comfort. The termination device must throw the air across the room far enough to enable it to blend with the room air, thus reducing velocity and temperature before it drops into the room's lower part. Engineers know that cool air is heavy and settles to the floor, but warm air is lighter and rises toward the ceiling. The location of windows must also be considered in a design because they can introduce a source of cold air that can produce occupant discomfort and cool circulating warm air.

Surface effect, another thing to be considered, is caused when the airstream from a duct outlet moves across a ceiling or contacts a wall. This creates a low pressure area along a surface, keeping the air in contact with it for most of the **throw**. Circular ceiling diffusers project an airstream across portions of a ceiling and create surface effect if they are long enough to cover the ceiling area. Grilles may also cause some ceiling effect (Fig. 46.36).

Smudging occurs with both slot diffusers and ceiling diffusers. Smudging is a band of discoloration that eventually appears on ceiling material around the edge of a diffuser. It is caused by dirt particles in the discharged airstream. They are held in air turbulence around a diffuser and can cause smudging around the outlet. This can be alleviated by placing anti-smudge rings around outlets.

Figure 46.36 Air surface effect keeps the incoming air from a ceiling diffuser moving across the ceiling.

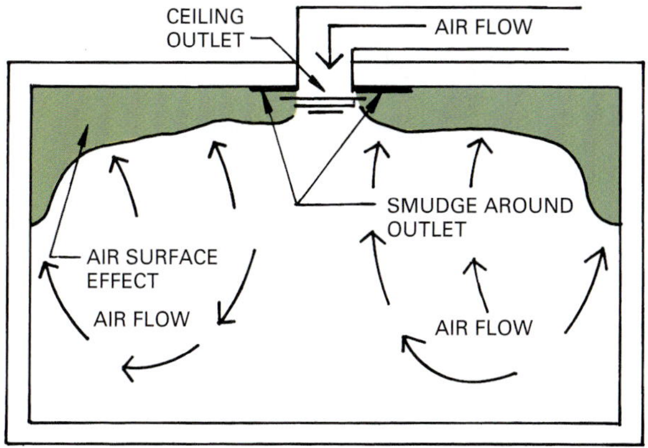

AIR SURFACE EFFECT WITH A CEILING DIFFUSER.

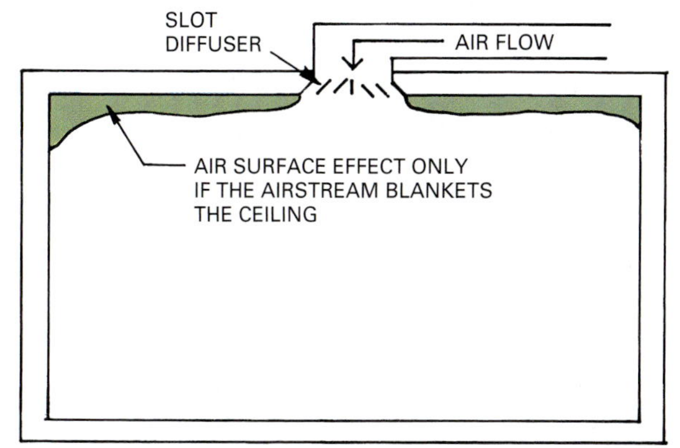

AIR SURFACE EFFECT WITH A SLOT DIFFUSER.

Another design consideration is an outlet's sound level. Outlets will transmit noise caused by mechanical equipment and high air velocities through a duct. High-pitched noise is frequently caused by air passing through an outlet, and a different design might need to be chosen. Recommended air velocities and decibel levels are found in publications of the American Society of Heating, Refrigerating, and Air-Conditioning Engineers.

Types of Supply Outlets

The basic types of supply outlets are grilles, slot **diffusers**, air distributing ceilings, and ceiling diffusers.

A grille is a rectangular opening with vanes that direct the airstream as it leaves an outlet. Adjustable grilles have adjustable vertical or horizontal vanes that deflect airstream in a horizontal or vertical direction. Some types have two sets of vanes at right angles to each other. Fixed bar grilles have vanes that cannot be moved. Variable area grilles (also called registers) have dampers that can vary the size of an outlet, thus varying the air that passes through (Fig. 46.37). Stamped grilles are simply a flat sheet with openings that permit air to pass with no control of direction or volume.

Slot diffusers are rectangular outlets with one or several slots, often installed in long lengths. Perpendicular slot diffusers discharge air perpendicular to or at a slight angle to the face of a diffuser. Parallel slot diffusers produce an airstream parallel with the face of a diffuser. Typical locations include in the sills of large windows and in the floor below glass doors or floor-length glass windows. They create an upflow of heated air over cold glass, producing increased comfort for occupants of a room.

Air distribution ceilings use the space between the bottom of the floor above and the top of a suspended ceiling as a plenum. The space is fed heated or cooled air by ducts located around its perimeter. The ceiling panels have perforations through which air enters the room below.

Ceiling diffusers may be round or square and contain a series of louvers that form air passages. They distribute the air supply in a uniform pattern around a diffuser (Fig. 46.38). Flush ceiling diffusers have concentric louvers that extend the same distance from the core. Stepped-down diffusers have louvers that project beyond the shell of a unit, as shown in Fig. 46.54. Perforated-face ceiling diffusers have a flat perforated face that is flush with and blends in with the surface of a suspended ceiling. They may have deflection vanes behind their exposed faces to direct airstream discharge. Variable ceiling diffusers have dampers that can be adjusted to control the amount of airflow from the diffuser.

Figure 46.37 Grilles may have fixed bars or adjustable vanes that vary the direction and amount of air flow from a duct.

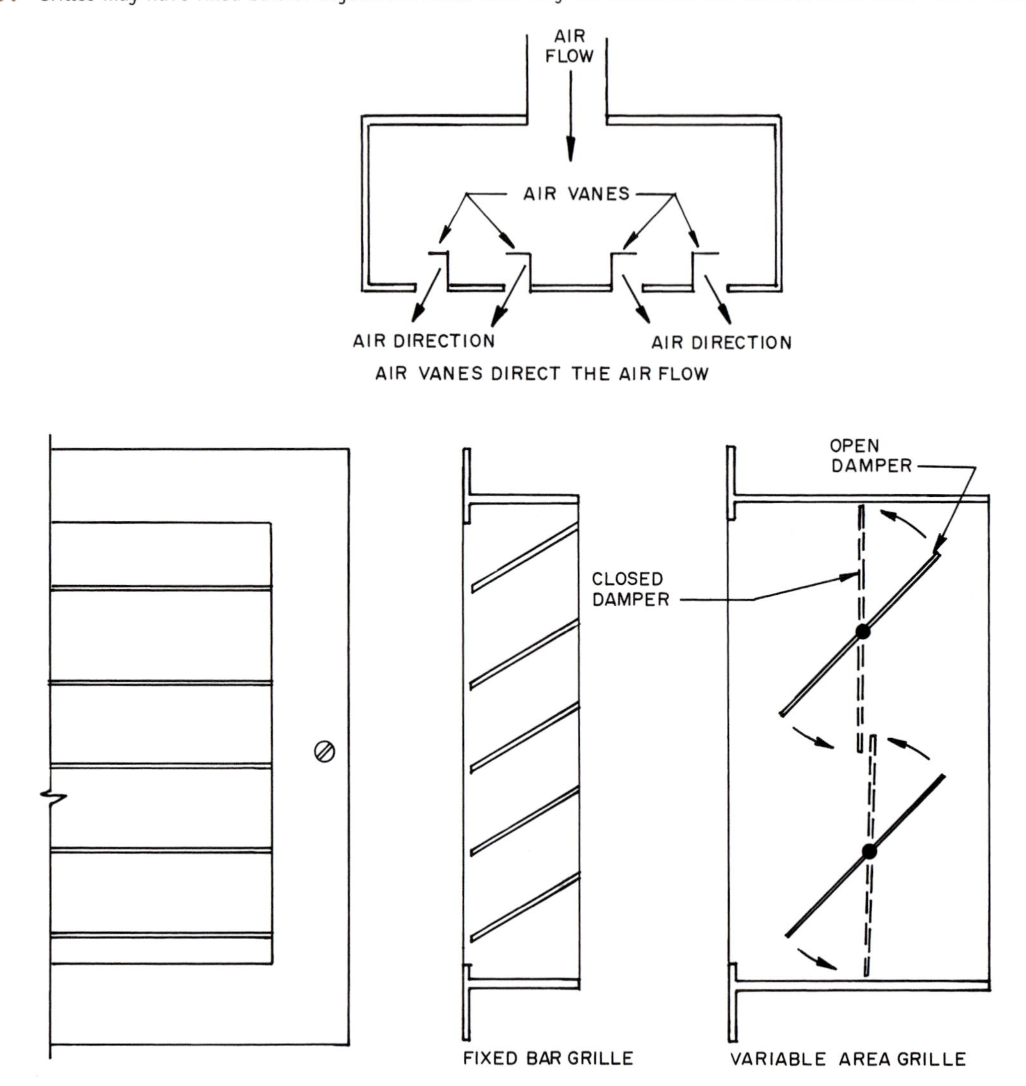

Figure 46.38 A round variable ceiling diffuser with dampers used to control the amount of air passing into the room.

Another type of ceiling air distribution diffuser is shown in Fig. 46.39. The supply duct runs above a suspended ceiling and supplies conditioned air to the air diffuser. The diffuser is a rectangular unit that rests upon the horizontal metal members of a suspended ceiling. Below this, parallel beams with spaces between them are used to distribute air into the room. Airflow direction is controlled in two directions by vanes. Diffusers are available in a variety of several sizes.

Diffuser Applications

Following are some general applications for the location of outlet and inlet diffusers. Outlet diffusers bring air into a space, and inlet diffusers are used with air returns or exhaust.

In general, in areas where air cooling needs are predominant, outlets will be most effective in the ceiling with air returns to the furnace on walls at the floor or in the

Figure 46.39 This diffuser fits into the grid of a suspended ceiling and is fed from a supply duct above the ceiling. *(Courtesy Chicago Metallic Corporation)*

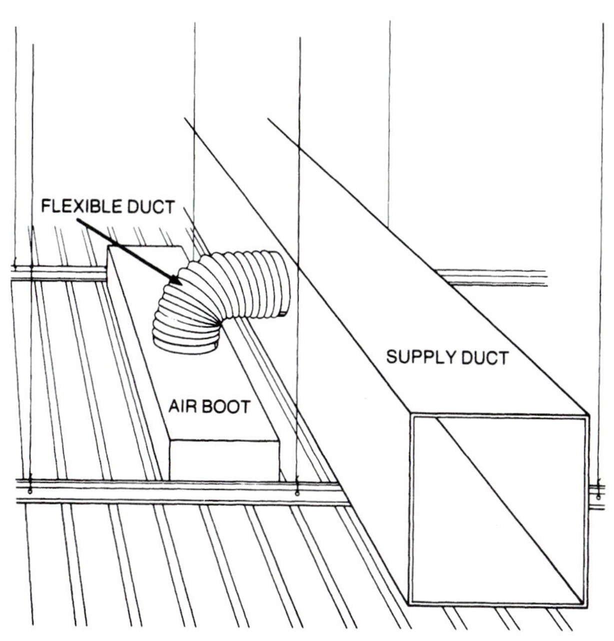

floor. This works because cool air sinks as it mixes with existing air in a room. In areas where the heating needs are predominant, outlets can be located floors or at the baseboard on walls. The return air can be high on the wall.

Outlet ducts are generally located on exterior walls and under windows and return air ducts on interior walls (Fig. 46.40). This permits a curtain of conditioned air to flow up the outside wall, warming or cooling its surface as needed to produce a more comfortable interior situation. A cold outside wall will pull body heat from an occupant even though the air temperature is adequate. Likewise a hot wall will radiate heat to an occupant.

Return Air Inlets and Exhaust Air Inlets

Return air inlets connect to ducts that return room air to a furnace for reconditioning. Exhaust air inlets move air from inside a building through ducts and discharge it outside. They typically do not require vanes to deflect or redirect the airstream. Some types do have dampers to control air flow amounts. Fixed bar grilles are commonly used. Bars may be situated on a slight angle or perpendicular to a grille's face (Fig. 46.41). Fixed bar louvered grilles used on the exterior of large commercial buildings for exhaust and fresh air intake can be from 30 to 60 feet in width and length.

Figure 46.40 Outlets under windows merge warm air with cold winter air or warm summer air with cool conditioned air to provide increased comfort to the occupants near the window.

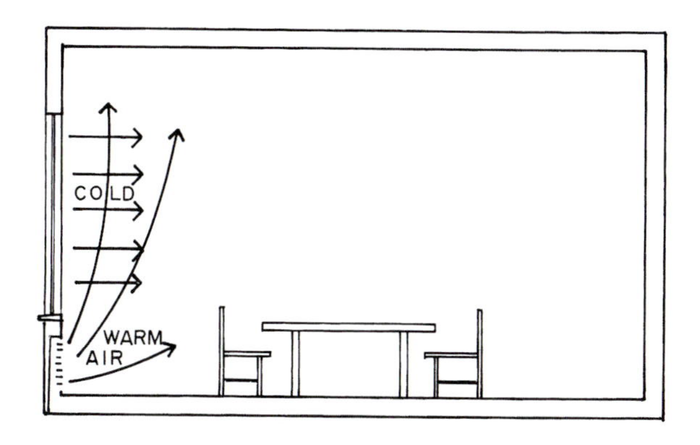

Figure 46.41 A return air grille installed in the ceiling.

Construction Methods

Building Commissioning

Building commissioning is the process of verifying that all HVAC, plumbing, electrical, and building fire and security systems are installed, calibrated, and operating as intended (by a building's owner) and as designed (by a building's architects and engineers). The process involves procuring the services of a qualified Commissioning Authority (C × A) to conduct and verify commissioning activities.

The basic procedure for building commissioning encompasses a comprehensive pre-construction review of design documents for compliance with Owner's Project Requirements (OPR), periodic site observations during the construction phase, and systems performance testing at project completion. The commissioning team begins by documenting the design requirements for systems to be commissioned prior to approval of contractor submittals for equipment. Commissioning requirements are then incorporated into the construction documents. These include detailed specifications for submittal review, construction verification, and functional performance testing procedures.

During the construction process, the commissioning authority inspects system installations to ensure all components are properly set up and any problems discovered are corrected prior to performance testing. Functional performance testing begins once all systems are in place, powered, and programmed. Testing procedures include a thorough review of all operations, including start up and shut down procedures, systems balancing, and emergency and failure modes of the various systems. A variety of methods are used to simulate operations and evaluate whether or not all systems perform as specified in construction documents. Any deficiencies discovered are reported to the owner in a summary commissioning report, and a plan of action is formulated to correct them.

While the practice of building commissioning is still fairly new in the construction industry, it has quickly become common practice. The LEED rating system mandates fundamental building commissioning as a prerequisite for certification and gives additional points for enhanced commissioning. The practice of commissioning results in a number of tangible benefits, including reduced energy use, improved occupant comfort, and reduced operating costs. The ultimate goal is to deliver for owners a project on schedule, under budget, and fully optimized at building occupancy.

HUMIDIFIERS AND DEHUMIDIFIERS

Control of a space's relative humidity is an important factor in the overall conditioning of its environment. Requirements vary depending on occupancy. **Humidity** describes the water vapor within a space. **Relative humidity** is a ratio of water vapor weight actually in the air to the maximum possible water vapor weight the air could contain at a given temperature; this ratio is expressed as a percentage. For example, if relative humidity is 100 percent, the air can hold no more water vapor, and an increase will cause moisture to condense and form water drops.

Human comfort depends a great deal on the relative humidity of air. Typically, indoor relative humidity should be kept between 30 to 60 percent. Low humidity causes drying of nose, throat, skin, and hair membranes. Furniture, cabinets, interior trim, and other wood products can shrink and check if the relative humidity is too low. Likewise, high relative humidity can cause doors and drawers to swell and stick. Heating air during winter removes moisture, so a humidifier is used to increase the relative humidity. In the summer, air in many geographic areas has a high relative humidity. An air-conditioning system must then dehumidify (remove water vapor from) the air.

Humidification Equipment

Humidifiers increase the amount of water vapor in air. They operate in two different ways. One type adds heat: as air flows through the humidifier's wet section, it picks up moisture, and the air is then distributed to the room. The other type takes heat from the air flow and evaporation occurs in the room. Typical types of humidifiers in residential and small commercial buildings are shown in Fig. 46.42.

A pan humidifier has water automatically fed into the pan. Vertical absorbent plates rest in the water and pass off moisture to the air as it passes over them, as shown in Fig. 46.42 part A. The water level is automatically controlled. The pan unit may have an electric heater that increases the rate of evaporation.

Wetted type humidifiers circulate air through or over a fibrous, porous material wetted by a spray, gravity water flow over it, or revolving through a water pan.

Figure 46.42 Types of humidification systems used in industrial buildings. *(Courtesy American Society of Heating, Refrigerating, and Air-Conditioning Engineers, Inc.)*

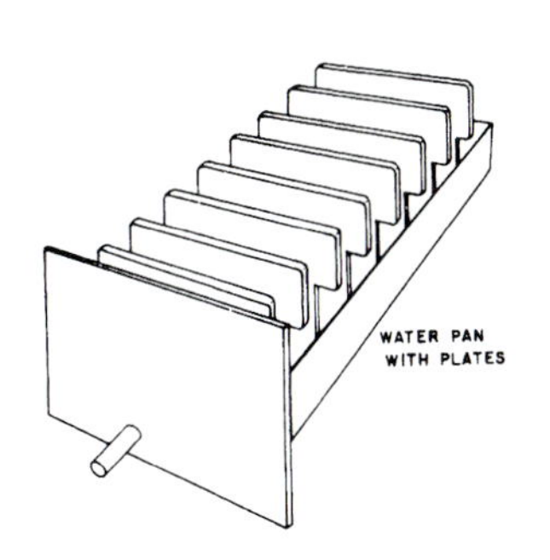

A. Pan Humidifier

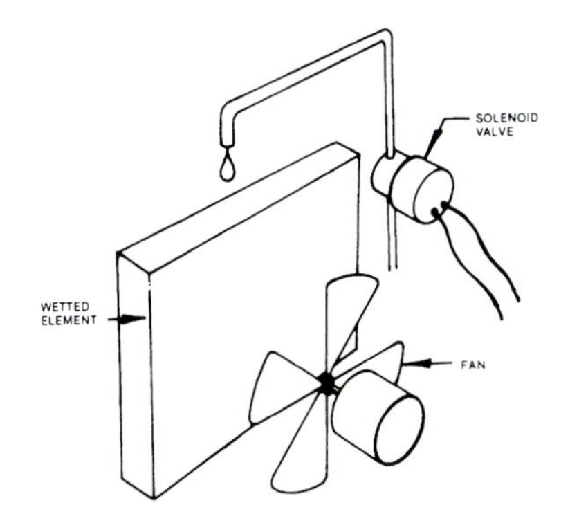

B. Power Wetted Element Humidifier

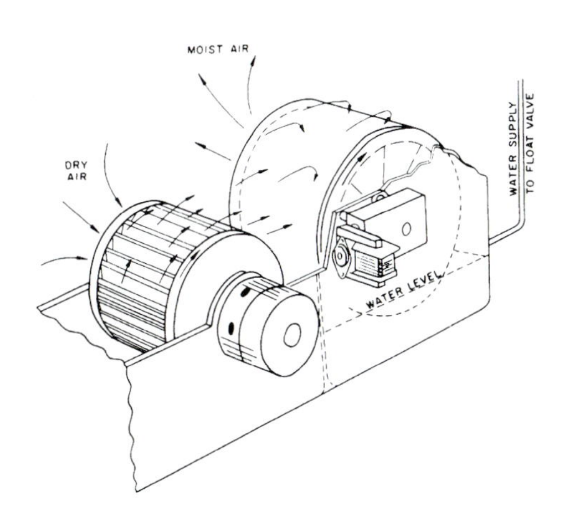

C. Wetted Drum Humidifier

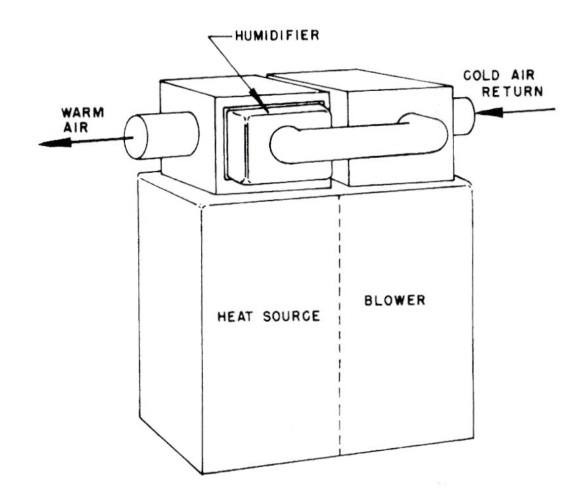

D. Bypass Wetted Element Humidifier

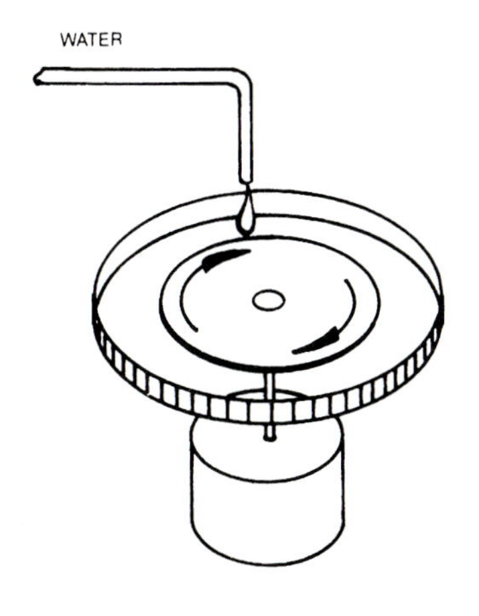

E. Atomizing Humidifier

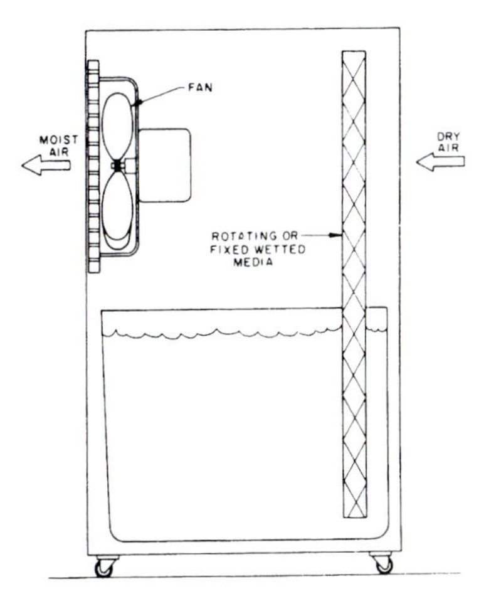

F. Appliance Portable Humidifier

One type uses a fan that moves air from the furnace plenum and blows it through the pad (Fig. 46.42 part B) or rotating drum (Fig. 46.42 part C) and back into the plenum.

A bypass humidifier is shown in Fig. 46.42 part D. The humidifier mounts on the furnace air supply and uses air moved by the furnace blower. Small humidifiers are available that can be mounted in ducts.

A spinning disc atomizing humidifier is shown in Fig. 46.42 part E. Small particles of water are injected into the airflow and absorbed, raising the humidity. Another type uses a spray nozzle to inject a mist into the airflow.

A small portable humidifier is illustrated in Fig. 46.42 part F. It has a rotating belt that moves through a reservoir of water. A fan blows air through the belt, where it picks up moisture, and distributes the air into a room.

Various industrial humidifiers are shown in Fig. 46.43. The pan humidifier shown can be heated by hot water, steam coils, or electric element. It may be installed on a duct as shown or operated in a remote location with hot water vapor carried to the main duct system by a connecting duct. Other types use an enclosed grid (Fig. 46.43 part B) or inject steam directly into the duct (Fig. 46.43 part C). The jacketed steam humidifier in Fig. 46.43 part D operates off a constant steam supply flowing into a steam trap in which any condensation drains off. The flow into the dispersing tube in the duct is controlled by a steam valve.

A self-contained steam humidifier is shown in Fig. 46.43 part E. It uses electric heaters to convert water to steam and then injects it into the airflow in the duct. Various types of atomizing humidifiers (Fig. 46.43 part G) are also used in industrial applications.

Desiccant Dehumidification

Dehumidification involves the removal of water vapor from air, gases, or other fluids by some mechanical or chemical means. A **dehumidifier** is a device used to remove moisture from the air, thus reducing the relative humidity.

Desiccation is the use of a desiccant to remove moisture from a material. A **desiccant** is any absorbent liquid or solid used to remove water or water vapor from a material. Dehumidification equipment uses both solid and liquid desiccant materials. **Absorbent** materials will extract substances from a liquid or gas medium with which they are in contact. **Adsorbent** materials have the ability to capture molecules of gases, liquids, or solids on their surfaces without changing the adsorbent material chemically or physically.

Dehumidification is typically accomplished in residences by the cooling of air during air-conditioning mode. For commercial and industrial applications, other methods are available.

Dehumidification is required for many industrial applications. They include maintaining a dry atmosphere in a warehouse, producing dry air to aid in the drying of a material in an industrial process, drying natural and liquified gas, and lowering the relative humidity in a plant manufacturing products using hygroscopic materials, such as wood.

Liquid desiccant dehumidification systems remove moisture by passing air through a mist-like spray of a liquid desiccant, such as a glycol solution. The desiccant spray absorbs moisture from the air and passes into a regeneration chamber, where it is heated and gives off the moisture. The moisture is discharged by outside air flow (Fig. 46.44).

Solid sorption dehumidification systems remove moisture by flowing air through a granular desiccant, such as silica gel or hygroscopic salts. Sorption is a general term used to include both absorption and adsorption. The granular desiccant has a vapor pressure below that of the vapor in the air. This difference in pressure drives water vapor in the air into the desiccant. When the desiccant becomes saturated, it is heated and dried in a reactivation chamber and can be reused (Fig. 46.45).

HYDRONIC (HOT WATER) HEATING SYSTEMS

Hydronic heating systems are used in residential and commercial installations. **Hydronics** refers to the use of water as the heat-transfer medium in heating and cooling systems. A system consists of a boiler fueled by oil or natural gas, and a system of pipes, radiators, pumps, and controls. Three types of systems are typically used in residential and small commercial buildings. These are one-pipe, two-pipe direct-return, and two-pipe reverse-return systems.

One-Pipe Systems

One-pipe systems use a single loop of pipe to circulate water at 180°F (82.9°C) to radiators and return cooler water to a boiler for reheating and recirculating, as shown in Fig. 46.46. Notice the water flows directly from the boiler to radiator 1, from which it flows into radiator 2. The water temperature at radiator 2 will be

Figure 46.43 Types of humidification equipment commonly used in residences. *(Courtesy American Society of Heating, Refrigerating, and Air-Conditioning Engineers, Inc.)*

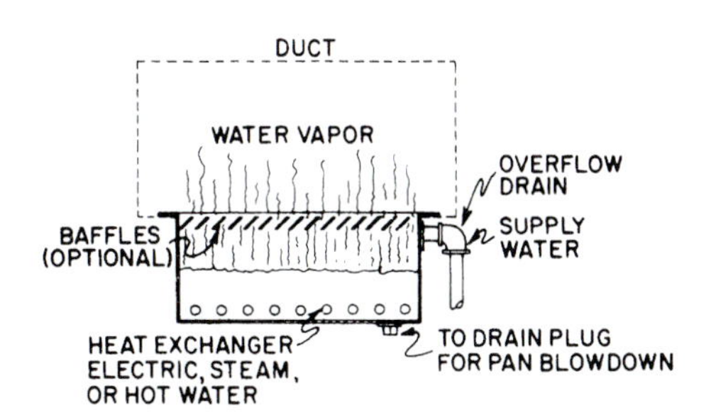

A. Heated Pan Humidifier

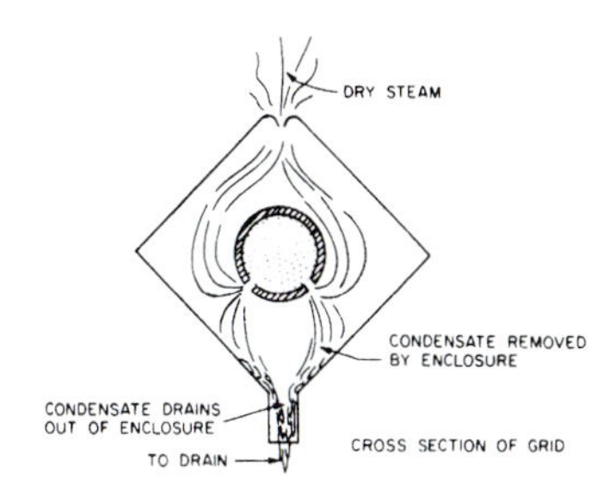

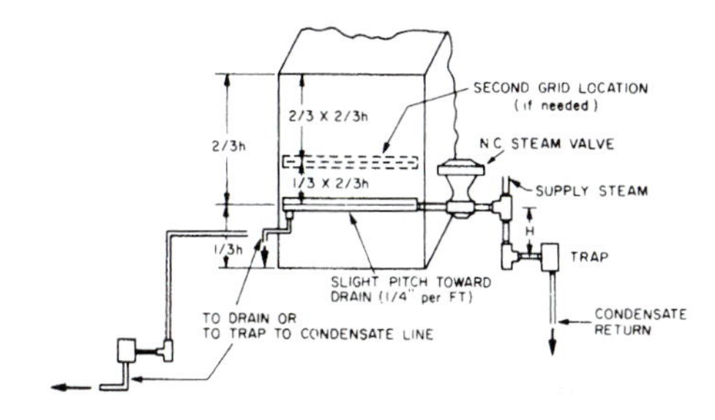

B. Enclosed Steam Grid Humidifier

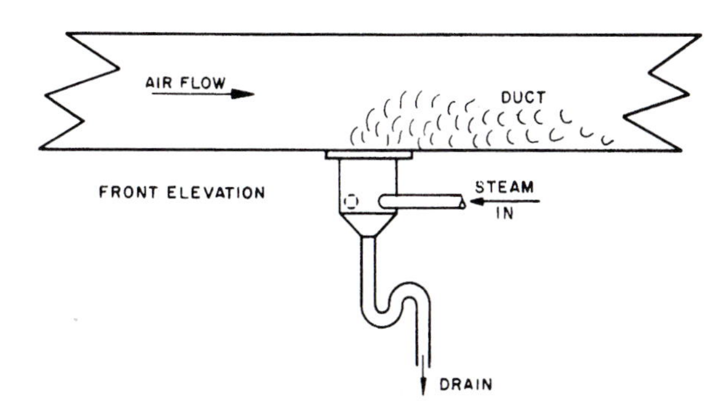

C. Cup Steam Humidifier

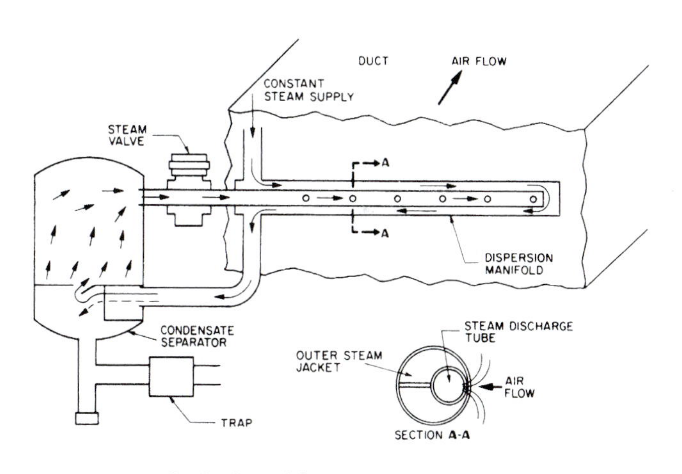

D. Jacketed Steam Humidifier

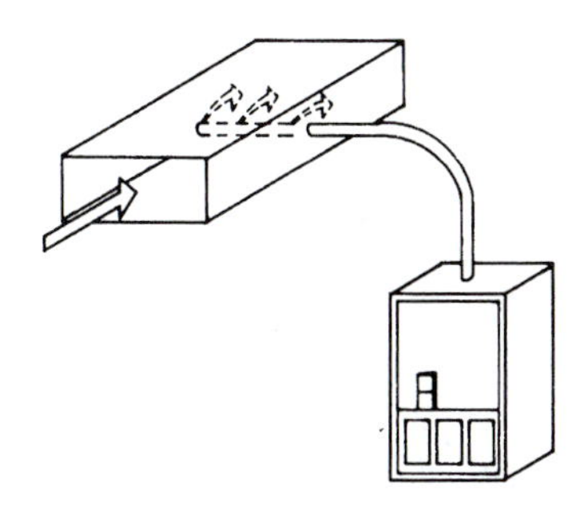

E. Self-contained Steam Humidifier

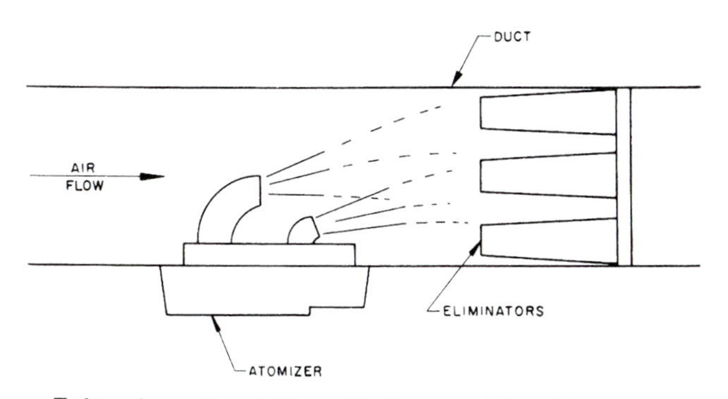

F. Atomizing Humidifier with Optional Filter Eliminator

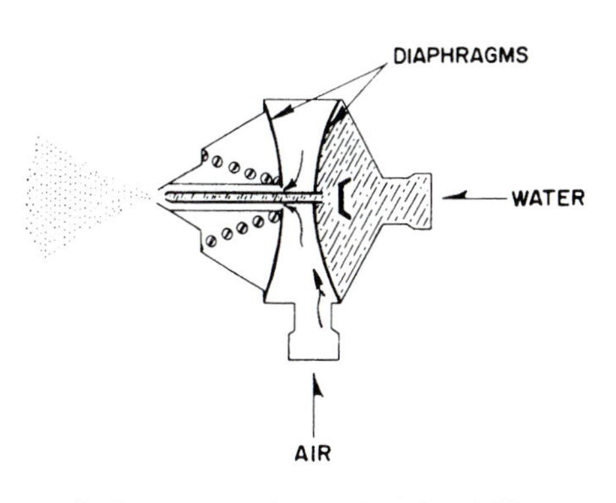

G. Pneumatic Atomizing Humidifier

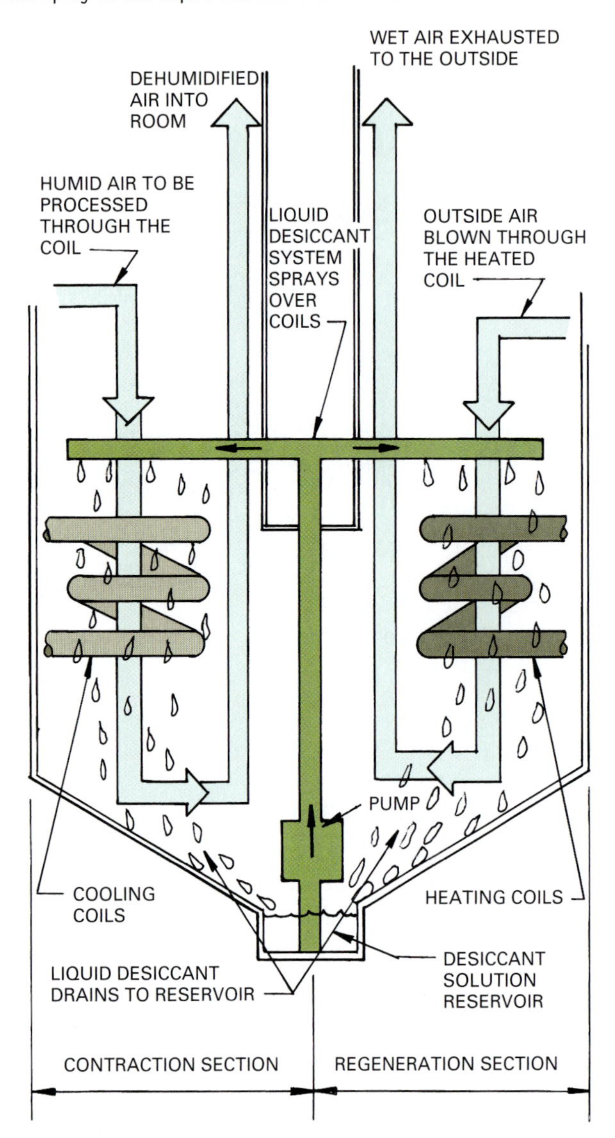

Figure 46.44 Liquid desiccant dehumidification uses a mist-like spray of the liquid desiccant to absorb moisture in the air.

lower than at radiator 1. This loss can be compensated for by having a larger No. 2 radiator, but the loss continues through the remainder of the radiators on the loop. This system is difficult to keep in balance.

Two-Pipe Direct-Return System

Each of the radiators in this system receive hot water directly from a boiler through the hot water supply line. The cool water from each radiator returns directly to the boiler through a separate loop of pipe. The last radiator tends to receive less water because of higher pipe resistance to water flow. It has the longest hot water supply and cool water return lines. This is overcome by increasing the size of the hot water supply line and sizing the circulation pump to provide needed flow

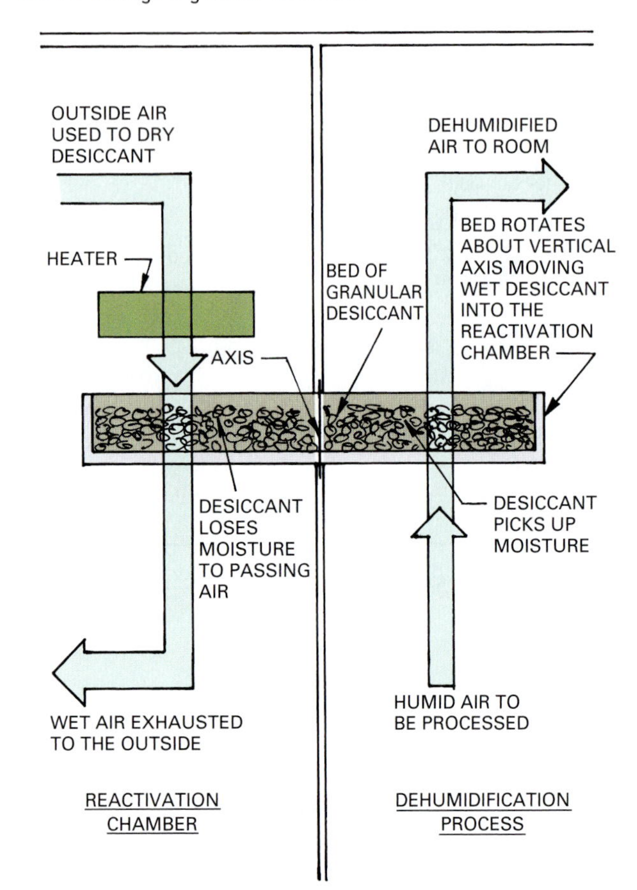

Figure 46.45 Solid sorption dehumidification flows the humid air through a granular desiccant.

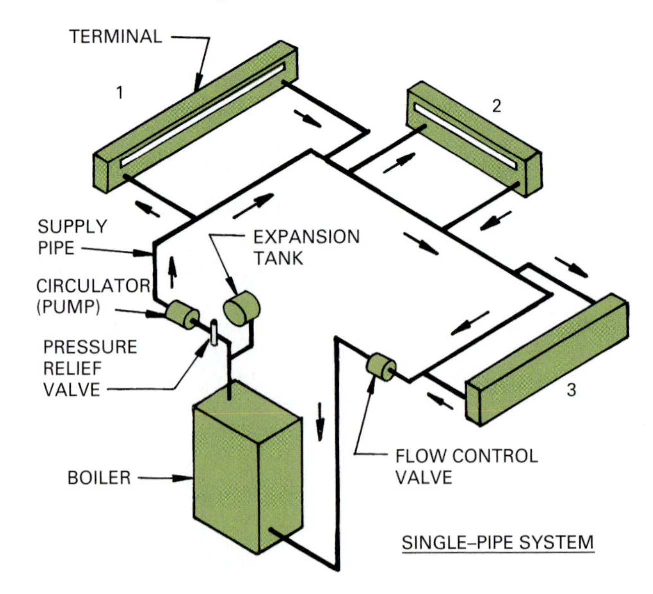

Figure 46.46 A single-pipe hydronic system carries water from the boiler through each terminal and back to the boiler.

at the radiators on the end of the circuit (Fig. 46.47). Balancing devices are used to regulate hot water flow through each radiator. In general, this system is not widely used because of the difficulty in getting a balanced distribution.

Figure 46.47 The two-pipe hydronic direct-return hot water system has a separate return line for the cooled water, which flows in a direction opposite the supply line.

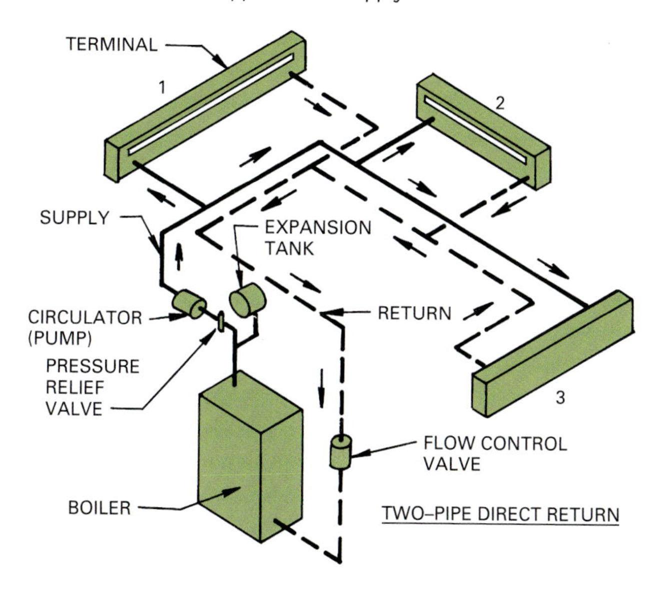

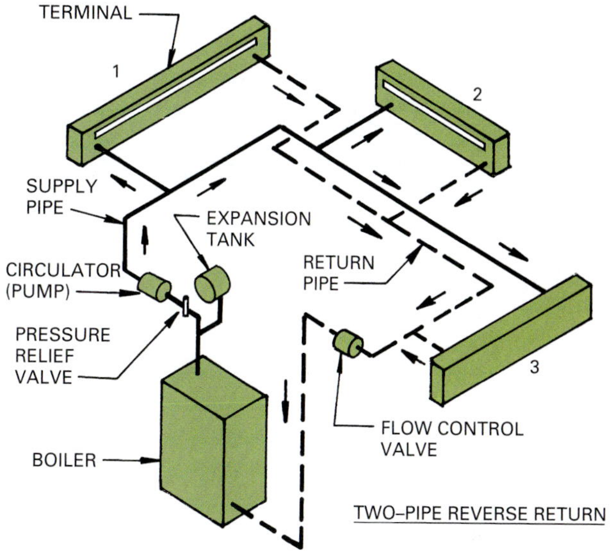

Figure 46.48 The two-pipe reverse-return hydronic hot water system has the return water flowing in the same direction as the hot supply water, providing about the same pipe resistance to water flow for each terminal.

Two-Pipe Reverse-Return System

This system is much like the two-pipe direct-return, except the return water flows in the same direction as the hot water. This provides a system having about the same pipe resistance on all radiators. For example, radiator 1 has the shortest hot water supply line but the longest return line. The reverse is true for radiator 3. This system is typically used in larger buildings with longer pipe runs (Fig. 46.48).

Multizone Two-Pipe Systems

A multizone system enables the temperatures in various parts of a building to be controlled separately. Each zone has a complete two-pipe system fed from a central boiler. The flow of hot water is individually controlled to each zone. This permits some zones to be kept at lower temperatures when not in use, resulting in a saving of energy costs. Large multistory and multi-use buildings will use multizone two-pipe hydronic systems (Fig. 46.49).

Hydronic Controls

The boiler's water temperature (180°F or 83°C) is controlled by a thermostat immersed in the boiler water. The water thermostat regulates the operation of the oil or gas burner. A room thermostat is used to start and stop the circulation pumps regulating the supply of water to the radiators. The boiler is often used to supply hot water for domestic use. In this case, it must be sized to meet both heating and domestic water demands.

Expansion Chambers

Hydronic heating systems require some form of expansion chamber (tank). This serves as a space into which uncompressible water can flow as it expands within the system due to temperature increases. Three types of expansion systems are commonly used: open tank, closed tank, and a diaphragm tank (Fig. 46.50).

The open expansion tank is located above the highest radiator and open to the atmosphere. It must be large enough to hold the maximum amount of water produced. They can cause poor performance and are not generally used.

The closed expansion system has a set volume of air that can be compressed and water the tank will hold as the air is compressed. The air pressure changes as the volume of water increases or decreases. The only time water can flow into or out of the tank is when the water expands or shrinks. The tank is sized to handle expansion due to the maximum temperature of the water. The size is determined by equations available in the ASHRAE publication *HVAC Systems and Equipment*.

The diaphragm expansion system has a flexible membrane in the tank to separate air and water. As the expansive water enters the tank from the top, the diaphragm moves down, compressing the air. This system is widely used.

Figure 46.51 shows a simplified illustration of a multistory hot water heating system showing both closed and diaphragm-type expansion systems. Each terminal unit has a vent that opens automatically to relieve excessive pressure.

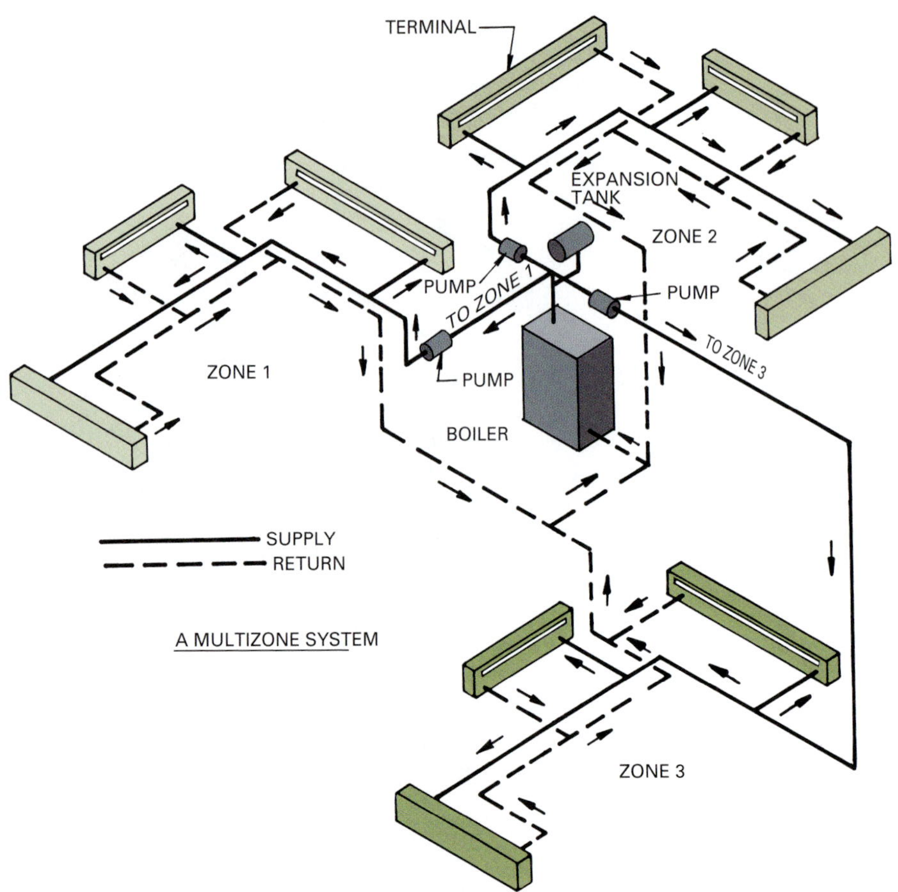

Figure 46.49 A multizone two-pipe hydronic system is used to provide different temperatures in each zone.

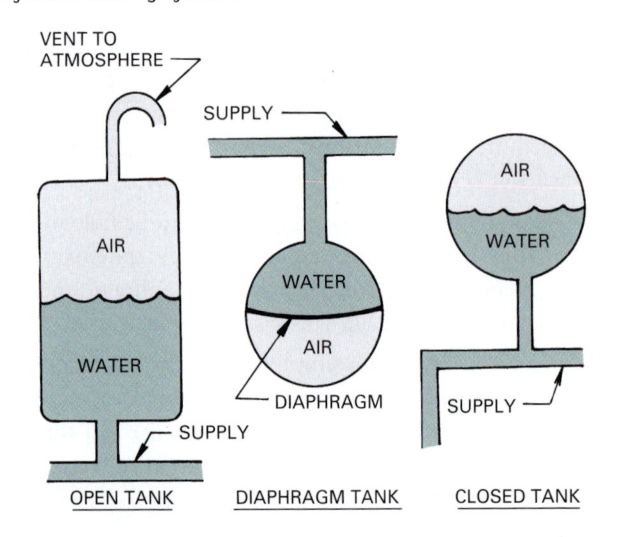

Figure 46.50 Types of expansion chambers used with hydronic heating systems.

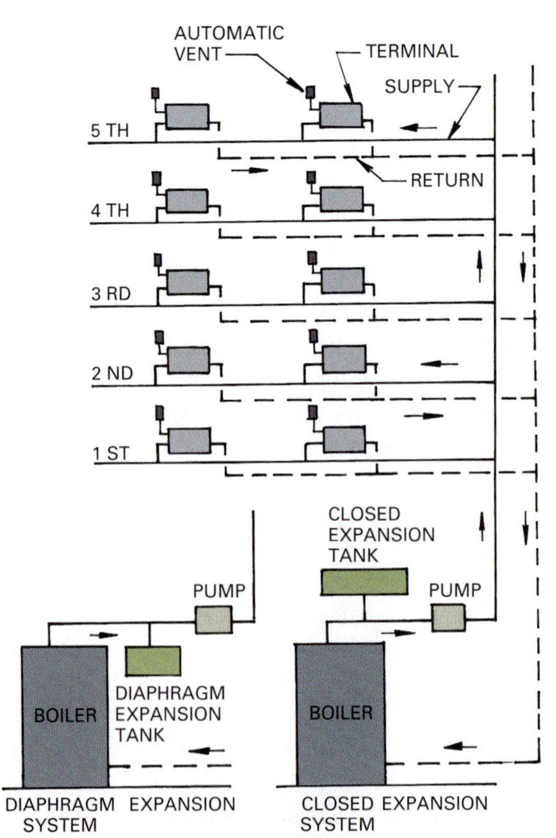

Figure 46.51 A simplified illustration of a multistory hot water heating system showing the diaphragm and closed expansion systems.

PIPE SYSTEMS FOR WATER HEATING AND COOLING

Water distribution systems can be used for heating with hot water and cooling with chilled water. The systems are usually closed and use circulators (pumps) to move water through the system and the terminal units. Hot water is supplied by a **boiler** and the cold water by a **chiller.** A chiller is a refrigerating machine used to remove heat from water that is circulated for cooling air in a building. Three-pipe and four-pipe distributions are used.

Three-Pipe Systems

Three-pipe systems provide heating and cooling supply to terminal units by running a heating supply pipe to each terminal. The third pipe holds the return in which the hot and chilled water are mixed and returned to the boiler and chiller, which results in warmed chilled water going into the chiller and cooled warm water going into the boiler. This results in increased costs to reheat and re-chill the water before it is recirculated.

Four-Pipe Systems

Four-pipe systems are used when a system provides both heating and cooling modes. In this system, each terminal receives separate hot water and chilled water supply and return lines, providing required heating and cooling any time either is needed. Systems that use the same coil for heating and cooling control the flow with two valves on each of the heating and cooling supplies at each terminal. Others put separate heating and cooling coils in the terminal unit (Fig. 46.52).

Figure 46.52 A four-pipe hydronic system providing both heating and cooling modes.

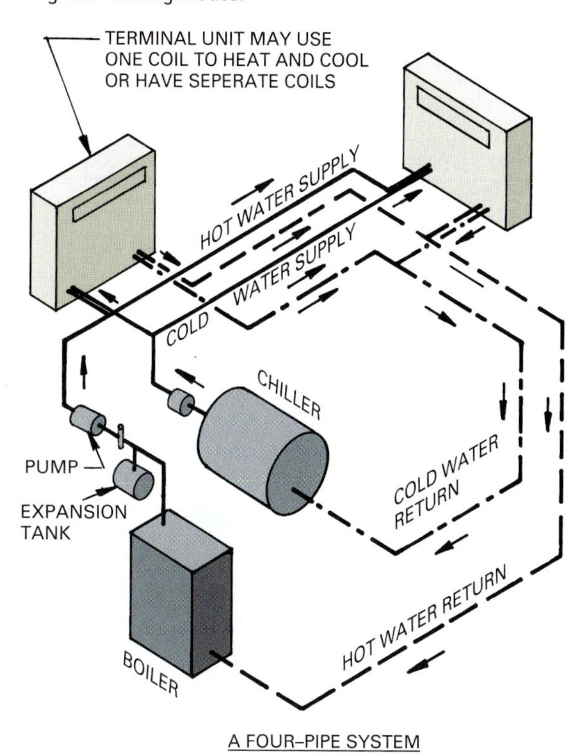

STEAM HEATING SYSTEMS

A steam heating system has a boiler or other steam-generating device, a piping system, radiators or **convectors**, and controls. The boiler is usually oil or gas fired, but coal, wood, waste products, solar, nuclear, electrical energy, or cogeneration sources can also be used. The two types of steam heating systems are one-pipe and two-pipe.

Steam delivers considerably more heat per pound (0.45 k) than a pound (0.45 k) of water, but when it becomes vapor, steam expands much more than hot water. Therefore, a steam system requires larger diameter pipes than do hot water systems. Steam produces high pressures that force it through a piping system without the pumps used in hot water systems. Pipes must also be sized to allow gravity flow of the condensed water back to the boiler without interfering with the steam flow. Engineers use the design data for pipe sizes available in the ASHRAE Handbook. Steam space heating systems are usually classified as low pressure.

It is more difficult to control the temperature of steam than hot water, so hot water is more widely used for space heating systems. If a building requires steam for an industrial process, the steam supply is generally used for space heating. This is often high-pressure steam requiring the use of pressure-reducing valves in series to achieve acceptable pressures for space heating.

Two-Pipe Steam Heating Systems

A two-pipe gravity return system is shown in Fig. 46.53. It has separate piping for the steam supply and the return of the **condensate** to the boiler. Thermostatic traps on the outlet line of each radiator or other terminal unit keep the steam within the unit until it has dispersed its latent heat. Then the trap opens, the condensate flows out the return line, and additional steam enters the terminal unit.

Two-pipe steam systems use either gravity or mechanical returns. Gravity flow two-pipe models are used only in small systems. Most systems use higher steam pressures to supply steam to terminal units and a condensate pump or vacuum pump to return condensate to the boiler.

Figure 46.53 Two-pipe steam heating systems have separate piping lines for the steam supply and the return condensate.

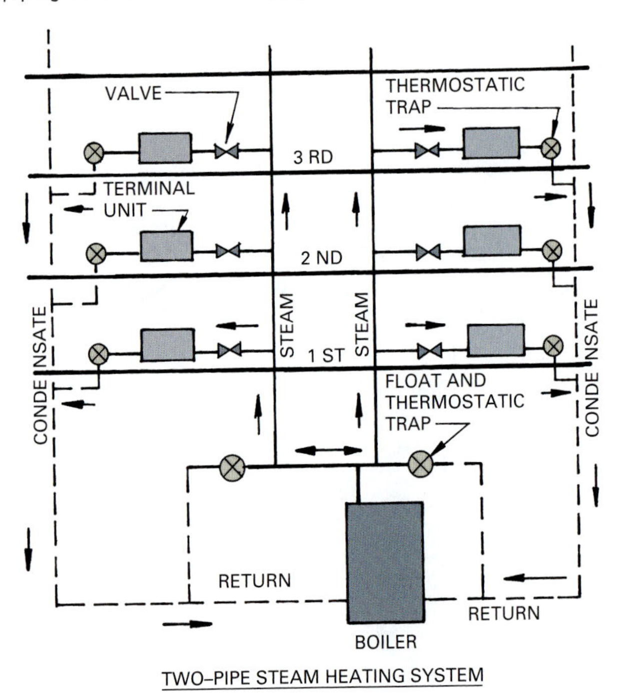

Figure 46.54 The two-pipe vacuum steam heating system uses a vacuum pump to increase the pressure difference and discharges gas that is not condensable to the atmosphere.

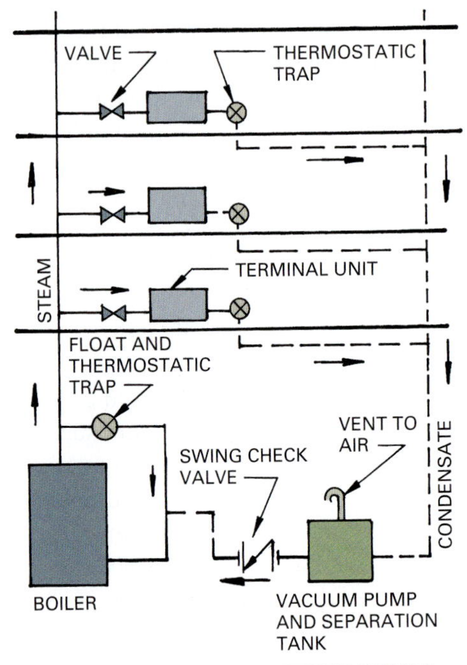

A two-pipe vacuum system is similar to the two-pipe mechanical return system but has a vacuum pump added to provide additional pressure difference. It circulates condensate, removing uncondensable gas and discharging it to the atmosphere, which creates a vacuum in both the steam supply line and the condensate return line. The vacuum return system is used on larger buildings because it requires a lower steam pressure and can fill a system with steam rapidly (Fig. 46.54).

BOILERS

Boilers are used to transfer heat from a fuel source to a fluid, such as water. The liquid is contained in a cast iron, steel, or copper pressure vessel that transfers heat to water to produce hot water or steam (Fig. 46.55). Boilers are constructed according to the American Society of Mechanical Engineers (ASME) *Boiler and Pressure Vessel Code*. The Hydronics Institute publishes the *Testing and Rating Standard for Heating Boilers*.

Boilers are classified by working pressure, temperature, fuel, size, and whether they are steam or water boilers. Steam boilers are used for space heating and auxiliary uses, such as in commercial laundries or industrial processes in which steam is required. Steam is typically used in large commercial and industrial buildings. Hot water boilers are used for space heating and domestic hot water supply. They are typically used in residential and small commercial buildings.

Boilers are classified as high-pressure or low-pressure. High-pressure boilers are referred to as power boilers. They produce steam pressures above 15 psi and hot water pressures and temperatures above 160 psi and 250°F (122°C). They are typically of steel construction and use firetube or watertube designs.

Low-pressure boilers, referred to as heating boilers, are limited to a maximum steam pressure of 15 psi, a maximum hot water pressure of 160 psi, and a temperature of 250°F (122°C). They are made from cast iron, steel, or copper.

Firetube boilers have hot combustion gases pass through tubes surrounded by water. They are used in both low- and high-pressure boilers. Watertube boilers have water inside tubes, and hot combustion gases pass over the tubes. Some boilers have the tubes in horizontal or nearly horizontal position, while some situate tubes vertically. Small, low-pressure boilers used for residential units may have helical coils. Watertube boilers are available ranging from small low-pressure units to large high-pressure steam units.

Figure 46.55 This is an oil-fired cast-iron wet base boiler. The letter "F" points to the cast-iron boiler sections that are assembled forming the boiler sections on all sides of the heat source. *(Courtesy Burnham Corporation, Hydronics Division, Lancaster, PA 17604)*

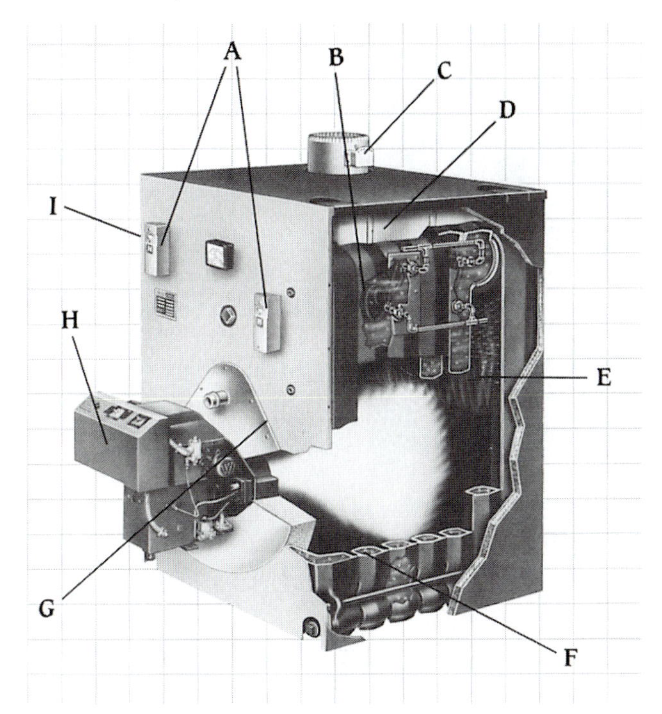

PF-5 Series Features

A. Front-mounted controls for easier adjustment and maintenance
B. Tankless heater for optimum domestic hot water
C. Adjustable lock-type damper for improved efficiency
D. Aluminized steel flue canopy for long life
E. Cast iron vertical flue design for maximum heat
F. Wet base thermal pump construction for improved circulation
G. Burner mounting plate with flame observation port
H. Four manufacturer burner options to best fit your needs
I. Left side cleanout for easy entrance to all flue surfaces

Boilers are available as condensing and non-condensing units. For many years, boilers operated so they did not condense flue gases in the boiler. This was done to prevent cast-iron and steel parts from corroding. Non-condensing boilers are operated with a high flue gas return temperature to prevent moisture in the flue gas from condensing on cast-iron and steel parts.

Condensing boilers salvage latent heat from the products of combustion by condensing the flue gas, increasing the efficiency of the boiler. This means lower flue gas temperatures can be used and the water vapor in the flue gas can condense and drain. The condensing section of the boilers must be made of materials that will withstand the required temperatures and resist corrosion. Certain types of stainless steel are used.

A pair of high-efficiency cast-iron wet base boilers is shown in Fig. 46.56. They operate on oil or natural/lpropane gas, and models are available for hot water and steam heatingsystems. Some types may have a combination gas/oil-fired burner/boiler unit—these are typically used in schools, apartments, and commercial buildings, where a switch in fuels may be essential for continued heating service.

Cast-iron boilers are made by assembling individual cast-iron boiler sections, called watertubes, in which water for heating flows. The size and rating of a boiler depends on the number of sections assembled. Tubes may be vertical or horizontal. The boiler may be dry base, which means the firebox is beneath the fluid-backed sections, or wet base, which has the firebox surrounded by fluid-backed sections on all sides. Wet leg boilers have the firebox surrounded on the top and sides by fluid-filled boiler sections.

Steel boilers are made by welding the parts forming the water chambers. One type of heat-exchange surface may be slanted, horizontal, or vertical firetubes. They may be dry base, wet leg, or wet base design.

Copper boilers use finned copper tube coils run from headers or serpentine copper tubing coil. These are usually fired by natural gas. Electric boilers consume no fuel and produce no exhaust gas, so no flue is required. The electric electrodes are immersed in the water.

Another type of boiler used more in recent times is the gas-fired pulse combustion boiler (Fig. 46.57). Pulse combustion provides efficient combustion and high heat transfer rates. These boilers reach operating temperature much faster than conventional boilers; they burn cleaner than conventional gas boilers and much cleaner than coal or oil.

The pulse combustion process has no power burner and no moving parts, and the combustion takes place in a sealed environment. The process begins with a charge of air and gas entering the burner/heat exchanger through metering valves. The charge is initially ignited by a spark plug. Once combustion starts, the spark plug shuts off. The positive pressure from combustion closes the metering valves and forces the hot gases created out the tailpipes into the exhaust decoupler. As this happens, the air and gas metering valves are sucked open, admitting a fresh charge of air and gas. This cycle repeats at a natural frequency of about thirty times a second.

The combustion process is so clean that the boiler does not require a chimney—it vents through the roof or side wall with small-diameter tubing. Pulse-combustion boilers may be condensing or non-condensing.

Large commercial hot water and steam heating systems use large boilers often installed in series, as shown in Fig. 46.58. Typical boiler connections for multiple boiler installations are controlled by local codes.

Figure 46.56 A pair of high-efficiency cast-iron wet base boilers. *(Courtesy H.B. Smith Company, Inc.)*

TERMINAL EQUIPMENT FOR HOT WATER AND STEAM HEATING SYSTEMS

Terminal units used on hot water and steam heating systems include natural convection units and forced convection units. Those used on high-pressure steam systems are more heavily constructed than those used for hot water and low-pressure steam. These include cast-iron and steel radiators, convectors, finned tube units, and baseboards.

Natural Convection Terminal Units

Natural convection heating devices include various types of terminal units that may be wall hung, recessed, or in the form of a baseboard. These devices distribute heat by using natural air circulation. This occurs because heated air rises and cool air settles, producing a natural circulation. Cool air enters a unit below the finned heating tube and rises as it heats, exiting through a grille or other opening at the top of the unit. These devices are available in various styles and sizes.

Finned tube units have metal fins secured to a metal tube and are a convection-type heater. However, some radiant heat is produced. If placed where human contact may be possible, they have an enclosing cover; otherwise they need not be covered. The fins are usually aluminum or copper and secured to copper tubing that conducts steam or hot water from the boiler. The fins become hot and air heats as it flows by. Several examples are shown in Fig. 46.59.

Baseboard units are located along a wall where it meets the floor. The heating element may be a finned tube, cast-iron, or aluminum unit. This unit is enclosed in metal, with openings at the floor and near the top of the enclosure, providing for natural air circulation.

Steel radiators are another device that heats by convection and radiation. These come in the form of panels typically mounted on a wall or ceiling, although they can be freestanding. They are heated with hot water circulating through tubes that feed flat hollow panels or a series of tubes (Fig. 46.60). Steel radiators work with low-temperature hydronic systems. Although they are not as hot as conventional hot water radiators, steel radiators are still somewhat hot to the touch. The panels are made of heavy-gauge steel or cast iron.

Forced Convection Heating Units

Forced convection units are used for spaces that are to be heated and cooled. However, they can be used for heating or cooling only as is done with natural convection units. Unit ventilators, unit heaters, induction units, fan coil units, and large central air handling units are available. Forced convection units use some form of fan or blower to produce air movement over a cooling or heating coil and through the space needing conditioning.

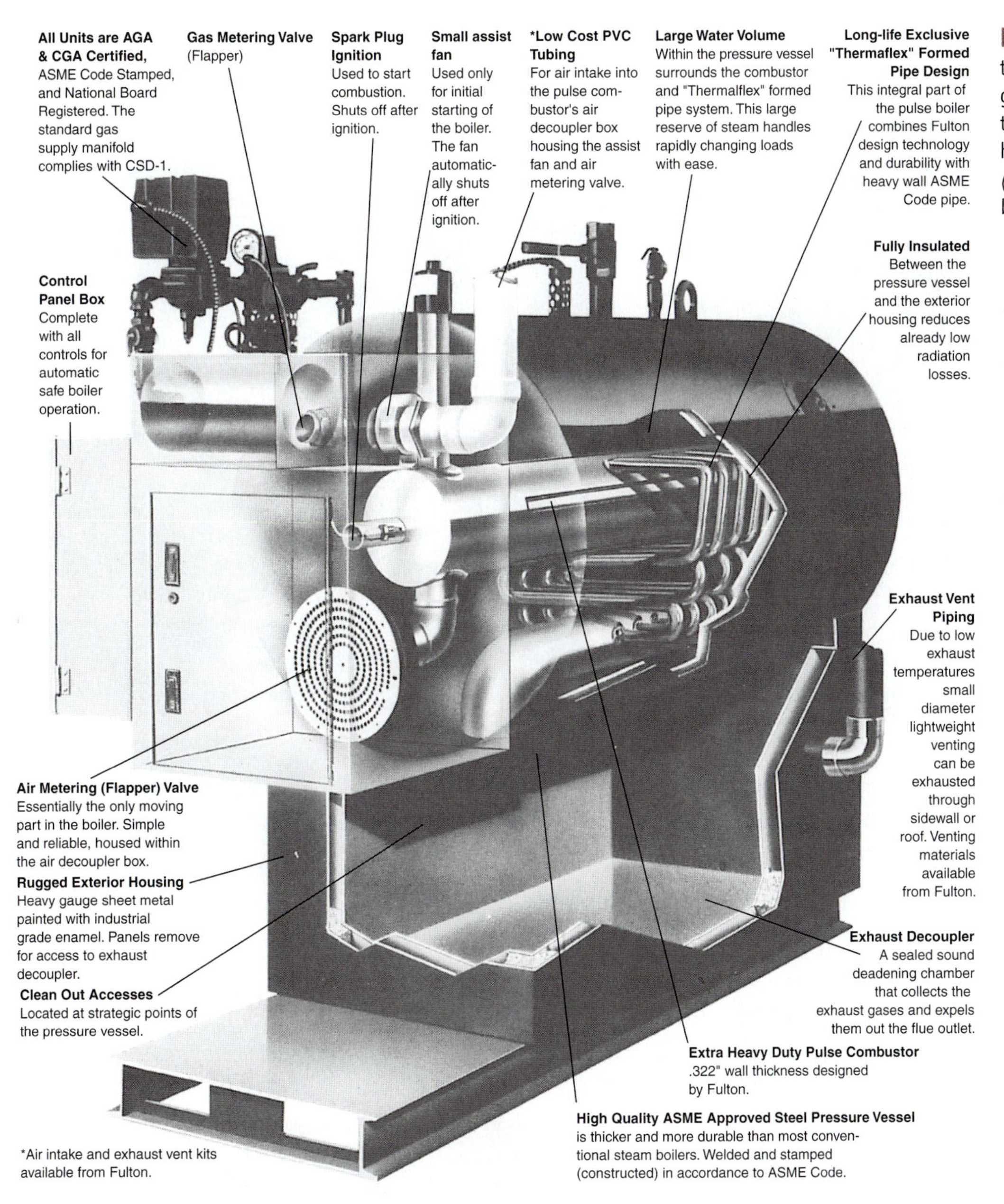

Figure 46.57 An internal view of a vertical gas-fired pulse combustion commercial/industrial hydronic heating boiler. *(Courtesy Fulton Boiler Works)*

Figure 46.58 Gas-fired boilers installed in series. *(Courtesy Weil-McLain)*

Fan coil units have a fan, a filter, and heating and cooling coils. Some have separate heating and cooling coils, but others use the same coil for heated and chilled water in winter and summer respectively (Fig. 46.61). If used on a two-pipe system, the entire system must be set for heating or cooling. Hot water in the pipes must be drained before chilled water is introduced. In a large building this can take several days. If used with three- or four-pipe systems, each fan coil unit has separate heating and cooling coils and can provide heating or cooling whenever required.

Fan coil units may be wall-mounted, ceiling-mounted or high-rise vertical units. The ceiling-mounted units may sit below or above the ceiling. If above, short ducts are used for the air return and supply. High-rise units are usually placed in room corners, but this is not mandatory.

Figure 46.59 Typical finned tube convection terminal units.

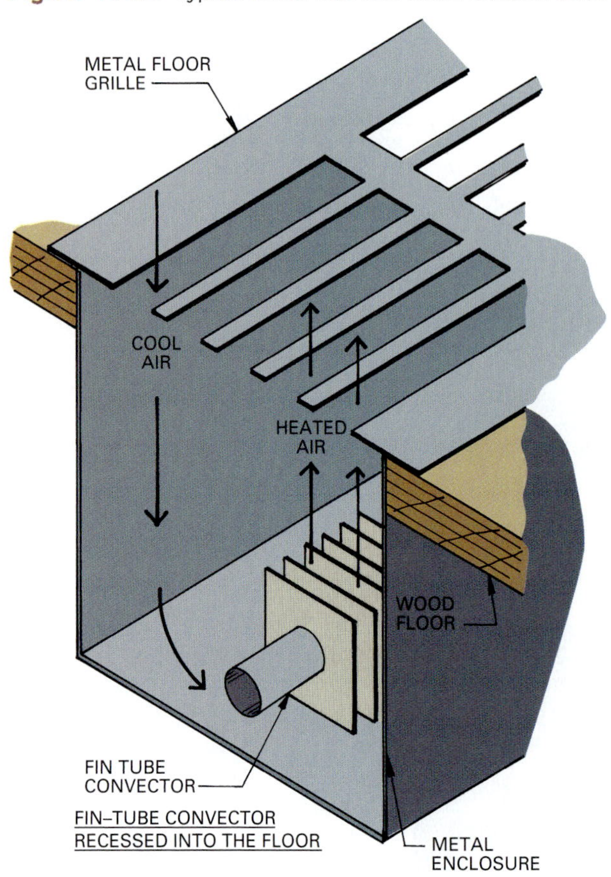

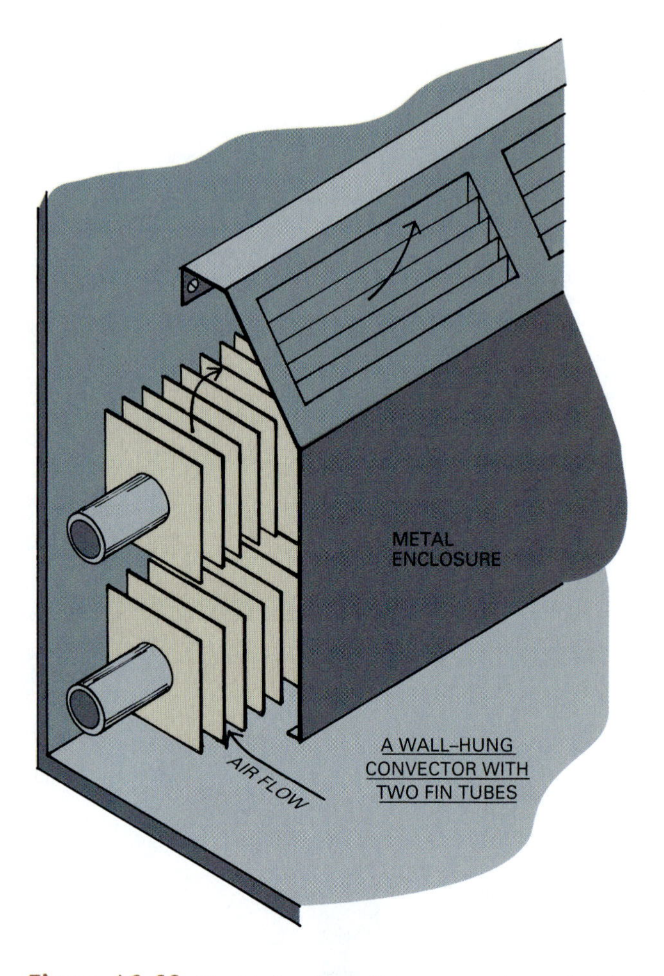

Figure 46.60 Steel hot water radiators are wall mounted, but baseboard units are available. They may be installed as freestanding units and serve as a room divider. *(Image copyright Maxim S. Sokolov, 2009. Used under license from Shutterstock.com)*

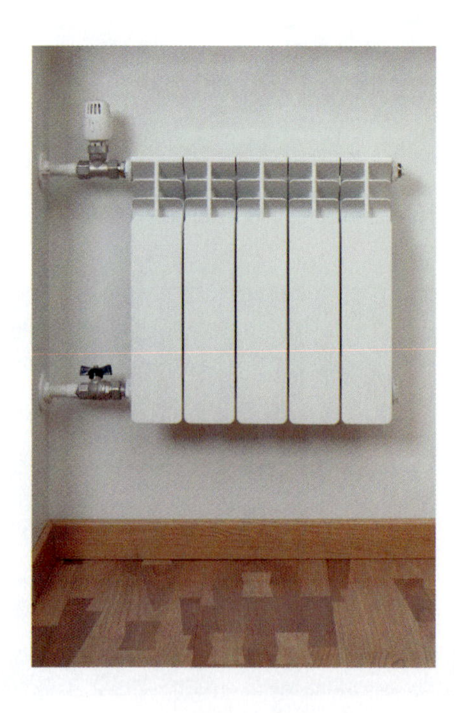

Figure 46.61 A typical wall-mounted fan coil. It is installed below a window so the conditioned air rises through a slot diffuser across the face of the window.

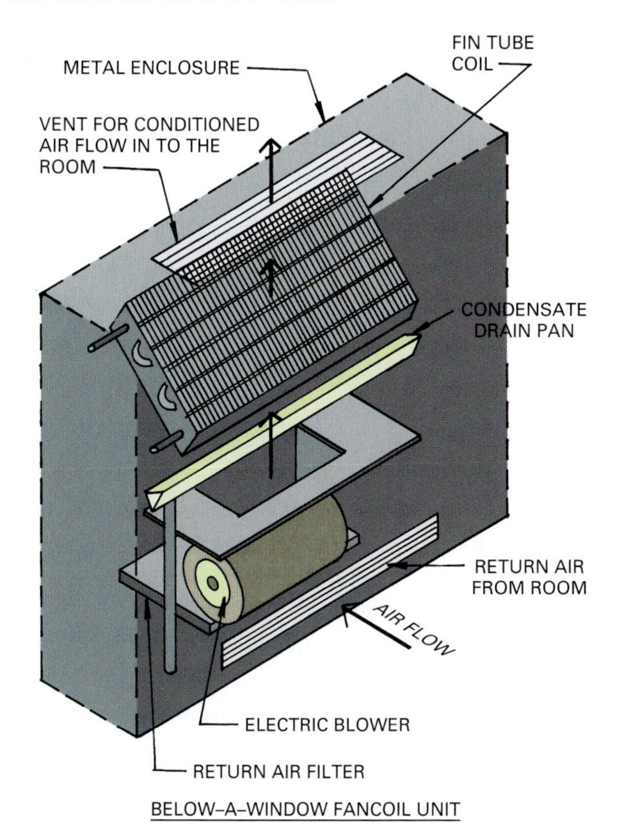

Figure 46.62 This illustrates a gas-fired heater designed for hanging from a ceiling. Heat also can be provided by electric coils, hot water, or steam.

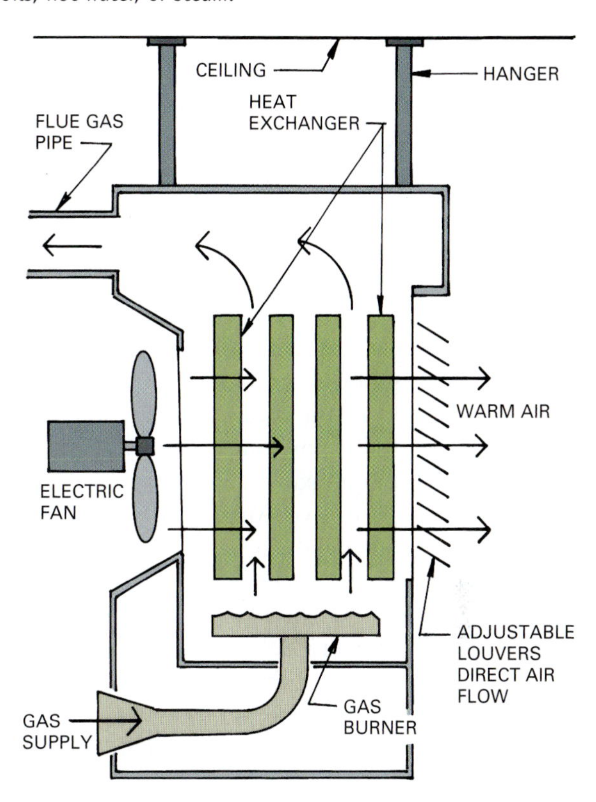

Unit ventilators are much like fan coil units except they have an opening through an outside wall that allows the bringing in of outside air. The opening is covered with a decorative louvered grille. The system regulates dampers to control the influx of outside air. Unit ventilators are typically used in large-occupancy rooms where frequent air changes may be necessary.

Induction units are much like fan coil units, but air movement in the room occurs via high-pressure air piped through a nozzle behind the coil. This causes room air to circulate through the coil and into the room. The nozzle and pressurized air replace the fan in the fan coil unit. They are usually mounted on an outside wall below a window.

Unit heaters have a fan that circulates air over some type of heat exchange surface, such as a hot water coil. They are enclosed in a metal case and usually suspended from the ceiling. Most common uses involve large open industrial plants or in businesses such as auto repair shops. They can use electric heating elements, natural gas or propane firing, or be part of a steam or hot water heating system. A typical gas-fired unit is shown in Fig. 46.62.

Unit heaters provide a rapid flow of heated air. They are also used to temper cold outside air introduced into a building through heavy-use openings, such as doors. They fill such areas with warm air, protecting occupants from sudden temperature changes. Unit heaters are controlled by thermostats and can be used to provide zoned heat by allowing one thermostat to control only the heaters in one part of a heated area.

Central air-handling units have a fan and a hot water or steam heating coil, which is used for cooling with a chilled water supply—a compressor is used if the water supply is not chilled. Heated and cooled air is distributed to various rooms through a duct system. Each space has terminal equipment, such as reheat coils, mixing boxes, and variable volume controls, to regulate air temperature. Reheat coils add additional heat to the air at the terminal device. Mixing boxes (also called blending boxes) are compartments in which two air supplies combine and then discharge into a room. Variable volume controls regulate airflow to control air temperature.

CHILLED WATER COOLING SYSTEMS FOR LARGE BUILDINGS

The most commonly used cooling system uses chilled water to remove heat from the air in a building. This involves the use of mechanical equipment, including a means of refrigeration, cooling coils, and heat exchangers. The chilled water system is a closed circuit of piping in which water recirculates through a chiller and other equipment used to remove heat. It then moves through terminal units in the cooling space and back to the chiller. The three types of liquid chillers are centrifugal, absorption, and positive displacement.

CENTRIFUGAL CHILLERS

Centrifugal chillers are driven by an electric motor, a steam or gas turbine, or an internal combustion engine, which may be diesel or natural gas fueled. They use a vapor compression refrigeration system. The major parts of a mechanical vapor-compression chilled water system include a compressor, a **condenser**, an evaporator, and an expansion valve. An air-cooled refrigerant condenser is a device in which heat removal from the refrigerant is accomplished entirely by heat absorption via air flowing over the condensing surfaces. A compressor is a machine that mechanically compresses the refrigerant. An evaporator is that part of a refrigerating system in which the refrigerant is evaporated to absorb heat from the contacting heat source. The expansion valve reduces the temperature and pressure of the refrigerant as it passes through the valve.

The vapor compression refrigeration cycle used by centrifugal chillers is illustrated in Fig. 46.63. This is a direct expansion (DX) cycle. The refrigerant in vapor form is compressed by the cylinder in the compressor. This causes it to become hot at a high pressure. The refrigerant moves to the condenser where a fan blows outside air through the condenser coil, removing much of the heat in the refrigerant and causing it to condense into a warm liquid, still at a high pressure. The pressure pushes the liquid refrigerant toward the evaporator. On the way, it passes through a thermal expansion valve that imposes a pressure drop, causing it to expand, which reduces the temperature. The cool liquid refrigerant passes through the cooling coil of the evaporator where it absorbs heat from air passing through the coil. When the refrigerant leaves the evaporator, it is a vapor. The vapor moves to the compressor, and the cycle repeats.

The centrifugal chiller system can also use a shell-and-coil heat exchanger and a cooling tower to dispose of heat from the refrigerant in the shell-and-coil condenser. A shell-and-coil heat exchanger has a coil of pipes within a shell or container. The pipes carry refrigerant through a second fluid held in the container. The shell-and-coil evaporator chills water in the shell or container and passes it through a cooling coil in the air handler. The air handler passes return air from the room and fresh outside air as required through the coil, conditioning the air supply to the rooms. The air flows through a system of ducts (Fig. 46.64).

The shell-and-coil condenser water absorbs heat from the refrigerant and moves it to a cooling tower on the roof. The hot condenser water drips or is sprayed in fine droplets. A fan pulls outside air through the spray, removing much of the heat. Other cooling tower designs are available.

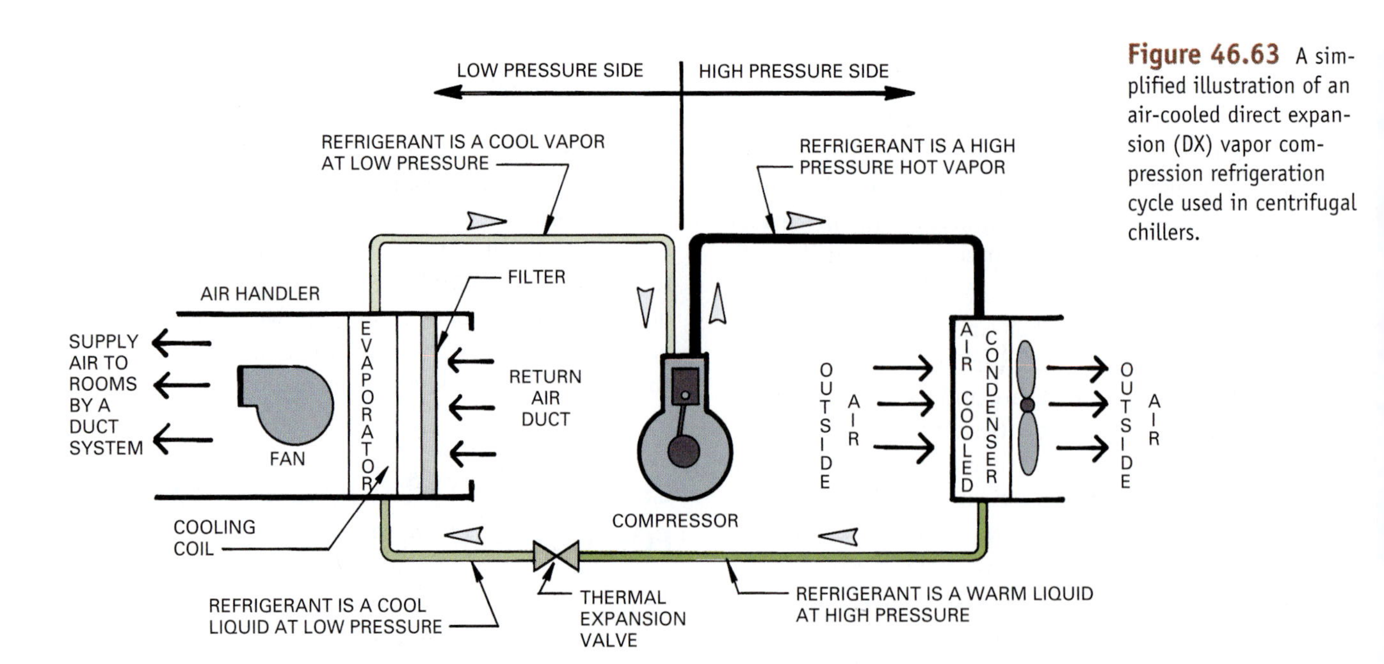

Figure 46.63 A simplified illustration of an air-cooled direct expansion (DX) vapor compression refrigeration cycle used in centrifugal chillers.

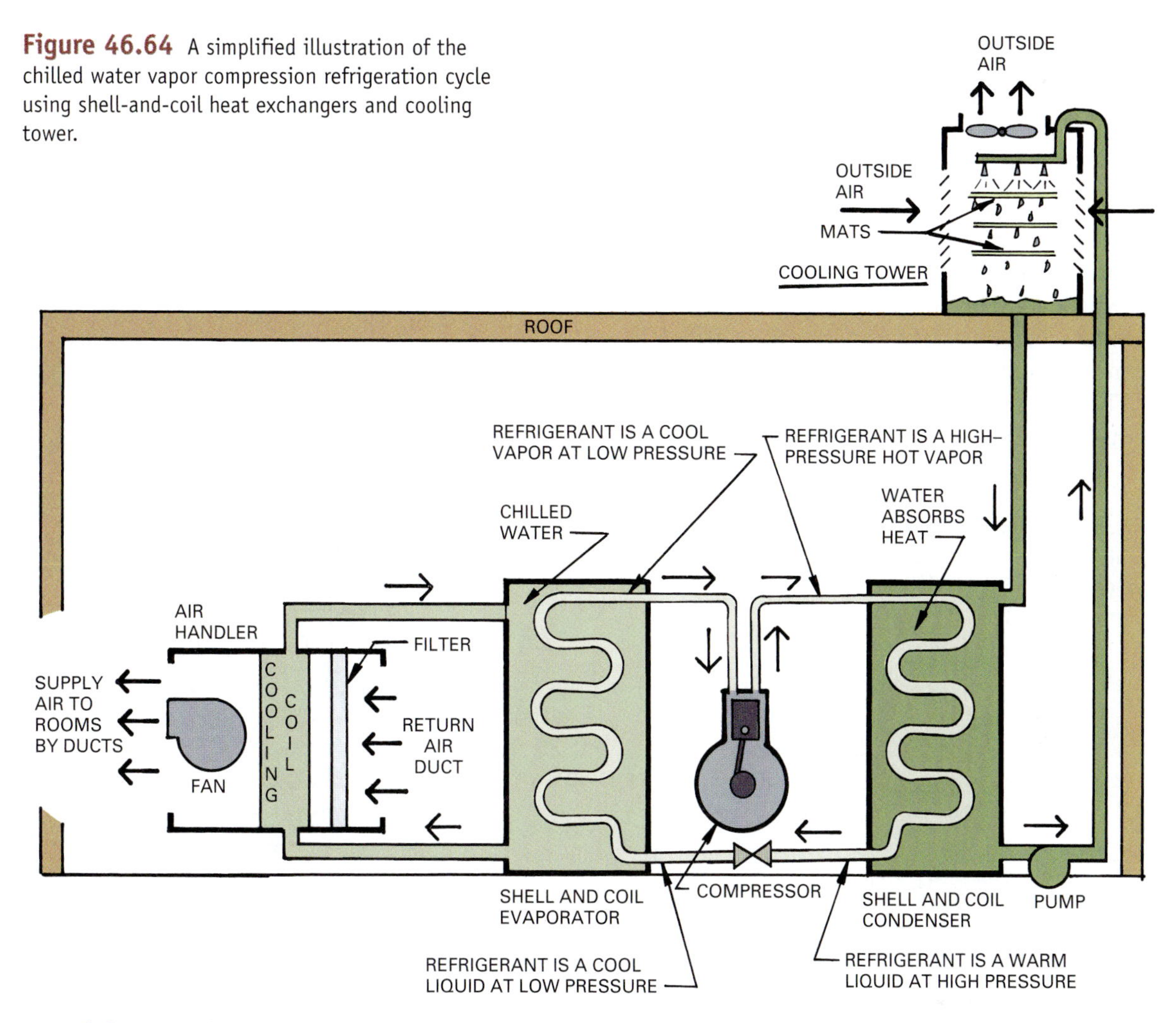

Figure 46.64 A simplified illustration of the chilled water vapor compression refrigeration cycle using shell-and-coil heat exchangers and cooling tower.

THE ABSORPTION CHILLER REFRIGERATION SYSTEM

An absorption chiller can be described as a refrigerating machine that uses heat energy and absorption input to generate chilled water. Absorption cooling uses an evaporated refrigerant (often water). Rather than the mechanical compression stage used in centrifugal chillers, it involves a process using an absorbent, such as lithium bromide solution. This solution pulls vapor off evaporator coils, creating a cooling reaction. A source of heat, such as a gas burner, low-pressure steam, or hot water, is used to regenerate the absorbent solution by separating it from the absorbed vapor. Since a source of heat is used to operate an absorption chiller, it is especially economical to operate when a building has waste steam or other heat sources that can be used to provide a major amount of the heat required.

The absorption chiller is a quiet operating unit having few moving parts and generating little vibration and is lightweight compared with compression type chillers. An absorption chiller is shown in Fig. 46.65.

Figure 46.65 A direct-fired double-effect absorption chiller/heater. *(Courtesy The Trane Company)*

Single-Effect Absorption Chillers

The basic absorption chiller refrigeration process for a water-cooled chiller is shown in Fig. 46.66. Following is a description of this single-effect water-cooled absorption process.

The absorbent material, lithium bromide, has a great affinity for the refrigerant, water. The absorbent material, which contains some refrigerant, is heated in the generator. The refrigerant vapor is then driven to the condenser where the process converts it to a liquid (water). The concentrated absorbent moves from the generator to the absorber.

The liquid refrigerant flows from the condenser through an expansion valve to a low-pressure evaporator where it flashes to a vapor, providing cooling for the chilled water system refrigerant circulated to cooling coils within the building. This is a refrigerant in a system separate from the chiller system refrigerant. The expansion valve reduces the refrigerant pressure as it enters the evaporator, which is on the low-pressure side of the cycle.

Next, the chiller refrigerant moves to the absorber, where it is reabsorbed by the concentrated absorbent solution. This results in reduced pressure in the system beyond the expansion valve. The absorber is cooled

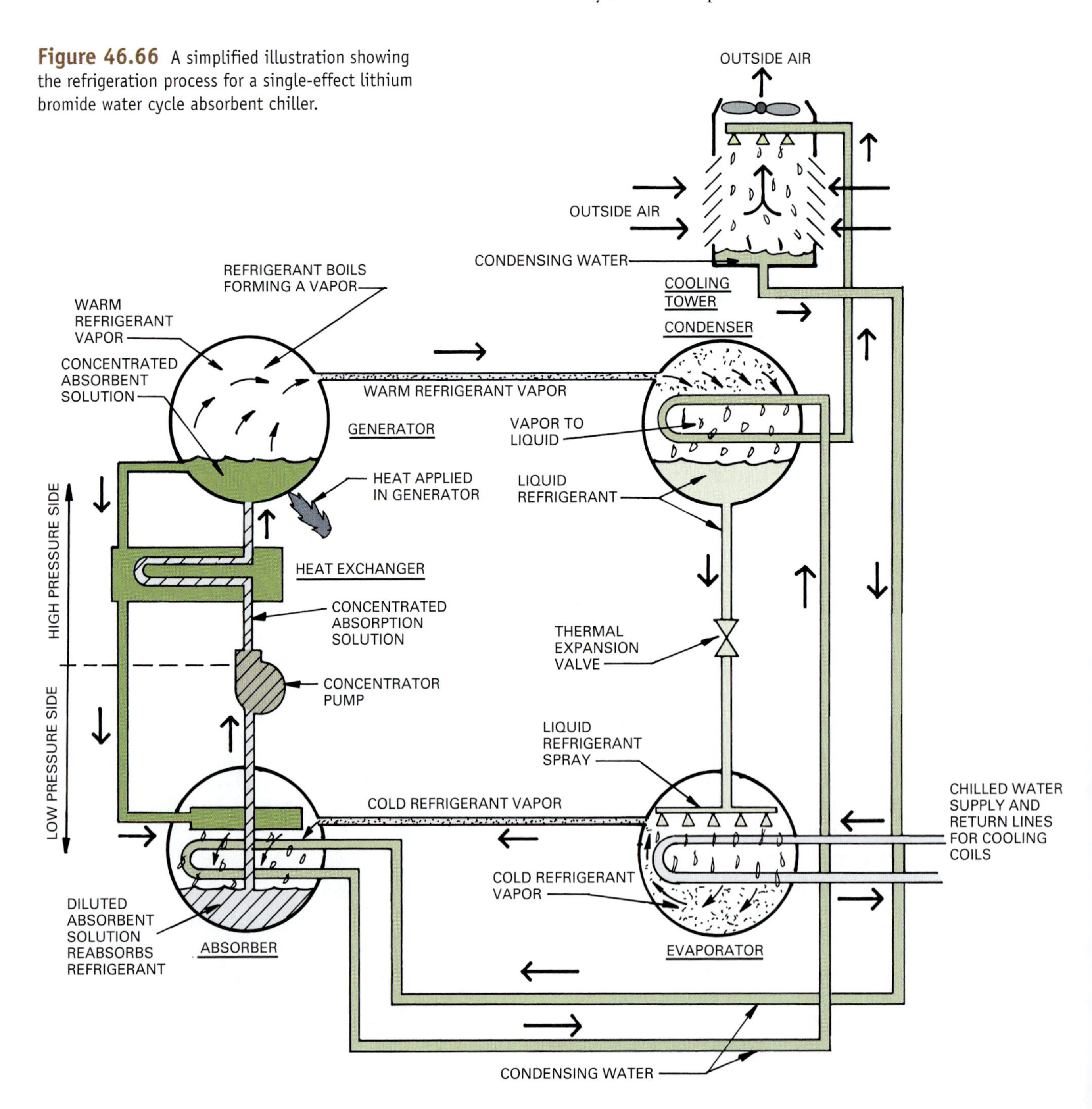

Figure 46.66 A simplified illustration showing the refrigeration process for a single-effect lithium bromide water cycle absorbent chiller.

to increase the rate of absorption of refrigerant into the concentrated absorption solution. This solution is pumped back into the generator, where the cycle repeats. The generator and condenser side operate at high pressure, and the evaporator and absorber operate at low pressure.

Double-Effect Absorption Chillers

Gas-fired double-effect absorption chillers/heaters are utilized in commercial applications with a central air-conditioning system using chilled water for cooling and hot water for heating. The condenser is water-cooled during the cooling period, and the heat is rejected through a cooling tower. The double-effect absorption cooling cycle has two generators. One is heated by natural gas and the other by the hot, semi-concentrated refrigerant vapor.

Cooling Cycle The cooling cycle for a double-effect chiller/heater is illustrated in Fig. 46.67. The gas burner heats a dilute absorbent solution (lithium bromide) in the high-temperature generator. The boiling process drives the refrigerant vapor and droplets of semi-concentrated solution to the primary separator. The semi-concentrated solution flows through a heat exchanger where it is pre-cooled before flowing to the low-temperature generator.

In the low-temperature generator, the hot refrigerant vapor flowing from the primary separator heats the semi-concentrated solution. This liberates refrigerant vapor, which flows to the condenser. The concentrated absorbent solution is pre-cooled in the condenser and flows to the absorber.

In the condenser, refrigerant vapor is condensed on the surfaces of the cooling coils, and the latent heat is

Figure 46.67 A simplified illustration of a gas-fired double-effect water-cooled absorption chiller/heater in the cooling mode. *(Developed from materials from American Yazaki Corporation)*

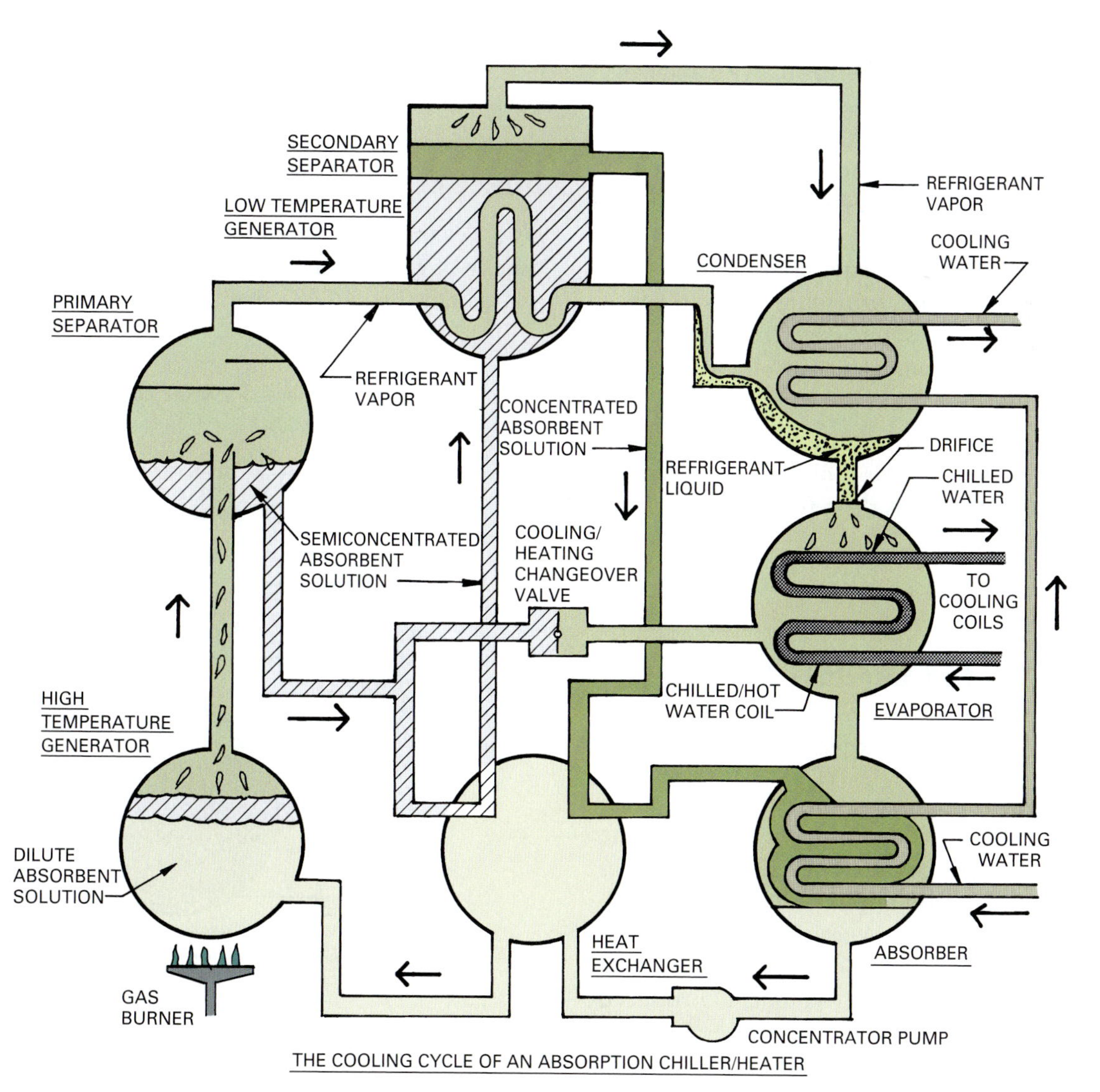

moved by the cooling water to a cooling tower where it is expelled to the outside air. Refrigerant liquid accumulates in the condenser and passes through an orifice into the evaporator.

The pressure in the evaporator is lower than that in the condenser. As the refrigerant liquid flows into the evaporator, it boils on the surface of the chilled/hot water coil. Heat, equivalent to the latent heat of the refrigerant, is removed from the recirculating water to the cooling coils, which chills the water to about 44°F (6.7°C). The refrigerant vapor flows to the absorber.

In the absorber, the concentrated lithium bromide solution (the absorbent) absorbs the refrigerant vapor as it flows across the absorber coil. The cooling water in the coil removes heat from the solution, after which the diluted lithium bromide flows through the heat exchanger, is preheated, and moves to the high-temperature generator to start through the cycle again.

Heating Cycle In the heating cycle for a double-effect absorption chiller/heater, the lithium bromide absorbent solution is brought to a boil in the high-temperature generator and the vapor with concentrated lithium bromide solution is lifted to the primary separator. The hot refrigerant vapor and droplets of concentrated absorbent solution flow through the open changeover valve into the evaporator and the absorber. The heat is transferred to recirculating water coils in the evaporator. This heated water flows to fan coil units within a building, providing required space heating. The refrigerant (water) vapor mixes with the concentrated absorbent and returns to the generator, where the cycle is repeated.

Absorption chillers may be water-cooled, as with a cooling tower, or air-cooled. Air-cooled absorption chillers generally use ammonia as the refrigerant and water as the absorbent.

POSITIVE DISPLACEMENT CHILLERS

Positive displacement chillers use scroll, reciprocating, and rotary screw compressors, typically with electric motor drives. A scroll compressor is a positive displacement compressor in which the reduction of the internal volume of the compression chamber is produced by a rotating scroll within a stationary scroll. A scroll is an involute spiral. A reciprocating compressor is a positive-displacement compressor in which the change in compression chamber volume is produced by the reciprocating movement of a piston. Positive-displacement compressors increase the pressure of the refrigerant vapor by reducing the volume of the compression chamber in which it has been injected using a mechanical means, such as a piston. A screw compressor uses a rotary motion to drive two intermeshing helical rotors to produce compression.

The capacity of the system is varied by having several compressors and turning them on and off as capacity requirements vary. Some reciprocating chiller compressors have an inlet suction valve. Reciprocating chillers use a compressive refrigeration cycle and are usually driven by electric motors. They are smaller than centrifugal chillers and usually release condenser heat into the air rather than through a cooling tower.

EVAPORATIVE AIR COOLERS

Evaporative cooling is sensible cooling obtained by the exchange of heat produced by water jets or sprays injected into an airstream. As the air passes through water vapor, the vapor retains some of its heat, thus cooling it. This cooling process is effective in hot and dry areas. There are two basic types, direct and indirect.

Direct Evaporative Air Coolers

Direct evaporative air coolers have evaporative pads, usually of a wood fiber, a water-circulating pump, and a large fan (Fig. 46.68). The fan pulls air through the

Figure 46.68 A direct evaporative air cooler pulls the air through wet evaporative pads.

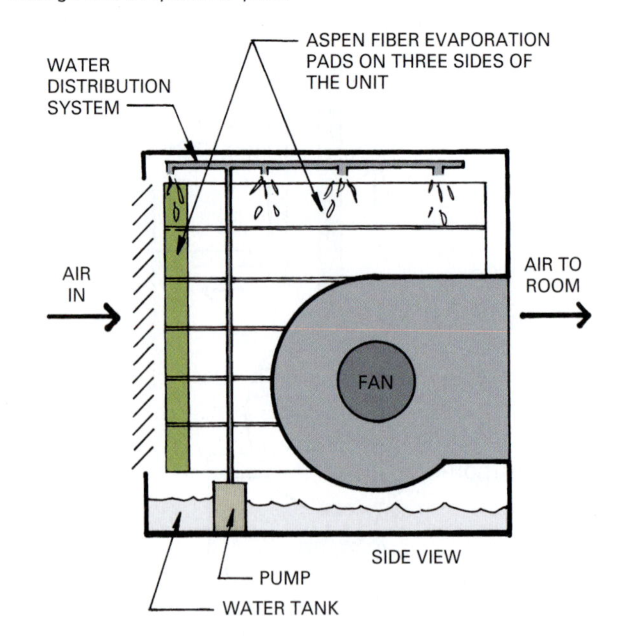

pads, which cool the air, and moves it to spaces within a building. The water that does not evaporate collects in a pan, and a pump lifts this water back through the cycle.

Indirect Evaporative Air Coolers

A variety of systems in use fall under this type. The following describes a package indirect evaporative air cooler. The unit contains a heat exchanger, a system to provide water vapor, a secondary fan, and a secondary air inlet. The heat exchanger typically is constructed with tubes, allowing one airstream to flow inside them and another over their exterior surface. Air filters are used to keep the system from becoming clogged with dust. The circulated water is usually treated to remove minerals, thus reducing corrosion in the heat exchanger.

The system is much like the direct evaporative system except the conditioned air is kept separate from the secondary air (which causes the water evaporation). Therefore, the conditioned air does not pick up moisture during the cooling process. This is why it is called an indirect system.

RADIANT HEATING AND COOLING

Radiant panel systems use panels whose surface temperatures can be controlled. They may be located in walls, ceilings, or floors. The panels are usually operated by electrical resistance units or by circulating water or air. Radiant energy is transmitted through the air in straight lines. It does not heat the air, but it does raise the temperature of solid objects upon which it falls. The objects obtain the heat by absorption, and the heat can be reflected off a surface.

There are a variety of piping systems used for radiant heating and cooling. Some use bronze or copper tubing while others utilize plastic tubing.

Piping in Ceilings

Several piping systems are used for radiant heating and cooling in ceilings. Hydronic ceiling panels can use a two-pipe or a four-pipe distribution system similar to those discussed for hot water systems. The design of the system is critical to a successful end result and must be done by a qualified engineer. Typically, panels are installed with the pipes embedded above or in a finished plaster ceiling, as shown in Fig. 46.69.

Figure 46.69 Typical radiant-heat ceiling panel piping installations.

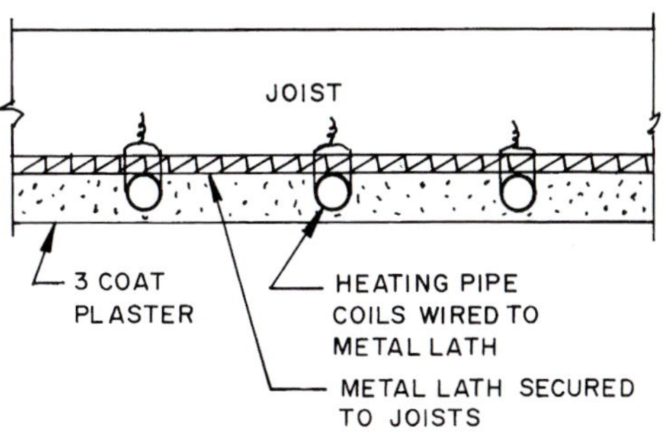

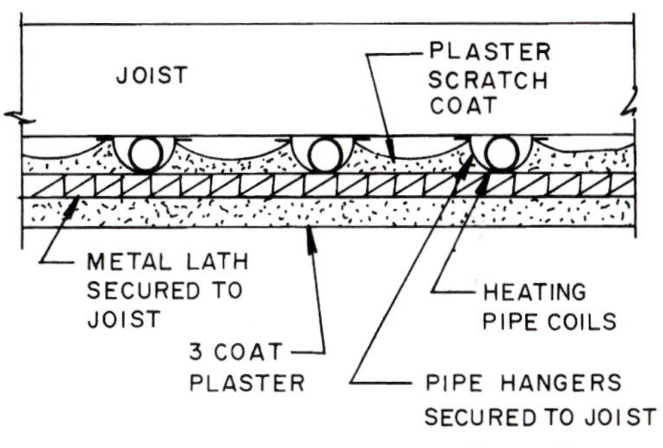

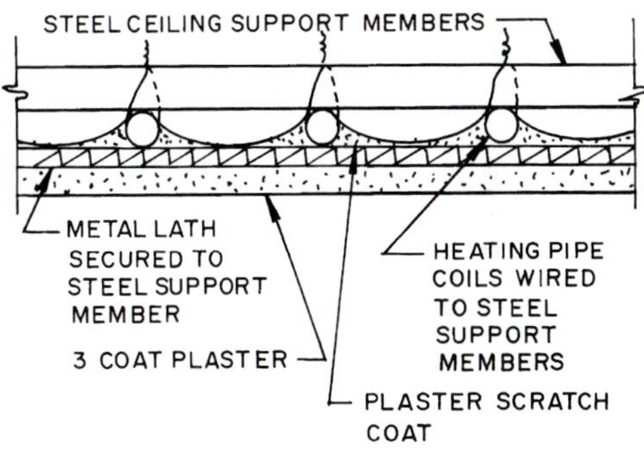

A suspended ceiling has the pipes tied to an overhead supporting member with metal lath and plaster below. When wood or metal joists are used, the pipes are secured to them with metal pipe hangers, and the metal lath and finished plaster are placed below. The coils may be embedded in the plaster coat by securing the metal lath above the pipes and plastering over them. Other types of finished ceilings can be used, but plaster is most common. Similar installations can be utilized on wall surfaces.

Another type of ceiling panel consists of tubing bonded to flat metal panels that act as the exposed finished ceiling. The individual panels are hung between the channels of a suspended ceiling. This system is used to heat and cool interior air.

Piping in Floors

Radiant heating piping may be embedded in concrete floors or placed above or below a wood subfloor. For concrete slab heating, ferrous and nonferrous pipe and tubing may be used. The piping may be arranged in continuous coils or have header coils (Fig. 46.70). Usually $1\frac{1}{2}$ to 4 in. (38 to 101 mm) of concrete covers the pipe. The edges of the slab must be fully insulated with rigid insulation. Sometimes insulating concrete is used. Piping may be placed on top of a wood subfloor and covered with 1 to 2 in. (25 to 50 mm) of concrete or gypsum underlayment (Fig. 46.71). Gypsum products designed specifically for floor heating are available, and concrete of structural quality should be used.

Piping under a subfloor is attached to the subfloor. Metal heat emission plates are used to improve heat transfer (Fig. 46.72).

Piping may also be intertwined with a subfloor consisting of spaced strips, which allows the piping to be between them and above the floor joists. Metal heat transfer plates are used. This is covered with wood subflooring and the finished floor covering.

Figure 46.70 Two arrangements for radiant heating pipes in a concrete slab.

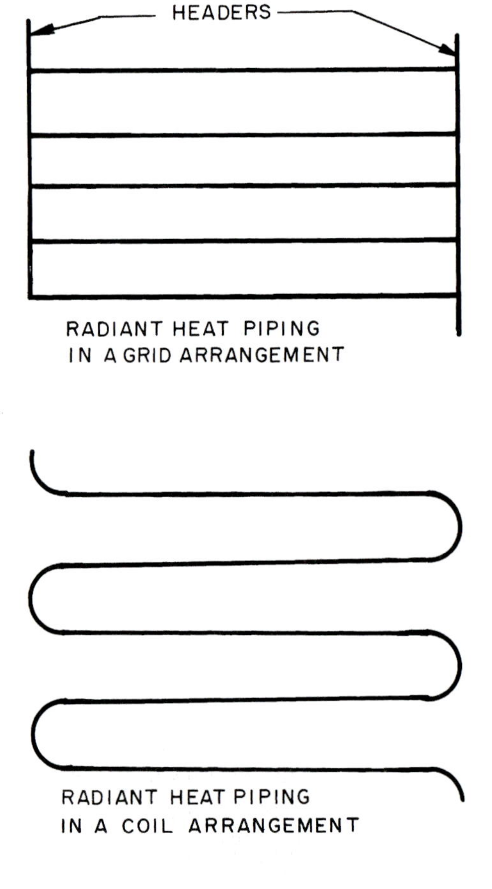

Figure 46.71 Radiant heating pipes can be installed over a wood subfloor and then embedded in a gypsum or concrete underlayment. *(Courtesy H. Raab, Wikipedia)*

Figure 46.72 This construction places the radiant heating pipes and metal heat emission plates below the subfloor.

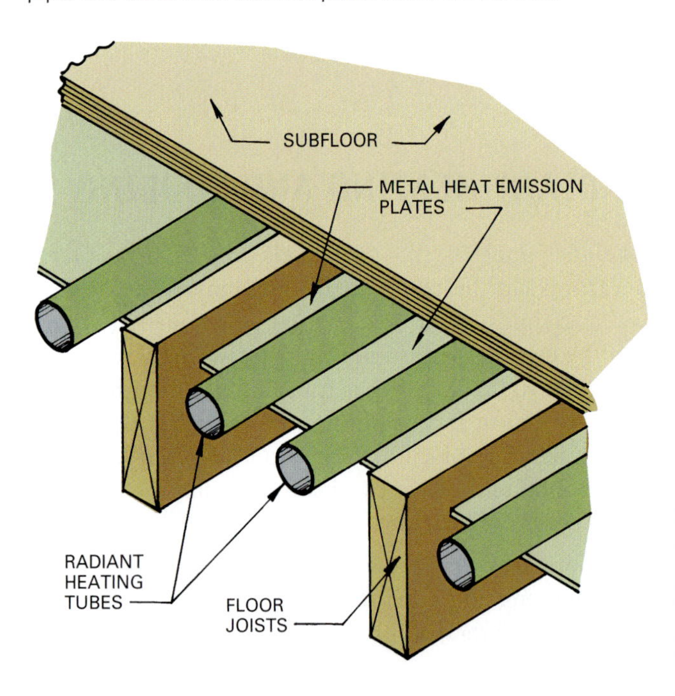

ELECTRIC HEATING SYSTEMS

Types of electrically heated systems include factory assembled panels that mount on walls or ceiling, fabrics and wall covering material containing resistance heating wires, and various types of electric resistance cables that may be embedded in concrete or laminated in drywall ceilings or in plaster.

One type of ceiling panel is made to fit into the grid formed by the structural members of suspended ceilings. The panels are available with various constructions, such as conductors embedded in a panel (maybe a gypsum panel) or some form of laminated panel.

Another ceiling heating system uses electric heating cable stapled to a ceiling covering material, such as gypsum board, plaster lath, or another fire-resistant material. The wires are covered by the various coats of plaster. If metal lath is used, it must first be covered with a brown coat of plaster to provide a nonelectrical conducting surface (Fig. 46.73).

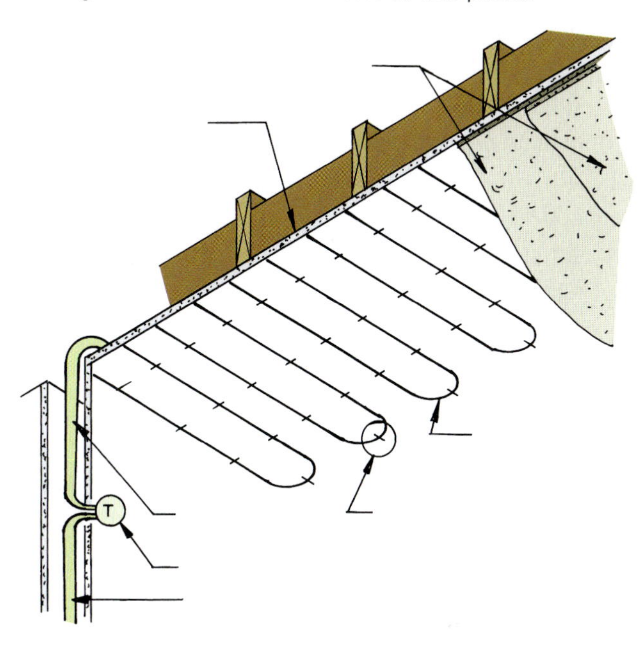

Figure 46.73 A ceiling electric heating system composed of heating cable secured to lath and covered with plaster.

Electrical heating cable can also be laid in concrete slabs. The floor is laid in two pours. The first pour is 3 in. (76 mm) or more of insulating concrete. The cable lays on this slab and is fastened in place by stapling or special nail anchors. These hold the cables so the spacing is maintained when pouring the second layer of concrete. The second layer is usually $1\frac{1}{2}$ to 2 in. (38 to 51 mm) thick and must not be insulating concrete. A finish floor can be laid over this slab.

Review Questions

1. What is the difference between thermal radiation, thermal conduction, and thermal convection?
2. What is a major source of information about thermal conditions necessary for human comfort?
3. What types of ventilation are used to assist with the conditioning of inside air?
4. What types of air cleaners are currently available?
5. What are the three basic directions of heat flow from warm air furnaces?
6. What fuels are used to fire warm air furnaces?
7. How does an air-to-air heat pump produce heat?
8. What is meant by a ground loop heat pump?
9. How can a gas-fired warm air furnace be used to cool a building?
10. What are the types of all-air distribution systems?
11. What factors must a designer consider as an all-air duct distribution system is being designed?
12. How does a reheat warm air distribution system differ from a variable air volume system?
13. How does a dual duct warm air system regulate air temperature?
14. What are the classes of commercial warm air ducts?
15. What is the difference between the perimeter loop and perimeter radial duct systems?
16. What is meant by the surface effect from a ceiling warm air diffuser?
17. What is a comfortable relative humidity for indoor air?
18. What types of residential and industrial humidifiers are available?
19. Why is the control of humidity in the air important?
20. How does a liquid desiccant dehumidification system remove moisture from the air?
21. What types of hydronic heating systems are in use?
22. What is the difference between a two-pipe direct-return and a two-pipe reverse-return hydronic system?
23. Why do hydronic heating systems need an expansion chamber?
24. What is the advantage of a four-pipe system for hydronic heating and cooling?
25. How does the condensate in a one-pipe steam heating system return to the boiler?
26. What purpose do thermostatic traps on each radiator of a steam heating system serve?

27. Why is a two-pipe vacuum steam heating system used in large buildings?
28. What is the major difference in the way water is heated in firetube and watertube boilers?
29. What is the difference in the operation of condensing and non-condensing boilers?
30. What is the unique feature identifying a wet-base boiler?
31. What are the commonly used natural convection terminal units?
32. What are the major components of a fan coil unit?
33. What types of refrigeration units are used to chill water for air-conditioning purposes?
34. How does a shell-and-coil condenser remove heat from the refrigerant?
35. How does a single-effect water-cooled absorption chiller cool the water for the air-conditioning system?

Key Terms

absorbent
adsorbent
boiler
Btu
chiller
compressor
condensate
condenser
convector
dehumidifier
desiccant
desiccation
diffuser
humidifier
humidity
hydronics
hydronic heating system
latent heat
makeup air
radiant heat
refrigerant
relative humidity
return air
sensible heat
steam
supply air
surface effect
thermal conduction
thermal convection
thermal radiation
thermostat
throw

Activities

1. Arrange visits to large multistory buildings and ask to see the equipment used for heating and cooling. Prepare a report describing what you saw.
2. Invite an engineer who designs multistory heating and cooling systems to address the class and review the steps of the design process and how decisions are reached.
3. Review the local building code and cite the general requirements relating to heating and air-conditioning systems for residential and commercial buildings. Report what the local building official checks as the building is under construction.

Additional Resources

ASHRAE Handbook—HVAC Applications; *ASHRAE Handbook*—Refrigeration; *ASHRAE Handbook*—HVAC Systems and Equipment; *ASHRAE Handbook*—Fundamentals, American Society of Heating, Refrigerating, and Air-Conditioning Engineers, Inc., Atlanta, GA.

International Mechanical Code, The, Plumbing and HVAC Collection, International Code Council, Falls Church, VA.

National Fuel Gas Code; Flammable and Combustible Liquids Code; Liquefied Petroleum Gas Code, National Fire Protection Association, Quincy, MA.

Pitchers, N., *Combined Heating, Cooling, and Power Handbook*, Taylor and Francis Books, Atlanta, GA.

Standard for the Installation of Oil-Burning Equipment, National Fire Protection Association, Quincy, MA.

Stein, B. and J. S. Reynolds, *Mechanical and Electrical Equipment for Buildings*, John Wiley and Sons, New York, NY.

See Appendix B for addresses of professional and trade organizations and other sources of technical information.

DIVISION

26

Electrical
CSI MasterFormat™

CHAPTER 47

Electrical Equipment and Systems

LEARNING OBJECTIVES

Upon completion of this chapter, the student should be able to:

- Understand the various sources of electrical power and how it is distributed from a generating source.
- Describe an electrical current and the difference between AC and DC current.
- Specify the types of electrical conductors for various applications.
- Identify and discuss the various units used to measure, control, and distribute electrical power within a building.
- Specify lighting installations for meeting needs in a wide range of situations.

The modern building depends heavily on electricity to make it functional and habitable. Electricity provides power for lighting; runs motors for heating, ventilating, and air-conditioning; powers elevators, escalators, and other conveying systems; supplies the power to operate the communications, fire, and security systems; and is used to operate a vast array of electrical devices. In case of power interruption, emergency systems can be utilized to provide a temporary source of electricity for critical operations.

ELECTRICAL LOADS

The loads put on an electrical system vary widely, depending on a building's occupancy. All buildings have extensive lighting requirements. Many functions, such as air-conditioning and heating, require the use of electric motors. Other systems, such as refrigeration and ventilation, have motors, compressors, and other electricity-consuming devices. The extensive range of appliances and other electrical devices produces varying loads as periods of demand fluctuate.

Industrial plants have heavy electrical demands to operate machinery that performs manufacturing operations, such as melting, fusing, and otherwise processing materials. Internal transportation (elevators, escalators, moving walks, conveyors) requires electric power. Internal communications and controls are electrically operated, and many building systems would not function without these systems. There are hundreds of special equipment items such as those found in hospitals, computer centers, and radio and television studios. The determination of electrical loads and internal systems is a major part of an adequately designed building.

BASICS OF ELECTRICITY

Electric current can be defined as the flow of electrons along a conductor, such as a copper wire. It is produced by a generator or battery that forces electrons to flow through the conductor to a consuming device—such as a light—and back to the producing source.

The flow of electric current in this continuing circuit resembles the flow of fluid in a hydraulic circuit (Fig. 47.1). In a hydraulic system, a pump puts the fluid under pressure (pounds per square inch). The fluid flow is measured in a quantity, such as gallons. The fluid meets some resistance as it enters a fixture, and the flow is controlled by a valve. An electric circuit has a battery or generator to produce electricity and the electromotive force (volts) to move it along the conductor in quantities measured in amperes. The flow of electricity finds **resistance** (ohms) when it enters a fixture, and the flow is controlled by a switch.

Figure 47.1 A comparison of electrical and hydraulic circuits reveals similarities. Electric switches and hydraulic valves control flow; electric current and hydraulic fluid flow in the circuit. Power is supplied by an electric device (a battery and a hydraulic pump); electric wire and hydraulic piping form the circuit; and resistance to flow occurs in both circuits.

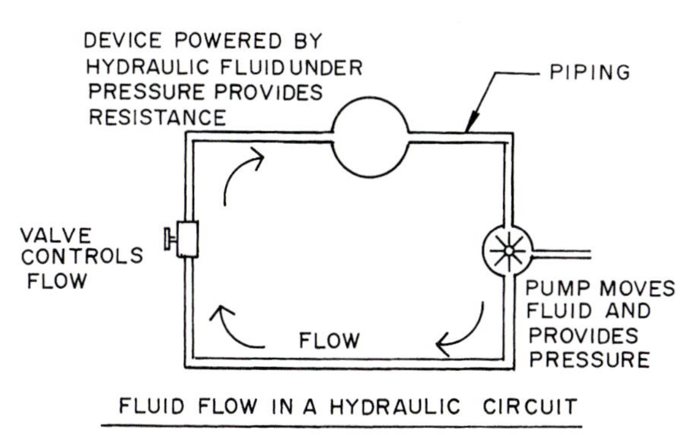

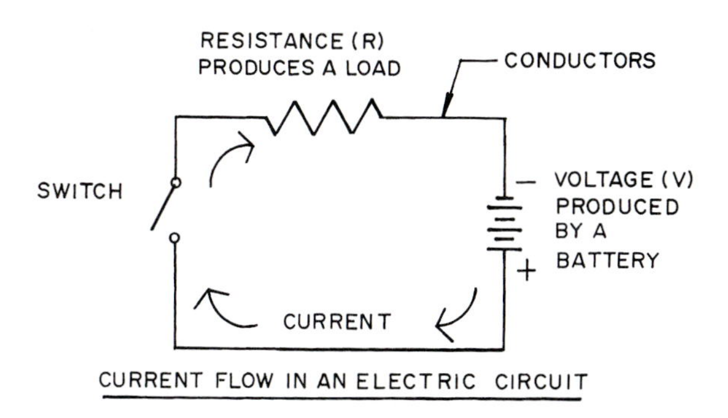

Units used to identify factors related to the flow of electric current are:

Ampere (A). A unit of the rate of flow of electric current. An electromotive force of 1 ohm results in a current flow of 1 ampere.

Volt (E). The unit of electromotive force (pressure) that causes electric current to flow along a conductor.

Ohm (Ω). The unit of electrical resistance of a conductor. The symbol for ohm is the Greek capital letter omega (Ω).

Ohm's Law

Amperes, volts, and ohms are related to each other, and a variance in one will affect the others. This relationship is identified as Ohm's Law. The relationship for each measure is shown by the following formula, in which I is the electric current or intensity of electron flow (measured in amperes), R is resistance (measured in ohms), and E is the electromotive force (measured in volts).

$I = E/R$	amperes = volts/ohms
$R = E/I$	ohms = volts/amperes
$E = I \bullet R$	volts = amperes • ohms

Figure 47.2 Electricity flows through a circuit from the positive connection to the negative connection.

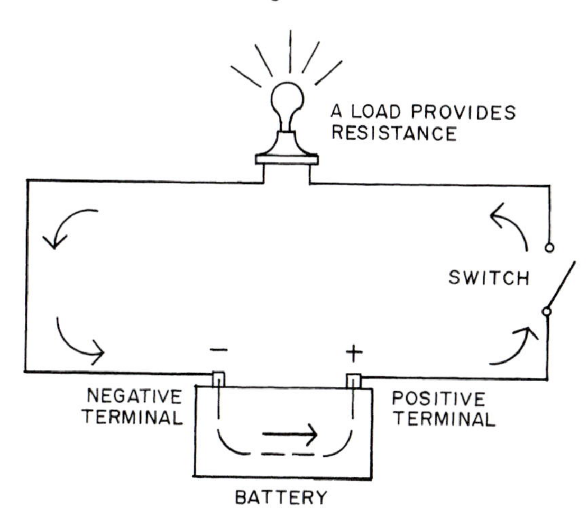

Conductors

Electric current flows along a material called a **conductor**. The commonly used conductors for electric wiring are copper and aluminum. Copper is a better conductor than aluminum, so if aluminum wire is used, it must have a larger diameter to carry the same amount of current. Good conductors have a low resistance to the flow of electricity. Materials with a high resistance to electrical flow get hot because of the friction generated by the flow of electrons. These materials, such as nichrome, are used in applications like electric heaters, in which the production of heat is desired.

Other materials, such as glass, ceramics, and plastics, do not conduct electricity and are called **insulators**. They are used on electrical devices to provide protection from the electricity. For example, the switch lever on a light switch is made of a nonconducting material.

For electricity to do work, it must flow through a circuit. It flows from the positive connection to the negative connection, as shown in Fig. 47.2. The negative electrons move to the positive pole, through the circuit to the consuming device, and back to the negative pole. A switch placed in the circuit can be opened to interrupt the flow of electrons and closed to complete the circuit.

AC AND DC CURRENT

The two types of electric current are direct current (DC) and alternating current (AC). Direct current has a constant flow in one direction. Alternating current varies periodically in value and directions, first flowing in one direction in the circuit and then flowing in the opposite direction. Each complete repetition is called a cycle. The number of repetitions per second is called the frequency and is measured in hertz (Hz). In the United States, the frequency for alternating current is 60 cycles per second or 60 hertz.

POWER AND ENERGY

Energy is the term used to express work. Energy is expressed in units of kilowatt-hours, foot-pounds, Btu's, joules, or calories. **Power** is the rate at which energy is used. Power is expressed in terms of watts, kilowatts, and other units shown in Table 47.1. Since power is the rate at which energy is used, time is a factor. The relationship between power and energy is shown by the following equation:

$$\text{Power} = \text{energy/time}$$
$$\text{Energy} = \text{power} \bullet \text{time}$$

The unit of electric power in electric circuits is expressed in watts (W) or kilowatts (kW). One kilowatt equals 1000 watts. One watt-hour of energy represents one watt of power used for one hour. A **watt** is one ampere flowing under an electromotive force of one volt. The power, W (watts), flowing into an electrical device having a resistance of R (ohms), in which the current is I (amperes), is found by the equation:

$$W = I^2 R$$
$$\text{watts} = \text{amperes}^2 \times \text{ohms}$$

It can be seen, therefore, that power is related to current (amperes), electromotive force (voltage), and resistance (ohms).

Table 47.1 Units of Power and Energy

English System	Metric System
Units of Power[a]	
Horsepower (hp)	Joule per second (J/sec)
Btu per second (Btu/sec)	Calorie per second (cal/sec)
Watt (W)	Watt (W)
Kilowatt (kW)	Kilowatt (kW)
Units of Energy[b]	
Btu	Calorie (cal)
Foot-pound (ft. • lb)	Joule (J)
Kilowatt-hour (kWh)	Kilowatt-hour (kWh)

[a] *The rate at which work is done*
[b] *The amount of work done*

ELECTRICAL CODES

Both the design and installation of electrical systems are carefully regulated by electrical codes. The National Fire Protection Association sponsors the model code, the *National Electrical Code Requirements for One- and Two-Family Dwellings*, as well as a series of Standards Publications related to electrical safety and equipment. The International Code Council offers the publication *International Code Council Electrical Code*, which details information related to the administration and enforcement of the NFPA National Electrical Code. Underwriters Laboratories, Inc. (UL), tests and certifies electrical devices. Approved electrical devices carry the UL label.

ELECTRIC POWER SOURCES

Most electrical current is produced by some type of generator. These include hydroelectric, fossil fuel, renewable energy, and nuclear powered electrical generators.

Alternating current is produced by an AC generator, also called an alternator. The alternator is powered by any of the four sources of energy mentioned above.

Hydroelectric Power Generation

Electric power is generated by hydroelectric generation plants using the energy of falling water to turn a generator and produce electricity. The water turns the turbine, which drives the generator that produces the power. Transformers step up voltage to requirements needed to transmit it over electric lines to various destinations. A typical installation has numerous turbines, generators, and transformers (Fig. 47.3).

Fossil Fuel Powered Generators

Fossil fuels used to produce electricity include coal, oil, and natural gas. Fossil fuel power plants use rotating machinery to convert the heat energy of combustion into mechanical energy, which operates an electrical generator. The process may utilize a steam turbine, gas turbine, or, in small isolated plants, a reciprocating internal combustion engine.

Generated waste heat must be released to the atmosphere, often using a cooling tower or river or lake water as a cooling medium. Fossil-fueled power stations are major emitters of greenhouse gases. Flue gases from combustion of fossil fuels are discharged into the air and contain carbon dioxide and water vapor, as well as other substances, such as nitrogen, nitrous oxides,

Figure 47.3 Hydropower uses water turbines to drive electric generators.

sulfur oxides, fly ash, and mercury. Since fossil fuels are nonrenewable resources—and are becoming increasingly expensive—alternate sources, such as solar, wind, and nuclear fission, are finding increasing use. Experiments aimed at generating electricity by burning waste materials are also underway.

Nuclear Powered Generators

Nuclear powered electric generation plants are similar to fossil fueled plants in that they produce steam to run a steam turbine that drives a generator. The energy is produced in a nuclear reactor by fission. Fission is a process in which the centers (or nucleuses) of certain atoms are split when struck by subatomic particles called neutrons. The products of fission fly apart at high speed and generate heat as they collide with surrounding matter. The fission reaction is controlled in the nuclear core of a reactor, which consists of fuel rods in a chemical form of plutonium or uranium and thorium. Heat energy is produced by the fission reaction of the nuclear fuel. The heat is removed by a coolant and used to produce steam to drive a steam turbine, which drives an electrical generator (Fig. 47.4).

Spent fuel from the nuclear fission process is highly radioactive and must be handled and stored following strict guidelines. Spent fuel rods are stored in shielded basins of water usually located on-site. The water provides both cooling for the continuously decaying fission products and shielding from radioactivity. To ensure safety

Figure 47.4 A simplified illustration showing how a nuclear electric generating plant operates. (*Courtesy Carolina Power and Light Company*)

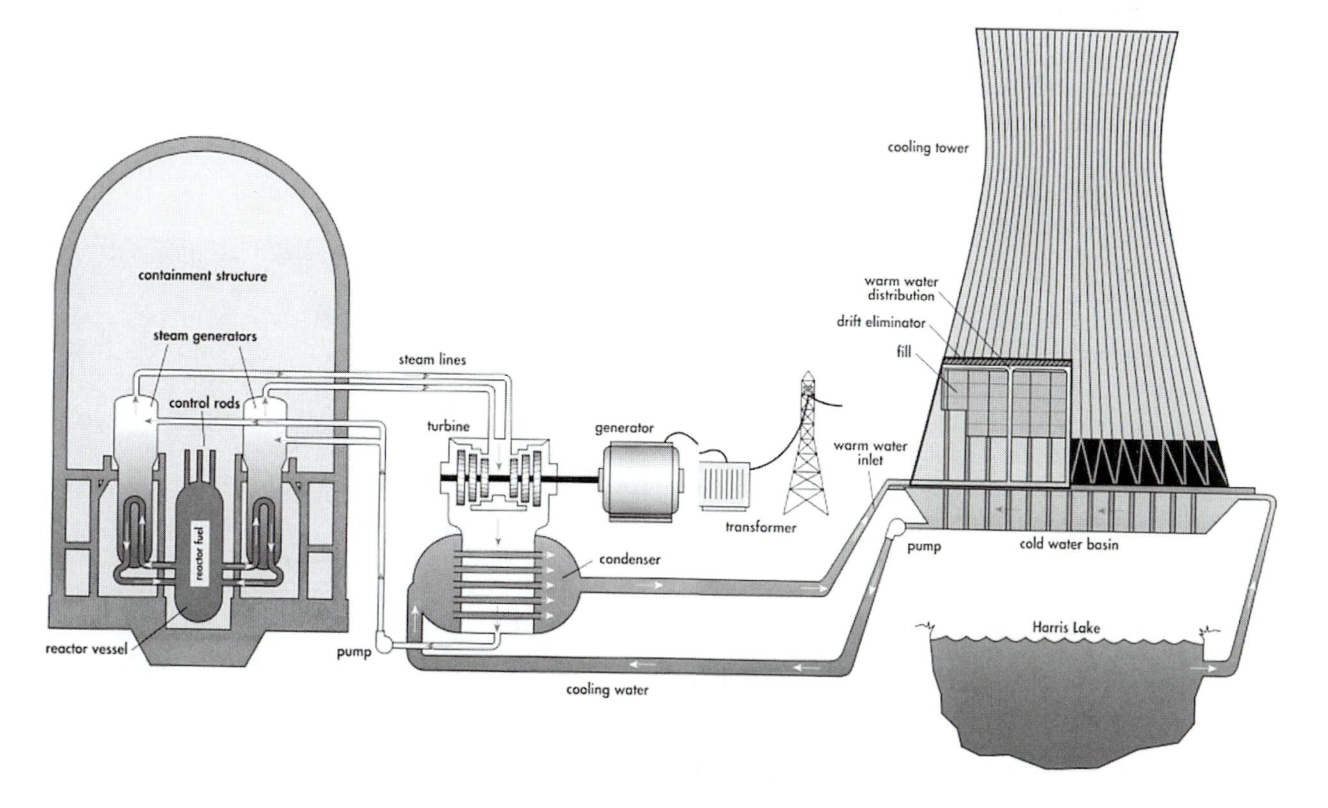

in a nuclear plant, the concept of "defense in depth" is employed. There are several layers of protection, each independent of the others, so if one failed others will continue protecting the plant, the workers, and the general public. Work is now underway to find methods of reprocessing spent fuel that could potentially recover up to 95 percent of remaining uranium and plutonium.

ON-SITE POWER GENERATION

On-site power generation is used when a utility system is not available or is unable to provide reliable service. Some systems provide additional service during peak periods. Certain facilities, such as hospitals, have on-site backup power generation systems that engage when the public utility system fails.

The methods used to generate electricity on-site include wind turbines, solar photovoltaic cells, thermal sources, gas- or steam-powered turbines, and cogeneration.

Thermal sources use an engine or turbine coupled to the shaft of a generator to produce electricity. Internal combustion engines run on fuels such as diesel fuel, methane gas, or natural gas. These are typically used as emergency backup sources of electricity for a vast range of commercial facilities, such as hospitals, banks, computer centers, stores, schools, and waste water treatment plants. Backup systems include the generator set, transfer switches, and paralleling switchgear.

Turbines are either gas- or steam-powered. Gas turbines burn various types of gaseous and liquid fuels, such as natural gas or fuel oil. Steam turbines are driven by a source that produces a large quantity of high-pressure steam. The revolving turbine drives the generator, producing electricity. The steam is produced by a boiler fired with a fuel, such as coal, solid waste, natural gas, or oil.

Solar Energy

Solar energy systems may be active or passive. A passive solar heating system is one in which thermal energy flows by radiation, conduction, or natural convection. Active solar systems use mechanical assistance to utilize solar energy with manufactured components that convert this energy to thermal energy and electrical power. The components include air and liquid flat plate collectors, concentrating collectors, and vacuum tube collectors.

Photovoltaic (PV) systems harness solar radiation and convert it to electricity. Photovoltaic semiconductor materials, including silicon, gallium, arsenide, copper indium diselenide, and cadmium telluride, exhibit a property that causes them to absorb photons of light and release electrons. When these free electrons are captured, an electric current that can be used as electricity is generated. Silicon is the most popular photovoltaic material in use today.

Components of a Photovoltaic System The fundamental building block of a photovoltaic system is the solar cell, a thin wafer consisting of an ultra-thin layer of phosphorus silicon on top of a thicker layer of boron silicon (Fig. 47.5). When sunlight strikes the surface of the cell, an electric field generate between these two materials and provides momentum and direction to light-stimulated electrons, resulting in a flow of current.

Solar cells are connected in series or parallel circuits to produce higher voltages, currents, and power levels. Photovoltaic modules consist of PV cells sealed in a protective enclosure. Photovoltaic panels include one or more PV modules assembled as a pre-wired, field-installable unit. A photovoltaic array is the complete power generating unit, consisting of any number of PV modules and panels (Fig. 47.6).

Additional components are required to make up an entire PV system. Mounting components support the panels and point them toward the sun. Inverters take the direct-current electricity produced by the modules and "condition" that electricity, usually by converting it to alternate-current electricity. Stand alone systems use batteries to store electricity for later use. All these items are referred to as "balance of system" (BOS) components.

Stand-alone PV systems are often used in places where utility-generated power is either unavailable (because the area is so remote from power plants) or too costly to hook up to because of the price of extending power lines. Batteries are used in stand-alone PV

Figure 47.5 A close up view of polycrystalline photovoltaic solar cells.

Figure 47.6 A complete photovoltaic array is composed of numerous connected solar panels.

Figure 47.8 Building-integrated photovoltaics may be used to provide shading while simultaneously generating electricity.

systems to store energy produced during the day and supply it to electrical loads as needed during nighttime and periods of cloudy weather.

PV Grid-Connected Systems are similar in components to stand-alone systems except a building is also connected to utility grid power (Fig. 47.7). The PV array can provide a portion of household energy demand while grid power supplies the remainder. When the batteries' voltage reaches a preset low point, an automatic transfer switch connects the grid to electrical loads.

With **net metering**, excess electricity produced from a photovoltaic system can be returned to the local utility, either by sale or crediting to an account. As the PV system produces electricity, kilowatts are first used to meet any electric requirements in the building. If the PV system produces more electricity than the building needs, the extra kilowatts are fed into the utility grid. An approved, utility-grade inverter converts the DC power from the PV modules into AC power that exactly matches the voltage and frequency of the electricity flowing in the utility line.

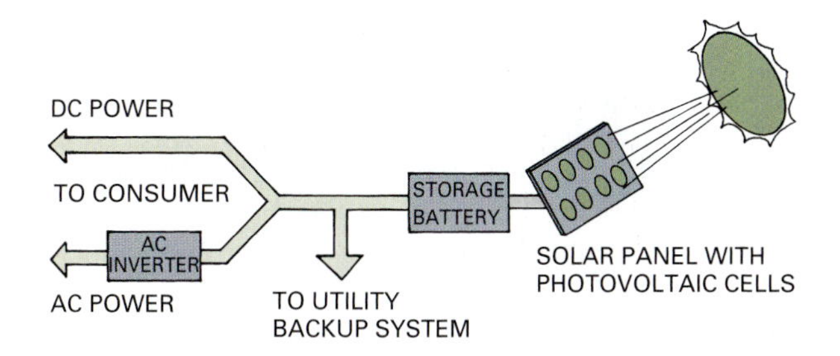

Figure 47.7 A schematic of a grid-connected photovoltaic power system.

Photovoltaic arrays must be mounted on a stable, durable structure that can support the array and withstand wind, rain, hail, and other adverse conditions. Sometimes, this mounting structure is designed to track the sun. The ideal orientation for a PV array is due south and tilted at an angle of 15 degrees higher than the latitude of the site.

Building-integrated PV (BIPV) provide an innovative and economical way to apply photovoltaics to a building. PV glazing modules can be integrated into a building as windows, skylights, curtain walls, roofing shingles and tiles, or shading elements (Fig. 47.8). Because solar products replace conventional materials while simultaneously producing electrical power from sunlight, material savings can be realized.

Wind Energy Systems

Another source of electrical energy is derived from prefabricated systems using wind to drive turbines. Wind turbines are available for a variety of applications, from residential stand alone or grid-connected to farm- or community-scale systems. Full-scale utility projects typically interconnect to existing local power distribution lines.

Horizontal axis turbines consist of four basic parts. A concrete or steel foundation is used to support the structure. The tower, commonly of lattice or tubular steel, contains the electrical conduits and supports the nacelle, which houses a generator and gearbox. The spinning blades are attached to the generator through a series of gears that increase the rotational speed of the blades and produce electricity. The most common tower is comprised of a white steel cylinder from 150 to 200 feet in height and 10 feet in diameter (Fig. 47.9). All towers require a ladder for access and a hoist for tools and equipment.

New designs for vertical axis turbines, including egg-beater and helical forms, are under development, although they are generally less efficient than horizontal axis systems. Similar to photovoltaic systems, wind turbines are integrated with an inverter to convert generated power to usable voltages. Battery backup is used if the system is not grid connected. Modern turbine technology is allowing architects and engineers to incorporate wind energy systems directly into buildings (Fig. 47.10).

Figure 47.9 The most commonly used wind turbine is comprised of a white steel cylinder ranging from 150 to 200 feet in height.

Figure 47.10 This high-rise tower in Dubai integrates wind turbines directly in the architecture.

Cogeneration

Cogeneration, also called combined heat and power (CHP), is the use of a heat engine or power station to simultaneously generate both electricity and useful heat. Conventional power plants emit heat, created as a by-product of electricity generation, into the environment through cooling towers or flue gases. CHP captures the waste heat and utilizes it for domestic or industrial heating purposes (Fig. 47.11). For example, if an internal combustion engine or a turbine is used to produce on-site electricity, the heat produced can be reclaimed and used to heat water or produce steam, which can be used to heat the building or provide cooling using an absorption chiller. An auxiliary conventionally fired boiler is used in the system to provide extra hot water or steam if needed.

Salvaged heat may also be used to produce steam used to drive a turbine, providing a supplemental on-site supply of electricity. This is especially effective in areas where high-pressure steam is required for an industrial process—such as food processing or pulp and paper manufacturing—that produces a steady, high-temperature source of wasted heat.

ELECTRICAL POWER CONDUCTORS

A major consideration in designing and installing an electric power system is the safe transmission of power to its end use. Electric power systems have the potential to cause fires, property damage, and human injury and death. Electrical codes are strict, and inspection during construction must be thorough. A crucial consideration

Figure 47.11 Cogeneration utilizes heat that normally would be wasted to produce on-site electricity or to heat or cool a building.

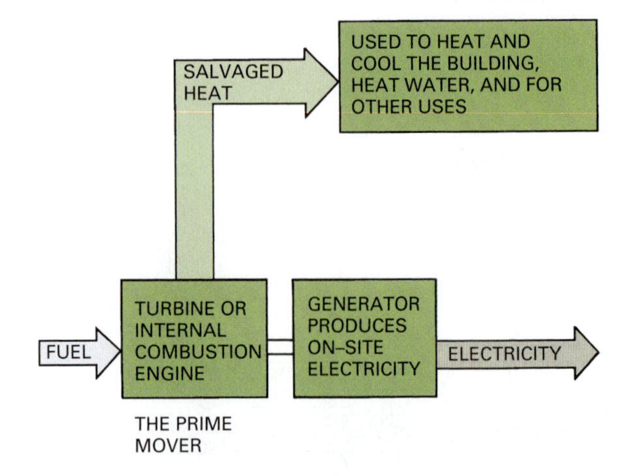

is whether or not to isolate electrical conductors as they pass through a building until they reach their point of use, such as an electric light.

Cables

Electric cable design and use is strictly regulated by codes such as the *National Electrical Code* published by the National Fire Protection Association. The types of conductors (Fig. 47.12) rated by this code are:

Type AC. Insulated conductors are wrapped in paper and enclosed in a flexible, spiral-wrapped, metal covering. An internal copper bonding strip in contact with the metal covering provides a means of grounding. Referred to as **BX cable**, it is used only in dry locations.

Type ACL. This type has the same insulation and covering as Type AC, but it uses lead-covered conductors that make it useful in wet applications.

Type ACT. The individual copper conductors have a moisture-resistant fibrous covering and run inside a spiral metal sheath.

Figure 47.12 Several of the commonly used electrical conductors.

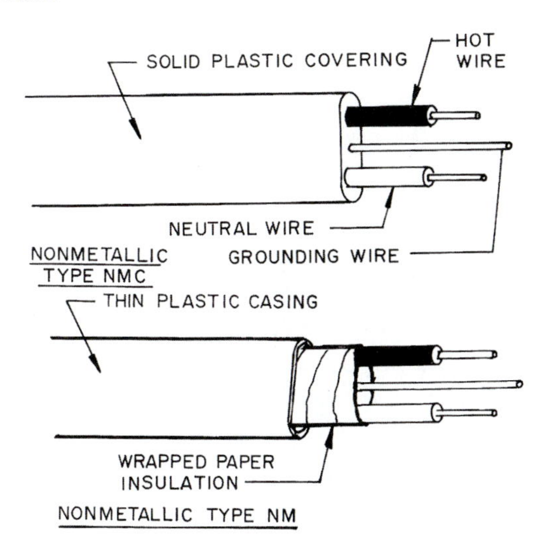

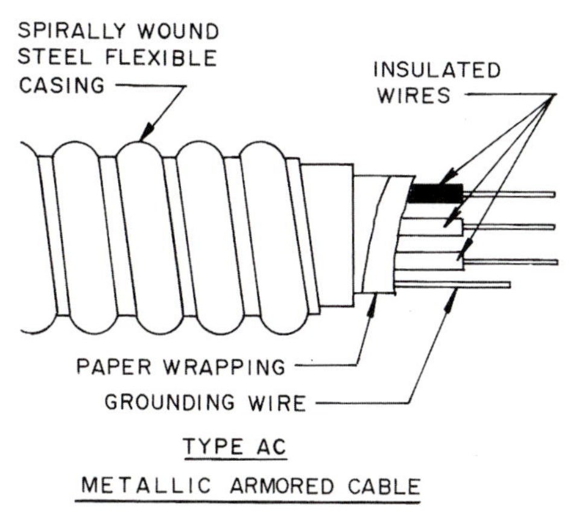

Type MC. Insulated copper conductors are sheathed in a flexible metal casing. If it has a lead sheath, it can be used in wet locations. It is a heavy-duty industrial feeder cable.

Type MI. Conductors are mineral insulated and sheathed in a gas-tight and watertight metal tube. This type can be fire-rated and is used in hazardous locations and underground.

Type NM or NMC. Non-metallic-sheathed cable used in protected areas. It is also called **Romex**. NM has a flame-retardant and moisture-resistant outer casing and is restricted to interior use. NMC is also fungus resistant and corrosion resistant and can be used on exterior applications.

Type SE or USE. Moisture-resistant, fire-resistant insulated cable with a braid of armor providing protection against atmospheric corrosion. Type USE has a lead covering, permitting use underground. It is used as underground service entrance cable. Type SE works for service entrance wiring or general interior use.

Type SNM. Conductors are situated in a core of moisture-resistant, flame-resistant, nonmetallic material. This assembly is covered with a metal tape and a wire shield and sheathed in an extruded non-metallic material impervious to oil, moisture, fire, sunlight, corrosion, and fungus. It can be used for hazardous applications.

Type UF. Conductors are enclosed in a sheath resistant to corrosion, fire, fungus, and moisture and can be directly buried in the earth.

Flat Conductor Cables

Flat conductor cables consist of copper cables formed flat and embedded in a plastic sheathing. A typical cable is 0.030 in. (0.78 mm) thick and around $2\frac{1}{2}$ in. (63.5 mm) wide (Fig. 47.13). The conductor is placed on the subfloor, a shielding material is placed over it, and carpet

Figure 47.13 A typical under-carpet wiring system. It uses a flat power cable that runs under the carpet. Outlets are placed as needed along the cable.

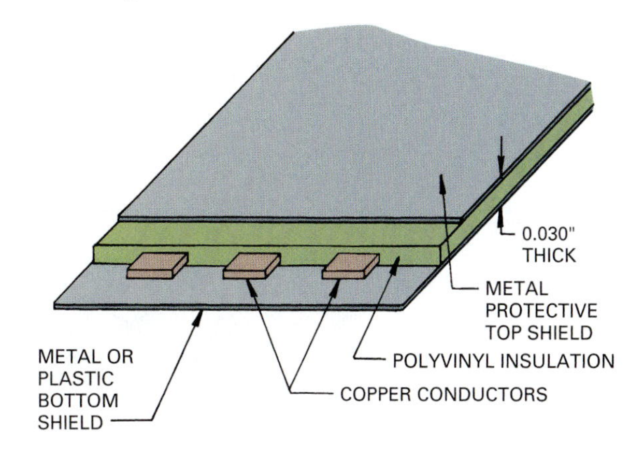

is installed. These cables are used for standard 120V electric power and in communication and data systems. Outlets are installed as required by making connections through the carpet (Fig. 47.14). Codes generally require that flat cable be covered with carpet squares to make it easily accessible.

Cable Bus and Busways

A **bus** is a bare conductor run in a metal trough, called a **busway**. The conductors are mounted on insulators to keep them clear of the trough. Electrical connections are made to the conductors with various types of plug-ins (Fig. 47.15).

RACEWAYS

Raceways are used to support, enclose, and protect electrical wires.

Cable Trays

Open raceways, or **cable trays**, are open-faced metal channels used to provide support for electric wires that have adequate insulation and do not require extra protection. Cable trays only support the wires. Wires in this system are open to inspection and modifications (Fig. 47.16).

Conduit

Conduit is a form of closed raceway. It supports insulated electric wire and provides protection. Conduit does not have conductors inside when installed. The conduit is installed first and the wires run through later (Fig. 47.17). Conduit can be run inside walls, ceilings, below and through floors, and in concrete slabs. Codes regulate the use and locations of the various types. Metal and non-metal conduit can be left exposed to view when appearance is not important.

Figure 47.14 Flat cable electric power, communications systems, and electronic systems are placed below the carpet and have outlets mounted on top of it.

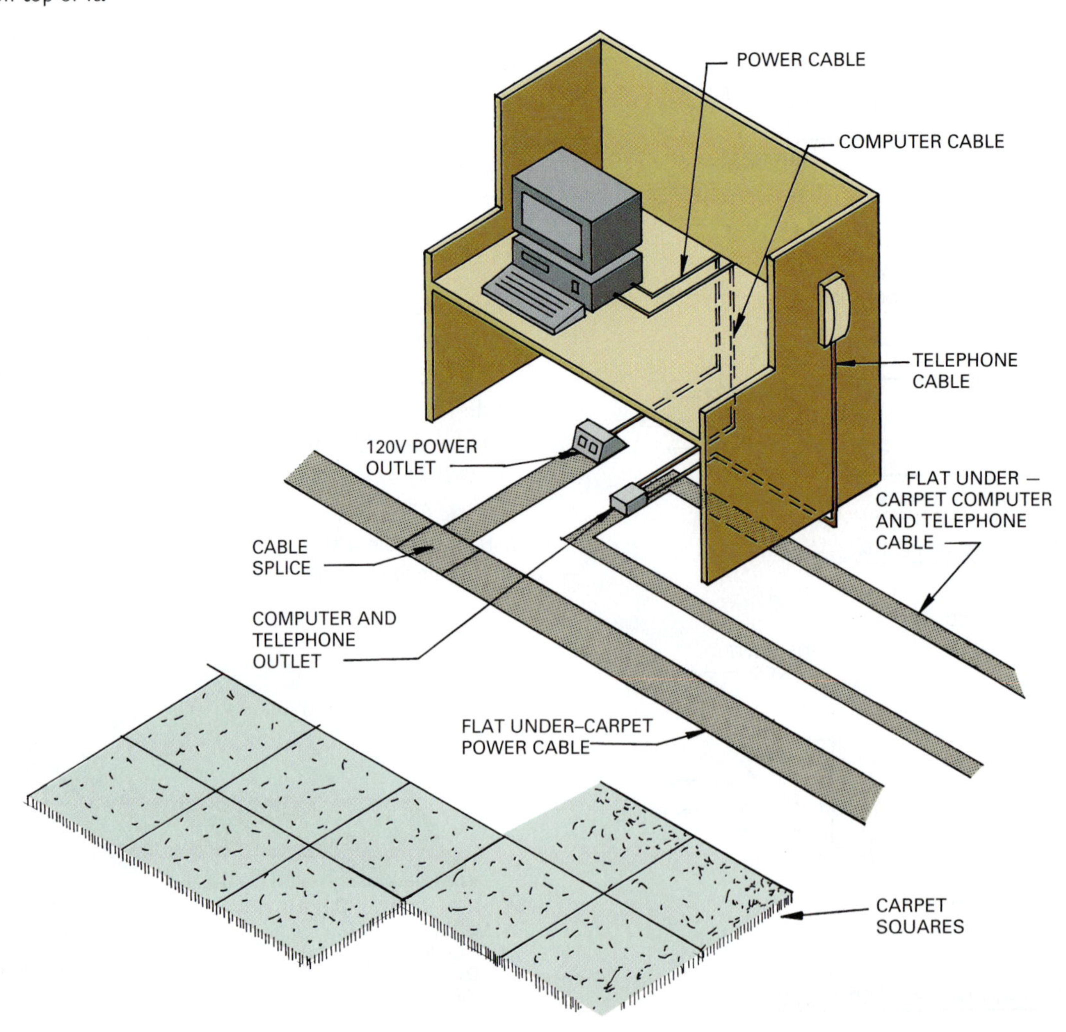

Figure 47.15 These lights are mounted on busways secured to the ceiling. The lights can be placed anywhere along the busway.

One type of conduit is a steel pipe available in three thicknesses: the heavy-wall conduit is referred to as rigid steel conduit (RSC); the intermediate-wall type is called intermediate metal conduit (IMC); and the thin-wall type is called electric metallic tubing (EMT) (Fig. 47.18).

Flexible metal conduit, called Greenfield, is used for short runs, such as connecting a furnace to its power source. It is available as a watertight conduit (Fig. 47.19). Rigid non-metallic conduit is also available in polyvinyl chloride (PVC) and high-density polyethylene.

Rigid metal conduit is available with inside diameters ranging from $\frac{1}{2}$ to 4 in. (12.7 to 101.6 mm). A thin-wall conduit can be bent to form curved corners. Junctions and sharp turns are made with metal fittings. Steel heavy-wall conduit is used in concrete slabs.

Conduit made from aluminum is also available in the same sizes. It is lightweight and non-sparking, weathers well, and is easy to work with. If embedded in concrete it may cause cracking. When used in underground locations it should be coated with asphalt or another type of protective coating.

Other raceways are made in rectangular shapes and are intended to be surface mounted and exposed to view. They are painted to match walls and ceilings and can incorporate electric outlets, telephone, computer and communications wiring connections (Fig. 47.20).

Metal raceways can be incorporated in a cellular steel floor deck. Metal decking serves as a substrate to support a cast-in-place concrete floor. Cells in the decking are used to carry electrical and communications wires (Fig. 47.21). Perpendicular to the cells in the steel decking are metal ducts spaced as required with outlets through which wires may be pulled to provide power for an area. These are typically spaced on a grid, permitting electric power to be available in a number of places in the floor.

METERS

The amount of electric power used is measured in watt-hours (Wh). Since the amount increases rapidly, it is reported in kilowatt-hours (kWh). One kWh is equal to 1,000 Wh. The amount used is measured by a kilowatt-hour **meter** (Fig. 47.22). Three-wire meters are typically used for residential applications. Commercial and industrial applications having heavy demands and higher voltage requirements typically use four-wire meters.

An electric utility supplies meters, which are usually placed outside buildings to provide ready access to utility staff. Meters can be installed inside a building if easy access is provided. The owner of a building provides meter pans and any current transformers needed to step down voltage within the building.

MOTOR CONTROL CENTERS

Motor control centers are used to start and stop motors and protect them from potential overloads. They have a disconnect switch that must be located within sight of the controller and the motor. The overload protection is typically a heat-operated relay that opens the circuit when line temperatures rise. The motor circuit is closed after the overload situation has been corrected by pressing a reset button (Fig. 47.23).

TRANSFORMERS

Transformers are used to change (transform) alternating current from one voltage to another. The voltage coming into a transformer is the primary voltage, and voltage leaving the transformer is the secondary voltage. For example, a transformer is used to step down a primary 4,160V current received from a utility distribution system to a secondary voltage, such as 480V, as it enters a building. Another transformer in the building's vault (or closet) could step this down to the 120V required for use within the building. The generally available primary

Figure 47.16 Some typical cable trays.

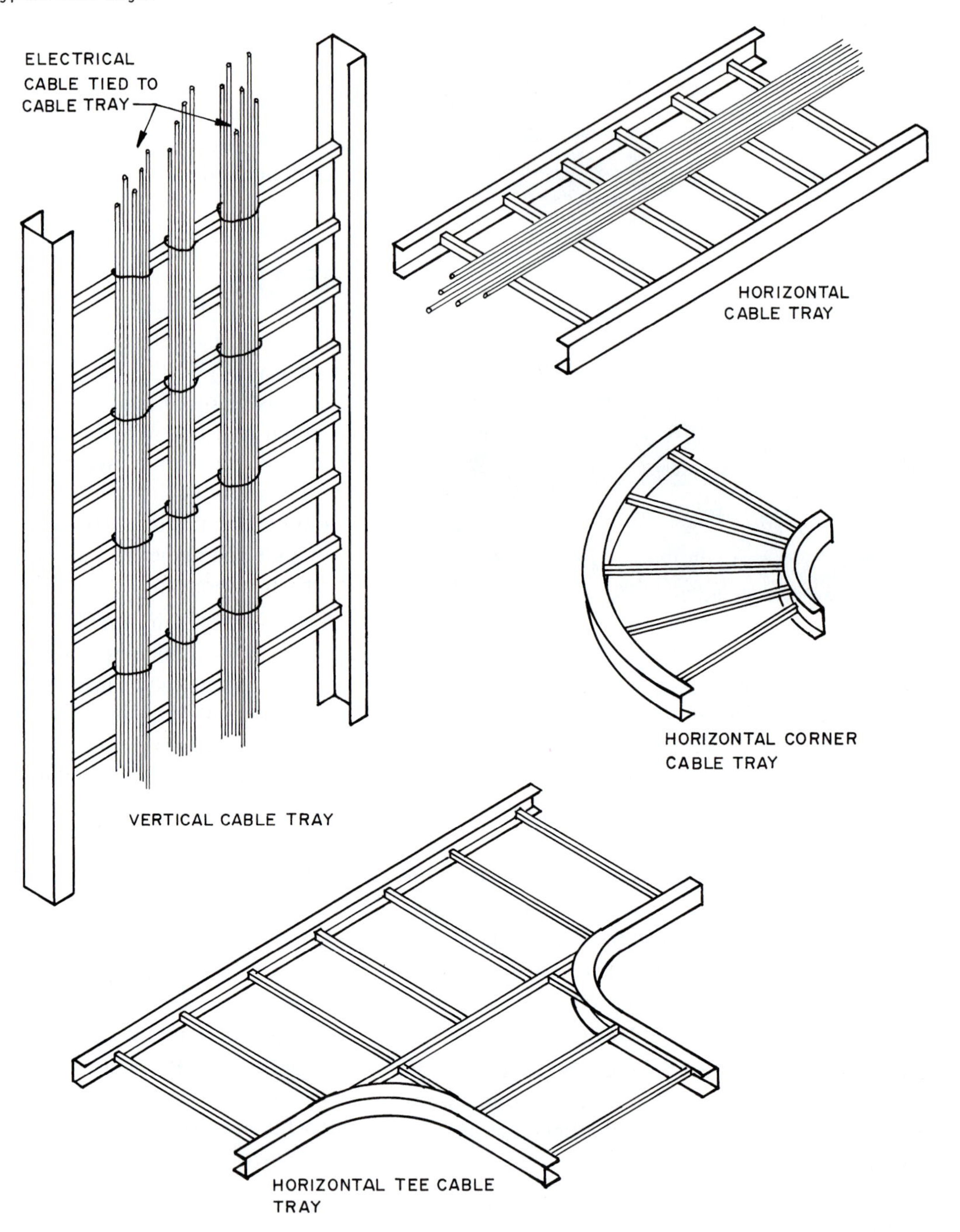

voltages include 2,400V, 4,160V, 7,200V, 12,470V and 13,200V. Secondary voltages include 480V, 277V, 240V, 208V, and 120V.

Transformers generate heat, which must be removed to prevent overheating. They may be dry (air cooled) or liquid cooled. Dry transformers remove heat by circulating air through spaces in the transformer. Liquid-cooled transformers circulate oil through coolers that absorb the heat and transfer it to the outside air. Mineral oil is the least expensive liquid, but it is flammable and cannot be used in all locations. A number of other liquid coolants, such as silicone liquids, have low flammability and are widely used. Liquid transformers are used on large installations.

Transformers are specified by the voltages in and out, the amount of power they can handle in kilovolt-amps (kVA) (thousands of volts times the rated maximum amperage), also called insulation class, and sound rating. Insulation specification pertains to electrical insulation temperature ratings in degrees

Figure 47.17 This surface-mounted steel pipe conduit protects insulated wire.

Figure 47.18 Types of rigid metal conduit.

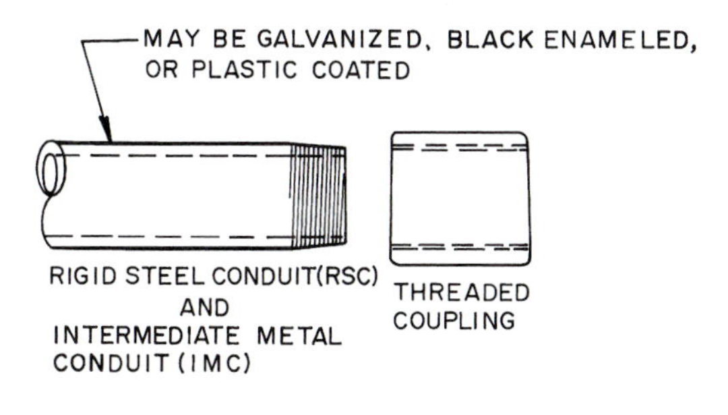

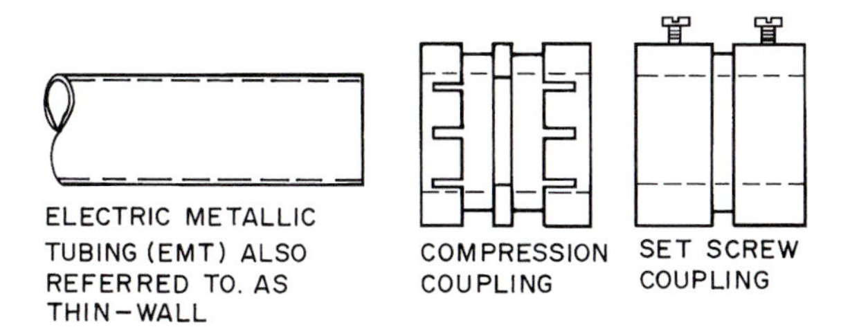

Figure 47.19 Flexible metal conduit is made with spiral-wrapped metal bands.

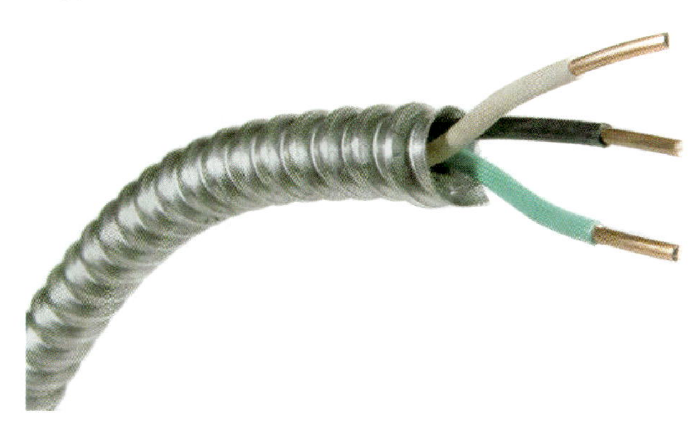

Figure 47.20 Rectangular raceways are used for surface-mounted wiring installations.

Figure 47.21 This cellular composite steel floor, Robertson Q-Floor, provides channels for running wiring and punch outs for the location of electrical outlets into the building. The outlet boxes are set in place before the concrete floor is poured. *(Courtesy H. H. Robertson)*

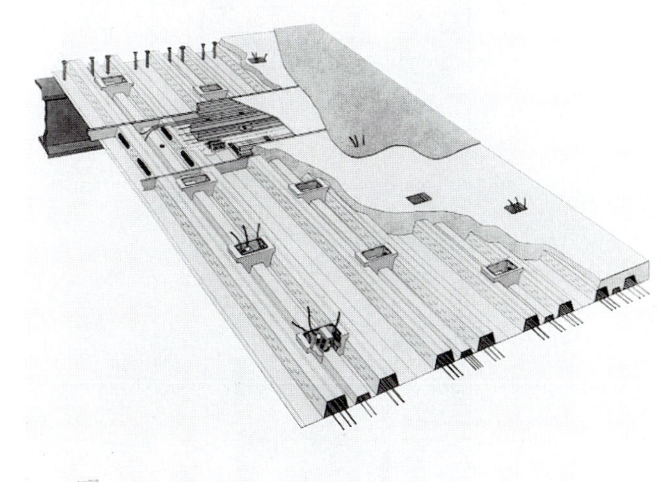

Figure 47.22 An electric meter is used to measure the kilowatt-hours of electricity used.

Figure 47.23 A motor control center utilizing electronic and programmable devices in addition to electromechanical controls. Motor controls start and stop motors and provide overload protection. *(Courtesy Allen-Bradley Co., A Rockwell Automation Business)*

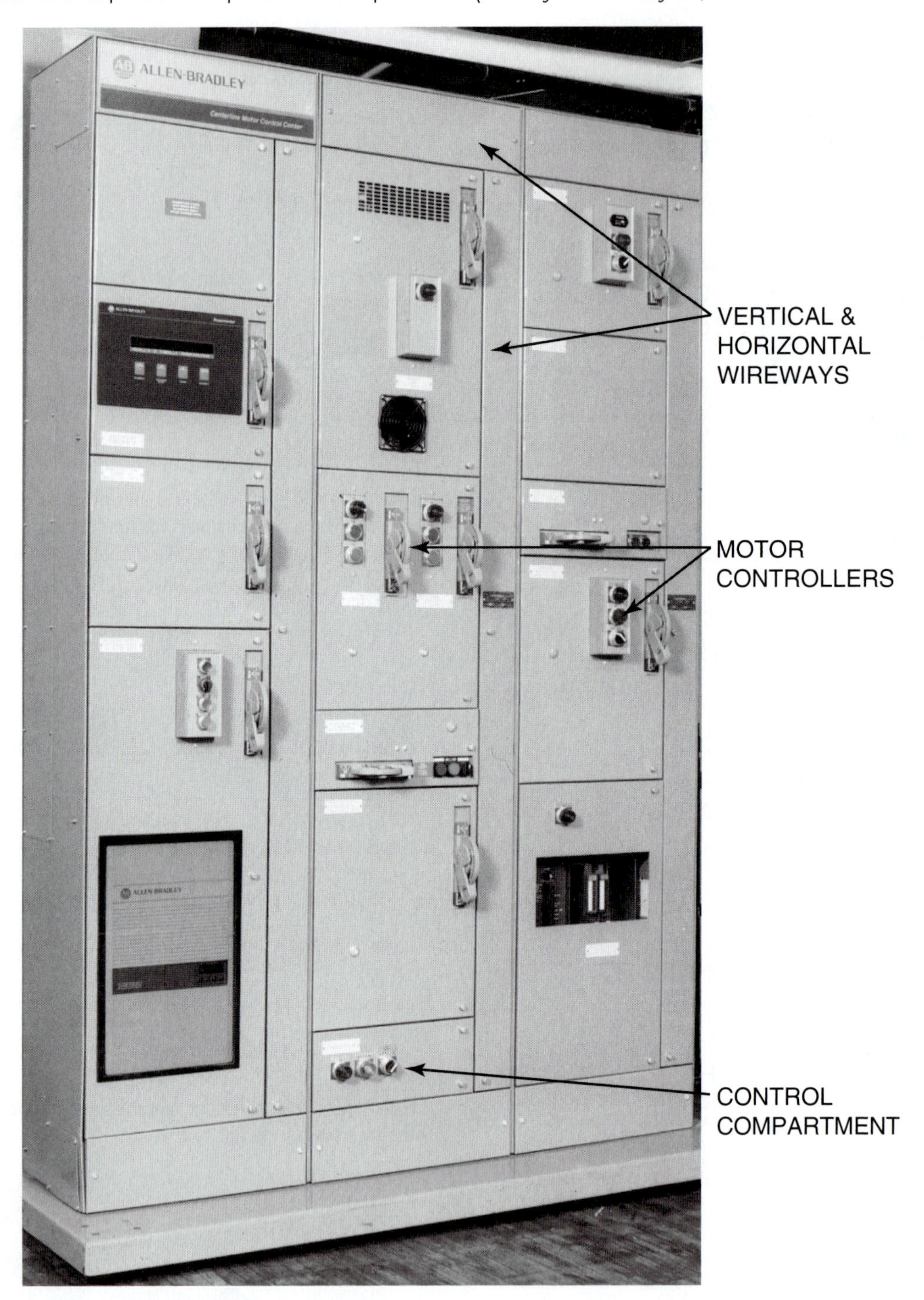

centigrade. Sound ratings have to do with the sound vibrations produced by the different types and sizes of transformers. Transformers are available in single-phase and three-phase construction. The term "phase" refers to the number of circuits, voltages, and currents in an alternating current system. Single-phase has only one circuit or path for the flow of current; three-phase has three paths of flow.

Transformers may be located indoors or outdoors. Dry-type outdoor installations must have weatherproof enclosures, and larger sizes must be kept away from combustible materials. Liquid-type transformers on or adjacent to buildings in which combustion is possible must have a means for protecting against fires caused by excess heat or oil leaks. This can be provided with fire-resistant enclosures, various types of barriers, or a sprinkler system.

SWITCHES

Switches are used to open electrical circuits and interrupt the flow of current. A typical switch used on low-power applications, such as lighting, has some means of physically opening and closing the circuit—manually moving a lever or pushing a button using an electric coil, a motor, or a spring. This action moves metal contacts that separate to open the circuit or close to complete the circuit. Solid-state switches also interrupt a circuit, but they do so by electronically creating a conducting or non-conducting condition. There are no moving parts. Switches are classified by the National Electric Manufacturers Association (NEMA) and Underwriters Laboratories (UL).

Different actions enable Switches to perform specific functions. These are illustrated by the simple knife switches shown in Fig. 47.24. A single-pole, single-throw (SPST) switch is moved to make or break a connection between two contact points. When the points are in contact, the circuit is complete, and when they separate or open, the circuit is broken.

A single-pole, double-throw (SPDT) action is used to control a unit, such as a light, from two locations. This requires an SPDT switch at both locations.

A double-pole, single-throw (DPST) action is similar to the SPST action. However, it opens both wires in the circuit. DPST switches are used when the wires are not grounded, for example, switches used on 240V motors and appliances. They can also be used to control two circuits at the same time. Both circuits may be open or closed simultaneously.

A double-pole, double-throw (DPDT) action can be used to control more than one circuit at a time. It can be used to reverse the direction of rotation of a DC motor by reversing the polarity.

Figure 47.24 Typical switch actions are illustrated by these knife switches.

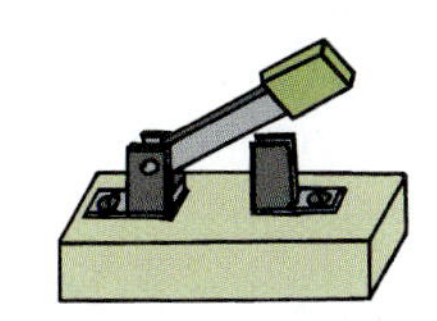

Switches that control circuits with loads up to 30A, such as lighting, are shown in Fig. 47.25. They are rated to carry various loads that typically run 15A, 20A, or 30A at 120V. They may be toggle, push, keyed, rocker, touch, tap-plate, or rotary type. They are available as SPST, DPST, DPDT, and SPDT and three-way and four-way switches. Three-way switches are used to control a light from two different locations. Four-way switches control a light from three different locations. Other switches can operate as timers that allow a unit, such as a fan, to operate for a set time after which it automatically shuts off. This type uses a spring-wound timer. A keyed switch provides security because a key is needed to operate it. Programmable switches are solid state switches that can be programmed to switch a circuit on and off at preset times.

The switches described above do not contain fuses. However, fusible switches are available (Fig. 47.26). The fuses and switches are enclosed in a metal box that can be closed and locked. The switch is manually operated with a handle on the outside of the box. Another type of switch using a solid state rectifier enables incandescent lights to have high, low, and off controls. A rectifier is an apparatus in which electric current flows more readily in one direction than in the reverse direction for the purpose of changing alternating current into direct current.

Other Switches

There are many other switches designed to serve special purposes. A service disconnect, disconnects all electric power to a building except for specific emergency equipment requirements. Generally, these are located outside a building. They may be part of a switchboard.

A contactor is a form of switch that uses an electromagnet to close the contact blocks. It is operated from a remote location and can be activated by various devices, such as pushbutton or thermostat. A remote-control switch is a form of contactor that remains latched until its electric circuit is energized. In this way, the coil energizes only when the circuit opens. It is used on applications where circuits must be kept closed (a light installation, for example) for long periods of time.

Time-controlled switches use some type of timer, such as an electronic timer or a miniaturized low-speed motor, that rotates a disc that contacts the switch to open and close the circuit.

An automatic transfer switch is a double-throw unit that switches to a source of emergency power when

Figure 47.25 Examples of switches typically used to control circuits with loads up to 30A.

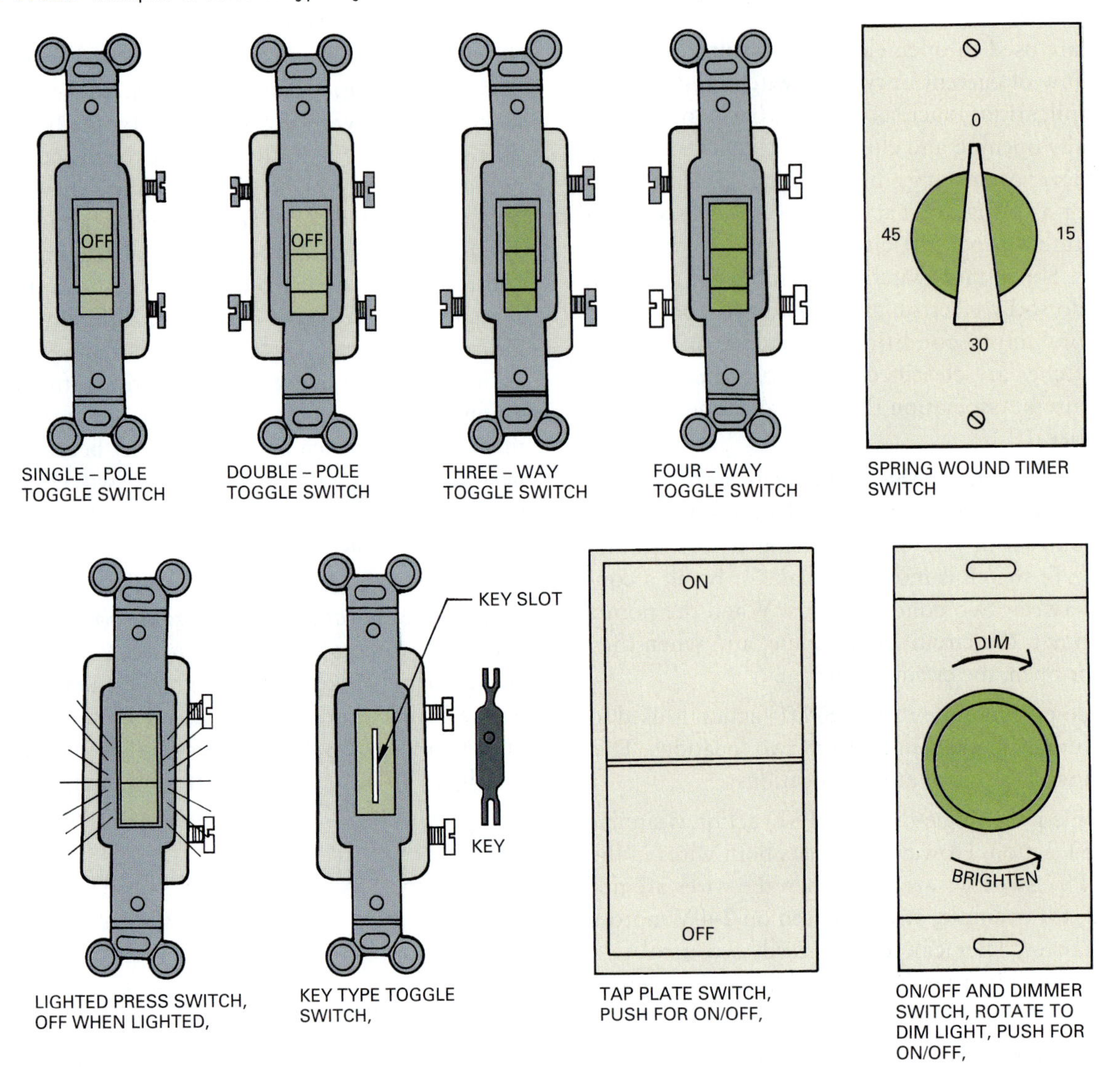

normal electrical service is interrupted (Fig. 47.27). A typical application would be in the power supply for a hospital operating room.

Isolation switches are opened only after current flow in a circuit has been interrupted by another regular-use switch. They are not used to interrupt the flow of current in a circuit.

OVERCURRENT PROTECTION DEVICES

Circuit breakers are electromechanical units that automatically disconnect a circuit when the current attains an established value that would cause overheating or a possible fire in the circuit. Examples of occurrences that could cause a circuit breaker to open are excess current flow caused by overloading or a short in the circuit. Circuit breakers are available in a molded case inserted in the circuit at the panel (Fig. 47.28) or a large air-type breaker not in a protective case. When a circuit breaker opens a circuit, the breaker is reset by moving the switch handle after the deficiency has been corrected.

Circuit breakers are mounted in a circuit breaker box or a panel board (Fig. 47.29).

A **fuse** is also used to protect circuits from overloads and shorts. Fuses have an internal metal link that melts when overheated, breaking the circuit. The two types of fuses are cartridge and plug (Fig. 47.30).

Cartridge fuses are either knife-blade or ferrule. The blades on the first type slip into metal clips on a panelboard, and the ferrule type's copper rings on each end fit into clips. This connects the incoming power at

Figure 47.26 This is a heavy-duty fused three-pole three-wire industrial switch. *(Courtesy Siemens Energy and Automation, Inc.)*

the service entrance to the bus bars in the panelboard. Most cartridge fuses are discarded after they blow, but some types have replaceable links.

Ferrule type cartridge fuses are rated from 10A to 60A and are generally used to protect currents for individual appliances, such as an electric stove. Knife-blade cartridge fuses work for service over 60A and are used in the service entrance between the incoming power line and the circuits in the panelboard.

Plug fuses are available in 15A, 20A, 25A, and 30A sizes. They are used to protect individual circuits requiring small current requirements, such as a series of lights. They are installed in a fuse box or panelboard and serve the same function as a circuit breaker. However, when plug fuses are blown, they are discarded and replaced with new fuses after the deficiency has been corrected.

There are several varieties of plug fuses. The standard plug fuse will blow when the fusible link overheats. The time delay plug fuse has a fusible link that will melt immediately only when a short circuit occurs. If there is an overload, the link softens but does not break. If the overload is quickly removed, such as when the momentary load that starts a large electric motor ceases, it will not break. However, if the overload continues the link will melt.

SWITCHGEAR AND SWITCHBOARDS

Switchgear and **switchboards** are freestanding units made up of an assembly of fuses and circuit breakers, switches, and other line components. They provide switching and feeder protection to circuits connected to a main power source. The switchboard distributes the incoming electrical power in smaller amounts within a building. Switchboards are manufactured with all components enclosed within a metal cabinet and controlled by insulated handles and push buttons in a front panel (Fig. 47.31).

Switchgear contains the same components and serves the same function as switchboards but handle higher voltages, usually above 600V. The term switchgear is also used to identify individual switching units not assembled in a panel.

Switchboards and switchgear used in commercial and industrial buildings are commonly placed in a basement vault designed specifically to house the switchgear. It must be adequately ventilated and meet requirements of the National Electrical Code. Typical space requirements are given in Table 47.2. The design of the vault should include adequate entrances and exits so equipment can be installed and removed and workers in the room can exit quickly in an emergency. Vaults must be located in permanently dry conditions, and any exceptions must meet specific code requirements. Switchgear installed outdoors is usually housed in a weatherproof metal case.

PANELBOARDS

A **panelboard** receives a large amount of electrical power from the public utility and distributes it in smaller amounts through a number of circuits. In small buildings,

Figure 47.27 An 800A three-pole automatic transfer switch with a control panel. When the normal flow of electricity fails, it will automatically transfer to the emergency service. *(Courtesy Automatic Switch Company)*

Figure 47.28 Ground fault (GFCI) and arc fault (AFCI) circuit interrupters.

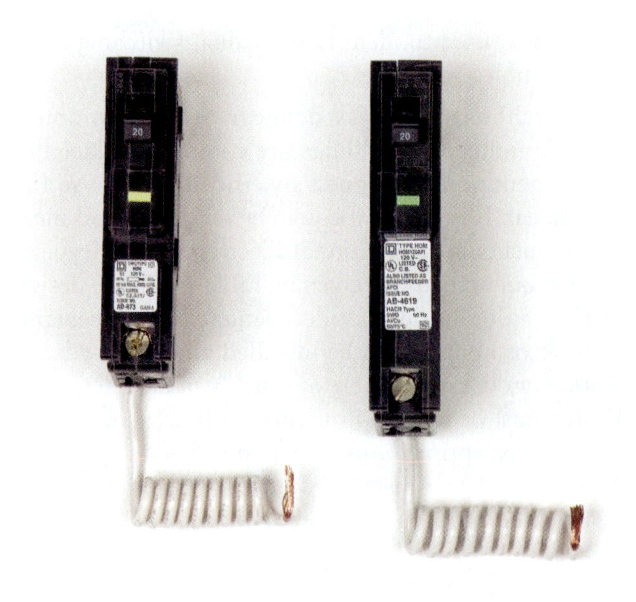

it serves the same basic purpose as a switchboard but on a smaller scale. A panelboard is typically used in residential and small commercial construction to distribute power to each of the circuits within a building after receiving input from the public utility line (Fig. 47.32).

A typical panelboard using circuit breakers is illustrated in Fig. 47.33. The incoming service entrance cable connects to the neutral bus bar and the main circuit breaker. The main controls the flow of power from the utility into the service panel. The white wire is neutral and connects to the neutral bus bar. The red and black wires connect to the main and through it to the snap-on connections for the circuit breaker. The circuit breakers control the power to each circuit.

Panelboards have locked metal covers so no live parts are exposed unless the door is open. Medium-duty panelboards are used for lighting and general purpose electrical outlets. Heavy-duty panels are used for distribution of power in industrial applications.

The panel may be surface-mounted, semi-recessed, or recessed into a wall. Each circuit is numbered, with a description identifying its location given on a chart, such as No. 6 – Kitchen wall outlets.

In large construction, panelboards are located on the various floors and fed from the service switches and switchgear. Lights, outlets, and other loads on a floor are controlled from a panelboard on that floor.

Figure 47.29 This circuit breaker box has one circuit breaker in place with blanks for six additional breakers.

Figure 47.30 Commonly used plug and cartridge fuses.

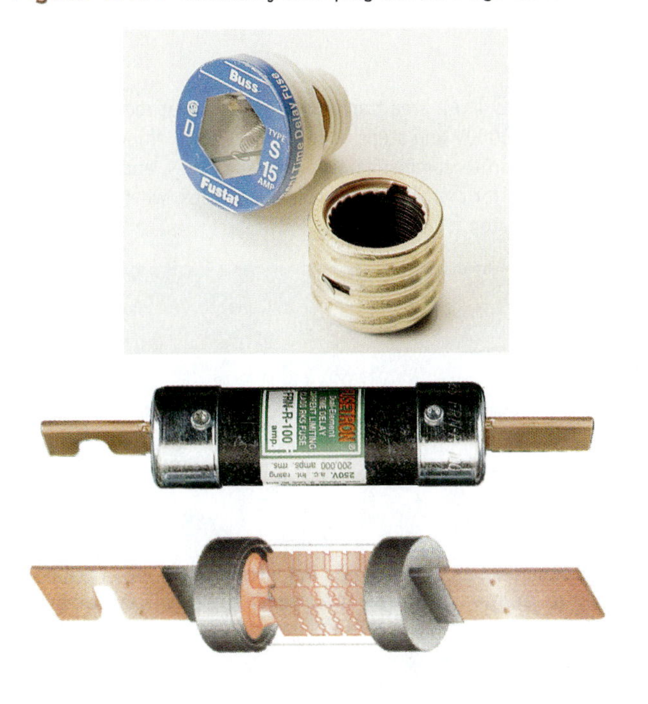

Figure 47.31 A switchboard is a single electric control panel or assembly of panels, enclosed in an insulated metal structure, on which switches, overcurrent devices, and instrumentation are mounted. The devices are controlled by handles on the front of the panel. *(Courtesy Square D Company)*

Table 47.2 Switchboard Working Clearances

Nominal Voltage to Ground	Minimum Clear Distance		
	Condition 1[a]	Condition 2[b]	Condition 3[c]
0–150	3 ft.	3 ft.	3 ft.
	(914 mm)	(914 mm)	(914 mm)
151–600	3 ft.	31/2 ft.	4 ft.
	(914 mm)	(1,066 mm)	(1,219 mm)

[a]Exposed live parts on one side and no live or grounded parts on the other side of the working space, or exposed live parts on both sides effectively guarded by suitable wood or other insulating materials. Insulated wire or insulated busbars operating at not over 300 volts shall not be considered live parts.
[b]Exposed live parts on one side and grounded parts on the other side.
[c]Exposed live parts on both sides of the work space (not guarded as provided in Condition 1) with the operator between.

ELECTRICAL SUPPLY

A public utility uses some form of energy or fuel to operate generators at their power plant that produce electricity for transmission to consumers. The electricity passes to a step-up transformer station, which increases voltage so the power can be transmitted with less current. This reduces line power losses and permits the use of smaller conductors. High-voltage transmission lines move the power to area transformer stations.

Area transformer stations step down the voltage of the power (Fig. 47.34), and it is transmitted with overhead and underground lines to different areas. Industries using large amounts of electricity often receive this power into their own transformer stations

Figure 47.32 This panelboard has the main breaker at the bottom and two vertical rows of circuit breakers.

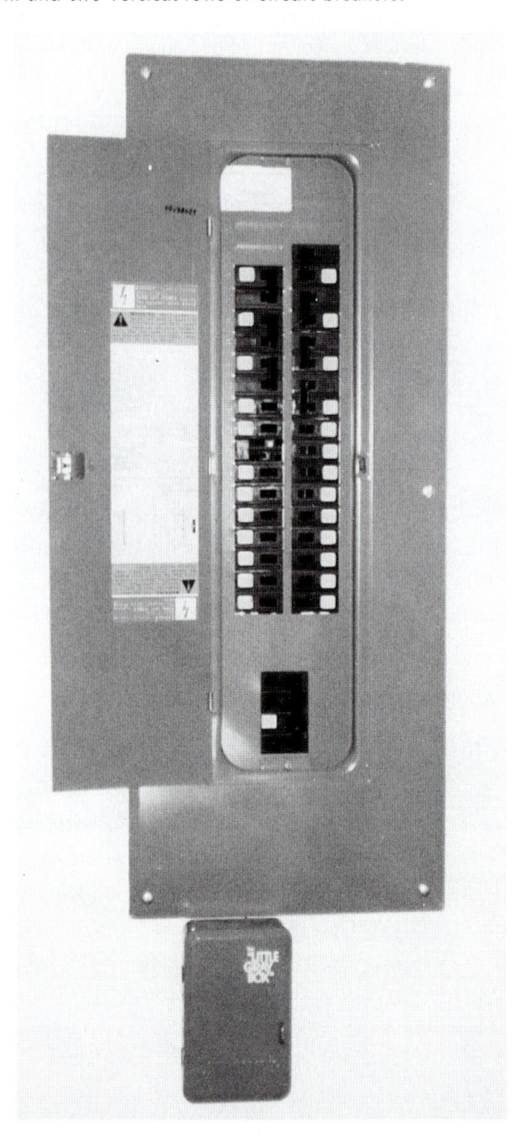

Figure 47.33 Assembly details for a typical panelboard using circuit breakers.

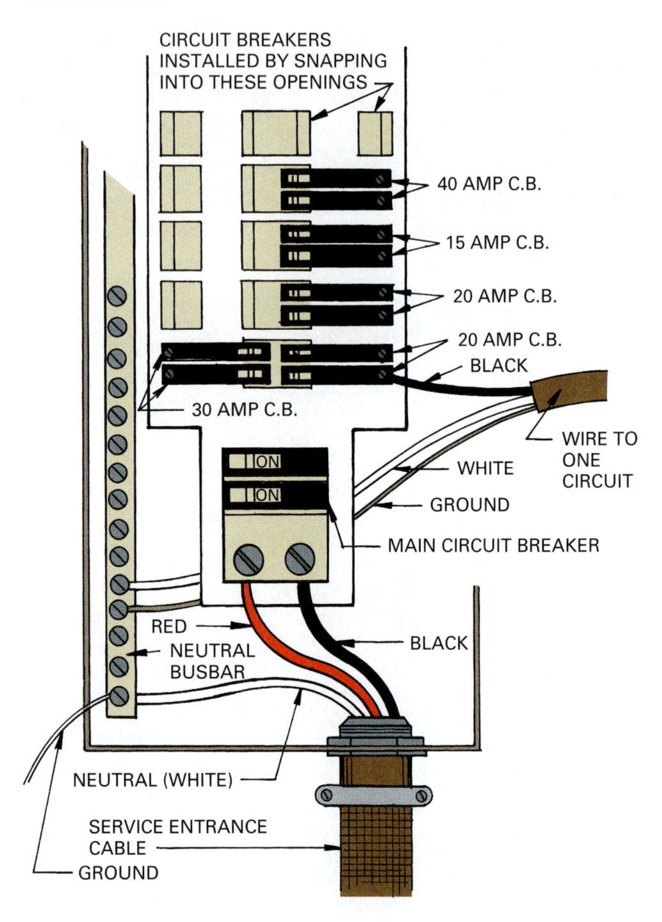

Figure 47.34 An area transformer station that receives power at 115,000V and steps it down to 23,000V three-phase, for distribution to commercial and industrial uses, and 13,200V single-phase, for distribution to residential uses, where on-site transformers step it down to 120/240V.

for distribution and voltage regulation within their plants. For most commercial and private consumers, the power is sent to a local transformer where it is stepped down to single-phase 120V or 120/240V power (Fig. 47.35). The 120/240V line is used to operate residential appliances (electric dryers, ranges, water heaters) and small industrial equipment requiring 240V. Larger commercial buildings may require either single-phase or three-phase at 120/208V, 120/240V, or 277/480V.

Power connects to a commercial building through a service entrance consisting of a disconnect switch, a meter, and distribution switchgear. Service entrances for residential and small commercial buildings consist of a meter and a panelboard.

ELECTRICAL DISTRIBUTION SYSTEMS IN BUILDINGS

Electrical distribution systems inside a building supply electrical power as needed to various sections and transmit information through an internal communications system. Standards for the design and installation of electrical distribution systems are maintained by the *National Electrical Code* published by the National Fire Protection Association. Underwriters Laboratory certifies electrical equipment and materials through the use of testing specifications and procedures.

Figure 47.35 Pole- and pad-mounted transformers are connected to high-voltage transmission lines. They step down the voltage for use by consumers.

The electrical power system of a commercial or industrial building includes the service entrance equipment, the interior distribution system, and the end equipment that uses the electricity. The service entrance incorporates equipment such as meters, switches, panelboards, switchgear, switchboards, fuses, and circuit breakers. The distribution system includes trays, wiring, raceways, wireways, conduit, and busways. The final part of the interior electrical system consists of the devices that use the electricity, including lighting, motors, heaters, and communications and industrial equipment. These constitute the loads that must be considered when designing a system.

The following discussion gives generalized examples. The actual design of a system varies considerably, depending on the requirements of the building and the equipment to be installed. The design of distribution systems requires the services of an experienced electrical engineer.

The Service Entrance

The *National Electrical Code* defines a **service entrance** as the conductors and equipment for delivering electricity from a utility to the wiring system of the building served. The service entrance includes the components that bring electrical power from the utility lines into the building. A typical service entrance for a residential or small commercial building is shown in Fig. 47.36. The meter is generally mounted on the exterior of the building with service supplied either overhead or underground. This enables the utility to read the meter and install or remove it without entering the building. From the service entrance, panelboard branch circuits run to various parts of the building for lights, outlets, appliances, furnace, and other required services. Each circuit has a circuit breaker overload device in the panelboard. Service can be run to a subpanel in an area some distance from the panelboard where branch circuits can be taken from it (Fig. 47.37). If the building has multiple occupancies, individual meters are set up for each unit.

A service entrance for commercial buildings can take various forms depending on occupancy needs. It requires a service switch or circuit breaker, which disconnects power to the entire building. This is typically located outside the building, but exceptions are made. The power enters through a meter installation that includes a meter pan, a meter cabinet, and a current transformer in a cabinet or vault. The service switch and

Figure 47.36 Typical above- and below-ground service entrances for residential and small commercial buildings.

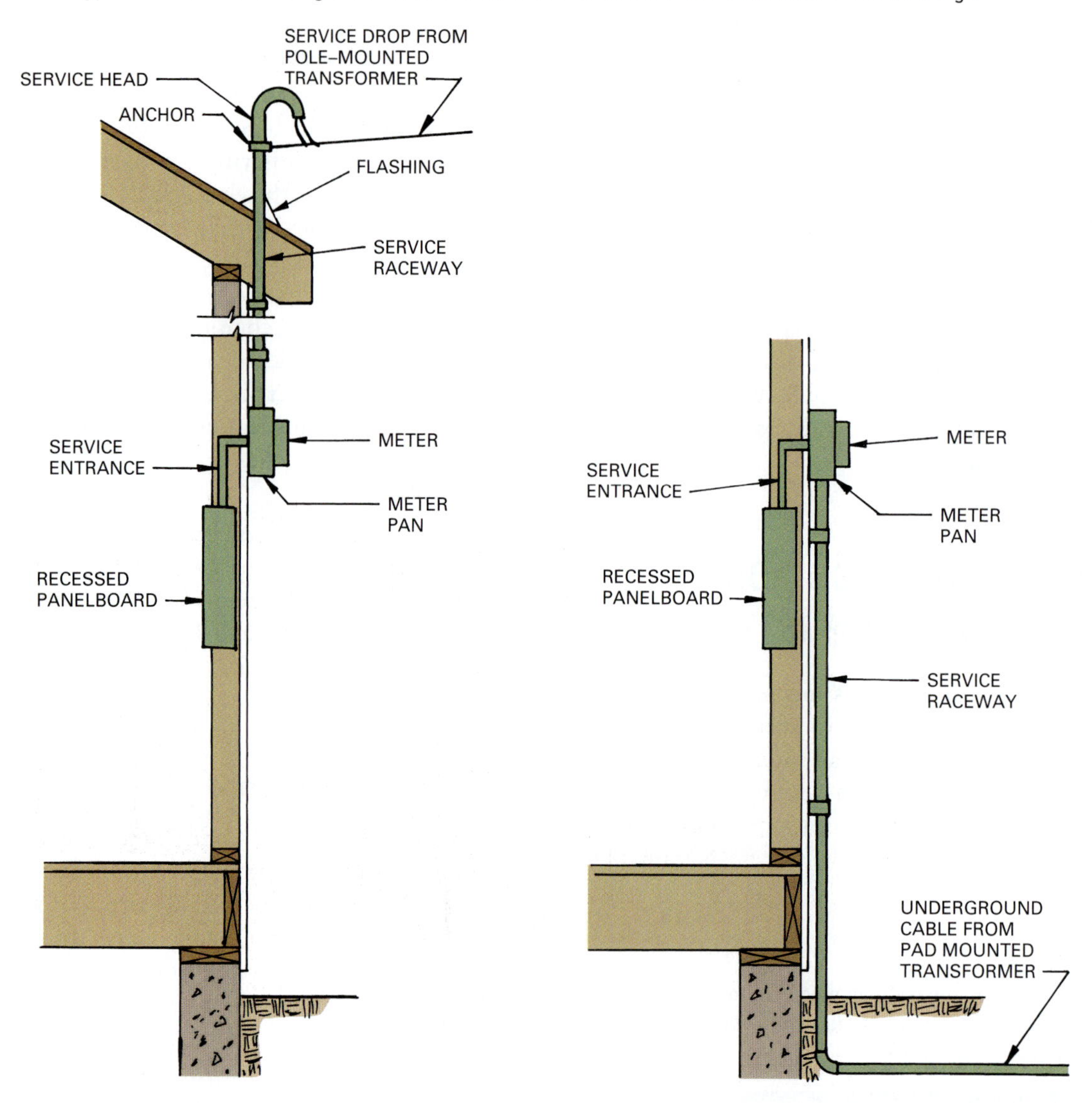

Figure 47.37 The service entrance panelboard is wired to distribute electricity through individual circuits to various parts of the building and to equipment.

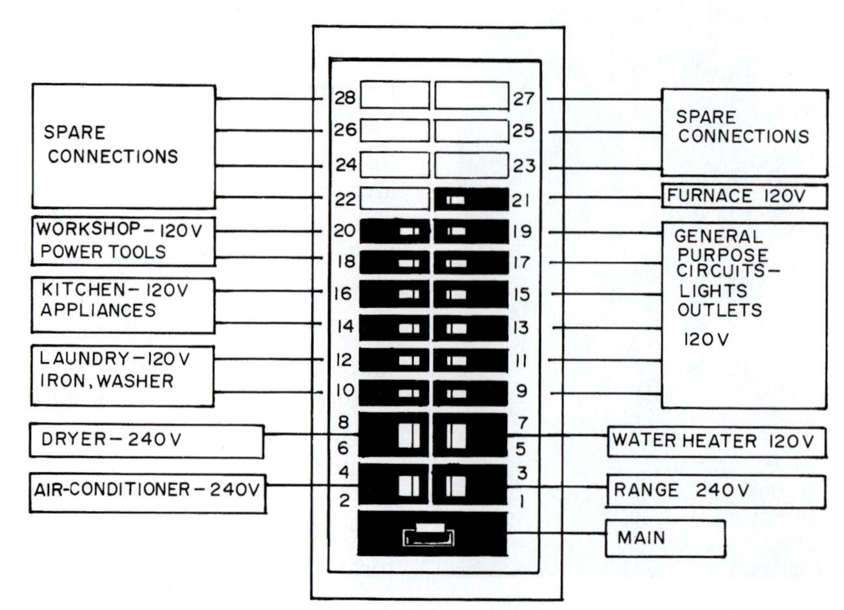

metering equipment may be assembled in one unit. The metering transformer steps down incoming electric current from the primary feeder to a voltage required for use in the building (120/240V, for example). It may sit outside the building on a concrete pad or inside or outside in a vault. Power feeds into the switchboard and is distributed to a number of individual circuits, as shown in Fig. 47.38.

LIGHTING

Sufficient light is essential for the proper functioning of a building. Standards for lighting design are written and maintained by the Illuminating Engineering Society of North America (IESNA). Most spaces integrate natural and artificial light to work together and provide the type and degree of illumination required. Natural

Figure 47.38 A generalized power riser diagram feeding from a switchboard to panelboards on the various floors and to equipment on the roof and in the basement.

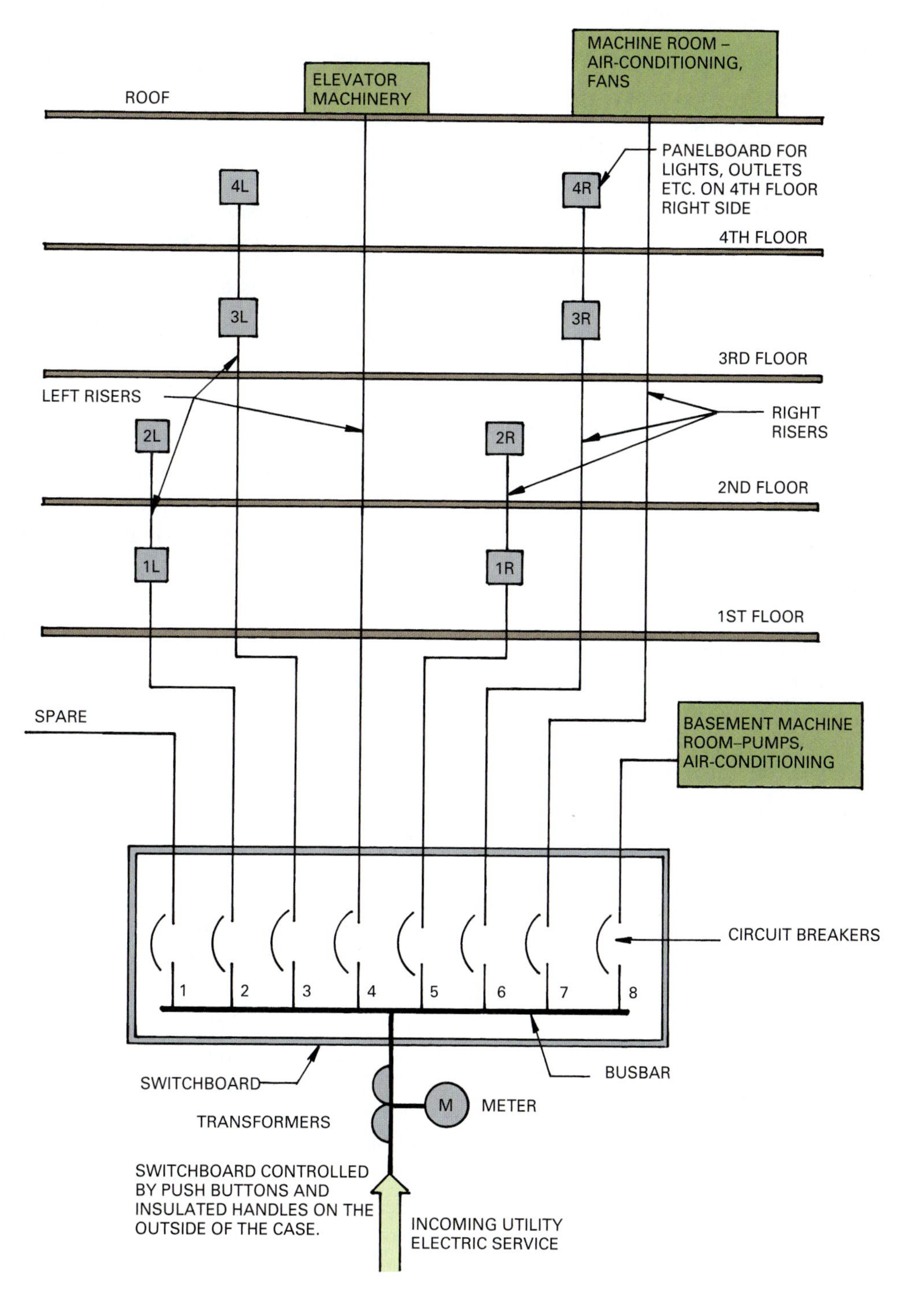

lighting can be controlled by various shading devices and the type of glazing used. Review Chapter 29 for information on glazing.

Lighting is a major source of energy use within buildings. In addition to the power used directly by light fixtures, many artificial light sources generate a substantial amount of heat within a building. This excess heat must be removed by the HVAC system, resulting in still higher energy consumption. The use of daylighting, energy efficient fixtures, and motion sensors to activate lighting should all be considered in an effort to reduce the energy consumption associated with artificial lighting.

Illumination Design Considerations

Lighting systems are designed to provide the specific lighting needs each building area requires. The purpose of a lighting system is to provide a pleasant atmosphere for occupants to and allow them sufficient visibility to perform required tasks. All of this must be accomplished using as little electrical power as possible.

When designing a lighting system, the ceiling height; the reflectance values of the ceiling, walls, floors, and furniture; and footcandle requirements all relate to one another. The choice of the type of **luminaire** (light fixture) and its spacing, mounting, and height above the area needing lighting must be considered.

Daylighting

Daylighting is the use of light from the sun and sky to complement or replace electric light. Daylight reduces the electric energy needed to power lights as well as the heat generated from lights, thereby also reducing cooling requirements. Lighting and its associated cooling energy use constitutes 30 to 40 percent of a commercial building's total energy expenditure.

Studies have shown that people are happier and more productive under daylight. Daylighting has been found to result in a reduced number of sick days in commercial offices and increased sales in retail spaces. Students learn better and day-lit schools have been proven to assist in higher levels of achievement.

The diffuse light of an overcast sky is similar in all orientations—it is soft and cool in both temperature and color. Every orientation provides useful daylight. North exposures give high-quality, consistent light with minimal heat gain, and minimal shading requirements. South orientations provide good access to strong illumination that varies throughout the day. Shading must be provided to prevent seasonal overheating (Fig. 47.39). East and West orientations must also incorporate proper shading to prevent glare and heat gain.

Figure 47.39 The large atrium space of the British Museum uses natural daylight to provide interior lighting.

Task Lighting

Task lighting seeks to provide an adequate level of light for a person to perform a specific activity without undue eyestrain. It is possible to provide targeted task lighting without having the same high levels of lighting in an entire space (Fig. 47. 40). To design task lighting, the activities to be performed, their location within a building, as well as the number of people performing tasks, their nearness to each other, and any patterns of movement must be specified. In many types of businesses, change in activities is normal, so the system must allow for easy revision, including relocating luminaires or redirecting light from existing luminaires. Since the level of light required may vary from time to time, luminaires must be individually switched so only those needed can be turned on or dimmer switches should be used. Windows near task activities must have blinds, shutters, or operable louvers to regulate natural light—available daylight can vary considerably over a period of hours.

The surface upon which a task is to be performed also influences lighting decisions. The most common situation involves some form of horizontal surface, but inclined and vertical work surfaces are frequently found. The widespread use of computers presents multiple problems: a horizontal keyboard, a vertical rack to hold the copy being set, and a monitor that produces light directly at the operator. The reflective values of a surface also require consideration because highly reflective work surfaces can cause uncomfortable light conditions.

Figure 47.40 This reception desk uses task lighting to provide a high level of illumination at the work surface.

General Illumination

General illumination provides lighting over a large area, including spaces in which specific task lighting luminaires are also present. It provides a way to ease visual discomfort caused when looking from a high-level lighted area (like a task area) out across a large space. General illumination reduces the difference in brightness levels and helps to maintain visual comfort.

Since the level of lighting intensity is less, general illumination luminaires can be widely spaced to reduce electric consumption. One example is in a library where study carrels have individual lights with levels required for reading, yet the entire reading room is equipped with luminaires providing a lower level of illumination—and visual comfort—when a person looks up from a book. General illumination also provides a warmer and more pleasant atmosphere by balancing brightness levels.

Selective Lighting

Selective lighting is used to focus light on a specific object or area, such as light focused on a painting or a retail display case (Fig. 47.41). The general illumination becomes the background, and the specific lighting focuses visual attention on an object.

Figure 47.41 Retail spaces make use of selective lighting to focus on specific merchandise.

Glare

Glare refers to light that is intense and uncomfortable to a viewer. It may emanate from luminaires that shine directly into a viewer's eyes. Glare can be reduced by lowering the brightness of luminaires or relocating and shielding them, if possible. Reflected glare is found more frequently. It occurs when light reflects off a surface. It can be controlled by selecting surfaces with low reflectance coefficients, reducing the brightness of the luminaire, or changing the angle of a reflecting surface.

Reflection

The **reflectance coefficient** is a measure of the percentage of incident light reflected from a surface. The **luminous intensity** (force) of a luminaire can be reduced or enhanced by the reflective potential of surrounding ceilings, walls, and floors. Dark colors absorb light, and light colors reflect light. Therefore, more light produced by a luminaire is available if light-colored ceilings and walls are used. Bright colors, such as red or yellow, are used infrequently because they tend to produce glare.

Manufacturers of wall and ceiling materials and paints generally list the reflectance coefficients of their products. For example, if half the incident light on a surface is reflected, the reflectance coefficient is 50 percent or 0.50. The other 50 percent is absorbed by or transmitted through the material. The amount transmitted is called **luminous transmittance**. The rate of flow of light energy through a surface is called **luminous flux**.

Construction **Methods**

Choosing Light and Color

Light quality is measured by color temperature and the color rendering index. Color temperature gives a measurement of the visual warmth or coolness of a light source, expressed in degrees Kelvin (°K). This measurement describes the quality of the actual light emitted by a lamp. The higher the color temperature, the cooler the light, and the lower the color temperature, the warmer the light. For example, typical inexpensive cool white fluorescent tube has a color temperature of about 4,100°K. Typical incandescent lamps have a color temperature of about 2,700°K. Light from an incandescent lamp is warmer, containing reds and yellows that contribute to the tones of surrounding materials, which people find warm and pleasing. And when high-temperature fluorescent tubes replace incandescent lamps, they cast a hot bluish to greenish light that gives surrounding materials a gray and flat appearance.

The color rendering index (CRI) provides a measure of light quality that indicates how natural an object looks when under a light source. This index is given as a percentage. The closer the CRI is to 100 percent, the more natural things appear. When a natural appearance is critical, as in an art museum, seldom is a lamp with less than 70 percent used. Incandescent lamps have CRI ratings close to 100 percent, but fluorescent tubes vary from below 50 to about 90 percent.

Following are some rules of thumb to help make color selection easier.

1. All space colors (wall and floor coverings, furniture, drapes, accents, et cetera) should be chosen under the lamp color specified for installation. Experience suggests that warm sources should be used at low lighting levels, cool sources at high levels. However, the choice may be influenced by space colors and by degree of luminaire brightness control.
2. Warm color schemes may appear overpowering if lighted with a warm source to relatively high levels—a cooler source should be used.
3. Cool color schemes may need warm sources, particularly at low lighting levels.
4. Well-shielded lighting systems, such as those using wedge louvers, or low brightness lenses will "cool off" a room. A slightly warmer lamp will counteract the effect.
5. Where color rendition is highly critical, high-CRI continuous-spectrum sources, such as Chroma 50 (C50) or Chroma 75 (C75), should be used.
6. Where both color and lighting level are important, the use of SP30, SP35, or SP41 GE Specification Series Color lamps, with three-peak rare earth phosphors, will provide good color rendering and high efficiency.
7. The Specification Series Colors tend to make spaces look more colorful because the three-peak phosphors compress all colors into the blue, green, and red-orange bands. This increases the contrast between colors. Three-peak lamps do not, however, increase the contrast of black-on-white tasks. Claims that less light is needed for typical office and industrial tasks when three-peak lamps are used are scientifically unfounded.
8. The color of a light source does not affect the visual performance of people doing black-on-white visual tasks. Vision and productivity studies, however, indicate that productivity may be affected by the color contrast and appearance of a visual environment and that color can contribute strongly to appearance.

**Principles adapted with permission from the General Electric publication* Specifying Light and Color.

The reflectance value of a ceiling used with a direct lighting system has little influence on the light projected, because direct lighting projects from a luminaire directly down into a room. Ceiling reflectance does greatly influence usable light if indirect lighting is used. Indirect lights project light up to a ceiling where it is reflected down into the room. The reflectance value of walls also influences reflected light. However, walls typically have doors, windows, cabinets, and other items that reduce the actual wall surface available. Floors have little influence on lighting; however, light-colored floors do reflect some light and provide visual comfort.

Room Cavity Ratio

The size of a space to be lighted affects the way light is distributed throughout the space. High ceilings reduce the efficiency of lighting. Open spaces can be more efficiently lighted than small rooms or partitioned space.

The relation between the space proportions, room height, room perimeter, and the height of work surfaces needing illumination divided by the floor area is called **the room cavity ratio.**

LIGHTING-RELATED MEASUREMENTS

Levels of Illumination

The illumination recommendations for residential and commercial buildings are detailed in Chapter 10 of the IESNA Lighting Handbook produced by the Illuminating Engineering Society of North America. The text outlines procedures for determining the quality and quantity recommendations for lighting design for various occupancies.

Luminous power of a light source is measured in either footcandles or candelas, the metric unit. The **candela** was adopted in 1948 as the international standard of luminous intensity. The unit used to measure luminous power at a predetermined distance from a light source is the lumen. The **lumen** is the luminous power on an area of one square foot at a distance of one foot from a one-footcandle light source—or on an area of one square meter at a distance of one meter from a one-candela light source. **Lux** is the metric unit used to describe the illumination produced by the luminous flux of one lumen falling perpendicular on a surface of one square meter. Lux is equivalent to 0.0929 footcandles. Lighting unit conversion factors are shown in Table 47.3. A related term, the **footlambert**, is a unit for measuring brightness or luminance. It is equal to 1 lumen per square foot when brightness is measured from a surface.

Measuring Illuminance Levels

The measure of **illuminance** is a measure of the quantity of illumination. One lumen of luminous flux uniformly imposed on one square foot of area results in an illuminance of one **footcandle.**

The level of illuminance is commonly measured with a **luminance meter**, often referred to as a light meter. It has a photoelectric panel connected to a microammeter and electronic control circuitry and is calibrated to read in footcandles or lux. Some types have a remote sensor. The meter is held so the photoelectric panel is parallel to the plane of the area being tested.

Measuring Luminance

To evaluate available illumination, it is necessary to get a measure of luminance, or the quality of illumination. **Luminance** (L) is a measure of brightness and brightness contrasts caused by photometric luminance. It is what is seen rather than illuminance (the quantity in footcandles), and it is an important part of the evaluation of the total illumination available. Luminance is measured by a luminance meter. Current luminance recommendations are developed following IESNA procedures.

LIGHTING SYSTEMS

Lighting systems fall into the categories of direct, indirect, semi-indirect, general-diffuse, and semi-direct (Fig. 47.42). The exact classification depends on the amount of light from the luminaire directed up and down.

Direct lighting is the most efficient system, projecting 90–100 percent of the light down to a floor or work surface. A light color reflective floor will reflect some light toward the ceiling, reducing ceiling darkness somewhat. Indirect lighting systems project most of the light up to the ceiling and some onto walls. The light is then reflected to the floor and work surfaces. This system requires that ceilings and walls have high reflectance coefficients. It produces an illumination that is uniform, free of glare, and generally diffuse. Indirect lighting uses baffles and requires that luminaires have more powerful lamps than direct lighting.

Semi-indirect lighting systems allow more light (10–40 percent) to project down than indirect systems do. They do this by using a translucent diffuser that permits some light to project down but reflects most of it up to the ceiling to produce a diffuse low-glare illumination.

General-diffuse systems (also called direct-indirect) project about equal amounts of light up and down. They produce a light ceiling and upper wall area but also project light down to the floor and work surfaces. This system typically uses a globe diffuser, which allows light to project in all directions.

Semi-direct lighting systems project most of their light down, with only a very small percentage projecting upward to light the ceiling. This produces efficient direct lighting while somewhat brightening dark ceiling areas (at least more so than when direct lighting is used).

TYPES OF LAMPS

A variety of **lamps** are in use, each with particular advantages and disadvantages. Some are useful only for special applications.

Table 47.3 Recommended Illumination Levels

Type of Space	Guideline (footcandles)[a]
Commercial and Institutional Interiors	
Art galleries	30–100
Auditoriums/assembly spaces	15–30
Banks	
Lobby	50
Customer areas	70
Teller stations and accounting areas	150
Hospitals	
Corridors, toilets, waiting rooms	20
Patient rooms	
General	20
Supplementary for reading	30
Supplementary for examination	100
Recovery rooms	30
Lab, exam, treatment rooms	
General lighting	50
Close work and examining table	100
Autopsy	
General lighting	100
Supplementary lighting	1,000
Emergency rooms	
General lighting	100
Supplementary lighting	2,000
Surgery	
General lighting	200
Supplementary on table	2,500
Hotels (rooms and lobbies)	10–50
Labs	50–100
Libraries	
Stacks	30
Reading rooms, carrels, book repair and binding, check-out	70
Catalogs, card files	100
Offices	
Corridors, stairways, washrooms	10–20[b]
Filing cabinets, bookshelves, conference tables	30
Secretarial desks (with task lighting as needed)	50–70
Routine work (reading, transcribing, filing, mail sorting, etc.)	100
Accounting, auditing, bookkeeping	150
Drafting	200
Post Offices	
Storage, corridors, stairways	20
Lobby	30
Sorting, mailing	100

Table 47.3 Recommended Illumination Levels *(Continued)*

Type of Space	Guideline (footcandles)[a]
Commercial and Institutional Interiors	
Restaurants	50
Schools	
Auditoriums	15–30
Reading or writing, libraries	70
Lecture halls	70–150
Drafting labs, shops	100
Sewing rooms	150
Stores	
Circulation areas, stock rooms	30
Merchandise areas	
Serviced	100
Self-service areas	200
Showcases and wall cases	
Serviced	200
Self-service	500
Feature displays	
Serviced	500
Self-service	1,000
Industrial Interiors (Manufacturing Areas)	
Storage areas	
Inactive	5
Active (rough, bulky)	10
Active (medium)	20
Active (fine)	50
Loading, stairways, washrooms	20
Ordinary tasks (rough bench and machine work, light inspection, packing/wrapping)	50
Difficult tasks (medium bench and machine work, medium inspection)	100
Very difficult tasks (fine bench and machine work, difficult inspection)	200–500
Extremely difficult tasks	500–1,000
Garages	
Active traffic areas	20
Service and repair	100
Exteriors	
Building security	1–5
Parking	
Self-parking	1
Attendant parking	2
Shopping centers (to attract customers)	5
Floodlighting	5–50
Bulletins and poster panels	20–100

[a]To convert to lux, multiply value by 10.76.
[b]Must be at least 20 percent of the level of the adjacent work space. Courtesy Illuminating Engineering Society of North America

Figure 47.42 The distribution of light can be varied depending on the type of fixture used.

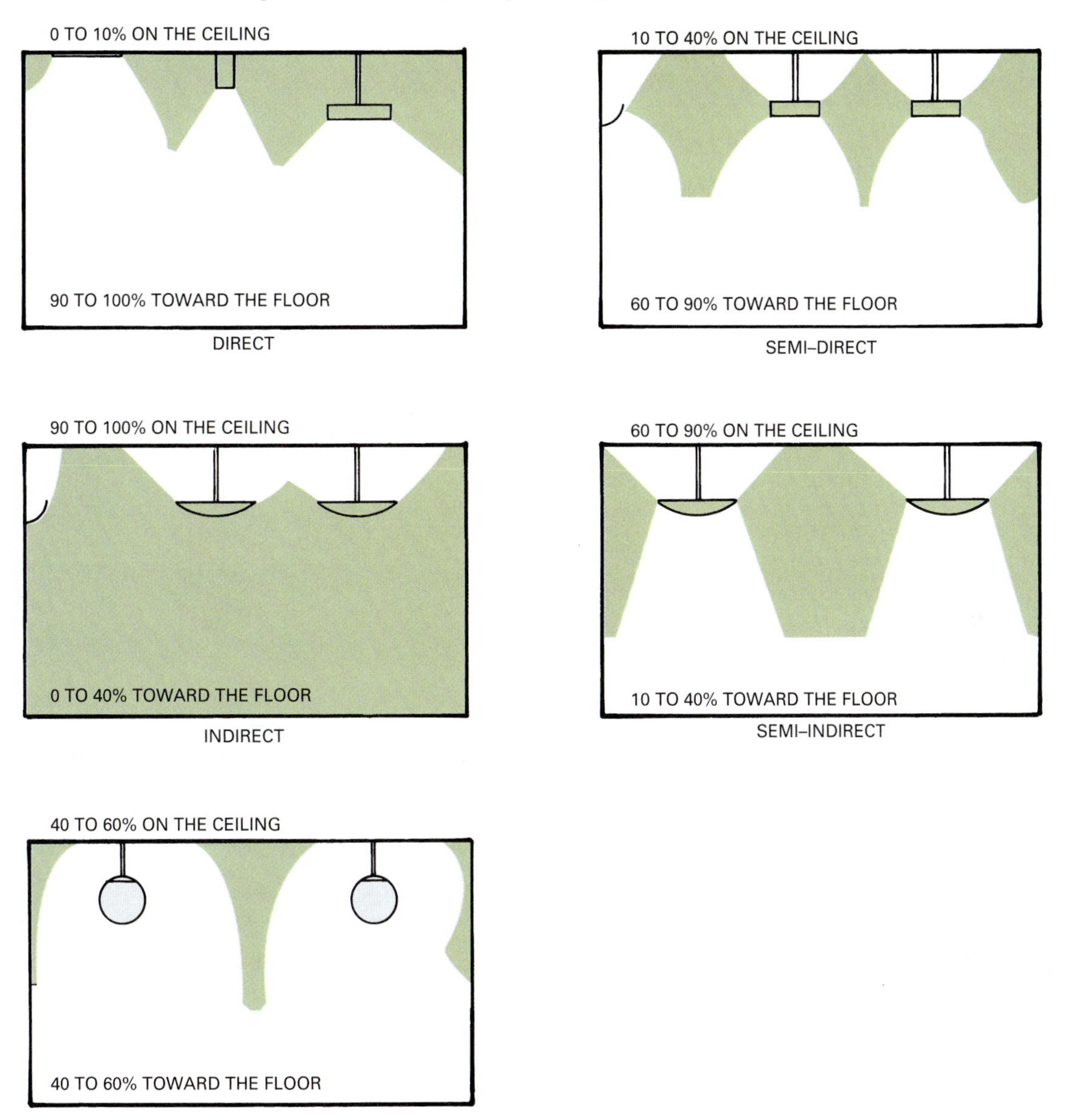

Incandescent Lamps

An **incandescent lamp** is a glass bulb containing a filament joined to a metal base with copper lead-in wires (Fig. 47.43). The filament is a tungsten wire that resists the flow of electricity. In doing so, the wire gets hot and glows, producing light. A lot of the energy used to create the heat that lights an incandescent bulb is waste, making them less energy efficient than other types of lamps. The metal base screws into a socket that is part of a luminaire, or lighting fixture. Most incandescent lamps are filled with argon and nitrogen, which makes the use of a high filament temperature possible. The melting point of tungsten is 3,655K (6,170°F). Lamps filled with krypton gas have a longer life than argon and nitrogen lamps and cost more.

Incandescent lamps are rated in watts, as shown in **Table 47.4**. They are made with several size bases. The larger bases are used on lamps with higher wattages. The voltage at which the lamp is used influences its life. If a 120V lamp is burned at a voltage higher than 120, its life will be shortened, but it will produce more lumens per watt. If operated at a voltage below this, it will have a prolonged life but produce fewer lumens per watt. Some of the commonly available incandescent lamps are shown in Fig. 47.44.

Table 47.4 Lighting Unit Conversion Factors

Unit	To convert from ...	Multiply by the correction factor ...	To convert to ...
Luminous Intensity (I)	Candlepower	1.00	Candela
Illuminance (E)	Footcandle	10.76	Lux
	Lux	0.09	Footcandle
Luminance (L)	Candela/m^2	0.29	Footlambert
	Candela/in.^{2}	1550.00	Candela/m^2
	Candela/ft.^{2}	10.76	Candela/m^2
	Footlambert	3.43	Candela/m^2

Tungsten-Halogen Lamps

Tungsten-halogen lamps (quartz-iodine) are a type of incandescent lamp. They have a tungsten filament and are filled with a halogen, such as iodine or bromine, and an inert gas that reduces the evaporation of the filament. The bulb is made from quartz because it will withstand higher temperatures than glass. The halogen additive in the lamp reacts chemically with any tungsten deposited on the bulb and redeposits it on the tungsten filament, improving the efficiency of the lamp. Tungsten-halogen lamps are available in a wide range of wattages, as shown in Table 47.4. Examples of some typical lamps are shown in Fig. 47.45.

Figure 47.43 The working components of an incandescent lamp.

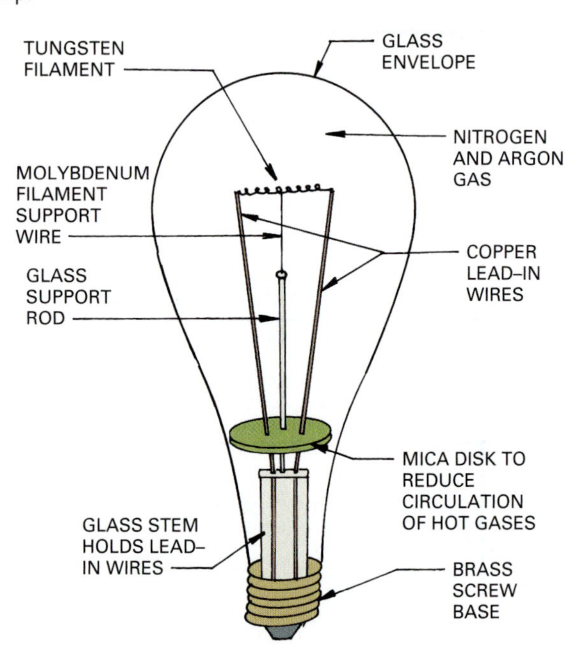

Two of the lamps shown have quartz tubes. One has a screw base and is installed vertically. The double-end lamp is installed horizontally. Both require reflectors, as shown in Fig. 47.46.

Figure 47.44 Some of the types of incandescent lamps available.

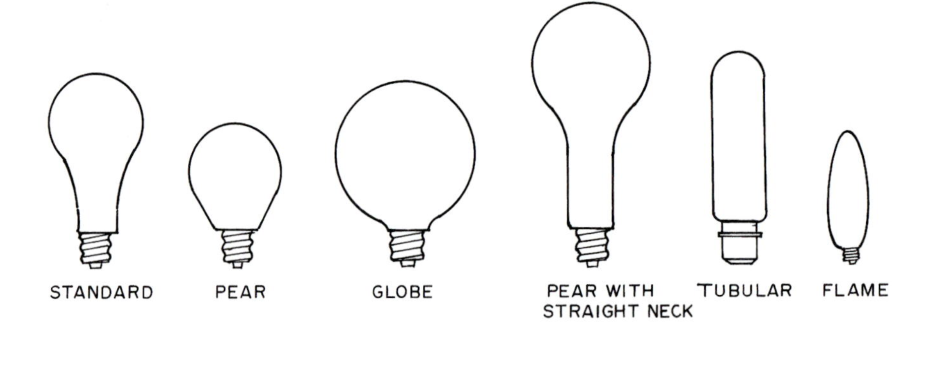

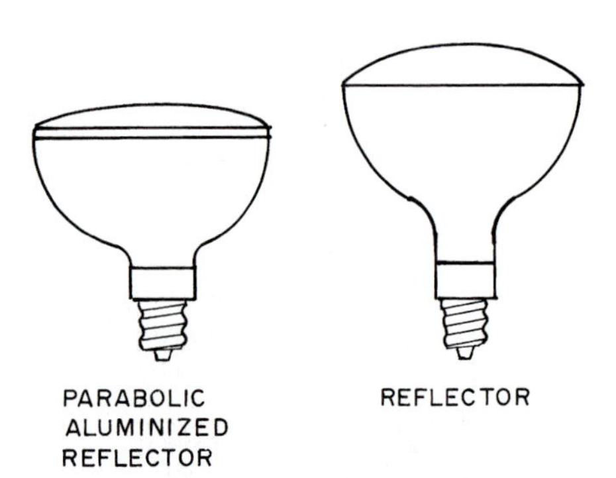

Figure 47.45 Typical tungsten-halogen lamps.

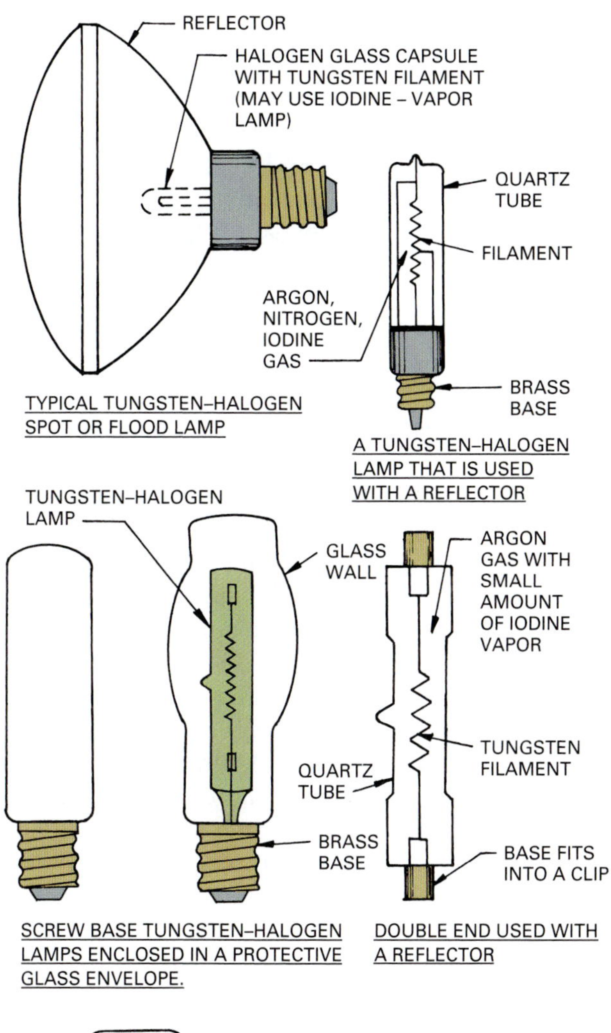

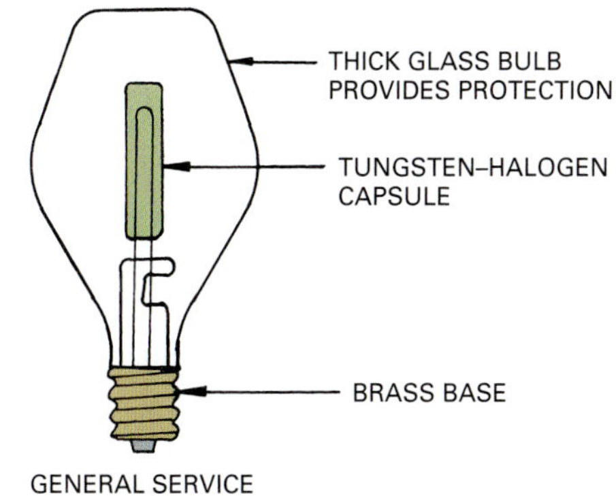

Tungsten-halogen lamps with parabolic reflectors are available from 15W to 1,500W. They come in a number of shapes. The slanted backside of the bulb is a reflector. Both flood and spot lamps are available. Tungsten-halogen lamps operate at high temperatures, and even though the quartz envelope can withstand high temperatures, the lamps can explode. Therefore, manufacturers recommend some type of shielding or protective screen be used to contain potential flying fragments.

Figure 47.46 A tungsten-halogen quartz-iodine lamp installed horizontally in a reflector to provide indirect lighting.

Fluorescent Lamps

Fluorescent lamps have a long glass tube sealed at each end. The tube is filled with a mixture of an inert gas, such as argon, and low-pressure mercury vapor. A cathode is placed in each end. The cathode produces electrons that start and maintain operation of a mercury arc, which produces an ultraviolet arc. The ultraviolet arc is absorbed by phosphors that coat the inside of the tube and cause it to fluoresce (radiate) light (Fig. 47.47). Because fluorescent bulbs don't use heat to create light, they are far more energy efficient than regular incandescent bulbs.

Fluorescent lamps will not operate directly off 120V alternating current because it will not cause the arc discharge. Each luminaire incorporates a ballast that

Figure 47.47 A generalized illustration showing the construction of a fluorescent lamp with bi-pin bases.

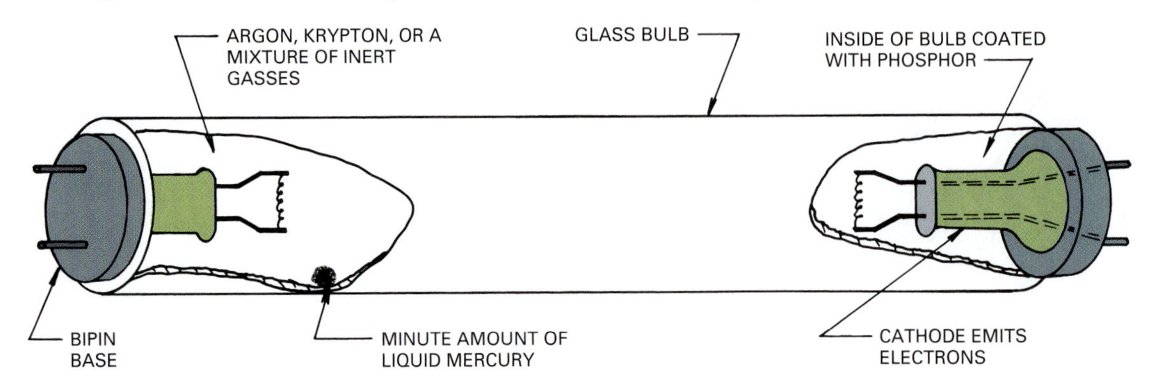

provides the starting and operating voltages. Standard lamps have a starter that preheats the cathodes, producing the high-voltage arc needed to start the lamp. Rapid-start lamps have the same basic construction as standard lamps, but a circuit keeps the lamp electrodes constantly preheated by means of low-voltage windings that are part of the ballast.

High-output (HO) and very-high-output (VHO) lamps require special ballasts. They are used where high output is required in a limited area, such as a merchandise display or an outdoor sign. They generate considerable heat that must be considered, but they function in cold situations where standard lamps will not light.

Instant-start fluorescent lamps use a high-voltage transformer to generate an arc, lighting the lamp without the preheating delay of standard lamps. They have a single pin at each end. The high-voltage start greatly reduces the life of the lamp, but it will start in lower temperatures than rapid-start lamps.

Fluorescent lamps are available in straight tubes, circles, and U-shapes. Small lamps are available that fit in table lamp sockets. Several different bases and lampholders are used, depending on type of lamp. Since fluorescent lamps are much more efficient than incandescent lamps, lower wattages can be used. Typical wattage ratings are shown in Table 47.4.

Standard fluorescent lamps contain mercury and are classified as hazardous waste by the Environmental Protection Agency. A new long-mercury fluorescent lamp is now available. It uses a chemical buffer to slow the absorption of mercury by the phosphor crystals that coat the inside of the tube, so less mercury is required in the lamp.

Each lamp receives a pre-measured amount of mercury encapsulated in a tiny glass bubble. As the lamp leaves the factory, a radio-frequency beam heats a wire wrapped around the bubble, which breaks, releasing the mercury in the tube.

Compact Fluorescents A **compact fluorescent light (CFL)** is a type of fluorescent lamp (Fig. 47.48) made in special shapes to fit standard household light sockets (like ceiling fixtures). Unlike fluorescent tubes, which require a separate ballast independent of the bulb, CFLs incorporate integral ballast built directly into the light bulb. Compared to incandescent lamps that provide the same amount of visible light, CFLs generally use less power, and have a longer rated life. Compared to an incandescent lamp, a CFL can save in electricity costs over its lifetime. It can also save 2000 times its own weight in greenhouse gases. Like all fluorescent lamps, CFLs contain mercury, which complicates their disposal. Both types offer energy-efficient light.

High-Intensity Discharge Lamps

High-intensity discharge lamps (HID) produce light by passing an electric arc through a metallic vapor confined in a sealed quartz or ceramic tube. The high-intensity discharge lamps available include mercury vapor, metal-halide, and high- and low-pressure sodium. With appropriate color correction, they can be used in many indoor and outdoor situations.

Mercury Vapor Lamps Mercury vapor lamps (M-V) are available in clear, white, white-deluxe, and color-corrected. The clear lamp produces a blue-green light that causes distortion of other colors. Color correction is accomplished by coating the outer bulb with phosphors to make the lamp useful for some indoor applications (Fig. 47.49).

Mercury vapor lamps use mercury and argon gas, which helps in starting, since mercury has a low vapor pressure at room temperature. A starting voltage energizes the circuit and moves voltage across the space between the starting electrode and the main electrode. This creates an argon arc that vaporizes the mercury.

Figure 47.48 The characteristic shape of a compact fluorescent light.

Figure 47.49 A simplified schematic of a typical mercury vapor lamp.

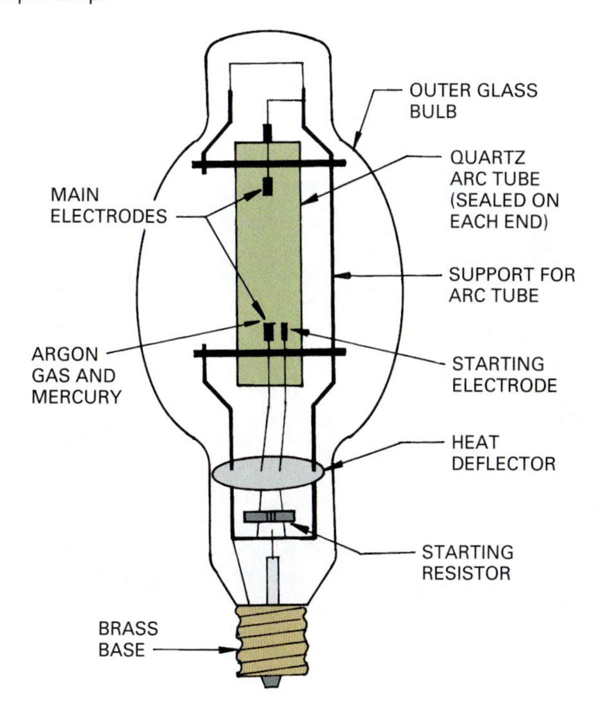

A mercury vapor lamp takes three to six minutes to reach full warm-up and maximum output. Ballast is used to start the lamp and control the arc. After the lamp is turned off, it must cool, and pressure must abate before it can be relit. This takes from three to eight minutes, depending on the construction of the lamp. As this wait time would leave an area in total darkness if a power failure of even a few seconds occurred, some form of emergency backup lighting is usually required. This could be a system using incandescent lamps connected to the electric system or a battery-powered emergency lighting system.

Mercury vapor lamps are not as efficient as fluorescent lamps, but they are more efficient than incandescent lamps. They are used on applications in which they remain in use for long periods, such as overnight in a parking lot. They should not be used if frequent switching on and off is required.

The mercury vapor lamp spectrum is in the ultraviolet (UV) range, but it is not harmful because the outer glass absorbs much of the UV. However, if the glass breaks, the lamp continues to function, and UV exposure could be hazardous; therefore, a safety warning appears on all mercury vapor lamps. They are available in various wattages, as shown in Table 47.4.

Metal-Halide Lamps Metal-halide lamps are another type of high-intensity discharge lamp (Fig. 47.50). They are designed much like mercury vapor lamps, but halides of metals, such as sodium, indium, or thallium, are added. These salts produce a light that radiates at frequencies different from colors radiated by mercury, producing better color rendition than mercury vapor lamps. It has about the same starting delay and restart times as the mercury vapor.

It should be noted that metal-halide lamps tend to explode and must be encased in an approved enclosing fixture. They carry the same safety warning as mercury vapor lamps. Some types have a plastic coating that makes them safe to use in open fixtures. They also are marked for their proper burning position, either base

Figure 47.50 Typical metal-halide lamps.

up, base down, or horizontally. If installed in the wrong position, their operating characteristics will be limited. Metal-halide lamps are available in various wattages, as shown in Table 47.4.

High- and Low-Pressure Sodium Lamps High-pressure sodium (HPS) lamps are arc discharge lamps containing sodium under high pressure in the glass arc tube. This produces a light with a yellow tint. They are highly efficient lamps, superior to all other types, and in areas where color is not important they can replace mercury vapor and metal-halide lamps (Fig. 47.51). These lamps are available in various wattages, as shown in Table 47.4.

Low-pressure sodium lamps, referred to as SOX, produce a deep yellow light that is not acceptable for general interior lighting. They are widely used for street, highway, and parking lighting. They have a long life and are the most economical type of lighting.

Figure 47.51 A simplified schematic of a high-pressure sodium lamp.

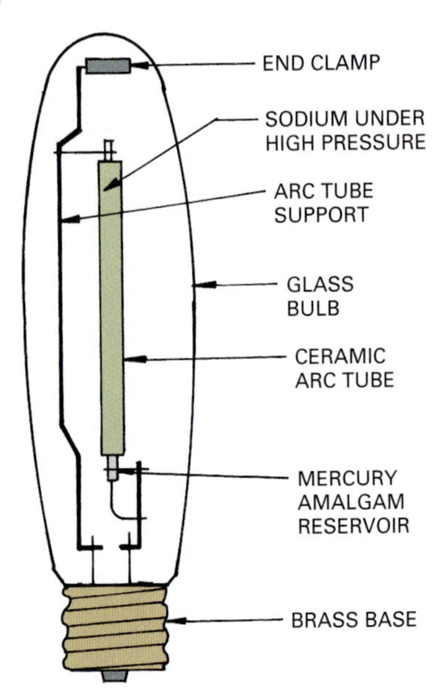

Light Emitting Diodes

Light Emitting Diodes (LED) are small light bulbs based on semiconductor technology. The electrons of a semiconductor diode are brought from a state of high energy to a state of low energy with the energy difference being emitted in the form of light. Until recently, LEDs were used only in single-bulb applications, such as in instrument panels and electronics. Today, LED bulbs are made by clustering many small bulbs encased in diffuser lenses that spread light in wider beams (Fig. 47.52). Bulbs are available with standard bases to fit common household light fixtures. Despite their high expense, LEDs provide a number of advantages over traditional light sources, including low energy consumption, a long lamp life, and small size.

Figure 47.52 Light Emitting Diodes (LED) are small light bulbs based on semiconductor technology.

Review Questions

1. What is the rate of flow of electric current if it is available at 120V and has a resistance of 15 ohms?
2. What is the difference between alternating current and direct current?
3. What are the various ways electric power can be produced?
4. What fuels are used to power fossil fuel electric-generating plants?
5. What power source is used to drive emergency generators?
6. How is electricity produced by utilizing normally wasted heat?
7. What are the types of electrical conductors recognized by codes?
8. What type of electrical conductor is used for underground applications?
9. What voltage current can be carried by flat conductor cables?
10. What types of conduit are in general use?
11. One kilowatt equals how many watt-hours?

12. What purpose do motor control centers serve?
13. What is the difference between the primary transformer voltage and the secondary voltage?
14. What are the commonly available switch actions?
15. What ways are various types of switches activated?
16. What devices are used to protect a circuit from overloads?
17. How does a type S plug fuse keep someone from putting the wrong size fuse in a panelboard?
18. What are the types of fuses in common use?

Key Terms

building-integrated PV (BIPV)
bus
busway
BX cable
cable tray
candela (cd)
circuit breaker
cogeneration
compact fluorescent light (CFL)
conductor, electric
conduit
current, electric
daylighting
footcandle (fc)
footlambert
fuse
illuminance (E)
incandescence
insulator, electric
lamp
lumen (lm)
luminaire
luminance (L)
luminance meter
luminous flux
luminous intensity
luminous transmittance
lux
meter, electric
motor control center
net metering
panelboard
Photovoltaic (PV)
power
raceway
reflectance coefficient
resistance, electric
romex
room cavity ratio
service entrance
switches
switchboard
transformer
watt

Activities

1. Invite a local building official to speak to the class about electrical code requirements.
2. Collect various electrical control and metering devices. Prepare a display and label each, giving the manufacturer's specifications.
3. Erect a section of a stud wall and install boxes for fuses, outlets, switches, lights, et cetera, and wire these. Bring power to the beginning end of the wiring and activate the system.
4. Walk through various buildings and note the types of lighting used. Prepare a report citing the location for each type found.
5. Arrange a visit to the electrical equipment room of a major building and have the building engineer explain the function of each unit. Describe the flow of power from the service entrance to the various floors of the building. View distribution equipment used on one of these floors.

Additional Resources

IESNA Lighting Handbook, and many other books on installation, standards, education, and design, Illuminating Engineering Society of America, New York, NY.

International Code Council Electrical Code, International Code Council, Falls Church, VA.

Kaufman, J. E., *IES Lighting Handbook Application*, Illuminating Engineering Society of America, New York, NY.

National Electrical Code®, National Fire Protection Association, Quincy, MA.

Other resources include:

Many electrical design, safety, and installation manuals, National Fire Protection Association, Quincy, MA.

See Appendix B for addresses of professional and trade organizations and other sources of technical information.

DIVISION

28

Electronic Safety and Security CSI MasterFormat™

CHAPTER 48

Electronic Safety and Security

LEARNING OBJECTIVES

Upon completion of this chapter, the student should be able to:

- Understand the signal systems used to provide electronic safety and security.
- Identify various fire detection and alarm strategies in common use.
- Describe the various devices used in the design of security systems.
- Understand the principles of intrusion protection.
- Understand the functions of a building automation system

Build Your Knowledge

For further information on these materials and methods, please refer to:

Chapter 44 Fire Suppression Systems

Electronic safety and security involve the design and operation of various control, signal, and communications systems. Topics listed in division 48 include fire detection and alarm, communications, access control, intrusion detection, and electronic monitoring and control systems. The signal systems used require a source, equipment that processes and transmits a signal, and a means of registering the signal through visual or auditory devices. Signals are typically conveyed via low voltage wiring through a variety of sensors. Modern buildings are increasingly using an integrated-systems design approach that incorporates security with other operational functions into a building automation system (BAS). These systems require monitoring and control to function reliably.

AUTOMATIC FIRE DETECTION AND ALARM SYSTEMS

Fire alarm systems are classified by the National Fire Alarm Code as either household fire warning systems, protected premises, or off premises systems. Household warning systems incorporate smoke and fire detectors to activate an auditory alarm. Protected premises alarms sound a warning only in the area of immediate danger and response is activated locally. Off premises systems are alarms that are directly connected to the municipal fire department. Large multi-building campuses may incorporate a central supervisory station that receives signals from a number of buildings.

A variety of fire detectors are able to sense the products of combustion at different stages of a fire's duration. Ionization detectors can detect microscopic particles emitted during the early stages of a fire (Fig. 48.1). Photoelectric detectors use beams of light to respond to smoke particles visible to the naked eye during development of a fire. Heat detectors sense the actual fire and immediately activate alarm and fire suppression systems. **Fire flame detectors** are used where combustible materials that may burn are present. They detect the presence of either infrared or ultraviolet radiation.

Fires may also be detected by building occupants with alarms activated via manually operated pull stations (Fig. 48.2). Various types of fire signal alarms are in use. The most common incorporate horns, bells, and flashing lights (Fig. 48.3). These are often combined into one unit, providing sound and light signals.

Fire-suppression systems include automatic sprinklers, water fog generators used on highly flammable solids or liquids, and a liquid foaming agent introduced

Figure 48.1 A common fire detector that senses smoke and sounds an alarm. *(Image copyright magicoven, 2009. Used under license from Shutterstock.com)*

Figure 48.2 Fire extinguishers are often placed in close proximity to manually activated fire alarm stations. *(Image copyright Jennie Book, 2009. Used under license from Shutterstock.com)*

into the water in a sprinkler system. Several types of automatic gas-suppression systems are available; carbon dioxide is commonly used. High expansion foams are sometimes used in small, compartmentalized areas. The foam must completely cover an area to put out the fire. Additional information on fire suppression systems can be found in Chapter 44.

Figure 48.3 Horns and bells are used to provide a loud warning when the fire-detection system senses a fire. *(Image copyright Joe Gough, 2009. Used under license from Shutterstock.com)*

SECURITY SYSTEMS

Electronic security systems provide surveillance and control for buildings to detect and prevent the occurrence of crime and vandalism. Systems range in size from small residential alarms to large central-control centers for multistory buildings. Large control centers can monitor different surveillance equipment and activate automatic response systems and voice communication with emergency personnel.

Visual surveillance systems use CCTV cameras mounted throughout a building to monitor problematic activity (Fig. 48.4). Typical locations for surveillance cameras include building entrances, hallways, elevators, exterior walks, as well as parking lots and garages. The cameras can be remote controlled to pan, tilt, and zoom for larger coverage areas. Small systems record information for review in case of an occurrence. Multiple cameras may be monitored continuously from a central command post where information is both recorded and stored. Emergency voice communications equipment is often incorporated to allow emergency personnel to remain in contact with the central control center.

A number of means are used to provide access control to a building. Locks may be operated manually or controlled electronically and programmed to permit or deny access on a timed schedule. Multi-family apartment buildings incorporate access control within the doorbell system. A two-way intercom provides communication from the access point to the apartment with a pushbutton activating the lock. In very large occupancies, the doorbell is replaced with an electronic tenant list and push-button telephone. For

Figure 48.4 Security cameras provide a wide coverage and record the activities within their area of surveillance. *(a) Image copyright Serggod, 2009. Used under license from Shutterstock.com (b) Image copyright Alistair Cotton, 2009. Used under license from Shutterstock.com*

office, hotel, or high risk occupancies, electronic coded key cards are used to gain access and monitor building occupancy (Fig. 48.5). The card transmits a signal that either opens a physical barrier or closes one in case of unauthorized access.

Intrusion detectors trigger an alarm when someone enters an unauthorized area. When an alarm is triggered, it sounds an auditory alarm, activates lights, and transmits information to the central control system where an operator can alert security staff. The most common detectors use magnetic devices mounted on doors and windows that transmit a signal when opened. Photoelectric devices use either light or infrared beams sent from a source to a receiver. When the beam of light is interrupted, the receiver signals the alarm to activate. Motion detectors use ultrasonic frequencies and microwaves to detect movement. Based on the Doppler Effect, the device detects changes in the frequency of waves reflected from a moving object. **Audio sensors** use controls that activate an alarm when a sound is detected.

Figure 48.5 Electronic coded key cards are used to gain access and monitor building occupancy. *(Image copyright Grandpa, 2009. Used under license from Shutterstock.com)*

One frequently used security system is found in banks and other businesses where valuable materials are present. Figure 48.6 illustrates a possible security arrangement for a bank. It uses audio sensors, waterproof outlets, and an alarm control cabinet, which is usually located in the vault. **Alarm actuators**, controls that sense a particular activity and signal an alarm, are located at the teller and drive-in windows, at the employees' desks, and surveillance cameras are placed in strategic locations.

ELECTRONIC MONITORING AND CONTROL

Large modern buildings typically incorporate a central control facility that combines the supervision and control of various building functions. Known as **building automation systems** (BAS), these computerized controls are able to process data from remote systems and calculate operational responses to optimize building performance. Fire protection, HVAC and electrical and lighting controls, elevator functions, communications, and security systems can all be monitored and regulated from this central command. Energy Management Control Systems that optimize HVAC functions, shut off systems when they are not needed, and cycle electrical loads are often integrated. The control center is staffed by trained personnel and includes automatic controls that respond electronically to preset signals (Fig. 48.7).

The core missions of a BAS system includes keeping the building climate within a specified range, providing lighting based on established schedules, and monitoring system performance and device failures. The data is processed and email or text notifications are transmitted to building engineering staff.

Lighting can be turned on and off with a building automation system based on time of day or occupancy sensors and timers. The monitoring system adjusts according to conditions and reduces building energy and maintenance costs when compared to a non-controlled building. A building controlled by a BAS is often referred to as an intelligent building system.

Figure 48.6 A schematic of security system devices and their configuration for a bank. *(Courtesy Mosler, Inc.)*

Specifications:

1. Audio sensors require junction box.
   Box shall be mounted flush with ceiling. Audio sensors shall not be mounted more than 25′-0″ from any wall surface or more than 50′-0″ apart. Minimum of one audio sensor per each enclosed area.
2. Waterproof floor outlet with bell nozzle.
3. Main junction box (12″ x 12″ x 4″) complete with cover and located in accessible plan in equipment room or work area.
4. Outlet box (2″W x 3″H x 2-1/4″D) for transformer by Mosler. Mount as illustrated in main junction box.
5. Status control.
6. Alarm control cabinet must be in vault whenever possible.

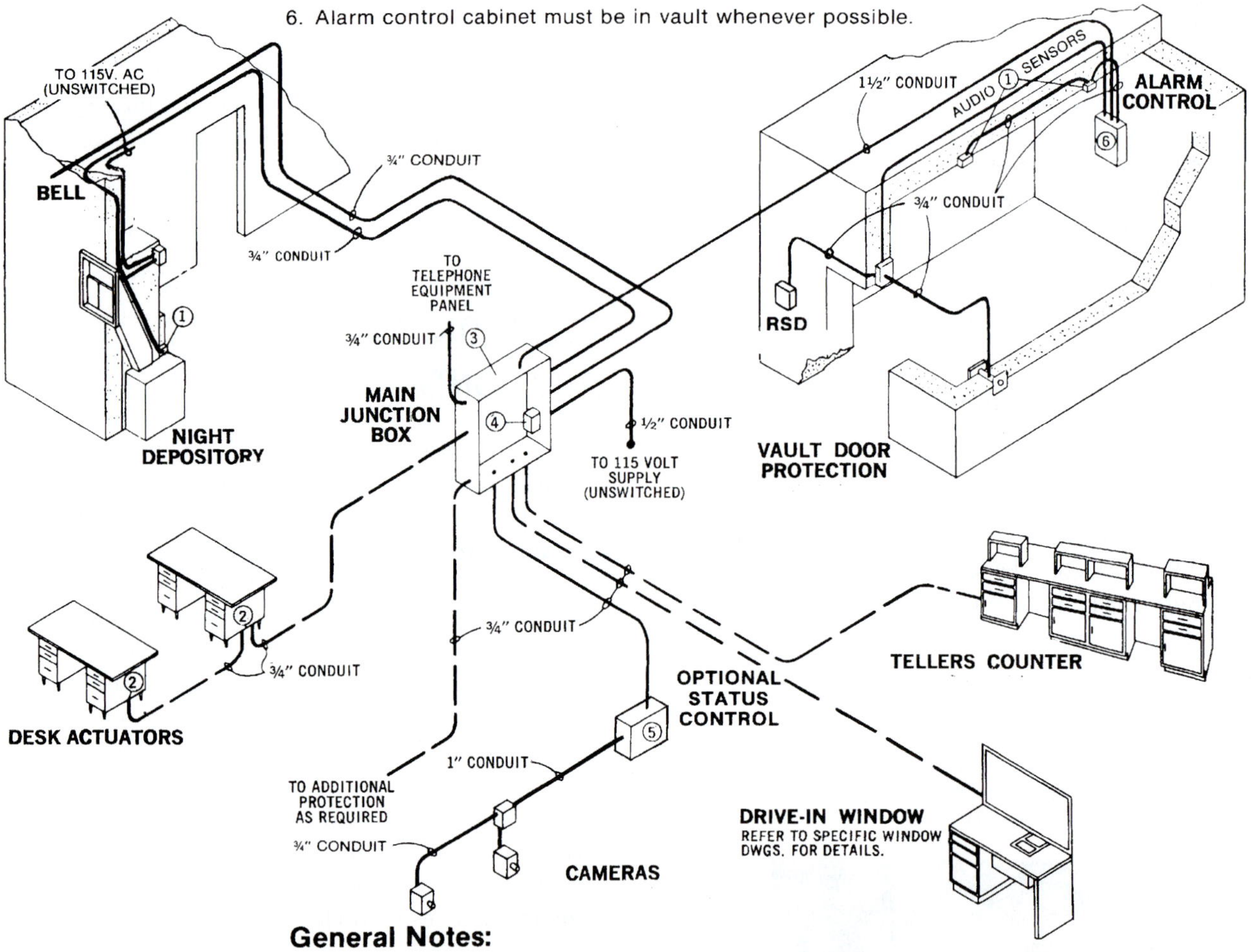

General Notes:

A. All conduit, outlet boxes and covers to be supplied and installed by others in accordance with national electric code and/or local codes without expense to Mosler.

B. No more than (2) 90° bends in a conduit run.

C. Conduit must have pull wires in place at time of installation.

——————— Must be separate conduit run.

——— ——— These conduits may be combined with each other wherever practical to eliminate and/or shorten conduit runs. Increase size as required.

Construction Techniques

Home Security Systems

The use of home security systems is increasing rapidly. A basic system uses sensors on doors and windows to activate an alarm. The system is armed and disarmed by a keypad located near the door that is most frequently used for entry by the occupant (Fig. A). When the system is armed and an intruder opens a door or window, the alarm is triggered. This may simply be an alarm inside and outside the building, or the system may be connected to a central monitoring station that verifies if it is a true entry and notifies local police. Another system is wireless with a central control unit. Each door or window has a battery-powered transmitter that sends a signal to the base station, which sounds the alarm. The system has a keypad that can be carried about the building.

Many types of protection sensors are available. Exterior sensors provide protection around the perimeter of a building. These can include pressure detectors that sense vehicles entering the driveway and photoelectric beams and motion detectors that can sense a person in the yard. Within the building, a system can have alarms on window screens that sound when the screen is cut, glass breakage detectors, and sensors on doors and windows that activate when they are opened, as well as interior motion sensors (Fig. B).

Figure A A close up of a security key pad. *(Image copyright Miles Boyer, 2009. Used under license from Shutterstock.com)*

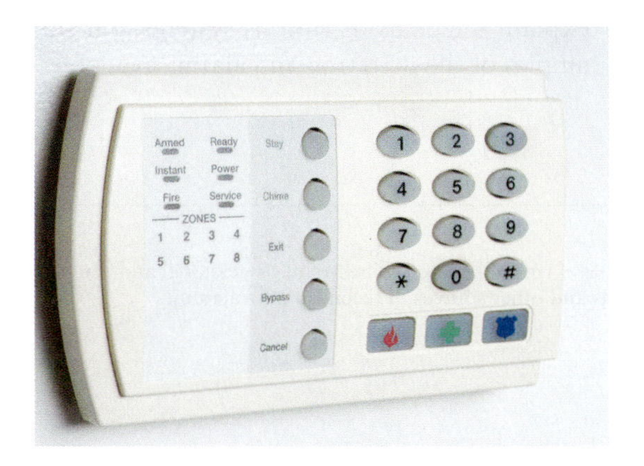

Figure B A passive infrared (PIR) motion detector. *(Image copyright Bill McKelvie, 2009. Used under license from Shutterstock.com)*

Figure 48.7 A building automation control center must be staffed by trained personnel. *(Image copyright Graham Taylor, 2009. Used under license from Shutterstock.com)*

Review Questions

1. What type of security units are used in areas requiring a very secure situation?
2. Where are fire flame detectors used?
3. What are the most commonly used type of fire alarms?
4. How are occupants alerted to the presence of carbon monoxide and natural and propane gas?
5. How can people entering the property near a building be detected?
6. Why are flashing light warning units used?

Key Terms

alarm actuators
audio sensors
building automation system
fire flame detector

Activities

1. Arrange a visit to a local business that has a security system. Ask to have security personnel demonstrate different system operations.
2. If there is a security firm in the area, try to arrange to have a sales representative visit the class and explain the goals of various systems and show examples of the detection and alarms available.

Additional Resources

Sweets Construction Catalog, Architects, Engineers, and Contractors Edition, Division 13, McGraw-Hill Construction, Two Penn Plaza, New York, NY 10121.

See Appendix B for addresses of professional and trade organizations and other sources of technical information.

DIVISION 1

GENERAL REQUIREMENTS

The level two titles within Division 1, General Requirements, of the CSI MasterFormat™ can be found in Chapter 2. A detailed listing of the level-three titles is available in the 2004 edition of MasterFormat™, published jointly by the Construction Specifications Institute (CSI), Alexandria, Virginia, and the Construction Specifications Canada (CSC), Toronto, Ontario, Canada. Following is descriptive information about the contents under each level-two title.

SUMMARY

The first level-two title is Summary, which includes Summary of Work, Multiple Contract Summary, Work Restrictions, and Project Utility Sources.

- Summary of Work identifies the work to be covered in a single contract and describes provisions for future construction. It may include work by an owner, work covered in the contract documents, and specific things, such as products ordered in advance, owner-furnished and installed products, and the use of salvaged material and products.
- Multiple Contract Summary describes the construction delivered under *more* than one construction contract, such as a construction management contract and a multiple-prime contract. It includes the sequence of construction, contract interface, any construction by the owner, and a summary of the contracts.
- Work Restrictions pertains to restrictions that affect construction operations, such as the location of construction, occupancy requirements, and the use of the building premises and site during construction. If the construction involves a building already occupied, procedures for coordination with the occupants are detailed.
- Project Utility Sources includes the identity of utility companies that are to provide permanent services to the project.

PRICE AND PAYMENT PROCEDURES

The second level-two title is Price and Payment Procedures. It details the allowance-adjusting procedures for cash and quantity allowances for products, installation, testing, and contingencies.

- Allowances includes the adjusting procedures for cash and quantity allowances for products, installation, inspection, and testing contingencies.
- Alternates provides for the submission and acceptance procedures for alternate bids, including, when desired, a list and description of each alternate.
- Value Analysis includes the procedures and submittal requirements for value analysis, value engineering, application for consideration, and consideration of proposals.
- Contraction Modification Procedures details the procedures for making clarifications and proposals for change and for changing the contract.
- Unit Prices establishes the procedures associated with unit prices and measurement and payment. It may include a list and descriptions of the actual unit price items.
- Payment Procedures describes the procedures for submitting schedules of values and applications for payment.

ADMINISTRATIVE REQUIREMENTS

The third level-two title, Administrative Requirements, encompasses the areas of Project Management and Coordination, Construction Progress Documentation, Submittal Procedures, and Special Procedures.

- Project Management and Coordination includes the administration of subcontractors and coordination with other contractors and the owner. This includes project meeting and on-site administration.
- Construction Progress Documentation covers the requirements for scheduling, recording, and reporting progress. This includes such things as

construction photographs, progress reports, site observation, purchase order tracking, scheduling of construction, and survey and layout data.

- Submittal Procedures includes the general procedures and requirements for submittals during the course of the construction. This can include items such as certificates, design data, field test reports, shop drawings, product data and samples, and source quality control reports.
- Special Procedures includes any special procedures that any of the project situations require, such as historic restoration, renovation or alteration; preservation; or hazardous material abatement.

QUALITY REQUIREMENTS

The fourth level-two title of Division 1 is Quality Requirements. This covers four areas: Regulatory Requirements, References, Quality Assurance, and Quality Control.

- Regulatory Requirements provides information necessary for conformance to requirements such as building codes, mechanical codes, electrical codes, and other regulations and to fee payments applicable to a project.
- References includes lists of reference standards cited in the contract documents and the organizations whose standards are cited. This includes items such as symbols used, abbreviations, and acronyms.
- Quality Assurance references are used to provide for procedures to ensure the quality of construction. This includes field observations and tests performed by manufacturers' representatives during installation. Quality assurance is provided by fabricators, installers, manufacturers, suppliers, and testing agencies.
- Quality Control establishes procedures to measure and report the quality and performance of construction. It may require field samples and mock-ups assembled at a site. The contractor provides quality control plans. Control can be provided by inspections, inspection services, and testing laboratories.

TEMPORARY FACILITIES AND CONTROLS

The fifth level-two title, Temporary Facilities and Controls, provides requirements for installation, maintenance, and removal of temporary utilities, controls, facilities, and construction aids during construction. This includes Temporary Utilities, Construction Facilities, Temporary Construction, Construction Aids, Vehicular Access and Parking, Temporary Barriers and Enclosures, Temporary Controls, and Project Identification.

- Temporary Utilities includes all such services used during construction, such as electrical, water, lighting, gas, telephone, fire protection, fuel oil, gasoline, and diesel fuel.
- Construction Facilities includes any temporary facilities built on-site for use during construction. Typical examples include field offices, storage buildings, sanitary facilities, and first aid stations.
- Temporary Construction includes those facilities built to provide access to various parts of a site and project so as to facilitate the construction process or to accommodate the needs of the owner and occupants. Typical examples include temporary bridges, ramps, overpasses, decking, and turnarounds.
- Construction Aids include all requirements and procedures related to tools and equipment used during construction, such as scaffolding, cranes, hoists, and construction elevators.
- Vehicular Access and Parking provides requirements for and procedures related to access to the site and parking facilities to meet needs of the construction and owner operations. Typically this can include access roads, haul routes, parking areas, temporary roads, and control of traffic.
- Temporary Barriers and Enclosures includes all facilities and procedures for the protection of the occupants or existing spaces during construction. This can include air barriers, barricades, dust barriers, fences, noise barriers, pollution control, security, and the protection of trees and other vegetation.
- Temporary Controls includes site or environment controls required to allow construction to proceed, including such things as erosion and sediment control and pest control.
- Project Identification includes any signs used to identify the construction site and areas on the site.

PRODUCT REQUIREMENTS

The sixth level-two title, Product Requirements, includes a comprehensive series of requirements.

- Basic Product Requirements includes the basic requirements for new, salvaged, and reused products used in construction.
- Product Options includes the basic requirements for options the contractor may have in selecting

products and how the determination is made for equal products.

- Product Substitution Procedures includes the basic requirements and procedures when proposals for the substitution of a product are made.
- Owner-Furnished Products includes the basic requirements for products that are to be furnished by the owner. This also involves scheduling, coordinating, handling, and storing owner-furnished products.
- Product Delivery Requirements specifies the basic requirements for packing, shipping, delivery, and acceptance of products at the site.
- Product Storage and Handling Requirements sets forth the basic requirements for storing and handling products on-site.

EXECUTION AND CLOSEOUT REQUIREMENTS

The seventh level-two title, Execution and Closeout Requirements, has eight sections.

- Examination involves the acceptance of conditions, existing conditions, and the basic requirements for determining acceptable conditions for installation.
- Preparation details the requirements for preparing to install, erect, or apply products, including activities such as field engineering, protection of adjacent construction, surveying, and construction layout.
- Execution includes the basic requirements for installing, applying, or erecting products that are new, prepurchased, salvaged, or owner furnished.
- Cleaning sets the requirements for maintaining the site in a neat condition during construction and the final cleaning in preparation for turning the project over to the owner.
- Starting and Adjusting involves establishing the initial checkout and startup procedures and any adjustments needed to ensure safe operation during the acceptance testing and commissioning.
- Protecting Installed Construction provides the requirements and procedures for protecting installed construction.
- Closeout Procedures includes the administrative procedures for substantial completion and final completion of the work.
- Closeout Submittals sets the procedures for closeout submittals, revised project documents, and delivery and distribution of spare parts and maintenance materials.

PERFORMANCE REQUIREMENTS

The eighth level-two title, Performance Requirements, has to do with the final requirements for preparing the facility for decommissioning.

- Commissioning requirements include commissioning summary, system performance evaluation, testing, adjusting, and balancing procedures.
- Demonstration and Training includes the requirements and procedures for the demonstration of the products and systems within the facility, including training the owner's operating and maintenance personnel.
- Operation and Maintenance sets forth the requirements and procedures for operating the facility after commissioning.
- Reconstruction involves any renovation or reconstruction of the existing facilities that may be required.

Life-Cycle Activities

The last and ninth level-two title, Life-Cycle Activities, includes the basic requirements for deactivating a facility or a portion of it from operation. This includes activities such as facility demolition and removal; hazardous materials abatement, removal, and disposal; and protection of deactivated facilities.

Appendix B

U.S. AND CANADIAN PROFESSIONAL AND TECHNICAL ORGANIZATIONS

DIVISION 3—CONCRETE

American Concrete Institute
PO Box 9094
Farmington Hills, MI 48333-9094

Concrete Reinforcing Steel Institute
933 North Plum Grove Road
Schaumberg, IL 60173-4758
www.crsi.org

American Society of Concrete Contractors
38800 Country Club Drive
Farmington Hills, MI 48331-3411
www.ascconline.org

Architectural Precast Association, Inc.
6710 Winkler Road, Suite 8
Fort Meyers, FL 33919
www.archprecast.org

Cement Association of Canada
Headquarters at 1500-60 Queen Street
Ottawa, Ontario
Canada K1P 5Y7
www.cement.ca
Regional Offices in Bedford, NB; Montreal, QC; Toronto, ON; Vancouver, BC; and Calgary, AB

National Concrete Masonry Association
13750 Sunrise Valley Drive
Herndon, VA 20171-4662
www.ncma.org

National Precast Concrete Association
10333 N. Meridian Street, Suite 272
Indianapolis, IN 46290
www.precast.org

National Ready Mixed Concrete Association
900 Spring Street
Silver Spring, MD 20910
www.nrmca.org

Portland Cement Association
5420 Old Orchard Road
Skokie, IL 60077-1083
www.cement.org

Post-Tensioning Institute
1717 W. Northern Avenue, Suite 114
Phoenix, AZ 85071-5471

Tilt-Up Concrete Association
PO Box 204
Mt. Vernon, IA 52314
www.tilt-up.org

DIVISION 4—MASONRY

Brick Industry Association
11490 Commerce Park Drive, Suite 300
Reston, VA 20191-1525
www.gobrick.com

Building Stone Institute
PO Box 507
Purdys, NY 10578

Cast Stone Institute
10 W. Kimball Street
Winder, GA 30680
www.caststone.org

Indiana Limestone Institute of America
Stone City Bank Building, Suite 400
1502J L Street
Bedford, IN 47421
www.iliai.com

International Masonry Institute
53 W. Jackson Boulevard, Suite 308
Chicago, IL 60604

The Masonry Society
3970 Broadway, Suite 201-D
Boulder, CO 80304-1135
www.masonrysociety.org

Marble Institute of America
28901 Clemens Road, Suite 100
Cleveland, OH 44145
www.marble-institute.com

National Concrete Masonry Association
13750 Sunrise Valley Drive
Herndon, VA 20171
www.ncma.org

National Lime Association
200 N. Glebe Road, Suite 806
Arlington, VA 22203
www.lime.org

Tile Council of North America, Inc.
100 Clemson Research Boulevard
Anderson, SC 29625
www.tileusa.com

Western States Clay Products Association
2550 Beverly Boulevard
Los Angeles, CA 90057-1019
www.brick-wscpa.org

DIVISION 5—METALS

The Aluminum Association
1525 Wilson Road, Suite 600
Arlington, VA 22209
www.aluminum.org

Aluminum Anodizers Council
1000 N. Eand, Suite 214
Wauconda, IL 60084
www.anodizing.org

American Galvanizers Association
6881 S. Holly Circle, No. 108
Englewood, CO 80112
www.galvanizeit.org

Aluminum Extruders Council
1000 N. Rand Road, Suite 214
Wauconda, IL 60084
www.aec.org

American Institute of Steel Construction
1 East Wacker Drive, Suite 3100
Chicago, IL 60601
www.aisc.org

Association for Iron & Steel Technology
186 Thorn Hill Road
Warrendale, PA 15086-7528
www.aist.org

Copper Development Association
260 Madison Avenue, 16th Floor
New York, NY 10016
www.copper.org

Metal Building Manufacturers Association
1300 Summer Avenue
Cleveland, OH 44115-2851
www.mbma.com

Metal Building Contractors and Erectors Association
242 East Main Street
Louisville, KY 40202
www.mbcea.org

Metal Construction Association
4700 W. Lake Avenue
Glenview, IL 60025
www.metalconstruction.org

National Association of Architectural Metal Manufacturers
8 S. Michigan Avenue, Suite 1000
Chicago, IL 60603
www.naamm.org

National Ornamental & Miscellaneous Metals Association
532 Forest Parkway, Suite A
Forest Park, GA 30297
www.nomma.org

Sheet Metal and Air-Conditioning Contractors National Association
4201 Lafayette Center Drive
Chantilly, VA 30151
www.smacna.org

Specialty Steel Industry of North America
3050 K Street NW, Suite 400
Washington, DC 20007
www.ssina.com

The Structural Engineering Institute of the American Society of Civil Engineers
1801 Alexander Bell Drive
Reston, VA 20191-4400
www.constructioninst.org

Steel Framing Alliance
1201 15th Street NW, Suite 320
Washington, DC 20005-2842
www.steelframing.org

Steel Erectors Association of America
PO Box 4891
Chapel Hill, NC 27515-4891
www.seaa.net

Steel Stud Manufacturers Association
8 S. Michigan Avenue, Suite 1000
Chicago, IL 60603
www.ssma.com

Steel Joist Institute
3127 Mr. Joe White Avenue
Myrtle Beach, SC 28577-6760
www.steeljoist.org

Wire Reinforcement Institute
942 Main Street, Suite 300
Hartford, CT 06103
www.wirereinforcementinstitute.org

Wire Association International
PO Box 578
Guilford, CT 06437
www.wirenet.org

Wire Fabricators Association
710 E. Ogden Avenue 600
Naperville, IL 66583

DIVISION 6—WOOD, PLASTICS AND COMPOSITES

American Fiberboard Association
853 N. Quentin Road
Palatine, IL 60067

American Forest and Paper Association
1111 Nineteenth Street NW, Suite 800
Washington, DC 20036
www.afandpa.org

American Institute of Timber Construction
7012 South Revere Parkway, Suite 140
Englewood, CO 80112
www.aitc-glulam.org

American Lumber Standards Committee
PO Box 210
Germantown, MD 20875-0210
www.alsc.org

American Wood Council
1111 Nineteenth Street NW, Suite 800
Washington, DC 20036
www.awc.org

Architectural Woodwork Institute
46179 Westlake
Potomac Falls, VA 20165
www.awinet.org

American Wood Preservers Association
PO Box 388
Selma, AL 36702
www.awpa.com

APA—The Engineered Wood Association
7011 South 19th Street
Tacoma, WA 91411-0700
www.apawood.org

Building Products Information
320-979 de Bourgogne
Ste-Foy, Quebec
Canada G1W 2L4

California Redwood Association
405 Enfrente Drive, Suite 200
Novato, CA 94940
www.calredwood.org

Canadian Wood Council
1400 Blair Place, Suite 210
Ottawa, Ontario
Canada K1P 6B9
www.cwc.ca

Canadian Plywood Association
735 West 15th Street
North Vancouver, British Columbia
Canada V7M 1T2
www.canply.org

Canadian Particleboard Association
27 Goulburn Avenue
Ottawa, Ontario
Canada K1N 8C7

Canadian Hardwood Plywood Association
27 Goulburn Avenue
Ottawa, Ontario
Canada K1N 8C7

Canadian Lumbermans Association
27 Goulbourn Avenue
Ottawa, Ontario
Canada K1N 8C7

COFI Plywood Technical Centre
735 West 15th Street
North Vancouver, British Columbia
Canada V7M 1T2

Composite Panel Association
18928 Premiere Court
Gaithersburg, MD 20879-1569
www.pbmdf.com

Composite Wood Council
18922 Premiere Court
Gaithersburg, MD 20879-1574
www.pbmdf.com

The Construction Institute of the American Society of Civil Engineers
1810 Alexander Bell Drive
Reston, VA 20191-4400
www.constructioninst.org

Cedar Shake and Shingle Bureau
PO Box 1176
Sumas, WA 98295-1178
www.cedarbureau.org

Forest Products Laboratory
One Gifford Pinchot Drive
Madison, WI 53726
www.fpl.fs.fed.us

Forest Products Society
2801 Marshall Court
Madison, WI 53705
www.forestprod.org

Forest Stewardship Council – U.S. (FSC-US)
212 Third Avenue North, Suite 280
Minneapolis, MN 55401
www.fscus.org

Hardwood Council
PO Box 525
Oakmont, PA 15139
www.americanhardwoods.org

Hardwood Manufacturers Association
400 Penn Center Boulevard, Suite 530
Pittsburgh, PA 15235
www.hardwoodinfo.com

Hardwood Plywood and Veneer Association
Laboratory and Testing Service
PO Box 2789
Reston, VA 20195-0789
www.hpva.org

Laminated Timber Institute of Canada
c/o Western Archrib Structures
PO Box 5648
Edmonton, Alberta
Canada T6C 4G1

Maple Flooring Manufacturers Association
60 Revere Drive, Suite 500
Northbrook, IL 60062
www.maplefloor.org

Modular Building Institute
413 Park Avenue
Charlottesville, VA 22902
www.mbinet.org

National Lumber Grades Authority
103-4400 Dominion Street
Burnaby, British Columbia
Canada K1N 8C7

National Hardwood Lumber Association
PO Box 34518
Memphis, TN 38184-0518
www.nhla.com

National Frame Builders Association
4840 Bob Billings Parkway
Lawrence, KS 66049
www.nfba.org

Quebec Wood Export Bureau
979 de Bourgagne, Bureau 450
Ste-Foy, Quebec
Canada G1W 2L4
www.quebecwoodexport.com

Structural Board Association
412-45 Sheppard Avenue East
Willowdale, Ontario
Canada M2N 5W9

Southern Forest Products Association
PO Box 641700
Kenner, LA 70064-1700
www.sfpa.org

Southern Pine Association
PO Box 641700
Kenner, LA 70064-1700
www.southernpine.com

Southern Pine Council
PO Box 641700
Kenner, LA 70064-1700

Structural Board Association
454 Sheppard Avenue E, Suite 412
Willowdale, Ontario
Canada M2N 5W9
www.osbguide.com

Structural Insulated Panel Association
PO Box 1699
Gig Harbor, WA 98335
www.sips.org

Truss Plate Institute
218 North Lee Street, Suite 312
Alexandria, VA 22314
www.tpinst.org

Truss Plate Institute of Canada
c/o M. Tek Canada, Inc.
100 Industrial Road
Bradford, Ontario
Canada L3Z 3G7
www.tpic.ca

The Wood Flooring Manufacturers Association
PO Box 3009
Memphis, TN 38173-0009
www.nofma.org

Wood Moulding & Millwork Producers Association
507 First Street
Woodland, CA 95695-4025
www.wmmpa.com

Western Wood Products Association
522 SW Fifth Avenue, Suite 600
Portland, OR 97204-2127
www.wwpa.org

Wood Truss Council of America
6300 Enterprise Lane
Madison, WI 53719
www.woodtruss.com

PLASTICS ORGANIZATIONS

Association of Postconsumer Plastic Recyclers
2000 L Street NW, Suite 835
Washington, DC 20006
www.plasticsrecycling.org

Extruded Polystyrene Foam Association
1223 Dale Boulevard
Woodbridge, VA 22193
www.XPSA.com

Plastics Institute of America
333 Aiken Street
Lowell, MA 01854
www.plasticsinstitute.org

Society of Plastics Engineers
PO Box 403
Brookfield, CT 06804-0403
www.4spe.org

The Society of Plastics Industry, Inc.
1667 K Street NW, Suite 1000
Washington, DC 20006-1620
www.plasticsindustry.org

Vinyl Siding Institute
1801 K Street NW, Suite 600K
Washington, DC 20006
www.vinylsiding.org

DIVISION 7—THERMAL AND MOISTURE PROTECTION

Adhesives Manufacturers Association
401 N. Michigan Avenue, Suite 2400
Chicago, IL 60611-4267
www.adhesives.org/ama

Asphalt Institute
2696 Research Park Drive
Lexington, KY 40511-8480
www.asphaltinstitute.org

Asphalt Roofing Manufacturers Association
1156 15th Street NW
Washington, DC 20005
www.asphaltroofing.com

Cellulose Insulation Manufacturers Association
1368 Keowee Street
Dayton, OH 45402
www.cellulose.org

Extruded Polystyrene Foam Association
4223 Dale Boulevard
Woodbridge, VA 22193
www.xpsa.com

Insulation Contractors of America
1321 Duke Street, Suite 303
Alexandria, VA 22314
www.insulate.org

Midwest Roofing Contractors Association
3840 Bob Billings Parkway
Lawrence, KS 66049
www.mrca.org

North American Insulation Manufacturers Association
44 Canal Center Plaza, Suite 310
Alexandria, VA 22314
www.naima.org

National Roofing Contractors Association
10255 W. Higgins Road, Suite 600
Rosemont, IL 60018
www.nrca.net

Roof Consultants Institute
7424 Chapel Hill Road
Raleigh, NC 27607
www.rci-online.org

Roof Tile Institute
PO Box 40337
Eugene, OR 97404-0049
www.rooftile.org

Sealant, Waterproofing, and Restoration Institute
14 W. Third Street, Suite 200
Kansas City, MO 64105
www.swrionline.org

Tile Roofing Institute
230 East Ohio, Suite 400
Chicago, IL 60611
www.tileroofing.org

Tile Roofing Institute Technical Office
PO Box 40337
Eugene, OR 97404-0048
www.tileroofing.org

Western States Roofing Contractors Association
1400 Marsten Road, Suite N
Burlingame, CA 94010-2422
www.wsrca.com

DIVISION 8—OPENINGS

American Architectural Manufacturers Association
1827 Walden Office Square, Suite 550
Schaumburg, IL 60173
www.aamanet.org

Builders Hardware Manufacturers Association
355 Lexington Avenue, 17th Floor
New York, NY 10017
www.buildershardware.com

Efficient Windows Collaborative
1850 M Street, NW, Suite 600
Washington, DC 20036
www.efficientwindows.org

Glass Association of North America
2945 SW Wanamaker Drive, Suite A
Topeka, KS 66614-5321
www.glasswebsite.com

Glass Technical Institute
12653 Portada Place
San Diego, CA 29130

Glazing Contractors Association
43636 Woodward Avenue
Bloomfield Hills, MI 48302

National Fenestration Rating Council
8484 Georgia Avenue, Suite 320
Silver Spring, MD 20910
www.nfrc.org

National Wood Window and Door Association
1400 E. Touhy Avenue, Suite 470
Des Plains, IL 60018-3337
www.wdma.com

Steel Window Institute
1300 Summer Avenue
Cleveland, OH 44145
www.steelwindows.com

DIVISION 9—FINISHES

Carpet and Rug Institute
PO Box 2048
Dalton, GA 30722
www.carpet-rug.com

Carpet Cushion Council
PO Box 546
Riverside, CT 06878
www.carpetcushion.org

Ceilings and Interior Systems Construction Association
1500 Lincoln Highway, Suite 202
St. Charles, IL 60174
www.cisca.org

Foundation of the Wall and Ceiling Industry
803 W. Broad Street, Suite 600
Falls Church, VA 22046
www.awci.org

Green Seal
1001 Connecticut Avenue, NW
Suite 827 Washington, DC
20036-5525 USA
www.greenseal.org

Gypsum Association
810 First Street NE, Suite 510
Washington, DC 20002
www.gypsum.org

International Institute for Lath and Plaster
3127 Los Fotir Boulevard
Los Angeles, CA 90039

National Oak Flooring Manufacturing Association
PO Box 3009
Memphis, TN 38103
www.nofma.org

National Paint and Coatings Association
1500 Rhode Island Avenue, NW
Washington, DC 20005
www.paint.org

National Terrazzo & Mosaic Association, Inc.
201 North Maple Avenue, Suite 208
Purcellville, VA 20132
www.ntma.com

National Tile Contractors Association
PO Box 13629
Jackson, MS 39236
www.tile-assn.com

National Wood Flooring Association
16388 Westwoods Business Plaza
Ellisville, MO 63021

Painting and Decorating Contractors of America
3913 Old Lee Highway
Fairfax, VA 22030-2401
www.pdca.com

Resilient Floor Covering Institute
401 E. Jefferson Street, Suite 102
Rockville, MD 20850

Rubber Manufacturers Association
1400 K Street NW, Suite 900
Washington, DC 20005
www.rma.org

Tile Council of America, Inc.
100 Clemson Research Boulevard
Anderson, SC 29625
www.tileusa.com

Wallcoverings Association
401 N. Michigan Avenue, Suite 2200
Chicago, IL 50611-4267
www.wallcoverings.org

THERE ARE NO REFERENCES FOR DIVISION 10—SPECIALTIES AND DIVISION 11—EQUIPMENT

DIVISION 12—FURNISHINGS

American Society of Furniture Designers
144 Woodland Drive
New London, NC 28127
www.asfd.com

Business and Institutional Furniture Manufacturers Association
2335 Burton Street NE
Grand Rapids, MI 49506

Contract Furnishings Council
1190 Merchandise Mart
Chicago, IL 60654

THERE ARE NO REFERENCES FOR DIVISION 13—SPECIAL CONSTRUCTION

DIVISION 14—CONVEYING EQUIPMENT

Conveyor Equipment Manufacturers Association
6724 Lone Oak Boulevard
Naples, FL 34109
www.cemanet.org

National Association of Elevator Contractors
1298 Wellbrook Circle NE, Suite A
Conyers, GA 30207-2872
www.naec.org

National Elevator Industry, Inc.
400 Frank W. Burr Boulevard
Teaneck, NJ 07666

THERE ARE NO DIVISIONS 15 THROUGH 20

DIVISION 21—FIRE SUPPRESSION

Building and Fire Research Laboratory
National Institute of Standards and Technology
100 Bureau Drive, MS 8600
Gaithersburg, MD 20899-8600
www.bfrl.nist.gov

Fire Suppression Systems Association
5024R Campbell Boulevard
Baltimore, MD 21236-5974
www.fssa.net

Fire Research Program
Institute for Research in Construction
Montreal Road Campus, M59
Ottawa, Ontario
Canada K1A 0R6
www.irc.nrc-cnrc.ga.ca/fr

Fire Equipment Manufacturers Association
1300 Summer Avenue
Cleveland, OH 44115-2851
www.yourfirstdefense.com

International Fire Code Institute
5360 S. Workman Mill Road
Whittier, CA 90501-22988
www.ifci.com

National Fire Protection Association
1 Batterymarch Park
Quincy, MA 02269-9101
www.nfpa.org

National Fire Laboratory Institute for Research in Construction
Montreal Road Campus
Ottawa, Ontario
Canada K1A 0R6

National Fire Sprinkler Association
40 Van Barrett Road
Patterson, NY 12563-1000
www.nfsa.org

Society of Fire Protection Engineers
7315 Wisconsin Avenue, Suite 620E
Bethesda, MD 20914
www.sfpe.org

DIVISION 22—PLUMBING

American Society of Mechanical Engineers
3 Park Avenue
New York, NY 10016-5990
www.asme.org

American Society of Plumbing Engineers
3617 Thousand Oaks Boulevard, Suite 210
Westlake Village, CA 91362-3649
www.aspe.org

American Society of Sanitary Engineers
29901 Clemens Road 100
West Lake, OH 44145
www.asse-plumbing.org

American Water Works
6666 W. Quincy A
Denver, CO 80235
www.awwa.org

Cast Iron Soil Pipe Institute
5959 Shallowford Road, Suite 419
Chattanooga, TN 37412
www.cispi.org

International Association of Plumbing
and Mechanical Officials
2001 Walnut Drive S
Walnut, CA 91789-7825

National Association of Plumbing, Heating,
and Cooling Contractors
PO Box 6808
Falls Church, VA 22040
www.phccweb.org

National Fire Sprinkler Association
40 Van Barrett Road
Patterson, NY 12563-1000
www.nfsa.org

Plastic Pipe Institute
1801 K Street NW, Suite 600K
Washington, DC 20006-1301
www.plasticpipe.org

Plumbing Manufacturers Institute
1340 Remington Road, Suite A
Schaumburg, IL 60173
www.pmihome.org

Plastic Pipe and Fittings Association
800 Roosvelt Road, Building C, Suite 312
Glen Ellyn, IL 60137
www.ppfahome.org

Waste Material Management Division
U.S. Department of Energy
1000 Independence Avenue SW
Washington, DC 20585

Water Quality Association
4151 Naperville Road
Lisle, IL 60532
www.wqa.org

DIVISION 23—HEATING, VENTILATING, AND AIR-CONDITIONING

Air-Conditioning and Refrigeration Institute
430 N. Fairfax Drive, Suite 425
Arlington, VA 22203
www.ari.org

Air Conditioning Contractors of America
2800 Shirlington Road, Suite 300
Arlington, VA 22206
www.acca.org

Air Movement and Control Association
International, Inc.
30 W. University Drive
Arlington Heights, IL 60004
www.amca.org

American Boiler Manufacturers Association
950 N. Glebe Road, Suite 160
Arlington, VA 22203-1824
www.abma.com

American Society of Heating
and Air-Conditioning Engineers
1791 Tullie Circle NE
Atlanta, GA 30329-2305
www.ashrae.org

American Solar Energy Society
2400 Central Avenue, Suite A
Boulder, CO 80301
www.ases.org

American Gas Association
400 North Capitol Street NW, 4th Floor
Washington, DC 20001
www.aga.org

Air Diffusion Council
1901 N. Roselle Road, Suite 800
Schaumburg, IL 60195

Building Commissioning Association
1400 SW 5th Ave, Suite 700
Portland, OR 97201
www.bcxa.org

Cooling Tower Institute
530 Wells Fargo Drive 218
Houston, TX 77090
www.cti.org

Florida Solar Energy Center
1679 Clearlake Road
Cocoa, FL 32922-5703
www.fsec.ucf.edu

Heat Exchange Institute
1300 Summer Avenue
Cleveland, OH 44115-2851
www.heatexchange.org

Heating, Refrigeration, and Air Conditioning Institute of Canada
5045 Orbitor Drive, Building 11, Ste. 300
Mississauga, Ontario
Canada L4W 4Y4
www.hrai.ca

The Hydronics Institute, Inc.
PO Box 218
Berkeley Heights, NJ 07922-0218
www.ahrinet.org

International Society for Indoor Air Quality and Climate
Box 22038, Sub 32
Ottawa, Ontario
Canada K1V 0W2

Institute of Heating and Air Conditioning Industries
454 W. Broadway
Glendale, CA 91204
www.ihaci.org

National Renewable Energy Laboratory
1617 Cole Boulevard
Golden, CO 80401
www.nrel.gov

Northeast Sustainable Energy Association
50 Miles Street, Suite 3
Greenfield, MA 01301

Plumbing, Heating, and Cooling Contractors National Association
180 S. Washington Street
Falls Church, VA 22046
www.phccweb.org

Sheet Metal and Air Conditioning Contractors National Association
4201 Lafayette Center Drive
Chantilly, VA 20151
www.smacna.org

U.S. Department of Energy
19901 Germantown Road
Germantown, MD 20874-1290
www.doe.gov

DIVISION 26—ELECTRICAL

American Lighting Association
PO Box 420288
Dallas, TX 75342
www.americanlightingassoc.com

American Public Power Association
2301 M Street NW, Suite 300
Washington, DC 20037
www.appanet.org

Edison Electric Institute
701 Pennsylvania Avenue, NW
Washington, DC 20004-2696
www.eei.org

Electric Power Research Institute
3412 Hillview Avenue
Palo Alto, CA 94034
www.epri.org

Electric Power Supply Association
1401 H Street NW, Suite 760
Washington, DC 20005
www.epsa.org

Illuminating Engineering Society of North America
120 Wall Street, 17th Floor
New York, NY 10005-4001
www.iesna.org

International Association of Lighting Designers
The Merchandise Mart, Ste. 11-114A
World Trade Center
Chicago, IL 60654
www.iald.org

Lighting Research Institute
120 Wall Street, 17th Floor
New York, NY 10005

Lighting Research Center
Watervliet Facility
Rensselaer Polytechnic Institute
110 8th Street
Troy, NY 12180
www.lrc.rpi.edu

National Lighting Bureau
8811 Colesville Road, Suite G106
Silver Spring, MD 20910
www.nlb.org

National Electric Manufacturers Association
1300 N 17th Street, Suite 1867
Rosslyn, VA 22209
www.nema.org

Underwriters Laboratories, Inc.
333 Pfingsten Road
Northbrook, IL 60062-2096
www.ul.com

DIVISION 28—ELECTRONIC SAFETY AND SECURITY

American Society for Industrial Security
1625 Prince Street
Alexandria, VA 22314
www.asis.online.org

National Burglar and Fire Alarm Association
7101 Wisconsin Avenue, Suite 901
Bethesda, MD 20814
www.alarm.org

National Crime Prevention Council
1700 K Street NW, 2nd Floor
Washington, DC 20006-3817
www.weprevent.org

National Crime Prevention Institute
University of Louisville
Burhans Hall
Louisville, KY 40292-0001
www.louisville.edu/a-s/ja

National Fire Protection Association
1 Batterymarch Park
Quincy, MA 02169-7471
www.nfpa.org

Security Hardware Distributors Association
100 N. 20th Street, 4th Floor
Philadelphia, PA 19103-1443
www.shda.org

Security Industry Association
635 Slaters Lane, Suite 110
Alexandria, VA 22314
www.siaonline.org

DIVISION 31—EARTHWORK

American Nursery and Landscape Association
1000 Vermont Avenue NW, Suite 300
Washington, DC 20005
www.anla.org

American Society of Civil Engineers
1801 Alexander Bell Drive
Reston, VA 20191-4400
www.asce.org

American Society of Landscape Architects
636 Eye Street NW
Washington, DC 20001
www.asla.org

ASFE: Professional Firms Practicing in the Gensciences
8811 Colesvill Road
Silver Spring, MD 20910
www.asfe.org

American Association of State Highway and Transportation Officials Systems
444 Capital Street NW, Suite 249
Washington, DC 20001
www.aashto.org

American Water Works Association
6660 W. Quincy Avenue
Denver, CO 80235
www.awwa.org

Canadian Society of Landscape Architects
1339 Fifteenth Avenue NW, Apartment 310
Calgary, Alberta
Canada T3C 3V3

Environmental Information Association
6935 Wisconsin Avenue, Suite 306
Chevy Chase, MD 20815-6112
www.eia-usa.org

The Geo-Institute of the American Society of Civil Engineers
1801 Alexander Bell Drive
Reston, VA 20191-4400
www.geoinstitute.org

Information Sources for Recycled and Energy-Efficient Materials

Athena Sustainable Materials Institute – Head Office
629 St. Lawrence St.
PO Box 189
Merrickville, Ontario, Canada, K0G 1N0
www.athenasmi.org

Alliance to Save Energy
1200 18th Street NW, Suite 800
Washington, DC 20036
www.ase.org

American Solar Energy Society
2400 Central Avenue, Suite G-1
Boulder, CO 80301
www.ases.org

American Wind Energy Association
122 C Street NW, Suite 400
Washington, DC 20001-2109
www.awea.org

Association of Energy Engineers
4025 Pleasantdale Road, Suite 420
Atlanta, GA 30340
www.aeecenter.org

American Council for an Energy-Efficient Economy
1001 Connecticut Avenue NW, Suite 801
Washington, DC 20036-5525
www.aceee.org

Environmental Building News – BuildingGreen, LLC
122 Birge Street, Suite 30
Brattleboro, VT 05301
www.buildinggreen.com

Canada Green Building Council
47 Clarence Street, Suite 202
Ottawa, ON K1N 9K1
www.cagbc.org

Center for Energy Policy and Research
New York Institute of Technology
Old Westbury, NY 11568

Center for Resourceful Building Technology
PO Box 100
Missoula, MT 59806
www.mountains.com/crbt

Construction Materials Recycling Association
P.O. Box 122
Eola, IL 60519
www.cdrecycling.org

Energy Efficiency and Renewable Energy Clearing House
PO Box 3048
Merrifield, VA 22136
www.eere.energy.gov

Energy Efficient Building Association
PO Box 22307
Eagan, MN 55122-0307

The Green Building Initiative
2104 SE Morrison,
Portland, Oregon 97214
www.greenglobes.com

The Geo-Institute of the American Society of Civil Engineers
1801 Alexander Bell Drive
Reston, VA 20191-4400
www.geoinstitute.org

NAHB Research Center
400 Prince George's Boulevard
Upper Marlboro, MD 20774-8731
www.nahbrc.org

Northeast Sustainable Energy Association
50 Miles Street
Greenfield, MA 01301
www.nesea.org

National Energy Management Institute
601 N. Fairfax Street, Suite 250
Alexandria, VA 22314
www.nemionline.com

National Renewable Energy Laboratory
1617 Cole Boulevard
Golden, CO 80401-3393
www.nrel.gov

National Research Council Canada and the Institute for Research in Construction
Montreal Road Campus, Building M-59
Ottawa, Ontario
Canada K1A 0R6
www.nrc.ca

National Energy Foundation
5225 Wiley Post Way, Suite 170
Salt Lake City, UT 84116
www.nef1.org

Sustainable Buildings Industry Council
1112 16th Street NW, Suite 240
Washington, DC 20036
www.SBICouncil.org

U.S. Green Building Council
2101 L Street, NW Suite 500
Washington, DC 20037
www.usgbc.org

Building Code Agencies

International Code Council
5203 Leesburg Pike, Suite 708
Falls Church, VA 22041-3401
www.iccsafe.org

National Fire Protection Association
1 Batterymarch Park
Quincy, MA 02169-7471
www.nfpa.org

International Association of Plumbing
and Mechanical Officials
20001 Walnut Drive South
Walnut, CA 91789-2825
www.iapmo.org

American Society for Testing and Materials
100 Barr Harbor Drive
West Conshohocken, PA 19428
www.astm.org

American Society of Mechanical Engineers
Three Park Avenue
New York, NY 10016-5990

American Concrete Institute
PO Box 9094
Farmington Hills, MI 48333
www.concrete.org

National Conference of States on Building Codes
and Standards
505 Hunter Park Drive, Suite 210
Herndon, VA 20170
www.ncsbcs.org

International Fire Code Institute
5360 S. Workman Mill Road
Whittier, CA 90501-2298
www.ifci.com

Other Organizations

American Water Works Association
6666 West Works Avenue
Denver, CO 80235-3098
www.awwa.org

American Architectural Manufacturers Association
1827 Walden Office Square, Suite 550
Schaumburg, IL 60173
www.aamanet.org

American Institute of Architects
1735 New York Avenue NW
Washington, DC 20006
www.aia.org

American National Metric Council
1735 N. Lynn Street, Suite 950
Arlington, VA 22209-2022

American National Standards Institute
25 West 42nd Street, 4th Floor
New York, NY 10036
www.ansi.org

American Society for Testing and Materials
100 Barr Harbor Drive
W. Conshohocken, PA 19428-2959
www.astm.org

American Society of Mechanical Engineers
3 Park Avenue
New York, NY 10016-5990
www.asme.org

American Society of Civil Engineers
World Headquarters
1801 Alexander Bell Drive
Reston, VA 20191-4400
www.asce.org

American Council for Construction Education
1717 N. Loop, 1604 East, Suite 320
San Antonio, TX 78232-1570
www.acce-hq.org

Association of Construction Inspectors
1224 N. Nokomis NE
Alexandria, MN 56308
www.iami.org

Building Research Council
University of Illinois
1 E. St. Mary's Road
Champaign, IL 61820

Canadian Housing Information Center
700 Chemin Montreal
Ottawa, Ontario
Canada K1A 0P7
www.cmhc-schl.gc.ca

Canadian Mortgage and Housing Corporation
700 Montreal Road
Ottawa, Ontario
Canada K1A 0P7
www.cmhc-schl.gc.ca

The Canadian Standards Association
5060 Spectrum Way
Mississauga, Ontario
Canada L4W 5N6
www.csa.ca

Canadian Construction Materials Centre
Institute for Research in Construction
National Research Council Canada
Building M-24, 1500 Montreal Road
Ottawa, Ontario
Canada K1A 0R6

Canadian Home Builders Association
150 Laurier Avenue West, Suite 200
Ottawa, Ontario
Canada K1P 5J4
www.chba.ca

The Construction Specifications Institute
99 Canal Center Plaza, Suite 300
Alexandria, VA 22314-1588
www.csinet.org

Construction Specifications Canada
100 Lombard Street, Suite 200
Toronto, Ontario
Canada M5C 1M3
www.csc-dcc.ca

The Coasts, Oceans, Ports, and Rivers Institute of the American Society of Civil Engineers
1801 Alexander Bell Drive
Reston, VA 20191-4400
www.coprinstitute.org

Environmental Hazards Management Institute
PO Box 932
Durham, NH 03824

The Environmental & Water Resources Institute of the American Society of Civil Engineers
1801 Alexander Bell Drive
Reston, VA 20191-4400
www.ewrinstitute.org

Federal Specifications
General Services Administration, Specifications Activity Office
Building 197
Washington Navy Yard
2nd and M Street SE
Washington, DC 20407

Home Builders Institute
1090 Vermont Avenue NW
Washington, DC 20005
www.hbi.org

Institute for Research in Construction
Montreal Road Campus, M59
Ottawa, Ontario
Canada K1A 0R6
www.irc.nrc-chrc.gc.ca

National Association of Home Builders
1201 15th Street NW
Washington, DC 20006
www.nahb.org

National Association of Home Builders Remodelers Council
1201 15th Street NW
Washington, DC 20005
www.nahb.org/remodelers

National Fire Protection Association
1 Batterymarch Park
Quincy, MA 02169-7471
www.nfpa.org

National Institute of Building Sciences
1090 Vermont Avenue NW, Suite 700
Washington, DC 20005-4905
www.nibs.org

National Association of the Remodeling Industry
780 Lee Street, Suite 200
Des Plaines, IL 60016
www.nari.org

National Institute of Science and Technology
U.S. Department of Commerce
Building 820, Room 306
Gaithersburg, MD 20899
www.nist.gov/metric

National Research Council of Canada
NRC Corporate Communications
1200 Montreal Road, Building M-58
Ottawa, Ontario
Canada K1A 0R6

National Institute of Standards and Technology
U.S. Department of Commerce
100 Bureau Drive MS2150
Gaithersburg, MD 20899-2150
www.nist.gov/srm

Occupational Safety and Health Administration
200 Constitution Avenue NW
Washington, DC 20210
www.osha.gov

Rubber Manufacturers Association
1400 K Street NW, Suite 900
Washington, DC 20005
www.rma.org

Sweets Construction Catalog
Architects, Engineers, and Contractors
McGraw-Hill Construction
Two Penn Plaza
New York, NY 10121
www.construction.com

The Transportation & Development Institute of the American Society of Civil Engineers
1801 Alexander Bell Drive
Reston, VA 20191-4400
www.tanddi.org

U.S. Metric Association
10245 Andasol Avenue
Northridge, CA 91325-1504
www.metric.org

Underwriters Laboratories, Inc.
333 Pfingsten Road
Northbrook, IL 60062-2096
www.ul.com

Underwriters Laboratories of Canada
7 Underwriters Road
Scarborough, Ontario
Canada M1R 3A9
www.ulc.ca

U.S. Environmental Protection Agency
Ariel Rios Building
1200 Pennsylvania Avenue NW
Washington, DC 20460
www.epa.gov

U.S. Department of Commerce
Building 820, Room 306
Gaithersburg, MD 20899
www.commerce.gov

U.S. Department of Housing and Urban Development
HUD USER
PO Box 23268
Washington, DC 20026-3268
www.huduser.org

U.S. Department of Energy
Energy Efficiency & Renewable Energy Center
1000 Independence Avenue
Washington, DC 20585-0121
www.eere.energy.gov

U.S. Environmental Protection Agency
Atmospheric Pollution Prevention Division
401 M Street, Mail Code 6202JSW
Washington, DC 20460

Air and Radiation Docket and Information Center
U.S. Environmental Protection Agency
401 M Street SW
Washington, DC 20460

U.S. EPA Main Library
Environmental Protection Agency
109 Alexander Drive
Research Triangle Park, NC 27711
www.epa.gov

U.S. Department of Energy
Office of Scientific Information
PO Box 62
Oak Ridge, TN 37831

The level-two numbers and titles used in this publication are from MasterFormat™ 2004 Edition, published by the Construction Specifications Institute (CSI) and Construction Specifications Canada (CS), and used with permission from CSI 2005.

Appendix C

METRIC INFORMATION

Tables giving metric equivalents for common fractions, two-place decimal inches, and millimeter-to-decimal-inches are printed inside the front and rear covers for ready reference.

Base SI Units		
Quantity	**Unit**	**Symbol**
Length	Meter	m
Mass	Kilogram	kg
Time	Second	s
Electric current	Ampere	A
Thermodynamic temperature	Kelvin	K
Amount of substance	Mole	mo
Luminous intensity	Candela	cd

Supplementary SI Units		
Quantity	**Unit**	**Symbol**
Plane angle	Radian	rad
Solid angle	Steradian	sr

Derived Metric Units with Compound Names		
Physical Quantity	**Unit**	**Symbol**
Area	Square meter	m^2
Volume	Cubic meter	m^3
Density	Kilogram per cubic meter	kg/m^3
Velocity	Meter per second	m/s
Angular velocity	Radian per second	rad/s
Acceleration	Meter per second squared	m/s^2
Angular acceleration	Radian per second squared	rad/s^2
Volume rate of flow	Cubic meter per second	m^3/s
Moment of inertia	Kilogram meter squared	$kg{\bullet}m^2$
Moment of force	Newton meter	N•m
Intensity of heat flow	Watt per square meter	W/m^2
Thermal conductivity	Watt per meter Kelvin	W/m•K
Luminance	Candela per square meter	cd/m^2

SI Prefixes		
Multiplication Factor	**Prefix**	**Symbol**
1 000 000 000 000 000 000 = 10^{18}	exa	E
1 000 000 000 000 000 = 10^{15}	peta	P
1 000 000 000 000 = 10^{12}	tera	T
1 000 000 000 = 10^{9}	giga	G
1 000 000 = 10^{6}	mega	M
1 000 = 10^{3}	kilo	k
100 = 10^{2}	hecto	h
10 = 10^{1}	deka	da
0.1 = 10^{21}	deci	d
0.01 = 10^{22}	centi	c
0.001 = 10^{23}	milli	m
0.000 001 = 10^{26}	micro	m
0.000 000 001 = 10^{29}	nano	n
0.000 000 000 001 = 10^{212}	pico	p
0.000 000 000 000 001 = 10^{215}	femto	f
0.000 000 000 000 000 001 = 10^{218}	atto	a

Metric Unit to Imperial Unit Conversion Factors[a]		
Metric Units		**Imperial Equivalents**
Length		
1 millimeter (mm)	=	0.0393701 inch
1 meter (m)	=	39.3701 inches
	=	3.28084 feet
	=	1.09361 yards
1 kilometer (km)	=	0.621371 mile
Length/Time		
1 meter per second (m/s)	=	3.28084 feet per second
1 kilometer per hour (km/h)	=	0.621371 mile per hour
Area		
1 square millimeter (mm^2)	=	0.001550 square inch
1 square meter (m^2)	=	10.7639 square feet
1 hectare (ha)	=	2.47105 acres
1 square kilometer (km^2)	=	0.386102 square mile

Metric Unit to Imperial Unit Conversion Factorsa (continued)		
Metric Units		**Imperial Equivalents**
	Volume	
1 cubic millimeter (mm^3)	=	0.0000610237 cubic inch
1 cubic meter (m^3)	=	35.3147 cubic feet
	=	1.30795 cubic yards
1 milliliter (mL)	=	0.0351951 fluid ounce
1 liter (L)	=	0.219969 gallon
	Mass	
1 gram (g)	=	0.0352740 ounce
1 kilogram (kg)	=	2.20462 pounds
1 tonne (t) (= 1,000 kg)	=	1.10231 tons (2,000 lb.)
	=	2204.62 pounds
	Mass/Volume	
1 kilogram per cubic meter (kg/m^3)	=	0.0622480 pound per cubic foot
	Force	
1 newton (N)	=	0.224809 pound-force
	Stress	
1 megapascal (MPa) (51 N/mm^2)	=	145.038 pounds-force per sq. in.
	Loading	
1 kilonewton per sq. meter (kN/m^2)	=	20.8854 pounds-force per sq. ft.
1 kilonewton per meter (kN/m) (51 N/mm)	=	68.5218 pounds-force per ft.
	Moment	
1 kilonewtonmeter (kNm)	=	737.562 pound-force ft.
	Miscellaneous	
1 joule (J)	=	0.00094781 Btu
1 joule (J)	=	1 watt-second
1 watt (W)	=	0.00134048 electric horsepower
1 degree Celsius (°C)	=	32 1 1.8 (°C) degrees Fahrenheit

Notes: 1. 1.0 newton = 1.0 kilogram – 9.80665 m/s^2 (International Standard Gravity Value)
2. 1.0 pascal = 1.0 newton per square meter

[a]*Multiply the number of metric units by the imperial (English) equivalent to convert a measurement from metric units to imperial (English) units.*

Brick and Block Masonry lb./ft.^{2} kg/m^2 Imperial Unit to Metric Conversion Factors[b]		
Imperial Units		**Metric Equivalents**
	Length	
1 inch	=	25.4 mm
	=	0.0254 m
1 foot	=	0.3048 m
1 yard	=	0.9144 m
1 mile	=	1.60934 km

Brick and Block Masonry lb./ft.^{2} kg/m^2 (continued)		
	Length/Time	
1 foot per second	=	0.3048 m/s
1 mile per hour	=	1.60934 km/h
	Area	
1 square inch	=	645.16 mm^2
1 square foot	=	0.0929030 m^2
1 acre	=	0.404686 ha
1 square mile	=	2.58999 km^2
	Volume	
1 cubic inch	=	16387.1 mm^3
1 cubic foot	=	0.0283168 m^3
1 cubic yard	=	0.764555 m^3
1 fluid ounce	=	28.4131 mL
1 gallon	=	4.54609 L
	Mass	
1 ounce	=	28.3495 g
1 pound	=	0.453592 kg
1 ton (2,000 lb.)	=	0.907185 t
1 pound	=	0.0004539 t
	Mass/Volume	
1 pcf	=	16.1085 kg/m^3
	Force	
1 pound	=	4.44822 N
	Stress	
1 psi	=	0.00689476 MPa
	Loading	
1 psf	=	0.0478803 kN/m^2
1 plf	=	0.0145939 kN/m
	Moment	
1 pound-force ft.	=	0.00135582 kNm
	Miscellaneous	
1 Btu	=	1055.06 J
1 watt-second	=	1 J
1 horsepower	=	746 W
1 degree Fahrenheit	=	(°F 2 32)/1.8 °C

Notes: 1. 1.0 newton = 1.0 kilogram – 9.80665 m/s^2 (International Standard Gravity Value)
2. 1.0 pascal = 1.0 newton per square meter

[b]*Multiply the number of imperial (English) units by the metric equivalent to convert a measurement from imperial (English) units to metric units.*

Appendix D

WEIGHTS OF BUILDING MATERIALS

Brick and Block Masonry		
4" brickwall	40	196
4" concrete brick, stone or gravel	46	225
4" concrete brick, lightweight	33	161
4" concrete block, stone or gravel	34	167
4" concrete block, lightweight	22	108
6" concrete, stone or gravel	50	245
6" concrete block, lightweight	31	152
8" concrete block, stone or gravel	55	270
8" concrete block, lightweight	35	172
12" concrete block, stone or gravel	85	417
12" concrete block, lightweight	55	270
Concrete	**lb./ft.^{3}**	**kg/m^3**
Plain, slag	132	2155
Plain, stone	144	2307
Reinforced, slag	138	2211
Reinforced, stone	150	2403
Lightweight Concrete	**lb./ft.^{3}**	**kg/m^3**
Concrete, perlite	35–50	561–801
Concrete, pumice	60–90	961–1442
Concrete, vermiculite	25–60	400–961
Wall, Ceiling, and Floor	**lb./ft.^{2}**	**kg/m^2**
Acoustical tile, $\frac{1}{2}$"	0.8	3.9
Gypsum wallboard, $\frac{1}{2}$"	2	9.8
Plaster, 2" partition	20	98
Plaster, 4" partition	32	157
Plaster, $\frac{1}{2}$"	4.5	22
Plaster on lath	10	49
Tile, glazed, $\frac{3}{8}$"	3	14.7
Tile, quarry, $\frac{1}{2}$"	5.8	28.4
Terrazzo, 1"	25	122.5
Vinyl composition floor tile	1.4	69
Hardwood flooring, $\frac{25}{32}$"	4	19.6
Flexicore 6", lightweight concrete	30	14.7
Flexicore 6", stone concrete	40	196
Plank, cinder concrete, 2"	15	73.5
Plank, gypsum, 2"	12	58.8
Concrete reinforced, stone, 1"	12.5	61.3

Wall, Ceiling, and Floor	lb./ft.^{2}	kg/m^2
Concrete reinforced, lightweight, 1"	6–10	29.4–49
Concrete plain, stone, 1"	12	58.8
Concrete plain, lightweight, 1"	3–9	14.7–44.1
Partitions	**lb./ft.^{2}**	**kg/m^2**
2 – 4 wood studs, gypsum wallboard 2 sides	8	39.2
4" metal stud, gypsum wallboard 2 sides	6	29.4
6" concrete block, gypsum wallboard 2 sides	35	171.5
Roofing	**lb./ft.^{2}**	**kg/m^2**
Built-up	6.5	31.9
Concrete roof tile	9.5	46.6
Copper	1.5–2.5	7.4–12.3
Steel deck alone	2.5	12.3
Shingles, asphalt	1.7–2.8	8.3–13.7
Shingles, wood	2–3	9.8–14.7
Slate, $\frac{1}{2}$"	14–18	68.6–88.2
Tile, clay	8–16	39.2–78.4
Stone Veneer	**lb./ft.^{3}**	**kg/m^3**
2" granite, $\frac{1}{2}$" parging	30	481
4" limestone, $\frac{1}{2}$" parging	36	577
4" sandstone, $\frac{1}{2}$" parging	49	785
1" marble	13	208
Structural Clay Tile	**lb./ft.^{2}**	**kg/m^2**
4" hollow	23	368
6" hollow	38	609
8" hollow	45	721
Structural Facing Tile	**lb./ft.^{2}**	**kg/m^3**
2" facing tile	14	68.6
4" facing tile	24	118
6" facing tile	34	167
Wood	**lb./ft.^{2}**	**kg/m^2**
Ash, white	40.5	198
Birch	44	202
Cedar	22	108
Cypress	33	162

(Continued)

Douglas fir	32	157
White pine	27	132
Pine, southern yellow	26	127
Redwood	26	127
Plywood, $\frac{1}{2}$"	1.5	7.4
Residential Assemblies	**lb./ft.^{2}**	**kg/m^2**
Wood framed floor	10	49
Ceiling	10	49
Frame exterior wall, 4" studs	10	49
Frame exterior wall, 6" studs	13	64
Brick veneer of 4" frame	50	245
Brick veneer over 4" concrete block	74	363
Interior partitions with gypsum both sides (allowance per sq. ft. of floor area—not weight of material)	20	320
Suspended Ceilings	**lb./ft.^{2}**	**kg/m^2**
Acoustic plaster on gypsum lath	10–11	49–54
Mineral fiberboard	1.4	6.9
Metals	**lb./ft.^{3}**	**kg/m^3**
Aluminum	165	2643
Copper	556	8907
Iron, cast	450	7209
Steel	490	7850
Steel, stainless	490–510	7850–8170

Glass	**lb./ft.^{2}**	**kg/m^2**
$\frac{1}{4}$" (6.3 mm) plate or float	3.3	16.2
$\frac{1}{2}$" (12.7 mm) plate or float	6.6	32.3
$\frac{1}{32}$" (0.79 mm) sheet	2.8	13.7
$\frac{1}{4}$" (6.3 mm) sheet	3.5	17.2
$\frac{1}{8}$" (3.2 mm) double strength	1.6	7.8
$\frac{7}{32}$" (5.6 mm) sheet	2.85	14.0
$\frac{1}{4}$" (6.3 mm) laminated	3.30	16.2
$\frac{1}{2}$" (12.7 mm) laminated	6.35	31.1
2" (50.2 mm) bullet resistant	26.2	128.4
1" (25.4 mm) insulating, $\frac{1}{2}$", (6.3 mm) air space	6.54	32.0
$\frac{1}{4}$" (6.3 mm) wired	3.5	17.1
$\frac{3}{8}$" (9.5 mm) wired	5.0	24.4
$3\frac{7}{8}$" = $5\frac{3}{4}$" square, (98.0 − 146 mm) glass block	16.0	78.4

NAMES AND ATOMIC SYMBOLS OF SELECTED CHEMICAL ELEMENTS

Name	Atomic Symbol
Aluminum	Al
Antimony	Sb
Argon	Ar
Arsenic	As
Barium	Ba
Beryllium	Be
Bismuth	Bi
Boron	B
Bromine	Br
Cadmium	Cd
Calcium	Ca
Carbon	C
Cerium	Ce
Cesium	Cs
Chlorine	Cl
Chromium	Cr
Cobalt	Co
Copper	Cu
Dysprosium	Dy
Erbium	Er
Europium	Eu
Fluorine	F
Gadolinium	Gd
Gallium	Ga
Germanium	Ge
Gold	Au
Hafnium	Hf
Helium	He
Holmium	Ho
Hydrogen	H
Illinium	Il
Indium	In
Iodine	I
Iridium	Ir
Iron	Fe
Krypton	Kr

Name	Atomic Symbol
Lanthanum	La
Lead	Pb
Lithium	Li
Lutetium	Lu
Magnesium	Mg
Manganese	Mn
Masurium	Ma
Mercury	Hg
Molybdenum	Mo
Neodymium	Nd
Neon	Ne
Nickel	Ni
Niobium	Nb
Nitrogen	N
Osmium	Os
Oxygen	O
Palladium	Pd
Phosphorus	P
Platinum	Pt
Polonium	Po
Potassium	K
Praseodymium	Pr
Protactinium	Pa
Radium	Ra
Radon	Rn
Rhenium	Re
Rhodium	Rh
Rubidium	Rb
Ruthenium	Ru
Samarium	Sm
Scandium	Sc
Selenium	Se
Silicon	Si
Silver	Ag
Sodium	Na
Strontium	Sr

Name	Atomic Symbol
Sulfur	S
Tantalum	Ta
Tellurium	Te
Terbium	Tb
Thallium	Tl
Thorium	Th
Thulium	Tm
Tin	Sn
Titanium	Ti
Tungsten	W
Uranium	U
Vanadium	V
Xenon	Xe
Ytterbium	Yb
Yttrium	Y
Zinc	Zn
Zirconium	Zr

Appendix F

COEFFICIENTS OF THERMAL EXPANSION[a] FOR SELECTED CONSTRUCTION MATERIALS

Material	Multiply by 10^{26} in./in./°F	Multiply by 10^{26} mm/mm/°C
Concrete		
Normal weight concrete	5.5	9.8
Gypsum		
Gypsum panels	9.0	16.2
Gypsum plaster	7.0	12.6
Wood fiber plaster	8.0	14.4
Masonry		
Brick (varies some)	3.0	5.6
Concrete masonry units	5.2	9.4
Marble	7.3	13.1
Granite	4.7	8.5
Limestone	4.4	7.9

Material	Multiply by 10^{26} in./in./°F	Multiply by 10^{26} mm/mm/°C
Metal		
Iron, gray cast	5.7	10.5
Iron, malleable	5.6	10.5
Steel, carbon (ASTM A285)	5.6	10.5
Steel, high strength (ASTM A141)	6.4	11.7
Steel, stainless (type 201)	8.7	15.7
Steel, stainless (type 405)	6.0	10.8
Nickel (211)	7.4	13.3
Copper (CA110)	9.4	16.5
Bronze, commercial	10.2	19.3
Brass, red	10.4	19.9
Aluminum, wrought	12.8	23.0

Material	Multiply by 10^{26} in./in./°F	Multiply by 10^{26} mm/mm/°C
Polymer, Thermosetting		
Phenolics	45.0	81.0
Urea-melamine	20.0	36.0
Polyesters	45.0	75.5
Epoxies	33.0	72.0

Material	Multiply by 10^{26} in./in./°F	Multiply by 10^{26} mm/mm/°C
Polymer, Thermoplastic		
Polyethylene, high density	70.0	120.0
Polypropylene	50.0	90.0
Polystyrene	38.0	68.5
Polyvinyl chloride (PVC)	30.0	54.0
Acrylonitrile-butadiene- styrene (ABS)	50.0	90.0
Acrylics	40.0	72.0
Wood		

For most hardwoods and softwoods the *parallel-to-grain* thermal expansion values range from 1.7×10^{26} to 2.5×10^{26} in./in./°F *or* 3.1×10^{26} to 4.5×10^{26} mm/mm/°C.

Linear expansion coefficients across the grain are proportional to wood density. They range from 5 to 10 times greater than the parallel-to-grain coefficients and, therefore, are of more concern.

Material	Multiply by 10^{26} in./in./°F	Multiply by 10^{26} mm/mm/°C
Foam Insulation		
Polystyrene	3.5	6.3
Polyurethane	2.7	4.9

Material	Multiply by 10^{26} in./in./°F	Multiply by 10^{26} mm/mm/°C
Glass		
Glass, soda lime window sheet	47.0	85.0
Glass, soda lime plate	48.0	87.0

[a] *The change in dimension of a material per unit of dimension per degree change in temperature.*

Appendix G

CONVERSION TABLES

Millimeters to Decimal Inches				
mm In.	mm In.	mm In.	mm In.	mm In.
1 = 0.0394	21 = 0.8268	41 = 1.6142	61 = 2.4016	81 = 3.1890
2 = 0.0787	22 = 0.8662	42 = 1.6536	62 = 2.4410	82 = 3.2284
3 = 0.1181	23 = 0.9055	43 = 1.6929	63 = 2.4804	83 = 3.2678
4 = 0.1575	24 = 0.9449	44 = 1.7323	64 = 2.5197	84 = 3.3071
5 = 0.1969	25 = 0.9843	45 = 1.7717	65 = 2.5591	85 = 3.3465
6 = 0.2362	26 = 1.0236	46 = 1.8111	66 = 2.5985	86 = 3.3859
7 = 0.2756	27 = 1.0630	47 = 1.8504	67 = 2.6378	87 = 3.4253
8 = 0.3150	28 = 1.1024	48 = 1.8898	68 = 2.6772	88 = 3.4646
9 = 0.3543	29 = 1.1418	49 = 1.9292	69 = 2.7166	89 = 3.5040
10 = 0.3937	30 = 1.1811	50 = 1.9685	70 = 2.7560	90 = 3.5434
11 = 0.4331	31 = 1.2205	51 = 2.0079	71 = 2.7953	91 = 3.5827
12 = 0.4724	32 = 1.2599	52 = 2.0473	72 = 2.8247	92 = 3.6221
13 = 0.5118	33 = 1.2992	53 = 2.0867	73 = 2.8741	93 = 3.6615
14 = 0.5512	34 = 1.3386	54 = 2.1260	74 = 2.9134	94 = 3.7009
15 = 0.5906	35 = 1.3780	55 = 2.1654	75 = 2.9528	95 = 3.7402
16 = 0.6299	36 = 1.4173	56 = 2.2048	76 = 2.9922	96 = 3.7796
17 = 0.6693	37 = 1.4567	57 = 2.2441	77 = 3.0316	97 = 3.8190
18 = 0.7087	38 = 1.4961	58 = 2.2835	78 = 3.0709	98 = 3.8583
19 = 0.7480	39 = 1.5355	59 = 2.3229	79 = 3.1103	99 = 3.8977
20 = 0.7874	40 = 1.5748	60 = 2.3622	80 = 3.1497	100 = 3.9371

Fractional Inches to Millimeters			
mm In.	mm In.	mm In.	mm In.
1/64 = 0.397	17/64 = 6.747	33/64 = 13.097	49/64 = 19.447
1/32 = 0.794	9/32 = 7.144	17/32 = 13.494	25/32 = 19.844
3/64 = 1.191	19/64 = 7.541	35/64 = 13.890	51/64 = 20.240
1/16 = 1.587	5/16 = 7.937	9/16 = 14.287	13/16 = 20.637
5/64 = 1.984	21/64 = 8.334	37/64 = 14.684	53/64 = 21.034
3/32 = 2.381	11/32 = 8.731	19/32 = 15.081	27/32 = 21.431
7/64 = 2.778	23/64 = 9.128	39/64 = 15.478	55/64 = 21.828
1/8 = 3.175	3/8 = 9.525	5/8 = 15.875	7/8 = 22.225
9/64 = 3.572	25/64 = 9.922	41/64 = 16.272	57/64 = 22.622
5/32 = 3.969	13/32 = 10.319	21/32 = 16.669	29/32 = 23.019
11/64 = 4.366	27/64 = 10.716	43/64 = 17.065	59/64 = 23.415
3/16 = 4.762	7/16 = 11.113	11/16 = 17.462	15/16 = 23.812
13/64 = 5.159	29/64 = 11.509	45/64 = 17.859	61/64 = 24.209
7/32 = 5.556	15/32 = 11.906	23/32 = 18.256	31/32 = 24.606
15/64 = 5.953	31/64 = 12.303	47/64 = 18.653	63/64 = 25.003
1/4 = 6.350	1/2 = 12.700	3/4 = 19.050	1 = 25.400

Design Services Of Carlisle Publishing Services

Decimal Equivalents of Common Fractions

4ths	8ths	16ths	32nds	64ths	To 4 Places	To 3 Places	To 2 Places	4ths	8ths	16ths	32nds	64ths	To 4 Places	To 3 Places	To 2 Places
				1/64	.0156	.016	.02					33/64	.5156	.516	.52
			1/32		.0312	.031	.03				17/32		.5312	.531	.53
				3/64	.0469	.047	.05					35/64	.5469	.547	.55
										9/16			.5625	.562	.56
		1/16			.0625	.062	.06								
				5/64	.0781	.078	.08					37/64	.5781	.578	.58
			3/32		.0938	.094	.09				19/32		.5938	.594	.59
												39/64	.6094	.609	.61
				7/64	.1094	.109	.11								
	1/8				.1250	.125	.12		5/8				.6250	.625	.62
				9/64	.1406	.141	.14					41/64	.6406	.641	.64
											21/32		.6562	.656	.66
			5/32		.1562	.156	.16								
				11/64	.1719	.172	.17					43/64	.6719	.672	.67
		3/16			.1875	.188	.19			11/16			.6875	.688	.69
												45/64	.7031	.703	.70
				13/64	.2031	.203	.20								
			7/32		.2188	.219	.22				23/32		.7188	.719	.72
				15/64	.2344	.234	.23					47/64	.7344	.734	.73
								3/4					.7500	.750	.75
1/4					.2500	.250	.25								
				17/64	.2656	.266	.27					49/64	.7656	.766	.77
			9/32		.2812	.281	.28				25/32		.7812	.781	.78
												51/64	.7969	.797	.80
				19/64	.2969	.297	.30								
		5/16			.3125	.312	.31			13/16			.8125	.812	.81
				21/64	.3281	.328	.33					53/64	.8281	.828	.83
											27/32		.8438	.844	.84
			11/32		.3438	.344	.34								
				23/64	.3594	.359	.36					55/64	.8594	.859	.86

(continued)

Design Services Of carlisle Publishing Services *(continued)*

Decimal Equivalents of Common Fractions

4ths	8ths	16ths	32nds	64ths	To 4 Places	To 3 Places	To 2 Places	4ths	8ths	16ths	32nds	64ths	To 4 Places	To 3 Places	To 2 Places
	3/8				.3750	.375	.38		7/8				.8750	.875	.88
												57/64	.8906	.891	.89
				25/64	.3906	.391	.39								
			13/32		.4062	.406	.41				29/32		.9062	.906	.91
				27/64	.4219	.422	.42					59/64	.9219	.922	.92
										15/16			.9375	.938	.94
		7/16			.4375	.438	.44								
				29/64	.4531	.453	.45					61/64	.9531	.953	.95
			15/32		.4688	.469	.47				31/32		.9688	.969	.97
												63/64	.9844	.984	.98
				31/64	.4844	.484	.48								
					.5000	.500	.50						1.0000	1.000	1.00

Metric Equivalents of Two-Place Decimal Inches									
		Millimeter Equivalent					Millimeter Equivalent		
		.02		.508			.52		13.208
			.03	.762				.53	13.462
		.04		1.016			.54		13.716
			.05	1.270				.55	13.970
	.06			1.524		.56			14.224
		.08		2.032			.58		14.732
			.09	2.286				.59	14.986
.10				2.540	.60				15.240
	.12			3.048		.62			15.748
		.14		3.556			.64		16.256
			.15	3.810				.65	16.510
		.16		4.064			.66		16.764
	.18			4.572		.68			17.272
			.19	4.826				.69	17.526
.20				5.080	.70				17.780
		.22		5.588			.72		18.288
	.24			6.096		.74			18.796
			.25	6.350				.75	19.050
		.26		6.604			.76		19.304
		.28		7.112			.78		19.812
.30				7.620	.80				20.320
			.31	7.874				.81	20.574
	.32			8.128		.82			20.828
		.34		8.636			.84		21.336
			.35	8.890				.85	21.590
		.36		9.144			.86		21.844
			.37	9.398				.87	22.098
	.38			9.652		.88			22.352
.40				10.160	.90				22.860
			.41	10.414				.91	23.114
		.42		10.668			.92		23.368
	.44			11.176		.94			23.876
			.45	11.430				.95	24.130
		.46		11.684			.96		24.384
			.47	11.938				.97	24.638
		.48		12.192			.98		24.892
.50				12.700	1.00				25.400

Millimeters to Decimal Inches

mm In.	mm In.	mm In.	mm In.	mm In.
1 = 0.0394	21 = 0.8268	41 = 1.6142	61 = 2.4016	81 = 3.1890
2 = 0.0787	22 = 0.8662	42 = 1.6536	62 = 2.4410	82 = 3.2284
3 = 0.1181	23 = 0.9055	43 = 1.6929	63 = 2.4804	83 = 3.2678
4 = 0.1575	24 = 0.9449	44 = 1.7323	64 = 2.5197	84 = 3.3071
5 = 0.1969	25 = 0.9843	45 = 1.7717	65 = 2.5591	85 = 3.3465
6 = 0.2362	26 = 1.0236	46 = 1.8111	66 = 2.5985	86 = 3.3859
7 = 0.2756	27 = 1.0630	47 = 1.8504	67 = 2.6378	87 = 3.4253
8 = 0.3150	28 = 1.1024	48 = 1.8898	68 = 2.6772	88 = 3.4646
9 = 0.3543	29 = 1.1418	49 = 1.9292	69 = 2.7166	89 = 3.5040
10 = 0.3937	30 = 1.1811	50 = 1.9685	70 = 2.7560	90 = 3.5434
11 = 0.4331	31 = 1.2205	51 = 2.0079	71 = 2.7953	91 = 3.5827
12 = 0.4724	32 = 1.2599	52 = 2.0473	72 = 2.8247	92 = 3.6221
13 = 0.5118	33 = 1.2992	53 = 2.0867	73 = 2.8741	93 = 3.6615
14 = 0.5512	34 = 1.3386	54 = 2.1260	74 = 2.9134	94 = 3.7009
15 = 0.5906	35 = 1.3780	55 = 2.1654	75 = 2.9528	95 = 3.7402
16 = 0.6299	36 = 1.4173	56 = 2.2048	76 = 2.9922	96 = 3.7796
17 = 0.6693	37 = 1.4567	57 = 2.2441	77 = 3.0316	97 = 3.8190
18 = 0.7087	38 = 1.4961	58 = 2.2835	78 = 3.0709	98 = 3.8583
19 = 0.7480	39 = 1.5355	59 = 2.3229	79 = 3.1103	99 = 3.8977
20 = 0.7874	40 = 1.5748	60 = 2.3622	80 = 3.1497	100 = 3.9371

Fractional Inches to Millimeters

In. mm	In. mm	In. mm	In. mm
1/64 = 0.397	17/64 = 6.747	33/64 = 13.097	49/64 = 19.447
1/32 = 0.794	9/32 = 7.144	17/32 = 13.494	25/32 = 19.844
3/64 = 1.191	19/64 = 7.541	35/64 = 13.890	51/64 = 20.240
1/16 = 1.587	5/16 = 7.937	9/16 = 14.287	13/16 = 20.637
5/64 = 1.984	21/64 = 8.334	37/64 = 14.684	53/64 = 21.034
3/32 = 2.381	11/32 = 8.731	19/32 = 15.081	27/32 = 21.431
7/64 = 2.778	23/64 = 9.128	39/64 = 15.478	55/64 = 21.828
1/8 = 3.175	3/8 = 9.525	5/8 = 15.875	7/8 = 22.225
9/64 = 3.572	25/64 = 9.922	41/64 = 16.272	57/64 = 22.622
5/32 = 3.969	13/32 = 10.319	21/32 = 16.669	29/32 = 23.019
11/64 = 4.366	27/64 = 10.716	43/64 = 17.065	59/64 = 23.415
3/16 = 4.762	7/16 = 11.113	11/16 = 17.462	15/16 = 23.812
13/64 = 5.159	29/64 = 11.509	45/64 = 17.859	61/64 = 24.209
7/32 = 5.556	15/32 = 11.906	23/32 = 18.256	31/32 = 24.606
15/64 = 5.953	31/64 = 12.303	47/64 = 18.653	63/64 = 25.003
1/4 = 6.350	1/2 = 12.700	3/4 = 19.050	1 = 25.400

Glossary

abrasion resistance - Resistance to being worn away by rubbing or friction.

absorbent - A material that extracts a substance for which it has an affinity from a liquid or gas and changes physically and chemically during the process.

accelerator - An admixture used to speed up the setting of concrete.

acceptable indoor air quality - Air containing no known contaminants at harmful concentrations.

access floor - A freestanding floor raised above the structural floor to provide access to mechanical and electrical services.

acoustical glass - A glazing unit used to reduce the transmission of sound through the glazed opening by bonding a soft interlayer between the layers of glass.

acoustical plaster - A plaster used to control sound made from calcined gypsum mixed with lightweight aggregates.

acoustics - The science of sound generation, sound transmission, and the effects of sound waves.

additive - Materials mixed with the basic plastic resin to alter its properties.

adhesion - The ability of a coating to stick to a surface.

adhesive - A substance used to hold materials together by surface attachment.

adhesive bonding agents - Bonding agents made from synthetic materials.

admixture - A material other than Portland cement, aggregate, and water that is added to concrete to alter its properties.

adsorbent - A material that extracts a substance for which it has an affinity from a liquid or gas and that changes physically and chemically during the process.

agglomeration - A process that bonds ground iron ore particles into pellets to facilitate handling.

aggregate - Inert granules such as crushed stone, gravel, and expanded minerals mixed with Portland cement and sand to form concrete.

air drying - Drying wood by exposing it to the air.

air-entrained Portland cement - A Portland cement with an admixture that causes a controlled quantity of stable, microscopic air bubbles to form in the concrete.

air gap - An unobstructed vertical distance between the lowest opening of pipe that supplies a plumbing fixture and the level at which the fixture will overflow.

air-supported structure - A structure that derives its structural integrity from the use of internal pressurized air to inflate a pliable material.

alarm actuators - Controls which sense a particular activity and signal an alarm or a pre-set process to control the situation.

alkali - A substance such as lye, soda, or lime that can be destructive to paint films.

alkyd - Synthetic resin modified with oil for good adhesion, gloss, color retention, and flexibility.

alloying elements - Any substance added to a molten metal to change its mechanical or physical properties.

alumina - A hydrated form of aluminum oxide from which aluminum is made.

anaerobic bonding agents - Bonding agents that set hard when not exposed to oxygen.

angle of repose - The steepest angle of a surface at which loose material will remain in place rather than sliding.

annealing - Heating a metal or glass to a high temperature followed by controlled cooling to relieve internal stresses.

anodizing - An electrolytic process that forms a permanent, protective oxide coating on aluminum.

APA performance-rated panels - Plywood manufactured to the structural specifications and standards of APA—The Engineered Wood Association.

arc resistance - The total elapsed time in seconds an electric current must arc to cause a part to fail.

arch - A curved construction spanning an opening and supported at the sides or ends.

architectural terra-cotta - Hard fired clay masonry units made with a textured or sculptured face.

ashlar stone - Stone cut into rectangular shapes.

asphalt - Dark brown to black hydrocarbon solids or semisolids having bituminous constituents that gradually liquefy when heated.

audio sensors - Controls which activate a security or safety system when a sound is detected.

Authority Having Jurisdiction - The governmental agency that regulates the construction process.

autoclave - A high-pressure steam room that rapidly cures green concrete units.

Axminster construction - Carpet formed by weaving on a loom that inserts each tuft of pile individually into the backing.

bagasse - Crushed sugarcane or beet refuse from sugar making used in paper production.

balloon framing - A system of framing a wood-framed building in which all vertical structural members (studs) of the exterior bearing walls and partitions extend the full height of the frame, from the bottom plate to the top plate, and support the floor joists and roof.

bank measure - The volume of soil in situ in cubic yards.

batch - The amount of concrete mixed at one time.

bauxite - Ore containing high percentages of aluminum oxide.

beam - A straight horizontal structural member whose main purpose is to carry transverse loads.

bending moment - The moment that produces bending on a beam or other structural member.

beneficiation - A process of grinding and concentration that removes unwanted elements from iron ore before the ore is used to produce steel.

binder - Film-forming ingredient in paint that binds the suspended pigment particles together.

bioremediation - The use of bacteria in site remediation that are able to digest and remove toxic chemicals.

bio-swale - A landscape element consisting of a linear drainage ditch with sloped sides covered in vegetation designed to remove silt and pollution from surface runoff water.

bitumen - A generic term describing a material that is a mixture of predominantly hydrocarbons in solid or viscous form derived from coal and petroleum.

bituminous coatings - Coatings formulated by dissolving natural bitumens in an organic solvent.

bleed - Excess water that rises to the surface of concrete shortly after it has been poured.

board foot - The measure of lumber having a volume of 144 $in.^3$

boards - Lumber less than 2 in. (50.8 mm) thick and 1 in. (25.4 mm) or more wide.

boiler - A closed vessel used to produce hot water or steam.

bonding agent - A compound that holds materials together by bonding joining surfaces.

branch soil pipe - A pipe in a plumbing system that discharges into a main or submain and into which no other branch pipes discharge.

brick - A solid or hollow masonry unit made from clay or shale, molded into a rectangular shape, and fired in a kiln to produce a hard, strong unit.

brick bond - The pattern of masonry units laid to form a multi-wythe wall.

brick veneer - A facing of brick that is laid against a wall but not structurally bonded to it.

British thermal unit (Btu) - The amount of heat required to raise the temperature of 1 lb. of water 1°F.

brown coat - The second coat of plaster in a three-coat plaster finish.

brownfield site - A property, the expansion, redevelopment, or reuse of which may be complicated by the presence or potential presence of a hazardous substance, pollutant, or contaminant.

buffer, elevator - Energy-absorbing units placed in an elevator pit.

building automation system - A central control system that combines the supervision and control of various building functions.

building code - A governmental ordinance that regulates the design, construction, and maintenance of buildings by enforcing minimum standards to safeguard the life, health, property, and general welfare of the public.

building construction - The addition of a building to a piece of real estate.

building drain - The lowest horizontal piping of a plumbing drainage system that receives the discharge from soil, waste, and other drainage pipes within a building and carries the waste to the building sewer.

building information modeling - A software program that administers a process for developing and managing all information of a building project throughout its life cycle.

building permit - A certificate issued by the authority having jurisdiction authorizing the construction of a project after review of the construction documents.

building section - A drawing that provides a representation of a building after a vertical plane has been cut through it and the front portion removed.

building sewer - Horizontal piping that carries the waste discharge from the building drain to the public sewer or septic tank.

building-integrated PV (BIPV) - photovoltaic materials used to replace conventional building materials in parts of the building envelope, such as the roof, skylights, or facades.

built-up roofing - A continuous, semi-flexible roof membrane consisting of plies of saturated felts, coated felts, fabrics, or mats that have surface coats of bitumens; the last ply is covered with mineral aggregates, bituminous materials, or a granular-surface roofing sheet.

bull float - A tool used to level concrete in a form after it has been screeded.

burning - Curing bricks by placing them in a kiln and subjecting them to a high temperature.

bus - A rigid electric conductor enclosed in a protective busway.

busway - A rigid conduit used to protect a bus running through it.

BX cable - A cable sheathed with spirally wrapped metal strip identified as Type AC.

cable tray - A ladder-like metal frame, open on the top, used to support insulated electrical cables.

caisson - A watertight structure within which work can be carried out below the surface of water.

caisson foundation - Foundations formed by drilling holes in the earth and filling them with concrete and reinforcing steel.

calcareous clays - Clays containing at least 15 percent calcium carbonate.

calcined gypsum - Ground gypsum that has been heated to drive off the water content.

camber - A slightly arched surface in a structural beam designed to resist applied loadings.

candela (cd) - A metric unit of luminous intensity that closely approximates candlepower.

candlepower (cp) - A term used to express the luminous intensity of a light source. It is the same magnitude as a candela.

capillary action - The movement of a liquid through small openings of fibrous material by the adhesive force between the liquid and the material.

car, elevator - The load-carrying unit of an elevator, consisting of a platform, walls, ceiling, door, and a structural frame.

car safeties, elevator - Devices used to stop a car and hold it in position should it travel at excessive speed or free fall.

carbohydrates - Organic compounds that form the supporting tissue of plants.

casing - The trim moldings used around doors, windows, and other openings.

cast alloys - Products made by pouring molten metal into either sand or permanent molds.

cast iron - A hard, brittle metal made of iron that contains a high percentage of carbon.

cast-in-place concrete - Concrete members formed and poured on the building site.

caulking compound - A resilient material used to seal cracks and prevent leakage of water.

cavity wall - A masonry wall made up of two wythes of masonry units separated by an air space.

cellulose - A complex carbohydrate that constitutes the chief part of the cell walls of plants, wood, and paper, and yields fiber for many products.

cement board - A panel product made with an aggregate Portland cement core reinforced with polymer-coated glass-fiber mesh on each side.

cementitious materials - Materials that have cementing properties.

cements - 1. A powder with adhesive and cohesive properties that sets into a hard, solid mass when mixed with water.
2. A bonding agent made from synthetic rubber suspended in a liquid.

certificate of occupancy - A certificate issued by the authority having jurisdiction certifying that a completed building is in compliance with all locally adopted building regulations and is in proper condition to be occupied.

chalkboard - A surface often made of slate for writing on with chalk.

chase - A recessed area in a wall for holding pipes and conduit that passes vertically between floors.

chemical strengthening - A process for strengthening glass that involves immersing it in a molten salt bath.

chiller - A refrigerating machine composed of a compressor, a condenser, and an evaporator, used to transfer heat from one fluid to another.

chromogenic glazing - Glazing materials that can change their optical properties in response to changes in solar radiation intensity, temperature, or through the introduction of an electric charge (photo chromic, thermo chromic, electro chromic).

circuit breaker - An electrical device used to open and close a circuit by non-automatic means or to open a circuit by automatic means at a predetermined overcurrent without damage to itself.

cladding - The external finish covering the structural frame of a building.

clay - A cohesive material made up of microscopic particles (less than 0.00008 in. or 0.002 mm).

clay tile - A unit made from fired and sometimes glazed clay and used as a finish surface on floors and walls.

cleanout - An opening in waste water piping systems that permits cleaning obstructions from the pipe.

clear coating - A transparent protective or decorative film.

coal tar pitch - A dark-brown-to-almost-black hydrocarbon solid or semisolid material derived by distilling coke-oven tar.

coating - A paint, varnish, lacquer, or other finish used to create a protective or decorative layer.

coefficient of thermal expansion - The total amount of heat, expressed in Btu, that passes by conduction through a 1 in. thickness of a homogeneous material per hr., per ft.^{2}, per °F, which is measured as the temperature difference between the two surfaces of the material.

cofferdam - A temporary watertight enclosure around an area of water-bearing soil or an area of water from which the liquid is removed (via pumping), allowing construction to take place in a water-free area.

cogeneration - The utilization of normally wasted heat energy to produce electricity or to heat or cool a building.

compact fluorescent light (CFL) - A type of fluorescent lamp made in special shapes to fit in standard household light sockets.

compaction - The act of compressing soil to increase its density.

compartment - A small area within a larger area enclosed by partitions.

competitive bidding - A method of selecting a general contractor by inviting bids on a competitive basis.

composite panels - Panels having a reconstituted wood core bonded between layers of solid veneer.

composting toilet - A system that converts human waste into organic compost and usable soil through the natural breakdown of organic matter into its essential minerals.

compressive stress - Stresses created when forces push on a member and tend to shorten it.

compressor - A mechanical device for increasing the pressure of a gas.

concentrated load - Any load that acts on a very small area of a structure.

concrete - A solid, hard material produced by combining Portland cement, aggregates, sand, and water (and sometimes admixtures).

concrete masonry units - Factory-manufactured concrete units, such as concrete brick or block.

concrete pump - A pump that moves concrete through hoses to the area where it is to be placed.

condensate - A liquid formed by the condensation of a vapor.

condenser - A heat exchange unit in which a vapor has some heat removed, causing it to form a liquid.

conduction - The process of heat transfer through a material to another part of that material or to a material touching it.

conductivity, electric - A measure of the ability of a material to conduct electric current.

conductor, electric - Wire through which electric current flows.

conduit - A steel or plastic tube through which electrical wires are run.

console - A decorative bracket used to support a cornice, window or other object.

consolidation - The process of compacting freshly placed concrete in a form.

construction bid - A A decorative bracket used to support a cornice, window or other object.

construction change directive - A document that describes a change to the original contract documents and specifies an adjustment in construction time and cost.

construction drawings - Drawings included in construction documents that graphically describe the dimensional relationships, form, sizes, and quantities of all building elements.

construction management - A project delivery method in which an owner hires a construction manager to act as his or her agent by overseeing both design and construction activities.

contract documents - The drawings and specifications that designate the exact requirements for the construction of a project.

control joint - A groove formed in concrete or masonry structures to allow a place where cracking can occur, thus reducing the development of high stresses.

convection - The process of carrying heat from one spot to another by movement of a liquid or gas. The heated liquid or gas expands and becomes lighter, causing it to rise, while the cooler, heavier dense liquid or air settles.

convector - A unit designed to transfer heat from hot water or steam to the air by by convection.

corona - The overhanging vertical member of a cornice.

corrosion - The deterioration of metal or concrete by chemical or electrochemical reaction caused by weather exposure.

covalent bonding - A process in which small numbers of atoms are bonded into molecules.

creep - Permanent dimensional deformation occurring over a period of time in a material subjected to constant stress at elevated temperatures.

cubicle - A small enclosed space often used in modular office partitions.

cupola - A small roofed structure built on a roof, usually to vent the area below the roof.

curing - Protecting concrete after placing so that proper hydration occurs.

current, electric - The flow of electrons along a conductor.

cushion - A layer of resilient material applied to a floor over which a carpet is to be laid.

daylighting - The use of natural daylight to offset electric lighting use.

dead load - The structural load on a building resulting from the weight of the actual materials of construction, such as walls, floors, roofs, and finishes.

decibel (dB) - A unit for measuring sound energy or power.

dehumidifier - A cooling, absorption, or adsorption device used for removing moisture from the air.

desiccant - Any absorbent, adsorbent, liquid, or solid that removes water or water vapor from a material.

desiccation - The process of evaporating or removing water vapor from a material.

design-bid-build - The traditional three step project delivery method in which the architect and contractor each sign separate contracts with the owner.

design-build - A project delivery method in which an owner signs a single contract with one entity to perform both design and construction services.

design drawings - Structural drawings prepared by the structural engineer that provide the necessary information for the detailing and fabrication of steel structural members.

design intent - A statement that defines the anticipated aesthetic, functional, and performance characteristics of a construction project.

dewatering - Pumping subsurface water from an excavation to maintain dry and stable working conditions.

dielectric strength - The maximum voltage a dielectric (nonconductor) can withstand without fracture.

diffuser - A circular, square, or rectangular air distributing outlet, usually in the ceiling, that has members to discharge supply air in several directions, mixing the supply air with the secondary air in a room.

dimension lumber - Lumber cut and dressed to standard sizes.

double tees - T-shaped precast floor and roof units that span long distances unsupported.

dressed lumber - Lumber having one or more sides planed smooth.

drypack - A stiff granular grout.

dry-press process - The process used to make bricks when the clay contains 10 percent or less moisture.

ductility - A measure of the ability of a material to be stretched or deformed without breaking.

durability (of a coating) - The ability of a coating to hold up against destructive agents, such as weathering and sunlight.

DWV - Drain, waste, vent pipe installation.

dynamic load - Any load that acts on a very small area of a structure.

eave - The part of a roof that extends beyond the side wall.

efflorescence - A white soluble salt deposit on the surface of concrete and masonry, usually caused by free alkalies leached from mortar by moisture moving through it.

elastomers - A macromolecular material that returns to its approximate initial dimensions and shape after being subjected to substantial deformation.

elevation - A drawing that provides a representation of the exterior face of a building, delineating geometries and the materials of construction.

elevator - A hoisting and lowering mechanism equipped with an enclosed car that moves in a hoistway between floors in a multi-story building.

embodied energy - The energy consumed by all processes associated with the production of a material or building.

enamel - A classification of paints that dry to a hard, flat semigloss or gloss finish.

environmentally preferable products (EPPs) - Materials that have a lesser or reduced effect on human health and the environment when compared to competing products that serve the same purpose.

epoxy varnish - A clear finish having excellent adhesion qualities, abrasion and chemical resistance, and water resistance.

equilibrium moisture content - The moisture content at which wood neither gains nor loses moisture when surrounded by air at a specified relative humidity and temperature.

erection plan - An assembly drawing showing where each structural steel member is located on a building frame.

escalator - A continuously moving stair used to move people between the floors of a multi-story building.

expansion joint - A joint used to separate two parts of a building to allow expansion and contraction movement of the parts.

faceted glass window - A window made by bonding glass pieces with an epoxy resin matrix or reinforced concrete.

fast-track project - A construction delivery method wherein design and construction activities occur simultaneously producing shorter project schedules.

fatigue strength - A measure of the ability of a material or structural member to carry a load without failure when the loading is applied repeatedly.

felt - A sheet material made using a fiber mat that has been saturated and topped with asphalt.

fenestration - The windows and openings in a building enclosure.

ferrous metals - Iron-based metallic materials.

fiber saturation point - The moisture content of wood at which the cell walls are saturated but there is no water in the cell cavities.

fibered gypsum - A neat gypsum with cattle hair or organic fillers added.

fillers - Inert material added to a plastic resin to alter its strength and working properties and to lower its cost.

finish coat - The third or final coat of gypsum plaster.

finish plaster - The topcoat of plaster on a wall or ceiling.

fire flame detector - A control that detects a flame that does not produce smoke when a fire first starts.

fire resistance rating - A partition assembly that has been tested and given a rating indicating the length of time it will resist a fire in hours.

fire resistant gypsum - A gypsum product that has increased fire-resistance properties due to the addition of fire-resistant materials in the gypsum core.

fire wall - A construction of noncombustible materials that subdivides a building or separates adjoining buildings to retard the spread of fire.

fireclays - Deep mined clays that are able to withstand heat.

fire-rated partition - A partition assembly that has been tested and given a rating indicating the length of time it will resist a fire in hours.

fire-resistant - The capacity of a material or assembly of materials to withstand fire or provide protection from it.

fixture unit - A measure of the potable discharge into a waste disposal system from its various plumbing fixtures expressed in cubic volume per minute.

flame spread rate - The rate at which flames will spread across the surface of a material.

flame spread rating - A numerical designation given to a material to indicate its comparative ability to restrict flame combustion over its surface.

flame-retardant - Having resistance to the propagation of flame and therefore not readily ignitable. Indicates a lower resistance than fire-resistant.

flammability - A measure of the extent to which a material is flammable.

flashing - A thin impervious material used to prevent water from penetrating the joints between building elements.

float process - A glass manufacturing process in which the molten glass ribbon flows through a furnace supported on a bed of molten metal.

flocked construction - Carpet formed by electrostatically spraying short strands onto an adhesive-coated backing material.

floor plan - A drawing that provides a representation of a building after a horizontal plane has been cut through it and the top portion removed.

flux - A mineral added to molten iron to cause impurities to separate into a layer of molten slag on top of the iron.

footcandle (fc) - The unit of illumination equal to 1 lumen per square foot.

footlambert - A unit for measuring brightness or luminance. It is equal to 1 lumen per square foot when brightness is measured from a surface.

framed connections - Connections joining structural steel members with a metal, such as an angle, that is secured to the web of the beam.

framing plan - A drawing showing the location, size, and spacing of structural members.

freeze cycle day - A day when the temperature of the air rises above or falls below 32°F or 0°C.

frequency - The number of cycles per second of current or voltage in alternating current, of a sound wave, or of a vibrating solid expressed in hertz.

fuel-contributed rating - A rating of the amount of combustible material in a coating.

fuse - An overcurrent protection device that opens an electric circuit when a fusible element breaks from heat due to overcurrent passing through it.

fusion-bonded carpet - Carpet formed by bonding pile yarn between two sheets of backing material and cutting the pile yarn in the center, forming two pieces of carpet.

galling - The wearing or abrading of one material against another under extreme pressure.

galvanic corrosion - Corrosion that develops by galvanic action when two dissimilar metals are in contact in the atmosphere.

galvanize - To coat steel or iron with zinc to prevent rust.

gearless traction elevator - An elevator with the traction sheave connected to a spur gear that is driven by a worm gear connected to the shaft of the electric motor.

general contractor - The primary contractor who is responsible for all the work on a construction site and oversees the work of subcontractors.

geodesic dome - A dome made of prefabricated short, straight, triangular sections interconnected to give stability in all directions.

glaze - A ceramic coating fused to the surface of bricks at high temperatures.

glued laminated lumber (glulam) - A structural wood member made by bonding laminas of dimension lumber.

glues - Bonding agents made from animal and vegetable products.

good indoor air quality - Indoor air in which there are no known contaminants at harmful concentrations.

grade - (1) Related to soil, the elevation or slope of the ground. (2) In relation to lumber, a means of classifying lumber or other wood products based on specified quality characteristics.

grade beam - A ground-level reinforced structural member that supports the exterior wall of a structure and bears directly upon columns or piers.

grading - The act of remodeling an existing land form to provide a level area for a structure and create circulation paths and drainage and landscape features.

gravel - Hard rock material in particles larger than an in. (6.4 mm) in diameter but smaller than 3 in. (76 mm).

graywater - Non-industrial wastewater generated from domestic processes such as dish washing, laundry, and bathing.

green lumber - Lumber having a moisture content greater than 19 percent.

Greenhouse Effect - A process by which the earth's atmosphere traps solar radiation by the existence of gases—such as carbon dioxide (CO2), water vapor, and methane—that allow incoming sunlight to pass through them but absorb the heat radiated back from the earth's surface.

grille - An open grate used to cover or conceal, protect, or decorate an opening.

groundwater - Water that exists below the surface of the earth and passes through the subsoil.

grout - A viscous mixture of Portland cement, water, and aggregate used to fill cavities in concrete. Also refers to a specially formulated mortar used under the baseplates of steel columns and in connections in precast concrete.

gypsum - Hydrous calcium sulfate.

gypsum board - A gypsum panel used for interior wall and ceiling surfaces. It contains a gypsum core and surfaces covered with paper.

gypsum lath - A panel having a gypsum core and a paper covering providing a bonding surface for plaster.

gypsum neat plaster - A gypsum plaster with no aggregates or fillers added. Sometimes called unfibered gypsum.

hardboard - A general term used to describe a panel made from interfelted ligno-cellulosic fibers consolidated under heat and pressure.

hardness - A measure of the ability of a material to resist indentation or surface scratching.

hardwood - A botanical group of trees that have broad leaves that are shed in the winter.

hardwood plywood - Plywood with various species of hardwoods used on the outer veneers.

haunch - A projection used to support a member, such as a beam.

heartwood - The wood extending from the pith to the sapwood.

heat capacity - The ability of a material to store and release heat

heat treating - Heating and cooling a solid metal to produce changes in physical and mechanical properties.

heat-treatable alloys - Aluminum alloys whose strength characteristics can be improved by heat treating.

heavy construction - The construction of large infrastructure projects, such as highways or bridges.

heavy timber construction - A type of wood-frame construction using heavy timbers for columns, beams, joists, and rafters.

hiding power - The ability of a paint to hide a previous color or substrate.

hinge joint - A joint that permits some bending action (similar to that of a hinge) and in which there is no appreciable separation of the joining members.

hoistway, elevator - A fire-resistant vertical shaft in which an elevator moves.

hollow clay masonry - A unit whose core area is 25 to 40 percent of the gross cross-sectional area of the unit.

hollow core door - A door with face veneers on the outer surfaces, wood stiles around the edges, and a hollow interior supported by a honeycomb grid.

horizontal shear - The tendency of top wood fibers to move horizontally in relationship to bottom fibers.

hot-melt adhesives - Adhesives that bond when heated to a liquid form.

humidifier - A device used to add moisture to the air.

humidity - The amount of water vapor within a given space.

hydrate - The capacity of lime to soak up water several times its weight.

hydration - A chemical reaction between water and cement that produces heat and causes the cement to cure or harden.

hydraulic elevator - An elevator having a car mounted on top of a hydraulic piston that moves by the action of hydraulic oil under pressure.

hydronic heating system - A system that circulates hot water through a series of pipes and convectors to heat a building.

hydronics - The science of cooling and heating water.

hydroxide of lime - The product produced by the chemical reaction during the slaking or hydrating of lime.

hygroscopic - The ability of a material to readily absorb and retain moisture from the air.

I joist - A wood joist made of an assembly of laminated veneer wood top and bottom flanges and a web of plywood or oriented strandboard.

ice dam - Ice that forms on an overhang, causing water buildup behind it to penetrate under roofing material.

igneous rock - Rock formed by the solidification of molten material.

illuminance (E) - The density of luminous power in lumens per a specified area.

impact isolation class (IIC) - An index of the extent to which a floor assembly transmits impact noise from a room above to a room below.

impact strength - The ability of a material to resist fracture when struck with a rapidly applied load.

incandescence - Emitting visible light as a result of being heated.

independent footing - Footings supporting a single structural element, such as a column.

industrial construction - The construction of large manufacturing or utility infrastructure projects.

instantaneous water heater - A water heating system without a storage tank that heats water instantaneously as taps are activated.

insulating glass - A glazing unit used to reduce the transfer of heat through a glazed opening by leaving an air- or gas-filled space between layers of glass.

insulator, electric - A material that is a poor conductor of electricity used on electrical devices to provide protection from electricity.

integrated design process - A project delivery approach that incorporates all stakeholders throughout the design and construction sequence. Used especially for sustainable building projects.

interceptor - A trapping device designed to collect materials a sewage treatment plant cannot handle, such as grease, glass or metal chips, and hair.

intermediate coat - A layer of coating between a primer coat and a topcoat whose main purpose is to provide structural strength.

intumescent coatings - Coatings that expand to form a thick fire-retardant coating when exposed to flame.

invitational bidding - A method of preselecting a general contractor based on experience and qualifications rather than cost.

joint compound - A plastic gypsum mixture similar to plaster used to seal joints and fasteners in gypsum wallboard installations.

jointing - Forming control, expansion, or construction joints in concrete construction.

joule - A meter-kilogram-second unit of work or energy.

jute - A coarse fiber, obtained from two East Indian plants, used in paper and carpet products.

kiln - (1) A chamber with controlled humidity, temperature, and airflow in which lumber is dried. (2) A low-pressure steam room in which green concrete units are cured.

kiln-dried - Wood products dried in a kiln.

kinetic energy - The energy possessed by a body because of its motion.

knitted carpet - Carpet formed by looping pile yarn, stitching, and backing together.

lacquer - A fast-drying clear or pigmented coating that dries by solvent evaporation.

laminated glass - Glass panels that have outer layers of glass laminated to an inner layer of transparent plastic.

laminated veneer lumber - A structural lumber manufactured from wood veneers so that the grain of all veneers runs parallel to the axis of the member.

lamp - A general term used to describe a source of artificial light. Often called a bulb or tube.

landing zone, elevator - The area 18 in. (5490 mm) above or below a landing floor.

latent heat - Heat involved with the action of changing the state of a substance, such as changing water to steam.

lateral load - Loads moving in a horizontal direction, such as those caused by wind.

latex - A water-based coating, such as styrene, butadiene, acrylic, and polyvinyl acetate.

leaded glass window - A window made of small glass pieces joined with lead cames.

LEED - The Leadership in Energy and Environmental Design is a third-party certification program that provides a nationally accepted standard for the design, construction, and maintenance of high-performance buildings.

life cycle analysis - A procedure for compiling and analyzing the inputs and outputs of resources and energy and their associated environmental impacts directly attributable to the functioning of a material or service system throughout its life cycle.

lightweight steel framing - Structural steel framing members made from cold-rolled lightweight sheet steel.

lignin - An amorphous substance that penetrates and surrounds the cellulose strands in wood, binding them together.

lime - A white-to-gray powder produced by burning limestone, marble, coral, or shells.

lintel block - A precast concrete structural member used to span an opening in a wall and carry the weight of the masonry units above it.

liquid limit (LL) - Related to soils, the water content expressed as a percentage of dry weight at which the soil will start to flow when tested by the shaking method.

live load - The structural load on a building resulting from occupancy and use—consists of people, furnishings, and environmental forces, such as wind, snow, and seismic activity.

loomed carpet - Carpet formed by bonding pile yarn to a rubber cushion.

low-E glass - Low emissivity glass with a thin metallic coating that selectively reflects ultraviolet and infrared wavelengths.

low-slope roofs - Roofs that are nearly flat.

lumber - A product produced by harvesting, sawing, drying, and processing wood.

lumber grader - Persons employed by sawmills and wood processing plants who inspect and grade lumber according to industry standards.

lumen (lm) - A unit for measuring the flow of light energy. See luminous flux.

luminaire - A complete lighting unit consisting of one or more lamps plus elements needed to distribute light, hold and protect the lamps, and connect power to the lamps. Also called a lighting fixture.

luminance - The luminous intensity of a surface of a given area viewed from a given direction.

luminance meter - A photoelectric instrument used to measure luminance. Also called a light meter.

luminescence - The emission of light not directly caused by incandescence.

luminous flux - The rate of flow of light energy through a surface, expressed in lumens.

luminous intensity - The force that generates visible light expressed by candela, lumens per steradian, or candlepower.

luminous transmittance - A measure of the capacity of a material to transmit incident light in relation to the total incident light striking it.

lux - A unit of illumination equal to 1 lumen per square.

makeup air - Air brought into a building from outside to replace interior exhausted air.

malleability - The characteristic of a material that allows plastic deformation in compression without rupture.

marker board - A surface that can be written on with a water-based or semi-permanent ink that can be removed.

MasterFormat - A standard developed by the Construction Specifications Institute for writing specifications using a system of descriptive titles and numbers to organize construction activities, products, and requirements.

mat foundation (also called spread foundation) - A large, single concrete footing equal in area to the area covered by the footprint of the building.

means of egress - A continuous and unobstructed path of vertical and horizontal egress from any occupied portion of a building or structure to a public way.

mechanical action - The bonding of materials by adhesives that enter pores and harden, forming a mechanical link.

melting temperature - The temperature at which a material turns from a solid to a liquid.

metal lath - Perforated sheets of thin metal secured to studs that serve as the base for a finished plaster wall.

metals - Materials refined from ores that have been extracted from the earth.

metamorphic rock - Rock formed by the action of pressure or heat on sedimentary soil or rock.

meter, electric - A device measuring and recording the amount of electricity passing through it in kilowatt-hours.

mildewcide - An agent that helps prevent the growth of mold and mildew on painted surfaces.

millwork - Any wood products that have been manufactured or preassembled, such as moldings, doors, windows, and built-in units.

mock-up - A full scale model of a construction component built on its construction site to judge appearance and test performance characteristics.

modified bitumens - A roofing membrane composed of a polyester or fiberglass mat saturated with a polymer-modified asphalt.

modillion - A stone scroll supporting a corona.

modulus of elasticity (E) - The property of a material that indicates its resistance to bending. It is the ratio of unit stress to unit strain.

moisture content - The amount of water contained in wood, expressed as a percentage of the weight of the wet wood to the weight of an oven-dry sample.

moment - A force that acts at a distance from a point and that tends to cause a body to rotate about that point.

monomer - An organic molecule that can be converted into a polymer by chemical reaction with similar molecules or organic molecules.

mortar - A plastic mixture of cementitious materials, water, and a fine aggregate.

mortar flow - A measure of the consistency of freshly mixed mortar related to the diameter of a molded truncated cone specimen after the sample has been vibrated a specified number of times.

mosaic - A decoration made up of small pieces of inlaid stone, glass, or tile.

motor control center - Controllers used to start and stop electric motors and protect them from overloads.

moving walk - A conveyor belt system operating at floor level used to move people in a horizontal direction.

natural fibers - Fibers found in nature, such as wool and cotton.

negotiated contract - A project delivery process in which the construction contractor is selected without the process of competitive bidding.

net-metering - A process wherein excess electricity generated by a photovoltaic system is sold back to a utility grid.

noise reduction coefficient (NRC) - A single number indicated by the amount of airborne sound energy absorbed into a material.

noncalcareous clays - Clays containing silicate of alumina, feldspar, and iron oxide.

nonferrous metals - Metallic materials in which iron is not a principal element.

non-heat-treatable alloys - Alloys that do not increase in strength when they are heat treated.

non-metallic inorganic materials - Materials extracted from soil and refined to produce a variety of products. Typical inorganic materials include sand, limestone, glass, brick, cement, gypsum, mortar, and mineral wool insulation.

non-potable water - Recycled water that can be used by plumbing fixtures, such as toilets, when there is no possibility of it becoming intermingled with potable water used for human consumption.

non-pre-stressed units - Concrete structural members in which the reinforcing steel is not subject to pre-stressing or post-tensioning.

oakum - A caulking material made from hemp fibers treated with tar.

off-gassing - A chemical process that occurs when solid but chemically unstable materials evaporate at room temperature and slowly release contaminants.

oil-based paints - Paint composed of resins requiring solvent for reduction purposes.

opaque coatings - Coatings that completely obscure the color and much of the texture of a substrate.

organic matter (or material) - A class of compounds comprising only those existing in plants and animals.

oriented strand board - A panel made from wood strands that has the strand face oriented in the long direction of the panel.

overlaid plywood - Plywood panels whose exterior surfaces are covered with a resin-impregnated fiber ply.

oxidation - A chemical reaction between a material and oxygen in the atmosphere.

oxide layer - In aluminum, a very thin protective layer formed naturally on aluminum due to its reaction to oxygen.

panelboard - A panel that includes fuses or circuit breakers used to protect the circuits in a building from overloads.

panelized construction - Construction that uses preassembled panels for walls, floors, and roof.

parallel strand lumber - Lumber made from lengths of wood veneer bonded to produce a solid member.

particleboard - A sheet product manufactured from wood particles and a synthetic resin or other binder.

photovoltaic (PV) - The use of semiconductor materials to convert solar radiation to electric current that can be used as electricity.

pig iron - A high-carbon-content iron produced by a blast furnace and used to manufacture cast iron and steel.

pigments - Paint ingredients mainly used to provide color and hiding power.

pilaster - A vertical projection from a masonry or concrete wall providing increased stiffening.

pile foundation - A foundation that uses long wood, concrete, or steel piles driven into the earth.

pile hammer - A machine for delivering blows to the top of a pile, driving it into the earth.

pit, elevator - The part of a hoistway that extends below the floor of the lowest landing to the floor at the bottom of the hoistway.

pitch - Related to carpets, the number of tufts in a 27 in. width of carpet.

plaster base - Any material suitable for the application of plaster.

plaster of Paris - A calcined gypsum mixed with water to form a thick, paste-like mixture.

plastic - An organic material that is solid in its finished state but is capable of being molded.

plastic behavior - The ability of a material to become soft and formed into desired shapes.

plastic limit (PL) - Related to soils, the percent moisture content at which soil begins to crumble when rolled into a thread $\frac{1}{8}$ in. (3 mm) in diameter.

plasticity index (PI) - The difference between the liquid limit and the plastic limit.

plasticizers - Liquid material added to some plastics to reduce their hardness and increase pliability. Also, an additive to concrete and mortar to increase plasticity.

platform framing - A wood structural frame for light construction with the studs extending only one floor high upon which the second floor is constructed.

plumbing chase - A hollow wall area accommodating piping used for drain waste or vent in plumbing systems.

plywood - All-veneer panels consisting of an odd number of cross-laminated layers, each layer consisting of one or more piles. Many such panels meet all of the prescriptive or performance provisions of U.S. Product Standard PS 1-83 / ANSI A199.1 for Construction and Industrial Plywood.

polymer - A chemical compound formed by the union of simple molecules to form more complex molecules.

Portland cement - A cementitious binder used in concrete, mortar, and stucco, obtained from pulverizing clinker consisting of hydraulic calcium silicates.

post-tensioned - A method used to place concrete under tension in which steel tenons are tensioned after the concrete has been poured and hardens.

potable water - Water that is safe to drink and meets standards of a local health authority.

power - The rate at which work is performed, expressed in watts or horsepower.

pozzolan - A siliceous or siliceous and aluminus material blended with Portland cement that chemically reacts with calcium hydroxide to form compounds possessing cementitious properties.

pre-design - The initial part of the design process in which a project team determines the goals and objectives of a project through design sketches and feasibility studies.

preservatives - Chemicals forced into wood to provide protection from decay, insects, and other damaging conditions.

pre-stressed units - Structural concrete units cast in molds in a factory that have stresses introduced as they are being cast (or post casting).

pre-tensioned - A method used to place a concrete member under tension by pouring concrete over steel tendons that are under tension before the concrete is poured.

primer coat - A base coat in a paint system. It is applied before the finish coats.

punch list - A listing of remaining items needing installation or repair near the completion of a construction project.

quartersawing - Sawing lumber so the hard annual rings are nearly perpendicular to the surface.

raceway - An enclosed channel designed to carry wires and cables.

radiant heat - Heat transferred by radiation.

radiation - The transmission of heat by electromagnetic waves.

rainscreen principle - The principle that states that wall cladding can be made watertight by placing wind-pressurized air chambers behind the joints, which reduces the air pressure differentials between the inside and outside that could cause water to move through the joints.

rake - The sloping portions at the gable end of a building.

rammed earth - A soil-cement mixture that is pneumatically rammed into forms to create walls generally 18 to 24 inches thick.

reduction - (1) A process in which iron is separated from oxygen with which it is chemically mixed by smelting the ore in a blast furnace. (2) With regard to aluminum, the electrolytic process used to separate molten aluminum from the alumina.

reflectance coefficient - A measure stated as a percentage of the amount of light reflected off a surface.

reflective glass - Glass having a thin layer of metal or metal oxide deposited on its surface to reflect heat and light.

refrigerant - A medium of heat transfer that absorbs heat by evaporating at low temperatures and pressure and releases heat when it condenses at a higher temperature and pressure.

reinforced masonry - Masonry construction that has steel reinforcing bars inserted to provide tensile strength.

relative humidity - The ratio of the amount of water vapor present in air to that which the air would hold if saturated at the same temperature.

resilient floor covering - Finished flooring made from a resilient material, such as polyvinyl chloride or rubber.

resin - A natural or synthetic material that is the main ingredient of paint and binds the ingredients together.

resistance, electric - The physical property of a conductor or electric-consuming device to resist the flow of electricity, reducing power and generating heat.

retarder - An admixture used to slow the setting of concrete.

return air - Air removed from a space and vented or reconditioned by a furnace, air conditioner, or other apparatus.

rigging - Hoisting equipment used to lift materials above grade.

rim joist - A type of sill used in frame construction in which the floor joists butt and are nailed to a rim joist and rest on the sill.

riser, plumbing - A water supply running vertically through a building to provide water to the various branches and fixtures on each floor.

rock - A solid mineral material found naturally in large masses.

romex - A non-metallic-sheathed electric cable of type NM or NMC.

room cavity ratio - A relationship between the height of a room, its perimeter, and the height of the work surface above the floor divided by the floor area.

rows - Related to carpets, the number of tufts per inch.

rubble - Rough stones in their natural irregular shapes and sizes.

safing - Fire-stopping material placed in spaces between a floor and curtain walls in high-rise construction.

sand - Fine rock particles from 0.002 in. (0.05 mm) in diameter to less than 0.25 in. (6.4 mm) in diameter.

sanitary piping - Pipe that carries liquid and water-borne waste from plumbing fixtures into which storm, surface, and groundwater are not admitted.

sanitary sewer - A sewer that receives sanitary sewage without the infusion of other water such as rain, surface water, or other clear water drainage.

sapwood - The wood near the outside of a log just under the bark.

scratch coat - The first coat of gypsum plaster that is applied to the lath.

screeding - The process of striking off the surface of freshly poured concrete with a screed so it is flush with the top of a form.

scupper - An outlet in a parapet wall for the drainage of overflow water from the roof to the outside of a building.

sealants - A mastic used to seal joints and seams.

sealer - A material used to seal the surface of a material against moisture.

seasoning - Removing moisture from green wood.

seated connections - Connections that join structural steel members with metal connectors, such as angles, upon which one member, such as a beam, rests.

sedimentary rock - Rock formed from the deposit of sedimentary materials on the bottom of a body of water or on the surface of the earth.

segregation - The tendency of large aggregate to separate from the sand-cement mortar in a concrete mix.

seismic area - A geographic area where earthquake activity may occur.

semi-transparent coatings - Coatings that allow some of the texture and color of a substrate to show through.

sensible heat - Heat that causes a detectable change in temperature.

service core - A fire-resistant vertical shaft through a multistory building used to route electrical, mechanical, and transportation systems.

service entrance - The point at which power is supplied to a building and where electrical equipment such as the service switch, meter, overcurrent devices, and raceways are located.

service temperature - The maximum temperature at which a plastic can be used without altering its properties.

shading coefficient (Sc) - The ratio of the solar heat gain of a window to the solar heat gain of clear float glass.

shales - Clays that have been subjected to high pressures, causing them to become relatively hard.

shear plate connector - A circular metal connector recessed into a wood member that is to be bolted to a steel member.

shear stress - The result of forces acting parallel to an area but in opposite directions, causing one portion of the material to "slide" past another.

shear studs - Metal studs welded to a steel frame that protrude up into a cast-in-place concrete deck.

sheave - A pulley over which an elevator wire hoisting rope runs.

sheeting - Wood, metal, or concrete members used to hold up the face of an excavation.

sherardize - To coat steel with a thin cladding of zinc by heating it in a mixture of sand and powdered zinc.

shop drawing - A drawing prepared by a contractor or material supplier that gives precise directives for the fabrication of a construction component.

shrinkage limit (SL) - Related to soils, the water content at which a soil volume is at its minimum.

Siamese connection - An outside connection to which firefighters connect an alternate source of water to boost the water used by a building's fire suppression system.

sick building syndrome - A medical condition caused by exposure to contaminants within a building resulting in headaches, dizziness, nausea, and respiratory and skin problems.

silt - Fine sand with particles smaller than 0.002 in. (0.05 mm) and larger than 0.00008 in. (0.002 mm).

single-ply roofing - A roofing membrane composed of a sheet of waterproof material secured to the roof deck.

sintering - A process that fuses iron ore dust with coke and fluxes into a clinker.

site plan - A drawing of a construction site, showing the proposed locations of buildings, site contours, and other improvements.

slag - A molten mass composed of fluxes and impurities removed from iron ore in a furnace.

slake - The process of adding water to quicklime, hydrating it and forming lime putty.

slump - A measure of the consistency of freshly mixed concrete, mortar, or stucco.

slurry - A liquid mixture of water and bentonite clay or Portland cement.

slurry wall - A wall built of a slurry used to hold up the sides of an area to be excavated.

smelting - A process in which iron ore is heated, separating the iron from the impurities.

smoke barriers - Continuous membranes used to resist the passage of smoke.

smoke developed rating - A relative numerical classification of the fumes developed by a burning material.

soffit - The underside of a cornice or other overhanging assembly.

soft-mud process - A process used to make bricks when the clay contains moisture in excess of 15 percent.

softwood - A botanical group of trees that have needles and are evergreen.

soil anchors - Metal shafts grouted into holes drilled into the sides of an excavation to stabilize it.

soil stack - A vertical plumbing pipe into which waste water flows through waste pipes from plumbing fixtures.

solid masonry - A unit whose core does not exceed 15 percent of the gross cross-sectional area of the unit.

solvent - Liquids used in paint and other finishing materials that give the coating workability and that evaporate, permitting the finish material to harden.

sound - A sensation produced by the stimulation of the organs of hearing by vibrations transmitted through the air. It is the movement of air molecules in a wavelike motion.

sound transmission class (STC) - A single number rating of the resistance of a construction to the passage of airborne sound.

space frame - A three-dimensional truss that forms a rigid, stable structure able to span long distances.

special units - Concrete masonry units that are designed and made for special applications.

specific adhesion - The bonding of dense materials through the attraction of unlike electrical charges.

specifications - Part of the construction documents that describes in writing the procedures, materials, methods, and performance requirements of a construction project.

spire - A tall pyramidal roof built upon a tower or steeple.

split ring connector - A ring-shaped metal insert placed in circular recesses cut in joining wood members that are held together with a bolt or lag screw.

spread foundation - A foundation that distributes the load over a large area.

spreading rate - The area over which a paint can be spread expressed in square feet per gallon.

stabilizers - Additives used to stabilize plastic by helping it resist heat, loss of strength, and the effect of radiation on the bonds between the molecular chains.

stain - A solution of coloring matter in a vehicle used to enhance the grain of wood during the finishing operation.

static load - Any load that does not change in magnitude or position with time.

steam - Water in a vapor state.

steeple - A tower-like ornamental construction, usually square or hexagonal, placed on the roof of a building and topped with a spire.

steep-slope roofs - Roofs with sufficient slope to permit rapid runoff of rain.

stiff-mud process - A process used to make bricks from clay that has 12 to 15 percent moisture.

stone - Rock selected or processed by shaping to size for building or other use.

story pole - A rod divided into equal parts with each part equal to the height of one course of masonry units.

strain - The deformation of a body or material when it is subjected to an external force.

strain hardening - Increasing the strength of a metal by cold-rolling.

strand-casting machine - A machine that casts molten steel into a continuous strand of metal that hardens and is cut into required lengths.

stress - The intensity of internally distributed forces that resist a change in the form of a body due to an applied force. It is measured in pounds per square inch (psi), mega-Newtons per square meter (MN/m^2) or mega-Pascals (MPa).

stress-rated lumber - Lumber that has its modulus of elasticity determined by actual tests.

structural clay tile - Rectangular masonry units made from clay and fired to harden them for use in load-bearing and non-load-bearing walls and partitions.

structural insulated panels - A structural panel consisting of two high-strength facing materials bonded to a rigid insulation core.

subcontractor - An entity that is contractually obligated to a general contractor to complete a portion of work on-site.

substrate - A subsurface to which another material is bonded.

suction - The rate at which clay masonry units absorb moisture.

suction rate - The weight of water absorbed when a brick is partially immersed in water for one minute—expressed in grams per minute or ounces per minute.

superintendent - A general contractor's on-site representative responsible for continuous field supervision, coordination, and completion of the work.

supply air - Conditioned air entering a space from an air conditioning, heating, or ventilating unit.

surface clays - Clays obtained by open-pit mining.

surface effect - The effect caused by the entrainment of secondary air against or parallel to a wall or ceiling when an outlet discharges air against or parallel to the wall or ceiling.

sustainable building - A building that provides the specified building performance requirements while minimizing disturbance to and improving the functioning of local, regional, and global ecosystems, both during and after its construction and specified service life.

sustainable development - Development that meets the needs of the present without compromising the ability of future generations to meet their own needs.

switchboard - A large single panel or an assembly of panels with switches, overcurrent protection devices, and buses that are mounted on the face or both sides of a panel.

switches - Devices to open and close electric circuits or to change connections within circuits.

synthetic fibers - Fibers formed by chemical reactions.

synthetic materials - Materials formed by the artificial building up of simple compounds.

tackboard - A surface used to display announcements and other materials, usually attached with tacks.

tapestry - A fabric upon which colored threads are woven by hand to produce a design.

temper designation - A specification of the temper or metallurgical condition of an aluminum alloy.

tempered glass - A process used to strengthen glass by raising its temperature to near the softening point and then suddenly blowing jets of cold air on both sides to chill it and create surface tension.

tensile bond strength - The ability of a mortar to resist forces tending to pull the masonry apart.

tensile stress - The stress per unit area of the cross section of a material that resists elongation.

terrazzo - A finish-floor material made up of concrete and an aggregate of marble chips that after curing is ground smooth and polished.

texture plaster - A finish plaster used to produce rough, textured finished surfaces.

thermal break - An insulating material or gap, placed between thermal conducting materials in contact to retard the passage of heat or cold.

thermal conductance (C) - Thermal conductance is the same as thermal conductivity except it is based on a specified thickness of a material rather than on the one inch used for conductivity.

thermal conduction - The process of heat transfer through a solid by transmitting kinetic energy from one molecule to the next.

thermal conductivity (k) - The rate of heat flow through one square foot of material one inch thick expressed in Btu per hour when a temperature difference of one degree Fahrenheit is maintained between the two surfaces.

thermal convection - Heat transmission by the circulation of a liquid or a heated air or gas.

thermal envelope - The exterior enclosure that separates conditioned space from the exterior.

thermal radiation - The transmission of heat from a hot surface to a cool one by means of electromagnetic waves.

thermal resistance - An index of the rate of heat flow through a material or assembly of materials. It is the reciprocal of thermal conductance (C).

thermal transmittance - A measure of heat transfer through an assembly of materials, such as an exterior wall—the reciprocal of the total resistance (R) of the assembly.

thermoplastic - Plastics that soften by heating and re-harden when cooled without changing chemical composition.

thermosets - Cured plastics that are chemically cross-linked and when heated will not soften but will degrade.

thermostat - A temperature-sensitive instrument that controls the flow of electricity to units used to heat and cool spaces in a building.

throw - The horizontal or vertical distance an airstream travels after leaving an air outlet before it loses velocity.

tieback anchors - Steel anchors grouted into holes drilled in an excavation wall to hold the sheeting, thus reducing the number of braces required.

timber - Wood structural members having a minimum thickness of 6 in. (140 mm).

timber joinery - The joining of structural wood members using wood joints, such as the mortise and tenon.

topcoat - The final coat of paint.

toughness - A measure of the ability of a material to absorb energy from a blow or shock without fracturing.

trade association - An organization whose membership comprises manufacturers or businesses involved in the production or supply of materials and services in a particular area.

transformer - An electrical device used to convert an incoming electric current from one voltage to another voltage.

trap - Device used to maintain a water seal against sewer gases that back up the waste pipe. Usually each fixture has a trap.

trowel - A hand tool with a flat, broad steel blade used to apply, shape, and spread mortar.

troweling - Producing a final smooth finish on freshly poured concrete with a steel-bladed tool after the concrete has been floated.

truss - A structural component made of a combination of members, usually in a triangular arrangement, to form a rigid framework often used to support a roof.

tufted construction - Carpet formed by stitching the pile yarn through the backing material.

underpinning - Placing a new foundation below an existing foundation

unfibered gypsum - A neat gypsum with no additives.

uniform load - Any load spread evenly over a large area.

universal design - A movement that advocates the design of products and environments to be usable by all people, to the greatest extent possible, without the need for adaptation or specialized design.

urban sprawl - The rapid and expansive growth of a city in an unplanned and often environmentally problematic fashion.

U-value - The overall coefficient of thermal expansion in Btu / (hr. ft2 °F) or W / m² K.

valence - The points on an atom to which valences of other elements can bond.

valve - A device used to regulate and stop the flow of water.

varnish - A transparent coating that dries on exposure to air, providing a protective coating.

vehicle - The liquid portion of a paint composed mainly of solvents, resins, or oils.

velvet construction - Carpet formed by joining the pile, stuffer, and weft yarns with double warp yarns.

veneer gypsum base - A gypsum board product designed to serve as the base for the application of gypsum veneer plaster.

veneer plaster - A thin layer of plaster applied over a special veneer gypsum base sheet.

vent stack - That part of the soil stack above the highest vent branch.

vents, plumbing - Pipes permitting a waste system to operate under atmospheric pressure while preventing sewer gases from entering a building. Plumbing vents allow air to enter and leave the system, preventing water in the traps from being siphoned off.

vertical shear - The tendency of one part of a member to move vertically in relationship to its adjacent part.

viscosity - The flow resistance of a liquid under an applied load or pressure.

volatile organic compound - Manmade chemicals used and produced in the manufacture of paints, composites, and refrigerants that can potentially harm humans.

volatile organic compounds (VOCs) - Compounds released to the atmosphere as a coating dries.

waferboard - A panel made of wood wafers, randomly arranged and bonded with a waterproof binder.

warp - A variation in a board from a flat, plane condition.

waste pipe - Horizontal plumbing pipes that connect a fixture to the soil pipe.

water retention - The property of a mortar that prevents the rapid loss of water by absorption into the masonry units.

water-based coatings - Coatings formulated with water as the solvent.

water-cement ratio - In a concrete or mortar mixture, the ratio of the amount of water (minus that held by the aggregates) to the amount of cement used.

water-vapor transmission rate (WVTR) - The steady-state vapor flow in a given time through a given area of a body, normal to specified parallel surfaces, under specific conditions of temperature and humidity at each surface.

watt - The unit of measurement of electrical power or rate of work. It is a pressure of one volt flowing at the rate of one ampere.

weatherability - The ability of a plastic to resist deterioration due to moisture, ultraviolet light, heat, and chemicals found in the air.

weathering index - A value that reflects the ability of clay masonry units to resist the effects of weathering.

weathering steel - A steel alloy that forms a natural self-protecting rust.

weep hole - Small openings at the bottom of exterior cavity walls to allow moisture in the cavities to drain out.

Wilton construction - Carpet formed on a loom capable of feeding yarns of various colors.

winning - A term used to describe the mining of clay.

workability - (1) Describes the ease or difficulty with which concrete can be placed and worked into its final location. (2) In relation to mortar, the property of freshly mixed mortar that determines the ease and homogeneity with which it can be spread and finished.

wrought alloys - Products formed by any of the standard manufacturing processes, such as drawing, rolling, forging, or extruding.

zoning ordinance - Local land use regulations written into municipal law that regulate the degree of building development according to land use.

Index